Antioxidants Effects in Health

Antioxidants Effects in Health

The Bright and the Dark Side

Edited by

Seyed Mohammad Nabavi

Baqiyatallah University of Medical Sciences, Iran

Ana Sanches Silva

National Institute of Agrarian and Veterinary Research (INIAV, I.P.) and
Center for Study in Animal Science (CECA), Porto, Portugal

ELSEVIER

Elsevier
Radarweg 29, PO Box 211, 1000 AE Amsterdam, Netherlands
The Boulevard, Langford Lane, Kidlington, Oxford OX5 1GB, United Kingdom
50 Hampshire Street, 5th Floor, Cambridge, MA 02139, United States

ISBN: 978-0-12-819096-8

For Information on all Elsevier publications visit our website at https://www.elsevier.com/books-and-journals

Publisher: Stacy Masucci
Acquisitions Editor: Andre G. Wolff
Editorial Project Manager: Zsereena Rose Mampusti
Production Project Manager: Maria Bernard
Cover Designer: Victoria Pearson

Typeset by Aptara, New Delhi, India

I dedicate this book to my mother, Seyed Maryam Nabavi
and my brother Seyed Fazel Nabavi
Seyed Mohammad Nabavi

To my beloved sons and husband, Maria Inês, João, and Ricardo
To Maria Lucília and Jacinto for their priceless support
and never-ending love
Ana Sanches Silva

Contents

PART 3 Synthetic antioxidants: bright and the dark side

Contributors

Amir Hossein Abdolghaffari
Medicinal Plants Research Center, Institute of Medicinal Plants, ACECR, Karaj, Iran; Toxicology and Diseases Group (TDG), Pharmaceutical Sciences Research Center (PSRC), The Institute of Pharmaceutical Sciences (TIPS), and Faculty of Pharmacy, Tehran University of Medical Sciences, Tehran, Iran; GI Pharmacology Interest Group (GPIG), Universal Scientific Education and Research Network (USERN), Tehran, Iran; Department of Toxicology & Pharmacology, Faculty of Pharmacy, Tehran Medical Sciences, Islamic Azad University, Tehran, Iran

Anjana Adhikari-Devkota
Graduate School of Pharmaceutical Sciences, Kumamoto University, Kumamoto, Japan

Salman Ahmad
Department of Pharmacognosy, Faculty of Pharmacy and Pharmaceutical Sciences, University of Karachi, Karachi, Pakistan

Reem Hasaballah Alhasani
Department of Biology, Faculty of Applied Science, Umm Al-Qura University, Makkah, Saudi Arabia

Mohammed Alqarni
Department of Pharmaceutical Chemistry, College of Pharmacy, Taif University, Taif, Saudi Arabia

Norah A. Althobaiti
Biology Department, College of Science and Humanities-Al Quwaiiyah, Shaqra University, Al Quwaiiyah, Saudi Arabia

Renata de Sousa Alves
Department of Pharmacy and Clinical Analysis, Federal University of Ceará (UFC), Fortaleza, Ceará, Brazil

Rajeshwar K.K. Arya
Department of Pharmaceutical Sciences, Kumaun University, Campus Bhimtal, Bhimtal, Uttarakhand, India

Mohammad Hossein Asghari
Department of Pharmacology and Toxicology, School of Medicine, Babol University of Medical Sciences, Babol, Iran

Chundoo B. Azeemah
Department of Health Sciences, Faculty of Medicine and Health Sciences, University of Mauritius, Réduit, Mauritius

Joseph I. Azzopardi
Department of Physiology and Biochemistry, Faculty of Medicine and Surgery, University of Malta, Msida, Malta

Nawshin Baureek
Department of Health Sciences, Faculty of Medicine and Health Sciences, University of Mauritius, Réduit, Mauritius

Tapan Behl
Chitkara College of Pharmacy, Chitkara University, Punjab, India

Tarun Belwal
College of Biosystems Engineering and Food Science, Zhejiang Key Laboratory for Agri-Food Processing, Zhejiang University, Hangzhou, People's Republic of China

Amira Y. Benmelouka
Faculty of Medicine, University of Algiers, Algiers, Algeria

Dhaka R. Bhandari
Institute of Inorganic and Analytical Chemistry, Justus Liebig University Giessen, Giessen, Germany

Dheeraj Bisht
Department of Pharmaceutical Sciences Sir J.C. Bose Technical Campus, Nainital, Uttarakhand, India

Arti Bisht
G.B. Pant National Institute of Himalayan Environment and Sustainable Development, Kosi-Katarmal, Almora, Uttarakhand, India

Renald Blundell
Department of Physiology and Biochemistry, Faculty of Medicine and Surgery, University of Malta, Msida, Malta; Centre for Molecular Medicine and Biobanking, University of Malta, Msida, Malta

Anna Blázovics
Department of Surgical Research and Techniques, The Heart and Vascular Center, Semmelweis University, Budapest, Hungary

Hasna Bouhenni
Faculty of Nature and Life Sciences, University of Tiaret, Algeria

Jacqueline Ramos Machado Braga
Federal University of Recôncavo of Bahia (UFRB), Cruz das Almas, Bahia, Brazil

Meriem Chafaa
Faculty of Nature and Life Sciences, University of Tiaret, Algeria

Sharmistha Chatterjee
Division of Molecular Medicine, Bose Institute, Kolkata, India

Zunera Chauhdary
Department of Pharmacology, Faculty of Pharmaceutical Sciences, Government College University, Faisalabad, Pakistan

Xiuping Chen
State Key Laboratory of Quality Research in Chinese Medicine, Institute of Chinese Medical Sciences, University of Macau, Macau, China

Lei Chen
College of Food Science and Technology, Guangdong Ocean University, Zhanjiang, China

Ericsson Coy-Barrera
Bioorganic Chemistry Laboratory, Faculty of Basic and Applied Science, Universidad Militar Nueva Granada, Cajicá, Colombia

Abhishek K. Das
Division of Molecular Medicine, Bose Institute, Kolkata, India

Hari P. Devkota
Graduate School of Pharmaceutical Sciences, Kumamoto University, Kumamoto, Japan

Prasanta Dey
School of Pharmacy, Sungkyunkwan University, Suwon, Republic of Korea

Arasana Dhariwal
Department of Pharmaceutical Sciences, Sir J.C. Bose Technical Campus, Nainital, Uttarakhand, India

Sanaa Dilmar A.
Department of Health Sciences, Faculty of Medicine and Health Sciences, University of Mauritius, Réduit, Mauritius

Koula Doukani
Faculty of Nature and Life Sciences, University of Tiaret, Algeria
Laboratory of Sciences and Technics of Animal Production, University of Abdelhamid Ibn Badis, Mostaganem, Algeria

Sumit Durgapal
Department of Pharmaceutical Sciences, Kumaun University, Campus Bhimtal, Bhimtal, Uttarakhand, India; Pharmaceutical Sciences Research Center, Health Institute, Kermanshah University of Medical Sciences, Kermanshah, Iran

Daphne Désiré A.-L.
Department of Health Sciences, Faculty of Medicine and Health Sciences, University of Mauritius, Réduit, Mauritius

Shahira M. Ezzat
Department of Pharmacognosy, Faculty of Pharmacy, Cairo University, Egypt
Department of Pharmacognosy, Faculty of Pharmacy, October University for Modern Sciences and Arts (MSA), Egypt

Sajad Fakhri
Pharmaceutical Sciences Research Center, Health Institute, Kermanshah University of Medical Sciences, Kermanshah, Iran

Mohammad Hosein Farzaei
Pharmaceutical Sciences Research Center, Health Institute, Kermanshah University of Medical Sciences, Kermanshah, Iran

Joomun B. Fatimah-Tuz-Zohra
Department of Health Sciences, Faculty of Medicine and Health Sciences, University of Mauritius, Réduit, Mauritius

Aakriti Garg
Department of Pharmacology, Indo-Soviet Friendship College of Pharmacy (ISFCP), Moga, Punjab, India
School of Pharmaceutical Sciences, Apeejay Stya University, Gurgaon, India

Noyel Ghosh
Division of Molecular Medicine, Bose Institute, Kolkata, India

Sumit Ghosh
Division of Molecular Medicine, Bose Institute, Kolkata, India

Jalaj K. Gour
Department of Biochemistry, Faculty of Science, University of Allahabad, Prayagraj, India

Daniel Cordeiro Gurgel
Federal Institute of Education, Science and Technology of Ceará, Limoeiro do Norte, Ceará, Brazil

Shokoufeh Hassani
Toxicology and Diseases Group (TDG), Pharmaceutical Sciences Research Center (PSRC), The Institute of Pharmaceutical Sciences (TIPS), and Faculty of Pharmacy, Tehran University of Medical Sciences, Tehran, Iran; Department of Toxicology and Pharmacology, School of Pharmacy, Tehran University of Medical Sciences, Tehran, Iran

Uday Hossain
Division of Molecular Medicine, Bose Institute, Kolkata, India

Subratty A. Hussein
Department of Health Sciences, Faculty of Medicine and Health Sciences, University of Mauritius, Réduit, Mauritius

Shabnoor Iqbal
Department of Zoology, Faculty of Life Sciences, Government College University, Faisalabad, Pakistan

Arvind Jantwal
Department of Pharmaceutical Sciences, Kumaun University, Campus Bhimtal, Bhimtal, Uttarakhand, India

Roberta Jeane Bezerra Jorge
Department of Physiology and Pharmacology, Drug Research and Development Center (NPDM), Federal University of Ceará (UFC), Fortaleza, Ceará, Brazil

Tanuj Joshi
Department of Pharmaceutical Sciences, Kumaun University, Campus Bhimtal, Bhimtal, Uttarakhand, India

Vijay Juyal
Department of Pharmaceutical Sciences, Kumaun University, Campus Bhimtal, Bhimtal, Uttarakhand, India

Rahul Kaldate
Department of Agrilcultural Biotechnology, Assam Agricultural University, Jorhat, Assam, India

Gökçe Ş. Karatoprak
Department of Pharmacognosy, Faculty of Pharmacy, Erciyes University, Kayseri, Turkey

Arnab Karmakar
Division of Molecular Medicine, Bose Institute, Kolkata, India

S. Khatoon Khadaroo
Department of Health Sciences, Faculty of Medicine and Health Sciences, University of Mauritius, Réduit, Mauritius

Haroon Khan
Department of Pharmacy, Abdul Wali Khan University, Mardan, Pakistan

Abdul H. Khan
Department of Pharmacy, Forman Christian College University (A Chartered University), Lahore, Pakistan

Ziyad Khan
Department of Pharmacy, University of Swabi, Swabi, Pakistan

Haroon Khan
Department of Pharmacy, Abdul Wali Khan University Mardan, Mardan, Pakistan

Anoop Kumar
Department of Pharmacology, Indo-Soviet Friendship College of Pharmacy (ISFCP), Moga, Punjab, India; Department of Pharmacology, Delhi Pharmaceutical Sciences and Research University (DPSRU), New Delhi, India

Prashant Kumar
Department of Pharmacy, Doon valley group of Institutions, Karnal, Haryana, India

Aadesh Kumar
Department of Pharmaceutical Sciences, Kumaun University, Campus Bhimtal, Bhimtal, Uttarakhand, India

Manish Kumar
Department of Biochemistry, Faculty of Science, University of Allahabad, Prayagraj, India

Ankit Kumar
Department of Pharmaceutical Sciences, Sir J.C. Bose Technical Campus, Nainital, Uttarakhand, India

Jankee T. Laxmi
Department of Health Sciences, Faculty of Medicine and Health Sciences, University of Mauritius, Réduit, Mauritius

Devina Lobine
Department of Health Sciences, Faculty of Medicine and Health Sciences, University of Mauritius, Réduit, Mauritius

Meeajan M. Irfaan
Department of Health Sciences, Faculty of Medicine and Health Sciences, University of Mauritius, Réduit, Mauritius

Filippo Maggi
School of Pharmacy, University of Camerino, Camerino, Italy

Nihal M. El Mahdy
Department of Pharmaceutics and Industrial Pharmacy, Faculty of Pharmacy, October University for Modern Sciences and Arts (MSA), Egypt

Marwa M. Mahfouz
Department of Pharmacology and Toxicology, Faculty of Pharmacy, Menoufia University, Menoufia, Egypt

Mohamad Fawzi Mahomoodally
Department of Health Sciences, Faculty of Medicine and Health Sciences, University of Mauritius, Réduit, Mauritius

Ankita Mandal
Division of Molecular Medicine, Bose Institute, Kolkata, India

Aline Diogo Marinho
Department of Physiology and Pharmacology, Drug Research and Development Center (NPDM), Federal University of Ceará (UFC), Fortaleza, Ceará, Brazil

Harikesh Maurya
M.G.B. Rajat College of Pharmacy. Gohila, Hanswar, Ambedkar Nagar, U.P., India

Lingchao Miao
Institute of Chinese Medical Sciences, University of Macau, Macao, China

Milad Moloudizargari
Department of Immunology, School of Medicine, Shahid Beheshti University of Medical Sciences, Tehran, Iran

Saeideh Momtaz
Medicinal Plants Research Center, Institute of Medicinal Plants, ACECR, Karaj, Iran; Toxicology and Diseases Group (TDG), Pharmaceutical Sciences Research Center (PSRC), The Institute of Pharmaceutical Sciences (TIPS), and Faculty of Pharmacy, Tehran University of Medical Sciences, Tehran, Iran;

GI Pharmacology Interest Group (GPIG), Universal Scientific Education and Research Network (USERN), Tehran, Iran

Ghulam Mujtaba Shah
Department of Botany, Faculty of Biological and Health Sciences, Hazara University, Mansehra, Pakistan

Seyed Mohammad Nabavi
Applied Biotechnology Research Center, Baqiyatallah University of Medical Sciences, Tehran, Iran

Nouzaifa Nabee
Department of Health Sciences, Faculty of Medicine and Health Sciences, University of Mauritius, Réduit, Mauritius

Francisco Assis Nogueira-Junior
Department of Physiology and Pharmacology, Drug Research and Development Center (NPDM), Federal University of Ceará (UFC), Fortaleza, Ceará, Brazil

Diana Célia Sousa Nunes-Pinheiro
Faculty of Veterinary Medicine, State University of Ceará, Fortaleza, Ceará, Brazil

Abhay K. Pandey
Department of Biochemistry, University of Allahabad, Prayagraj, India

Kiran Patni
Graphic Era Hill University Bhimtal, Nainital, Uttarakhand, India

Pooja Patni
School of Pharmaceutical Sciences, Lovely Professional University, Phagwara, Punjab, India

Lokesh Patni
Shree Dev Dental Clinic, Nainital, Uttarakhand, India

Vinay Pratap
Department of Biochemistry, Faculty of Science, University of Allahabad, Prayagraj, India

Govind Rajpal
Department of Pharmaceutical Sciences, Sir J.C. Bose Technical Campus, Nainital, Uttarakhand, India

Márcia Maria Vieira Ramos
University Center Estácio of Ceará, Fortaleza, Ceará, Brazil

Harvesh Kumar Rana
Department of Biochemistry, University of Allahabad, Prayagraj, India

Mahendra Rana
Department of Pharmaceutical Sciences, Kumaun University, Campus Bhimtal, Bhimtal, Uttarakhand, India

Amita J. Rana
Department of Pharmaceutical Sciences, Kumaun University, Campus Bhimtal, Bhimtal, Uttarakhand, India

Shahid Rasool
College of Pharmacy, University of Sargodha, Sargodha, Pakistan

Azhar Rasul
Department of Zoology, Faculty of Life Sciences, Government College University, Faisalabad, Pakistan

Akhtar Rasul
Department of Pharmaceutics, Faculty of Pharmaceutical Sciences, Government College University, Faisalabad, Pakistan

Elodie Rosette M. A.-L.
Department of Health Sciences, Faculty of Medicine and Health Sciences, University of Mauritius, Réduit, Mauritius

Archana N. Sah
Department of Pharmaceutical Sciences, Faculty of Technology, Bhimtal Campus, Kumaun University, Nainital, Uttarakhand, India

Mohamed A. Salem
Department of Pharmacognosy, Faculty of Pharmacy, Menoufia University, Menoufia, Egypt

Kasturi Sarkar
Department of Microbiology, St. Xavier's College, Kolkata, India

Ammar S.M. Selles
Institute of Veterinary Sciences, University of Tiaret, Algeria

Muhammad Ajmal Shah
Department of Pharmacognosy, Faculty of Pharmaceutical Sciences, Government College University, Faisalabad, Pakistan

Ghulam Mujtaba Shah
Department of Botany, Faculty of Biological and Health Sciences, Hazara University, Mansehra, Pakistan

Shahid Shah
Department of Pharmacy Practice, Faculty of Pharmaceutical Sciences, Government College University, Faisalabad, Pakistan

Ruchika Sharma
Department of Biotechnology, Indo-Soviet Friendship College of Professional Studies (ISFCPS), Moga, Punjab, India

Parames C. Sil
Division of Molecular Medicine, Bose Institute, Kolkata, India

Ana Sanches Silva
National Institute for Agricultural and Veterinary Research (INIAV), I.P., Vairão, Vila do Conde, Portugal; Center for Study in Animal Science (CECA), University

of Oporto, Oporto, Portugal; University of Coimbra, Faculty of Pharmacy, Coimbra, Portugal

João Alison de Moraes Silveira
Department of Physiology and Pharmacology, Drug Research and Development Center (NPDM), Federal University of Ceará (UFC), Fortaleza, Ceará, Brazil

Amit Kumar Singh
Department of Biochemistry, University of Allahabad, Prayagraj, India

Anita Singh
Department of Pharmaceutical Sciences, Kumaun University, Campus Bhimtal, Bhimtal, Uttarakhand, India

Manoj K. Singh
Centre for Non Communicable Diseases (NCD), National Centre for Disease Control (NCDC), Ministry of Health & Family Welfare-Government of India, Delhi, India

Sushil Kumar Singh
DBT NECAB, Assam Agricultural University, Jorhat, Assam, India

Laxman Singh
Centre of Biodiversity Conservation & Management, G.B. Pant National Institute of Himalayan Environment, Kosi-Katarmal, Almora, Uttarakhand, India

Leila Soudani
Faculty of Nature and Life Sciences, University of Tiaret, Algeria

Ipek Süntar
Department of Pharmacognosy, Faculty of Pharmacy, Gazi University, Ankara, Turkey

Devesh Tewari
Department of Pharmacognosy, School of Pharmaceutical Sciences, Lovely Professional University, Phagwara, Punjab, India

Nidhi Tiwari
Department of Pharmaceutical Sciences, Kumaun University, Campus Bhimtal, Bhimtal, Uttarakhand, India

Malik Saad Ullah
Department of Pharmaceutical Chemistry, Faculty of Pharmaceutical Sciences, Government College University, Faisalabad, Pakistan

Hammad Ullah
Department of Pharmacy, School of Medicine and Surgery, University of Naples Federico II, Via Domenico Montesano, Naples, Italy

Jyoti Upadhyay
School of Health Science and Technology, Department of Pharmaceutical Sciences, University of Petroleum and Energy Studies, Dehradun, Uttarakhand, India

Mirele da Silveira Vasconcelos
Federal Institute of Education, Science and Technology of Ceará (IFCE), Baturité, Ceará, Brazil

Sandeep Visvarma
Deep Dental Clinic, Nainital, Uttarakhand, India

Jianbo Xiao
Universidade de Vigo, Nutrition and Bromatology Group, Department of Analytical and Food Chemistry, Faculty of Sciences, Ourense, Spain

Yixi Xie
Key Laboratory for Green Organic Synthesis and Application of Hunan Province, Xiangtan University, Xiangtan, China

Ömer F. Yakıncı
National Poisons Information Service, Ministry of Health, Ankara, Turkey
Department of Pharmacognosy, Faculty of Pharmacy, Gazi University, Ankara, Turkey

Li Yang
Institute of Chinese Medical Sciences, University of Macau, Macao, China

Toorabally B. Zaynab
Department of Health Sciences, Faculty of Medicine and Health Sciences, University of Mauritius, Réduit, Mauritius

Haolin Zhang
Institute of Chinese Medical Sciences, University of Macau, Macao, China

Francisco Rogênio da Silva Mendes
Northeast Biotechnology Network (RENORBIO), Center of Experimental Biology (Nubex), University of Fortaleza (UNIFOR), Fortaleza, Ceará, Brazil

Dirce Fernandes de Melo
Department of Biochemistry and Molecular Biology (DBBM), Federal University of Ceará (UFC), Fortaleza, Ceará, Brazil

Paulo Carvalho de Paula
Department of Biochemistry and Molecular Biology, Federal University of Ceará, Fortaleza, Ceará, Brazil

Luciana de Siqueira Oliveira
Department of Food Technology, Federal University of Ceará, Fortaleza, Ceará, Brazil

Felipe Domingos de Sousa
Department of Physics, Federal University of Ceará, Fortaleza, Ceará, Brazil

Tamiris de Fátima Goebel de Souza
Department of Physiology and Pharmacology, Nucleus of Drug Research and Development - NPDM, Federal University of Ceará, UFC, Ceará, Brazil

Juan M. Álvarez-Caballero
Chemistry and Bioprospecting of Natural Products Group, Universidad del Magdalena, Santa Marta, Colombia

About the editors

Short Biography—Seyed Mohammad Nabavi

Seyed Mohammad Nabavi is a biotechnologist and senior researcher in Applied Biotechnology Research Center, Baqiyatallah University of Medical Science and member of Iran's National Elites Foundation. His research focused on the health-promotion effects of natural products, namely antioxidant compounds. He is author/co-author of 300 publications in peer-reviewed international journals with a high impact factor. He is a referee of several international journals. His Scopus h-index is 51 (August 2021).

Short Biography—Ana Sanches Silva

Ana Sanches Silva received the degree in pharmaceutical sciences from the Pharmacy Faculty, University of Coimbra (FFUC), Portugal, and received her European PhD degree in pharmacy from the University of Santiago de Compostela (USC), Spain, with honors. In addition, she was awarded with two awards for best PhD thesis. She is a member of the executive board of Animal Science Studies Center and invited professor at the FFUC. Ana has a remarkable track record, namely as co-author of papers in peer-reviewed journals with high impact factor, book chapters, and as co-editor of scientific books in the food science field. Ana's research has focused on antioxidants, namely natural antioxidants, with potential to be used in active food packaging. In addition, she has a special interest in the development and validation of analytical methodologies, especially mass spectrometry related, to determine food and food packaging components and contaminants.

Preface

Antioxidants are among the compounds that contribute for the body's redox homeostasis. They are being intensively studied for their health benefits, mainly due to their ability to scavenge free radicals. However, many questions can be raised related to their safety and effectiveness:

- May antioxidants have a more preventive than curative effect on diseases?
- Can antioxidants be detrimental for human health? In what extent?
- What are the optimal antioxidants' dosages? At what level the antioxidants behave as pro-oxidants?
- What is the ideal route of administration?
- Can (food/food supplements) antioxidants reach specific compartments of the cells (with high levels of free radicals)?
- Supplementation with exogenous antioxidants can lead to decrease antioxidants production within the body?

The book, *Antioxidants Effects in Health: The Bright and the Dark Side*, aims to evaluate the current scientific evidence on these topics. The book will be useful for everyone who is interested in antioxidants, namely researchers, health professionals, industry and government regulatory agencies; for students in phytochemistry, pharmacognosy, and natural product synthesis; and for experts in the formulation of herbal and natural pharmaceuticals.

The target audience have to face every day new challenges in a field that is in rapid growth, with continuous increase information on antioxidants, concerning their properties, content in foodstuffs, as well as in the health promoting or detrimental properties of these compounds. This book will introduce recent and updated information on antioxidants in a systematic way which will allow to easily compare the antioxidants and find a conductive line among different chapters.

The book is composed of six parts. Part 1 addresses the evolution of antioxidants over time (Chapter 1.1), the oxidative stress (causes, free radicals, targets, mechanisms, affected organs, effects, indicators) (Chapter 1.2) and food autooxidation (Chapter 1.3).

Part 2 is dedicated to endogenous antioxidants, namely bilirubin, catalase, coenzyme Q, ferritin, glucose-6-phosphate-dehydrogenase, melatonin, superoxide dismutase, and uric acid. It addresses endogenous sources, mechanisms of action, beneficial, and detrimental effects on health, in vitro evidence, animal studies, and clinical studies of these antioxidants.

Part 3 is devoted to the bright and dark side of synthetic antioxidants. It addresses the following topics in each chapter: sources, chemistry, bioavailability, legal status, mechanisms of action, beneficial and detrimental effects on health, in vitro evidence, animal studies, and clinical studies. Ascorbyl palmitate, butylated hydroxyanisole (BHA), butylated hydroxytoluene (BHT), erythorbic acid (D-ascorbic acid), nordihydroguaiaretic acid (NDGA), octyl gallate (OG), propyl gallate (PG), and tert-butylhydroxyquinone (TBHQ) are the compounds focused in this part of the book.

Part 4 of the book addresses natural occurring antioxidants in order to give answer to the above-mentioned questions. Amino acids, carnosine, carnosol, carotenoids, citric acid, coenzyme Q, curcumin, flavonoids, lecithin, lignans, organosulfur compounds, phenolic acids, phytic acid, protein hydrolysates, selenium, sterols, tartaric acid, turmeric, uric acid, vanillin, vitamin A (retinol), vitamin C, vitamin E (tocopherols and tocotrienols), vitamin K, and zinc are focused in this part of the book.

Part 5 explores the relation between antioxidants and several diseases or disorders, namely cancer, cardiovascular diseases, cataracts/age-related macular degeneration, cognitive decline in elderly, dentistry, gastric lesions, immune functions, infertility, kidney, and liver diseases.

Part 6 discusses the actual and future perspectives on antioxidants. Chapter 6.1 devotes to the duality of antioxidants/prooxidants and chapter 6.2 to "Food and food supplements antioxidants: targets in human antioxidant system and effects on the production of endogenous antioxidants." Finally, Chapter 6.3 presents the concluding remarks and future perspectives.

Throughout the book, the relationship of antioxidants with different beneficial and detrimental effects are examined, and the current controversies and future perspectives are addressed and explored.

We sincerely acknowledge and thank all the authors for their valuable contributions for this book.

We are aware of other antioxidants that could have been addressed, as well as the relation of antioxidants with other diseases or disorders. We look forward *Antioxidants Effects in Health: The Bright and the Dark Side* being useful and well-received by readers. We hope, in the future, another edition of the book can address other antioxidants.

Ana Sanches Silva and Seyed Mohammad Nabavi

PART

Introduction

1

CHAPTER

1.1 Evolution of antioxidants over times (including current global market and trend)

Noyel Ghosh[a], Sharmistha Chatterjee[a], Parames C. Sil

Division of Molecular Medicine, Bose Institute, Kolkata, India

[a]NG and SC contributed to this manuscript equally

1.1.1 Introduction

Adaptations are "traits that arise and are maintained under selection" (Hochachka, 1998). In the course of evolution, these are the characters in biological entities that help them course through the changing ambient conditions. Interaction of the biological systems with the external environment is crucial for maintaining an internal environment favoring growth, reproduction, and overall survival. The early earth had a highly reducing atmosphere, with a major portion of carbon dioxide and methane, apart from ammonia. Probably, the net process that might have occurred was, water dissociated into hydrogen, which escaped from the earth, and into oxygen. This oxygen then oxidized the reduced carbon compounds to carbon dioxide. Similarly, ammonia was oxidized to nitrogen, and reduced iron was converted to more oxidized states. Free oxygen only appeared when the methane and ammonia were oxidized, and the cyanobacteria started releasing oxygen into the environment; and hence, the present atmospheric conditions were established on earth (Urey, 1952).

Approximately 150 million years ago the level of molecular oxygen approached present atmospheric level of 21% (Graham et al., 1995). If the interaction of biological systems with external environment is studied, it could be seen that oxygen is an essential inorganic chemical, needed for maintaining favorable life conditions. This is true for all living systems, except some obligate anaerobes, which get killed even by normal atmospheric concentrations of oxygen. But, though mostly all living beings essentially need oxygen for survival, it is paradoxical that oxidative damage also does occur in their bodies at key biological sites, and as a result, threatens their structure and functionality. In defense, this oxygenic threat is encountered by a biological antioxidant system that has evolved over time in the organisms (Benzie, 2000; Fridovich, 1998), in parallel with the evolution of our oxygenic atmosphere. Plants have evolved potent antioxidants which accumulate near and around the active

Antioxidants Effects in Health. DOI: https://doi.org/10.1016/B978-0-12-819096-8.00011-2

sites of photosynthesis, which happens to be a light-harvesting and oxygen-releasing process. For the early plants, oxygen was a toxic waste product of photosynthesis and thus they devised methods of its removal from their internal system. Also, they have evolved mechanisms of quenching the reactive oxygen radicals so that oxidative damage can be avoided. Plants, therefore, are known to produce quite an impressive array of antioxidant compounds which includes flavonoids, polyphenols, carotenoids, benzoic acids, cinnamic acids, ascorbic acid, folic acid, tocotrienols, and tocopherols (Benzie and Strain, 1999; Hollman, 2001). All of these have been found to be concentrated in the oxidation-prone sites of the plant, with high rates of oxidative turnover. In a similar fashion, we have certain antioxidant enzymes like superoxide dismutase (SOD) and catalase, and molecules like glutathione, which are dedicated to solely prevent oxidative stress in our body. Although very effective, these internal antioxidants sometimes fall insufficient in providing protection against continuous external oxidative assault. This is where the plant-derived dietary antioxidants come to the rescue (Fig. 1.1.1).

In simple terms, an antioxidant can be defined as anything that inhibits or prevents oxidation of a susceptible substrate (Benzie, 2003). However, the antioxidant systems found in the biological world are complex, and all of them act in sync to

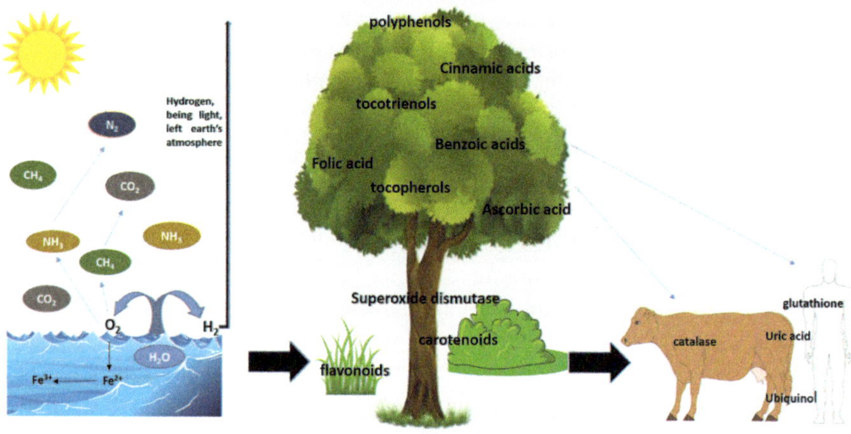

FIG. 1.1.1 The evolution of antioxidants in the history of earth.

The early earth had mostly a reducing environment. The first of the blue-green algae, which evolved in the sea, released oxygen into the atmosphere. Water split up into oxygen which oxidized the methane, nitrogen, and lesser oxidized metals, and into hydrogen, which being light, left the atmosphere of earth. With time, green plants evolved which released oxygen into the atmosphere as a by-product of photosynthesis, and themselves devised an array of antioxidants to prevent oxidation to key sites, which could lead to fatal oxidative damage. Animals evolved later, and arranged their inherent enzymatic antioxidants to meet their oxidation preventing needs. They also consumed plants, which were rich sources of antioxidants, and helped escalate the antioxidant defense put up by the animals.

avert the imminent oxidative damage. This is done by decreasing the reactive oxygen species (ROS) load, diverting ROS to other biological pathways, which have less reactive products, selectively rendering transition metal ions inactive (in redox terms), and, when everything else fails, providing sacrificial molecules which would act as replaceable or recyclable "buffers" and absorb incoming oxidative hits and any excess energy if present in the system. As has been discussed earlier, antioxidant defenses present in the human body are effective but are not impeccable, and hence, oxidative damage to key biological sites does occur. These accumulate with age, and finally contribute and lead to senescence and age-related diseases (Ames et al., 1993; Beckman and Ames, 1998; Finkel and Holbrook, 2000; Halliwell, 1999).

1.1.1.1 Oxidative damage and antioxidants

Besides opening huge opportunity for aerobic catabolic pathways, increased contact with molecular oxygen imposed a great threat of oxygen toxicity (Benzie, 2000). Here the key questions are: how does oxygen become toxic and what is the method of protection against such toxicity? Now the toxicity is due to the formation of free radicals. A free radical is any molecular species containing an unpaired electron in an atomic orbital and capable of independent existence (Fridovich, 1974; Fridovich, 1975). Presence of such solitary electron makes them highly reactive and unstable as it requires another molecule either for electron donation or acceptance. Thus, it results in oxidation (due to electron loss) and reduction (due to electron addition).

Molecular oxygen is a biradical molecule which requires four electrons and hydrogen atoms for its complete reduction to water. However, there is a large energy barrier for complete reduction to happen. Moreover, addition of four spontaneous electrons to the molecular oxygen is largely restricted to cytochrome oxidase complex at the end of electron transport system (ETS). However, partial reduction of oxygen by single electron transfer can be formed easily. Such partially reduced oxygen intermediates are known as ROS (Fridovich, 1998).

Both molecular oxygen and ROS are oxidizing agents which mean they are capable of taking electrons from other species. Oxidation of biological macromolecules such as DNA, lipid and proteins change their structure and function, resulting in mutation, damage and cell death. Oxidation power of ground-state molecular oxygen is somewhat restricted as another species having antiparallel electron spin to that of the unpaired parallel-spin electrons of diatomic oxygen, can only act as the electron donor. However, reactivity of molecular oxygen can be increased by removal of spin restriction, that is, by energy transfer through photosensitisers or by electron addition. Photosensitizers, such as chlorophyll, flavins, or porphyrin containing compounds are capable of light harvesting and energizing molecular oxygen, generating singlet oxygen ($^1O_2$) (Halliwell et al., 2000). $^1O_2$ has an energy level of 92 kJ above the ground state energy level of oxygen (Fridovich, 1974). It is capable of doing an easy transfer of energy to another molecule through which it can impose change in the structure of the target molecule. Notable target molecules are amino acids like tryptophan, cysteine, methionine and histidine, and lipids containing carbon-carbon covalent double bonds (C=C).

Besides, reduction of ground state molecular oxygen by one electron generates superoxide radical ($O_2^{\cdot-}$) (Miller et al., 1990). Superoxide anion radical formation from electron leakage during mitochondrial electron transport is inevitable. The process is also mediated by nicotine adenine dinucleotide phosphate [NAD(P)H] oxidase or xanthine oxidase. NAD(P)H oxidase is found in neutrophils, macrophages and in monocytes where a burst of $O_2^{\cdot-}$ leads to bactericidal activity. Thus, rate of generation of $O_2^{\cdot-}$ increases in inflammation, post ischemic reperfusion and also during exercise. Further, one electron reduction of $O_2^{\cdot-}$ by SOD generates hydrogen peroxide (H_2O_2). H_2O_2 can also be generated by xanthine oxidase, NAD(P)H oxidase, amino acid oxidase and also in peroxisome through molecular oxygen consumption (Dupuy et al., 1991; Granger, 1988). Being uncharged, H_2O_2 can easily diffuse across the plasma membrane. As H_2O_2 is not highly reactive, it may accumulate. Concentrations of 100 μmol/L or more have been found in biological fluids (Halliwell et al., 2000). H_2O_2 and $O_2^{\cdot-}$ together can form fiercely reactive hydroxyl radical ($OH^{\cdot}$) through Haber- Weiss reaction (Liochev and Fridovich, 2002).

$$H_2O_2 + O_2^{-} = OH + OH^{-} + O_2$$

This reaction proceeds very slowly in aqueous solution unless there is the presence of free transition metal ions like Fe^{2+} or Cu^{+}. In such conditions, reaction (known as Fenton reaction (Fenton, 1894), when iron-catalyzed) proceeds quickly. $OH^{\cdot}$ is the most potent and damaging ROS that can cause damage to proteins, carbohydrates, lipids and also to DNA.

ROS can be also generated by myeloperoxidase (MPO)-halide-H_2O_2 generating system. MPO, an enzyme present in activated neutrophilic granules, can convert H_2O_2 to hypochlorous acid (HOCl) in presence of chloride (Cl^{-}) ions (Klebanoff, 2005).

HOCl is highly oxidative and effective to kill the invading pathogens. However, HOCl is also able to react with DNA and generate pyrmidine oxidation products. Moreover, inducible nitric oxide synthase is capable of producing a large amount of reactive nitrogen species, such as $NO^{\cdot}$, which can function as a $O_2^{\cdot-}$ quencher. The $NO^{\cdot}$ and $O_2^{\cdot-}$ react together to form a highly strong oxidizer, peroxynitrite ($ONOO^{-}$) (Zhu et al., 1992).

$$\mathbf{NO}^{\cdot} + \mathbf{O}_2^{-} = ONOO^{-}$$

Other ROS include peroxyl radicals (ROO-.); simplest form of which is the hydroperoxyl radical (HOO-.). Lipid peroxidation through abstracting hydrogen atom from side chain methylene carbon can generate lipid radicals, which further reacts with oxygen to produce peroxyl radicals. Apart from aforesaid endogenous source of oxidants, cigarette smoke, ozone exposure, hyperoxia, ionizing radiation, and even heavy metal ions of lead, arsenic, etc., are considered as potent exogenous sources of ROS production. Though ROS play important role in a number of physiological processes such as microbial killing, gene transcription, cell division, apoptosis, etc., excessive ROS generation causes oxidative stress. Importance of ROS induced oxidative stress has been implicated in a number of chronic diseases, such as cancer, osteoporosis, coronary heart disease, etc. ROS attacks important biomolecules from fatty acid to DNA. Oxidative damage to these biomolecules can lead to enzyme inactivation, membrane disruption, mitochondrial dysfunction, mutation and cell death, etc. Hence, limitation of harmful interaction between ROS and their vulnerable macromolecules is of utmost importance. The molecules which play pivotal role in the prevention of oxidation are simply designated as the antioxidants.

1.1.2 Antioxidants in early human use

Originally, the term "antioxidant" specifically referred to a chemical, which could prevent oxygen consumption. The antioxidants were extensively studied in the latter half of 19th and early 20th centuries, the era of industrial revolution, to understand their roles mainly in key industrial processes, like vulcanization of rubber, preventing metal corrosion, and how they affected fuel polymerization during fouling of internal combustion engines (Mattill, 1947).

In biology though, early antioxidant research focused mainly on their role in prevention of rancidity that is, studying their effects on oxidation of unsaturated fatty acids (German, 1999). Back then, the fat was simply placed inside a closed container with oxygen and the rate of oxygen consumption was measured to study the antioxidant activity. However, the field got revolutionized when the antioxidant activity of vitamins A, C, and E, as well as SOD was discovered (Lobo et al., 2010). Earlier, they were only identified as important intermediate metabolites and physiological modulators in our body. But the ground-breaking discoveries one after the other led the scientific community to realize the importance of these molecules as antioxidants in the physiology and biochemistry of living organisms, and thus, antioxidant research accomplished a new interest worldwide (Jacob, 1996; Knight, 1998).

As the free radicals are highly reactive, the scientific community was oblivious to their existence in the biological system. Their recognition probably came after it was discovered that water underwent homolytic dissociation under the effect of ionizing radiation (Gerschman et al., 1954). Overall, their general acceptance in the biological systems came only when SOD was discovered as an enzyme having antioxidant function by McCord and Fridovich (McCord and Fridovich, 1969) in 1969.

The plausible mechanisms of antioxidant action started to be discovered henceforth, as the possible mechanisms of free radical generation inside the biological system had already been reported (speculated to be Fenton and Haber-Weiss reactions), and scientists conjectured that any substance having antioxidant activity should be one that could itself be readily oxidized (Société de, 1849). This hypothesis was confirmed by the discovery of the mechanism of how vitamin E prevented the lipid peroxidation process, and thus, the antioxidants were identified as reducing agents which prevented oxidative reactions, mostly by scavenging the reactive oxygen species, before they could inflict any damage upon cells (Wolf, 2005).

1.1.3 Types of antioxidants and their mode of action

Antioxidants inhibit and protect the system from cellular damage caused by free radicals mostly through their free radical scavenging property. These low molecular weight molecules are stable enough to donate electrons and thus able to terminate chain reaction before vital biomolecules get damaged. Some of these antioxidants are entirely physiological in origin, such as glutathione, uric acid, and ubiquinol; whereas, others must be supplied from outside through regular diet. Such dietary antioxidants can be exemplified by vitamins like ascorbic acid, vitamin E, plant-based antioxidants, such as curcumin, β-carotene, etc. Still, antioxidants can be broadly classified into two categories: enzymatic and nonenzymatic. Some of the nonenzymatic antioxidants, like vitamins, carotenoids, and flavonoids, are widely used as dietary supplements and as prescription medicine in current times.

1.1.3.1 Enzymatic antioxidants

Cells are equipped with a variety of antioxidant enzymes that serve to counterbalance the dreadful effects of cellular oxidative stress. Table 1.1.1 portrays a quick look through them. The major enzymatic antioxidants are SOD, catalase (CAT) and glutathione system, including glutathione, glutathione reductase, glutathione peroxidases, and glutathione-S-transferase. In addition, heme oxygenase-1, and other redox proteins

Table 1.1.1 List of enzymatic antioxidants.

Name of antioxidants	Acronym	Enzyme commission (EC) number	Catalyzed reaction
Superoxide dismutase	SOD	EC 1.15.1.11	$M^{(n+1)+} - SOD + O_2^- = M^{n+} - SOD + O_2$ $M^{n+} - SOD + O_2^- + 2H^+ = M^{(n+1)+} - SOD + H_2O_2$
Catalase	CAT	EC 1.11.1.6	$2H_2O_2 = O_2 + H_2O$
Glutathione peroxidase	GSH-Px	EC 1.11.1.9	$2GSH + H_2O_2 = GSSG + 2H_2O$ $2GSH + ROOH = GSSG + ROH + H_2O$
Thioredoxin	TRX	EC 1.8.4.10	Adenosine monophosphate + sulfite + thioredoxin disulfide = 5′-adenylyl sulfate + thioredoxin Adenosine 3′,5′-bisphosphate + sulfite + thioredoxin disulfide = 3′-phosphoadenylyl sulfates + thioredoxin
Peroxiredoxin	PRX	EC 1.11.1.15	$2R'\text{-}SH + ROOH = R'\text{-}S\text{-}S\text{-}R' + H_2O + ROH$
Glutathione transferase	GST	EC 2.5.1.11	$GSSG + NADPH + H^+ = 2GSH + NADP^+$

such as thioredoxin, peroxiredoxins have also been reported to play vital role in pulmonary antioxidant defense mechanisms (Table 1.1.1). Major ones are discussed below.

1.1.3.1.1 ***Superoxide dismutase***

SODs are a group of closely related antioxidant enzymes that are capable of breakdown of O2·- into O_2 and H_2O_2. Superoxides are moderately reactive but due to their charged nature they cannot readily diffuse out of the cell. Thus, dismutation of superoxides is of utmost importance. However, at physiological pH, spontaneous dismutation occurs very slowly which speaks of the importance of SODs. There are three distinct families of SODs depending upon the type of cofactor they use for their functions: Cu/Zn-type (which uses both copper and zinc), Fe, and Mn-type (which either uses iron or manganese) and lastly Ni-type, which acts with nickel. Mn-SODs are predominant in mitochondrial matrix and peroxisomes; whereas, Cu/Zn-types are mostly found in cytosol, peroxisome, apoplast, and chloroplasts (Wuerges et al., 2004). Again, though Fe-SODs are mostly detected in chloroplasts, but can also be found in peroxisomes (Corpas et al., 2001; Corpas et al., 2006).

1.1.3.1.2 ***Catalase***

Dismutation of superoxide to H_2O_2 seems to be the principal antioxidant strategy and thus, being harmful, removal of the H_2O_2 from the cell is highly recommended. Being uncharged, H_2O_2 can easily diffuse out of the cell. But this only happens in case of prokaryotes and single-celled eukaryotes. However, in multicellular eukaryotes with structural complexity, it is highly necessary to have a proper system to flush out H_2O_2 for prevention of possible toxicity. This is done through the action of catalase which catalyzes decomposition of H_2O_2 into water and O_2 (Gaetani et al., 1996). Degradation is achieved through the conversion between catalase-ferricatalase (iron coordinated to H_2O) and compound I (iron complexed with oxygen atom). Catalase is one among the highly conserved enzymes through the evolutionary course, which exists as a tetramer composed of four identical subunits with a heme group at each of four active sites.

1.1.3.1.3 ***Glutathione system***

Though not exclusively, but catalase activity is largely restricted to peroxisomes. In other organelles, such as in chloroplasts and mitochondria, peroxidases are the key enzymatic antioxidants for H_2O_2 removal. In animals, glutathione peroxidase (GSH-Px) acts in cooperation with catalase. GSH-Px is a unique enzyme that holds selenocysteine in the active site and uses endogenous tripeptide glutathione (GSH) for reduction of H_2O_2 and lipid peroxides to their corresponding alcohols. In animals, four different isoforms of GSH-Px are available. Cellular GSH-Px or GSH-Px-1 is ubiquitous in nature (Arthur, 2001). It reduces H_2O_2 and fatty acid peroxides except the esterified peroxyl lipids. On the other hand, membrane-bound GSH-Px-4 is responsible for reduction of the esterified peroxides. GSH-Px-2 serves to reduce the dietary peroxides in gastrointestinal epithelial cells (Chu, 1993). Lastly GSH-Px 3 is the only member within this group of antioxidant enzymes that is found in the extracellular milieu (Comhair, 2001).

1.1.3.2 Nonenzymatic antioxidants

Such antioxidants are extremely valuable in defense against oxidants. Most of them, including vitamin E, ascorbic acid, flavonoids, carotenoids, etc., are derived from the dietary sources. However, cell itself synthesizes some of the nonenzymatic antioxidants, which can be grouped as physiological antioxidants, such as uric acid, glutathione, etc. A quick look at the following table summarizes the sources of these dietary antioxidants, their bioavailability, and chemical structures, along with their concentrations in the plasma (Table 1.1.2).

1.1.3.2.1 Vitamins

Vitamin E, a highly effective lipid-soluble antioxidant, is a collective description for all tocopherol and trocotrienol derivatives. Tocopherols possess phytyl chain in their structures whereas; tocotrienols bear the same chain with three double bonds at the positions 3′, 7′, and 11′ (Gaßmann, 1991; Machlin, 1991). Tocopherols can be found in polyunsaturated vegetable oils and in the germ of cereal seeds, whereas tocotrienols are present in the aleurone and subaleurone layers of cereal seeds and in palm oils Both of these derivates remain in four possible isoforms designated from α to δ that differ in the number and position of methyl groups. However, all of them act as chain breakers during lipid peroxidation.

$$\text{LOO}^{\cdot} + \alpha - \text{tocopherol} - \text{OH} = \text{LOOH} + \alpha - \text{tocopherol} - \text{O}^{\cdot}$$

Amongst all such derivates, α –tocopherol (the major vitamin E *in vivo*) is found to be most efficient (Schneider, 2005; Yoshida et al., 2003). Such efficacy might be due to three possible reasons: firstly, it intercepts lipid peroxidation by reacting with the fatty acid peroxyl (LOO·) radicals (a product of lipid peroxidation) extremely fast which doesn't allow these radicals to oxidize other target components; secondly, it takes away the reactive character of peroxyl radicals and lastly, such reaction generates stable tocopheroxyl radical (α-tocopherol–O·) which itself doesn't initiate lipid peroxidation (Burton and Ingold, 1981). Vitamin E is prescribed by doctors to patients who have digestive complications that make it difficult for them to absorb vitamin E. It may also be prescribed to treat a movement disorder, tardive dyskinesia. It is also speculated to have beneficial effects on hair and skin, whether applied topically or taken orally. The Tolerable Upper Intake Level (UL) for men and women over 18 years old, as per the National Institutes of Health guidelines, is 1000 mg a day.

Ascorbic acid or vitamin C is a water-soluble antioxidant that functions mostly as a free radical scavenger. Vitamin C is a potential electron donor, which is capable of regeneration of oxidized vitamin E along with GSH or other components capable of electron donation (Oh et al., 2010). During reaction, vitamin C donates one electron to lipid peroxyl radical to end the chain reaction of lipid peroxidation and itself convert to ascorbate radical. Two such ascorbate radicals together react promptly to regenerate one molecule of ascorbate and one hydroascorbate molecule. Hydroascorbate lacks antioxidant property and thus reverts back to ascorbate by

Table 1.1.2 Sources of nonenzymatic antioxidants, bioavailability, plasma concentration, and chemical structures.

Antioxidants	Sources	Human plasma concentration (per liter)	Bioavailability	Structure
Vitamin C	Fruits and vegetables, especially in strawberries, kiwi, citrus, sprouts, cauliflower, and in other green vegetables	25–80 μM	Though unstable at neutral pH at low doses (<100 mg): 100% decreasing to 15% (>10g)	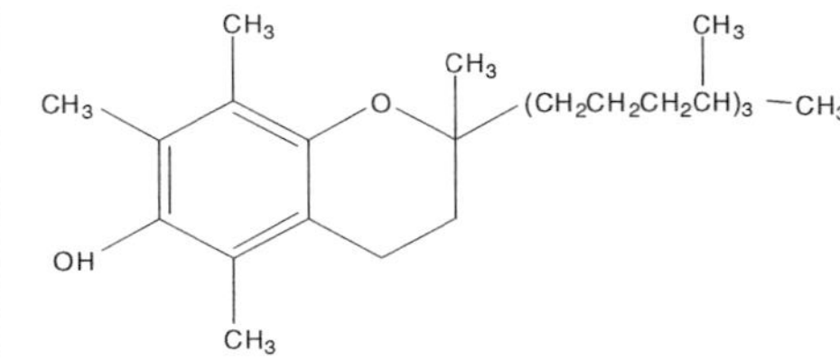 i. Vitamin C
Vitamin A	Liver and fish oils are best sources of preformed vitamin A; apart from these, eggs and milk also contain preformed vitamin A as well as provitamin A. Provitamin A can be widely found in green leafy vegetables like broccoli and spinach, orange and yellow vegetables like carrots, tomatoes, fruits, and vegetable oils.	2.0–4.0 μM at fast	Retinol obtained from provitamin A carotenoids via oxidative cleavage in intestine are absorbed with high efficacy, as is the nature of fat-soluble vitamins getting easily absorbed in aqueous formulations. Retinol is stored in the liver (70% of total retinol in body); released from the liver into the bloodstream.	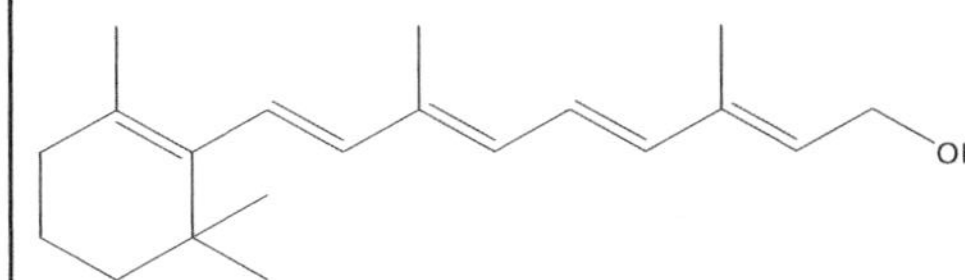 ii. Vitamin A (Retinol)

(continued)

Table 1.1.2 Sources of nonenzymatic antioxidants, bioavailability, plasma concentration, and chemical structures. *Continued*

Antioxidants	Sources	Human plasma concentration (per liter)	Bioavailability	Structure
Vitamin E	Green leafy vegetables, particularly in spinach; seeds, nuts, such as in wheat germ oil, vegetable oil, specially sunflower oil	15–40 μM	Hepatic uptake is preferential process; bioavailability 15%–95%. Though γ-isoform is chiefly present in diet, liver uptake of α-isoform occurs most preferentially	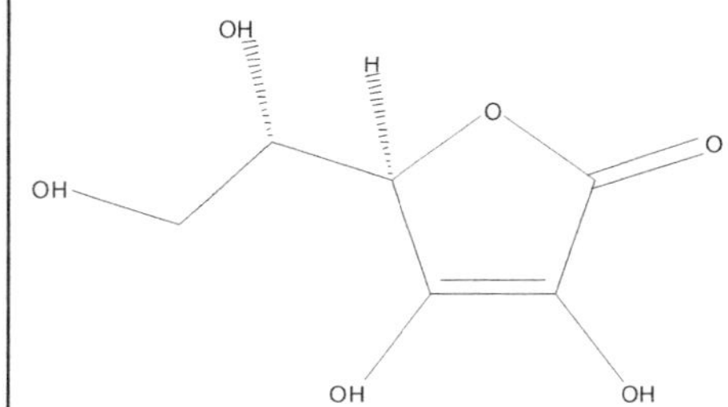 iii. Vitamin E
Carotenoids	Fruits and vegetables having orange or red color, such as carrot, tomato, broccoli, melon, green leafy vegetables, etc.	< 1 μM	Dependent upon the dose and forms; bioavailability <15%. Zeaxanthin and leutin are concentrated at macula layer of eye	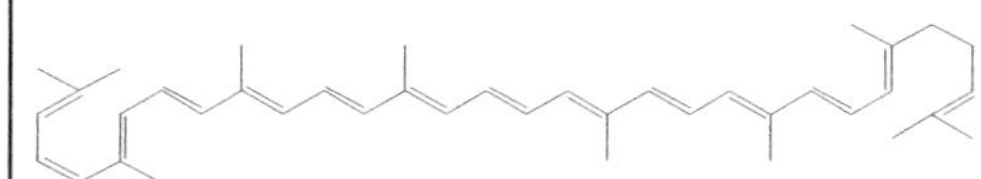 iv. Lycopene 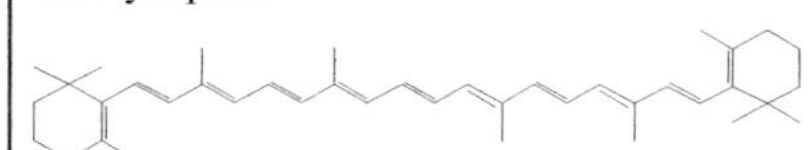 v. B-carotene
Flavonoids	Berries, cherries, citrus fruits, apples, tea, onion, herbs (thyme, parsley)	Never above 3 μM	Dose and form dependent; quercetin shows 20%–50%, whereas for catechin bioavailability is <2%	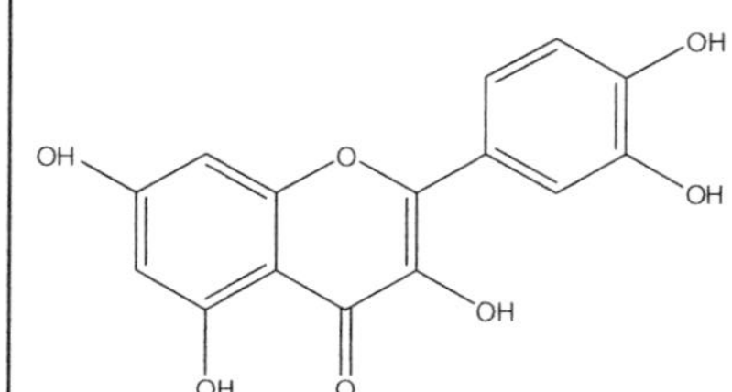 vi. Quercetin

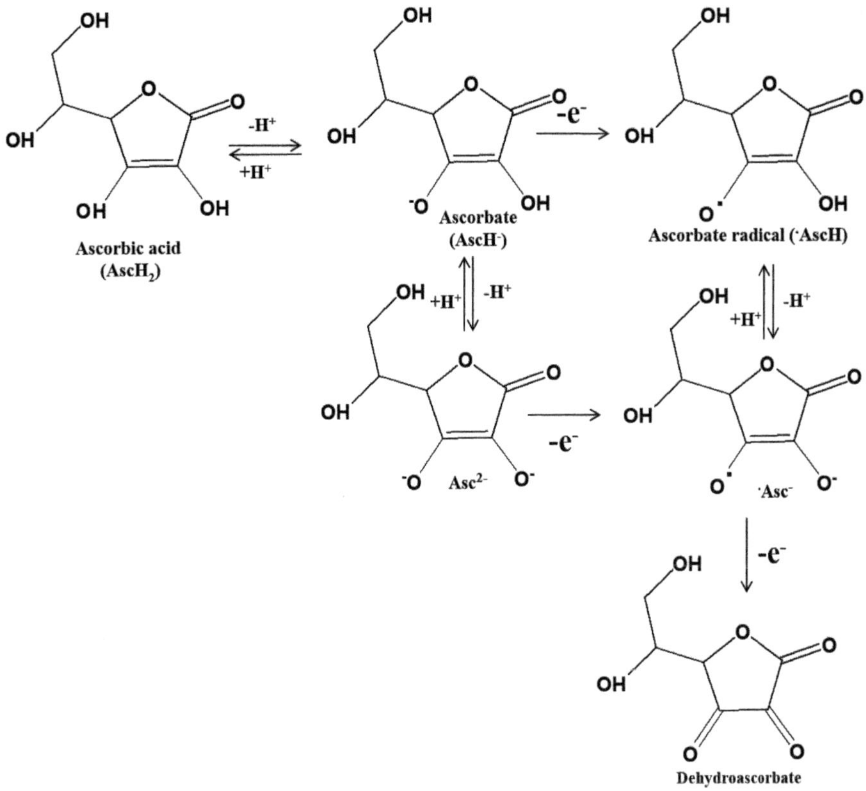

FIG. 1.1.2 Radical scavenging mechanism of ascorbic acid (vitamin C).

Ascorbic acid changes to dehydroascorbate; along the way it donates two protons to reactive oxygen species, and other reactive oxygen radicals, thus neutralizing them.

receiving two electrons (Fig. 1.1.2). Vitamin C is presently medically recommended for the treatment and prevention of the disease scurvy. It is parenterally administered to patients with an acute deficiency of it, or for those patients who face uncertainties or problems in absorbing orally ingested ascorbic acid (vitamin C). It is also prescribed to patients for enhanced wound healing. A daily dose of 70 to 150 mg daily for adults is the average protective dose of vitamin C, but in the presence of scurvy, patients are recommended doses of 300 mg to 1 g daily. However, parenteral administration of as much as 6 g has been given to normal adults without evidence of significant toxicity.

Vitamin A was first identified as a fat-soluble fraction (McCollum and Davis, 1913) which prevented rancidity of lipids (Monaghan and Schmitt, 1932). Some carotenoids, like β and α-carotene, etc., do not only act as antioxidants themselves, but also undergo hydrolysis to retinal (Krinsky et al., 1997; Olson, 1989), which is the visually active form of Vitamin A (Moore, 1930). Thus, they are also classified as

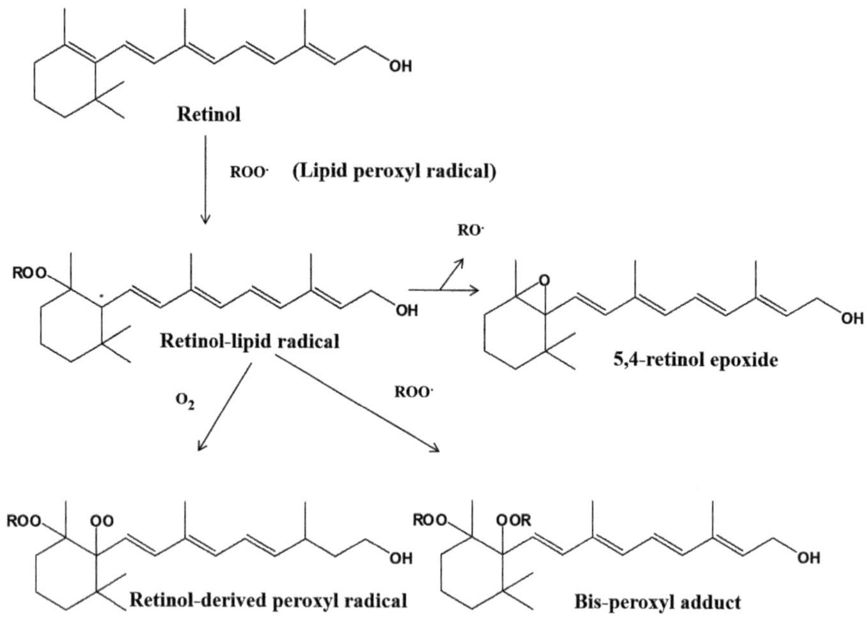

FIG. 1.1.3 Radical scavenging mechanism of vitamin A.

Retinol halts the lipid peroxidation chain by reacting with the lipid peroxyl radical, and ultimately forming a stable adduct.

provitamin A. The physiologically active forms of Vitamin A are retinol, retinal, and retinoic acid; their antioxidant activities could be compared as retinol ≥ retinal >> retinoic acid (Das, 1989). Vitamin A cannot be synthesized in the body, so it must be acquired from outside by consumption of food containing vitamin A and provitamin A (Palace et al., 1999). Vitamin A acts as an antioxidant by acting as a chain breaker via combination with peroxyl radicals, thus preventing the cell's lipid phase peroxidation which would eventually generate hydroperoxides (Tesoriere et al., 1993) (Fig. 1.1.3).

Medically, it is prescribed to treat vitamin A deficiency, which could result in xerophthalmia, immune weakness, pregnancy complications, etc. It is also recommended for treating acne and other skin infections. It is available over the counter as a nutritional supplement in the form of capsules, but when high doses are needed, it is injected. The tolerable upper intake level of vitamin A for men and women is 3 mg a day. Vitamin A causes substantial toxicity when taken in high doses, including bone loss and increased risk of osteoporosis, dry and cracked skin, severe headaches, etc.

Vitamin K is a well-known biological ubiquinone. The oxidized form of ubiquinone, ubiquinol, has been extensively studied in recent years and has been found to be effective in inhibiting lipid peroxidation in the biological systems. Suggestively, ubiquinol (QH) functions as a potent antioxidant in two ways – one, by scavenging the lipid peroxyl radical (LOO·) and two, by regenerating the tocopheroxyl radical to tocopherol (Kagan et al., 1990; Landi et al., 1984; Mukai et al., 1988; Scarpa et al., 1984; Stocker et al., 1991; Takayanagi et al., 1980). The biological activities of the quinones, mostly vitamin K hydroquinone and plastoquinone, have been recently started to be unfolded. The common function of these biological quinones is acting as redox components of the extensive transmembrane ETS inside the cells. The vitamin K cycle in particular has been shown to exhibit potent antioxidant effects. Vitamin K firstly aids in converting glutamic acid to γ-carboxy glutamic acid, by acting as the cofactor. Next, this vitamin K is reduced to vitamin K hydroquinone. This vitamin K hydroquinone gets oxidized to vitamin K epoxide and subsequently provides the energy necessary for the carboxylation of glutamic acid. Again, vitamin K epoxide gets reduced, thereby regenerating vitamin K, and the cycle continues (Vervoort et al., 1997). Vitamin K hydroquinone thus, is a potential radical scavenger (Mukai et al., 1993). Vitamin K has till now been prescribed for patients with problems in blood clotting, as the K vitamins are all established bioactive compounds affecting thrombogenesis positively, and its dosage depends upon the type of K vitamin being prescribed. For Vitamin K4 (or phylloquinone or phytomenadione or phytonadione) it is 2.5 mg to 25 mg orally for teenagers and adults, and 0.5 mg to 1 mg intramuscular or subcutaneous injection in newborns. For menadiol (a derivative of p-hydroquinone, it is an intermediate in the synthesis of vitamin K4. It is oxidized to menadione, i.e., vitamin K3), the oral dose for children as well as adults is 5 mg to 10 mg a day. The injectable dose for the same is 5 mg to 15 mg, once or twice per day in adults and teenagers, injected subcutaneously or intramuscularly, and 5 mg to 10 mg in children.

1.1.3.2.2 ***Physiological antioxidants***

This category includes Uric acid, GSH etc. Uric acid is the most potent antioxidant found in human beings. In hydrophilic environment it acts as the scavenger of peroxyl and carbon-centred radicals. Moreover, it is an exceptional scavenger of peroxynitrite radicals ($ONOO^-$) in association with ascorbic acid and thiols (Squadrito et al., 2000).

Glutathione or GSH is a cysteine containing peptide (Meister and Anderson, 1983), which is highly abundant in most of the aerobic cell compartments. GSH possesses the antioxidant capacity as it owns thiol group in its cysteine moiety that undergoes reversible oxidation and reduction (Masella et al., 2005). GSH donates one electron for detoxification of peroxynitrite radicals and lipid peroxides. Oxidized GSH or GSSG reverts back to its reduced form by glutathione peroxidase enzyme using NAD(P)H as the electron donor. Besides, it also plays important role in conversion of vitamin C and vitamin E into their active forms.

Another notable physiological antioxidant is melatonin, chemically denoted as N-acetyl-5-methoxytryptamine (Nassar et al., 2007). Melatonin is a hormone, found in living organisms, including several algae (Caniato et al., 2003). It is a powerful antioxidant as it can scavenge several free radicals and pass across the cell membrane and blood-brain barrier readily (Reiter et al., 1997). However, unlike other conventional antioxidants, it cannot go through repeated rounds of oxidation-reduction. Hence, it is designated as the terminal or suicidal antioxidant.

1.1.3.2.3 ***Carotenoids***

Carotenoids are one of the most common lipophilic phytonutrients. Carotenoids function as antioxidants in low partial pressures of oxygen, but in high oxygen partial pressures they act as pro-oxidant molecules (El-Agamey, 2004). This group is well exemplified by antioxidants like β-carotene, lycopene etc. Lycopene, the most commonly found antioxidant in fruits and vegetables, is a powerful singlet oxygen quencher due to possession of high number of conjugated double bonds. Coming next to β-carotene, another potential antioxidant, which is able to inhibit the oxidant-induced NF-kB activation and interleukin (IL)-6 and tumor necrosis factor-α (TNF-α) production. Though β-carotene is able to react with hydroxyl ($OH^{\cdot}$) as well as superoxide radical ($O_2^{\cdot -}$), it is most potent to scavenge peroxyl radicals generated from lipid peroxidation (Rice-Evans, 1997). β-carotene deactivates the peroxyl radicals to form resonance stabilized carbon-centered radical adducts.

1.1.3.3 Other natural antioxidants

1.1.3.3.1 ***Bioflavonoids***

Bioflavonoids are a class of natural benzo-γ-pyran derivatives that are widely distributed in fruits and vegetables. It has been found that bioflavonoids exert protective effects on DNA damage caused by hydroxyl radicals with the involvement of chelating ions, such as iron or copper. Bioflavonoids include flavonol (e.g., quercetin, myricetin), isoflavone (e.g., genistein, daidzein), flavan-3-ols (e.g., catechin, epigallocatechin), flavonone (e.g., naringenin, hisperidin), anthocyanidin (e.g., cynidin), etc., all of which are known to have strong antioxidant activities.

Quercetin has been reported to protect DNA from the oxidative damage conferred by $OH^{\cdot}$, $O_2^{\cdot}$, and H_2O_2 (Krishnamachari, 2002). On the contrary, quercetin has also been shown to exert opposite reaction on DNA damage depending on concentration of cupric ions. Thus, although at low concentrations (<25 mM) of cupric ion, quercetin shows protective nature, at higher concentrations, quercetin exerts damage to DNA by ROS (Galaris and Evangelou, 2002; Jun, 2007).

1.1.3.3.2 ***Anthocynidines***

Anthocynidines are also potential antioxidants, which inhibits lipid oxidation by their free radical scavenging activity and ion-chelating activity (Van Acker, 1996). Cynidin, a well-known anthocynidin, can donate one electron to generate a free radical from –OH groups attached to phenolic rings which ultimately stabilizes the free radical. During this process, reducing polyphenolic agents change to a comparatively stable aroxyl radical.

1.1.3.3.3 ***Hydroxycinnamates***

Hydroxycinnamates are another widely accepted group of antioxidants. This group includes compounds such as ferulic acid, sinapic acid, coumaric acid etc. This group effectively prevents oxidative damage of LPL (Meyer, 1998). Antioxidant capacity of such compounds is clearly due the hydroxylation and methylation of aromatic rings in their structure; for which they are capable of hydroxyl electron donation and formation of resonance-stabilized antioxidant radicals.

Apart from aforesaid antioxidant groups, there are also a number of compounds that are reported to be potent antioxidants. One of them is curcumin. Lipid soluble curcumin is an excellent free radical scavenger. It has the chain breaking antioxidant ability comparable to that of vitamin E. The free radical scavenging nature of curcumin is due to the phenolic OH group and the CH_2 group of the β-diketone moiety (Salem, 2014). However, solely phenolic OH has been credited for its antioxidant capacity by pulse radiolysis and other biochemical methods. It has been also reported that copper complex of curcumin has excellent SOD activity. During the reaction with cucurmin-Cu^{2+} complex, most of the superoxide radicals reacts with copper rather than curcumin. This results in the reduction of Cu^{2+} to Cu^{+}. Cu^{+} reacts with another superoxide radical that finally results in the regeneration of the parent complex (Barik, 2005).

1.1.4 Current research focus and trends

Once the antioxidant research attained full swing, pharmaceutical industries worldwide initiated programs under natural product discovery. It focused on a wide range of drug classes ranging from antibacterial and antifungal to infectious and communicable diseases. This worldwide spurge led to the development of numerous lead compounds for a large array of diseases, from microbial infections to cancer, and from metabolic disorders to graft rejection (Baker, 2007; Ojima, 2008). However, these natural product discovery programs reached their greatest momentum with the advent of automated techniques of high throughput screening, which led to creation of huge libraries of lead molecules based on combinatorial chemistry. Many natural molecules were rediscovered by screening of extracts. Nowadays, this classical natural product chemistry has been replaced with more specific molecular target-based drug discovery, by utilization of those lead molecule combinatorial libraries which are purported to obtain effective molecular hits inside the body system (Dias, 2012).

Combinatorial chemistry has been revolutionizing the field of natural product drug discovery and antioxidant research by providing active leads for synthesis of structural analogues. After stacking of the libraries in this manner by synthetic compounds, when the synthetic chemists realized that the libraries lacked the intricate complexity of the natural products found in the environmental network, they adopted the concept of diversity-oriented synthesis, where they started synthesizing compounds that resembled or mimicked the natural products themselves (Newman, 2008). These are the compounds which are currently under research and are undergoing extensive biological screening for being established as novel drug moieties.

In this whole process of isolating and identifying natural products as drug leads, there arise numerous possibilities of isolation of known compounds, which are known to be responsible for the activity of a particular natural extract (a process termed "dereplication") (Sarker, 2006). But there now exist numerous advanced methods that can discriminate the novel compounds from the known ones, and at a very early stage. Isolation of novel natural compounds was very frequent in the 1970s, but the rate is declining now; although, the natural sources are still considered as limitless sources for novel chemicals which could act as drug leads (Ramakrishna, 1999).

1.1.5 Current global market

Antioxidants have also been playing a significant role in the global economy. The FDA approved antioxidants (Table 1.1.3), as well as unregistered antioxidants, have acquired a significant market share in recent times. Based on market trends from the past 5 years, it could be seen that antioxidants have been growing in popularity as they are widely accepted and used for improvement in health and well-being. Apart from this, increased use of anti-aging serums, the inclination of the consumers toward better and organic personal care products, and growing demand for food preservatives as well as fuel additives (used in the plastic and rubber industry) have provided impetus to its growth. So mainly, the changes in consumption patterns as well as changing lifestyle patterns in the modern health-conscious and result-

Table 1.1.3 List of antioxidants registered with the American FDA.

List of status of food additives (antioxidants) as licensed by the American FDA (as on 24th October 2019)

Antioxidant name	Uses listed by the US FDA
Anoxomer	Food additive; nondigestible polymeric antioxidant
Butylated hydroxyanisole (BHA)	Food additive, preservative, and industrial antioxidant
Butylated hydroxytoluene (BHT)	Food additive and industrial antioxidant
Tert-butylhydroquinone (TBHQ)	Unsaturated vegetable oils and edible animal fats preservative; often used in combination with BHA.
2,4,5-trihydroxybutyrophenone (TBHP)	Potent antioxidant; often used in combination with other antioxidants
Dilauryl thiodipropionate	Secondary industrial antioxidant
Ethoxyquin	Food preservative
Propyl gallate	Food preservative against rancidity of oils
Thiodipropionic acid	Antioxidant preservative in cosmetics and personal care products

List of antioxidants deemed unfit for consumption by American FDA

Antioxidant name	Current status
Nordihydroguaiaretic acid (NDGA)	Illegal; not for use in foods
Thiourea	Banned; use illegal

driven society have led to the overgrowth of this section of the market. Globally, the natural antioxidants market is currently segmented into North America, Europe, Asia Pacific, and RoW, that is, rest of the world. Statistical reports from various agencies suggest that the growth outlook for antioxidants is expected to be highest in Asia Pacific, followed by North America and Europe. In the USA, the growth of natural antioxidants is being driven by a growing inclination toward functional food and beverages. Emerging economies developing at a high pace, particularly China, India, and Australia, have created a high demand for antioxidants; and hence, this particular sector has seen remarkable growth, more so with the advent of various start-ups and setting up of other small companies particularly in the Asia-Pacific region, and is expected to be steady (over the forecast period 2017–23). Mostly, the massive expansion of pharmaceutical as well as food industry in this region has been the supportive factor behind the growth of the natural antioxidant market in this region. On top of that, China is the prime producer as well as consumer of vitamin C, that is, ascorbic acid, which has also substantially contributed to the positive growth of the market. Even with the danger lurking because of the presence of cheaper substitutes of natural antioxidants which might hamper their market expansion, all the positive factors direct to a compound annual growth rate of 9.8% for the natural antioxidants in the Asia-Pacific market during the said forecast period.

Categorically, the natural antioxidants could be classified according to their application as, fats and oils, frozen desserts and dairy, confectionery and bakery, sweet and savory snacks, sports nutrition and beverages, meat products and other processed foods, etc. Out of these, the meat product category has been dominating the market as these are easily degradable, and hence need a generous helping of preservatives for processing and storage. Although, market analysis suggests that in the near future, this spot is likely to be occupied by the sports nutrition segment, followed by substantial growth in the dairy and frozen desserts market. Apart from this, based on their source, the natural antioxidant market could be classified into vitamin C, vitamin E, carotenoids, polyphenols, and others. Market surveys have revealed that among all these, vitamin C is dominating the market, closely followed by vitamin E. The domination of vitamin C in the market comes from its easy availability besides its extensive application in a wide range of industries and arguable beneficial effect on patients suffering from deadly diseases like cancer, etc. However, another class of natural antioxidants is projected to take a substantial dive to grow over the said forecast period, owing to its application in extensive research and healthcare. These are the polyphenols like curcumin, quercetin, resveratrol, etc.

As antioxidants are predominantly used as food preservatives, hence, with the processed food market conquering the industry, the antioxidant sector saw a concomitant growth. This is particularly so in the case of frozen and processed meat, as antioxidants are the preferred choice for preservation and as additives for the manufacturers. However, with the increased awareness about health and well-being, customers are gradually shifting to natural antioxidants from synthetic ones. For example, rosemary extracts are in heavy use nowadays for their ability to efficiently prevent fat oxidation processes. This growth is expected to remain steady amidst increased concerns about health and food safety, as, natural antioxidants are considered to be comparatively safe and devoid of any side-effects.

Apart from the increasing awareness regarding derogatory effects of eating processed food containing artificial additives which could lead to chronic and sometimes terminal diseases, consumers are also choosing natural antioxidants for their another important property. This is their antiaging attribute – with the advent of recurring proofs that consumption of antioxidants can slow down the aging process, a plethora of cosmetic products have been launched that cites the presence of natural antioxidants as their unique selling point, and hence, the global antioxidant market is expected to be driven by increased demand. Also, the increasing occurrence of chronic and lethal diseases like amyotrophic lateral sclerosis, diabetes, neurological disorders, and cancer are intensifying the demand for natural antioxidants in consumable products. In India, consumers are gradually getting inclined toward organic personal care products and this has been providing a boost towards the application of natural antioxidants in the personal care industry as well.

Manufacturing synthetic antioxidants, as well as their availability, are easy. These are the main factors that support their growth. The natural antioxidant segment is also expected to see steady growth, but it will be slower in comparison due to the higher extraction and manufacturing costs. As discussed above, a steadily increasing interest in natural ingredients and green food would act to stimulate its demand.

In the wake of such a spurt in the antioxidant market, consumers find it difficult to look out for authentic antioxidant supplements, amidst a heap of companies marketing their products under various brands, which are sometimes not even registered with the FDA. As recently as in 2019, the FDA of the USA in the far west as well as the Philippines in the far east, have released multiple circulars warning people to refrain from buying specific products which are not registered with the FDA. Interestingly, all these products claim to have multiple benefits as they are seemingly constituted of a mix-up of most of the established beneficial dietary antioxidants. Consumers are easily fooled by the sheer marketing strategies employed by these companies amidst the rush to secure a healthy lifestyle composed of the right kind of nutritious food. The FDA of various countries under such dire circumstances has created an exclusive search tool on their websites for the consumers to be aware of the fact whether the antioxidant supplement they are buying has been approved and licensed by the authorities, or they are simply being duped. As recently as in late 2019, the FDA in the Philippines has blacklisted quite a few brands selling "high power antioxidants" as food supplements in the promise of a healthy life. Some of those being Nature's Plus Animal Parade Kids Immune Booster Children's Chewable Antioxidant Supplement With Whole Food Concentrate, ABW Antioxidant Supplement Alfalfa Barley Wheatgrass, and Dok Apo Ax5 Antioxidant Times Five. If stringent measures are not taken as such, the unbridled sale of these so-called food supplements would have dire consequences on the consumer population.

While this might be the story of registered and nonregistered antioxidants available in marketable packaging, current research focused on extracts and natural ingredients of easily available fruits and vegetables has led to a spurt of the fruit and vegetable market, cashing specifically on the fruits and vegetables containing established active ingredients, like guava for quercetin and pomegranate for a strong mixture of powerful antioxidants. This is an indirect market of antioxidants, where the pharmaceutical sector doesn't gain any revenue from, but small industries and agricultural farmers benefit heavily. Consumers are now more driven to buy and eat healthy to live a longer and more fulfilling life. In the process, they contribute to the development of the unorganized antioxidant industry. To give an overview of some of the registered and once registered but now illegal antioxidants in the US market, a table has been attached herewith.

1.1.6 Economic burden and cost benefit of antioxidants

According to Drummond et al., cost-effectiveness analysis (CEA) is of three types – cost minimization, cost benefit, and cost effectiveness, where cost benefit approach is defined as one in which the costs and benefits are measured in terms of dollar, and all those treatments are selected for which the net benefit is positive, that is, greater than zero. This approach needs to place a dollar value on outcomes of public health (Drummond, 2015).

Cost-benefit analysis or CBA has been historically the most widely-used approach to evaluate economic impacts of the major health and safety regulations in the U.S. As of 2010, the U.S. Office of Dietary Supplements (ODS) charted out in its strategic plan of 2010–14 that nearly half of the total American population takes at least one dietary supplement on a daily basis and approximately one-third of them buy dietary supplements including multivitamins/minerals (MVMs) worth about $25 billion each year. Further, Huang et al. noted that in addition to the above statistics, a huge number of people, that is, nearly 65% of the U.S. population also uses fortified foods/beverages amounting to a heaping $36 billion annually (Office of Dietary Supplements, 2010) (Huang, 2006). Studies by Tice et al. noted that folic acid supplementation alongwith cyanocobalamin in older men (>45 years) and women (>55 years) would aid in the prevention of 300,000 deaths over a period of 10 years, which would eventually end up saving nearly $2 billion (Tice, 2001). Their updated study by Bentley et al. found that fortification of folic acid would reduce incidences of myocardial infarction, neural tube defects, and colon cancer. A 700 μg folic acid fortification would result in 266,649 quality-adjusted life-year (QALY) increase and an eventual cost decrease of $3.6 billion. The same study however, noted that the said fortification would in turn increase the risk of masking or precipitating vitamin B_{12} deficiency (Bentley, 2009). Overall, each year, the Americans spend about $61 billion on multivitamins/minerals and fortification; although, fact remains that significant evidence is lacking for drawing any firm conclusion about whether MVM use really effects disease prevention (Wong, 2011). Economic studies all over the world provide a similar trend, concominant with the increase in antioxidant research as well as the growth of antioxidant market at a very high rate. Hence, until now, it could be considered that the cost-benefit analysis tilts more on the benefit side. However, more research is required to delve deeper into the hidden costs and collateral damage incurred in the process.

1.1.7 Adverse effect of antioxidants

As discussed above, antioxidants potentially scavenge ROS to decrease oxidative stress. Thus, antioxidant supplementation makes sense as unhealthy lifestyle encourages oxidative-stress condition. Moreover, obesity results in inflammatory response with high level of ROS (Jia, 2018); and exposure to different toxicants such as smoke, environmental pollutants, alcohol also does the same (Mathur and D'cruz, 2011; Sharma, 2013). In such a scenario, antioxidants are highly advertised; at the same time, there is regulation available too. However, there is a possibility that patients can unintentionally take very high doses of these antioxidants which can evoke detrimental effects presumably because of an imbalance between oxidative and reductive stress of body. For instance, use of excessive vitamin A supplementation in smokers can increase risk of cancer (Group, 1994). Even high concentration of vitamin C can impose oxidative stress in turn and can cause DNA damage (Aruoma, 1991; Fraga, 1991). A condition, where the redox equilibrium shifts toward reductive side, is known as "reductive stress" (Wendel, 1987). Similar to oxidative stress, reductive stress can be also harmful for health. High level of antioxidants prevents sperms from fertilizing oocytes (as high level of ROS is required for capacitation and acrosomal reaction for mammalian fertilization), resulting in male infertility (De Lamirande and Gagnon, 1995; Henkel, 2011; Kothari, 2010). From developmental arena, high level of antioxidants is prone to cause teratogenic effects (Wang and Rogers, 2007). Thus, maintenance of proper redox balance is of utmost importance. However, the problem persists because we do not know what the body's antioxidant tolerance level is, and where the actual equilibrium lies. Therefore, excessive intake of antioxidants can worsen redox balance. In this respect, next paragraphs describe harmful effect of both dietary antioxidants and antioxidant supplements.

a. **Adverse effects of dietary antioxidants**
   Fruits and vegetables containing high concentration of tannin, oxalic acid and phytic acid may have antinutrient effect. Inside the gut such strong reducing acids can bind to dietary minerals and prevent their proper absorption.
   Tannin, a flavonoid family antioxidant, includes both proanthrocyanidins and processed tannins (Beecher, 2003). Proanthrocyanidins are widely available in apples, berries, nuts, chocolates, and red wine, whereas black and oolong tea, coffee, and red wine are rich in processed tannins. In herbivores, condensed tannins prevent protein digestion by interferring the activity of digestive enzymes and protein absorption across the gut. It is reported that high intake of tannin may also cause bowel irritation, liver damage as well as kidney irritation (Brune, 1989).

Oxalic acid is another such strong antioxidant acid which is found highly concentrated in spinach and also occurs in sweet potatoes and peanuts (Mosha, 1995). Consumption of oxalate at a high concentration often results in a deficit in calcium absorption (Kelsay, 1985). Oxalic acid reacts with dietary calcium to form insoluble calcium oxalate and thus, high level of oxalate imposes health risks for growing infants and metabolically disposed adults.

In developing countries, iron deficiency is widespread in infants and young children where vegetable protein sources are often associated with cereals. This is partly due to high concentration of phytic acid in different kinds of legumes and cereals. Phytic acid can be found in legumes such as soybean, black bean, moong bean, lentils and also in chickpeas. Being a strong inhibitor of iron absorption (Hallberg, 1989), phytic acid can promote iron deficiency causing mental and psychomotor underdevelopment. Reducing phytic acid consumption by 90% (approximately 100 mg/100 g dried product) seems to enhance iron absorption about twofold. Hence, for at-risk population, complete enzymatic degradation of phytic acid by several cooking procedures such as blanching has been recommended (Hurrell, 2003; Mosha, 1995).

In this context, it must be noted that high dose of polar antioxidants is less harmful than that of the nonpolar antioxidants. For example, eugenol, a nonpolar antioxidant, present in clove oil is more toxic than ascorbate which can be excreted through urination due to its water solubility.

b. Adverse effects of antioxidant supplementation

For normal functioning of body, a daily dose of vitamins, minerals as well as different antioxidants are required in diet. However, during supplementation, their exact dosing, proper requirement and associated risks remain largely unknown. Most of us unconsciously intake them in far high doses than their actual recommended daily allowance (RDA). To investigate potential risk factors several studies were made which proves that instead of being helpful, unnecessary supplementation does more harm. Table 1.1.4 enlists reported side effects of antioxidant supplementation.

β-Carotene and Retinol Efficacy Trial (CARET) was carried out in 18,314 men and women at high risk of developing lung cancer (Omenn, 1996). This study was based on the finding that people with high concentration of serum β-carotene had lower risk of developing lung cancer. However, in this study, the smokers who were recommended to take a combination of 30 mg β-carotene and 25,000 IU retinyl palmitate (vitamin A) daily had 28% more lung cancer and 17% more deaths than placebo subjects.

In another cancer prevention study, the effect of α-tocopherol (vitamin E) and β-carotene was made among 29,133 smokers of age 50–69 from south-west Finland (Group, 1994). Individuals were randomly assigned to one of four groups: α-tocopherol (50 mg/day) alone, β-carotene (20 mg/ day) alone, both α-tocopherol and β-carotene, or placebo; followed for 5–8 years. However, there was no reduction in the occurrence of lung cancer in the group, treated with vitamin E alone. Moreover, an increased incidence of lung cancer and

Table 1.1.4 Side effects of antioxidant supplementation.

Antioxidants	Recommended daily allowance (RDA)	Acute side effects	Chronic side effects
β-Carotene	15–30 mg/day	Change in skin coloration	Increased risk of certain types of cancer
Ascorbic acid (vitamin C)	75–90 mg/day	Diarrhea	Iron overload, hyperoxaluria, renal stone formation
α-Tocopherol (vitamin E)	22.4 IU/day	Fatigue, headache, muscle weakness, creatinuria	Increased risk of hemorrhage, impaired bone mineralization, cardiovascular disorder
Zinc	8–11 mg/day	Intranasal anosmia, gastrointestinal disturbances	Copper deficiency, anemia, suppression of immune responses, and increased risk of prostate cancer
Selenium	55 μg/day	Skin rashes, irritability, gastrointestinal disturbances, fatigue	Hair loss, nail disorder, neuropathy, increased risk of diabetes
Glutathione	250 mg/day or 600 mg IM QOD for male infertility	Gastrointestinal disturbances	-
Ubiquinol (coenzyme Q10)	60–90 mg/day	Bowel discomfort, gastrointestinal disorders, heart burn	Hemorrhagic toxicity
Melatonin	10 mg/day at bedtime	Dizziness, headache, nausea, diarrhea, heartburn, skin rash	Disturbance in sleeping cycle

hemorrhagic stroke was reported compared to placebo. Besides, the group treated with β-carotene alone displayed 18% higher incidence of lung cancer as well as death due to ischemic heart disease.

Similar observation was made from Bjelakovic's meta-analysis from 2007 which included 68 randomized trials with 2,32,606 participants. It was concluded that the treatment with β-carotene, vitamin A, and vitamin E may increase all-cause mortality. However, the potential roles of selenium and ascorbate on mortality were recommended for further inspection. Later these results were confirmed by the same researchers with the help of cochrane collaboration methodology (Bjelakovic, 2007). The key findings were: (1) β-carotene, vitamin A, and vitamin E supplementation either alone or in combination with other antioxidants significantly increase mortality, (2) selenium showed a tendency to reduce mortality, and (3) there was no evidence of the effect of vitamin C toward increasing longevity (Bjelakovic, 2006; Gluud, 2008; Moher, 1998; Schulz, 1995).

Still, the mechanism of the possible disadvantage of antioxidant supplementation is poorly understood. First of all, it is well established that oxidative stresses are a part of the pathogenesis of different chronic diseases; however, could the oxidative stress be the cause of the chronic diseases? (Halliwell, 1994) Secondly, some essential cellular defense mechanisms including detoxification, phagocytosis, apoptosis etc. require free radical generation. If impaired, cellular homeostasis may be disturbed (Kimura, 2005; Salganik, 2001; Simon, 2000). And thirdly, unlike prescription drugs, antioxidant supplements must go through toxicity studies in order to be sold to consumers (Bast and Haenen, 2002). Thus, a clear understanding of the mechanisms of action of the antioxidants towards particular diseases is highly essential.

Conclusion

Oxygen is a basic requirement for all biological systems, excepting few. But early in biological research, it has come to light how oxygen could be harmful as well. Molecular oxygen, and its set off chain-reactions generate numerous highly reactive radicals, which are fatal in large concentrations. So, very early in history of life on earth, antioxidants have evolved to counter the oxidative damage and prevent any harm due to oxygen or its reactive radicals, collectively termed as reactive oxygen species. Plants were the first to design extensive antioxidant systems in their bodies, by developing highly effective antioxidant compounds like flavonoids, polyphenols, carotenoids, and many more, which prevent the oxidative harm to the plant tissues even after producing large volumes of molecular oxygen as a photosynthetic by-product and leaked radicals during the photophosphorylation reactions. Radicals are effectively neutralized and molecular oxygen is efficiently flushed out. The animals, not very far behind, developed their own antioxidant systems which not only neutralize and flush out the oxidants from the body, but also maintain critical balances between the oxidants and antioxidants in the system. They also derived the plant-sourced antioxidants by consuming vegetation. The oxidants actually prove to be of great help in killing invading pathogens or transformed cells, but a large concentration of these wreak havoc inside the system, often proving fatal, inducing DNA damage, lipid peroxidation, and playing a huge role in senescence and associated complications. So, keeping their levels under control is of utmost importance to living beings.

Humans knew of antioxidants primarily in the industrial sector in the beginning, but with advanced research in biology, the toxic effects of oxidants in the system, and the effect of antioxidant compounds against the oxidative damage came into awareness. The extensive search for antioxidants that followed led to the compilation of a huge library of molecules having antioxidant properties, which was later used by combinatorial chemists to study the synergistic effects of these antioxidants and synthetic chemists to synthesize new chemical derivatives which mimicked the naturally sourced antioxidants.

Antioxidants nowadays have a huge role in our diet and supplementations. They are medically prescribed for proven therapeutic effects. Research is still underway to uncover newer antioxidant molecules from various sources, and their effects are being studied extensively on a wide array of diseases. The future of this research looks promising, as antioxidants have been proven to be effective as therapeutics even in deadly diseases like cancer (Kundu, 2019) and diabetes (Manna, 2009). Concomitant growth has been recorded in the market and sale of various antioxidants, with consumers also being sometimes fooled into buying unregistered antioxidants from the open market. The global population incurs a huge cost in the process, research shows, which eventually benefits them in the long run. Although many loopholes and lack of proper evidence could be found, it needs further research to cement findings if such an investment actually works in the population's favor. The FDA regularly lists unregistered antioxidants available freely in the market, and warns people against buying them. With continuous and combined efforts of scientists working in the medical and economic field, we can surely hope for better answers in the near future.

Conflict of interest

The authors declare no conflicts of interest.

Abbreviations

ROS	Reactive oxygen species
ETS	Electron transport system
DNA	Deoxyribonucleic acid
SOD	Superoxide dismutase
CAT	Catalase
NAD(P)H	Nicotinamide adenine dinucleotide phosphate
iNOS	Inducible nitric oxide synthase
RNS	Reactive nitrogen species
GSH	Glutathione
GSSG	Reduced glutathione
GSH-Px	Glutathione peroxidase
LOO·	Lipid peroxyl radical
$ONOO^-$	Peroxynitrite radicals
OH·	Hydroxyl radical
$O_2^{\cdot -}$	Superoxide radical
UL	Upper intake level
QH	Ubiquinol
H_2O_2	Hydrogen peroxide
TNF-α	Tumor necrosis factor

CARET	β-Carotene and retinol efficacy trial
RDA	Recommended daily allowance

References

Ames, B.N., Shigenaga, M.K., Hagen, T.M., 1993. Oxidants, antioxidants, and the degenerative diseases of aging. Proc. Natl. Acad. Sci. U S A 90, 7915–7922.

Arthur, J.R., 2001. The glutathione peroxidases. Cell. Mol. Life Sci. 57, 1825–1835.

Aruoma, O.I., Halliwell, B., Gajewski, E., Dizdaroglu, M., 1991. Copper-ion-dependent damage to the bases in DNA in the presence of hydrogen peroxide. Biochem. J. 273, 601–604.

Baker, D.D., Chu, M., Oza, U., Rajgarhia, V., 2007. The value of natural products to future pharmaceutical discovery. Nat. Prod. Rep. 24, 1225–1244.

Barik, A., Mishra, B., Shen, L., Mohan, H., Kadam, R., Dutta, S., Zhang, H.-Y., Priyadarsini, K.I., 2005. Evaluation of a new copper (II)–curcumin complex as superoxide dismutase mimic and its free radical reactions. Free Radic. Biol. Med. 39, 811–822.

Bast, A., Haenen, G.R., 2002. The toxicity of antioxidants and their metabolites. Environ. Toxicol. Phar. 11, 251–258.

Beckman, K.B., Ames, B.N., 1998. The free radical theory of aging matures. Physiol. Rev. 78, 547–581.

Beecher, G.R., 2003. Overview of dietary flavonoids: nomenclature, occurrence and intake. J. Nutr. 133, 3248S–3254S.

Bentley, T.G., Weinstein, M.C., Willett, W.C., Kuntz, K.M., 2009. A cost-effectiveness analysis of folic acid fortification policy in the United States. Public Health Nutr. 12 (4), 455–467.

Benzie, I.F., 2000. Evolution of antioxidant defence mechanisms. Eur. J. Nutr. 39, 53–61.

Benzie, I.F., 2003. Evolution of dietary antioxidants. Comp. Biochem. Phys. A 136, 113–126.

Benzie, I.F., Strain, J.J., 1999. [2]Ferric reducing/antioxidant power assay: direct measure of total antioxidant activity of biological fluids and modified version for simultaneous measurement of total antioxidant power and ascorbic acid concentration. Method Enzymol, 299. Academic Press, Cambridge, Massachusetts, pp. 15–27.

Bjelakovic, G., Nagorni, A., Nikolova, D., Simonetti, R., Bjelakovic, M., Gluud, C., 2006. Meta-analysis: antioxidant supplements for primary and secondary prevention of colorectal adenoma. Aliment. Pharmacol. Ther. 24, 281–291.

Bjelakovic, G., Nikolova, D., Gluud, L.L., Simonetti, R.G., Gluud, C., 2007. Mortality in randomized trials of antioxidant supplements for primary and secondary prevention: systematic review and meta-analysis. JAMA 297, 842–857.

Brune, M., Rossander, L., Hallberg, L., 1989. Iron absorption and phenolic compounds: importance of different phenolic structures. Eur. J. Clin. Nutr. 43, 547–557.

Burton, G., Ingold, K., 1981. Autoxidation of biological molecules. 1. Antioxidant activity of vitamin E and related chain-breaking phenolic antioxidants in vitro. J. Am. Chem. Soc. 103, 6472–6477.

Caniato, R., Filippini, R., Piovan, A., Puricelli, L., Borsarini, A., Cappelletti, E.M., 2003. Melatonin in plants. Developments in Tryptophan and Serotonin Metabolism. Springer, Manhattan, New York City, pp. 593–597.

Chu, F.-F., Doroshow, J., Esworthy, R., 1993. Expression, characterization, and tissue distribution of a new cellular selenium-dependent glutathione peroxidase, GSHPx-GI. J. Biol. Chem. 268, 2571–2576.

Comhair, S.A., Bhathena, P.R., Farver, C., Thunnissen, F.B., Erzurum, S.C., 2001. Extracellular glutathione peroxidase induction in asthmatic lungs: evidence for redox regulation of expression in human airway epithelial cells. FASEB J. 15, 70–78.

Corpas, F.J., Barroso, J.B., del Río, L.A., 2001. Peroxisomes as a source of reactive oxygen species and nitric oxide signal molecules in plant cells. Trends Plant Sci. 6, 145–150.

Corpas, F.J., Fernandez-Ocana, A., Carreras, A., Valderrama, R., Luque, F., Esteban, F.J., Rodríguez-Serrano, M., Chaki, M., Pedrajas, J.R., Sandalio, L.M., 2006. The expression of different superoxide dismutase forms is cell-type dependent in olive (Olea europaea L.) leaves. Plant Cell Physiol. 47, 984–994.

Das, N.P., 1989. Effects of vitamin A and its analogs on nonenzymatic lipid peroxidation in rat brain mitochondria. J. Neurochem. 52, 585–588.

De Lamirande, E., Gagnon, C., 1995. Impact of reactive oxygen species on spermatozoa: a balancing act between beneficial and detrimental effects. Mol. Hum. Reprod. 10, 15–21.

Dias, D.A., Urban, S., Roessner, U., 2012. A historical overview of natural products in drug discovery. Metabolites 2, 303–336.

Drummond, M.F., Sculpher, M.J., Claxton, K., Stoddart, G.L., Torrance, G.W., 2015. Methods for the Economic Evaluation of Health Care Programmes. Oxford University Press, Oxford, England.

Dupuy, C., Virion, A., Ohayon, R., Kaniewski, J., Deme, D., Pommier, J., 1991. Mechanism of hydrogen peroxide formation catalyzed by NADPH oxidase in thyroid plasma membrane. J. Biol. Chem. 266, 3739–3743.

El-Agamey, A., Lowe, G.M., McGarvey, D.J., Mortensen, A., Phillip, D.M., Truscott, T.G., Young, A.J., 2004. Carotenoid radical chemistry and antioxidant/pro-oxidant properties. Arch. Biochem. Biophys. 430, 37–48.

Fenton, H.J.H., 1894. LXXIII.—Oxidation of tartaric acid in presence of iron. J. Chem. Soc., Trans. 65, 899–910.

Finkel, T., Holbrook, N.J., 2000. Oxidants, oxidative stress and the biology of ageing. Nature 408, 239.

Fraga, C.G., Motchnik, P.A., Shigenaga, M.K., Helbock, H.J., Jacob, R.A., Ames, B.N., 1991. Ascorbic acid protects against endogenous oxidative DNA damage in human sperm. Proc. Natl. Acad. Sci. U S A 88, 11003–11006.

Fridovich, I., 1974. Superoxide and evolutionHorizons in Biochemistry and Biophysics, 1. Adison-Wesley, pp. 1–37.

Fridovich, I., 1975. Superoxide dismutases. Annu. Rev. Biochem. 44, 147–159.

Fridovich, I., 1998. Oxygen toxicity: a radical explanation. J. Exp. Biol. 201, 1203–1209.

Gaetani, G.F., Ferraris, A., Rolfo, M., Mangerini, R., Arena, S., Kirkman, H., 1996. Predominant role of catalase in the disposal of hydrogen peroxide within human erythrocytes. Blood 87, 1595–1599.

Galaris, D., Evangelou, A., 2002. The role of oxidative stress in mechanisms of metal-induced carcinogenesis. Crit. Rev. Oncol. Hemat. 42, 93–103.

Gaßmann, B., 1991. Handbook of Vitamins. 2. Aufl. Herausgegeben von L.J. Machlin. 595 Seiten, zahlr. Abb. und Tab. Marcel Dekker, Inc., New York, Basel 1991. Preis: 150,- $ (USA und Canada 125,- $; bei Abnahme von mindestens 5 Exemplaren "for classroom use only" 49,75 $). Food /Nahrung 35, 1079.

German, J.B., 1999. Food processing and lipid oxidation. Adv. Exp. Med. Biol. 459, 23–50.

Gerschman, R., Gilbert, D.L., Nye, S.W., Dwyer, P., Fenn, W.O., 1954. Oxygen poisoning and x-irradiation: a mechanism in common. Science 119, 623–626.

Gluud, L.L., Thorlund, K., Gluud, C., Woods, L., Harris, R., Sterne, J.A., 2008. Correction: reported methodologic quality and discrepancies between large and small randomized trials in meta-analyses. Annals of internal medicine, 149(3), p.219.

Graham, J.B., Aguilar, N.M., Dudley, R., Gans, C., 1995. Implications of the late Palaeozoic oxygen pulse for physiology and evolution. Nature 375, 117.

Granger, D.N., 1988. Role of xanthine oxidase and granulocytes in ischemia-reperfusion injury. Am. J. Physiol. Heart C 255, H1269–H1275.

Group, A.-T.B.C.C.P.S., 1994. The effect of vitamin E and beta carotene on the incidence of lung cancer and other cancers in male smokers. N. Engl. J. Med. 330, 1029–1035.

Hallberg, L., Brune, M., Rossander, L., 1989. Iron absorption in man: ascorbic acid and dose-dependent inhibition by phytate. Am. J. Clin. Nutr. 49, 140–144.

Halliwell, B., 1994. Free radicals, antioxidants, and human disease: curiosity, cause, or consequence? Lancet (British edition) 344, 721–724.

Halliwell, B., 1999. Antioxidant defence mechanisms: from the beginning to the end (of the beginning). Free Radic. Res. 31, 261–272.

Halliwell, B., Clement, M.V., Ramalingam, J., Long, L.H., 2000. Hydrogen peroxide. Ubiquitous in cell culture and in vivo? IUBMB Life 50, 251–257.

Henkel, R.R., 2011. Leukocytes and oxidative stress: dilemma for sperm function and male fertility. Asian J. Androl. 13, 43.

Hochachka, P.W., 1998. Mechanism and evolution of hypoxia-tolerance in humans. J. Exp. Biol. 201, 1243–1254.

Hollman, P.C.H., 2001. Evidence for health benefits of plant phenols: local or systemic effects? J. Sci. Food Agric. 81, 842–852.

Huang, H.Y., Caballero, B., Chang, S., Alberg, A.J., Semba, R.D., Schneyer, C.R., Wilson, R.F., Cheng, T.Y., Vassy, J., Prokopowicz, G., Barnes 2nd, G.J., Bass, E.B., 2006. The efficacy and safety of multivitamin and mineral supplement use to prevent cancer and chronic disease in adults: a systematic review for a National Institutes of Health state-of-the-science conference. Ann. Intern. Med. 145 (5), 372–385.

Hurrell, R.F., 2003. Influence of vegetable protein sources on trace element and mineral bioavailability. J. Nutr. 133, 2973S–2977S.

Jacob, R.A., 1996. Introduction: Three Eras of Vitamin C Discovery. Ascorbic Acid: Biochemistry and Biomedical Cell Biology. Springer, Manhattan, New York City.

Jia, Y.-F., Feng, Q., Ge, Z.-Y., Guo, Y., Zhou, F., Zhang, K.-S., Wang, X.-W., Lu, W.-H., Liang, X.-W., Gu, Y.-Q., 2018. Obesity impairs male fertility through long-term effects on spermatogenesis. BMC Urol. 18, 42.

Jun, T., Liancai, Z., Bochu, W., 2007. Effects of quercetin on DNA damage induced by copper ion. Int. J. Pharmacol. 3, 19–26.

Kagan, V.E., Serbinova, E.A., Koynova, G.M., Kitanova, S.A., Tyurin, V.A., Stoytchev, T.S., Quinn, P.J., Packer, L., 1990. Antioxidant action of ubiquinol homologues with different isoprenoid chain length in biomembranes. Free Radic. Biol. Med. 9, 117–126.

Kelsay, J., 1985. Effect of oxalic acid on bioavailability of calciumNutritional Bioavailability of Calcium. American Chemical Society, Washington, DC.

Kimura, H., Sawada, T., Oshima, S., Kozawa, K., Ishioka, T., Kato, M., 2005. Toxicity and roles of reactive oxygen species. Curr. Drug Targets Inflamm. Allergy 4, 489–495.

Klebanoff, S.J., 2005. Myeloperoxidase: friend and foe. J. Leukoc. Biol. 77, 598–625.

Knight, J.A., 1998. Free radicals: their history and current status in aging and disease. Ann. Clin. Lab. Sci. 28, 331–346.

Kothari, S., Thompson, A., Agarwal, A., du Plessis, S.S., 2010. Free radicals: their beneficial and detrimental effects on sperm function. Indian J. Exp. Biol. 48 (5), 425–435.

Krinsky, N., Wang, O.-D., Tang, G., Russell, R., 1997. Mechanism of carotenoid cleavage to retinoids. Scientifur 21, 68.

Krishnamachari, V., Levine, L.H., Paré, P.W., 2002. Flavonoid oxidation by the radical generator AIBN: a unified mechanism for quercetin radical scavenging. J. Agric. Food Chem. 50, 4357–4363.

Kundu, M., Sadhukhan, P., Ghosh, N., Chatterjee, S., Manna, P., Das, J., Sil, P.C., 2019. pH-responsive and targeted delivery of curcumin via phenylboronic acid-functionalized ZnO nanoparticles for breast cancer therapy. J. Adv. Res. 18, 161–172.

Landi, L., Cabrini, L., Sechi, A.M., Pasquali, P., 1984. Antioxidative effect of ubiquinones on mitochondrial membranes. Biochem. J. 222, 463–466.

Liochev, S.I., Fridovich, I., 2002. The Haber-Weiss cycle—70 years later: an alternative view. Redox Rep. 7, 55–57.

Lobo, V., Patil, A., Phatak, A., Chandra, N., 2010. Free radicals, antioxidants and functional foods: Impact on human health. Pharmacogn. Rev. 4, 118–126.

Machlin, L.J., 1991. Handbook of vitamins. New York: M. Dekker, 40.

Manna, P., Sinha, M., Sil, P.C., 2009. Protective role of arjunolic acid in response to streptozotocin-induced type-I diabetes via the mitochondrial dependent and independent pathways. Toxicology 257, 53–63.

Masella, R., Di Benedetto, R., Varì, R., Filesi, C., Giovannini, C., 2005. Novel mechanisms of natural antioxidant compounds in biological systems: involvement of glutathione and glutathione-related enzymes. J. Nutr. Biochem. 16, 577–586.

Mathur, P.P., D'cruz, S.C., 2011. The effect of environmental contaminants on testicular function. Asian J. Androl. 13, 585.

Mattill, H.A., 1947. Antioxidants. Annu. Rev. Biochem. 16, 177–192.

McCollum, E.V., Davis, M., 1913. The necessity of certain lipins in the diet during growth. J. Biol. Chem. 15, 167–175.

McCord, J.M., Fridovich, I., 1969. Superoxide dismutase. An enzymic function for erythrocuprein (hemocuprein). J. Biol. Chem. 244, 6049–6055.

Meister, A., Anderson, M.E., 1983. Glutathione. Annu. Rev. Biochem. 52, 711–760.

Meyer, A.S., Donovan, J.L., Pearson, D.A., Waterhouse, A.L., Frankel, E.N., 1998. Fruit hydroxycinnamic acids inhibit human low-density lipoprotein oxidation in vitro. J. Agric. Food Chem. 46, 1783–1787.

Miller, D.M., Buettner, G.R., Aust, S.D., 1990. Transition metals as catalysts of "autoxidation" reactions. Free Rad. Biol. Med. 8, 95–108.

Moher, D., Jones, A., Cook, D.J., Jadad, A.R., Moher, M., Tugwell, P., Klassen, T.P., 1998. Does quality of reports of randomised trials affect estimates of intervention efficacy reported in meta-analyses? Lancet North Am. Ed. 352, 609–613.

Monaghan, B.R., Schmitt, F.O., 1932. The effects of carotene and of vitamin A on the oxidation of linoleic acid. J. Biol. Chem. 96, 387–395.

Moore, T., 1930. Vitamin A and carotene: the absence of the liver oil vitamin A from carotene. VI. The conversion of carotene to vitamin A in vivo. Biochem. J. 24, 692.

Mosha, T., Gaga, H., Pace, R., Laswai, H., Mtebe, K., 1995. Effect of blanching on the content of antinutritional factors in selected vegetables. Plant Food Hum. Nutr. 47, 361–367.

Mukai, K., Fukuda, K., Tajima, K., Ishizu, K., 1988. A kinetic study of reactions of tocopherols with a substituted phenoxyl radical. J. Org. Chem. 53, 430–432.

Mukai, K., Morimoto, H., Kikuchi, S., Nagaoka, S.-i., 1993. Kinetic study of free-radical-scavenging action of biological hydroquinones (reduced forms of ubiquinone, vitamin K and tocopherol quinone) in solution. BBA Gen. Subjects 1157, 313–317.

Nassar, E., Mulligan, C., Taylor, L., Kerksick, C., Galbreath, M., Greenwood, M., Kreider, R., Willoughby, D.S., 2007. Effects of a single dose of N-Acetyl-5-methoxytryptamine (Melatonin) and resistance exercise on the growth hormone/IGF-1 axis in young males and females. J. Int. Soc. Sport Nutr. 4, 14.

Newman, D.J., 2008. Natural products as leads to potential drugs: an old process or the new hope for drug discovery? J. Med. Chem. 51, 2589–2599.

Office of Dietary Supplements. Strategic Plan 2010–2014, 2010. Office of Dietary Supplements, National Institutes of Health, U.S. Department of Health & Human Services; Bethesda. NIH Publication Number, pp. 10–7527.

Oh, C.-W., Li, M., Kim, E.-H., Park, J.-S., Lee, J.-C., Ham, S.-W., 2010. Antioxidant and radical scavenging activities of ascorbic acid derivatives conjugated with organogermanium. B. Korean Chem. Soc. 31, 3513–3514.

Ojima, I., 2008. Modern natural products chemistry and drug discovery. J. Med. Chem. 51, 2587–2588.

Olson, J.A., 1989. Provitamin A function of carotenoids: the conversion of β-carotene into vitamin A. Nutr. J. 119, 105–108.

Omenn, G.S., Goodman, G.E., Thornquist, M.D., Balmes, J., Cullen, M.R., Glass, A., Keogh, J.P., Meyskens Jr, F.L., Valanis, B., Williams Jr, J.H., 1996. Risk factors for lung cancer and for intervention effects in CARET, the beta-carotene and retinol efficacy trial. J. Natl. Cancer. I. 88, 1550–1559.

Palace, V.P., Khaper, N., Qin, Q., Singal, P.K., 1999. Antioxidant potentials of vitamin A and carotenoids and their relevance to heart disease. Free Radic. Biol. Med. 26, 746–761.

Ramakrishna, N., Nadkarni, S., Bhat, R., Naker, S., Kumar, E., Lal, B., 1999. Screening of natural product extracts for antibacterial activity: early identification and elimination of known compounds by dereplication. Indian J. Chem., Sect. B 38B (12), 1384–1387.

Reiter, R., Carneiro, R.C., Oh, C.-S., 1997. Melatonin in relation to cellular antioxidative defense mechanisms. Horm. Metab. Res. 29, 363–372.

Rice-Evans, C.A., Sampson, J., Bramley, P.M., Holloway, D.E., 1997. Why do we expect carotenoids to be antioxidants in vivo? Free Rad. Res. 26, 381–398.

Salem, M., Rohani, S., Gillies, E.R., 2014. Curcumin, a promising anti-cancer therapeutic: a review of its chemical properties, bioactivity and approaches to cancer cell delivery. RSC Adv. 4, 10815–10829.

Salganik, R.I., 2001. The benefits and hazards of antioxidants: controlling apoptosis and other protective mechanisms in cancer patients and the human population. J. Am. Coll. Nutr. 20, 464S–472S.

Sarker, S.D., Latif, Z., Gray, A.I., 2006. Natural Product Isolation. In: Sarker, S.D., Latif, Z., Gray, A.I. (Eds.), Natural Products Isolation. Humana Press, Totowa, NJ, pp. 1–25.

Scarpa, M., Rigo, A., Maiorino, M., Ursini, F., Gregolin, C., 1984. Formation of α-tocopherol radical and recycling of α-tocopherol by ascorbate during peroxidation of phosphatidylcholine liposomes: an electron paramagnetic resonance study. BBA - Gen Subjects 801, 215–219.

Schneider, C., 2005. Chemistry and biology of vitamin E. Mol. Nutr. Food Res. 49, 7–30.

Schulz, K.F., Chalmers, I., Hayes, R.J., Altman, D.G., 1995. Empirical evidence of bias: dimensions of methodological quality associated with estimates of treatment effects in controlled trials. JAMA 273, 408–412.

Sharma, R., Biedenharn, K.R., Fedor, J.M., Agarwal, A., 2013. Lifestyle factors and reproductive health: taking control of your fertility. Reprod. Biol. Endocrin. 11, 66.

Simon, H.-U., Haj-Yehia, A., Levi-Schaffer, F., 2000. Role of reactive oxygen species (ROS) in apoptosis induction. Apoptosis 5, 415–418.

Société de, b., 1849. Comptes rendus hebdomadaires des séances et mémoires de la Société de biologie et de ses filiales. Comptes rendus hebdomadaires des séances et mémoires de la Société de biologie et de ses filiales.

Squadrito, G.L., Cueto, R., Splenser, A.E., Valavanidis, A., Zhang, H., Uppu, R.M., Pryor, W.A., 2000. Reaction of uric acid with peroxynitrite and implications for the mechanism of neuroprotection by uric acid. Arch. Biochem. Biophys. 376, 333–337.

Stocker, R., Bowry, V.W., Frei, B., 1991. Ubiquinol-10 protects human low density lipoprotein more efficiently against lipid peroxidation than does alpha-tocopherol. Proc. Natl. Acad. Sci. U. S. A 88, 1646–1650.

Takayanagi, R., Takeshige, K., Minakami, S., 1980. NADH- and NADPH-dependent lipid peroxidation in bovine heart submitochondrial particles. Dependence on the rate of electron flow in the respiratory chain and an antioxidant role of ubiquinol. Biochem. J. 192, 853–860.

Tesoriere, L., Ciaccio, M., Bongiorno, A., Riccio, A., Pintaudi, A.M., Livrea, M.A., 1993. Antioxidant activity of all-trans-retinol in homogeneous solution and in phosphatidylcholine liposomes. Arch. Biochem. Biophys. 307, 217–223.

Tice, J.A., Ross, E., Coxson, P.G., Rosenberg, I., Weinstein, M.C., Hunink, M.G., Goldman, P.A., Williams, L., Goldman, L., 2001. Cost-effectiveness of vitamin therapy to lower plasma homocysteine levels for the prevention of coronary heart disease: effect of grain fortification and beyond. JAMA 286 (8), 936–943.

Urey, H.C., 1952. On the early chemical history of the earth and the origin of life. Proc. Natl. Acad. Sci. U. S. A. 38, 351.

Van Acker, S.A., Tromp, M.N., Griffioen, D.H., Van Bennekom, W.P., Van Der Vijgh, W.J., Bast, A., 1996. Structural aspects of antioxidant activity of flavonoids. Free Rad. Biol. Med. 20, 331–342.

Vervoort, L.M., Ronden, J.E., Thijssen, H.H., 1997. The potent antioxidant activity of the vitamin K cycle in microsomal lipid peroxidation. Biochem. Pharmacol. 54, 871–876.

Wang, C.C., Rogers, M.S., 2007. Oxidative stress and fetal hypoxia. Reactive oxygen species and diseases. Trivandrum: Research Signpost, pp. 257–282.

Wendel, A., 1987. Measurement of in vivo lipid peroxidation and toxicological significance. Free Radic. Biol. Med. 3, 355–358.

Wolf, G., 2005. The discovery of the antioxidant function of vitamin E: the contribution of Henry A. Mattill. Nutr. J. 135, 363–366.

Wong, J.B., Coates, P.M., Russell, R.M., Dwyer, J.T., Schuttinga, J.A., Bowman, B.A., Peterson, S.A., 2011. Economic analysis of nutrition interventions for chronic disease prevention: methods, research, and policy. Nutr. Rev. 69 (9), 533–549.

Wuerges, J., Lee, J.-W., Yim, Y.-I., Yim, H.-S., Kang, S.-O., Carugo, K.D., 2004. Crystal structure of nickel-containing superoxide dismutase reveals another type of active site. Proc. Natl. Acad. Sci. U. S. A. 101, 8569–8574.

Yoshida, Y., Niki, E., Noguchi, N., 2003. Comparative study on the action of tocopherols and tocotrienols as antioxidant: chemical and physical effects. Chem. Phys. Lipids. 123, 63–75.

Zhu, L., Gunn, C., Beckman, J.S., 1992. Bactericidal activity of peroxynitrite. Arch. Biochem. Biophys. 298, 452–457.

CHAPTER

1.2 The oxidative stress: Causes, free radicals, targets, mechanisms, affected organs, effects, indicators

Amit Kumar Singh, Harvesh Kumar Rana, Abhay K. Pandey
Department of Biochemistry, University of Allahabad, Prayagraj, India

1.2.1 Introduction

Oxidative stress (OS), a disorderd situation where the generation of reactive oxygen species (ROS) inside the system surpasses the antioxidant defense system of the body (Kumar et al., 2013a). This disparity may be a result of altered antioxidant capacity because of compromised production, distribution of antioxidant molecule, or the overproduction of ROS from endogenous sources or external stresses. Excess ROS has a detrimental effect on cellular macromolecules such as lipid, proteins, or DNA, which results in improper signal transduction pathways. Therefore the OS has been occupied a place in the growing list of human diseases such as cancer, hepatotoxicity, neurodegeneration, cardiovascular as well as in aging symptoms (Kumar et al., 2013b).

Free radicals are very reactive in nature because of the availability of an unpaired electron. They are produced inside cells either as metabolic by products or because of mitochondrial dysfunction. Oxygen species, such as ROS, are nonreactive; however, they have the potential to generate free radicals. They are also the by-product of metabolism generated mainly from the electron leakage during the process of cellular respiration. Instead of this, the enzyme/enzyme complexes sited in or linked with cell membrane or organelles were also involved in the process of generating ROS (Singh et al., 2018).

External agents, such as xenobiotics, radiation also help in the production of ROS. Scientific findings have already established the involvement of free radicals in the initiation and progression of diseases; however, it is still complicated to assess its role due to the very short lifetime of the free radical species; hence, it is still not clear that they are the cause of the disease or produced as a result of the diseases. Therefore, the evidence is still circumstantial regarding the molecular mechanism of free radical-induced cellular damage. Consequently, the current need is to search for OS sensitive biomarkers, which

Antioxidants Effects in Health. DOI: https://doi.org/10.1016/B978-0-12-819096-8.00012-4

can be used as the diagnostic parameter to check their association in disease pathogenesis or the xenobiotic toxicity (Kumar and Pandey 2015).

The biological system has developed a comprehensive set of antioxidant defense mechanisms to circumvent the free radical generation and further limit their damaging effects. They performed this task by enzymes that inactivate the peroxide radicals, protein molecules to seize the transition metals or a variety of compound to scavenge the free radicals. Antioxidants are natural or synthetic compounds, which prevent the generation of the radicals or inhibits their reaction to the biological macromolecules. This defense mechanism classified into enzymatic and non-enzymatic defense. Enzymatic defense involves the enzymes, such as superoxide dismutase (SOD), catalase, thioredoxin reductase, glutathione transferase, glutathione peroxidase, etc., and the nonenzymatic antioxidants defense mechanism involved the tocopherols, ascorbate, glutathione, selenium, zinc, ubiquinols, carotenes, and polyphenols (Klotz et al., 2003).

1.2.2 Oxidative stress

Oxidative stress may be defined as "an imbalance between oxidants and antioxidants in favor of the oxidants, leading to a disruption of redox signaling and control and/or molecular damage." A healthy redox system requires the balance between the oxidation and reduction; hence within the biological system, the net flow of electron from reductants to different classes of molecules must be zero at the end of the pathway. This shift in redox steady-state mediates several molecular changes, which leads to the cell and tissue damages depending on the extent of the redox shift. Cell and organ face continuous exposure to OS, either from the endogenous sources or from the exogenous sources (Kumar and Pandey, 2013).

1.2.2.1 Free radicals and their sources

The occurrence of an unpaired electron in the free radical makes them highly reactive molecules. In biology/medicine, free radical is generally known as ROS, such as hydroxide radical, superoxide radical, peroxide radicals, singlet oxygen or reactive nitrogen species (RNS), such as nitric oxide, peroxynitrite, nitrogen dioxide (Valko et al., 2007). These radicals play either a detrimental or beneficial effect in the living system. The beneficial effect of free radicals usually occur at lower concentration and participate in the various physiological process as well as in cellular signaling pathways (Wu et al., 2015). The harmful effect of free radicals happens when there is an overproduction or an impaired antioxidant defense mechanism.

Free radicals can be produced by both endogenous and exogenous sources (Halliwell, 2011). The source of endogenous generation is the mitochondrial respiratory chain enzymes, such as NADPH oxidase, cytochrome oxidase, xanthine oxidase, impaired endothelial nitric oxide synthase myeloperoxidase, etc., in addition to this unbound metals such as iron (Fe) and copper (Cu) can generate free radicals

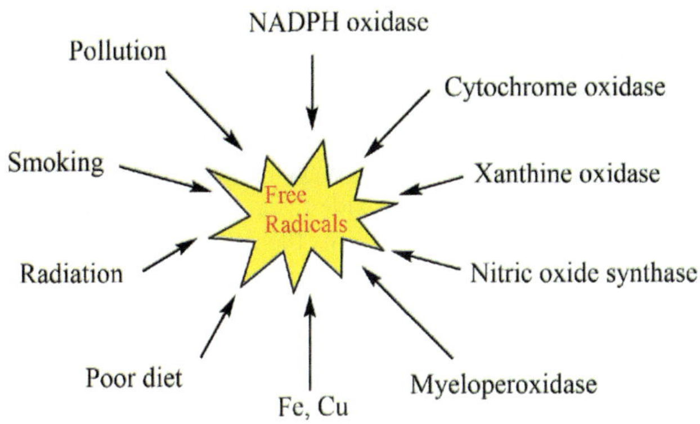

FIG. 1.2.1 Exogenous and endogenous sources of free radical generation.

through their catalytic decomposition of hydrogen peroxide known as Fenton reaction (Valko et al., 2007). Oxidative insult caused by the external sources is air pollution, smoking, radiation (ionizing or solar), or poor diet (fatty and processed food), as given in Fig. 1.2.1.

1.2.3 Targets of free radicals

Free radicals can damage all three crucial classes of macromolecules, that is, protein, lipids, and nucleic acid, whenever there is an imbalance between the free radical (ROS/RNS) generation and antioxidant defenses mechanism.

1. **Deoxyribonucleic acid (DNA):** Elevated ROS/RNS level has the potential to damage the DNA oxidatively. However, the mitochondrial DNA is more susceptible to oxidative damage as compared to the nuclear DNA on account of its closer proximity to the generation site. Hydroxyl radical attacks to all the constituent of DNA such as purines and pyrimidine bases and to the sugar backbone, which leads to the number of modifications including single and double-strand breaks in DNA. It also leads to epigenetic alterations, telomere shortening, and Y chromosomal microdeletions (Bui et al., 2018).
2. **Ribonucleic acid (RNA):** It is well known that thee RNA are more vulnerable to the oxidative damage compared to DNA because of its proximity to the mitochondria where lots of ROS are formed, its single-stranded nature, absence of active repair mechanism to the oxidative damages in RNA, less protection by the proteins as compared to DNA molecule (Hofer et al., 2005). The most extensively studied RNA damage biomarkers are 7, 8-dihydro-8-deoxyguanosine (8-oxoG) which reported to be increased during the disease condition, for example, hemochromatosis (Broeddbaek et al., 2009), Alzheimer's disease (Abe et al., 2002), and atherosclerosis (Martinet et al., 2004).

3. **Lipids:** Higher free radical concentration inside the cell causes peroxidation of the lipid molecule. Lipid peroxidation starts whenever a free radicals attack methylene group of fatty acid and abstract hydrogen atom from it. This process also results in the formation of carbon-centered lipid radical (L•), which in contact with molecular oxygen forms lipid peroxide radicals (LOO•).Consequently, this lipid peroxide undergoes a cyclization reaction to form endoperoxide which finally forms the end product malondialdehyde (MDA), 4-oxy-2-nonenal (ONE), and 4-hydroxyl nonenal (4-HNE). This end product is toxic and causes damage to protein and DNA (Marnett, 1999). This lipid peroxidation causes loss of membrane functioning, such as decreased membrane fluidity as well as inactivation of membrane-bound receptors and enzymes (Bast, 1993).
4. **Protein:** The carbonylation of protein residue been contemplated as the biomarkers of ROS induced protein damage. ROS mediated damage of amino acid residue, namely lysine, threonine, proline, and arginine, results in the formation of carbonyl derivatives (Chevion et al., 2000; Singh et al., 2019). O-tyrosine (a marker for hydroxyl radical) and 3-nitrotyrosine (a marker for RNS) are the other protein oxidation biomarkers. Elevated protein carbonyl content has been observed in various disease condition such as diabetes (Jones and Hothersall 1993), aging (Smith et al., 1991) Rheumatoid arthritis (Chapman et al., 1989), alzheimer's disease (Smith et al., 1991), muscular dystrophy, and Parkinson's disease (Floor and Wetzel, 1998).

1.2.4 Free radicals and their damaging effect on organs

The higher concentration of radicals imparts damaging effects on the organ system of the body. Generally the oxidative stress is the result of alcohol, drugs, environmental pollutants, temperature, radiation, as well as routine lifestyle. Scientific studies have linked the role of oxidative stress in hepatic, renal, brain as well as lung damages.

1. **Lung:** A significant number of experimental studies establish an association between the activities of pulmonary antioxidant enzymes and the levels of nonenzymatic lung antioxidants (Farzaei et al., 2019), demonstrating that these defense systems are vital, and by inference that free radicals are involved in pulmonary oxygen toxicity. Several experimental findings further mark a strong correlation between oxidative stress in lung diseases, for example, asthma, chronic obstructive pulmonary disease, acute lung inflammation, pulmonary fibrosis, and lung cancer. Oxidative stress also plays a critical role in inflammatory responses in lung diseases by increasing the expression of redox-sensitive transcription factors and thus upregulated proinflammatory gene expression. Inflammation itself causes oxidative stress in the lungs. ROS seems to be key regulatory factors in the molecular pathways leading to the induction of lung diseases (Park et al., 2009).

2. **Liver:** The liver, with its comprehensive metabolic activity, is both a source and a target of free radicals. In addition to this, Kupffer cells also play a critical role in free radical associated hepatic damage. These phagocytic cells get activated by the presence of various inflammatory mediators, such as TNF-α, IL-1 IL-6, IL-8, MCP-1, etc., that might contribute towards hepatic cancer development as well as several other damages linked with xenobiotic exposures (Muriel, 2009). ROS/RNS leads to hepatocyte damage because of varied conditions, for example, excessive alcohol consumption, fibrosis/cirrhosis of the liver, hepatocellular carcinoma, radiation-induced liver injury, paracetamol overdose, and viral hepatitis (Singh et al., 2018). Both ROS and RNS might disturb the energetic, respiratory, and regenerative pathways in hepatic cells. The disparity of proinflammatory and anti-inflammatory cytokine in immune and inflammatory cells, angiogenesis in endothelial and stellate cells, and the expression of collagen genes exacerbate the hepatic damage (Paublo, 2009).
3. **Kidney:** Several other organs such as the kidney and the brain also affected by free radical-mediated damage. This damage occurs because of the organ-specific bioactivation, such as has been shown for quinone–glutathione conjugates formed in the kidney (Monks et al., 1990), or the occurrence of specific compounds that are freely oxidized, for instance, catecholamines in the brain (Bondy, 1992). In fact, oxidative stress is also being blamed for the initiation and progression of numerous neurological disorders (Reynolds et al., 2007).

1.2.5 Biomarkers

There are various oxidative stress-related biomarkers employed to study the damaging effect of ROS on organ systems (Griffiths et al., 2002; Frijhoff et al., 2015). The usual ROS induced biomarkers are protein carbonyl content, advanced glycation end products, oxidized LDL, 3-nitrotyrosine, an oxidation product of genetic elements, such as 8-oxoguanine, the end product of lipid peroxidation (MDA, 4-hydroxy-nonenal), protein thiol content, glutathione, etc. Other specialized biomarkers also exist, such as the length of telomere varies with the extent of oxidative stress (Von Zglinicki, 2002). Comet assay is used to assay the DNA damage in cells (Collins, 2014). The ratio of reduced to oxidized glutathione (GSH/GSSG) and cysteine/cystine in the plasma sample has been used as a biomarker of oxidative stress. Higher cystine levels have been linked to death as an outcome in liver diseases (Patel et al., 2016). Polyunsaturated lipid moiety is more prone to oxidative insult as they have more double bonds in the molecular structure (Porter et al., 1995), hence the lipid peroxidation indicators are important biomarkers to study the damaging effect of free radicals. Experimental studies have proved that lipid peroxidation is linked to cardiovascular diseases. MDA, 4-HNE is the most widely used lipid peroxidation biomarker. MDA, a ketoaldehyde, is the final product of lipid peroxidation and its level increased during the tissue damage. Thiobarbituric acid-reactive substance (TBARS) assay is used to assay the lipid peroxidation as MDA reacts with thiobarbituric acid to give a red color pigment. The reduced form of glutathione (GSH) inhibits the activity of lipid peroxidase enzyme thus inhibits the lipid peroxidation. However, during oxidative stress, GSH oxidized to glutathione disulfide (GSSG); therefore, decrease in the level of GSH and an increase in the level of GSSG occur (Mieyal et al., 2008; Jones, 2002). Protein carbonyl (PCO) content usually increases during the oxidative insult. It formed due to the oxidative cleavage of the protein backbone or deamination of glutamic acid and lysine amino acid residue. Another widely used biomarkers are advanced glycation end (AGE) products formed by the non-enzymatic reaction between reducing sugar and protein, DNA, and lipids. This glycation reaction also occurs during the normal metabolism, however, during hyperlipidemia, hyperglycemia, and oxidative stress condition it becomes more pronounced. In humans, the elevated PCO and AGE content are associated with diabetes, aging, neurodegenerative disorder, and hepatic diseases (Jeroen et al., 2015).

The measurement of oxidized LDL (ox-LDL) as a biomarker of oxidative stress in atherosclerosis has been stated due to its potential to promote lipid deposition. Ox-LDL is supposed to generate by the activated platelets. It is usually assayed in plasma or isolated LDL by immunological methods (Massberg et al., 2005). F2-isoprostanes are formed by the oxidation of amino acid. They are considered as the most reliable biomarker of measuring oxidative stress *in vivo*. Like F2-isoprostanes, isolevuglandins are also the product of amino acid oxidation and are sensitive to the change in redox balance (Frijhoff, et al., 2015). 3-nitrotyrosine is a stable marker of oxidative stress. It is formed by the replacement of the C-3 hydrogen atom of tyrosine with the nitro group. This reaction can occur in a free tyrosine molecule or in a polypeptide sequence. Serum C-reactive protein is a widely used clinical biomarker of ROS induced inflammatory insult in cells (Frijhoff, et al., 2015). ROS induced oxidation of uric acid results in the formation of allantoin, which is a significant biomarker to assess oxidative stress both *in vitro* and *in vivo* (Kandar et al., 2006).

The elevated level of allantoin has been reported in relation to cardiovascular diseases in the plasma of people with type 2 diabetes, obesity, and smoking habit (Seet et al., 2011). Oxidative stress-related heart failure cases reported having an elevated plasma level of allantoin (Doehner et al., 2002). Myeloperoxidase a heme oxidase found in inflammatory cells such as macrophages and neutrophils. It catalyzes the production of hypochlorous acid (HOCl) via a reaction between hydrogen peroxide and chloride molecule. HOCl is a primary oxidant that causes the oxidation of LDL molecule (Hazen and Heinecke 1997). Myeloperoxidase is a potential prognostic biomarker of inflammation, atherosclerosis, and coronary heart disease (Zhang et al., 2001).

Conclusion

Oxidative stress has been occupied a place in the growing list of human diseases, such as cancer, hepatotoxicity, neurodegeneration, cardiovascular, as well as in aging symptoms. Imbalance in redox system of body resulted in several harmful effects on the cellular macromolecules (protein, lipid, DNA, RNA) and tissue system of the body. There are several sensitive biomarkers discovered related to the oxidative stress-mediated damages. However, research is still needed to fill the gap where the exact mechanism of oxidative stress-mediated damage is not understood.

Acknowledgments

Amit Kumar Singh acknowledges CSIR-New Delhi India for providing financial support in the form of the senior research fellowship. All the authors also acknowledge DST-FIST facilities provided to the Department of Biochemistry, University of Allahabad, Allahabad, India.

References

Abe, T., Tohgi, H., Isobe, C., Murata, T., Sato, C., 2002. Remarkable increase in the concentration of 8-hydroxyguanosine in cerebrospinal fluid from patients with Alzheimer's disease. J. Neurosci. Res. 70 (3), 447–450.

Bast, A., 1993. Oxidative stress and calcium homeostasis. In: Halliwell, B., Aruoma, OI (Eds.), DNA and Free Radicals. Ellis Horwood, London, pp. 95–108.

Bondy, S.C., 1992. Reactive oxygen species: relation to aging and neurotoxic damage. Neurotoxicology 13, 87–100.

Broedbaek, K., Poulsen, H.E., Weimann, A., Kom, G.D., Schwedhelm, E., Nielsen, P., Boger, R.H., 2009. Urinary excretion of biomarkers of oxidatively damaged DNA and RNA in hereditary hemochromatosis. Free Radical Biol. Med. 47 (8), 1230–1233.

Bui, A.D., Sharma, R., Henkel, R., Agarwal, A., 2018. Reactive oxygen species impact on sperm DNA and its role in male infertility. Andrologia 50 (8), e13012.

Chapman, M.L., Rubin, B.R., Gracy, R.W., 1989. Increased carbonyl content of proteins in synovial fluid from patients with rheumatoid arthritis. J. Rheumatol. 16 (1), 15–18.

Chevion, M., Berenshtein, E., Stadtman, E.R., 2000. Human studies related to protein oxidation: protein carbonyl content as a marker of damage. Free Radic. Res. 33, S99–108.

Collins, A.R., 2014. Measuring oxidative damage to DNA and its repair with the comet assay. Biochim. Biophys. Acta 1840, 794–800.

Doehner, W., Schoene, N., Rauchhaus, M., Leyva-Leon, F., Pavitt, D.V., Reaveley, D.A., Schuler, G., Coats, A.J., Anker, S.D., Hambrecht, R., 2002. Effects of xanthine oxidase inhibition with allopurinol on endothelial function and peripheral blood flow in hyperuricemic patients with chronic heart failure: results from 2 placebo-controlled studies. Circulation 105 (22), 2619–2624.

Farzaei, M.H, Singh, A.K., Kumar, R., Croley, C.R., Pandey, A.K., Coy-Barrera, E., Patra, J.K., Das, G., Kerry, R.G., Annunziata, G., Tenore, G.C., Khan, H., Micucci, M., Budriesi, R., Momtaz, S., Nabavi, S.M., Bishayee, A., 2019. Targeting inflammation by flavonoids: novel therapeutic strategy for metabolic disorders. Int. J. Mol. Sci. 20, 4957.

Floor, E., Wetzel, M.G., 1998. Increased protein oxidation in human substantia nigra pars compacta in comparison with basal ganglia and prefrontal cortex measured with an improved dinitrophenylhydrazine assay. J. Neurochem. 7091, 268–275.

Frijhoff, J., Winyard, P.G., Zarkovic, N., Davies, S.S., Stocker, R., Cheng, D., Knight, A.R., Taylor, E.L., Oettrich, J., Ruskovska, T., Gasparovic, A.C., Cuadrado, A., Weber, D.,

Poulsen, H.E., Grune, T., Schmidt, H.H., Ghezzi, P., 2015. Clinical relevance of biomarkers of oxidative stress. Antioxid. Redox Signal 23, 1144 –70.

Griffiths, H.R., Møller, L., Bartosz, G., Bast, A., Bertoni-Freddari, C., Collins, A., Cooke, M., Coolen, S., Haenen, G., Hoberg, A.M., Loft, S., Lunec, J., Olinski, R., Parry, J., Pompella, A., Poulsen, H., Verhagen, H., Astley, S.B., 2002. Biomarkers. Mol. Aspects Med. 23, 101–208.

Halliwell, B., 2011. Free radicals and antioxidants – Quo vadis? Trends Pharmacol. Sci. 32, 125–130.

Hazen, S.L., Heinecke, J.W., 1997. 3-Chlorotyrosine, a specifc marker of myeloperoxidase-catalyzed oxidation, is markedly elevated in low density lipoprotein isolated from human atherosclerotic intima. J. Clin. Invest. 99 (9), 2075–2081.

Hofer, T., Badouard, C., Bajak, E., Ravanat, J.L., Mattsson, A., Cotgreave, I.A., 2005. Hydrogen peroxide causes greater oxidation in cellular RNA than in DNA. Biol. Chem. 386 (4), 333–337.

Jeroen, F., Paul, G.W., Neven, Z., Sean, S.D., Roland, S., David, C., Annie, R.K., Emma, L.T., Jeannette, O., Tatjana, R., Ana, C.G., Antonio, C., Daniela, W., Henrik, E.P., Tilman, G., Harald, H.H.W.S., Pietro, G., 2015. Clinical relevance of biomarkers of oxidative stress. Antioxid. Redox Signal. 23 (14), 1144–1170.

Jones, D.P., 2002. Redox potential of GSH/GSSG couple: assay and biological signifcance. Methods Enzymol. 348, 93–112.

Jones, R.H., Hothersall, J.S., 1993. The effect of diabetes and dietary ascorbate supplementation on the oxidative modification of rat lens beta L crystallin. Biochem. Med. Metab. Biol. 50 (2), 197–209.

Kand'ar, R.P.Z., Muzakova, V., 2006. Monitoring of antioxidant properties of uric acid in humans for a consideration measuring of levels of allantoin in plasma by liquid chromatography. Clin. Chim. Acta 365 (1–2), 249–256.

Klotz, L.O., Kröncke, K.D., Buchczyk, D.P., Sies, H., 2003. Role of copper, zinc, selenium and tellurium in the cellular defense against oxidative and nitrosative stress. J. Nutr 133, 1448S–1451S.

Kumar, S., Chashoo, G., Saxena, A.K., Pandey, A.K., 2013. Parthenium hysterophorus: a probable source of anticancer, antioxidant and anti-HIV agents. Hindawi Publishing Corporation BioMed Research International, p. 11.

Kumar, S., Pandey, A.K., 2013. Chemistry and biological activities of flavonoids: an overview. 2013, 162750.

Kumar, S., Pandey, A.K., 2015. Free radicals: Health implications and their mitigation by herbals. Brit. J. Med. Med. Res. 7 (6), 438–457.

Marnett, L.J., 1999. Lipid peroxidation—DNA damage by malondialdehyde. Mutat. Res. 424 (1–2), 83–95.

Martinet, W., de Meyer, G.R., Herman, A.G., Kockx, M.M., 2004. Reactive oxygen species induce RNA damage in human atherosclerosis. Eur. J. Clin. Invest. 34 (5), 323–327.

Massberg, S., Schürzinger, K., Lorenz, M., Konrad, I., Schulz, C., Plesnila, N., Kennerknecht, E., Rudelius, M., Sauer, S., Braun, S., Kremmer, E., Emambokus, N.R., Frampton, J., Gawaz, M., 2005. Platelet adhesion via glycoprotein IIb integrin is critical for atheroprogression and focal cerebral ischemia: an in vivo study in mice lacking glycoprotein IIb. Circulation 112 (8), 1180–1188.

Mieyal, J.J., Gallogly, M.M., Qanungo, S., Sabens, E.A., Shelton, M.D., 2008. Molecular mechanisms and clinical implications of reversible protein S-glutathionylation. Antioxid. Redox. Signal. 10 (11), 1941–1988.

Monks, T.J., Anders, M.W., Dekant, W., Stevens, J.L., Lau, S.S., van Bladeren, P.J., 1990. Contemporary issues in toxicology. Glutathione conjugate mediated toxicities. Toxicol. Appl. Pharmacol. 106, 1–19.

Muriel, P., 2009. Role of free radicals in liver diseases. Hepatol. Int. 3, 526–536.

Park, H.S., Kim, S.R., Lee, Y.C., 2009. Impact of oxidative stress on lung diseases. Respirology 14, 27–38.

Patel, R.S., Ghasemzadeh, N., Eapen, D.J., Sher, S., Arshad, S., Ko, Y.A., Veledar, E., Samady, H., Zafari, A.M., Sperling, L., Vaccarino, V., Jones, D.P., Quyyumi, A.A., 2016. Novel biomarker of oxidative stress is associated with risk of death in patients with coronary artery disease. Circulation 133, 361–369.

Paublo, M., 2009. Role of free radical in liver diseases. Hepatol Int. 3 (4), 526–536.

Porter, N.A., Caldwell, S.E., Mills, K.A., 1995. Mechanisms of free radical oxidation of unsaturated lipids. Lipids 30 (4), 277–290.

Reynolds, A., Laurie, C., Mosley, R.L., Gendelman, H.E., 2007. Oxidative stress and the pathogenesis of neurodegenerative disorders. Int. Rev. Neurobiol. 82, 297–325.

Seet, R.C., Lee, C.Y., Loke, W.M., Huang, S.H., Huang, H., Looi, W.F., Chew, E.S., Quek, A.M., Lim, E.C., Halliwell, B., 2011. Biomarkers of oxidative damage in cigarete smokers: which biomarkers might reflect acute versus chronic oxidative stress? Free Radic. Biol. Med. 50 (12), 1787–1793.

Singh, A.K, Rana, H., Tshbalala, T., Kumar, R., Gupta, A., Nadhlala, A.R., Pandey, A.K., 2019. Phytochemical, nutraceutical and pharmacological attributes of a functional crop Moringa oleifera Lam: an overview. S. Afr. J. Bot. 129, 209–220. doi:10.1016/j.sajb.2019.06.017.

Singh, A.K., Kumar, R., Pandey, A.K., 2018. Hepatocellular carcinoma: cause, mechanism of progression and biomarkers. Curr. Chem. Genom. Transl. Med. 12, 9–26.

Smith, C.D., Carney, J.M., Starke-Reed, P.E., Oliver, C.N., Stadtman, E.R., Floyd, R.A., Markesbery, W.R., 1991. Excess brain protein oxidation and enzyme dysfunction in normal aging and in Alzheimer disease. Proc. Natl. Acad. Sci. USA. 88 (23), 10540–10543.

Valko, M., Leibfritz, D., Moncol, J., Cronin, M.T., Mazur, M., Telser, J., 2007. Free radicals and antioxidants in normal physiological functions and human disease. Int. J. Biochem. Cell Biol. 39, 44–84.

von Zglinicki, T., 2002. Oxidative stress shortens telomeres. Trends Biochem. Sci. 27, 339–344.

Wu, P., Nielsen, T.E., Clausen, M.H., 2015. FDA-approved small-molecule kinase inhibitors. Trends Pharmacol. Sci. 36, 422–439.

Zhang, R., Brennan, M.L., Fu, X., Aviles, R.J., Pearce, G.L., Penn, M.S., Topol, E.J., Sprecher, D.L., Hazen, S.L., 2001. Association between myeloperoxidase levels and risk of coronary artery disease. JAMA 286 (17), 2136–2142.

CHAPTER

Food auto-oxidation: An overview

1.3

Aakriti Garg[a,b], Ruchika Sharma[c], Prasanta Dey[d], Anoop Kumar[a,e]

[a]*Department of Pharmacology, Indo-Soviet Friendship College of Pharmacy (ISFCP), Moga, Punjab, India*

[b]*School of Pharmaceutical Sciences, Apeejay Stya University, Gurgaon, India*

[c]*Department of Biotechnology, Indo-Soviet Friendship College of Professional Studies (ISFCPS), Moga, Punjab, India*

[d]*School of Pharmacy, Sungkyunkwan University, Suwon, Republic of Korea*

[e]*Department of Pharmacology, Delhi Pharmaceutical Sciences and Research University (DPSRU), New Delhi, India*

1.3.1 Introduction

Auto-oxidation is defined as the spontaneous oxidation of a substance at ambient temperature in the presence of oxygen. Lipids form an essential component in food products due to nutrition, satiety, mouthfeel, and health promotions. However, lipids are the main targets for the auto-oxidation process due to their low stability. Thus, unsaturated fatty acids containing food products, such as meat, emulsions, low moisture foods, nuts, and oils are prone to auto-oxidation. However, due to prolonged storage, compounds, such as sterols, and highly processed products, such as spray-dried food products, may also undergo auto-oxidation. The various factors involved in food auto-oxidation are humidity, radiations, oxygen, high temperature, and microorganisms, which results in the change in flavor, odor, and the texture of food products (Kanner, 2007), also known as "rancidity" as shown in Fig. 1.3.1. Food processing such as deep-frying and cooking can be problematic as it combines high temperature and water, which speed up the process of auto-oxidation (Nayak et al., 2016). Also, cooking of meat releases protein-bound metals (Papuc et al., 2017; Yu et al., 2017), which inactivates antioxidants (Kerry, 2012) and disrupts the lipid membrane, finally accelerating the process of auto-oxidation (Cheng, 2016).

Food auto-oxidation results in the loss of nutritional value and a decrease in the shelf life of food products (Ahmed et al., 2016), making them unfit for consumption. The utilization of vitamins in the body is also adversely affected due to the presence of rancid fats (Bickford et al., 1948). The auto-oxidation of unsaturated fatty acids generates toxic degradation products, such as unsaturated aldehydes which conjugate with protein, cell membranes, and DNA, resulting in their denaturation (Antolovich et al., 2002).

Antioxidants Effects in Health. DOI: https://doi.org/10.1016/B978-0-12-819096-8.00013-6

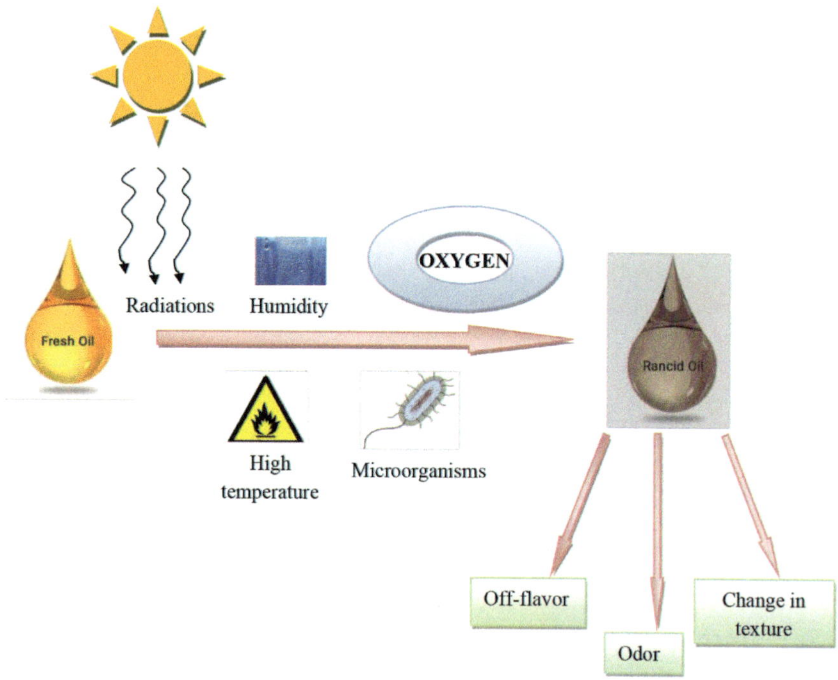

FIG. 1.3.1 Food auto-oxidation.

Since the products originating from the oxidation can present toxicity, they can also affect food safety. In this chapter, we have discussed the mechanism and methods for the determination of auto-oxidation of food. Further, the various factors involved in auto-oxidation and toxic effects of food auto-oxidation have been discussed. The measures which can be taken to prevent the auto-oxidation process and the antioxidants used in food industries have been also discussed. Finally, this chapter concludes with challenges faced during the prevention of food auto-oxidation.

1.3.2 Mechanism of auto-oxidation

The auto-oxidation process involves mainly three steps, that is, initiation, propagation, and termination (Gryn'ova et al., 2011) as described below (Fig. 1.3.2).

1.3.2.1 Initiation

Initiation of auto-oxidation, that is, the formation of free radicals can be triggered by various factors such as singlet oxygen ($^1O_2$), radiations, an excited state of photosensitizers, environmental pollutants, heat and/or light (Ahmed et al., 2016).

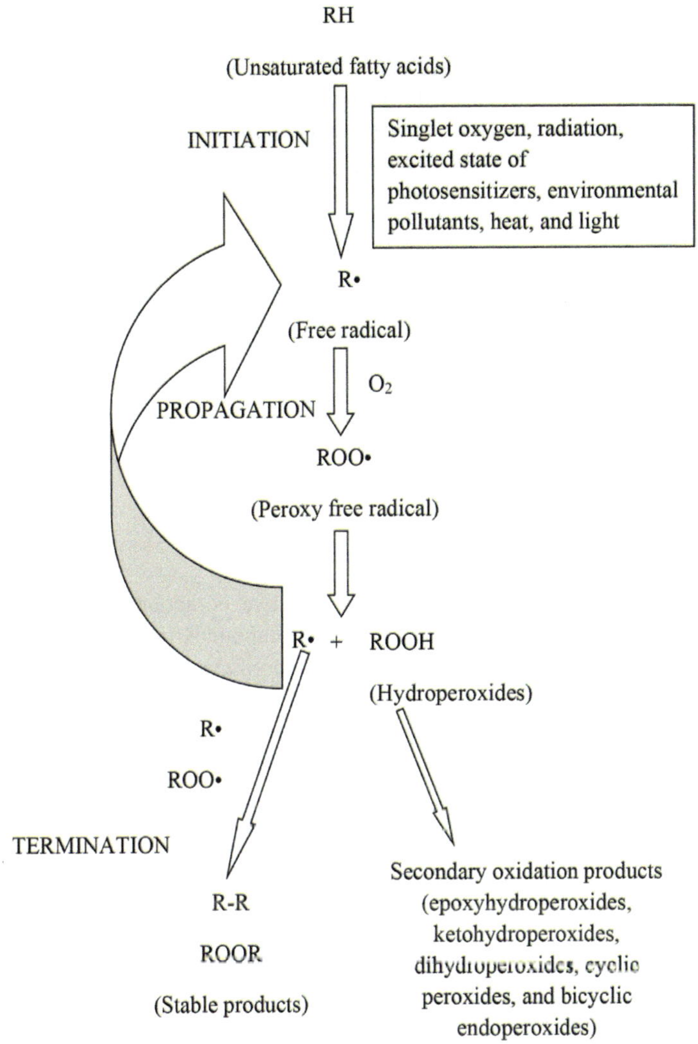

FIG. 1.3.2 Mechanism of auto-oxidation.

These radicals, such as oxygen or hydroxyl radical, possess one or more unpaired electrons which are highly reactive. In unsaturated fatty acid (RH), these radicals can combine with a hydrogen atom which results in the formation of a free radical (R•).

$$X\bullet + RH \xrightarrow{\text{Heat or Light}} R\bullet$$

This is the step where measures can be taken to delay or prevent the oxidation of food as once the oxidation starts and reaches to propagation step; it cannot be delayed or stopped (Ayala et al., 2014).

1.3.2.2 Propagation

The major auto-oxidation process occurs through a free radical chain propagation reaction. This reaction results in the generation of peroxides and hydroperoxides which are quite unstable. Thus, the splitting of these compounds results in the formation of more free radicals and propagation of chain reactions as mentioned below.

$$R\bullet + O_2 \rightarrow RO_2\bullet$$

$$RO_2\bullet + RH \rightarrow ROOH + R\bullet$$

$$ROOH \rightarrow RO\bullet + HO\bullet$$

$$2ROOH \rightarrow RO_2\bullet + RO\bullet + H_2O$$

$$RO\bullet + RH \rightarrow ROH + R\bullet$$

$$HO\bullet + RH \rightarrow R\bullet + H_2O$$

1.3.2.3 Termination

The propagation step increases the highly reactive compounds as mentioned in Section 1.3.2.2. Further, these highly reactive compounds interact with each other and result in the formation of stable nonradical products (Antolovich et al., 2002). These stable nonradical products result in the termination of the auto-oxidation process. However, auto-oxidation is a circular process in which stable products may also combine with free radicals and lead to continuous degradation of lipids.

$$2RO_2\bullet \rightarrow \text{Inert Products}$$

$$RO_2\bullet + R\bullet \rightarrow ROOH$$

$$2R\bullet \rightarrow R\text{-}R$$

After the destruction of the fatty acids, secondary products of oxidation such as epoxyhydroperoxides, dihydroperoxides, cyclic peroxides, etc., are generated which further break up to form volatile compounds and are responsible for the rancidity (Choe, 2006).

1.3.3 Methods for the determination of food auto-oxidation

During the process of auto-oxidation, the various properties, such as peroxide value, oxygen absorption, the content of volatile compounds, color, etc., are changed. These properties can be utilized for the determination of the presence and/or extent of auto-oxidation occurred in the food product. Over the past few years, various methods are developed for the determination of food auto-oxidation. Some of the methods are discussed below.

1.3.3.1 Determination of primary products

The extent of auto-oxidation can be determined by measuring the number of primary products. The main primary products of food auto-oxidation are peroxides and hydroperoxides. The methods for the determination of these products are discussed below.

1.3.3.1.1 Peroxide value

In the initial stage of lipid oxidation, hydroperoxides are formed which can be measured by calculating the peroxide value. The most commonly used method for the determination of peroxide value is the iodometric titration method. The principle of this method is titrating produced iodine by the reaction of potassium iodide and peroxide with sodium thiosulfate. Thus, the amount of peroxide present in the food product is directly proportional to the amount of iodine produced. The content of iodine is expressed in terms of milliequivalents of oxygen per kilogram of lipid (Osawa et al., 2007). The reaction equation is shown below:

$$ROOH + 2H^+ + 2KI \rightarrow ROH + I_2 + 2K^+ + H_2O$$

$$I_2 + Na_2S_2O_3 \rightarrow NaI + Na_2S_2O_4$$

However, there are various limitations of this method. The most common limitation is the oxidation of iodide in the presence of atmospheric oxygen. In the presence of atmospheric oxygen, iodide results in the high peroxide value whereas the absorption of iodine by lipids results in the underestimation of hydroperoxides. Moreover, this test is extremely sensitive to changes in temperature (Popa et al., 2017) thus, the sensitivity and accuracy of this method are questionable. The iodometric titration method also requires a large amount of lipid (Mureşan et al., 2010), therefore this method cannot be applied for the microanalysis of hydroperoxides. Thus, a new method was developed for the spectrometric determination of triiodide at 357 nm. This method is specific for the quantification of hydroperoxides.

Peroxides are unstable compounds; thus, the level of peroxides reaches the maximum and then decreases to disappear. Similarly, PV also reaches the maximum and then decreases (Popa et al., 2017).

1.3.3.1.2 Iron thiocyanate method

The iron thiocyanate method is a spectrophotometric method for the determination of peroxides (Shantha and Decker, 1994). The basic principle of this method is the conversion of ferrous ions to ferric ions in the presence of peroxides. Further, the addition of thiocyanate solution results in the formation of colored compound, that is, flammulated iron (III) thiocyanate. The color produced by this reaction is obtained with calorimetry and the intensity of the color is directly proportional to the number of peroxides present in the sample (Navas et al., 2004).

Another method known as the modified ferrous oxidation-xylenol orange method can also be used for the determination of peroxides. The basic principle behind this method is the oxidation of ferrous ions to ferric ions by hydroperoxides in acidic medium (Bou et al., 2008). Ferric ions thus formed react with xylenol orange to form a blue-purple colored complex, which has UV maximum around 550–600 nm (Meisner et al., 2009). This method is simple, fast, and easy to perform, however, the method is sensitive to oxygen present in solution, hence may give a false-positive result.

1.3.3.1.3 Conjugated dienes value

Conjugated dienes are measured as the indicator of free radicals that are produced during the auto-oxidation process of lipids. The most common method used to detect conjugated unsaturation is ultraviolet spectroscopy. Conjugated dienes have UV maxima around 230–240 nm with strong absorption at 234 nm (Farhoosh et al., 2009). The main limitation of this method is interference in results due to the presence of unoxidized lipids and peroxides of material other than lipid. Thus, this method is suitable only for the determination of lipid auto-oxidation.

1.3.3.1.4 Active oxygen method

The active oxygen method (AOM) is useful for the determination of peroxides and hydroperoxides by calculating the AOM value. The AOM value is defined as the number of hours required for the peroxide concentration to reach 100 meq/kg of lipid. In this method, the air is bubbled through the lipid solution to predict the stability of lipids at a specific flow rate, concentration, and temperature (Läubli and Bruttel, 1986). Further, the iodine titration method is used for the determination of formed peroxides and hydroperoxides. The disadvantages of this method are expensive, nonreproducible, and time consuming (Anwar et al., 2003) as stable lipids may take 48 h or more for the production of the desired concentration of peroxides.

1.3.3.1.5 High-performance liquid chromatography

High-performance liquid chromatography (HPLC) is a widely used and well-known technique for the separation and determination of components from the mixture. This technique is also applied for the determination of lipid peroxides with different volatility, polarity, and/or molecular weight. Different samples can be determined by using different detectors in HPLC, such as evaporative light-scattering detector, ultraviolet detector, diode array detector, and electrochemical detector (Kamal-Eldin et al.,2003). Various oxidation products can be detected using HPLC coupled with electron spin resonance chromatography (HPLC-ESR) and high-performance liquid chromatography-electron spin resonance-mass spectrometry (HPLC/ESR/MS).

1.3.3.1.6 Infrared spectroscopy method

Infrared (IR) spectroscopy is a nondestructive, easy, simple, economical, and sensitive technique, which is applied for the quantitative and qualitative determination of auto-oxidation products such as hydroperoxyl and lipid hydroxyl groups. Near-infrared spectroscopy (NIR) is used for the routine analysis of various functional groups such as –CH, –OH, –NH, and other chemical bonds having major absorption in the NIR region. Mid-infrared spectroscopy can also be applied for the determination of food product oxidation such as edible oils and canned tomato juice (Barriuso et al., 2013).

IR method can be used for the determination of auto-oxidation of samples whose peroxide value falls in the range of 0 to 100 meq/kg.

1.3.3.2 Determination of secondary products

The actual extent of oxidation of the products can be measured by the determination of secondary products. These products are formed due to decomposition or reaction of primary products (unstable) with other compounds to form stable products (Fenaille et al., 2001).

1.3.3.2.1 Thiobarbituric acid reactive substances assay

The auto-oxidation products which react with thiobarbituric acid (TBA) to produce color are known as thiobarbituric acid reactive substances (TBARS). These substances can be detected by measuring the malondialdehyde (MDA), that is, end product of lipid peroxidation using thiobarbituric acid. This method is widely used for the determination of food auto-oxidation because of its simplicity and low cost. The intensity of color produced by the reaction of auto-oxidation products with thiobarbituric acid is measured with the help of UV spectroscopy. For example, the reaction of malondialdehyde (protein and sugar degradation product) with two molecules of TBA produces red/pink color (Bull et al., 1985) while the reaction of TBA with 2-alkenals and 2,4-alkadienals turns the solution yellow in acidic medium. The maximum absorption of red and yellow products is 532–535 nm and 450 nm respectively (Jardine et al., 2002). The TBA value expresses lipid oxidation in milligrams of MDA equivalents per kilogram sample or as micromoles MDA equivalents per gram sample (Pikul et al., 1989; Salih et al., 1987).

This method is criticized because of its operational limit, low sensitivity at low MA concentration, and nonspecificity (Csallany et al., 1984). The presence of other TBARS such as aldehyde and saccharides may give false-positive results for the determination of the MA-TBA complex. Regardless of these limitations, the TBA test is commonly used for the determination of food auto-oxidation.

1.3.3.2.2 Anisidine value

Hydroperoxides formed due to the auto-oxidation of lipids breaks to form volatile aldehydes such as hexanal, leaving behind nonvolatile compounds, such as 2-alkenlas and 2,4-dienals. The methods discussed above are used for the determination of volatile compounds. However, volatile aldehydes may get removed during the process, giving a false estimation of auto-oxidation. Thus, nonvolatile compounds are also necessary to measure. The anisidine value (AV) is calculated to measure these compounds, which are defined as 100 times the absorbance (at 350 nm) of a solution resulting from the reaction of 1 g of fat in 100 mL of solvent (Gunstone and Harwood, 2007). The basic principle behind this method is the formation of pink-red color by the reaction of carbonyl compounds and anisidine which is measured at 350 nm. However, this method is not suitable for highly colored oil with absorbance at 350 nm (Steele, 2004). This method is convenient to determine the auto-oxidation of oils and edible fats; however, if the lipid is extracted from a product, the absorbance of the product may produce false results.

Peroxide value (PV) and anisidine value (AV) can be used for the determination of total oxidation (totox) (Steele, 2004) which is given by the following equation:

$$\text{Totox} = 2 \times \text{PV} + \text{AV}$$

1.3.3.2.3 Oxidative stability index method/Rancimat method

Oxidative stability index/Rancimat method is used to determine the stability of the oil. More the oxidative stability index, more is the stability of the oil. The basic principle behind this test is the same as that of AOM (described earlier), however, OSI is relatively fast and automated. In the OSI process, under strict temperature control, lipid degradation is accelerated by passing air through the sample. The volatile acids are flushed into a reservoir containing deionized water. In this method, volatile acids dissociate and produce ions formed to alter the conductivity of water. The OSI value is defined as the hours needed for the rate of conductivity change to reach a predetermined value. The automated instrument, that is, racimat is used to measure the conductivity of low molecular weight fatty acids produced during the auto-oxidation process at 100° or above. This instrument is automated, reproducible, and can test multiple samples simultaneously, thus, reducing time and labor (Läubli, 1986).

However, the main limitation of this method is that the mechanism of oxidation changes at elevated temperature (Amorati et al., 2013) as the solubility of oxygen decreases at elevated temperature, thus making the reaction dependent on oxygen concentration.

1.3.3.2.4 Acid value

The lipid/fat auto-oxidation leads to the production of free fatty acids. These free fatty acids break in smaller molecules which degrades the food quality. The presence of free fatty acids in food products can be estimated by titrating the product with standard potassium hydroxide/sodium hydroxide solution in an alcoholic medium (Henry, 2016). The acid value is defined as the number of milligrams of potassium hydroxide required to neutralize the free fatty acids present in 1 g of fat (Koczoñ et al., 2008).

1.3.3.2.5 Fluorescence spectroscopy method

As the name suggests, fluorescence spectroscopy is used for the detection of fluorescent compounds. Malondialdehyde, a carbonyl compound is produced during the lipid oxidation process. Malondialdehyde reacts with proteins or amino acids to produce a fluorescent compound known as Schiff base. Due to characteristic excitation and emission spectra of the Schiff base, it is used for the determination of oxidation using fluorescent spectroscopy. Aldehyde itself polymerizes to form a fluorescent compound that can be detected using fluorescent spectroscopy. The extent of lipid oxidation is usually expressed as relative fluorescence intensity compared with a standard fluorescent substance (Sun et al., 2011). This method is highly sensitive.

1.3.3.2.6 Gas chromatography method

Gas chromatography (GC) is an analytical technique used for the separation and/or identification of compounds present in minute quantity. GC can be used for the detection of volatile compounds produced during the oxidation of food products. The quantity of volatile compounds increases with the increase in the oxidation level. Additionally, a higher quantity of smaller molecular weight aldehyde indicates the greater extent of oxidation. This method is highly sensitive and indicates a change in the flavor of the product.

1.3.3.3 Determination of oxygen absorption

The consumption of oxygen is high in the initial stages of the auto-oxidation process which increases the weight of fats and oils. Thus, oxygen absorption of the samples can be used to detect food auto-oxidation. In the oxygen absorption method, the sample is kept in the oven with no air circulation and is periodically tested for change in weight. The oxygen absorption is detected by mass change (Santos-Fandila et al., 2014). This method can be applied to detect oxygen absorptions in samples containing highly unsaturated fatty acids.

Another method known as the headspace oxygen method can also be employed for the determination of oxygen absorption. In this method, the food product is kept in a close vial containing a fixed amount of oxygen under high temperatures (100°C). Then the rate of decrease in oxygen is used for the determination of the stability of the food product. If the rate of decrease in oxygen level is faster, the food product is less stable.

1.3.4 Factors influencing auto-oxidation

Factors influencing food auto-oxidation may be intrinsic, that is, regarding the product or extrinsic, that is, due to applied technology. Some of the factors influencing food auto-oxidation process are mentioned below:

Temperature: The rate of auto-oxidation is directly proportional to temperature. It can affect not only the auto-oxidation rate but also the reaction mechanisms (Choe et al., 2006).

Light: The presence of light does not play great importance in auto-oxidation as fatty acids and their peroxides do not absorb visible light. However, the presence of sensitizer or ultraviolet light accelerates the auto-oxidation process (Bekbölet, 1990).

Oxygen: The rate of auto-oxidation process increases with an increase in oxygen pressure until its concentration reaches a steady state.

Humidity: The rancidity of food products occurs at high as well as very low moisture content. Thus, the optimum level of moisture is required to produce a protective effect in the form of a monolayer.

Ionizing radiation: Ionizing radiations increase the susceptibility of auto-oxidation in food.

Catalysts: Various catalysts, such as heavy metals increase the rate of auto-oxidation in food.

Type of oil: Rate of oxidation is also influenced by the type of oil present and increases with the increase of unsaturated groups (Ahmed et al., 2016; Flick et al., 1992), for example, murine oils are highly susceptible to auto-oxidation due to the presence of a high amount of polyunsaturated fatty acids. The polyunsaturated fatty acids contain reactive double bonds which make them more prone to auto-oxidation.

Microorganism: Certain microorganisms can produce the hydrolytic enzyme called lipase, which directly interferes the hydrolysis of triglycerides and produce glycerols and fatty acid. These fatty acids undergo auto-oxidation to form rancid. The microbial lipase requires suitable pH and other conditions for its activity upon fats and oil.

Processing and storage conditions: Various factors involved in processing and storage, such as reducing particle size, heating, oxygen exposure, additives, storage time can determine the rate of auto-oxidation (Amaral et al., 2018).

Thus, the auto-oxidation of food products can be prevented by inhibiting the effect of the above-mentioned factors.

1.3.5 Toxic effects of food auto-oxidation

Auto-oxidation of food products results in loss of nutritional value, characteristic unpalatable, and off-flavor in the product. Apart from this, the oxidized products may have toxic effects on the human body. For example, MDA, a major secondary product of oxidation is highly cytotoxic and mutagenic to mammalian cells (Rajinder et al., 2001; Niederhofer et al., 2003; Bont and Larebeke, 2004; Marquez-Ruiz et al., 2008).

The lipid oxidation products may react with nitrogenous compounds present in biological systems, such as amino acids, proteins, bases of phospholipids, and DNA, forming fluorescent and pigmented compounds that are related to tissue injury. Other toxicity effects that can be produced by oxidized products in the cell include the destruction of cytosolic enzyme activities, swelling, and lysis of cell membranes. Oxidized products also produce pro-inflammatory responses, peptic ulcers, teratogenicity (Grootveld et al., 2001; Estévez et al., 2017; Kanner, 2007), thrombosis/spasm, atherosclerotic plaque, and arterial injury which are dangerous to human health (Esterbauer, 1993). The symptoms of oxidized fat toxicity are poor growth rate, diarrhea, myopathy, hepatomegaly, steaties or yellow fat disease, secondary deficiencies of vitamin A and E, and hemolytic anemia.

In a study, it was observed that under *in vivo* conditions, administration of oxidized lipids and protein leads to oxidative stress (OS) in tissues, blood, and urine (Estevez and Luna, 2016). Chiang et al. (2011) have reported an increase in reactive oxygen species (ROS) and lipid hydroperoxides and a decrease in glutathione peroxidase, vitamin E in pancreatic islets of mice after the administration of 20% oxidized soybean oil diet. Various studies have indicated an increase in ROS level and a decrease in antioxidant capacity of the gastrointestinal tract (GIT) after stomach perfusion of oxidized soybean protein or casein in experimental animals (Fang et al., 2012; Li et al., 2013; Li et al., 2014). The cholesterol oxidative product like oxysterols resulted in a cytotoxic effect on human monocytic blood cell line (O'Callaghan et al., 2001), promote cancer, atherosclerosis, and cell membrane damage (Staprans et al., 1998; Phillips et al., 2001; Pietras et al., 2012). Xie and colleagues (2014) have reported the induction of intestinal epithelial death due to food oxidative products. The harmful effects of oxidized products on the human body are summarized in Table 1.3.1.

Table 1.3.1 Effect of oxidized food products on human health.

Oxidized product	Effect on human health	Reference
Oxidized lipids	Risk of atherosclerosis	Khan-Merchant et al., 2002
Lipid hydroperoxides	May enhance tumor	Bull et al., 1984; 1988
Advanced lipid oxidation end products (carbonyls, alcohols, hydrocarbons, and furans), especially MDA	Cytotoxic	Niederhofer et al., 2003
	Highly mutagenic and carcinogenic compounds	De et al., 2004 Marquez-Ruiz et al., 2008
	Activate inflammatory response	Kanner, 2007
	Damage hepatocytes	Kanazawa et al., 1985
	Increase the risk of atherogenesis	Esterbauer, 1993
	Dysfunction of red blood cells	Tesoriere et al., 2002
Cholesterol oxidative products	Promote atherosclerosis	Staprans et al., 1998
	Carcinogenic	Phillips et al., 2001
	Damage cell membranes	Pietras et al., 2012
	Cytotoxic	O'Callaghan et al., 2001
Digestion-resistant oxidized proteins	Affect colon epithelial renewal and homeostasis	Kim et al., 2013
	Intestinal tumorigenesis	Le Leu et al., 2007
Oxidized amino acids	Ortho- and meta-tyrosine – may replace phenylalanine, thus forming damaged protein in mammalian cells	Rodgers et al., 2002 Gurer-Orhan et al., 2006 Klipcan et al., 2009 Dunlop et al., 2013
	Aminoadipic acid (AAA) – toxic effects on retinal glial cells and cerebral astrocytes	Ishikawa and Mine, 1983 Brown and Kretzschmar, 1998
	Kynurenine and N-formylkynurenine – potential carcinogens	Friedman and Cuq, 1988

1.3.6 Prevention of food auto-oxidation

The auto-oxidation of food products can be prevented by controlling various factors responsible for the auto-oxidation process. The most common factor involved in the auto-oxidation process is temperature. Thus, the temperature can be reduced while processing, manufacturing, packaging, and shipping of food products to prevent auto-oxidation. The contact of food products with oxygen, which acts as a catalyst for the production of free radicals can be minimized by excluding oxygen from the system, reducing the headspace to a minimum, and/or filling the headspace with an inert gas, such as nitrogen. Oxygen can also be eliminated by adding oxygen scavengers, such as catechol, ascorbic acid, and glucose oxidase (Cruz et al., 2012). Radiations such as UV light accelerates the process of auto-oxidation, thus, the exposure from UV light can be prevented by packaging food product in brown glass/plastic containers or black plastic bags. The humidity level should also be controlled as moisture in combination with other factors accelerates the auto-oxidation process.

Food containing fats and oils tend to absorb foreign odor, thus should not be placed in the vicinity of strong-smelling food products. Food products should be packed in aluminum containers as iron or copper rusted containers act as pro-oxidants and hasten the process of auto-oxidation. Fats and oils can be hydrogenated to prevent oxidation and prolong the shelf life of food. The most common method used to prevent auto-oxidation in the industry is the addition of antioxidants which are discussed in detail in the next section. The preventive measures of food auto-oxidation are shown in Fig. 1.3.3.

1.3.7 Antioxidants used in the food industry

Antioxidants are the common ingredients used by the food industry to prevent oxidative damage to food products. Antioxidants slow down the rate of oxidation process by accepting or donating electrons and by directly interacting with free radicals, forming new free radicals, which are comparatively less reactive and dangerous (Lü et al., 2010). Antioxidants either protect target lipids from oxidation initiators or the stall propagation phase.

Based on mechanism, antioxidants can be classified into **primary antioxidants,** that is, chain-breaking antioxidants and **secondary antioxidants,** that is, preventive antioxidants (Antolovich et al., 2002). Primary antioxidants act by two mechanisms. First, they delay or inhibit the initiation step by accepting free radicals. Second, they interrupt the propagation step by interacting with peroxy radicals to convert them into stable compounds. These antioxidants are more useful if added at the initial step of the auto-oxidation process when the propagation process has not occurred. Examples of primary antioxidants are butylated hydroxyanisole (BHA), tertiary butylhydroquinone (TBHQ), propyl gallate (PG), butylated hydroxytoluene (BHT), tocopherols, carotenoids, and flavonoids.

Secondary antioxidants prevent auto-oxidation through various mechanisms such as to hinder reactive oxygen species formation, scavenging species, which are

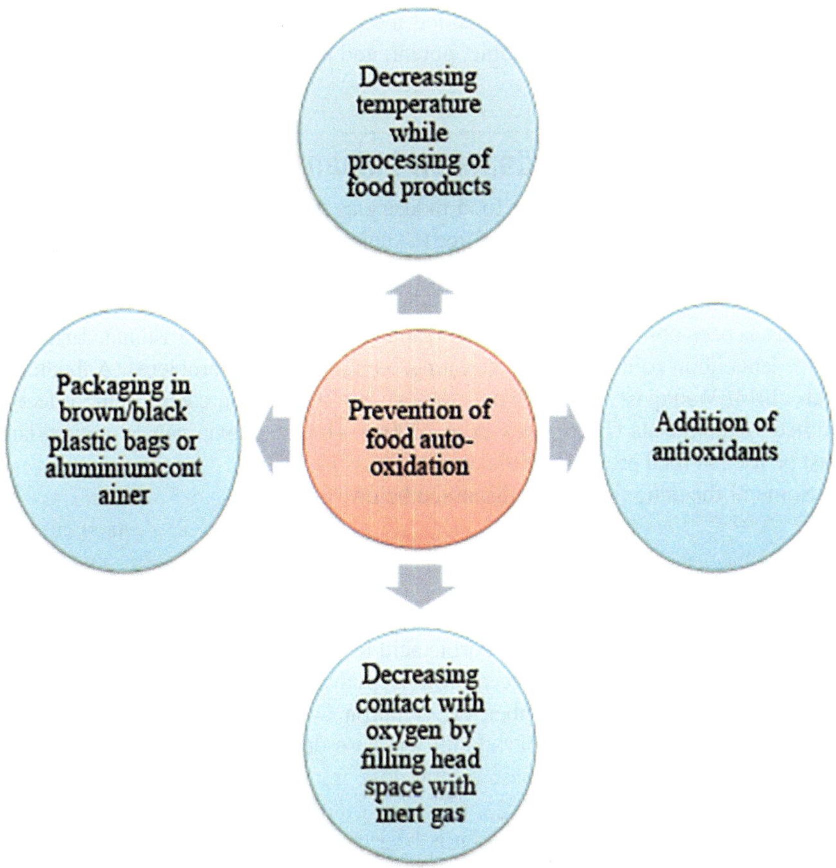

FIG. 1.3.3 Preventive measures of food auto-oxidation.

responsible for oxidation initiation, transition metals chelation, UV filtration, singlet oxygen deactivation, and inhibiting prooxidant enzyme or antioxidant enzyme cofactor. Additionally, they may also act as oxygen scavengers, reducing agents, and can decompose hydroperoxides to nonradical species. Secondary antioxidants enhance the activity of primary antioxidants, hence are also known as synergists. Examples of secondary antioxidants are citric acid, ascorbyl palmitate, ascorbic acid, tartaric acid, and lecithin, etc.

A combination of antioxidants is added to produce a synergistic effect. However, the antioxidant ability varies markedly as per the food products. Food products are predominantly multiphase systems. The effectiveness of antioxidants depends on several factors, such as lipid substrate, the polarity of the antioxidants, nonlipid constituents, ionic strength, pH, temperature, presence of metal ions, the concentration of antioxidants, and the physical properties of the food (Huang et al., 1996), for example, in bulk lipids, polar antioxidants are more effective, whereas nonpolar antioxidants are more active in emulsified media. Antioxidants that are used in the food industry should be cheap, stable, potent, and nontoxic (Schuler, 1990).

1.3.8 Effect of antioxidants on human health

Antioxidants are important in the food industry as well as for human beings as they help to prevent the damaging effect of free radicals either by donating hydrogen atom or scavenging them (Pérez and Aguilar, 2013; Réblová, 2012; Hamid et al., 2010). However, some studies indicate the harmful effect of antioxidants (mainly synthetic) on the human body. It has been observed that BHA, BHT, and/or TBHQ may cause asthma, dermatitis, angioedema, joint pains, excessive sweating, stomach and eye problems (Anbudhasan et al., 2014; Wahlqvist, 2013). Some studies also indicate the carcinogenic effect of synthetic antioxidants (Botterweck et al., 2000). Thus, nontoxic natural antioxidants must be used in food products (Muluken and Shimelis, 2019).

Some of the antioxidants are discussed below.

1.3.8.1 Ascorbic acid

The ascorbic acid, also known as vitamin C, which is a monosaccharide with molecular formula $C_6H_8O_6$. The ascorbic acid has antioxidant activity and is mainly obtained from poblan chili, yucca flower, guava, cauliflower, red bell pepper, garlic, orange, pumpkin, cucumber, etc. Vitamin C is administered orally (Frikke-Schmidt et al., 2011) and its bioavailability is dose-dependent, that is, the absorption fraction decreases with an increase in oral dose fraction (Kubler and Gehler, 1970; Mayersohn, 1972).

In food products, such as jam, gelatin, beer, bread, fruit juices, etc., ascorbic acid can be used as an antioxidant either in free form or as a salt, such as calcium ascorbate, sodium ascorbate, potassium ascorbate, etc. (Morrissey et al., 1998; Bauernfield et al., 1970; Liao et al., 1988). Ascorbic acid arrests the chain reaction by reacting with free radicals, thus preventing the deleterious effect of free radicals (Cerutti, 2006). Ascorbic acid inhibits the oxidative modifications of low-density lipoproteins (LDL) and prevention of the formation of atherosclerotic plaque. It was shown that the adverse reaction of vitamin C was rare at dose <4 g/d (Thorat et al., 2013), however, at higher doses, it could result in harmful effects.

1.3.8.2 Butylated hydroxyanisole and butylated hydroxytoluene

Butylated hydroxyanisole (BHA) and butylated hydroxytoluene (BHT) are the structurally related compounds generally used as antioxidants in food products. They donate a hydrogen atom to oxygen radicals derived from fatty acids, thus scavenging ROS. Apart from this, BHA also increases the activity of hepatic cytosolic gamma-glutamylcysteine synthetase which synthesizes glutathione, an antioxidant.

Several studies have reported both beneficial and deleterious effects of BHA and BHT. Festjens et al. (2006) have reported the strong antinecrotic property of BHA as compared to BHT. In literature, some studies have reported the anticarcinogenic property of BHA and BHT (Hocman, 1988; Wattenberg, 1972, 1986; Wattenberg et al., 1980; Williams, 1986; Williams and Iatropoulos, 1996), whereas some studies have reported the carcinogenic effect of BHA and BHT in experimental animals (Botterweck et al., 2000; Hocman, 1988; Ito et al., 1983; Williams, 1986). The BHA and BHT (0.75%) induced 50% hemolysis in the rat (Jayalakshmi and colleagues, 1986). Recently, Park et al. (2019) have shown the potent cytotoxic activity of BHA on human astrocytes which may cause impairment of brain and nerve development. The BHA (300 mg/kg BW) and BHT (75 mg/kg b.w.) also result in the disruption of the endocrine system of Wistar rats (Pop et al., 2013). Moreover, several studies reported that BHA and BHT lead to urticaria (Thune and Granholt, 1975; Tosti et al., 1987; Fisherman and Cohen, 1977; Juhlin, 1981). Thus, the use of synthetic antioxidants may cause a detrimental effect on the body and hence should be used wisely in food products.

1.3.8.3 Gallates

Propyl gallate, octyl gallate, and dodecyl gallate are the commonly used antioxidants in the gallate group. Propyl gallate (PG) acts as an antioxidant in both aqueous and lipid medium scavenging peroxy radical (Medina et al., 2013). Currently, 25 to 1000 mg/kg of PG is authorized in the EU for food products. Currently, propyl gallate is an authorized antioxidant in the European Union (EU) with concentration ranging from 25 to 1000 mg/kg in foods. The nongenotoxic and noncarcinogenic of propyl gallate has been reported under *in vivo* conditions (Aguilar et al., 2014). The adverse effects of PG (405 mg/kg BW) on the endocrine system of Wistar rats was reported by (Pop et al., 2013). The gallates induced skin and/or stomach irritation has also been reported in the literature (Winter, 1972; Dastychová et al., 2008).

1.3.8.4 Tocopherols

Tocopherols, also known as vitamin E, are phenolic antioxidants which act by scavenging free radicals and/or reacting with singlet oxygen, thus inhibiting lipid auto-oxidation. Tocopherol is commonly found in vegetable oils, oilseeds, nuts, cereals, whole grains, margarine, etc. (Surai et al., 2003). Vitamin E protects DNA, proteins, and cell membranes from oxidation and thereby contributes to cellular health. The antioxidative properties of vitamin E have been found to play a vital role in the battle against various diseases such as OS, atherosclerosis, cancer, Alzheimer's disease (AD), and cataract (Rizvi et al., 2014).

1.3.8.5 Tertiary butylhydroquinone

Tertiary butylhydroquinone is a lipophilic antioxidant used in food containing high-fat content due to its free radical scavenging activity (Ooi et al., 2013). In a study it was reported that 1 g of THBQ caused vomiting, nausea, ringing in the ears, a sense of suffocation, delirium, and collapse and 5 g of THBQ lead to death (Winter, 1972).

It is also found that contact with THBQ results in skin problems (Aalto-Korte, 2000; Orton and Shaw, 2001). Gharavi et al. (2007) have reported THBQ induced cancer after chronic exposure. The food allergy due to the administration of THBQ has also been reported (Jin et al. 2018).

1.3.9 Regulatory guidelines/aspects

In India, Food Safety and Standards (Food Product Standards and Food Additives) Regulation, 2011 describe the standard of various products and limits of additives used (FDA, 2011). Although many synthetic and natural compounds have antioxidant properties, only a few are permitted as "generally recognized as safe (GRAS)" substances for use in food products by international bodies such as the Joint FAO/WHO Expert Committee on Food Additives (JECFA) and the European Community's Scientific Committee for Food (SCF) as compiled in Table 1.3.2 (Miková, 2001). Food and Drug Administration (FDA) and U.S. Department of Agriculture (USDA) has also specified the limits of antioxidants which can be used in food products (compiled in Table 1.3.3) (Shahidi et al., 2005; Directorate of Food and Drugs Administration, 2011).

1.3.10 Current challenges

The major challenge faced nowadays is to develop a tool for testing the antioxidant capacity and its efficacy in preventing the auto-oxidation process (Laguerre et al., 2007). Another challenge in food auto-oxidation is the presence of more polyunsaturated fatty acids. More the polyunsaturated fatty acids in the food, more it prone to auto-oxidation and difficult to select suitable antioxidants. Currently, the use of natural products is encouraging by the regulatory authority which restricted the use of synthetic antioxidants used for the prevention of auto-oxidation

Table 1.3.2 Antioxidants permitted in food products.

Ascorbic acid, sodium, calcium salts, glycine	Glycine
Ascorbyl palmitate and stearate anoxomer	Gum guaiac
Butylated hydroxyanisole (BHA)	Lecithin
Butylated hydroxytoluene (BHT)	Polyphosphates
Tert-butyl hydroquinonea (TBHQ)	Propyl, octyl, and dodecyl gallates
Citric acid, stearyl, and isopropyl esters	Tartaric acid
Erythorbic acid and sodium salt	Thiodipropionic acid, dilauryl and distearyl esters
Ethylenediaminetetraacetic acid (EDTA)	Trihydroxy butyrophenone
Calcium disodium salt	Glycine

Table 1.3.3 Maximum limit of antioxidants used in various food products.

		Maximum limits (ppm)			
S. No.	Food product	*Butylated hydroxy anisole (BHA)*	*Butylated hydroxy-toluene (BHT)*	*Propyl gallate (PG)*	*Tertiary butyl hydro quinone (TBHQ)*
1.	Active dry yeast	1000	–	–	–
2.	Beverages from dry mixes	2	–	–	–
3.	Dehydrated potato shreds	50	50	–	–
4.	Dried meat	100	100	100	100
5.	Dry breakfast cereals	50	50	–	–
6.	Dry diced fruits	32	–	–	–
7.	Dry mixed for beverages and desserts	50	–	–	–
8.	Dry sausage	30	30	30	30
9.	Emulsion stabilizers for shortenings	200	200	–	–
10.	Fresh sausage	100	100	100	100
11.	Potato flakes	50	50	–	–
12.	Snacks/savories (fried products):- chiwda, bhujia, dalmoth, kadubale, kharaboondi, spiced & fried dals, banana chips and similar fried products sold by any name	200	–	–	200
13.	Chewing gum	250	–	–	–
14.	Sweets (carbohydrates based and, milk product based):- halwa, mysore pak, boondi ladoo, jalebi, khoya burfi, peda, gulab jamun, rasogolla, and similar milk product based sweets sold by any name	200	–	–	200
15.	Chocolate – white, milk, plain, composite, filled	200		200	200

(Waraho et al., 2011). However, the natural antioxidants need to be extracted, purified, and dried before use in food products, thus increasing its cost. Moreover, natural products affect the sensory properties of food products (Raikos, 2017). Thus, all the above-mentioned challenges need to be handled in further studies.

Conclusion

In conclusion, food products, such as meat, vegetable oil, dairy products, and bakery products containing unsaturated fatty acids are more prone to auto-oxidation which could result in the formation of unwanted as well as toxic products. This can be prevented by controlling various factors responsible for the auto-oxidation process. Further, it can be prevented by the addition of suitable antioxidants in food products. Natural antioxidants are effective but their cost is high as compared to synthetic antioxidants. The efficacy of synthetic antioxidants in the prevention of food auto-oxidation is still questionable. Thus, further studies should be carried out for the development of safe and effective antioxidants to prevent food auto-oxidation.

Acknowledgments

Authors are thankful to Professor R.K. Goyal, Vice Chancellor, Delhi Pharmaceutical Sciences and Research University (DPSRU), New Delhi and Mr. Parveen Garg, Chairman, Dr. G.D. Gupta, Director, ISF College of Pharmacy, Moga for their constant support and motivation.

Abbreviations

AOM	Active oxygen method
AD	Alzhiemer's disease
AV	Anisidine value
BHA	Butylated hydroxyanisole
BHT	Butylated hydroxytoluene
DNA	Deoxyribonucleic acid
EU	European Union
FDA	Food and Drug Administration
GC	Gas chromatography
GRAS	Generally recognized as safe
GSHPx	Glutathione peroxidase
HPLC	High-performance liquid chromatography
HPLC-ESR	High-performance liquid chromatography-electron spin resonance
HPLC/ESR/MS	High-performance liquid chromatography-electron spin resonance-mass spectrometry
IR	Infrared
JECFA	Joint FAO/WHO Expert Committee on Food Additives
LDL	Low-density lipoprotein
MA	Malonaldehyde
MA-TBA	Malonaldehyde-thiobarbituric acid

MDA	Malondialdehyde
NIR	Near-infrared
OSI	Oxidative stability index
OS	Oxidative stress
PV	Peroxide value
PG	Propyl gallate
ROS	Reactive oxygen species
SCF	Scientific Committee for Food
TBHQ	Tertiary butylhydroquinone
TBA	Thiobarbituric acid
TBARS	Thiobarbituric acid reactive substances
USDA	U.S. Department of Agriculture
UV	Ultraviolet

References

Aalto-Korte, K., 2000. Allergic contact dermatitis from tertiary-butylhydroquinone (TBHQ) in a vegetable hydraulic oil. Case Reports 43 (5), 303.

Aguilar, F., Crebelli, R., Dusemund, B., Galtier, P., Gott, D., Gundert-Remy, U., Konig, J., Lambre, C., Leblanc, J.C., Mortensen, A., Mosesso, P., 2014. Scientific opinion on the re-evaluation of propyl gallate (E 310) as a food additive. EFSA J. 14, 3642.

Ahmed, M., Pickova, J., Ahmad, T., Liaquat, M., Farid, A., Jahangir, M., 2016. Oxidation of lipids in foods. Sarhad J. Agric. 32 (3), 230–238.

Amaral, A.B., Silva, M.V.D., Lannes, S.C.D.S., 2018. Lipid oxidation in meat: mechanisms and protective factors–a review. Food Sci. Technol. 38, 1–15.

Amorati, R., Foti, MC., Valgimigli, L., 2013. Antioxidant activity of essential oils. J. Agric. Food Chem. 61 (46), 10835–10847.

Anbudhasan, P., Surendraraj, A., Karkuzhali, S., Sathishkumaran, P., 2014. Natural antioxidants and its benefits. Int. J. Food Sci. Nutr 3, 225–232.

Antolovich, M., Prenzler, P.D., Patsalides, E., McDonald, S., Robards, K., 2002. Methods for testing antioxidant activity. Analyst 127, 183–198.

Anwar, F., Bhanger, M.I., Kazi, T.G., 2003. Relationship between rancimat and active oxygen method values at varying temperatures for several oils and fats. J. Am. Oil Chem' Soc. 80, 151–155.

Ayala, A., Muñoz, M.F., Argüelles, S., 2014. Lipid Peroxidation: Production, Metabolism, and Signaling Mechanisms of Malondialdehyde and 4-hydroxy-2-nonenal. Oxidative medicine and cellular longevity, Egypt.

Barriuso, B., Astiasarán, I., Ansorena, D., 2013. A review of analytical methods measuring lipid oxidation status in foods: a challenging task. Eur. Food Res. Technol. 6 (1), 1–5.

Bauernfeind, J.C., Pinkert, D.M., 1970. Food processing with added ascorbic acid. Adv. Food. Res. 18, 219–315.

Bekbölet, M., 1990. Light effects on food. J. Food Prot. 53, 430–440.

Bickford, W.G., Fisher, G.S., Dollear, F.G., Swift, C.E., 1948. Autoxidation of fats. I. Preparation and oxidation of alkylbenzene-maleic anhydride adducts. J. Am. Oil Chem' Soc 25, 251–254.

Bont, Rinne De., Larebeke, Nik Van., 2004. Endogenous DNA damage in humans: a review of quantitative data. Mutagenesis 19, 169–185.

Botterweck, A., Verhagen, H., Goldbohm, R., Kleinjans, J., Van Brandt, P., 2000. Intake of butylated hydroxyanisole and butylated hydroxytoluene and stomach cancer risk: Results from analyses in the Netherlands cohort study. Food Chem. Toxicol. 38, 599–605.

Bou, R., Codony, R., Tres, A., Decker, E.A., Guardiola, F., 2008. Determination of hydroperoxides in foods and biological samples by the ferrous oxidation–xylenol orange method: a review of the factors that influence the method's performance. Anal. Biochem 377, 1–15.

Brown, D.R., Kretzschmar, H.A., 1998. The glio-toxic mechanism of α-aminoadipic acid on cultured astrocytes. J. Neurocytol 27, 109–118.

Bull, A.W, Nigro, N.D., Marnett, L.J., 1988. Structural requirements for stimulation of colonic cell proliferation by oxidized fatty acids. Cancer Res. 48, 1771–1776.

Bull, A.W., Marnett, L.J., 1985. Determination of malondialdehyde by ion-pairing high-performance liquid chromatography. Anal. Biochem 149, 284–290.

Bull, A.W., Nigro, N.D., Golembieski, W.A., 1984. In vivo stimulation of DNA synthesis and induction of ornithine decarboxylase in rat colon by fatty acid hydroperoxides, autoxidation products of unsaturated fatty acids. Cancer Res. 44, 4924–4928.

Cerutti, G., 2006. Residui, additivi e contaminanti degli alimenti. Tecniche nuove, Italy.

Cheng, J.H., 2016. Lipid oxidation in meat. J Nutr Food Sci 6, 494.

Chiang, Y.F., Shaw, H.M., Yang, M.F., Huang, C.Y., Hsieh, C.H., Chao, P.M., 2011. Dietary oxidised frying oil causes oxidative damage of pancreatic islets and impairment of insulin secretion, effects associated with vitamin E deficiency. Br. J. Nutr. 105, 1311–1319.

Choe, E., Min, D.B., 2006. Mechanisms and factors for edible oil oxidation. Compr. Rev. Food Sci. Food Saf 5, 169–186.

Cruz, R.S., Camilloto, G.P., dos Santos Pires, A.C., 2012. Oxygen scavengers: an approach on food preservationStructure and Function of Food Engineering. InTech, Austria.

Csallany, A.S., Guan, M.D., Manwaring, J.D., Addis, P.B., 1984. Free malonaldehyde determination in tissues by high-performance liquid chromatography. Anal. Biochem. 142, 277–283.

Dastychová, E., Necas, M., Vasku, V., 2008. Contact hypersensitivity to selected excipients of dermatological topical preparations and cosmetics in patients with chronic eczema. Acta Dermatovenerol Alp Pannonica Adriat 17 (2), 61–68.

Directorate of Food and Drugs Administration, 2011. https://dfda.goa.gov.in/images/PDF-DOCUMENTS/Food-product-standards-and-food-additives-regulation2011partII.pdf. (Accessed 7 May, 2019).

Dunlop, R.A., Cox, P.A., Banack, S.A., Rodgers, K.J., 2013. The non-protein amino acid BMAA is misincorporated into human proteins in place of L-serine causing protein misfolding and aggregation. PLoS One 8, e75376.

Esterbauer, H., 1993. Cytotoxicity and genotoxicity of lipid-oxidation products. Am. J. Clin. Nutr. 57, 779S–786S.

Estévez, M., Li, Z., Soladoye, O.P., Van-Hecke, T., 2017. Health risks of food oxidation. Adv. Food. Nutr. Res. 82, 45–81.

Estevez, M., Luna, C., 2016. Dietary protein oxidation: A silent threat to human health? Crit. Rev. Food Sci. Nutr. 57, 3781–3793.

Fang, W., Sun, J., Li, Z.L., Le, G., Shi, Y., 2012. Effect of oxidated food protein on mice gut flora and redox state. Chin. J. Microecol 24, 193–196.

Farhoosh, R., Moosavi, S.M.R., 2009. Evaluating the performance of peroxide and conjugated diene values in monitoring quality of used frying oils. J. Agric. Sci. Technol. 11, 173–179.

Fenaille, F., Mottier, P., Turesky, R.J., Ali, S., Guy, P.A., 2001. Comparison of analytical techniques to quantify malondialdehyde in milk powders. J. Chromatogr. A 921, 237–245.

Festjens, N., Kalai, M., Smet, J., Meeus, A., Van Coster, R., Saelens, X., Vandenabeele, P., 2006. Butylated hydroxyanisole is more than a reactive oxygen species scavenger. Cell Death Differ. 13, 166–169.

Fisherman, E.W., Cohen, G.N., 1977. Chronic and recurrent urticaria: new concepts of drug-group sensitivity. Ann Allergy 39 (6), 404–414.

Flick, Jr., G.J., Hong, G.P., Knobl, G.M., 1992. Lipid Oxidation of Seafood During Storage. ACS SympSerAm Chem Soc, USA.

Food and Drug Administration, 2011. http://fda.up.nic.in/Legislation/regulations_2011/Food%20safety%20and%20standards%20(Food%20product%20standards%20and%20Food%20Additives)%20regulation,%202011.pdf (accessed 7 May, 2019).

Friedman, M., Cuq, J.L., 1988. Chemistry, analysis, nutritional value, and toxicology of tryptophan in food. A review. J. Agric. Food Chem. 36, 1079–1093.

Frikke-Schmidt, H., Tveden-Nyborg, P., Lykkesfeldt, J., 2011. Vitamin C in human nutrition. In: Herrmann, W., Obeid, R. (Eds.), Vitamins in the prevention of human disease. De Gruyter: Berlin, Germany, pp. 323–347.

Gharavi, N., Haggarty, S., S El-Kadi, A.O., 2007. Chemoprotective and carcinogenic effects of tert-butylhydroquinone and its metabolites. Curr. Drug Metab. 8 (1), 1–7.

Grootveld, M., Silwood, C.J., Addis, P., Claxson, A., Serra, B.B., Viana, M., 2001. Health effects of oxidized heated oils 1. Food Res. Int. 13, 41–55.

Gryn'ova, G., Hodgson, J.L., Coote, M.L., 2011. Revising the mechanism of polymer autooxidation. Org. Biomol. Chem 9, 480–490.

Gurer-Orhan, H., Ercal, N., Mare, S., Pennathur, S., Orhan, H., Heinecke, J.W., 2006. Misincorporation of free m-tyrosine into cellular proteins: a potential cytotoxic mechanism for oxidized amino acids. Biochem. J. 395, 277–284.

Gunstone, F.D., Harwood, J.L., 2007. Occurrence and characterisation of oils and fats. The lipid handbook. CRC press, United States, pp. 51–156.

Hamid, A.A., Aiyelaagbe, O.O., Usman, L.A., Ameen, O.M, Lawal, A., 2010. Antioxidants: its medicinal and pharmacological applications. Afr. J. Pure Appl. Chem. 4, 142–151.

Henry, L.N., 2016. Effect of light and air on the quality and stability of selected vegetable oils. Int. J. Innov. Res. Sci. Eng. Technol 5, 6609–6616.

Hocman, G., 1988. Chemoprevention of cancer: phenolic antioxidants (BHT, BHA). Int. J. Biochem. Cell B 20, 639–651.

Huang, S-W., Frankel, EN., Schwarz, K., Aeschbach, R., German, J.B., 1996. Antioxidant Activity of Carnosic Acid and Methyl Carnosate in Bulk Oils and Oil-in-Water Emulsions. J. Agric. Food Chem. 44 (10), 2951–2956.

Ishikawa, Y., Mine, S., 1983. Aminoadipic acid toxic effects on retinal glial cells. Jpn. J. Ophthalmol. 27, 107–118.

Ito, N., Fukushima, S., Hagiwara, A., Shibata, M., Ogiso., T., 1983. Carcinogenicity of butylated hydroxyanisole in F344 rats. J. Natl. Cancer I 70, 343–352.

Jardine, D., Antolovich, M., Prenzler, P.D., Robards, K., 2002. Liquid chromatography–mass spectrometry (LC-MS) investigation of the thiobarbituric acid reactive substances (TBARS) reaction. J. Agric. Food Chem. 50 (6), 1720–1724.

Jayalakshmi, C.P., Sharma, J.D., 1986. Effect of butylated hydroxyanisole (BHA) and butylated hydroxytoluene (BHT) on rat erythrocytes. Environ. Res. 41, 235–238.

Jin, Y., Bursley, J., Dover, H., Gangur, V., Rockwell Elizabeth, C., 2018. The effect of the common food additive tBHQ in OVA-elicited food allergy. J. Immunol. 200 (1), 104 .4.

Juhlin, L., 1981. Recurrent urticaria: clinical investigation of 330 patients. Br J Dermatol 104 (4), 369–381.

Kamal-Eldin, A., 2003. Analytical Methods Used to Evaluate Oxidation in Microencapsulated Oils. Lipid oxidation pathways. AOCS Publishing, Illinois, pp. 252–267.

Kanazawa, K., Kanazawa, E., Ntake, M., 1985. Uptake of secondary autoxidation products of linoleic acid by the rat. Lipids 20, 412–419.

Kanner, J., 2007. Dietary advanced lipid oxidation endproducts are risk factors to human health. Mol. Nutr. Food Res. 51, 1094–1101.

Kerry, J.P., 2012. Advances in Meat, Poultry and Seafood Packaging. Woodhead Publishing, Cambridge.

Khan-Merchant, N., Penumetcha, M., Meilhac, O., 2002. Oxidized fatty acids promote atherosclerosis only in the presence of dietary cholesterol in low density lipoprotein receptor knockout mice. J. Nutr. 132, 3256–3262.

Kim, E., Coelho, D., Blachier, F., 2013. Review of the association between meat consumption and risk of colorectal cancer. Nutr. Res. 33, 983–994.

Klipcan, L., Moor, N., Kessler, N., Safro, M.G., 2009. Eukaryotic cytosolic and mitochondrial phenylalanyl-tRNA synthetases catalyze the charging of tRNA with the meta-tyrosine. Proc. Natl. Acad. Sci. USA. 106, 11045–11048.

Koczoñ, P., Gruczyñska, E., Kowalski, B., 2008. Changes in the acid value of butter during storage at different temperatures as assessed by standard methods or by FT-IR spectroscopy. Am. J. Food Technol. 3, 154–163.

Kubler, W., Gehler, J., 1970. Kinetics of intestinal absorption of ascorbic acid. Calculation of non-dosage-dependent absorption processes. Int. Z. Vitaminforsch. 40, 442–453.

Laguerre, M., Lecomte, J., Villeneuve, P., 2007. Evaluation of the ability of antioxidants to counteract lipid oxidation: existing methods, new trends and challenges. Prog. Lipid Res. 46, 244–282.

Läubli, M.W., Bruttel, P.A., 1986. Determination of the oxidative stability of fats and oils: comparison between the active oxygen method (AOCS Cd 12-57) and the rancimat method. J. Am. Oil. Chem. Soc. 63, 792–795.

Le, Leu, R.K., Brown, I.L., Hu, Y., Morita, T., Esterman, A., Young, G.P., 2007. Effect of dietary resistant starch and protein on colonic fermentation and intestinal tumourigenesis in rats. Carcinogenesis 28, 240–245.

Li, Z.L., Mo, L., Le, G., Shi, Y., 2014. Oxidized casein impairs antioxidant defense system and induces hepatic and renal injury in mice. Food Chem. Toxicol. 64, 86–93.

Li, Z.L., Wu, L., Le, G., Shi, Y., 2013. Effect of oxidized casein on the oxidative damage of blood and digestive organs in mice. Acta Nutrimenta Sinica 35, 39–43.

Liao, M.L., Seib, P.A., 1988. Chemistry of L-ascorbic acid related to foods. Food Chem. 30, 289–312.

Lü, J.M., Lin, P.H., Yao, Q., Chen, C., 2010. Chemical and molecular mechanisms of antioxidants: experimental approaches and model systems. J. Cell. Mol. Med. 14, 840–860.

Marquez-Ruiz, G., Garcia-Martinez, M.C., Holgado, F., 2008. Changes and effects of dietary oxidized lipids in the gastrointestinal tract. Lipids Insights 2, 11–19.

Mayersohn, M., 1972. Ascorbic acid absorption in man—Pharmacokinetic implications. Eur. J. Pharmacol. 19, 140–142.

Medina, M.E., Iuga, C., Alvarez-Idaboy, J.R., 2013. Antioxidant activity of propyl gallate in aqueous and lipid media: a theoretical study. PCCP 15, 13137–13146.

Meisner, P., Gebicki, J.L., 2009. Determination of hydroperoxides in aqueous solutions containing surfactants by the ferrous oxidation-xylenol orange method. Acta Biochim. Pol. 56 (3), 523–527.

Miková, K., 2001. The regulation of antioxidants in food Handbook of Food Preservation. CRC Press, United States, pp. 267–284.

Morrissey, P.A., Sheehy, P.J., Galvin, K., Kerry, J.P., Buckley, D.J., 1998. Lipid stability in meat and meat products. Meat Sci. 49S1, S73–S86.

Muluken, K., Shimelis, E., 2019. Application of antioxidants in food processing industry: options to improve the extraction yields and market value of natural products. Adv. Food Technol. Nutr. Sci. 5, 38–49.

Mureşan, V., Muste, S., RACOLtA, E., Semeniuc, C.A., Man, S., Birou, A., Chircu, C., 2010. Determination of peroxide value in sunflower halva using a spectrophotometric method. Bull. UASVM Agric. 67, 334–339.

Navas, J.A., Tres, A., Codony, R., Boatella, J., Bou, R., Guardiola, F., 2004. Modified ferrous oxidation-xylenol orange method to determine lipid hydroperoxides in fried snacks. Eur. J. Lipid Sci. Technol. 106, 688–696.

Nayak, P.K., Dash, U., Rayaguru, K., Krishnan, K.R., 2016. Physio-chemical changes during repeated frying of cooked oil: a review. J. Food Biochem. 40, 371–390.

Niederhofer, L.J., Daniels, J.S., Rauzer, C.A., Greene, R.E., Marnett, L.J., 2003. Malonaldehyde, a product of lipid peroxidation is mutagenic in human cells. J. Biol. Chem. 278, 31426–31433.

O'Callaghan, Y.C., Woods, J.A., O'Brien, N.M., 2001. Comparative study of the cytotoxicity and apoptosis-inducing potential of commonly occurring oxysterols. Cell Biol. Toxicol. 17, 127–137.

Ooi, N., Chopra, I., Eady, A., Cove, J., Bojar, R., O'Neill, A.J., 2013. Antibacterial activity and mode of action of tert-butylhydroquinone (TBHQ) and its oxidation product, tert-butylbenzoquinone (TBBQ). J. Antimicrob. Chemother. 68 (6), 1297–1304.

Orton, D.I., Shaw, S., 2001. Allergic contact dermatitis from pharmaceutical grade BHA in Timodine, with no patch test reaction to analytical grade BHA. Case Reports 44 (3), 191–192.

Osawa, C.C., Gonçalves, L.A., Ragazzi, S., 2007. Determination of hydroperoxides in oils and fats using kits. J. Sci. Food Agric. 87, 1659–1666.

Papuc, C., Goran, G.V., Predescu, C.N., Nicorescu, V., 2017. Mechanisms of oxidative processes in meat and toxicity induced by postprandial degradation products: A review. Compr. Rev. Food Sci. Food Saf 16, 96–123.

Park, S., Lee, J.Y., Lim, W., You, S., Song, G., 2019. Butylated hydroxyanisole exerts neurotoxic effects by promoting cytosolic calcium accumulation and endoplasmic reticulum stress in astrocytes. J. Agric. Food Chem. 67, 9618–9629.

Pérez, J.A.M., Aguilar, T.A.F., 2013. Chemistry of natural antioxidants and studies performed with different plants collected in Mexico Ed. In: Morales-González, JA (Ed.), Oxidative Stress and Chronic Degenerative Diseases - A Role for Antioxidants. Intech, London, UK, pp. 59–81.

Phillips, J.E., Geng, Y., Mason, R.P., 2001. 7-Ketocholesterol forms crystalline domains in model membranes and murine aortic smooth muscle cells. Atherosclerosis 159, 125–135.

Pikul, J., Leszczynski, D.E., Kummerow, F.A., 1989. Evaluation of three modified TBA methods for measuring lipid oxidation in chicken meat. J. Agric. Food Chem. 37, 1309–1313.

Pietras, M.P., Pieszka, M., Bederska-Łojewska, D., Orczewska-Dudek, S., 2012. Cholesterol and lipid peroxides in animal products and health implications—a review. Ann. Anim. Sci 12, 25–52.

Pop, A.N., Berce, C., Bolfa, P.O., Nagy, A.N., Catoi, C., Dumitrescu, I.B., Silaghi-Dumitrescu, L., Loghin, F.E., 2013. Evaluation of the possible endocrine disruptive effect of butylated hydroxyanisole, butylated hydroxytoluene and propyl gallate in immature female rats. Farmacia 61, 202–211.

Popa, M., Glevitzky, I., Dumitrel, G.A., Glevitzky, M., Popa, D., 2017. Study on peroxide values for different oils and factors affecting the quality of sunflower oil. Scientific papers-series e-land reclamation earth observation & surveying environmental engineering 6, 137–140.

Raikos, V., 2017. Natural antioxidants for food applications: challenges and recent developments. EC Nutrition 8 (2), 33–34.

Rajinder, S., Chiara, L., Shylaja, J., Marek, A.S., Bhalchandra, A.D., Kazimierz, S.K., Herman, A.J., Lawrence, J.M., Lucy, M.A., David, E.G.S., 2001. Lobe specific increases in malondialdehyde DNA adduct formation in the liver of mice following injection of Helicobacter hepaticus. Carcinogenesis 22, 1281–1287.

Réblová, Z., 2012. Effect of temperature on the antioxidant activity of phenolic acids. Czech J. Food Sci. 30, 171–177.

Rizvi, S., Raza, S.T., Faizal, Ahmed, A.A., Abbas, S., Mahdi, F., 2014. The role of vitamin E in human health and some diseases. Sultan Qaboos Univ Med J. 14, e157.

Rodgers, K.J., Wang, H., Fu, S., Dean, R.T., 2002. Biosynthetic incorporation of oxidized amino acids into proteins and their cellular proteolysis. Free Radic. Biol. Med. 32, 766–775.

Salih, A.M., Smith, D.M., Price, J.F., Dawson, L.E., 1987. Modified extraction 2- thiobarbituric acid method for measuring lipid oxidation in poultry. Poult. Sci. 66, 1483–1488.

Santos-Fandila, A., Camino-Sánchez, F., Zafra-Gómez, A., 2014. Degradation markers in nutritional products a review. Austin J. Anal. Pharm. Chem 1, 1005.

Schuler, P., 1990. Natural antioxidants exploited commercially. Food Antioxidants. Springer, Dordrecht, pp. 99–170.

Shahidi, F., Zhong, Y., 2005. Antioxidants: regulatory status. Bailey's Industrial Oil and Fat Products. Memorial University of Newfoundland, Wiley, Canada.

Shantha, N.C., Decker, E.A., 1994. Rapid, sensitive, iron-based spectrophotometric methods for determination of peroxide values of food lipids. J. AOAC Int. 22 (2), 421–424.

Staprans, I., Pan, X.M., Joseph, H.R., Feingold, K.R., 1998. Oxidized cholesterol in the diet accelerates the development of aortic atherosclerosis in cholesterol-fed rabbits. Arterioscler. Thromb. Vasc. Biol. 18, 977–983.

Steele, R., 2004. Understanding and measuring the shelf-life of food. Woodhead Publishing, Cambridge.

Sun, Y.E., Wang, W.D., Chen, H.W., Li, C., 2011. Autoxidation of unsaturated lipids in food emulsion. Crit. Rev. Food Sci. Nutr 51, 453–466.

Surai, K.P., Surai, P.F., Speake, B.K., Sparks, N.H., 2003. Antioxidant-prooxidant balance in the intestine: food for thought. 1. Prooxidants. Nutritional Genomics & Functional Foods 1, 51–70.

Tesoriere, L., D'Arpa, D., Butera, D., Pintaudi, A.M., Allegra, M., Livrea, M.A., 2002. Exposure to malondialdehyde induces an early redox unbalance preceding membrane toxicity in human erythrocytes. Free Radic. Res. 36, 89–97.

Thorat, I.D., Jagtap, D.D., Mohapatra, D., Joshi, D.C., Sutar, R.F., Kapdi, S.S., 2013. Antioxidants, their properties, uses in food products and their legal implications. Int. J. Food Stud 2.

Thune, P., Granholt, A., 1975. Provocation tests with antiphlogistica and food additives in recurrent urticaria 1975. Dermatology 151 (6), 360–367 151 (6), 360–367.

Tosti, A., Bardazzi, F., Valeri, F., Russo, R., 1987. Contact dermatitis from butylated hydroxyanisole. Contact Dermatitis 17 (4), 257–258.

Wahlqvist, M.L., 2013. Antioxidant relevance to human health. Asia Pac. J. Clin. Nut 22, 171–176.

Waraho, T., McClements, D.J., Decker, E.A., 2011. Mechanisms of lipid oxidation in food dispersions. Trends Food Sci. Technol. 22, 3–13.

Wattenberg, L.W., 1972. Inhibition of carcinogenic and toxic effects of polycyclic hydrocarbons by phenolic antioxidants and ethoxyquin. J. Natl. Cancer I 48, 1425–1430.

Wattenberg, L.W., 1986. Protective effects of 2(3)-tertbutyl-4-hydroxyanisole on chemical carcinogenesis. Food Chem. Toxicol. 24, 1099–1102.

Wattenberg, L.W., Coccia, J.B., Lam, L.K., 1980. Inhibitory effects of phenolic compounds on benzo(a)pyreneinduced neoplasia. Cancer Res. 40, 2820–2823.

Williams, G.M., 1986. Epigenetic promoting effects of butylated hyroxyanisole. Food Chem. Toxicol. 24, 1163–1166.

Williams, G.M., Iatropoulos, M.J., 1996. Inhibition of the hepatocarcinogenicity of a¯atoxin B1 in rats by low levels of the phenolic antioxidants butylated hydroxyanisole and butylated hydroxytoluene. Cancer Lett. 104, 49–53.

Winter, R., 1972. A consumer's dictionary of food additives. Crown Publishers Inc., NY, pp. 143.

Xie, F., Sun, S., Xu, A., Zheng, S., Xue, M., Wu, P., Zeng, J.H., Bai, L., 2014. Advanced oxidation protein products induce intestine epithelial cell death through a redox-dependent, c-jun N-terminal kinase and poly (ADP-ribose) polymerase-1-mediated pathway. Cell Death. Dis. 5, e1006.

Yu, T.Y., Morton, J.D., Clerens, S., Dyer, J.M., 2017. Cooking-induced protein modifications in meat. Compr. Rev. Food Sci. Food Saf 16, 141–159.

PART

Endogenous antioxidants

2

CHAPTER

Bilirubin

2.1

Mohamad Fawzi Mahomoodally, Jankee T. Laxmi

Department of Health Sciences, Faculty of Medicine and Health Sciences, University of Mauritius, Réduit, Mauritius

2.1.1 Origin and structure

A crucial heme metabolite originally isolated from a German scientist, Rudolf Virchow and his colleagues (American Chemical Society, 2019), bilirubin aids in the placement of iron in numerous proteins complexes (Kalakonda and John, 2018). With a molecular formula of $C_{33}H_{36}N_4O_6$ this brownish yellow endogenous compound, has an opened-ring porphyrin structure, with a chain of substituted pyrrole and 1, 3-dihydro-2-pyrrol-2-one rings (Chambers, 2019). It is synthesized in the bone marrow cells within the liver as the end-product of red blood cell (hemoglobin) degradation and is responsible for the coloration of feces (Augustyn et al., 2017).

2.1.2 Bilirubin synthesis

The synthesis of bilirubin occurs through two major pathways, where 80% are *via* the breakdown of hemoglobin in senescent red blood cells together with shattered erythroid cells within the bone and the rest of 20% originates from different heme-containing proteins such as myoglobin, cytochrome, catalase, peroxidase, and tryptophan pyrrolase found in other tissues, principally the liver and muscles (Ponka, 1999). Heme which is a small 4 pyrroles pentagon-shaped structure made up from 4 carbons and 1 nitrogen are found in these proteins (Bhagavan and Ha, 2015). The four pyrroles of heme together shape a tetrapyrrole and presence of substitutions on the side chains of tetrapyrrole, permits the structure to hold metals such as iron, forming a porphyrin (Blackstone, 2001). Heme are acted upon by the enzyme heme oxygenase which frees the chelated iron by catalysing the oxidation of the alpha carbon bridge. This response produces an equimolar sum of carbon monoxide which is excreted by the lungs and leads to the development of the green pigment biliverdin. This green pigment is then converted to bilirubin via biliverdin reductase with the assistance of nicotinamide adenine dinucleotide (Perlman and Volpe, 2018).

Antioxidants Effects in Health. DOI: https://doi.org/10.1016/B978-0-12-819096-8.00021-5

2.1.3 Bilirubin metabolism and excretion

Once synthesized, bilirubin is discharged within the plasma and taken up by albumin (Wang and Chowdhury, 2006). Bilirubin has a high binding affinity to the albumin transporter. The binding limits the evasion of bilirubin from the vascular spaces, minimises glomerular filtration and thus prevents its precipitation and deposition in tissues. When the albumin-bilirubin complex reaches the liver, the exceedingly penetrable hepatic circulation permits the complex to reach the sinusoidal surface of hepatocytes. This allow the dissociation of bilirubin from albumin and its entrance to the hepatocytes either as conjugated or unconjugated form. Unconjugated or indirect bilirubin do not dissolve in water and travel through the bloodstream to the liver, where they changed into soluble form (direct or conjugated) (Erlinger and Dhumeaux, 2014). A larger part of the unconjugated bilirubin entering the hepatocytes is extricated within the periportal locale (Perlman and Volpe 2018). Moreover, a fraction of the conjugated and unconjugated bilirubin inside the hepatocytes is transported back into the sinusoidal space while the conjugated bilirubin that gets away from being re uptake, is excreted within the urine.

Conjugated bilirubin and other substances predetermined to be excreted in the bile are effectively transported over the bile canalicular membrane of the hepatocytes (Kalakonda and John, 2018). The conjugated bilirubin, at that point, effectively discharges into the canalicular bile and channels into the little digestive tract. Bacteria, within the intestinal lumen, metabolise bilirubin and produce a compound known as urobilinogen. The urobilinogen, which is colourless, is then excreted in the form of yellow colour in the urine termed as urobilin (Anon, 2013). The urobilin gets reduced in the intestine to stercobilin and this confers the brown colour of stool. Eventually some of the urobilinogen reaches back the liver via the hepatic portal vein and get metabolised and excreted in bile (Greenberg, 2014).

2.1.4 Bilirubin as an antioxidant

Bilirubin is considered as a harmful waste product responsible for liver infection and other illnesses, including hepatitis, kernicterus and jaundice (Sedlak and Snyder 2004). Quite a few mechanisms have been anticipated in how bilirubin can act as impending endogenous antioxidants. Apart from having antioxidant bilirubin has shown to be effective in preventing other diseases as shown in Table 2.1.1. Table 2.1.2 provides some in vitro, in vivo, and clinical studies that support the fact about the beneficial effect of bilirubin as an antioxidant.

2.1.5 Bilirubin and its potential detrimental effects

Bilirubin is known to be a toxic metabolite that ought to be eliminated from the body *via* urination or defecation. Bilirubin that circulates within the body is either in conjugated or unconjugated form. Impairment can occur which leads to

Table 2.1.1 In vitro, in vivo, and clinical studies on bilirubin as endogenous antioxidant.

Illness	Benefits	References
Cardiovascular		
Relationship between bilirubin levels and the risk of ischemic heart disease was studied in 7685 British men.	Men with lowest concentration of bilirubin were seen to have the greatest chance of ischemic heart diseases. Thus, illustrating the fact that bilirubin aid in preventing ischemic heart diseases.	(Gupta et al., 2016)
A case control study was conducted on 1240 Utah adults that had an early onset of coronary artery disease.	It was found that those who had a higher bilirubin level were less likely to develop an early onset of coronary artery disease compared to those that had a lower level.	(Hunt et al., 1996)
A prospective cohort study was conducted using Veterans Aging Cohort Study where cardiovascular diseases as well as acute myocardial infarction, heart failure, and ischemic stroke events were assessed.	It was found that those VACS participants that had an elevated bilirubin levels had a lower risk of incident of total CVD acute myocardial infarction, heart failure, and ischemic stroke events after adjusting for known risk factors.	(Marconi et al., 2018)
A study was conducted with 1741 subjects who underwent screening for carotid artery disease. A link between amount of bilirubin and plaque formation were establish by measuring the level of bilirubin in those subjects.	The study concluded that bilirubin levels in the highest quartile were linked to a 32% reduction in risk of formation of plaques.	(Ishizaka et al., 2001)
A prospective study among Korean men and women were carried out to observe the benefits of bilirubin and its role in decrease and managing CVD	Findings suggested that serum bilirubin confer protective function against stroke risk in men	(Kimm et al., 2009)
Retinopathy		
Relationship between low bilirubin level and retinopathy were studied in 45 premature babies in a clinical study.	It was reported that low level of bilirubin was associated with the worsening of retinopathy, hence indicating that bilirubin is an eye protector substance.	(Sedlak and Snyder 2004)
Diabetes		
A randomised, controlled clinical study was conducted on a large sample size of diabetic patients to establish a possible relationship between bilirubin and diabetes.	The study concluded that there was a possible causal association between an increase amount of total bilirubin and the decrease of type 2 diabetes.	(Abbasi et al., 2014)

Table 2.1.2 Health benefits of bilirubin.

Design of study	Bilirubin's effects	References
In vitro		
Re-crystallised bilirubin was dissolved in 0.05M NaOH solution prior to its addition to a phosphate-buffered solution of essentiall fatty acid. Purified linoleic acid was added to an albumin solution as an aqueous dispersion and stirred until the solution was clear. Cold AAPH was then added and the reaction was initiated by placing the reaction tube in a 37^0 water bath. Aliquots were evacuated, and the bile pigments extricated by addition of 1 volume of reaction blend to 4 volume of cold methanol. The protein was pelleted, supernatant evacuated, and an aliquot was examined and quantitated for bilirubin and its oxidation item by HPLC. Uric acid and ascorbic acid were utilised along the bilirubin to compare the peroxyl radical binding abilities.	Albumin-bound bilirubin was seen to oxidised at the same rate of peroxyl radicals formation and biliverdin was synthesised stoichiometrically as the oxidation item. On an equimolar preface, it was observed that Alb-BR effectively competes with uric acid for peroxyl radicals but is less proficient in scavenging these radicals than vitamin C. The results illustrated that 1 mol of Alb-BR can scavenge 2 mol of peroxyl radicals. It demonstrated that small amount of plasma bilirubin was enough to anticipate oxidation of albumin-bound fatty acids. The data depicted a portion for Alb-BR as a physiological antioxidant in plasma and in vascular spaces.	(Stocker et al., 1987)
10 nM of bilirubin were given to treat HeLa cells that were damaged by the addition 100 µM of hydrogen peroxide.	It was found that treatment with bilirubin increased the viability of HeLa cells in the presence of toxic concentrations of hydrogen peroxide.	(Baranano et al., 2002)
The role of bilirubin and biliverdin in inhibiting peroxynitrite-mediated protein tyrosine nitration was tested. It was conducted using bovine serum egg, where tyrosine accumulation was made either by nitration via peroxynitrite (ONOO–) or *in situ* produced from the thermal decay of 3-morpholino-sydnonimine.	Peroxynitrite-mediated tyrosine nitration involves tyrosyl-radical-intermediates as a result of homolytic bond cleavage in ONO-OH yielding hydroxyl (HO•) and nitrogen dioxide (NO_2•) radicals. These radicals are known to cause DNA damage and lead to several point mutations. Based on the results obtained, both bilirubin and biliverdin were capable of inhibiting peroxynitrite-mediated protein tyrosine nitration, with bilirubin being threefold more potent than biliverdin.	(Jansen et al.,2010)

Design of study	Bilirubin's effects	References
The ability of bilirubin to scavenge superoxide radicals was tested via two diverse frameworks. In the primary one superoxide was ceaselessly being generated by xanthine oxidase (XO) and hypoxanthine, while for the second, authentic superoxide was utilised to demonstrate any inhibitory effect of bilirubin on enzymatic activity. The results were obtained via an HPLC.	A decline in superoxide genesis by XO was observed on treatment with bilirubin. Furthermore, bilirubin showed inhibitory effects on enzyme activity.	(Jansen et al.,2010)
The effect of bilirubin was tested on the enzyme superoxide dismutase, catalase, and glucose-6-phosphate dehydrogenase and on cumene hydroperoxide-treated erythrocytes. Bilirubin were added at different concentrations and thiobarbituric acid-reactive substance, reduced glutathione levels, and the enzyme activities were measured.	The results showed that bilirubin had moderately barred the oxidant effects of cumene hydroperoxide and a greater increase in glucose-6-phosphate dehydrogenase and reduced glutathione activity were shown in comparison to the rest when treated with bilirubin	(Yeşilkaya et al., 1998)
The antioxidant level of bilirubin was determined using a ROS-sensitive fluorophore and dichlorofluoresin diacetate. Bilirubin was incubated with ROS *Escherichia coli* or isolated human neutrophils and added to a buffer solution where its concentration was monitored spectrophotometrically. Its effect was determined by counting viable *E. coli* after it was incubated on a nutrient agar.	It was found that bilirubin had the ability to attenuate the bacterial activity of the *E. coli* or of isolated neutrophils through scavenging ROS.	(Arai et al., 2001)
In vivo		
The health benefits of an increase bilirubin level on NADPH oxidase complex were investigated in rat lab. Bilirubin was intraperitoneally administered to rats that were fed a high fat and sugar diet daily for 14 days. The effect of bilirubin on the weight of the rats were examined and noted.	Leptin resistance is known to be present in obese rodents, due ingestion of high caloric food. This resistance leads to the activation and proliferation of microglia in the arcuate nucleus, which can produce cytokines as well as activate NADPH oxidase complexes, responsible for pro-inflammatory mediators release. It was found that leptin insensitivity was drastically reduced, along secretion of pro-inflammatory molecule secretion when bilirubin was injected to the rats.	(DiNicolantonio et al., 2019)

(Continued)

Table 2.1.2 Health benefits of bilirubin. *Continued*

Design of study	Bilirubin's effects	References
The potency of bilirubin to scavage free radical were investigated in freshly isolated hepatocytes of rats. 100 μmol/l of glycochenodeoxycholate (GCDC) were added to the hepatocytes together with unconjugated and conjugated bilirubin separately. It was then incubated for 4 hours and the amount of ROS generated were tested using fluorescence at 490 nm and 520 nm.	It was found that both UCB and CB inhibited GCDC induced apoptosis.	(Granato et al., 2003)
The benefit of bilirubin was tested for CD4+ T cells infiltration and pro-inflammatory cytokine expression in the brain tissue of rodent model having multiple sclerosis in experimental autoimmune encephalomyelitis.	When bilirubin was injected to the rodent model, a reduction in CD4+ T cells infiltration and pro-inflammatory cytokine expression was seen.	(Liu et al., 2008)
An animal model study was conducted on rats to investigate the antioxidant benefit of bilirubin on tubular injuries and intestinal fibrosis caused by oxidative stress. The animals were injected intraperitoneal bilirubin and the level of generated superoxide in the vascular endothelial cells and renal tubular cells were investigated.	Bilirubin demonstrated to lower the level of ROS in the aanimal models providing protection against oxidative stress as well as cell apoptosis.	(Ratliff et al., 2016)
Benefit of bilirubin against ischemia-reperfusion injury in rat kidney were investigated where rat models were injected with 10 μM bilirubin and histological examination were performed.	It was observed that bilirubin injected to the isolated perfused kidney model conferred renal protection.	(Adin et al.,2005)
The antihypertensive benefit of bilirubin was investigated on mice model where HO-1 enzyme activity was supressed in superimposed uni-nephrectomy mice via the administration of DOCA.	The development of systolic hypertension was observed in mice with having the HO-1 knocked off compared to those with the normal enzymatic function.	(Nath et al., 2007)
Clinical studies		
Respiratory distress, circulatory failure, sepsis, aspiration, and asphyxia was seen to be due to ROS in 44 infants. ROS generation in those infant lead to a decline in serum bilirubin in comparison with those neonates who were ill from non-oxidative disease. Hence to investigate the efficacy of bilirubin against ROS, bilirubin was injected to those 44 infants.	An increase in serum bilirubin was seen to manage those free radicals causing infant illnesses.	(Rogers et al., 2000; Gupta et al., 2004)

Design of study	Bilirubin's effects	References
It was seen that 25 preterm infants that were suffering from oxygen-radical diseases such as intraventricular haemorrhage, retinopathy, bronchopulmonary dysplasia, and necrotising enterocolitis had significantly low level of bilirubin in comparison to 57 control cases.	It was concluded that low level of bilirubin plays a major role in oxygen radical diseases in preterm infants.	(Hegyi et al., 2018)
ROS generation is found to be high in people suffering from sepsis. Study conducted showed that bilirubin has the ability to form bilirubin oxidative metabolites that gets excreted via the urine. A clinical study was engineered on 47 people with 19 septic as case and 28 non septic people as control to investigate the amount of bilirubin oxidative metabolites in the urine.	On the conduction of the investigation an increase in bilirubin oxidative metabolites in the urine of 19 septic patients was observed in comparison with the 28 control subjects. Bilirubin oxidative metabolites are generated from bilirubin due to its scavenging action against free radicals. The genesis of BOM evinced the fact that bilirubin has the ability to bind with ROS and thus prevent further damages caused by it. This can prove to be beneficial to sepsis suffers	(Otani et al., 2001)
Grafts tissues were being taken from mice model and its acceptance and tolerance level as well as the pro-inflammatory substances were measured.	Bilirubin treatment has shown to induce expansion of Foxp3-positive regulatory T cells and suppress the expression of pro-inflammatory and pro-apoptotic genes in islet grafts.	(Rocuts et al., 2010) (Tananka et al., 2009)
A clinical study was conducted on 637 participants to establish a link between coronary artery calcification and low level of serum bilirubin via tomography method.	It was observed that low concentration of bilirubin is strongly associated with the genesis of coronary artery calcification, and when the concentration increases the amount of calcification demonstrated a decline.	(Bhuiyan et al., 2008)
The relationship between bilirubin and radial artery pulse pressure was conducted via a clinical study with 777 individuals where the pulsatile arterial function was investigated in terms of small and large artery via radical artery pressure pulse contour analysis.	Results concluded the fact that bilirubin had a positive correlation with radial artery pulse pressure.	(Yoshino et al., 2011)
A clinical study was conducted on 107 obese patients without coronary heart disease to investigate relationship between coronary endothelial function and bilirubin.	The results depicted the fact that high bilirubin level lead to favourable coronary endothelium function as well as bilirubin depicted anti-inflammatory properties as well.	(Han et al., 2010)
In a Korean population of 93,909 individuals a cross-sectional study was carried out to investigate the relation between diabetes mellitus, CKD originated from diabetes mellitus and serum bilirubin.	Based on the results obtained it could be concluded that a high serum bilirubin level was strongly associated with a decline in diabetes prevalence as well as diabetes induced CKD.	

an overproduction or accumulation of bilirubin resulting in hyperbilirubinemia or a decrease in its concentration leading to hypobilirubinemia (William, 2019). Unconjugated hyperbilirubinemia is reported to cause haemolytic jaundice, Cigler-Najjar syndrome, and Gilbert syndrome predominately, while the conjugated one results in Dubin-Johnson syndrome, rotor syndrome and recurrent intrahepatic cholestasis (Mesquita and Casartelli, 2017). A retrospective cohort analysis was conducted by Hodgson et al. in 2018 on neonatal with jaundice, and it was seen that those that had an elevated level of bilirubin was more susceptible to neonatal liver disease. Another study conducted by Lang et al. in 2014, showed that conjugated bilirubin triggered anaemia by inducing erythrocyte death both *in vitro* and *in vivo*. Unconjugated bilirubin was observed to activate and damage microglia cells via morphological changes and releasing high levels of TNF-α, IL-1β, and IL-6 in a concentration-dependent manner. It was also noted that unconjugated bilirubin triggered extracellular accumulation of glutamate and cell death through apoptosis and necrosis (Gordon et al., 2006).

Conclusion

The use and exploitation of bilirubin, as an endogenous antioxidant, is experiencing a new momentum in the medical field. Its potential benefits have been demonstrated via several *in vitro*, *in vivo* studies and clinical studies as. Whether bilirubin is detrimental or beneficial to human this is still a key question that should be answered with more scientific and clinical studies.

References

Abbasi, A., Deetman, P.E., Corpeleijn, E., Gansevoort, R.T., Gans, R.O.B., Hillege, H.L., van der Harst, P., Stolk, R.P., Navis, G., Alizadeh, B.Z., Bakker, S.J.L., 2014. Bilirubin as a potential causal factor in type 2 diabetes risk: a Mendelian randomization study. Diabetes 64 (4), 1459–1469.

Adin, C.A., Croker, B.P., Agarwal, A., 2005. Protective effects of exogenous bilirubin on ischemia-reperfusion injury in the isolated, perfused rat kidney. Am. J. Physiol. - Renal Physiol. 288 (4), F778–F784.

American Chemical Society., 2019. Bilirubin. https://www.acs.org/content/acs/en/molecule-of-the-week/archive/b/bilirubin.html (accessed 9 April 2019).

Arai, T., Yoshikai, Y., Kamiya, J., Nagino, M., Uesaka, K., Yuasa, N., Oda, K., Sano, T., Nimura, Y., 2001. Bilirubin impairs bactericidal activity of neutrophils through an antioxidant mechanism in vitro. J. Surg. Res. 96 (1), 107–113.

Augustyn, A., Brian, P., Duignan Eldridge, A., Gregersen, E., Amy., Petruzzello, M., John, P., Ray, M., Rogers, K., Tikkanen, A., Wallenfeldt, J., Zeidan, A., Zelazko, A., 2017. Bilirubin biochemistry. Encyclopædia Britannica. https://www.britannica.com/science/bilirubin accessed 8 April 2019.

Baranano, D.E., Rao, M., Ferris, C.D., Snyder, S.H., 2002. Biliverdin reductase: a major physiologic cytoprotectant. Proc. Natl. Acad. Sci. 99 (25), 16093–16098.

Bhagavan, N.V., Ha, C.-E., 2015. Metabolism of iron and HemeEssentials of Medical Biochemistry, 2nd ed. Academic Press, USA.

Bhuiyan, A.R., Srinivasan, S.R., Chen, W., Sultana, A., Berenson, G.S., 2008. Association of serum bilirubin with pulsatile arterial function in asymptomatic young adults: the Bogalusa Heart Study. Metabolism 57 (5), 612–616.

Blackstone, N.W., Lodish, H., Berk, A., Zipursky, L., Matsudaira, P., Baltimore, D., Darnell, J., 2001. Molecular cell biology. Q. Rev. Biol. 76 (1), 76.

Chambers, M., 2019. Bilirubin - similar structures search, synonyms, formulas, resource links, and other chemical information. https://chem.nlm.nih.gov/chemidplus/sid/0000635654.

DiNicolantonio, J.J., McCarty, M., OKeefe, J., 2019. Does elevated bilirubin aid weight control by preventing development of hypothalamic leptin resistance? Open Heart 6 (1), e000897.

Erlinger, S., Arias, I.M., Dhumeaux, D., 2014. Inherited disorders of bilirubin transport and conjugation: New insights into molecular mechanisms and consequences. Gastroenterology 146 (7), 1625–1638.

Gordo, A.C., Falcão, A.S., Fernandes, A., Brito, M.A., Silva, R.F.M., Brites, D., 2006. Unconjugated bilirubin activates and damages microglia. J. Neurosci. Res. 84 (1), 194–201.

Granato, A., Gores, G., Vilei, M.T., Tolando, R., Ferraresso, C., Muraca, M., 2003. Bilirubin inhibits bile acid induced apoptosis in rat hepatocytes. Gut 52 (12), 1774–1778.

Greenberg, A., 2014. Urinalysis and urine microscopy. National Kidney Foundation. Primer on Kidney Diseases, 33–41.

Gupta, N., Singh, T., Chaudhary, R., Garg, S.K., Sandhu, G.S., Mittal, V., Gupta, R., Bodin, R., Sule, S., 2016. Bilirubin in coronary artery disease: cytotoxic or protective? World J. Gastrointest. Pharmacol. Ther. 7 (4), 469.

Han, S.S., Na, K.Y., Chae, D.W., Kim, Y.S., Kim, S., Chin, H.J., 2010. High serum bilirubin is associated with the reduced risk of diabetes mellitus and diabetic nephropathy. Tohoku J. Exp. Med. 221 (2), 133–140.

Hegyi T., Goldie E., Hiatt M., 1994. The protective role of bilirubin in oxygen-radical diseases of the preterm infant. J. Perinatol. 14 (4), 296–300. PMID: 7965225.

Hodgson, J.M., Someren, V.H., Smith, C., Goyale, A., 2018. Direct bilirubin levels observed in prolonged neonatal jaundice: a retrospective cohort study. BMJ Paediatr. Open 2 (1), e000202. doi: 10.1136/bmjpo-2017-000202. PMID: 29637188; PMCID: PMC5843007.

https://www.semanticscholar.org/paper/The-protective-role-of-bilirubin-in-oxygen-radical-Hegyi-Goldie/e3e60fe8de48aa7c359a5ebb17867171e925bacb (accessed 9 April 2019).

Hunt, S.C., Wu, L.L., Hopkins, P.N., Williams, R.R., 1996. Evidence for a major gene elevating serum bilirubin concentration in Utah pedigrees. Arterioscler. Thromb. Vasc. Biol. 16 (8), 912–917.

Ishizaka, N., Ishizaka, Y., Takahashi, E., Yamakado, M., Hashimoto, H., 2001. High serum bilirubin level is inversely associated with the presence of carotid plaque. Stroke 32 (2), 580–583.

Jansen, T., Hortmann, M., Oelze, M., Opitz, B., Steven, S., Schell, R., Knorr, M., Karbach, S., Schuhmacher, S., Wenzel, P., Münzel, T., Daiber, A., 2010. Conversion of biliverdin to bilirubin by biliverdin reductase contributes to endothelial cell protection by heme oxygenase-1—evidence for direct and indirect antioxidant actions of bilirubin. J. Mol. Cell Cardiol. 49 (2), 186–195.

Kalakonda, A. and John, S., 2018. *Physiology, bilirubin.* https://www.ncbi.nlm.nih.gov/books/NBK470290/ (accessed 8 April 2019).

Kimm, H., Yun, J.E., Jo, J., Jee, S.H., 2009. Low serum bilirubin level as an independent predictor of stroke incidence. Stroke 40 (11), 3422–3427.

Labpedia.net, 2013. Urine for urobilinogen and ehrlich reagent. http://www.labpedia.net/test/185. (Accessed on: 8 April 2018).

Lang, E., Gatidis, S., Freise, N.F., Bock, H., Kubitz, R., Lauermann, C., Orth, H.M., Klindt, C., Schuier, M., Keitel, V., Reich, M., Liu, G., Schmidt, S., Xu, H.C., Qadri, S.M., Herebian, D., Pandyra, A.A., Mayatepek, E., Gulbins, E., Lang, F., Häussinger, D., Lang, K.S., Föller, M., Lang, P.A., 2014. Conjugated bilirubin triggers anemia by inducing erythrocyte death. Hepatology 61 (1), 275–284.

Liu, Y., Li, P., Lu, J., Xiong, W., Oger, J., Tetzlaff, W., Cynader, M., 2008. Bilirubin possesses powerful immunomodulatory activity and suppresses experimental autoimmune encephalomyelitis. J. Immunol. 181 (3), 1887–1897.

Marconi, V.C., Duncan, M.S., So-Armah, K., Re, V.L., Lim, J.K., Butt, A.A., Goetz, M.B., Rodriguez-Barradas, M.C., Alcorn, C.W., Lennox, J., Beckman, J.A., Justice, A., Freiberg, M., 2018. Bilirubin is inversely associated with cardiovascular disease among HIV-positive and HIV-negative individuals in VACS (Veterans Aging Cohort Study). J. Am. Heart Assoc. 7 (10).

Mesquita, M., Casartelli, M., 2017. Neonatal hyperbilirubinemia, acute bilirubin encephalopathy and Kernicterus: The continuum is still valid in the 21st century. Pediatría (Asunción) 44 (2), 153–158.

Nath, K.A., d'Uscio, L.V., Juncos, J.P., Croatt, A.J., Manriquez, M.C., Pittock, S.T., Katusic, Z.S., 2007. An analysis of the DOCA-salt model of hypertension in HO-1−/− mice and the Gunn rat. Am. J. Physiol. - Heart Circ. Physiol. 293 (1), H333–H342.

Otani, K., Shimizu, S., Chijiiwa, K., Yamaguchi, K., Kuroki, S., Tanaka, M., 2001. Increased urinary excretion of bilirubin oxidative metabolites in septic patients: a new marker for oxidative stress in vivo. J. Surg. Res. 96 (1), 44–49.

Perlman, J.M., Volpe, J.J., 2018. BilirubinVolpe's Neurology of the Newborn, 4th ed, 730–762. https://www.elsevier.com/books/volpes-neurology-of-the-newborn/volpe/978-0-323-42876-7.

Ponka, P., 1999. Cell biology of heme. Am. J. Med. Sci. 318 (4), 241.

Ratliff, B.B., Abdulmahdi, W., Pawar, R., Wolin, M.S., 2016. Oxidant Mmechanisms in renal injury and disease. Antioxid. Redox Signaling 25 (3), 119–146.

Rocuts, F., Zhang, X., Yan, J., Yue, Y., Thomas, M., Bach, F.H., Czismadia, E., Wang, H., 2010. Bilirubin promotes de novo generation of T regulatory cells. Cell Transplant. 19 (4), 443–451.

Rogers, S., Witz, G., Anwar, M., Hiatt, M., Hegyi, T., 2000. Antioxidant capacity and oxygen radical diseases in the preterm newborn. Arch. Pediatr. Adolesc. Med. 154 (6), 544.

Sedlak, T.W., Snyder, S.H., 2004. Bilirubin benefits: cellular protection by a biliverdin reductase antioxidant cycle. Pediatrics 113 (6), 1776–1782.

Stocker, R., Glazert, A., Ames, B., 1987. Antioxidant activity of albumin-bound bilirubin (reactive oxygen species/plasma antioxidants/biliverdin/evolution). Med. Sci. 84, 5918–5922.

Wang, X., Chowdhury, J.R., Chowdhury, N.R., 2006. Bilirubin metabolism: Applied physiology. Current Paediatrics 16 (1), 70–74.

Yeşilkaya, A., Yeğin, A., Özdem, S., Aksu, T.A., 1998. The effect of bilirubin on lipid peroxidation and antioxidant enzymes in cumene hydroperoxide-treated erythrocytes. Int. J. Clin. Lab. Res. 28 (4), 230–234.

Yoshino, S., Hamasaki, S., Ishida, S., Kataoka, T., Yoshikawa, A., Oketani, N., Saihara, K., Okui, H., Shinsato, T., Ichiki, H., Kubozono, T., 2011. Relationship between bilirubin concentration, coronary endothelial function, and inflammatory stress in overweight patients. J. Atheroscler. Thromb. 18 (5), 403–412. DOI https://doi.org/10.5551/jat.6346.

CHAPTER

Catalase

2.2

Mohamad Fawzi Mahomoodally, Daphne Désiré A.-L., Elodie Rosette M. A.-L.

Department of Health Sciences, Faculty of Medicine and Health Sciences, University of Mauritius, Réduit, Mauritius

2.2.1 Introduction

Catalase, a key clinical enzyme, is involved in the breakdown of hydrogen peroxide (H_2O_2) to water and molecular oxygen by using iron or manganese as cofactor. It is a common antioxidant enzyme which presents in most living tissues that use oxygen (Ighodaro and Akinloye, 2018). In the 19th century, the origin of catalase (EC 1.11.1.6) was reported by Louis Jacques Thénard who discovered H_2O_2 and questioned that tissue deterioration in living organisms was due to a "special" substance activity. In 1863, Schobein demonstrated that the "ferment" could detoxify H_2O_2 and after sometimes the name "catalase" was given by Oscar Loew to the enzyme (Glorieux and Calderon, 2017). Catalase is a tetrameric enzyme which consists of four interchangeable subunits of 60 KDa arranged in a tetrahedral position where each subunit contains an iron heme group and NADPH in its active center. The enzyme has one of the highest turnover rate of all enzymes where the ultimate turnover rate for hydrogen peroxide is approximately 16,000,000–44,000,000 s^{-1} per folded tetra-heme molecule with each chain comprising of over 500 amino acids (Karadag and Fadil, 2015).

From an old-fangled study conducted by Morgulis et al. in 1926, it was found that independently of the concentration of H_2O_2 present, the activity of catalase is at its peak between 0 to 10°C. Thus, increasing temperature will cause a decrease in catalase activity. A decrease in catalase activity has been linked to mutations in the catalase genes leading to the development of various diseases. This mutation can alter the expression or activity of this antioxidant enzyme. There are two main categories of genetic deficiencies of erythrocyte catalase; (1) hypocatalasemia: reduced catalase activity about 50% and (2) acatalasemia: characterised by very low catalase activity, that is, less than 10%. However, this condition is relatively rare.

Gene alterations as well as the deletion of chromosome 11p causes a decrease in catalase's reaction. Deletion of chromosome 11p is more common among children suffering from ailments like Wilms' tumour and retardation syndrome amongst many others (Glorieux and Calderon, 2017).

Antioxidants Effects in Health. DOI: https://doi.org/10.1016/B978-0-12-819096-8.00022-7

2.2.2 Endogenous and exogenous sources

Catalase is commonly found in most living organisms. Catalase is normally obtained from microorganisms and the bovine liver (Singh et al., 2019). Good sources of catalase include plants like *Malva sylvestris* L. in particular the leaves which showed significant antioxidant properties and radical scavenging actions (Gasparetto et al., 2011). Kandukuri et al., have studied, purified and characterised the enzyme from certain plants, for example, the black gram (Vignamungo) seeds. Other endogenous sources from plants include cotton, sunflower and pumpkin. In mammals, the enzyme is more prominent in erythrocytes, the liver tissues where it is mainly found in the peroxisomes and sometimes in the kidney (Kodydkova et al., 2014). Catalase can also be extracted from numerous sources like bovine liver, *Aspergillus niger*, sweet potatoes and this antioxidant enzyme is used commercially in the food and dairy industries. In countries where proper refrigeration equipment are lacking, H_2O_2 is often used to cold-sterilize milk, therefore catalase is a cost-effective and environmentally friendly method adopted to remove the residual H_2O_2 (McSweeney, 2011). There are several types of catalase.

2.2.3 Catalase: Importance, benefits, and activity

Free radicals and reactive oxygen species (ROS) are formed from normal metabolic processes, namely from both enzymatic and non-enzymatic reactions (Lobo et al., 2010), which contribute towards oxidative stress. Oxidative stress occurs when there is an imbalance between the free radicals production and the antioxidant defences (Burton, 2011). Some ailments associated with oxidative stress are carcinogenesis, inflammatory conditions (Lobo et al., 2010), age-associated degenerative diseases like anaemia, vitiligo, Alzheimer's disease, Parkinson's disease, bipolar disorder, schizophrenia, inflammatory conditions (Nandi et al, 2019) and metabolic syndrome which is in turn linked with problems such as hypertension, diabetes, insulin resistance, dyslipidemia, and obesity (Cambray Guerra et al, 2014).

Catalase main function is to eradicate free radicals like H_2O_2 from the body. It decomposes H_2O_2 into oxygen and water molecules. H_2O_2 is potentially toxic as it can form ROS, but it also acts as a secondary messenger in various biological reactions, including proliferation, signalling or even apoptosis (Glorieux and Calderon, 2017).

Glorieux and Calderon (2017) reported that catalase can "act in its peroxidatic mode" by decomposing small substrates like methanol. On the other hand it was also found that this enzyme can additionally oxidize ethanol into acetaldehyde rendering it more available for metabolism in the liver (Glorieux and Calderon, 2017). Furthermore, catalase may decompose peroxynitrite and oxidize nitric oxide (a free radical) into nitrate. Other roles which are fulfilled by catalase are the detoxification of toxic substances and the activation of anti-tumour compounds (Glorieux and Calderon, 2017).

2.2.3.1 Catalase therapy in diabetic retinopathy

In a study conducted by Giordano et al. (2015), it was noted that retinal oxidative stress is a major contributor to diabetic retinopathy. Diabetic mice were treated with a cell penetrating catalase molecule (CAT-SKL) which can significantly regulate hyperglycaemia–induced oxidative stress in retinal cells and nonretinal tissue. Although the study comprises of several limitations, like the technical method used which did not accurately measure the effect of this treatment on retinal oxidative stress, it has revealed a new scope for further research on this topic.

2.2.3.2 Catalase therapy in cancer treatment

Catalase plays a major role by detoxifying hydrogen peroxide under stress like conditions. Any change in the expression of catalase in cancer cells promotes cell proliferation by inducing genetic instability and activation of oncogenes. The regulation of catalase expression seems to be mainly controlled at transcriptional levels despite other mechanisms may also be involved. Moreover transcription factors like Sp1 and NF-Y, JunB and RARα transcription factors are important regulators in breast cancer cells by recruiting proteins involved in transcriptional complexes and chromatin re-modelling. Thus, catalase can be a future therapeutic target in the context of cancer by using pro-oxidant approaches (Glorieux and Calderon, 2017).

2.2.3.3 Catalase therapy in cardiac patients

Reactive protein cysteine thiolates are involved in redox regulation. Oxidants, such as H_2O_2 react with thiolates to form oxidative post-translational modifications, enabling physiological redox signalling. Cardiac disease and aging are linked with oxidative stress which can impair redox signalling by changing essential cysteine thiolates. It was found that cardiac-specific overexpression of catalase conferred protection from oxidative stress and delayed cardiac aging in mice. Thus, catalase may help to alleviate cardiac disease and aging by moderating global protein cysteine thiol oxidation (Yao et al, 2015).

2.2.4 Impact of physiological, behavioral and environmental factors on catalase activity

Human behavior, one's health condition and lifestyle greatly influence catalase activity in the body (Thimraj et al., 2018). ROS can be obtained from both exogenous sources like fuel combustion and from vehicles and endogenous sources like cellular and subcellular metabolic reactions. Cigarette or tobacco smoke is found to be a potentially high source of oxidants which cause the activation of inflammatory cells. Upon activation, those inflammatory cells produce free radicals which greatly contribute to oxidative stress. From a study conducted by (Thimraj et al., 2018), it was found that people suffering from asthma have a decrease in catalase protein content as well as its activity. Additionally, they also observed that smokers suffering from chronic obstructive pulmonary disease (COPD) have a decrease in both the transcript and protein levels of catalase in their bronchiolar epithelium. Therefore, a decline in catalase activity can lead to the pathogenesis of diseases, like type II diabetes, vitiligo and hypertension (Glorieux and Calderon, 2017).

2.2.5 Role of catalase as a biomarker for oxidative stress

According to Nandi et al. (2019) our body has a defence mechanism against oxidative stress and catalase forms an integral part of this antioxidant defence system which catalytically removes reactive species. H2O2 is a freely diffusible and long-lived reactive species which plays an indirect role in the development of numerous cellular injuries. Even though this molecule is not very reactive, it is responsible for the production of more reactive free radicals like OH radical. Generally, certain enzymes like glutathione peroxidase can easily offset H2O2's activity, but catalase plays a fundamental role in the neutralization of hydrogen peroxide. Therefore catalase is an important biomarker for oxidative stress and the pathogenesis of many diseases and infections (Nandi et al., 2019).

2.2.6 Mechanism of action of catalase

The first step in the mechanism of action of catalase starts by:

1. Decomposition of H_2O_2: catalase reacts with H_2O_2 to form compound I, known as oxoferryl porphyrin cation radical, and water.
2. Two-electron redox process: the compound I reacts with a second H_2O_2 molecule and is reduced back into catalase, water and oxygen molecules. Glorieux and Calderon, 2017 reported that the same molecule of H_2O_2 was used to produce the water and oxygen molecules. (Gebicka and Krych-Madej, 2019) stated that only a few compounds can reduce compound II back into the original catalase.
3. Formation of compound II from compound I: compound I undergoes the one-electron reduction to form compound II, only when there is a low level of H_2O_2 and in the presence of one-electron donors, such as phenols, salicylic acid or ferrocyanide. Compound II is an inactive intermediate, which can be transferred back into its resting state by another one-electron reduction reaction. During the conversion of compound I to compound II, a proton is usually released. Both compound I and compound II are described as an oxoferryl-heme species but one difference between those two is that in compound I, the porphyrin holds a cation radical, whereas compound II lacks this porphyrin cation radical.
4. Formation of compound III: when there is a high level of H_2O_2, NADPH prevents the formation of compound III by participating in a two-electron reduction process.

5. Compound II may be subjected to two different reactions depending on the conditions present: either (1) it will return to a resting state because of the presence of another one-electron donor or (2) compound II will be converted into compound III, which is an inactive intermediate, in the presence of a H_2O_2 molecule. Gebicka and Krych-Madej (2019) attest that compound III "structurally resembles oxy forms of myoglobin and haemoglobin." In this inactive intermediate state, iron exists at an oxyferrous state. At last, compound III will go back to a resting state or it will cause the inactivation of the catalase. Therefore, compound III does not directly form part of the normal catalytic cycle of catalase (Glorieux and Calderon, 2017; Gebicka and Krych-Madej, 2019). A brief overview of the steps is represented below:
   - Catalase + H_2O_2 →H_2O + compound I (oxoferryl porphyrin cation radical)
   - Compound I + H_2O_2 → H_2O + O_2 + catalase
   - Compound I + [one-electron donor] → compound II (inactive intermediate)
   - Compound II + [one-electron donor] → resting state
   - Compound II + H_2O_2 → compound III
   - Compound III → resting state
   - Compound III → inactivation of catalase (Glorieux and Calderon, 2017; Gebicka and Krych-Madej, 2019).

2.2.7 In vitro and in vivo studies

A preclinical study carried out by Giordano et al. displayed the link between retinal oxidative stress and diabetic retinopathy. The aim of the study was to evaluate whether a treatment designed to enhance cellular catalase reduces oxidative stress in retinal cells cultured in high glucose and in diabetic mice corrects an imaging biomarker responsive to antioxidant therapy (manganese-enhanced magnetic resonance imaging [MEMRI]). The levels of H_2O_2 were firstly calculated in the cell lines of the retina exposed to normal or high glucose levels and treated with a cell-penetrating product of the peroxisomal enzyme catalase (CAT-SKL). The amount of H_2O_2 were calculated using a quantitative fluorescence-based assay. Following catalase transduction, an elevated amount of glucose-induced peroxide production diminished significantly in both human retinal cell lines. For the *in vivo* studies, 2 month old male mice weighing 20 g were put into different groups. The mice were inferred diabetes C57Bl/6 with streptozotocin and treated with CAT-SKL once a week for a period of 3 to 4 months. As control, non-treated age-matched non-diabetic mice and non-treated mice with diabetes were studied. MEMRI was used to evaluate the effectiveness of the treatment of CAT-SKL on diabetes evoked oxidative stress-related pathophysiology. Same analyses were done using difluoromethylornithine (DFMO). In the group of mice suffering from diabetes, insufficient intra-retinal uptake of manganese was enhanced in a significant way by catalase supplementation. Furthermore, in the peroxisome-rich liver of treated mice, the enzyme activity of catalase increased and there was a decrease in oxidative damage. However, difluoromethylornithine (DFMO) was ineffective in the assay. It was then concluded that increasing catalase is a therapy for treating the retinal oxidative stress associated with diabetic retinopathy.

Another study carried by Akateh et al. in 2019 investigated the effect of pegylated catalase in preventing ischemia/reperfusion injury (IRI) during liver surgery. Pegylation of catalase (PEG-CAT) was tested both *in vitro* and *in vivo*. *In vitro* studies consisted of populations of enriched rat liver cells which were exposed to H_2O_2 so as to cause oxidative stress injury. The health and activity of the cells were noted. *In vivo* studies consisted of rats which experienced segmental (70 %) hepatic warm ischemia for one hour followed by six hours of reperfusion and the evaluation of aminotransferase, plasma alanine, aspartate aminotransferase, tissue malondialdehyde, adenosine triphosphate, glutathione and histology. Pre-treatment of the liver cells with PEG-CAT indicated an important uptake and protection against injury due to oxidative stress, *in vitro*. *In vivo*, the level of alanine and aspartate aminotransferase were lowered in a significant way with respect to time compared to the control following the direct intrahepatic delivery of PEG-CAT. Similarly tissue malondialdehyde, adenosine triphosphate and glutathione all showed significant results. Also, the degree of centrilobular necrosis were enhanced by intrahepatic compared to systemic delivery of PEG-CAT. It was therefore concluded that the direct intrahepatic administration of PEG-CAT protected the liver cells against ischemia reperfusion injury (Akateh et al. in 2019).

2.2.8 Clinical Study

Góth et al. conducted a study in 2001 aimed to determine if there was a relationship between blood catalase deficiency and diabetes among the Hungarian people. The authors attested that a low catalase activity has been linked to long term exposure to oxidative stress which may contribute to the development of late-onset disorders like type II diabetes. In their study, they analysed families suffering from genetic deficiencies of catalase, that is, one family was suffering from acatalasemia and twelve other families were suffering from hypocatalasemia.

A total of 26,680 subjects were involved in this study from which 21,750 were hospital patients, 1,630 were clinic patients and 3,300 were healthy control subjects. In this study, several biomarkers were observed and carefully examined, namely:

- The frequency of catalase deficiency in Hungarian subjects
- The blood catalase activity of randomly selected diabetic patients
- Indicators of glycemic control, insulin, and C-peptide concentrations in nondiabetic hypocatalasemic and normocatalasemic of the members of five Hungarian families with catalase deficiency and finally,
- Insulin and C-peptide concentrations in diabetic patients suffering from acatalasemia and hypocatalesemia.

Góth et al. (2001) deducted that there is indeed a relationship between low catalase activity and an increase in the frequency of diabetes. It was perceived that elevated concentrations of hydrogen peroxide, which resulted from a catalase deficiency, may damage pancreatic cells. This will then influence insulin release and signalling which may afterwards affect glucose metabolic processes. In healthy subjects, it was noted that there was a boost up in "catalytic" activity when there was an increase in oxidative stress. This mechanism was an effective means used to protect the pancreatic β-cells. Nonetheless, in patients suffering from inherited catalase deficiency, their body lack this ability to protect their β-cells and they eventually amplify their risk of developing type II diabetes.

Ultimately, it was concluded that this decrease in catalase activity causes an accumulation in hydrogen peroxide. This has a direct impact on the oxidative destruction of the pancreatic β-cells which finally diminish both insulin secretion and effectiveness leading to the onset of diabetes (Góth et al., 2001)

A study carried out by Dhanapal et al. (2010) investigated the possible link of the activity of catalase and genetic polymorphism in patients suffering from diabetes of type 2. The blood sample of 20 patients consisting of both males and females aged between 40-65 years were collected. These patients were analysed for the presence of superoxide dismutase in their catalase gene and studies of restricted fragment length polymorphism demonstrated that 80 % of the patients were homozygous (TT genotype) and the remaining were of the heterozygous genotype. It was noted that patients with T allele had a lower fasting plasma glucose, red blood cell catalase activity and HbA1c compared to the heterozygous patients. This suggests a possible link between being heterozygous with a poor glycaemic control and a higher activity of catalase. Therefore, it can be concluded that humans having lower level of catalase are more likely to be predisposed to type 2 diabetes mellitus (Dhanapal et al., 2010).

An analytical overview performed by Rubio-Requelme et al. (2020) highlighted a significant association between increased catalase activity and seminal quality improvements. To further prove it, Hajizadeh and Tartibian (2018) performed three independent randomized controlled trials on 1228 sedentary men who registered in an infertility clinic. Following 24 weeks of chronic resistance exercise, a substantial increase in catalase enzymatic activity was detected and this lead to a reduction of peroxides in the seminal plasma. The results therefore showed that an increase in catalase will promote fertility in male adults.

Several authors came across the beneficial effects of the Indian herb *Withania somnifera* which is used as a common treatment for male reproductive health complications like infertility and reproductive endocrinological problems. After administration of this herb, a significant decrease in lipid peroxidation was observed and this might be explained by the synergistic effect between the enzymes catalase and superoxide dismutase along with the inherent antioxidant activity of the herb itself (Rubio- Riquelme et al., 2020).

Conclusion

Catalase has been found to be a key enzyme in the breakdown of H_2O_2 to water and molecular oxygen. The enzyme has one of the highest turnover rate and it is found in all living organisms. Catalase plays an important role in eradicating free radicals from the body, and thus preventing carcinogenesis, neurological disorders and inflammatory conditions. In addition, it helps to detoxify toxic substances and activate anti-tumour compounds. A decrease in the enzyme might lead to a rare condition known as acatalasemia which plays a role in the pathogenesis of type 2 diabetes, vitiligo and increased blood pressure.

References

Akateh, C., Beal, E.W., Kim, J.L., Reader, B.F., Maynard, K., Zweier, J.L., Whitson, B.A., Black, S.M., 2019. Intrahepatic delivery of pegylated catalase is protective in a rat ischemia/reperfusion injury model, Ohio. J. Surg. Res. 238, 152–163.

Burton., G.J., 2011. Oxidative stress, United Kingdom. Best practice & research. Clin. Obstet. Gyneacol. 25 (3), 287–299.

Dhanapal, K., Selvan, N., Dhananjeyan., V., 2010. A study on catalase activity and its genetic polymorphism in diabetes mellitus patients. J. Biol. Sciences 10, 653–657.

Gasparetto, J.C., Martins, C.A., Hayashi, S.S., Otuky, M.F., Pontarolo, R., 2011. Ethnobotanical and scientific aspects of *Malva Sylvestris* L.: a millennial herbal medicine, Brazil. J. Pharm. Pharmacol. 64 (2), 172–189.

Gebicka, L., Krych-Madej, J., 2019. The role of catalase in the prevention/promotion of oxidative stress, Poland. J. Inorg. Biochem. 197, 110–699.

Giordano, C.R., Roberts, R., Krentz, K.A., Bissig, D., Talreja, D., Kumar, A., Terlecky, S.R., Berkowitz, B.A., 2015. Catalase therapy corrects oxidative stress-induced pathophysiology in incipient diabetic retinopathy, United States. Invest. Ophthalmol. Vis. Sci. 56, 3095–3102.

Glorieux, C., Calderon, P.B., 2017. Catalase, a remarkable enzyme: targeting the oldest antioxidant enzyme to find a new cancer treatment approach. Biol. Chem. 398 (10), 1095–1108. doi:10.1515/hsz-2017-0131. PMID: 28384098.

Góth, L., Lenkey, A., Bigler, W.N., 2001. Blood catalase deficiency and diabetes. Hungary. Diabetes Care 24, 1839–1840.

Guerra, R.C., Zuñiga-Muñoz, A., Guarner Lans, V., Díaz-Díaz, E., Tena Betancourt, C.A., Pérez-Torres, I., 2014. Modulation of the activities of catalase, Cu-Zn, Mn superoxide dismutase, and glutathione peroxidase in adipocyte from ovariectomised female rats with metabolic syndrome. Int. J. Endocrinol. 2014, 175080. doi:10.1155/2014/175080. PMID: 24987414; PMCID: PMC4060422.

Hajizadeh Maleki, B., Tartibian, B., 2018. Resistance exercise modulates male factor infertility through anti-inflammatory and antioxidative mechanisms in infertile men: a RCT, Germany. Life Sci. 203, 150–160.

Ighodaro, O.M., Akinloye., O.A., 2018. First line defence antioxidants-superoxide dismutase (SOD), catalase (CAT) and glutathione peroxidase (GPX): Their fundamental role in the entire antioxidant defence grid, Nigeria. Alexandria J. Med. 54, 287–293.

Karadag, H., Ozhan, F., 2015. Effect of Cyprodinil and Fludioxonil pesticides on bovine liver catalase activity. Biotechnol. Biotechnol. Equip. 29 (1), 40–44.

Kodydkova, J., Vavrova, L., Kocik, M., Zak, A., 2014. Human catalase its polymorphisms regulation and changes of activity in different diseases, Czech Republic. Folia Biologica (Praha) 60, 153–167.

Lobo, V., Patil, A., Phatak, A., Chandra, N., 2010. Free radicals antioxidants and functional foods: impact on human health, India. Pharmacogn. Rev. 4, 421 -304.

McSweeney, P.L.H., 2011. Enzymes exogenous to milk in dairy technology catalase, glucose oxidase, glucose isomerase and hexose oxidase. Encyclopaedia of Dairy Sciences, 301–303. https://www.elsevier.com/books/encyclopedia-of-dairy-sciences/mcsweeney/978-0-12-818766-1.

Morgulis, S., Berber, M., Rabkin, I., 1926. Studies on the effect of temperature on the catalase reaction, Omaha. J. Biol. Chem. 68, 521–533.

Nandi, A., Yan, L.J., Jana, C.K., Das, N., 2019. Role of catalase in oxidative stress- and age-associated degenerative diseases. Oxid. Med. Cell. Longev. 2019, 9613090. doi:10.1155/2019/9613090. PMID: 31827713; PMCID: PMC6885225.

Rubio-Riquelme, N., Huerta-Retamal, N., Gómez-Torres, M., J., M.-.E., R, M., 2020. Catalase as a molecular target for male infertility diagnosis and monitoring: an overview, Spain. Antioxidants (Basel) 9 (1), 78.

Singh, R., Singh, A., Sachan, S., 2019. Enzymes in Food Biotechnology. Academic Press, pp. 237–843. https://doi.org/10.1016/B978-0-12-813280-7.00048-7.

Thimraj, T.A., George, L., Asrafuzzaman, S., Upadhyay., S., Ganguly, K., 2018. Chapter 7-oxidative signaling in chronic obstructive airway diseases. In: Chatterjee, S., Jungraithmayr, W., Bagchi, D. (Eds.). Immunity and Inflammation in Health and Disease, Academic Press, Boston, MA, USA, 79–98.

Yao, C., Behring, J.B., Shao, D., Sverdlov, A.L., Whelan, S.A., Elezaby, A., Yin, X., Siwik, D.A., Seta, F., Costello, C.E., Cohen, R.A., Matsui, R., Colucci, W.S., McComb, M.E., Bachschmid, M.M., 2015. Overexpression of catalase diminishes oxidative cysteine modifications of cardiac proteins. PLoS One 10 (12), e0144025. doi:10.1371/journal.pone.0144025. PMID: 26642319; PMCID: PMC4671598.

CHAPTER

Coenzyme Q: An endogenous antioxidant 2.3

Rajeshwar K.K. Arya[a], Prashant Kumar[b], Anita Singh[a], Mahendra Rana[a], Amita J. Rana[a], Aadesh Kumar[a]

[a]*Department of Pharmaceutical Sciences, Kumaun University, Campus Bhimtal, Bhimtal, Uttarakhand, India*

[b]*Department of Pharmacy, Doon valley group of Institutions, Karnal, Haryana, India*

2.3.1 Introduction

Coenzyme Q10 (CoQ10) is a nonproteinous, lipid-soluble antioxidant synthesized inside the body, and chemically known as Ubiquinone (Ernster and Dallner, 1995). Coenzyme Q is a benzoquinone derivate and chemically coenzyme Q is 2,3-dimethoxy, 5-methyl, 6-polyisoprene parabenzoquinone (Crane, 2001). It is abundantly found in aerobic organisms and mammalians (Saini, 2011). In 1940, coenzyme Q was initially recognized by Moor and coworker, and in 1957, F. Crane isolated it from the mitochondria of the beef heart (Crane, 2001). CoQ10 obtained from animal source has 10 isoprene units (five carbons each) (Crane, 2001). The organs having higher metabolic activity contains a higher amount of CoQ10, for example, heart, kidney, and liver, and works as energy transferring molecule. Fig. 2.3.1 represents the chemical structure of ubiquinone (Griffin et al., 2007).

Sources: CoQ10 is also found in many foods, for example, cold water fish, vegetable oils, meats, etc. Table 2.3.1 (Weber et al., 1997). The natural food source of CoQ10 contains comparatively very low CoQ10 content rather than supplements. The various food sources of CoQ10 are given in Table 2.3.1 (Rodick et al., 2018; Muztaba et al., 2018; Garrido et al., 2014).

Table 2.3.1 Various Sources of CoQ10 (Saini, 2011; Weber et al., 1997; Rodick et al., 2018; Muztaba et al., 2018; Garrido et al., 2014).

CoQ10 is synthesized from tyrosine and mevalonate in the cells (Bank et al., 2011). The genetic mutation is responsible for CoQ10 synthesis failure (Doimo et al., 2014). Various factors that can affect the CoQ10 level in the tissue or serum like aging, disease condition, drugs belonging to inhibitors of isoprene synthesis (HMG CoA reductase) (Garrido et al., 2014; Barcelos and Haas, 2019). The low level of CoQ10 in mitochondria causes various cardiac, neurological, muscle disorders, diabetes, tumors and encephalomyopathies, muscle coordination impairment,

Antioxidants Effects in Health. DOI: https://doi.org/10.1016/B978-0-12-819096-8.00023-9

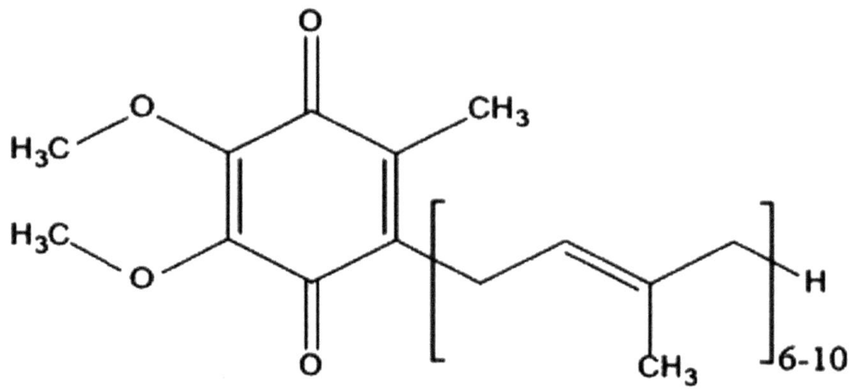

FIG. 2.3.1 Structure of ubiqunone (Bentinger et al., 2007).

hearing impairment, retinitis pigmentosa, and a steroid-resistant nephrotic syndrome (Garrido et al., 2014). Various factors affecting CoQ10 level in body were discussed in Table 2.3.2 (Crane, 2001; Garrido et al., 2014; Doimo et al., 2014).

Table 2.3.2 Factors affecting the level of Coq10 in the body (Crane, 2001; Garrido et al., 2014; Doimo et al., 2014; Barcelos and Haas, 2019).

CoQ10 prevents cellular impairment and plays a substantial contribution to the metabolic function. CoQ10 distributed in all the cells of the body like the mitochondrial respiratory chain and various inner membranes (Weber et al., 1997). The main function of CoQ10 is carrying electron in the respiratory chain, helping in coupling respiratory chain to oxidative phosphorylation and synthesis of ATP (Barcelos and

Table 2.3.1 Various food source of CoQ10.

Organ meats	Heart, liver, and kidney (Saini, 2011; Rodick et al., 2018)
Some muscle meats	Pork, beef, and chicken (Weber et al., 1997; Rodick et al., 2018)
Fatty fish	Herring, trout, tuna, salmon, mackerel, and sardines (Saini, 2011; Muztaba et al., 2018)
Vegetables:	Cauliflower, broccoli, spinach, parsley, and potato (Muztaba et al., 2018)
Fruit	Oranges and strawberries, avocado, apple, grape (Weber et al., 1997)
Legumes: Soybeans	Lentils and peanuts (Muztaba et al., 2018)
Nuts and seeds	Sesame seeds and pistachios (Muztaba et al., 2018)
Oils	Soybean and canola oil (Weber et al., 1997; Muztaba et al., 2018)
Eggs and dairy product	Butter, egg yolk, Swiss cheese (Muztaba et al., 2018)

Table 2.3.2 Factors affecting CoQ10 level in body.

Genetic defects (Crane, 2001; Garrido et al., 2014; Doimo et al., 2014)	Affecting CoQ10 synthesis or its utilization by cells
Mitochondrial diseases (Garrido et al., 2014)	Increases CoQ10 demands by tissues
Aging (Garrido et al., 2014; Barcelos and Haas, 2019)	Increases oxidative stress
HMG Co-A reductase inhibitors of isoprene synthesis (Garrido et al., 2014; Barcelos and Haas, 2019)	For example, lovastatin, simvastatin, pravastatin
Nutritional deficiencies	Vitamin B6 deficiency (Garrido et al., 2014; Doimo et al., 2014; Barcelos and Haas, 2019)

Haas, 2019). It participated in the generation of bioenergy in mitochondria (Freye and Strobel, 2018), therefore, CoQ10 is very essential for human health because it energies all the cells of the entire body (Saini, 2011). CoQ10 is the only endogenous antioxidant that participates in redox reaction and neutralizes the free radicals (Saini, 2011; Freye and Strobel, 2018). It protects DNA, protein, and lipid from oxidation. CoQ10 uses two mechanisms for lipid peroxidation: (1) it sequesters free radical and (2) it reduces α–tocopheryl radical to α-tocopherol (Muztaba et al., 2018; Bentinger et al., 2007) CoQ10 prevents the calcium overload and thus stabilizes the calcium channel (Hano et al., 1994).

2.3.2 Mechanism of action of coenzyme Q10

The CoQ10 exists in three redox forms, oxidize form (ubiquinone), semiquinone (ubisemiquinone), and deoxidized form (ubiquinol) (Ernster and Dallner, 1995). Endogenous CoQ10 protects protein oxidation, the enzymes impede with the peroxidation process at two steps: (1) when the chain reaction is being initiated and (2) the peroxidation process is propagated.

In the membrane perferryl radical ($Fe_3^+ O_2^-$) interact with the nearby lipids and protein molecules (LH), produce lipid radical (L*). The lipid radical (L*) interact with O_2 and peroxyl radical (LOO^-) is formed, with the formation of peroxyl radicals (LOO*) the peroxidation process is started. Due to high lipophillicity and high distribution CoQ10 remains in the vicinal membrane where the peroxyl radical is generated. In the presence of NADPH, CoQ10 (ubiquinol) reduces the perferryl radical and formation of ubisemiquinone, and H_2O_2 takes place, that blocks the lipid oxidation at the initial step. The ubisemiquinone also remove LOO* and blocks the propagation step. The reduced lipid converts tocopheroxyl radical into vitamin E. Apart from the lipid oxidation CoQ10 also prevents protein oxidation by interrupting and the formation of perferryl radical at the initial step and ceasing the propagation. It also intervenes with DNA oxidation (Garrido et al., 2014; Bentinger et al., 2007).

2.3.3 Coenzyme Q10 as pro-oxidant

The coenzyme Q10 also works as a pro-oxidant. During the metabolic process in mitochondria, electrons escaped from the electron transport chain and superoxide radicals are formed, the continuous production of superoxide results in the formation of H_2O_2 and ROS or free radicals (Bentinger et al., 2007). The accumulation of ROS in the cells leads to lipid peroxidation, oxidative cell damage, and DNA damage. The oxidative damage is responsible for various diseases, for example, infertility, cancer, aging and neurological disorder and cardiac disorder, etc. (Sumien et al., 2009). The respiration process generates ubisemiquinone, which interacts with oxygen that antimycin stimulates the electron removal and that leaching of CoQ10 from mitochondria inhibits H_2O_2 generation (Bentinger et al., 2007). Thus, it is assumed that CoQ10 can promote oxidation in mitochondria. The ubisemiquinone is generated in the external part of the tissue where auto-oxidation may occur and superoxide radicals are produced and converted into H_2O_2. This H_2O_2 converted into OH^- radicals by reductive homolytic cleavage. CoQ10 binds with proteins of mitochondrial complex, which encounters with auto-oxidation process. This show CoQ10 served as a pro-oxidant (Bentinger et al., 2007).

2.3.4 Beneficial effects

Several investigations have been done for estimating the beneficial effect of CoQ10 supplement, the high potent antioxidant property prevents the lipoproteins oxidation and generates high energy that improve cardiac muscle contraction and hence heart function. A study shows a long time supplementation of CoQ10 is good for heart diseases (Sumien et al., 2009). Another investigation revealed cardiac failure patients have low CoQ10 in cardiac muscles and tissue (Langisjoen et al., 1994; Folkers et al., 1985). The first-ever CoQ10-deficient patient's case of two sisters was reported in 1989, where a 3 month 50 mg thrice a day CoQ10 supplement regimen improved patient health (Ogasahara et al., 1989). The supplementation is also very beneficial in various diseases like Parkinson, migraine, diabetes, cancer, and skin problems.

2.3.4.1 Can endogenous antioxidants be detrimental for human health; in what extant?

Endogenous antioxidants may have some detrimental effect on human health when its supplement is given in excess amount, or when produced in excess, As the coenzyme CoQ10 is an endogenous antioxidant we should know in what extent of can be harmful to humans? A study on mice revealed a long time oral administration of CoQ10 at low dose does not show any evident effect on mental and motor functioning, whereas high dose impaired the mental and motor functioning (Sumien et al., 2009). CoQ10 supplements frequently given orally in many diseases and provide symptomatic benefits. Supplementing with CoQ10 has rare to mild side effects in humans. They include diarrhea, nausea, upper abdominal pain, rashes, and insomnia (Hidaka et al., 2008). Precaution should be taken while supplementing with CoQ10 to the patient suffering from heart disease, nephritic disorder, hepatic dysfunction, or diabetes due to its hypoglycemic and antihypertensive property. A dose greater than 300 mg affects the level of the hepatic enzymes (Hidaka et al., 2008; DiNicolantonio et al., 2015). The CoQ10 enzyme shows interaction with some drugs, for example, anticoagulants, antihypertensive, or antidiabetic. The scientific evidence regarding the safety data is not available, the CoQ10 supplementations should be cautiously administered to the patient with anticancer treatment, and supplementations should be avoided for the pregnant women or lactating mothers.

2.3.4.2 Beneficial effect of CoQ10 as evidenced by clinical studies

Many researchers have evaluated the effectiveness of CoQ10 supplementation on patients. Various clinical evidences are there, regarding the supplementation benefits. Various factors are responsible for declining the CoQ10 level in the, for example, age, disease, etc. Several clinical investigations has been done on animal and humans so far, those indicate supplementation could improve the disease condition. The prescribed dose of supplement is for adult 1200 mg/day and children 10 mg/kg (Garrido et al., 2014). Some major function of CoQ10 supplementation are listed below.

2.3.4.2.1 Heart failure

Oxidative stress could be the main cause of cardiac disease, such as cardiac failure and high blood pressure. The low level of CoQ10 in mitochondria produces low energy and hence the cardiac muscle contractility is suppressed that resulted in cardiac failure. The low level of CoQ10 was observed in cardiomyopathy patients, and the low level also correlated with the severity of the disease. CoQ10 supplementation enhances the myocardial contractility and improves patient life (Folkers et al., 1985) other research also showed CoQ10 supplementation can be used effectively in oxidative stress-induced coronary artery associated heart failure or high blood pressure, where the heart becomes unable to contract, relax and pump the blood efficiently (DiNicolantonio et al., 2015; Sharma et al., 2016). The researcher noticed, supplementation improves symptoms with lowering the patient death risk (Morstensen et al., 2014). The CoQ10 supplementation also maintains the optimum level of ATP production and reduces the cell damage (Hidaka et al., 2008).

2.3.4.2.2 Fertility

The low CoQ10 level affects both the male and female fertility, a study shows that the female fertility decreased with the age. The CoQ10 level is declined with the age and the female body becomes inefficient to protect the egg from oxidative damage

(Ben-Meir et al., 2015). Another study shows that the lower level of CoQ10 also results in low sperm count and bad sperm quality in males (Walczak et al., 2013; Lafuente et al., 2013).

2.3.4.2.3 Effect on skin

The antioxidant property of CoQ10 helps in protecting the skin from oxidative damage and protects skin from external and internal factors (Knott et al., 2015). The topically applied CoQ10 has antiwrinkle property and can be used as an antiwrinkle agent in creams (Hoppe et al., 1999). Research also revealed that people with low levels of CoQ10 are more prone to skin cancer (Rusciani et al., 2006).

2.3.4.2.4 Migraine

Some researchers suggested that oxidative stress is responsible for low energy in the brain and causes migraines (Yorns and Hadison, 2013). Another study revealed that the CoQ10 supplement improves the mitochondrial function and reduces the migraines associated with inflammation (Slater et al., 2011). A research performed on a large number of patients, the low level of CoQ10 was observed in those patients (Hershey et al., 2007) and supplementation can reduce migraine (Sandor et al., 2005).

2.3.4.2.5 Diabetes

A study shows that oxidative stress-induced cell damage disturbs the functioning of mitochondria and makes them insulin resistant (Xu et al., 2015). The supplementation overcome this insulin resistivity and modulates glucose level (El-ghoroury et al., 2009; Ericksson et al., 1999). The supplementation maintain the glucose level in type 2 diabetic patient without changing the lipid profile but stimulates fat decomposition and prevents obesity (Zahedi et al., 2014; Alam and Rahman, 2014).

2.3.4.2.6 Cancer treatment

The researcher estimated the relationship between CoQ10, lung cancer, and lipid peroxidation level in cancer patients. The researcher estimated malondialdehyde for assessing lipid and DNA damage, high DNA damage rate with lipid peroxidation was found in patients with lower CoQ10 levels (Cobanoglu et al., 2011). The oxidative stress-induced cell damage increases the possibility of carcinoma formation (Gupta et al., 2014; Chai et al., 2011). Researchers also found that the CoQ10 supplementation improves the cellular health and life of the patient by protecting cells from oxidative stress and enhances the synthesis of cellular energy (Zahedi et al., 2014; Alam and Rahman, 2014). The researchers also observed the low level of CoQ10 in cancer patients, the low level of CoQ10 can increase about 53% chances of development of various types of cancers (Folker et al., 1997; Chai et al., 2011). The supplementation can diminish the risk of cancer recurrence (Rusciani et al., 2007).

2.3.4.2.7 Brain

The main function of mitochondria is the production of energy in brain cells. These functions of mitochondria decreased with age. Increased dysfunction of mitochondria results in the death of brain cells and causes neurological disorder, for example, Alzheimer's and Parkinson's (Cassarino and Bennett, 1999). The brain contains fatty acid in higher concentrations which makes the brain prone to oxidative damage which aggravates the synthesis of harmful products which can alter memory, cognition, and physical functions (Hyun et al., 2010; Kones, 2010). The supplementation can slow down the amelioration of Alzheimer's and Parkinson's disease (Wadsworth et al., 2008; Shults et al., 2002).

2.3.4.2.8 ***Lungs***

The lungs are very vulnerable for oxidative damage, because oxygen directly comes in contact with the lungs. A study on lung cancer patients revealed that the damaged lungs with insufficient antioxidant function and lower CoQ10 concentration resulted in lungs disorder such as asthma and other pulmonary disorders (Wada et al., 2006). Other studies also show pulmonary disorder patients possess a low CoQ10 level (Tanrikulu et al., 2011; Gazdik et al., 2002). The researcher also found that the CoQ10 supplementation in asthma patients can reduce the swelling and required steroid dose (Gvozdjakova et al., 2005). The researcher found, CoQ10 also improves the physical performance of COPD patients by enhancing tissue oxygen level and cardiac rate (Fujimoto et al., 1993).

2.3.5 *In-vitro* and *in-vivo* studies

Some animal and human trials are summarized in the Table 2.3.3.

Table 2.3.3 Summary of animal and human study.

Table 2.3.3 Summary of animal and human study.

Zahrooni et al., 2019 conducted 60 days study on breast cancer, the people were divided into two groups, group one with 29 patient, received CoQ10 at a dose of 100 mg/day and the second group of 29 healthy individuals received the placebo, a remarkable decline in cytokine level was observed that reduces the inflammation caused breast cancer.	(Zahrooni et al., 2019)
Zmitek et al., 2017 performed 49 days study to know the effect of CoQ10 on aging; the CoQ10 has a significant effect on wrinkles, skin smoothing, and firmness. The supplementation is done at two different dose levels, both have shown good antiwrinkle property in almost all parts of the face, but the placebo group has not shown any change.	(Zmitek et al., 2017)
Li et al., 2017 collected the preconducted study data from various database regarding the effect of CoQ10 on cardiac disorder and reconducted the study on 2149 patient with supplement and placebo. The mortality, physical performance, ejected fraction from heart, and NYHA classification were evaluated. The decline in mortality rate was observed with improved physical performance on CoQ10 supplementation and but the supplementation has no effect on ejected fraction from heart and NYHA Classification	(Lei and Liu, 2017)

(continued)

Table 2.3.3 Summary of animal and human study. *Continued*

Aponte et al., 2015 conducted a literature review based survey and evaluated the evidence for the beneficiary effect of CoQ10 supplement for the treatment of breast cancer, and set up a relation between CoQ10 levels and breast cancer. The supplementation enhances the antioxidant level, which declines the cytokine (cancer biomarker) level.	(Aponte et al., 2015)
Zahedi et al., 2014 performed 12 weeks study on type 2 diabetic patient, the patient was treated with CoQ10, the supplementation reduce hyperglycemia by protecting β-cell from oxidative stress, but it does not show any significant effect on the patient's lipid profile.	(Zahedi et al., 2014)
Buric et al., 2019 evaluated the effectiveness of CoQ10 supplement against glioblastoma using *in-vitro* and *in-vivo* model, the RC6 cell was targeted in the study, both the study revealed, the CoQ10 supplementation has a potential effect on cancer cells, the CoQ10 supplement with temozolomide has increased death of cancer cell and also ceases the infiltration of RC16 into the other cells.	(Buric et al., 2019)
X. Zhang et al., 2018 evaluated the benefits of CoQ10 on cardiac cell damage in apolipoprotein deficient mice, the study was conducted for 16 weeks, the blood sample was analyzed for metabolic parameter, and the heart tissue was excised for histopathological study and immune-histochemical analysis. The total cholesterol, LDL-c, and triglyceride level were analyzed and a decrease in all parameters was observed. The supplementation also decreased the cardiac damage.	(Zhang et al., 2018)
Vetvicka et al., 2018 studied the combinatory effect of CoQ10 and β-glucan on immunity and anticancer activity. They performed *in-vitro* and *in-vivo* study on mice, the study revealed that the CoQ10 supplementation was found less noticeable anticancer activity but when given in combination with β-glucan, the synergistic effect enhances anticancer activity.	(Vetvicka and Vetvicka, 2018)
Ulla et al., 2017 performed a study on the rats, the isoprenaline generated remodeled heart was used in the study, the oxidative stress marker was analyzed to check lipid oxidation and the histopathological analysis was done to know the effect of CoQ10 supplementation on the inflammation and fibrosis the study confirmed the 100 mg/kg supplementation reduces the oxidative marker level and also avert the inflammation and fibrosis	(Ulla et al., 2017)
Jang et al., 2017 performed an *in-vitro* study on colon cancer cell lines; they performed MTT assay, nitric oxide assay, and western blotting method. The study revealed that nitric oxide generation is increased by an enhanced apoptotic signal with increasing the dose of the supplement. All the combining effects show good repression action on the development of the cancerous cell.	(Jang et al., 2017)
Mortensen et al., 2014 performed a randomized double-blind trial on 420 heart failure patients and discussed that when patients were treated with CoQ10 for 2 years, the symptoms were improved and the risks of patient death were decreased.	(Morstensen et al., 2014)

Conclusion

Antioxidants protect from various free radical-induced damage of tissue and organs. This can be reversible or irreversible and will depend upon external factors such as environmental or biochemical agents. The damage to the cell is reversible depending on the levels of antioxidant stress. The CoQ10 is believed as a good antioxidant for lipids, protein, and DNA due to its high lipophilicity, wide distribution in tissues, and efficacious reactivity. The CoQ10 counterbalance the free radicals and thus preserve the organs. CoQ10 antioxidants act like the first line of defense for the protection of cell damage. Increased intake of CoQ10 in the diet helps in maintaining cellular integrity and also the normal physiological and biochemical functions of the living system. CoQ10 supplementation is beneficial in various disease, for example, Parkinson, cancer, cardiac disorder, migraine, and diabetes. Thus, we can conclude that CoQ10 shows more beneficial effects with lesser reported harmful effects but there is the need for exhaustive research in this field to know more about the benefit and harm ratio.

Conflict of interest

No interest.

Abbreviations

CoQ10	Coenzyme Q10
ROS	Reactive oxygen species
NYHA	New York Heart Association System
NADPH	Nicotinamide adenine dinucleotide phosphate
ATP	Adenosine triphosphate
COPD	Chronic obstructive pulmonary disease
MTT	(3-(4, 5-dimethylthiazolyl-2)-2, 5-diphenyltetrazolium bromide)

References

Alam, M.A., Rahman, M.M., 2014. Mitochondrial dysfunction in obesity: potential benefits and mechanism CoQ10 supplementation in metabolic syndrome. J. Diabetes Metab. Disord. 13 (60), 1–11.

Aponte, B., Carlos, D.J., Hernandez, J., 2015. Role of coenzyme Q10 in breast cancer. Pharm. Pharm. Int. J. 3 (2), 261–267.

Bank, G., Kagan, D., Madhavi, D., 2011. Coenzyme Q10: Clinical update and bioavailability, J. Evid. Based Complement. Altern. Med. 16 (2)129–137.

Barcelos, I.P.D., Haas, RH., 2019. CoQ10 and aging. Biology 8 (2), 1–22.

Ben-Meir, A., Burstein, E., Borrego-Alvarez, A., Chong, J., Wong, E., Yavorska, T., 2015. Coenzyme Q10 restores oocyte mitochondrial function and fertility during reproductive aging. Aging Cell 14 (5), 887–895.

Bentinger, M., Kerstin, B., Gustav, D., 2007. The antioxidant role of coenzyme Q. Mitochondrion 7S7, S41–S50.

Buric, S.S., Podolski-Renic, A., Dinic, J., Stankovic, T., Jovanovic, M., Hadzic, S., Ayuso, J.M., Virumbrales-Munoz, M., Fernandez, L.J., Ochoa, I., Perez-García, V.M., Pesic, M., 2019. Modulation of Antioxidant Potential with Coenzyme Q10 Suppressed Invasion of Temozolomide-Resistant Rat Glioma *in-vitro* and *in-vivo*. Oxid. Med. Cell Longev. 12, 1–14. doi: 10.1155/2019/3061607. PMID: 30984333; PMCID: PMC6432727.

Cassarino, D.S., Bennett, J.P., 1999. An evaluation of the role of mitochondria in neurodegenerative diseases: mitochondrial mutations and oxidative pathology, protective nuclear responses, and cell death in neurodegeneration. Brain Res. Brain Res. Rev. 29 (1), 1–25.

Chai, W., Cooney, R.V., Franke, A.A, Caberto, C.P., Wilkens, L.R., Le et al., M.L., 2011. Plasma coenzyme Q level and prostate cancer risk: the multiethenic cohort study. Cancer epidemiology biomarkers prevention 20 (4), 708–710.

Chai, W., Cooney, R.V., Franke, A.A, Caberto, C.P., Wilkens, L.R., Le, M.L., 2011. Plasma coenzyme Q level and prostate cancer risk: the multiethnic cohort study. Cancer Epidemiol. Biomark. Prev. 20 (4), 708–710.

Cobanoglu, U., Demir, H., Cebi, A., Sayir, F., Alp, H.H., Akan, Z., 2011. Lipid peroxidation, DNA damage and coenzyme Q10 in lung cancer patients-markers for risk assessment? Asian Pac. J. Cancer Prev. 12 (6), 1399–1403.

Crane, L.F., 2001. Biochemical functions of coenzyme Q10. J. Am. Coll. Nutr. 20, 591–598.

DiNicolantonio, J.J., Bhutani, J., McCarty, M.F., & O'Keefe, J.H., 2015. Coenzyme Q10 for the treatment of heart failure: a review of the literature. Open heart, 2 (1), e000326. https://doi.org/10.1136/openhrt-2015-000326.

Doimo, M., Desbats, M.A., Cerqua, C., Cassina, M., Trevisson, E., Salviati, L., 2014. Genetics of coenzyme Q10 deficiency. Mol. Syndromol. 5 (3-4), 156–162.

El-ghoroury, H.M., Raslan, E.A., El-Saaid, G.S., Agybi, M.H., Siam, I., Salem, S.I., 2009. Malondialdehyde and coenzyme Q10 in platelets and serum in type2 diabetes mellitus: correlation with glycemic control. Blood Coagul. Fibrin. 20 (4), 248–251.

Ericksson, J.G., Forsen, T.J., Motensen, S.A., Rohde, M., 1999. The effect of coenzyme Q10 administration on metabolic control in patients with type 2 diabetes mellitus. Biofactors 9 (2-4), 315–318.

Ernster, L., & Dallner, G., 1995. Biochemical, physiological and medical aspects of ubiquinone function. Biochim. Biophys. Acta. 1271 (1), 195–204. https://doi.org/10.1016/0925-4439(95)00028-3.

Folker, K.O., Nylander, M., Morita, M., Mellstedt, H., 1997. Activities of vitamin Q10 in animal models and a serious deficiency in patients with cancer. Biochim. Biophys. Res. Commun. 234 (2), 296–299.

Folkers, K., Wolaniuk, J., Simonsen, R., Morishita, M., Vadhanavikit, S., 1985. Biochemical rationale and the cardiac response of patients with muscle disease to therapy with Coenzyme Q10, Proceedings of the national academy of sciences of the USA America, 82, 4513–4516.

Freye, E.C.K., Strobel, H.P., 2018. Supplements which increases ATP synthesis within the mitochondria and protect against toxic sodium azide. Act. Sci. Nutr. Health 2 (7), 02–09.

Fujimoto, S., Kurihara, N., Hirata, K., Taeda, T., 1993. Effects of coenzyme Q10 administration on pulmonary function and exercise performance in patients with chronic lung diseases. Clin. Investig. 71 (8), S162–S166.

Garrido, J.D., Cordero, M.D., Oropesa, A.M., Fernandez, V.A., Mata, M.D., Minguel, P.A.M., Perez, C.C., Villanueva, P.M., Cotan, D., Sanchez-Alcazar, J.A., 2014. Coenzyme Q10 therapy. Mol. Syndromol. 5 (3-4), 187–197.

Gazdik, F., Gvozdjakova, A., Nadvornikova, R., Repocka, L., Jahnova, E., Kucharska, J., 2002. Decreased levels off coenzyme Q10 in patients with bronchial asthma. Allergy 57 (9), 811–814.

Griffin, C., Gueorguieva, D., Burgess, A.M., Nitu, M.S., Pandey, S., 2007. Artificial Cells, Cell Engineering and Therapy. Woodhead Publishing Series In Biomaterials, United Kingdom, pp. 117–142.

Gupta, R.K., Patel, A.K., Shah, N., Chaudhary, A.K., Jha, U.K., Yadav, U.C., 2014. Oxidative stress and antioxidants in disease and cancer: a review. Asian Pac. J. Cancer Prev. 15 (11), 4405–4409.

Gvozdjakova, A., Kucharska, J., Bartkovjakova, M., Gazdikova, K., Gazdik, F.E., 2005. Coenzyme Q10 supplementation reduces corticosteroids dosage in patients with bronchial asthma. Biofactors 25 (1-4), 235–240.

Hano, O., Thompson-Gorman, S.L., Zweier, J.L., Lakatta, E.G., 1994. Am. J. Physiol. Circ. Physiol. 266 (6), H174–H218.

Hershey, A.D., Powers, S.W., Vockl, A.L., Lecates, S.L., Ellinor, P.L., Segers, A., 2007. Q10 response to supplementation in pediatric and adolescent migraine. Headache 47 (1), 73–80.

Hidaka, T., Fujii, K., Funahashi, I., Fukutomi, N., Hosoe, K., 2008. Safety assessment of coenzyme Q10 (CoQ10). Biofactors 32 (1-4), 199–208.

Hoppe, U., Bergemann, J., Diembeck, W., Ennen, J., Gohla, S., Harris, I., 1999. Coenzyme Q10, a cutaneous antioxidant and energizer. Biofactors 9 (2-4), 371–378.

Hyun, D.H., Mughal, M.R., Yang, H., Lee, J.H., Ko, E.J., Hunt, N.D., 2010. The plasma membrane redox system is impaired by amyloid β-peptide and in the hippocampus and cerebral cortex of 3xTgAD mice. Exp. Neurol. 225 (2), 423–429.

Jang, S., Lee, L., Ryu, S.M., Park, H.J., Kim, H., Kim, D., 2017. Effect of coenzyme Q10 via nitric acid production and growth arrest of human colon cancer HCT116 cells. J. Prev. Med. 41 (2), 59–65.

Knott, A., Achterberg, V., Smuda, C., Mieke, H., Sperling, G., Dunkelmann, K., 2015. Topical treatment with coenzyme Q10-containing formulas improves skin's Q10 level and provides antioxidative effects. Biofactors 41 (6), 383–390.

Kones, R., 2010. Parkinson's disease: mitochondrial molecular pathology, inflammation, statins, and therapeutic neuroprotective nutrition. Nutr. Clin. Pract. 25 (4), 371–389.

Lafuente, R., Gonazalez-Comadran, M., Sola, I., Lopez, G., Brassesco, M., Carreras, R., 2013. Coenzyme Q10 and male infertility: a meta-analysis. J. Assist. Reprod. Genet. 30 (9), 1147–1156.

Langisjoen, H., Langisjoen, P., Willis, R., Folkers, K., 1994. Usefullnes of coenzyme Q10 in clinical cardiology: a long term study. Mol. Aspects Med. 15, 165–175.

Lei, L., Liu, Y., 2017. Efficacy of coenzyme Q10 in patients with cardiac failure: a meta-analysis of clinical trials. BMC Cardiovasc. Disord. 17 (196).

Morstensen, S.A., Rosenfeldt, F., Kumar, A., Dolliner, P., Filipiak, K.J., Alehagen, U., 2014. The effect of coenzyme Q10 on morbidity and mortality in chronic heart failure Q-symbio: a randomized double-blind trial. Study Investig. 2 (6), 641–649.

Muztaba, M., Badruddeen, M.I., Khan, M., Khalid, Ghazanfer, M., Tarique, J., Akhtar, M., 2018. Ahmad CoQ10: A miraculous and clinically vital coenzyme for normal body functions, Prevention and treatment of diseases. J. Pharm. Sci. Res. 10 (5), 976–982.

Ogasahara, A.G., Engel, D., Mack, D.F., 1989. Muscle coenzyme Q deficiency in familial mitochondrial encephalomyopathy. Proceedings of the national academy of sciences of the USA America 86 (7), 2379–2382.

Rodick, C.T., Seibels, D.R., Babu, J.R., Huggins, K.W., Ren, G., Mathews, S.T., 2018. Nutr. Diet. Suppl. 10, 1–11.

Rusciani, L., Proietti, I., Paradisi, A., Rusciani, A., Guerriero, G., Mammone, A., 2007. Recombinant interferon alfa-2b and coenzyme Q10 as a postsurgical adjuvant therapy for melanoma: a 3-year trial with recombinant interferon alfa and 5-year follow up. Melenoma Res. 17 (3), 177–183.

Rusciani, L., Proietti, I., Rusciani, A., Paradisi, A., Sbordoni, G., Alfano, C., 2006. Low plasma coenzyme Q10 levels as an independent prognostic factor for melanoma progression. J. Am. Acad. Dermatol. 54 (2), 234–241.

Saini, R., 2011. Coenzyme Q10: the essential nutrient. J. Pharm. Bioallied Sci. 3 (3), 466–467.

Sandor, P.S., Di Clemente, L., Coppola, G., Saenger, U., Fumal, A., Magis, D., 2005. Efficacy of coenzyme Q10 in migraine prophylaxis: a randomized controlled trial. Neurology 64 (4), 713–715.

Sharma, A., Fonarow, G.C., Butler, J., Ezekowitz, J.A., Felker, G.M., 2016. Coenzyme Q10 and heart failure: a state of the art review. Circ. Heart Fail. 9 (4).

Shults, C.W., Oakes, D., Kieburtz, K., Beal, M.F., Haas, R., Plumb, S., 2002. Effects of coenzyme Q10 in early Parkinson disease: evidence of slowing of the functional decline. Arch. Neurol. 59 (10), 1541.

Slater, S.K., Nelson, T.D., Kabbouche, M.A., LeCates, S.L., Horn, P., Segers, A., 2011. A randomized, double-blinded, placebo-controlled, crossover, add-on study of coenzyme Q10 in the prevention of pediatric and adolescent migraine. Cephalalgia 31 (8), 879–905.

Sumien, N., Heinrich, K.R., Shetty, R.A., Sohal, R.S., Forster, M.J., 2009. Prolonged intake of coenzyme Q10 impairs cognitive function in mice. J. Nutr. 139 (10), 1926–1932.

Tanrikulu, A.C., Abakay, A., Evliyaoglu, O., Palanci, Y., 2011. Coenzyme Q10, copper, zinc, andlipid peroxidation levels in serum of patients with chronic obstructive pulmonary disease. Biol. Trace Elem. Res. 143 (2), 659–667.

Ulla, A., Mohammad, M.K., Sikder, B., 2017. Coenzyme Q10 prevents oxidative stress and fibrosis in isoprenaline induced cardiac remodeling in aged rats. BMC Pharmacol. Toxicol. 18 (29).

Vetvicka, V., Vetvicka, J., 2018. Combination therapy with glucan and coenzyme Q10 in murine experimental autoimmune disease and cancer. Anticancer Res. 38 (6), 3291–3297.

Wada, H., Hagiwara, S., Saitoh, E., Leki, P., Okamuara, T., Ota, T., 2006. Oxidative stress in patients with chronic obstructive pulmonary disease (COPD) as measured by redox status of plasma coenzyme Q10. Pathophysiol. 13 (1), 29–33.

Wadsworth, T.L., Bishop, J.A., Pappu, A.S., Woltjer, R.L., Quinn, J.F., 2008. Evaluation of coenzyme Q as an antioxidant strategy for alzheimer's disease. J. Alzheimers Dis. 14 (2), 225–234.

Walczak, R., Wolski, J.K., Slowikowska, H., 2013. the role of oxidative stress and antioxidants in male fertility. Cent. European J. Urol. 66 (1), 60–67.

Weber, C., Bysted, A., Holmer, G., 1997. Intestinal absorption of coenzyme Q10 administered in a meal or as capsule to healthy subjects. Nutr. Res. 17, 941–945.

Xu, Z., Huo, J., Ding, X., Yang, M., Li, L., Dai, J., 2015. Sawashita, coenzyme Q10 improves lipid metabolism and ameliorates obesity by regulating CaMKII-mediated PDE4 inhibition. Sci. Rep. 7 (1).

Yorns, W.R., Hadison, H.H., 2013. Mitochndrial dysfunction in migraine. Semin. Pediatr. Neurol. 20 (3), 188–195.

Zahedi, H., Eghtesadi, S., Seifirad, N., Rezaee, N., F.Shidfar, I.H., 2014. Effects of CoQ10 supplementation on lipid profiles and glycemic control in patients with type 2 diabetes: a randomized, double-blind, placebo-controlled trial. J. Diabetes Metab. Disord. 13 (81).

Zahrooni, N., Hosseini, S.A., Ahmadzadeh, A., Angali, K.A., Assarehzadega, M.A., 2019. The effect of coenzyme Q10 supplementation on vascular endothelial growth factor and Serum level of interleukin 6 and 8 in women with breast cancer: double-blind, placebo-control, randomized clinical trial. Clin. Risk Manag. 15, 1043–1410.

Zhang, X., Liu, H., Hao, Y., Xu, L., Zhang, T., Liu, Y., 2018. Coenzyme Q10 protects against hyperlipidemia-induced cardiac damage in apolipoprotein E-deficient mice. Lipids Health Dis. 17 (279).

Zmitek, K., Pogacnic, T., Mervic, L., Zmitek, J., Pravst, I., 2017. The effect of dietary intake of coenzyme Q10 on skin parameters and condition: results of the randomized placebo-controlled, double-blind study. Res. Commun. 43 (1), 132–140.

CHAPTER

Ferritin

2.4

Mohamad Fawzi Mahomoodally, Joomun B. Fatimah-Tuz-Zohra

Department of Health Sciences, Faculty of Medicine and Health Sciences, University of Mauritius, Réduit, Mauritius

2.4.1 Introduction

Ferritin is a water-soluble protein that can be produced by the liver, the spleen and the bone marrow Macara et al., 1973 suggested that ferritin molecules consist of two major components: (1) a protein shell called apoferritin (Fig. 2.4.1) that is the Fe free form of the protein and (2) a microcrystalline core situated within the hollow shell containing up to about 4500 Fe atoms (Schnell et al., 2019).

The protein ferritin is made up of 24 subunits that are mainly of two types, H and L subunits. It is denoted H as these isoforms of ferritin were originally isolated from the human heart, where it was found to be in large amount. It also presents itself as the heavier subunit when their electrophoretic migration is compared (Goodsell, 2002). Similarly, L refers to ferritin as it has been isolated from the human liver, which is rich in the lighter L subunit. The proportion of the presence of L subunit to H subunit within a ferritin protein greatly depends on tissue type and the developmental stage reached (Wang et al., 2010).

The protein coat also termed as the apoferritin consists of 20–24 subunits, which are symmetrically arranged which allow access to the inner cavity through inter-subunit channels (Macara et al., 1973). The H-ferritin acts as a rapid detoxifier of Fe while the L-ferritin is involved in continuing storage of Fe along with nucleation and mineralization (Harrison and Arosio, 1996). The ferritin protein particular shape and alignment cause it to have two types of conformation: (1) the three-fold channels and (2) the four-fold symmetric channels that are considered to provide pathways for transferring Fe^{2+} ions between inside and outside of the protein. According to a study done by Takahashi and Kuyucak 2003, while the three-fold channel was found to be in charge of the movement of Fe^{2+} ions, the four-fold channel was established as being possibly permeate to protons only.

Transferrin is another protein that is produced locally in the testes and the central nervous system (CNS). It is a specific type of transporter that captures Fe release into the plasma by either enterocytes or reticuloendothelial macrophages and delivers it to other tissues in the body (Shiel, 2018). Endocytosis and the uptake of Fe occurs from binding of the ferro-transferrin to cell surface transferrin specific receptors known as

Antioxidants Effects in Health. DOI: https://doi.org/10.1016/B978-0-12-819096-8.00024-0

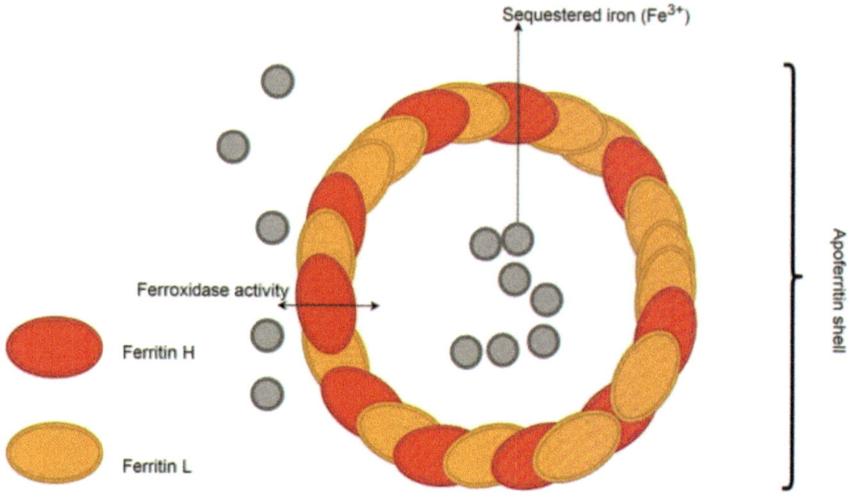

FIG. 2.4.1 The 24 subunits of ferritin.

TfR. Once absorbed, the Fe is transported to the mitochondria to synthesise heme or to be stored and detoxified in cytosolic ferritin (Abbaspour et al., 2014). Availability of excessive Fe in cells has been proven to be dangerous as it speeds up the rate of oxidation of DNA, proteins, and some cell membranes due to the production of free radicals as by-products of Fenton reactions (Giardi et al., 2011). Hence, it is imperative to control the concentration of cellular Fe. Cellular iron balance is maintained by the post-transcriptional mechanism that involves two particular regulatory proteins, IRP-1 and IRP-2. These proteins control both the regulation of ferritin translations and of the activity of transferrin receptor (Giardi et al., 2011).

2.4.2 Serum ferritin level as a diagnostic biomarker

The ferritin test measures the amount of ferritin present in the blood. The ferritin test can be used to understand how much iron is stored in the body and also help in diagnosis of various diseases (Knovich et al., 2009). According to Mosby's manual, if the test shows a higher level of ferritin, it could be indicative of certain conditions that cause the body to store excess levels of Fe. Furthermore, it could also indicate the presence of liver diseases, renal dysfunction, rheumatoid arthritis, hyperthyroidism or some other inflammatory conditions. Some types of cancer may also cause blood ferritin level to be elevated (Pagana and Pagana, 2014). The normal range for blood ferritin level, for men, is 20–500 and for women is 20–200 ng/mL. Low levels of ferritin are indicative of an iron deficiency, which can lead to anemia, that is, a reduction in the number of oxygen-carrying red blood cells (Soppi, 2018). A study found that patients with iron deficiency anemia (IDA) were recorded to

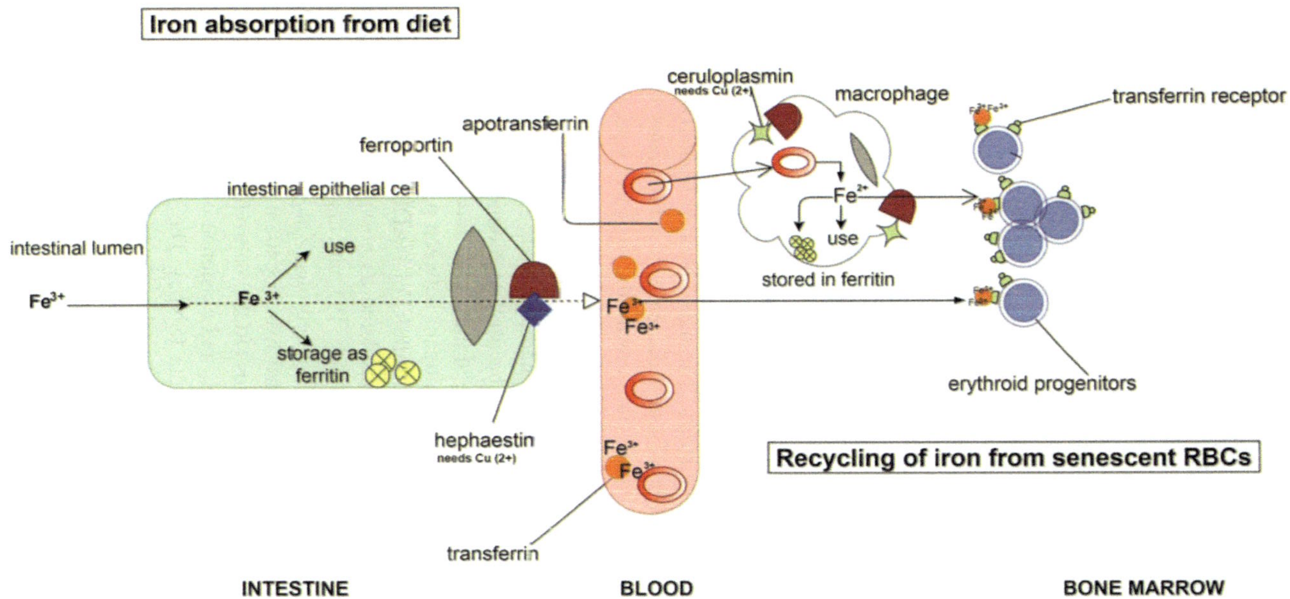

FIG. 2.4.2 Illustration of iron pathway in the body.

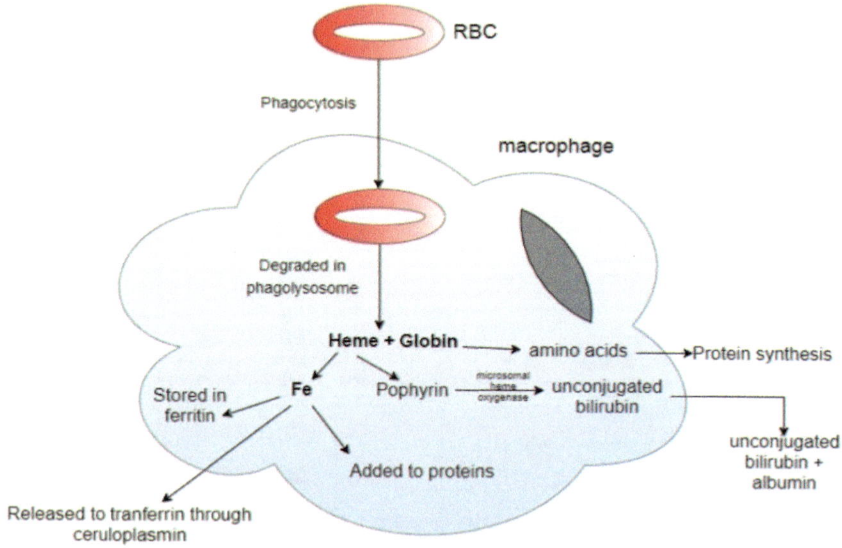

FIG. 2.4.3 Red blood cell degradation in macrophage during iron absorption.

have an elevated oxidative stress. The data obtained from IDA patients illustrated an increase in oxidants level and a decrease in antioxidants levels. This, in turn, resulted in a shift in the equilibrium toward the oxidative side, and hence increasing oxidative stress which consequently contributes to pathogenesis of diseases in IDA cases (Yoo et al., 2009).

In another study by Hemmati-Dinarvand et al., 2017, it was found that level of serum ferritin concentration could be used as biomarker for the diagnosis of Parkinson's disease (PD). Results showed that the difference of serum concentration of ferritin and selenium (Se) was significant between patients suffering from PD and healthy ones. Hence, serum concentrations of NADPH oxidase 1 (and superoxide radical), ferritin, and Se can be considered as possible diagnostic biomarkers for diagnosis and monitoring of PD patients. PD patients are mostly affected due to the degeneration of dopamine synthesising neurons (Kaufman et al., 2017). Researchers suggested that dopaminergic neuronal loss in patients suffering from PD is significantly due to reactive oxygen species (ROS) as products of dopamine metabolism, decreased gluthathione (GSH) level, and existence of large amounts of Fe and Ca in the substantia nigra pars compacta (SNpc). The generation of reactive oxygen species (ROS) is attributed to various mechanisms and sources in the body: metabolism of dopamine, mitochondrial dysfunction, neuroinflammation, and aging. Even though PD is closely linked to oxidative stress, it is unclear whether the accretion of ROS is its primary cause or it is the consequence of a series of cellular dysfunctions (Dias et al., 2013).

A cohort clinical study on newly diagnosed patients (n = 305) with acute myeloid leukemia (AML) concluded that an elevated ferritin level might indicate tumor burden in patients with AML and also predict worsening event-free survival (EFS) in the high-risk group (Tachibana et al., 2018). The median follow-up period of 58 months among survivors established that high ferritin group (≥400 ng/mL) demonstrated inferior EFS at the 5-year interval (30% vs 40%; P = 0.033) compared to the low ferritin group. Analysis of multiple variables in the high-risk group karyotype revealed that high ferritin levels predicted worse EFS (hazard ratio = 2.07; 95%; confidence interval, 1.28–3.33; P = 0.003) (Tachibana et al., 2018).

2.4.3 Role of ferritin in Fe homeostasis

Ferritin plays an important role during Fe homeostasis. It has as primary role of acting as a ferroxidase converting Fe^{2+} to Fe^{3+}, when it is sequestered into the ferritin mineral core (Andrews and Schmidt, 2007). Ferritin was found to have a maximum storage capacity of 4500 Fe^{3+} ions per molecule (Arosio et al., 2009). Fe is able to enter the ferritin molecules through the three-fold channels in its ferrous state; this transfer is facilitated by the chaperone poly r(C)-binding protein 1 (PCBP1) (Arosio et al., 2009). It requires an oxidation reaction caused by the ferroxidase activity of the H chain. Subsequently, the oxidised Fe is accumulate in the inner cavity of the ferritin molecule in its mineralised form, where the nucleation process is assisted by the L chain. Exiting Fe from ferritin requires the reduction of mineralized Fe^{3+} to Fe^{2+}. This exit can also occur through electron transfer or even through direct Fe release mechanism (Bresgen and Eckl, 2015).

2.4.4 Ferritin and oxidative stress

Several studies have been conducted to elaborate on the antioxidant properties of ferritin. A study using the super coil assay (Surguladze et al., 2004) showed that ferritin is able can act on Fe in order to reduce its capacity to catalyse oxidative damages even when DNA is present as a competing ligand to iron. Additionally, comparing other proteins like catabolite activator protein (CAP; a trans-acting transcriptional activator that exists as a homodimer in solution.) or transferrin did not yield the same protective result as ferritin (Surguladze et al., 2004).

It was also hypothesised that ferritin/iron content can be related to the activation state of microglia. Therefore, Rodríguez-Callejas et al., 2019 endeavoured to delineate the possible role of ferritin in microglia activation in a non-human primate model. They concluded that ferritin offers protection to microglia in adult and old marmosets. They also found that, in aged subjects, the decrease in ferritin and the increased amount of Fe in the tissues of the brain, which was argued to be due to increased number of cells with oxidized RNA, most probably precluding the onset of neurodegeneration.

A study performed on lab cultured endothelial cells effectively demonstrated the antioxidant properties of ferritin (Balla et al., 1992). They observed that when endothelial cells were momentarily pulsed with heme and then incubated for 16 h, those cells exhibit high resistance to oxidant mediated injury and to the build-up of endothelial lipid peroxidation products. The protective effect was linked to the activation (within 4 h) of mRNAs for both heme oxygenase and ferritin. Once the 16 h mark passed, they recorded an increase in heme oxygenase and ferritin, which was approximately 50 and 10 fold, respectively. Therefore, it was suggested that ferritin inhibits oxidant-mediated cytolysis in direct relation to its intracellular concentration (Balla et al., 1992).

Another study aimed to test the hypothesis that ferritin regulates Fe metabolism in a way that might impact energy balance and thermal homeostasis in any organism (Blankenhaus et al., 2019). A mouse strain of ferritin heavy chain ($FtH^{R26\ fl/fl}$) was developed. Tamoxifen was administered to induce global deletion of ferritin heavy chain (FtH) in adult $Fth^{R26\Delta/\Delta}$ mice, hence testing whether FtH is necessary for the maintenance of organismal homeostasis. It was found that under normal nutritional Fe supply, FtH deletion in adult $Fth^{R26\Delta/\Delta}$ mice led to an intense misbalance of the organism Fe metabolism, oxidative stress, inflammation, and multi-organ damage, resulting in death. Surprisingly, FtH deletion was also linked to the acute atrophy of white and brown adipose tissues in addition to the disintegration of energy expenditure and thermogenesis. Henceforth, it was argued that the FtH constituent of ferritin operates as a master regulator of organismal Fe homeostasis and oxidative stress, together with thermoregulation (Blankenhaus et al., 2019).

Sepsis is a severe, active, and unstable immune response to an infection (Hotchkiss et al., 2016). An *in vivo* research done on the effects of myeloid ferritin heavy chain (FtH) in controlling the pathogenic sequelae of sepsis found that the deletion of FtH in the myeloid cells impacts cecal ligation puncture and induced inflammation and organ injury. It was also revealed that deletion of FtH was linked to significant disease protection against sepsis. This was proved by the evident of lower mortality rate, better maintained kidney function and blood pressure, reduced liver injury and serum cytokine levels and decreased organ inflammation in the myeloid FtH deficient mice (Zarjou et al., 2019).

Human ferritin H subunits in addition to murine ferritin H, when compared to the subunits L have been found to have a ferroxidase activity. It was also suggested that ferritin might act as a cytoprotective protein and help in reducing the formation of oxygen free radicals by sequestering intracellular Fe. Moreover, studies done on endothelial, tumor, and leukemia cells after pretreatment with hemin (ferric chloride heme) following introduction of ferritin have demonstrated anti-oxidative properties (Orino et al., 2001). It was found that increasing the amount of ferritin decreased the build-up of reactive oxygen species (ROS) in cells affected by H_2O_2. Nonetheless, it was also found that when ferritin was injected into cells through transient transfection, it was unable to shield them from oxidative stress. Due to the variability of transient transfection, shielding from oxidative stress is unlikely unless the preponderant cells express that particular transfected gene (Orino et al., 2001).

A study by Lee et al., 2009 established that the mechanism of the ROS-induced activation of p53 was found to also involve the protein ferritin. Physical interaction between ferritin and ferroxidase activity was shown to have an effect on the levels of p53 protein. P53 have two main dose dependent responses: (1) low level of p53 reduces ROS formation and (2) high level encourages the accretion of ROS. Additionally, the protein ferritin reduced ROS formation and was acting as an anti-apoptotic protein regardless of the ferroxidase activity manner. It was also found to bind with p53 protein and keep it at a constant level.

Ferritin itself, as a protein, does not seem to play detrimental role in the body. Nonetheless, an abnormally high level of serum ferritin can be indicative of an iron overload in the body and can also be termed as hemochromatosis (Lewis, 2019). A study done by Bester et al., 2013 add on to how high levels of ferritin can accelerate the pathogenesis of Alzheimer's disease (AD). During a random selection of group of AD patients, 60% of them were found to have an elevated serum ferritin level and consequently disfigured red blood cells structure. This change in structure, impacted the oxygen carrying capacity of the red blood cells causing a tension in those AD affected individuals.

Free iron is able to generate oxidative stress and DNA damage. Animal based studies have shown that excess free iron is carcinogenic. A study done by Kabat and Rohan reviewed the impact of the overload of free iron in the progression of breast cancer. They found that Fe^{3+} released from both ferritin and hemosiderin were reduced to Fe^{2+} and once catalysed was able to form hydroxyl radicals which are powerful oxidizing agents that were promoting mutagenesis, activation of oncogenes, tumor suppressor gene inhibition along with lipid peroxidation (Kabat and Rohan, 2007).

Conclusion

The level of serum ferritin can act as a biomarker of health in order to detect certain pathologies. Ferritin is commonly studied for its primary role of iron sequestration by entrapping labile Fe and its ability to coordinate cellular defence against inflammation and stress. Additionally, ferritin protein was also observed to have antioxidative attributes, whereby cells treated with heme were reported to induce high resistance to oxidant mediated injuries as a result of the presence of ferritin. Moreover, ferritin and free iron in the body were reported to have a positive correlation in the pathogenesis of breast cancer. Functions of ferritin beyond iron storage are of potential research interest.

References

Abbaspour, N., Hurrell, R., Kelishadi, R., 2014. Review on iron and its importance for human health. J. Res. Med. Sci. 19 (2), 164–174.

Andrews, N., Schmidt, P., 2007. Iron homeostasis. Annu. Rev. Physiol. 69 (1), 69–85.

Arosio, P., Ingrassia, R., Cavadini, P., 2009. Ferritins: a family of molecules for iron storage, antioxidation and more. Biochim. Biophys. Acta Gen. Subj. 1790 (7), 589–599.

Balla, G., Jacob, H., Balla, J., Rosenberg, M., Nath, K., Apple, F., Eaton, J., Vercellotti, G., 1992. Ferritin: a cytoprotective antioxidant strategem of endothelium. J. Biol. Chem. 267 (25), 18148–18153.

Bester, J., Buys, A., Lipinski, B., Kell, D., Pretorius, E., 2013. High ferritin levels have major effects on the morphology of erythrocytes in Alzheimer's disease. Front. Aging Neurosci. 5, 88.

Blankenhaus, B., Braza, F., Martins, R., Bastos-Amador, P., González-García, I., Carlos, A., Mahu, I., Faisca, P., Nunes, J., Ventura, P., Hoerr, V., Weis, S., Guerra, J., Cardoso, S., Domingos, A., López, M., Soares, M., 2019. Ferritin regulates organismal energy balance and thermogenesis. Mol. Metab. 24, 64–79.

Bresgen, N., Eckl, P., 2015. Oxidative stress and the homeodynamics of iron metabolism. Biomolecules 5 (2), 808–847.

Dias, V., Junn, E., Mouradian, M., 2013. The role of oxidative stress in Parkinson's disease. J. Parkinson's Dis. 3 (4), 461–491.

Giardi, M., Rea, G., Berra, B., 2011. Bio-Farms for Nutraceuticals. Landes Bioscience, Boston.

Goodsell, D., 2002. Ferritin and transferrin. *RCSB Protein Data Bank*. https://pdb101.rcsb.org/motm/35.

Harrison, P., Arosio, P., 1996. The ferritins: molecular properties, iron storage function and cellular regulation. Biochimi. Biophys. Acta Bioenerg 1275 (3), 161–203.

Hemmati-Dinarvand, M., Taher-Aghdam, A., Mota, A., Zununi Vahed, S., Samadi, N., 2017. Dysregulation of serum NADPH oxidase1 and ferritin levels provides insights into diagnosis of Parkinson's disease. Clin. Biochem. 50 (18), 1087–1092.

Hotchkiss, R., Moldawer, L., Opal, S., Reinhart, K., Turnbull, I., Vincent, J., 2016. Sepsis and septic shock. Nat. Rev. Dis. Primers 2 (1).

Kabat, G., Rohan, T., 2007. Does excess iron play a role in breast carcinogenesis? an unresolved hypothesis. Cancer Causes Control. 18 (10), 1047–1053.

Kaufman, D., Geyer, H., Milstein, M., 2017. Neurotransmitters and drug abuse. Kaufman's Clinical Neurology for Psychiatrists, 495–517.

Knovich, M., Storey, J., Coffman, L., Torti, S., Torti, F., 2009. Ferritin for the clinician. Blood Rev. 23 (3), 95–104.

Lee, J., Jang, H., Cho, E., Youn, H., 2009. Ferritin binds and activates p53 under oxidative stress. Biochem. Biophys. Res. Commun. 389 (3), 399–404.

Lewis, E., 2019. Ferritin level in iron overload [online]. Hemochromatosis Help. https://hemochromatosishelp.com/ferritin-level/ (Accessed 6 July, 2019).

Macara, I., Hoy, T., Harrison, P., 1973. The formation of ferritin from apoferritin. Catalytic action of apoferritin. Biochem. J. 135 (2), 343–348.

Orino, K., Lehman, L., Tsuji, Y., Ayaki, H., Torti, S., Torti, F., 2001. Ferritin and the response to oxidative stress. Biochem. J. 357 (1), 241.

Pagana, K., Pagana, T., 2014. Mosby's Manual of Diagnostic and Laboratory Tests, 5th ed. St Louis, Missouri.

Rodríguez-Callejas, J., Cuervo-Zanatta, D., Rosas-Arellano, A., Fonta, C., Fuchs, E., Perez-Cruz, C., 2019. Loss of ferritin-positive microglia relates to increased iron, RNA oxidation, and dystrophic microglia in the brains of aged male marmosets. Am. J. Primatol. 81 (2).

Schnell, Z., Van Leeuwen, A., Kranpitz, T., 2019. Davis's Comprehensive Handbook of Laboratory and Diagnostic Tests – With Nursing Implications, 8th ed. F.A. Davis, Philadelphia.

Shiel, W., 2018. Definition of transferrin [online]. MedicineNet. https://www.medicinenet.com/script/main/art.asp?articlekey=15591 (Accessed 14 June, 2019).

Soppi, E., 2018. Iron deficiency without anemia–a clinical challenge. Clin. Case Rep. 6 (6), 1082–1086.

Surguladze, N., Thompson, K., Beard, J., Connor, J., Fried, M., 2004. Interactions and Reactions of Ferritin with DNA. J. Biol. Chem. 279 (15), 14694–14702.

Tachibana, T., Andou, T., Tanaka, M., Ito, S., Miyazaki, T., Ishii, Y., Ogusa, E., Koharazawa, H., Takahashi, H., Motohashi, K., Aoki, J., Nakajima, Y., Matsumoto, K., Hagihara, M., Hashimoto, C., Taguchi, J., Fujimaki, K., Fujita, H., Fujisawa, S., Kanamori, H., Nakajima, H., 2018. Clinical significance of serum ferritin at diagnosis in patients with acute myeloid leukemia: a YACHT multicenter retrospective study. Clin. Lymphoma Myeloma Leuk. 18 (6), 415–421.

Takahashi, T., Kuyucak, S., 2003. Functional properties of threefold and fourfold channels in ferritin deduced from electrostatic calculations. Biophys. J. 84 (4), 2256–2263.

Wang, W., Knovich, M., Coffman, L., Torti, F., Torti, S., 2010. Serum ferritin: past, present and future. Biochim. Biophys. Acta Gen. Subj. 1800 (8), 760–769.

Yoo, J., Maeng, H., Sun, Y., Kim, Y., Park, D., Park, T., Lee, S., Choi, J., 2009. Oxidative status in iron-deficiency anemia. J. Clin. Lab. Anal. 23 (5), 319–323.

Zarjou, A., Black, L., Mccullough, K., Hull, T., Esman, S., Boddu, R., Varambally, S., Chandrashekar, D., Feng, W., Arosio, P., Poli, M., Balla, J., Bolisetty, S., 2019. Ferritin light chain confers protection against sepsis-induced inflammation and organ injury. Front. Immunol. 10 (131). doi:10.3389/fimmu.2019.00131.

CHAPTER

Glucose-6-phosphate-dehydrogenase 2.5

Mohamad Fawzi Mahomoodally, Chundoo B. Azeemah
Department of Health Sciences, Faculty of Medicine and Health Sciences, University of Mauritius, Réduit, Mauritius

2.5.1 Introduction

In a damp, chilly laboratory in Boston on a December morning in 1926, German investigators assayed the first synthetic antimalarial drug, Pamaquine. Albeit a strategic move in curbing the malarial pandemic, Peter Mühlens, its inventor claimed that the latter bore no side effects (Beutler, 2007). Soon, however, sporadic episodes of severe toxicity were recorded among West Indian immigrants of the African lineage complaining of nausea, anemia, and cyanosis, amounting with the death of two patients when administered the drug. Startled, Cordes wrote: "*One is tempted to think of an idiosyncrasy in certain persons. Or there may be an action of pamaquine [in which] it activates a haemolytic mechanism already prepared in a malaria-infected patient; it is, metaphorically, the match which sets the house on fire* (Comfort, 2009)."

Unequivocally, Cordes's postulations were far sighted. Decades later, the idiosyncratic hemolytic anemia caused by synthetic antimalarial, mainly in blacks, became a vigorous area of research, trail blazing the path for further investigations. This upheaval consequently elucidated the mechanism of hemolysis three decades later, heralded by the discovery of a glucose-6-phosphate-dehydrogenase (G6PD) deficiency being the root of the disease (Slater, 2004). Well established as a ubiquitous enzyme, G6PD is present in all contemporary organisms and tissues and is conspicuously primeval in evolution. The enzyme is X-linked cytosolic in nature and has thus become a popular subject to the phenomenon of X chromosome inactivation (Luzzatto, 1987). Located near the telomeric region of the X chromosome, the *g6pd* gene is ascribed a size of 18.5 Kb (Szabo et al., 1984). Additionally, the concluding sequence of the *g6pd* gene consists of 13 exons and 12 introns encoding a product of 1545 bp with the *g6pd* gene comprising of 514 amino acids (Pai et al., 1980).

2.5.2 Mechanism of action

The metabolic role of G6PD has been extensively studied. In a nudge, it is a significant regulatory enzyme in the hexose monophosphate shunt (HMS), catalyzing

Antioxidants Effects in Health. DOI: https://doi.org/10.1016/B978-0-12-819096-8.00025-2

the oxidation of glucose-6-phosphate (G6P) to 6-phosphogluconolactone concluding with the subsequent generation of reduced nicotinamide adenine dinucleotide phosphate (NADP) required for various biosynthetic reactions as well as shielding the cell from oxidative stress. In assonance, various investigations have highlighted the essential role of G6PD in regulating hypertrophy by maintaining the cellular redox status as well as providing an alternative pathway to the main glycolytic mechanism for glucose production (Luzzatto, 1987).

While the mechanistic understanding of the G6PD enzyme remains an evolving ground for research, much accent has been laid on its disparity of action in red blood cells (RBCs). It has been widely applauded that the enzyme thrives as a tetramer or dimer in erythrocytes (Gómez-Manzo et al., 2016). Of particular interest, the dimer structure of the two subunits in the enzyme have been found to be symmetrically confided across the intricate interface of β-sheets while each subunit binds to a $NADP^+$ molecule for their structural strength (Au et al., 1999).

What advantage does the enzyme confer by being the rate-determining step of the pentose phosphate pathway (PPP) for the RBCs? Part of this answer lies in the complexity of the pathway itself while part of it is attributed to the structure of the RBCs-albeit it being an important twist. Though conspicuous, it is worth highlighting that in contrast to other cells, mature erythrocytes do not contain mitochondria and ergo, the PPP pathway is the only provenance of NADPH; the latter being a critical regulator for protecting endogenous cells against oxidative damage stemming from the generation of reactive oxygen species (ROS) (Manganelli et al., 2010).

Proclaimed as a subsequent imbalance between two antagonistic forces (pro- and antioxidant species), and culminating with molecular and cellular damage, oxidative stress has been ascribed as an aetiology for distinct pathologies (Nóbrega-Pereira et al., 2016). Ergo, organisms are endowed with several antioxidant mechanisms that maintain ROS levels below an immaculate threshold under homoeostatic conditions (D'Autréaux and Toledano, 2007), the dark side of these species stems from the damage induced to macromolecules. Additionally, numerous studies have substantiated the 'free radical theory of ageing' outlining that ROS-derived atrophy culminates in the functional decline of a multitude of organ systems predisposing to pathologies, such as neurodegenerative diseases and cancer (Finkel and Holbrook, 2000). In support of the former, it has been noted that in the case of transgenic mice, bearing an overexpression of antioxidant enzymes, protection from a number of pathologies is generally conferred and in remarkable cases, an increased longevity is observed.

Typically, NADPH functions as a mandatory scavenger of cellular ROS, with its involvement being central to three antioxidant pathways: the glutaredoxin, glutathione, and thioredoxin cycles (Cappellini and Fiorelli, 2008). Reduced glutathione plays a pivotal role in the antioxidant defence system by becoming oxidized, thereby neutralizing the damaging oxidative process. To assert its role as a detoxifying agent glutathione must be kept in its reduced form and must be continuously replenished from its oxidized form. The regeneration mechanism uses hydrogen ions obtained from NADPH, which is formed from NADP with the G6PD enzyme catalyzing the first step in the hexose monophosphate pathway (Kaplan and Hammerman, 2000).

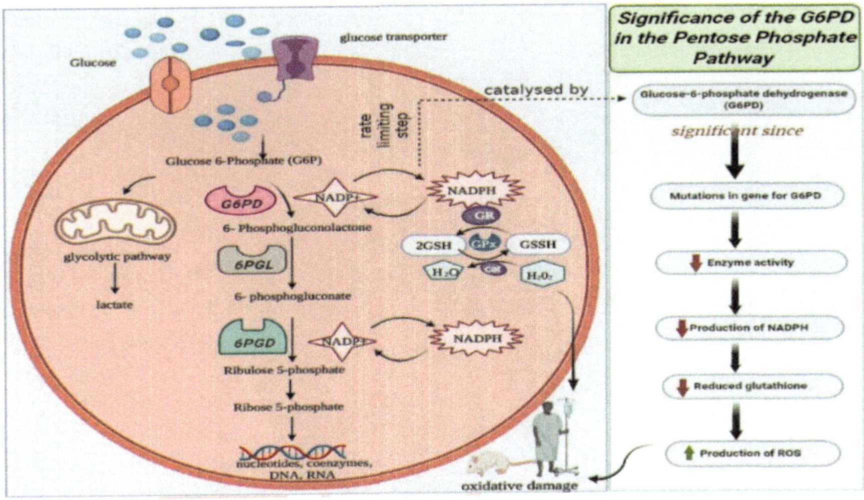

FIG. 2.5.1 An illustration of the pentose phosphate pathway (PPP) in an erythrocyte highlighting the significance of the glucose 6-phosphate dehydrogenase (G6PD) enzyme in the process.

ADP, adenosine diphosphate; ATP, adenosine triphosphate; cat, catalase; G6P, glucose 6-phosphate; GR, glutathione reductase; GSH, reduced glutathione; GSSG, oxidized glutathione; GPx, glutathione peroxidase; 6PGD, 6-phosphogluconate dehydrogenase; 6PGL, 6-phosphogluconolactonase.

The cascades of steps involved in the glutathione pathway are tightly regulated in RBCs to provide a first line of defence against the generation of ROS (Fig. 2.5.1). The electrons from the NADPH conveniently passes to the glutathione dimers (GSSG), the reaction catalysed by glutathione reductase enzyme amount in the production of two reduced glutathione monomers (GSH) asserting their antioxidant properties on the ROS. In consonance, the elimination of peroxides from the RBCs is brought about using glutathione peroxidases (GPX) coupled with the GSH as substrates. Besides, the NADPH is essential to reduce the GSSG and the sulfhydryl groups of some proteins to protect the cell against oxidative stress. The RBCs that succumb to this stress ultimately undergoes hemolysis (Cappellini and Fiorelli, 2008). Consequently, G6PD can be regarded legitimately as a typical and generally mandatory household or housekeeping enzyme (Luzzatto, 1987).

2.5.3 Beneficial effects of glucose-6-phosphate on health

The beneficial effects of G6PD on health is connected to its function. Numerous studies have shed light on the necessity of G6PD for cell growth. The notion was substantiated by Raineri and Levy, 1970, who illustrated that suppression of the enzyme activity led to educed cell lines differentiation when treated with a putative

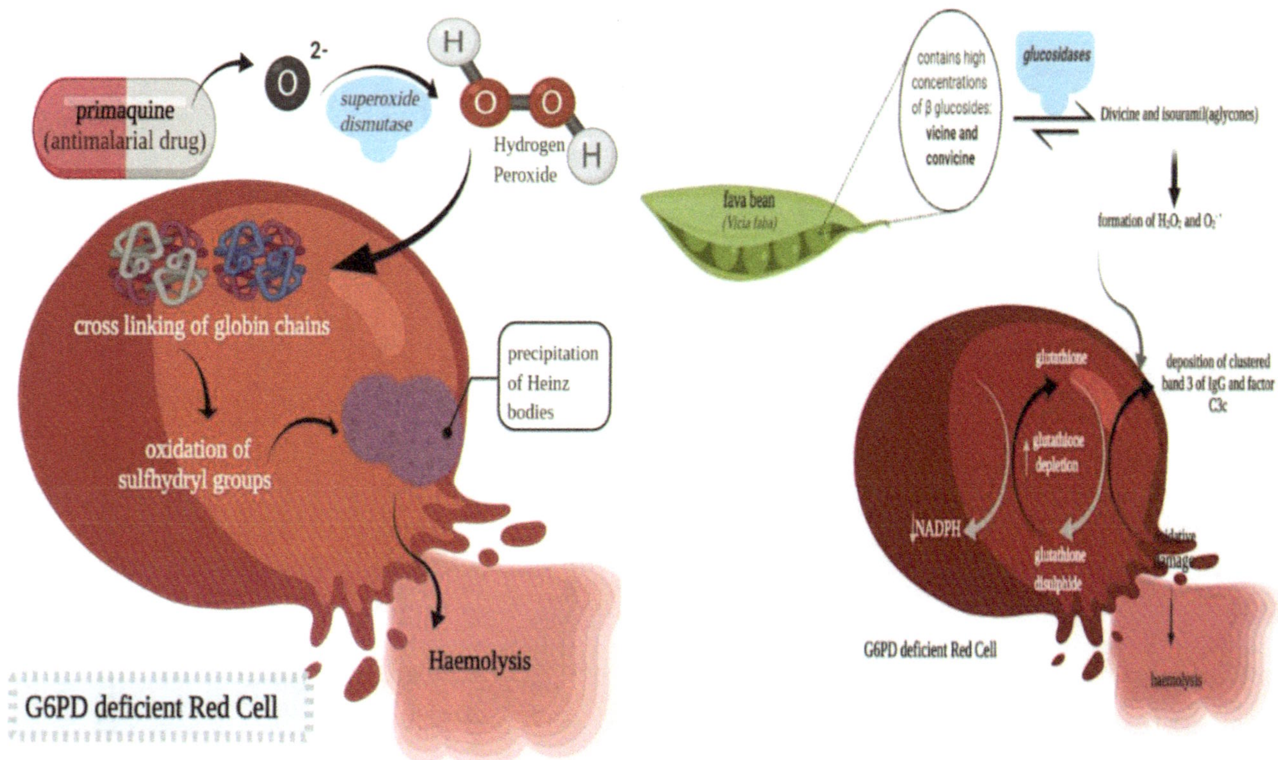

FIG. 2.5.2

Schematic representation of (A) drug induced hemolysis and (B) fava induced hemolysis in G6PD deficient erythrocytes.

inhibitor of dehydroepiandrosterone (DHEA) (Raineri and Levy, 1970). It is however, disputable whether DHEA actually inhibited the enzyme's activity in the cultured cells. While precedent research ventures have shown that DHEA had the potency to inhibit the activity of a purified G6PD preparation, yet, it seems that DHEA and its analogues do not sustain a prolonged inhibitory effect on G6PD in the cultured cells heralded by a transient decreased G6PD activity after undergoing DHEA treatment bouncing back several hours later. Similar inhibition kinetics was observed with the enzyme in erythrocytes (Oertel and Rebelein, 1970).

The key role of G6PD activity in cellular hypertrophy was further illustrated by the potential of G6PD to recover deficient cells from growth arrest premature senescence. The mechanism responsible has been hypothesised to be due to heightened oxidative stress instead of telomere shortening. These findings establish the significance of G6PD in cellular proliferation and senescence (Ho et al., 2006). Besides, the participation of ROS in cellular senescence is not surprising. ROS produced during cellular metabolism causes progressive damage to the cell can thereby accelerate senescence. As a result of the imperfect nature of respiration, about 1%–2% of electron flow catalyses the formation of oxygen free radicals, which is subsequently permuted to other ROS such as hydrogen peroxide or hydroxy radicals. These ROS consequently destroy protein, lipids as well as the genomic material of the cells. The damage incurred ultimately averts the cells potential to grow and stimulates senescence as witnessed in G6PD deficient cells.

Coupled with the above, the enzyme also plays an important role in death signalling. Human fibroblasts lacking the enzyme exhibited altered response to nitric oxide. These deficient cells sustained apoptosis after being administered nitric oxide (Cheng et al., 2000). This was contrasting to normal cells, which showed distinct proliferation after treatment with nitric oxide. In assonance, it has been shown that G6PD deficient fibroblasts are more likely to undergo peroxynitrite-induced apoptosis, further stimulated by glutathione depletion and the activation of p53, emphasizing the necessity of redox control in the apoptotic pathways. Ergo, the level of G6PD activity clearly dictates the cell's potential in eliminating ROS and conspicuously their abundance within the cell (Ho et al., 2006). Ultimately, it can be inferred that cells with normal G6PD levels conserve a constant GSH/GSSG ratio and that an inverse correlation of G6PD activity can be established with the quantity of ROS.

2.5.4 Detrimental effects of glucose-6-phosphate on health

While the G6PD enzyme served as a prototype in correlating enzymatic deficiencies to pathological disorders, it also offered an unparalleled opportunity for various scientific strides. Convergent with the unveiling of the biochemical pathways through which erythrocytes metabolize sugar and eventually culminated with the development of isotopic methods offering precise estimations about the red cell survival. Whilst most of the detrimental effects on GD6P on health arise from a deficiency or an overexpression of the enzyme (Table 2.5.1), the most common and

Table 2.5.1 Association of G6PD with distinct pathologies.

Disease	Study type/design	Results	References
Composite cardiovascular diseases	Retrospective, cross sectional/737 G6PD deficient United States military personnel matched with counterpart	↑ 39.6% risk of developing cardiovascular diseases in G6PD deficient individuals ↑ Vascular oxidant stress among G6DP patients	(Thomas et al., 2018)
Hypertension	Cross sectional/154,917 prepregnant females with abnormal blood pressure participated in check-ups programs	↑ Odd ratios for G6PD deficient patients compared with parallel group ↑ Systolic blood pressure ↔ Diastolic blood pressure	(Zhao et al., 2018)
Atherosclerosis	In vivo/80% G6PD deficient mutant mice compared with G6PD hemizygous and wild-type mice	↑ Blood pressure in G6DP deficient mice ↓ Serum cholesterol levels in G6DP deficient mice ↓ Superoxide anion & iNOS in hemizygous mice	(Matsui et al., 2006)
Cancer	In vivo/human epidermal melanocyte cells with four G6PD gene variants (wild-type, deficiency, overexpression, and mutant) injected in five groups of mice.	↓ Growth and formation of tumor tissues in G6PD deficient mice ↓ Downregulation of cell cycle proteins (cyclin D1, E, and p53) ↑ Apoptosis-inhibited factors Bcl-2	(Hu et al., 2013)
Gastric cancer	In vivo/expression of G6PD in 167 patients with pathology detected by immunohistochemistry	↑ Expression of G6PD associated with progression of gastric cancer ↑ Tumor size, metastasis, cancer survival rate with G6PD overexpression.	(Wang et al., 2011)
Type II diabetes	Cross sectional/52,371 diabetic and G6DP deficient patients matched with parallel cohort.	↑ Incidence of diabetes in G6DP deficient population ↓ Enzymic expression in hyperglycemic patients	(Heymann et al., 2012)
Diabetes mellitus	Case control of patients with hemolytic crises	↓ Levels of insulin ↔ Acidosis	(Carette et al., 2011)
Amyotrophic lateral sclerosis (ALS)	Cross sectional/20 sporadic ALS patients matched with parallel group	↑ Lipid peroxidation in ALS patients ↓ Expression of G6PD enzyme in course of ALS progression	(Babu et al., 2008)
Huntington disease	In vivo/overexpression of G6PD *Drosophila* model injected with the exon 1 of the human *huntingtin* gene	↑ Lifespan of flies with Huntington disease ↑ Arrest of eye neurodegeneration ↑ Tolerance to oxidative stress	(Besson et al., 2015)

↑, increase; ↓, decrease; ↔, no change; ALS, Amyotrophic lateral sclerosis; G6PD, glucose-6 phosphate dehydrogenase; iNOS, inducible nitric oxide synthase.

widely researched consequence is that of haemolytic anaemia. In a nutshell, triggers for haemolytic anaemia in G6PD deficient subjects can be due to (1) drugs, (2) fava beans, and (3) infections.

2.5.4.1 Drug induced hemolysis

Numerous clinical investigations have been brought forward in order to elucidate the link between primaquine and hemolysis. Effectuated during the rigid circumstances bordering the Korean War and World War II, the clinical trials were essentially carried out on prisoner volunteers sentenced in the Illinois State Penitentiary at Joliet, funded by the United States Army (Beutler, 2007).

Paradoxically to the notion of how research was performed before the rise of the Institutional Review Boards (IRBs), the prisoner's participation was truly benevolent, with each signing informed consent forms defining clearly the details of the trial (Ebaugh and Ross, 1985). The aim of the trial was conspicuous. When normal volunteers were administered 30 mg of primaquine daily, acute hemolytic anemia was noted as an outcome in some subjects while the majority didn't. Did those who developed the pathology metabolize primaquine differently or were their red cells any different from the others? The answer to the paradigm was rendered possible by the development of the ^{51}Cr technique (Gray and Sterling, 1950). The trial concluded with the findings that when ^{51}Cr-labeled cells from a primaquine-sensitive subject were transfused into a non-sensitive subject, primaquine administration resulted in rapid destruction of the labeled erythrocytes, however, when ^{51}Cr-labeled cells from a nonsensitive subject were transfused into a primaquine-sensitive subject, they survived normally even when the host's red cells were being rapidly destroyed highlighting clearly that sensitivity to the hemolytic effect of primaquine was due to an intrinsic defect of the erythrocyte (Beutler, 2007).

Few morphological changes were noted in subjects experiencing primaquine-induced hemolysis. However, the detection of Heinz bodies before the onset of hemolysis was remarkable (Vásquez Vivar and Augusto, 1994). The emergence of these Heinz bodies both *in vitro* and *in vivo* in G6PD deficient cells emphasized the cell's inefficacy to protect GSH against probable drug-induced hemolysis due to the cell's inability to avert oxidative insults at the sulfhydryl functional groups (Beutler et al., 1957).

2.5.4.1.1 *Fava induced hemolysis (favism)*

The correlation of dry grains or frozen fava beans (*Vicia faba*) consumption and its subsequent pathological consequence (Favism) were noted mostly in the Mediterranean countries. The syndrome is most predominant among males than females heralded by an elevated incidence during 2 and 6 years. Favism is also preponderant in breast fed infant whose mother has previously consumed fava beans (Mehta et al., 2000). From an epidemiological perspective, fava beans consumption is mostly common in Iran, as a consequence, the disease is mostly concentrated in the area, the causality being the high manifestation of G6PD deficiency in Iran (Noori-Daloii et al., 2007).

Favism is explained as the acute haemolysis following the consumption of fava beans. Symptoms include chills, headache, chills, and fever within 24 h of ingestion, followed by hemoglobinuria and jaundice. An example of this type of clinical manifestation was witnessed in an 18-year-old male conferring the mutant G6PD variant which established that hemolytic anemia was triggered by repeated consumption of fava beans. Reported with a neonatal jaundice in infancy, biochemical data of the patient revealed an under expression of enzyme activity during the hemolytic crisis (Benmansour et al., 2013). Although favism may be quite mild, yet occasionally; the anemia may be very severe. Kattamis et al. (1969) reported a series of hospitalized patients whereby blood hemoglobin concentration was found to be lower than 6 g/100 rnL in 81%, and below 4 g/100 rnL in 30% of the patients while in one patient a modest hemoglobin level of less than 1 g/100 rnL was recorded (Kattamis et al., 1969). According to Luisida (1941), acute renal failure, one of the consequences of the pathology is apparently common in adults. Of 62 cases of favism reviewed, 16 patients developed the complication, 10 required dialysis, and 3 terminated in death (Luisida, 1941).

Besides, the hemolytic crisis has been attributed to be similar to that triggered by drugs albeit it is worth highlighting that though patients with favism are G6PD deficient, not all of them develop favism upon the consumption of the beans (Yoshida, 1967). However, emphasizing that though G6PD deficiency is a mandatory requirement for favism, the latter is not a sufficient cause for the pathology. Ultimately, oxidizing compounds such as vicine, convicine and isouramil found in fava beans have been proposed to be the causative mechanism of hemolysis (Noori-Daloii et al., 2004).

2.5.4.1.2 *Infection-induced hemolysis*

Though the frequent trigger of hemolytic anemia in GD6P subjects is induced by the administration of drugs, hemolysis precipitated by concomitant infectious disease is by far the most common source of morbidity in the patients. Yet, despite infection induced hemolysis may range from a continuum of mild to occasionally very severe, the exact mechanism has not been uncovered. Howbeit, it has been hypothesized that the release of oxidants by leucocytes during an infection subsequently prompting oxidative stress in erythrocytes might be a plausible explanation in explaining the infections (Kattamis and Tjortjatou, 1970).

Panich and Sungnate (1973) have reported a number of cases of acute renal failure resulting from infection-induced hemolytic episodes in GD6P deficient subjects in Thailand, they highlighted that in many instances, the coupling of hemolytic drugs administered during the infection might have also played a role (Panich and Sungnate, 1973). Mengel (1967), on the other hand, in his review noted that at least 19 of 102 G6PD deficient patients admitted to a New York hospital developed hemolytic episodes complicating such an infection (Mengel, 1967).

Albeit in one survey it has been noted that half of the G6PD deficient patients became anemic following up an infection and suffered from pneumonia (Siler et al., 2017), several studies have highlighted the precipitation of anemia due to typhoid fever (Hersko and Vardy, 1967; Lampe et al., 1975). In assonance several infectious agents such as *Salmonella*, *Proteus*, *Escherichia* coli, as well as *Streptococci* have been associated with causing hemolysis (Beutler, 2007). Finally, hemolytic anemia appears to be common in infection hepatitis whereby in a study by Kattamis and Tjortjatou in 1970, in a cohort of 125 children with viral hepatitis 23% of those with normal G6PD activity experienced hemolysis while 87% of the group with G6PD deficiency developed hemolytic anemia (Kattamis and Tjortjatou, 1970).

Conclusion

70 years ago, the enzyme G6PD was still foreign to the scientific community. It is conspicuous that studying the enzyme has probed a critical eye on the world history, unraveling myriad of discoveries along the way. Well established as a ubiquitous enzyme in all contemporary organisms, besides being the rate limiting enzyme of the pentose phosphate pathway, the enzyme is eventually indispensable in maintaining the cytosolic pool of NADPH and averting potential oxidative insults. While much interest have been geared towards underpinning the enzyme's role in mitigating several pathologies, it is feasible to assert that future research ventures coupled with novel analytical technologies such as proteomics and functional genomics, we might fully unveil the latter's status in determining cell physiology and human health.

References

Au, S., Naylor, C., Gover, S., Vandeputte-Rutten, L., Scopes, D., Mason, P., Luzzatto, L., Lam, V., Adams, M., 1999. Solution of the structure of tetrameric human glucose 6-phosphate dehydrogenase by molecular replacement. Acta Crystallogr. D Biol. Crystallogr. 55 (4), 826–834. doi:10.1107/s0907444999000827.

Babu, G., Kumar, A., Chandra, R., Puri, S., Singh, R., Kalita, J., Misra, U., 2008. Oxidant–antioxidant imbalance in the erythrocytes of sporadic amyotrophic lateral sclerosis patients correlates with the progression of disease. Neurochem. Int., 52 (6), 1284–1289. doi:10.1016/j.neuint.2008.01.009.

Benmansour, I., Moradkhani, K., Moumni, I., Wajcman, H., Hafsia, R., Ghanem, A., Abbès, S., Préhu, C., 2013. Two new class III G6PD variants [G6PD Tunis (c.920A>C: p.307Gln>Pro) and G6PD Nefza (c.968T>C: p.323 Leu>Pro)] and overview of the spectrum of mutations in Tunisia. Blood Cells Mol. Dis. 50 (2), 110–114. doi:10.1016/j.bcmd.2012.08.005.

Besson, M., Alegría, K., Garrido-Gerter, P., Barros, L., Liévens, J., 2015. Enhanced neuronal glucose transporter expression reveals metabolic choice in a HD Drosophila model. PLoS One 10 (3), e0118765. doi:10.1371/journal.pone.0118765.

Beutler, E., 2007. Glucose-6-phosphate dehydrogenase deficiency: a historical perspective. Blood, 111 (1), 16–24. doi:10.1182/blood-2007-04-077412.

Beutler, E., Robson, M., Buttenwieser, E., 1957. The mechanism of glutathione destruction and protection in drug-sensitive and non-sensitive erythrocytes; in vitro studies.. J. Clin. Investig., 36 (4), 617–628. doi:10.1172/JCI103461.

Cappellini, M., Fiorelli, G., 2008. Glucose-6-phosphate dehydrogenase deficiency. Lancet, 371 (9606), 64–74. doi:10.1016/S0140-6736(08)60073-2.

Carette, C., Dubois-Laforgue, D., Gautier, J., Timsit, J., 2011. Diabetes mellitus and glucose-6-phosphate dehydrogenase deficiency: from one crisis to another. Diabetes Metab. 37 (1), 79–82. doi:10.1016/j.diabet.2010.09.004.

Cheng, M., Ho, H., Liang, C., Chou, Y., Stern, A., Lu, F., Chiu, D., 2000. Cellular glucose-6-phosphate dehydrogenase (G6PD) status modulates the effects of nitric oxide (NO) on human foreskin fibroblasts. FEBS Lett. 475 (3), 257–262. doi:10.1016/s0014-5793(00)01687-2.

Comfort, N., 2009. The prisoner as model organism: malaria research at Stateville Penitentiary.. Stud. Hist. Philos. Sci. Biomed. Sci. 40 (3), 190–203. doi:10.1016/j.shpsc.2009.06.007.

D'Autréaux, B., Toledano, M., 2007. ROS as signalling molecules: mechanisms that generate specificity in ROS homeostasis. Nat. Rev. Mol. Cell Biol. 8 (10), 813–824. doi:10.1038/nrm2256.

Ebaugh Jr., F., Ross, J., 1985. The radioactive sodium chromate method for erythrocyte survival. J. Lab. Clin. Med. 49 (4), 304–307. doi:10.1159/000466392.

Finkel, T., Holbrook, N., 2000. Oxidants, oxidative stress and the biology of ageing. Nature 408 (6809), 239–247. doi:10.1038/35041687.

Gómez-Manzo, S., Marcial-Quino, J., Vanoye-Carlo, A., Serrano-Posada, H., Ortega-Cuellar, D., González-Valdez, A., Castillo-Rodríguez, R., Hernández-Ochoa, B., Sierra-Palacios, E., Rodríguez-Bustamante, E., Arreguin-Espinosa, R., 2016. Glucose-6-phosphate dehydrogenase: update and analysis of new mutations around the world. Int. J. Mol. Sci. 17 (12), 2069. doi:10.3390/ijms17122069.

Gray, S., Sterling, K., 1950. Determination of circulating red cell volume by radioactive chromium. J. Clin. Invest. 112 (2902), 179–180. doi:10.1172/JCI102404.

Hersko, C., Vardy, P., 1967. Haemolysis in typhoid fever in children with G-6-PD deficiency. Brit. Med. J. 1 (5534), 214–215. doi:10.1136/bmj.1.5534.214.

Heymann, A., Cohen, Y., Chodick, G., 2012. Glucose-6-phosphate dehydrogenase deficiency and type 2 diabetes. Diabetes Care, 35 (8), e58. doi:10.2337/dc11-2527.

Ho, H., Wei, T., Cheng, M., Chiu, D., 2006. Green tea polyphenol Epigallocatechin-3-gallate protects cells against peroxynitrite-induced cytotoxicity: modulatory effect of cellular g6pd status. J. Agric. Food Chem. 54 (5), 1638–1645. doi:10.1021/jf0524372.

Hu, T., Zhang, C., Tang, Q., Su, Y., Li, B., Chen, L., Zhang, Z., Cai, T., Zhu, Y., 2013. Variant G6PD levels promote tumor cell proliferation or apoptosis via the STAT3/5 pathway in the human melanoma xenograft mouse model. BMC Cancer 13 (1), 1–11. doi. https://doi.org/10.1186/1471-2407-13-251.

Kaplan, M., Hammerman, C., 2000. Glucose-6-phosphate dehydrogenase deficiency: a worldwide potential cause of severe neonatal hyperbilirubinemia. Neo Rev. 1 (2), e32–e39. doi:10.1053/siny.2002.0099.

Kattamis, C., Kyriazakou, M., Chaidas, S., 1969. Favism: clinical and biochemical data. J. Med. Genet. 6 (1), 34–41. doi:10.1136/jmg.6.1.34.

Kattamis, C., Tjortjatou, F., 1970. The haemolytic process of viral hepatitis in children with normal or deficient glucose-6-phosphate dehydrogenase activity. J. Pediatr. 77 (3), 422–430. https://doi.org/10.1016/S0022-3476(70)80010-5.

Lampe, R., Kirdpon, S., Mansuwan, P., Benenson, M., 1975. Glucose-6-phosphate dehydrogenase deficiency in Thai children with typhoid fever. J. Pediatr. 87 (4), 576–578. doi:10.1016/S0022-3476(75)80826-2.

Luisada, A., 1941. A singular disease chiefly affecting the red blood cells. Medicine, 20 (2), 229–250.

Luzzatto, L., 1987. Glucose 6-phosphate dehydrogenase: genetic and haematological aspects. Cell Biochem. Funct. 5 (2), 101–107. doi:10.1007/s00439-005-0126-5.

Manganelli, G., Fico, A., Martini, G., Filosa, S., 2010. Discussion on pharmacogenetic interaction in g6pd deficiency and methods to identify potential hemolytic drugs. Cardiovasc. Hematol. Disord. Drug Targets, 10 (2), 143–150. doi: 10.2174/187152910791292547.

Matsui, R., Xu, S., Maitland, K., Mastroianni, R., Leopold, J., Handy, D., Loscalzo, J., Cohen, R., 2006. Glucose-6-Phosphate Dehydrogenase Deficiency Decreases Vascular Superoxide and Atherosclerotic Lesions in Apolipoprotein E–/–Mice. Arterioscler. Thromb. Vasc. Biol. 26 (4), 910–916. doi:10.1161/01.ATV.0000205850.49390.3b.

Mehta, A., Mason, P., Vulliamy, T., 2000. Glucose-6-phosphate dehydrogenase deficiency. Best Pract. Res. Clin. Haematol. 13 (1), 21–38. doi:10.1053/beha.1999.0055.

Mengel, C., 1967. Anemia during acute infections. Role of glucose-6-phosphate dehydrogenase deficiency in Negroes. Arch. Intern. Med. 119 (3), 287–290. doi:10.1001/archinte.1967.00290210119011.

Nóbrega-Pereira, S., Fernandez-Marcos, P., Brioche, T., Gomez-Cabrera, M., Salvador-Pascual, A., Flores, J., Viña, J., Serrano, M., 2016. G6PD protects from oxidative damage and improves healthspan in mice. Nat. Commun. 7 (1), 1–9. doi:10.1038/ncomms10894.

Noori-Daloii, M., Hajebrahimi, Z., Najafi, L., Mesbah-Namin, S., Mowjoodi, A., Mohammad Ganji, S., Yekaninejad, M., Sanati, M., 2007. A comprehensive study on the major mutations in glucose-6-phosphate dehydrogenase-deficient polymorphic variants identified in the coastal provinces of Caspian Sea in the north of Iran. Clin. Biochem. 40 (9-10), 699–704. doi:10.1016/j.clinbiochem.2007.02.008.

Noori-Daloii, M., Najafi, L., Ganji, S., Hajebrahimi, Z., Sanati, M., 2004. Molecular identification of mutations in G6PD gene in patients with favism in Iran. J. Physiol. Biochem. 60 (4), 273–277. doi:10.1007/BF03167073.

Oertel, G., Rebelein, I., 1970. Effects of dehydroepiandrosterone and its conjugates upon the activity of glucose-6-phosphate dehydrogenase. J. Steroid Biochem. 1 (2), 93–97. doi:10.1016/0022-4731(70)90004-X.

Pai, G., Sprenkle, J., Do, T., Mareni, C., Migeon, B., 1980. Localization of loci for hypoxanthine phosphoribosyltransferase and glucose-6-phosphate dehydrogenase and biochemical evidence of non-random X chromosome expression from studies of a human X-autosome translocation. Proc. Natl. Acad. Sci. 77 (5), 2810–2813. doi:10.1073/pnas.77.5.2810.

Panich, V., Sungnate, T., 1973. Characterization of glucose-6-phosphate dehydrogenase in Thailand. Hum. Genet. 18 (1).

Raineri, R., Levy, H., 1970. Specificity of steroid interaction with mammary glucose 6-phosphate dehydrogenase. Biochemistry, 9 (11), 2233–2243. doi:10.1021/bi00813a003.

Siler, U., Romao, S., Tejera, E., Pastukhov, O., Kuzmenko, E., Valencia, R., MedaSpaccamela, V., Belohradsky, B., Speer, O., Schmugge, M., Kohne, E., Hoenig, M., Freihorst, J., Schulz, A., Reichenbach, J., 2017. Severe glucose-6-phosphate dehydrogenase deficiency leads to susceptibility to infection and absent NETosis. J. Allergy. Clin. Immunol. 139 (1), 212–219.e3. doi:10.1016/j.jaci.2016.04.041.

Slater, L., 2004. Malaria chemotherapy and the "kaleidoscopic" organisation of biomedical research during World WarII. Ambix, 51 (2), 107–134. doi:10.1179/amb.2004.51.2.107.

Szabo, P., Purrello, M., Rocchi, M., Archidiacono, N., Alhadeff, B., Filippi, G., Toniolo, D., Martini, G., Luzzatto, L., Siniscalco, M., 1984. Cytological mapping of the human glucose-6-phosphate dehydrogenase gene distal to the fragile-X site suggests a high rate of meiotic recombination across this site. Proc. Natl. Acad. Sci. 81 (24), 7855–7859. doi:10.1073/pnas.81.24.7855.

Thomas, J., Kang, S., Wyatt, C., Kim, F., Mangelsdorff, A., Weigel, F., 2018. Glucose-6-phosphate dehydrogenase deficiency is associated with cardiovascular disease in U.S. military centers. Tex. Heart Inst. J. 45 (3), 144–150. doi:10.14503/THIJ-16-6052.

Vásquez-Vivar, J., Augusto, O., 1994. Oxidative activity of primaquine metabolites on rat erythrocytes in vitro and in vivo. Biochem. Pharmacol. 47 (2), 309–316. doi:10.1016/0006-2952(94)90022-1.

Wang, J., Yuan, W., Chen, Z., Wu, S., Chen, J., Ge, J., Hou, F., Chen, Z., 2011. Overexpression of G6PD is associated with poor clinical outcome in gastric cancer. Tumor Biol. 33 (1), 95–101. doi:10.1186/1471-2407-13-251.

Yoshida, A., 1967. Human glucose 6-phosphate dehydrogenase: Purification and characterization of negro type variant (A+) and comparison with normal enzyme (B+). Biochem. Genet. 1 (2), 81–99. doi:10.1007/BF00486511.

Zhao, J., Zhang, X., Guan, T., Wang, X., Zhang, H., Zeng, X., Dai, Q., Wang, Y., Zhou, L., Ma, X., 2018. The association between glucose-6-phosphate dehydrogenase deficiency and abnormal blood pressure among prepregnant reproductive-age Chinese females. Hypertens. Res. 42 (1), 75–84. doi:10.1038/s41440-018-0118-1.

CHAPTER

Melatonin

2.6

Mohammad Hossein Asghari[a], Milad Moloudizargari[b]

[a]*Department of Pharmacology and Toxicology, School of Medicine, Babol University of Medical Sciences, Babol, Iran*

[b]*Department of Immunology, School of Medicine, Shahid Beheshti University of Medical Sciences, Tehran, Iran*

2.6.1 Endogenous sources

Melatonin (N-acetyl-5-methoxy tryptamine), an evolutionarily conserved molecule, is endogenously produced from tryptophan. It was initially considered as an exclusive hormone synthetized in the pineal gland to regulate the circadian rhythm. However, it is now clear that this indoleamine is produced by many cells within different organs, including lymphocytes, astrocytes, retinal cells, glial cells, cells of the gut, testes, placenta, ovary, skin, etc. (Tan et al., 2010). The synthesis of melatonin in the pineal gland consists of L-tryptophan uptake from the circulation into the pinealocytes, the subsequent conversion of L-tryptophan into 5-hydroxytryptophan by 5-monooxygenase/hydrolase and its decarboxylation by the aromatic amino acid decarboxylase to form 5-hydroxytryptamine (or serotonin). In the following step, serotonin undergoes N-acetylation by the action of aralkyl amine-N-acetyltransferase (AANAT), giving rise to N-acetylserotonin. Finally, N-acetyl serotonin is methylated by acetylserotonin-O-methyltransferase (ASMT), giving rise to melatonin (Acuña-Castroviejo et al., 2014; Gonzalez-Arto et al., 2016; Reiter et al., 2010). Among different sources, the gastrointestinal (GI) tract has been revealed to be a main site of melatonin abundance. The cells of the gut are perhaps the leading cells in melatonin synthesis and secretion considering the finding that the GI can store up to 400 times the melatonin stored in the pineal gland, indicating the importance of this molecule in the regulation of GI function and health. The melatonin produced within the cells of the GI tract and the intestinal microbial flora as well as the melatonin content of the ingested foods rich in melatonin altogether constitute the melatonin storage of the GI tract (Asghari et al., 2017). Melatonin has also been detected in almost all biological fluids, including saliva, cerebrospinal fluid, urine, follicular fluid, bile, amniotic fluid, synovial fluid, seminal plasma, and milk (Casao et al., 2010; Claustrat et al., 2005; Gonzalez-Arto et al., 2016). The widespread range of organs and bodily fluids in which melatonin exists is another evidence of its importance as a regulator of different mechanisms throughout the whole body beside

Antioxidants Effects in Health. DOI: https://doi.org/10.1016/B978-0-12-819096-8.00026-4

its regulation of the day/night cycle (Tan et al., 2015). However, melatonin from tissues other than the Pineal gland is not responsible for circadian rhythm regulation and cannot replace the function and role of pineal melatonin, and studies have shown that the extra pineal melatonin functions primarily as an antioxidant. Studies are still underway on the safety of melatonin, and according to the currently available data, melatonin has proven highly safe. However, there are still some considerations regarding the use of melatonin. A critical consideration is the use of the inhibitors of CYP1A2 and CYP2C19, two enzymes by which melatonin is mainly metabolized. The inhibition of these enzymes could result in considerably high concentrations of melatonin. Melatonin also possesses hypotensive and glucose-lowering effects, which necessitates caution to be taken by the patients that take antihypertensive or glucose-lowering medicines for their conditions. Melatonin is also not recommended for pregnant and lactating women. The absorption of melatonin depends on the site and the rectum has the highest absorption. Melatonin appears to reach a maximum plasma concentration 30–45 min after intravenous administration and 30–60 min after oral administration. Melatonin has a low bioavailability as well as a high first-pass effect such that solely 10%–15% of its total ingested amount reaches the circulation following oral consumption (Andersen et al., 2016; Pourhanifeh et al., 2020; Tan et al., 2018).

2.6.2 Mechanisms of action

Melatonin is a major role-player in the network of biochemical and signaling pathways. This modulatory role of melatonin is mediated through its direct and indirect effects on a range of proteins and free radicals. Melatonin also induces its effects via its receptors including MT1 and MT2, both belonging to the family of G-protein coupled receptors (GPCR), which dampens cAMP and, as a result, mitigate the activity of CREB and PKA (Asghari et al., 2018). The interaction of melatonin with MT1 increases the concentration of cytoplasmic calcium, while its interaction with MT2 decreases cGMP formation and activates PKC. Phospholipase C (PLC) is present downstream of both receptors. The MAPK/MEK/ERK pathway is triggered following the activation of PLC and the PI3K/AKT signaling pathway may also play a role (Hardeland, 2009). Dysregulation of melatonin production is found to be associated with aging and many disorders, including malignancies, immune-mediated diseases, and type 2 diabetes (Goradel et al., 2017). On the other hand, the melatonin has proven effective in a number of diseases such as the Alzheimer's disease, insomnia, Huntington's disease, Parkinson's disease, migraine, amyotrophic lateral sclerosis (ALS), peptic ulcer, gastroesophageal reflux disease (GERD), irritable bowel syndrome (IBS), and inflammatory bowel disease (IBD). Moreover, the effectiveness of melatonin in reducing myocardial injury and modulating reproductive activity has been already approved in animal models (Bai et al., 2018; Galano et al., 2013; Zhang and Zhang, 2014). It has been also shown that melatonin can mitigate heavy metal- and pesticide-induced toxicity through the reduction of oxidative stress, apoptosis and inflammatory responses (Asghari et al., 2017; Haghi-Aminjan et al., 2018; Romero et al., 2014).

2.6.3 Beneficial effects of melatonin on health

2.6.3.1 Melatonin as an antioxidant

Oxidative stress occurs as a result of a disequilibrium between free radical generation and the related biological defenses (Haghi-Aminjan et al., 2018; Zhang and Zhang, 2014). Our knowledge concerning the detrimental effects of excess reactive oxygen species dates back as far as 1950s as numerous experiments prompted researchers postulate that reactive oxygen radicals are implicated in damaging vital cellular components, such as enzymes, fatty acids, and nucleic acids. In addition, the protective role of several molecules with anti-oxidant properties against oxygen radical injury have also been elucidated (Di Meo et al., 2016). Since then, extensive experiments have been conducted to explain the involvement of reactive oxygen species (ROS) in health and diseases. The astonishing outcomes of these experiments explain how low concentrations of such reactive radicals and oxidants play a crucial role in physiologic processes, such as immune system responses, cellular signaling pathways, transcription factor activation, and programmed cell death (Rossignol and Frye, 2012). This is while oxidative stress generated by high concentrations of ROS is implied in the etiology of numerous disorders, such as cancer, cardiovascular disorders, rheumatoid arthritis, diabetes, various neurodegenerative diseases, and aging (Essick and Sam, 2010). This has led to an upsurge in using antioxidants for prevention and treatment of a variety of disorders such that the dietary supplementation of antioxidants is becoming an ever-increasing trend, especially in industrial countries. As a result, a growing body of studies has focused on the efficacy and safety of antioxidants as well as comparing them under different conditions (Pham-Huy et al., 2008).

To protect against free radicals, various free radical scavenging mechanisms such as direct neutralization of free radicals or indirect upregulation of antioxidant enzymes have been evolved in the cells, which convert toxic metabolites into nontoxic molecules (David et al., 2007; Moore, 2008). Melatonin has recently drawn the attentions of many researchers due to its promising antioxidant properties. The mechanisms of melatonin's antioxidant effects seem to involve both the receptor-mediated and direct pathways (Kim et al., 2013; Rahimi et al., 2005).

It has been documented that melatonin alters both the gene expression and function of antioxidant enzymes. It was first reported that melatonin increases the gene expression of antioxidant enzymes, including Cu/Zn-SOD (copper zinc superoxide dismutase, or SOD1) and Mn-SOD (manganese superoxide dismutase, or SOD2), which are involved in detoxification of ROS (Antolín et al., 1996). Not only does melatonin upregulate the expression of genes related to detoxifier agents, but also decreases the activity and/or expression of agents involved in free radical generation. Interaction with calmodulin has been shown as a main mechanism of melatonin in modulating the activity of antioxidant enzymes. This interaction also leads to the inactivation of RORα, which in turn releases the inhibition of NF-κB-induced expression of antioxidant enzymes via down regulating the upstream NF-κB inhibitor, IκB. Thus, the blockade of the RORα pathway plays a role in the increased activity of antioxidant enzymes induced by melatonin (Tomás-Zapico and Coto-Montes, 2005; Zhang and Zhang, 2014).

2.6.3.2 Direct reactive oxygen species scavenging

Aerobic organisms permanently produce free radicals as byproducts of their oxidative reactions. A great deal of evidence highlights the involvement of two classes of toxic agents, which also include some free radicals, in various diseases and aging: Reactive oxygen and nitrogen species (ROS and RNS, respectively). The first class constitutes a group of oxygen-based molecules possessing a single electron, known as free radicals, as well as some other strong oxidizing molecules such as hydrogen peroxide (H_2O_2) and singlet oxygen. Singlet oxygen, superoxide ($O_2^{\bullet-}$), hydrogen peroxide (H_2O_2), and the hydroxyl radical ($^{\bullet}OH$) are among the most common ROS of which the hydroxyl radical is the most reactive and responsible for a great proportion of ROS-induced damage in living cells (R Ramis et al., 2015; Valko et al., 2007; Zhang and Zhang, 2014).

Melatonin have been recognized as a potent direct radical scavenger for 25 years (Galano et al., 2011, 2013; Reiter et al., 2016; Reiter et al., 2001; Tan et al., 2007). The first report on the direct scavenging effect of melatonin was made by Tan et al. who showed through *in vitro* study that the highly toxic $^{\bullet}OH$ can be neutralized by melatonin. Subsequent studies further confirmed this report in a number of well-controlled and occasionally cell-free experiments. The first *in vivo* evidence of melatonin as a direct neutralizer of $^{\bullet}OH$ was provided by Li et al. They showed that melatonin administration can decrease $^{\bullet}OH$ generation during a cerebral ischemia/reperfusion model, which was also shown by Bromme et al (Brömme et al., 2000; Li et al., 1997; Tan, 1993; Zhang and Zhang, 2014).

Melatonin reacts with free radicals mainly via three major mechanisms including the transfer of a single electron, a hydrogen atom or the formation of a radical adduct. Nitric oxide (NO) has an extremely high affinity for the superoxide anion radical ($O_2^{\bullet-}$) making it almost impossible for many conventional pharmacological and enzymatic antioxidants to overcome the generation of high amounts of the peroxynitrite ($ONOO^-$) radical. The discovery of $ONOO^-$ as a powerful causative agent in oxidative stress as turned this field into a new era. In this regard, the two elements of elevated NO and excess $O_2^{\bullet-}$ along with $ONOO^-$ production as the outcome of such a combination constitute a dangerous triad which is referred to as the "devil's triangle" (Galano et al., 2011; Korkmaz et al., 2009; Royano and Reiter, 2010). Melatonin metabolites, just like the melatonin itself, are capable of scavenging free radicals; this makes melatonin a powerful neutralizer of oxidative stress. Reportedly, every single melatonin molecule is capable of scavenging up to 10 free radicals, which is way more effective than many other antioxidants (Reiter et al., 2014).

2.6.4 Effects of melatonin on diseases

Medicinal administration of melatonin has been recommended for the treatment of several diseases in which melatonin has shown encouraging results. Thanks to its dual hydrophilic and lipophilic nature, which endows it the ability to mobilize between almost all tissues, melatonin can reach subcellular organelles such as mitochondria, where the highest levels of ROS is produced, and therefore, exert potent antioxidant effects. Subsequently, melatonin prevents cell damage caused by ROS by neutralizing them. The two facts that (1) melatonin is almost cheap and easily synthesized and (2) that it stays in the body for a long time, emphasize the benefits of melatonin for treatment (Pohanka, 2011). Studies that show the role of melatonin in the improvement of gastrointestinal diseases like IBS, IBD and necrotizing enterocolitis indicate its key role in the physiological function of the gastrointestinal tract. These include the effects of melatonin on immune response and microbiota. Gastrointestinal microbiota (the microbial community assemblage) plays an important role with regard to melatonin absorption from its dietary sources. The function of microbiota involves releasing absorbable melatonin from indigestible nutritional components and therefore, turning it usable for the host (Kim et al., 2020). Melatonin can reduce the severity of inflammatory bowel pathologies such as colitis in animal models. Owing to its anlagesic and motility regulating effects in the gastrointestinal tract that can enhance bowel habits, melatonin relieves abdominal pain and also due to its beneficial effects on sleep, it helps relieve sleep disorders. Melatonin's antianxiety effects have also been shown to be useful to improve psychological parameters in IBS patients who suffer from anxiety (Siah et al., 2014). Melatonin may also play a role in mucosal regeneration. Melatonin appears to inhibit the release of stomach acid, improve gastric circulation and enhance the regeneration of mucosal membranes. In individuals with less melatonin release, it takes longer for gastric ulcer to improve (Vaughn et al., 2014).

Neurodegenerative diseases are another group of disorders in which extensive studies have shown melatonin to be of benefit. The potential effectiveness of melatonin in patients and animal models of diseases, such as AD, PD, ALS, epilepsy, and stroke has been demonstrated. These include age-associated conditions in which oxidative stress, mitochondrial disturbances, and apoptosis can be observed in various parts of the brain (M Escribano et al., 2014). Neurodegeneration is a prominent feature of many chronic progressive neurological conditions, representing certain clinical, biochemical and morphological characteristics. Due to the presence of large amounts of unsaturated fatty acids, high oxygen demand, high amounts of ascorbate and iron in some areas, as well as low concentrations of antioxidants, the deteriorative effects of ROS could be extremely pronounced in the central nervous system. High NO production is an important indicator of neurodegeneration. Increased concentrations of NO lead to increased calcium and sodium ions within the cell and may cause disruptions in the mitochondrial function (Miller et al., 2015). Increasing evidence suggests mitochondrial involvement in neurological diseases. The disease status caused by mitochondrial dysfunction is due to the lack of natural biological energy, increased radical production, reduced ATP production, and stimulated cell death. For this reason, mitochondria are now the target of new therapies for neurological disorders. Melatonin can interact with mitochondrial complexes and modulate electron transfer activities. In addition, melatonin reacts specifically with electron transport chain complexes I and IV, and increases their activity (M Escribano et al., 2014). Experiments have shown that melatonin can neutralize the exagerated production of free radicals by mitochondrial complex I through cardiolipin (Pohanka, 2011). Melatonin use in AD animal models dramatically dampens cognitive impairment progression and reduces brain atrophy. Melatonin also hampers the deposition of amyloid plaques and improves learning and memory loss (Escames et al., 2010; Rudnitskaya et al., 2015). Furthermore, owing its reversing effects on the elevated lipid peroxidation levels and nigrostriatal dopaminergic degradation, melatonin has been shown to be effective in maneb and paraquat induced animal models of PD through the promotion of locomotor activity (Singhal et al., 2011). Metal complexes can accelerate the precipitation of β-amyloid (Aβ), and studies have shown that aluminum, a neurotoxic element, is an underlying etiological factor in neurodegenerative diseases like AD and ALS. Melatonin could be useful in the treatment of AD by reducing the production of Aβ and mitochondria-related enhancement of cell viability, and could be used as a supplement for the treatment of such neurological diseases (Esparza et al., 2019).

2.6.4.1 Clinical studies

Most of the melatonin studies over the years have been related to the study of its effects on circadian rhythm and sleep disorders, while recent years of research has revealed the role of melatonin in other physiological mechanisms. Due to properties such as its anti-inflammatory and antioxidant effects, human studies have been conducted on its beneficial effects in other disorders and diseases (Moghaddam et al., 2020). Melatonin was studied for its neuroprotective effects in a double-blind randomized controlled trial (RCT) where participants who suffered from ischemia and reperfusion (I/R) injury received 6 mg/day melatonin orally for a period of 6 days starting 3 days before the carotid endarterectomy (CEA) to 3 days post surgery. Based on the results, melatonin dampened the expression of NF-κB, tumor necrosis factor-α (TNF-α), interleukin-6 (IL-6), while it induced the expression of Nrf2, SOD, CAT, and GPx following CEA. It was pointed out that melatonin, owing to its antioxidant and anti-inflammatory properties, can be beneficial in ameliorating the brain I/R injury following CEA (Zhao et al., 2018). Another study aimed at investigating the efficacy of simultaneous administration of N-acetyl cysteine (NAC) and melatonin in subsiding acute oxidative damage and early reperfusion injury in patients undergoing coronary artery bypass grafting (CABG). The serum concentration of several factors, including cardiac troponin I, lactate, malondealdehyde (MDA), and TNF-α was then measured. The results showed that both the melatonin and NAC have potent antioxidant properties possessing similar efficacies in terms of subsiding

CABG-associated cardiac injury and oxidative damage with the dosages used in the intervention (Shafiei et al., 2018).

In a RCT study, Raygan et al. studied the antioxidant effects of melatonin and its effectiveness in reducing cardiometabolic risk in type 2 diabetes patients, who had coronary heart disease (CHD). The study allocated 60 of these patients with concurrent CHD into two groups of 30. For 12 weeks, one group received 10 mg melatonin and the other group received placebo. At the end of the study, the beneficial effects of melatonin on plasma levels of total glutathione (GSH), NO, malondialdehyde (MDA), protein carbonyl (PCO), high-sensitivity C-reactive protein (hs-CRP), HDL-cholesterol, blood pressure, and mental health parameters were observed (Raygan et al., 2019). Migraine is another condition in which the clinical effectiveness of melatonin has been also investigated. In a double-blind study, men and women between the ages of 18 and 65 with migraines who had attacked 2–8 times a month were examined. Participants in the study were accidentally assigned to three study groups. The first group received 25 mg amitriptyline, one received 3 mg melatonin and the other group received placebo. The results of this study showed that 3 mg melatonin was better than that of placebo in preventing migraines and was as effective as amitriptyline, but at the same time more tolerable and had fewer side effects (Gonçalves et al., 2016). Mania is another disorder for which clinical studies of melatonin have been performed. In another RCT study, 54 acute mania patients with bipolar disorder were randomly divided into two groups. The study found that in the group that 6 mg/day melatonin was added to the common used medications, impressive improvement in manic symptoms and the general clinical condition was observed (Moghaddam et al., 2020).

2.6.4.2 Melatonin and cancer

It has been shown that ROS are more abundant in cancer cells than in normal cells. ROS are primarily thought to be oncogenic due to their detrimental effects on lipids, proteins and specially DNA, which make the cellular genome unstable and promote the development of tumors. ROS also contribute to cancer cell survival, proliferation, metastasis, and angiogenesis, playing roles as signaling molecules. Cancer cells tend to over activate the signaling pathways involved in tumorigenesis by increasing their rates of ROS production. This is mainly done as tumors acquire oncogenic mutations, lose tumor suppressor genes and increase their metabolism to adapt to hypoxic conditions. However, the role of ROS in cancer should not be looked at superficially since conflicting results have been achieved so far. In contrast to the tumor promoting roles for ROS, they have been shown to induce cancer cell death by increasing oxidative stress (Nogueira et al., 2008; Reczek and Chandel, 2017; Sabharwal and Schumacker, 2014). The fact that antioxidant agents are widely taken by many healthy individuals and cancer patients in hopes of fighting cancer seems quite alarming, since clinical trials in this regard have given rise to conflicting results; some of which indicate that antioxidants increase cancer risk rather than inducing protection against it (Godic et al., 2014; Goodman et al., 2011; Westerlund et al., 2010).

The observation of increased colorectal cancer risk among night shift workers whose circadian rhythms have been disrupted has drawn the attentions of researchers to melatonin as a protective agent in cancer (Asghari et al., 2017; Rodriguez et al., 2013). In line with these findings, it has been observed that the plasma concentrations of melatonin are altered in cancer patients compared to the healthy individuals. The increased likelihood of night shift workers to be affected by cancer was further revealed by a series of other studies on several types of cancer, including ovarian, lung, endometrial breast, prostate, and gastric cancer (Asghari et al., 2018).

While it is generally believed that only the antiproliferative effects of melatonin are present at its low concentrations, its apoptotic effects can be observed at its higher concentrations in many tumor cells. While nanomolar concentrations of melatonin have been shown to have no significant effects on normal cells, its higher concentrations can inhibit and/or cause differentiation of especially neuronal cells. Previous studies have also shown that melatonin can alter neither the viability nor the proliferation of normal cells (Bizzarri et al., 2013).

A biphasic or dual role (antioxidant vs pro-oxidant) for melatonin has been suggested by several studies based on its concentration and exposure time. Osseni et al. showed that high concentrations of melatonin can induce pro-oxidant effects in the hepatocellular carcinoma cell line (HepG2), while its low concentrations showed antioxidant effects. Long exposure (96 h) to the lower concentrations could also decrease cell viability significantly. The highest concentrations of melatonin were able to decrease cell viability even when ROS production was increased and GSH levels were low (Osseni et al., 2000). Therefore, the pro-oxidant effect of melatonin depends not only on the cell type but also on the concentration used. More importantly, since these effects of melatonin are solely supported by *in vitro* data with cancer cells, it is of great importance to check whether or not these effects can be also achieved *in vivo*. Accordingly, melatonin may regarded as a conditional pro-oxidant.

Conclusion/future prospects

Melatonin, a neurohormone involved in circadian rhythm regulation, is a powerful antioxidant with a wide range of activities. This indoleamine and its metabolites with direct and indirect antioxidant properties have an important role in controlling oxidative stress and various diseases and their antiapoptotic, antioxidant, and anti-inflammatory effects have been extensively studied. Importantly, the antioxidant mechanisms of melatonin appear to be more complex than conventional antioxidants. Some areas of melatonin research, such as its effects on mitochondrial function and metabolism, and the use of melatonin in cancer, require further attention due to the inconsistent reports. Further clinical investigations will also help elucidate the effectiveness of melatonin in various diseases.

References

Acuña-Castroviejo, D., Escames, G., Venegas, C., Díaz-Casado, M.E., Lima-Cabello, E., López, L.C., Rosales-Corral, S., Tan, D.-X., Reiter, R.J., 2014. Extrapineal melatonin: sources, regulation, and potential functions. Cell. Mol. Life Sci. 71, 2997–3025.

Andersen, L.P.H., Gögenur, I., Rosenberg, J., Reiter, R.J., 2016. The safety of melatonin in humans. Clin. Drug Investig. 36, 169–175.

Antolín, I., Rodríguez, C., Saínz, R.M., Mayo, J.C., Uría, H., Kotler, M., Rodriguez-Colunga, M., Tolivia, D., Menendez-Pelaez, A., 1996. Neurohormone melatonin prevents cell damage: effect on gene expression for antioxidant enzymes. FASEB J. 10, 882–890.

Asghari, M.H., Abdollahi, M., de Oliveira, M.R., Nabavi, S.M., 2017. A review of the protective role of melatonin during phosphine-induced cardiotoxicity: focus on mitochondrial dysfunction, oxidative stress and apoptosis. J. Pharm. Pharmacol. 69, 236–243.

Asghari, M.H., Ghobadi, E., Moloudizargari, M., Fallah, M., Abdollahi, M., 2018. Does the use of melatonin overcome drug resistance in cancer chemotherapy? Life Sci. 196, 143–155.

Asghari, M.H., Moloudizargari, M., Bahadar, H., Abdollahi, M., 2017. A review of the protective effect of melatonin in pesticide-induced toxicity. Expert Opin. Drug Metab. Toxicol. 13, 545–554.

Asghari, M.H., Moloudizargari, M., Ghobadi, E., Fallah, M., Abdollahi, M., 2017. Melatonin as a multifunctional anti-cancer molecule: implications in gastric cancer. Life Sci. 185, 38–45.

Bai, C., Gao, Y., Zhang, X., Yang, W., Guan, W., 2018. Melatonin promotes self-renewal of nestin-positive pancreatic stem cells through activation of the MT2/ERK/SMAD/nestin axis. Artificial Cells Nanomed. Biotechnol. 46, 62–74.

Bizzarri, M., Proietti, S., Cucina, A., Reiter, R.J., 2013. Molecular mechanisms of the pro-apoptotic actions of melatonin in cancer: a review. Expert Opin. Ther. Targets 17, 1483–1496.

Brömme, H., Mörke, W., Peschke, E., Ebelt, H., Peschke, D., 2000. Scavenging effect of melatonin on hydroxyl radicals generated by alloxan. J. Pineal Res. 29, 201–208.

Casao, A., Cebrián, I., Asumpção, M.E., Pérez-Pé, R., Abecia, J.A., Forcada, F., Cebrián-Pérez, J.A., Muiño-Blanco, T., 2010. Seasonal variations of melatonin in ram seminal plasma are correlated to those of testosterone and antioxidant enzymes. Reprod. Biol. Endocrinol. 8, 59.

Claustrat, B., Brun, J., Chazot, G., 2005. The basic physiology and pathophysiology of melatonin. Sleep Med. Rev. 9, 11–24.

David, S.S., O'shea, V.L., Kundu, S., 2007. Base-excision repair of oxidative DNA damage. Nature 447, 941.

Di Meo, S., Reed, T.T., Venditti, P., Victor, V.M., 2016. Role of ROS and RNS sources in physiological and pathological conditions. Oxid. Med. Cell. Longev. 2016, 1245049. doi:10.1155/2016/1245049. 27478531.

Escames, G., López, A., Antonio Garcia, J., García, L., Acuña-Castroviejo, D., Joaquin Garcia, J., Carlos Lopez, L., 2010. The role of mitochondria in brain aging and the effects of melatonin. Curr. Neuropharmacol. 8, 182–193.

Escribano, B.M., Colin-Gonzalez, A.L., Santamaria, A., Túnez, I., 2014. The role of melatonin in multiple sclerosis, Huntington's disease and cerebral ischemia. CNS Neurol. Disord. Drug Targets 13, 1096–1119.

Esparza, J.L., Gómez, M., Domingo, J.L., 2019. Role of melatonin in aluminum-related neurodegenerative disorders: a review. Biol. Trace Elem. Res. 188, 60–67.

Essick, E.E., Sam, F., 2010. Oxidative stress and autophagy in cardiac disease, neurological disorders, aging and cancer. Oxid. Med. Cell. Longev. 3, 168–177.

Galano, A., Tan, D.X., Reiter, R.J., 2011. Melatonin as a natural ally against oxidative stress: a physicochemical examination. J. Pineal Res. 51, 1–16.

Galano, A., Tan, D.X., Reiter, R.J., 2013. On the free radical scavenging activities of melatonin's metabolites, AFMK and AMK. J. Pineal Res. 54, 245–257.

Godic, A., Poljšak, B., Adamic, M., Dahmane, R., 2014. The role of antioxidants in skin cancer prevention and treatment. Oxid. Med. Cell. Longev. 2014, 860479. doi:10.1155/2014/860479. 24790705.

Gonçalves, A.L., Ferreira, A.M., Ribeiro, R.T., Zukerman, E., Cipolla-Neto, J., Peres, M.F.P., 2016. Randomised clinical trial comparing melatonin 3 mg, amitriptyline 25 mg and placebo for migraine prevention. J. Neurol. Neurosurg. Psychiatry 87, 1127–1132.

Gonzalez-Arto, M., Hamilton, T.R.d.S., Gallego, M., Gaspar-Torrubia, E., Aguilar, D., Serrano-Blesa, E., Abecia, J., Pérez-Pé, R., Muiño-Blanco, T., Cebrián-Pérez, J., 2016. Evidence of melatonin synthesis in the ram reproductive tract. Andrology 4, 163–171.

Goodman, M., Bostick, R.M., Kucuk, O., Jones, D.P., 2011. Clinical trials of antioxidants as cancer prevention agents: past, present, and future. Free Radic. Biol. Med. 51, 1068–1084.

Goradel, N.H., Asghari, M.H., Moloudizargari, M., Negahdari, B., Haghi-Aminjan, H., Abdollahi, M., 2017. Melatonin as an angiogenesis inhibitor to combat cancer: Mechanistic evidence. Toxicol. Appl. Pharmacol. 335, 56–63.

Haghi-Aminjan, H., Asghari, M.H., Farhood, B., Rahimifard, M., Hashemi Goradel, N., Abdollahi, M., 2018. The role of melatonin on chemotherapy-induced reproductive toxicity. J. Pharm. Pharmacol. 70, 291–306.

Hardeland, R., 2009. Melatonin: signaling mechanisms of a pleiotropic agent. Biofactors 35, 183–192.

Kim, S.W., Kim, S., Son, M., Cheon, J.H., Park, Y.S., 2020. Melatonin controls microbiota in colitis by goblet cell differentiation and antimicrobial peptide production through toll-like receptor 4 signalling. Sci. Rep. 10, 1–10.

Kim, T.-K., Kleszczyński, K., Janjetovic, Z., Sweatman, T., Lin, Z., Li, W., Reiter, R.J., Fischer, T.W., Slominski, A.T., 2013. Metabolism of melatonin and biological activity of intermediates of melatoninergic pathway in human skin cells. FASEB J. 27, 2742–2755.

Korkmaz, A., Reiter, R.J., Topal, T., Manchester, L.C., Oter, S., Tan, D.-X., 2009. Melatonin: an established antioxidant worthy of use in clinical trials. Mol. Med. 15, 43–50.

Li, X., Zhang, L., Gu, J., Zhang, A., Sun, F., 1997. Melatonin decreases production of hydroxyl radical during cerebral ischemia-reperfusion. Zhongguo yao li xue bao. Acta Pharmacol. Sin. 18, 394–396.

Miller, E., Morel, A., Saso, L., Saluk, J., 2015. Melatonin redox activity. Its potential clinical applications in neurodegenerative disorders. Curr. Top. Med. Chem. 15, 163–169.

Moghaddam, H.S., Bahmani, S., Bayanati, S., Mahdavinasa, M., Rezaei, F., Akhondzadeh, S., 2020. Efficacy of melatonin as an adjunct in the treatment of acute mania: a double-blind and placebo-controlled trial. Int. Clin. Psychopharmacol. 35, 81–88.

Moore, M.N., 2008. Autophagy as a second level protective process in conferring resistance to environmentally-induced oxidative stress. Autophagy 4, 254–256.

Nogueira, V., Park, Y., Chen, C.-C., Xu, P.-Z., Chen, M.-L., Tonic, I., Unterman, T., Hay, N., 2008. Akt determines replicative senescence and oxidative or oncogenic premature senescence and sensitizes cells to oxidative apoptosis. Cancer Cell 14, 458–470.

Osseni, R.A., Rat, P., Bogdan, A., Warnet, J.-M., Touitou, Y., 2000. Evidence of prooxidant and antioxidant action of melatonin on human liver cell line HepG2. Life Sci. 68, 387–399.

Pham-Huy, L.A., He, H., Pham-Huy, C., 2008. Free radicals, antioxidants in disease and health. Int. J. Biomed. Sci. 4, 89.

Pohanka, M., 2011. Alzheimer's disease and related neurodegenerative disorders: implication and counteracting of melatonin. J. Appl. Biomed. 9, 185–196.

Pourhanifeh, M.H., Hosseinzadeh, A., Dehdashtian, E., Hemati, K., Mehrzadi, S., 2020. Melatonin: new insights on its therapeutic properties in diabetic complications. Diabetol. Metab. Syndr. 12, 1–20.

Rahimi, R., Nikfar, S., Larijani, B., Abdollahi, M., 2005. A review on the role of antioxidants in the management of diabetes and its complications. Biomed. Pharmacother. 59, 365–373.

Ramis, M.R., Esteban, S., Miralles, A., Tan, D.-X., Reiter, R.J., 2015. Protective effects of melatonin and mitochondria-targeted antioxidants against oxidative stress: a review. Curr. Med. Chem. 22, 2690–2711.

Raygan, F., Ostadmohammadi, V., Bahmani, F., Reiter, R.J., Asemi, Z., 2019. Melatonin administration lowers biomarkers of oxidative stress and cardio-metabolic risk in type 2 diabetic patients with coronary heart disease: a randomized, double-blind, placebo-controlled trial. Clin. Nutr. 38, 191–196.

Reczek, C.R., Chandel, N.S., 2017. The two faces of reactive oxygen species in cancer. Ann. Rev. Cancer Biol. 1, 79–98.

Reiter, R.J., Mayo, J.C., Tan, D.X., Sainz, R.M., Alatorre-Jimenez, M., Qin, L., 2016. Melatonin as an antioxidant: under promises but over delivers. J. Pineal Res. 61, 253–278.

Reiter, R.J., Tan, D.-X., Fuentes-Broto, L., 2010. Melatonin: a multitasking molecule. Prog. Brain Res. 181, 127–151.

Reiter, R.J., Tan, D.-X., Galano, A., 2014. Melatonin reduces lipid peroxidation and membrane viscosity. Front. Physiol. 5, 377.

Reiter, R.J., Tan, D.-x., Manchester, L.C., Qi, W., 2001. Biochemical reactivity of melatonin with reactive oxygen and nitrogen species. Cell Biochem. Biophys. 34, 237–256.

Rodriguez, C., Martín, V., Herrera, F., García-Santos, G., Rodriguez-Blanco, J., Casado-Zapico, S., Sánchez-Sánchez, A., Suárez, S., Puente-Moncada, N., Anítua, M., 2013. Mechanisms involved in the pro-apoptotic effect of melatonin in cancer cells. Int. J. Mol. Sci. 14, 6597–6613.

Romero, A., Ramos, E., de Los Ríos, C., Egea, J., del Pino, J., Reiter, R.J., 2014. A review of metal-catalyzed molecular damage: protection by melatonin. J. Pineal Res. 56, 343–370.

Rossignol, D.A., Frye, R.E., 2012. A review of research trends in physiological abnormalities in autism spectrum disorders: immune dysregulation, inflammation, oxidative stress, mitochondrial dysfunction and environmental toxicant exposures. Mol. Psychiatry 17, 389.

Royano, S.P., Reiter, R., 2010. Melatonin: Helping cells cope with oxidative disaster. Cell Membr. Free Radic. Res. 2, 99–111.

Rudnitskaya, E.A., Muraleva, N.A., Maksimova, K.Y., Kiseleva, E., Kolosova, N.G., Stefanova, N.A., 2015. Melatonin attenuates memory impairment, amyloid-β accumulation, and neurodegeneration in a rat model of sporadic Alzheimer's disease. J. Alzheimers Dis. 47, 103–116.

Sabharwal, S.S., Schumacker, P.T., 2014. Mitochondrial ROS in cancer: initiators, amplifiers or an Achilles' heel? Nat. Rev. Cancer 14, 709.

Shafiei, E., Bahtoei, M., Raj, P., Ostovar, A., Iranpour, D., Akbarzadeh, S., Shahryari, H., Anvaripour, A., Tahmasebi, R., Netticadan, T., 2018. Effects of N-acetyl cysteine and melatonin on early reperfusion injury in patients undergoing coronary artery bypass grafting: A randomized, open-labeled, placebo-controlled trial. Medicine (Baltimore) 97 (30), e11383.

Siah, K.T.H., Wong, R.K.M., Ho, K.Y., 2014. Melatonin for the treatment of irritable bowel syndrome. World J. Gastroenterol. 20, 2492.

Singhal, N.K., Srivastava, G., Patel, D.K., Jain, S.K., Singh, M.P., 2011. Melatonin or silymarin reduces maneb-and paraquat-induced Parkinson's disease phenotype in the mouse. J. Pineal Res. 50, 97–109.

Tan, D.-X., 1993. Melatonin: a potent, endogenous hydroxyl radical scavenger. Endocr. J. 1, 57–60.

Tan, D.-X., Manchester, C.L., Sanchez-Barcelo, E., Mediavilla, D.M., Reiter, J.R., 2010. Significance of high levels of endogenous melatonin in mammalian cerebrospinal fluid and in the central nervous system. Curr. Neuropharmacol. 8, 162–167.

Tan, D.-X., Manchester, L., Esteban-Zubero, E., Zhou, Z., Reiter, R., 2015. Melatonin as a potent and inducible endogenous antioxidant: synthesis and metabolism. Molecules 20, 18886–18906.

Tan, D.X., Manchester, L.C., Terron, M.P., Flores, L.J., Reiter, R.J., 2007. One molecule, many derivatives: a never-ending interaction of melatonin with reactive oxygen and nitrogen species? J. Pineal Res. 42, 28–42.

Tan, D.X., Xu, B., Zhou, X., Reiter, R.J., 2018. Pineal calcification, melatonin production, aging, associated health consequences and rejuvenation of the pineal gland. Molecules 23, 301.

Tomás-Zapico, C., Coto-Montes, A., 2005. A proposed mechanism to explain the stimulatory effect of melatonin on antioxidative enzymes. J. Pineal Res. 39, 99–104.

Valko, M., Leibfritz, D., Moncol, J., Cronin, M.T., Mazur, M., Telser, J., 2007. Free radicals and antioxidants in normal physiological functions and human disease. Int. J. Biochem. Cell Biol. 39, 44–84.

Vaughn, B.V., Rotolo, S., Roth, H.L., 2014. Circadian rhythm and sleep influences on digestive physiology and disorders. Chronophysiol. Ther. 4, 67.

Westerlund, A., Steineck, G., Bälter, K., Stattin, P., Grönberg, H., Hedelin, M., 2010. Dietary supplement use patterns in men with prostate cancer: the cancer prostate Sweden study. Ann. Oncol. 22, 967–972.

Zhang, H.M., Zhang, Y., 2014. Melatonin: a well-documented antioxidant with conditional pro-oxidant actions. J. Pineal Res. 57, 131–146.

Zhao, Z., Lu, C., Li, T., Wang, W., Ye, W., Zeng, R., Ni, L., Lai, Z., Wang, X., Liu, C., 2018. The protective effect of melatonin on brain ischemia and reperfusion in rats and humans: In vivo assessment and a randomized controlled trial. J. Pineal Res. 65, e12521.

CHAPTER

Superoxide dismutase

2.7

Arnab Karmakar*, Abhishek K. Das*, Noyel Ghosh*, Parames C. Sil

Division of Molecular Medicine, Bose Institute, Kolkata, India

**Authors contributed equally*

2.7.1 Introduction

Our body works by maintaining redox homeostasis within all its intricate systems to sustain its stability and flexibility and to survive even in a very stressful environment. Such a balance is opted by fine tuning between the generation of free radicals along with other oxidants and their neutralization by antioxidants. Free radicals are unstable molecules having an extra unpaired electron in their outermost shells, making them highly reactive (Genestra, 2007). Reactive oxygen species (ROS) and reactive nitrogen species (RNS) are the two prominent types of free radicals found in biological systems. Few examples regarding ROS are – superoxide radical ($O_2^•$), hydroxyl radical ($OH^•$), hydrogen peroxide (H_2O_2), etc., and RNS are – nitric oxide radical ($NO^•$), nitrogen dioxide radical ($NO_2^•$), peroxynitrite anion ($ONOO^−$), etc.

Synthesis of free radicals is highly essential for our body. Macrophages kill invading microorganisms by releasing $O_2^{•−}$ through NADPH oxidase-dependent catalysis (Droge, 2002). Role of ROS and RNS in many intracellular signaling cascades that take place in fibroblasts, myocytes, and endothelial cells has been discovered too (Genestra, 2007). This can be exemplified by the activity of $NO^•$, which acts as an intracellular messenger in numerous signaling events occurring in different organs and organelles (Pacher et al., 2007).

Besides their advantageous character, free radicals have detrimental sides as well. Upon generation, free radicals tend to uptake an electron from their nearest donor molecule or release their own unpaired electron to become stable and eventually converting the acceptor molecule into another free radical. The newly generated radical now imparts the same action on other molecules to make a third radical and thus, the chain of reaction continues. Antioxidants avert free radical induced injury by preventing the synthesis of radicals, assisting disintegration of radicals to lesser reactive products or scavenging them. Deficiency of antioxidants results in excess free radical-load inside the cell and subsequently reaction of the excess free radicals with the cellular macromolecules such as proteins,

Antioxidants Effects in Health. DOI: https://doi.org/10.1016/B978-0-12-819096-8.00027-6

lipids and DNA. Such reactions are hazardous for the cellular homeostasis in various aspects as they can cause structural deformity of many proteins leading to their loss of activity. Upon excess $OH^{\bullet}$ and ONOO- generation, membrane lipids undergo peroxidation by initiating a radical chain reaction producing malondialdehyde (MDA) as a by-product, which is toxic to the cell (Halliwell, 2007). Free radicals also react with DNA and cause structural changes leading to the formation of oxidative DNA lesions which are mutagenic as well. Excess amount of ROS leads to oxidative stress to the cell, which after a certain limit can initiate programmed cell death (McCord, 2016).

Superoxide radical generates enzymatically via NADPH oxidase, xanthine oxidase, cytochrome p450 system, etc. (Halliwell, 2007) and reacts with various other molecules to generate more of free radicals like H_2O_2, $OH^{\bullet}$, $ONOO^-$, HOCl, etc. The most common source of free radical generation in aerobic organisms is the electron transport chain in the inner mitochondrial membrane where the terminal electron acceptor molecule oxygen is reduced into water. This is a nonenzymatic process of ROS generation (Valko et al., 2007) as shown below:

$$O_2 + e^- = O_2^{\bullet -} \tag{2.7.1}$$

$$O_2^{\bullet -} + 2H^+ + e^- = H_2O_2 \tag{2.7.2}$$

$$H_2O_2 + H^+ + e^- = H_2O + OH^{\bullet} \tag{2.7.3}$$

$$OH^{\bullet} + H^+ + e^- = H_2O \tag{2.7.4}$$

(Conversion of O_2 into H_2O after accepting the electron)

From these reactions, it can be seen that the intermediate molecules are $O_2^{\bullet -}$ via reaction (I), H_2O_2 via reaction (II), and $OH^{\bullet}$ via reaction (III), which are the common constituents of ROS. Among them, H_2O_2 is comparatively less reactive than other short-lived radicals and ions (Collin, 2019).

To counteract the excess free radicals and to maintain a balance between their generation and neutralization, importance of antioxidants is unparalleled. Antioxidants produced by our body itself are called endogenous antioxidants; whereas, antioxidants that are obtained through food and supplements, as our body cannot produce them, are called exogenous or nutrient-derived antioxidants. Endogenous antioxidants can be classified into two types – nonprotein antioxidants like glutathione, ferritin, bilirubin, uric acid, coenzyme Q, alpha-lipoic acid, etc., and protein antioxidants or antioxidant enzymes like superoxide dismutase (SOD), catalase (CAT), and glutathione peroxidase (GPX). Nonprotein antioxidants act as ligand molecules that directly plays a role in scavenging or stabilizing free radicals whereas antioxidant enzymes catalyze the reactions that convert free radicals to relatively stable molecules (Pham-Huy et al., 2008). SOD forms the first line of defense against ROS as it neutralizes the $O_2^{\bullet -}$, which is the first free radical to be generated via the electron transport chain in aerobic organisms.

2.7.2 Classifications of different types of superoxide dismutase

2.7.2.1 General classification of superoxide dismutase

Superoxide dismutase, E.C. number 1.15.1.1, is found to be present throughout all six realms of life. Depending upon the metal cofactor, SOD is divided into, Cu/Zn-SOD (Cu and Zn as metal cofactors), Mn-SOD (Mn as the metal cofactor), Fe-SOD (Fe as the metal cofactor) and Ni-SOD (Ni as the metal cofactor). The evolutionary origin of different SOD was established based on the homology modelling study. Based on the outcome, it is thought that Mn-SOD and Fe-SOD have been evolved in prokaryotes (Smith and Doolittle, 1992), whereas Cu/Zn-SOD has been evolved in eukaryotes (Steinman, 1978) separately. However, eukaryotes have Mn-SOD in their mitochondria, and Cu/Zn-SOD has been found in some bacteria as well, leading to controversies.

In animals, Cu/Zn-SOD is primarily found in the cytosol, the intermembrane space of mitochondria, nucleus, peroxisome and lysosome (Crapo et al., 1992; Keller et al., 1991). In plants, the majority of Cu/Zn-SOD is found in the cytosol and chloroplast (Gill et al., 2015). Mn-SOD is found in prokaryotes and eukaryotes only in the mitochondrial matrix (Fridovich, 1995). Fe-SOD is present in bacteria and some chloroplasts of plants (Bridges and Salin, 1981). Recently, Ni-SOD was identified and has been isolated from Streptomyces and Cyanobacteria (Abreu and Cabelli, 2010) (See Fig. 2.7.1).

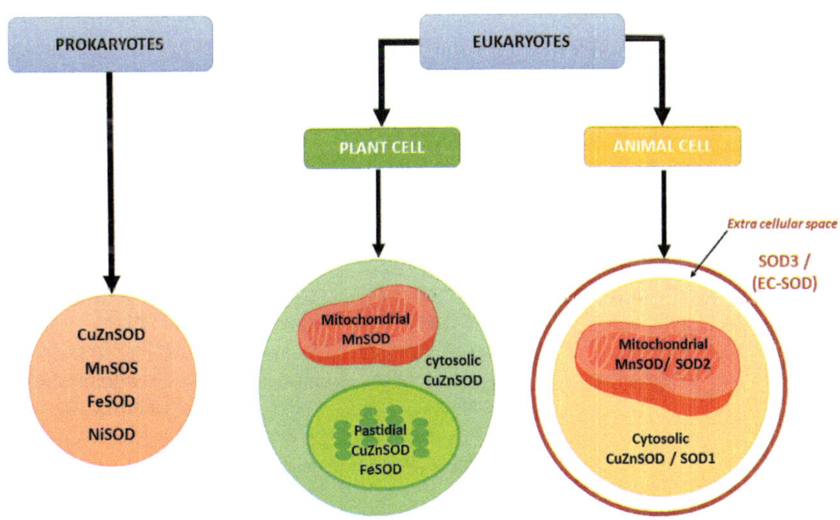

FIG. 2.7.1 Availability of superoxide dismutase in a diverse form in aerobic organisms, including both prokaryotes and eukaryotes.

2.7.2.2 Classification of mammalian superoxide dismutase

In mammals, especially in humans, three forms of SOD have been identified – SOD1, SOD2 and SOD3 (See Table 2.7.1). Among the three types, SOD1 constitutes 85% or more of the total SOD enzyme activity (Marklund, 1984) and was first discovered by McCord and Fridovich in 1969 (McCord, 2016). It is a 32 kDa homodimeric protein, found primarily in the cytosol. It contains Cu and Zn in the catalytic centre (Cu/Zn-SOD). The human SOD1 gene is present in chromosome no. 21 (Zelko et al., 2002) containing five exons and four introns which finally codes for a monomeric protein. Although SOD1 has an omnipresent and high expression level and thus, regarded as a housekeeping gene; yet, it was found to be finetuned by several positive and negative regulatory elements (Milani et al., 2011). After the discovery of the SOD1 gene construct, several regulatory domains, along with the promoter region was identified having binding sites for nuclear localizing proteins like nuclear factor kappa B (NF-κB), activator protein 1 (AP-1) and 2 (AP-2), specificity protein 1 (Sp-1), heat shock transcription factor (HSF), NF-1, metal responsive element, etc. (Kim et al., 1994; Yoo et al., 1999). Generally, enhanced oxidative stress results in the elevation of SOD1 expression. Upon translation, apo-SOD1 is modified by the copper chaperons for SOD (CCS) protein by delivering the Cu ion to the apo-SOD1, converting it to mature SOD1 in the presence of an oxidative insult (Brown et al., 2004).

The counterpart of Mn-SOD in mammals is SOD2, a homotetrameric protein with a molecular mass of 23 kDa for each subunit. SOD2, which is a mitochondrial protein (Karnati et al., 2013), accounts for 10%–15% of the total cellular SOD activity in mammals (Williams et al., 1998). SOD2 gene is located in chromosome 6 of the human genome and its promoter region has putative binding sites for proteins like Ap-2, Sp-1 and NF-κB (Zelko et al., 2002). Similar to the SOD1 gene, the coding region of SOD2 is interrupted by four introns. In response to various stresses like enhanced oxidative tension, exposure to environmental toxicants like cigarette smoke,

Table 2.7.1 Endogenous source of SODs in mammals.

Mammalian SOD forms	Characterization	Occurrence	Gene location	SOD activity%
SOD1	Cu/Zn-SOD, dimeric protein	Cytosol of all mammalian cells types	On chromosome 21, region- 21q22	Almost 85% of total SOD activity
SOD2	Mn-SOD, tetrameric protein	Mitochondria of all mammalian cells types	On chromosome 6, region-6q25	10%–15% of the total SOD activity
SOD3	60 % homology with Cu/Zn-SOD, minimal or no homology with Mn-SOD; tetrameric protein	Cell surface and extracellular matrix of few cell types like fibroblast cells and endothelial cells	On chromosome 4, region-4p-q21	Only 1% or less of total SOD activity

ionizing radiations, ozone, etc., the tetrameric protein expression is upregulated, providing a compensatory response to the damages (Gilks et al., 1998). Enhanced level of cytokines like TNF-α, IL-6, IL-1, IFN-γ, etc., is also linked with the increase in expression of SOD2 (Warner et al., 1996; Weller et al., 1997). Further research revealed that Mn-SOD activation in response to cytokines, is very rapid (within 2 h), and a very low concentration of cytokines is enough for its activation (Shull et al., 1991). The metal insertion system of SOD2 is found to be quite different from SOD1 (Yamakura and Kawasaki, 2010). The precursor polypeptides are imported into the mitochondrial matrix, followed by an N-terminus pre-sequence elimination, resulting in a conformational change of apo-SOD2 which is compulsory for Mn incorporation. This "close" to "open" conformational change of apo-SOD2 is driven by increased temperature and pH inside the mitochondrial environment (Voelkl et al., 2018; Whittaker, 2010). Few metal transporters have been identified as facilitators of Mn uptake and localizer of Mn in the mitochondrial matrix. Inactivation of SOD2 by misincorporation of iron is common in the bacterial system as mitochondrial Fe concentration is higher than Mn, both having a similar affinity to the active site of SOD2. Though, such a phenomenon is not observed in eukaryotic system reasons yet to be known (Culotta et al., 2006).

SOD3, also known as extracellular SOD (EC-SOD), identified in the 1980s, is a 135 kDa tetrameric protein having Cu and Zn in the active site. Although EC-SOD and Cu/Zn-SOD share a similar catalytic trait, they are structurally dissimilar (Marklund, 1982). EC-SOD is chiefly expressed on the cell surface and extracellular matrix of fibroblast cells and endothelial cells. EC-SOD is also found with smaller quantities in the lymph, blood plasma, cerebrospinal fluid, etc. SOD3 activity accounts for only ≤1% of the total SOD activity in most of the tissues of mammalian cells (Marklund, 1984) (See Fig. 2.7.1). In contrast to *SOD1* and *SOD2*, *SOD3* genomic construct is not highly conserved among different mammalian species (Zelko et al., 2002). *SOD3* is localized in chromosome 4 of the human genome (Hendrickson et al., 1990) and shows 40%–60% homology with Cu/Zn-SOD and negligible homology with Mn-SOD at the coding region (Zelko et al., 2002). In human fibroblast cells, SOD3 expression is increased in response to IL-1 and IFN-γ (Marklund, 1992). Angiotensin II upregulates the synthesis of SOD3 in vascular smooth muscle cells (Fukai et al., 2002). In order to become functional, SOD3 undergoes proteolytic cleavage by furin and carboxypeptidase before the secretion into the extracellular matrix (Bowler et al., 2002).

2.7.3 Mechanism of action

Dismutation of superoxide without the interference of any catalytic enzymes:

$$2O_2^{\bullet -} + 2H^+ = H_2O_2 + O_2 \tag{2.7.5}$$

(See Fig. 2.7.2)

With the presence of SOD, the reaction becomes four times faster. The mechanism of action of SODs is based on the "ping-pong" method, where sequential

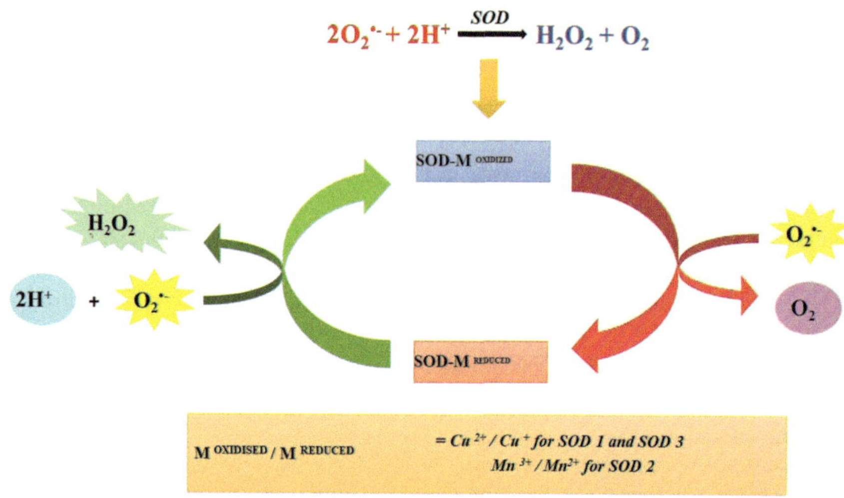

FIG. 2.7.2 Generalized mechanism of SOD-catalyzed superoxide radical scavenging.

This involves sequential reduction and reoxidation of catalytic metal (Cu or Mn) at the enzyme's active site.

reduction and oxidation of the metal center takes place (Abreu and Cabelli, 2010). The general mechanism of enzymatic dismutation of superoxide:

$$E\text{-}M^{(n+1)+} + O_2^{\bullet-} = E\text{-}M^{n+} + O_2 \quad (2.7.6)$$

$$E - M^{n+} + O_2^{\bullet-} + 2H^{+} = E - M^{(n+1)+} + H_2O_2 \quad (2.7.7)$$

Here $E\text{-}M^{(n+1)+}$ refers to the metal-bound enzyme and M can be any of the metals-Cu, Mn, Fe, or Ni, and its oxidized state or charge is designated by "(n+1)+." The enzyme oxidizes the $O_2^{\bullet-}$ (reaction VI) into molecular oxygen, reducing the metal ion to M^{n+}. In the next step (reaction VII) M^{n+}-enzyme complex reduces one $O_2^{\bullet-}$ to H_2O_2 and oxidizes itself to gain the native $M^{(n+1)+}$-enzyme state.

Cu/Zn-SOD or SOD1 is a dimeric enzyme in which each monomeric subunit contains one Zn^{2+} and one Cu^{2+} metal in their active site, linked by a histidine imidazole. Three additional histidine residues bind the Cu^{2+} making a square planar structure (Abreu and Cabelli, 2010; Hough and Hasnain, 2003). On the other hand, Zn^{2+} stabilizes the system by binding to one aspartate and two histidines along with the bridging imidazole (Aminake and Pradel, 2013). The transition of Cu^{2+} to Cu^{+} takes place with the help of relaxation and distortion in the complex zinc-ligand geometry where His61 bridges copper and zinc ions. The reactions for Cu/Zn-SOD catalysis are as follows:

$$E - Cu^{2+} + O_2^{\bullet-} = E - Cu^{+} + O_2 \quad (2.7.8)$$

$$E - Cu^{+} + O_2^{\bullet-} + 2H^{+} = E - Cu^{2+} + H_2O_2 \quad (2.7.9)$$

Cu/Zn-SOD has a net negative charge with the pH range between 4.6 and 6.8 (Salin and Wilson, 1981). Since, the substrate $O_2^{\bullet-}$ is also negatively charged, electrostatic repulsion between the enzyme and the substrate is avoided by the presence of the positively charged local amino acids, distributed near the active site of the enzyme. Near the copper centred active site, a deep channel was identified which is comprised of positively charged amino acids like lysine and arginine. This positively charged region guides the substrate $O_2^{\bullet-}$ to bind the enzyme near the copper centre, facilitating the reaction to take place. This mechanism is called an electrostatic guidance mechanism (Getzoff et al., 1983).

Human Cu/Zn-SOD has Glu132, Glu133 and Lys136 residues in the active site channel rim that forms a hydrogen bond network. The nitrogen atom of Lys136 with the help of its positive charge forms hydrogen bonds with negatively charged carboxylate groups of Glu132 and Glu133. It has been observed that if Glu132 and Glu133 are mutated site-specifically to Gln, the mutated enzyme has higher activity than wild-type SOD (Getzoff et al., 1992). Pulse radiolysis techniques were used to study the reaction rate of various mutants of positively charged arginine in Cu/Zn-SOD and it was found that arginine was one of the responsible residues for electrostatic guidance (Fisher et al., 1994).

Mn-SOD was first discovered in 1970 by Fridovich and coworkers from *E. coli* (Keele et al., 1970).The active site Mn^{3+} is bound to four amino acids acting as ligands. A fifth ligand binding site is present, where the superoxide anion sits. This looks like a trigonal bipyramidal geometry (Abreu and Cabelli, 2010). The dismutation reaction takes place by alternating cycles of oxidized form- Mn^{3+} and reduced form- Mn^{2+} of manganese:

$$\text{E-Mn}^{3+} + O_2^{\bullet-} = \text{E-Mn}^{2+} + O_2 \quad (2.7.10)$$

$$\text{E-Mn}^{2+} + O_2^{\bullet-} + 2H^+ = \text{E-Mn}^{3+} + H_2O_2 \quad (2.7.11)$$

Fe-SOD shows similar homology with the Mn-SOD in the protein level as well as in the oxidized and reduced state of its corresponding metal centers; thus, it has a similar kind of mechanism of action as well. Ni-SOD is a hexamer with a molecular weight of 80 kDa, whose subunits contain a "Nickel hook". Ni is a transition metal and involved in dismutation reaction, although the details of its mechanism are not clear (Abreu and Cabelli, 2010).

2.7.4 Beneficial roles of superoxide dismutases

In gist, the enzyme SOD catalyzes the conversion of $O_2^{\bullet-}$ to oxygen molecule and H_2O_2. Pivotally they act to maintain the fine balance between production and neutralization of free radicals (See Fig. 2.7.3).

$O_2^{\bullet-}$ may not be considered to be a highly oxidizing agent, but it is potentially toxic to our system. $O_2^{\bullet-}$ oxidizes iron-sulfur clusters found in protein molecules and leads to disrupt their enzymatic activity (Liochev and Fridovich, 1999). Moreover,

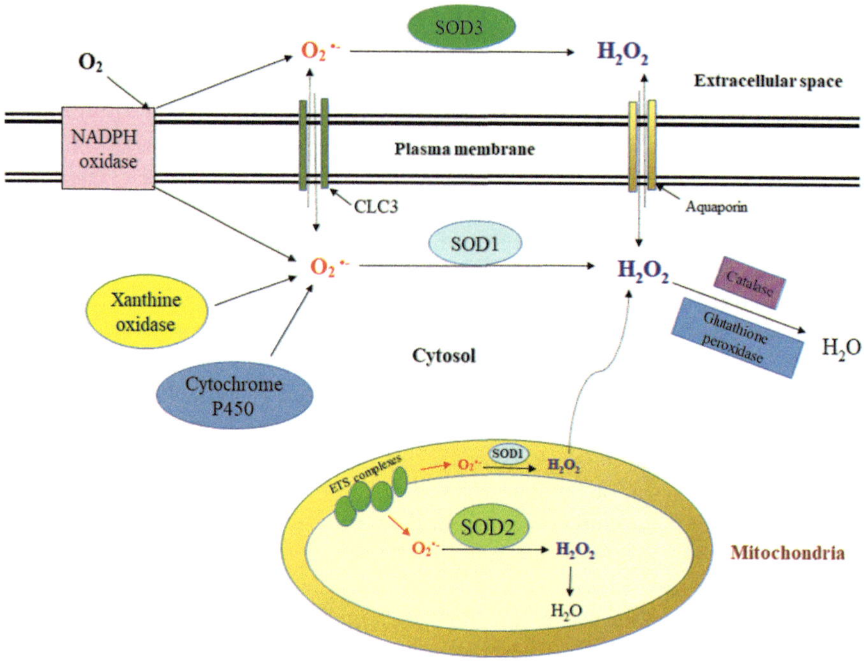

FIG. 2.7.3 Beneficial roles of different SODs in subcellular compartments of animal cells.

the iron atom which is released in the overall process reduces H_2O_2 and generates highly toxic $OH^{\bullet}$ (Liochev and Fridovich, 1999). In all organisms listed from bacteria to humans, SOD plays a vital role. Loss of catalytic efficiency of SOD results in oxidative stress that further responsible for harmful oxidative damages, such as DNA breakage, protein carboxylation as well as peroxidation of membrane lipids (Dukan and Nyström, 1999; Van Remmen et al., 2001). Besides, $O_2^{\bullet-}$ reacts with $NO^{\bullet}$ to produce highly detrimental $ONOO^-$, which takes part in the generation of other reactive species like carbonate anion radical ($CO_3^{\bullet-}$) (Pacher et al., 2007). SOD prevents the formation of $ONOO^-$ by the removal of $O_2^{\bullet}$. Apart from this, SOD plays a diverse role in the maintenance of the regulation of various redox signaling. The H_2O_2 generated in process of dismutation plays an important role in cellular signaling. This can be clarified by studies revealing the result of SOD inhibition. Tetrathiomolybdate, a SOD1 inhibitor, prevents PTPs oxidation and this in turn, ceases the growth factor-mediated phosphorylation of the ERK 1/2 in tumor and endothelial cells (Juarez et al., 2008). Here, inhibiting SOD1 prevents generation H_2O_2 through $O_2^{\bullet-}$ dismutation.

Alternation of SOD activity results in dysregulation of a number of crucial signaling pathways associated with several pathological conditions like cancer and familial amyotrophic lateral sclerosis (FALS) (Robberecht and Philips, 2013). Few mutations in SOD1 gene, have been detected for FALS to be responsible for pathophysiology caused by the disease (Cleveland and Rothstein, 2001). Hence, it is very clear that SOD plays a big role in maintaining a stable system within our body and any alteration in its activity is harmful to the body.

RNS like $NO^{\bullet}$, $NO_2^{\bullet}$, $ONOO^-$ and many others are functionally significant, but they are also hazardous in many ways. $NO^{\bullet}$ acts as an important molecule in signaling cascades associated with vasodilation, immune response, neurotransmission etc. $O_2^{\bullet-}$ and $NO^{\bullet}$ react with each other in a three times faster rate than SOD-mediated dismutation of $O_2^{\bullet-}$ (Beckman and Koppenol, 1996). SODs control the level of $O_2^{\bullet-}$ and thus, reduce the probability of the reaction with $NO^{\bullet}$ by several folds. In this way, SODs also maintain the presence and physiological function of $NO^{\bullet}$ (Zhao et al., 2015). In the extracellular matrix, SOD3 scavenges $O_2^{\bullet-}$ and inhibits the inactivation of vascular $NO^{\bullet}$. This, in turn, maintains a proper vascular tone and regulates blood pressure (Fukai et al., 2002). Thus, SODs are vital for the proper maintenance of physiological and developmental balance in the aerobic organism.

2.7.5 Superoxide dismutases and diseases

To explore the diverse effect of SOD in the physiological and developmental aspects of human, studies are carried out extensively.

Though different SODs perform similar catalysis, they peruse distinct roles in different human cancer due to their distinct biological localizations. The role of SOD1 is highly paradoxical in cancer. *In vitro* studies by Yamanaka and Deamer were the first to report that SOD activity in the malignant cells was different from the activity in normal cells (Yamanaka, 1974). In SV-40 transformed WI-38 cells, it was observed that the band corresponding Mn-SOD was absent; still the net SOD activity was a bit higher than that of the normal cells. This observation indicates that in transformed cells the activities of Cu/Zn-SOD are uplifted by several folds in order to protect cells against oxidative stress. Despite extensive DNA-damage, SOD1-null mice were observed to develop only hepatocellular carcinoma (Elchuri et al., 2005). However, overexpression of SOD1 was reported in both lung (Glasauer et al., 2014) and breast cancers (Papa et al., 2014). Moreover, induced inhibition of SOD1 was found to be effective for K-Ras and epidermal growth factor receptor (EGFR) -driven lung carcinoma (Somwar et al., 2011). Keeping all the evidence in mind, it can be said that in the later stage of cancer, SOD1 is oncogenic rather being tumor suppressive. The role of SOD2 is again controversial. Reduced SOD2 expression is evident early-stage tumor progression whereas, SOD2 overexpression is often reported to be associated with late-stage tumor invasion and metastasis. Studies reveal that SOD2 expression is crucial for Warburg's effect associated with tumor progression. SOD2 maintains H_2O_2 generation to activate AMP-activated protein kinase (AMPK) in order to cause a metabolic shift from Kreb's cycle to glycolysis (Koppenol et al., 2011). SOD3 being extracellular in localization, seems to affect cancer by changing the tumor environment. SOD3 downregulation was reported to be linked with breast

(Teoh-Fitzgerald et al., 2014) and lung cancer (Teoh-Fitzgerald et al., 2012) due to a change in DNA copy number or DNA hypermethylation. Besides, in pancreatic ductal adenocarcinoma, SOD3 overexpression was found to decrease further tumor invasiveness (O'Leary et al., 2015).

In a study, human endothelial cells, cultured in high glucose (5 mM to 20 mM) containing medium; glycol-oxidative stress was found to upregulate free radical-mediated cellular injury markers, such as MDA, conjugated dienes, etc. When treated with external SOD, glucose induced oxidative stress and consequent cellular damage were ameliorated to a great extent, suggesting that SOD can be used as a therapeutic agent against diabetes-based complications such as vascular pathophysiology (Curcio et al., 1995). In addition, endothelial cells, cultured in high glucose containing media, were found to express hyperglycemia-induced increase in protein kinase C (PKC) activity, TGF-β1 expression and overproduction of advanced glycation end products (AGEs) etc. However, endothelial cells overexpressing Mn-SOD or SOD2, cultured in a similar condition, were observed to be protected from this injury (Nishikawa et al., 2000). *In vitro* studies also showed that dysregulation of SOD activity in diabetes can impair $NO^{\bullet}$ functioning on renal arterioles and enhances endothelial albumin permeability (Dileepan et al., 1993; Ohishi and Carmines, 1995).

Plenty of studies describe the connection between uplifted oxidative stress markers and irreversible neuronal damage (Labbadia and Morimoto, 2013). SOD1 is indicated as a potential biomarker of oxidative damage. In a study to evaluate the effect of neurorehabilitation exercise in Huntington's disease, it was found that the level of SOD1 remained high during the whole experimental period and even after completion of the experiment. Such a finding seems to describe the protective role of SOD1 in Huntington's disease (Ciancarelli et al., 2015). Amyotrophic lateral sclerosis (ALS), which is characterized by progressive loss of motor neurons in the anterior horn of the spinal cord (Kiernan et al., 2011), can be of two types: sporadic and familial. Sporadic ALS develops between the age of 50 and 60 (Ingre et al., 2015), but its underlying cause yet remains to be elusive. However, misfolded or oxidized SOD1 has been reported to cause mitochondrial dysfunction contributing to SALS (D'Amico et al., 2013). Whereas, 20% of cases of FLAS emerge due to mutation in the SOD1 gene (Gamez et al., 2006). A study by Bastow et al. has suggested that mutant SOD1 can impair amino acid biosynthesis and results in neurodegeneration similar in ALS (Bastow et al., 2016). It has been also concluded that the gain of function mutant in the SOD1 gene rather than its functional loss seems to be highly effective to generate motor neuron dysfunction characterizing ALS (Hensley et al., 2006). Being an important component in signaling cascade mutant SOD1 can harm motor neurons as well as the activity of glial cells in FALS (Lee et al., 2016). Under normal physiological conditions, SOD1 activates Nox via regulating Rac1. Nox-containing endosomes also known as redoxosomes are essential for redox balance as they produce $O_2^{\bullet-}$ which in turn gets converted by SODs. SOD1 mutation is thought to enhance Nox-dependant ROS production which in turn, is responsible for neuronal death in ALS (Li et al., 2011).

2.7.6 Superoxide dismutase as a therapeutic target against various diseases

SOD has been examined diversely as a protective, curative and ameliorative agent against different pathophysiological conditions through many *in vitro*, *in vivo*, and clinical studies (See Fig. 2.7.4). Few recent studies have been put under the Table 2.7.2.

2.7.6.1 Superoxide dismutases and cancer

In carcinogenesis, the role of oxidative stress is revealed by many studies. ROS has been reported as an endogenous class of cancer stimulating agents (Feig et al., 1994; Guyton and Kensler, 1993). Researchers have shown that malignant cells increase the production of ROS to induce some alteration in cellular signaling pathways like the PI3L/AKT pathway which implies to inhibit cell proliferation and cell mobility. Therefore, SODs which provide the first line defense against ROS mediated oxidative stress can be targeted meaningfully as an anticancer therapeutic agent. Cu/Zn-SOD has been shown as a therapeutic target in treating multiple myeloma (Salem et al., 2015). A report has shown cancer prevention through antioxidants by dietary supplement-based SOD (Robbins and Zhao, 2014). SOD mimetics are reported to be beneficial in the treatment of castration-resistant prostate cancer, where the expression of SOD2 is suppressed (Thomas and Sharifi, 2012). In addition, implications of SODs in radiation therapy are also important (Holley et al., 2014). SOD-antagonists have also been described as anti-cancer agents. Recently, ATN-224, a bis-choline

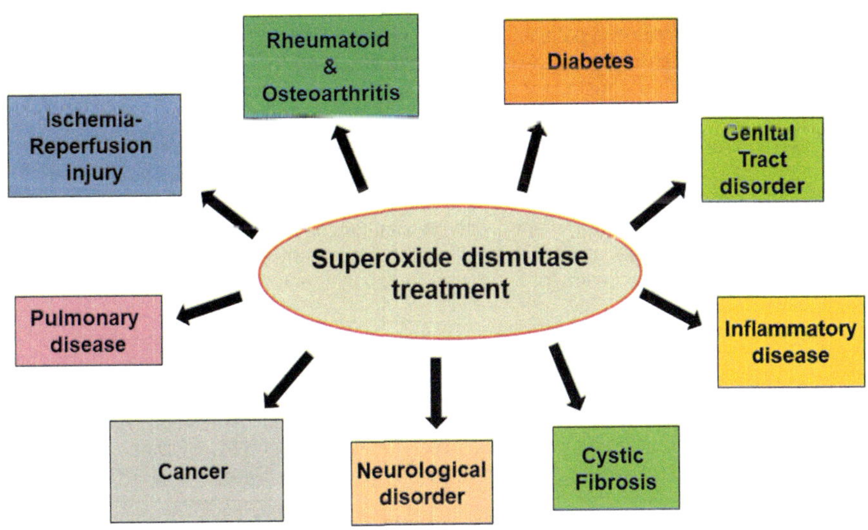

FIG. 2.7.4 SOD in therapeutic use against various diseases: an overview.

Table 2.7.2 Recent in vivo and clinical studies.

In vivo experimental study

- The recombinant Mn-SOD (rMn-SOD), a human recombinant protein, has shown to have significant effect as a radioprotective agent against radiotherapy for normal cells and radiosensitizing agent for cancer cells in mice. Thus, it has been suggested for clinical trial as well for the cancer patients during radiotherapy as it would be a safe tool to amplify the radiation effect specifically on cancer cells as well as giving protection for healthy cells (Borrelli et al., 2009).
- The use of a recombinant Mn-SOD (rMn-SOD) has been reported to prevent acute kidney injury induced by contrast media (CM) in a rat model. CM administration causes increased ROS level in kidney, leading to oxidative stress mediated tissue damage, and the animal develops CIN or contrast media induced nephropathy. Upon treatment with rMn-SOD, the O2− anion mediated oxidative stress was reduced in CM administered rats, and the consequences of CM administration like GFR reduction and renal tissue damage were prevented as well (Pisani et al., 2014).
- A lacithinized version of SOD, called PC-SOD, has been suggested to have preventive role against the mechanical ventilation induced injury in ARDS patients. In this study, the lungs of mice were subjected to cecal ligation and puncture (CLP) injuries. Mice which were administered with intravenous injection of PC-SOD, had increased survival rate after getting CLP (Tanaka et al., 2017).
- A nonreducible Cu/Zn-SOD nanozyme has been shown to have protective role against angiotensin II mediated hypertension in mice model. PLL_{50}-PEG Cu/Zn-SOD nanozyme was intracerebroventricularly injected in mice that were subcutaneously infused with angiotensin II before. The nanozyme successfully delivered active Cu/Zn-SOD to neurons and was able to counteract excess superoxide followed by decreasing blood pressure in those hypertensive mice (Savalia et al., 2014).

Clinical study

- Lecithinized SOD or PC-SOD has been clinically trialled in patients with idiopathic interstitial pneumonia or IIP, at 10 medical institutes in Japan. The administration was safe and well tolerated in patients with moderate to advanced IIPs. The PC-SOD administration also improved the levels of serum markers for IIP as well in patients with advanced IIP, and suggested for a large-scale clinical study to confirm its role against IIP (Kamio et al., 2014).
- Chronic neck pain (CNP) is linked with oxidative stress, and so, a combination of SOD and alpha-lipoic acid (ALA) was thought to relieve pain in CNP patients. In an open study, random patients were chosen and administered with 600 mg of ALA and 140 IU SOD in combination, daily, along with physiotherapy for 2 months. There was no drug reaction reported, along with significant improvement reports for patients receiving this combinatorial treatment compared to patients receiving only physiotherapy (Mauro et al., 2014).
- Breast cancer patients receiving anthracycline treatment are prone to heart failure due to anthracycline mediated increased free radicals and cardiotoxicity. PC-SOD has been thought to protect this cardiotoxicity by scavenging superoxide, and was administered in anthracycline treated breast cancer patients. The results showed no improvement in biomarkers for cardiac effect assessment, and PC-SOD has been reported to have no role in cardio protection in this context (Broeyer et al., 2014).

tetrathiomolybdate, has been reported to act by chelating copper ions. When treated with ATN-224, some promising results were obtained in the phase II trial in relapsed prostate cancer patients (Lin et al., 2009). In the case of solid tumors, ATN-224 was found to preferentially attack tumor cells rather than normal endothelial cells. Besides, in the A549 cell line as well as tumor xenograft model ATN-224 was found to cause apoptotic cell death similar to SOD1 knock-down condition. It was studied that ATN-224 mediated constant elevation of H_2O_2 directs activation of p38 and subsequent downregulation of antiapoptotic Mcl1 in order to execute apoptosis of tumor cells. Another compound, 4,5-dichloro-2-m-tolylpyridazin-3(2H)-1 (LCS-1) was also been studied to bind with and inactivate SOD1 *in vitro* (Somwar et al., 2009). In addition, an oestrogen-mimetic 2-methoxyoestradiol was reported to target SOD1 indirectly and to kill preferentially human leukemia cells by p53 accumulation and concomitant release of mitochondrial cytochrome c in order to address apoptosis (Glasauer et al., 2014).

2.7.6.2 Superoxide dismutase and diabetes

The hyperglycemia-induced oxidative stress is very common in diabetes which causes various pathophysiological complications including nephropathy, hepatic and vascular dysfunction etc. SOD administration has been reported to be a promising treatment against many diabetic complications. Treatment with exogenous Cu/Zn-SOD has shown to significantly cure liver oxidative stress in streptozotocin (STZ)-induced diabetic male Wister rats (Di Naso et al., 2011). Another study was done to investigate whether chemically modified SOD treatment can reduce lipid peroxidation in diabetic animals. Administration of PMVE/MA-SOD (poly-methyl vinyl ether-co-malcic anhydride SOD) in STZ induced diabetic rats had shown to reduce MDA level compared to diabetic control rats in the liver, kidney and brain tissues. GSH and SOD activity was found to be higher as well in the treated animals (Mansuroğlu et al., 2015). These studies suggested that SOD has a promising therapeutic advantage in clinical studies against diabetes.

2.7.6.3 Superoxide dismutase in bronchopulmonary dysplasia

Bronchopulmonary dysplasia (BPD) is a chronic lung disease in newborns, suffering from respiratory distress syndrome. It occurs due to the long-term use of ventilators in premature infants after birth. This disease was one of the very first against which the advantage of SOD therapy was accepted. Due to the fact that superoxide radicals could easily be formed in these neonates, SOD therapy was thought to be a way for the prevention of this disease. Rosenfeld et al. conducted a trial on 45 neonates suffering from respiratory distress syndrome, on whom he administered Bovine SOD after every 12 hours through injection, and monitored them continuously (Rosenfeld et al., 1984). The trial seemed to have a notable effect on the health of the neonates and appeared to be effective.

2.7.6.4 Superoxide dismutase in cystic fibrosis

Cystic fibrosis (CF) is a chronic inflammatory disease where activated neutrophils are recruited in huge amount at the site of inflammation. In cystic fibrosis patients, SOD activity was found to be low in their blood plasma (Madarasi et al., 2000). Also, in their RBC, monocytes and polymorphonuclear cells, the Cu/Zn-SOD activity was reported to be lower than that of a normal healthy person (Percival et al., 1995). It has been reported that Cu/Zn-SOD has antifibrotic action which is mediated by TGF-β1 repression (Vozenin-Brotons et al., 2001).

An *in vitro* study showed that CF cells have higher oxidative stress and a higher rate of apoptosis compared to normal cells. The expression of Cu/Zn-SOD and Mn-SOD were found to be much lower in these cells than normal cells, though their activity were unaltered. However, EC-SOD had a similar expression level but lower activity compared to normal cells (Rottner et al., 2011). In this study, SOD mimetics were used to treat to proapoptotic agent induced CF cells after which the apoptosis rate was found to be reduced (Rottner et al., 2011). These findings suggest that SOD can be used as a protective agent in cystic fibrosis patients.

2.7.6.5 Superoxide dismutase in rheumatoid arthritis and osteoarthritis

Chronic inflammation in the synovial membrane or synovium of bone joints is seen in patients with rheumatoid arthritis, which results in the degeneration of cartilage. The pathogenesis of this disease is also linked with enhanced oxidative stress or impaired antioxidant defense system in the patients (Hitchon and El-Gabalawy, 2004; Mahajan and Tandon, 2004). A study has found that in the patients suffering from rheumatoid arthritis have low SOD and glutathione peroxidase activity (Karatas et al., 2003).

Patients suffering from osteoarthritis were injected with bovine SOD intraarticularly into their Synovial cleft. This study was first reported by Lund-Olesen and Menander (Marberger et al., 1974). 2–3 mg SOD, 3 times daily, was given to the patients, and 16 out of 19 patients showed signs of improvement at a steady rate. Later on, the curative effect of this treatment was carried out on placebo-controlled, randomized trials in three different regions namely, Europe, Scandinavia and UK (Marberger et al., 1974). Promising results were seen in this trial also and hence, SOD therapy is considered as a way out for the treatment of osteoarthritis by many investigators.

2.7.6.6 Superoxide dismutase in ischemia-reperfusion injury

An interruption in arterial blood supply to organs or tissues is caused by ischemia, which leads to local cell death over time. After an interrupted period, reperfusion of blood to the oxygen lacking regions can cause a sudden burst of ROS, leading to oxidative stress-mediated injuries. An *in vivo* study, demonstrating the role of

$O_2^{\bullet-}$ in the pathogenesis of ischemic cerebral infarction and reperfusion has shown that transgenic mice overexpressing Cu/Zn-SOD were resistant to this ischemia-reperfusion injury (Yang et al., 1994). Another study has revealed these focal cerebral ischemic infarction injuries are more severe in transgenic mice lacking Mn-SOD, revealing the role of $O_2^{\bullet-}$ in the mitochondria as well for this pathophysiology (Kim et al., 2002). SOD mimetics have been reported as *in vivo* protective agents for ischemia-reperfusion led tissue damages (Salvemini and Cuzzocrea, 2002).

2.7.6.7 Superoxide dismutase and neurological disorders

In the case of neurological disorders, oxidative stress has always been an underlying cause. To combat enhanced oxidative stress, SOD trials were carried out on patients with Parkinson's disease. Both cytosolic and mitochondrial SOD plays an important role in lowering the overproduction and adverse effect of $O_2^{\bullet-}$ on Parkinson's disease (Henchcliffe and Beal, 2008). Notably, transgenic SOD1 And SOD2 transgenic mice were found to be resistant against 1-methyl-4-phenyl-1,2,3,6-tetrahydropyridine (MPTP)-induced neurotoxicity. In light of such findings role of SODs were evaluated in paraquat-treated *Drosophila melanogaster*. Acute treatment with high concentration of paraquat was responsible to elicit SOD2 over expression, whereas chronic treatment with paraquat with a sub-lethal dose of paraquat was able to upregulate SOD1 expression in dopaminergic neurons to fight neurotoxicity (Filograna et al., 2016). Among porphyrin-based SOD mimetics, Mn-porphyrins are found to be beneficial to reduce oxidative stress related to Parkinson's disease by scavenging $ONOO^{\bullet}$ radical (Batinić-Haberle et al., 2010).

The oxidative injury affects the cytoskeletal structures and membrane of brain cells, and thus, can cause Alzheimer's disease (AD). AD is characterized by the progressive disintegration of behavior, cognition and functionality where daily activities are hampered to a great extend (Zuo et al., 2015). Here the pathophysiology is mainly associated with the extracellular deposition of amyloid-beta (Aβ) plaques and the accumulation of intracellular tau neurofibrillary tangles (NFT). Interestingly, Cu^{2+}/Zn^{2+}-bound Aβ has been shown to possess a structure similar to SOD, with potential antioxidant properties. Hence, Cu^{2+} and Zn^{2+} supplementation have been suggested as a novel strategy to decrease Aβ-induced ROS generation and metal-catalyzed Aβ deposition (Curtain et al., 2001). A study has reported that the activity of SOD1 in Alzheimer's cells was found to be higher by 30% than that of the normal cells. Another study revealed that Cu/Zn-SOD is modified by oxidative stress in the AD brain and forms proteinaceous aggregations that are associated with the characteristic amyloid plaques and tangles of AD brain (Choi et al., 2005). In a study, with Mn-SOD overexpressing mice, it was found that overexpression of Mn-SOD significantly reduced hippocampal $O_2^{\bullet-}$ and decreased amyloid-beta plaques resulting in the prevention of AD related learning and memory loss (Massaad et al., 2009). Supplementation of SOD in the AD mice model has been shown curative results (Persichilli et al., 2015).

2.7.6.8 Superoxide dismutase in gastrointestinal disorders

The level of different SODs differs in GI diseases. Increased level of Mn-SOD was observed in oesophageal and gastric carcinomas whereas the Cu/Zn-SOD level was decreased (Miranda et al., 2000). When colitis was induced in the mice model, transgenic mice with overexpression of SOD showed disease induction than wild type mice (Krieglstein et al., 2001). Intrarectal acetic acid stimulates SOD activity was observed on patients suffering from Crohn's disease and ulcerative colitis (Beltran et al., 2010). SOD level in the peripheral bloodstream is used as an oxidative stress biomarker of inflammatory bowel disease (Guan et al., 2018).

2.7.6.9 Superoxide dismutase in inflammatory disorders

Most of the inflammatory diseases are linked with the activation of neutrophils at the site of inflammation. Activated neutrophils adhere to vascular endothelium in order to migrate within extracellular milieu where they damage surrounding tissue structures by releases ROS and different types of proteolytic enzymes. $O_2^{\bullet-}$ have been reported to facilitate the infiltration of neutrophils in endothelial cells and cause enhanced inflammation (Masini et al., 2002; Salvemini et al., 1999).

Particulate matters, such as fly ash have been postulated to produce oxidative stress and neutrophil-mediated inflammation in the lower respiratory tract. Transgenic mice overexpressing EC-SOD were treated with a particulate matter-ROFA (residual oil fly ash) and had been shown to prevent neutrophil infiltration; thereby, reducing inflammation and oxidative stress (Ghio et al., 2002).

It is well documented that SOD is involved in the regulation of neutrophil apoptosis, which can be targeted for the resolution of inflammation. Patients with Down syndrome, where Cu/Zn-SOD is overexpressed, show a high rate of neutrophil apoptosis (Yasui et al., 1999). H_2O_2 production and Caspase-3 activation are reported to be directly involved in neutrophil apoptosis (Yasui et al., 2005). Excess SOD administration from outside has been reported to lower H_2O_2-mediated apoptosis of neutrophils *in vitro* (Yasui et al., 2005).

SOD is extensively used in treating sports injuries including frozen shoulder, tennis elbow, tendon synovitis, etc. SOD is commonly injected into the site locally in tennis elbow, which results in a substantially increased relief of pain and re-establishes working capability in patients (Muller and Moll, 1983).

The idea behind SOD treatment in some human diseases is sometimes based on the fact that the disease shows resistance to all known therapies and hence SOD trials are given a shot. In the case of Peyronie's disease (characterized by chronic inflammation of penis due to fibrosis and induration), as a curative therapy, SOD has been widely used by injecting it directly into the fibrotic plaques (Schneider et al., 1985). Iontophoresis techniques were also used to transfer the SOD across the surface of the skin by the usage of electric current. Here also, positive results from these trials of the SOD therapy were reported (Verges and Chateau, 1988).

2.7.7 Adverse effects of superoxide dismutase

SOD converts two $O_2^{\bullet-}$ to molecular oxygen and H_2O_2. This H_2O_2 is a potent oxidant that often generates $OH^\bullet$, which is far more reactive than $O_2^{\bullet-}$. This leads to oxidative damage to a great extent. Sometimes, molecules other than $O_2^{\bullet-}$ can also act as an electron donor for Cu/Zn-SOD and reduce copper from Cu^{2+} to Cu^+. The resultant reduced enzyme then can act as superoxide reductase (SOR) on some other substrates by reducing them and oxidizing itself back to natural Cu^{2+} form. If an electron acceptor takes one electron back from SOR to favor Cu^+ to Cu^{2+} transition, the enzyme can act as superoxide oxidase (SOO) (Liochev and Fridovich, 2000). Yet, Mn-SOD does not show such kinds of (SOR or SOO) activities. Cu/Zn-SOD when acts as SOR, causes the generation of one molecule of H_2O_2 in order to convert one $O_2^{\bullet-}$ and thus, increases the chance of H_2O_2-mediated oxidative stress.

In contrast to the aforementioned discussion, superoxide is sometimes beneficial to prevent oxidative damages to our body. For example, it acts as the chain breaking molecule for lipid peroxidation as it scavenges reactive lipid peroxyl radical ($LOO^\bullet$) generated in lipid peroxidation and converts them into lipid hydroperoxide (LOOH).

$$LOO^\bullet + O_2^{\bullet-} + H^+ = LOOH \tag{2.7.12}$$

(Nelson et al., 1994)

In a system of lipid peroxidation, increased SOD activity can lead to uplifted $O_2^{\bullet-}$ scavenging and subsequent an excess incidence of lipid peroxidation (Groner et al., 1986). In the case of Down's syndrome, due to trisomy in chromosome 21, this enzyme is overexpressed as the gene for SOD is present in chromosome 21. Lipid peroxidation level and overall oxidative stress are seen to be elevated in the individuals having Down's syndrome than normal persons, maybe due to over-scavenging of $O_2^{\bullet-}$ (Brooksbank and Balazs, 1984; Sinet, 1982).

Other than dismutation, Cu/Zn-SOD catalyzes certain other kinds of reactions as well, depending on the availability of different redox molecules that can cause the transition of copper cofactor at the active site of the enzyme. In the presence of H_2O_2 it shows peroxidase activity as described below:

$$\text{E-Cu}^{2+} + H_2O_2 = \text{E-Cu}^+ + HO_2^- + H^+ \tag{2.7.13}$$

$$\text{E-Cu}^+ + HO_2^- + H^+ = \text{E-Cu}^{2+}\text{-OH} + OH^- \tag{2.7.14}$$

Here, H_2O_2 first attacks on Cu^{2+} centre of the Cu/Zn-SOD and reduces it to Cu^+ while itself gets oxidized to hydroperoxyl anion (HO_2^-). In the second step, the reduced copper ion of the enzyme breaks the peroxide bond (O-O) of the HO_2^- and itself become oxidized, generating a hydroxide anion (OH^-) and an $OH^\bullet$; both of which eventually become attached to the enzyme's active site. The second step of this reaction is a kind of 'Fenton reaction'. The enzyme bound $OH^\bullet$ is reactive and has been reported to attack the amino acids of the enzyme itself, and can lead to serious damage (Hodgson and Fridovich, 1975).

In presence of bicarbonate ions, the rate of the reaction (XIII) increases many folds as $OH^{\bullet}$ attached to the enzyme (shown as $E\text{-}Cu^{2+}\text{-}OH$) oxidizes bicarbonates to carbonate radical anions ($CO_3^{\bullet -}$), which are potent oxidants (Zhang et al., 2000) (See Fig. 2.7.5).

$$E\text{-}Cu^{2+}\text{-}OH + HCO_3^- = E\text{-}Cu^{2+} + H_2O + CO_3^{\bullet -} \qquad (2.7.15)$$

Cu/Zn-SOD can also catalyze the reaction mediated by $ONOO^-$ that causes nitration of the proteins at their free tyrosine and tyrosyl residues (Ischiropoulos et al., 1992).

There are reports revealing that Cu/Zn-SOD catalyzes the breakdown of nitrosothiols resulting in the release of $NO^{\bullet}$ (Jourd'Heuil et al., 1999). The pro-oxidant activity of a SOD mimetic, EUK-8 was observed in starving *E. coli* cells upon administration. These 'side reactions' of Cu/Zn-SOD occurring in various peculiar conditions can lead to various pathophysiological conditions if increased beyond the limit. That is the reason why overexpressing SOD or over injecting SOD into the body as antioxidant treatments against various pathophysiological conditions does not always give expected results; rather sometimes becomes detrimental as well.

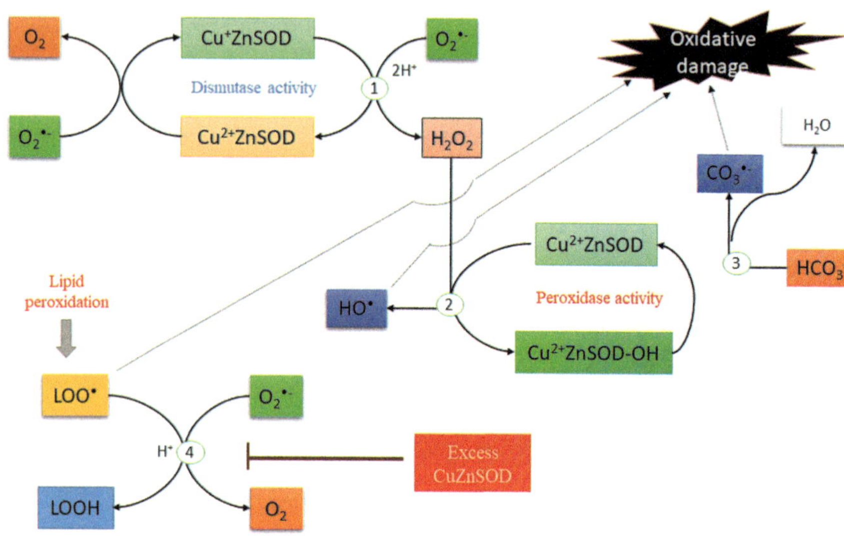

FIG. 2.7.5 Reactions describing the detrimental effects of superoxide dismutase (Cu/Zn-SOD shown here).

H_2O_2 is produced via reaction 1 by the dismutase activity of CuZnSOD. H_2O_2 is involved in the formation of $OH^{\bullet}$ via reaction 2 by the help of peroxidase activity of Cu/Zn-SOD, which results in formation of Cu^{2+} Zn-SOD-OH. In presence of HCO_3^-, Cu^{2+} ZnSOD-OH generates $CO_3^{\bullet -}$ via reaction 3. $OH^{\bullet}$ and $CO_3^{\bullet -}$ are potential radicals to cause oxidative damage. $O_2^{\bullet -}$ can act as a chain breaker of lipid peroxyl radicals ($LOO^{\bullet}$) via reaction 4, as it converts $LOO^{\bullet}$ it to LOOH (lipid hydroperoxyl molecule). Excess Cu/Zn-SOD, if present at the site of lipid peroxidation, can inhibit this step by over scavenging $O_2^{\bullet -}$, thus can enhance lipid peroxidation.

2.7.8 Optimum dose, route of administration, and limitations of therapeutic use of superoxide dismutase

The optimal dose of SOD is not well established and depends on various factors like gender, age and other health conditions. In one study, 16 mg SOD injection twice was found to be most effective for treating osteoarthritis (McIlwain et al., 1989). Another study was done with rowing team members with a dosage of 500 mg of a specific plant SOD (Skarpanska-Stejnborn et al., 2011). It has been seen that high doses of SOD is detrimental for health but the exact dose of SOD beyond which it exerts toxicity is still unknown (McCord, 2008). The route of administration of SOD is either orally or through injections (intraperitoneal, intramuscular, intravenous and subcutaneous) (Carillon et al., 2013). SOD administration has certain disadvantages, as SOD has a very low half-life *in vivo*, along with instability, low cellular uptake, low bioavailability, and high immunogenicity. Due to its low half-life and rapid renal excretion, the accumulation of SOD at the site of inflammation is very low, which makes it less effective against inflammation. That is why the clinical administrations of SOD against various pathophysiological conditions are very limited, even though they are suggested to have a strong therapeutic potentiality. To overcome these limitations, SOD administrations are being carried out via incorporating them in highly loaded SA-liposomes or PEG-liposomes with long circulation capacity (Cruz et al., 2005). SOD has been coated with wheat-derived gliadin to increase its bioavailability (Romao, 2015). Various SOD mimetics have been designed and shown to have the dismutation capacity of SOD but higher stability and other advantages over SOD itself. This makes them effective against diseases for which the native SOD remained ineffective (Salvemini et al., 2002).

Conclusion

Ever since life began and the environment became oxidative, SOD has been playing a vital role in the defense against ROS-mediated oxidative injuries. Whenever $O_2^{\bullet}$ generation increases, this enzyme comes into play to neutralize and reduce the level of ROS back to normal and thus, maintaining healthy redox homeostasis. As most of the diseases are reported to cause oxidative stress to the body and damage various organs, it is a worthy decision to take SOD as the protective agent against disease mediated enhanced ROS production and oxidative stress. After various cell and animal-based studies, it has been proved that SOD surely has therapeutic roles against various diseases and many clinical trials have been done successfully as well. But again, SOD administration has limitations due to its instability and lower life span, for which the SOD mimetics are now being designed and examined extensively against complications where SOD treatment failed. These mimetics are the future of targeted antioxidant-based therapy against different oxidative stress-related diseases and their proper scientific use may ameliorate various pathophysiological conditions.

However, overexpression of SOD or overdose of SOD administration leads to over-scavenging of $O_2^{\bullet -}$ which cause deleterious effects to the system, as a certain amount of $O_2^{\bullet -}$ is necessary for the body. Few diseases are related to the overexpression of SOD; for example – in Down's syndrome, SOD1 is overexpressed and linked with enhanced lipid peroxidation. Thus, SOD can be beneficial as well as detrimental for the body depending on its amount of presence, which breaks the myth that antioxidants are always good for health, on the contrary states that they have both the bright and the dark side.

Abbreviations

SOD	Superoxide dismutase
ROS	Reactive oxygen species
RNS	Reactive nitrogen species
Cu/Zn-SOD	Copper and zinc-containing superoxide dismutase
Mn-SOD	Manganese -containing superoxide dismutase
Fe–SOD	Iron-containing superoxide dismutase
Ni-SOD	Nickel-containing superoxide dismutase
EC-SOD	Extracellular superoxide dismutase
HOCl	Hypochlorous acid
$ONOO^-$	Peroxynitrite
MDA	Malondialdehyde
GPX	Glutathione peroxidase
$O_2^{\bullet -}$	Superoxide
$OH^{\bullet}$	Hydroxide radical
$CO_3^{\bullet -}$	Carbonate anion radical
H_2O_2	Hydrogen peroxide
$NO^{\bullet}$	Nitric oxide radical
$NO_2^{\bullet}$	Nitrogen dioxide radical
IL-1	Interleukin 1
IFN-γ	Interferon gamma
Glu	Glutamate
Lys	Lysine
Gln	Glutamine
PTP	Protein tyrosine phosphatase
ERK	Extracellular signal-regulated kinase
FALS	Familial amyotrophic lateral sclerosis
SOO	Superoxide oxidase
SOR	Superoxide reductase
BPD	Bronchopulmonary dysplasia
CF	Cystic fibrosis

References

Abreu, I.A., Cabelli, D.E., 2010. Superoxide dismutases-a review of the metal-associated mechanistic variations. Biochim. Biophys. Acta 1804, 263–274.

Aminake, M.N., Pradel, G., 2013. Antimalarial drugs resistance in Plasmodium falciparum and the current strategies to overcome them. Microbial pathogens and strategies for combating them. Formatex 2013, 1–774.

Bastow, E.L., Peswani, A.R., Tarrant, D.S., Pentland, D.R., Chen, X., Morgan, A., Staniforth, G.L., Tullet, J.M., Rowe, M.L., Howard, M.J., 2016. New links between SOD1 and metabolic dysfunction from a yeast model of amyotrophic lateral sclerosis. J. Cell Sci. 129, 4118–4129.

Batinić-Haberle, I., Rebouças, J.S., Spasojević, I., 2010. Superoxide dismutase mimics: chemistry, pharmacology, and therapeutic potential. Antioxid. Redox Signaling 13, 877–918.

Beckman, J.S., Koppenol, W.H., 1996. Nitric oxide, superoxide, and peroxynitrite: the good, the bad, and ugly. Am. J. Physiol. Cell Physiol. 271, C1424–C1437.

Beltran, B., Nos, P., Dasí, F., Iborra, M., Bastida, G., Martínez, M., O'Connor, J.E., Saez, G., Moret, I., Ponce, J., 2010. Mitochondrial dysfunction, persistent oxidative damage, and catalase inhibition in immune cells of naive and treated Crohn's disease. Inflamm. Bowel Dis. 16, 76–86.

Borrelli, A., Schiattarella, A., Mancini, R., Morrica, B., Cerciello, V., Mormile, M., d'Alesio, V., Bottalico, L., Morelli, F., D'Armiento, M., D'Armiento, F.P., 2009. A recombinant MnSOD is radioprotective for normal cells and radiosensitizing for tumor cells. Free Radic. Biol. Med. 46, 110–116.

Bowler, R.P., Nicks, M., Olsen, D.A., Thøgersen, I.B., Valnickova, Z., Højrup, P., Franzusoff, A., Enghild, J.J., Crapo, J.D., 2002. Furin proteolytically processes the heparin-binding region of extracellular superoxide dismutase. J. Biol. Chem. 277, 16505–16511.

Bridges, S.M., Salin, M.L., 1981. Distribution of iron-containing superoxide dismutase in vascular plants. Plant Physiol. 68, 275–278.

Broeyer, F.J., Osanto, S., Suzuki, J., de Jongh, F., van Slooten, H., Tanis, B.C., Bruning, T., Bax, J.J., Ritsema van Eck, H.J., de Kam, M.L., Cohen, A.F., 2014. Evaluation of lecithinized human recombinant super oxide dismutase as cardioprotectant in anthracycline-treated breast cancer patients. Br. J. Clin. Pharmacol. 78, 950–960.

Brooksbank, B.W., Balazs, R., 1984. Superoxide dismutase, glutathione peroxidase and lipoperoxidation in Down's syndrome fetal brain. Dev. Brain Res. 16, 37–44.

Brown, N.M., Torres, A.S., Doan, P.E., O'Halloran, T.V., 2004. Oxygen and the copper chaperone CCS regulate posttranslational activation of Cu, Zn superoxide dismutase. Proc. Natl. Acad. Sci. 101, 5518–5523.

Carillon, J., Rouanet, J.M., Cristol, J.P., Brion, R., 2013. Superoxide dismutase administration, a potential therapy against oxidative stress related diseases: several routes of supplementation and proposal of an original mechanism of action. Pharm. Res. 30, 2718–2728.

Choi, J., Rees, H.D., Weintraub, S.T., Levey, A.I., Chin, L.-S., Li, L., 2005. Oxidative modifications and aggregation of Cu, Zn-superoxide dismutase associated with Alzheimer and Parkinson diseases. J. Biol. Chem. 280, 11648–11655.

Ciancarelli, I., De Amicis, D., Di Massimo, C., Sandrini, G., Pistarini, C., Carolei, A., Ciancarelli, M.G.T., 2015. Influence of intensive multifunctional neurorehabilitation on neuronal oxidative damage in patients with Huntington's disease. Funct. Neurol. 30, 47.

Cleveland, D.W., Rothstein, J.D., 2001. From Charcot to Lou Gehrig: deciphering selective motor neuron death in ALS. Nat. Rev. Neurosci. 2, 806–819.

Collin, F., 2019. Chemical basis of reactive oxygen species reactivity and involvement in neurodegenerative diseases. Int. J. Mol. Sci. 20, 2407.

Crapo, J.D., Oury, T., Rabouille, C., Slot, J.W., Chang, L.-Y., 1992. Copper, zinc superoxide dismutase is primarily a cytosolic protein in human cells. Proc. Natl. Acad. Sci. 89, 10405–10409.

Cruz, M.E., Manuela Gaspar, M., Barbara, M., Martins, F., Luisa Corvo, M., 2005. Liposomal superoxide dismutases and their use in the treatment of experimental arthritis. Methods Enzymol. 391, 395–413.

Culotta, V.C., Yang, M., O'Halloran, T.V., 2006. Activation of superoxide dismutases: putting the metal to the pedal. Biochim. Biophys. Acta. 1763, 747–758.

Curcio, F., Pegoraro, I., Russo, P.D., Falleti, E., Perrella, G., Ceriello, A., 1995. SOD and GSH inhibit the high glucose-induced oxidative damage and the PDGF increased secretion in cultured human endothelial cells. Thromb. Haemost. 73, 969–973.

Curtain, C.C., Ali, F., Volitakis, I., Cherny, R.A., Norton, R.S., Beyreuther, K., Barrow, C.J., Masters, C.L., Bush, A.I., Barnham, K.J., 2001. Alzheimer's disease amyloid-β binds copper and zinc to generate an allosterically ordered membrane-penetrating structure containing superoxide dismutase-like subunits. J. Biol. Chem. 276, 20466–20473.

D'Amico, E., Factor-Litvak, P., Santella, R.M., Mitsumoto, H., 2013. Clinical perspective on oxidative stress in sporadic amyotrophic lateral sclerosis. Free Radic. Biol. Med. 65, 509–527.

Dileepan, K.N., Sharma, R., Stechschulte, D.J., Savin, V.J., 1993. Effect of superoxide exposure on albumin permeability of isolated rat glomeruli. J. Lab. Clin. Med. 121, 797–804.

Droge, W., 2002. Free radicals in the physiological control of cell function. Physiol. Rev. 82, 47–95.

Dukan, S., Nyström, T., 1999. Oxidative Stress Defense and Deterioration of Growth-arrestedEscherichia coli Cells. J. Biol. Chem. 274, 26027–26032.

Elchuri, S., Oberley, T.D., Qi, W., Eisenstein, R.S., Roberts, L.J., Van Remmen, H., Epstein, C.J., Huang, T.-T., 2005. CuZnSOD deficiency leads to persistent and widespread oxidative damage and hepatocarcinogenesis later in life. Oncogene 24, 367–380.

Feig, D.I., Reid, T.M., Loeb, L.A., 1994. Reactive oxygen species in tumorigenesis. Cancer Res. 54, 1890S–1894S.

Filograna, R., Godena, V.K., Sanchez-Martinez, A., Ferrari, E., Casella, L., Beltramini, M., Bubacco, L., Whitworth, A.J., Bisaglia, M., 2016. Superoxide dismutase (SOD)-mimetic M40403 is protective in cell and fly models of paraquat toxicity implications for parkinson disease. J. Biol. Chem. 291, 9257–9267.

Fisher, C.L., Cabelli, D.E., Tainer, J.A., Hallewell, R.A., Getzoff, E.D., 1994. The role of arginine 143 in the electrostatics and mechanism of Cu,Zn superoxide dismutase: computational and experimental evaluation by mutational analysis. Proteins 19, 24–34.

Fridovich, I., 1995. Superoxide radical and superoxide dismutases. Annu. Rev. Biochem. 64, 97–112.

Fukai, T., Folz, R.J., Landmesser, U., Harrison, D.G., 2002. Extracellular superoxide dismutase and cardiovascular disease. Cardiovasc. Res. 55, 239–249.

Gamez, J., Corbera-Bellalta, M., Nogales, G., Raguer, N., García-Arumí, E., Badia-Canto, M., Lladó-Carbó, E., Álvarez-Sabín, J., 2006. Mutational analysis of the Cu/Zn superoxide dismutase gene in a Catalan ALS population: should all sporadic ALS cases also be screened for SOD1? J. Neurol. Sci. 247, 21–28.

Genestra, M., 2007. Oxyl radicals, redox-sensitive signalling cascades and antioxidants. Cell. Signal. 19, 1807–1819.

Getzoff, E.D., Cabelli, D.E., Fisher, C.L., Parge, H.E., Viezzoli, M.S., Banci, L., Hallewell, R.A., 1992. Faster superoxide dismutase mutants designed by enhancing electrostatic guidance. Nature 358, 347–351.

Getzoff, E.D., Tainer, J.A., Weiner, P.K., Kollman, P.A., Richardson, J.S., Richardson, D.C., 1983. Electrostatic recognition between superoxide and copper, zinc superoxide dismutase. Nature 306, 287–290.

Ghio, A.J., Suliman, H.B., Carter, J.D., Abushamaa, A.M., Folz, R.J., 2002. Overexpression of extracellular superoxide dismutase decreases lung injury after exposure to oil fly ash. Am. J. Physiol. Lung Cell. Mol. Physiol. 283, L211–L218.

Gilks, C.B., Price, K., Wright, J.L., Churg, A., 1998. Antioxidant gene expression in rat lung after exposure to cigarette smoke. Am. J. Pathol. 152, 269.

Gill, S.S., Anjum, N.A., Gill, R., Yadav, S., Hasanuzzaman, M., Fujita, M., Mishra, P., Sabat, S.C., Tuteja, N., 2015. Superoxide dismutase—mentor of abiotic stress tolerance in crop plants. Environ. Sci. Pollut. Res. 22, 10375–10394.

Glasauer, A., Sena, L.A., Diebold, L.P., Mazar, A.P., Chandel, N.S., 2014. Targeting SOD1 reduces experimental non–small-cell lung cancer. J. Clin. Invest. 124, 117–128.

Groner, Y., Elroy-Stein, O., Bernstein, Y., Dafni, N., Levanon, D., Danciger, E., Neer, A., 1986. Molecular Genetics of Down's Syndrome: Overexpression of Transfected Human Cu/Zn-Superoxide Dismutase Gene and the Consequent Physiological Changes. Cold Spring Harb. Symp. Quant. Biol. 51, 381–393.

Guan, G., Lan, S., 2018. Implications of antioxidant systems in inflammatory bowel disease. BioMed Res. Int. 2018, 1290179.

Guyton, K., Kensler, T.W., 1993. Oxidative mechanisms in carcinogenesis. Br. Med. Bull. 49, 523–544.

Halliwell, B., 2007. Biochemistry of Oxidative Stress. Portland Press Limited. Biochem. Soc. Trans. 35 (5), 1147–1150.

Henchcliffe, C., Beal, M.F., 2008. Mitochondrial biology and oxidative stress in Parkinson disease pathogenesis. Nat. Rev. Neurol. 4, 600.

Hendrickson, D.J., Fisher, J.H., Jones, C., Ho, Y.-S., 1990. Regional localization of human extracellular superoxide dismutase gene to 4pter-q21. Genomics 8, 736–738.

Hensley, K., Mhatre, M., Mou, S., Pye, Q.N., Stewart, C., West, M., Williamson, K.S., 2006. On the relation of oxidative stress to neuroinflammation: lessons learned from the G93A SOD1 mouse model of amyotrophic lateral sclerosis. Antioxid. Redox Signal. 8, 2075–2087.

Hitchon, C.A., El-Gabalawy, H.S., 2004. Oxidation in rheumatoid arthritis. Arthritis Res. Ther. 6, 265.

Hodgson, E.K., Fridovich, I., 1975. Interaction of bovine erythrocyte superoxide dismutase with hydrogen peroxide. Inactivation of the enzyme. Biochemistry 14, 5294–5299.

Holley, A.K., Miao, L., St. Clair, D.K., St. Clair, W.H., 2014. Redox-modulated phenomena and radiation therapy: the central role of superoxide dismutases. Antioxid. Redox Signal. 20, 1567–1589.

Hough, M.A., Hasnain, S.S., 2003. Structure of fully reduced bovine copper zinc superoxide dismutase at 1.15 A. Structure 11, 937–946.

Ingre, C., Roos, P.M., Piehl, F., Kamel, F., Fang, F., 2015. Risk factors for amyotrophic lateral sclerosis. Clin. Epidemiol. 7, 181.

Ischiropoulos, H., Zhu, L., Chen, J., Tsai, M., Martin, J.C., Smith, C.D., Beckman, J.S., 1992. Peroxynitrite-mediated tyrosine nitration catalyzed by superoxide dismutase. Arch. Biochem. Biophys. 298, 431–437.

Jourd'Heuil, D., Laroux, F.S., Miles, A.M., Wink, D.A., Grisham, M.B., 1999. Effect of superoxide dismutase on the stability of S-nitrosothiols. Arch. Biochem. Biophys. 361, 323–330.

Juarez, J.C., Manuia, M., Burnett, M.E., Betancourt, O., Boivin, B., Shaw, D.E., Tonks, N.K., Mazar, A.P., Doñate, F., 2008. Superoxide dismutase 1 (SOD1) is essential for H2O2-mediated oxidation and inactivation of phosphatases in growth factor signaling. Proc. Natl. Acad. Sci. 105, 7147–7152.

Kamio, K., Azuma, A., Ohta, K., Sugiyama, Y., Nukiwa, T., Kudoh, S., Mizushima, T., 2014. Double-blind controlled trial of lecithinized superoxide dismutase in patients with idiopathic interstitial pneumonia–short term evaluation of safety and tolerability. BMC Pulm. Med. 14, 86.

Karatas, F., Ozates, I., Canatan, H., Halifeoglu, I., Karatepe, M., Colak, R., 2003. Antioxidant status & lipid peroxidation in patients with rheumatoid arthritis. Indian J. Med. Res. 118, 178–181.

Karnati, S., Lüers, G., Pfreimer, S., Baumgart-Vogt, E., 2013. Mammalian SOD2 is exclusively located in mitochondria and not present in peroxisomes. Histochem. Cell Biol. 140, 105–117.

Keele, B.B., McCord, J., Fridovich, I., 1970. Superoxide dismutase from Escherichia coli BA new manganese-containing enzyme. J. Biol. Chem. 245, 6176–6181.

Keller, G.-A., Warner, T.G., Steimer, K.S., Hallewell, R.A., 1991. Cu, Zn superoxide dismutase is a peroxisomal enzyme in human fibroblasts and hepatoma cells. Proc. Natl. Acad. Sci. 88, 7381–7385.

Kiernan, M.C., Vucic, S., Cheah, B.C., Turner, M.R., Eisen, A., Hardiman, O., Burrell, J.R., Zoing, M.C., 2011. Amyotrophic lateral sclerosis. Lancet North Am. Ed. 377, 942–955.

Kim, G.W., Kondo, T., Noshita, N., Chan, P.H., 2002. Manganese superoxide dismutase deficiency exacerbates cerebral infarction after focal cerebral ischemia/reperfusion in mice: implications for the production and role of superoxide radicals. Stroke 33, 809–815.

Kim, H.T., Kim, Y.H., Nam, J., Lee, H.J., Rho, H.M., Jung, G., 1994. Study of 5′-flanking region of human Cu/Zn superoxide dismutase. Biochem. Biophys. Res. Commun. 201, 1526–1533.

Koppenol, W.H., Bounds, P.L., Dang, C.V., 2011. Otto Warburg's contributions to current concepts of cancer metabolism. Nat. Rev. Cancer 11, 325–337.

Krieglstein, C.F., Cerwinka, W.H., Laroux, F.S., Salter, J.W., Russell, J.M., Schuermann, G., Grisham, M.B., Ross, C.R., Granger, D.N., 2001. Regulation of murine intestinal inflammation by reactive metabolites of oxygen and nitrogen: divergent roles of superoxide and nitric oxide. J. Exp. Med. 194, 1207–1218.

Labbadia, J., Morimoto, R.I., 2013. Huntington's disease: underlying molecular mechanisms and emerging concepts. Trends Biochem. Sci. 38, 378–385.

Lee, J., Hyeon, S.J., Im, H., Ryu, H., Kim, Y., Ryu, H., 2016. Astrocytes and microglia as non-cell autonomous players in the pathogenesis of ALS. Exp. Neurobiol. 25, 233–240.

Li, Q., Spencer, N.Y., Pantazis, N.J., Engelhardt, J.F., 2011. Alsin and SOD1G93A proteins regulate endosomal reactive oxygen species production by glial cells and proinflammatory pathways responsible for neurotoxicity. J. Biol. Chem. 286, 40151–40162.

Lin, J., Beer, T., Ryan, C., Mathew, P., Wilding, G., Morris, M., Callahan, J., Gordon, G., Reich, S., Carducci, M., 2009. A randomized, phase II study of ATN-224 in patients with biochemically relapsed, hormone-naive prostate cancer: a DOD/PCF prostate cancer clinical trials consortium trial. J. Clin. Oncol. 27, 5135.

Liochev, S.I., Fridovich, I., 1999. Superoxide and iron: partners in crime. IUBMB Life 48, 157–161.

Liochev, S.I., Fridovich, I., 2000. Copper-and zinc-containing superoxide dismutase can act as a superoxide reductase and a superoxide oxidase. J. Biol. Chem. 275, 38482–38485.

Madarasi, A., Lugassi, A., Greiner, E., Holics, K., Biro, L., Mozsary, E., 2000. Antioxidant status in patients with cystic fibrosis. Ann. Nutr. Metab. 44, 207–211.

Mahajan, A., Tandon, V.R., 2004. Antioxidants and rheumatoid arthritis. J. Indian Rheumatol. Assoc. 12, 139–142.

Mansuroğlu, B., Derman, S., Yaba, A., Kızılbey, K., 2015. Protective effect of chemically modified SOD on lipid peroxidation and antioxidant status in diabetic rats. Int. J. Biol. Macromol. 72, 79–87.

Marberger, H., Huber, W., Bartsch, G., Schulte, T., Swoboda, P., 1974. Orgotein: a new antiinflammatory metalloprotein drug evaluation of clinical efficacy and safety in inflammatory conditions of the urinary tract. Int. Urol. Nephrol. 6, 61–74.

Marklund, S., 1992. Regulation by cytokines of extracellular superoxide dismutase and other superoxide dismutase isoenzymes in fibroblasts. J. Biol. Chem. 267, 6696–6701.

Marklund, S.L., 1982. Human copper-containing superoxide dismutase of high molecular weight. Proc. Natl. Acad. Sci. 79, 7634–7638.

Marklund, S.L., 1984. Extracellular superoxide dismutase and other superoxide dismutase isoenzymes in tissues from nine mammalian species. Biochem. J. 222, 649–655.

Masini, E., Cuzzocrea, S., Mazzon, E., Marzocca, C., Mannaioni, P.F., Salvemini, D., 2002. Protective effects of M40403, a selective superoxide dismutase mimetic, in myocardial ischaemia and reperfusion injury in vivo. Br. J. Pharmacol. 136, 905–917.

Massaad, C.A., Washington, T.M., Pautler, R.G., Klann, E., 2009. Overexpression of SOD-2 reduces hippocampal superoxide and prevents memory deficits in a mouse model of Alzheimer's disease. Proc. Natl. Acad. Sci. 106, 13576–13581.

Mauro, G.L., Cataldo, P., Barbera, G., Sanfilippo, A., 2014. α-Lipoic acid and superoxide dismutase in the management of chronic neck pain: a prospective randomized study. Drugs R. D. 14, 1–7.

McCord, J.M., 2008. Superoxide dismutase, lipid peroxidation, and bell-shaped dose response curves. Dose-Response 6, 223–238.

McCord, J.M., 2016. The discovery of superoxide dismutase and its role in redox biology. Redox-Active Ther. 5–10

McIlwain, H., Silverfield, J.C., Cheatum, D.E., Poiley, J., Taborn, J., Ignaczak, T., Multz, C.V., 1989. Intra-articular orgotein in osteoarthritis of the knee: a placebo-controlled efficacy, safety, and dosage comparison. Am. J. Med. 87, 295–300.

Milani, P., Gagliardi, S., Cova, E., Cereda, C., 2011. SOD1 transcriptional and posttranscriptional regulation and its potential implications in ALS. Neurol. Res. Int. 2011, 458427.

Miranda, A., Janssen, L., Bosman, C.B., van Duijn, W., Oostendorp-van de Ruit, M.M., Kubben, F.J., Griffioen, G., Lamers, C.B., Han, J., Van Krieken, J.M., van de Velde, C.J., 2000. Superoxide dismutases in gastric and esophageal cancer and the prognostic impact in gastric cancer. Clin. Cancer Res. 6, 3183–3192.

Muller, U., Moll, G., 1983. Uber die Behandlung der Epicondylitis mitlokalinjiziertemOrgotein (Doppelblindstudie). Z. Rheumatol. 42, 21–24.

Naso, D., Cangeri, F., Simões Dias, A., Porawski, M., Marroni, N.A.P., 2011. Exogenous superoxide dismutase: action on liver oxidative stress in animals with streptozotocin-induced diabetes. Exp. Diabetes Res. 2011, 754132.

Nelson, S.K., Bose, S.K., McCord, J.M., 1994. The toxicity of high-dose superoxide dismutase suggests that superoxide can both initiate and terminate lipid peroxidation in the reperfused heart. Free Radic. Biol. Med. 16, 195–200.

Nishikawa, T., Edelstein, D., Du, X.L., Yamagishi, S.-i., Matsumura, T., Kaneda, Y., Yorek, M.A., Beebe, D., Oates, P.J., Hammes, H.-P., 2000. Normalizing mitochondrial superoxide production blocks three pathways of hyperglycaemic damage. Nature 404, 787.

Ohishi, K., Carmines, P.K., 1995. Superoxide dismutase restores the influence of nitric oxide on renal arterioles in diabetes mellitus. J. Am. Soc. Nephrol. 5, 1559–1566.

O'Leary, B.R., Fath, M.A., Bellizzi, A.M., Hrabe, J.E., Button, A.M., Allen, B.G., Case, A.J., Altekruse, S., Wagner, B.A., Buettner, G.R., 2015. Loss of SOD3 (EcSOD) expression promotes an aggressive phenotype in human pancreatic ductal adenocarcinoma. Clin. Cancer Res. 21, 1741–1751.

Pacher, P., Beckman, J.S., Liaudet, L., 2007. Nitric oxide and peroxynitrite in health and disease. Physiol. Rev. 87, 315–424.

Papa, L., Manfredi, G., Germain, D., 2014. SOD1, an unexpected novel target for cancer therapy. Genes Cancer 5, 15.

Percival, S.S., Bowser, E., Wagner, M., 1995. Reduced copper enzyme activities in blood cells of children with cystic fibrosis. Am. J. Clin. Nutr. 62, 633–638.

Persichilli, S., Gervasoni, J., Di Napoli, A., Fuso, A., Nicolia, V., Giardina, B., Scarpa, S., Desiderio, C., Cavallaro, R.A., 2015. Plasma thiols levels in alzheimer's disease mice under diet-induced hyperhomocysteinemia: effect of s-adenosylmethionine and superoxide-dismutase supplementation. J. Alzheimers Dis. 44, 1323–1331.

Pham-Huy, L.A., He, H., Pham-Huy, C., 2008. Free radicals, antioxidants in disease and health. Int. J. Biomed. Sci. 4, 89.

Pisani, A., Sabbatini, M., Riccio, E., Rossano, R., Andreucci, M., Capasso, C., De Luca, V., Carginale, V., Bizzarri, M., Borrelli, A., Schiattarella, A., 2014. Effect of a recombinant manganese superoxide dismutase on prevention of contrast-induced acute kidney injury. Clin. Exp. Nephrol. 18, 424–431.

Robberecht, W., Philips, T., 2013. The changing scene of amyotrophic lateral sclerosis. Nat. Rev. Neurosci. 14, 248.

Robbins, D., Zhao, Y., 2014. Manganese superoxide dismutase in cancer prevention. Antioxid. Redox. Signal. 20, 1628–1645.

Romao, S., 2015. Therapeutic value of oral supplementation with melon superoxide dismutase and wheat gliadin combination. Nutrition 31, 430–436.

Rosenfeld, W., Evans, H., Concepcion, L., Jhaveri, R., Schaeffer, H., Friedman, A., 1984. Prevention of bronchopulmonary dysplasia by administration of bovine superoxide dismutase in preterm infants with respiratory distress syndrome. J. Pediatr. 105, 781–785.

Rottner, M., Tual-Chalot, S., Mostefai, H.A., Andriantsitohaina, R., Freyssinet, J.-M., Martínez, M.C., 2011. Increased oxidative stress induces apoptosis in human cystic fibrosis cells. PLoS One 6, e24880.

Salem, K., McCormick, M.L., Wendlandt, E., Zhan, F., Goel, A., 2015. Copper–zinc superoxide dismutase-mediated redox regulation of bortezomib resistance in multiple myeloma. Redox. Biol. 4, 23–33.

Salin, M.L., Wilson, W.W., 1981. Porcine superoxide dismutase. Isolation and characterization of a relatively basic cuprozinc enzyme. Mol. Cell. Biochem. 36, 157–161.

Salvemini, D., Cuzzocrea, S., 2002. Superoxide, superoxide dismutase and ischemic injury. Curr. Opin. Invest. Drugs 3, 886–895.

Salvemini, D., Riley, D.P., Cuzzocrea, S., 2002. SOD mimetics are coming of age. Nat. Rev. Drug Discov. 1, 367–374.

Salvemini, D., Wang, Z.Q., Zweier, J.L., Samouilov, A., Macarthur, H., Misko, T.P., Currie, M.G., Cuzzocrea, S., Sikorski, J.A., Riley, D.P., 1999. A nonpeptidyl mimic of superoxide dismutase with therapeutic activity in rats. Science 286, 304–306.

Savalia, K., Manickam, D.S., Rosenbaugh, E.G., Tian, J., Ahmad, I.M., Kabanov, A.V., Zimmerman, M.C., 2014. Neuronal uptake of nanoformulated superoxide dismutase and attenuation of angiotensin II-dependent hypertension after central administration. Free Radic. Biol. Med. 73, 299–307.

Schneider, H.-J., Rugendorff, E., Röhrborn, C., 1985. Pathogenesis, diagnosis and therapy of Induratio penis plastica (IPP). Int. Urol. Nephrol. 17, 235–244.

Shull, S., Heintz, N.H., Periasamy, M., Manohar, M., Janssen, Y., Marsh, J.P., Mossman, B.T., 1991. Differential regulation of antioxidant enzymes in response to oxidants. J. Biol. Chem. 266, 24398–24403.

Sinet, P.M., 1982. Metabolism of oxygen derivatives in Down's syndrome. Ann. N. Y. Acad. Sci. 396, 83–94.

Skarpanska-Stejnborn, A., Pilaczynska-Szczesniak, L., Basta, P., Deskur-Smielecka, E., Woitas-Slubowska, D., Adach, Z., 2011. Effects of oral supplementation with plant superoxide dismutase extract on selected redox parameters and an inflammatory marker in a 2,000-m rowing-ergometer test. Int. J. Sport. Nutr. Exerc.Metab. 21 (2), 124–134.

Smith, M.W., Doolittle, R.F., 1992. A comparison of evolutionary rates of the two major kinds of superoxide dismutase. J. Mol. Evol. 34, 175–184.

Somwar, R., Erdjument-Bromage, H., Larsson, E., Shum, D., Lockwood, W.W., Yang, G., Sander, C., Ouerfelli, O., Tempst, P.J., Djaballah, H., 2011. Superoxide dismutase 1 (SOD1) is a target for a small molecule identified in a screen for inhibitors of the growth of lung adenocarcinoma cell lines. Proc. Natl. Acad. Sci. 108, 16375–16380.

Somwar, R., Shum, D., Djaballah, H., Varmus, H., 2009. Identification and preliminary characterization of novel small molecules that inhibit growth of human lung adenocarcinoma cells. J. Biomol. Screen 14, 1176–1184.

Steinman, H.M., 1978. The amino acid sequence of mangano superoxide dismutase from Escherichia coli B. J. Biol. Chem. 253, 8708–8720.

Tanaka, K.I., Tamura, F., Sugizaki, T., Kawahara, M., Kuba, K., Imai, Y., Mizushima, T., 2017. Evaluation of lecithinized superoxide dismutase for the prevention of acute respiratory distress syndrome in animal models. Am. J. Respir. Cell Mol. Biol. 56, 179–190.

Teoh-Fitzgerald, M., Fitzgerald, M., Zhong, W., Askeland, R., Domann, F., 2014. Epigenetic reprogramming governs EcSOD expression during human mammary epithelial cell differentiation, tumorigenesis and metastasis. Oncogene 33, 358–368.

Teoh-Fitzgerald, M.L., Fitzgerald, M.P., Jensen, T.J., Futscher, B.W., Domann, F.E., 2012. Genetic and epigenetic inactivation of extracellular superoxide dismutase promotes an invasive phenotype in human lung cancer by disrupting ECM homeostasis. Mol. Cancer Res. 10, 40–51.

Thomas, R., Sharifi, N., 2012. SOD mimetics: a novel class of androgen receptor inhibitors that suppresses castration-resistant growth of prostate cancer. Mol. Cancer Ther. 11, 87–97.

Valko, M., Leibfritz, D., Moncol, J., Cronin, M.T., Mazur, M., Telser, J., 2007. Free radicals and antioxidants in normal physiological functions and human disease. Int. J. Biochem. Cell Biol. 39, 44–84.

Van Remmen, H., Williams, M.D., Guo, Z., Estlack, L., Yang, H., Carlson, E.J., Epstein, C.J., Huang, T.T., Richardson, A., 2001. Knockout mice heterozygous for Sod2 show alterations in cardiac mitochondrial function and apoptosis. Am. J. Physiol. Heart Circ. Physiol. 281, H1422–H1432.

Verges, J., Chateau, A., 1988. Nouveau traitement de la maladie de La Peyronie: la superoxide dismutase par ionisations: comparaison avec une série ancienne classique. Ann. urol. 22 (2), 143–144.

Voelkl, B., Vogt, L., Sena, E.S., Würbel, H., 2018. Reproducibility of preclinical animal research improves with heterogeneity of study samples. PLoS Biol. 16, e2003693.

Vozenin-Brotons, M.-C., Sivan, V., Gault, N., Renard, C., Geffrotin, C., Delanian, S., Lefaix, J.-L., Martin, M., 2001. Antifibrotic action of Cu/Zn SOD is mediated by TGF-β1 repression and phenotypic reversion of myofibroblasts. Free Radic. Biol. Med. 30, 30–42.

Warner, B.B., Stuart, L., Gebb, S., Wispe, J.R., 1996. Redox regulation of manganese superoxide dismutase. Am. J. Physiol. Lung Cell. Mol. Physiol. 271, L150–L158.

Weller, B.L., Crapo, J.D., Slot, J., Posthuma, G., Plopper, C.G., Pinkerton, K.E., 1997. Site- and cell-specific alteration of lung copper/zinc and manganese superoxide dismutases by chronic ozone exposure. Am. J. Respir. Cell Mol. Biol. 17, 552–560.

Whittaker, J.W., 2010. Metal uptake by manganese superoxide dismutase. Biochim. Biophys. Acta (BBA)-Proteins Proteom. 1804, 298–307.

Williams, M.D., Van Remmen, H., Conrad, C.C., Huang, T.T., Epstein, C.J., Richardson, A., 1998. Increased oxidative damage is correlated to altered mitochondrial function in heterozygous manganese superoxide dismutase knockout mice. J. Biol. Chem. 273, 28510–28515.

Yamakura, F., Kawasaki, H., 2010. Post-translational modifications of superoxide dismutase. Biochim. Biophys. Acta (BBA)-Proteins Proteom. 1804, 318–325.

Yamanaka, N., 1974. Superoxide dismutase activity in WI-38 cell cultures: effects of age, trypsinization, and SV-40 transformation. Physiol. Chem. Phys. 6, 95–106.

Yang, G., Chan, P.H., Chen, J., Carlson, E., Chen, S.F., Weinstein, P., Epstein, C.J., Kamii, H., 1994. Human copper-zinc superoxide dismutase transgenic mice are highly resistant to reperfusion injury after focal cerebral ischemia. Stroke 25, 165–170.

Yasui, K., Kobayashi, N., Yamazaki, T., Agematsu, K., Matsuzaki, S., Ito, S., Nakata, S., Baba, A., Koike, K., 2005. Superoxide dismutase (SOD) as a potential inhibitory mediator of inflammation via neutrophil apoptosis. Free Radic. Res. 39, 755–762.

Yasui, K., Shinozaki, K., Nakazawa, T., Agematsu, K., Komiyama, A., 1999. Presenility of granulocytes in Down syndrome individuals. Am. J. Med. Genet. 84, 406–412.

Yoo, H., Chang, M., Rho, H., 1999. Heavy metal-mediated activation of the rat Cu/Zn superoxide dismutase gene via a metal-responsive element. Mol. Gen. Genet. 262, 310–313.

Zelko, I.N., Mariani, T.J., Folz, R.J., 2002. Superoxide dismutase multigene family: a comparison of the CuZn-SOD (SOD1), Mn-SOD (SOD2), and EC-SOD (SOD3) gene structures, evolution, and expression. Free Radic. Biol. Med. 33, 337–349.

Zhang, H., Joseph, J., Felix, C., Kalyanaraman, B., 2000. Bicarbonate enhances the hydroxylation, nitration, and peroxidation reactions catalyzed by copper, zinc superoxide dismutase intermediacy of carbonate anion radical. J. Biol. Chem. 275, 14038–14045.

Zhao, Y., Vanhoutte, P.M., Leung, S.W., 2015. Vascular nitric oxide: beyond eNOS. J. Pharmacol. Sci. 129, 83–94.

Zuo, L., Hemmelgarn, B.T., Chuang, C.-C., Best, T.M., 2015. The role of oxidative stress-induced epigenetic alterations in amyloid-β production in Alzheimer's disease. Oxid. Med. Cell. Longev. 2015, 604658.

CHAPTER

Uric acid

2.8

Mohamad Fawzi Mahomoodally, Daphne Désiré A.-L., Sanaa Dilmar A., Elodie Rosette M. A.-L.

Department of Health Sciences, Faculty of Medicine and Health Sciences, University of Mauritius, Réduit, Mauritius

2.8.1 Introduction

Uric acid can be categorized as an antioxidant, a powerful peroxynitrite as well as reactive oxygen species (ROS) scavenger (El Ridi and Tallima, 2017). This weak acid, a $C_5H_4N_4O_3$ heterocyclic compound (Bartoli et al., 2018), is formed as the end product of the metabolism of purine nucleotides. Ingestion, endogenous synthesis of purine from non-purine precursors and the reutilization of preformed compounds are all potential sources of purine nucleotides in the body (Pasilic et al., 2012).

Uric acid is the final product of the breakdown of purine nucleotide by enzymes in the body. The major source of purines is from its ingestion from foods like meat, fish and alcoholic beverages besides those endogenously produced by the body from non-purines precursors and those that can be reused in the form of preformed purine compounds (Maiuolo et al., 2019).

After ingestion, nucleoproteins are broken down by pepsin in the stomach producing nucleins which will be further broken down in the small intestine by the enzyme trypsin into nucleic acid complexes. Depolymerase and ribonucleases are present in the small intestine which act on the nucleic acid complexes till the formation of adenosine monophosphate (AMP) and guanosine monophosphate (GMP) which are mononucleotides (Maiuolo et al., 2019).

Nucleotidase, first, removes a phosphate group from AMP forming adenosine which, in turn, is deaminated to form inosine. Nucleotidase also acts on GMP converting it to guanosine. By purine nucleoside phosphorylase, inosine and guanosine are then converted into purine base, which are hypoxanthine and guanine (Maiuolo et al., 2019). Hypoxanthine undergoes oxidation by xanthine oxidase, whereas guanine is deaminated by guanine deaminase, both forming xanthine. Furthermore, xanthine oxidase oxidises xanthine resulting in the formation of uric acid which is the final product of purine degradation (Maiuolo et al., 2019).

Antioxidants Effects in Health. DOI: https://doi.org/10.1016/B978-0-12-819096-8.00028-8

2.8.2 Antioxidant effect of uric acid

The antioxidant effect of uric acid was studied by many scientists but more in-depth by Ames et al. (1981) under certain conditions, where they found that exogenously added uric acid is effective in protecting cells against oxidants. From these studies we could infer that uric acid can scavenge radicals formed from various reaction in the blood like those produced by the self-oxidation of hemoglobin or peroxide production by macrophages. However, even in the plasma uric acid can prevent lipid peroxidation only as long as ascorbic acid is present (Sautin and Johnson, 2008).

High amount of uric acid was found to have a protective effect on the central nervous system against conditions like multiple sclerosis, Parkinson's disease, and acute stroke. In an experiment involving a mouse model of allergic encephalomyelitis, the administration of uric acid ceased the action of peroxynitrite (ONOO–) mediated nitrosylation of neuronal proteins and halted their increase in the blood-brain barrier, thus diminishing the infiltration of leukocytes compared to ascorbate,which did not block this action (Sautin and Johnson, 2008).

Kuzkaya et al. (2005) found that uric acid cannot scavenge superoxide radicals, whereas peroxynitrite can be scavenged in a unique way in extracellular space; however, this action needs the presence of ascorbic acid and thiols to be completed or else peroxynitrite will react with tetrahydrobiopterin and other compounds resulting in the uncoupling of NO synthase and thus making uric acid lose its antioxidant effect. Another mechanism responsible for preventing oxidative damage of proteins is the nitrosylation of tyrosine. It was also found that the antioxidant effect of bicarbonate deactivates the antioxidant effect of uric acid. (Sautin and Johnson, 2008).

In hydrophilic environment, uric acid can strongly scavenge carbon-centered and peroxyl radicals but not lipophilic radicals .This is because it cannot stop the radical chain reaction occurring within lipid membranes, which causes oxidative damage. The antioxidant effect of uric acid can occur only in hydrophilic environment like plasma and can be explained by favourability of this condition for tyrosine nitration and the ability of peroxynitrous radicals to be extremely diffusible across membranes (Sautin and Johnson, 2008).

2.8.3 Pro-oxidant activity of uric acid

The major limitation of the antioxidant ability of uric acid is that it occurs only in hydrophilic environment and is restricted to the plasma, whereas its pro-oxidant activity happens within the cells. The pro-oxidant activity of uric acid arises when it reacts with oxidants resulting in the formation of other radicals that could amplify the propagation of radical chain reaction and cause oxidative damage to cells. These formed radicals are more prone to target lipids like low density lipids (LDL) and membranes rather than components of cells. Uric acid, in the presence of copper ions and lipid hydroperoxide, can further oxidises LDL. However, the oxidation of these lipids can create a hydrophobic medium which is not favourable for the antioxidant activity of uric acid and at the same time convert uric acid to an oxidant (Sautin and Johnson, 2008).

The antioxidant function of uric acid was found to be effective at physiological concentration, which is 300 μM, but at 500 μM concentration of urate, oxidation of LDL and liposomes occurs due to peroxynitrite as a result of oxidative damage. High concentration of uric acid was also found to be strongly related to insulin resistance, type 2 diabetes, hypertension, kidney diseases, dyslipidaemia, visceral obesity and other cardiovascular and cerebrovascular complications. Although the pathogenesis of these diseases is complex in nature, it was discerned that they were all accompanied with oxidative stress and proteins and lipids that were modified due to oxidation (Sautin and Johnson, 2008). It should be noted that that there is a thin line between the antioxidant and pro-oxidant activity of uric acid which is determined by the micro-environment where its reaction occurs and availability of lipid hydro peroxides (Kang and Ha, 2014).

2.8.4 Beneficial effects of uric acid

2.8.4.1 Antioxidant properties

Uric acid forms part of the most important antioxidants found in the human biological fluids. High amount of serum uric acid passes through the kidney glomeruli where it is filtered and about 90% of the filtered uric acid is absorbed again which demonstrates that it has an ample physiological role. More than half of the capacity of blood plasma to act as an antioxidant comes from uric acid. Indeed, it is a good antioxidant, peroxynitrite scavenger and a powerful reactive oxygen species (ROS) scavenger. Elevated amount of uric acid can be easily found in the human cytosol and cells of mammals, particularly in the liver, nasal secretions of humans and vascular endothelial cells, where it acts as an antioxidant (El Ridi and Tallima 2017).

2.8.4.2 Endothelial functions

A study carried out by Oberbach et al. (2014) have documented the capacity of uric acid to cause harm to the vascular endothelial cells integrity. However, a current report pointed out that low levels of serum uric acid can lead to the loss of function mutations of SLC22A12, which encodes blood vessels and URAT1 in the kidney proximal tubular cells transporters, resulting in the malfunctioning of the endothelial cells *in vivo* (Sugihara et al., 2015). Uric acid may play a crucial role in healing tissues by initiating inflammatory processes needed for repairing of tissues, scavenging oxygen free radicals and mobilizing endothelial cells precursors (Nery et al., 2015).

2.8.4.3 Booster of type 2 immune responses

Allergens, such as cysteine proteases, cysteine peptidases papain and bromelin are capable to stimulate barrier epithelial cells to produce type 2 cytokines such as thymic stromal lymphopoietin (TSLP), interleukin (IL)-25, and IL-33, which are responsible for directing the immune environment to the type 2 axis and hypersensitive inflammation. Hara et al. reported in 2014 that allergens and cysteine peptidases, such as papain can cause stress and damage tissue cells leading to the release of uric acid which, in turns, activates epithelial cells to release thymic stromal lymphopoietin (TSLP) and IL-33. Uric acid has been recognized as the principal element that monitors the development of type 2 immune responses to cysteine peptidase allergens.

2.8.4.4 Defence against neurological and autoimmune diseases

In patients diagnosed with multiple sclerosis, a low level of uric acid in plasma may result in a decrease in the amount of antioxidant molecules. There is a belief that peroxynitrites and ROS play an important role in the degradation of myelin in multiple sclerosis which can be prevented by high levels of uric acid despite it has been reported that patients suffering from gout rarely have multiple sclerosis. Many studies have shown the positive correlation of having a low uric acid serum level with an increased risk of having multiple sclerosis disease (Drulovie et al., 2001; Sotgiu et al., 2002; Rentos et al., 2006).

In addition, a low level of uric acid in plasma has been correlated with disorders of the neurological system, Parkinson's disease and Alzheimer's disease. It has been observed that a diet rich in purine content lowers the risk of Parkinson's disease and helps to slow down the progression of symptoms (Álvarez-Lario and MacArrón-Vicente, 2011). Furthermore, it has been linked to pemphigus vulgaris, an auto immune disorder which can be identified by the presence of blisters and sores of the skin and mucous membrane (Yousefi et al., 2011). Bakhtiari et al. reported in 2017 that low levels of uric acid in the saliva leads to an autoimmune disorder known as lichen planus which results in the inflammation of the mucocutaneous tissue.

2.8.5 Increasing the bioavailability of uric acid

Uric acid is the end product of purine metabolism, mostly derived from endogenous synthesis, but a minor part also arises from exogenous sources such as foods with purine content, alcohol, and fructose drinks. Uric acid is synthesized mainly in the liver and intestines but is also synthesized in other tissues, such as muscles, kidneys, and the vascular endothelium (Jakse et al., 2019). The bioavailability of uric acid can be influenced by the intake of certain compounds obtained naturally from certain food items, such as olives (hydroxytyrosol) (Wan et al., 2018), lemons (Wang et al., 2017) and some medicinal plants such as *Barringtonia racemosa* L (Osman et al., 2016).

2.8.5.1 Hydroxytyrosol

In vivo and in vitro studies carried by Wan et al. (2018) investigated effect of hydroxytyrosol, a phenolic compound naturally present in the form of esterified oleuropein in olives, as an inhibitor of xanthine oxidase (XO). The results showed that hydroxytyrosol lowered the level of uric acid found in rat serum, inhibited XO activity and changed mRNA transcription of certain uric acid transporters towards their normal transcription level. It was, therefore, concluded that hydroxytyrosol could decrease the level of uric acid in rats suffering from hyperuricemia and it helps in adjusting the mRNA genes of the transporter of renal uric acid.

2.8.5.2 Lemons

In addition, the extent to which lemon fruit juice and/or water-soluble extracts can be used to lower the level of blood uric acid was determined in humans. During six weeks, humans suffering from elevated levels of uric acid were given 30 ml/day of the juice of one freshly squeezed lemon. The serum samples of the participants were taken at different time intervals for biochemical assessments and fasting blood samples were collected for blood tests at the end of the clinical study. Results showed that the levels of serum uric acid were considerably lowered after the intake of lemon fruit juice and/or water-soluble extracts. Furthermore, no cases of renal or liver malfunction was accounted. It was, therefore, concluded that lemon might play a role in decreasing the serum level of uric acid independent of XO inhibition (Wang et al., 2017).

2.8.5.3 *Barringtonia racemosa* L

In vitro studies carried out by Osman et al in 2016 assessed the anti-inflammatory effects and total phenolic content (TPC) of different parts of *B. racemosa*, a plant commonly found in certain tribes. The study was carried out by determining its anti-inflammatory potential through denaturation inhibition assays of albumin and XO. The TPC in the plant extracts were determined by Folin-Ciocalteu assay. As a result, the in vitro anti-inflammatory activities of *B. racemose* were confirmed and its potential to be used in alleviating gouty arthritis and XO-related diseases was confirmed. It was, then, concluded that *B. racemosa* has *in vitro* anti-inflammatory activities and can help to relieve gouty arthritis and diseases related to XO.

2.8.6 Detrimental effects of uric acid

Even though uric acid has many beneficial antioxidant effects, in an excessive amount, it can also have destructive effects on the body, which are discussed below.

2.8.6.1 Life expectancy in breast cancer patients

Yue et al. (2017) investigated the serum uric acid concentration in patients suffering from breast cancer .They concluded that breast cancer patients with elevated serum uric acid concentration have a higher mortality rate. Those researchers also discovered that many factors are responsible for this increase in serum uric acid concentration, notably obesity, an elevated BMI in conjunction with increasing age. They suspected that a raise in uric acid may encourage both carcinogenesis and tumorigenesis.

2.8.6.2 Eclampsia

Paula et al. (2018) studied women suffering from both pre-eclampsia and eclampsia.

For this retrospective cohort study, a total of 733 pregnant women suffering from hypertension were considered, from which 329 women were diagnosed with pre-eclampsia without eclampsia associated and 45 subjects had developed eclampsia during the course of the study. The majority of patients were suffering from pre-eclampsia symptoms, notably an increased blood pressure and proteinuria after twenty weeks of pregnancy. They were all admitted to the São Lucas Hospital or the Pontifical Catholic University of Rio Grande de Sul and this study was carried out from January 2005 till September 2010.

Paula et al. (2018) concluded that a raise in the level of uric acid, usually above 5.9 mg/dl along with proteinuria are strongly linked to the increase risk of developing eclampsia. From another study done by Thangaratinan et al in 2006, it was identified that an increased serum uric acid can almost double the risk of eclampsia in pregnant women.

2.8.6.3 Gout

Hyperuricaemia is a condition in which there is an imbalance between the production and excretion of uric acid in the body. This excess uric acid will supersaturate the body fluids causing the deposition of monosodium urate crystals (MSU) in the tissues and joints. Those MSU crystals will lead to the formation of tophi (deposits of MSU crystals in the joints of hyperuricemia patients) which will cause inflammation, damage to soft tissues and joints along with the debut of acute gouty attacks. Gout is a painful ailment associated with a rich diet which occurs mainly between the ages of 30 to 50 years old. However, after 80 years old, women are more affected by gout than men.

From a previous study, it was seen that male participants who have a serum uric acid level greater than 9.0 mg/dL were more at risk of developing gout. Alternatively, it was discovered that people older than 75 years of age have a 4.1% prevalence of suffering from hyperuricaemia.

In 2014, Pineda et al., 2015 analyzed the effect of MSU crystals on the knees of 21 male adult, white New Zealand rabbits with a weight of approximately 2 kg. The environment and diet of those subjects were carefully controlled prior to the study. Synthetic MSU and allopurinol crystals were prepared according to the Denko and Whitehouse method. Those MSU crystals were in a needle-shape, uniform size incorporated in a 1 ml suspension and they were injected in 42 rabbit knees. Consequently, Pineda et al. found that those MSU crystal undoubtedly induced an acute joint inflammatory process in the rabbit knee model. But also, early morphostructural changes in the rabbit knee model were observed during an acute gouty attack.

2.8.6.4 Behavioral and mental disorders

According to the results obtained from three studies previously conducted, Bartoli et al. (2018) attested that at univariate level, a connection was observed between serum uric acid and suicidal ideation, psychological distress, verbal aggression as well as a history of violence. Nonetheless, another research demonstrated verbal aggression as the sole factor responsible for predicting future violent behaviours (de Girolamo et al., 2016).

Along with all the destructive consequences of uric acid on one's body, it can also be held responsible for an increase in aggressivity among youths (Mrug and Mrug, 2016). Previous studies have demonstrated the relationship between elevated levels of uric acid and the development of impulsive and disinhibited behaviour (Lorenzi et al., 2010; Sutin et al., 2014). A longitudinal study was performed on African American adolescents who presented symptoms of hyperuricaemia so as to test the hypothesis mentioned above. Furthermore, genetic disorders of purinergic metabolism, leading to an overproduction of uric acid, may incite the development of behavioural abnormalities like impulsive, aggressive and self-injurious behavior (Fu et al., 2014).

In patients suffering from bipolar disorders, it was observed that uric acid levels were higher during manic episodes and psychotropic medications were not the cause. Mrug and Mrug, 2016 deducted from their study that "higher 12-h excretion of uric acid predicted more frequent last-month aggression reported 1.5 years later." Even if this effect did not vary between genders, it was seen that this association was slightly stronger in females.

This finding can, therefore, be used as a new scope to help identify adolescents who are at risk of promoting increased aggressive behavior overtime. Controlling one's diet and medications can help lower uric acid level, thus reducing aggressive and other disinhibited behavior in youngsters. However, more research should be performed on this study to better understand the contributing factors and causal mechanisms influencing the connections between the level of uric acid and aggression in adolescents.

Conclusion

Uric acid has been reported to be a peroxynitrite and a reactive oxygen species scavenger formed at the end of purine metabolism. This antioxidant can be obtained mainly from food sources and its bioavailability might be altered by the intake of certain compounds obtained naturally from certain food items, such as olives (hydroxytyrosol), lemons and certain plants as *Barringtonia racemosa*. Uric acid can also act as a pro-oxidant whereby it mediates the generation of radicals leading to oxidation of molecules such as proteins and lipids. A high amount of uric acid has been linked to positive effect on the neurological system and provide defence against autoimmune problems, such as multiple sclerosis. In addition, it boosts up type two immune responses and enhances endothelial functions. However, uric acid is also known to have detrimental effects on health, such as increasing the risk of eclampsia, gout, affective and aggressive disorders.

References

Álvarez-Lario, B., MacArrón-Vicente, J., 2011. Is there anything good in uric acid? Int. J. Med. 104, 1015–1024.

Ames, B., Cathcart, R., Schwiers, E., Hochstein, P., 1981. Uric acid provides an antioxidant defense in humans against oxidant- and radical-caused aging and cancer: a hypothesis. Proc. Natl. Acad. Sci. 78, 6858–6862.

Bakhtiari, S., Toosi, P., Samadi, S., Bakhsi, M., 2017. Assessment of uric acid level in the saliva of patients with oral lichen planus. Med. Princ. Pract. 26, 57–60.

Bartoli, F., Crocamo, C., Bava, M., Castagna, G., Dibrita, C., Ribolidi, I., Trotta, G., Verrengia, E., Clerici, M., Carrà, G., 2018. Testing the association of serum uric acid levels with behavioral and clinical characteristics in subjects with major affective disorders: A cross-sectional study, Italy. Psychiatry Res. 269, 118–123.

de Girolamo, G., Buizza, C., Sisti, D., Ferrari, C., Bulgari, V., Iozzino, L., Boero, M.E., Cristiano, G., De Francesco, A., Giobbio, G.M., Maggi, P., Rossi, G., Segalini, B., Candini, V., Andreose, S., Basso, P., Beneduce, R., Bertolotti, P., Braida, V., Bonelli, M., Bongiorno, F., Bussi, R., Castagro, E., Dominicis, F., Ghersi, L., Greppo, S., Sodano, A.J., Lo Presti, E., Milone, V., Panigada, F., Pasquadibisceglie, L., Rigamonti, D., Rillosi, L., 2016. Monitoring and predicting the risk of violence in residential facilities. No difference between patients with history or with no history of violence, Italy. J Psychiatr. Res. 80, 5–13.

Drulovic, J., Dujmovic, I., Stojsavljevic, N., Mesaros, S., Andjelkovic, S., Miljkoovic., D., 2001. Uric acid levels in sera from patients with multiple sclerosis. J. Neurol. 248, 121–126.

El Ridi, R., Tallima, H., 2017. Physiological functions and pathogenic potential of uric acid: a review. Egypt. J. Adv. Res. 8, 487–493.

Fu, R., Chen, C., Jinnah, H.A., 2014. Genotypic and phenotypic spectrum in attenuated variants of Lesch- Nyhan disease, United States of Amercia. Mol. Genet. Metab. 112, 280–285.

Hara, K., Iijima, K., Elias, M.K., Seno, S., Tojima, I., Kobayashi, T., 2014. Airway uric acid is a sensor of inhaled protease allergens and initiates type 2 immune responses in respiratory mucosa. J. Immunol. 192, 4032–4042.

Jakše, B., Jakše, B., Pajek, M., Pajek, J., 2019. Uric acid and plant-based nutrition. Nutrients 11 (8), 1736.

Kang, D., Ha, S., 2014. Uric acid puzzle: dual role as anti-oxidantand pro-oxidant. Electrolyte Blood Press. 12, 1–6.

Kuzkaya, N., Weissmann, N., Harrison, D.G., Dikalov, S., 2005. Interactions of peroxynitrite with uric acid in the presence of ascorbate and thiols: implications for uncoupling endothelial nitric oxide synthase. Biochem. Pharmacol. 70, 343–354.

Lorenzi, T.M., Borba, D.L., Dutra, G., Lara, D.R., 2010. Association of serum uric acid levels with emotional and affective temperaments, Brazil. J. Affect. Disord. 121, 161–164.

Maiuolo, J., Oppedisano, F., Gratteri, S., Muscoli, C., Mollace, V., 2019. Regulation of uric acid metabolism and excretion, Italy. Int. J. Cardiol. 213, 8–14.

Mrug, S., Mrug, M., 2016. Uric acid excretion predicts increased aggression in urban adolescents, United States. Physiol. Behav. 163, 144–148.

Nery, R.A., Kahlow, B.S., Skare, T.L., Tabushi, F.I., Amarale Castro, A., 2015. Uric acid and tissue repair. Brazil. Arq. Bras. Cir. Dig. 28, 290–292.

Oberbach, A., Neuhaus, J., Jehmlich, N., Schlichting, N., Heinrich, M., Kullnick., Y., 2014. A global proteome approach in uric acid stimulated human aortic endothelial cells revealed regulation of multiple major cellular pathways. Int. J. Cardiol. 176, 746–752.

Osman, N.I., Sidik, N.J., Awal, A., Adam, N.A.M., Rezali, N.I., 2016. *In vitro* xanthine oxidase and albumin denaturation inhibition assay of *Barringtonia racemosa* L. and total phenolic content analysis for potential anti-inflammatory use in gouty arthritis, Malaysia. J. Intercult. Ethnopharmacol. 5, 343–349.

Pasalic, D., Marinkovic, N., Feher-Turkovic, L., 2012. Uric acid as one of the important factors in multifactorial disorders – facts and controversies, Croatia. Biochem. Med. 22, 63–75.

Paula, G., L, P.C., E, B., Hentschke, R., M., A., C, I., Luz, H.J., Filho, C., V, E., Poli-de-Figueiredo, E.C., 2018. Increased proteinuria and uric acid levels are associated with eclamptic crisis, Brazil. Pregnancy Hypertens. 15, 93–97.

Pineda, C., Fuentes-Gómez, A.J., Hernández-Dias, C., Zamudio-Cuevas, Y., Fernández-Torres, J., Ventura, L., Gómez-Quiróz, L.E., Gutiérrez, M., López-Reyes, A., 2015. Animal model of acute gout reproduces the inflammatory and ultrasonographic joint changes of human gout, Mexico. Arthritis Res. Ther 17, 37.

Rentzos, M., Nikolaou, C., Anagnostouli, M., Rombos, A., Tsakanikas, K., Economou, M., Dimitrakopoulos, A., Karouli, M., Vassilopoulous, D., 2006. Serum uric acid and multiple sclerosis. Clin. Neurol. Neurosurg. 108, 527–531.

Sautin, Y., Johnson, R., 2008. Uric acid: the oxidant-antioxidant paradox. Nucleosides. Nucleotides. Nucleic. Acids. 27, 608–619.

Sotgiu, S., Pugliatti, M., Samma, A., Sotguiu, A., Fois, M.L., Arru, G., Rosati, G., 2002. Serum uric acid and multiple sclerosis. Neurol. Sci. 23, 183–188.

Sugihara, S., Hisatome, J., Kuwabara, M., Niwa, K., Kato, M., 2015. Depletion of uric acid due to SLC22A12 (URAT1) loss-of-function mutation causes endothelial dysfunction in hypouricemia. Circ. J. 79, 1125–1132.

Sutin, A.R., Cutler, R.G., Camandola, S., Uda, M., Feldman, N.H., Cucca, F., Zonderman, A.B., Mattson, M.P., Ferrucci, L., Schlessinger, D., Terracciano, A., 2014. Impulsivity is associated with uric acid: Evidence from humans and mice, Florida. Biol. Psychiatry 75, 31–37.

Wan, Y., Liang, Y.X., Zal, B., Fu, G.M., Xie, M.Y., 2018. The possible mechanism of hydroxytyrosol on reducing uric acid levels, China. J. Funct. Foods 42, 319–326.

Wang, H., Cheng, L., Lin, D., Ma, Z., Deng, X., 2017. Lemon fruits lower the blood uric acid levels in humans and mice. Sci. Hort. 220, 4–10.

Yousefi, M., Rahimi, H., Barikbin, B., Toossi, P., Lotfi, S., Hedayati, M., Younespour, S., 2011. Uric acid: a new antioxidant in patients with pemphigus vulgaris, Iran. Indian J. Dermatol. 56, 278–281.

Yue, C., Feng, P., Yao, Z., Yu, X., Lin, W., Qian, Y., Guo, Y., Li, L., Liu, M., 2017. High serum uric acid concentration predicts poor survival in patients with breast cancer, China. Clin. Chim. Acta. 473, 160–165.

PART

3 Synthetic antioxidants: bright and the dark side

CHAPTER

Ascorbyl palmitate

3.1

Renald Blundell[a,b], Muhammad Ajmal Shah[c], Joseph I. Azzopardi[a], Amira Y. Benmelouka[d], Mohammed Alqarni[e], Haroon Khan[f]

[a]*Department of Physiology and Biochemistry, Faculty of Medicine and Surgery, University of Malta, Msida, Malta*

[b]*Centre for Molecular Medicine and Biobanking, University of Malta, Msida, Malta*

[c]*Department of Pharmacognosy, Faculty of Pharmaceutical Sciences, Government College University, Faisalabad, Pakistan*

[d]*Faculty of Medicine, University of Algiers, Algiers, Algeria*

[e]*Department of Pharmaceutical Chemistry, College of Pharmacy, Taif University, Taif, Saudi Arabia*

[f]*Department of Pharmacy, Abdul Wali Khan University, Mardan, Pakistan*

3.1.1 Chemistry

Ascorbyl palmitate ([(2S)-2-[(2R)-3,4-dihydroxy-5-oxo-2H-furan-2-yl]-2-hydroxyethyl]hexadecanoate) is the product of an esterification reaction between the carboxylic group of palmitic acid and the primary alcohol group of ascorbic acid. The chemical structure is shown in Fig. 3.1.1. This compound is found as with a white or yellowish white powder with an odour of citrus and a melting range of 107–117°C. It has a high solubility in ethanol and in water but it is insoluble in fat (EFSA, 2016; Igoe, 2011).

3.1.2 Synthesis

Ascorbyl palmitate can be synthesised enzymatically such as through a reaction of esterification mediated by a lipase, such as *Bacillus stearothermophilus* SB 1 lipase or via a reaction of esterification between palmitic acid and ascorbic acid using a catalyst like concentrated sulfuric acid, after which the product can be extracted and then purified by recrystallization (EFSA, 2016). Industrial synthesis can be also done using another catalyst called hydrofluoric acid (Ji, 2011). Enzyme biocatalysis, which is the more recent approach, can offer many benefits in comparison to the use of catalysts. One of these benefits is the lower formation of byproduct (Humeau et al., 1995). However, this method imposes a major burden which is related to the difficulty of attaining a balance between the enzymatic activity and the solubility of the substrate (Tufiño et al., 2019).

Antioxidants Effects in Health. DOI: https://doi.org/10.1016/B978-0-12-819096-8.00031-8

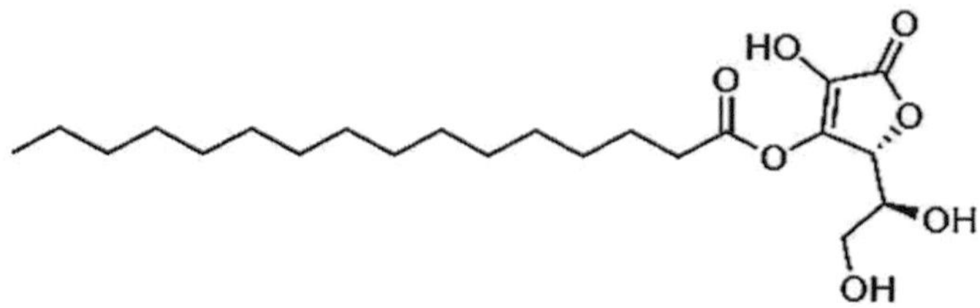

FIG. 3.1.1 The chemical structure of ascorbyl palmitate.

3.1.3 Legal status

In the EU, ascorbyl palmitate – given the E number E 304 – is allowed for use as a food additive at maximum permitted level (MPL) of *quantum satis* (i.e., as much is sufficient) except in foods targeted to young children and infants. For other categories of food, the MPL is based on the fat basis of the said food product (EFSA, 2016); the MPL for all the different foods covered in "Annex II of Regulation (EC) No 133/2008) are shown in Table 3.1.3. In the US, the FDA allows the administration of ascorbyl palmitate as a preservative in foods and lists it as 'generally recognized as safe' (GRAS) (FDA, 2018).

Table 3.1.3 The maximum permitted levels of ascorbyl palmitate in different categories of foods as set out in the Annex II of Regulation (EC) No 1333/2008 and its amendments.

Food category	Maximum permitted level (mg/L or mg/kg as appropriate)
Unflavored fermented products, heat treated after fermentation Products made of flavored fermented milk, including heat-treated ones Dried milk Other creams Unripened cheese (except mozzarella)[a] Processed cheese Cheese products[a] Dairy analogues; beverage whiteners included	*Quantum satis*
Oils and fats which are essentially free from water (except virgin oils, anhydrous milk fat, and olive oils) Other oil and fat emulsions; spreads and liquid emulsions included Vegetable oil pan spray Edible ices Dried vegetables and fruit Vegetables and fruits in vinegar, brine, or oil	*Quantum satis*

(continued)

Table 3.1.3 The maximum permitted levels of ascorbyl palmitate in different categories of foods as set out in the Annex II of Regulation (EC) No 1333/2008 and its amendments. *Continued*

Food category	Maximum permitted level (mg/L or mg/kg as appropriate)
Vegetable and fruit preparations, except compote	
Nut butters and but spreads	
Processed potato products	
Chocolate products and cacao (as included in Directive 2000/36/EC)	
Other confectioneries	
Chewing gum	
Decorations, fillings and coatings (excluding fruit-based fillings that fall in category 4.2.4)	
Starches	
Breakfast cereals	
Dry pasta	
Potato gnocchi	
Fillings of stuffed pasta	
Noodles	
Batter	
Processed or pre-cooked cereals	
Bread prepared using only: salt, water, wheat four, leaven or yeast	
Pain courant francais; fehér és félbarna kenyerek, Friss búzakenyér	*Quantum satis*
Rolls and bread	
Fine bakery wares	
Non heat treated processed meat	
Heat treated processed meat	
Casings, coatings and decorations for meat	
Fishery products and processed fish, crustaceans included	
Processed fish roe	
Egg products and processed eggs	
Other syrups and sugars	
Salt substitutes	
Condiments and seasonings	
Vinegars	
Mustard	
Broths and soups	
Sauces	
Savoury and salads – based sandwich spreads	
Yeast and yeast products	
Protein products (exceot category 1.8 products)	

Table 3.1.3 The maximum permitted levels of ascorbyl palmitate in different categories of foods as set out in the Annex II of Regulation (EC) No 1333/2008 and its amendments. *Continued*

Food category	Maximum permitted level (mg/L or mg/kg as appropriate)
Dietary foods reserved for special medical use (as defined in Directive 1999/21/EC; categories 13.1.5 excluded) Dietary foods that replace total daily food take or a person's meal and that are produced for weight control Foods suitable produced for patients with gluten intolerance (as defined by Regulation EC 41/2009)	
Fruit juices[b] and vegetable juices Vegetable and fruit[b] nectars and related products Flavored drinks Other Perry and cider Fruit wine and made wine Spirit beverages (as defined in Regulation (EC) No 110/2008) Aromatized wines Aromatized wine based beverages Aromatized wine cocktails Other alcoholic beverages, including mixtures of alcoholic drinks with spirits containing less than 15% of alcohol and nonalcoholic drinks Snacks made from cereal, potato, starch, or flour Processed nuts Desserts (except products from categories 1, 3, and 4) Solid food supplements, including capsules and tables and similar fors (chewable forms are excluded) Liquid food supplements Food supplements with a chewable or a syrup-type form Processed foods that are not included in categories 1 to 17 (except foods for young children and infants)	*Quantum satis*
Infant formulae[c] Follow-on formulae[c]	10
Baby foods and processed cereal-based foods for infants and young children[c] Other food products designed for young children Infants dietary foods that are designed for special medical uses and special infants formulae	100

[a]*excluding category 16 products*
[b]*defined by Council Directive 2001/112/EC*
[c]*defined by Directive 2006/141/EC*

3.1.4 Mechanism of action

Ascorbyl palmitate is able to function as a chain-breaking antioxidant because of its hydrogen-donating potential (Perricone et al., 1999). While it can act independently as a free radical scavenger, it can also regenerate tocopherol, another antioxidant, from its oxidized state (Cort, 1974; Frankel et al., 1994; Let et al., 2007; Marinova and Yanishlieva, 1992). It can limit lipid peroxidation via blocking the activity of lipoxygenases (Mohamed et al., 2014). Additionally, it decreases ionizing radiation-induced lipid peroxidation and protein carbonylation. It also reduces radiation-induced caspase-3 upregulation and stops pycnosis and karyorrhexis (Xiao and Miwa, 2017). Moreover, it has an anticarcinogenic capacity, which is mediated by the downregulation of DNA production in proliferating malignant cells (Kageyama et al., 1999). Further, it alters the production of a broad spectrum of reactive oxygen species and increases cytotoxicity in tumor cells (Miwa et al., 1988). Finally, it can interact with cytochrome P4503A4 (CYP3A4) and inhibit the oxidation of some medications (Dresser et al., 2002).

3.1.5 Effects on health

Reports about the health effects of ascorbyl palmitate are scarce. Toxicity studies showed that the medium lethal dose (LD50) of ascorbyl palmitate was in the range of 4700 to more than 2×10^4 mg/kg b.w and 5150 to more than 10^4 mg/kg b.w in mice and rats, respectively (Bächtold, 1972, 1973a, 1973b). In a more recent assessment, the LD50 was found to be 2×10^3 mg/kg b.w and $>5 \times 10^3$ mg/kg b.w in mice and rats, respectively (EFSA, 2016). A bacterial mutation assay found no mutagenic activity in palmitoyl ascorbic acid at a dose level fluctuating between 0.033 and 3.3 mg (Priva et al., 1991). Also, no reports were published of dermal irritation or sensitization caused by ascorbyl palmitate in clinical studies (Elmore, 2005).

In male MF1 mice fed 600 mg/kg of acetaminophen with the intention of inducing hepatotoxicity, the coadministration of ascorbyl palmitate prevented the 35% mortality at 24 h which was previously observed in rats that were treated with acetaminophen alone. Ascorbyl palmitate protected the liver of the mice by eliminating reactive acetaminophen metabolites and by decreasing hepatic glutathione levels (Jonker et al., 1988). In another study, ascorbyl palmitate showed the same hepatoprotective activities as shown in the aforementioned study, as well as an antipyretic effect, 15 and 30 min after the administration of the dose (Mitra et al., 1988). Furthermore, it can decrease the deamidation of gliadin by transglutaminase 2. Therefore, it can be administered as an additive component in foods delivered to celiac disease patients (Engstrom et al., 2017). In immunology, the use of liquid crystals made of ascorbyl palmitate as a vaccine adjuvant was found to provide a slow release of the antigen both in vitro and in vivo. It can also enhance the efficacy of the vaccine *via* multiple pathways, including the activation of the MyD88 signalling mechanism and the improvement of cytokine release (Sanchez Vallecillo et al., 2015). In addition to that, channel blockers are widely used in the treatment of cardiac and extracardiac diseases. Their extended use is associated with the disruption of fibrillary collagen deposition in the vascular wall structure (Ivanov et al., 2016; Roth et al., 1996). Ascorbyl palmitate could reduce the adverse effects of calcium, potassium, and sodium channel blockers on the formation of the extracellular matrix (Ivanov et al., 2016). These findings highlight the benefits of ascorbate palmitate supplementation to minimize the vascular adverse effects of these medications, especially in patients who are taking them chronically.

In vitro, ascorbyl palmitate was also shown to inhibit the human term placental and the activity of fetal liver glutathione S transferase (Mitra et al., 1992). High glutathione S transferase activation is seen in rapidly growing tumour cells in both humans and rodents, and it has also been implicated in tumor-cell drug-resistance (Chatterjee and Gupta, 2018).

Recent studies are focusing on the application of ascorbyl palmitate-based nanotechnology in delivering multiple drugs, such as astaxanthin, Paclitaxel, and docetaxel (Fratter et al., 2019; Li et al., 2016; Sohrabi et al., 2017; Zhou et al., 2017). In antitumor therapies, the use of ascorbyl palmitate does not serve as a carrier only, it also plays a role in treating cancer since it stops malignant cell proliferation and can control the synthesis of DNA in a variety of cancer cells, including glioblastoma, breast, skin, and colon malignant cells (Naidu, 2003). Finally, this antioxidant can significantly decrease sebum secretions in women. It can thus be used in treating undesired oily skin and in reducing acne (Khan et al., 2017).

Conclusions

Taking into consideration all the studies mentioned in this chapter as well as the conclusions reached by the European Food Safety Authority, palmitoyl ascorbic acid can be used without harms for human consumption at the levels recommended by the food safety authorities. Toxicity and genotoxic studies have shown that this chemical has a very low acute toxicity and is not genotoxic, respectively. Furthermore, current studies show ascorbyl palmitate to be a promising agent in preventing hepatotoxicity due to acetaminophen overdoses, and in tackling the phenomenon of drug-resistance in tumour cells.

Authors' contribution

RB - Designed, supervised, and reviewed the contents of the chapter, MAS - Preparing the draft of the chapter, JIA & AYB - Helped in writing the draft of the chapter, MA & HK - Helped to revise the chapter.

References

Bächtold, 1972. Akute Toxizitätsversuche von vitamin C, dessen Vorstufen und einigen möglichen Zersetzungsprodukten. Interne mitteilung No 4155. Unpublished study report submitted by DSM, July 2010.

Bächtold, 1973a. Prüfung im 5-tage-toxizitätsversuch an mäusen und ratten. Interne mitteilung No 3996. Unpublished study report submitted by DSM, July 2010.

Bächtold, 1973b. Akute Toxizitätsversuche von vitamin C, dessen Vorstufen und einigen möglichen Zersetzungsprodukten und metaboliten. Interne mitteilung No 4881. Unpublished study report submitted by DSM, July 2010.

Chatterjee, A., Gupta, S., 2018. The multifaceted role of glutathione S-transferases in cancer. Cancer Lett. 433, 33–42. https://doi.org/10.1016/j.canlet.2018.06.028.

Cort, W.M., 1974. Antioxidant activity of tocopherols, ascorbyl palmitate, and ascorbic acid and their mode of action. J. Am. Oil Chem. Soc. 51, 321. https://doi.org/10.1007/BF02633006.

Dresser, G.K., Wacher, V., Wong, S., Wong, H.T., Bailey, D.G., 2002. Evaluation of peppermint oil and ascorbyl palmitate as inhibitors of cytochrome P4503A4 activity in vitro and in vivo. Clin. Pharmacol. Ther. 72, 247–255. https://doi.org/10.1067/mcp.2002.126409.

EFSA, 2016. Scientific Opinion on the re-evaluation of ascorbyl palmitate (E 304(i)) and ascorbyl stearate (E 304(ii)) as food additives. EFSA J 13, 1–57. https://doi.org/10.2903/j.efsa.2015.4289.

Elmore, A.R., 2005. Final report of the safety assessment of L-ascorbic acid, calcium ascorbate, magnesium ascorbate, magnesium ascorbyl phosphate, sodium ascorbate, and sodium ascorbyl phosphate as used in cosmetics1. Int. J. Toxicol. 24, 51–111. https://doi.org/10.1080/10915810590953851.

Engstrom, N., Saenz-Mendez, P., Scheers, J., Scheers, N., 2017. Towards celiac-safe foods: decreasing the affinity of transglutaminase 2 for gliadin by addition of ascorbyl palmitate and ZnCl2 as detoxifiers. Sci. Rep. 7, 77. https://doi.org/10.1038/s41598-017-00174-z.

FDA, 2018. Food Additive Status List [WWW Document].

Frankel, E.N., Huang, S.-W., Kanner, J., German, J.B., 1994. Interfacial phenomena in the evaluation of antioxidants: bulk oils vs emulsions. J. Agric. Food Chem. 42, 1054–1059. https://doi.org/10.1021/jf00041a001.

Fratter, A., Biagi, D., Cicero, A.F.G., 2019. Sublingual delivery of astaxanthin through a novel ascorbyl palmitate-based nanoemulsion: preliminary data. Mar. Drugs 17, 494–508. https://doi.org/10.3390/md17090508.

Humeau, C., Girardin, M., Coulon, D., Miclo, A., 1995. Synthesis of 6-O-palmitoyl L-ascorbic acid catalyzed by Candida antartica lipase. Biotechnol. Lett. 17, 1091–1094. https://doi.org/10.1007/BF00143107.

Igoe, R.S., 2011. Dictionary of Food Ingredients, 5th ed. Choice Reviews Online. Springer, San Diego, CA, USA. https://doi.org/10.5860/choice.49-1813.

Ivanov, V., Ivanova, S., Kalinovsky, T., Niedzwiecki, A., Rath, M., 2016. Inhibition of collagen synthesis by select calcium and sodium channel blockers can be mitigated by ascorbic acid and ascorbyl palmitate. Am. J. Cardiovasc. Dis. 6, 26–35.

Ji, J.M., 2011. Chemical synthesis of ascorbyl palmitate in [BMIM] BF4, in: Advanced Materials Research. Trans Tech Publ 236–238, 1962–1965.

Jonker, D., Lee, V.S., Hargreaves, R.J., Lake, B.G., 1988. Comparison of the effects of ascorbyl palmitate and L-ascorbic acid on paracetamol-induced hepatotoxicity in the mouse. Toxicology 52, 287–295. https://doi.org/10.1016/0300-483X(88)90133-3.

Kageyama, K., Bradbury, M.J., Zhao, L., Blount, A.L., Vale, W.W., 1999. Urocortin messenger ribonucleic acid: tissue distribution in the rat and regulation in thymus by lipopolysaccharide and glucocorticoids. Endocrinology 140, 5651–5658. https://doi.org/10.1210/endo.140.12.7223.

Khan, H., Akhtar, N., Ali, A., 2017. Assessment of combined ascorbyl palmitate (AP) and sodium ascorbyl phosphate (SAP) on facial skin sebum control in female healthy volunteers. Drug Res. (Stuttg). 67, 52–58. https://doi.org/10.1055/s-0042-118171.

Let, M.B., Jacobsen, C., Meyer, A.S., 2007. Ascorbyl palmitate, γ-tocopherol, and EDTA affect lipid oxidation in fish oil enriched salad dressing differently. J. Agric. Food Chem. 55, 2369–2375. https://doi.org/10.1021/jf062675c.

Li, J., Guo, C., Feng, F., Fan, A., Dai, Y., Li, N., Zhao, D., Chen, X., Lu, Y., 2016. Co-delivery of docetaxel and palmitoyl ascorbate by liposome for enhanced synergistic antitumor efficacy. Sci. Rep. 6, 38787. https://doi.org/10.1038/srep38787.

Marinova, E.M., Yanishlieva, N.V.L., 1992. Inhibited oxidation of lipids III: on the activity of ascorbyl palmitate during the autoxidation of two types of lipid systems in the presence of α-tocopherol. Lipid /Fett 94, 448–452. https://doi.org/10.1002/lipi.19920941203.

Mitra, A., Govindwar, S., Joseph, P., Kulkarni, A., 1992. Inhibition of human term placental and fetal liver glutathione-S-transferases by fatty acids and fatty acid esters. Toxicol. Lett. 60, 281–288. https://doi.org/10.1016/0378-4274(92)90286-S.

Mitra, A., Ravikumar, V.C., Bourn, W.M., Bourcier, D.R., 1988. Influence of ascorbic acid esters on acetaminophen-induced hepatotoxicity in mice. Toxicol. Lett. 44, 39–46. https://doi.org/10.1016/0378-4274(88)90127-0.

Miwa, N., Yamazaki, H., Nagaoka, Y., Kageyama, K., Onoyama, Y., Matsui-Yuasa, I., Otani, S., Morisawa, S., 1988. Altered production of the active oxygen species is involved in enhanced cytotoxic action of acylated derivatives of ascorbate to tumor cells. Biochim. Biophys. Acta (BBA)-Molecular Cell Res. 972, 144–151.

Mohamed, R., Tarannum, S., Yariswamy, M., Vivek, H.K., Siddesha, J.M., Angaswamy, N., Vishwanath, B.S., 2014. Ascorbic acid 6-palmitate: a potent inhibitor of human and soybean lipoxygenase-dependent lipid peroxidation. J. Pharm. Pharmacol. 66, 769–778. https://doi.org/10.1111/jphp.12200.

Naidu, K.A., 2003. Ascorbyl esters for the treatment of cancer.

Perricone, N., Nagy, K., Horváth, F., Dajkó, G., Uray, I., Zs.-Nagy, I., 1999. The hydroxyl free radical reactions of ascorbyl palmitate as measured in various in vitro models. Biochem. Biophys. Res. Commun. 262, 661–665. https://doi.org/10.1006/bbrc.1999.1277.

Priva, M.J., Simmon, V.F., Mortelmans, K.E., 1991. Bacterial mutagenicity testing of 49 food ingredients gives very few positive results. Mutat. Res. Toxicol. 260, 321–329. https://doi.org/10.1016/0165-1218(91)90017-G.

Roth, M., Eickelberg, O., Kohler, E., Erne, P., Block, L.H., 1996. Ca2+ channel blockers modulate metabolism of collagens within the extracellular matrix. Proc. Natl. Acad. Sci. U. S. A. 93, 5478–5482. https://doi.org/10.1073/pnas.93.11.5478.

Sanchez Vallecillo, M.F., Escalera, Minguito de la, M.M., Aguirre, M., V, Gamboa, Ullio, G., V, Palma, S.D., Gonzalez-Cintado, L., Chiodetti, A.L., Soldano, G., Moron, G., Allemandi, D.A., Ardavin, C., Pistoresi-Palencia, M.C., Maletto, B.A., 2015. A liquid crystal of ascorbyl palmitate, used as vaccine platform, provides sustained release of antigen and has intrinsic pro-inflammatory and adjuvant activities which are dependent on MyD88 adaptor protein. J. Control. Release 214, 12–22. https://doi.org/10.1016/j.jconrel.2015.07.008.

Sohrabi, Y., Panahi-Azar, V., Barzegar, A., Ezzati Nazhad Dolatabadi, J., Dehghan, P., 2017. Spectroscopic, thermodynamic and molecular docking studies of bovine serum albumin interaction with ascorbyl palmitate food additive. Bioimpacts 7, 241–246. https://doi.org/10.15171/bi.2017.28.

Tufiño, C., Bernal, C., Ottone, C., Romero, O., Illanes, A., Wilson, L., 2019. Synthesis with immobilized lipases and downstream processing of ascorbyl palmitate. Molecules 24, 3227. https://doi.org/10.3390/molecules24183227.

Xiao, L., Miwa, N., 2017. The lipophilic vitamin C derivative, 6-O-palmitoylascorbate protects human keratinocytes and 3D-human skin equivalents against X-ray-induced oxidative stress and apoptosis more markedly than L-ascorbic acid. J. Cell. Biochem. 118, 318–329. https://doi.org/10.1002/jcb.25639.

Zhou, M., Li, X., Li, Y., Yao, Q., Ming, Y., Li, Z., Lu, L., Shi, S., 2017. Ascorbyl palmitate-incorporated paclitaxel-loaded composite nanoparticles for synergistic anti-tumoral therapy. Drug Deliv. 24, 1230–1242. https://doi.org/10.1080/10717544.2017.1370619.

CHAPTER

Butylated hydroxyanisole 3.2

Renald Blundell[a,b], Muhammad Ajmal Shah[c], Joseph I. Azzopardi[a], Amira Y. Benmelouka[d], Azhar Rasul[e], Norah A. Althobaiti[f]

[a]Department of Physiology and Biochemistry, Faculty of Medicine and Surgery, University of Malta, Msida, Malta

[b]Centre for Molecular Medicine and Biobanking, University of Malta, Msida, Malta

[c]Department of Pharmacognosy, Faculty of Pharmaceutical Sciences, Government College University, Faisalabad, Pakistan

[d]Faculty of Medicine, University of Algiers, Algiers, Algeria

[e]Department of Zoology, Faculty of Life Sciences, Government College University, Faisalabad, Pakistan

[f]Biology Department, College of Science and Humanities-Al Quwaiiyah, Shaqra University, Al Quwaiiyah, Saudi Arabia

3.2.1 Chemistry

Butylated hydroxyanisole (BHA) is a monohydric phenolic antioxidant (Burdock, 2005) found as a mixture of two isomeric compounds, 2-tert-butyl-4-hydroxyanisole (2-BHA) as shown in Fig. 3.1.2A and 3-tert-butyl-4-hydroxyanisole (3-BHA) as shown in Fig. 3.2.1B; in commercial mixtures, the latter isomer is found in greater quantities than the former at a ratio of 9:1 and is preferred for its antioxidant properties (Burdock, 2005; Rajadhyaksha et al., 2006). BHA is a crystalline or waxy solid that can be white or slightly yellow in colour and has a faint, characteristic odor. It is freely soluble in alcohols, such as ethanol, and in petroleum ether, chloroform, fats, oils, and propylene glycol, but it is insoluble in water. It melts at temperatures ranging from 48 to 63 degrees Celsius (EFSA, 2016).

3.2.2 Synthesis

There are different methods by which BHA can be manufactured, such as:

1. Through an alkylation of 4-hydroxyanisole with isobutylene (Kirk-Othmer, 2007).
2. Through a reaction between tert-butyl alcohol and 4-methoxyphenol over alumina or silica at a temperature of 150°C (Kirk-Othmer, 2007).
3. Through a reaction between hydroquinone and either tert-butyl alcohol or isobutene, which is catalyzed by an acid and then methylated (Kirk-Othmer, 2007).

Antioxidants Effects in Health. DOI: https://doi.org/10.1016/B978-0-12-819096-8.00032-X

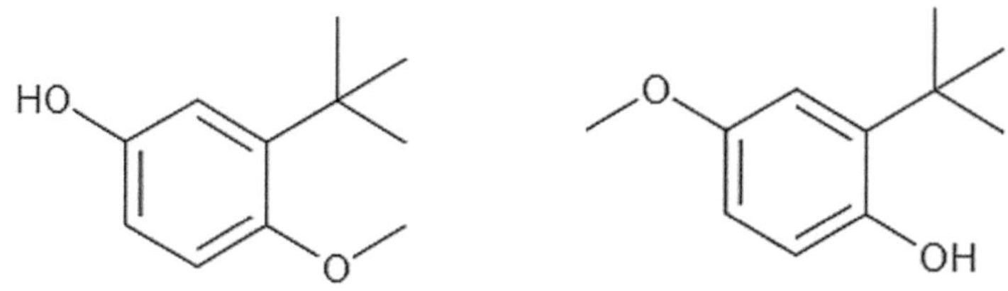

FIG. 3.2.1 The chemical structure of 2-tert-butyl-4-hydroxyanisole (A) and 3-tert-butyl-4-hydroxyanisole (B).

4. Through the methylation of hydroquinone, which yields a mixture of 2-BHA and 3-BHA upon treatment with phosphoric acid and tert-butyl alcohol (Burdock, 2005).

3.2.3 Legal status

In the EU, BHA – which has the E number E 320 – is authorized for use alone or in combination with other antioxidant components like butylated hydroxytoluene (BHT) or gallate, with a maximum permitted level (MPL) of up to 400 mg/kg depending on the type of food in which the additives are added to. BHA is authorized for use in beverages (EFSA, 2016). BHA is classified by the FDA as "Antioxidant/Substances permitted as optional ingredients in a standardized food" and "GRAS" (Food Additive Status List, 2018), and is listed as a carcinogen in California under "Proposal 65" (OEHHA, 19AD).

3.2.4 Mechanisms of action

The antioxidant effect of BHA, like that of other phenolic compounds such as BHT and tertiary butyl hydroquinone (tBHQ), is due to its ability to scavenge free radicals through donating hydrogen to the free radicals causing them to become more stable and therefore less likely to cause oxidative damage as the cycle of new radical formation is broken (Pereira et al., 2009). How effective the phenolic antioxidants are in scavenging free radicals depends on different factors, such as the bond dissociation energy between a phenolic hydrogen and oxygen, the reduction potential and delocalization of the antioxidant radicals and the pH related to the acid dissociation constant. The bond dissociation energy of the bond in the hydroxyl group of the phenolic compounds is also a predictor of the potency of the antioxidant properties of the compounds. This is because the lower the bond dissociation energy for the hydroxyl group is, the more stable the resulting antioxidant radical is (Amorati et al., 2003; Pereira et al., 2009).

Other mechanisms by which phenolics exert their antioxidant effect is through the chelation of metal ions; and the chemical quenching of singlet oxygen, becoming themselves oxidized in the process (Soobrattee et al., 2005).

3.2.5 Effects on health

In vitro, BHA was found to inhibit cellular growth in a time- and dose-dependent fashion, being associated with both apoptosis and fragmentation of single-strand DNA of the A549 cells being studied (Vandghanooni et al., 2013). *In vivo* experiments showed that BHA lead to an increase the incidence of obesity in mice due its effects on adipogenesis (Sun et al., 2020, 2019). At concentrations of 1.5%–2%, BHA might promote tumour formation in the forestomach of rodents (Haqlwara et al., 1983; Verhagen et al., 1991); but this was later confirmed to be species specific (Williams et al., 1999), and it is therefore not relevant for human health. Furthermore, at the levels found in food, that is, between 0.25% and 0.75%, BHA is able to prevent the acute toxicity engendered by monocrotaline in female mice (Miranda et al., 1981). Also, it could successfully prevent the nephrotoxicity caused by ferric nitrilotriacetate (Ansar and Iqbal, 2016).

In a 2015 study, BHA was found to have a protective effect against hydrogen peroxide-induced cellular death in hepatocytes (Hwang et al., 2015). Moreover, BHA could significantly reduce hepatic enzymes levels in a dose-dependent manner and it could also increase antioxidant enzymes level and prevent lipid peroxidation and hydrogen peroxide generation in ferric nitrilotriacetate treated samples (Ansar and Iqbal, 2015). Similar results have been obtained when carbon tetrachloride was used to induce the injury instead of nitrilotriacetate (Dassarma et al., 2017). BHA is a well-characterized Nrf2 activator which has protective effects against inflammation and colitis-associated colon tumorigenesis (Zheng et al., 2019).

Recent studies have raised concerns about the possibility of BHA being an endocrine disrupter where it was shown to have weak oestrogenic and anti-androgenic properties *in vitro* and antiestrogenic properties *in vivo* (Pop et al., 2018, 2013). In addition, BHA can also act as a selective inhibitor of 11β-hydroxysteroid dehydrogenase isoform 2 (HSD11B2) – a key regulator of the local level of glucocorticoids. Therefore, it may enhance glucocorticoid action in local tissues including the placenta and kidneys (Li et al., 2016). Hence, exposure to this molecule may potentially lead to adverse events, such as hypertension, hypokalemia, and retardation of fetal growth. BHA has been show to also affect the biosynthesis of testosterone in Leydig cells (Li et al., 2016). Additionally, the prolonged exposure to BHA may cayse abnormal calcium levels, endoplasmic reticulum stress, and mitochondrial dysfunction in the testes eventually resulting in their dysfunction (Ham et al., 2020).

BHA is also a powerful inhibitor of 5α-reductase (SRD5A1) and 3α-hydroxysteroid dehydrogenase (AKR1C14) in rats, and hence it reduces the formation of neurosteroids (Guo et al., 2017). Therefore, this compound has the potential for pharmacological exploitation in the future, since inhibitors of SRD5A1 can provide a protection for dopaminergic neurons in mouse models of Parkinson's disease.

Conclusions

In conclusion to the studies and their results discussed above, while BHA is likely not a direct carcinogen in humans at the recommended doses, it can still act as an endocrine disrupter and/or produce DNA damage. On the other hand, other studies have shown BHA to act as a protective agent in mouse hepatocytes. However, it may come with potent negative effects that can be species-specific. Taking this into consideration, the recommended levels of BHA in food should be maintained and further studies on order to understand the impact of its use in health and to prevent its negative effects are needed.

Authors' contribution

RB - Designed, supervised, and reviewed the contents of the chapter, MAS - Preparing the draft of the chapter, JIA & AYB - Helped in writing the draft of the chapter, AR & NAA - Helped to revise the chapter.

References

Amorati, R., Ferroni, F., Pedulli, G.F., Valgimigli, L., 2003. Modeling the co-antioxidant behavior of monofunctional phenols. Applications to some relevant compounds. J. Org. Chem. 68, 9654–9658. https://doi.org/10.1021/jo0351825.

Ansar, S., Iqbal, M., 2016. Antioxidant and nephroprotective potential of butylated hydroxyanisole against ferric nitrilotriacetate-induced oxidative stress and early tumor events. Hum. Exp. Toxicol. 35, 448–453. https://doi.org/10.1177/0960327115591378.

Ansar, S., Iqbal, M., 2015. Ameliorative effect of butylated hydroxyanisole against ferric nitrilotriacetate-induced hepatotoxicity and oxidative stress in rats. Toxicol. Ind. Health 31, 967–973. https://doi.org/10.1177/0748233714554409.

Burdock, G.A., 2005. Fenaroli's Handbook of Flavor Ingredients, 6th ed. CRC Press, New York.

Dassarma, B., Nandi, D.K., Gangopadhyay, S., Samanta, S., 2017. Hepatoprotective effect of food preservatives (butylated hydroxyanisole, butylated hydroxytoluene) on carbon tetrachloride-induced hepatotoxicity in rat. Toxicol. Reports 5, 31–37. https://doi.org/10.1016/j.toxrep.2017.12.009.

EFSA, 2016. Scientific opinion on the re-evaluation of butylated hydroxyanisole BHA (E 321) as a food additive. EFSA J 10, 2588. https://doi.org/10.2903/j.efsa.2012.2588.

Food Additive Status List, 2018. https://www.fda.gov/Food/IngredientsPackagingLabeling/FoodAdditivesIngredients/ucm091048.htm.

Guo, J., Li, L., Zhou, S., Su, Y., Li, X., Sun, J., Ge, R.S., 2017. Butylated hydroxyanisole alters rat 5α-reductase and 3α-hydroxysteroid dehydrogenase: Implications for influences of neurosteroidogenesis. Neurosci. Lett. 653, 132–138. https://doi.org/10.1016/j.neulet.2017.05.034.

Ham, J., Lim, W., You, S., Song, G., 2020. Butylated hydroxyanisole induces testicular dysfunction in mouse testis cells by dysregulating calcium homeostasis and stimulating endoplasmic reticulum stress. Sci. Total Environ. 702, 134775.

Haqlwara, A., Shibata, M., Ito, N., Fukushima, S., Ogiso, T., 1983. Carcinogenicity of butylated hydroxyanisole in F344 Rats234. JNCI J. Natl. Cancer Inst. 70, 343–352. https://doi.org/10.1093/jnci/70.2.343.

Hwang, G.H., Jeon, Y.J., Han, H.J., Park, S.H., Baek, K.M., Chang, W., Kim, J.S., Kim, L.K., Lee, Y.-M., Lee, S., Bae, J.-S., Jee, J.-G., Lee, M.Y., 2015. Protective effect of butylated hydroxylanisole against hydrogen peroxide-induced apoptosis in primary cultured mouse hepatocytes. J. Vet. Sci. 16, 17–23. https://doi.org/10.4142/jvs.2015.16.1.17.

Kirk-Othmer, 2007. Kirk-Othmer Encyclopedia of Chemical Technology, 5th ed. Wiley, New York, NY.

Li, L., Wu, Y., Wu, X., Wang, H., Bai, Y., Wang, X., Akingbemi, B.T., Huang, P., Ge, R.-S., 2016. Butylated hydroxyanisole potently inhibits rat and human 11beta-hydroxysteroid dehydrogenase type 2. Pharmacology 97, 10–17. https://doi.org/10.1159/000441034.

Li, X., Cao, S., Mao, B., Bai, Y., Chen, X., Wang, X., Wu, Y., Li, L., Lin, H., Lian, Q., Huang, P., Ge, R.-S., 2016. Effects of butylated hydroxyanisole on the steroidogenesis of rat immature Leydig cells. Toxicol. Mech. Methods 26, 511–519. https://doi.org/10.1080/15376516.2016.1202367.

Miranda, C.L., Reed, R.L., Cheeke, P.R., Buhler, D.R., 1981. Protective effects of butylated hydroxyanisole against the acute toxicity of monocrotaline in mice. Toxicol. Appl. Pharmacol. 59, 424–430. https://doi.org/10.1016/0041-008X(81)90294-5.

OEHHA, 19 March 2021. The proposition 65 list. https://oehha.ca.gov/proposition-65/proposition-65-list

Pereira, D.M., Valentão, P., Pereira, J.A., Andrade, P.B., 2009. Phenolics: from chemistry to biology. Molecules 14, 2202–2211. https://doi.org/10.3390/molecules14062202.

Pop, A., Drugan, T., Gutleb, A.C., Lupu, D., Cherfan, J., Loghin, F., Kiss, B., 2018. Estrogenic and anti-estrogenic activity of butylparaben, butylated hydroxyanisole, butylated hydroxytoluene and propyl gallate and their binary mixtures on two estrogen responsive cell lines (T47D-Kbluc, MCF-7). J. Appl. Toxicol. 38, 944–957.

Pop, A., Kiss, B., Loghin, F., 2013. Endocrine disrupting effects of butylated hydroxyanisole (BHA - E320). Clujul Med 86, 16–20.

Rajadhyaksha, M.N., Sharma, A.K., Gupta, R.S., Shirwaikar, A.S., 2006. Synthesis of butylated hydroxyanisole from tertiary butyl hydroquinone.

Soobrattee, M.A., Neergheen, V.S., Luximon-Ramma, A., Aruoma, O.I., Bahorun, T., 2005. Phenolics as potential antioxidant therapeutic agents: Mechanism and actions. Mutat. Res. Mol. Mech. Mutagen. 579, 200–213. https://doi.org/10.1016/j.mrfmmm.2005.03.023.

Sun, Z., Tang, Z., Yang, X., Liu, Q.S., Liang, Y., Fiedler, H., Zhang, J., Zhou, Q., Jiang, G., 2020. Perturbation of 3-tert-butyl-4-hydroxyanisole in adipogenesis of male mice with normal and high fat diets. Sci. Total Environ. 703, 135608.

Sun, Z., Yang, X., Liu, Q.S., Li, C., Zhou, Q., Fiedler, H., Liao, C., Zhang, J., Jiang, G., 2019. Butylated hydroxyanisole isomers induce distinct adipogenesis in 3T3-L1 cells. J. Hazard. Mater. 379, 120794.

Vandghanooni, S., Forouharmehr, A., Eskandani, M., Barzegari, A., Kafil, V., Kashanian, S., Ezzati Nazhad Dolatabadi, J., 2013. Cytotoxicity and DNA Fragmentation Properties of Butylated Hydroxyanisole. DNA Cell Biol. 32, 98–103. https://doi.org/10.1089/dna.2012.1946.

Verhagen, H., Schilderman, P.A.E.L., Kleinjans, J.C.S., 1991. Butylated hydroxyanisole in perspective. Chem. Biol. Interact. 80, 109–134. https://doi.org/10.1016/0009-2797(91)90019-4.

Williams, G.M., Iatropoulos, M.J., Whysner, J., 1999. Safety assessment of butylated hydroxyanisole and butylated hydroxytoluene as antioxidant food additives. Food Chem. Toxicol. 37, 1027–1038. https://doi.org/10.1016/S0278-6915(99)00085-X.

Zheng, Z., Chen, Y., Huang, J., Deng, H., Tang, X., Wang, X.J., 2019. Mkp-1 is required for chemopreventive activity of butylated hydroxyanisole and resveratrol against colitis-associated colon tumorigenesis. Food Chem. Toxicol. 127, 72–80.

CHAPTER

Butylated hydroxytoluene

3.3

Renald Blundell[a,b], Muhammad Ajmal Shah[c], Joseph I. Azzopardi[a], Shabnoor Iqbal[d], Akhtar Rasul[e], Ghulam Mujtaba Shah[f]

[a]*Department of Physiology and Biochemistry, Faculty of Medicine and Surgery, University of Malta, Msida, Malta*

[b]*Centre for Molecular Medicine and Biobanking, University of Malta, Msida, Malta*

[c]*Department of Pharmacognosy, Faculty of Pharmaceutical Sciences, Government College University, Faisalabad, Pakistan*

[d]*Department of Zoology, Faculty of Life Sciences, Government College University, Faisalabad, Pakistan*

[e]*Department of Pharmaceutics, Faculty of Pharmaceutical Sciences, Government College University, Faisalabad, Pakistan*

[f]*Department of Botany, Faculty of Biological and Health Sciences, Hazara University, Mansehra, Pakistan*

3.3.1 Chemistry

Butylated hydroxytoluene (BHT; IUPAC name: 4-methyl-2,6-ditertiarybutylphenol) is a monohydric phenolic antioxidant and used in food to enhance the antioxidants capacity while its excretion process is slow in experimental animals after consumption (Burdock, 2005). Its chemical structure is illustrated in Fig. 3.3.1. It is a white crystalline solid with a faint aromatic odour, having a melting point of 70°C. BHT is sparingly soluble in water (1.1 mg/L 20°C) and completely soluble in ethanol and fatty oils (EFSA, 2016).

3.3.2 Synthesis

The commercial production of BHT is obtained through an alkylation reaction between 4-methylphenol (*para*-cresol) and 2-methylpropene (isobutylene) in the presence of an acid catalyst. The product is then neutralised by the addition of sodium carbonate followed by crystallization, filtration, and washing with isopropanol. The product is then dried and sieved (Burdock, 2005; EFSA, 2016). BHT is also synthesized by using p-cresol obtained from coal tar (25%) and also used in numerous synthetic processes. The synthetic route involves sulfonation of toluene then heating with sodium hydroxide. The isobutylene gas is used for alkylation

Antioxidants Effects in Health. DOI: https://doi.org/10.1016/B978-0-12-819096-8.00033-1

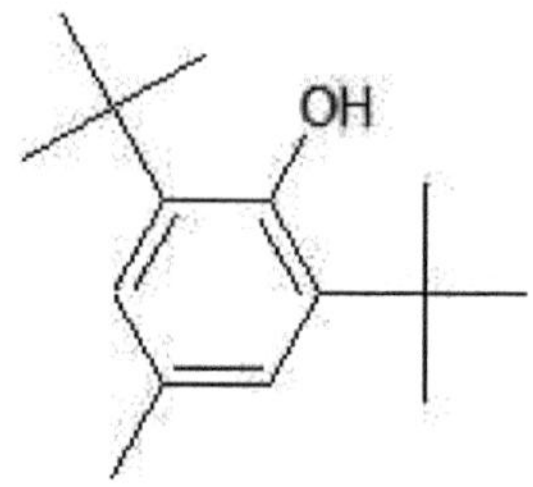

FIG. 3.3.1 The chemical structure of butylated hydroxytoluene.

of p-cresol in a catalyzed reaction and the products are sensitive to catalyst. The isobutylene gas was bubbled through the solution of p-cresol and 5% phosphoric acid while heating at 70°C. The sodium hydroxide is used to wash the product. The crystals of BHT have 46% yield (Stillson, 1947). Another process involves heating of p-cresol with 5% methane disulfonic acid at 40°C and bubbled through isobutylene for about 6–70 h. The crystals are washed with sodium hydroxide. The crystals (BHT) have 71% to 88% yields and methanol is used for recrystallization (McConnell and Davis, 1963).

3.3.3 Legal status

In the EU, BHT (E 321) is permitted to be used in fats and oils either alone or in combination with other antioxidants at levels of up to 100 mg/kg expressed as fat. It is also authorized for use, either alone or in combination with other antioxidants in the chewing gum manufacture at concentration of 400 mg/kg of chewing gum as shown in Table 3.3.1 (EFSA, 2016). BHT is classified as "GRAS" by the FDA and its use is permitted in other foodstuff apart from fat and oils, such as dry breakfast cereals and certain potato products, including sweet potato flakes and potato granules (FDA, 2018).

Table 3.3.1 The optimum levels of butylated hydroxytoluene in different food categories according to Directive No 95/2/EC.

Food category	Maximum permitted level (mg/kg or mg/L as appropriate)
Fats and oils for the commercial manufacturing of heat-treated foodstuffs Frying fat and frying oil and (w olive pomace oil) Lard; fish oil; beef, poultry, and sheep fat	100
Chewing gum Food supplements as defined in Directive 2002/46/EC	400

3.3.4 Mechanisms of action

BHT performs antioxidant activity either by hindering the ROS generation or by scavenging the free radicals that involves various mechanisms (Fig. 3.3.2): (1) BHT scavenges reactive species that induce peroxidation, (2) metal chelating effect, particular iron ions that catalyze oxidative processes and release radicals of hydroxyl, and theses radicals chelate hydroperoxides by Fenton reactions, (3) BHT also scavenges $\cdot O^{-}_2$ mediated peroxides formation, (4) BHT assists to hinder the chain reactions of autoxidation, (5) lower the localized level of O_2 (Brewer et al., 2011; Stochs et al., 1995). The phenolic ring of BHT provides H to the reactive oxygen species and BHT itself becomes radical. It involves two pathways through which usually BHT and its derivatives performed antioxidative activity. At first step free radical R is formed while in second and third steps, a chain reaction yields various lipid molecules (ReH) that further transform to lipid hydroperoxides (ROOH). The BHT as an antioxidant (ArOH) interrupts the chain reaction. However, BHT radicals (ArO) are relatively stable, they need time to react with the substrate RH but react rapidly with ROO. BHT also neutralizes the free radicals by electron transfer that involve reversible deprotonation and hydroperoxide formation (Minishi et al., 1997).

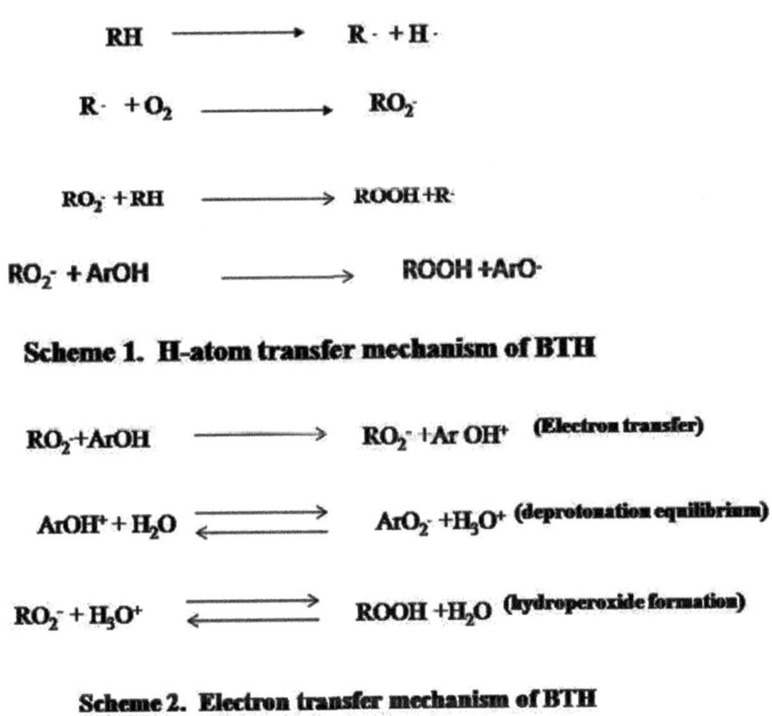

FIG. 3.3.2 The schematic representation of antioxidative mechanism of BHT.

3.3.5 Effects on health

A donepezil-BHT hybrid has been formed with promising results obtained from animal models as a potential future treatment for Alzheimer's disease with antioxidant, cholinergic, and neuroprotective properties (Yang et al., 2018). BHT can be used for cryopreservation to protect human semen from HIV virus as it increased the spermatozoa membrane fluidity to prevent the virus interaction and transmission (Merino et al., 2015). BHT has been reported to ameliorate the ferric nitrilotriacetate (Fe-NTA)-induced liver injuries in mice by reducing the oxidative stress and enhancing the levels of the antioxidative enzymes (Ansar et al., 2016).Various animal studies in the 1980's reported polarizing results about the carcinogenic and anticarcinogenic properties of BHA and BHT (Haqlwara et al., 1983; Hocman, 1988; Wattenberg, 1985; Williams, 1986), a cohort study took place between September 1986 and December 1992. The study recruited 62,573 women and 58,279 men aged between 55 and 69 years. This study failed to find any significant relationship between the incidence of stomach cancer in the intake of BHT and BHA (Botterweck et al., 2000). Beside that an *in vitro* study observed, the symmetric S-BHT components effectively potentiated the oxidative stress in human cancer cell lines MCF7 (breast cancer), and HT29 (colon cancer) to reduce the cancer cell progression and found anticancer capacity (Ahmad et al., 2019). An acute toxicity studies depicted that BHT has less acute toxicity with an acute oral LD_{50} of 1700–1930 mg BHT/kg body weight in rats and 940–2100 mg BHT/kg body weight in cats (Madhavi et al., 1995).

Conclusions

BHT has shown low acute toxicity and is therefore safe at the recommended levels in food. Furthermore, clinical trials have failed to show any carcinogenic effect that BHT might have in the human body. Furthermore, it would be of great academic and commercial interest if newer studies on the antiviral properties that BHT exhibits against the herpes simplex virus are carried out. Overall therefore, this antioxidant is not only safe for use in food but it may also be a potential agent in the treatment of Alzheimer's and herpes infections, amongst others.

Authors' contribution

RB - Designed, supervised, and reviewed the contents of the chapter, MAS - Preparing the draft of the chapter, JIA & SI - Helped in writing the draft of the chapter, AR & GMS - Helped to revise the chapter.

References

Ahmad, M.H., et al., 2019. Design and synthesis of sulfur-containing butylated hydroxytoluene: antioxidant potency and selective anticancer agent. J. Chem. Sci. 131 (10), 110. doi:10.1007/s12039-019-1682-x.

Ansar, S., et al., 2016. Amelioration of ferric nitrilotriacetate-induced hepatotoxicity in Wistar rats by diallylsulfide. Hum. Exp. Toxicol. 35 (3), 259–266. doi:10.1177/0960327115583362.

Botterweck, A.A.M., Verhagen, H., Goldbohm, R.A., Kleinjans, J., 2000. Intake of butylated hydroxyanisole and butylated hydroxytoluene and stomach cancer risk: results from analyses in the Netherlands cohort study. Food. Chem Toxicol. 38, 599–605.

Brewer, M.S., 2011. Natural antioxidants: sources, compounds, mechanisms of action, and potential applications. Compr. Rev. Food Sci. Food Saf. 10 (4), 221–247.

Burdock, G.A., 2005. Fenaroli's Handbook of Flavor Ingredients, 6th ed. CRC Press, New York.

EFSA, 2016. Scientific opinion on the re-evaluation of butylated hydroxytoluene BHT (E 321) as a food additive. EFSA J. 10, 2588. https://doi.org/10.2903/j.efsa.2012.2588.

FDA, 2018. Food additive status list. https://www.fda.gov/food/ingredientspackaginglabeling/foodadditivesingredients/ucm091048.htm (Accessed 24 February, 2019).

Haqlwara, A., Shibata, M., Ito, N., Fukushima, S., Ogiso, T., 1983. Carcinogenicity of butylated hydroxyanisole in F344 Rats234. J. Natl. Cancer Inst. 70, 343–352. https://doi.org/10.1093/jnci/70.2.343.

Hocman, G., 1988. Chemoprevention of cancer: phenolic antioxidants (BHT, BHA). Int. J. Biochem. 20, 639–651. https://doi.org/10.1016/0020-711X(88)90158-9.

Madhavi, D.L., Singhal, R.S., Kulkarni, P.R., 1995. Technological aspects of food antioxidants. In: Madhavi, D.L., Deshpande, S.S., Salunke, D.K. (Eds.), Food Antioxidants: Technological; Toxicological and Health Perspectives. CRC Press, Boca Raton, pp. 159–167.

McConnell, W.V., Davis, H.E., 1963. Alkylation of phenols. U.S. Patent No. 3,082,258. March 19, 1963. Eastman 637 Kodak, Rochester, NY.

Merino, O., Aguagüiña, W.E., Esponda, P., Risopatrón, J., Isachenko, E., Isachenko, V., Sánchez, R., 2015. Protective effect of butylated hydroxytoluene on sperm function in human spermatozoa cryopreserved by vitrification technique. Andrologia 47 (2), 186–193.

Minisci, F., 1997. Free Radicals in Biology and Environment. Springer, Netherlands.

Stillson, G.H. 1947., Alkylation of phenols.U.S. Patent #2,428,745. Assigned to Gulf Oil.

Wattenberg, L.W., 1985. Chemoprevention of cancer. Cancer Res. 45, 1 LP-8.

Williams, G.M., 1986. Epigenetic promoting effects of butylated hydroxyanisole. Food Chem. Toxicol. 24, 1163–1166. https://doi.org/10.1016/0278-6915(86)90303-0.

Yang, X.-L., Cai, P., Wang, X., Fang, S., Liu, Q., Yang, H.-L., Kong, L., 2018. Donepezil-butylated hydroxytoluene (BHT) hybrids as Anti-Alzheimer's disease agents with cholinergic, antioxidant, and neuroprotective properties. Eur. J. Med. Chem. 157, 161–176. https://doi.org/10.1016/j.ejmech.2018.08.005.

CHAPTER

Erythorbic acid (D-ascorbic acid)

3.4

Renald Blundell[a,b], Muhammad Ajmal Shah[c], Joseph I. Azzopardi[a], Shabnoor Iqbal[d], Tapan Behl[e], Abdul H. Khan[f]

[a]*Department of Physiology and Biochemistry, Faculty of Medicine and Surgery, University of Malta, Msida, Malta*

[b]*Centre for Molecular Medicine and Biobanking, University of Malta, Msida, Malta*

[c]*Department of Pharmacognosy, Faculty of Pharmaceutical Sciences, Government College University, Faisalabad, Pakistan*

[d]*Department of Zoology, Faculty of Life Sciences, Government College University, Faisalabad, Pakistan*

[e]*Chitkara College of Pharmacy, Chitkara University, Punjab, India*

[f]*Department of Pharmacy, Forman Christian College University (A Chartered University), Lahore, Pakistan*

3.4.1 Chemistry

Erythorbic acid (EA), also known as D-ascorbic acid (IUPAC name: (5R)-5-[(1R)-1,2-dihydroxyethyl]-3,4-dihydroxy-5-methylfuran-2(5H)-one) is a diprotic acids and an epimer of vitamin C (L-ascorbic acid). As compared to L-ascorbic acid its activity is 5% and its chemical structure is shown in Fig. 3.4.1. It is a crystalline solid of white to slightly pale color, that gradually blackens upon subjection to light. Its melting point is 164°C and it is freely soluble in water and ethanol. While in solution, it quickly, degrades in the presence of air, it is relatively stable in air, but when it is in the dry state (Burdock, 2005; EFSA, 2016).

3.4.2 Synthesis

EA is manufactured by the acidification of calcium 2-keto-D-gluconate producing 2-keto-D-gluconic acid. Through a cation exchange resin, the resulting acidified mixture is filtered and decalcified. Methyl 2-keto-D-gluconate is produced when filtrate was concentrated and esterified with methanol. The resultant ester is then formed into crystal, isolated, rinsed with methanol and then turned to sodium erythorbate by heating the solution following the incorporation of sodium carbonate or sodium bicarbonate (EFSA, 2016). The resulting salt is solidified, isolated and washed with methanol and then suspended in a water/methanol mixture and

Antioxidants Effects in Health. DOI: https://doi.org/10.1016/B978-0-12-819096-8.00034-3

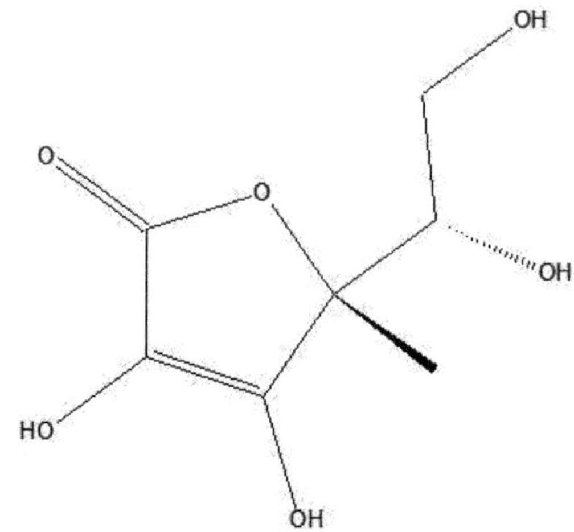

FIG. 3.4.1 The chemical structure of octyl gallate.

converted to erythorbic acid by the addition of sulfuric acid. The resulting sodium sulfate is removed by filtration and the solution is then concentrated, deionized by means of ion exchange resins and decolourized using activated carbon. The solution is then crystallized, separated, washed, and dried (EFSA, 2016). EA can also be synthesized from sucrose. It also occurs in nature by *Saccharomyces cerevisiae*, a species of yeast (Burdock, 2005).

3.4.3 Legal status

Erythorbic acid is an approved food additive both in the EU, where it is referred to by the E number E 315, and in the US where it is listed as "substances generally recognized as safe in foods but limited in standardized foods where the standard provides for its use" by the FDA (EFSA, 2016; FDA, 2018). In the EU, the MPL (maximum permissible level) of combined erythorbic acid and its sodium salt ranges from 250 to 1500 mg/kg or mg/L (as appropriate), as shown in Table 3.4.1 below. Erythorbic acid is used in combination with sodium erythorbate (sodium salt of erythorbic acid, E316) as an antioxidant and food preservative. The corresponding MPL values in food categories showed the combined value of erythorbic acid and sodium erythorbate.

3.4.4 Mechanisms of action

Erythorbic acid is a metal-chelating agent where it partly deactivates trace metals that are often present as salts of fatty acid, which would otherwise cause oxidative damage (Mukhopadhyay, 2006). This compound is also a strong reducing agent where it reduces nitrite to nitric oxide and also inhibits the formation of nitrosamines

Table 3.4.1 The maximum permitted levels of erythorbic acid and its sodium salt in some of the food categories as set out in the Annex II of Regulation (EC) No 1333/2008 and its amendments.

Food categories	Maximum permitted level (mg/kg or mg/L as appropriate)
Filling of stuffed pasta	250
Nonheat-treated meat products Heat-treated meat products	500
Unprocessed fish Processed fish and fishery products, including molluscs and crustaceans Fish roe	1500

(Toldra, 2002). It acts as a scavenger of free radicals and stimulates the hydroxylation of peptidyl proline to induce collagen synthesis (Kutnink et al., 1969). The underlying mechanism behind antitumor, anti-inflammatory activities is the down regulation of transcriptional factors (hypoxia inducible factor 1) (Traber and Stevens, 2011).

3.4.5 Effects on health

EA has chemical and antioxidant properties parallel to its stereoisomer ascorbic acid, but the former has only 1/20th of the *in vivo* antiscorbutic activity in guinea pigs, and *in vitro* 1/8th of the collagen synthesis stimulation activity when compared to that of the latter. This difference in the activity between the two stereoisomers is due to the difference in quantity of the intracellular uptake (Kipp and Schwarz, 1990; Suzuki et al., 1995).

A 2004 study on 10 healthy women aged between 20 to 26 years demonstrated that EA enhances the absorption of nonheme iron from iron sulfate more than ascorbic acid does. Previous studies, one carried out on healthy men and another on rats found no effect of EA on the absorption of iron. Fidler et al. claim that the reason for these results is because of the low sensitivity of the method used to evaluate the iron absorption in the former study and because of the fact that iron absorption in humans cannot be predicted from studies on rats as the iron absorption in both study subjects are not comparable (Fidler et al., 2018).

When applied to pig skin, 10% erythorbic acid was found to reduce ultraviolet B-induced phototoxicity (Andersen, 1999). A previous study had reported that, when EA is administered into a human mammary tumor xenograft mouse at a low dose by itself fails to inhibit tumor growth. However, when used in conjunction with cupric sulfate, tumor growth was found to be depressed (Tsao, 1991).

In another study, high-dose administration of EA and its sodium salt combination showed antitumor activity both *in vitro* (on colon-26 cells) and *in vivo* on mice model of cancer by damaging the cancerous cells by the reactive oxygen species produced by its auto-oxidation (Miura et al., 2015).

Abe et al. (1984) investigated the carcinogenic potential of sodium erythorbate on 306 F344/DuCrj rats of both gender. Sodium erthorbate in drinking water marginally decreased the aggregate tumor incidences and tumor development in female rats with reduced weight gain. Few and far between scientific studies reported the pharmacological effects of sodium erythorbate in combination with erythorbic acid, although its non-injurious health status is declared by WHO (Miková, 2001). The antiscorbutic activity erythorbic acid (250 mg/day) was investigated in guinea pigs fed on vitamin C depleted diet. Erythobic acid administration has no beneficial effect to slow down acute vitamin C deficit. However, animal fed on suboptimal dose of vitamin C along with erythorbic acid showed the protective effect against scurvy *via* preserving vitamin C in body (Reiff and Free, 1959).

Fabianek and Herp (1967) investigated the anti-scorbutic activity of this compound in adult male guinea pigs fed on vitamin C lacking diet. Daily dose of erythorbic acid 1, 2, 10, 50, 100, and 200 mg was administered for 38 days and up to 115 days to three animals of 10 mg dose group. Their findings showed that animals survived after 115 days treated with 10 mg erythorbic acid with slightly reduced weight gain. It has been concluded that erythorbic acid compensated the vitamin C antiscorbutic effect.

The effect of erythorbic acid on wound healing, weight gain and alkaline phosphatases was investigated in a study on female guinea pigs treated with scorbutagenic diet for six days. Erythorbic acid at dose of 100 mg cure the wound healing, weight loss, and serum alkaline phosphatase level $1/20^{th}$ compared to ascorbic acid. This study concluded that erythorbic acid has poor tissue retention and uptake, otherwise might have equal efficacy to vitamin C (Goldman et al., 1981). Robertson (1963) explored the stimulatory potential of erythorbic acid equivalent to vitamin C on collagen synthesis in carrageenan induced scorbutic granulomas on vitamin C deficient guinea pigs. Kutnink et al., 1969 revealed the stimulatory potency of erythorbic acid comparable to ascorbic acid in hydroxylation of peptidyl proline. Erythorbic acid showed restorative effect on lipid peroxidation induced by CCl_4 similar to ascorbic acid in animal models, concluding equal *in vivo* antioxidant potential to ascorbic acid (Kunert and Tappel, 1983).

In another study, a partial ascorbic acid sparing effect of erythorbic acid was reported in eleven adult healthful nongravid matron volunteers. In this study volunteers were kept in metabolic unit and vitamin C deficient formula diet was provided for 54 days. Ascorbic acid supplements administered after 24 days did not significantly recovered scurvy signs. Although, erythorbic acid preserved the effects of vitamin C at 30, 60, and 90 mg/day dose (Sauberlich et al., 1989).

Conclusions

Taking into account the conclusions that were made by the European Food Safety Assessment, there is no reason why erythorbic acid should not be considered as safe for use by the food industry. Furthermore, erythorbic acid has shown promising results in the treatment for different cancers as well as in reducing phototoxicity and increasing the absorption of nonheme iron, the latter providing a new modality of treatment for anemia.

Authors' contribution

RB - Designed, supervised, and reviewed the contents of the chapter, MAS - Preparing the draft of the chapter, JIA & ZC - Helped in writing the draft of the chapter, AR & US - Helped to revise the chapter.

References

Andersen, F.A., 1999. Final report on the safety assessment of ascorbyl palmitate, ascorbyl dipalmitate, ascorbyl stearate, erythorbic acid, and sodium erythorbate. Int. J. Toxicol. 18, 1–26. https://doi.org/10.1177/109158189901800303.

Abe, I., Saito, S., Hori, K., Suzuki, M., Sato, H., 1984. Sodium erythorbate is not carcinogenic in F344 rats. Exp. Mol. Pathol. 41 (1), 35–43.

Burdock, G.A., 2005. Fenaroli's Handbook of Flavor Ingredients, 6th ed. CRC Press, New York.

EFSA, 2016. Scientific opinion on the re-evaluation of erythorbic acid (E 315) and sodium erythorbate (E 316) as food additives. EFSA J. 14, 4360. https://doi.org/10.2903/j.efsa.2016.4360.

Fabianek, J., Herp, A., 1967. Anti-scorbutic activity of D- araboascorbic acid. Proc. Soc. Exp. Biol. Med. 125, 462–465.

FDA, 2018. Food additive status list. https://www.fda.gov/food/ingredientspackaginglabeling/foodadditivesingredients/ucm091048.htm (Accessed 24 February, 2019).

Fidler, M.C., Davidsson, L., Zeder, C., Hurrell, R.F., 2018. Erythorbic acid is a potent enhancer of nonheme-iron absorption. Am J. Clin. Nutr. 1–3, 99–102.

Goldman, H.M., Gould, B.S., Munro, H.N., 1981. The antiscorbutic action of L-ascorbic acid and D-isoacorbic acid (erythorbic acid) in the guinea pig. Am J. Clin. Nutr. 34, 24–33.

Kipp, D.E., Schwarz, R.I., 1990. Effectiveness of Isoascorbate versus Ascorbate as an Inducer of Collagen Synthesis in Primary Avian Tendon Cells. J. Nutr. 120, 185–189. https://doi.org/10.1093/jn/120.2.185.

Kutnink, M.A., Tolbert, B.M., Richmond, V.L., Baker, E.M., 1969. Efficacy of the ascorbic acid stereoisomers in proline hydroxylation *in vitro*. Proc. Soc. Exp. Biol. Med. 132, 440–442.

Kunert, K.J., Tappel, A.L., 1983. The effect of vitamin C on *in vivo* lipid peroxidation in guinea pigs as measured by pentane and ethane production. Lipids 18, 271–274.

Miková, K., 2001. The regulation of antioxidants in food. Handbook of Food Preservation. Illinois, United States, 267–284.

Miura, K., Yazama, F., Tai, A., 2015. Oxidative stress-mediated antitumor activity of erythorbic acid in high doses. Biochem. Biophys. Reports 3, 117–122. https://doi.org/10.1016/j.bbrep.2015.07.018.

Mukhopadhyay, A.K., 2006. Antioxidants – Natural and Synthetic. Amani International Publishers, Kiel.

Reiff, J.S., Free, A.H., 1959. Nutritional studies with isoascorbic acid in the guinea pig. J. Agric. Food Chem. 7, 55–56.

Robertson, W., Van, B., 1963. D-ascorbic acid and collagen synthesis. Biochem. Biophys. Acta. 74, 137–139.

Sauberlich, H.E., Kretsch, M.J., Taylor, P.C., Johnson, H.L., Skala, J.H., 1989. Ascorbic acid and erythorbic acid metabolism in nonpregnant women. Am. J. Clin. Nutr. 50, 1039–1049.

Suzuki, E., Kurata, T., Tomita, M., Arakawa, N., 1995. Vitamin C activity of erythorbic acid in ascorbic acid-deficient guinea pigs. J. Nutr. Sci. Vitaminol. 41, 17–24.

Toldra, F., 2002. Dry Cured Meats. Food & Nutrition PRess Inc., United States.

Traber, M.G., Stevens, J.F., 2011. Vitamins C and E: Beneficial effects from a mechanistic perspective. Free Radic. Biol. Med. 51 (5), 1000–1013. 10.1016/j.freeradbiomed.2011.05.017.

Tsao, C.S., 1991. Inhibiting effect of ascorbic acid on the growth of human mammary tumor xenografts. Am. J. Clin. Nutr. 54, 1274S–1280S. https://doi.org/10.1093/ajcn/54.6.1274s.

CHAPTER

Nordihydroguaiaretic acid

3.5

Renald Blundell[a,b], Muhammad Ajmal Shah[c], Joseph I. Azzopardi[a], Shabnoor Iqbal[d], Akhtar Rasul[e], Zunera Chauhdary[f]

[a]*Department of Physiology and Biochemistry, Faculty of Medicine and Surgery, University of Malta, Msida, Malta*

[b]*Centre for Molecular Medicine and Biobanking, University of Malta, Msida, Malta*

[c]*Department of Pharmacognosy, Faculty of Pharmaceutical Sciences, Government College University, Faisalabad, Pakistan*

[d]*Department of Zoology, Faculty of Life Sciences, Government College University, Faisalabad, Pakistan*

[e]*Department of Pharmaceutics, Faculty of Pharmaceutical Sciences, Government College University, Faisalabad, Pakistan*

[f]*Department of Pharmacology, Faculty of Pharmaceutical Sciences, Government College University, Faisalabad, Pakistan*

3.5.1 Chemistry

Nordihydroguaiaretic acid (NDGA), 4,4′-(2,3-Dimethylbutane-1,4-diyl)dibenzene-1,2-diol, is a phenolic lignan and a lipoxygenase inhibitor (Gowri et al., 2000). It is a greyish-white crystalline solid in appearance (Perry et al., 1972). It has a melting in the range of 184–185°C; it is freely soluble both in ether and ethanol, in propylene glycol at a temperature of 116°C and slightly soluble in hot water (Madhavi et al., 1995). The chemical structure of NDGA is shown in Fig. 3.5.1.

3.5.2 Synthesis

While NGDA can be found naturally in the resin of many plants, especially *Larreadivaricata*, a creosote bush found in the southwestern US, it can also be obtained through different methods of chemical synthesis. One such technique, which is claimed to give the highest yield of the product, involves the acylation of 1,2-dimethoxybenzene with propionyl chloride in chloroform resulting in the formation of 3,4-dimethoxypropiophenone which then undergoes bromination to give α-bromo-3,4-dimethocypropiophenone which is then heated in the presence of copper powder and refluxed in a xylene solution (Perry et al., 1972). Another method by which NDGA can be synthesized is through the hydrogenation, and subsequent demethylation of guaiaretic acid dimethyl ether (Ramis-Ramos, 2003).

Antioxidants Effects in Health. DOI: https://doi.org/10.1016/B978-0-12-819096-8.00035-5

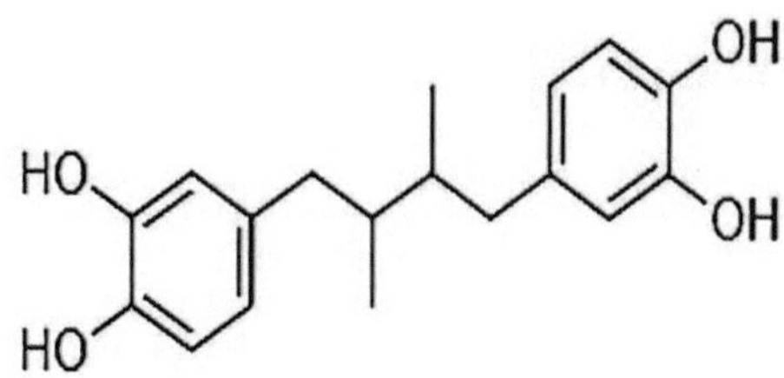

FIG. 3.5.1 The chemical structure of nordihydroguaiaretic acid.

3.5.3 Legal status

The use of NDGA as a food additive is prohibited both by the European Food Safety Agency (EFSA) in the EU and by the FDA in the US (EFSA, 2017; FDA, 2018; Parvez et al., 2009).

3.5.4 Mechanisms of action

The antioxidant properties of NDGA are due to its action as a potent scavenger of reactive oxygen species, such as the hydroxyl radical (•OH), peroxynitrite ($ONOO^-$), amongst others. In a particular study, NDGA was found to be more efficient in scavenging certain reactive oxygen species compared to the reference compounds used (mannitol, glutathione, penicillamine, and uric acid). The antioxidative attribute of NDGA is owing to the occurrence of the four reducing species derivative of two catechol groups in its structure (Lü et al., 2010; Norris and Carr, 2013).

3.5.5 Effects on health

Animal studies have shownthat NDGA can cause renal proximal tubular damage and cyst formation, while chaparral (which contains NDGA) has been reported to cause hepatotoxicity in humans; as a result of such findings, NDGA chronic use might cause deleterious effects.

Starting in 2008, studies found that NDGA can enhance the median lifespan of mice when given at a dose of 2500 ppm, but interestingly the increase in lifespan was only observed in males. This difference in the extension of the lifespan between the two sexes was hypothesized to be due to the difference in the pharmacokinetic properties between males and females as the blood levels of NDGA were found to be higher in the former compared to the latter (Harrison et al., 2014; Strong et al., 2008). In a more recent study, Strong et al. (2008) noticed that NDGA failed to increase the lifespan of female mice even at concentrations of 5000 ppm, therefore the pharmacokinetics relating to the blood levels between the different sexes could not account to the difference in the effect of NDGA on mice. Furthermore, the effect of

NDGA on the lifespan in male mice was found to be dose dependent, with the lifespan increasing as the dose of the NDGA increased. NDGA was also noted to attenuate age-related deficits in rotarod performance in both male and female mice (Javors et al., 2016). Similar results were obtained in a study involving the use of transgenic *Drosophila* model of alzheimer's disease where it was found that exposure to NDGA resulted in a significant dose dependent increase in the life span of these fruit flies. The symptoms associated with alzheimer's disease were also found to be attenuated in the transgenic fruit flies that were exposed to NDGA (Siddique and Ali, 2017).

In their study, Bergren and Valentine found that NDGA exhibited anti-bronchoconstrictive action when given as an aerosol or intravenous injection in ovalbumin-sensitized guinea pigs which were challenged with an ovalbumin aerosol. This antibronchoconstrictive was reported both in the anesthetized and the conscious, nonsedated guinea pigs. It was hypothesized by the authors of this study that this effect might be due to the blockage of 5-lipoxygenase and therefore the synthesis of leukotrienes (Bergren and Valentine, 2016). In a similar study, it was reported that there was a reduction in protein extravasation following the administration of ovalbumin aerosol in ovalbumin-sensitised rats that were exposed to NDGA.The mucosal mast cell morphology in these rats was noticed to be preserved suggesting that NDHA has a mast cell stabilizing property (Damazo et al., 2001).

In vitro, NDGA has also been shown to inhibit the multiplication of a vast range of viruses, including the hepatitis C virus, West Nile virus, Zika virus and dengue virus, amongst others (Craigo et al., 2000; Merino-Ramos et al., 2017; Soto-Acosta et al., 2014; Villareal et al., 2015). NDGA brings about antiviral activity through different mechanisms, including the inhibition of virion assembly, a reduction in the viral genome replication and also through the modification of the host's lipid synthesis (Soto-Acosta et al., 2014; Villareal et al., 2015).

In their study, Huang et al. established that NDGA analogues could minimize brain injury during reperfusion in rats who were subjected to transient middle cerebral artery occlusion through the attenuation of oxidative stress, both directly by scavenging reactive oxygen species and indirectly by activating the keap1/Nrf2/ARE pathway (Huang et al., 2018).

NDGA has also been found to be an anti-parasitic agent, being active against *Naegleriafowleri*, *Entamoeba histolytica* and *Giardia lamblia* (Bashyal et al., 2017). Other in vitro studies have found NDGA to be able to induce growth inhibition and apoptosis in human cancer cells through different mechanisms depending on the type of cancer (Mcdonald et al., 2002; Seufferlein et al., 2002). Indeed, NDGA has been marketed as an antineoplastic drug under the generic name of "masoprocol" for the treatment of solar keratosis (Gupta et al., 2012; Olsen et al., 1991; Schmitt and Miot, 2012).

A clinical trial where daily 2000 mg doses of NDGA were administered orally in 28-day cycles in patients with nonmetastatic hormone-sensitive prostate cancer reported that NDGA is ineffective in treating this disease as there wereno significant decrease in theprostate-specific antigen (PSA) levels after three cycles (Friedlander et al., 2012). Another study found that in mice and *in vitro*, NDGA can inhibit prostate cancer cell migration and tumour metastasis through the suppression of neuropilin 1 (Li et al., 2016). NDGA was also found to downregulate a receptor tyrosine kinase, fibroblast growth factor receptor 3 (FGFR3) which provokes human skeletal dysplasias and mutations that lead to the onset of certain malignancies such as multiple myeloma, cervical cancer, and bladder carcinoma. NDGA was reported to block downstream signaling for deactivation of FGFR3 expression related to multiple myeloma cell lines (Meyer et al., 2008). NDGA-P21, a derivative of NDGA was also shown to inhibit glioma cell progression as well as the self-renewal of glioma stem-like cell *in vitro* (Zhao et al., 2017).

Of interest to note is that though *in vitro* studies have shown that NDGA is an inhibitor of glucose-stimulated β-cells (Yamamoto et al., 1982) and is therefore expected to be diabetogenic, recent animal studies have demonstrated that NDGA can lower the glucose levels in the plasma of noninsulin-dependent diabetic mouse (Luo et al., 1998). Furthermore, NDGA was found to protect diabetic rats from the development of diabetic nephropathy; one of the important microvascular complications of diabetes (Anjaneyulu and Chopra, 2004).

Conclusions

Although NDGA cannot be used in the food industry as an antioxidant due to its nephrotoxicity and hepatotoxicity, it has very promising results in the treatment of a diverse array of diseases if administered as a pharmaceutical product. Indeed, NDGA has already been marketed as "masoprocol" in the treatment of solar keratosis. Its antitumor effects have also been reported against gliomas, multiple myeloma and prostatic cancer. Furthermore, NDGA has shown promising results as an antiviral, anti-Alzheimer's and anti-bronchoconstrictive agent, amongst others. As a result, further larger-scale studies are warranted to employ NDGA or its derivatives as a treatment option for different diseases in the real world.

Authors' contribution

RB - Designed, supervised, and reviewed the contents of the chapter, MAS - Preparing the draft of the chapter, JIA & SI - Helped in writing the draft of the chapter, AR & ZC - Helped to revise the chapter.

References

Anjaneyulu, M., Chopra, K., 2004. Nordihydroguairetic acid, a lignin, prevents oxidative stress and the development of diabetic nephropathy in rats. Pharmacology 72, 42–50. https://doi.org/10.1159/000078631.

Bashyal, B., Li, L., Bains, T., Debnath, A., LaBarbera, D.V., 2017. Larrea tridentata: A novel source for anti-parasitic agents active against Entamoeba histolytica, Giardia lamblia and Naegleria fowleri. PLoS Negl. Trop. Dis. 11, 1–19. https://doi.org/10.1371/journal.pntd.0005832.

Bergren, D.R., Valentine, J.L., 2016. Anti-anaphylactic action of nordihydroguaiaretic acid in antigen sensitized guinea pigs. Respir. Physiol. Neurobiol. 234, 26–31. https://doi.org/10.1016/j.resp.2016.09.003.

Craigo, J., Callahan, M., Chih, R., Huang, C., Delucia, A.L., 2000. Inhibition of human papillomavirus type 16 gene expression by nordihydroguaiaretic acid plant lignan derivatives. Antiviral Res. 47, 19–28.

Damazo, A.S., de Lima, W., Perretti, M., Oliani, S.M., 2001. Pharmacological modulation of allergic inflammation in the rat airways and association with mast cell heterogeneity. Eur. J. Pharmacol. 426, 123–130. https://doi.org/10.1016/s0014-2999(01)01093-7.

EFSA, 2017. EFSA Compendium of Botanicals That Have Been Reported to Contain Toxic, Addictive, Psychotropic or Other Substances of Concern. EFSA Supporting Publications, Parma, Italy. https://doi.org/10.2903/j.efsa.2009.281.

FDA, 2018. Food additive status list. https://www.fda.gov/food/ingredientspackaginglabeling/foodadditivesingredients/ucm091048.htm (Accessed 24 February, 2019).

Friedlander, T.W., Weinberg, V.K., Huang, Y., Mi, J.T., Formaker, C.G., Small, E.J., Harzstark, A.L., Lin, A.M., Fong, L., Ryan, C.J., 2012. A phase II study of insulin-like growth factor receptor inhibition with nordihydroguaiaretic acid in men with non-metastatic hormone-sensitive prostate cancer. Oncol. Rep. 27, 3–9. https://doi.org/10.3892/or.2011.1487.

Gowri, M.S., Azhar, R.K., Kraemer, F.B., Reaven, G.M., Azhar, S., 2000. Masoprocol decreases rat lipolytic activity by decreasing the phosphorylation of HSL. Am. J. Physiol. Metab. 279, E593–E600. https://doi.org/10.1152/ajpendo.2000.279.3.E593.

Gupta, A.K., Paquet, M., Villanueva, E., Brintnell, W., 2012. Interventions for actinic keratoses. Cochrane Database Syst. Rev. https://doi.org/10.1002/14651858.CD004415.pub2.

Harrison, D.E., Strong, R., Allison, D.B., Ames, B.N., Astle, C.M., Atamna, H., Fernandez, E., Flurkey, K., Javors, M.A., Nadon, N.L., Nelson, J.F., Pletcher, S., Simpkins, J.W., Smith, D., Wilkinson, J.E., Miller, R.A., 2014. Acarbose, 17-α-estradiol, and nordihydroguaiaretic acid extend mouse lifespan preferentially in males. Aging Cell 13, 273–282. https://doi.org/10.1111/acel.12170.

Huang, L., Wang, J., Chen, L., Zhu, M., Wu, S., Chu, S., Zheng, Y., Fan, Z., Zhang, J., Li, W., Chen, D., Yang, X., Wang, S., Qiu, P., Wu, J., 2018. Design, synthesis, and evaluation of NDGA analogues as potential anti-ischemic stroke agents. Eur. J. Med. Chem. 143, 1165–1173. https://doi.org/10.1016/j.ejmech.2017.09.028.

Javors, M.A., Astle, C.M., Denzel, M.S., Nadon, N.L., Nelson, J.F., Ndukum, J., Richardson, A., Salmon, A.B., Rainger, G.E., Antebi, A., Steegenga, W.T., Simpkins, J.W., Martinez, P.A., Boguc, M., Lamming, D.W., Fernandez, E., Strong, R., Harrison, D.E., Flurkey, K., de Magalhães, J.P., Miller, B.F., Miller, R.A., Sabatini, D.M., McCord, J.M., Hamilton, K.L., Müller, M., 2016. Longer lifespan in male mice treated with a weakly estrogenic agonist, an antioxidant, an α-glucosidase inhibitor or a Nrf2-inducer. Aging Cell 15, 872–884. https://doi.org/10.1111/acel.12496.

Li, X., Fan, S., Pan, X., Xiaokaiti, Y., Duan, J., Shi, Y., Pan, Y., Tie, L., Wang, X., Li, Y., Li, X., 2016. Nordihydroguaiaretic acid impairs prostate cancer cell migration and tumor metastasis by suppressing neuropilin 1. Oncotarget 7, 86225–86238. https://doi.org/10.18632/oncotarget.13368.

Lü, J., Nurko, J., Weakley, S.M., Jiang, J., Kougias, P., Lin, P.H., Yao, Q., Chen, C., 2010. Molecular mechanisms and clinical applications of nordihydroguaiaretic acid (NDGA) and its derivatives: an update. Med. Sci. Monit. 16, 93–100.

Luo, J., Chuang, T., Cheung, J., Quan, J., Tsai, J., Sullivan, C., Hector, R.F., Reed, M.J., Meszaros, K., King, S.R., Carlson, T.J., Reaven, G.M., 1998. Masoprocol (nordihydroguaiaretic acid): a new antihyperglycemic agent isolated from the creosote bush (Larrea tridentata). Eur. J. Pharmacol. 346, 77–79. https://doi.org/10.1016/S0014-2999(98)00139-3.

Madhavi, D.L., Singhal, R.S., Kulkarni, P.R., 1995. Technological aspects of food antioxidants. In: Madhavi, D.L., Deshpande, S.S., Salunke, D.K. (Eds.), Food Antioxidants: Technological; Toxicological and Health Perspectives. CRC Press, pp. 159–167.

Mcdonald, R.W., Bunjobpon, W., Liu, T., Fessler, S., Pardo, O.E., Freer, I.K.A., Glaser, M., Seckl, M.J., Robins, D.J., 2002. Synthesis and anticancer activity of nordihydroguaiaretic acid (NDGA) and analogues, Anti-Cancer Drug Des. 16 (6), 261–270.

Merino-Ramos, T., Jiménez de Oya, N., Saiz, J.-C., Martín-Acebes, M.A., 2017. Antiviral activity of nordihydroguaiaretic acid and its derivative tetra- O -methyl nordihydroguaiaretic acid against West Nile virus and Zika virus. Antimicrob. Agents Chemother. 61, 1–10. https://doi.org/10.1128/aac.00376-17.

Meyer, A.N., McAndrew, C.W., Donoghue, D.J., 2008. Nordihydroguaiaretic acid inhibits an activated fibroblast growth factor receptor 3 mutant and blocks downstream signaling in multiple myeloma cells. Cancer Res. 68, 7362–7370. https://doi.org/10.1158/0008-5472.CAN-08-0575.

Norris, D.O., Carr, J.A., 2013. Chapter 3 - Synthesis, Metabolism, and Actions of Bioregulators. In: Norris, D.O., Carr, J.A.B.T.-V.E. (Eds.). Academic Press, San Diego, pp. 41–91. https://doi.org/10.1016/B978-0-12-394815-1.00003-3.

Olsen, E.A., Lisa Abernethy, M., Kulp-Shorten, C., Callen, J.P., Glazer, S.D., Huntley, A., McCray, M., Monroe, A.B., Tschen, E., Wolf, J.E., 1991. A double-blind, vehicle-controlled study evaluating masoprocol cream in the treatment of actinic keratoses on the head and neck. J. Am. Acad. Dermatol. 24, 738–743. https://doi.org/10.1016/0190-9622(91)70113-G.

Parvez, S., Raisuddin, S., Rehman, H., Rahman, S., Ansari, R.A., 2009. Nordihydroguaiaretic acid from creosote bush (Larrea tridentata) mitigates 12-O-tetradecanoylphorbol-13-acetate-induced inflammatory and oxidative stress responses of tumor promotion cascade in mouse skin. Evid. Based Complement. Altern. Med. 2011, 1–10. https://doi.org/10.1093/ecam/nep076.

Perry, C.W., Kalnins, M.V, Deitcher, K.H., 1972. Synthesis of Lignans. I. nordihydroguaiaretc acid. J. Org. Chem. 37, 4371–4376. https://doi.org/10.1021/jo00799a605.

Ramis-Ramos, G., 2003. ANTIOXIDANTS | Synthetic Antioxidants. In: Caballero, B.B.T.-E (Ed.), Encyclopedia of Food Sciences and Nutrition (Second Edition) Academic Press, Oxford, pp. 265–275. https://doi.org/10.1016/B0-12-227055-X/00054-7.

Schmitt, J.V., Miot, H.A., 2012. Actinic keratosis: a clinical and epidemiological revision. An. Bras. Dermatol. 87, 425–434.

Seufferlein, T., Seckl, M.J., Schwarz, E., Beil, M., Wichert, G., Baust, H., Lu, H., 2002. Mechanisms of nordihydroguaiaretic acid-induced growth inhibition and apoptosis in human cancer cells, 1188–1196. https://doi.org/10.1038/sj/bjc/6600186

Siddique, Y.H., Ali, F., 2017. Protective effect of nordihydroguaiaretic acid (NDGA) on the transgenic Drosophila model of Alzheimer's disease. Chem. Biol. Interact. 100 (269), 59–66. https://doi.org/10.1016/j.cbi.2017.04.005.

Soto-Acosta, R., Bautista-Carbajal, P., Syed, G.H., Siddiqui, A., Del Angel, R.M., 2014. Nordihydroguaiaretic acid (NDGA) inhibits replication and viral morphogenesis of dengue virus. Antiviral Res. 109, 132–140. https://doi.org/10.1016/j.antiviral.2014.07.002.

Strong, R., Miller, R.A., Astle, C.M., Floyd, R.A., Flurkey, K., Hensley, K.L., Javors, M.A., Leeuwenburgh, C., Nelson, J.F., Ongini, E., Nadon, N.L., Warner, H.R., Harrison, D.E., 2008. Nordihydroguaiaretic acid and aspirin increase lifespan of genetically heterogeneous male mice. Aging Cell 7, 641–650. https://doi.org/10.1111/j.1474-9726.2008.00414.x.

Villareal, V.A., Rodgers, M.A., Costello, D.A., Yang, P.L., 2015. Targeting host lipid synthesis and metabolism to inhibit dengue and hepatitis C viruses. Antiviral Res. 124, 110–121. https://doi.org/10.1016/j.antiviral.2015.10.013.

Yamamoto, S., Nakadate, T., Nakaki, T., Ishii, K., Kato, R., 1982. Prevention of glucose-induced insulin secretion by lipoxygenase inhibitor. Eur. J. Pharmacol. 78, 225–227. https://doi.org/10.1016/0014-2999(82)90240-0.

Zhao, Q.-W., Lin, Y., Xu, C.-R., Yao, Y.-L., Cui, Y.-H., Zhang, X., Bian, X.-W., 2017. NDGA-P21, a novel derivative of nordihydroguaiaretic acid, inhibits glioma cell proliferation and stemness. Lab. Investig. 97, 1180.

CHAPTER

Octyl gallate 3.6

Renald Blundell[a,b], Muhammad Ajmal Shah[c], Joseph I. Azzopardi[a], Shabnoor Iqbal[d], Reem Hasaballah Alhasani[e], Ghulam Mujtaba Shah[f]

[a]*Department of Physiology and Biochemistry, Faculty of Medicine and Surgery, University of Malta, Msida, Malta*

[b]*Centre for Molecular Medicine and Biobanking, University of Malta, Msida, Malta*

[c]*Department of Pharmacognosy, Faculty of Pharmaceutical Sciences, Government College University, Faisalabad, Pakistan*

[d]*Department of Zoology, Faculty of Life Sciences, Government College University, Faisalabad, Pakistan*

[e]*Department of Biology, Faculty of Applied Science, Umm Al-Qura University, Makkah, Saudi Arabia*

[f]*Department of Botany, Faculty of Biological and Health Sciences, Hazara University, Mansehra, Pakistan*

3.6.1 Chemistry

OG (IUPAC name: *n*-octyl 3,4,5-trihydroxybenzoate) is the octyl ester of 3,4,5-trihydroxybenzoic acid and its chemical structure is illustrated in Fig. 3.6.1. It is a white to creamy-white crystalline odourless solid and melting point is in the range of 99–102°C. OG is insoluble in water while completely soluble in ether, propane-1,2-diol, and ethanol. Because of its longer chain, OG is stable at higher temperatures and more lipid-soluble as compared to propyl gallate (Reische et al., 2008).

3.6.2 Synthesis

OG is synthesized through the esterification of gallic acid, and octanol (molar ratio of 3:1) in the presence of sulfuric acid which acts as a catalyst. The resultant azeotropic solution distilled and mixed with petroleum ether (Sas et al., 2001). The mixture is then allowed to cool at room temperature and the OG crystals are separated by filtration, then washed with petroleum ether and dried for 24 h at a temperature of 60°C under vacuum. This results in approximately 95% pure OG at a yield of approximately 75% (Sas et al., 2001).

Antioxidants Effects in Health. DOI: https://doi.org/10.1016/B978-0-12-819096-8.00036-7

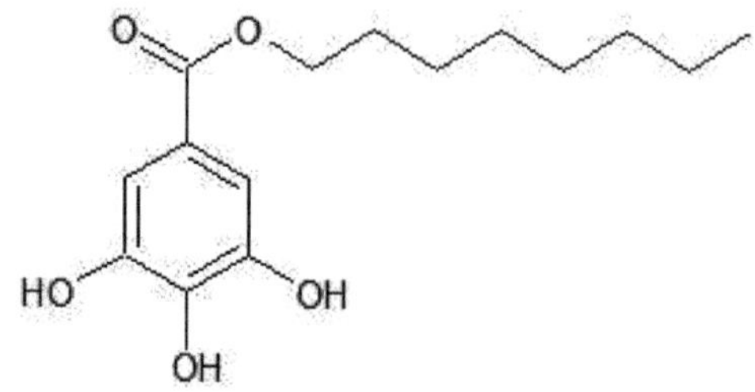

FIG. 3.6.1 The chemical structure of OG.

3.6.3 Legal status

In the EU, OG–given the E number E 311–was a food additive with MPL of between 25 to 400 mg/kg as shown in Table 3.6.1 (EFSA, 2016). Its use in the food industry as an additive in the EU was banned in 2018 in view of insufficient data regarding its toxicology profile (European Commission Directorate-General for Health and Food

Table 3.6.1 The maximum permitted levels of OG in some of the food categories as set out in Annex II of Regulation (EC) No 1333/2008 and its amendments before being banned by the EU in 2018.

Food categories	Maximum permitted level (mg/kg or mg/L as appropriate)
Processed potato products (only dehydrated potatoes)	25
Nut butters and spreads Dehydrated milk Essential oil and fat (excluding anhydrous milk fat) Oil emulsions, including spreads and liquid emulsions Breakfast cereals	200
Precooked or processed cereals Fine bakery wares Nonheat-treated processed meat Seasonings and condiments Soups and broths Sauces Processed nuts Potato-, cereal-, flour-, or starch-based snacks	200
Chewing gum Solid food supplementations like tablets and capsules (excluding chewable forms) Liquid food supplementations Syrup-type food supplementations	400

Safety, 2018). Meanwhile, in the US, its use is permitted as a chemical preservative, and for which a petition has been filed and a regulation issued (FDA, 2018).

3.6.4 Mechanisms of action

Like other alkyl gallates, OG performs its antioxidant activities through a variety of mechanisms, such as by inhibiting the enzymes involved in the production of reactive oxygen species, such as xanthine oxidase and lipoxygenases; quenching reactive oxygen species; chelating transition metal ions, such as Cu^{2+} and Fe^{2+} ions, which cause free radical damage. In their study, Kubo et al. claim that these antioxidative mechanisms are largely associated with the head and tail structural functions of these compounds (Kubo et al., 2010).

3.6.5 Effects on health

Acute toxicity studies reported LD_{50} levels of OG, that is, 4700 mg/kg per body weight of each albino rat. Other studies reported the LD50 of OG at 2710 mg/kg per body weight and 2300 mg/kg per body weight of male and female Sprague-Dawley rats, respectively. No effects were reported in pigs in whom OG was administered at a single dose (2000–6000 mg/kg per body weight) (EFSA, 2016; Van Esch, 1955). Its ban in the EU in 2018 was the result of insufficient data regarding its safety as a food additive being available at the time, including the presence of impurities, such as lead and arsenic. Thus, further safety assessments are needed to warrant the reintroduction of OG in the food industry within the EU (European Commission Directorate-General for Health and Food Safety, 2018).

Apart from slight hypochromic anemia, no tissue or organ damage was observed in rats that were fed a diet containing 180 mg/kg per body weight of OG per day for 3 months (van Esch, 1955). OG was reported to have higher sensitizing potential than other chemicals, including other alkyl gallates, such as propyl gallate and dodecyl gallate. The commonest mechanism of sensitization was found to be through the use of lipstick. Despite this, rarely have reactions due to oral intake of OG been reported, and therefore, the use of this antioxidant as an additive in food does not warrant concerns regarding allergenicity, intolerance, and hypersensitivity (Giordano-Labadie et al., 2000; Pemberton et al., 1993; Schnuch et al., 1998).

OG was shown to have antifungal activity against *Candida albicans, Aspergillus niger, Saccharomyces cerevisiae,* and *Zygosaccharomyces bailii* (Fujita and Kubo, 2002a; Hsu, 2007; Kubo et al., 2001). Apart from its antifungal properties, OG was also shown to have a broad antimicrobial spectrum, including bactericidal activity against MRSA (Rúa et al., 2013) and *Enterococcus facealis* (Rúa et al., 2013).

Furthermore, OG was exhibited an antiviral effect against RNA viruses such as herpes simplex virus type 1 (HSV-1), stomatitis virus, and poliovirus. Studies on the effect of OG on HSV-1 infection was revealed that its antiviral activity is due to the direct attenuation of the virus, as well as suppressed both the intracellular multiplication of the virus and its release besides that OG selectively accelerating the death of the virus-infected cells (Uozaki et al., 2007).

Conclusions

Although the current studies on the toxicology of OG show that it has low toxicity, and the current the toxicology results show that OG has low acute toxicity; furthermore, reports of sensitization due to its oral intake are sparse. Therefore, it is safe to assume that the use of OG in the food industry as an antioxidant does not carry any foreseeable harm. However, further studies are required to ensure the validity of such results. In the meantime, the EU does not allow OG to be used anymore as a food additive. Further, re-evaluation is needed not only to allow evaluation of its safety as a food additive but also for other potential uses, such as in the development of antiviral and antifungal medications (European Commission Directorate-General for Health and Food Safety, 2018; Fujita and Kubo, 2002b; Yamasaki et al., 2007).

Authors' contribution

RB - Designed, supervised, and reviewed the contents of the chapter, MAS - Preparing the draft of the chapter, JIA & SI - Helped in writing the draft of the chapter, RHA & GMS - Helped to revise the chapter.

References

EFSA, 2016. Scientific opinion on the re-evaluation of octyl gallate (E 311) as a food additive. EFSA J. 13 (10), 4248. https://doi.org/10.2903/j.efsa.2015.4248.

European Commission Directorate-General for Health and Food Safety, 2018. Commission Regulation (EU) 2018/1481 of 4 October 2018 amending Annexes II and III to Regulation (EC) No 1333/2008 of the European Parliament and of the Council and the Annex to Commission Regulation (EU) No 231/2012 as regards octyl gallate (E 311) and d. Official Journal of the European Union, L 251, 13–18.

FDA, 2018. *Food additive status list*. https://www.fda.gov/food/ingredientspackaginglabeling/foodadditivesingredients/ucm091048.htm. (Accessed 9 September 2018).

Fujita, K., Kubo, I., 2002a. Antifungal activity of octyl gallate. Int. J. Food Microbiol. 79 (3), 193–201. https://doi.org/10.1016/S0168-1605(02)00108-3.

Fujita, K., Kubo, I., 2002b. Antifungal activity of octyl gallate. Int. J. Food Microbiol. 79 (3), 193–201.

Giordano-Labadie, F., Schwarze, H., Bazex, J., 2000. Allergic contact dermatitis from octyl gallate in lipstick. Contact Dermatitis 42 (1), 51.

Hsu, F., 2007. Evaluation of antifungal properties of octyl gallate and its synergy with cinnamaldehyde. Bioresour. Technol. 98, 734–738. https://doi.org/10.1016/j.biortech.2006.04.002.

Kubo, I., Nihei, K., Shimizu, K., Joung Ha, T., Masuoka, N., 2010. Multifunctional antioxidant activities of alkyl gallates. Open Bioact. Compd. J. 3, 1–11. https://doi.org/10.2174/1874847301003010001.

Kubo, I., Xiao, P., Fujita, K., 2011. Antifungal activity of octyl gallate: structural criteria and mode of action, 11, 347–350.

Pemberton, M., Yeoman, C.M., Clark, A., Craig, G.T., Franklin, C.D., Gawkrodger, D.J., 1993. Allergy to octyl gallate causing stomatitis. Br. Dent. J. 175, 106. https://doi.org/10.1038/sj.bdj.4808229.

Reische, D.W., Lillard, D.A., Eitenmiller, R.R., 2008. Antioxidants. In: Akoh, C.C., Min, D.B. (Eds.), Food Lipids, Chemistry, Nutrition and Biotechnology. 3rd ed. CRC Press, New York, p. 418.

Rúa, J., García-Armesto, M.R., del Valle, P., de Arriaga, D., Gutiérrez-Fernández, J., Álvarez-Alonso, R., 2013. Antimicrobial activity of binary combinations of natural and synthetic phenolic antioxidants against Enterococcus faecalis. J. Dairy Sci. 96 (8), 4912–4920. https://doi.org/10.3168/jds.2013-6643.

Sas, B., Coppens, B., Van Hemel, J., 2001. Method of Synthesizing Alkyl Gallates (Patent No. WO 01/30299 A3). World Intellectual Property Organisation.

Schnuch, Geier, Uter, & Frosch, 1998. Patch testing with preservatives, antimicrobials and industrial biocides. Results from a multicentre study. Br. J. Dermatol. 138 (3), 467–476. https://doi.org/10.1046/j.1365-2133.1998.02126.x.

Uozaki, M., Yamasaki, H., Katsuyama, Y., Higuchi, M., 2007. Antiviral effect of octyl gallate against DNA and RNA viruses, 73, 85–91. https://doi.org/10.1016/j.antiviral.2006.07.010

Van Esch, G.J., 1955. The toxicity of the antioxidants propyl, octyl, and dodecyl gallate. Voeding 16, 683–686.

Yamasaki, H., Uozaki, M., Katsuyama, Y., Utsunomiya, H., Arakawa, T., Higuchi, M., Higuti, T., Koyama, A.H., 2007. Antiviral effect of octyl gallate against influenza and other RNA viruses. Int. J. Mol. Med. 19 (4), 685–688.

CHAPTER

Propyl gallate

3.7

Renald Blundell[a,b], Muhammad Ajmal Shah[c], Joseph I. Azzopardi[a], Shabnoor Iqbal[d], Malik Saad Ullah[e], Shahid Rasool[f]

[a]*Department of Physiology and Biochemistry, Faculty of Medicine and Surgery, University of Malta, Msida, Malta*

[b]*Centre for Molecular Medicine and Biobanking, University of Malta, Msida, Malta*

[c]*Department of Pharmacognosy, Faculty of Pharmaceutical Sciences, Government College University, Faisalabad, Pakistan*

[d]*Department of Zoology, Faculty of Life Sciences, Government College University, Faisalabad, Pakistan*

[e]*Department of Pharmaceutical Chemistry, Faculty of Pharmaceutical Sciences, Government College University, Faisalabad, Pakistan*

[f]*College of Pharmacy, University of Sargodha, Sargodha, Pakistan*

3.7.1 Chemistry

Propyl gallate (IUPAC name: 3,4,5-trihydroxybenzoate) is the propyl ester of 3,4,5-trihydroxybenzoic acid. It is an odorless, white or creamy-white, crystalline solid. It is soluble in water, propane-1,2-diol, ethanol, and ether and the melting point is 146– 150°C (EFSA, 2016). The chemical structure of propyl gallate is shown in Fig. 3.7.1.

3.7.2 Synthesis

Propyl gallate can be synthesized using the same method as that used for the synthesis of octyl gallate and other alkyl gallates described in Chapter 3.6, with the only difference being that propanol is used instead of octanol. Propyl gallate can be synthesized by using tannase of *Lactobacillus plantarum* to catalyze the transesterification of tannic acid with 1-propanol in aqueous media. The synthetic yield is depended on the pH and amount of 1-propanol used. The synthetic yields of 45% were obtained by 30% of 1-propanol at pH 5.0. The resultant product was pure chromatographically. Another noncommercial method used immobilized lipase of *Staphylococcus xylosus* to catalyze the esterification of propanol and gallic acid. The maximum yield was 90% ± 3.5 at a molar ratio of 160 at 52°C and used 400 IU of immobilized lipase and propanol/gallic acid (Bouaziz et al., 2010)

Antioxidants Effects in Health. DOI: https://doi.org/10.1016/B978-0-12-819096-8.00037-9

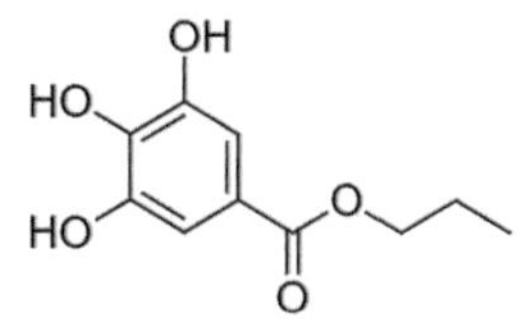

FIG. 3.7.1 The chemical structure of propyl gallate.

3.7.3 Legal status

In the EU, given the E number of E 310, propyl gallate is an authorized antioxidant with MPLs ranging between 25 and 400 mg/kg or mg/L as appropriate, depending on the food, as shown in Table 3.7.1 (EFSA, 2016). It is given the GRAS status by the FDA (FDA, 2018).

Table 3.7.1 The maximum permitted levels of propyl gallate in some of the food categories as set out in the Annex II of Regulation (EC) No 1333/2008 and its amendments.

Food categories	Maximum permitted level (mg/kg or mg/L as appropriate)
Processed potato products	25
Dehydrated milk	200
Fat and oils essentially (excluding anhydrous milk fat)	
Oil emulsions and other fat products (either individual or in combination of other food additives)	
Nut butters and nut spreads	
Breakfast cereal	
Precooked or processed cereals	
Fine bakery wares	
Nonheat-treated processed meat	
Soups and broths	
Sauces	
Potato-, cereal-, flour-, or starch-based snacks	
Processed nuts	
Chewing gum	400
Solid food supplementations, including capsules and tablets and similar forms (excluding chewable forms)	
Liquid foods	
Syrup-type food	

3.7.4 Mechanisms of action

Three reaction mechanisms: hydrogen transfer (HT), radical adduct formation (RAF), and single electron transfer (SET), are involved in the antioxidant activity of propyl gallate. Overall, the mechanism involves hydrogen removal from the hydroxyl groups. PG has three reactive sites. Two of them are located at hydroxyl equivalent 3(OH) positions, while the other one is at the hydroxyl on 4(OH) position. In medium, the PG has only two reactive sites at the hydroxyls on the 3(OH) position as the 4(OH) hydroxyl is deprotonated. The neutral and deprotonated form of PG tend to more reactive towards ·OOH radical. While the deprotonated forms is more reactive than neutral forms (Medina et al., 2013).

3.7.5 Effects on health

Exposure to propyl gallate in fertilized chicken eggs was found to have an inhibitory effect on the angiogenesis in the chorioallantoic membrane of the chicks through the upregulation of endostatin, although the mechanism by which this is upregulated on an mRNA level is still not understood (Jung and Lim, 2011). An antiangiogenic property is desired in the treatment of cancer amongst other diseases (Ng and Adamis, 2005; Petrovic, 2016). Propyl gallate was also found to be anti-nociceptive (Jung and Lim, 2011).

The growth of A549 lung cancer cells was inhibited when treated with propyl gallate. This inhibition of growth was a result of propyl gallate inducing apoptosis through chromatin and DNA fragmentation (Hamishehkar et al., 2014). Similarly, propyl gallate was also associated with the inhibition of HeLa cell growth in a dose-dependent manner by inducing apoptosis through the regulation of glutathione (Han et al., 2009). The proteasome inhibitor MG132 was upregulated the propyl gallate-induced apoptosis in HeLa cells (You and Park, 2011).

Recently, it was determined that the antitumor effect of propyl gallate extends also to hepatocellular carcinoma where it inhibits its growth both *in vitro* and in zebrafish models in a time- and dose-dependent manner. This is done through the increase in the intracellular levels of superoxide and reactive oxidative stress in the cancer cells as well as through the activation of autophagy via the formation of autophagosomes and lysosomes (Wei et al., 2019). While propyl gallate might show antitumor effects on its own, it was documented that in the presence of copper, it causes extensive DNA damage in the form of strand breaks in the isolated DNA of the bacteriophage PM2; the same results were obtained at a cellular level in human fibroblasts. Therefore, the antioxidative and cytoprotective properties are commonly associated with propyl gallate that may easily change to prooxidative, genotoxic and cytotoxic properties on exposure to copper (Hamishehkar et al., 2014; Jacobi et al., 1998). Another study also identified propyl gallate as the disrupter of the DNA-damage response due to inhibition of camptothecin-induced mediator of DNA damage checkpoint 1, leading to chromosomal aberrations (Matsuda et al., 2016).

At concentrations of 0.0032% and 0.266%, it was found that propyl gallate inhibited the growth of 27 strains of mostly Gram-positive bacteria from the human oral cavity, possibly being able to inhibit the formation of caries (Jordan et al., 1961). Indeed, a subsequent study found that the teeth of hamsters that were supplied with water containing propyl gallate exhibited significantly fewer caries than the controls (Lisanti and Eichel, 1963). Propyl gallate was also showed anaesthetic activity when applied to the lumbar plexus of frogs or intradermally in rabbits and guinea pigs; the anaesthetic effect was found to be potentiated by adrenaline (Modak and Rao, 1971).

When used at a concentration of 300 mg/mL, propyl gallate showed little antibacterial activity. However, when added to meclocyline, propyl gallate was shown to potentiate the antibacterial effect of the former, especially against resistant strains. The bacterial strains against which the meclocyline/propyl gallate mixture was effective against were *Proteus, Klebsiella, Pseudomonas,* and *Escherichia coli* (Retico et al., 1981). Propyl gallate was also shown to potentiate the antifungal activity of imidazole, itraconazole, fluconazole, azoxystrobin, and amphotericin B (D'Auria et al., 2001; Miguez et al., 2004; Strippoli et al., 2000; Teodoro et al., 2015).

Administration of 2.5% propyl gallate in gravid rats caused both maternal toxicity and slight retardation of fetal development, but no teratogenic effects were reported (Tanaka et al., 1979). In 2011, a case of contact depigmentation induced by propyl gallate in a 41-year-old woman was reported (Pandhi et al., 2011). In another case report, propyl gallate was regarded as a possible cause of allergic contact cheilitis (Özkaya et al., 2007). Propyl gallate has also been shown to exert anti-inflammatory activity in animal studies using mice and rats through the downregulation of the NF-κB and JNK pathways (Jung et al., 2011).

Conclusions

Propyl gallate is suggested as a safe food additive for commercial utilization. Propyl gallate has also shown potential use in the treatment of cancer as it can both induce apoptosis in cancerous cells and exhibit an antiangiogenic effect. Propyl gallate can also be used in the prevention of caries and, when used in conjunction with meclocyline, can also be used in the treatment of those bacterial strains which are difficult to treat with meclocyline alone. However, caution must be taken, however, to avoid exposing propyl gallate to copper, as this can cause extensive damage to the DNA.

Authors' contribution

RB - Designed, supervised, and reviewed the contents of the chapter, MAS - Preparing the draft of the chapter, JIA & SI - Helped in writing the draft of the chapter, MSU & SR - Helped to revise the chapter.

References

Bouaziz, A., et al., 2010. Enzymatic propyl gallate synthesis in solvent-free system: optimization by response surface methodology. J. Mol. Catal. B Enzym. 67 (3-4), 242–250. doi:10.1016/j.molcatb.2010.08.013.

D'Auria, F.D., Tecca, M., Strippoli, R., Simonetti, N., 2001. In vitro activity of propyl gallate–azole drug combination against fluconazole- and itraconazole-resistant Candida albicans strains. Lett. Appl. Microbiol. 32, 220–223. https://doi.org/10.1046/j.1472-765X.2001.00893.x.

EFSA, 2016. Scientific opinion on the re-evaluation of propyl gallate (E 311) as a food additive. EFSA J. 13, 4248. https://doi.org/10.2903/j.efsa.2015.4248.

FDA, 2018. Food additive status list. https://www.fda.gov/food/ingredientspackaginglabeling/foodadditivesingredients/ucm091048.htm (Accessed 24 February, 2019).

Hamishehkar, H., Khani, S., Kashanian, S., Ezzati Nazhad Dolatabadi, J., Eskandani, M., 2014. Geno- and cytotoxicity of propyl gallate food additive. Drug Chem. Toxicol. 37, 241–246. https://doi.org/10.3109/01480545.2013.838776.

Han, Y.H., Park, W.H., Pg, M., 2009. Propyl gallate inhibits the growth of HeLa cells via regulating intracellular GSH level. Food Chem. Toxicol. 47, 2531–2538. https://doi.org/10.1016/j.fct.2009.07.013.

Jacobi, H., Eicke, B., Witte, I., 1998. DNA strand break induction and enhanced cytotoxicity of propyl gallate in the presence of copper(II). Free Radic. Biol. Med. 24, 972–978. https://doi.org/10.1016/S0891-5849(97)00400-0.

Jordan, H.V., Bowler, A.E., Berger, N.D., 1961. Testing of antioxidants against experimental caries in rats. J. Dent. Res. 40, 878–883. https://doi.org/10.1177/00220345610400051201.

Jung, H.-J., Su-Jung, K., Jeon, W.-K., Kim, B.-C., Ahn, K., Kim, K., Kim, Y.-M., Park, E.-H., Lim, C.-J., 2011. Anti-inflammatory activity of n-propyl gallate through down-regulation of NF-κB and JNK pathways. Inflammation 34, 352–361.

Jung, H.J., Lim, C.J., 2011. The antiangiogenic and antinociceptive activities of n-propyl gallate. Phytother. Res. 1573, 1570–1573.

Lisanti, V.F., Eichel, B., 1963. Antioxidant inhibition of experimentally induced caries in hamsters. J. Dent. Res. 42, 1030–1035. https://doi.org/10.1177/00220345630420040501.

Matsuda, S., Matsuda, Y., Yanagisawa, S.Y., Ikura, M., Ikura, T., Matsuda, T., 2016. Disruption of DNA damage-response by propyl gallate and 9-aminoacridine. Toxicol. Sci. 151, 224–235. https://doi.org/10.1093/toxsci/kfw039.

Miguez, M., Reeve, C., Wood, P.M., Hollomon, D.W., 2004. Alternative oxidase reduces the sensitivity of Mycosphaerella graminicola to QOI fungicides. Pest Manag. Sci. 60, 3–7. https://doi.org/10.1002/ps.837.

Modak, A.T., Rao, M.R., 1971. Propyl gallate as a local anesthetic agent. Indian J. Med. Res. 59, 795–798.

Medina, M.E., et al., 2013. Antioxidant activity of propyl gallate in aqueous and lipid media: a theoretical study. Phys. Chem. Chem. Phys. 15 (31), 13137–13146. doi:10.1039/c3cp51644j.

Ng, E.W.M., Adamis, A.P., 2005. Targeting angiogenesis, the underlying disorder in neovascular age-related macular degeneration. Can. J. Ophthalmol. 40, 352–368. https://doi.org/10.1016/S0008-4182(05)80078-X.

Özkaya, E., Topkarcı, Z., Özarmağan, G., 2007. Allergic contact cheilitis from a lipstick misdiagnosed as herpes labialis: Subsequent worsening due to Zovirax® contact allergy. Australas. J. Dermatol. 48, 190–192. https://doi.org/10.1111/j.1440-0960.2007.00380.x.

Pandhi, D., Vij, A., Singal, A., 2011. Contact depigmentation induced by propyl gallate. Clin. Exp. Dermatol. 36, 366–368. https://doi.org/10.1111/j.1365-2230.2010.03930.x.

Petrovic, P., 2016. Targeting angiogenesis in cancer treatments: Where do we stand? J. Pharm. Pharm. Sci. 19, 226–238.

Retico, A., Simonetti, N., Torosantucci, A., Strippolo, V., 1981. Effect of propyl gallate on the antibacterial activity of meclocycline sulfosalicylate. Farm. Ed. Sci. 36, 817–826.

Strippoli, V., Dauria, F.D., Tecca, M., Callari, A., Simonetti, G., 2000. Propyl gallate increases in vitro antifungal imidazole activity against Candida albicans. Int. J. Antimicrob. Agents 16, 73–76.

Tanaka, S., Kawashima, K., Nakaura, S., Nagao, S., Omori, Y., 1979. Effects of dietary administration of propyl gallate during pregnancy on the prenatal and postnatal developments of rats. Food Hyg. Saf. Sci. (Shokuhin Eiseigaku Zasshi) 20, 378–384. https://doi.org/10.3358/shokueishi.20.378.

Teodoro, G.R., Ellepola, K., Seneviratne, C.J., Koga-Ito, C.Y., 2015. Potential use of phenolic acids as anti-candida agents: a review. Front. Microbiol. 6, 1410–1420.

Wei, P., Huang, C., Id, Y.C., 2019. Propyl gallate inhibits hepatocellular carcinoma cell growth through the induction of ROS and the activation of autophagy. Plos One 14 (1), 1–18.

You, B.O.R.A., Park, W.O.O.H., 2011. The enhancement of propyl gallate-induced apoptosis in HeLa cells by a proteasome inhibitor. Oncol. Rep. 25 (3), 871–877. https://doi.org/10.3892/or.2010.1121.

CHAPTER

Tert-butylhydroquinone 3.8

Renald Blundell[a,b], Muhammad Ajmal Shah[c], Joseph I. Azzopardi[a], Zunera Chauhdary[d], Shahid Shah[e], Ghulam Mujtaba Shah[f]

[a]*Department of Physiology and Biochemistry, Faculty of Medicine and Surgery, University of Malta, Msida, Malta*

[b]*Centre for Molecular Medicine and Biobanking, University of Malta, Msida, Malta*

[c]*Department of Pharmacognosy, Faculty of Pharmaceutical Sciences, Government College University, Faisalabad, Pakistan*

[d]*Department of Pharmacology, Faculty of Pharmaceutical Sciences, Government College University, Faisalabad, Pakistan*

[e]*Department of Pharmacy Practice, Faculty of Pharmaceutical Sciences, Government College University, Faisalabad, Pakistan*

[f]*Department of Botany, Faculty of Biological and Health Sciences, Hazara University, Mansehra, Pakistan*

3.8.1 Chemistry

Tert-butylhydroquinone (tBHQ, IUPAC name: 2-(1,1-Dimethylethyl)-1,4-benzenediol) is a white to brownish crystalline matter with a faint aromatic odor. It has a melting range of 126.5–128.5°C and is insoluble in water at 25°C, but soluble in fats, alcohol, acetone, and ethyl acetate (Furia, 1980; Larranaga et al., 2016; Lewis, 2012). The chemical structure of tBHQ is shown in Fig. 3.8.1.

3.8.2 Synthesis

Mostly tBQH is synthesized by reacting hydroquinone with either tert-butyl alcohol or isobutylene in the presence of a Friedel-Crafts catalyst (Kirk-Othmer, 2007).

In typical method of tertiary butyl-hydroquinone synthesis, 0.6 mole (66 g) of hydroquinone and 240 mL of phosphoric acid are mixed in toluene solvent. Tert-butanol 0.6 mole (60 mL) poured into above solution through dropping funnel at 110°C for 1–2 h. The reaction mixture is stirred for 1 h at 110°C, until layer of phosphoric acid got separated and cooled to obtain solid substance following removal of solvent under reduce pressure. This solid is further purified with toluene and dioxane. tBHQ can be synthesized by heptane solvent procedure, in which heptane 300 mL mixed with 110 g hydroquinone and 85% phosphoric acid (400 mL),

Antioxidants Effects in Health. DOI: https://doi.org/10.1016/B978-0-12-819096-8.00038-0

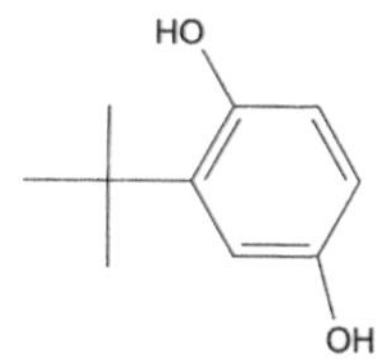

FIG. 3.8.1 The chemical structure of tert-butylhydroquinone.

refluxed at 88°C. Tertiary butyl alcohol 74 g is mixed for half an hour. The muddy layer of heptane is hardened and filtered via dilution with three volume distilled water, cooled at 2.5°C and suspended in hot distilled water. In hot suspension solvent is removed by steam distillation, filtered to obtain 60 g di-tertiary butyl hydroquinone. The filtrate is crystallized to isolate mono tertiary butyl hydroquinone (De Walt & Rodgers, 1955).

Preparation of tBHQ by hexane solvent required phosphoric acid 15 g mixed with 55 g of hydroquinone and 300 cc hexane in an autoclave equipped with stirring mode. Isobutylene passed though stirring in closed autoclave and mixture is heated at 74°C up to 14 h and cooled. The manure is surged into hot water and hexane is removed through steam distillation. This mixture is filtered to isolate 58 g of di-tertiary butyl hydroquinone, filtrate is cooled to obtain monotertiary butyl hydroquinone less than 1 g (De Walt and Rodgers, 1955).

3.8.3 Legal status

In the EU, tBHQ, with E number E 319, is authorized for use as a food additive with an acceptable daily intake of 0.7 mg/kg body weight (EFSA, 2016). tBHQ is also authorized for use in food products in combination with gallates, butylated hydroxyanisole, and butylated hydroxytoluene by European Commission (Commission, 2008) and the FDA in the US (FDA, 2018). The maximum permitted levels of antioxidants are listed in Tale 3.8.1.

tBHQ (tertiary-butyl hydroquinone), gallates (propyl gallate, octyl gallate, and dodecyl gallate), BHA (butylated hydroxyanisole), and BHT (butylated hydroxytoluene).

3.8.4 Mechanisms of action

Tertiary butyl hydroquinone (tBHQ) conferred the antioxidant potential by direct free radical scavenging potential and through singlet oxygen quenching by charge transfer mechanism (Bekdeşer et al., 2011; Kim et al., 2009). tBHQ in combination with BHA markedly decreased oxidative stress via its modulatory action on lipid

Table 3.8.1 The maximum permitted levels (MPLs) of Tbhq in combination of gallates, butylated hydroxyanisole and butylated hydroxytoluene in some of the food categories as set out in the Annex II of Regulation (EC) No 1333/2008 and its amendments.

Food additives	Food categories	MPLs as mg/kg
tBHQ, Gallates, and BHA	Dried milk powder only for vending machines	200
tBHQ, Gallates, and BHA	Water free fats and oils for manufacturing heat-treated foods. Except animal fat and olive pomace oil	200
tBHQ, Gallates, and BHA	Spreading emulsions of fats and oil	200
tBHQ, Gallates, and BHA	Processed nuts spreads	200
tBHQ, Gallates, and BHA	Processed products of dehydrated potatoes	25
tBHQ, Gallates, BHA and BHT	Chewing gum	400
tBHQ, Gallates, and BHA	Processed cereals for breakfast and cake mixes	200
tBHQ, Gallates, and BHA	Non-heat treated dehydrated meat items	200
tBHQ, Gallates, BHA and BHT	Condiments and seasonings	200
tBHQ, Gallates, and BHA	Flour, cereals, and potato based snacks	200
tBHQ, Gallates, BHA and BHT	Food supplements in solid, liquid, syrup type, and chewable forms	400

peroxidation and NADPH oxidase 2 (Sugioka and Nakano, 1976). Antioxidant molecular mechanisms comprise the upregulation of glutathione synthesizing enzymes, activation of nuclear factor erythroid related factor 2 (Nrf-2), and c-Jun transcription factors, their translocations and DNA binding to ARE (antioxidant response element) (Lee et al., 2001). It is reported that tBHQ mitigated oxidative insult by activation of ERK2, JNK1, and p38 MAP kinase signaling cascades (Park and Kim, 2014). Its antiAlzheimer's activity is granted by dramatic inhibition of plasminogen activator inhibitor-1 (PAI-1), critically associated with amyloid beta peptide deposition (Akhter et al., 2011). tBHQ upregulated the mRNA expression and proteins of hemeoxygenase-1 antioxidant enzyme through ARE dependent manner (Park and Kim, 2014),

3.8.5 Effects on health

tBHQ is an activator of Nrf2 transcription factor which is effective against inflammation and oxidative injury. As a result, tBHQ reduces the extent of hepatic ischemia/reperfusion injury that occurs in certain surgeries such as liver transplantation. This could pave the way for the therapeutic use of tBHQ to prevent hepatic ischemia/reperfusion injury in liver surgeries (Li and Xiong, 2017).

In vitro, tBHQ has also shown promising results as a potential therapeutic agent via the Nrf2/ARE pathway for those neurodegenerative diseases caused by oxidative and inflammatory injury due to iron toxicity (Xu et al., 2017). tBHQ can also attenuate neonatal hypoxic-ischemic brain damage in rats, as well as having a protective potential on the cerebral inflammatory response subsequent to traumatic brain injury in mice; both effects being attributed to the activation of the Nrf2 signaling cascades (Jin et al., 2011; Zhang et al., 2018). In rat model of hypertension, tBHQ administration has also been shown to reduce hypertension by regulating Nrf2 signaling in the hypothalamic paraventricular nucleus (Bai et al., 2017).

The healing process after a tooth extraction involves an inflammatory phase where copious amounts of reactive oxygen species are generated, causing oxidative stress leading to delay in wound healing and tissue injury. In their study, Tusi et al. found that treating the tooth sockets of rats with tBHQ following a tooth extraction resulted in enhanced healing of the hard tissues (Tusi et al., 2017).

In human monocytic leukemia U937 cells, both tBHQ and its metabolite tert-butylquinone (tBQ) were cytotoxic (Okubo et al., 2003).

tBHQ altered the thymocytes membrane potential, impairing intracellular signaling pathways by creating hyperpolarization through activation of calcium dependent potassium channels and depolarization due to increased ionic membrane permeability. Therefore may exert immunotoxic effects in neonates and adolescents through cytotoxic effect on lymphocytes (Takeda et al., 2017). Like BHA (butylated hydroxyanisole), tBHQ is cytotoxic to human lymphocytes in a dose-dependent manner where at levels of 17 mg/L reduced the lymphocytes' viability to 72% after 50 h of incubation (Schilderman et al., 1995). For comparison, an intake of 125 mg tBHQ by human volunteers resulted in serum levels of 31–37 mg/L at 3 h, which fell to 15–16 mg/L at 25 h (van Esch, 1986).

Oral administration of tBHQ has antidiabetic effect on streptozotocin induced diabetic rat model. Such findings show the potential use of this antioxidant as a way to prevent the diabetogenic effect of free-radical producers (Nishizono et al., 2000). tBHQ has also been shown to provide a protective action on the kidneys in rats against cisplatin, a chemotherapeutic agent associated with serious side effects, including nephrotoxicity (Pérez-rojas et al., 2011).

To investigate hyperplastic activity of tBHQ and related phenols, Wistar rats were divided into 10 groups and treated with tBHQ 2%, BHA 2%, and BHT 1%, mixed with their regular diet for 4 weeks. BHA significantly induced hyperkeratosis, hyperplasia, and acanthosis in forestomach mucosa and near glandular stomach. tBHQ 2% treatment resulting in mild hyperplasia and brownish discoloration of forestomach mucosa, although no stomach lesion was induced by BHA 2%. Concluding, hyperplasiogenic potential of tBHQ is due to methoxy group at para position (Altmann et al., 1985).

Basal diet containing 0.5% tBHQ for 20 weeks did not produced hyperplasia and tumorous lesion in forestomach, urinary bladder, esophagus, and glandular stomach in Syrian male hamsters. Labeling indexes analysis manifested no increase in labeling index of forestomach and urinary bladder sections (Hirose et al., 1986).

Nakagawa and Moldéus (1992) reported that F344 rats' hepatocytes incubation with 0.5 mM tBHQ as10^6 cells/mL at ambient temperature induced 100% cell death within 1–2 h. Loss of ATP and decreased protein thiol amount were noticed after initiation of cell death in medium. However, the level of malondialdehyde was not significantly increased in spite of auto-oxidation of tBHQ.

Nakagawa et al. (1994) in another study investigated the cytotoxic potential of tBHQ on hepatocytes isolated from F344 male rats. Time dependent exhaustion in the level of adenine nucleotides, GSH and protein thiols were observed at 0.5 mM concentration of tBHQ. *N*-acetylcysteine pretreatment mitigated the toxic effect of tBHQ on GSH. Oxidative phosphorylation coupled respiration in isolated mitochondria was damaged by tBHQ.

The analysis of CASE structure-activity relational system explored that tBHQ was not anticipated to prompt hepatic GST-positive preneoplastic lesions emergence in rat model (Sakai et al., 1994).

Dicoumarol, an inhibitor of DT-diaphorase increased the cytotoxic potential of tBHQ on hepatocytes isolated from F344 rats through hampering the redox cycling of hydroquinone (Nakagawa, 1996).

In male F334 rats 2% tBHQ in basal diet for 4–8 weeks, elevated the pH, and decreased osmolarity via decreasing potassium and phosphate contents in urine. tBHQ promoted DNA synthesis in urothelium and altered the surface morphology of urinary bladder epithelium (Shibata et al., 1991). Peters et al. (1996) reported the tumor promoting and nephrotoxic effect of tBHQ and BHA triggered by GSH conjugates of tBHQ in kidney and bladder tissues of male F344 rats.

The protective effect of tBHQ on mammary gland, forestomach, and ear duct carcinogenesis intitiated by 7, 12 dimethylbenz anthracene (25 mg/kg) was investigated. It was noticed that 0.8% tBHQ in basal diet of rats for 51 weeks decreased the rate mammary tumor development initiated by 7,12 dimethylbenz anthracene. Histopathological analysis revealed that tBHQ did not produce protective against tumor development in ear duct and forestomach (Hirose et al., 1988).

Hepatic necrosis was induced by nitrosamines forming mixture as sodium nitrite (125 mg/kg) and dimethylamine (1g/kg) in rats. tBHQ was administered at doses 0, 25, 75, and 225 mg/kg along sodium nitrite and dimethylamine. Ascorbic acid was administered to positive control group. Nitrosamine forming mixture alone produced marked hepatic necrosis in disease control group. tBHQ only at highest dose level and ascorbic acid significantly decreased necrosis development and suppressed the induction of ALT, AST, and ornithine carbamoyl transferase enzymes (Astill and Mulligan, 1977).

Diethyl nitrosamine single dose was administered to a group of Fischer rats, after 2 weeks followed by tBHQ 1% in basal diet for 6 weeks. Partial hepatectomy was performed at completion of 3 weeks, showed that preneoplastic GST placental form positive foci number and area was remarkably decreased in tBHQ 1% treated group compared to disease control group treated with alone diethyl nitrosamine (Ito et al., 1988).

Conclusions

tBHQ is safe for use as a food additive as established by the EFSA. Furthermore, it has also shown promising results as an adjuvant with the chemotherapeutic agent cisplatin to reduce risks of nephrotoxicity amongst other serious side effects. Further studies are needed regarding its use in the hastening the healing process following a tooth extraction through its antioxidant effect. Its cytotoxicity to lymphocytes, however, is the major drawback in its widespread use.

Authors' contribution

RB - Designed, supervised, and reviewed the contents of the chapter, MAS - Preparing the draft of the chapter, JIA & ZC - Helped in writing the draft of the chapter, SS & GMS - Helped to revise the chapter.

References

Akhter, H., Katre, A., Li, L., Liu, X., Liu, R.-M., 2011. Therapeutic potential and anti-amyloidosis mechanisms of tert-butylhydroquinone for alzheimer's disease. J. Alzheimer's D. 26, 767–778.

Altmann, H.-J., Grunow, W., Wester, P.W., Mohr, U., 1985. Induction of forestomach lesions by butylhydroanisole and structurally related substances. Arch. Toxicol. 8, 114–116.

Astill, B.D., Mulligan, L.T., 1977. Phenolic antioxidants and the inhibition of hepatotoxicity from N-dimethylnitrosamine formed *in situ* in the rat stomach. Food Cosmet. Toxicol. 15, 167–171.

Bai, J., Yu, X., Liu, K., Wang, F., Jing, G., Li, H., Zhang, Y., Huo, C., Li, X., Gao, H., Qi, J., Kang, Y., 2017. Central administration of tert -butylhydroquinone attenuates hypertension via regulating Nrf2 signaling in the hypothalamic paraventricular nucleus of hypertensive rats. Toxicol. Appl. Pharmacol. 333, 100–109. https://doi.org/10.1016/j.taap.2017.08.012.

Bekdeşer, B., Özyürek, M., Güçlü, K., Apak, R., 2011. tert-Butylhydroquinone as a spectroscopic probe for the superoxide radical scavenging activity assay of biological samples. Anal. Chem. 83, 5652–5660.

De Walt, S.Y., Rodgers, G.F., 1955. Preparation of tertiary butyl hydroquinone: Google Patents.

European Commission, 2008. Regulation (EC) No 1333/2008 of the European Parliament and of the Council of 16 December 2008 on food additives. OJEC 50, 18.

EFSA, 2016. Statement on the refined exposure assessment of tertiary-butyl hydroquinone (E 319). EFSA J. 14, 4363. https://doi.org/10.2903/j.efsa.2016.4363.

FDA, 2018. Food additive status list. https://www.fda.gov/food/ingredientspackaginglabeling/foodadditivesingredients/ucm091048.htm (Accessed 24 February, 2019).

Furia, T.E., 1980. CRC Handbook of Food Additives, 2nd ed. CRC Press, Boca Raton, FL.

Hirose, M., Inoue, T., Asamoto, M., Tagawa, Y., Ito, N., 1986. Comparison of the effects of 13 phenolic compounds in induction of proliferative lesions of the forestomach and increase in the labellin indices of the glandular stomach and urinary bladder epithelium of Syrian golden hamsters. Carcinogenesis 7, 1285–1289.

Hirose, M., Masuda, A., Fukushima, S., Ito, N., 1988. Effects of subsequent antioxidant treatment on 7,12-dimethylbenz[a]anthracene initiated carcinogenesis of the mammary gland, ear duct and forestomach in Sprague-Dawley rats. Carcinogenesis 9, 101–104.

Ito, N., Tsuda, H., Tatematsu, M., Inoue, T., Tagawa, Y., Aoki, T., Uwagawa, S., Kagawa, M., Ogiso, T., Masui, T., Imaida, K., Fukushima, S., Asamoto, M., 1988. Enhancing effect of various hepatocarcinogens on induction of preneoplastic glutathione S-transferase placental form positive foci in rats – an approach for a new medium-term bioassay system. Carcinogenesis 9, 387–394.

Jin, W., Kong, J., Wang, H., Wu, J., Lu, T., Jiang, J., Ni, H., Liang, W., 2011. Protective effect of tert-butylhydroquinone on cerebral inflammatory response following traumatic brain injury in mice 42, 714–718. https://doi.org/10.1016/j.injury.2011.03.009.

Kim, J.I., Lee, J.H., Choi, D.S., Won, B.M., Jung, M.Y., Park, J., 2009. Kinetic study of the quenching reaction of singlet oxygen by common synthetic antioxidants (tert-butylhydroxyanisol, tert-di-butylhydroxytoluene, and tert-butylhydroquinone) as compared with α-tocopherol. J. Food Sci. 74, 362–369.

Kirk-Othmer, 2007. Kirk-Othmer Encyclopedia of Chemical Technology, 5th ed. Wiley, New York, NY.

Larranaga, M.D., Lewis, R.J.S., Lewis, R.A., 2016. Hawley's Condensed Chemical Dictionary, 16th ed. John Wiley & Sons, Ltd, Hoboken, NJ.

Lee, J.-M., Moehlenkamp, J.D., Hanson, J.M., Johnson, J.A., 2001. Nrf2-dependent activation of the antioxidant responsive element by tert-butylhydroquinone is independent of oxidative stress in IMR-32 human neuroblastoma cells. Biochem. Biophys. Res. Commun 280, 286–292.

Lewis, R.J.S., 2012. Sax's Dangerous Properties of Industrial Materials, 12th ed. John Wiley & Sons, Inc, Hoboken, NJ.

Li, L., Xiong, Y., 2017. Tert-butylhydroquinone protects liver against ischemia/reperfusion injury in rats through Nrf2-activating anti-oxidative activity 372, 366–372. https://doi.org/10.1016/j.transproceed.2016.12.008.

Nakagawa, Y., 1996. Effects of dicoumarol on cytotoxicity caused by *tert* butylhydroquinone in isolated rat hepatocytes. Toxicol. Lett 84, 63–68.

Nakagawa, Y., Moldéus, P., 1992. Cytotoxic effects of phenyl-hydroquinone and some hydroquinones on isolated rat hepatocytes. Biochem. Pharmacol. 44, 1059–1065.

Nakagawa, Y., Nakajima, K., Moore, G., Moldeus, P., 1994. On the metabolism of 3- *tert*-butyl-4-hydroxyanisole- and its metabolites-induced cytotoxicities in isolated rat hepatocytes. Eur. J. Pharmacol. Environ. Toxicol. Pharmacol. Sect. 270, 341–348.

Nishizono, S., Hayami, T., Ikeda, I., Imaizumi, K., 2000. Protection against the diabetogenic effect of feeding tert-butylhydroquinone to rats prior to the administration of streptozotocin. Biosci. Biotechnol. Biochem. 64, 1153–1158. https://doi.org/10.1271/bbb.64.1153.

Okubo, T., Yokoyama, Y., Kano, K., Kano, I., 2003. Cell death induced by the phenolic antioxidant tert-butylhydroquinone and its metabolite tert -butylquinone in human monocytic leukemia U937 cells. Food Chem. Toxicol. 41, 679–688. https://doi.org/10.1016/S0278-6915(03)00002-4.

Park, J.-S., Kim, H.-S., 2014. Regulation of hemeoxygenase-1 gene expression by Nrf2 and c-Jun in tertiary butylhydroquinone-stimulated rat primary astrocytes. Biochem. Biophys. Res. Commun 447, 672–677.

Pérez-rojas, J.M., Guerrero-beltrán, C.E., Cruz, C., Sánchez-gonzález, D.J., Martínez-martínez, C.M., Pedraza-chaverri, J., 2011. Preventive effect of tert -butylhydroquinone on cisplatin-induced nephrotoxicity in rats 49, 2631–2637. https://doi.org/10.1016/j.fct.2011.07.008.

Sakai, T., Klopman, G., Rosenkranz, H.S., 1994. Structural basis for the induction of preneoplastic glutathione S-transferase positive foci by hepatocarcinogens. Teratog. Carcinog. Mutagen. 14, 219–237.

Schilderman, P.A.E.L., Rhijnsburger, E., Zwingmann, I., Kleinjans, J.C.S., 1995. Induction of oxidative DNA damage and enhancement of cell proliferation in human lymphocytes in vitro by butylated hydroxyanisole. Carcinogenesis 16, 507–512. https://doi.org/10.1093/carcin/16.3.507.

Shibata, M.-A., Asakawa, E., Hagiwara, A., Kurata, Y., Fukushima, S., 1991. DNA synthesis and scanning electron microscopic lesions in renal pelvic epithelium of rats treated with bladder cancer promoters. Toxicol. Lett. 55, 263–272.

Sugioka, K., Nakano, M., 1976. A possible mechanism of the generation of singlet molecular oxygen in NADPH-dependent microsomal lipid peroxidation. BBA 423, 203–216.

Takeda, M., Oyama, K., Kamemura, N., Kanemaru, K., Yuasa, K., Yokoigawa, K., Oyama, Y., 2017. Change in plasma membrane potential of rat thymocytes by tert-butylhydroquinone, a food additive: Possible risk on lymphocytes. Food Chem. Toxicol. 109, 296–301.

Tusi, S.K., Manesh, T.E., Fathollahi, M.S., Bagherian, A., 2017. Can tert-butylhydroquinone improve the healing of extracted tooth socket in rats? Dent Res J (Isfahan) 14 (1), 8–12. doi:10.4103/1735-3327.201134

van Esch, G.J., 1986. Toxicology of tert-butylhydroquinone (TBHQ). Food Chem. Toxicol. 24, 1063–1065. https://doi.org/10.1016/0278-6915(86)90289-9.

Xu, W., Li, F., Xu, Z., Sun, B., Cao, J., Liu, Y., 2017. Tert-butylhydroquinone protects PC12 cells against ferrous sulfate-induced oxidative and inflammatory injury via the Nrf2/ARE pathway. Chem. Biol. Interact. https://doi.org/10.1016/j.cbi.2017.05.021.

Zhang, J., Donovan, L., Lu, Y., Yang, L., Wu, C., Li, Y., Zhang, Q., 2018. Neurochemistry international tert-butylhydroquinone post-treatment attenuates neonatal hypoxic-ischemic brain damage in rats. Neurochem. Int. 116, 1–12. https://doi.org/10.1016/j.neuint.2018.03.004.

PART

Natural occurring antioxidants: bright and the dark side

4

CHAPTER

4.1 Amino acid: Essential builiding blocks for Human body

Dheeraj Bisht[a], Harikesh Maurya[b], Tanuj Joshi[d], Tarun Belwal[c], Arasana Dhariwal[a], Aadesh Kumar[d]

[a]*Department of Pharmaceutical Sciences Sir J.C. Bose Technical Campus, Nainital, Uttarakhand, India*

[b]*M.G.B. Rajat College of Pharmacy. Gohila, Hanswar, Ambedkar Nagar, U.P., India*

[c]*College of Biosystems Engineering and Food Science, Zhejiang Key Laboratory for Agri-Food Processing, Zhejiang University, Hangzhou, People's Republic of China*

[d]*Department of Pharmaceutical Sciences, Kumaun University, Campus Bhimtal, Bhimtal, Uttarakhand, India*

4.1.1 Introduction

Amino acid defined as an accumulation of organic compounds that comprises an amino group (-NH_2) as a basic moiety along with an acidic carboxyl group (-COOH), and winged with an alkyl group (-R) that considered as a side chain, which is exclusive to every amino acid. The α-amino carboxylic acid is an actual amino acid. Every molecule possesses a central carbon (C) atom, known as α-carbon, to that an amino, as well as carboxyl group, adhere. The remaining two bonds of the α-carbon atom are usually satisfied with a hydrogen (H) atom and the al moiety (Akram et al., 2011).

The amino acids alter one another in the specified chemical structure of the R group. Amino acids are made up of monomers that make contrive proteins (Azad, 2017).

4.1.2 Sources

The National Institutes of Health reported that amino acids contain all nine essential amino acids associated with dietary proteins including animal sources, such as meats, milk, fish, beef, pork, poultry, eggs, the main food category of meat products is a source of 6 essential amino acids and plant sources such as whole sources of soy, beans, legumes, nut butter, and grains are called a complete protein (Hou and Wu, 2018). It is well known that the plant protein contents lower essential amino acids as compared to animal proteins but health professionals should encourage

Antioxidants Effects in Health. DOI: https://doi.org/10.1016/B978-0-12-819096-8.00041-0

a vegetarian protein diet having all essential amino acids that would be sufficient without consuming animal products (Warsewicz et al., 2018).

4.1.3 Chemistry

A fundamental structure of proteins comprises Amino acids as monomers in the basic frame. Every amino acid has an innermost carbon atom (α-carbon) bonded with an amino group (-NH_2) and carboxyl group (-COOH) as well their hydrogen atom (Azad, 2017). Under physiological conditions, the aqueous position of the final structure is ionized via amino (-NH_3^+) and carboxyl (-COO^-) groups respectively. Each amino acid also bears one another -R group (side chain) attached to the central atom that is admitted for atom or class of atoms. This alkyl group is responsible for specific characteristics of each protein include size, polarity, and pH (Reddy, 2019).

The chemical structure of a common amino acid is Fig. 4.1.1.

4.1.4 Bioavailability

The comparative bioavailability of two evaluated products of protein-containing amino acids was evaluated after single oral administration under the circadian plasma concentration rhythms has been considered before and after processing of the same protein. Bioavailability is also affected by gastric emptying time. Lactalbumin on enzymatic hydrolysis lowers 12% of the bioavailability of the amino acids quantitatively. Concerning the nutrition of patients and baby's liquid products, it is reported around 10% better bioavailability than identical dry products. The sterilization and ultrahigh heat treatment of milk protein products also reported as slight improvements (1%) in bioavailability. The dried green peas intended for use in simple processing food were observed as raise bioavailability of about 20% of the total protein (Moch and Kübler, 1993).

4.1.5 Mechanisms of action

Amino acid-like arginine, cysteine, glutamine, leucine, and methionine are the main components of protein synthesis that regulate the immune system and metabolism

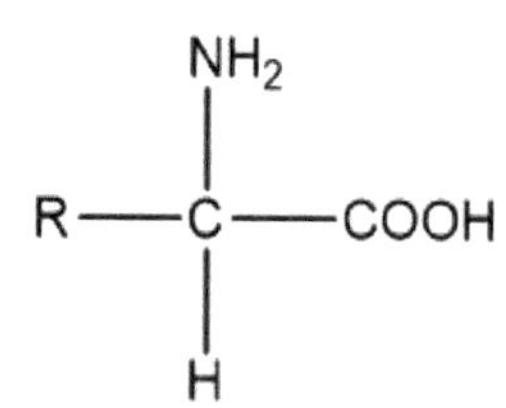

FIG. 4.1.1 Fundamental arrangement of amino acid (Azad, 2017).

in the living body. Glutamines responsible for immunity boost up, minimize the inflammatory reaction, and develop the intestinal barrier. The catabolism of arginine may produce NO, alleviate inflammatory reaction, generate ornithine, and promoting wound healing, while the muscle protein synthesis was promoted by leucine through the mTOR pathway. Cysteine and methionine can facilitate the formation of collagen and connective tissue that beneficial for wound healing and leads to a favourable impact on lipid metabolism (Tian and Wang, 2013).

4.1.6 Pro-oxidant activity

The pro-oxidant and antioxidant activities of universal food additives and broccoli's amino acids were evaluated in a study, which shows that there was none of the hydrogen peroxide-scavenging effects was observed with broccoli's amino acids. In another study, it was observed that the descending order of Asp, SMC, GABA, Glu, Gln, Pro, Phe, Leu, Lys, Arg, Asn, Val, Ile, His, Ser, Gly, Orn, and Ala moderately reserved deoxyribose destruction in the existence of ferric-EDTA and H_2O_2 after dissolving in water at a concentration of 0.5 and 0.05 mM. Eventually, it was observed that the Tyr and Thr acted as pro-oxidants in this method (Martinez-Tome et al., 2001).

4.1.7 Beneficial and detrimental effects on health

All the essential amino acids perform numerous significant activities that have beneficial as well as detrimental effects shown in Table 4.1.1.

4.1.8 *In-vitro* evidence of amino acids

An in vitro study designed as the nuclear fractions isolated from rat liver were capable to incorporate labelled amino acids, where the DNA containing a fraction of proteins were very smaller amount. Finally, it was reported that the active RNP-particles are presented in rat liver nuclei (Rendi, 1960).

Several *in-vitro* studies analyze the growth of human embryos achieved after IVF blastocyst phase with or without the arrangement of two to four cell juncture of amino acids. The study was designed on 129 human embryos that arbitrarily dispersed among three serum-free chronological medium separately for subsequent periods of 24 h with 20 essential AA that responsible for blastocyst formation. All embryos were fully developed on the sixth day; the blastocysts stage differentially labeled and examined with the number of trophectoderm and inner cell mass, mitoses, and dead cells. The study reviles that the arrangement of AA significantly ($P = 0.005$) increases blastocyst cells such as Earle's + Gln (61.8 ± 4.2), Earle's + AA (99.3 ± 8.4), and K-SCIM (100.2 ± 9.4). It was also present in both trophectoderm and inner cell mass lineages ($P < 0.02$). While the dead cell index was significantly ($P = 0.047$) lesser with Earle's + AA (Devreker et al., 2001).

Table 4.1.1 Beneficial and detrimental effect of amino acids.

Amino acids	Safe dose	Beneficial effects	Detrimental effects
Phenylalanine	50–100 mg/kg	Help alleviate symptoms of alcohol withdrawal, pain relief, treating Parkinson's disease, attention deficit hyperactivity disorder (ADHD). Precursor for neurotransmitters likes tyrosine, dopamine, epinephrine, and norepinephrine (Akram et al., 2020).	It is recommended by the experts to take or reduce phenylalanine diet for at least 20 weeks before getting pregnant. This should reduce the risk of birth defects. A movement disorder (tardive dyskinesia) was reported in people with schizophrenia while taking phenylalanine (George Kapalka, 2010).
Valine	220–240 mg/kg	Valine helpful in decreasing fatigue due to extensive exercise, building muscles, protecting against liver cancer, serious muscle injuries, stimulate muscle growth and regeneration, maple syrup urine disease (Holecek, 2018).	Very high intake of protein and mixed amino acids (more than three times the RDA or 2.4 g/kg) is thought to increase the risk of renal glomerular sclerosis and accelerate osteoporosis (Kohlmeier, 2003).
Threonine	20–30 mg/kg	Useful in spinal spasticity, multiple sclerosis, familial spastic paraparesis, and amyotrophic lateral sclerosis. Components of collagen and elastin, useful in fat metabolism and immune function (Borgonha et al., 2002).	The detrimental effects of high temperatures are exacerbated by high RH. Further, heat stress negatively affects the immune functions. The effect of supplemental l-threonine on laying performance, serum free amino acids, and immune function of laying hens under high-temperature and high-humidity environmental climates (Azzam et al., 2011).
Tryptophan	50–80 mg/kg	Useful in attention deficit-hyperactivity disorder, seasonal depression, anxiety, gout, premenstrual syndrome, Tourette syndrome, myofascial pain syndrome. Maintain proper nitrogen balance and precursor to serotonin, a neurotransmitter that regulates appetite, sleep, and mood (Markus, 2008).	Higher dose of L-tryptophan can produces some side effects like acidity, stomach ache, nausea, belching, vomiting, and diarrhea, and loss of appetite. And sometimes it causes headache, drowsiness, visual blurring, muscle weakness, and dry mouth (Anonymous, n.d.).

Table 4.1.1 Beneficial and detrimental effect of amino acids. *Continued*

Amino acids	Safe dose	Beneficial effects	Detrimental effects
Methionine	5–10 grams per day	Useful in liver disorders, recover wound healing, depression, alcoholism, allergies, asthma, copper poisoning, schizophrenia, drug withdrawal, and Parkinson's disease. Essential for metabolism and detoxification, tissue growth and absorption of zinc, selenium, and minerals (Holecek, 2018).	Methionine was labeled as most toxic amino acid in animal growth. In human at a higher dose methionine work as a precursor of homocysteine that plays a major role in vascular damage and cardiovascular disease. A 10-fold higher dose, give by mistake, results in death (Garlick, 2006).
Leucine	2–5 grams per day	Helpful to regulate blood sugar levels, stimulates wound healing and produces growth hormones. Essential for protein synthesis and muscle repair (Balage and Dardevet, 2012).	Very high dose of leucine may cause hypoglycemia (low blood sugar level). Pregnant woman should not use leucine supplememnts (Anonymous, n.d.).
Isoleucine	48–72 mg/kg	Help to control blood sugar, boost energy, endurance, muscles injuries, muscle development, and lean body mass. Vital for immune function, hemoglobin production, and energy regulation (Kurpad et al., 2006).	After prolonged use of isoleucine (upto 2 year), some side effects, such as fatigue and loss of co-ordination may occur (Anonymous, n.d.).
Lysine	1–3 grams per day	Useful in red eyes, fatigue, hair loss, anorexia, anemia, reproductive system, cold sores, canker sores, diabetes, stress. Essential for protein synthesis, energy production, hormone, antibodies, enzyme production, immune function, production of collagen and elastin (Wolfe, 2017).	When excess (more than 61g/day) amount of lysine was administration in 140–150 kg bull calves, severe but transit diarrhea was occurred, reported by Abe et al. in their study (Abe et al., 2001).
Histidine	4–5 grams per day	Useful in rheumatoid arthritis, ulcers, cholera, anemia associated with kidney failure or kidney dialysis, metabolic syndrome. Essential to produce histamine, a neurotransmitter help in immune response, digestion, sexual function, sleep-wake cycles and maintain the myelin sheath (Spillane et al., 2012).	Linear decline in whole body protein turnover was found after excess use of histidine, in which significant decrease in albumin, transferrin, and hemoglobin was noted over the histidine depletion period (Kriengsinyos et al., 2002).

In another study, it was observed that the brain interstitial fluid concentration of excitotoxic AA and L-glutamate closely regulated via uptake transporters and metabolism in astrocytes and neurons. The study suggested that the blood-brain barrier itself may contribute to regulating the concentration of brain L-glutamate (Helms et al., 2012).

An *in-vitro* study designed as rabbit optic nerve divisions represents the concentrative accumulation of neutral, acidic, and basic AA, especially of glycine, proline, and aspartic acid incubated in Na^+ deficient medium in cold (0°C). The anoxia, cyanide, di-nitrophenol (DNP), ouabain present in media inhibited saturated proline uptake. The study reveals that the rabbit optic nerve restrains carrier systems for neutral and acidic amino acids, whether the neutral group has similar systems as the specific carrier system and the Ehrlich cell to substantial degrees by system A (Cotlier et al., 1971).

4.1.8.1 Animal studies

Animal studies play a major role in defining the biological activity of any compound or plant material. That could be better understood with Table. 4.2.

4.1.9 Clinical studies

Nishitani et al. (2004) reported that the combination of BCAA: Leu, Ile, and Val (branched-chain amino acid) were used to treat hypoalbuminemia in Japanese patients with liver cirrhosis. It is well known that during protein synthesis, leucine stimulates mammalian target of rapamycin (mTOR) signals and inhibits the protein degradation and also activates glycogen synthase via mTOR signals in L6 cell, but not hepatocyte. The study reported that leucine improved glucose metabolism in normal and liver cirrhosis rats (Nishitani et al., 2004).

The phase I and II clinical trials in the National Cancer Institute evaluated a new nitrosourea, that is, diethyl-l-(3-(2-chloroethyl)-3-nitrosoureido] ethyl phosphonate (S 10036) because of its high potency. The phase I study has been performed on 22 cancer patients of advanced stage. The drug was administered at a dose range of 25–200 mg/m^2/week in slow i.v. infusion for 60 min on days 1, 8, 15, and 22 followed by four consecutive weeks using the modified Fibonacci scheme. The dose-limiting toxicity like acute thrombocytopenia at a dose of 100 mg/m^2/week or above, while hematological toxicity was observed as dose dependent. Nausea and vomiting were observed as moderate to severe.

Phase II study designed for S 10036 at a dose of 100 mg/m^2/week for four successive weeks (induction therapy) in the group of patients exclusive of prior therapy and another group as 100 mg/m^2/week for three successive weeks of patients with prior chemotherapy or radiotherapy. The recommended dose for the second cycle of S 10036 chemotherapy (maintenance therapy) is 100 mg/m^2/week every 3 weeks seems to be legitimate due to cumulative toxicity (Khayat et al., 1987).

Scherbakov et al. (2016) observed a randomized placebo-controlled dual-blinded clinical trial as a supplement of essential amino acids (EAAs) in patients with stroke who are at a major risk for long-term disability. Physical and functional performances

Table 4.1.2 Summary of animal safety studies with amino acids.

Amino acid	Animal used/ dose/ duration	Inference
Monosodium glutamate	Wistar rats – daily oral dose 30 and 60 g/kg b.w. for 2 months	Monosodium glutamate may produce deleterious effects on the testes of Wistar rats and by extension may contribute to the causes of male infertility. Thus, it is important to reconsider the usage as a flavor enhancer (Alalwani, 2014).
L-glutamine	Sprague-Dawley rats. Doses 1.25%, 2.5%, and 5.0% (*w/w*) for 13 weeks	The elevated urine parameters were observed with the dose of 2.5% and 5.0% groups at the end of study, while little increase in hematology parameters were observed with 5.0% group and nonadverse effect was estimated at 1.25% (w/w) dose (Tsubuku et al., 2004).
L-arginine	Sprague-Dawley rats – doses 1.25%, 2.5%, and 5.0% (*w/w*) for 13 weeks	It was found an increase in hemoglobin, together with a tendency toward an increase in the red blood cell counts, but the change was toxicologically insignificant and non adverse effect was estimated at 5.0% (w/w) dose (Tsubuku et al., 2004).
L-valine	Sprague-Dawley rats – doses 1.25%, 2.5%, and 5.0% (*w/w*) for 13 weeks	Valine at the dose of 5.0% (w/w) showed slightly decreased body weight at the end of study, while hepatology and ophthalmology were shows nontoxicological effects and also estimate that no adverse effect observed at the same dose (Tsubuku et al., 2004).
L-leucine	Sprague-Dawley rats – doses 1.25%, 2.5%, and 5.0% (*w/w*) for 13 weeks	The rats fed with leucine supplemented diet showed nonsignificant dose-related effects on body weight and also hepatology and ophthalmology were shows nontoxicological effects further it was also estimate with no adverse effect at the dose of 5.0% w/w (Tsubuku et al., 2004).
L-isoleucine	Sprague-Dawley rats – doses 1.25%, 2.5%, and 5.0% (*w/w*) for 13 weeks	Isoleucine at the dose of 5.0% w/w shows nonsignificant effects on body weight, and also hepatology and ophthalmology were shows nontoxicological effects at the end of study. It slightly affected the urine electrolytes, protein, ketone bodies, urine glucose, and urobilinogen. Isoleucine at the dose of 2.5% w/w observed no adverse effect during and after the study (Tsubuku et al., 2004).
L-lysine	Sprague-Dawley rats – doses 1.25%, 2.5%, and 5.0% (*w/w*) for 13 weeks	Lysine treated group of animal were not showed any changes in clinical signs, body weights, diet consumption, water intake, ophthalmology, histopathology, organ weights, and also none of the functional, biochemical, or renal function changes were observed at the end of study. It was estimate with no adverse effect found at the dose of 5.0% w/w (Tsubuku et al., 2004).

(Continued)

Table 4.1.2 Summary of animal safety studies with amino acids. *Continued*

Amino acid	Animal used/ dose/ duration	Inference
L-alanine	Sprague-Dawley strain SPF rats – daily oral dose 2000 mg/kg for 4 weeks.	L-alanine showed nonsignificant toxicological changes in body weight, food consumption, ophthalmology, hematology, blood chemistry, organ weight. The urinalysis was observed as positive urine protein or phosphate salt with increased urine volume. Further it is suggest that repeated oral administration of L-alanine at 2000 mg/kg/day well tolerated in animal (Mami Aoki et al., 2014).
L-asparagine	Rats – at dose levels of 0%, 1.25%, 2.5%, 5% in diet for 13 weeks	L-asparagine showed significant decrease in body weight of rats treated with 5% and 1.25% groups. While, the rats treated with 5% showed significant increases in relative organ weights like brain, kidney, and testis. The biochemical test, that is, GLU, PL, K, and ALT were increased significantly in 5% treated group. The histopathology shows nonsignificant variation in development of lesions among the groups (Yokohira et al., 2008).
L-asparatic acid	Rats – at dose levels of 0%, 0.05%, 1.25%, 2.5%, 5.0% in diet for 13 weeks	L-aspartic acid groups showed decrease in BUN, creatinine and uric acid levels but the levels of urinary ketone and protein were significantly increased, while relative kidney weight was significantly increased in the 5.0%, and regenerative renal tubules with tubular dilation were observed in the 2.5% or greater groups. Finally, it indicates that 2.5% and/or higher doses of L-aspartic acid cause toxic effects on kidneys and salivary glands at high dose levels (Tada, et al., 2008).
L-cystein	Sprague-Dawley rats – daily i.v. dose of 0, 100, 300, and 1000 mg/kg b.w. for 4 weeks	L-cysteine 1000 mg/kg group showed significant gain in body weight even food consumption was reduced on day 3 of study. The salivation, stereotypy, ptosis, tremor, hemoglobin, hematocrit, and mean corpuscular volume were decreases, while the slight increase in the count of reticulocyte was observed. Histopathological examination showed sperm granulomas in the epididymis and necrosis of the Purkinje cells and granular layer in the cerebellum with 1000 mg/kg group. Slight tubular basophilia with blood or hyaline casts, proteinuria, or occult blood in urinalysis was observed in the kidney with 300 mg/kg and 1000 mg/kg groups (Sawamoto, et al., 2003).

Table 4.1.2 Summary of animal safety studies with amino acids. *Continued*

Amino acid	Animal used/ dose/ duration	Inference
Glycine	Rats – daily oral doses of 500, 1000, and 2000 mg/kg b.w. in a volume of 10 mL/kg for 4 weeks	All the glycine-treated groups did not showed any toxicological changes at a dose of 2000 mg/kg/day during the study. There were no histopathological changes in the kidneys or urinary bladder and non changes in other biochemical parameters (Shibui et al., 2013).
L-histidine	Rats – dose levels of 0%, 0.31%, 0.62%, 1.25%, 2.5%, and 5.0% in the diet for 13-week sub-chronic toxicity study.	L-histidine treated group showed increased levels of BUN and creatinine and sperm granulomas in the epididymis were found in half of the 5.0% group. It was concluded that the dose of 2.5%, 5.0% in diet proved to exert significant toxicity reflected in suppression of body weight gain and formation of sperm granulomas (Ikezaki et al., 1994).
L-methionine	Rats – dose levels of 0%, 0.1%, 0.3%, 0.9%, 2.7% (w/w) in the diet for 4-week toxicity study	The maximum methionine ingestion group was showed decrease in WBC count, thymus atrophy, and histological abnormalities in the adrenal gland and testis. Significant growth suppression and minor changes in biochemical parameters were observed with rat containing 0.9% (w/w) group. None of adverse effects were observed at 0.3% and 0.9% (w/w), corresponding to 236 and 705 mg/kg/d body weight, respectively (Chin et al., 2015).

were examined by exercise reports before additives of EAAs and discharge from the in-patient rehabilitation, at 12 weeks to 1 year. Muscle weakness was observed after giving the stroke in the first week of treatment. The physical and functional performance was directed to improve independently before the post-stroke rehabilitation plan. Finally, it was observed that the rehabilitation plan and supplementation of EAAs have prevented muscle wasting and augmenting to improve the accumulation of muscle strength and functional independence after stroke (Scherbakov et al., 2016).

Xu et al. (2014) observed nine meta-analyses in elderly people while potentially relative clinical trials to compare the efficacy of AA supplement with placebo control for the enhancement of lean body mass (LBM), leg muscle strength, and/or reduction in sarcopenia. The studies investigating this issue have been incompatible and unclear since the assistance of nutritional intervention, intervention time, and physical exercise that improve muscle loss. The overall difference was not significant in changes of LBM from baseline to the end of the study compared to the treatment and placebo groups. In contrast, the mean changes from baseline in double leg press and leg extension were also shown non-significant differences respectively. Finally, it was indicated that the AA supplements did not increase LBM and muscle strength as compared to placebo control in a range of aged people (Xu et al., 2014).

4.1.10 Effect of antioxidants on the gastrointestinal tract

The superior quantity of antioxidants present in convinced foods, such as fruits, vegetables, grains, and green tea plays an important role in defending gastrointestinal tract (GIT) damage to oxidative stress, and interruption in the development of stomach, colon, and rectal cancer. It is familiar that the carotenoids and flavonoids didn't absorb vitamin C and E so that their concentrations are greatly higher in the lumen of the GIT than plasma or other body tissues. Finally, it was concluded that the development of antioxidant action in the GIT is more prone to the gastrointestinal tract (Halliwell et al., 2000).

The dietary nutrients are digested and absorbed by the GIT, which protects the body against physical and chemical damage from lumen contents, immunity against antigens, and optimum environment for the gut microbiota. In the case of inflammatory bowel disease and irritable bowel syndrome that involves ulcerative colitis and Crohn's disease. It works as an antioxidant as obtained from diet but unable to digest adequately. As it is clear that these kinds of ailments are involved with oxidative stress, a range of antioxidant additives are employed for the conservation and rehabilitation of gut functions. A few studies reveal that the supplements administered in Crohn's disease, the apparent benefits occurred with such as allopurinol, omega-3 fatty acids, Boswellia serrata, Tripterygium wilfordii, and Artemesia species. In the case of ulcerative colitis, the related useful supplements were as follows; allopurinol, omega-3 fatty acids, Matricaria chamomilla, Curcuma longa, Chinese multiherbal preparation, and Ayurvedic preparation were also used with beneficial effects. The secondary benefits of several antioxidant supplements were useful during irritable bowel syndrome. Finally, it was concluded that the various antioxidant supplements could be favorable at certain stages of precise ailments by inhibiting oxidative stress pathways in tissues and by a natural process, not by merely acting like scavengers (Khan et al., 2017).

Conclusion

Amino acids are indispensable chemicals, working as essential building blocks of peptides and proteins, that are needed by the body for the most favourable metabolism and appropriate body functioning. It is known as essential, nonessential, and conditionally essential, that plays vital roles such as protein synthesis and act as precursors in the production of secondary metabolism molecules. The components of amino acid contain glutamine, arginine, leucine, methionine, and cysteine regulates the immune system and metabolism, and also participating in body protein synthesis. In which, glutamine promotes recovery of immunity, improves the intestinal barrier, but reduces inflammatory reaction. The catabolism of arginine may generate ornithine which leads to producing NO^-, and alleviate inflammatory reaction to promote wound healing. The leucine through the mTOR pathway promotes muscle protein synthesis; however, the influence on metabolism is still disputed. Methionine and cysteine promote the synthesis of connective tissue and collagen beneficial to wound healing, and their valuable effects help in lipid biotransformation. Excess intake of some specific antioxidants may produce harmful effects on human health. Many folds of dose may cause death. So amino acids are essential for health and body functioning but with limited concentration, higher and prolonged use would lead to serious complications.

Abbreviations

AA	Amino acids
Ala	Alanine
Arg	Arginine
BCAA	Branched-chain amino acids
Gly	Glycine
Glu	Glutamine
Leu	Leucine
Met	Methionine
mTOR	Mammalian target of rapamycin
mRNA	Messenger ribose nucleic acid
NO	Nitric oxide
Val	Valine

References

Abe, M., Iriki, T., Kaneshige, K., Kuwashima, K., Wantanabe, S., Sato, H., Funaba, M., 2001. Adverse effects of excess lysine in calves. J. Anim. Sci. 79 (5), 1337–1345.

Akram, M., Asif, H.M., Uzair, M., Akhtar, N., 2011. Amino acids: a review article. J. Med. Plant. Res. 5 (17), 3997–4000.

Akram, M., Daniyal, M., Ali, A., Zainab, R., Muhammad Ali Shah, S., Munir, N., Mahmood Tahir, I., 2020. Role of phenylalanine and its metabolites in health and neurological disorders. Synucleins Biochemistry and Role in Diseases. doi:10.5772/intechopen.83648.

Alalwani, A.D., 2014. Monosodium glutamate-induced testicular lesions in rats (histological study). Middle East Fertil. Soc. J. 19 (4), 274–280.

Anonymous, L-tryptophan, 2021. https://www.rxlist.com/l-tryptophan/supplements.htm.

Anonymous, 2021. Leucine, Health Encyclopedia. University of Rochester Medical Centre, New York. https://www.urmc.rochester.edu/encyclopedia/content.aspx?contenttypeid=19&contentid=Leucine. (Accessed on May 20, 2021).

Anonymous, 2021. Branched-chain amino acids. Vitamin and supplements. https://www.webmed.com/vitamins/ai/ingredientmono-1005/branched-chain-aminino-acids-bcaas. (Accessed on May 20, 2021).

Aoki, M., Mochizuki, M., Okamura, T., Hatayama, K., Nakamura, A., Morishita, K., 2014. A 4-week oral toxicity study of L-alanine in rats with a recovery period of 2 weeks. Fundam. Toxicol. Sci. 1 (2), 63–72.

Azad, S., 2017. Amino acids: its types and uses. Int. J. Clin. Diagn. Pathol. 1 (1), 13–16.

Azzam, M.M.M., Dong, X.Y., Xie, P., Wang, C., Zou, X.T, 2011. The effect of supplemental l-threonine on laying performance, serum free amino acids, and immune function of laying hens under high-temperature and high-humidity environmental climates. J. Appl. Poult. Res. 20 (3), 361–370.

Balage, M., Dardevet, D., 2012. Long-term effects of leucine supplementation on body composition. Curr. Opin. Clin. Nutr. Metab. Care 13 (3), 265–270.

Borgonha, S., Regan, M.M., Seung, H.O., Condon, M., Young, V.R., 2002. Threonine requirement of healthy adults, derived with a 24-h indicator amino acid balance technique. Am. J. Clin. Nutr. 75 (4), 698–704.

Chin, K., Toue, S., Kawamata, Y., Watanabe, A., Miwa, T., Smriga, M., Sakai, Ryosei, 2015. A 4-week toxicity study of methionine in male rats. Int. J. Toxicol. 34 (3), 233–241.

Cotlier, E., Becker, B., Beaty, C., Coben, L.A., 1971. Amino acid transport by rabbit optic nerve *in-vitro*. Invest. Ophthalmol. Vis. Sci. 10, 876–886.

Devreker, F., Hardy, K., Bergh, M.V.D., Vannin, A.S., Emiliani, S., Englert, Y., 2001. Amino acids promote human blastocyst development *in vitro*. Hum. Reprod. 16 (4), 749–756.

Garlick, P.J., 2006. Toxicity of methionine in humans. J. Nutr. 136 (6), 1722S–1725S.

George Kapalka, G.M., 2010. Depression. Nutritional and Herbal Therapies for Children and Adolescents, 141–187.

Halliwell, B., Zhao, K., Whiteman, M., 2000. The gastrointestinal tract: a major site of antioxidant action? Free Radic. Res. 33 (6), 819–830.

Helms, H.C., Madelung, R., Waagepetersen, H.S., Nielsen, C.U., Brodin, B., 2012. *In-vitro* evidence for the brain glutamate efflux hypothesis: brain endothelial cells cocultured with astrocytes display a polarized brain-to-blood transport of glutamate. Glia 60 (6), 882–893.

Holecek, M., 2018. Branched-chain amino acids in health and disease: metabolism, alterations in blood plasma, and as supplements. Nutr. Metab. (Lond) 15, 33.

Hou, Y., Wu, G., 2018. Nutritionally essential amino acids. Adv. Nutr. 9 (6), 849–851.

Ikezaki, S., Nishikawa, A., Furukawa, F., Imazawa, T., Enami, T., Mitsui, M., Takahashi, M., 1994. 13-week subchronic toxicity study of L-histidine monohydrochloride in F344 rats. Eisei Shikenjo Hokoku 112, 57–63.

Khan, I., Samson, S.E., Grover, A.K., 2017. Antioxidant supplements and gastrointestinal diseases: a critical appraisal. Med. Princ. Pract. 26 (3), 201–217. doi:10.1159/000468988.

Khayat, D., Lokiec, F., Bizzari, J.P., Weil, M., Meeus, L., Sellami, M., Rouesse, J., Banzet, P., Jacquillat, C., 1987. Phase I clinical study of the new amino acid-linked nitrosourea, S 10036, administered on a weekly schedule. Canc. Res. 47, 6782–6785.

Kohlmeier, M.V., 2003. Valine, Nutrient Metabolism In Food Sciences and Technology. Academic Press, p. 370–377. https://www.sciencedirect.com/topics/medicine-and-dentistry/valine.

Kriengsinyos, W., Raffi, M., Wykes, L.J., Ball, R.O., Pencharz, P.B., 2002. Long term effect of histidine depletion on whole body protein metabolism in healthy adults. J. Nutr. 132 (11), 3340–3348.

Kurpad, A.V., Regan, M.M., Raj, T., Gnanou, J.V., 2006. Branched chain amino acid requirements in healthy adult human subjects. J. Nutr. 136 (1), 2565–2635.

Markus, C.R., Firk, C., Gerhardt, C., Kloek, J., Smolders, G.F., 2008. Effect of different tryptophan sources on amino acids availability to the brain and mood in healthy volunteers. Psychopharmacology (Berl.) 201 (1), 107–114.

Martinez-Tome, M., Garcia-Carmona, F., Murcia, MA., 2001. Comparison of the antioxidant and pro-oxidant activities of broccoli amino acid with those of common food additives. J. Sci. Food Agric. 81 (10), 1019–1026.

Milan, H., 2018. Branched-chain amino acids in health and disease: metabolism, alterations in blood plasma, and as supplements. Nutr. Metab. (Lond) 15, 33.

Moch, K.J., Kübler, W., 1993. Bioavailability of amino acids from several industrially-prepared protein-containing products. Z. Ernahrungswiss. 32 (1), 2–20.

Nishitani, S., Ijichi, C., Takehana, K., Fujitani, S., Sonaka, I., 2004. Pharmacological activities of branched-chain amino acids: specificity of tissue and signal transduction. Biochem. Biophys. Res. Commun. 313 (2), 387–389.

Reddy M.K., 2019. Amino acid, Encyclopædia Britannica. https://www.britannica.com/science/amino-acid. (Accessed on 12.06.2021).

Rendi, R., 1960. *In vitro* incorporation of labeled amino acids into nuclei isolated from rat liver. Exp. Cell Res. 19 (3), 489–498.

Sawamoto, O., Kyo, S., Kaneda, S., Harada, M., Kishimoto, S., Koshitani, O., Kurisu, K., Nakashima, Y., 2003. Four-week intravenous repeated dose toxicity study of L-systeine in male rats. J. Toxicol. Sci. 28, 95–107.

Scherbakov, N., Ebner, N., Sandek, A., Meisel, A., Haeusler, K.G., Haehling, S.V., Anker, S.D., Dirnagl, U., Joebges, M., Doehner, W., 2016. Influence of essential amino acids on muscle mass and muscle strength in patients with cerebral stroke during early rehabilitation: protocol and rationale of a randomized clinical trial (AMINO-Stroke Study). BMC Neurol. 16, 10.

Shibui, Y., Miwa, T., Yamashita, M., Chin, K., Kodama, T., 2013. A 4-week repeated dose toxicity study of glycine in rats by gavage administration. J. Toxicol. Pathol. 26 (4), 405–412.

Spillane, M., Emerson, C., Willoughby, D.S., 2012. The effects of 8 weeks of heavy resistance training and branched-chain amino acid supplementation on body composition and muscle performance. Nutr. Health 21, 263–273.

Tada, Y., Yano, N., Takahashi, H., Yuzawa, K., Ando, H., Kubo, Y., Nagasawa, A., Uehara, S., Ogata, A., Nakae, D., 2008. Toxic effects of l-aspartic acid at high dose levels on kidneys and salivary glands in Fischer 344 rats detected in a 90-day feeding study. Food Chem. Toxicol. 46, 2789–2795.

Tian, F., Wang, X.Y., 2013. Pharmacological mechanisms and effects of amino acid nutrients. Zhonghua Wei Chang Wai Ke Za Zhi 16 (11), 1123–1126.

Tsubuku, S., Hatayama, K., Katsumata, T., Nishimura, N., Mawatari, K., Smriga, M., Kimura, T., 2004. Thirteen-week oral toxicity study of branched-chain amino acids in rats. Int. J. Toxicol. 23, 119–126.

Tsubuku, S., Hatayama, K., Mawatari, K., Smriga, M., Kimura, T., 2004. Thirteen-week oral toxicity study of L-glutamine in rats. Int. J. Toxicol. 23, 107–112.

Tsubuku, S., Hatayama, K., Mawatari, K., Smriga, M., Kimura, T., 2004. Thirteen-week oral toxicity study of L-arginine in rats. Int. J. Toxicol. 23 (2), 101–105.

Tsubuku, S., Mochizuki, M., Mawatari, K., Smriga, M., Kimura, T., 2004. Thirteen-week oral toxicity study of L-lysine hydrochloride in rats. Int. J. Toxicol. 23, 113–118.

Warsewicz, H.G., Laskowski, W., Kulykovets, O., Chylak, A.K., Czeczotko, M., Rejman, K., 2018. Food products as sources of protein and amino acids—the case of Poland. Nutrients 10 (12), 1977.

Wolfe, R.R., 2017. Branched-chain amino acids and muscle protein synthesis in humans: myth or reality? J. Int. Soc. Sports Nutr. 14, 30.

Xu, Z.R., Tan, Z.J., Zhang, Q., Gui, Q.F., Yang, Y.M., 2014. Clinical effectiveness of protein and amino acid supplementation on building muscle mass in elderly people: a meta-analysis. PLoS One 9 (9), e109141.

Yokohira, M., Hosokawa, K., Yamakawa, K., Hashimoto, N., Suzuki, S., Matsuda, Y., Saoo, K., Kuno, T., Imaida, K., 2008. A 90-day toxicity study of L-asparagine, a food additive, in F344 rats. Food Chem Toxicol. 46, 2568–2572.

CHAPTER

Carnosine

4.2

Mohamad Fawzi Mahomoodally, Sanaa Dilmar A., S. Khatoon Khadaroo
Department of Health Sciences, Faculty of Medicine and Health Sciences, University of Mauritius, Réduit, Mauritius

4.2.1 Introduction

Carnosine, also referred to as β-alanyl-L-histidine or L-carnosine, is a naturally occurring dipeptide having the molecular formula $C_9H_{14}N_4O_3$ and possessing a molecular weight of 226.236 g/mol (Bildyrev, 1992). It was first discovered in 1900 by the Russian chemist, V.S. Gulewitch, during his quest for unidentified nitrogen-containing not-protein compound(s) in Liebig's meat extracts (Boldyrev, 2013). Carnosine also has a melting point within the range of 253 to 256°C (Bildyrev, 1992) and is a naturally occurring dipeptide, neuropeptide found in the brain, stomach, kidneys, olfactory bulbs, cardiac muscles, and in abundant amount in skeletal muscles (Artioli et al., 2009). It is one of the most abundant small molecule compounds in skeletal muscles with similar concentration to creatinine and adenosine triphosphate (ATP).

Being a water-soluble and endogenously synthesised dipeptide, it composes of two different amino acids (Bildyrev, 1992), namely β-alanine and L-histidine. β-alanine is a dispensable and non-proteogenic amino acid (does not take part in protein structures), that is endogenously produced by the liver (Boldyrev, 2013). It is formed from the degradation of uracil and once produced, it is taken by several tissues, one of which can be the skeletal muscles (Harris, 2010). It should be noted that carnosine synthesis is highly dependent on the availability of β-alanine as studies have shown that L-carnosine synthesis *in vivo* is limited by β-alanine availability (Artioli et al., 2009). Ultimately, this can be an effective way of increasing endogenous carnosine synthesis.

L-histidine is an indispensable or non-essential amino acid in globular proteins (Harris, 2010). It is not essential in adults but needed by children for growth and people with uraemia, where there is an elevated concentration of urea, creatinine and other nitrogenous compounds. Synthesized from 5-phosphoribosyl-l-pyrophosphate and ATP, it is regarded as an ancient molecule being derived from an RNA metabolite. Miflin described it as the '*Cinderella*' of all the amino acids as little work is carried out for its synthesis in plants (Lea and Azevelo, 2017). Thus, β-alanine is created internally by our body and is not obtained from dietary sources while histidine

Antioxidants Effects in Health. DOI: https://doi.org/10.1016/B978-0-12-819096-8.00042-2

is found in plentiful amount in dietary sources like meat, eggs, sea foods, quinoa and dairy products (Artioli et al., 2009). The other related compounds which can act as antioxidants *in vivo*, are believed to possess the ability to protect neuronal cells against ischemic injury and oxidative stress (Kiso, 2003).

Carnosine has shown its effectiveness in various processes in the body, one of which can be the contribution to the physicochemical buffering of lactate by exercising and it has anti-fatigue effects (Barger, 1918). In addition, it possesses several beneficial characteristics like anti-glycation, anti-stress, antioxidant, hydroxy radical scavenging and chelation properties (Artioli et al., 2009), which are because of reducing risk of developing inflammation and reducing oxidative and glycaemic stress in the body (Kiso, 2003).

4.2.1.1 Sources of carnosine

β-alanyl-L-histidine is a cytoplasmic intracellular dipeptide located in high concentration in the skeletal muscle of both vertebrates and non-vertebrates, produced by the enzyme carnosine synthase (Kiso, 2003). It comes primarily from meat sources due to its ability to get accumulated in the muscles (Table 4.2.1). Carnosine is predominantly found in the muscle and brain cells and the dietary sources of this compound is known to only come from animals (Furuse, 2007). Belitz et al. mentioned that beef contains around 1500 mg of carnosine per pound. The average daily intake of carnosine for an individual ranges from 50 to 250 mg (Furuse, 2007).

4.2.1.2 Factors determining carnosine content in muscles

Carnosine content in the muscles of human beings is affected by various factors. The following can be some of the major internal and external factors which cause a difference in presentence of β-alanyl-L-histidine in muscles.

Table 4.2.1 Carnosine content in different types of raw meat extract from different species.

Species raw extract	Carnosine content (mg/g)	References
Beef	5.8–7.12	(Mora et al., 2008) (Jones et al., 2011) (Jung et al., 2013)
Pork	0.13–4.19	
Lamb	7.06	
Chicken	0.66–1.83	
Turkey	0.86–7.9	
Seafood		
Mackerel	7.78	(Jones et al., 2011) (Preedy, 2015) (Jones et al., 2011)
Prawn	9.25–11.6	
Rainbow trout	0.11	
Tuna	5.29	

4.2.1.2.1 Gender/sex

From several studies conducted regarding carnosine content with gender (Penafiel et al., 2004), it has been found from biopsy studies that males have higher carnosine level (22%-82%) compared with females (Wilson, 1938). This theory was further scrutinised by the testosterone administration to female mice. When injecting the male hormone, carnosine content was increased by 268% while castration of mice reduced by 40% (Williams and Krehl, 1952). This can be attributed to the ability of testosterone to increase the expression of taurine and β-alanine transporter rather than the enhancing the transcription of carnosine synthase enzyme. In addition, a similar study on orchidectomy, with or without testosterone replacement in mice, confirmed this mechanism (Harris et al., 1990). The physiological effect that carnosine exerts should not be omitted. It provides protection against exercise-induced acidosis and it has been reported that high carnosine content in muscle could be advantageous for high intensity exercise (Penafiel et al., 2004), and this could be justified by the fact that men are more likely to perform strenuous exercises like weight-lifting (Harris, 2010).

4.2.1.2.2 Age

Data from cross-sectional studies demonstrate that muscle carnosine content negatively correlates with age. As such, in a study conducted on 293 participants aged between 9 and 83 years old, Baguet et al. (2009) reported that the reason behind the low carnosine level in old age individuals is attributed mainly to a decrease in muscle carnosine that occurs during adulthood or even shortly after puberty compared to the decline that takes place from adulthood to old age (Baguet et al., 2009). The exact cause for this reduction is still unclear but could possibly be attributed to the reduction in androgen/oestrogen level in elderly individuals, especially in the postmenopausal stage (Harris, 2010) or even the loss of degenerative skeletal muscle mass, also refreed as sarcopenia (Xing et al., 2019).

4.2.1.2.3 Exercise

Graef et al. (2009) reported that exercise training could change the muscle carnosine content. However, a short-term training intervention suggested that a few weeks of training does not stimulate carnosine accumulation in muscles (Graef et al., 2009).

4.2.1.2.4 Diet

What we eat has been seen to have an effect on the carnosine content in our muscles. Evidences have shown that vegetarians possess > 20% lower muscle carnosine compared to omnivores (Penafiel et al., 2004). Nevertheless, longitudinal studies may be required to further show the cause of this percentage in carnosine content. Regular ingestion of chicken breast extract which is a popular dietary product in Asia has shown to cause elevated level of carnosine in rats. Harris et al. (2010) have shown, in 2006, that daily ingestion of very high dose of carnosine, around 13 mg a day during a period of four weeks, could elevate carnosine content in healthy individual as much as 65% (Furuse, 2007).

4.2.1.3 Synthesis of carnosine

Carnosine is an intracellular dipeptide molecule made from two bonded amino acids which are β-alanine and histidine. Apart from the dietary sources of carnosine, it may also be found in human tissues whereby it is especially concentrated in tissues such as the brain and muscle which requires greater energy demand (Artioli et al., 2009).

Carnosine obtained from the diet is hydrolysed, when it passes the gastrointestinal tract, by the enzyme carnosinase into β-alanine and histidine, which then go through the bloodstream. Since carnosine has been broken down, it cannot reach its target tissues. Fortunately, skeletal muscles can produce it in muscle cells from β-alanine and histidine found in the bloodstream (Artioli et al., 2009).

Histidine is an essential amino acid and is easily available from diet; making it abundant in the body. However, β-alanine is a non-essential amino acid which has to be produced by the body and only part of the required amount can be obtained from the hydrolysis of carnosine in the gastrointestinal tract. Therefore, β-alanine need to be manufactured in the liver by the degradation of uracil which is transported by the blood (Artioli et al., 2009).

Muscle cells then uptake β-alanine and histidine from the bloodstream in the presence of the enzyme carnosine synthase. It should be noted that the amount of produced carnosine is dependent on the concentration of β-alanine because histidine is usually found in greater concentration than β-alanine in plasma, and also carnosine synthase has a higher affinity to histidine (Artioli et al., 2009).

4.2.1.4 Bioavailability of carnosine

Bioavailability is a key process determining the extent and rate at which a specific metabolite or drug (also referred to as an active moiety) enters the systemic circulation of the human body where it can have access to the site of action (Bildyrev, 1992). In other words, the bioavailability of carnosine will help to control its uptake and the amount that reaches the bloodstream after its synthesis and also the fraction that is absorbed and escapes from first-pass elimination. Carnosine is different from many other phytonutrients as it has an extremely short half-life (Barger, 1918). As such, a person can have a meal loaded with meat for breakfast, which is high in carnosine but has hardly any carnosine which left by lunch (Baguet et al., 2009). Several factors take part in changing the bioavailability of carnosine in the body, which are discussed below.

4.2.1.5 Intestinal transport

Gardner et al. (1991) investigated the intestinal absorption of this molecule in human beings. From the study, it was reported that carnosine crosses the intestine to a huge extent. It was also seen that up to 14% of ingested L-carnosine was found in the urine over 5 hours after eating (Furuse, 2007). However, the rapid post absorptive hydrolysis was a hindrance for the proper quantification of intact peptide absorption.

In a study on the administration of chicken breast extract (CBEX) to increase carnosine level in the brain, it was found that indeed single oral administration of CBEX to male Wistar rats boosted both anserine and carnosine level in the hippocampus and hypothalamus. This increase is due to the passage of these dipeptides through the blood-brain barrier. Carnosine was reported to increase until 120 minutes after administration (Tomonaga et al., 2007).

It should however be pointed out that even though carnosine is regarded as a bioavailable dipeptide capable of crossing the blood brain barrier, its availability in the body depends greatly on β-alanine. As such, it was found that supplementation of the latter produces up to 80% increase in carnosine levels in the skeletal muscle tissues (Xing et al., 2019). In addition, oral ingestion of a dietary supplement containing carnosine only was found not to increase carnosine levels in muscles compared to when a supplement containing equal amount of β-alanine (Derave et al., 2010).

4.2.1.6 Carnosine transporters (PEPT 1 and PEPT 2)

Carnosine can be transported across the cellular membrane through a number of transporters from the proton-oligopeptide transported family (POT-family). Mammalia members of this family are PEPT 1 (Peptide transporter 1) and PEPT 2 (peptide transporter 2) (also referred to as oligopeptide transporter 1 and 2) (Barger, 1918).

PEPT 1 is defined as a high capacity, low affinity transporter and plays a fundamental nutritive role in intestinal absorption of peptides (Barger, 1918). β-alanyl-L-histidine can be absorbed in the small intestine where part of it enters the blood without being hydrolyzed upon ingestion. Carnosine uptake across the brush-border membrane is accomplished by PEPT 1.

When the molecule β-alanyl-L-histidine enters the enterocytes, it is either broken down by carnosinase or further transported across the basolateral membrane. PEPT 1, a peptide-proton cotransporter, has increased activity in lower luminal pH (Barger, 1918). On the other hand, PEPT 2 is a low capacity, high affinity transporter and contributes to reabsorption of filtered peptides in the renal tubules. Its ablation markedly impedes the tubular reabsorption of carnosine. Once in the epithelial cytosol, carnosine either accumulates or is hydrolysed by carnosinase (Artioliet al., 2009).

The extent to which these transporters carry carnosine has an effect on its bioavailability depending on how effectively they work to make that compound available to the body for exerting its effects (Barger, 1918).

4.2.1.7 Carnosinase activity

The presence of the enzyme carnosinase in plasma can affect the bioavailability of carnosine (Penafiel et al., 2004). Carnosinase quickly degrades carnosine in the body and may prevent it to provide that sustained protective effect (Graef, 2009).

Once the compound hits the digestive tract, dietary carnosine is broken down by the enzyme carnosinase into the its components, β-alanine and L-histidine, as the body uses these amino acids for protein synthesis (Graef, 2009).

In addition, Gardner et al. (1991) found a highly significant negative correlation between serum carnosinase activity and urine recovery of intact carnosine after ingestion by human subjects. Hydrolysis of carnosine by the liver, kidney and other tissues plays a role in limiting the bioavailability of carnosine (Hipkiss et al., 1995). Specific drugs may be designed to inhibit that enzyme carnosinase but the response to these remain uncertain.

Serum carnosinase is synthesized in the brain where it is secreted into the cerebrospinal fluid and then into the systemic circulation (Boldyrev et al., 2013). It exists in two isoforms, firstly, tissue carnosinase which is found in the liver, kidney, spleen and secondly serum carnosinase found in plasma, the brain and the spinal fluid (Barger, 1918).

Under appropriate conditions in the body, both isoforms catalyse the hydrolysis of the dipeptide carnosine. The rate of breakdown by the enzyme carnosinase determines the availability of carnosine in the body for absorption and function (Barger, 1918).

4.2.1.8 Mechanism of action

Carnosine exhibit antioxidant activity mainly due to its biological properties (Table 4.2.2). As it is a water-soluble molecule, it allows reactions with water-soluble oxidation mediators (e.g. transition metals and reactive oxygen species) in the cytoplasm of cells (Gariballa, 2000). It should be noted that the antioxidant activity depends on the chemical structure, bioavailability and on the level of severity of oxidative stress which can change the mechanism of action (Prokopieva et al., 2016). Since 1980, both direct and indirect antioxidant activity mechanisms of carnosine have been investigated (Boldyrev et al., 2013). The two main mechanisms are the chelation of metal ion and reactive oxygen radicals scavenging which are explained below.

4.2.1.8.1 Chelation of metal ions

Carnosine can form complexes with the elements of the first transition metals series and its activity depends on the type of metal ion. Chelation is the process by which compounds (organic molecules) bond to metal ions by forming coordinate bonds between a multiple ligand and a single central atom, which leads to the sequestration of the metal (Tandon et al., 1985). Copper and zinc carnosine complexes have been mostly studied.

4.2.1.8.1.1 Copper-carnosine complex

Two six-membered chelate ring structures of Cu^{2+}-carnosine complex were first proposed by Dobbie and Kermack, where Cu^{2+} bonds to imidazole, amino and peptide nitrogen atom. Later on, other authors confirmed that the monomeric Cu^{2+} binds to imidazole when the pH increases, resulting in amino acid and peptide nitrogen to deprotonate which then bind to the metal (Boldyrev et al., 2013).

With the advance of technology, Freeman and Szymanski used x-ray crystallography to study this complex in solid state, where they found that it is a dimer and not a monomer. Actually, both monomer and dimer are found in solution but when the equilibrium shifts in a neutral pH, the dimer $[Cu_2H_{-2}L_2]^0$ becomes more prominent.

Table 4.2.2 Antioxidative properties of carnosine.

Type of study	Description	Outcome	References
Clinical trial	Placebo controlled trail in neurodegenerative patients - 42 patient with chronic discirculatory encephalopathy Carnosine daily dose of 0.75 g/g or 2g + basic therapy After 21 days: cognitive functions of patient's brain before and after treatment were compared	Higher dose of carnosine (2g) daily was used Latency of cognitive spikes ↓ from 378±21 to 345± 12 m sec Blood lipoproteins were protected from Fe^{2+}-induced oxidation ↑ duration of acidic haemolysis of Red blood cells from 134 ± 4 s to 15[illegible] ± 6 s	Kawahara et al., 2018
In vitro	Assay: MDA Peroxide assay	Carnosine dose-dependently inhibited TBARS formation in LDL incubated with 1.25 μM copper for 1.5 h Control: 0.0617 ±0.0067 Carnosine: 0.0237 ±0.0085 Carnosine effectively prolonged the lag time and reduced maximum rate of LDL oxidation	Bogardus et al., 2000
In vitro	Protective effect of carnosine on Cu(II)-induced damage in human blood cells	Incubation of RBC with carnosine, prior to treatment with Cu(II) protected effects by: - ↓MetHb level and increased the activity of MetHb reductase - ↓ the carbonyl and malondialdehyde content - ↓ protein and lipid hydroperoxide levels - restored activity of hexokinase Protective effect of carnosine was dose dependent	Husain and Mahmood, 2020
In vitro	Protective effects of carnosine to human peritoneal mesothelial cells in both acute and chronic conditions with PDF and GDP	↑ cell viability, preservation of protein from modification, ↓ ROS Carnosine enhanced HPMC viability against toxic effect of GDPs by protecting cellular protein from modification and from ROS-mediated damage	Alhamdani et al., 2007
In vitro	96 h of carnosine treatment was provided to human glioblastoma cells with 20 mM and 50 mM carnosine	↓ ATP and dehydrogenase 50 mM reduced ATP to 42.7± 13.5 % and dehydrogenase to 41.0± 19.3 % Inhibit proliferation of cells ↓ ROS, ↑ in mitochondrial superoxide dismutase in tumour cells	Renner et al., 2008

Type of study	Description	Outcome	References
In vivo	Cataract induced in rats by sodium selenite L-carnosine eye drops (50 g/L or 20 g/L) and 0.9% normal saline L-carnosine was added to the medium of cultured rat cataract lens at 1.00, 0.10, 0.01 g/L for 1 week	Lesions in 50 g/L L-carnosine was lighter than that of control group Protein number 182 in carnosine group 161 in control group (high molecular weight protein ↓ and low molecular protein weight ↑) At 50 g/L carnosine could restrain the development of early stage rat cataract	Liu et al., 2009
In vivo	Diabetic rats induced with streptozotocin Carnosine treatment	Lens opacification progressed in a biphasic manner in diabetic rats Initial slow increase in the first 8 weeks of diabetes followed by steep increase in the next 5 weeks Carnosine delayed cataract progression in diabetic rats on week 4 ↓ CAT and GSH-Px in lens of untreated diabetic rats at 13th weeks after injection Lower glycated lens in treated grp Carnosine prevents AGEs formation	Yan et al., 2008
In vivo	Experimental model of focal cerebral ischemia/reperfusion in Wistar rats Carnosine provided with a diet-daily dose 150 mg/kg for 7 d before temporary occlusion of the middle cerebral artery performed for 60 min After 24 h of onset of ischemia, effect of carnosine on area of necrotic core was evaluated in animals	↓ in infarct size, ↑ total antioxidant capacity ↓ MDA and isoprostanes in brain tissue Carnosine consumed prophylactically with the diet for 7 days before induction of ischemia by means of MCA occlusion in rats retains high antioxidant activity of brain tissue ↓ in MDA	Devyator et al., 2018
In vivo	4 groups: Regular chow diet(control) 0.5% α-lipoic acid (ALA) 0.5% α-lipoic acid+ 0.25% L-carnosine 0.5% L-carnosine (LC)	Skin and liver SOD of ALA and LC grp greater than control grp Liver GSH-Px activity greater in LC group than chow diet Skin MDA and serum in ALA and LC groups lower than in chow diet HDL-C level in LC group greater ↑ in antioxidant activity ↓ in LPO in serum, liver, skin of rats	Kim et al., 2011
In vivo	90 guinea pigs	Accelerates reparative process in injured lungs by activation of fibroblast proliferation, connective tissues, and intracellular regeneration	(Seidman and Castor, 1981)

The monomer is usually found in low pH environment. Bonding with the dimer form then results in a square-planar coordination sphere complex (Boldyrev et al., 2013). However, the chance of the existence of the dimer in living organism is very low due to the very small concentration of copper and competing ligands which will decrease the probability of the monomeric complexes from forming ligands (Boldyrev et al., 2013).

4.2.1.8.1.2 Zinc-carnosine complex

This complex has gain in popularity due to its pharmacological application as it has proven that it protects gastric mucosa from ulceration *in vivo* and is effective against *Helicobacter pylori* associated gastritis. The complex is known as polaprezinc (Z-103) which can be dissolved in acid and its stability in the stomach leading to better adhesion to ulcerous sites (Boldyrev et al., 2013).

Chelation of zinc and carnosine occurs in a one to one ratio forming a quadridentate complex of polymeric nature (Mahmood et al., 2007). This zinc-carnosine complex is less stable than the copper-carnosine one and interestingly the zinc complexation has the ability to reverse the tautomeric preference between the $N_{\pi}OH$ and $N_{T}OH$ tautomeric forms of the imidazole ring, as compared to the free ligand whereby the $N_{T}OH$ tautomer is predominant (Baran, 2000).

Formation of zinc-carnosine complex depend on the available nitrogen and pH. Two neutral complexes are formed in basic environment, a water-insoluble monomer, $[ZnH_1L]^0$ involving the $N_{T}OH$ form and a dimer, $[Zn_2H_2L_2]^0$ involving the $N_{\pi}OH$ (Torreggiani et al., 2000).

4.2.1.8.2 Scavenging ROS

4.2.1.8.2.1 Reaction of carnosine with ROS, peroxy-nitrite, and hypochlorous acid

Carnosine can react with superoxide anions (e.g., superoxide dismutase) in aqueous medium and this mechanism of action was investigated by Pavlov et al. This reaction is based on principle of carnosine being able to form a charge-transfer complex with the radical which affects the oxygen species (O^{2-}). Also, by using pulse radiolysis it was found that carnosine effectively quenches ROS like hydroxyl radicals (OH^-) which results in a stable intermediate obtained from a base catalysed water elimination reaction and is less reactive than OH^-.This quenching mechanism occurs when imidazole part of carnosine reacts singlet oxygen and forms endoperoxide products (Boldyrev et al., 2013).

Carnosine can counteract peroxy-nitrite dependent reaction which gives rise to a protective effect. Fontana et al. (2002) found out that it protects tyrosine against nitration, α1-antiproteinase against inactivation and it prevents modification of low-density lipoprotein. Hipkiss et al. (1995) reported that carnosine acts as protective agent against hypochlorite (HOCl) depending on the dosage of carnosine which inhibits protein cross linking and high molecular weight oligomers caused by HOCl. This mechanism involves a chlorine shift in the carnosine from the imidazole ring to the amine terminal due to its unique relationship of the three adjacent functional groups (imidazole ring, carboxylic group and the β-alanyl group) (Boldyrev et al., 2013).

4.2.1.8.2.2 Reaction of carnosine with peroxyl radicals

The L-histidine part of the molecule is very important for its antioxidant activity and the mechanism involves the reduction of the radical whereby Decker et al. found that millimolar concentration range of purified carnosine can scavenge peroxyl radicals with similar potency of that of L-histidine and no effect was seen with β-alanine. However, later this reducing ability was investigated and it was found that carnosine was unable to reduce Fe (III) in myoglobin. As well, carnosine cannot scavenge 2, 2-diphenyl-1-picrylhydrazyl radicals which confirm that carnosine does not act as an electron transfer agent (Boldyrev et al., 2013).

4.2.2 Possible pro-oxidant activity

Carnosine can be considered as an ideal antioxidant as it is naturally synthesised in the body, easily absorbed in the gastrointestinal tract, can penetrate blood-brain barrier, has increased bioavailability and has membrane stabilising action. It is low molecular water-soluble compound and is not addictive. An overdose of carnosine is not fatal as the enzyme carnosinase act on the excess carnosine and breaks it down into products that would be excreted, thus it does not accumulate in the body (Prokopieva et al., 2016).

However, depending on the cases, carnosine has pro-oxidant action which is the reverse of an antioxidant effect. Shi et al reported that it did not counteract the excess production of hydroxyl radical in the presence of nickel ions and H_2O_2. It was also found that it amplifies DNA oxidative damage to free 2'-deoxyguanosine. In the certain conditions, it can enhance manufacturing of hydroxyl radicals in the presence of H_2O_2 due to the reduction of certain transitional ions. In an *in vivo* study carried out on the concentration of carnosine was concluded that it reduces Fe^{3+} to Fe^{2+} ions which demonstrate the reducing activity of carnosine and is more effective in the presence of copper (Mozdzan et al., 2005).

4.2.3 Beneficial effects of carnosine on health

4.2.3.1 Carnosine in diabetes and associated complications

Carnosine has the ability to affect glycaemic control and help in alleviating or preventing diabetic complications like nephropathy and ocular damage. Lee et al. first studied the protective role of carnosine in diabetic animal models using BALB/c mice, where oral supplementation of carnosine was given to them for four weeks dose-dependently and it was found that the level of plasma glucose and fibronectin were decreased, level of insulin was increased and the oxidative damage was reduced (Boldyrev et al., 2013).

Some years later on, Soliman et al. (2007) confirmed the beneficial effect of carnosine on diabetes using a streptozotocin diabetic-induced model where reduced hyperglycaemia, normalized dyslipidaemia, and reduced liver damage were seen using a carnosine dose- dependent treatment (100 and 200 mg/kg) (Boldyrev et al., 2013).

Carnosine also exhibited hypoglycaemic effect. Nagai et al. studied this effect and proposed that the ability of carnosine to control blood glucose level is due to a receptor-mediated mechanism that regulates the autonomic nervous system via the histamine H3 receptor (Boldyrev et al., 2013).

Carnosine works by lowering the activity of sympathetic nerves and allows better transmission of parasympathetic nerves which leads to an increased secretion of insulin as carnosine can maintain or increase β-cells in the pancreas and also prevent glucagon secretion resulting in the hypoglycaemic effect (Boldyrev et al., 2013).

Carnosine has the ability to cause a reduction in diabetes associated diseases or complications like nephropathy and ocular diseases. It was found that in nephropathy caused by diabetes carnosine acts by preventing glomerular apoptosis, restraining podocytes loss and reducing expression of apoptotic proteins like cytochrome c (Boldyrev et al., 2013). Carnosine has a protective effect on the capillaries of the retina for diabetic retinopathy in rats (Prokopieva et al., 2016).

4.2.3.2 Combating and preventing neurological issues

Neurological issues constitute a group of both brain and spinal cord diseases characterized by a progressive deterioration of the structure and function of the neuronal cells (Farooqui, 2016) .The cells and tissues of the nervous system are sensitive to free radical oxidation due to high intensity metabolic processes, high level of oxygen consumption, large amount of lipids with polyunsaturated fatty acids (Hardy and Higgins, 1992) and the induction of oxidative stress along with neuroinflammation exacerbates this condition. These processes initiate excitotoxic chain reactions in which neurons continuously experience a great amount of extracellular glutamate level (Hardy and Higgins, 1992). The protection of nervous tissues against free radicals is of major importance (Barger, 1918). Since carnosine exhibits antioxidant properties by directly interacting with single oxygen as well as scavenging peroxyl and superoxide radicals, it can prevent the negative impacts resulting from ROS. The direct interaction of carnosine with ROS also prevents he production of lipid peroxidation products like malondialdehyde (MDA) which deems delirious to the organism (Xing et al., 2019). The ability to cross the blood-brain barrier also makes carnosine a potential compound for neuronal issues.

Glycosylation, oxidative stress, inflammation secondary to oxidative stress can lead to the progression of Alzheimer's disease (AD) (Bae et al., 2013). Carnosine can reduce the toxic effect of beta-amyloid protein, an abnormal protein built up in the brain of people with AD which damages the surrounding brain tissue (Hardy, 1992). The possible protective effect of L-carnosine in AD was first introduced in the late 1990 when *in vivo* studies demonstrated the ability of this natural antioxidant to inhibit the formation of the beta-amyloid polymerisation and cell toxicity (Bae et al., 2013). Preston et al. reported that carnosine provided some protection against the effect of the neurotoxic amyloid peptide on rat brain vascular endothelial cells.

Parkinson's disease (PD) is the second most common neurodegenerative disease after AD and is a progressive disorder associated with aging that affects both motor and cognitive function (Frohlich, 2016). The pathological hallmark of PD is the loss of dopaminergic neurons which results in the initiation and execution of voluntary movements (tremors) (Bae et al., 2013) and occurrence of Lewy bodies in specific brain stem areas (Chand and Litvan, 2007).Considerable improvement of clinical state of PD patients was seen during administration of carnosine at doses 1.5 g/day for 30 days in addition to traditional therapy for PD. Improved coordination of movements was observed (Hardy, 1992). Furthermore, a randomized, double-blind, placebo-controlled study revealed that carnosine inclusion (2.0 g/day), in addition to basic therapy, improved cognitive function (Hardy, 1992).

Antiglycation activity, regulation of glutamate release were some hypotheses proposed for the therapeutic action induced by carnosine. In vivo and in vitro studies indicated that its possible mechanism of neuroprotection can be due to metal ion chelating activity.

Observations from these preclinical and clinical studies demonstrated the potency of carnosine as an antioxidant against neuronal degradation and improvement of symptoms in patients suffering from neurodegenerative diseases.

4.2.3.3 Carnosine against cancer and side effects

The antiproliferative properties that carnosine possesses makes it of particular interest in cancer management and prevention (Hipkiss et al., 1995). Pointed by Hipkiss in a perspective study of carnosine and cancer, the antineoplastic effects were described in 1986 by Nagai and Suda. It was reported that carnosine treatment (50 mg/kg/day) inhibited tumour growth and mortality in mice.

The antiproliferative activity of carnosine was demonstrated in HCT 16, human colon cancer cells where the naturally-occurring antioxidant, carnosine (Hipkiss et al., 1995) was found to inhibit ATP production, from 5mM to300 mM. Since it can target ATP, L-carnosine has the ability to inhibit mitochondrial reactive oxygen species generation which is induced by Kirsten Rat Sarcoma (KRAS) virus. Renner et al. demonstrated the ability of carnosine to inhibit tumours growth *in vivo* in nude mouse model, in which NIH3T3 fibroblasts expressing the human epidermal growth receptor 2 was implanted in the dorsal skin. Further studies on inhibition of tumour proliferation was confirmed in an animal model, whereby human colon cancer (HCT 116) cells were implanted in BALB/c athymic mice.

Carnosine can provide protection in lung injury caused by radiation treatment (Bildyrev, 1992). Since this type of treatment is shown to be effective in cancer, it is widely used among cancer patients. To support this statement, Severin et al. reported that carnosine administered in doses ranging from 50-200 mg/kg/day during a period of 20 days prior to irradiation increased survival rates in albino mice subjected to whole body irradiation. The formation of reactive oxygen species can lead to the development of cell injury and the use of L-carnosine may protect healthy tissues from inflammation due to its anti-inflammatory and anti-oxidant properties.

4.2.3.4 Carnosine and its effect on aging

The presence of carnosine can suppress fibroblast senescence (Hipkiss, 2009) by switching the cell phenotype from senescent to juvenile and this effect can be reversed by removing carnosine (Boldyrev et al., 2013). It also causes a protective effect on telomere shortening in fibroblast medium which could lead to expanded lifespan (Hipkiss, 2009). It was shown that carnosine can retard ageing in senescence accelerated mice (SAM) and Drosophila (Hipkiss, 2009). Boldyrev et al. conducted a study on SAM by supplementing carnosine to their standard diet and found out a reduction in senile features and a reduction in aging by 20 % which was due to the increase in the antioxidant activity (Boldyrev et al., 2013).

Carnosine antioxidant, anti-glycation and anti-crosslinking properties contributes to its geroprotective effect. Carnosine affects altered polypeptides metabolism by neutralising (Hipkiss, 2009) the accumulation of products of carbonylation, glycation and cross-linking process which characterises the senescence phenotype (Boldyrev et al., 2013).

4.2.4 Application of carnosine

Muscle contractions during exercise is associated with oxidative stress due to an increase in oxidative metabolism (Begum et al., 2005). Increase in respiration rate along with an increase in the flow of electrons in the electron transport chain is also associated with elevated reactive oxygen species. The increase in the production of ROS has been implicated as a principal cause of muscle fatigue and dysregulation in muscle homeostasis (Boldyrev et al., 2013).

4.2.4.1 Ergogenic aid in sports

Ergogenic aids can be nutritional, mechanical and physiological tools that are used by athletes to help boost sports performance. Although several endogenous antioxidative systems are present in the human body, carnosine can be used as an ergogenic aid in high intensity exercise as it can help improve performance by behaving as an antioxidant (Boldyrev et al., 2013). As such, in 2007, Hill et al. denoted the capacity to cycle at 110% maximal power which was improved by 13 and 16% respectively when β-alanine supplementation was provided (Hill et al., 2007). Furthermore, in a meta-analysis composing of 15 experimental studies on β-alanine and exercise, Hobston et al. postulated that supplementation of this amino acid helped in exercise with a duration of less than for minutes.

Carnosine can also act as a neurotransmitter by abating exercise-induced fatigue and can inhibit protein and fat oxidation (Caruso et al., 2012). By its scavenging and chelating properties, β-alanyl-L-histidine can chelate copper and iron which decrease muscle damage and in doing so, they are prevented from reacting with peroxides in the Fenton reaction (Trexler et al., 2015). A reduction in carnosine content in caused suppression of muscle strength and endurance while its supplementation has been associated with quicker recovery, strength and resistance (Matthews et al., 2019).

4.2.5 Studies demonstrating antioxidative properties of carnosine

4.2.5.1 Detrimental effects of carnosine on health

The naturally occurring dipeptide does not seem to prove any side effects as it is synthesized in the body and quickly degraded by the enzyme carnosinase (Barger, 1918). The safety and toxicity of carnosine is evaluated using supplementation either orally or intravenously. Overall, carnosine supplements are regarded as being possibly safe for most of children and adults.

To support this statement, a study has been conducted on the safety and tolerability of carnosine in rats (Krehl, 1952). In this *in vivo* study, based on the FDA guidelines on preclinical acute toxicity studies, daily assessment was done for systemic signs of toxicity. The safety and efficacy of this treatment was evaluated in rats with permanent or transient middle cerebral artery occlusion. Experiments also included a control group for comparison the effectiveness of the treatment (Bae et al., 2013).

Body weight, food consumption, activity and mortality were monitored over a period of 14 days after single intravenous L-carnosine treatment at doses of 100, 500, 1000, 2000 mg/kg. Observations regarding weight and mortality were done and no significant difference was observed between the control groups (saline treated) and the carnosine treated group and no rat died. In addition, histopathologic evaluations were done on bone marrow, cerebellum, hippocampus, lung, liver, heart and kidneys in randomly selected animals to examine organ toxicity during administration (Bae et al., 2013).

Carnosine (2000 mg/kg) did not induce any sign of toxicity in organs. Histological evaluation, complete blood count, coagulation tests and serum chemistry did not reveal any abnormalities. This is while, drug interaction has been observed with carnosine supplements and blood pressure medications, which have resulted in low blood pressure in patients taking both and in some cases this combination has acted as a stimulant (Kiso, 2003). No evidence has shown detrimental effects on health from carnosine as it is easily degraded by the enzyme carnosinase (Bildyrev, 1992). Chez et al. (2002) demonstrated in a human clinical trial assessing the health benefit of carnosine that oral administration of β-alanine in doses of 1.6-6.4 g per day showed good tolerance and no side effects (Chez et al., 2002).

Although carnosine has been regarded as a safe bioactive component in both dietary and supplemental forms and no significant side effects have been documented till now in studies conducted, safety considerations have been investigated on the supplementation of β-alanine, especially among athletes. One of the prominent effects observed was paraesthesia, which is characterised by a benign sensation of tingling and numbness when supplementation of β-alanine exceeded 100 μM (Xing et al.,2019). In addition, the congenital lack of degradative enzyme carnosinase results in serious conditions like carnosinaemia whereby carnosine cannot be broken down to their respective constituents (Boldyrev et al., 2013).

Conclusion

Carnosine is considered as one of the most powerful, natural antioxidant which is synthesised by the body and also obtained from animal food sources. The literature refers the molecule β-alanyl-L-histidine as being an antioxidant due to its ability to reduce both oxidative and glycaemic stress and inflammation through several mechanisms as discussed above. These positive effects entail several health benefits associated with carnosine like the management of different non-communicable diseases (prevalence of cancer and neurological disorders) which are now among global health issues. However, saying that carnosine is completely beneficial and safe is still questionable, based on the data from both preclinical and clinical studies. There is still a dearth of scientific evidence from studies to demonstrate the negative side of carnosine as almost all studies carried out focused on the antioxidative properties as well as other physiological benefits. Studies assessing safety and toxicity of carnosine require longitudinal trials as well as various types of exercises among athletes to be able to conclude the drastic effects of carnosine. As a conclusive remark, the molecule β-alanyl-L-histidine can be regarded as both a powerful endogenous and exogenous antioxidant as illustrated from in vitro, in vivo and clinical studies which demonstrated the ability to chelate metal ions and scavenge free radicals which are the main contributor to oxidative stress which consequently increases the prevalence of ailments.

References

Alhamdani, M.S., Al-Kassir, A.H., Abbas, F.K., Jaleel, N.A., Al-Taee, M.F., 2007. Antiglycation and antioxidant effect of carnosine against glucose degradation products in peritoneal mesothelial cells. Nephron. Clin. Pract. 107 (1), c26–34. doi:10.1159/000106509. PMID: 17664891.

Artioli, G.G., Gualano, B., Smith, A., Stout, J., Herbert Lancha, A., 2009. The role of β-alanine supplementation on muscle carnosine and exercise performance. Med. Sci. Sports Exerc. 42, 1162–1173.

Bae, O.N., Serfozo, K., Baek, S.H., Young Lee, K., Dorrance, A., Rumbeiha, W., Fitzgerald, S.D., Farooq, M.U., Naravela, B., Bhatt, A., Majd, A., 2013. Safety and efficacy evaluation of carnosine an endogenous neuroprotective agent for Ischemic stroke. Stroke 44, 205–212. https://doi.org/10.1161/STROKEAHA.112.673954.

Baguet, A., Reyngoudt, H., Pottier, A., 2009. Carnosine loading and washout in muscles. J. Appl. Physiol 106, 837–842.

Baran, E., 2000. Metal complexes of Carnosine. Biochemistry 65, 789–797.

Barger, G., Tutin, E., 1918. Carnosine constitution and synthesis. Biochem. J. 12, 402–407.

Begum, G., Cunliffe, A., Leveritt, M., 2005. Physiological role of carnosine in contracting muscle. Int. J. Sport Nutr. Exerc. Metab. 5 (5), 493–514.

Bildyrev, A.A., Formazuyk, V.E., Quinn, P.J., 1992. Carnosine, its properties, function and potential therapeutic application. Mol. Aspects Med. 13, 379–444.

Bogardus, S.L., Boissonneault, G.A., 2000. Carnosine inhibits *in vitro* low-density lipoprotein oxidation. Nutr. Res. 20 (7), 967–976.

Boldyrev, A., Aldini, G., Derave, W., 2013. Physiology and pathophysiology of carnosine. Physiol. Rev. 93, 1803–1845.

Caruso, J., Charles, J., Unruh, K., Giebel, R., Learnmonth, L., Potter, W., 2012. Ergogenic effects of β-alanine and carnosine- proposed future research to quantify their efficacy. Nutrients 4 (7), 585–601.

Chand, P., Litvan, I., 2007. Parkinson's disease. Encyclopedia of Gerontology 2nd ed. Elsevier, New York, pp. 322–333. https://doi.org/10.1016/B0-12-370870-2/00146-3.

Chez, M.G., Buchanan, C.P., Aimonovitch, M.C., Becker, M., Schaefer, K., Black, C., Komen, J., 2002. Double-blind, placebo-controlled study of L-carnosine supplementation in children with autistic spectrum disorders. J. Child Neurol. 17 (11), 833–837.

Derave, W., Everaert, I., Beeckman, S., Baguet, A., 2010. Muscle carnosine metabolism and β-alanine supplementation in relation to exercise and training. Sports Med. 40 (3), 247–263.

Devyatov, A.A., Fedorova, T.N., Stvolinsky, S.L., Ryzhkov, I.N., Riger, N.A., Tutelyan, V.A., 2018. Issledovanie neĭroprotektornykh mekhanizmov deĭstviia karnozina pri éksperimental'noĭ fokal'noĭ ishemii/reperfuzii [Study of the neuroprotective effects of carnosine in an experimental model of focal cerebral ischemia/reperfusion]. Biomed. Khim. 64 (4), 344–348. Russian. doi:10.18097/PBMC20186404344. PMID: 30135281.

Farooqui, A.A., 2016. Neurochemical aspects of neurological disorders. Trace Amine and Neurological Disorders, 237–256. ISBN: 9780128036167, Hardcover ISBN: 9780128036037.

Fontana, M., Pinnen, F., Lucente, G., Pecci, L., 2002. Prevention of peroxynitrite-dependent damage by carnosine and related sulphonamido pseudodipeptides. Cell Mol. Life Sci. 59, 546–551.

Furuse, M., Hayakwa, T., Maemura, H., Morimatsu, F., Sato, M., Takahata, Y., Tomonaga, S., Yamane, H., 2007. Oral administration of chicken breast extract increases brain carnosine and anserine concentration in rats. Nutr. Neuro. 10, 181–186.

Gardner, M.L., Illingworth, K.M., Kelleher, J., Wood, D., 1991. Intestinal absorption of the intact carnosine in man and comparison with intestinal permeability to lactulose. J. Physiol. 439, 411–422.

Gariballa, S., 2000. Review. Carnosine: physiological properties and therapeutic potential. Age Ageing 29, 207–210.

Graef, J.L., Smith, A.E., Walter, A.A., 2009. Effects of β-alanine supplementation on high intensity interval endurance performance and body composition in men, a double-blind trial. J. Int. Soc. Sports Nutr. 6, 116–125.

Hardy, J.A., Higgins, G.A., 1992. Alzheimer's disease: the amyloid cascade hypothesis. Science 256, 184–185.

Harris, R.C., Marlic, D.J., Dunnett, M., Snow, D.H., Hultman, E., 1990. Muscle buffering capacity and dipeptide content in man. Comp. Biochem. Physiol. 97, 249–251.

Harris, R.C., Sale, C., Saunders, B., 2010. Effect of beta-alanine supplementation on muscle carnosine concentration and exercise performance. Amino Acids 39, 321–333.

Hill, C.A., Harris, R.C., Kim, H.J., Harris, B.D., Sale, C., Boobis, L.H., Kim, C.K., Wise, J.A., 2007. Influence of beta-alanine supplementation on skeletal muscle carnosine concentration and high intensity cycling capacity. Amino Acids 32 (2), 225–233.

Hipkiss, A., 2009. On the enigma of carnosine's anti-ageing actions. Exp. Gerontol. 44, 237–242.

Hipkiss, A.R., Michaelis, J., Syrris, P., 1995. Non-enzymatic glycosylation of the dipeptide carnosine. FEBS Lett. 37, 81–85.

Husain, N., Mahmood, R., 2020. Mitigation of Cu(II)-induced damage in human blood cells by carnosine: an in vitro study. Toxicol. in Vitro 68, 104956.

Intarapichet, KO., Maikhunthod, B., 2005. Heat and ultrafiltration extraction of broiler meat carnosine and its antioxidant activity. Meat. Sci. 71, 364–374.

Jones, G., Smith, M., Harris, R., 2011. Imidazole dipeptide content of dietary sources commonly consumed within the British diet. Pro. Nutr. Soc. 70, 363–377.

Jung, S., Bae, Y.S., Kim, H.J., Jayasena, D., Lee, J.H., Park, H.B., Heo, K.N., Jo, C., 2013. Carnosine, anserine, creatinine and inosine 5'-monophosphate contents in breast and thigh meats from 5 lines of Korean native chicken. Poultry Sci. 92 (12), 3275–3282.

Kawahara, M., Tanaka, K.I., Kato-Negishi, M., 2018. Zinc, Carnosine, and Neurodegenerative Diseases. Nutrients 10 (2), 147. doi:10.3390/nu10020147.

Kim, M.Y., Kim, E.Y., Kim, Y.N., Choi, C., Lee, B.H., 2011. Effects of α-lipoic acid and L-carnosine supplementation on antioxidative activities and lipid profiles in rat. Nutr. Res. Pract. 5 (5), 421–428.

Kiso, Y., Kurata, H., Matsumura, Y., Muraoka, T., Ono, H., Takaota, M., Tanaka, T., 2003. Preventive effect of L-carnosine on ischemia/reperfusion-induced acute renal failure in rats. Eur. J. Pharmacol. 474, 261–267.

Krehl, W.A., Williams, H.M., 1952. The microbiological determination of carnosine and its formation by rat liver slices. Biol. Chem. 196, 443–448.

Lea, P.J., Azevelo, R.A., 2017. Encyclopedia of Applied Plant Sciences, 2nd ed.

Liu, Y.F., Liu, H.W., Peng, S.L., 2009. Effects of L-carnosine in preventing and treating rat cataract induced by sodium selenite. Zhonghua Yan Ke Za Zhi. 45 (6), 533–536. Chinese. PMID: 19957677.

Mahmood, A., FitzGerald, A., Marchbank, T., Ntatsaki, E., Murray, D., Ghosh, S., Playford, R., 2007. Zinc carnosine, a health food supplement that stabilises small bowel integrity and stimulates gut repair processes. Gut 56, 168–175.

Matthews, J.J., Artioli, G.G., Turner, M.D., Sale, C., 2019. The physiological roles of carnosine and β-alanine in exercising human skeletal muscle. Med. Sci. Sports Exerc. 51 (10), 2098–2108.

Mora, L., Sentandrem, M.A., Toldra, F., 2008. Contents of creatine, creatinine and carnosine in porcine muscles of different metabolic types. Meat Sci. 79 (4), 709–715.

Mozdzan, M., Szemraj, J., Rysz, J., Nowak, D., 2005. Antioxidant properties of carnosine re-evaluated with oxidizing systems involving iron and copper ions. Basic Clin. Pharmacol. Toxicol. 96, 352–360.

Penafiel, R., Ruzafa, C., Monserrat, F., Cremades, A., 2004. Gender-related differences n carnosine, anserine and lysine content of murine skeletal muscles. Amino Acid 20, 53–58.

Preedy, V.R., 2015. Imidazole Dipeptides: Chemistry, Analysis, Function and Effects. Royal Society of Chemistry, London, UK, pp. 43–60.

Prokopieva, V., Yarygina, E., Bokhan, N., Ivanova, S., 2016. Use of carnosine for oxidative stress reduction in different pathologies. Oxid. Med. Cell Longev. 2016, 1–8.

Renner, C., Seyffarth, A., Garcia de Arriba, S., Meixensberger, J., Gebhart, R., Gaunitz, F., 2008. Carnosine inhibits growth of cells isolated from Human Glioblastoma multiform. Int. J. Pep. Res. Ther. 14, 127–135.

Seidman, J.C., Castor, C.W., 1981. Connective tissue activation in guinea pig lung fibroblast cultures: regulatory effects of glucocorticoids. In Vitro 17 (2), 133–138. doi:10.1007/BF02618070. PMID: 6765898.

Soliman, K., Mohamed, A., Metwally, N., 2007. Attenuation of some metabolic deteriorations induced by diabetes mellitus using carnosine. J. Appl. Sci. 7, 2252–2260.

Tandon, S., Flora, S., Singh, S., 1985. Chelation in metal intoxication. Int. J. Environ. Res. Public Health 79, 204–210.

Tomonaga, S., Hayakawa, T., Yamane, H., Maemura, H., Sato, M., Takahata, Y., Furuse, M., 2007. Oral administration of chicken breast extract increases brain carnosine and anserine concentration in rats. Nutr. Neurosci. 10 (3-4), 181–186.

Torreggiani, A., Bonora, S., Fini, G., 2000. Raman and IR spectroscopic investigation of zinc (II)-carnosine complexes. Biopolymers 57, 352–364.

Trexler, E.T., Smith-Ryan, A.E., Stout, J.R., Hoffman, J.R., Wilborn, C.D., Sale, C., Kreider, R.B., Jäger, R., Eranest, C.P., Bannock, L., Campbell, B., Kalman, D., Ziegenfus, T.N., Antonio, J., 2015. International society of sports nutrition position stand: Beta-Alanine. J. Int. Soc Sport Nutr. 12, 30.

Williams, H.M., Krehl, W.A., 1952. The microbiological determination of carnosine and its formation by rat liver slices. J. Biol. Chem. 196, 443.

Wilson, D., Zapp, J., 1938. Quantitative studies of carnosine and anserine in mammalian muscle. J. Biol. Chem. 126, 19–27.

Xing, L., Chee, M.E., Zhang, H., Zhang, W., Mine, Y., 2019. Carnosine- a natural bioactive dipeptide: bioaccessibility, bioavailability and health benefits. JFB 5, 8–17.

Yan, H., Gou, Y., Zhang, J., Ding, Z.H., Ha, W.J., Harding, J.J., 2008. Inhibition of cataracts in diabetic rats by carnosine, aminoguanidine and aspirin drops. Mol Vis. 14, 2282–2291.

CHAPTER

Carnosol

4.3

Xiuping Chen

State Key Laboratory of Quality Research in Chinese Medicine, Institute of Chinese Medical Sciences, University of Macau, Macau, China

4.3.1 Introduction

Medical herbs have a long history in application in almost all civilizations worldwide. Many herbs remain widely used as traditional medicines in some areas, especially in China and India. The history of modern drug discovery clearly shows that natural plants, especially medical herbs, are an important source of new drugs. Artemisinin, paclitaxel, camptothecin, quinine, and digoxin are few representative examples of these drugs. Approximately 80% of clinically available drugs are directly or indirectly obtained from natural plants (Pan et al., 2013). Terpenes and terpenoids, the largest class of natural products, defend many species of plants, animals, and microorganisms against predators, pathogens, and competitors (Gershenzon and Dudareva, 2007). They also demonstrate various biological activities, such as antcancer, anti-inflammatory, antioxidant, antibacterial, and antifungal properties (Cor et al., 2018; Gonzalez-Burgos and Gomez-Serranillos, 2012; Nwodo et al., 2016; Wu et al., 2018; Zacchino et al., 2017). Carnosic acid and carnosol (Fig. 4.3.1) are the two major polyphenolic diterpenes found in *Rosmarinus officinalis* (rosemary), a perennial herb native to the Mediterranean region; these two compounds account for over 90% of the antioxidant activity of rosemary leaves (de Oliveira, 2016). Carnosol has anticancer, antioxidant, and anti-inflammatory effects (Kashyap et al., 2017).

4.3.2 Source and chemistry

The main natural source of carnosol is the plant family *Lamiaceae*. Among the family members, rosemary has the highest carnosol content. Dried rosemary leaves contain approximately 0.2%–0.4% of carnosol (Kashyap et al., 2017). Other plants from the family *Lamiaceae* that also contain carnosol include *Lepechinia mutica* (Benth.) Epling (Ramirez et al., 2018), *L. radula* (Benth.) Epling, *L. paniculata* (Kunth) Epling (Gilardoni et al., 2018), *L. hastata* (A. Gray) Epling (Dimayuga et al., 1991), *Origanum majorana* L. (marjoram; Vagi et al., 2005), *Salvia officinalis* L. (sage), and

Antioxidants Effects in Health. DOI: https://doi.org/10.1016/B978-0-12-819096-8.00043-4

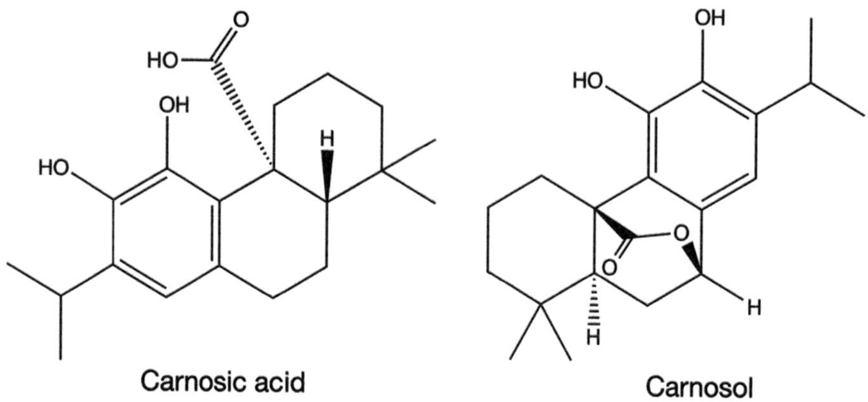

FIG. 4.3.1 Chemical structures of carnosic acid and carnosol.

Thymus vulgaris L. (thyme; Miura et al., 2002; Rutherford et al., 1992). Carnosol is also found in other plants, such as *Psidium guajava* L (family Myrtaceae; Ouyang et al., 2015). The distribution of the polyphenolic profile (including carnosol) in rosemary leaves is significantly modified by the inoculation of arbuscular mycorrhizal fungi (Sero et al., 2018).

Carnosol is an orthodiphenol of abietane-type diterpene lactone that consists of an abietane carbon skeleton with two hydroxyl groups. Its molecular formula was assigned in 1942 as a diphenolic, ester-containing $C_{19}H_{26}O_4$ hydrophenanthrene (White and Jenkins, 1942). However, a following study showed that the structure of picrosalvin, a diterpene-o-diphenollactone, is a $C_{20}H_{26}O_4$ *o*-diphenolic hydrophenanthrene lactone (Brieskorn and Fuchs, 1962). Carnosol and picrosalvin were eventually found to be the same compound (Brieskorn et al., 1964). The crystal structure, absolute configuration, and spectroscopic properties of carnosol were investigated by X-ray analyses in 1990 (Gajhede et al., 1990), and the total synthesis of carnosol dimethyl ether was reported in 1976 (Meyer et al., 1976).

Some of the physical and chemical properties of carnosol are listed as follows:

Formula: $C_{20}H_{26}O_4$
IUPAC name: (1R,8S,10S)-3,4-dihydroxy-11,11-dimethyl-5-propan-2-yl-16-oxatetracyclo [6.6.2.01,10.02,7] hexadeca-2,4,6-trien-15-one
Other names: picrosalvin, NSC 39143
CAS no: 5957-80-2
PubChem CID: 442009
Molecular weight: 330.424 g/mol
Solubility: 1425 mg/L at ~25°C (in water)
pKa (predicted) value: 8.91 ± 0.40
Density (predicted) value: 1.26 ± 0.1 g/cm^3

In general, pure carnosol (>95%) in solution has very poor stability, which is dependent on the solvents used. The stability of carnosol stored in clear vials in an autosampler at room temperature is in the following sequence: ethyl acetate–acetonitrile (10:90) > dimethyl sulfoxide (DMSO)–acetonitrile (10:90) > DMSO > methanol (Thorsen and Hildebrandt, 2003). Its stability is also affected by temperature and light. Temperature is the major factor affecting its degradation in solution, and light exposure further accelerates this process. The degradation rate of carnosol occurs in the following sequence: −10°C in the dark < 4°C in the dark < room temperature in the dark < room temperature with light exposure at < 40°C in the dark < 40°C with light exposure (Zhang et al., 2012).

4.3.3 Bioavailability and toxicity

The bioavailability of carnosol alone has been seldomly investigated. Few reports of extracts containing carnosol mentioned this issue (Arranz et al., 2017; Arranz, Santoyo et al., 2015; Fernandez-Ochoa et al., 2017; Perez-Sanchez et al., 2017; Romo Vaquero et al., 2013; Soler-Rivas et al., 2010). Among these studies, several applied Caco-2, a human epithelial colorectal adenocarcinoma cell line, to evaluate the absorption, transportation, and permeability of these phenolic compounds (Arranz et al., 2017; Mes et al., 2015; Arranz et al., 2015; Perez-Sanchez et al., 2017; Soler-Rivas et al., 2010). Other studies involved an in situ intestine perfusion assay (Fernandez-Ochoa et al., 2017) or a rat model of obesity (Romo Vaquero et al., 2013). Arrazn et al. (Arranz et al., 2015) reported that only 3% of the initial carnosol in an extract mainly containing carnosic acid and carnosol (256 and 38 mg/g, respectively) was taken up after 12 h in a Caco-2 model, suggesting the low absorption of carnosol. However, the absorption of carnosol may be higher than that in previous reports and can be improved in encapsulated form (Perez-Sanchez et al., 2017). Fernandez-Ochoa et al. (Fernandez-Ochoa et al., 2017) compared the absorption coefficients of several diterpenes in a rosemary leaf extract. The results showed that the absorption coefficients of epiisorosmanol, carnosol, and rosmadial is 2.9 h^{-1}, which is lower than that of carnosic acid and rosmanol (3.7 h^{-1}). Romo et al. (Romo Vaquero et al., 2013) performed a metabolic study of a rosemary leaf extract containing carnosic acid (38.9%), carnosol (6.5%), and methyl carnosate (6.9%) by using female Zucker rats. The results showed that carnosol glucuronide can be detected in multiple tissues, including small intestine, cecum, colon, liver, brain, and plasma. The time to reach maximum peak concentration, maximum peak concentration, the time of the last measurable concentration, area under the curve, and mean residence time for carnosol after the oral administration of 100 mg extract were 13.3, 18.2, 13.3, 137.4, and 8.5, respectively.

No specific study focused on the toxicity of carnosol in animals. However, studies of its anticancer (Johnson et al., 2010), liver-protective (Zhao et al., 2018), renal-protective (Zheng et al., 2018), antidiabetic (Samarghandian et al., 2017), and neuroprotective (Samarghandian et al., 2017) effects in mice and rats suggest its possible minor toxicity.

4.3.4 Antioxidant and pro-oxidant activities

The antioxidant properties of carnosol have been well investigated (Aruoma et al., 1992). Carnosol and carnosic acid are strong antioxidants that account for more than 90% of the antioxidant properties of rosemary extracts. Carnosol is a powerful inhibitor of Fe^{3+}/ascorbic acid-triggered lipid peroxidation in microsomal and liposomal systems and is more effective than propyl gallate (Aruoma et al., 1992). It also shows an inhibitory effect on copper-induced low-density lipoprotein oxidation and oxidized apo B formation, showing half maximal inhibitory concentration (IC_{50}) values of 10 and 4.5 μM, respectively (Zeng et al., 2001). $CCl_3O_2^{\bullet}$ is an active organic radical widely used to determine the ability of a compound to react with peroxyl radicals. Carnosol is a good $CCl_3O_2^{\bullet}$ scavenger with a calculated rate constant of 3.0×10^6 $M^{-1}S^{-1}$ (Aruoma et al., 1992; Aruoma et al., 1996). In a 2,2-diphenyl-1-picrylhydrazyl assay, carnosol is more effective than butylated hydroxytoluene, a most commonly used lipophilic antioxidant (Escuder et al., 2002). Carnosol scavenges superoxide anions generated in a xanthine/xanthine oxidase system (Aruoma et al., 1992; Zeng et al., 2001). It is a strong hydroxyl radical (•OH) scavenger, and the calculated rate constant for reaction with •OH in a deoxyribose system is 8.7×10^{10} $M^{-1}S^{-1}$ (Aruoma et al., 1992). In summary, carnosol is a potent antioxidant and can react with various types of free radicals. It shows no effect on xanthine oxidase activity (Zeng et al., 2001) but weakly inhibits lipoxygenase (Laughton et al., 1991). Thus, carnosol serves as a reactive oxygen species (ROS) scavenger without affecting the sources and processes of ROS generation, such as the mitochondria, mitochondrial respiratory chain, and nicotinamide adenine dinucleotide phosphate oxidase.

Many antioxidants can also induce oxidative stress and may be pro-oxidant. Ascorbic acid is a good example. High concentrations of ascorbic acid triggers the generation of high levels of ROS, thereby causing cancer cell death (Shenoy et al., 2018). However, the pro-oxidant activities of carnosol have been rarely reported. Carnosol stimulates Fe^{3+}/bleomycin-dependent DNA degradation and damage (Aruoma et al., 1992; Laughton et al., 1991). It kills cancer cells by inducing ROS generation (see below).

4.3.5 Pharmacological effects and underlying mechanisms

4.3.5.1 Anticancer

The anticancer effect of carnosol has been well established and summarized in recent publications (Chun et al., 2014; Johnson, 2011; Kashyap et al., 2017; Samarghandian et al., 2018). Here, I briefly summarize its effects in cancer therapy. First, carnosol has a broad anticancer spectrum. It demonstrates anticancer effects in various types of cancers, including breast cancer, prostate cancer, skin cancer, gastric cancer, leukemia, melanoma, colon cancer, osteosarcoma, glioblastoma, and fibrosarcoma.

Second, similar to that of widely investigated natural products, such as curcumin, resveratrol, and berberine, the efficacy of carnosol is relatively low. Their IC_{50} levels are several or dozens of micromoles on most cancer cell lines, which is much higher than that of most clinical anticancer drugs. Third, the anticancer mechanisms involve many aspects of cancer biology. The anticancer effects of carnosol are related to the inhibition of cell proliferation, induction of cell cycle arrest and apoptosis, inhibition of metastasis and angiogenesis, and suppression of epithelial–mesenchymal transition and stemness. These effects are mediated by multiple signaling pathways, which include the PI3K/AKT/mTOR, JAK/STAT3, Hedgehog, p53/MDM2, and intrinsic and extrinsic apoptotic pathways. Fourth, the exact target for carnosol remains to be clarified. Similar to other natural products, it may nonselectively affect many targets. Fifth, all anticancer data regarding these natural products have been obtained from laboratory experiments, and clinical evidence remains lacking. Although carnosol itself is a potent antioxidant, it can inhibit cancer proliferation by the induction of ROS. Carnosol induces ROS-mediated apoptosis and autophagy in MG-63 human osteosarcoma cells and MDA-MB-231 breast cancer cells (Al Dhaheri et al., 2014; Lo et al., 2017). A similar effect has been observed in HCT116 human colon cancer cells (Park et al., 2014). These reports reveal that ROS may play a role in the anticancer effect of carnosol.

4.3.5.2 Neuroprotective

The neuroprotective effect of carnosic acid and carnosol has been reviewed recently (de Oliveira, 2016). Carnosol protects rotenone-induced neurotoxicity in cultured dopaminergic cells (Kim et al., 2006). It prevent hydrogen peroxide (H_2O_2)-induced PC12 cell injury by the activation of extracellular signal-regulated protein kinase, c-Jun N-terminal kinase, p38MAPK, PI3K/Akt, Nrf2, and heme oxygenase-1 (HO-1) and the inhibition of ROS generation (Martin et al., 2004). Carnosol also increases ARE binding activity, leading to increased glutathione levels in HT22 neuronal cells (Tamaki et al., 2010). Furthermore, carnosol inhibits sodium nitroprusside-induced nitric oxide production and apoptosis in rat C6 glial cells, possibly through the modulation of apoptosis and induction of HO-1 (Kim et al., 2010). The neuroprotective effect of carnosol has been confirmed in an animal model of depression (Machado et al., 2013). Similar to the anticancer effect of carnosol, its role in ROS generation remains to be clarified.

4.3.6 Clinical studies

The evaluation of carnosol is at the stage of preclinical studies. No related clinical study was retrieved by searching the clinical trial website (https://clinicaltrials.gov/) with the keyword “carnosol,” “picrosalvin,” or “NSC 39143.” This result suggests that although various studies on this compound are being conducted, much work remains to be done before it can be clinically available.

Conclusion

Carnosol is a natural potent antioxidant with various biological activities. However, these effects require verification from additional solid evidence. Importantly, its toxicity data are lacking. Numerous studies that have investigated extracts containing carnosol have provided only limited information. Pure carnosol must be investigated in future works to determine its pharmacodynamic and pharmacokinetic profiles accurately.

Acknowledgments

This study was supported by the Science and Technology Development Fund, Macao SAR (FDCT, 078/2016/A2, 175/2017/A3) and the National Natural Science Foundation of China (no. 81774200). I thank Miss Bingling Zhong for drawing Fig. 4.3.1.

Abbreviations

OH	Hydroxyl radical
AUC	Area under the curve
BHT	Butylated hydroxytoluene
C_{max}	Maximum peak concentration
DMSO	Dimethyl sulfoxide
DPPH	2,2-diphenyl-1-picrylhydrazyl
EMT	Epithelial-mesenchymal transition
ERK	Extracellular signal-regulated protein kinase
GSH	Glutathione
H_2O_2	Hydrogen peroxide
HO-1	Heme oxygenase-1
IC_{50}	Half maximal inhibitory concentration
JNK	c-Jun N-terminal kinase
LDL	Low-density lipoprotein
MRT	Mean residence times in the body
NADPH	Nicotinamide adeninc dinucleotide phosphate
ROS	Reactive oxygen species
SNP	Sodium nitroprusside
T_{max}	Time to reach maximum peak concentration
T_{last}	Time of the last measurable concentration

References

Al Dhaheri, Y., Attoub, S., Ramadan, G., Arafat, K., Bajbouj, K., Karuvantevida, N., AbuQamar, S., Eid, A., Iratni, R., 2014. Carnosol induces ROS-mediated beclin1-independent autophagy and apoptosis in triple negative breast cancer. PLoS One 9, e109630.

Arranz, E., Guri, A., Fornari, T., Mendiola, J.A., Reglero, G., Corredig, M., 2017. In vitro uptake and immune functionality of digested Rosemary extract delivered through food grade vehicles. Food Res. Int. 97, 71–77.

Arranz, E., Mes, J., Wichers, H.J., Jaime, L., Mendiola, J.A., Reglero, G., Santoyo, S., 2015. Anti-inflammatory activity of the basolateral fraction of Caco-2 cells exposed to a rosemary supercritical extract. J. Funct. Foods 13, 384–390.

Arranz, E., Santoyo, S., Jaime, L., Fornari, T., Reglero, G., Guri, A., Corredig, M., 2015. Improved bioavailability of supercritical rosemary extract through encapsulation in different delivery systems after in vitro digestion. Food Dig. 6, 30–37.

Aruoma, O.I., Halliwell, B., Aeschbach, R., Loligers, J., 1992. Antioxidant and pro-oxidant properties of active rosemary constituents: carnosol and carnosic acid. Xenobiotica 22, 257–268.

Aruoma, O.I., Spencer, J.P., Rossi, R., Aeschbach, R., Khan, A., Mahmood, N., Munoz, A., Murcia, A., Butler, J., Halliwell, B., 1996. An evaluation of the antioxidant and antiviral action of extracts of rosemary and Provençal herbs. Food Chem. Toxicol. 34, 449–456.

Brieskorn, C.H., Fuchs, A., 1962. The constituents of Salvia officinalis. XIII. The structure of picrosalvin, a diterpene-o-diphenollactone from the sage leaf. Chem. Ber. 95, 3034–3041.

Brieskorn, C.H., Fuchs, A., Bredenberg, J.B.s., McChesney, J.D., Wenkert, E., 1964. The structure of carnosol. J. Org. Chem. 29, 2293–2298.

Chun, K.S., Kundu, J., Chae, I.G., Kundu, J.K., 2014. Carnosol: a phenolic diterpene with cancer chemopreventive potential. J. Cancer Prev. 19, 103–110.

Cor, D., Knez, Z., Knez Hrncic, M., 2018. Antitumour, antimicrobial, antioxidant and antiacetylcholinesterase effect of Ganoderma lucidum terpenoids and polysaccharides: a review. Molecules 23 (3), 649.

de Oliveira, M.R., 2016. The dietary components carnosic acid and carnosol as neuroprotective agents: a mechanistic view. Mol. Neurobiol. 53, 6155–6168.

Dimayuga, R.E., Garcia, S.K., Nielsen, P.H., Christophersen, C., 1991. Traditional medicine of Baja California Sur (Mexico). III. Carnosol: a diterpene antibiotic from Lepechinia hastata. J. Ethnopharmacol. 31, 43–48.

Escuder, B., Torres, R., Lissi, E., Labbé, C., Faini, F., 2002. Antioxidant capacity of abietanes from Sphacele salviae. Nat. Prod. Lett. 16, 277–281.

Fernandez-Ochoa, A., Borras-Linares, I., Perez-Sanchez, A., Barrajon-Catalan, E., Gonzalez-Alvarez, I., Arraez-Roman, D., Micol, V., Segura-Carretero, A., 2017. Phenolic compounds in rosemary as potential source of bioactive compounds against colorectal cancer: In situ absorption and metabolism study. J. Funct. Foods 33, 202–210.

Gajhede, M., Anthoni, U., Halfdan Nielsen, P., Pedersen, E.J., Christophersen, C., 1990. Carnosol. Crystal structure, absolute configuration, and spectroscopic properties of a diterpene. J. Crystallogr. Spectrosc. Res. 20, 165–171.

Gershenzon, J., Dudareva, N., 2007. The function of terpene natural products in the natural world. Nat. Chem. Biol. 3, 408–414.

Gilardoni, G., Ramirez, J., Montalvan, M., Quinche, W., Leon, J., Benitez, L., Morocho, V., Cumbicus, N., Bicchi, C., 2018. Phytochemistry of Three Ecuadorian Lamiaceae: Lepechinia heteromorpha (Briq.) Epling, Lepechinia radula (Benth.) Epling and Lepechinia paniculata (Kunth) Epling. Plants (Basel) 8 (1), 1.

Gonzalez-Burgos, E., Gomez-Serranillos, M.P., 2012. Terpene compounds in nature: a review of their potential antioxidant activity. Curr. Med. Chem. 19, 5319–5341.

Johnson, J.J., 2011. Carnosol: a promising anti-cancer and anti-inflammatory agent. Cancer Lett. 305, 1–7.

Johnson, J.J., Syed, D.N., Suh, Y., Heren, C.R., Saleem, M., Siddiqui, I.A., Mukhtar, H., 2010. Disruption of androgen and estrogen receptor activity in prostate cancer by a novel dietary diterpene carnosol: implications for chemoprevention. Cancer Prev. Res. (Phila.) 3, 1112–1123.

Kashyap, D., Kumar, G., Sharma, A., Sak, K., Tuli, H.S., Mukherjee, T.K., 2017. Mechanistic insight into carnosol-mediated pharmacological effects: Recent trends and advancements. Life Sci. 169, 27–36.

Kim, S.J., Kim, J.S., Cho, H.S., Lee, H.J., Kim, S.Y., Kim, S., Lee, S.Y., Chun, H.S., 2006. Carnosol, a component of rosemary (Rosmarinus officinalis L.) protects nigral dopaminergic neuronal cells. Neuroreport 17, 1729–1733.

Kim, S.Y., Park, E., Park, J.A., Choi, B.S., Kim, S., Jeong, G., Kim, C.S., Kim, D.K., Kim, S.J., Chun, H.S., 2010. The plant phenolic diterpene carnosol suppresses sodium nitroprusside-induced toxicity in c6 glial cells. J. Agric. Food Chem. 58, 1543–1550.

Laughton, M.J., Evans, P.J., Moroney, M.A., Hoult, J.R., Halliwell, B., 1991. Inhibition of mammalian 5-lipoxygenase and cyclo-oxygenase by flavonoids and phenolic dietary additives. Relationship to antioxidant activity and to iron ion-reducing ability. Biochem. Pharmacol. 42, 1673–1681.

Lo, Y.C., Lin, Y.C., Huang, Y.F., Hsieh, C.P., Wu, C.C., Chang, I.L., Chen, C.L., Cheng, C.H., Chen, H.Y., 2017. Carnosol-induced ROS inhibits cell viability of human osteosarcoma by apoptosis and autophagy. Am. J. Chin. Med. 45, 1761–1772.

Machado, D.G., Cunha, M.P., Neis, V.B., Balen, G.O., Colla, A., Bettio, L.E., Oliveira, A., Pazini, F.L., Dalmarco, J.B., Simionatto, E.L., Pizzolatti, M.G., Rodrigues, A.L., 2013. Antidepressant-like effects of fractions, essential oil, carnosol and betulinic acid isolated from Rosmarinus officinalis L. Food Chem. 136, 999–1005.

Martin, D., Rojo, A.I., Salinas, M., Diaz, R., Gallardo, G., Alam, J., De Galarreta, C.M., Cuadrado, A., 2004. Regulation of heme oxygenase-1 expression through the phosphatidylinositol 3-kinase/Akt pathway and the Nrf2 transcription factor in response to the antioxidant phytochemical carnosol. J. Biol. Chem. 279, 8919–8929.

Meyer, W.L., Manning, R.A., Schindler, E., Schroeder, R.S., Shew, D.C., 1976. Diterpenoid total synthesis, an A .far. B .far. C approach. VIII. Introduction of oxygen at carbon-11. Total synthesis of (+-)-carnosic acid dimethyl ether and (+-)-carnosol dimethyl ether. J. Org. Chem. 41, 1005–1015.

Miura, K., Kikuzaki, H., Nakatani, N., 2002. Antioxidant activity of chemical components from sage (Salvia officinalis L.) and thyme (Thymus vulgaris L.) measured by the oil stability index method. J. Agric. Food Chem. 50, 1845–1851.

Nwodo, J.N., Ibezim, A., Simoben, C.V., Ntie-Kang, F., 2016. Exploring cancer therapeutics with natural products from African medicinal plants, part II: alkaloids, terpenoids and flavonoids. Anticancer Agents Med. Chem. 16, 108–127.

Ouyang, W., Zhu, X.A., He, C.X., Chen, X.X., Ye, S.M., Peng, S., Cao, Y., 2015. Chemical constituents from ethyl acetate extract of Psidium guajava leaves (II). Zhong Yao Cai 38, 1649–1652.

Pan, S.-Y., Zhou, S.-F., Gao, S.-H., Yu, Z.-L., Zhang, S.-F., Tang, M.-K., Sun, J.-N., Ma, D.-L., Han, Y.-F., Fong, W.-F., Ko, K.-M., 2013. New perspectives on how to discover drugs from herbal medicines: CAM's outstanding contribution to modern therapeutics. Evid-Based Complement. Altern. Med, 627375.

Park, K.W., Kundu, J., Chae, I.G., Kim, D.H., Yu, M.H., Kundu, J.K., Chun, K.S., 2014. Carnosol induces apoptosis through generation of ROS and inactivation of STAT3 signaling in human colon cancer HCT116 cells. Int. J. Oncol. 44, 1309–1315.

Perez-Sanchez, A., Borras-Linares, I., Barrajon-Catalan, E., Arraez-Roman, D., Gonzalez-Alvarez, I., Ibanez, E., Segura-Carretero, A., Bermejo, M., Micol, V., 2017. Evaluation of the intestinal permeability of rosemary (Rosmarinus officinalis L.) extract polyphenols and terpenoids in Caco-2 cell monolayers. PLoS One 12, e0172063.

Ramirez, J., Gilardoni, G., Ramon, E., Tosi, S., Picco, A.M., Bicchi, C., Vidari, G., 2018. Phytochemical study of the Ecuadorian sspecies Lepechinia mutica (Benth.) Epling and high antifungal activity of carnosol against Pyricularia oryzae. Pharmaceuticals (Basel), 11.

Romo Vaquero, M., Garcia Villalba, R., Larrosa, M., Yanez-Gascon, M.J., Fromentin, E., Flanagan, J., Roller, M., Tomas-Barberan, F.A., Espin, J.C., Garcia-Conesa, M.T., 2013. Bioavailability of the major bioactive diterpenoids in a rosemary extract: metabolic profile in the intestine, liver, plasma, and brain of Zucker rats. Mol. Nutr. Food Res. 57, 1834–1846.

Rutherford, D.M., Nielsen, M.P., Hansen, S.K., Witt, M.R., Bergendorff, O., Sterner, O., 1992. Isolation and identification from Salvia officinalis of two diterpenes which inhibit t-butylbicyclophosphoro[35S]thionate binding to chloride channel of rat cerebrocortical membranes in vitro. Neurosci. Lett. 135, 224–226.

Samarghandian, S., Azimi-Nezhad, M., Borji, A., Samini, M., Farkhondeh, T., 2017. Protective effects of carnosol against oxidative stress induced brain damage by chronic stress in rats. BMC Complement. Altern. Med. 17, 249.

Samarghandian, S., Azimi-Nezhad, M., Farkhondeh, T., 2018. Anti-carcinogenic effects of carnosol-an updated review. Curr. Drug Discov. Technol. 15, 32–40.

Samarghandian, S., Borji, A., Farkhondeh, T., 2017. Evaluation of antidiabetic activity of carnosol (phenolic diterpene in rosemary) in streptozotocin-induced diabetic rats. Cardiovasc. Hematol. Disord. Drug. Targets 17, 11–17.

Sero, R., Nunez, N., Nunez, O., Camprubi, A., Grases, J.M., Saurina, J., Moyano, E., Calvet, C., 2018. Modified distribution in the polyphenolic profile of rosemary leaves induced by plant inoculation with an arbuscular mycorrhizal fungus. J. Sci. Food Agric.

Shenoy, N., Creagan, E., Witzig, T., Levine, M., 2018. Ascorbic Acid in Cancer Treatment: Let the Phoenix Fly. Cancer Cell 34, 700–706.

Soler-Rivas, C., Marin, F.R., Santoyo, S., Garcia-Risco, M.R., Senorans, F.J., Reglero, G., 2010. Testing and enhancing the in vitro bioaccessibility and bioavailability of Rosmarinus officinalis extracts with a high level of antioxidant abietanes. J. Agric. Food Chem. 58, 1144–1152.

Tamaki, Y., Tabuchi, T., Takahashi, T., Kosaka, K., Satoh, T., 2010. Activated glutathione metabolism participates in protective effects of carnosic acid against oxidative stress in neuronal HT22 cells. Planta Med. 76, 683–688.

Thorsen, M.A., Hildebrandt, K.S., 2003. Quantitative determination of phenolic diterpenes in rosemary extracts. Aspects of accurate quantification. J. Chromatogr. A 995, 119–125.

Vagi, E., Rapavi, E., Hadolin, M., Vasarhelyine Peredi, K., Balazs, A., Blazovics, A., Simandi, B., 2005. Phenolic and triterpenoid antioxidants from Origanum majorana L. herb and extracts obtained with different solvents. J. Agric. Food Chem. 53, 17–21.

White, A.I., Jenkins, G.L., 1942. Salvia carnosa (Dougl.). II. Carnosol. J. Am. Pharm. Assoc. 31 (1912-1977), 37–43.

Wu, Q., Sun, J., Chen, J., Zhang, H., Guo, Y.W., Wang, H., 2018. Terpenoids from marine soft coral of the genus Lemnalia: chemistry and biological activities. Mar. Drugs 16.

Zacchino, S.A., Butassi, E., Liberto, M.D., Raimondi, M., Postigo, A., Sortino, M., 2017. Plant phenolics and terpenoids as adjuvants of antibacterial and antifungal drugs. Phytomedicine 37, 27–48.

Zeng, H.H., Tu, P.F., Zhou, K., Wang, H., Wang, B.H., Lu, J.F., 2001. Antioxidant properties of phenolic diterpenes from Rosmarinus officinalis. Acta Pharmacol. Sin. 22, 1094–1098.

Zhang, Y., Smuts, J.P., Dodbiba, E., Rangarajan, R., Lang, J.C., Armstrong, D.W., 2012. Degradation study of carnosic acid, carnosol, rosmarinic acid, and rosemary extract (Rosmarinus officinalis L.) assessed using HPLC. J. Agric. Food Chem. 60, 9305–9314.

Zhao, H., Wang, Z., Tang, F., Zhao, Y., Feng, D., Li, Y., Hu, Y., Wang, C., Zhou, J., Tian, X., Yao, J., 2018. Carnosol-mediated Sirtuin 1 activation inhibits Enhancer of Zeste Homolog 2 to attenuate liver fibrosis. Pharmacol. Res. 128, 327–337.

Zheng, Y., Zhang, Y., Zheng, Y., Zhang, N., 2018. Carnosol protects against renal ischemia-reperfusion injury in rats. Exp. Anim. 67, 545–553.

CHAPTER

Carotenoids (Xanthophylls and Carotenes)

4.4

Koula Doukani[a,b], Ammar S.M. Selles[c], Hasna Bouhenni[a], Meriem Chafaa[a], Leila Soudani[a]

[a]*Faculty of Nature and Life Sciences, University of Tiaret, Algeria*

[b]*Laboratory of Sciences and Technics of Animal Production, University of Abdelhamid Ibn Badis, Mostaganem, Algeria*

[c]*Institute of Veterinary Sciences, University of Tiaret, Algeria*

4.4.1 Carotenoids

Carotenoids are one of the most frequent groups of natural colors. They are produced by plants and some species of archaea and fungi as well as photosynthetic bacteria and algae. However, animals must obtain them through diet. The red pea aphid (*Acyrthosiphon pisum*) and spider mite (*Tetranychus urticae*) are the only animals that make carotenoids from fungi via gene transfer. There are over 1100 known carotenoids which have been identified and characterized and are responsible for the red, orange, and yellow colors (Yabuzaki, 2017).

4.4.2 Chemical composition

Xanthophylls and carotenes are two types of carotenoids that vary only in their oxygen content. The base structure of carotenoids is identical, consisting of eight isoprene molecules. Isoprene molecules have 5 carbons each, making a total of 40 carbons. Tetraterpenoids are a group of carotenoids that all have the same structure (Rao and Rao, 2007).

Carotenoids are classified into two groups based on their chemical composition (Fig. 4.4.1): (1) carotenes, which are composed of only carbon and hydrogen molecules, for example, α-carotene, β-carotene, γ-carotene, lycopene, phytoene, and phytofluene and (2) xanthophylls, which contain oxygen functionality at the cyclic end groups in the form of methoxy, hydroxy, keto, carboxy, and epoxy positions. Examples of these compounds are zeaxanthin, lutein, β-cryptoxanthin, astaxanthin, fucoxanthin, spirilloxanthin, echinenone, and antheraxanthin (Goodwin, 1980).

Because of the several conjugated double bonds in the polyene backbone, many geometric isomeric forms are possible. The most common isomer found in nature is the linear all-trans isomer, and many physicochemical and biological

Antioxidants Effects in Health. DOI: https://doi.org/10.1016/B978-0-12-819096-8.00044-6

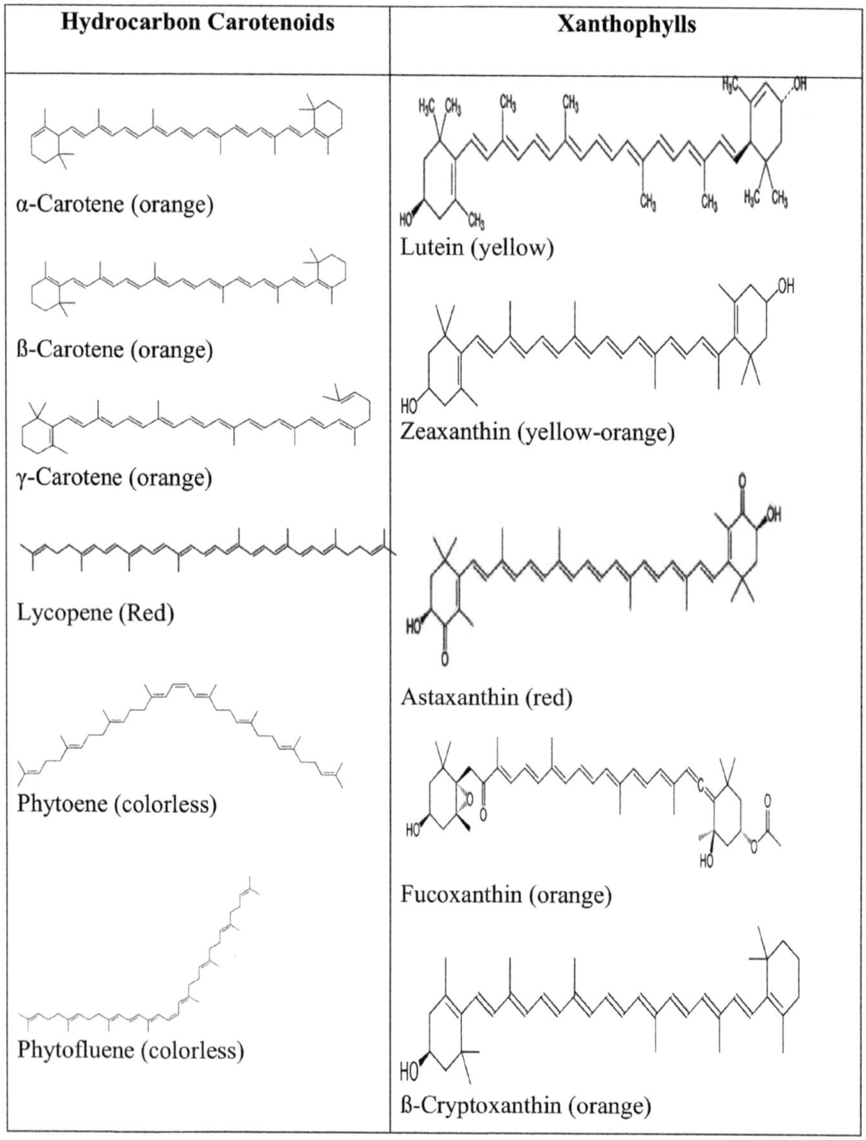

FIG. 4.4.1 Classification of carotenoids.

properties of these compounds are due to their electron-rich polyene backbone (Britton, 1995).

Carotenoids are generated in plants through a complex process that combines C5-isoprenoid units which are biosynthesized into C20 molecules (geranylgeranyl pyrophosphate GGPP) by the geranylgeranyl pyrophosphate synthase; however, carotenoids with 40 carbon atoms are the most common, which is the product of phytoene synthase mediated condensation of two molecules of 20 carbon atoms (Liang et al., 2006).

Carotenoids are naturally synthesized by all photosynthetic and non-photosynthetic species, including bacteria, fungi and archaea, which have a complex carotenogenic metabolism (Britton et al., 2004). They are categorized as C30, C40, C45, and C50 carotenoids based on the number of carbons in their structure, but only the C40 carotenoids are present in nature in greater abundance and, as a result, are more studied in the literature. Furthermore, species such as eukaryotes, archaea, and bacteria biosynthesize C40 carotenoids, and their chemical structures are made up of a variety of terminal classes. C30 and C50 carotenoids, on the other hand, are generated by archaea and bacteria and contain only 6 and 10 isoprenoid units, respectively. On the contrary, only bacteria are capable of producing C45 carotenoids, which are made up of nine isoprenoid units (Yabuzaki, 2017).

The acyclic C40 isoprenoid (tetraterpene) lycopene is the source of all C40 carotenoids. Lycopene is made up of eight C5 isoprene units in total. Initially, four C5 units combine to form geranylgeranyl diphosphate, a C20 intermediate. The C40 intermediate phytoene, a more saturated precursor of lycopene, is formed when two C20 precursors merge head-to-head. By stepwise enzymatic dehydrations, the other double bonds are introduced in phytoene. Various modifications, such as cyclizations, oxidative functionalizations, rearrangements, and oxidative degradations, are used to produce other carotenoids from lycopene (Zechmeister, 1962).

Carotenoids geometric isomers, as well as carotenoid epoxides, can be found in all-trans and cis forms. While the all-trans isomer is the most common type of carotenoid, the cis isomers are only found in trace amounts. The amount of cis carotenoid isomers could be increased by heating and thermal processing. Due to carotenoid deficiency, the depletion of carotenoids in fruits and vegetables is a major concern. Controlling carotenoid geometry isomer degradation and improving the efficiency of dietary carotenoids should be prioritized. Color variations in carotenoids during geometric isomerization, thermal processing, and even fruit ripening have all been studied extensively. However, the chemical and kinetic mechanisms of color changes during carotenoid degradation and isomerization are still unknown. More research into the isomerization of carotenoids in relation to the colorful pigments found in the biodiversity of fruits and vegetables is required in the future (Khoo et al., 2011).

4.4.3 Sources of carotenoids

Vegetables and fruits are the most major sources of carotenoids in the human diet (Elvira-Torales et al., 2019). Yellow-orange fruits and vegetables, as well as dark green, leafy vegetables, are high in carotenoids (Britton and Khachik, 2009). Animal-derived foods, such as dairy products, eggs, and some fish and seafood, can also contain large amounts of carotenoids (Mezzomo and Ferreira, 2016).

Carotenoids are found in a wide range of fruits and vegetables, and their composition and content are influenced by a number of factors, including variety, genotype, season, geographical location/climatic conditions, soil, maturation stage, type of processing, and storage conditions (Yahia and Ornelas-Paz, 2010; Alonso, 2017).

In addition, food cooking and processing techniques and methods of cultivation differ widely across the world and can have a significant impact on the stability and hence the quality of carotenoids (Maiani et al., 2009).

4.4.3.1 Carotenes

Beta carotene is abundantly found in foods that has the highest provitamin A activity. It is mostly found in yellow-orange and dark green fruits and vegetables (Gul et al., 2015) with large amounts in buriti (*Mauritia vinifera* Mart.), camu-camu (*Myrciaria dubia*), tucumã (*Astrocaryum aculeatum*), bocaiuva (*Acrocomia mokayayba* Barb. Rodr.), some varieties of pumpkin, acerola, mango, carrot, spinach, squash, papaya, sweet potatoes, apricots, nuts, rose hip fruits carrot noodles, and oil palm (Mezzomo and Ferreira, 2016).

Lycopene is a red color that exists naturally only in the tissues of vegetables and algae. It is deficient in provitamin A activity and is responsible for the red to pink colors in fruits and vegetables, such as tomatoes, red grapefruit, watermelon, guava, and papaya (Kong et al., 2010). Tomatoes and their derivatives, such as soups, juices, sauces, and ketchup are commonly the most commonly cited sources of lycopene. However, even greater concentration in cherry, guava and guava products, similar concentrations in watermelon and Thai papaya, and smaller amounts in Solo and Formosa papaya can be found (Mezzomo and Ferreira, 2016).

Phytoene (PT) and phytofluene (PTF) are carotenes found in a wide variety of fruits and vegetables, for example, in tomatoes, carrots, and light orange apricots (Meléndez-Martínez et al., 2015).

4.4.3.2 Xanthophylls

Lutein and zeaxanthin are found in green and dark green leafy vegetables, such as parsley, spinach, broccoli, and Brussels sprouts (Mezzomo and Ferreira, 2016). Egg yolk is a highly bioavailable source of zeaxanthin and lutein (Handelman et al., 1999).

Lutein is a non provitamin A that is distributed in a wide variety of vegetables, such as kale, some varieties of pumpkin, acerola, spinach, and winter squash, caja (*Spondias lutea*) and camu-camu (*Myrciaria dubia*) (Mezzomo and Ferreira, 2016) and fruits, such as mango, papaya, peaches, plums, and oranges (Elvira-Torales et al., 2019). It is also produced by some microalgae, such as *Chlorella sorokiniana* MB-1, *Chlorella vulgaris*, and *Scenedesmus obliquus* CNW-N (Mezzomo and Ferreira, 2016).

Zeaxanthin is found in high amounts in pequi (*Caryocar villosum*) and even the native marine microalgae *Chlorella saccharophila* (Mezzomo and Ferreira, 2016).

Astaxanthin is a characteristic marine carotenoid in crustaceans (lobster, crab, and shrimp). Also, high amounts of this pigment are produced by the microalgae *Chlorella vulgaris, Phaffia rhodozyma,* and *Haematococcus pluvialis* (Ambati et al., 2014).

Undaria pinnatifida (Wakame), *Laminaria japonica* (Ma-Kombu), *Phaeodactylum tricornutum*, and *Cylindrotheca closterium* are examples of macroalgae and microalgae that contain also a marine carotenoid called **fucoxanthin** (Zhang et al., 2015a).

β-Cryptoxanthin is a less well- known carotenoids. It is found in carrots, peppers, tangerines, pumpkins, peaches, oranges, and in tropical fruits, such as papaya (Elvira-Torales et al., 2019).

Britton and Khachik (2009) suggested a classification system for carotenoids in fruits and vegetables based on the following levels:

Low (0–0.1 mg/100 g FW)
Moderate (0.1–0.5 mg/100 g FW)
High (0.5–2 mg/100 g FW)
Very high (>2 mg/100 g FW)

Table 4.4.1 shows examples of foods having high, high-very high levels of a certain carotenoid.

4.4.4 Carotenoids accumulation and bioavailability

Plant physiological, genetic, and biochemical characteristics, as well as environmental growth factors, including light, temperature, and fertility, tend to influence carotenoids accumulation in plant tissues (Kurilich et al., 1999; Kopsell et al., 2003). There have been reports of significant differences in carotenoids accumulation among different vegetable crop species (Klein and Perry, 1982; Kimura and Rodriguez-Amaya, 2003). Carrot, corn, kale, lettuce, potato (*Solanum tuberosum* subsp. *tuberosum*), pepper, and soybean all have substantial genetic variation within species (Glycine max). Vegetable improvement programs can benefit from genetic variation in carotenoids concentrations within species (Nicolle et al., 2004).

Environmental growing conditions have an effect on the accumulation of carotenoids in plant foods. Carotenoids levels have been shown to rise and fall in response to environmental changes, with varying results for different plant species. Plant carotenoid accumulation is affected by changes in rising air temperature, irradiance level, irradiance photoperiod, and nutritional fertility (Lefsrud et al., 2005; Lefsrud et al., 2006a; Lefsrud et al., 2006b). Increased carotenoids concentrations are also associated with increased coloration in vegetable and fruit tissues as they mature (Russo and Howard, 2002). Carotenoids concentrations in leaf tissues rise with maturity but fall as the plant ages. Manipulation of cultural growing conditions and harvest time will thus influence the concentrations of carotenoids in fruit and vegetable crops (Gross, 1991).

The species and structures of carotenoids present in food, the composition and release of carotenoids from the food matrix, ingested quantities and absorption in the intestinal tract, transportation inside the lipoprotein fractions, the nutritional status of the ingesting host, as well as biochemical conversions and tissue-specific depositions; all affect the bioavailability of carotenoids from plant foods (Faulks and Southon, 2005). Carotenes are fully lipophilic molecules found in plant membranes' hydrophobic cores. Xanthophylls, on the other hand, are often hydrophobic

Table 4.4.1 Some examples of foods having high, high very high levels of a given carotenoid (Britton and Khachik, 2009).

High content	High-very high content
ß-carotene	
Brussels sprouts (*Brassica oleracea* [*Gemmiera*])	Apricot (*Prunus ameniaca*)
Karat banana (*Musa troglodytarum*)	Broccoli (*Brassica oleracea*)
Peach (*Prunus persica*)	Buriti (*Mauritia vinifera*)
Pepper (red, orange, green) (*Capsicum annuum*)	Carrot (*Daucus carota*)
West Indian cherry (*Malpighia glabra*)	Gac oil (*Momordica cochinchinnensis*)
	Kale (*Brassica oleracea* (*Acephala*))
	Mango (*Mangifera indica*)
	Red palm oil (*Elaesis guineensis*)
	Spinach (*Spinacia oleracea*)
	Sweet potato (*Ipomoea batatas*)
	Tomato (*Lycopersicum sculentum*), high-beta
ß-cryptoxanthin	
Permisimmon (*Diospyros kaki*)	
Pitanga (*Eugenia uniflora*)	
Lutein	
Broccoli (*Brassica oleracea* [*Italica*])	
Green leafy vegetables	
Pepper (yellow, green) (*Capsicum annuum*)	
Zeaxanthin	
Buriti (*Mauritia vinifera*)	
Chinese wolfberry (*Lycium chinensis*)	
Pepper (orange, red) (*Capsicum annuum*)	
Lycopene	
Carrot (red) (*Daucus carota*)	
Guava (*Psidium guajava*)	
Tomato (*Lycopersicon esculentum*)	
Water melon (*Citrullus lanatus*)	

molecules with polar groups on opposite ends of a non-polar carbon skeleton. Since carotenoids are lipophilic, any biotic or abiotic activity that exposes them to possible oxidation, degradation, or isomerization would have an effect on their biochemistry and bioavailability (Gruszecki, 1999).

The release of carotenoids from the food matrix is the first step in their bioavailability. Carotenoids chemistry shifts as a result of food processing practices such as thermal processing, mincing, or liquefying, most likely due to isomerization or oxidation reactions (Kopas-Lane and Warthesen, 1995). However, by minimizing possible enzymatic oxidation, freezing or low-temperature storage normally preserves carotenoids concentrations. Processing activities normally increase bioavailability by increasing the release of bound carotenoids from the food matrix; however, in some food crops, thermal degradation of carotenoids chemistry can reduce bioavailability (Rodriguez-Amaya, 1999).

Carotenoids are absorbed passively in humans and follow digestive processes similar to lipids. Carotenoids that are bound to proteins or membranes must first be released from tissues and dissolved in a hydrophobic domain (oils, fats or bulk lipid emulsions). They must be moved to bulk lipids or intestinal micelles in the digesta because they are hydrophobic (Faulks and Southon, 2005). The presence of dietary fat in the small intestine, which promotes the release of emulsifying bile acids by the gallbladder, is needed for carotenoids absorption (Zaripheh and Erdman, 2002). Recent research has found that when carotenoids are consumed with dietary lipids, their absorption increases (Brown et al., 2004). Carotenoids are absorbed into chylomicrons after being released into the enterocyte, and then delivered to the bloodstream and, finally, the liver. Until eventual tissue specific deposition, carotenoids compounds can stay in the liver or be transferred to low-density or high-density lipoproteins (Brown et al., 2004).

In dietary trials or cohort studies, carotenoids bioavailability is normally measured in blood serum after ingestion. The relatively simple study measures changes in serum carotenoids levels over time after consumption of whole foods or supplements. The following are several caveats to consider when interpreting serum carotenoids bioavailability: (1) serum responses to single oral doses of carotenoids are greatly variable, (2) carotenoids levels measured in serum indicate an equilibrium between intestinal absorption, breakdowns, tissue uptake, and tissue release, and (3) serum already contains high levels of endogenous carotenoids (i.e. -carotene, -carotene, lycopene, and lutein) (Yeum and Russell, 2002). *In vitro* Caco-2 human intestinal cell lines have also been used in recent studies to determine carotenoid bioavailability (Chitchumroonchokchai et al., 2004).

4.4.5 Beneficial and detrimental effects of carotenoids on health

4.4.5.1 Beneficial effects

A diet containing carotenoids is essential for normal behavior and good health of humans and animals (Nisar et al., 2015). The activities of carotenoids as well as provitamin A and antioxidants (reduction of oxidative stress by capture of free radicals) increase interest toward these compounds. However, plants are rich sources of carotenoids (Rao and Agarwal, 1999).

4.4.5.1.1 Provitamin A activity

In humans, carotenoids regulate a number of genes related to developmental and physiological processes. Moreover, they are involved in the visual cycle as precursors of vitamin A related retinoids, such as retinol, retinal, and retinoic acid (Bonet et al., 2015). Divided into provitamin A and non provitamin compound (Olson and Krinsky,

1995). Three carotenoids classes (β-carotene, α-carotene, and β-cryptoxanthin) possess this provitamin A activity (Cazzonelli et al., 2010). They help prevent vitamin A deficiency (Stahl and Sies, 2005).

A toxic effect of vitamin A can be seen with diets rich in it. While, an excess of dietary β-carotene is not, therefore plants with a high β-carotene level constitute a safe and effective means as being a source of vitamin A (Cazzonelli et al., 2010), these β-carotenes are cleaved to form two retinals (vitamin A aldehyde) in the human organism by β-carotene 15,15′-dioxygenase (Von Lintig and Vogt, 2000; Wyss et al., 2000).

4.4.5.1.2 *Antioxidant activity*

Several epidemiological and clinical studies have examined the antioxidant capacity of carotenoids to prevent formation of various ROS in pathological conditions, such as cancer, inflammation, retinal degeneration and neurodegeneration (Cho et al., 2018). Likewise, these antioxidant properties are due to the various molecules of carotenoids such as lutein and zeaxanthin, ketocarotenoids and which form part of the beneficial human dietary components of the man (Jyonouchi et al., 1995; Zhu et al., 2008).

Antioxidant activity is probably due to trapping of singlet oxygen ($^1O_2$) and peroxyl radicals. In addition, carotenoids inhibit radicals generation and singlet oxygen by effectively deactivating the electronically excited sensitizing molecules (Truscott, 1990; Young and Lowe, 2001).

However, Miller et al. (1996) show that the presence of carbonyl and hydroxyl groups, in the terminal rings, as well as the number of conjugated double bonds increases the relative capacities of carotenoids to scavenge the ABTS ·+ radical cation.

Nevertheless, Edge et al. (1997) considered carotenoids as the most powerful natural scavengers of singlet oxygen, with a rapid extinction rate (10^{10} $M^{-1}s^{-1}$).

Moreover, El-Agamey et al. (2004) and Milani et al. (2017) report the three stages by which carotenoids scavenge radicals:

First step: an oxidation-reduction reaction involving a transfer of electrons:

CAR + ROO → CAR++ ROO–, this electron acceptance is favored by the presence of conjugated double bonds which makes it possible to neutralize free radicals (Rutz et al., 2016; Milani et al., 2017).

Second step: hydrogen abstraction: CAR + ROO– → CAR + ROOH

Third step: addition:CAR + ROO → ROOCAR (El-Agamey et al., 2004; Milani et al., 2017).

Besides, cooperative synergistic effect of vitamins E, C and β-carotene can trap reactive nitrogen species and inhibit lipid peroxidation (Bfhm et al., 1998; Stahl and Sies, 2005; Milani et al., 2017).

Wang et al. (2000) concluded that antioxidants might be a novel strategy for treating *H. pylori* infection in humans. This conculsion was made on the basis of the *in vivo* study carried out on Balbe / cA mice infected with *H. pylori* and treated with algal meal rich in astaxanthin (0.4, 2, and 4 g/kg of body weight, with the astaxanthin content at 10, 50, and 100 mg/kg respectively). The results of this study showed significantly lower colonization rates, significantly decreased lipid peroxidation and lower inflammation scores than animals not treated or treated with a control meal.

El-Akabawy and El-Sherif (2019) reported that zeaxanthin exerts potent anti-inflammatory and antioxidant effects by lowering myeloperoxidase and malondialdehyde and increasing the enzymatic activity of superoxide dismutase and catalase as well as glutathione levels. Additionnelly, zeaxanthin suppressed levels of tumor necrosis factor alpha, interferon-gamma, interleukin-6, interleukin-1 beta, and nuclear transcription factor kappa B, and inhibited the expression of nitric oxide synthase and cyclooxygenase-2.

Under some circumstances carotenoids may act as cellular antioxidants. For example, β-carotene can suppress the upregulation of hem oxygenase-1 gene expression provoked by UVA exposure in human dermal fibroblasts (FEK4) in a dose-dependent manner (Trekli et al., 2003; Elliott, 2005). This is consistent with a direct antioxidant (singlet oxygen quenching) effect as is the observation that UVA exposure caused the depletion of cellular β-carotene and the accumulation of apocarotenal. However, the activation of retinoid signaling via retinoic acid receptors RARs and RXRs could not be ruled out as a possible alternative mechanism. Curiously, the authors of this study also reported that at the lowest level used (0.2 AM) the presence of β-carotene actually augmented UVA induced hem oxygenase-1 induction. Furthermore, others have reported that β-carotene can act as a pro-oxidant augmenting the induction of hem oxygenase-1 in human skin fibroblasts (HFP-1) (Obermuller-Jevic et al., 1999; Elliott, 2005).

Several enzymatic and nonenzymatic antioxidants are involved in the body's antioxidant defense system (Sies, 1993; Stahl and Sies, 2005). Carotenoids protect cells against oxidative stress by activation of endogenous antioxidant enzymes and reduction of DNA damage, helps protect cells against oxidative stress (Cho et al., 2018).

While, several studies have shown that high concentrations carotenoid increase DNA damage (Woods et al., 1999; Palozza, 2005). These same authors reported that β-carotene increase H_2O_2-induced oxidative DNA damage has been reported in HepG2 cells (Woods et al., 1999; Palozza, 2005), whereas, Lowe et al. (1999) have observed that a failure of β-carotene and other carotenoids, such as lycopene, protect human cells against free-radical-induced DNA damage.

Another study noted oxidation of β-carotene resulted in a greater increase in DNA oxidation in human Hs68 fibroblasts (Yeh and Hu, 2001).

4.4.5.1.3 ***Carotenoids effect on immune function***

Carotenoids have a stimulating effect on the immune system mediated by their vitamin A metabolism and the subsequent mediation of the RAR/RXR response pathways. However, even carotenoids without provitamin A, such as lutein, canthaxantine, and lycopene have marked effects on the immune system (Chew and Park, 2004; Hughes, 2001; Rühl, 2007).

Several studies have reported the stimulatory effects of β-carotene on immune function. Many clinical trials with various doses of β-carotene have been used (from 15 mg/day to 180 mg/day, administered over periods ranging from 14 days to one year). These studies reported an increase in the helper T lymphocytes number (CD4 +) or CD4 +: CD8 + T cells (T suppressor/cytotoxic cells) and in lymphocytes percentage of expressing the activation markers, interleukin (IL) 2 receptor, and transferrin receptor (Alexander et al., 1985; Watson et al., 1991; Murata et al., 1994). In the same context, Fryburg et al. (1995) suggests that β-carotene appears an immuno-enhancing agent in the management of HIV infections by potentiating the increase in CD4 + cells.

Likewise, dietary carotenoids have a stimulating effect on the immune system in inflammatory diseases or human immunodeficiency (Austin et al., 2006). This stimulation results in an increase in proliferation of lymphocytes induced by mitogens, an increase in myeloperoxidase and phagocytic activity, an increased antibody response and an increase in cytochrome oxidase and peroxidase activities in macrophages due to an increase in respiratory burst, which leads to stimulation of blood neutrophil killing activity (Chew and Park, 2004; Elliott, 2005). Moreover, the administration of β-carotene in patients with acquired immunodeficiency syndrome at 60 mg /day for four weeks leads to a slight increase in the number of CD4 + cells (Fryburg et al., 1995; Hughes, 1999).

Additionally, Chew and Park (2004) reported a stimulating effect of β-carotene on the growth of the thymus gland and a strong increase in the number of thymic small lymphocytes. Likewise, *in vivo* studies show the stimulatory effect of β-carotene on lymphocytic blastogenesis in rats, pigs, and cattle. Also, oral β-carotene supplementation in adult humans has a stimulatory effect by increasing the number of T helper and T inducing lymphocytes.

Kang and Kim (2017) report that among carotenoids astaxanthin is the best immune stimulator by stimulating production of antibodies directed against T-dependent antigens and the activity of T helper cells. These authors found that the T cell response of mice infected with *H. pylori* was different in the astaxanthin-treated group with a predominant T-helper (Th) 2 cell-like response and IL-4 release.

4.4.5.1.4 Carotenoids effect on gastrointestinal tract

People with inflammatory bowel disease commonly suffer from vitamin A deficiency (Bousvaros et al., 1998). The study by Głąbska et al. (2019) on 56 individuals with ulcerative colitis in remission with gastrointestinal symptoms (daily number of bowel movements, and the presence of painful tenesmus, flatulence, and constipation) allowed reduction of the incidence of constipation to be observed following consumption of higher doses of lutein and zeaxanthin (varying from 289.0–13221.3 μg and 432.7–1309.0 μg, respectively). However, higher doses of the retinoids, such as lycopene, lutein, and zeaxanthin in individuals with ulcerative colitis in remission lead to a decrease in fecal blood, mucus, and pus but not abdominal pain. While, higher carotene intake in individuals with ulcerative colitis in remission may contribute to higher incidence of fecal mucus (Głąbska et al., 2016).

Wang et al. (2000) found that astaxanthin-rich algae meal posses an inhibitory effect on *H. pylori* growth *in vitro*. Likewise, El-Akabawy and El-Sherif (2019) reported that zeaxanthin can be used for the treatment of ulcerative colitis induced in rats from day 15 by transrectal administration of 3% acetic acid and pretreated with zeaxanthin (50 mg/kg/day) orally for 14 days as it causes a significant reduction in disease activity index, wet weight of the colon, ulcer area, macroscopic scores, and histological changes.

4.4.5.1.5 Carotenoids effect on neurological system

Antineuroinflammatory effects of carotenoids have been highlighted (Cho et al., 2018). Several studies have shown the suppression of inflammation murine retinal cells by lutein (Sasaki et al., 2009; Li et al., 2012). Additionally, lutein in the presence of a variety of oxidative stressors suppresses activation of the nuclear factor-κB (NF-κB) pathway causing reduction lipid peroxidation and release pro-inflammatory cytokines (Kim et al., 2012; Liu et al., 2017). Likewise, in brain the inhibition of NF-κB activity and related expression of pro-inflammatory cytokines (Sachdeva and Chopra, 2015) that can contribute to the suppression of Aβ formation (Katayama et al., 2011) and improvement of memory retention (Prakash and Kumar, 2014; Sachdeva and Chopra, 2015).

Furthermore, Lee et al. (2012); and Marcotorchino et al. (2012) reported that lycopene reduces proinflammatory cytokine and chemokine expression in macrophages. In addition, lycopene increases the permeability of the blood-brain barrier. Whereas, Parkinson's disease and vascular dementia have been observed with significantly low lycopene levels. Also, lycopene can confer protection against amyotrophic lateral sclerosis (ALS) in humans (Rao and Rao, 2007).

Moreover, Nam et al. (2010) report that crocin and crocetin, by sitmulating lipopolysaccharides, interferon γ, and β-amyloids (Aβ) in microglial cells, suppress the production of proinflammatory cytokines and nitric oxide. Though, crocin was shown to be beneficial in both Alzheimer's disease (Ghahghaei et al., 2013; Asadi et al., 2015) and Parkinson disease (Zhang et al., 2015b; Rao et al., 2016).

In addition, Zhou et al. (2015) and Zhou et al. (2017) found that astaxanthin reduced hippocampal and retinal inflammation in diabetic rats induced by streptozotocin. Also, it attenuated cognitive deficits, retinal oxidative stress, and depression. Furthermore, astaxanthin has been shown to protect neurons in various neurodegenerative diseases, including Alzheimer's disease (Chang et al., 2010; Lobos et al., 2016), Parkinson disease (Ikeda et al., 2008; Ye et al., 2013), and amyotrophic lateral sclerosis (Isonaka et al., 2011).

As well, Lin et al. (2016) reported that the consumption of fucoxanthin can inhibit acetylcholinesterase and improve the expression of neurotrophic factors derived from the brain, hence these advantages in the treatment of Alzheimer's disease.

4.4.5.1.6 Anticarcinogenic effects

Several studies have reported the beneficial effects of carotenoids in the treatment of various cancers (Milani et al., 2017). Acting themselves or their metabolites, carotenoids influence the expression of certain genes, inhibit regulatory enzymes, which probably enables preventive properties of cancer (Stahl and Sies, 2005). In addition, a correlation between a high intake of carotenoids in the diet and the reduced risk of breast, cervical, ovarian and colorectal cancers have been mentioned in epidemiological studies (Milani et al., 2017).

Furthermore, several mechanisms are involved in cancer chemoprevention by dietary carotenoids. These have effects on gap junctional intercellular communication, on the growth factor signaling, on cell cycle progression, on differentiation-related proteins, on retinoid-like receptors, on the antioxidant response element, on nuclear receptors, on the AP-1 transcriptional complex, on the Wnt/β-catenin pathway and on inflammatory cytokines. Moreover, carotenoids can stimulate the proliferation of B- and T-lymphocytes, the activity of macrophages and cytotoxic T-cells, effector T-cell function, and the production of cytokines (Milani et al., 2017).

Giovannucci (2002) reported that increased blood levels of lycopene were associated with a decreased risk of prostate cancer. These high lycopene levels are due to consumption of tomatoes and tomato products. Based on the results of previous studies, Kucuk et al. (2001) and van Chen et al. (2001) suggested that supplementation with lycopene or diet rich in lycopene may slow the growth of prostate cancer.

Furthermore, Wang et al. (2015) report, "Neither dietary β-carotene intake nor its blood levels was associated with reduced prostate cancer risk. Dietary α-carotene intake and lycopene consumption (both dietary intake and its blood levels) were all associated with reduced risk of prostate cancer. However, neither blood α-carotene levels nor blood lycopene levels could reduce the risk of advanced prostate cancer". These authors concluded, "α-carotene and lycopene, but not β-carotene, were inversely associated with the risk of prostate cancer. However, both α-carotene and lycopene could not lower the risk of advanced prostate cancer".

β-cryptoxanthin acts as a chemopreventive agent against lung cancer. It negatively regulates the α7 / PI3K signaling pathway of nicotinic neuronal acetylcholine receptors (Iskandar et al., 2016). Similarly, this molecule enhances the action of a chemotherapeutic agent, oxaliplatin, in the treatment of colon cancer (San Millan et al., 2015). Likewise, a diet with high levels of lycopene and β-cryptoxanthin could protect against aggressive prostate cancer (Antwi et al., 2016).

Astaxanthin by reactivating the expression of Nrf2 and Nrf2-target genes through epigenetic modification and chromatin remodeling possess a beneficial health effects against the formation and tumor progression of prostate cancer (Yang et al., 2016). Likewise, astaxanthin by inhibiting ERK1/2 activity exerts antitumorigenic and anti-inflammatory effects on human lung cancer cell lines (Liao et al., 2016). Additionally, high concentrations of astaxanthin may suppress mammary carcinoma (Yuri et al., 2016).

In *vitro* studies indicate that fucoxanthin stimulates apoptosis and decreases proliferation and migration in glioma cancer cell lines U87 and U251 through Akt/mTOR and p38 pathway inhibition (Liu et al., 2016; Merhan, 2017).

4.4.5.1.7 Carotenoids effect on skin

β-Carotenes have anti-inflammatory properties and have the ability to prevent the formation of reactive oxygen species protects the skin from the harmful effects of UV light, the formation of erythema, premature aging of the skin, the development of photodermatitis and skin cancer (Stahl and Sies, 2007; Cazzonelli, 2011; Merhan, 2017). Moreover, several studies in humans have shown that carotenoids levels in plasma and skin decrease upon UV irradiation; lycopene is lost preferentially as compared to other carotenoids (Ribaya-Mercado et al., 1995).

However, a protective effect on human skin of dehydrated molecules and accumulated phytoene and phytofluene was due to its UV absorption properties and antioxidant and anti-inflammatory activities (Aust et al., 2005; Hsu et al., 2012; Merhan, 2017).

Several studies have investigated the protective effect of supplementation with *β*-carotene alone at 24 mg/day or associated with supplementation with vitamin E or other carotenoids (lutein and lycopene), for erythema caused by UV or induced by illumination with a solar simulator. The results showed a decrease and an attenuation of the intensity of the erythema (Stahl et al., 2000; Heinrich et al., 2003; Stahl and Sies, 2005).

4.4.5.1.8 Carotenoids and cardiovascular diseases

Many studies have indicated the protective effect of carotenoids against cardiovascular diseases (Stahl and Sies, 2005; Li and van Eck, 2007; Lu and Li, 2008). Likewise, several clinical trials have shown that carotenoids can reduce the risk of developing cardiovascular disease via several mechanisms such as: (1) lowering blood pressure, (2) reducing proinflammatory cytokines, (3) decreasing markers inflammation (e.g., reactive protein C), and (4) improving the sensitivity of the liver, muscle, and fatty tissue to insulin. In addition, it can modulate the expression of specific genes involved in cell metabolism (Gammone et al., 2015; Milani et al., 2017).

4.4.5.1.9 Eye diseases

Between 1987 and 1998, a number of studies indicated that dietary carotenoids (lutein, zeaxanthin, lycopene, carotene, and carotene) from fruits and vegetables could help prevent eye diseases (Sommerburg, 1998). Given the especially high concentrations and exclusive presence of both xanthophylls in some ocular tissues, lutein, and zeaxanthin can play a role in visual health, demonstrating their contributions to vision enhancement or delaying the onset of ocular diseases (Alves-Rodrigues and Shao, 2004). Lutein has been shown to play a central role in reducing the incidence of eye diseases, such as AMD, cataract, and retinitis pigmentosa (Aleman et al., 2001). It is impractical to specifically measure the effect of eye lutein concentration on the occurrence of ocular diseases in living subjects due to the fragile nature of the eye and the intrusive nature of techniques for assessing and quantifying metabolic products in the retina and lens. Direct measurement of the actual concentration of lutein and zeaxanthin in the macular pigment of donor eyes with and without AMD was published in a key study. This study concluded that control subjects with the highest levels of lutein are 82% less likely to produce AMD than those with the lowest levels of xanthophylls after analyzing 56 retinas from AMD and control subjects (Bone et al., 2001). Observational studies are appropriate for demonstrating relations between nutrient supplementation and tissue concentrations of a particular compound with disease risk, but fall short of establishing a clear cause-and-effect relationship between nutrient intake and a specific benefit (Alves-Rodrigues and Shao, 2004).

4.4.5.2 Detrimental effects

We use the term detrimental to refer to both direct and indirect effects (Zahavi and Zahavi, 1997). The theory of direct toxicity is intriguing because, while the toxicity of carotenoids has yet to be thoroughly assessed, certain other chemicals, such as tannins and alkaloids, are considered to be harmful to animals. Toxicity may be especially problematic for herbivorous species, whose diets are typically rich in carotenoids. Detoxifying the components of plant metabolism may be energy intensive for herbivores (Bendich, 1993; Olson, 1993). Furthermore, carotenoids may be poisonous to any organism that consumes them, based on the degree and type of toxicity. This is true regardless of the quantity consumed. The vast majority of human observational studies indicate that carotenoids can minimize the risk of some chronic diseases (Bendich, 1993; Olson, 1993). However, experimental carotenoids supplementation has not always been shown to be successful, and in one review, supplementation in smokers was linked to an increased risk of lung cancer (Mayne, 1996), smoking is an uncommon animal behavior, and associations between disease incidence and carotenoids in smokers cannot actually be extrapolated to other species or non smokers. Despite the evidence to the contrary, Zahavi and Zahavi (1997) proposed that carotenoids may contribute to the disintegration of cell membranes when transported into cells. More research is required to establish carotenoids possible harmful effects.

4.4.6 Toxicity of carotenoids

The enhancement of the immune response observed in animal models, which may be attributed to the synthesis of tumor specific antigens, is one of the main effects of carotenoids that can be linked to cancer prevention (International Agency for Research on Cancer, 1998). Furthermore, carotenoids have been shown to influence cytochrome P450 metabolism, to inhibit arachidonic acid metabolism, chromosome destruction and instability, to influence apoptosis, and to affect a variety of other biological processes (Mathews-Roth, 1993).

For several years, patients with erythropoietic protoporphyria have been treated with doses of 20–180 mg/day ß-carotene with no signs of toxicity or abnormally elevated serum vitamin A concentrations (Mathews-Roth, 1993). Then, Sies and Krinsky (1995) discovered that low dietary intake and plasma concentrations of carotenoids are often linked to an increased risk of cervical dysplasia, cardiovascular disease, lung cancer, cortical cataract, and AMD.

Some studies have shown that people who consume more fruits and vegetables rich with carotenoids and have elevated levels of serum ß-carotene have a reduced risk of cardiovascular disease and cancer. Although no clinical experiments of ß-carotene as a single agent has indicated a decrease in cancer risk at any particular site; contrarily, the risk of lung cancer among smokers and asbestos employees who received elevated doses of ß-carotene supplements (concentrations that were 10–15 times higher than usual in blood) was higher (Bohlke et al., 1999). According to Olson (1994), carotenoids are non toxic when consumed in foods, however, a very high dose of oxocarotenoid and canthaxanthin used for medicinal reasons can induce retinopathy.

Increased consumption of vegetables and fruits rich with carotenoids has generally been related to a lower risk of lung cancer in observational trials, whether prospective or retrospective. Furthermore, in prospective trials, elevated levels of ß carotene in blood were linked to a lower risk of lung cancer (Hennekens et al., 1996). Also, a high dose of ß carotene (in supplement form, 20–30 mg/day) is contraindicated for smokers due to the high risk of lung and stomach cancer, according to a study conducted by a group of researchers in the United States, who tested the effects of a mixture of 30 mg/day ß carotene and 25.000 IU/day retinol vitamin A in more than 18.000 smokers (men and women), former smokers, or others who have been exposed to asbestos on the jobsite. In groups of smokers who took more than 20 cigarettes/day and consume elevated doses of ß carotene supplements (5 to 10 times the recommended dose), the prevalence of lung cancer was higher (16%) after 6 years in the ATBC participants and (28%) after 4 years in the CARET participants (Omenn et al., 1996a).

Furthermore, other studies written about 1990 summarized the literature on diet and lung cancer conducted over the previous 25 years. The consensus was that increased consumption of carotenoids from vegetables and fruits decreased the risk of lung cancer in clinical trials of diet and lung cancer, whether prospective or retrospective. While in prospective trials, elevated levels of ß-carotene in blood were reliably linked to a lower risk of lung cancer. The most basic reason for the epidemiology was that ß-carotene was defensive, despite the fact that other carotenoids or other substances from vegetables and fruits, as well as related dietary habits, had not been thoroughly investigated. Human chemoprevention studies conducted in the last decade have shown that ß-carotene raises the prevalence of lung cancer and death in human smokers. As well, excessive use of certain carotenoids can also result in carotenemia, a reversible skin yellowing (Steinmetz and Potter, 1996). As a result, recent data from human testing suggests that supplementing with ß-carotene (20 mg or more per day) is not recommended for heavy smokers (Hennekens et al., 1996).

Wang and his colleagues at Tufts University attempted to understand the negative effects of high dose ß-carotene supplementation on a molecular degree. They affirmed that a high dose of ß-carotene disrupts retinoid signalling in lung cells, resulting in changes in the expression levels of genes included in tumorigenesis (Liu et al., 2000).

Other researchers looked at the impact of ß-carotene supplements (50 mg/day) on cancer incidence in over 22.000 male doctors in the United States, with 11% of them currently smoking, supplementing of ß-carotene for longer than 12 years was not related to an elevated risk of lung cancer (Scientific Committee on Food on the safety of use of beta carotene from all dietary sources, 2000).

More recently, the Women's Health Study (WHS) found no elevated lung cancer risk in 40.000 women, including 13% smokers, who received 50 mg ß-carotene a day over two years and two years of follow-up (Lee et al., 1999).

According to Rao and Rao (2007), women with higher circulating levels of *a*-carotene, ß-carotene, lutein + zeaxanthin, lycopene, and total carotenoids could have a lower risk of breast cancer.

Carotenoids, such as lutein, are adequate to preserve wellness when consumed as part of a well-balanced diet. However, supplementation is needed, in cases of chronic disease or insufficient carotenoids absorption (Ravikrishnan et al., 2011).

According to some reports, lutein intake has no effect on cytochrome P450 enzyme function, implying that lutein may not affect the metabolism of endogenous or exogenous molecules (Zheng et al., 2013). Also, in an animal study, mice deficient in carotene oxygenase 2 developed pathologic carotenoid aggregation as well as mitochondrial dysfunction and oxidative stress (Amengual et al., 2011). This result suggested that, under some cases, a high carotenoid consumption could cause toxicity. Lutein supplementation also raised the incidence of crystalline maculopathy in elderly people (Satia et al., 2009).

While previous research has shown a positive interaction between lutein and the risk of some diseases, the EFSA survey concluded that the data collected were inadequate to prove a negative outcome (European Food Safety Authority (EFSA), 2008).

4.4.7 *In-vitro* evidence, animal studies, and clinical studies of carotenoids

Total carotenoid concentrations and overall mortality rate in older adults have been studied in several observational trials (Akbaraly et al., 2009; Ray et al., 2006) as well as adults of a variety of ages (Shardell et al., 2011). The majority of studies indicate inverse correlations, while a separate pattern of interaction by gender has been proposed (Ray et al., 2006). Overall mortality for individual plasma carotenoids has been studied in a smaller number of experiments (Bates et al., 2011), but it's too early to make conclusion about the association for individual carotenoids.

CVD is the leading cause of death worldwide, and diet plays a critical part in its progression (Perk et al., 2012). The evidence for dietary factors and CVD risk was classified as high for vegetables and was rated as moderate for fruit and dietary ß-carotene (Mente et al., 2009). Despite the fact that a recent meta-analysis affirmed the relation between FV intake and CVD risk (Wang et al., 2014). A variety of retrospective trials have been performed in relation to overall or human carotenoid intake or status and CVD outcomes (Goyal et al., 2013; Sesso et al., 2005).

Any detected inconsistencies may be attributed to demographic variability and carotene consumption ratios, as well as the degree of correction for possible confounders within individual studies and the subsequent probability of residual confounding. Most experiments have used a single carotenoid status measurement, and it's likely that participant's carotenoid status improved during follow-up, causing certain participant's long-term carotenoid status to be misclassified, affecting the reported correlations (Aune et al., 2012).

Despite the relatively stable relation between increased carotene intake and status and the risk of CVD, carotene intake has been shown to have little effect on cardiovascular disease (CVD) or lung cancer in randomized controlled trials (RCTs) (Voutilainen et al., 2006). According to meta-analyses, carotene supplements have been linked to an increased risk of death, especially in smokers (Bjelakovic et al., 2013), as a result, the data from retrospective trials and RCTs on the impact on CVD results has been inconsistent. The RCTs' design has been criticized (Riccioni et al., 2012, Stanner et al., 2004). Supplements were used in the trials and doses were within the pharmacological spectrum, the observed findings may have been influenced by genetic differences in supplementation response, the wellbeing of study subjects was also taken into consideration, participants were normally at high risk of CVD and present a considerable degree of atherosclerosis, as a result, the chance of the intervention minimizing risk is decreased. Furthermore, it has been hypothesized that supplementation would only help those with low nutritional status in RCTs (Goyal et al., 2013).

Indeed, results were found in those with originally lower nutrient status in the only two carotene-containing, as conclusion, trials demonstrate a beneficial influence of supplementation (Blot et al., 1993, Hercberg et al., 2004). Bjelakovic et al. (2013) studies found that at larger doses, vitamins have little benefit or have a negative effect, however, as a result of these carotene trial results, health care organizations have advised against taking carotene supplements for the treatment of CVD or cancer. To better evaluate the impact of carotenoids on CVD risk, RCTs that consider the health and the nutritional status of participants entering trials are needed. Despite the scholarly interest in doing such research, it is unlikely that further supplementation trials will be performed.

Depending on the cancer site, the evidence that increased FV intake was associated with reduced cancer risk was graded as "probable" or "limited-suggestive" as reported by the World Cancer Research Fund (2007). More recent findings have also shown that there is a poor association (Boffetta et al., 2010). Although it's still possible that one kind of FV, or a certain compound within certain FV, is linked to a lower risk of cancer, or that FV in general is protective at certain cancer sites, and those who consume very little FV may still benefit from raising their intake (Key, 2011).

The WCRF reported the evidence for carotenoid containing foods at various locations in terms of carotenoids and cancer risk. It was classified as probable that foods containing carotenoids protected against cancer in the mouth, pharynx, and larynx, as well as lung cancer, whereas, this association is improbable for prostate cancer and nonmelanoma skin cancer. On the other hand, foods rich in lycopene are thought to protect against prostate cancer (Bjelakovic et al., 2013).

A number of other studies, including of lycopene supplementation, have looked into the correlation between lycopene and prostate cancer risk. Meta-analyses found no impact of lycopene supplementation on benign prostate hyperplasia (BPH) or prostate cancer danger, while a meta-analysis of two studies found a decrease in prostate specific antigen levels in men diagnosed with prostate cancer who obtained lycopene. Given the small number of RCTs and their variable quality, the researchers found that it is currently impossible to endorse or deny the use of lycopene for the prevention or treatment of BPH or prostate cancer (Ilic and Misso, 2012).

Another topic that has gotten a lot of attention since the WCRF study was released in 2007 is the relation between carotenoid intake and status and breast cancer risk. A meta-analysis found a relation between higher levels of carotene, cryptoxanthin, lutein + zeaxanthin, lycopene, and total carotenoids in the blood and a lower risk of breast cancer (Eliassen et al., 2012), whereas another meta-analysis supports this relation for carotenoid status, it also indicates that blood carotenoids concentrations are more closely correlated with reduced breast cancer risk than carotenoid consumption as measured by a dietary questionnaire (Aune et al., 2012).

According to epidemiological research, there is a positive correlation between higher dietary intake and tissue carotenoids concentrations and a lower risk of chronic diseases (Johnson, 2002; Agarwal and Rao, 2000), ß-carotene, and lycopene have been linked to a lower incidence of coronary disease and some tumors, while lutein and zeaxanthin have been linked to ocular disorders (Ribaya-Mercado and Blumberg, 2004). Carotenoids' antioxidant properties have been proposed as the primary mechanism by which they achieve their beneficial effects. Carotenoids can also exert their effects by other pathways such as cell growth control, gap junction connectivity, modulators of phase I and II drug metabolizing enzymes, modulating gene expression, and immune response (Astrog, 1997; Jewell and O'Brien, 1999; Bertram, 1999).

Carotenoids, including *a*- and ß-carotene, as well as ß-cryptoxanthin, have the additional benefit of being able to be converted to vitamin A, which has a role in disease growth and prevention. Carotenoids, including ß-carotene and lycopene have been shown to have antioxidant effects in various *in vitro*, animal, and human studies. ß-carotene suppressed the upregulation of heme oxygenase-1 gene expression in human dermal fibroblasts (FEK4) exposed to UVA in a dose-dependent manner (Elliott, 2005). Worth noting that ß-carotene has been documented to behave as a pro-oxidant in some circumstances. UVA-induced heme oxygenase-1 induction was enhanced by ß-carotene at a concentration of 0.2 mM, suggesting a pro-oxidant function (Obermuller-Jevic et al., 1999). In another study, 10 mM ß-carotene increased the generation of reactive oxygen species (ROS) and the levels of cellular oxidized glutathione in leukemia and colon adenocarcinoma cell lines *in vitro* (Palozza et al., 2003). In rats, the pro-oxidant role of ß-carotene was also shown, with increased phase I enzyme production in the liver, kidney, and intestine, as well as increased oxidative stress (Paolini et al., 2001). Human experiments confirm also the pro-oxidant properties of ß-carotene. The alpha-tocopherol ß-carotene (ATBC) trial found that supplementing ß-carotene at pharmacological levels improved lung cancer incidence in smokers (Alpha Tocopherol beta Carotene Cancer Prevention Study Group, 1994). The ß-carotene and retinol efficiency trial (CARET) found elevated CVD mortality in a population of users, former smokers, and asbestos exposed people (Omenn et al., 1996b). These findings point to a potential biphasic reaction of ß-carotene, which may encourage wellbeing when consumed in small quantities but may have negative consequences when consumed in larger amounts. While scientists are still debating the validity of the results of human trials and the specifics of research designs, questions about ß pro-oxidant carotene's properties which could increase the risk of lung cancer and CVD in smoking, have prompted a moratorium on further intervention studies using ß-carotene. ß-carotene has been used as a "gold standard" model to research the association between oxidative stress and chronic diseases for a long time. The subject of study has now moved to lycopene, a different carotenoid antioxidant (Ma et al., 2018).

Conclusion

Today, scientific research has demonstrated a range of carotenoids health benefits in food. These antioxidant, immune stimulating, pro-inflammatory, and provitamin A activities have enabled it to have beneficial effects on several organs. These compounds are recognized, as safe and side effects are very rare and generally mild when they do occur, which allows their are use in many treatments. However, their overall effects are variable depending on the total food intake of carotenoids. Further research must be carried out to establish adequate doses used in the various pathologies studied as well as a clarification of the synergistic or antagonistic effect of these compounds with the other antioxidants contained in foods to better understand their therapeutic effects.

Abbreviations

GGPP	Geranyl geranyl pyrophosphate
FW	Fresh weight
ROS	Reactive oxygen species
UVA	Ultraviolet A
RARs	Retinoic acid receptors
RXRs	Retinoid X receptors
ALS	Amyotrophic lateral sclerosis
DNA	Deoxyribonucleic acid
Aβ	β-amyloid
Nrf2	Nuclear factor erythroid-2-related factor 2
ERK	Extracellular signal-regulated kinases
CVD	Cardiovascular disease
AMD	Age-related macular degeneration

NF-KB	Nuclear factor-Kappa B
α7 / PI3K	Phosphatidyl inositol 3-kinase
Akt /mTOR	Protein kinase B/mammalian target of rapamycin
FV	Fruits and vegetables
RCTs	Randomized controlled trials
WCRF	World Cancer Research Fund
BPH	Benign prostatic hyperplasia
FEK4	Normal human dermal fibroblasts
ATBC	Alpha-tocopherol beta-carotene
CARET	Carotene and Retinol Efficiency Trial

References

Agarwal, S., Rao, AV., 2000. Carotenoids and chronic diseases. Drug Metab. Drug Interact. 17 (1–4), 189–210.

Akbaraly, T.N., Favier, A., Berr, C., 2009. Total plasma carotenoids and mortality in the elderly: results of the epidemiology of vascular ageing (EVA) study. Br. J. Nutr. 101 (1), 86–92.

Aleman, T.S., Duncan, J.L., Bieber, M.L., De Castro, E., Marks, D.A., Gardner, L.M., Steinberg, J.D., Cideciyan, A.V., Maguire, M.G., Jacobson, S.G., 2001. Macular pigment and lutein supplementation in retinitis pig-mentosa and usher syndrome. Invest. Ophthalmol. Vis. Sci. 42 (8), 1873–1881.

Alexander, M., Newmark, H., Miller, G., 1985. Oral beta-carotene can increase the number of OKT4 positive cells in human blood. Immunol. Lett. 9 (4), 221–224.

Alonso, B., 2017. Carotenoids: content in foods, in diet and bioavailability. Sci. News Lett. 2, 1–9.

Alpha-Tocopherol beta Carotene Cancer Prevention Study Group, 1994. The effects of vitamin E and beta-carotene on the incidence of lung cancer and other cancers in male smokers. N. Engl. J. Med. 330, 1029–1035.

Alves-Rodrigues, A., Shao, A., 2004. The science behind lutein. Toxicol. Lett. 150 (1), 57–83.

Ambati, R.R., Phang Siew, M., Sarada, R., Ravishankar Gokare, A., 2014. Astaxanthin: sources, extraction, stability, biological activities and its commercial applications. Mar. Drugs 12, 128–152.

Amengual, J., Lobo, G.P., Golczak, M., Li, H.N., Klimova, T., Hoppel, C.L., Wyss, A., Palczewski, K., vonLintig, J., 2011. A mitochondrial enzyme degrades carotenoids and protects against oxidative stress. FASEB J. 25, 948–959.

Antwi, S.O., Steck, S.E., Su, L.J., Hebert, J.R., Zhang, H., Craft, N.E., 2016. Carotenoid intake and adipose tissue carotenoid levels in relation to prostate cancer aggressiveness among African-American and European-American men in the North Carolina-Louisiana prostate cancer project (PCaP). Prostate 76 (12), 1053–1066.

Asadi, F., Jamshidi, A.H., Khodagholi, F., Yans, A., Azimi, L., Faizi, M., Vali, L., Abdollahi, M., Ghahremani, M.H., Sharifzadeh, M., 2015. Reversal effects of crocin on amyloid β-induced memory deficit: modification of autophagy or apoptosis markers. Pharmacol. Biochem. Behav. 139 (Pt A), 47–58.

Astrog, P., 1997. Food carotenoids and cancer prevention: an overview of current research. Trends Food Sci. Technol. 8, 406–413.

Aune, D., Chan, D.S., Vieira, A.R., 2012. Dietary compared with blood concentrations of carotenoids and breast cancer risk: a systematic review and meta-analysis of prospective studies. Am. J. Clin. Nutr. 96, 356–373.

Aust, O., Stahl, W., Sies, H., Tronnier, H., Heinrich, U., 2005. Supplementation with tomato-based products increases lycopene, phytofluene, and phytoene levels in human serum and protects against UV-light-induced erythema. Inter. J. Vitam. Nutr. Res. 75 (1), 54–60.

Austin, J., Singhal, N., Voigt, R., Smaill, F., Gill, M.J., Walmsley, S., Salit, I., Gilmour, J., Schlech, W.F., Choudhri., S., Rachlis, A., Cohen, J., Trottier, S., Toma, E., Phillips, P., Ford, P.M., Woods, R., Singer, J., Zarowny, D.P., Cameron, D.W., 2006. A community randomized controlled clinical trial of mixed carotenoids and micronutrient supplementation of patients with acquired immunodeficiency syndrome. Eur. J. Clin. Nutr. 60, 1266–1276.

Bates, C.J, Hamer, M., Mishra, G.D., 2011. Redox-modulatory vitamins and minerals that prospectively predict mortality in older British people: the National Diet and Nutrition Survey of people aged 65 years and over. Br. J. Nutr. 105, 123–132.

Bendich, A., 1993. Biological functions of dietary carotenoids. Ann. New York Acad. Sci. 691 (1), 61–67.

Bertram, J.S., 1999. Carotenoids and gene regulation. Nutr. Rev. 57, 182–191.

Bfhm, F., Edge, R., McGarvey, D.J., Truscott, T.G., 1998. Beta-carotene with vitamins E and C offers synergistic cell protection against NOx. FEBS Lett. 436, 387–389.

Bjelakovic, G., Nikolova, D., Gluud, C., 2013. Antioxidant supplements to prevent mortality. JAMA 310 (11), 1178–1179.

Blot, W.J., Li, J.Y., Taylor, P.R., 1993. Nutrition intervention trials in Linxian, China: supplementation with specific vitamin/mineral combinations, cancer incidence, and disease-specific mortality in the general population. J. Natl. Cancer Inst. 85, 1483–1492.

Boffetta, P., Couto, E., Wichmann, J., 2010. Fruit and vegetable intake and overall cancer risk in the European Prospective Investigation into Cancer and Nutrition (EPIC). J. Natl. Cancer Inst. 102, 529–537.

Bohlke, K., Spiegelma, D., Trichopoulou, A., Katsouyanni, K., Trichopoulos, D., 1999. Vitamins A, C and E and the risk of breast cancer: results from a case-control study in Greece. Br. J. Cancer 79, 23–29.

Bone, R.A., Landrum, J.T., Mayne, S.T., Gomez, C.M., Tibor, S.E., Twaroska, E.E., 2001. Macular pigment in donor eyes with and without AMD: a case-control study. Invest. Ophthalmol. Vis. Sci. 42 (1), 235–240.

Bonet, M.L., Canas, J.A., Ribot, J., Palou, A., 2015. Carotenoids and their conversion products in the control of adipocyte function, adiposity and obesity. Arch. Biochem. Biophys. 572, 112–125.

Bousvaros, A., Zurakowski, D., Duggan, C., Law, T., Rifai, N., Goldberg, N.E., 1998. Vitamins A and E serum levels in children and young adults with inflammatory bowel disease: Effect of disease activity. J. Pediatr. Gastroenterol. Nutr. 26, 129–135.

Britton, G., Khachik, F., 2009. Carotenoids in Food. In: Britton, G., Pfander, H., Liaaen-Jensen, S. (Eds.). Carotenoids, vol 5. Birkhäuser, Basel. pp. 45–66.

Britton, G., 1995. Structure and properties of carotenoids in relation to function. FASEB J. 15, 1551–1558.

Britton, G., Liaaen-Jensen, S., Pfander, H., 2004. Carotenoids: Handbook. Springer, Basel, Switzerland AG.

Brown, M.J., Ferruzzi, M.G., Nguyen, M.L., Cooper, D.A., Eldridge, A.L., Schwartz, S.J., White, W.S., 2004. Carotenoid bioavailability is higherfrom salads ingested with full-fat than with fat-reduced salad dressings as measured with electrochemical detection. Am. J. Clin. Nutr. 80, 396–403.

Cazzonelli, C.I., 2011. Carotenoids in nature:insights from plants and beyond. Funct. Plant Biol. 38, 833–847.

Cazzonelli, C.I., Nisar, N., Hussain, D., Carmody, M.E., Pogson, B.J., 2010. Biosynthesis and regulation of carotenoids in plants – micronutrients, vitamins and health benefits. In: Pua, E.C., Davey, M.R. (Eds.), Plant Developmental Biology- Biotechnological Perspectives. Volume 2, Springer, Berlin Heidelberg, 117–137.

Chang, C.H., Chen, C.Y., Chiou, J.Y., Peng, R.Y., Peng, C.H., 2010. Astaxanthine secured apoptotic death of PC12 cells induced by β-amyloid peptide 25–35: its molecular action targets. J. Med. Food 13 (3), 548–556.

Chen, L., Stacewicz-Sapuntzakis, M., Duncan, C., Sharifi, R., Ghosh, L., van Breemen, R., Ashton, D., Bowen, P.E., 2001. Oxidative DNA damage in prostate cancer patients consuming tomato sauce-based entrees as a whole-food intervention. J. Natl. Cancer Inst. 93, 1872–1879.

Chew, B.P., Park, J.S., 2004. Carotenoid action on the immune response. J. Nutr. 134 (1), 257–261.

Chitchumroonchokchai, C., Schwartz, S.J., Failla, M.L., 2004. Assessment of lutein bioavailability from meals as a supplement using simulated digestion and Caco-2 human intestinal cells. J. Nutr. 134, 2280–2286.

Cho, K.S., Shin, M., Kim, S., Lee, S.B., 2018. Recent advances in studies on the therapeutic potential of dietary carotenoids in neurodegenerative diseases. Oxid. Med. Cell Longev., 412–458 2018.

Edge, R., McGarvey, D., Truscott, T., 1997. The carotenoids as anti-oxidants. A review. J. Photochem. Photobiol. B 41 (3), 189–200.

El-Agamey, A., Lowe, G.M., McGarvey, D.J., Mortensen, A., Phillip, D.M., Truscott, T.G., 2004. Carotenoid radical chemistry and antioxidant/pro-oxidant properties. Arch. Biochem. Biophys. 430, 37–48.

El-Akabawy, G., El-Sherif, N.M., 2019. Zeaxanthin exerts protective effects on acetic acid-induced colitis in rats via modulation of pro-inflammatory cytokines and oxidative stress. Biomed. Pharmacother. 111, 841–851.

Eliassen, A.H., Hendrickson, S.J., Brinton, L.A., 2012. Circulating carotenoids and risk of breast cancer: pooled analysis of eight prospective studies. J. Natl. Cancer Inst. 104, 1905–1916.

Elliott, R., 2005. Mechanisms of genomic and non-genomic actions of carotenoids. Biochim. Biophys. Acta 1740, 147–154.

Elvira-Torales, L.I., García-Alonso, J., Periago-Castón, M.J., 2019. Nutritional importance of carotenoids and their effect on liver health: a review. Antioxidants 8, 229.

European Food Safety Authority (EFSA), 2008. Safety, bioavailability and suitability of lutein for the particularnutritional use by infants and young children—scientific opinion of the panel on dietetic products, nutrition and allergies. EFSA J. 823, 1–24.

Faulks, R.M., Southon, M., 2005. Challenges to understanding and measuring carotenoid bioavailability. Biochim. Biophys. Acta Mol. Basis Dis. 1740, 95–100.

Fryburg, D.A., Mark, R.J., Griffith, B.P., Askenase, P.W., Patterson., T.F., 1995. The effect of supplemental beta-carotene on immuno-logical indices in patients with AIDS: a pilot study. Yale J. Biol. Med. 68 (1-2), 19–23.

Gammone, M.A., Riccioni, G., D'Orazio, N., 2015. Carotenoids: potential allies of cardiovascular health? Food Nutr. Res. 59, 26762.

Ghahghaei, A., Bathaie, S.Z., Kheirkhah, H., Bahraminejad, E., 2013. The protective effect of crocin on the amyloid fibril formation of Aβ42 peptide in vitro. Cell Mol. Biol. Lett. 18 (3), 328–339.

Giovannucci, E., 2002. A review of epidemiologic studies of tomatoes, lycopene, and prostate cancer. Exp. Biol. Med. 227, 852–859.

Głąbska, D., Guzek, D., Zakrzewska, P., Lech, G., 2019. Intake of Lutein and Zeaxanthin as a possible factor influencing gastrointestinal symptoms in Caucasian individuals with ulcerative colitis in remission Phase. J. Clin. Med. 8 (1), 77.

Głąbska, D., Guzek, D., Zakrzewska, P., Włodarek, D., Lech, G., 2016. Lycopene, Lutein and Zeaxanthin may reduce faecal blood, mucus and pus but not abdominal pain in individuals with ulcerative colitis. Nutrients 8 (10), 613.

Goodwin, T.W., 1980. The biochemistry of the carotenoids, second ed. Plants, Vol.1. Chapman and Hall, New York.

Goyal, A., Terry, M.B., Siegel, A.B., 2013. Serum antioxidant nutrients, vitamin a, and mortality in US Adults. Cancer Epidemiol. Biomarkers Prev. 22 (12) 2202–2211.

Gross, J., 1991. Pigments in Vegetables: Chlorophylls and Carotenoids, AVI/Van Nostrand Reinhold. Springer, US.

Gruszecki, W.I., 1999. Carotenoids in membranes. In: Frank, H.A., Young, A.J., Britton, G., Cogdell, R.J. (Eds.), The Photochemistry of Carotenoids. Advances in Photosynthesis and Respiration, 8. Springer publisher, Dordrecht. Netherlands, pp. 363–379.

Gul, K., Tak, A., Singh, A.K., Singh, P., Yousuf, B., Wani, A.A., 2015. Chemistry, encapsulation, and health benefits of β-carotene - A review. Cogent Food Agric. 1 (1), 101869.

Handelman, G.J., Nightingale, Z.D., Lichtenstein, A.H., Schaefer, E.J., Blumberg, J.B., 1999. Lutein and zeaxanthin concentrations in plasma after dietary supplementation with egg yolk. Am. J. Clin. Nutr. 70, 247–251.

Heinrich, U., Gärtner, C., Wiebusch, M., Eichler, O., Sies, H., Tronnier, H., Stahl, W., 2003. Supplementation with beta-carotene or a similar amount of mixed carotenoids protects humans from UV-induced erythema. J. Nutr. 133, 98–101.

Hennekens, C.H., Buring, J.E., Manson, J.E., 1996. Lack of effect of long-term supplementation with beta carotene on the incidence of malignant neoplasm's and cardiovascular disease. N. Engl. J. Med. 334, 1145–1149.

Hercberg, S., Galan, P., Preziosi, P., 2004. The SU.VI MAX Study: A randomized, placebo controlled trial of the health effects of antioxidant vitamins and minerals. Arch. Intern. Med. 164 (21), 2335–2342.

Hsu, B.Y., Pu, Y.S., Inbaraj, B.S., Chen, B.H., 2012. An improved high performance liquid chromatography diode array detection-mass spectrometry method for determination of carotenoids and their precursors phytoene and phytofluene in human serum. J. Chromatogr. B 899, 36–45.

Hughes, D., 1999. Effects of carotenoids on human immune function. Proc. Nutr. Soc. 58 (3), 713–718.

Hughes, D.A., 2001. Dietary carotenoids and human immune function. Nutrition 17 (10), 823–827.

Ikeda, Y., Tsuji, S., Satoh, A., Ishikura, M., Shirasawa, T., Shimizu, T., 2008. Protective effects of astaxanthin on 6-hydroxydopamine-induced apoptosis in human neuroblas- toma SH-SY5Y cells. J. Neurochem. 107 (6), 1730–1740.

Ilic, D., Misso, M., 2012. Lycopene for the prevention and treatment of benign prostatic hyperplasia and prostate cancer: a systematic review. Maturitas 72 (4), 269–276.

International Agency for Research on Cancer, 1998. IARC Handbooks of Cancer Prevention, Vol.2. Carotenoids. IARC, Lyon, France.

Iskandar, A.R., Miao, B., Li, X., Hu, K.Q., Liu, C., Wang, X.D., 2016. β-cryptoxanthin reduced lung tumor multiplicity and inhibited lung cancer cell motility by down regulating nicotinic acetylcholine receptor α7 signaling. Cancer Prev. Res. 9 (11), 875–886.

Isonaka, R., Hiruma, H., Katakura, T., Kawakami, T., 2011. Inhibition of superoxide dismutase selectively suppresses growth of rat spinal motor neurons: comparison with pho phorylated neurofilament-containing spinal neurons. Brain Res. 1425, 13–19.

Jewell, C., O'Brien, N.M., 1999. Effect of dietary supplementation with carotenoids on xenobiotic metabolizing enzymes in the liver, lung, kidney and small intestine of the rat. Br. J. Nutr. 81, 235–242.

Johnson, E.J., 2002. The role of carotenoids in human health. Nutr. Clin. Care 5 (2), 47–49.

Jyonouchi, H., Sun, S., Tomita, Y., Gross, M.D., 1995. Astaxanthin, a carotenoid without vitamin A activity, augments antibody responses in cultures including T-helper cell clones and suboptimal doses of antigen. J. Nutr. 125, 2483–2492.

Kang, H., Kim, H., 2017. Astaxanthin and β-carotene in *Helicobacter pylori*-induced gastric inflammation: a mini-review on action mechanisms. J. Cancer Prev. 22 (2), 57–61.

Katayama, S., Ogawa, H., Nakamura, S., 2011. Apricot carotenoids possess potent anti-amyloidogenic activity *in vitro*. J. Agri. Food Chem. 59 (23), 12691–12696.

Key, T.J., 2011. Fruit and vegetables and cancer risk. Br. J. Cancer 104, 6–11.

Khoo, H.E., Prasad, K.N., Kong, K.W., Jiang, Y., Ismail, A., 2011. Carotenoids and their isomers: color pigments in fruits and vegetables. Molecules 16, 1710–1738.

Kim, J.E., Clark, R.M., Park, Y., Lee, J., Fernandez, M.L., 2012. Lutein decreases oxidative stress and inflammation in liver and eyes of guinea pigs fed a hypercholesterolemic diet. Nutr. Res. Pract. 6 (2), 113–119.

Kimura, M., Rodriguez-Amaya, D.B., 2003. Carotenoid composition of hydroponic leafy vegetables. J. Agric. Food Chem. 51, 2603–2607.

Klein, B.P., Perry, A.K., 1982. Ascorbic acid and vitamin A activity in selected vegetables from different geographic areas of the United States. J. Food Sci. 47, 941–948.

Kong, K.-W., Khoo, H.-E., Prasad, K.N., Ismail, A., Tan, C.-P., Rajab, N.F., 2010. Revealing the power of the natural red pigment lycopene. Molecules 15 (2), 959–987.

Kopas-Lane, L.M., Warthesen, J.J, 1995. Carotenoid photostability in raw spinach and carrots during cold storage. J. Food Sci. 60, 773–776.

Kopsell, D.E., Kopsell, D.A., Randle, W.M., Coolong, T.W., Sams, C.E., Celentano, J.C., 2003. Kale carotenoids remain stable while flavor compounds respond to changes in sulfur fertility. J Agric. Food Chem. 51 (18), 5319–5325.

Kucuk, O., Sarkar, F.H., Sakr, W., Djuric, Z., Pollak, M.N., Khachik, F., Li, Y.W., Banerjee, M., Grignon, D., Bertram, J.S., Crissman, J.D., Pontes, E.J., Wood Jr, D.P., 2001. Phase II randomized clinical trial of lycopene supplementation before radical prostatectomy. Cancer Epidemiol. Biomark. Prev. 10, 861–868.

Kurilich, A.C., Tsau, G.J., Brown, A., Howard, L., Klein, B.P., Jeffery, E.H., Kushad, M., Wallig, M.A., Juvik, J.A., 1999. Carotene, tocopherol, and ascorbate in subspecies of *Brassica oleracea*. J. Agric. Food Chem. 47, 1576–1581.

Lee, I-M., Cook, N.R., Manson, J.E., Buring, J.E., Hennekens, C.H., 1999. Beta-carotene supplementation and incidence of cancer and cardiovascular disease: the Womens' Health Study. J. Natl. Cancer. Inst. 91, 2102–2109.

Lee, W., Ku, S.K., Bae, J.W., Bae, J.S., 2012. Inhibitory effects of lycopene on HMGB1-mediated pro-inflammatory responses in both cellular and animal models. Food Chem. Toxicol. 50, 1826–1833.

Lefsrud, M.G., Kopsell, D.A., Augé, R.M., 2006a. Biomass production and pigment accumulation in kale grown under increasing photoperiods. Hort. Sci. 41 (3), 603–606.

Lefsrud, M.G., Kopsell, D.A., Kopsell, D.E., Celentano, J.C., 2006b. Irradiance levels affect growth parameters and carotenoid pigments in kale and spinach grown in a controlled environment. Physiol. Plant. 127, 624–631.

Lefsrud, M.G., Kopsell, D.A., Kopsell, D.E., Celentano, J.C., 2005. Air temperature affects biomass and carotenoid pigment accumulation in kale and spinach grown in a controlled environment. Hort. Sci. 40 (7), 2026–2030.

Li, L., van Eck, J., 2007. Metabolic engineering of carotenoid accumulation by creating a metabolic sink. Transgenic Res. 16, 581–585.

Li, S.Y., Fung, F.K., Fu, Z.J., Wong, D., Chan, H.H.L., Lo, A.C.Y., 2012. Anti-inflammatory effects of lutein in retinal ischemic/hypoxic injury: *in vivo* and *in vitro* studies. Invest Ophtalmol. Vis. Sci. 53 (10), 5976–5984.

Liang, C., Zhao, F., Wei, W., Wen, Z., Qin, S., 2006. Carotenoid biosynthesis in Cyanobacteria: Structural and evolutionary scenarios based on comparative genomics. Int. J. Biol. Sci. 2 (4), 197–207.

Liao, K.S., Wei, C.L., Chen, J.C., Zheng, H.Y., Chen, W.C., Wu, C.H., 2016. Astaxanthin enhances pemetrexed-induced cytotoxicity by down regulation of thymidylate synthase expression in human lung cancer cells. Regul. Toxicol. Pharmacol. 81, 353–361.

Lin, J., Huang, L., Yu, J., Xiang, S., Wang, J., Zhang, J., 2016. Fucoxanthin, a marine carotenoid, reverses scopolamine-induced cognitive impairments in mice and inhibits acetylcholin- esterase *in vitro*. Mar. Drugs 14 (4), 67.

Liu, C., Wang, X.D., Bronson, R.T., Smith, D.E., Krinsky, N.I., Russell, R.M., 2000. Effects of physiological versus pharmacological beta-carotene supplementation on cell proliferation and histopathological changes in the lungs of cigarette smoke-exposed ferrets. Carcinogenesis 21 (12), 2245–2253.

Liu, Y., Zheng, J., Zhang, Y., Wang, Z., Yang, Y., Bai, M., Dai, Y., 2016. Fucoxanthin activates apoptosis via inhibition of PI3K/Akt/mTOR pathway and suppresses invasion and migration by restriction of p38-MMP-2/9 pathway in human glioblastoma cells. Neurochem. Res. 41, 2728–2751.

Liu, T., Liu, W.H., Zhao, J.S., Meng, F.Z., Wang, H., 2017. Lutein protects against β-amyloid peptide-induced oxidative stress in cerebrovascular endothelial cells through modulation of Nrf-2 and NF-κB. Cell Biol. Toxicol. 33 (1), 57–67.

Lobos, P., Bruna, B., Cordova, A., Barattini, P., Galáz, J.L., Adasme, T., Hidalgo, C., Muñoz, P., Paula-Lima, A., 2016. Astaxanthin protects primary hippocampal neurons against noxious effects of Aβ-oligomers. Neural Plast. 34, 567–583.

Lowe, G.M., Booth, L.A., Bilton, R.F., Young, A.J., 1999. Lycopene and β-carotene protect against oxidative damage in HT29 cells at low concentrations but rapidly lose this capacity at higher doses. Free Radic. Res. 30, 141–151.

Lu, S., Li, L., 2008. Carotenoid metabolism: biosynthesis, regulation, and beyond. J. Integr. Plant Biol. 50 (7), 778–785.

Ma, E., Iso, H., Yamagishi, K., Ando, M., Wakai, K., Tamakoshi, A., 2018. Dietary antioxidant micronutrients and all-cause mortality: the Japan collaborative cohort study for evaluation of cancer risk. J. Epidemiol. 28 (9), 388–396.

Maiani, G., Periago Castón, M.J., Catasta, G., Toti, E., Cambrodón, I.G., Bysted, A., Böhm, V., 2009. Carotenoids: actual knowledge on food sources, intakes, stability and bioavailability and their protective role in humans. Mol. Nutr. Food Res. 53, S194–S218.

Marcotorchino, J., Romier, B., Gouranton, E., Riollet, C., Gleize, B., Malezet-Desmoulins, C., Landrier, J.F., 2012. Lycopene attenuates LPS-induced TNF-alpha secretion in macrophages and inflammatory markers in adipocytes exposed to macrophage-conditioned media. Mol. Nutr. Food Res. 56, 725–732.

Mathews-Roth, M.M., 1993. Carotenoids in erytropoietic protoporphyria and other photosensitivity diseases. Ann. N Y. Acad. Sci. 691, 127–138.

Mayne, S.T., 1996. Beta-carotene, carotenoids, and disease prevention in humans. FASEB J. 10, 690–701.

Meléndez-Martínez, A.J., Mapelli-Brahm, P., Benítez-González, A., Stinco, C.M., 2015. A comprehensive review on the colorless carotenoids phytoene and phytofluene. Arch. Biochem. Biophys. 572, 188–200.

Mente, A., de Koning, L., Shannon, H.S, Anand, S.S., 2009. A systematic review of the evidence supporting a causal link between dietary factors and coronary heart disease. Arch. Intern. Med. 169 (7), 659–669.

Merhan, O., 2017. The biochemistry and antioxidant properties of carotenoids. In: Cvetković, D.J., Nicolić, G.S. (Eds.), Carotenoids. InTechOpen, Rijeka, pp. 51–66.

Mezzomo, N., Ferreira, S.R.S., 2016. Carotenoids functionality, sources, and processing by supercritical technology: a review. J. Chem. 3, 164–312.

Milani, A., Basirnejad, M., Shahbazi, S., Bolhassani, A., 2017. Carotenoids: biochemistry, pharmacology and treatment. Br. J. Pharmacol. 174 (11), 1290–1324.

Miller, N.J., Sampson, J., Candeias, L.P., Bramley, P.M., Rice-Evans, C.A., 1996. Antioxidant activities of carotenes and xanthophylls. FEBS Lett. 384 (3), 240–242.

Murata, T., Tamai, H., Morinobu, T., Manago, M., Takenaka, H., Hayashi, K., Mino, M., 1994. Effect of long-term administration of beta-carotene on lymphocyte subsets in humans. Am. J. Clin. Nutr. 60 (4), 597–602.

Nam, K.N., Park, Y.M., Jung, H.J., 2010. Anti-inflammatory effects of crocin and crocetin in rat brain microglial cells. Eur. J. Pharmacol. 648 (1–3), 110–116.

Nicolle, K., Simon, G., Rock, E., Amouroux, P., Rémésy, C., 2004. Genetic variability influences carotenoid, vitamin, phenolic, and mineral content in white, yellow, purple, orange, and dark-orange carrot cultivars. J. Am. Soc. Hortic. Sci. 129, 523–529.

Nisar, N., Li, L., Lu, S., Khin, N.C., Pogson, B.J., 2015. Carotenoid metabolism in plants. Mol. Plant 8, 68–82.

Obermuller-Jevic, U.C., Francz, P.I., Frank, J., Flaccus, A., Biesalski, H.K., 1999. Enhancement of UVA induction of haem oxygenase-1 expression by beta-carotene in human skin fibroblasts. FEBS Lett. 460, 212–216.

Olson, J.A., 1993. Molecular actions of carotenoids. Ann. New York Acad. Sci. 691, 156–166.

Olson, J.A., 1994. Vitamin A, Retinoids and Carotenoids. In: Shils, M.E., Olson J.A. and Shike M. (Eds.). 8th Edn. Modern Nutrition in Health and Disease, Pa, Lea and Febiger, Philadelphia, pp: 287–307.

Olson, J.A., Krinsky, N.I., 1995. Introduction: the colorful fascinating world of the carotenoids: important physiologic modulators. FASEB J. 9, 1547–1550.

Omenn, G.S., Goodman, G.E., Thornquist, M.D., 1996a. Effects of a combination of beta-carotene and vitamin A on lung cancer and cardiovascular disease. N. Eng. J. Med. 334, 1150–1155.

Omenn, G.S., Goodman, G.E., Thornquist, M.D., 1996b. Risk factors for lung cancer and for intervention effects in CARET, the Beta-carotene and retinol efficacy Trial. J. Natl. Cancer. Inst 88 (21), 1550–1559.

Palozza, P., 2005. Can β-carotene regulate cell growth by a redox mechanism? An answer from cultured cells. Biochim. Biophys. Acta Mol. Basis Dis. 1740, 215–221.

Palozza, P., Serini, S., Torsello, A., 2003. Beta-carotene regulates NF-kappaB DNA-binding activity by a redox mechanisms in human leukemia and colon adenocarcinoma cells. J. Nutr. 133, 381–388.

Paolini, M., Antelli, A., Pozzetti, L., 2001. Induction of cytochrome P450 enzymes and over-generation of oxygen radicals in beta-carotene supplemented rats. Carcinogenesis 22 (9), 1483–1495.

Perk, J., De Backer, G., Gohlke, H., 2012. European association for cardiovascular prevention & rehabilitation. European guidelines on cardiovascular disease prevention in clinical practice. Int. J. Behav. Med. 19, 403–488.

Prakash, A., Kumar, A., 2014. Implicating the role of lycopene in restoration of mitochondrial enzymes and BDNF levels in β-amyloid induced Alzheimer's disease. Eur. J. Pharmacol. 741, 104–111.

Rao, A.V., Agarwal, S., 1999. Role of lycopene as antioxidant carotenoid in the prevention of chronic diseases: a review. Nutr. Res. 19, 305–323.

Rao, A.V., Rao, L.G., 2007. Carotenoids in human health. Pharmacol. Res. 55, 207–216.

Rao, S.V., Yenisetti, S.C., Rajini, P.S., 2016. Evidence of neuro-protective effects of saffron and crocin in a Drosophila model of Parkinsonism. Neurotoxicology 52, 230–242.

Ravikrishnan, R., Rusia, S., Ilamurugan, G., Salunkhe, U., Deshpande, J., Shankaranarayanan, J., Shankaranarayana, M.L., Soni, M.G., 2011. Safety assessment of lutein and zeaxanthin (Lutemax 2020): subchronic toxicity and mutagenicity studies. Food Chem. Toxicol. 49, 2841–2848.

Ray, A.L., Semba, R.D., Walston, J., 2006. Low serum selenium and total carotenoids predict mortality among older women living in the community: the women's health and aging studies. J. Nutr. 136, 172–176.

Ribaya-Mercado, J.D., Blumberg, J.B., 2004. Lutein and zeaxanthin and their potential roles in disease prevention. J. Am. Coll. Nutr. 23 (6), 567S–587S.

Ribaya-Mercado, J.D., Garmyn, M., Gilchrest, B.A., Russell, R.M., 1995. Skin lycopene is destroyed preferentially over β-carotene during ultraviolet irradiation in humans. J. Nutr. 125, 1854–1859.

Riccioni, G., D'Orazio, N., Salvatore, C., Franceschelli, S., Pesce, M., Speranza, L., 2012. Carotenoids and vitamins C and E in the prevention of cardiovascular disease. Int. J. Vitam. Nutr. Res. 82, 15–26.

Rodriguez-Amaya, D.B., 1999. A Guide to Carotenoid Analysis in Foods. ILSI, Washington DC.

Rühl, R., 2007. Effects of dietary retinoids and carotenoids on immune development. Proc. Nutr. Soc. 66 (3), 458–469.

Russo, V.M., Howard, L.R., 2002. Carotenoids in pungent and non-pungent peppers at various developmental stages grown in the field and glasshouse. J. Sci. Food Agric. 82, 615–624.

Rutz, J.K., Borges, C.D., Zambiazi, R.C., da Rosa, C.G., da Silva, M.M., 2016. Elaboration of microparticles of carotenoids from natural and synthetic sources for applications in food. Food Chem. 202, 324–333.

Sachdeva, A.K., Chopra, K., 2015. Lycopene abrogates Aβ (1–42)-mediated neuroinflammatory cascade in an experimental model of Alzheimer's disease. J. Nutr. Biochem. 26 (7), 736–744.

San Millan, C.S, Soldevilla, B., Martin, P., Gil-Calderon, B., Compte, M., Perez-Sacristan, B., Donoso, E., Pena, C., Romero, J., Granado-Lorencio, F., et al., 2015. β-Cryptoxanthin synergistically enhances the antitumoral activity of oxaliplatin through ΔNP73 negative regulation in colon cancer. Clin. Cancer Res. 21, 4398–4409.

Sasaki, M., Ozawa, Y., Kurihara, T., 2009. Neuroprotective effect of an antioxidant, lutein, during retinal inflammation. Invest. Ophthalmol. Vis. Sci. 50 (3), 1433–1439.

Satia, J.A., Littman, A., Slatore, C.G., Galanko, J.A., White, E., 2009. Long-term use of beta-carotene, retinol, lycopene, and lutein supplements and lung cancer risk: Results from the Vitamins and Lifestyle (VITAL) study. Am. J. Epidemiol 169, 815–828.

Scientific Committee on Food on the safety of use of beta carotene from all dietary sources (Opinion adopted by the SCF on 7 September 2000).

Sesso, H.D., Buring, J.E., Norkus, E.P, Gaziano, J.M., 2005. Plasma lycopene, other carotenoids, and retinol and the risk of cardiovascular disease in men. Am. J. Clin. Nutr. 81, 990–997.

Shardell, M.D., Alley, D.E., Hicks, G.E., 2011. Low-serum carotenoid concentrations and carotenoid interactions predict mortality in US adults: the Third National Health and Nutrition Examination Survey. Nutr. Res. 31, 178–189.

Sies, H., 1993. Strategies of antioxidant defense. Eur. J. Biochem. 215 (2), 213–219.

Sies, H., Krinsky, N.L., 1995. Antioxidant vitamins and b-carotene in disease prevention. Am. J. Clin. Nutr. 62, 1299–1540.

Sommerburg, O., 1998. Fruits and vegetables that are sources for lutein and zeaxanthin: the macular pigment in human eyes. Br. J. Ophthalmol. 82, 907–910.

Stahl, W., Heinrich, U., Jungmann, H., Sies, H., Tronnier, H., 2000. Carotenoids and carotenoids plus vitamin E protect against ultraviolet light-induced erythema in humans. Am. J. Clin. Nutr. 71, 795–798.

Stahl, W., Sies, H., 2005. Bioactivity and protective effects of natural carotenoids. Biochim. Biophys. Acta 1740, 101–107.

Stahl, W., Sies, H., 2007. Carotenoids and flavonoids contribute to nutritional protection against skin damage from sunlight. Mol. Biotechnol. 37, 26–30.

Stanner, S.A., Hughes, J., Kelly, C.N., Buttriss, J., 2004. A review of the epidemiological evidence for the antioxidant hypothesis. Public Health Nutr. 7, 407–422.

Steinmetz, K.A., Potter, J.D., 1996. Vegetables, fruit and cancer prevention. J. Am. Diet. Assoc. 96, 1027–1039.

Trekli, M.C., Riss, G., Goralczyk, R., Tyrrell, R.M., 2003. Beta-carotene suppresses UVA-induced HO-1 gene expression in cultured FEK4. Free Radic. Biol. Med. 34, 456–464.

Truscott, T.G., 1990. The photophysics and photochemistry of the carotenoids. J. Photochem. Photobiol. B Biol. 6, 359–371.

Von Lintig, J., Vogt, K., 2000. Molecular identification of an enzyme cleaving b-carotene to retinal. J. Biol. Chem. 275, 11915–11920.

Voutilainen, S., Nurmi, T., Mursu, J., Rissanen, T.H., 2006. Carotenoids and cardiovascular health. Am. J. Clin. Nutr. 83, 1265–1271.

Wang, X., Ouyang, Y., Liu, J., 2014. Fruit and vegetable consumption and mortality from all causes, cardiovascular disease, and cancer: systematic review and dose–response meta-analysis of prospective cohort studies. BMJ 349, g4490.

Wang, X., Willén, R., Wadström, T., 2000. Astaxanthin-rich algal meal and vitamin C inhibit *Helicobacter pylori* infection in BALB/cA mice. Antimicrob. Agents Chemother. 44 (9), 2452–2457.

Wang, Y., Cui, R., Xiao, Y., Fang, J., Xu, Q., 2015. Effect of carotene and lycopene on the risk of prostate cancer: a systematic review and dose–response meta-analysis of observational studies. PLoS One 10 (9), e0137427.

Watson, R.R., Prabhala, R.H., Plezia, P.M., Alberts, D.S., 1991. Effect of beta-carotene on lymphocyte subpopulations in elderly humans: evidence for a dose-response relationship. Am. J. Clin. Nutr. 53 (1), 90–94.

Woods, J.A., Young, A.J., Bilton, R.F., 1999. Carotene enhances hydrogen peroxide-induced DNA damage in human hepatocellular HepG2 cells. FEBS Lett. 449, 255–258.

World Cancer Research Fund, 2007. Food, nutrition, physical activity, and the prevention of cancer: a global perspective. AICR, Washington, DC.

Wyss, A., Wirtz, G., Woggon, W., Brugger, R., Wyss, M., Friedlein, A., Bachmann, H., Hunziker, W., 2000. Cloning and expression of beta, beta-carotene 15,150-dioxygenase. Biochem. Biophys. Res. Commun. 271, 334–336.

Yabuzaki, J., 2017. Carotenoids database: structures, chemical fingerprints and distribution among organisms. Database 17, bax004.

Yahia, E.M., Ornelas-Paz, J.J., 2010. Chemistry, stability, and biological actions of carotenoid. In: de la Rosa, L.A., Álvarez-Parrilla, E., González-Aguilar, G.A. (Eds.), Fruit and Vegetable Phytochemicals: Chemistry, Nutritional Value, and stability. Wiley-Blackwell, Ames, IA, USA, pp. 177–222.

Yang, Y., Fuentes, F., Shu, L., Wang, C., Pung, D., Li, W., 2016. Epigenetic CpG methylation of the promoter and reactivation of the expression of GSTP1 by astaxanthin in human prostate LNCaP Cells. AAPS J. 30, 795–800.

Ye, Q., Zhang, X., Huang, B., Zhu, Y., Chen, X., 2013. Astaxanthin suppresses MPP+-induced oxidative damage in PC12 cells through a Sp1/NR1 signaling pathway. Mar. Drugs 11 (12), 1019–1034.

Yeh, S., Hu, M., 2001. Induction of oxidative DNA damage in human foreskin fibroblast Hs68 cells by oxidized β-carotene and Lycopene. Free Radic. Res. 35, 203–213.

Yeum, K-J., Russell, R.M., 2002. Carotenoid bioavailability and bioconversion. Annu. Rev. Nutr. 22, 483–504.

Young, A.J., Lowe, G.M., 2001. Antioxidant and prooxidant properties of carotenoids. Arch. Biochem. Biophys. 385, 20–27.

Yuri, T., Yoshizawa, K., Emoto, Y., Kinoshita, Y., Yuki, M., Tsubura, A., 2016. Effects of dietary xanthophylls, canthaxanthin and astaxanthin on N-methyl-N-nitrosourea-induced rat mammary carcinogenesis. In Vivo 30 (6), 795–800.

Zahavi, A., Zahavi, A., 1997. The Handicap Principle: A Missing Piece of Darwin's Puzzle. Oxford University Press, UK.

Zaripheh, S., Erdman, J.W. Jr, 2002. Factors that influence the bioavailablity of xanthophylls. J. Nutr. 132, 531S–534S.

Zechmeister, L., 1962. Cis-Trans Isomeric Carotenoids, Vitamins A and Aryl-polyenes. Academic Press, New York.

Zhang, B., Liu, C., Wang, Y., Yao, X., Wang, F., Wu, J., King, G.J., Liu, K., 2015a. Disruption of a carotenoid cleavage dioxygenase 4 gene converts flower colour from white to yellow in Brassica species. New Phytol. 206 (4), 1513–1526.

Zhang, G.F., Zhang, Y., Zhao, G., 2015b. Crocin protects PC12 cells against MPP+-induced injury through inhibition of mitochondrial dysfunction and ER stress. Neurochem. Int. 89, 101–110.

Zheng, Y.F., Bae, S.H., Kwon, M.J., Park, J.B., Choi, H.D., Shin, W.G., Bae, S.K., 2013. Inhibitory effects of astaxanthin, b-cryptoxanthin, canthaxanthin, lutein, and zeaxanthin on cytochrome P450 enzyme activities. Food Chem. Toxicol. 59, 78–85.

Zhou, X., Zhang, F., Hu, X., Chen, J., Wen, X., Sun, Y., Liu, Y., Tang, R., Zheng, K., Song, Y., 2015. Inhibition of inflammation by astaxanthin alleviates cognition deficits in diabetic mice. Physiol. Behav. 151, 412–420.

Zhou, X.Y., Zhang, F., Huetal, X.T., 2017. Depression can be prevented by astaxanthin through inhibition of hippocampal inflammation in diabetic mice. Brain Res. 1657, 262–268.

Zhu, C., Naqvi, S., Capell, T., Christou, P., 2008. Metabolic engineering of ketocarotenoid biosynthesis in higher plants. Arch. Biochem. Biophys. 483, 182–190.

CHAPTER

Citric acid, antioxidant effects in health

4.5

Sushil Kumar Singh[a], Rahul Kaldate[b], Arti Bisht[c]

[a]*DBT NECAB, Assam Agricultural University, Jorhat, Assam, India*

[b]*Department of Agrilcultural Biotechnology, Assam Agricultural University, Jorhat, Assam, India*

[c]*G.B. Pant National Institute of Himalayan Environment and Sustainable Development, Kosi-Katarmal, Almora, Uttarakhand, India*

4.5.1 Introduction

The citric acid ($C_6H_8O_7$) is also known as tricarboxylic acid (TCA) metabolite, synthesized by condensation of acetate with oxaloacetate in the cellular oxidative metabolism of mitochondria. Its nomenclature originally comes from the Latin word *citrus*, a plant with phenotypically similar fruit and taste (Max et al., 2010). It is naturally acidic in taste and acts as a good preservative. It can also easily manufacture and converted into a soluble form. It is widely used as an acidifier, flavoring agent, chelating agent, and utilized for the stability of the fruit (Apelblat., 2014; Vandenberghe et al., 1999; Hotha et al., 2014). The annual production of citric acid is exciding million tons every year. In 2007, the total production of citric acid was 1.6 million tones around the world, with a 3.5 – 4.0% annual rise in demand and expenditure (Anastassiadis et al., 2008). In the last 20 years, the availability of citric acid has been increased globally, from 0.5 to 2 million tons (Ciriminna et al., 2017).

In a human mitochondrial cell, citric acid is synthesized in-between the metabolic pathways of TCA cycle where pyruvate manufactured through glycolysis is breakdown into CO_2, $4H^+$, and NADH, provides energy to the cell by utilizing $4H^+$ by electron transport chain. It is naturally available in tissues of many plants, animals, and physiological fluids. It can be originated from natural sources like citrus fruits or industrial production (e.g., chemical reaction and microbial fermentation). Citrus fruits (lemons, oranges, tomatoes, beets, etc.) contain a higher concentration of citric acid than other fruits, and hence, they are grouped into acid fruit. It is most concentrated in lemons and limes (Muller et al., 1996), lemons contain 4.0%–8.0%, black currants 1.5%–3%, grapefruits 1.2%–2.1%, oranges, tangerines, red currents, raspberries, and strawberries carry a range between 0.6% and 1.3% of the dry weight basis (Apelblat, 2014).

The world production of citric acid by industry-based microbial fermentation technology is rapidly enhanced with the help of various molds, yeasts, and bacteria. With the advancement in metagenomics technologies, numerous different species of

Antioxidants Effects in Health. DOI: https://doi.org/10.1016/B978-0-12-819096-8.00045-8

microorganisms have been identified and studied in the last few years. Swain et al., 2011 enlisted these microorganisms capable of producing citric acid in which 26 native strains of bacteria, mainly belonging to *Aspergillus* and *Penicillium* genus, 14 strains of yeasts and three strains of bacteria are mentioned. Genetically modified microorganisms are developed in such a way that commercially enhances the production of citric acid. Moreover, *Aspergillus niger* was mostly used in industry for large scale production of citric acid due to its superior production yield as compared to other microorganisms (Pau et al., 2015).

Nowadays, the production of citric acid is carried out by solid-state fermentation, utilizing agricultural residues like sugarcane and cassava bagasse, and food-producing solid wastes, such as grape and apple pomace with a range of 22%–155% yield (Swain et al., 2011). In 1880, Grimaux and Adam, firstly, chemically produced citric acid by purely chemical reactions using glycerol as a starting material.

4.5.2 Chemistry

The IUPAC nomenclature of citric acid is 2-hydroxy-1, 2, 3-tricarboxylic acid with a chemical formula of $C_6H_8O_7$ and structural formula as represented in Fig. 4.5.1 (PubChem Identifier: CID 311, URL: https://pubchem.ncbi.nlm.nih.gov/compound/Citric-acid).

In 1934, the anhydrous crystalline structure of citric acid was first explained by Bennett and Yuill with the help of X-ray diffraction (Bennett and Yuill, 1935). Citric acid is synthesized as a critical metabolic output in the mitochondrial TCA cycle. Citric acid monohydrate has 210.14 g/mol concentration (NCBI, PubChem Database), which contains three similar COOH- functional groups at different positions contributing 3.1, 4.7, and 6.4 values of pKa (Papagianni, 2007). It is odorless, colorless, slightly hygroscopic, and readily dissolvable in water (62.07% at 25°C) (Dalman, 1937). The absence of asymmetric carbon atoms in its structure makes them optically inactive.

The anhydrous citric acid has a density of 1.665 g cm^{-3} at a temperature of 18°C with a melting point of 156–157°C (Apelblat, 2014) and density of its monohydrate form is 1.542 g cm^{-3} at 25°C (Laguerie et al., 1976). In 1991, Trask-Morrell and Kottes Andrews reported that its anhydrous form melts at 152–154°C

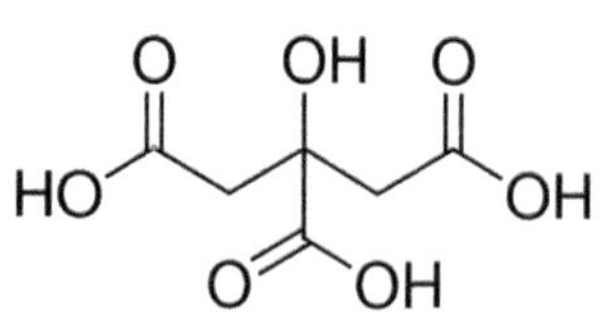

FIG. 4.5.1 The structural formula of citric acid.

and disintegrate at 228–242°C, which is solid at room temperature. It can occur mainly as the trivalent anion at physiologic pH and smaller concentration extent in the urine.

4.5.3 Bioavailability

The fresh juice of citrus fruit like lemon, orange, and lime supply more citric acid per liter than cane-packed juice of grapefruit, and other extracted juice from the citrus fruit (Penniston et al., 2008). Citric acid helps in increasing the accumulation of Ca, Mg, P, and Zn. In contrast, that of ascorbic acid only ameliorates the concentration of Fe and assist in improving the bioaccumulation of other micronutrients of chicken eggshells (Siddique et al., 2016). The bioaccumulation of calcium and phosphorus upon intake of citric acid in the food were studied by modifying the dietary habits of a rat. This study indicates that taking citric acid supplementation together with a Ca-rich diet enhances the bioavailability of Ca and P in the bone. Continues intake of Ca citrate may, therefore, aid in improving the deposition of Ca in the bone (Lacour et al., 1997). Moreover, it has been reported that the accumulation of citric acid reduces the uptake of both lead and cadmium in the body. It was observed that the accumulation of citric acid to the soil altered the concentration and relative abundance of other forms of acids (Chen et al., 2003). It may reduce the phytoremediation of heavy metal ions lead and cadmium due to the decline of soil pH after the addition of citric acid.

4.5.4 Mechanisms of action

Citric acid acts as an oxygen scavenger and delays unsaturated lipid/free fatty acid oxidation by depletion of essential agents like metals-based oxygen scavenger (Rostamzad et al., 2011). It plays an active role in limiting the degree of lipids degradation due to oxidation of food materials (Gordon, 1990) and chelates photo-oxidative metal ions by combining metal ions with free radicals of carboxyl or hydroxyl groups thereby quench catalytic oxygen (Anon, 1985). The concentration of available dissolved oxygen is a vital factor for lipid oxidation, and reducing it can help to enhance the stability of lipids. Chelate complexes have inhibited oxidation process due to the following regions: (1) formation of unstable complexes, (2) reducing the activity of chelate metals ions, or (3) providing steric hindrance between metals and free triplet and singlet electronic states of oxygen or other constituents of food products (Polumbryk et al., 2013).

Ethylene diamine tetra acetic acid (EDTA) is the well-known chemically synthesized chelating agent that can bind to metal by forming carbonate and amine bonds. This water-soluble EDTA, in combination with lipid-soluble citric acid, can act as an effective antioxidant but not a true antioxidant. The synergistic effect of citric acid is due to its consumption of free radicals when it is associated with a metal chelator like

EDTA as Ca-EDTA or Na-EDTA form. This partial antioxidant activity of EDTA basically reduces the oxidation catalyzed by metal, and decrease the amount of free radicals (Phaniendra et al., 2015; Palmer et al., 1997). It catalyzed the production of reactive oxygen species act as a scavenger in the step of propagation while that of chelator at the initial stage (Choe et al., 2009).

4.5.5 Possible proxidant activity

Citric acid was applied to decrease the growth of microorganisms, add sour flavor, and enhance the quality as well as stability of food components in the beverage industry. The aqueous solution of citric acid and other exclusive hydrogen peroxide equalizer is lower down the conversion of H_2O_2 to H_2O and O_2 gas. This type of microbial regulator outcome is because of the development of per-organic acids, as shown in the following reaction.

$$H_2O_2 \rightarrow H_2O + O_2 \uparrow$$

$$H_2O_2 + R-COOH \rightarrow R-COOOH + H_2O$$

For packaged food products, citric acid, along with other ingredients, can act as an antioxidant preservative and helps in maintaining quality, enhance taste, masking effect, texture improver, color, and freshens of industrially produced packed food. In the case of freshly cut fruit like apple citric acid proved to be the best protective agent that increases the shelf-life of food (Rossle et al., 2009).

The presence of citric acid reduced the pH, and break the lipid membrane of microorganism with the help of H_2O_2 and other organic acids with free –COOH functional group. Due to membrane disruption property of citric acid, it protects from various foodborne pathogens that can cause severe diseases in human beings, for example, *Clostridium botulinum, Salmonella typhimurium, Staphylococcus aureus, Shigella, Listeria monocytogenes,* Campylobacter, etc. The treatments of citric acid solution to the raw food before packaging wash out the entire toxic components.

4.5.6 Safety profile or toxicity studies

The safety concern related to citric acid depends on the effect of toxicity at diverse concentrations. The FDA enlisted citric acid in categories of generally recognized as safe (GRAS), direct food additives based on the report of dermal toxicity, which proved to be safe for ingestion. However, FDA makes it compulsory to label as a warning note in the cosmetic product containing AHA, which may increase the skin sensitivity under the sun, causing sunburn (Fiume et al., 2014).

It shows low irritation at a concentration of 10% and mild irritation at 30% concentration in rabbit eyes. Similarly, up to 10% of citric acid in skin cream is safe to use. Still, the concentration above that shows a significant increase in epidermal thickness along with Langerhans cells and glycosaminoglycan (GAG) content.

Spermicidal activity determined that a concentration of 0.1% and 1% citric acid show slight and complete immortality of sperm, respectively. However, citric acid does not show any sign of genotoxicity, both *in vivo* and *in vitro* (Fiume et al., 2014). It has been reported that oral intake of citric acid is easily absorbed by the gastrointestinal tract and metabolized approximately 2 kg of citric acid a day to provide energy. The physiological level of citric acid in the body is about 25 mg/L (Dzik and Kirkley, 1988). It does not act as a mutagenic, carcinogen, or teratogenic substance *in vivo* or *in vitro*. It can be easily filterable from the glomerulus of the kidney. The NOAEL toxicity dose rate is 1200 mg/kg/d. Due to the low toxicity and biodegradable nature of citric acid, it can be considered as safe for the environment.

4.5.7 Beneficial and detrimental effects on health

Citric acid and citrates are extensively used as a food additive for additional protection against oxidation and show no side effects on human health (Lawrence, 1998; Voss, 2002; Lidon et al., 2014; Inetianbor et al., 2015). The composition of citric acid with K-citrate and Na-citrate is utilized for the treatment of kidney diseases such as kidney stones or gout or metabolic acidosis. A skin cream containing citric acid helps to remove skin infection. It can also help in retarding the acidic content of urine. It helps in recovering damaged tissue in the body, protects the brain and liver from oxidative damage, and reduces aging, stress, and acid in the urine due to its antioxidant properties. In our body it is present in every tissue, take part in oxidative metabolism, play a role for reducing lipid peroxidation and inflammation through antioxidant redox homeostasis mechanism of decreasing the level of PMN degranulation and release of IL-1beta, platelet factor-4, elastase and myeloperoxidase in the cell (Gabutti et al., 2004). It can also help in the recovery of CCl-4 induced hepatocellular necrosis injury in rats (Abdel et al., 2009).

The fresh juice of lemon or lime is regarded as the most abundant source of citric acid, and its intake substantiates as useful for treating oxalate stones (Halliwell, 1992; Kang et al., 2007; Haleblian et al., 2008). It can be quickly broken down in the water surface and readily biodegradable to the environment, declared as safe for human consumption (Ciriminna et al., 2017). It is now well known that citric acid, along with their salts, are metal chelators and create an acidic buffer for preservation (Oktar et al., 2001; Kim et al., 2006). Due to its antioxidant properties, citric acid acts as an inert substance in pharmaceutical drug development. It reduced the delivery of drug *in vitro*, when combining with multiple tablets and delayed the absorption of drug *in vivo*. It enhances the acidity while adjusting the pH and provides stability of food ingredients, chelate blood Ca ion, and prevents blood clotting. Due to its unique property of pH adjustment, chelation, masking agent, preservative, antioxidant and dissolution, citric acid has widely been utilized in the food and pharmaceutical industry. Hence, daily intake of citric acid increases in the body through both in the form of processed food, drugs, and natural sources.

4.5.8 Animal studies and clinical studies

There are various *in vivo/in vitro* clinical and animal trials are investigated to assess the effectiveness of citric acid, which is summarised with details in Table 4.5.1. Multiple researchers have studies clinical clearance and established safety concerns of citric acid. In support of this *in vivo* animal studies, Omer et al. (2013) observed the effect of citric acid under endotoxin-induced oxidative damage in the brain and liver tissue of mice. They found that 1–2 g/kg dose of citric acid retard brain lipid-peroxidation and inflammation, liver damage, and DNA-fragmentation upon LPS (lipopolysaccharide) induced oxidative stress. The study of antioxidative activity of citric and ascorbic acids helps in lipid oxidation, was observed in the experiment of frozen Persian sturgeon fillets, which revealed significant differences in biochemical parameters when added with citric acid compared with control. It might be due to oxygen scavenger property, which delays lipid oxidation by reducing oxygen and metals. In different studies, the detailed result was observed by Aubourg et al. (2004) that acetic acid and citric acid have a synergistic outcome on each other. It can also change the fat improperly due to the reduction of lipid oxidation. Hence, ascorbic acid, along with citric acid, demonstrated an active role in avoiding lipid oxidation in freeze stored fillets (Rostamzad et al., 2011).

Similarly, Abdelrazek et al. (2016), experimented with studying the effect of citric acid and acetic acid in water acidification on broilers performance concerning thyroid hormones levels. The investigation reveals the further accumulation of citric acid to broilers water, boost broilers' development by controlling the health of gut mucosal, liver, and thyroid hormones (T3 and T4) after observing lipid summary. Citric acid shows a substantial effect on balancing internal body homeostasis. Nagoba et al. (2011) used citric acid for controlling the growth of microorganisms related to chronic wound infection in animals. There were 38 cases of this infection which was not reacting to the conventional treatment. They are divided into two groups. In group-I, 3%, and in group-II, 5% of citric acid solutions were applied to check its efficiency. They found that the application of citric acid provides faster recovery of wounds in all 38 animals achieving a 100% success rate. Moreover, the use of citric acid was substantiated to be an effective and economical method for the successful recovery of continuously infected wounds in animals.

Significant emphasis has also been given to the safety of citric acid in animals. In this regard, Bonting et al. (1956) identify the result of citric acid in combination with phosphoric acid for its prolonged intake in rats. It was found that a concentration of 1.20% citric acid shows no harmful effect on developmental stages in rats. Moreover, multiple studies display enormous consumption of citric acid composition of beverages and fruits prove to be supportive in treating kidney stones (Seltzer et al., 1996; Penniston et al., 2008; Gul and Monga, 2014). However, it is interesting to note that citric acid displays potential antimicrobial effects on the growth of *Helicobacter pylori* strains associated with peptic ulcers in human beings (Zazgornik and Mittermayer, 2011). It is now well known that citric acid is used as a sole antimicrobial agent for successful treatment of lepromatous ulcers and chronic oral ulcers as compared to

Table 4.5.1 Summary of *in-vitro*/*in-vivo* clinical and animal studies to assess the effect of citric acid.

S. No.	Organ/ tissue	Model	Dose and route of administration/ method/treatment	Investigation/ objective	Results/outcome	References
1	-	Sprague-Dawley rats	100 mg/kg/d; orally	Access protective effect of CA against hepatic I/R injury	The I/R-citric acid group shows more expression of catalase, superoxide dismutase, antioxidants, nitric oxide, and lower expression of aspartate aminotransferase and alanine aminotransferase compared to the control group	Kim et al., 2019
2	Brain	Rat	1.150 mg/kg malathion; intraperitoneal (i.p.) injection of – 200–400 mg/kg citric acid; orally 2. 1 mg/kg atropine; i.p./ 200 mg/kg citric acid + 1 mg/kg atropine	Access the effect of CA given only or along with atropine on organ or tissue or cells understudy	CA shows beneficial effect upon oxidative stress in the brain, neuronal injury, liver and damage to DNA	Abdel-Salam et al., 2016
3	-	One thirty-two, Cobb broiler chicks	Four treatment groups: 1. Control group 2. CA group 3. Acetic acid (AA) group 4. Combination group	Access the effect of H_2O acidification by CA on broilers' performance	Enhanced poultry performance by modulating gut and liver health as well as thyroid hormones	Abdelrazek et al., 2016

(continued)

Table 4.5.1 Summary of *in-vitro*/*in-vivo* clinical and animal studies to assess the effect of citric acid. *Continued*

S. No.	Organ/ tissue	Model	Dose and route of administration/ method/treatment	Investigation/ objective	Results/outcome	References
4	Thirty dental implants with anodized surface	Beagle dogs	Four treatment groups: 1. Er:YAG laser 2. PDT 3. Titanium bur alone 4. Titanium bur with CA	Assess the effects of 4 different treatment modalities	It has been evaluated that the titanium bur along with CA group shows more significant improvement on vertical bone height	Htet et al., 2016
5	Bovine kidney cell line (LFBK) with foot-and-mouth disease (FMD) virus O & A serotypes	-	-	FMD virus inactivated by CA & Na_2CO_3 along with Deicers	With all the deicers, 0.2% CA could lower the virus titer 4 logs @20°C	Hong et al., 2015
6	Brain & liver	Mice	1, 2, or 4 g/kg CA; orally	Access the effect of CA on oxidative stress induced by endotoxin	The dose of 1–2 g/kg CA shows reduced lipid peroxidation and inflammation in brain, liver damage, and fragmentation of DNA	Abdel-Salam et al., 2014

S. No.	Organ/ tissue	Model	Dose and route of administration/ method/treatment	Investigation/ objective	Results/outcome	References
7	-	Nine *Helicobacter pylori* strains	1. H_2O_2 - 3% 2. $NaHCO_3$ - 8.4% 3. $C_6H_8O_6$ - 2% 4. CA combination with sodium citrate 5. 7% & 14% CA solutions	Evaluate the inhibitory effect of different chemicals or compounds understudy on *H. pylori* strains	The potent inhibitory activity showed by CA an on *H. pylori* strains growth	Zazgornik and Mittermayer, 2011
8	-	Broiler	1. CA - 0.5% 2. Flavomycin (FA) - 0.001% 3. Combination of 0.5% CA + 0.001% flavomycin; orally through diet	To study the effect of CA, FL and their combination	0.5% dietary supplementation of CA boost nonspecific immunity and increase feed efficiency as well as enhance the yield of broiler	Haque et al., 2010
9	-	Diabetic rats	2 g/L CA; orally	Access the effect of CA on diabetic rats	CA protects against the development of diabetic complications and ameliorates ketosis	Nagai et al., 2010
10	Human peripheral blood cells	-	50, 100, 200, & 3,000 μg ml^{-1} of CA for 24 h and 48 h	Analyze the clastogenic effects of CA (as a food additive) in human peripheral lymphocytes	CA with compared to negative control significantly enhanced the micronuclei frequency	Yılmaz et al., 2008

other therapies or treatments (Nagoba et al., 2012; Jamadar et al., 2019). In the case of a patient with anogenital warts, treatment with a composition of 9% citric acid along with 6% sodium nitrite shows promising effects (Ormerod et al., 2015). In another therapeutic approach for the treatment of a 6-year-old girl with propionic acidaemia, an acute life-threatening metabolic disorder, citric acid shows a positive effect in curing the diseases (Siekmeyer et al., 2013). Many methods have been explored for the treatment of wrinkles in which treatment of citric acid was proved to be a better solution or remedy and got patented for this method with U.S. patent No. 5,470,880 in the United States (Yu and Van Scoott, 1995).

Conclusion

The present study indicates that citric acid is a weak acid found in citrus fruit, mostly in lemons and limes. It has been used as a natural preservative and takes part in foods and soft drinks and has an acidic sour taste. The study of clinical and preclinical trail demonstrated the involvement of antioxidative activity in citric acid against several disorders including metabolic, neurological, and cardiological, among others.

In various research studies, citric acid has been picked as a possible drug agent for AD. With these studies, we can say that citric acid administration is helpful in the inhibition or reversal of Alzheimer's disease. Citric acid also increased bioavailability.

Consequently, due to the high national and international demand of the citrus fruit and production of citric acid under the Pharmaceutical sector, these have got much attention and need to be reintroduced for further advanced research.

References

Abdel-Salam, O.M.E., Shaffie, N.M., Sleem, A.A., 2009. Hepatoprotective effects of citric acid and aspartame on carbon tetrachloride-induced hepatic damage in rats. EXCLI J. 8, 41–49.

Abdelrazek, H.M.A., Abuzead, S.M.M., Ali, S.A., El-Genaidy, H.M.A., Abdel-Hafez, S.A., 2016. Effect of citric and acetic acid water acidification on broiler's performance with respect to thyroid hormones levels. Adv. Anim. Vet. Sci. 4 (5), 271–278.

Abdel-Salam, O.M., Youness, E.R., Mohammed, N.A., Morsy, S.M.Y., Omara, E.A., Sleem, A.A., 2014. Citric acid effects on brain and liver oxidative stress in lipopolysaccharide-treated mice. J. Med. Food 17 (5), 588–598.

Abdel-Salam, O.M., Youness, E.R., Mohammed, N.A., Yassen, N.N., Khadrawy, Y.A., El-Toukhy, S.E., Sleem, A.A., 2016. Novel neuroprotective and hepatoprotective effects of citric acid in acute malathion intoxication. Asian Pac. J. Trop. Med. 9 (12), 1181–1194.

Anastassiadis, S., Morgunov, IG., Kamzolova, SV., Finogenova, TV., 2008. Citric acid production patent review. Recent Pat. Biotechnol 2, 107–123.

Anon, 1985. Ullmann's Encyclopedia of Industrial Chemistry, fifth ed. VCH, Weinheim.

Apelblat, A., 2014. Citric Acid Springer, Cham, Switzerland.

Aubourg, S.P., Perez-Alonso, F., Gallardo, J.M., 2004. Studies on rancidity inhibition in frozen horse mackerel (*Trachurus trachurus*) by citric acid and ascorbic acids. Eur. J. Lipid Sci. Technol. 106, 232–240.

Bennett, G.M., Yuill, J.L., 1935. The crystal form of anhydrous citric acid. J. Chem. Soc. 130, 130.

Bonting, S.L., Jansen, B.C.P., 1956. The effect of a prolonged intake of phosphoric acid and citric acid in rats. Voeding 17, 137–148.

Chen, Y.X., Lin, Q., Luo, Y.M., He, Y.F., Zhen, S.J., Yu, Y.L., Tian, G.M., Wong, M.H., 2003. The role of citric acid on the phytoremediation of heavy metal contaminated soil. Chemosphere 50 (6), 807–811.

Choe, E., Min, D.B., 2009. Mechanisms of antioxidants in the oxidation of foods. Compr. Rev. Food Sci. Food Saf. 8 (4), 345–358.

Ciriminna, R., Meneguzzo, F., Delisi, R., 2017. Citric acid: emerging applications of key biotechnology industrial product. Chem. Cent. J. 11 (1), 22.

Dalman, LH., 1937. The solubility of citric and tartaric acids in water. J. Am. Chem. Soc. 59, 2547–2549.

Dzik, W.H., Kirkley, S.A., 1988. Citrate toxicity during massive blood transfusion. Transfus. Med. Rev. 2, 76–94.

Fiume, M.M., Heldreth, B.A., Bergfeld, W.F., Belsito, D.V., Hill, R.A., Klaassen, C.D., Liebler, D.C., Marks Jr, J.G., Shank, R.C., Slaga, T.J., Snyder, P.W., 2014. Safety assessment of citric acid, inorganic citrate salts, and alkyl citrate esters as used in cosmetics. Int. J. Toxicol. 33 (2_suppl), 16S–46S.

Gordon, M.H., 1990. The Mechanism of Antioxidant Action in Vitro. In: Hudson B.J.F. (Ed.), Food antioxidants. Elsevier, New York, pp. 1–18.

Gabutti, L., Nicola, F., Giorgio, M., Mombell, K., Claudio, M., 2004. The favorable effect of regional citrate anticoagulation on interleukin-1beta release is dissociated from both coagulation and complement activation. J. Nephrol. 6 (17), 819–825.

Grimaux, C., Adam, P., 1880. Synthèse de l'acide citrique. C R Acad. Sci. 90, 1252–1255.

Gul, Z., Monga, M., 2014. Medical and dietary therapy for kidney stone prevention. Korean J. Urol. 55 (12), 775–779.

Haleblian, GE., Leitao, VA., Pierre, SA., 2008. Assessment of citrate concentrations in citrus fruit-based juices and beverages: implications for management of hypocitraturic nephrolithiasis. J. Endourol. 22, 1359–1366.

Halliwell, B., 1992. Reactive oxygen species and the central nervous system. J. Neurochem. 59 (5), 1609–1623.

Haque, M.N., Islam, K.M., Akbar, M.A., Chowdhury, R., Khatun, M., Karim, M.R., Kemppainen, B.W., 2010. Effect of dietary citric acid, flavomycin and their combination on the performance, tibia ash and immune status of broiler. Can. J. Anim. Sci. 90 (1), 57–63.

Hong, J.K., Lee, K.N., You, S.H., Kim, S.M., Tark, D., Lee, H.S., Ko, Y.J., Seo, M.G., Park, J.H., Kim, B., 2015. Inactivation of foot-and-mouth disease virus by citric acid and sodium carbonate with deicers. Appl. Environ. Microbiol. 81 (21), 7610–7614.

Hotha, K.K., Patel, T., Roychowdhury, S., Subramanian, V., 2014. Development of better-quality assay method for the citric acid and sodium citrate in ophthalmic/oral solutions and their application to deformulation studies. Am. J. Anal. Chem. 5 (17), 1249.

Htet, M., Madi, M., Zakaria, O., Miyahara, T., Xin, W., Lin, Z., Aoki, K., Kasugai, S., 2016. Decontamination of anodized implant surface with different modalities for peri-implantitis treatment: lasers and mechanical debridement with citric acid. J. Periodontol. 87 (8), 953–961.

Inetianbor, J.E., Yakubu, J.M., Ezeonu, S.C., 2015. Effects of food additives and preservatives on man – a review. Asian J. Sci. Technol. 6 (2), 11181135.

Jamadar, N., Nagoba, B., Davane, M., Ahmed, A., Tangsal, A., 2019. Citric acid treatment of oral ulcers refractory to conventional treatment: a case study. J. Wound Care 28 (7), 461–463.

Kang, D.E., Sur, R.L., Haleblian, G.E., 2007. Long-term lemonade based dietary manipulation in patients with hypocitraturic nephrolithiasis. J. Urol. 177, 1358–1362.

Kim, S.K., Kang, S.W., Jin, S.A., Ban, J.Y., Hong, S.J., Park, M.S., 2019. Protective effect of citric acid against hepatic ischemia reperfusion injury in sprague-dawley rats. Transplant. Proc. 51 (8), 2823–2827.

Kim, S., Lee, K., Park, J., Lee, H., Hwang, I., 2006. Effect of natural antioxidants on stored freeze dried food product formulated using horse mackerel (*Trachurus trachurus*). J. Sci. Technol. 41, 90–95.

Lacour, B., Tardivel, S., Drüeke, T., 1997. Stimulation by citric acid of calcium and phosphorus bioavailability in rats fed a calcium-rich diet. Miner. Electrolyte Metab. 23 (2), 79–87.

Laguerie, C., Aubry, M., Couderc, J.P., 1976. Some physicochemical data on monohydrate citric acid in water: solubility, density, viscosity, diffusivity, pH of standard solution, and refractive index. J. Chem. Eng. Data 21, 85–87.

Lawrence, A., 1998. In: Stephenson, M., Thomas, G. (Eds.), Understanding Food Additives, FAIA – Food Additives and Ingredient Association, University of York, p. 148.

Lidon, F.C., Almeida, A.S., Leitão, A.E., Pinheiro, M.M.N., Macas, B., Costa, R., 2014. A synoptic overview of durum wheat production in the Mediterranean region and processing following the European Union requirements. Emirates J. Food Agric. 26 (8), 693705.

Max, B., Salgado, J.M., Rodríguez, N., Cortés, S., Converti, A., Domínguez, J.M., 2010. Biotechnological production of citric acid. Braz. J. Microbiol. 41 (4), 862–875.

Muller, M., Irkens-Kiesecker, U., Rubinstein, B., 1996. On the mechanism of hyper acidification in lemon: comparison of the vacuolar H(+)-ATPase activities of fruits and epicotyls. J. Biol. Chem. 271 (4), 1916–1924.

Nagai, R., Nagai, M., Shimasaki, S., Baynes, J.W., Fujiwara, Y., 2010. Citric acid inhibits development of cataracts, proteinuria and ketosis in streptozotocin (type 1) diabetic rats. Biochem. Biophys. Res. Commun. 393 (1), 118–122.

Nagoba, B.S., Wadher, B.J., Rao, A., Selkar, S.P., Gandhi, R.C., 2012. Treatment of lepromatous ulcers using citric acid as a sole antimicrobial agent. Int. Wound J. 9 (5), 553–556.

Nagoba, B.S., Wadher, B.J., Selkar, S.P., 2011. Citric acid treatment of chronic wounds in animals. Int. J Animal Vet. Adv. 3 (1), 26–28.

Oktar, G.L., Sinci, V., Kalaycioğlu, S., Soncul, H., Gökgöz, L., Halit, V., Ersöz, A., 2001. Biochemical and hemodynamic effects of ascorbic acid and alpha-tocopherol in coronary artery surgery. Scand. J. Clin. Lab. Inv. 61 (8), 621–629.

Omer, E.A., Said-Al Ahl, H.A.H., El Gendy, A.G., Shaban, Kh.A., Hussein, M.S., 2013. Effect of amino acids application on production, volatile oil and chemical composition of chamomile cultivated in saline soil at Sinai. J. Appl. Sci. Res. 9 (4), 3006–3021.

Ormerod, A.D., van Voorst Vader, P.C., Majewski, S., Vanscheidt, W., Benjamin, N., Van der Meijden, W., 2015. Evaluation of the efficacy, safety, and tolerability of 3 dose regimens of topical sodium nitrite with citric acid in patients with anogenital warts: a randomized clinical trial. JAMA Dermatol. 151 (8), 854–861.

Palmer, H.J., Paulson, K.E., 1977. Reactive oxygen species and antioxidants in signal transduction and gene expression. Nutr. Rev. 55, 353–361.

Papagianni, M., 2007. Advances in citric acid fermentation by *Aspergillus niger*: biochemical aspects, membrane transport and modelling. Biotechnol. Adv. 25, 244–263.

Pau, L.S., Kehinde, O.O., Qi, Y.S., Fitri, A.A.Z., John, C.W.L., Tau, C.L., 2015. Overview of citric acid production from *Aspergillus niger*. Front Life Sci. 8 (3), 271–283.

Penniston, K.L., Nakada, S.Y., Holmes, R.P., Assimos, D.G., 2008. Quantitative assessment of citric acid in lemon juice, lime juice, and commercially-available fruit juice products. J. Endourol. 22 (3), 567–570.

Phaniendra, A., Jestadi, D.B., Periyasamy, L., et al., 2015. Free radicals: properties, sources, targets, and their implication in various diseases. Indian J. Clin. Biochem. 30 (1), 11–26.

Polumbryk, M., Ivanov, S., Polumbryk, O., 2013. Antioxidants in food systems, Mechanism of action. Ukr. J. Food Sci. 1, 15–40.

Rossle, C., Gormley, T.R., Butler, F., 2009. Efficacy of Nature seal AS1 browning inhibitor in fresh cut fruit salads applications, with emphasis on apple wedges. J. Hortic. Sci. Biotechnol. 84, 62–67.

Rostamzad, H., Shabanpour, B., Shabani, A., Shahiri, H., 2011. Enhancement of the storage quality of frozen Persian sturgeon fillets by using ascorbic acid. Int. Food Res. J. 18 (1), 109–116.

Seltzer, M.A., Low, R.K., McDonald, M., Shami, G.S., Stoller, M.L., 1996. Dietary manipulation with lemonade to treat hypocitraturic calcium nephrolithiasis. J. Urol. 156 (3), 907–909.

Siddique, S., Firdous, S., Durrani, A.I., Khan, S.J., Saeed, A., 2016. Hesperidin, a citrus flavonoid, increases the bioavailability of micronutrients of *Gallus domesticus* (chicken) eggshell: in vitro study. Chem. Speciation Bioavailability 28 (1-4), 88–94.

Siekmeyer, M., Petzold-Quinque, S., Terpe, F., Beblo, S., Gebhardt, R., Schlensog-Schuster, F., Kiess, W., Siekmeyer, W., 2013. Citric acid as the last therapeutic approach in an acute life-threatening metabolic decompensation of propionic acidaemia. J. Pediatr. Endocrinol. Metab. 26 (5-6), 569–574.

Swain, M.R., Ray, R.C., Patra, J.K., 2011. Citric acid: microbial production and applications in food and pharmaceutical industries. In: Vargas, D.A., Medina, J.V. (Eds.), Citric Acid. Nova Science Publisher, New York, pp. 97–118.

Trask-Morrell, B.J., Kottes, A.B.A., 1991. Thermoanalytical characteristics of polycarboxylic acids investigated as durable press agents for cotton textiles. J. Appl. Polym. Sci. 42, 511–521.

Vandenberghe, L.P., Soccol, C.R., Pandey, A., Lebeault, J.M., 1999. Microbial production of citric acid. Braz. Arch. Biol. Technol. 42 (3), 263–276.

Voss, C., 2002. Veneno no seu prato? Utilidades e Riscos dos Aditivos Alimentares. EDIDECO – Editores para a Defesa do Consumidor Lda, Lisboa.

Yılmaz, S., Ünal, F., Yüzbaşıoğlu, D., Aksoy, H., 2008. Clastogenic effects of food additive citric acid in human peripheral lymphocytes. Cytotechnology 56 (2), 137–144.

Yu, J.R., Van Scott, J.E., 1995. Method of using citric acid for the treatment of wrinkles. U.S. Patent No. 5,470,880.

Zazgornik, J., Mittermayer, H., 2011. Citric acid inhibits growth of *Helicobacter pylori* in vitro: a new strategy for eradication. Wien. Klin. Wochenschr. 123 (1-2), 38–40. https://doi.org/10.1007/s00508-010-1524-9.

CHAPTER

Antioxidant activity of coenzyme-Q; bright and dark side

4.6

Ziyad Khan[a], Haroon Khan[b], Marya[b], Muhammad Ajmal Shah[c]

[a]*Department of Pharmacy, University of Swabi, Swabi, Pakistan*

[b]*Department of Pharmacy, Abdul Wali Khan University Mardan, Mardan, Pakistan*

[c]*Department of Pharmacognosy, Faculty of Pharmaceutical Sciences, Government College University, Faisalabad, Pakistan*

4.6.1 Introduction

4.6.1.1 Origin and sources of coenzyme-Q10

In normal healthy individuals coenzyme-Q10 (Co-Q10) is synthesized in all cells from tyrosine (or phenylalanine) and mevalonate (Fig. 4.6.1) (Farsi et al., 2018). Co-Q10 exists in the body of many organisms where there are two forms of Co-Q10, the oxidized form is ubiquinone and the reduced form is ubiquinol. Co-Q10 serves as an essential carrier for the electron transfer in the mitochondrial respiratory chain for the synthesis of ATP. In 1957, Co-Q10 was first isolated from beef mitochondria (Bonakdar and Guarneri, 2005). Co-Q10 is a lipid-soluble antioxidant compound found in all organisms (Dhanasekaran and Ren, 2005). Dr. Folkers in 1958 discovered the chemical structure of Co-Q10 (Littarru and Tiano, 2008). In humans the endogenous Co-Q10 is mainly found in the most energetic organs like kidney, heart, and liver where its level decreases inside these tissues after the age of 21. Inside the cell, 50% of Co-Q10 is present in mitochondria; that is why it is readily accessible to deal with the free radicals produced during oxidative phosphorylation. Also, 5%–10% of Co-Q10 is located in cytosol (Sastry et al., 1961). Human beings can also obtain this important antioxidant from exogenous sources in their daily diet like fishes, vegetables (spinach and brocolli), meat, and dairy products (Mattila and Kumpulainen, 2001; Strazisar, et al., 2005). The daily intake of these foods provides between 3 and 5 mg of Co-Q10 (Pravst et al., 2010).

4.6.1.2 Chemistry

Chemically, Co-Q10 is a quinone molecule, it is found in almost all cells of the body; hence, the term ubiquinones is given to it (Bonakdar and Guarneri, 2005). Structurally, Co-Q10 contains benzoquinone nucleus having 6 to 10 polyisoprenoid side chain units, based upon the number of side chains it is named as Q6, Q7, and so Q10 for Co-Q10. Chemically, Co-Q10 is 2, 3-dimethoxy-5-methyl-6-decaprenyl

Antioxidants Effects in Health. DOI: https://doi.org/10.1016/B978-0-12-819096-8.00046-X

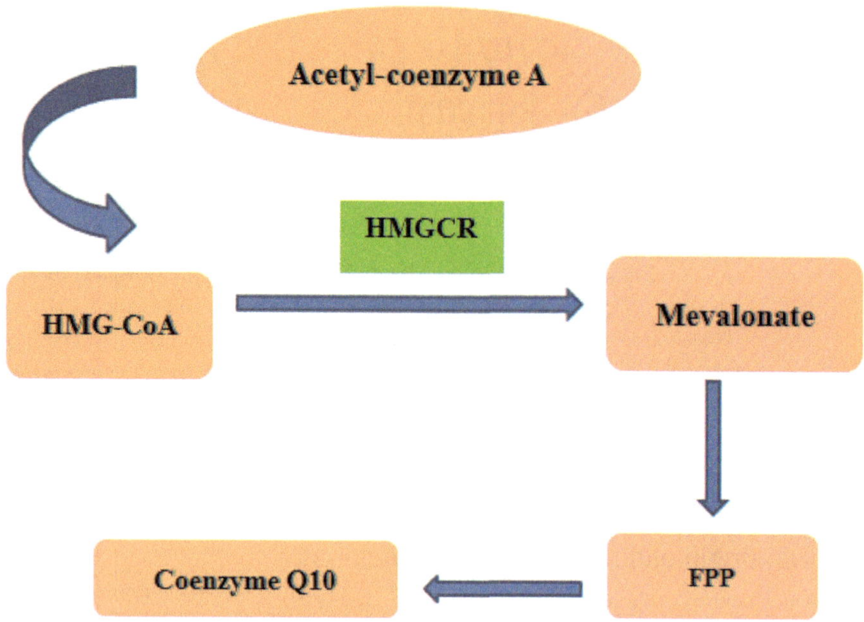

FIG. 4.6.1 Pathway of coenzyme-Q10 synthesis.

HMG-CoA, 3-hydroxy-3-methyl-glutaryl-CoA; HMGCR, 3-hydroxy-3-methyl-glutaryl-CoA Reductase; FPP, farnesyl diphosphate.

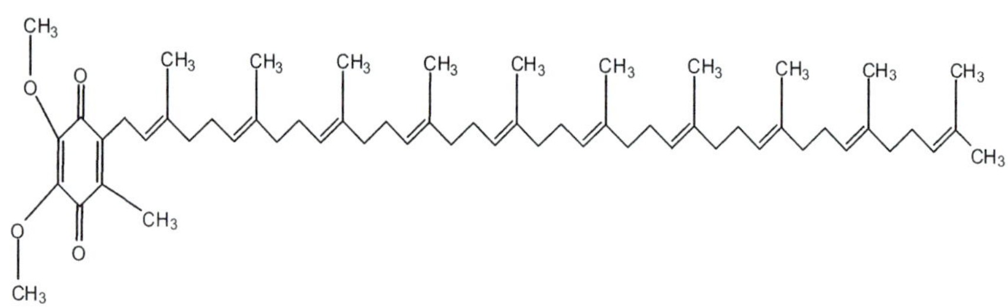

FIG. 4.6.2 Structure of coenzyme Q10.

benzoquinone. These side chains attribute hydrophobic nature to this Co-Q10 (Fig. 4.6.2) (Orsucci et al., 2011).

4.6.1.3 Recommended oral dose and bioavailability

Co-Q10 is absorbed from the diet mainly in the small intestine and its absorption is enhanced in presence of food rich in lipids. Co-Q10 is transported to the liver where a complex of lipoprotein is produced and then it is deposited in tissues.

The recommended oral dosage of Co-Q10 generally range from 100 to 200 mg a day for cardiovascular patients (Alam and Rahman, 2014). Tissues with high-energy requirements and metabolic rates, such as the heart and the skeletal muscle contain relatively high concentrations of Co-Q10. About 95% of Co-Q10 in human circulatory system exists in its reduced form as ubiquinol (Bhagavan and Chopra, 2006). The safety of high and prolong oral doses of Co-Q10 is well acknowledged in human and also in animals (López-Lluch et al., 2019).

Co-Q10 is mostly available in oral dosage forms in the form of tablet and capsule. Due to its low solubility in water, the oral absorption of Co-Q10 is poor. In a study conducted by Kaikkonen et al. (Parkinson et al., 2013) the oral administration of 30 mg of Co-Q10, in the form of granules, showed marginal quantity in plasma; however, with increasing in oral dose to 200 mg per day, plasma level of Co-Q10 was increased 6.1 times. The solubility of Co-Q10 is also affected by formulation and nowadays efforts are made to increase its bioavailability by increasing the solubility through different techniques (Bhagavan and Chopra, 2006). It is very difficult to determine the exact amount of Co-Q10 that passes the blood-brain barrier because it is formed endogenously. Increased levels of Co-Q10 inside the brain of rats and mice after oral administration of Co-Q10 was recorded compared to controls. When given intraperitoneal, level of Co-Q10 was found much lower in cerebrospinal fluid (CSF) than in other tissues like spleen, kidney and liver. To the best of our knowledge, in humans, no determination of Central nervous system (CNS) penetration of Co-Q10 is recorded while CSF level of Co-Q10 can be found (Duberley et al., 2013).

In vivo study was conducted to determine the oral bioavailability of Co-Q10. Due to high molecular weight and poor solubility of Co-Q10, it was mixed with beta-cyclodextrin and the resultant complex formed was Q10vital, which is more water soluble. Q10vital in liquid and powder form were given orally to healthy humans and its plasma levels were compared to its reference standard softgel capsule (Co-Q10 mixed in soyabean oil). The mean plasma concentrations obtained for Q10vital liquid (120%) and powder (79%) was higher than its reference standard. This increase in plasma level is attributed to enhance water solubility of novel Q10vital (Žmitek, et al., 2008).

4.6.1.4 Antioxidant mechanisms of action

The antioxidant activity of Co-Q10 is attributed to its benzoquinone group because it can both donate and accept electrons, which is a very important property of a good antioxidant. Co-Q10 is a free radical scavenger and prevents the protein and lipid peroxidation. Co-Q10 is equally potent as Vitamin E in the prevention of lipid peroxidation (Stocker et al., 1991). Another study suggests that the antioxidant activity of Co-Q10 in preventing LDL oxidation is higher than β-carotene and α-tocopherol (Shekelle, et al., 2003). Besides the antioxidant activity of Co-Q10, it also increases the bioavailability of other antioxidants like ascorbic acid, beta-carotene, and vitamin E (Sohal and Forster, 2007).

It has been reported that supplementation of Co-Q10 readily eliminates markers of oxidative stress and inflammation (Fig. 4.6.3) (Sohal and Forster, 2007). *In vivo*, Co-Q10 showed outstanding antioxidant activity. It is activator of mitochondrial

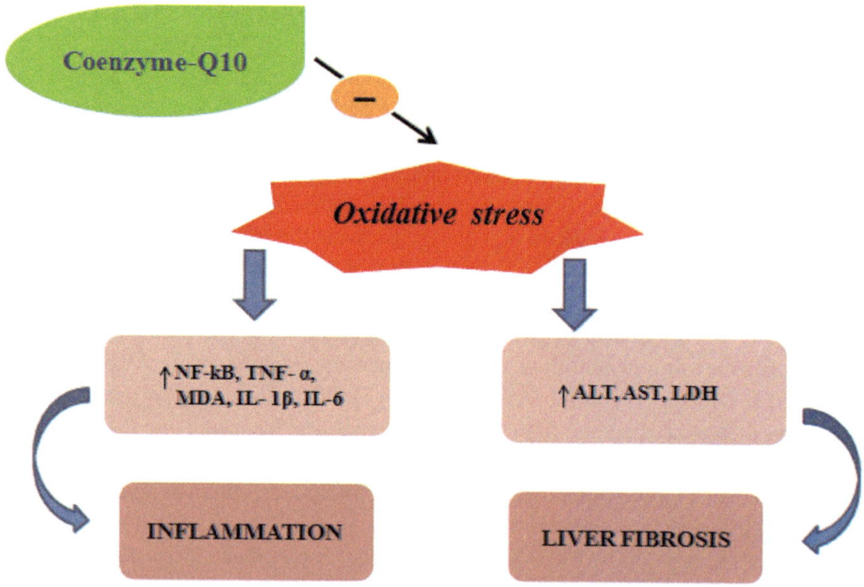

FIG. 4.6.3 Co-Q10 inhibits markers of oxidative stress and inflammation.

ALT, alanine aminotransaminase; AST, aspartate aminotransferase; IL-1β, interleukin-1 beta; IL-6, interleukin-6; LDH, lactic acid dehydrogenase; MDA, malondialdehyde; NF-kB, nuclear factor kappa-light-chain-enhancer of activated B cells; TNF-α, tumor necrosis factor alpha.

uncoupling proteins and as cofactor, which ultimately decreased the generation of free radicals. Ubiquinol-10, the reduced form of Co-Q10, protects the hydrogen peroxide-induced DNA strand breaks in lymphocytes. With supplementation of Co-Q10, *in vivo*, a remarkable recovery of human lymphocytes due to oxidative DNA damage was reported by (Tomasetti et al., 1999). Enhanced cellular antioxidant enzyme activity was also reported for Co-Q10 supplementation. The antioxidant activity of enzymes like glutathione, superoxide dismutase and catalase is increased by treatment of Co-Q10. This study was conducted on rats liver homogenates which resulted in decreased lipid peroxidation (Modi et al., 2006). Also, the antioxidant activity of enzyme catalase and concentration of growth stimulating hormone was increased in rats treated with Co-Q10. Recently, a study has proved that Co-Q10, at a dose of 150mg/day, facilitated other antioxidant enzymes activity and decreased the oxidative stress in patients with coronary artery diseases (Lee et al., 2012).

The antioxidant action of Co-Q10 is attributed to its phenol ring structure. Ubiquinone react with reactive oxygen species to prevent lipid peroxidation in membranes and other lipid structures in the cell. During oxidative stress, Co-Q plays an important role in recycling of vitamin E through NADPH-dependent system (Powers et al., 2004). Decreased mRNA expression of the lipogenic enzymes, fatty acid synthase (FAS) and acetyl-CoA carboxylase 1(ACC1), and the glycerogenic enzyme, phosphoenolpyruvate carboxykinase (PEPCK), is responsible for the lipid lowering effect of Co-Q10 (Alam and Rahman, 2014).

Co-Q10 is well found in membranes in close proximity to the unsaturated lipid chains where it acts as a primary scavenger of free radicals. The amount of Co-Q10 in many membranes is from 3 to 30 times more than the tocopherol content (Crane, 2001). A direct revelation of the effectiveness of Co-Q10 as an antioxidant can be related with Co-Q10 deficient yeast. A yeast mutant deficient in Co-Q10 synthesis showed more lipid peroxide production than normal yeast (Poon et al., 1997). Another direct revelation of elimination of free radicals was shown by Co-Q10 treatment of skin in older people. Luminescence from free radicals was eliminated when skin cream containing Co-Q10 was applied (Hoppe et al., 1999).

4.6.1.5 Possible Pro-oxidant activity

Co-Q10 plays an important role in the regulated pro-oxidant formation of the superoxide anion/H_2O_2 second messenger system. We have a general concept that superoxide anion/H_2O_2 usually are harmful and cause damage to tissues and macromolecules. On the other hand, there is an important physiological role of these pro-oxidants as second messenger in intracellular signaling against the protection of pathogens (Linnane et al., 2007). In healthy organisms, there is a balance in the production of reactive oxygen species (ROS) and antioxidants. An imbalance between these antioxidants and oxidants in the favor of oxidants leads to cells/tissue damage, known as oxidative stress. Oxidants are produced as a by-product of aerobic metabolism at controlled levels but their production is elevated under diseased conditions. ROS damages biomolecules like amino acids, carbohydrates, fatty acids, and nucleotides leading to different diseases like diabetes mellitus, cancer, inflammations, and neurological diseases. Despite these harmful effects, these oxidants play an important role in mitochondrial electron transport in the defense process against infections (Silvestri et al., 2015).

4.6.1.6 Beneficial and detrimental effects on health

The antioxidant properties of Co-Q10 decrease the chances of cardiovascular diseases by enhancing the endothelial cell functions and decreasing the oxidative damage of low-density lipoproteins (LDL) caused by free radicals (Hughes et al., 2002). Studies also suggest that Co-Q10 is also beneficial in the prevention and treatment of certain diseases like neurodegenerative diseases, Parkinson's disease, cancer, acquired immunodeficiency syndrome (AIDS). Some diseases are caused by deficiency of Co-Q10 like ataxic syndrome, which is a neurodegenerative disease, also called Friedreich's ataxia. It is associated with cardiomyopathy. The main cause of this disease is the deficiency of frataxin protein which leads to deposition of iron, impaired production of adenosine triphosphate, and thus increasing the chance of oxidative stress. Co-Q10 acts as a potent antioxidant which prevents the oxidation of DNA, proteins, lipids and lipoproteins, and protects other antioxidants, such as vitamin C and vitamin E. Co-Q10 is the only antioxidant which is endogenously synthesized and it is lipid-soluble. It also prevents the opening of the mitochondrial membrane transition pores which permit passage of enzymes and other molecules. These enzyme and molecules can further contribute to the depolarization of the mitochondrial membrane, apoptotic events, and DNA fragmentation. Co-Q10 also protects the mitochondrial membrane from oxidation and depolarization from enzymes. Co-Q10 is known to have anti-inflammatory and antiatherosclerotic properties (Bentinger et al., 2010).

Co-Q10 is found to decrease the gastric ulceration by increasing the mucus secretion in gastric juices and this protective activity is attributed to the increased level of nitric oxide and glutathione in the gastric mucosa (Malash et al., 2012). Co-Q10 shows promising hepatoprotective effects by reducing oxidative stress and inflammation (Saleh et al., 2017).

Co-Q10 is responsible for preventing the impairment in left ventricular function and sparing ATP by lowering the degree of oxidation of LDL. Co-Q10 is also known for protecting mitochondria, platelets, lipids and smooth muscles from oxidative stress during plaque disruption as well as during ischemia and reperfusion damage (Singh et al., 2003).

Co-Q10 deficiency can cause metabolic disorder. Evidences suggest that oxidative stress is responsible in the pathogenesis of type 2 diabetes mellitus and its complications (Chew and Watts, 2004). Lower plasma concentration of Co-Q10 was found when blood samples of patients having active diabetic complications and deranged glycemic index were tested. The anti-inflammatory activity of Co-Q10 is extensively reported. It stimulates cells having lipopolysaccharides which results in a special release of tumor necrosis factor-α (TNF-α), monocyte chemo attractant protein-1 (MCP-1) and macrophage inflammatory protein-1 alpha (MIP-1α). Such factors were considerably reduced by pre-incubation of cells with the reduced form of Co-Q10 (Schmelzer et al., 2007). Supplementation with Co-Q10 also decreased the raised plasma lipid profiles and also the level of the pro-inflammatory cytokine, TNF-α, in adipose tissues of obese (ob/ob) mice. Co-Q10 treatment also cured the inflammatory state in liver of rats fed with high fructose diet. Recent studies proved that Co-Q10 can serve as an agonist for peroxisome proliferator activated receptor (PPAR) and activates the anti-inflammatory response mediated by PPAR. Antiadipogenic activity of Co-Q10 was exhibited *in vitro*, in 3T3-F442A cell line. Prevention of Co-Q10 synthesis strongly increased adipocyte differentiation while allowing synthesis of Co-Q10 strongly inhibited adipocyte differentiation. Treatment with Co-Q10 showed an increase in energy consumption in inguinal adipose tissues and also increased oxidation of fats (Alam and Rahman, 2014).

4.6.1.7 *In vitro* antioxidant evidence

Supplementation with Co-Q10 decreased the release of proinflammatory cytokines in oxidized low density lipoprotein (Ox-LDL) induced inflammatory process in Tamm Horsefall Protein 1 (THP-1) cells. Extracellular superoxide dismutase (ecSOD) activity and endothelium-dependent vasodilatation were also enhanced astonishingly after Co-Q10 supplementation through decreasing local vascular oxidative stress. Significant evidence showed that Ox-LDL induced endothelial dysfunction is linked with down-regulation of endothelial nitric oxide synthase (eNOS) and upregulation of inducible nitric oxide synthase (iNOS). Nuclear factor kappa-light-chain-enhancer of activated B cells (NF-kB) plays a key role in mediating caspase-3 protein activation which is responsible for programmed cell death. Study showed that Co-Q10 inhibited the apoptosis in human umbilical vein endothelial cells (HUVEC) due to lowering of Ox-LDL and inhibiting NF-kB from activating caspase-3 protein. Co-Q10 also weakened the Ox-LDL-mediated down regulation of eNOS and up-regulation of inducible iNOS (Alam and Rahman, 2014). The expression of lipogenic genes and fatty acid synthesis are linked to the activation of an enzyme, adenosine monophosphate activated protein kinase (AMPK) (Foretz et al., 1999). The phosphorylation of AMPK is increased by Co-Q10 by increasing the level of calcium in cytoplasm through calcium-calmodulin-dependent protein kinase kinase (CaMKK). In addition, Co-Q10 was found to enhance the oxidation of fatty acids in 3T3-L1 pre-adipocytes and also uplifts the levels of PPARα in protein and mRNA. This process is responsible for suppression of adipocyte differentiation (Alam and Rahman, 2014).

4.6.2 Animal studies

The details of animal studies are presented in Table 4.6.1.

4.6.3 Clinical studies

The details of clinical studies are presented in Table 4.6.2.

4.6.4 Side effects

In clinical study conducted on 29 human subjects, it was reported that Co-Q10 exhibited side effects of nausea and vomiting at dose of 300 mg/day, 1200 mg/day, and 1800 mg/day in three patients respectively. Stomach upset was reported in one patient at a dose of 1800 mg/day (Yeung et al., 2015). Studies conducted to reveal side effects of oral Co-Q10 in human subjects. No serious side effects were found with the supplementation of Co-Q10. Only 1% of 5000 patients experienced diarrhea, loss of appetite and epigastric discomfort. He also noticed that higher doses of Co-Q10 (>100 mg) causes insomnia (Heck et al., 2000; Al-Hasso, 2001).

Table 4.6.1 Showing detail *in vivo* animal studies of Co-Q10.

Study design	Animals (numbers)	Dose/route/duration	Mechanism/effect	References
Experimental model of diabetic neuropathy (DN)	Rats (n = 70)	10 mg/kg PO for 5 weeks	Co-Q10 treatment decreased DN induced motor neuron deficiency, attenuated oxidative stress and suppressed degeneration of dorsal root ganglionic neurons	(Galeshkalami et al., 2019)
Morris water maze task design	Male Wistar rats (n =6 0)	600 mg/kg intragastric lavage for 90 days	High dose of Co-Q10 improves cognitive function in rats by increasing level of adenosine triphosphate (ATP) and inhibiting oxidative stress	(Omidi et al., 2019)
Mouse model of restraint Stress	Adult male mice (n = 72)	500 mg/kg/day PO for 5 days	Co-Q10 decreased marker of neuroinflammation, NF-kB, and marker of apoptosis like caspase-3. While enhanced the total antioxidant capacity (TAC) in brain of mice	(Salehpour et al., 2019)
Allergic rhinitis model	Sprague-Dawley rats (n = 30)	20 mg/kg intragastric lavage for 7 days	Co-Q10 reduced the elevated levels of plasma inflammatory markers like IL-13, IL-4, NO, and malondialdehyde (MDA). Co-Q10 relieves symptoms of allergic rhinitis in rats	(Sakat et al., 2018)
Hepatotoxicity model of rats	Albino rats (n = 60)	200 mg/kg/day PO for 4 weeks	Co-Q10 is hepatoprotective supplement, it significantly decreased the level of hepatic enzymes like alanine transaminase (ALT), serum lactate dehydrogenase (LDH), aspartate transaminase (AST), and proinflammatory cytokines tumor necrosis factor alpha (TNF-α), interleukin-1β (IL-1β), and interleukin-6 (IL-6)	(Saleh, et al., 2017)
Rat model of diabetes encephalopathy	Male Wistar rats (n = 90)	10 mg/kg/day PO for 2 weeks	Co-Q10 intake decreased oxidative stress markers like MDA and superoxide dismutase (SOD). Levels of apoptosis marker caspase-3 and inflammatory cytokine, TNF-α were also decreased. Co-Q10 is helpful in treating diabetic encephalopathy.	(Motawi et al., 2017)

Study design	Animals (numbers)	Dose/route/duration	Mechanism/effect	References
Morris water maze	Young mice (n = 134)	50 mg/kg i.p before anesthesia	Co-Q10 increased mitochondrial membrane potential, up-regulated levels of excitatory postsynaptic density protein-95 and synaptophysin in brain. Co-Q10 alleviated the sevoflurane induced cognitive deficiency in young mice	(Xu et al., 2017)
Mouse model of neurotoxicity	Mice (n = 30)	10 mg/kg Intraperitoneal (i.p) for 7 days	Co-Q10 treatment, decreased the level of ROS and increased the activity of protective enzymes like glutathione dismutase and catalase in methamphetamine induced neurotoxicity in mice brain	(Thanh et al., 2016)
Mouse model of Parkinson disease (PD)	15-month-old mices (n = 6–9/experiment)	150 μM PO for 1 week	Co-Q10 ameliorates the onset of motor impairment in mice by inhibiting accumulation of phosphorylated α-synuclein in motor neurons	(Takahashi et al., 2016)
Model of rat colitis	Adult male Wistar rats (n = 32)	30 mg/kg/day PO for 7 days	Co-Q10 alleviated ulcerative colitis in rats by decreasing inflammatory markers like NO and MDA while increased antioxidant marker GSH and activity of catalase.	(Ewees et al., 2016)
Rats gastropathy model	Male Wistar rats (n = 63)	40, 200, 400 mg/kg PO	Co-Q10 inhibits peptic acivity, increases mucin sectretion due to glutathione (GSH), nitric oxide (NO) replenishment in the gastric mucosa	(Malash et al., 2012)

Table 4.6.2 Showing detailed clinical studies of Co-Q10.

Study design	Subject (numbers)	Dose/route/duration	Mechanism/outcomes	References
Randomized double blind, placebo control study	Women with polycystic ovary syndrome (PCOS) (n = 86)	200 mg/day PO for 8 weeks	Co-Q10 significantly decreased total testosterone level, improved plasma fasting blood sugar and insulin level in women with PCOS	(Izadi et al., 2019)
Randomized controlled clinical study	Smokers with chronic periodontitis (n = 40)	200 mg/5 mL oral gel for 3 months	Co-Q10 having radical scavenging effect enhanced the recovery of periodontitis in smokers	(Raut et al., 2019)
Randomized double blind, placebo control study	Dyslipidemic patients (n = 101)	120 mg/day PO for 24 weeks	Co-Q10 intake attenuates atherothrombosis by inhibiting the thrombotic modulater, integrin αllbβ3 and upregulation of cyclic adenosine phosphate (cAMP) and protein kinase A (PKA) pathway	(Ya et al., 2019)
Randomized double blind, placebo control study	Diabetic nephropathy patients (n = 50)	100 mg/day PO for 12 weeks	Co-Q10 decrease the level of plasma markers MDA and advanced glycation end products (AGEs) responsible for oxidative stress and inflammation respectively	(Gholnari et al., 2018)
Randomized double blind, placebo controlled clinical study	Women with type 2 diabetes (T2D) (n = 80)	100 mg/day PO for 12 weeks	Co-Q10 significantly increases level of catalase, TAC and increases the insulin sensitivity in T2DM patients	(Zarei et al., 2018)
Randomized double blind, clinical study	Hemodialysis patients (n = 65)	600 to 1200 mg/day PO for 04 months	Co-Q10 intake decreased the plasma concentration of F_2-isoprostanes a biomarker of oxidative stress	(Rivara et al., 2017)
Randomized double blind, placebo controlled clinical study	Patients with Huntington's disease (HD) (n = 609)	2400 mg/day PO for 60 months	Study suggests that Co-Q10 has no significant relation to improve progressive functional decline in HD	(McGarry et al., 2017)
Randomized placebo control clinical study	Nonalcoholic fatty liver disease patients (n = 41)	100 mg/day PO for 12 weeks	Co-Q10 treatment lowered the inflammatory biomarkers like C-reactive protein (CRP), IL-6, and TNF-α	(Farsi et al., 2016)

Study design	Subject (numbers)	Dose/route/ duration	Mechanism/outcomes	References
Randomized double blind, placebo controlled clinical study	Chronic fatigue syndrome (CFS) patients (n = 80)	200 mg/day PO for 8 weeks	Co-Q10 notably decreased the heart rate and fatigue after exercise. No effect on pain and sleep was noted.	(Castro-Marrero et al., 2016)
Open-label add-on controlled study	Acute migraine headache patients (n = 80)	100 mg/day PO for 1 month	Co-Q10 decreased the severity and frequency of attack of acute migraines to 50% per month	Coq migraine 2016
Randomized double blind, placebo controlled clinical study	Patients with multiple sclerosis (n = 48)	500 mg/day PO for 12 weeks	Co-Q10 treatment notably improved the symptoms of fatigue and depression	(Sanoobar et al., 2016)
Randomized double blind, controlled clinical study	Rheumatoid arthritis patients (n = 44)	100 mg/day PO for 02 months	Co-Q10 lowered the level of inflammatory cytokines like IL-6 and TNF-α and oxidative stress marker MDA	(Abdollahzadet al., 2015)
Randomized placebo control clinical study	Hyperlipidemic patients with myocardial infarction (MI) (n = 52)	200 mg/day PO for 12 weeks	Co-Q10 intake has beneficial effects on hyperlipidemic patients with MI. It reduced the inflammatory cytokine IL-6 and intercellular adhesion molecule-1 and enhanced level of high density lipoprotien (HDL)	(Mohseni et al., 2015)
Randomized placebo controlled clinical study	Mild hypertensive patients (n = 60)	100 mg/day PO for 12 weeks	Co-Q10 significantly reduced the inflammatory biomarkers like CRP, IL-6, and TNF-α	(Bagheri Nesami et al., 2015)

4.6.5 Safety profile

Co-Q10 is used a dietary supplement and published data suggest that Co-Q10 is safe and does not exhibit any serious side effects. In open label dose escalation study, Co-Q10 at a dose of 1600 mg/day was safe and well tolerated in hemodialysis patients (n = 29) (Yeung et al., 2015). In animal studies the maximum dose of Co-Q10 was 4000 mg orally for both rats and mice. While the lethal dose for Co-Q10 was higher than 5000 mg/kg for rats of either sex. In a study conducted on rats, there is no significant effects of Co-Q10 (600 mg/kg/day) on fetus nor any postnatal toxicity or abnormality is attributable (Hidaka et al., 2008). Co-Q10 do not possess any genotoxic effects (Kitano et al., 2006). No toxicity of Co-Q10 intake at an oral dose of 100 mg/day is reported in clinical studies. The maximum tolerated dose of Co-Q10 in human was 1200 mg/day, with plasma concentration reached up to 4 μg/m (Hidaka et al., 2008).

Conclusion

Co-Q10 has numerous beneficial health effects and the most important of all is its antioxidant property. Co-Q10 readily eliminates free radicals like lipid peroxyl or alkoxyl radicals both *in vitro* and *in vivo*. Co-Q10 is used in the prevention of many diseases like diabetes mellitus, cancer, hepatoprotective, migraine, cardiovascular diseases, ulcerative colitis, autoimmune and neurodegenerative disorders. Based on evidence, the safety of high and prolonged oral doses of Co-Q10 is well acknowledged in human. Co-Q10 is readily available in the market in the form of tablet and capsule and is well tolerated by human body. Animal and clinical studies also revealed the free radical scavenging activity of Co-Q10, which plays a vital role in protecting tissues against the harmful effects of oxidative stress. Further research is required to evaluate the useful effects of Co-Q10 in AIDs, male infertility, and periodontal diseases.

Abbreviations

Co-Q10	Coenzyme-Q10
CNS	Central nervous system
CSF	Cerebrospinal fluid
Ox-LDL	Oxidized low density lipoprotein
ACC1	Acetyl-CoA carboxylase 1
NF-kB	Nuclear factor kappa-light-chain-enhancer of activated B cells
PEPCK	Phosphoenolpyruvate carboxykinase
ROS	Reactive oxygen species
AIDS	Acquired immunodeficiency syndrome
ALT	Alanine transaminase

LDH	Serum lactate dehydrogenase
AST	Aspartate transaminase
IL-1 β	Interleukin-1β
TNF-α	Tumor necrosis factor-α
IL-6	Interleukin-6
MCP-1	Monocyte chemo attractant protein-1
MIP-1α	Macrophage inflammatory protein-1
PPAR	Peroxisome proliferator activated receptor
ecSOD	Extracellular superoxide dismutase
eNOS	Endothelial nitric oxide synthase
iNOS	Inducible nitric oxide synthase
HUVEC	Human umbilical vein endothelial cells
AMPK	Adenosine monophosphate activated protein kinase
CaMKK	Calcium-calmodulin-dependent protein kinase.
T1D	Type 1 diabetes
T2D	Type 2 diabetes
MD	Malondialdehyde
GSH	Glutathione
TAC	Total antioxidant activity
PO	Orally
i.p	Intraperitoneally
HD	Hungtington's disease
PD	Parkinson's disease.
NO	Nitric oxide
PCOS	Polycystic ovary syndrome.

References

Abdollahzad, H., Aghdashi, M.A., Jafarabadi, M.A., Alipour, B., 2015. Effects of coenzyme Q10 supplementation on inflammatory cytokines (TNF-α, IL-6) and oxidative stress in rheumatoid arthritis patients: a randomized controlled trial. Arch. Med. Res. 46 (7), 527–533.

Al-Hasso, S., 2001. Coenzyme Q10: A review. Hosp. Pharm. 36 (1), 51–66.

Alam, M.A., Rahman, M.M., 2014. Mitochondrial dysfunction in obesity: potential benefit and mechanism of Co-enzyme Q10 supplementation in metabolic syndrome. J. Diabetes Metab. Disord. 13 (1), 60–67.

Bagheri Nesami, N., Mozaffari-Khosravi, H., Najarzadeh, A., Salehifar, E., 2015. The effect of coenzyme Q10 supplementation on pro-inflammatory factors and adiponectin in mildly hypertensive patients: a randomized, double-blind, placebo-controlled trial. Int. J. Vitam. Nutr. Res. 85 (3-4), 156–164.

Bentinger, M., Tekle, M., Dallner, G., 2010. Coenzyme Q–biosynthesis and functions. Biochem. Biophys. Res. Commun. 396 (1), 74–79.

Bhagavan, H.N., Chopra, R.K., 2006. Coenzyme Q10: absorption, tissue uptake, metabolism and pharmacokinetics. Free Radic. Res. 40 (5), 445–453.

Bonakdar, R.A., Guarneri, E., 2005. Coenzyme Q10. Am. Fam. Physician 72 (6), 1065–1070.

Castro-Marrero, J., Sáez-Francàs, N., Segundo, M.J., Calvo, N., Faro, M., Aliste, L., et al., 2016. Effect of coenzyme Q10 plus nicotinamide adenine dinucleotide supplementation on maximum heart rate after exercise testing in chronic fatigue syndrome–A randomized, controlled, double-blind trial. Clin. Nutr. 35 (4), 826–834.

Chew, G., Watts, G., 2004. Coenzyme Q10 and diabetic endotheliopathy: oxidative stress and the 'recoupling hypothesis'. QJM 97 (8), 537–548.

Crane, F.L., 2001. Biochemical functions of coenzyme Q10. J. Am. Coll. Nutr. 20 (6), 591–598.

Dhanasekaran, M., Ren, J., 2005. The emerging role of coenzyme Q-10 in aging, neurodegeneration, cardiovascular disease, cancer and diabetes mellitus. Curr. Neurovasc. Res. 2 (5), 447–459.

Duberley, K.E., Hargreaves, I.P., Chaiwatanasirikul, K.A., Heales, S.J., Land, J.M., Rahman, S., et al., 2013. Coenzyme Q10 quantification in muscle, fibroblasts and cerebrospinal fluid by liquid chromatography/tandem mass spectrometry using a novel deuterated internal standard. Rapid Commun. Mass Spectrom. 27 (9), 924–930.

Ewees, M.G., Messiha, B.A.S., Abo-Saif, A.A., El, H.A.E.-T.A., 2016. Is coenzyme Q10 effective in protection against ulcerative colitis? An experimental study in rats. Biol. Pharm. Bull. 39 (7), 1159–1166.

Farsi, F., Mohammadshahi, M., Alavinejad, P., Rezazadeh, A., Zarei, M., Engali, K.A., 2016. Functions of coenzyme Q10 supplementation on liver enzymes, markers of systemic inflammation, and adipokines in patients affected by nonalcoholic fatty liver disease: a double-blind, placebo-controlled, randomized clinical trial. J. Am. Coll. Nutr. 35 (4), 346–353.

Farsi, F., Nezhad, P.A., Morshedzadeh, N., 2018. The Hepatoprotective Effects of Coenzyme Q10 Against Oxidative Stress *The Liver*. Elsevier, Amsterdam, Netherlands, pp. 281–294.

Foretz, M., Pacot, C., Dugail, I., Lemarchand, P., Guichard, C., le Lièpvre, X., et al., 1999. ADD1/SREBP-1c is required in the activation of hepatic lipogenic gene expression by glucose. Mol. Cell. Biol. 19 (5), 3760–3768.

Galeshkalami, N.S., Abdollahi, M., Najafi, R., Baeeri, M., Jamshidzade, A., Falak, R., et al., 2019. Alpha-lipoic acid and coenzyme Q10 combination ameliorates experimental diabetic neuropathy by modulating oxidative stress and apoptosis. Life Sci. 216, 101–110.

Gholnari, T., Aghadavod, E., Soleimani, A., Hamidi, G.A., Sharifi, N., Asemi, Z., 2018. The effects of coenzyme Q10 supplementation on glucose metabolism, lipid profiles, inflammation, and oxidative stress in patients with diabetic nephropathy: a randomized, double-blind, placebo-controlled trial. J. Am. Coll. Nutr. 37 (3), 188–193.

Heck, A.M., DeWitt, B.A., Lukes, A.L., 2000. Potential interactions between alternative therapies and warfarin. Am. J. Health Syst. Pharm. 57 (13), 1221–1227.

Hidaka, T., Fujii, K., Funahashi, I., Fukutomi, N., Hosoe, K., 2008. Safety assessment of coenzyme Q10 (CoQ10). Biofactors 32 (1-4), 199–208.

Hoppe, U., Bergemann, J., Diembeck, W., Ennen, J., Gohla, S., Harris, I., et al., 1999. Coenzyme Q_{10}, a cutaneous antioxidant and energizer. Biofactors 9 (2-4), 371–378.

Hughes, K., Lee, B.L., Feng, X., Lee, J., Ong, C.-N., 2002. Coenzyme Q10 and differences in coronary heart disease risk in Asian Indians and Chinese. Free Radic. Biol. Med. 32 (2), 132–138.

Izadi, A., Ebrahimi, S., Shirazi, S., Taghizadeh, S., Parizad, M., Farzadi, L., et al., 2019. Hormonal and metabolic effects of coenzyme Q10 and/or vitamin E in patients with polycystic ovary syndrome. J. Clin. Endocrinol. Metab. 104 (2), 319–327.

Kitano, M., Hosoe, K., Fukutomi, N., Hidaka, T., Ohta, R., Yamakage, K., et al., 2006. Evaluation of the mutagenic potential of ubidecarenone using three short-term assays. Food Chem. Toxicol. 44 (3), 364–370.

Lee, B.-J., Huang, Y.-C., Chen, S.-J., Lin, P.-T., 2012. Coenzyme Q10 supplementation reduces oxidative stress and increases antioxidant enzyme activity in patients with coronary artery disease. Nutrition 28 (3), 250–255.

Linnane, A.W., Kios, M., Vitetta, L., 2007. Coenzyme Q10–Its role as a prooxidant in the formation of superoxide anion/hydrogen peroxide and the regulation of the metabolome. Mitochondrion 7, S51–S61.

Littarru, G.P., Tiano, L., 2008. Clinical aspects of coenzyme Q 10 in relationship with its bioenergetic and antioxidant properties, *Mitochondrial Medicine*. Springer, Amsterdam, Netherlands, pp. 303–321.

López-Lluch, G., del Pozo-Cruz, J., Sánchez-Cuesta, A., Cortés-Rodríguez, A.B., Navas, P., 2019. Bioavailability of coenzyme Q10 supplements depends on carrier lipids and solubilization. Nutrition 57, 133–140.

Malash, A.M., Abdallah, D.M., Agha, A.M., Kenawy, S.A., 2012. Gastroprotective efficacy of coenzyme Q10 in indomethacin-induced gastropathy: other potential mechanisms. Ulcers 2012.

Mattila, P., Kumpulainen, J., 2001. Coenzymes Q9 and Q10: contents in foods and dietary intake. J. Food Compos. Anal. 14 (4), 409–417.

McGarry, A., McDermott, M., Kieburtz, K., de Blieck, E.A., Beal, F., Marder, K., et al., 2017. A randomized, double-blind, placebo-controlled trial of coenzyme Q10 in Huntington disease. Neurology 88 (2), 152–159.

Modi, K., Santani, D., Goyal, R., Bhatt, P., 2006. Effect of coenzyme Q10 on catalase activity and other antioxidant parameters in streptozotocin-induced diabetic rats. Biol. Trace Elem. Res. 109 (1), 25–33.

Mohseni, M., Vafa, M., Zarrati, M., Shidfar, F., Hajimiresmail, S.J., Forushani, A.R., 2015. Beneficial effects of coenzyme Q10 supplementation on lipid profile and intereukin-6 and intercellular adhesion molecule-1 reduction, preliminary results of a double-blind trial in acute myocardial infarction. Int. J. Prev. Med. 6, 73–77.

Motawi, T.K., Darwish, H.A., Hamed, M.A., El-Rigal, N.S., Naser, A.F.A., 2017. A therapeutic insight of niacin and coenzyme Q10 against diabetic encephalopathy in rats. Mol. Neurobiol. 54 (3), 1601–1611.

Omidi, G., Karimi, S.A., Rezvani-Kamran, A., Monsef, A., Shahidi, S., Komaki, A., 2019. Effect of coenzyme Q10 supplementation on diabetes induced memory deficits in rats. Metab. Brain Dis. 34 (3), 833–840.

Orsucci, D., Mancuso, M., Ienco, E.C., LoGerfo, A., Siciliano, G., 2011. Targeting mitochondrial dysfunction and neurodegeneration by means of coenzyme Q10 and its analogues. Curr. Med. Chem. 18 (26), 4053–4064.

Parkinson, M.H., Schulz, J.B., Giunti, P., 2013. Co-enzyme Q10 and idebenone use in Friedreich's ataxia. J. Neurochem. 126, 125–141.

Poon, W.W., Do, T.Q., Marbois, B.N., Clarke, C.F., 1997. Sensitivity to treatment with polyunsaturated fatty acids is a general characteristic of the ubiquinone-deficient yeast coq mutants. Mol. Aspects Med. 18, 121–127.

Powers, S.K., Deruisseau, K.C., Quindry, J., Hamilton, K.L., 2004. Dietary antioxidants and exercise. J. Sports Sci. 22 (1), 81–94.

Pravst, I., Žmitek, K., Žmitek, J., 2010. Coenzyme Q10 contents in foods and fortification strategies. Crit. Rev. Food Sci. Nutr. 50 (4), 269–280.

Raut, C.P., Sethi, K.S., Kohale, B., Mamajiwala, A., Warang, A., 2019. Subgingivally delivered coenzyme Q10 in the treatment of chronic periodontitis among smokers: A randomized, controlled clinical study. J. Oral Biol. Craniofac. Res. 9 (2), 204–208.

Rivara, M.B., Yeung, C.K., Robinson-Cohen, C., Phillips, B.R., Ruzinski, J., Rock, D., et al., 2017. Effect of coenzyme Q10 on biomarkers of oxidative stress and cardiac function in hemodialysis patients: the CoQ10 biomarker trial. Am. J. Kidney Dis. 69 (3), 389–399.

Sakat, M.S., Kilic, K., Kandemir, F.M., Yildirim, S., Sahin, A., Kucukler, S., et al., 2018. The ameliorative effect of berberine and coenzyme Q10 in an ovalbumin-induced allergic rhinitis model. Eur. Arch. Otorhinolaryngol. 275 (10), 2495–2505.

Saleh, A.A.S., Shahin, M.I., Kelada, N.A., 2017. Hepatoprotective effect of taurine and coenzyme Q10 and their combination against acrylamide-induced oxidative stress in rats. Trop. J. Pharm. Res. 16 (8), 1849–1855.

Salehpour, F., Farajdokht, F., Cassano, P., Sadigh-Eteghad, S., Erfani, M., Hamblin, M.R., et al., 2019. Near-infrared photobiomodulation combined with coenzyme Q10 for depression in a mouse model of restraint stress: reduction in oxidative stress, neuroinflammation, and apoptosis. Brain Res. Bull. 144, 213–222.

Sanoobar, M., Dehghan, P., Khalili, M., Azimi, A., Seifar, F., 2016. Coenzyme Q10 as a treatment for fatigue and depression in multiple sclerosis patients: a double blind randomized clinical trial. Nutr. Neurosci. 19 (3), 138–143.

Sastry, P.S., Jayaraman, J., Ramasarma, T., 1961. Distribution of coenzyme Q in rat liver cell fractions. Nature 189 (4764), 577–585.

Schmelzer, C., Lorenz, G., Rimbach, G., Döring, F., 2007. Influence of Coenzyme Q_ {10} on release of pro-inflammatory chemokines in the human monocytic cell line THP-1. Biofactors 31 (3, 4), 211–217.

Shekelle, P., Hardy, M., Coulter, I., Udani, J., Spar, M., Oda, K., et al., 2003. Effect of the supplemental use of antioxidants vitamin C, vitamin E, and coenzyme Q10 for the prevention and treatment of cancer. Evid. Rep. Technol. Assess. (Summ) 75, 1–3.

Silvestri, S., Orlando, P., Armeni, T., Padella, L., Brugè, F., Seddaiu, G., et al., 2015. Coenzyme Q10 and α-lipoic acid: antioxidant and pro-oxidant effects in plasma and peripheral blood lymphocytes of supplemented subjects. J. Clin. Biochem. Nutr. 57 (1), 21–26.

Singh, R.B., Neki, N.S., Kartikey, K., Pella, D., Kumar, A., Niaz, M.A., et al., 2003. Effect of coenzyme Q10 on risk of atherosclerosis in patients with recent myocardial infarction. Mol. Cell. Biochem. 2012, 75–82.

Sohal, R.S., Forster, M.J., 2007. Coenzyme Q, oxidative stress and aging. Mitochondrion 7, S103–S111.

Stocker, R., Bowry, V.W., Frei, B., 1991. Ubiquinol-10 protects human low density lipoprotein more efficiently against lipid peroxidation than does alpha-tocopherol. Proc. Natl. Acad. Sci. 88 (5), 1646–1650.

Strazisar, M., Fir, M., Golc-Wondra, A., Milivojevic, L., Prosek, M., Abram, V., 2005. Quantitative determination of coenyzme Q10 by liquid chromatography and liquid chromatography/mass spectrometry in dairy products. J. AOAC Int. 88 (4), 1020–1027.

Takahashi, K., Ohsawa, I., Shirasawa, T., Takahashi, M., 2016. Early-onset motor impairment and increased accumulation of phosphorylated α-synuclein in the motor cortex of normal aging mice are ameliorated by coenzyme Q. Exp. Gerontol. 81, 65–75.

Thanh, H.N., Minh, H.P.T., Duc, L.V., Thanh, T.B., 2016. Protective effect of coenzyme Q10 on methamphetamine-induced neurotoxicity in the mouse brain. Trends Med. Res. 11 (1), 1–10.

Tomasetti, M., Littarru, G., Stocker, R., Alleva, R., 1999. Coenzyme Q10 enrichment decreases oxidative DNA damage in human lymphocytes. Free Radic. Biol. Med. 27 (9-10), 1027–1032.

Xu, G., Lu, H., Dong, Y., Shapoval, D., Soriano, S., Liu, X., et al., 2017. Coenzyme Q10 reduces sevoflurane-induced cognitive deficiency in young mice. Br. J. Anaesth. 119 (3), 481–491.

Ya, F., Xu, X.R., Shi, Y., Gallant, R.C., Song, F., Zuo, X., et al., 2019. Coenzyme Q10 upregulates platelet cAMP/PKA pathway and attenuates integrin αIIbβ3 signaling and thrombus growth. Mol. Nutr. Food Res. 63 (23), 1900662.

Yeung, C.K., Billings, F.T., Claessens, A.J., Roshanravan, B., Linke, L., Sundell, M.B., et al., 2015. Coenzyme Q 10 dose-escalation study in hemodialysis patients: safety, tolerability, and effect on oxidative stress. BMC Nephrol. 16 (1), 183–189.

Zarei, P., Rezvanfar, M.R., Ansarihadipour, H., Delavar, M., Abdollahi, M., Khosrowbeygi, A., 2018. Effects of coenzyme Q10 supplementation on the serum levels of amylase, adenosine deaminase, catalase, and total antioxidant capacity in women with type 2 diabetes mellitus: a randomized, double-blind placebo-controlled trial. J. Res. Med. Sci. 23, 9–14.

Žmitek, J., Šmidovnik, A., Fir, M., Prošek, M., Žmitek, K., Walczak, J., et al., 2008. Relative bioavailability of two forms of a novel water-soluble coenzyme Q10. Ann. Nutr. Metab. 52 (4), 281–287.

CHAPTER

Curcumin

4.7

Hari P. Devkota[a], Anjana Adhikari-Devkota[a], Dhaka R. Bhandari[b]

[a]*Graduate School of Pharmaceutical Sciences, Kumamoto University, Kumamoto, Japan*
[b]*Institute of Inorganic and Analytical Chemistry, Justus Liebig University Giessen, Giessen, Germany*

4.7.1 Introduction

Curcumin is the main curcuminoid polyphenol present in the rhizomes of turmeric (*Curcuma longa* L.) (Fig. 4.7.1), a plant of Zingiberaceae family. The content of curcuminoids in dried turmeric rhizomes is reported to be around 1% to 5% and is also responsible for the yellow colour of the rhizome (Li et al., 2019; Prasad et al., 2014; The Ministry of Health Labor and Welfare of Japan, 2016). Turmeric is used as spice and medicine all over the world which is reported to be originated from India, where it is also known as "Indian Saffron" or "Golden Spice." Turmeric is widely used in the traditional medicine systems in India for hundreds of years for the treatment of cold and cough, wounds, inflammation, and liver diseases (Prasad et al., 2014). Modern medicine has also found various health beneficial effects of turmeric, for which curcumin is reported to be the principal bioactive compound.

Curcumin, also known as diferuloylmethane is a member of the diarylheptanoid class of chemical compounds having two aryl (phenyl) rings attached to heptane (consisting seven carbon) derivative. Chemically, having bis-α, β-unsaturated β-diketone moiety, it exists as keto-enol tautomers depending upon pH. Two other common natural derivatives of curcumin are demethoxycurcumin (DMC) and bisdemethoxycurcumin (BDMC) (Fig. 4.7.2). The usual content of curcumin and its two derivatives in commercial curcuminoid mixture is curcumin (77%), demethoxycurcumin (17%), and bisdemethoxycurcumin (3%) (Anand et al., 2008). Curcumin was firstly reported by Vogel and Pelletier in 1815. Since then the structure confirmation and syntheses have explored various chemical and physiological properties of curcumin in 200 years from the first discovery (Anand et al., 2008). Curcumin and its derivatives have been widely studied for antioxidant, anti-inflammatory, anticancer, antimicrobial, hepatoprotective, neuroprotective, and many other biological activities (Anand et al., 2008; Basnet and Skalko-Basnet, 2011; Gupta et al., 2012; Lee et al., 2013; Li et al., 2019; Maheshwari et al., 2006; Ohori et al., 2006; Prasad et al., 2014; Singh and Khar, 2008; Weber et al., 2005). Curcumin is also one of the most widely studied

Antioxidants Effects in Health. DOI: https://doi.org/10.1016/B978-0-12-819096-8.00047-1

FIG. 4.7.1 Photographs of plant, flower, rhizomes, and rhizome powder of turmeric (Curuma longa L.).

natural products for its beneficial effects in human and animal health (Yeung et al., 2018). Scopus and PubMed search (on August 13, 2019) with the keyword "curcumin" were found to have 24,071 and 13,598 documents, respectively. A Scopus database search with the term "curcumin" for the past 20 years (from 2000–2018) showed constant growth in the number of publications each year (Fig. 4.7.3). Many of these publications are also focused on the *in vitro* or *in vivo* antioxidant activity of curcumin or its natural/synthetic derivatives, which are discussed in sections below along with the bioavailability issues.

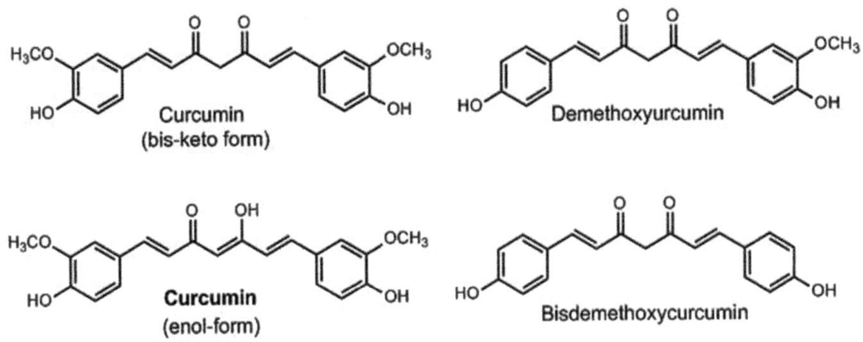

FIG. 4.7.2 Structures of curcumin, demethoxycurcumin, and bisdemethoxycurcumin.

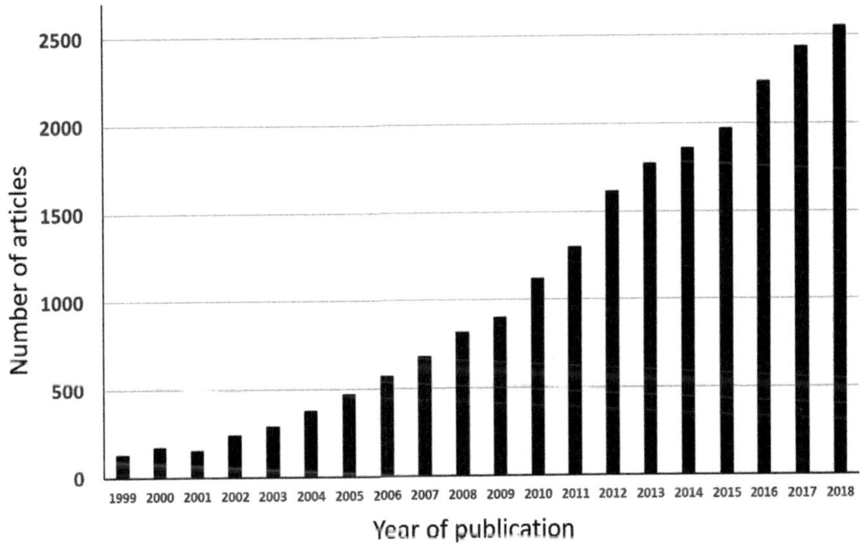

FIG. 4.7.3 Trends in scientific publications regarding curcumin from the year 2000 to 2018 (Retrieved on August 10, 2019).

4.7.2 Antioxidant activities of curcumin

4.7.2.1 *In-vitro* studies

Free radical scavenging activities of curcumin and its natural/synthetic derivatives in *in vitro* models are well studied. Somparn et al. studied the antioxidant activities of curcumin, DMC, BDMC, tetrahydrocurcumin, hexahydrocurcumin, and octahydrocurcumin by 1,1-diphenyl-2-picrylhydrazyl (DPPH), 2,2′-azobis(2-amidinopropane)dihydrochloride (AAPH) induced linoleic oxidation and AAPH induced red blood cell hemolysis assays (Somparn et al., 2007). All the hydroxylated derivatives of curcumin showed stronger antioxidant activity as compared to curcumin, DMC, and BDMC and the activity of DCM and BDCM were weaker as compared to curcumin. The study suggested the hydrogenation of the double bonds in the central heptanoid chain increased the antioxidant activity and the ortho-methoxy groups in the phenyl ring of curcumin were involved in the antioxidant activity. Kunchandy and Rao reported the potent hydroxyl radical and superoxide anion scavenging activities of curcumin, however, at lower concentration, it generated a higher amount of hydroxyl radicals (Kunchandy and Rao, 1990). Ak and Gülçin evaluated the free radical scavenging activity of curcumin using various assays such as DPPH, 2,2′-azino-bis(3-ethylbenzothiazoline-6-sulfonic acid) (ABTS) *N,N*-dimethyl-*p*-phenylenediamine dihydrochloride (DMPD) radical scavenging activities, superoxide anion and hydrogen peroxide scavenging activities, total antioxidant activity (TAA) determination by ferric thiocyanate, total reducing ability determination by the Fe^{3+}–Fe^{2+} transformation method, and ferrous ions (Fe^{2+}) chelating activities. Curcumin showed potent activities in all of these assay systems (Ak and Gülçin, 2008). Barzegar and Moosavi-Movahedi studied the free radical scavenging activity of curcumin in polar solvents by using electron spin resonance (ESR), reduction of ferric iron in aqueous medium and intracellular ROS/toxicity assays. Curcumin showed potent scavenging activity against intracellular hydrogen peroxide and hydroxyl and peroxy radicals (Barzegar and Moosavi-Movahedi, 2011). Fujisawa et al. studied the ROS generation and radical scavenging activities of curcumin and other related compounds where curcumin showed superior activity to eugenol and its derivatives. Curcumin showed potent scavenging activity against peroxy radicals and thus generated curcumin radical formed stable dimer, which resulted in the less ROS generation from the scavenging process (Fujisawa et al., 2004). Various synthetic derivatives have also been studied for their *in vitro* antioxidant activities. Li and Leu synthesized ferrocenylidene curcumin, an organometallic derivative of curcumin which showed potent scavenging activities against DPPH and ABTS radicals and DNA protective activity against Cu^{2+}/GSH induced oxidation and AAPH-induced oxidation (Li and Liu, 2011). Li et al. synthesized various mono-carbonyl analogues of curcumin and evaluated their Nrf2-activating anti-oxidative activity. Among the screened compounds, (1E,4E)-1-(3-allyl-4-methoxyphenyl)-5-(2-chlorophenyl)penta-1,4-dien-3-one, named as 14p, attenuated H_2O_2-induced oxidative stress, inhibited ROS-induced cell death and inhibited the tert-butyl hydroperoxide induced apoptosis by activating nuclear factor erythroid 2-related factor 2 (Nrf2) in H9C2 cells (Li et al., 2015). Ahsan et al. evaluated the activities of curcumin, DMC and BDMC for cleavage of plasmid DNA by Fe(II)-EDTA system (hydroxyl radicals) and the generation of singlet oxygen by riboflavin. Curcumin was found to be the most effective in the DNA cleavage reaction and a reducer of Cu(II) followed by DMC and BDMC (Ahsan et al., 1999).

The phenolic hydroxyl group of curcumin is the active chemical group in curcumin that is responsible for the antioxidant activity by donating a hydrogen atom. Jovanovic et al. studied the antioxidant mechanism of curcumin by laser flash photolysis and pulse radiolysis and revealed that the phenolic hydroxyl group acts as an H-atom donor (Jovanovic et al., 1999). This was further confirmed by Barclay et al. (Barclay et al., 2000) that synthesized curcumin derivatives without free phenolic hydroxyl groups showed no antioxidant activity. Curcumin and its derivatives also exert potent metal chelating activities that result in increased antioxidant activity (Ferrari et al., 2014; Priyadarsini, 2014). In a recent study, Lin et al. evaluated the antioxidant activity of curcumin in RAW264.7 cells, where oxidative stress was introduced by hydrogen peroxide. Curcumin treatment increased the activity of catalase, SOD, and glutathione peroxidase. Lower doses of curcumin decreased the ROS and malondialdehyde (MDA) levels, however, higher doses increased the ROS and MDA (Lin et al., 2019). Low and middle dose curcumin also activated the Nrf2-Keap1 signaling pathway.

4.7.2.2 *In-vivo* (animal) studies

The potent antioxidant activity of curcumin and its derivatives is reported in many preclinical studies. For example, Ataie et al. evaluated the neuroprotective activity of curcumin against homocysteine-induced cognitive impairment and oxidative stress in the adult male Wister rats. Intraperitoneal injection of curcumin (5 and 50 mg/kg) attenuated the homocysteine-induced increase in MDA and superoxide anion levels in rat hippocampi (Ataie et al., 2010). Sahoo et al. evaluated the antioxidative activity of curcumin in l-thyroxine-induced testicular oxidative stress in Wistar male rats. Curcumin increased the SOD level in postmitochondrial fraction (PMF) and a mitochondrial fraction (MF) and catalase in PMF, with decreased GPx activity in MF (Sahoo et al., 2008). Jagetia and Rajanikant evaluated the effects of curcumin (100 mg/kg body weight) in antioxidant status in Swiss albino mice skin exposed to a total dose of 10, 20 and 40 Gy γ-radiation. The irradiation of mouse skin resulted in dose-dependent decline in glutathione concentration, glutathione peroxidase, and superoxide dismutase activities and increased lipid peroxidation. Pretreatment of curcumin before irradiation reversed these oxidative stress parameters induced by radiation (Jagetia and Rajanikant, 2015).

4.7.2.3 Clinical studies

Few studies have also reported antioxidant activity of curcumin in humans. Durgaprasad et al. evaluated the antioxidative activity of curcumin in 20 topical pancreatitis patients. Patients were treated with curcumin (500 mg) with piperine (5 mg) or placebo for 6 weeks. Curcumin combined with piperine treatment resulted in decreased erythrocyte MDA levels and increased GSH levels as compared to placebo (Durgaprasad et al., 2005). Kalpravidh et al. evaluated the effects of curcuminoids treatment (12 months) in hematological profile, oxidative stress, and antioxidant parameters in β-thalassemia/Hb E patients. The ratio of curcumin, DMC and BDMC in curcuminoid mixtures was 1:0.3:0.1. Patients were treated with two capsules of 250 mg curcuminoid mixtures (500 mg daily dose). Curcuminoids treatment attenuated the increased oxidative stress in β-thalassemia/Hb E patients represented by higher levels of MDA, SOD, GSH-Px in RBC, serum NTBI, and lower level of RBC GSH. After 3 months of treatment, all these parameters were reversed to the baseline level (Kalpravidh et al., 2010).

4.7.3 Bioavailability of curcumin

Curcumin has very low solubility in water but it is readily soluble in dimethylsulfoxide (DMSO), acetone, and ethanol (Basnet and Skalko-Basnet, 2011). Being highly lipophilic and presenting low aqueous solubility makes curcumin a poorly bioavailable molecule (Anand et al., 2008). In Sprague-Dawley rats, Wahlström and Blennow reported that when 1g/kg curcumin was administered orally, 75% of it was excreted in feces with a negligible amount in the urine suggesting poor absorption after oral administration. Even after intravenous injection, curcumin was soon metabolized (Wahlström and Blennow, 1978). Holder et al. prepared deuterium and tritium labeled curcumin and administered by oral, intravenous, and intraperitoneal routes. Orally and intraperitoneally administered curcumin was mostly excreted in feces, and intravenous and intraperitoneal doses of labeled curcumin were well excreted in the bile of cannulated rats. The main metabolites in bile were glucuronides of tetrahydrocurcumin and hexahydrocurcumin and a minor amount of dihydroferulic acid with traces of ferulic acid. Similarly, when curcumin was added to isolated hepatocytes or liver microsomes, about 90% was metabolized within 30 min (Holder et al., 1978). In 1980, Ravindranath and Chandrasekhara reported that after oral administration of 400 mg curcumin to rats, about 60% of the dose was absorbed as about 38% was remained in lower parts of the large intestine. The absorbed curcumin was excreted in urine as conjugated glucuronides and sulfates (Ravindranath and Chandrasekhara, 1980). Yang et al. developed a validated liquid chromatography-tandem mass spectrometry (LC-MS/MS) for the quantification of curcuminoids in *Curcuma longa* extract and rat plasma. After oral administration of curcumin (500 mg/kg), the maximum concentration (C_{max}) was 0.06 ± 0.01 μg/mL and the time to reach maximum concentration (T_{max}) was 41.7 ± 5.4 min. The elimination half-life was found to be 28.1 ± 5.6 min for oral administration (500 mg/kg) and 44.5 ± 7.5 min for intravenous administration (10 mg/kg). The oral bioavailability was found to be only about 1% (Yang et al., 2007). These studies suggest that the oral bioavailability of curcumin is very low for therapeutic effects and even the systemic administration will result in fast metabolism. However, a review of clinical trials by Kunnumakkara et al. (2019) debated that low bioavailability of curcumin can be addressed by using either higher concentrations of curcumin and/or by using curcumin in combination with other compounds or as formulations that enhance bioavailability. Various studies have reported that showed the increase in the bioavailability of curcumin using novel drug formulations, such as nanoformulations, nanoemulsions, liposomes, micelles, microemulsions, etc. (Jiang et al., 2020; Lee et al., 2014; Mock et al., 2015; Sanidad et al., 2019; Yu and Huang, 2012).

4.7.4 The pro-oxidant activity of curcumin

Many antioxidant compounds have also been reported to possess prooxidant capacity depending upon various factors such as concentration, physiological conditions, among others (Carocho and Ferreira, 2013; Galati et al., 2002). Many dietary polyphenols including curcumin containing phenol rings were reported to be metabolized by the enzyme peroxidase to produce prooxidant phenoxy radicals (Galati et al., 2002). Various researchers have reported that curcumin at lower concentration (e.g., 10 μM) acts as an antioxidant to protect cells from oxidative stress. However, at higher concentration (e.g., 50 μM), it acts as a pro-oxidant thus generating free radicals having an important role in cellular growth and apoptosis (Sandur et al., 2007; Shuto et al., 2010). Curcumin is reported to exert its anticancer activity through a prooxidant mechanism. For example, Yoshino et al. reported that curcumin caused a copper-dependent DNA damage and the induction of apoptosis through the prooxidant mechanism. Curcumin along with copper ion treatment to DNA from plasmid pBR322 and calf thymus caused DNA strand scission and the formation of 8-hydroxy-2′-deoxyguanosine, which was protected by adding catalase in reaction mixture suggesting the role of hydroxyl radicals in DNA damage. Curcumin also caused apoptotic cell death of HL60 cells in a dose- and time-dependent manner which was related to an increase in intracellular ROS (Yoshino et al., 2004). Similarly, Banarjee et al. also reported the concentration-dependent antioxidant and pro-oxidant activity through the AAPH induced hemolysis of RBCs (Banerjee et al., 2008). Lin et al. also reported that the lower doses of curcumin decreased the ROS and malondialdehyde (MDA) levels; however, higher doses increased the ROS and MDA (Lin et al., 2019). However, such antioxidant/prooxidant balance should be considered in detail for therapeutic activities of curcumin.

4.7.5 Beneficial and detrimental effects of curcumin in human health

As mentioned in the above sections, curcumin exerts various health beneficial effects. However, the strong concern should also be taken regarding the possible health detrimental effects as there are not many reports on the high dose toxicity study of turmeric powder or curcumin. In Nepal, India and many South Asian countries, turmeric powder is used daily as a spice for making curry recipes and the daily intake is estimated 0.5–1.5 g (Eigner and Scholz, 1999). Similarly, in India, the daily intake of turmeric powder is estimated to be 2–2.5 g (Basnet and Skalko-Basnet, 2011). Considering the curcumin content to be 3%, the maximum intake of curcumin in Nepal and India for an average of 60 kg human would be around 0.75 mg/kg/day and 1.25 mg/kg/day, respectively. Curcumin has been granted an acceptable daily intake (ADI) of 0–3 mg/kg-BW by the Joint FAO/WHO Expert Committee on Food Additives (WHO, 2004). BCM-95®, a reconstituted, purified and standardized turmeric extract containing no less than 85% curcuminoids and 5–7% volatile oils was approved as Generally Recognized As Safe (GRAS) by the US FDA (https://www.fda.gov/media/130730/download). The European Parliament And The Council Of The European Union has approved curcumin (E100) as food additive (https://eur-lex.europa.eu/legal-content/EN/TXT/?uri=celex:32012R0231). In clinical trials, high daily doses of curcumin up to 8000 mg also showed no apparent adverse effects (Basnet and Skalko-Basnet, 2011).

Conclusions

Turmeric, the main source of curcumin, is widely used as spice for hundreds of years in many parts of the world. Various studies have supported the strong free radical scavenging and antioxidant activity of curcumin. However, the limitation related to the low oral bioavailability and rapid metabolism in the human body is the main concern. Newer curcumin derivatives have been developed with improved bioavailability and also improved drug delivery formulations, which in the future will explore new applications of curcumin as a potent therapeutic agent. However, the issues related to the interferences of curcumin in various biological activity assays and difficulties in obtaining good results in clinical studies as compared to preclinical or *in vitro* studies remain a big concern among researchers.

References

Ahsan, H., Parveen, N., Khan, N.U., Hadi, S.M., 1999. Pro-oxidant, anti-oxidant and cleavage activities on DNA of curcumin and its derivatives demethoxycurcumin and bisdemethoxycurcumin. Chem. Biol. Interact. 121, 161–175. https://doi.org/10.1016/S0009-2797(99)00096-4.

Ak, T., Gülçin, I., 2008. Antioxidant and radical scavenging properties of curcumin. Chem. Biol. Interact. 174, 27–37. https://doi.org/10.1016/j.cbi.2008.05.003.

Anand, P., Thomas, S.G., Kunnumakkara, A.B., Sundaram, C., Harikumar, K.B., Sung, B., Tharakan, S.T., Misra, K., Priyadarsini, I.K., Rajasekharan, K.N., Aggarwal, B.B., 2008. Biological activities of curcumin and its analogues (Congeners) made by man and Mother Nature. Biochem. Pharmacol. 76, 1590–1611. https://doi.org/10.1016/j.bcp.2008.08.008.

Ataie, A., Sabetkasaei, M., Haghparast, A., Moghaddam, A.H., Kazeminezhad, B., 2010. Neuroprotective effects of the polyphenolic antioxidant agent, Curcumin, against homocysteine-induced cognitive impairment and oxidative stress in the rat. Pharmacol. Biochem. Behav. 96, 378–385. https://doi.org/10.1016/j.pbb.2010.06.009.

Banerjee, A., Kunwar, A., Mishra, B., Priyadarsini, K.I., 2008. Concentration dependent antioxidant/pro-oxidant activity of curcumin. Studies from AAPH induced hemolysis of RBCs. Chem. Biol. Interact. 174, 134–139. https://doi.org/10.1016/j.cbi.2008.05.009.

Barclay, L.R.C., Vinqvist, M.R., Mukai, K., Goto, H., Hashimoto, Y., Tokunaga, A., Uno, H., 2000. On the antioxidant mechanism of curcumin: classical methods are needed to determine antioxidant mechanism and activity. Org. Lett. 2, 2841–2843. https://doi.org/10.1021/ol000173t.

Barzegar, A., Moosavi-Movahedi, A.A., 2011. Intracellular ROS protection efficiency and free radical-scavenging activity of curcumin. PLoS One 6 (10), e26012. https://doi.org/10.1371/journal.pone.0026012.

Basnet, P., Skalko-Basnet, N., 2011. Curcumin: an anti-inflammatory molecule from a curry spice on the path to cancer treatment. Molecules 16 (6), 4567–4598. https://doi.org/10.3390/molecules16064567.

Carocho, M., Ferreira, I.C.F.R., 2013. A review on antioxidants, prooxidants and related controversy: natural and synthetic compounds, screening and analysis methodologies and future perspectives. Food Chem. Toxicol. 51, 15–25. https://doi.org/10.1016/j.fct.2012.09.021.

Durgaprasad, S., Pai, C.G., Vasanthkumar, A.J.F., Namitha, S., 2005. A pilot study of the antioxidant effect of curcumin in tropical pancreatitis. Indian J. Med. Res. 122, 315–318.

Eigner, D., Scholz, D., 1999. Ferula asafoetida and Curcuma longa in traditional medical treatment and diet in Nepal. J. Ethnopharmacol. 67 (1), 1–6. https://doi.org/10.1016/S0378-8741(98)00234-7.

Ferrari, E., Benassi, R., Sacchi, S., Pignedoli, F., Asti, M., Saladini, M., 2014. Curcumin derivatives as metal-chelating agents with potential multifunctional activity for pharmaceutical applications. J. Inorg. Biochem. 139, 38–48. https://doi.org/10.1016/j.jinorgbio.2014.06.002.

Fujisawa, S., Atsumi, T., Ishihara, M., Kadoma, Y., 2004. Activity of curcumin and related compounds. Anticancer Res. 24, 563–570.

Galati, G., Sabzevari, O., Wilson, J.X., O'Brien, P.J., 2002. Prooxidant activity and cellular effects of the phenoxyl radicals of dietary flavonoids and other polyphenolics. Toxicology 177, 91–104. https://doi.org/10.1016/S0300-483X(02)00198-1.

Gupta, S.C., Patchva, S., Koh, W., Aggarwal, B.B., 2012. Discovery of curcumin, a component of golden spice, and its miraculous biological activities. Clin. Exp. Pharmacol. Physiol. 39 (3), 283–299. https://doi.org/10.1111/j.1440-1681.2011.05648.x.

Holder, G.M., Plummer, J.L., Ryan, A.J., 1978. The metabolism and excretion of curcumin (1,7-bis-(4-hydroxy-3-methoxyphenyl)-1,6-heptadiene-3,5-dione) in the rat. Xenobiotica 8, 761–768. https://doi.org/10.3109/00498257809069589.

Jagetia, G., Rajanikant, G., 2015. Curcumin stimulates the antioxidant mechanisms in mouse skin exposed to fractionated γ-Irradiation. Antioxidants 4, 25–41. https://doi.org/10.3390/antiox4010025.

Jiang, T., Liao, W., Charcosset, C., 2020. Recent advances in encapsulation of curcumin in nanoemulsions: a review of encapsulation technologies, bioaccessibility and applications. Food Res. Int. 132, 109035. https://doi.org/10.1016/j.foodres.2020.109035.

Jovanovic, S.V., Steenken, S., Boone, C.W., Simic, M.G., 1999. H-atom transfer is a preferred antioxidant mechanism of curcumin. J. Am. Chem. Soc. 121, 9677–9681. https://doi.org/10.1021/ja991446m.

Kalpravidh, R.W., Siritanaratkul, N., Insain, P., Charoensakdi, R., Panichkul, N., Hatairaktham, S., Srichairatanakool, S., Phisalaphong, C., Rachmilewitz, E., Fucharoen, S., 2010. Improvement in oxidative stress and antioxidant parameters in β-thalassemia/Hb E patients treated with curcuminoids. Clin. Biochem. 43, 424–429. https://doi.org/10.1016/j.clinbiochem.2009.10.057.

Kunchandy, E., Rao, M.N.A., 1990. Oxygen radical scavenging activity of curcumin. Int. J. Pharm. 58, 237–240. https://doi.org/10.1016/0378-5173(90)90201-E.

Kunnumakkara, A.B., Harsha, C., Banik, K., Vikkurthi, R., Sailo, B.L., Bordoloi, D., Gupta, S.C., Aggarwal, B.B., 2019. Is curcumin bioavailability a problem in humans: lessons from clinical trials. Expert Opin. Drug Metab. Toxicol. https://doi.org/10.1080/17425255.2019.1650914.

Lee, W.-H., Loo, C.-Y., Bebawy, M., Luk, F., Mason, R., Rohanizadeh, R., 2013. Curcumin and its derivatives: their application in neuropharmacology and neuroscience in the 21st century. Curr. Neuropharmacol. https://doi.org/10.2174/1570159x11311040002.

Lee, W.H., Loo, C.Y., Young, P.M., Traini, D., Mason, R.S., Rohanizadeh, R., 2014. Recent advances in curcumin nanoformulation for cancer therapy. Expert Opin. Drug Deliv. https://doi.org/10.1517/17425247.2014.916686.

Li, H., Sureda, A., Devkota, H.P., Pittalà, V., Barreca, D., Silva, A.S., Tewari, D., Xu, S., Nabavi, S.M., 2019. Curcumin, the golden spice in treating cardiovascular diseases. Biotechnol. Adv. https://doi.org/10.1016/j.biotechadv.2019.01.010.

Li, P.Z., Liu, Z.Q., 2011. Ferrocenyl-substituted curcumin: Can it influence antioxidant ability to protect DNA? Eur. J. Med. Chem. 46, 1821–1826. https://doi.org/10.1016/j.ejmech.2011.02.041.

Li, W., Wu, M., Tang, L., Pan, Y., Liu, Z., Zeng, C., Wang, J., Wei, T., Liang, G., 2015. Novel curcumin analogue 14p protects against myocardial ischemia reperfusion injury through Nrf2-activating anti-oxidative activity. Toxicol. Appl. Pharmacol. 282, 175–183. https://doi.org/10.1016/j.taap.2014.12.001.

Lin, X., Bai, D., Wei, Z., Zhang, Y., Huang, Y., Deng, H., Huang, X., 2019. Curcumin attenuates oxidative stress in RAW264.7 cells by increasing the activity of antioxidant enzymes and activating the Nrf2-Keap1 pathway. PLoS One 14. https://doi.org/10.1371/journal.pone.0216711.

Maheshwari, R.K., Singh, A.K., Gaddipati, J., Srimal, R.C., 2006. Multiple biological activities of curcumin: a short review? Life Sci. 78 (18), 2081–2087. https://doi.org/10.1016/j.lfs.2005.12.007.

Mock, C.D., Jordan, B.C., Selvam, C., 2015. Recent Advances of Curcumin and Its Analogues in Breast Cancer Prevention and Treatment. RSC Adv 5 (92), 75575–75588. https://doi.org/10.1039/c5ra14925h.

Ohori, H., Yamakoshi, H., Tomizawa, M., Shibuya, M., Kakudo, Y., Takahashi, A., Takahashi, S., Kato, S., Suzuki, T., Ishioka, C., Iwabuchi, Y., Shibata, H., 2006. Synthesis and biological analysis of new curcumin analogues bearing an enhanced potential for the medicinal treatment of cancer. Mol. Cancer Ther. 5 (10), 2563–2571. https://doi.org/10.1158/1535-7163.mct-06-0174.

Prasad, S., Gupta, S.C., Tyagi, A.K., Aggarwal, B.B., 2014. Curcumin, a component of golden spice: from bedside to bench and back. Biotechnol. Adv. 32 (6), 1053–1064 https://doi.org/10.1016/j.biotechadv.2014.04.004.

Priyadarsini, K.I., 2014. The chemistry of curcumin: From extraction to therapeutic agent. Molecules 19 (12), 20091–20112. https://doi.org/10.3390/molecules191220091.

Ravindranath, V., Chandrasekhara, N., 1980. Absorption and tissue distribution of curcumin in rats. Toxicology 16, 259–265. https://doi.org/10.1016/0300-483X(80)90122-5.

Sahoo, D.K., Roy, A., Chainy, G.B.N., 2008. Protective effects of vitamin E and curcumin on l-thyroxine-induced rat testicular oxidative stress. Chem. Biol. Interact. 176, 121–128. https://doi.org/10.1016/j.cbi.2008.07.009.

Sandur, S.K., Ichikawa, H., Pandey, M.K., Kunnumakkara, A.B., Sung, B., Sethi, G., Aggarwal, B.B., 2007. Role of pro-oxidants and antioxidants in the anti-inflammatory and apoptotic effects of curcumin (diferuloylmethane). Free Radic. Biol. Med. 43, 568–580. https://doi.org/10.1016/j.freeradbiomed.2007.05.009.

Sanidad, K.Z., Sukamtoh, E., Xiao, H., McClements, D.J., Zhang, G., 2019. Curcumin: Recent advances in the development of strategies to improve oral bioavailability. Annu. Rev. Food Sci. Technol. 10, 597–617. https://doi.org/10.1146/annurev-food-032818-121738.

Shuto, T., Ono, T., Ohira, Y., Shimasaki, S., Mizunoe, S., Watanabe, K., Suico, M.A., Koga, T., Sato, T., Morino, S., Sato, K., Kai, H., 2010. Curcumin decreases toll-like receptor-2 gene expression and function in human monocytes and neutrophils. Biochem. Biophys. Res. Commun. 398, 647–652. https://doi.org/10.1016/j.bbrc.2010.06.126.

Singh, S., Khar, A., 2008. Biological Effects of curcumin and its role in cancer chemoprevention and therapy. Anticancer. Agents Med. Chem. 6 (3), 259–270. https://doi.org/10.2174/187152006776930918.

Somparn, P., Phisalaphong, C., Nakornchai, S., Unchern, S., Morales, N.P., 2007. Comparative antioxidant activities of curcumin and its demethoxy and hydrogenated derivatives. Biol. Pharm. Bull. 30, 74–78. https://doi.org/10.1248/bpb.30.74.

The Ministry of Health Labor and Welfare of Japan, 2016. The Japanese Pharmacopoeia, 17th Edition. The Ministry of Health Labor and Welfare of Japan, Tokyo.

Wahlström, B., Blennow, G., 1978. A study on the fate of curcumin in the rat. Acta Pharmacol. Toxicol. (Copenh). 43, 86–92. https://doi.org/10.1111/j.1600-0773.1978.tb02240.x.

Weber, W.M., Hunsaker, L.A., Abcouwer, S.F., Deck, L.M., Vander Jagt, D.L., 2005. Antioxidant activities of curcumin and related enones. Bioorganic Med. Chem. 13 (11), 3811–3820. https://doi.org/10.1016/j.bmc.2005.03.035.

WHO, 2004. WHO Technical Report Series 922. Evaluation of certain food additives and contaminants. Sixty-first report of the Joint FAO/WHO Expert Committee on Food Additives. WHO.

Yang, K.Y., Lin, L.C., Tseng, T.Y., Wang, S.C., Tsai, T.H., 2007. Oral bioavailability of curcumin in rat and the herbal analysis from *Curcuma longa* by LC-MS/MS. J. Chromatogr. B Anal. Technol. Biomed. Life Sci. 853, 183–189. https://doi.org/10.1016/j.jchromb.2007.03.010.

Yeung, A.W.K., Aggarwal, B.B., Barreca, D., Battino, M., Belwal, T., Horbanczuk, O.K., ... Atanasov, A.G., 2018. Dietary natural products and their potential to influence health and disease including animal model studies. Anim. Sci. Pap. Rep. 36 (4), 345–358.

Yoshino, M., Haneda, M., Naruse, M., Htay, H.H., Tsubouchi, R., Qiao, S.L, Li, W.H., Murakami, K., Yokochi, T., 2004. Prooxidant activity of curcumin: Copper-dependent formation of 8-hydroxy-2′-deoxyguanosine in DNA and induction of apoptotic cell death. Toxicol. Vitr. 18, 783–789. https://doi.org/10.1016/j.tiv.2004.03.009.

Yu, H., Huang, Q., 2012. Improving the oral bioavailability of curcumin using novel organogel-based nanoemulsions. J. Agric. Food Chem. 60, 5373–5379. https://doi.org/10.1021/jf300609p.

CHAPTER

Flavonoids

4.8

Lingchao Miao[a], Haolin Zhang[a], Li Yang[a], Lei Chen[b], Yixi Xie[c], Jianbo Xiao[d]

[a]Institute of Chinese Medical Sciences, University of Macau, Macao, China

[b]College of Food Science and Technology, Guangdong Ocean University, Zhanjiang, China

[c]Key Laboratory for Green Organic Synthesis and Application of Hunan Province, Xiangtan University, Xiangtan, China

[d]Universidade de Vigo, Nutrition and Bromatology Group, Department of Analytical and Food Chemistry, Faculty of Sciences, Ourense, Spain

4.8.1 Introduction

Flavonoids, with a C_6–C_3–C_6 structure, are well known for wide distribution in nature and various beneficial bioactivities. Dietary flavonoids widely found in vegetables and fruits are the most important phytochemicals in diets. Flavonoids usually occur as their O/C-glycosidic form in plant food. The most common flavonoids are the glucosides, rutinosides, rhamnosides, and galactosides of flavones (apigenin/luteolin), flavonols (keampferol/quercetin/myricetin), flavanones (hesperetin/naringenin/eriodictyol), isoflavones (genistein/daidzein), and anthocyanidins (cyanidin/delphinidin/malvidin). Cereals (wheat/barley/oat/rye/spelt/buckwheat/maize/millet), citrus fruits (bergamot/grapefruits/kumquats/lemons/oranges/limes/pummel), legumes (peas/Fava beans/chick peas/Mung beans), and others (passion fruits/Cayenne pepper/dates/tomatoes/chayote/cucumber/carambola fruit/fenugreek seeds/cocoa seeds/bamboo shoots/beet root/green tea) are main dietary resources of flavonoids (Fig. 4.8.1).

4.8.2 Chemistry

Flavonoids are composed of two benzene rings (A- and B-rings) with phenolic hydroxyl groups linked by the central three carbon atoms (C6-C3-C6). The basic parent nucleus is 2-phenylchromone (Fig. 4.8.2). Natural flavonoids are usually biosynthesized from derivatives of acetic acids/phenylalanine via shikimic acid pathway (Wang et al., 2018). The structures of flavonoids are usually linked with hydroxyl, methoxyl, and glycosidic moieties. According to the oxidation degree of the central three carbon chain, the position of B-ring connection (2 - or 3-position) and whether the three carbon chain forms a ring, the main natural flavonoids can be roughly classified as several subclasses, such as flavones, flavonols, flavanones,

Antioxidants Effects in Health. DOI: https://doi.org/10.1016/B978-0-12-819096-8.00048-3

FIG. 4.8.1 Dietary resources of flavonoids.

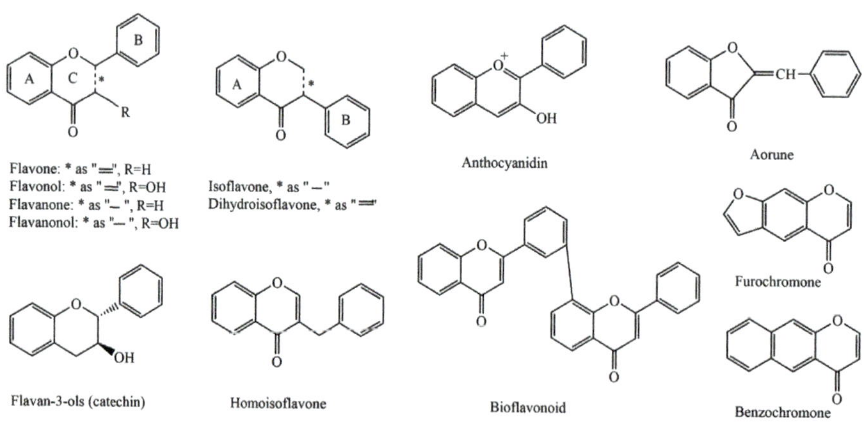

FIG. 4.8.2 Structural classifications of flavonoids.

flavanonols, anthocyanidins, flavan-3, flavan-3,4-diols, xanthones, chalcones, and so on (Xiao, 2017).

Some flavonoids do not conform to the basic skeleton of C6-C3-C6. Biflavonoids, as the name suggests, refer to the flavonoids consisting of two monomeric subunits. The linkage sites of the flavonoid subunits are either on the backbone or on the substituents of flavonoid units. Biflavonoids are widely distributed in plant derived foods and beverages. Homoisoflavone can be regarded as the isoflavone with one more CH_2 between the B ring and C ring. Furochromone has a furan ring at the C6-C7 position of chromone molecule. Benzochromone refers to chromone with a benzene ring the at the C6-C7 position.

4.8.3 Physical and chemical properties

Most of flavonoids are crystalline solids; but flavone glycosides are amorphous powder. Most flavonoids are yellow, which are linked to the existence of cross conjugation system in the molecule. Besides, the type, number and substitution position of the chromophores (OH, OCH_3, etc.) also have a certain influence on their color. Colored flavonoids are widely distributed in plant tissues such as flowers, fruits, leaves, and stems, which exhibit various colors ranging from red to blue and even purple (Yoshida et al., 2012). Flavonoid glycosides have optical activity, and most of them are left-handed. Only free flavonoids with chiral carbon atoms in the molecule have optical activity. Free flavonoids are generally soluble in methanol, ethanol, ethyl acetate, chloroform, ether, and other organic solvents and dilute lye, but insoluble or insoluble in water (Zhao et al., 2019). Since the presence of a sugar moiety usually increases the solubility of flavonoids in water, the flavonoid glycosides are generally soluble in high polar solvents such as water, methanol and ethanol, but insoluble in lipophilic organic solvents such as benzene, chloroform, and ether. Anthocyanins existed in the form of ions have the property of salts, so they are usually water soluble. Besides, it has been revealed that acylation of anthocyanins with aromatic acids tended to decrease the solubility of anthocyanins in cell cultures, which are hydrophilic environments (Plaza et al., 2014). Most flavonoids are acidic due to the presence of phenolic hydroxyl groups. The acidity is related to the number and position of phenolic hydroxyl groups (Musialik et al., 2009). The different acidity of individual flavonoid determines its separation method is alkali extraction or acid precipitation.

Flavonoid glycosides are mainly phenolic glycosides formed by dehydration of phenolic hydroxyl groups on the benzene ring and hemiacetal hydroxyl groups of sugars. Since the aromatic ring certainly donating electron on the glycoside atoms, so the hydrolysis of phenolic glycosides is easier than that of alcohol glycosides. Hence, flavonoid glycosides are easy to be hydrolyzed, even without acid. Flavonoid glycosides can also be hydrolyzed under the action of enzymes. During fermentation, microorganisms will produce enzymes to promote the hydrolysis of flavonoid glycosides. For example, in the process of soybean fermentation into soya-bean milk, fermentation microorganisms produce beta-glucosidase, which hydrolyzes daidzein and genistein to corresponding aglycones daidzein and genistein (Hur et al., 2014). Dietary flavonoid glycosides are hydrolyzed both in the oral cavity and intestine to produce more bioactive and absorbable aglycones, which is meaningful to the health benefits of flavonoids (Walle et al., 2005).

4.8.4 Bioavailability

4.8.4.1 Absorption and metabolism of flavonoids in gastrointestinal tract and liver

It has been clearly shown that the gastrointestinal tract plays a very important role in the absorption and metabolism of flavonoids before entering the circulatory system of human body. The jejunum and ileum cells in the small intestine mainly transport flavonoids in the form of glycosides from the lumen to the portal vein (flavonoids containing the B ring may also be transported in the form of *O*-methylation). In general, it is believed the fact that the solubility of flavonoids dependent on their molecular mass. Flavonoids with large molecular mass always showing low solubility and is difficult to pass through the small intestinal epithelial cells by passive diffusion. Therefore, flavonoid glycosides need to be metabolized into aglycones by enzymes located in the intestinal cells, intestinal lumen or intestinal flora before they can be absorbed. At present, the absorption mechanism of flavonoid glycosides mainly has the following possibilities: (1) active transport, that is, the sodium dependent glucose transporter (SGLT) located on the epithelial cell membrane of the small intestinal wall may mediate flavonoid. For example, the absorption of quercetin 4-*O*-β-glucoside and quercetin-3-glucoside is reported to be related to SGLT1 and (2) absorption after being hydroxylation into their aglycones. For example, the lactase-phloridin hydrolase (LPH) at the edge of the small intestine can participate in the hydrolysis of flavonoid glycosides in the intestinal flora, and can also hydrolyze flavonoid glycosides into aglycones.

4.8.4.2 Phase I metabolism of flavonoids

It is well known that phase I metabolism always linked to oxidation, reduction and hydrolysis reaction. Some flavonoid glycosides will be hydrolyzed into aglycones during the phase I metabolism. In the pharmacokinetic study of scutellarin, the average plasma concentration of scutellarin reached a peak at about 1.5 h, and the two peak mass concentrations were 26.57 ± 10.89 μg/L and 18.86 ± 9.70 μg/L, indicating that this bimodal phenomenon may be caused by the hepatoenteric circulation of scutellarin. In another word, flavonoids could be hydrolyzed into aglycone by the flora or relative metabolic enzymes in the intestine through phase I metabolism. After absorption, the remaining aglycone undergoes glucuronic acid combination to form scutellarin/isoscutellarin, excreted into the small intestine via bile and absorbed again.

4.8.4.3 Phase II metabolism of flavonoids

Phase II metabolism is known to affect the bioavailability of flavonoids in human. Usually, most flavonoids undergo sulfation, methylation, and glucuronidation in the small intestine and liver and conjugated metabolites can be found in plasma after flavonoid ingestion. For example, quercetin can be metabolized by phase II metabolic enzymes and inhibit its dehydrogenase activity. Similar to the binding of glucuronic acid, sulfation can also be inhibited by some flavonoids, such as lacustrine, galangin, quercetin, kaempferol, and genistein. In addition, the polarity of flavonoids increases after phase II metabolism in the body, which is conducive to their excretion. Some metabolites of flavonoids will continue to be beneficial to their excretion. As an attempt to increase the circulatory and organ levels of quercetin, several ester derivatives of quercetin were formulated to bypass the phase II metabolism during absorption. Quercetin ester derivatives were synthesized and transepithelial transport across monolayers of three cell lines (canine MDCK-1, -2 and human Caco-2) were tested (Biasutto et al., 2007). Certain esters crossed some monolayers (MDCK and some Caco-2 clones) without phase II metabolism whereas for certain Caco-2 clones, complete deacylation occurred followed by metabolism of quercetin. Since elimination of residual acyl groups are expected *in vivo*, this method may increase the systemic levels of quercetin.

4.8.4.4 ATP-dependent efflux transporter

There are various drug transporters in the human body to control the absorption, distribution, metabolism and excretion of flavonoids. Among them, ATP-dependent efflux transporter (ATP-binding cassette [ABC] transporter) P-glycoprotein (P-gp/MDR1/ABCB1) has attracted widespread attention. P-gp mainly mediates the normal secretion of drugs into the intestinal lumen, bile, urine and blood. Distributed on the top membrane side of the small intestine, the efflux function of P-gp is one of the main reasons for the low bioavailability of flavonoids. Lohner et al. (2007) used Bcap37/MDR1 cells transfected with the P-gp gene as a model to investigate the effects of P-gp on the transport of flavonoids such as quercetin, kaempferol and isorhamnetin in Ginkgo extract. The results showed that when the P-gp inhibitor verapamil was added, the absorption of the above three flavonoids increased significantly, indicating that the P-gp efflux pump is one of the reasons that limit the bioavailability of the flavonoids (Lohner et al., 2007). Meanwhile, numerous studies have pointed out that flavonoids can bind to P-gp and inhibit P-gp-mediated transport including genistein, biochanin A, phloretin, silymarin, chrvsin, eriodictyol, myricetin, taxifolin, naringenin, hubutin, EGCG, daidzein, catechin, genistein, chrysin, quercetin, and cyaniding. In terms of function regulation, flavonoids not only act as substrates or inhibitors of ABC transporters, but also regulate the expression of P-gp.

4.8.5 Stability

The chemical instability of flavonoids usually leads to poor practical applications. For instance, health-beneficial green tea beverage may occur browning through a multistep oxidation reaction of rich catechins during shelf life (Tan et al., 2020). Therefore, the stability of flavonoids plays a crucial role in the practical value of flavonoids in food and medicine fields. Generally, the stability of flavonoids depends on their naturally chemical structure as well as external factors. The external factors include the initial concentration of flavonoids in solvents, solvents, pH, temperature, light, oxygen, enzymes, metal ions, proteins (Xiao and Högger, 2015), polysaccharide (Liu et al., 2020) as well as other components such as nitrite salt, sulfur dioxide, ascorbic acid, etc. (Cao et al., 2021; Deng et al., 2018).

The structure-stability relationships of flavonoids are summarized as following points (Fig. 4.8.3):

(1) Hydroxylation of flavonoids always decreases the stability, with some exceptions. The degree and position of hydroxylation influence the stability of flavonoids. The hydroxyl groups on ring B dramatically reduced the stability of flavonoids and the stability of flavones and flavonols was found to be in order: resorcinol-type > catechol-type > pyrogallol-type (most instable) in Dulbecco's modified Eagle's medium (DMEM) at 37°C (Xiao and Högger, 2015). In phosphate buffered saline (PBS, pH = 7.4) at 4°C, the flavonols with catechol or pyrogallol groups were also highly instable to form stable dimers and oxidized products within 5 s (Cao et al., 2020). Besides, hydroxylation of C-3 on ring C dramatically decreased the stability of flavones and flavanones in DMEM as well as in PBS. Particularly, myricetin, and dihydromyricetin is extremely instable in DMEM (Xiao and Högger, 2015; Cao et al., 2020). However, the contribution of hydroxyl groups on ring A to the stability of flavonoids is not so certain in distinct solutions. Baicalein (5,6,7-OH on ring A) showed extremely instable in PBS compared to apigenin (5,7-OH on ring A, 4′-OH on ring B) (Cao et al., 2020). Separately, it was found that the hydroxylation on ring A of flavones decreased the stability of flavones in DMEM (Xiao and Högger, 2015).

(2) C or O-glycosylation dramatically improved the stability of flavonols in various solutions (Xiao, 2017). For example, rutin, a quercetin *O*-glycoside, exhibited remarkably higher stability than quercetin in alkaline PBS (containing Fe^{2+} or Cu^{2+}) (Makris and Rossiter, 2000) and in aqueous solution of different pH values at 100°C with air perfusion (Buchner et al., 2006), as well as in DMEM (Xiao and Högger, 2015). C-glycosylated flavonoids was thought to be more

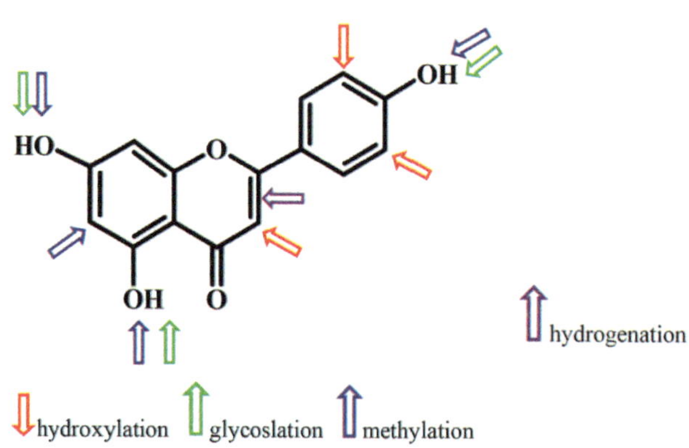

FIG. 4.8.3 The structure-stability relationships of flavonoids.

stable than aglycones or O-glycosylated flavonoids (Xiao, 2017; Rawat et al., 2009), but there were not many reports about that. Moreover, the size of sugar moiety has an important effect on the stability of the flavonoid glycosides and the increasing number of conjugated sugar molecules stabilizes the flavonoids. For example, the stability of anthocyanin and its glycosides is in order: cyanidin 3-*O*-β-rutinoside (two sugar molecules) > cyanidin 3-*O*-β-glucoside (one sugar molecules) > anthocyanin (Levy et al., 2019).

(3) Most methylation of flavonoids improves their stability. Methoxylation on ring B enhanced the stability of multi-hydroxy flavonols at 37°C in DMEM (Xiao and Högger, 2015). The methoxyflavones (5,7-dimethoxyflavone, 5,7,4′-trimethoxyflavone) also had apparently higher metabolic stability than original flavones (chrysin, apigenin) in human liver homogenate 9000 g supernatant (S9 fraction) (Walle, 2007). However, the methoxylation may cause a slight decrease in the stability of flavonols possessing less hydroxyl groups (Xiao and Högger, 2015).

(4) Hydrogenation of C_2=C_3 double bond slightly decreases the stability of flavones and flavonols. For example, the stability of dihydromyricetin and naringenin were slightly higher than myricetin and apigenin, respectively.

(5) The acylation could further enhance the stability of flavonoid glycosides (Ishihara and Nakajima, 2003).

Furthermore, a few researches about the degradation kinetics of flavonoids have found that flavonoids rapidly degraded and formed many degradation products *via* dimerization, further oxidation and ring-cleavage. Myricetin was extremely unstable in DMEM at 37°C which largely converted to dimers within 1 min and further formed their ring-cleavage product (Cao et al., 2020). Flavonols with a catechol-type or pyrogallol-type groups on ring B formed dimers and oxidized products in neutral PBS at 4°C within 5 s (Cao et al., 2020).

Finally, low temperature (Cao et al., 2020) and addition of ascorbic acids (Sun et al., 2021) could notably enhance the stability of flavonoids, as well as many other promising methods have been developed, such as cocrystallization, solid dispersion, nanotechnology, microemulsion, inclusion complexes, and chemical or enzymatic acylation (Liu et al., 2019).

4.8.6 Antioxidant activity of flavonoids

The capacity of flavonoids to scavenging free radicals mainly depends on the difficulty of hydrogen extraction of the hydroxyl groups on their structures (especially ring B and C). Theoretical and experimental studies on the relationship between the structure of flavonoids and antioxidant activity have been reported wildly. These reports provide some explanations for the structure-activity relationships of these compounds, such as the suggestion that the activity of the B ring is higher than that of the A ring, and that the ortho hydroxyl groups on the B ring are necessary for effective antioxidants, etc. However, there is not enough evidence on other influencing factors, such as

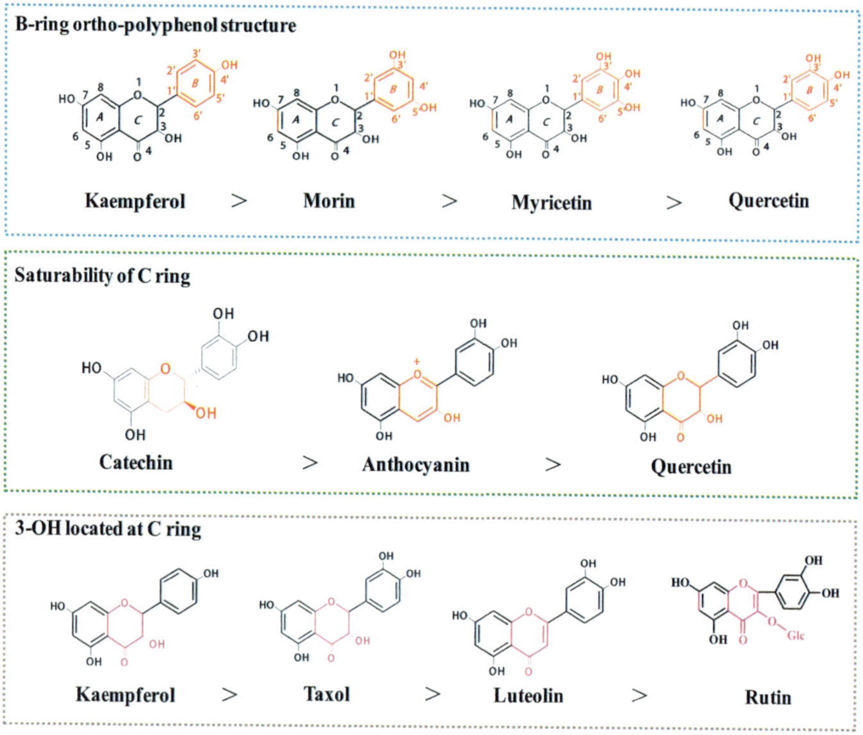

FIG. 4.8.4 Structure-antioxidant activity relationship of some flavonoids.

molecular planarity, the relationship between C ring and A ring and B ring, and the contribution of 3 hydroxylation to the antioxidant activity. Followed, the structure-activity relationships of antioxidant activities of flavonoids are discussed (Fig. 4.8.4).

4.8.6.1 Ortho-polyphenol structure and double bond between C3 and C4

The most important factors affecting the antioxidant activity of flavonoids are the degree of hydroxylation and the position of hydroxyl. Among them, the *O*-dihydroxy group in B ring plays an important role in the antioxidant activity of flavonoids. Rice-Evans and other studies found that the B ring linked with ortho phenolic hydroxyl group, which showed higher activity than that of morin with meta phenolic hydroxyl group in B ring. The reason might be due the fact that the ortho-phenolic hydroxyl flavonoids could react with free radicals to form a more stable conjugated semiquinone free radical. This interrupts the chain reaction of free radicals, so it has strong antioxidant properties. In addition, the antioxidant activity of quercetin is much greater than that of morin may also be due to the fact that the quercetin B cyclosemiquinone free radical and the ortho position. Meanwhile, researchers experimentally proved that the order of antioxidant activity of some flavonoids: myricetin > quercetin > kaempferol. These three substances have the same A and C ring structure only differs on the B ring, myricetin has 3 phenolic hydroxyl groups at C-3′, 4′, and 5′ positions, quercetin has 2 phenolic hydroxyl groups at C-3′ and 4′ positions, and kaempferol has only one hydroxyl group at 4′ position. These results indicated that the B-ring ortho polyphenol structure affects the antioxidant activity of flavonoids. In addition, the double bond between C-3 and C-4 and the carbonyl group at C-4 could also affect the antioxidant activity of flavonoids. For example, the C ring of anthocyanin does not contain a carbonyl group, but contains an unsaturated double bond between the C-3 and C-4 positions, which also makes it shows an antioxidant activity similar to that of quercetin (Fig. 4.8.4).

4.8.6.2 Hydroxylation

The 3-position hydroxyl of the C ring of flavonoids is a very important structure to determine their antioxidant activities. Compared with the corresponding aglycon, the glycosylation of the 3-position hydroxyl of the flavonoids may reduce its antioxidant activity. For example, rutin converted from quercetin after the glycosylation of the 3-position hydroxyl (still retaining the 3′ and 4′ ortho hydroxyl structure of the B ring), the TEAC value is reduced by almost 50%. To luteolin (compared with quercetin), the TEAC value is reduced by 55%. Moreover, the double bond located to the 3-position hydroxyl group or the 3-position hydroxyl group without the double bond will largely lose its antioxidant activity. This may be due to the hydroxyl group at the 3 position and the double bond between C2 and C3 could be isomerized and form a diketo group and further to affect their antioxidant activity.

4.8.6.3 Glycosylation

Glycosylation generally performs in the metabolism of flavonoids, and flavonoid glycosides have been shown to possess more-hydrophilic properties than aglycones. Generally, flavonoid glycosides could be metabolized to aglycones by human intestinal microflora, producing α-rhamnosidase, exo-β-glucosidase, endo-β-glucosidase, ord β-glucuronidase. Accordingly, rutin, hesperidin, naringin, and poncirin were transformed to their respective aglycones with arhamnosidase and β-glucosidase produced by intestinal bacteria. The formation of flavonoid glycosides inevitably involves the substitution of phenolic hydroxyl groups and the formation of hydroxyl glycosides, which will have an effect on the antioxidant activity. According to the reported data, it is found that the substitution position of phenolic hydroxyl groups and the formation of hydroxyl glycosides have no adverse effect on the antioxidant activity of flavonoids after glycosidation. For example, Chung et al. (2006) found that the antioxidant activity of genistein is similar to genistein after glycosidation (geniposide). The free radical scavenging activity of geniposide was slightly lower than that of genistein at low concentration, but the free radical scavenging activity of geniposide was significantly improved at high concentration due to its low water solubility. Kosina et al. (2002) studied the antioxidant activity and free radical scavenging effect of four silymarin glycosylated products. The results revealed that these silymarin glycosylated products had better cytoprotective effect against oxidative stress than silymarin in rat erythrocytes. In addition to antioxidant activity, the bioavailability of flavonoid glycosides was also improved. Nevertheless, Rüfer and co-workers through a randomized, double-blind, crossover study found that daidzein had better bioavailability than daidzein through oral administration. In addition, it was found that flavonoid glycosides had stronger binding effect on DNA than their aglycones (Rüfer et al., 2008).

4.8.7 Mechanisms of bioactivities in cell levels

4.8.7.1 Anti-inflammation activity

Studies have found that flavonoids have good anti-inflammatory activities, and its anti-inflammatory mechanisms mainly involve cytokines and the receptors, arachidonic acid metabolic pathways, and other pathways.

Flavonoids could exert anti-inflammatory effects by inhibiting the production of some inflammatory cytokines. In the study of flavonoids, it was found that quercetin, naringenin, kaempferol, and glycyrrhizin have a significant inhibitory effect on inflammatory cytokines, therapy exerting an anti-inflammatory effect (Yasui et al., 2015; Wang et al., 2015). In the study of Yilma et al. (2013), it was found that naringenin could significantly inhibit the increase of IL-10 in macrophages caused by Chlamydia trachomatis infection. Similarly, Chung et al. (2015) found that kaempferol and its glycosides could inhibit the Th2 cytokines (such as IL-, IL-5, and IL-13).

Meanwhile, the anti-inflammatory effect of flavonoids is also related to the metabolic pathway of arachidonic acid. It is reported that the arachidonic acid pathway includes cyclooxygenase pathway, lipoxygenase pathway, prostaglandin pathway, thromboxane A2 pathway, phospholipase A2 pathway, and leukotriene pathway (Patel et al., 2008). Ispquercitrin and alflavin could inhibit the expression of COX1, COX2, and COX5, thus play an anti-inflammatory effect (Yasui et al., 2015). Li et al. (2012) found that baicalein could inhibit the biosynthesis of LTB4 in inflammatory exudates by inhibiting 5-LOX activity. Kim et al. (2015) found that kaempferol could significantly inhibit the production of PGE2 in macrophages RAW 264.7 cells induced by lipopolysaccharide. In addition, flavonoids could not only inhibit the production of TXA2 by inhibiting AA and COX, but they could also act as TXA2 receptor antagonists to exert anti-inflammatory effects (Guglielmone et al., 2005).

Except these pathways mentioned above, flavonoids could also exert anti-inflammatory effects by inhibiting the production of NO, reducing the activation of NF-κB signaling pathway, as well as inhibiting the expression of ICAM-1.

4.8.7.2 Antioxidant activity

When the oxidative balance in the body is disrupted, it will cause oxidative stress, causing structural and functional damage such as lipids and proteins, leading to the occurrence of many diseases (Alvarez-Suarez et al., 2013). It's reported that flavonoids are a natural anti-oxidant, which contain phenolic hydroxyl groups and could capture oxygen free radicals (Agati et al., 2012). The ability to scavenge free radicals is related to its chemical structure and the more phenolic hydroxyl groups of flavonoids, the stronger antioxidant activity (Li, 2016). The phenolic hydroxyl in the structure of flavonoids could clear free radicals by decarboxylation, also could reduce mitochondrial damage through a redox-dependent mechanism, and it play a key role in the process of protecting mitochondria from oxidative stress (Oliveira, 2015). In addition, the antioxidative stress effects of flavonoids also related to decrease the ROS generation by inhibiting the enzymes, including glutathione, S-transferase, NADH oxidase, mitochondrial succinoxidase, cyclooxygenase, microsomal monooxygenase, and lipoxygenase (Pietta, 2000). Meanwhile, flavonoids could improve the antioxidant capacity through the activities of SOD, GSH-Px and CAT in the body (Zhang et al., 2014). In summary, the antioxidant effect of flavonoids is mainly related to scavenging free radicals, reducing ROS levels, and exerting antioxidant effects by regulating signal pathways and some metabolic pathways. And its structure is the key to its antioxidant effect.

4.8.7.3 Anticancer activity

In the process of research, it was found that flavonoids have good anti-tumor activity. The mechanism mainly involves directly killing tumor cells and preventing their division and reproduction. For example, apigenin could make the HL-60 cell cycle in the G2/M phase, thereby inhibiting tumors (Lepley et al., 1996). In addition, flavonoids could reduce or eliminate the carcinogenic toxicity of some chemical carcinogens. That is, flavonoids have a certain antagonistic effect on some carcinogens and mutagenic agents. For example, studies have found that quercetin could inhibit the proliferation of human lung cancer cells by inhibiting the activity of extracellular signal-regulated kinases (Nguyen et al., 2004). Meanwhile, the antitumor effect of flavonoids is also related to regulating or enhancing the activity of other substances to indirectly kill tumor cells. For example, the total flavonoids of *Pueraria lobata* could enhance the activity of NK cells, SOD, and CYP450 enzymes in the body, which could kill the cancer cells. Also, the flavonoids could promote the secretion of Ca^{2+} in the endoplasmic reticulum and the infiltration of Ca^{2+} in the extracellular space, so that intracellular Ca^{2+} are significantly increased and calcium-dependent apoptotic channels are activated to cause breast cancer cells to apoptosis (Lee et al., 2008).

4.8.7.4 Antidiabetic activity

With the improvement of people's living standards, people have gradually formed an unhealthy eating habit. According to statistics, it is reported that there are currently more than 422 million diabetic patients in the world, which has become one of the diseases that seriously endanger human health (Xu et al., 2018). Studies have shown that many flavonoids play an important role in the prevention and treatment of diabetes.

The key to the treatment of diabetes by flavonoids is to regulate some glucose metabolism-related enzymes, such as aldose reductase, sodium-glucose co-transporter and α-glucosidase (Nicolle et al., 2011). By inhibiting these enzymes, the body's intake of glucose is reduced, while the excretion of glucose is increased, thereby reducing blood sugar levels. Schwanck and Blaschek found that the flavonoids in kale have a strong inhibitory effect on sodium-glucose co-transporter (Schwanck and Blaschek, 2010). In addition, studies have also found that oxidative stress is closely related to the occurrence and development of diabetes and its chronic complications, and has become a primary and independent damage factor of diabetes. Anti-oxidant therapy could reduce oxidative stress in diabetic patients and help improve glucose and lipid metabolism and prevent the occurrence and development of chronic complications of diabetes (Nicolle et al., 2011). Zhao et al. (2007) found that *sweet potato leaf* flavonoids could significantly reduce the level of malondialdehyde in diabetic rats and increase the activity of SOD.

4.8.8 Pharmacology in animal studies

4.8.8.1 Anticancer activity

An *in-vivo* experimental investigation adopting skin cancers model on mice was performed by George et al. (2011). In the study, it was found that the synergistic effect of black tea polyphenols (BTP) and resveratrol on inhibiting two-stage mouse skin carcinoma through significantly regressing the number and volume of tumor and decreasing incidence of tumor, was owing to down-regulation on expression of phosphorylated mitogen-activated protein kinase family proteins, extracellular signal regulated kinase 1/2 c-Jun N-terminal kinase 1/2, P38. Apart from this, this combinatorial treatment of natural products also could help to upregulate the total P53 and p-P53. This study provided some promising results on suppressing mouse skin tumor growth by the natural polyphenols. Furthermore, as one kind of the most effective natural products in anti-cancer drug discovery, taxanes, are of great significance to show the anticancer bioactivity based on years of research worldwide, still serve as the first lines medicine for breast, ovary, lung and other metastatic cancers. And synergistic effects of the green tea extract epigallocatechin-3-gallate (EGCG) and taxane in extinguishing the malignant human prostate carcinomas *in-vivo* using the severe combined immunodeficient mice has was also illustrated by Stearns and Wang (2011). And the potential mechanism of the combination of EGCG and taxane anticancer bioactivity was due to the increase in expression of apoptotic genes like P53, P21, P73, and caspase-3. In addition, by 7, 12-dimethylbenz (a) anthracene (DMBA) to induce mammary tumorigenesis in Wistar rats, Roy et al. (2011) investigated the antitumorigenic activity of tea polyphenols, owing to the scavenging of reactive oxygen species through suppressing the expression of cyclooxygenase-2 (Cox-2), inactivating phosphorylated forms of NF-κB and Akt, and downregulating mutant type P53 by tea polyphenols of which the ability of tumor suppression was lost. Also, in a rat bladder carcinogenesis model, the inhibitory effect of intravesical fisetin against bladder cancer by activation of P53 and downregulation of NF-KB pathway has been documented (Li et al., 2014). Being efficacious and safe product, intravesical fisetin was suggested by author as a novel and promising therapeutic approach for bladder cancer.

In a word, through plenty of animal experiments, the natural occurring antioxidants exhibit relatively decent anticancer activity.

4.8.8.2 Antidiabetic activity

Classified as anthocyanins, tea catechins, procyanidins, flavonoids, nonflavonoid phenolic and polyphenols-rich extract, natural polyphenols have shown great antidiabetic bioactivity according to the current animal studies. For instance, the famous traditional medicinal plants like purslane, mulberry leaves, fenugreek seeds, and so on, which are abundant of polyphenols and other effective constituents, have shown excellent antidiabetic effects on animal or clinical investigation. C57BL/Ksj-db/db mice were taken by Lee et al. (2020) to illustrate that the purslane extract could reduce hyperglycemia via PI3K/Akt and AMPK pathways in the skeletal muscles of mice. In addition, the purslane extract could activate the phosphorylation of TBC1D1 Ser 231 and ACC Ser 79, as well as upregulating plasma membrane-glucose transporter type 4 (GLUT4) expression. Apart from this, polysaccharides purified from purslane has also be reported to exhibit excellent anti-oxidant and anti-diabetic pharmacological results using diabetic Sprague Dawley rats model induced by STZ injection (Bai et al., 2016). Glucose tolerance, FINS, and ISI levels could be significantly improved while FBG level decreased obviously. What's more, MDA, SOD, and the proinflammatory cytokines levels of IL-6 and TNF α were reduced distinctly, which revealed that the antidiabetic activity might be associated with the intrinsic antioxidant and anti-inflammatory effects. For fenugreek seeds, there are even many clinical human studies showing its anti-diabetic activity.

Very early in 1986, defat fenugreek seeds, fenugreek gum, fenugreek seeds, and ripe fenugreek were discovered that the blood sugar level of diabetics could be effectively lowered and the insulin response could be improved by them (Sharma, 1986). Consisting of a large amount of diosgenin, soluble dietary fiber (galactomannan), 4-hydroxyisoleucine, flavone C-glycosides, trigonelline and other constituents, fenugreek seeds exhibited hypoglycemic activity on animal studies. Pancreatic islet β-cells could be protected by diosgenin, as well as downregulating hepatic glucose heteroplasia, upregulating hepatic glucose kinase, and the antioxidase activity also could be enhanced by diosgenin (Fuller and Stephen, 2015). Named galactomannan, the soluble dietary fiber could reduce the blood sugar level in diabetic adult male Wistar rats which were induced by nicotinamide and STZ solution through inhibiting digestive enzymes existing in the small intestine to make the gastric emptying of carbohydrates delayed (Hamden et al., 2010). To flavone C-glycosides, digestive enzymes and the formation of advanced glycation end products could be inhibited, activating insulin signaling (Xiao et al., 2016). What's more, ω-3 fatty acids abundant in the essential oils from fenugreek seeds, could inhibit maltase activity and pancreas α-amylase in diabetic rats to protect pancreatic islet β-cells (Hamden et al., 2011). And there are so many animals research on the antidiabetic activity of natural plants where we will not go into details here.

In conclusion, the anti-diabetic bioactivity of the occurring antioxidants is of great significance and deserved to be investigated for the potential of anti-diabetic bioactivity.

4.8.8.3 Anti-inflammation activity

Among so many differential kinds of natural occurring antioxidants, flavonoids, with the chemical entities of benzo-pyrone derivatives, are widely distributed in the natural plant Kingdom. They are mainly classified as flavan-3-ols, chalcones, flavones, flavanones, isoflavones, flavonols, and bioflavonoids (Fig. 4.8.1). With more than 4000 derivatives have been reported from nature which indicates their chemical diversities. Of their bioactivities, the anti-inflammatory activity of flavonoids has long been utilized in the cosmetic industry and Chinese medicine as a form of crude plant extracts. It has been proven that varieties of flavonoid molecules possess anti-inflammatory activity on various inflammatory animal models. Some flavonoids were specifically found to suppress chronic inflammation of several experimental animal models. Thus, it is worthy to constantly investigate the anti-inflammatory activity of flavonoids, not only for figuring out the anti-inflammatory mechanisms, but also for exploiting a new kind of anti-inflammatory medicine.

Using arachidonic acid or croton oil as an inflammagen to build the mouse ear edema model, Lee et al. (1993) fund out that flavone and flavanols glycosides possess significant anti-inflammatory activities (15%–19% inhibition), while the flavonoid derivatives detected exhibited weaker anti-inflammatory effect than indomethacin or hydrocortisone. And glycosides of kaempferol-type possessed a higher oral anti-inflammatory activity than that of quercetin-type flavanols glycosides in mice. Apart from this, apigenin, belong to flavone, has been found to downregulate the response of Th1 and Th17 cells to major lupus autoantigen, and subsequently suppress the ability of lupus B cells to generate pathogenic autoantibodies that limit the inflammatory state in SFN-1 mice. In apigenin therapy of SNF1 mice with established lupus mice model, the serum levels of pathogenic autoantibodies to nuclear antigens could be inhibited almost up to 97% and the development of severe glomerulonephritis could be markedly delayed. Also, the downregulation of COX-2 expression in lupus T cells, B cells, and antigen-presenting cells (APCs) was involved, causing their apoptosis (Kang et al., 2009). Moreover, intake of apigenin also could modulate immune reaction triggered by TNF-α in rheumatoid arthritis mouse model (Yoshio et al., 2006). Besides, luteolin, quercetin, kaempferol and the catechin compounds including epigallocatechin-3-gallate (EGCG), epigallocatechin (EGC), epicatechin-3-gallate (ECG), and epicatechin (EC) all show the excellent anti-inflammatory bioactivity according to numerous animal research.

4.8.8.4 Antiobesity activity

Early research has reported that oolong tea which belongs to the natural anti-oxidant possesses the anti-obesity bioactivity. L-K Han et al. (1999) took high fat diet mice to test the capacity of oolong tea in preventing obesity. They found that the obesity and fatty liver induced by a high-fat diet could be suppressed by oolong tea administration via enhancing noradrenaline-induced lipolysis with the active substance identified as caffeine. Caffeine enhanced noradrenaline-induced lipolysis in fat cells, accelerating the hormone-induced lipolysis in a cell-free system consisting of lipid droplets and HSL, but not with sonicated lipid droplets and HSL. Also, the pancreatic lipase activity could be inhibited by oolong tea extract. Apart from this, green tea of which the bioactive compounds containing catechins, and epigallocatechin gallate (EGCG) has been demonstrated to be anti-obesity using animal obesity models via reducing adipocyte proliferation and differentiation, fat mass lipogenesis, fat absorption, body weight, plasma levels of triglycerides, free fatty acids, cholesterol, glucose, as well as to upregulate beta-oxidation and thermogenesis. Green tea put liver, adipose tissue, intestine and skeletal muscle all as target organs, playing its role of antiobesity (Wolfram et al., 2006). Apart from tea, there are many other anti-oxidant products existing in the nature showing the anti-obesity bioactivity based on animal studies.

4.8.9 Clinical studies

Flavonoids have been considered as prospective anticancer agents due to their potential to arrest cell cycle, downregulate mutant p53 protein, and inhibit Ras protein expression. The monomers, extracts and derivatives of flavonoids (Fig. 4.8.5)

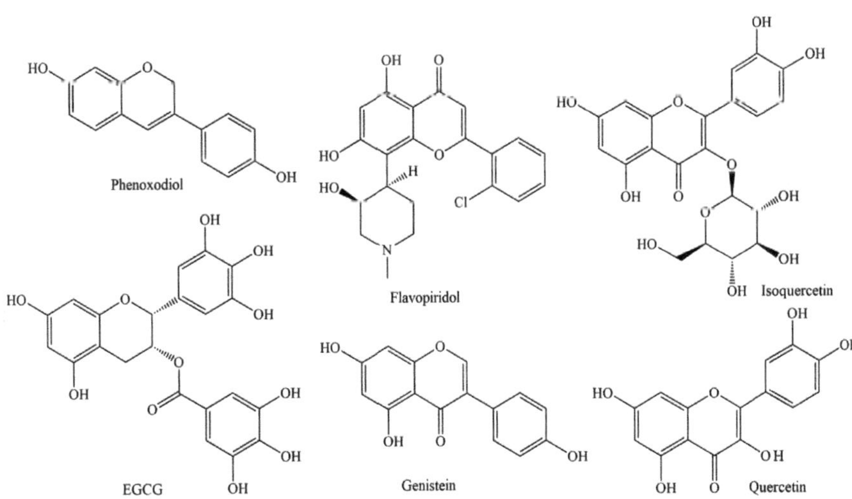

FIG. 4.8.5 Flavonoids have completed phase II or III clinical studies.

were applied to phase II or phase III clinical trial studies (Bisol et al., 2020; Kay et al., 2012).

Flavopiridol is a flavonoid alkaloid for the treatment of acute myeloid leukemia and the first cyclin-dependent kinase inhibitor in clinical trials (Deep et al., 2018). Phenoxodiol is a synthetic isoflavone derivative with synergistic interaction with cisplatin and carboplatin in phase II studies showing good tolerance for the cancer treatment (Fotopoulou et al., 2014). Polyphenon E is a standardized extract from *Camellia sinensis* (L.) mainly consisting of (-)-EGCG (65%), (-)-EGC (4%), (-)-epicatechin (9%), (-)-epicatechin-3-gallate (6%) (Shin et al., 2020). As it is the first complex botanical extract approved by the FDA, polyphenon E was approved for genital warts in 2006 is a milestone for herb medicines. The green tea extract rich in EGCG increased the insulin resistance and glucagon-like peptide 1 in patients with type 2 diabetes and lipid abnormalities (Liu et al., 2014). EGCG has been tested for the treatment of multiple system atrophy, basal cell carcinoma, multiple sclerosis, traumatic brain injury in phase III clinical studies.

A combination of quercetin, rutin, bromelain, and L-carnosine can be useful in reducing uricemia in patients with borderline uricemia in phase IV clinical study (Derosa, D'Angelo and Maffioli, 2020). The combination of luteolin and quercetin was found to effectively reduce the symptoms in children with autism spectrum disorders (Taliou et al., 2013). Seventeen clinical trials from a period of 20 years in six countries showed that quercetin showed hypotensive, anti-obesity, and hypoglycemic activities in humans (Huang et al., 2020). Recent clinical studies showed that quercetin was safe up to 2000 mg/day (Han et al., 2020).

A multicenter phase II trial found that isoquercetin improves hypercoagulability in cancer patients at high risk for thrombosis by targeting protein disulfide isomerase (Zwicker et al., 2019). A recent randomized phase 2 pilot trial reported that sodium nitrite and isoquercetin together did not effectively ameliorate endothelial dysfunction on the patients with CKD (Chen et al., 2020).

Genistein displayed a possible bimodal effect on bladder cancer (Messing et al., 2012). Genistein can treat osteoporosis in a clinical study. Genistein (24 months) showed positive effects on bone mineral density in osteopenic postmenopausal women (Marini et al., 2007). Genistein (6-month treatmen) significantly increases brachial artery flow-mediated vasodilation in postmenopausal women with metabolic syndrome (Irace et al., 2013).

4.8.10 Future perspectives

Natural flavonoids with different structure have been widely used as antioxidants. However, their stability and oxidized products are rarely studied. Although natural flavonoids displayed various bioactivities in cell and animal levels; however, their clinical outcomes are still limited. It is really important to carry out clinical trial and human studies for flavonoids benefits.

References

Agati, G., Azzarello, E., Pollastri, S., et al., 2012. Flavonoids as antioxidants in plants: location and functional significance. Plant Sci. 196, 67–76.

Alvarez-Suarez, J.M., Giampieri, F., Battino, M., 2013. Honey as a source of dietary antioxidants: structures, bioavailability and evidence of protective effects against human chronic diseases. Curr. Med. Chem. 20 (5), 621–638.

Bai, Y., et al., 2016. Anti-diabetic effect of *Portulaca oleracea* L. polysaccharide and its mechanism in diabetic rats. Int. J. Mol. Sci. 17 (8), 1201.

Biasutto, L., Marotta, E., De Marchi, U., Zoratti, M., Paradisi, C., 2007. Ester-based precursors to increase the bioavailability of quercetin. J. Med. Chem. 50 (2), 241–253.

Bisol, Â., de Campos, P.S., Lamers, M.L., 2020. Flavonoids as anticancer therapies: a systematic review of clinical trials. Phytother. Res. 34 (3), 568–582.

Buchner, N., Krumbein, A., Rohn, S., Kroh, L.W., 2006. Effect of thermal processing on the flavonols rutin and quercetin. Rapid Commun. Mass Spectrom. 20 (21), 3229–3235.

Cao, H., Högger, P., Arroo, R., Xiao, J., 2020. Flavonols with a catechol or pyrogallol substitution pattern on ring B readily form stable dimers in phosphate buffered saline at four degrees celsius. Food Chem. 311, 125902.

Cao, H., Saroglu, O., Karadag, A., Diaconeasa, Z., Zoccatelli, G., Conte-Junior, C.A., Gonzalez-Aguilar, G.A., Ou, J., Bai, W., Zamarioli, C.M., de Freitas, L.A.P., Shpigelman, A., Campelo, P.H., Capanoglu, E., Hii, C.L., Jafari, S.M., Qi, Y., Liao, P., Wang, M., Xiao, J., 2021. Available technologies on improving the stability of polyphenols in food processing. Food Front 2, 109-139.

Cao, H., Yi, L., Zhong, J., Högger, P., Wang, M., Prieto, M., Simal-Gandara, J., Xiao, J., 2020. Investigation of new products and reaction kinetics for myricetin in DMEM via an in situ UPLC–MS–MS analysis. Food Front. 1 (3), 243–252.

Chen, J., Hamm, L.L., Bundy, J.D., Kumbala, D.R., Bodana, S., Chandra, S., Chen, C.S., Starcke, C.C., Guo, Y., Schaefer, C.M., Lustigova, E., Mahone, E., Vadalia, A.M., Livingston, T., Obst, K., Hernandez, J., Bokhari, S.R., Kleinpeter, M., Alper, A.B., Lukitsch, I., He, H., Nieman, D.C., He, J., 2020. Combination treatment with sodium nitrite and isoquercetin on endothelial dysfunction among patients with CKD: A Randomized phase 2 pilot trial. Clin. J. Am. Soc. Nephrol. 15 (11), 1566–1575.

Chung, M.J., Kang, A.Y., Lee, K.M., Oh, E., Jun, H.J., Kim, S.Y., Park, K.H., 2006. Water-soluble genistin glycoside isoflavones up-regulate antioxidant metallothionein expression and scavenge free radicals. J. Agric. Food Chem. 54 (11), 3819–3826.

Chung, M.J., Pandey, R.P., Choi, J.W., et al., 2015. Inhibitory effects of kaempferol-3-O-rhamnoside on ovalbumin-induced lung inflammation in a mouse model of allergic asthma. Int. Immunopharmacol. 25 (2), 302–310.

Deng, J., Yang, H., Capanoglu, E., Cao, H., Xiao, J.B., 2018. 9 - Technological aspects and stability of polyphenols. In: Galanakis, C.M.B.T.-P.P. (Ed.), Recovery, and Applications. Woodhead Publishing, Netherland, pp. 295–323.

Deep, A., Marwaha, R.K., Marwaha, M.G., Jyoti, N.R., Sharma, A.K., 2018. Flavopiridol as cyclin dependent kinase (CDK) inhibitor: a review. New J. Chem. 42 (23), 18500–18507.

Derosa, G., D'Angelo, A., Maffioli, P., 2020. Evaluation of the effects of a supplement composed by quercetin, rutin, bromelain and l-carnosine in patients with borderline uricemia. J. Food Nutr. Res. 8 (10), 550–555.

Fotopoulou, C., Vergote, I., Mainwaring, P., Bidzinski, M., Vermorken, J.B., Ghamande, S.A., Gabra, H., 2014. Weekly AUC2 carboplatin in acquired platinum-resistant ovarian cancer with or without oral phenoxodiol, a sensitizer of platinum cytotoxicity: The phase III OVATURE multicenter randomized study. Ann. Oncol. 25 (1), 160–165.

Fuller, S., Stephen, J.M., 2015. Diosgenin, 4-hydroxyisoleucine, and fiber from fenugreek: Mechanisms of actions and potential effects on metabolic syndrome. Adv. Nutr. 6 (2), 189–197.

George, J., Singh, M., Srivastava, A.K., Bhui, K., Roy, P., Chaturvedi, P.K., Shukla, Y., 2011. Resveratrol and black tea polyphenol combination synergistically suppress mouse skin tumors growth by inhibition of activated MAPKs and p53. PLoS One 6, e23395.

Guglielmone, H.A., Agnese, A.M., Nunez Montoya, S.C., et al., 2005. Inhibitory effects of sulphated flavonoids isolated from flaveria bidentis on platelet aggregation. Thromb. Res. 115 (6), 495–502.

Han, M.K., Barreto, T.A., Martinez, F.J., Comstock, A.T., Sajjan, U.S., 2020. Randomised clinical trial to determine the safety of quercetin supplementation in patients with chronic obstructive pulmonary disease. BMJ Open Respir Res. 7 (1), e000392.

Han, L.K., Takaku, T., Li, J., et al., 1999. Anti-obesity action of oolong tea. Int. J. Obes. 23, 98–105.

Huang, H.H., Liao, D., Dong, Y., Pu, R., 2020. Effect of quercetin supplementation on plasma lipid profiles, blood pressure, and glucose levels: a systematic review and meta-analysis. Nutr. Rev. 78 (8), 615–626.

Hamden, K., Jaouadi, B., Carreau, S., Bejar, S., Elfeki, A., 2010. Inhibitory effect of fenugreek galactomannan on digestive enzymes related to diabetes, hyperlipidemia, and liver-kidney dysfunctions. Biotechnol. Bioprocess Eng. 15 (3), 407–413.

Hamden, K., Keskes, H., Belhaj, S., Mnafgui, K., Feki, A., Allouche, N., 2011. Inhibitory potential of omega-3 fatty and fenugreek essential oil on key enzymes of carbohydrate-digestion and hypertension in diabetes rats. Lipids Health Dis. 10, 226.

Hur, S.J., Lee, S.Y., Kim, Y.C., Choi, I., Kim, G.B., 2014. Effect of fermentation on the antioxidant activity in plant-based foods. Food Chem. 160, 346–356.

Ishihara, K., Nakajima, N., 2003. Structural aspects of acylated plant pigments: stabilization of flavonoid glucosides and interpretation of their functions. J. Mol. Catal. B Enzym. 23 (2), 411–417.

Irace, C., Marini, H., Bitto, A., Altavilla, D., Polito, F., Adamo, E.B., Arcoraci, V., Minutoli, L., Di Benedetto, A., Di Vieste, G., de Gregorio, C., Gnasso, A., Corrao, S., Licata, G., Squadrito, F., 2013. Genistein and endothelial function in postmenopausal women with metabolic syndrome. Eur. J. Clin. Invest. 43 (10), 1025–1031.

Kang, H.K., Ecklund, D., Liu, M., et al., 2009. Apigenin, a non-mutagenic dietary flavonoid, suppresses lupus by inhibiting autoantigen presentation for expansion of autoreactive Th1 and Th17 cells. Arthritis Res. Ther. 11, R59.

Kay, C.D., Hooper, L., Kroon, P.A., Rimm, E.B., Cassidy, A., 2012. Relative impact of flavonoid composition, dose and structure on vascular function: a systematic review of randomised controlled trials of flavonoid-rich food products. Mol. Nutr. Food Res. 56 (11), 1605–1616.

Kim, S.H., Park, J.G., Lee, J., et al., 2015. The dietary flavonoid kaempferol mediates anti-inflammatory responses via the Src, Syk, IRAK1 and IRAK4 molecular targets. Mediat. Inflamm 2015, 904142.

Kosina, P., Křen, V., Gebhardt, R., Grambal, F., Ulrichová, J., Walterová, D., 2002. Antioxidant properties of silybin glycosides. Phytother. Res. 16 (S1), 33–39.

Kumazawa, Y., Kawaguchi, K., Takimoto, H., 2006. Immunomodulating effects of flavonoids on acute and chronic inflammatory responses caused by tumor necrosis factor α. Curr. Pharm. Des. 12, 4271–4279.

Lee, J.H., Li, Y.C., Ip, S.W., et al., 2008. The role of Ca^{2+} in baicalein-induced apoptosis in human breast MDA-MB-231 cancer cells through mitochondria- and caspase-3-dependent pathway. Anticancer Res. 28 (3A), 1701–1711.

Lee, J.H., et al., 2020. Portulaca oleracea L. extract reduces hyperglycemia via PI3k/Akt and AMPK pathways in the skeletal muscles of C57BL/Ksj-db/db mice. J. Ethnopharmacol. 260, 112973.

Lee, S.J., Son, K.H., Chang, H.W., et al., 1993. Anti-inflammatory activity of naturally occurring flavone and flavonol glycosides. Arch. Pharm. Res. 16, 25.

Lepley, D.M., Li, B., Birt, D.F., et al., 1996. The chemopreventive flavonoid apigenin induces G2/M arrest in keratiocytes. Carcinogenesis 17 (11), 2367–2375.

Levy, R., Okun, Z., Shpigelman, A., 2019. The influence of chemical structure and the presence of ascorbic acid on anthocyanins stability and spectral properties in purified model systems. Foods 8 (6), 207.

Li, L., Zeng, H.W., Shan, L., et al., 2012. The different inhibitory effects of Huang-lian-Jie-Du-Tang on cyclooxygenase 2 and 5- lipoxygenase. J. Ethnopharmacol. 143 (2), 732–739.

Li, J., Qu, W., Cheng, Y., Sun, Y., Jiang, Y., Zou, T., Wang, Z., Xu, Y., Zhao, H., 2014. The inhibitory effect of intravesical fisetin against bladder cancer by induction of p53 and down-regulation of NF-kappa B pathways in a rat bladder carcinogenesis model. Basic Clin. Pharmacol. Toxicol. 115, 321–329.

Li, H.D., 2016. Optimal enzyme-assisted ethanol extaction of flavonoids from B roccoli by RSM and research on antioxidant effect. Chem. Biomol. Eng. 1 (1), 12–20.

Liu, D., Mao, Y., Ding, L., Zeng, X.A., 2019. Dihydromyricetin: A review on identification and quantification methods, biological activities, chemical stability, metabolism and approaches to enhance its bioavailability. Trends Food Sci. Technol. 91, 586–597.

Liu, F., Sun, L., You, G., Liu, H., Ren, X., Wang, M., 2020. Effects of Astragalus polysaccharide on the solubility and stability of 15 flavonoids. Int. J. Biol. Macromol. 143, 873–880.

Liu, C.Y., Huang, C.J., Huang, L.H., Chen, I.J., Chiu, J.P., Hsu, C.H., 2014. Effects of green tea extract on insulin resistance and glucagon-like peptide 1 in patients with type 2 diabetes and lipid abnormalities: a randomized, double-blinded, and placebo-controlled trial. PLoS One 9 (3), e91163.

Lohner, K., Schnäbele, K., Daniel, H., Oesterle, D., Rechkemmer, G., Göttlicher, M., Wenzel, U., 2007. Flavonoids alter P-gp expression in intestinal epithelial cells in vitro and *in vivo*. Mol. Nutr. Food Res. 51 (3), 293–300.

Makris, D.P., Rossiter, J.T., 2000. Heat-induced, metal-catalyzed oxidative degradation of quercetin and rutin (quercetin 3-O-rhamnosylglucoside) in aqueous model systems. J. Agric. Food Chem. 48 (9), 3830–3838.

Marini, H., Minutoli, L., Polito, F., Bitto, A., Altavilla, D., Atteritano, M., Gaudio, A., Mazzaferro, S., Frisina, A., Frisina, N., Lubrano, C., Bonaiuto, M., D'Anna, R., Cannata, M.L., Corrado, F., Adamo, E.B., Wilson, S., Squadrito, F., 2007. Effects of the phytoestrogen genistein on bone metabolism in osteopenic postmenopausal women: a randomized trial. Ann. Intern. Med. 146 (12), 839–847.

Messing, E., Gee, J.R., Saltzstein, D.R., Kim, K., diSant'Agnese, A., Kolesar, J., Harris, L., Faerber, A., Havighurst, T., Young, J.M., Efros, M., Getzenberg, R.H., Wheeler, M.A., Tangrea, J., Parnes, H., House, M., Busby, J.E., Hohl, R., Bailey, H., 2012. A phase 2 cancer chemoprevention biomarker trial of isoflavone G-2535 (genistein) in presurgical bladder cancer patients. Cancer Prev. Res. (Phila.) 5 (4), 621–630.

Musialik, M., Kuzmicz, R., Pawłowski, T.S., Litwinienko, G., 2009. Acidity of hydroxyl groups: an overlooked influence on antiradical properties of flavonoids. J. Org. Chem. 74 (7), 2699–2709.

Nguyen, T.T.T., Tran, E., Nguyen, T.H., et al., 2004. The role of activated MEK-ERK pathway in quercetin-induced growth inhibition and apoptosis in A549 lung cancer cells. Carcinogenesis 25 (5), 647–659.

Nicolle, E., Souard, F., Faurc, P., et al., 2011. Flavonoids as promising lead compounds in type 2 diabetes mellitus: molecules of interest and structure-activity relationship. Curr. Med. Chem. 18, 2661–2672.

Oliveira, M.R.D., Nabavi, S.F., Habtemariam, S., et al., 2015. The effects of baicalein and baicalin on mitochondrial function and dynamics: a review. Pharmacol. Res. 100, 296–308.

Patel, M.I., Kurek, C., Dong, Q.H., 2008. The arachidonic acid pathway and its role in prostate cancer development and progression. J. Urol. 179 (5), 1668–1675.

Pietta, P.G., 2000. Flavonoids as antioxidants. J. Nat. Prod. 63 (7), 1035–1042.

Plaza, M., Pozzo, T., Liu, J., Gulshan Ara, K.Z., Turner, C., Nordberg Karlsson, E., 2014. Substituent effects on *in vitro* antioxidizing properties, stability, and solubility in flavonoids. J. Agric. Food Chem. 62 (15), 3321–3333.

Rawat, P., Kumar, M., Sharan, K., Chattopadhyay, N., Maurya, R., 2009. Ulmosides A and B: flavonoid 6-C-glycosides from Ulmus wallichiana, stimulating osteoblast differentiation assessed by alkaline phosphatase. Bioorg. Med. Chem. Lett. 19 (16), 4684–4687.

Roy, P., George, J., Srivastava, S., Tyagi, S., Shukla, Y., 2011. Inhibitory effects of tea polyphenols by targeting cyclooxygenase-2 through regulation of nuclear factor kappa B, Akt and p53 in rat mammary tumors. Investig. New Drugs 29, 225–231.

Rüfer, C.E., Bub, A., Möseneder, J., Winterhalter, P., Stürtz, M., Kulling, S.E., 2008. Pharmacokinetics of the soybean isoflavone daidzein in its aglycone and glucoside form: a randomized, double-blind, crossover study. Am. J. Clin. Nutr. 87 (5), 1314–1323.

Schwanck, B., Blaschek, W., 2010. HPLC-DAD/ESI-MS identification and quantification of flavonoids and hydroxycinamic derivatives in kale. Planta Med. 76, 512–519.

Sharma, R.D., 1986. Effect of fenugreek seeds and leaves on blood glucose and serum insulin responses in human subjects. Nutr. Res. 6 (12), 1353–1364.

Shin, D.M., Nannapaneni, S., Patel, M.R., Shi, Q., Liu, Y., Chen, Z., Chen, A.Y., El-Deiry, M.W., Beitler, J.J., Steuer, C.E., Roser, S.M., Klein, A.M., Owonikoko, T.K., Ramalingam, S.S., Khuri, F.R., Chen, Z.G., Saba, N.F., 2020. Phase Ib study of chemoprevention with green tea polyphenon E and erlotinib in patients with advanced premalignant lesions (APL) of the head and neck. Clin. Cancer Res. 26 (22), 5860–5868.

Stearns, M.E., Wang, M., 2011. Synergistic effects of the green tea extract epigallocatechin-3-gallate and taxane in eradication of malignant human prostate tumors. Transl. Oncol. 4, 147–156.

Sun, C.C., Li, Y., Yin, Z.P., Zhang, Q.F., 2021. Physicochemical properties of dihydromyricetin and the effects of ascorbic acid on its stability and bioavailability. J. Sci. Food Agric. 101, 3862–3869.

Taliou, A., Zintzaras, E., Lykouras, L., Francis, K., 2013. An open-label pilot study of a formulation containing the anti-inflammatory flavonoid luteolin and its effects on behavior in children with autism spectrum disorders. Clin. Ther. 35 (5), 592–602.

Tan, J., de Bruijn, W.J.C., van Zadelhoff, A., Lin, Z., Vincken, J.P., 2020. Browning of epicatechin (EC) and epigallocatechin (EGC) by auto-oxidation. J. Agric. Food Chem. 68 (47), 13879–13887.

Walle, T., 2007. Methylation of dietary flavones greatly improves their hepatic metabolic stability and intestinal absorption. Mol. Pharm. 4 (6), 826–832.

Walle, T., Browning, A.M., Steed, L.L., Reed, S.G., Walle, U.K., 2005. Flavonoid glucosides are hydrolyzed and thus activated in the oral cavity in humans. J. Nutr. 135 (1), 48–52.

Wang, T.Y., Li, Q., Bi, K.S., 2018. Bioactive flavonoids in medicinal plants: structure, activity and biological fate. Asian J. Pharm. Sci. 13 (1), 12–23.

Wang, Z., Zhang, Z., Du, N., et al., 2015. Hepatoprotective effects of grape seed procyanidin B2 in rats with carbon tetrachloride-induced hepatic fibrosis. Altern. Ther. Health. Med. 21 (Suppl 2), S12–S21.

Wolfram, S., Wang, Y., Thielecke, F., 2006. Anti-obesity effects of green tea: from bedside to bench. Mol. Nutr. Food Res. 50, 176–187 2006.

Xiao, J.B., 2017. Dietary flavonoid aglycones and their glycosides: which show better biological significance? Crit. Rev. Food Sci. Nutr. 57 (9), 1874–1905.

Xiao, J.B., Capanoglu, E., Jassbi, A.R., Miron, A., 2016. Advance on the flavonoid C-glycosides and health benefits. Crit. Rev. Food Sci. Nutr. 56 (Suppl 1), S29–S45.

Xiao, J.B., Högger, P., 2015. Stability of dietary polyphenols under the cell culture conditions: avoiding erroneous conclusions. J. Agric. Food Chem. 63 (5), 1547–1557.

Xu, L.N., Li, Y., Dai, Y., et al., 2018. Natural products for the treatment of type 2 diabetes mellitus: pharmacology and mechanisms. Pharmacol. Res. 130, 451–465.

Yasui, M., Matsushima, M., Omura, A., 2015. The suppressive effect of Quercetin on Toll-like Receptor 7-mediated activation in alveolar macrophages. Pharmacology 96 (5-6), 201–209.

Yilma, A.N., Singh, S.R., Morici, L., et al., 2013. Flavonoid naringenin: a potential immunomodulatro for chalamydia trachomatis inflammation. Mediat. Inflamm 2013, 102457.

Yoshida, K., Oyama, K.I., Kondo, T., 2012. Chemistry of flavonoids in color development. Recent Advances in Polyphenol Research. Wiley, USA, 99–129.

Zhang, T., Wu, W.L., Li, D.Y., et al., 2014. Anti-oxidant and anti-apoptotic effect of luteolin on mice peritoneal macrophages stimulated by angiotensin II. Int. Immunopharmacol 20 (2), 346–351.

Zhao, R., Li, Q., Long, L., et al., 2007. Anti-diabetic activity of flavone from Ipomoea Batatas leaf in non-insulin dependent diabetic rats. Int. J. Food Sci. Technol. 42, 80–85.

Zhao, J., Yang, J., Xie, Y., 2019. Improvement strategies for the oral bioavailability of poorly water-soluble flavonoids: an overview. Int. J. Pharm. 570, 118642.

Zwicker, J.I., Schlechter, B.L., Stopa, J.D., Liebman, H.A., Aggarwal, A., Puligandla, M., Caughey, T., Bauer, K.A., Kuemmerle, N., Wong, E., Wun, T., McLaughlin, M., Hidalgo, M., Neuberg, D., Furie, B., Flaumenhaft, R., CATIQ Investigators11, 2019. Targeting protein disulfide isomerase with the flavonoid isoquercetin to improve hypercoagulability in advanced cancer. JCI Insight 4 (4), 125851.

CHAPTER

4.9

Lecithin

Shahira M. Ezzat[a,b], Mohamed A. Salem[c], Nihal M. El Mahdy[d], Marwa M. Mahfouz[e]

[a]*Department of Pharmacognosy, Faculty of Pharmacy, Cairo University, Egypt*

[b]*Department of Pharmacognosy, Faculty of Pharmacy, October University for Modern Sciences and Arts (MSA), Egypt*

[c]*Department of Pharmacognosy, Faculty of Pharmacy, Menoufia University, Menoufia, Egypt*

[d]*Department of Pharmaceutics and Industrial Pharmacy, Faculty of Pharmacy, October University for Modern Sciences and Arts (MSA), Egypt*

[e]*Department of Pharmacology and Toxicology, Faculty of Pharmacy, Menoufia University, Menoufia, Egypt*

4.9.1 Introduction

Lecithin is the primary fat constituents that build up the membranes surrounding each cellular structure. It is found in the cell membranes of all living things. Lecithin helps to control the way the cells function and support intercellular communication (Williams and Chapman, 1971).

Lecithin was first discovered by the French scientist Maurice Gobley in 1806 who named it lekithos after the Greek word that means egg yolk, which has been the only commercial source for lecithin till the discovery of soybean processing in the 1930s (Williams and Chapman, 1971).

Lecithin or phosphatidylcholine is a type of fat called phospholipid, it consists of saturated and unsaturated fatty acids, phosphoric acid, and choline (Juneja, 1996).

Lecithin is produced not only by plants and animals but also by the human body as a constituents of the bile juice (Singh et al., 2009). Lecithin is an amphiphilic molecule that attracts both water and fat so it helps in fat digestion (Wilde and Chu, 2011). Lecithin made by the body and the natural lecithin in food usually provide the required amount for human health, thus there is no need to take lecithin supplements (van Nieuwenhuyzen and Tomás, 2008). Although no official recommended daily intake for lecithin, some dietary studies recommend a dose of 550 mg per day for men and 425 mg per day for women.

Antioxidants Effects in Health. DOI: https://doi.org/10.1016/B978-0-12-819096-8.00060-4

4.9.2 Sources of lecithin

Lecithin is the yellow to brown fatty substances that can be obtained from both plants and animals. Although eggs are still the best whole food source of lecithin, soy lecithin, and sunflower lecithin are the most common varieties (Szuhaj, 2005). Lecithin can be also found in many other sources, such as cauliflower, whole grains, chickpeas (garbanzo beans), cabbage, organic meat, seeds, liver, split peas, peanuts, milk, and nuts (Szuhaj, 1989).

Commercial lecithin, which is commonly derived from soybeans is often used in food industry as emulsifier in processed goods, like ice creams and salad dressings (Kralova and Sjöblom, 2009).

Lecithin supplements are usually used for controlling the high cholesterol levels, aid for lactation, and as antiulcer (Küllenberg et al., 2012). These lecithin supplements most probably prepared from eggs, sunflower seeds, or soybeans. Although, animal derived lecithin is now used for preparation of supplements but still soybeans is the most important source for lecithin supplements (Küllenberg et al., 2012).

Sunflower lecithin is not as common as soy lecithin but in many case it is preferred because soybeans are sometimes subjected to genetic modification in mass production but sunflower seeds never subjected to such modification. Moreover, the process of extraction of sunflower lecithin is also gentler and doesn't require harsh chemicals (Ma et al., 2003).

4.9.3 Chemistry of lecithin

The term "lecithin" refers to a fatty mixture rather than a single compound. Lecithins can be obtained from either plant or animal origin. Lecthins have amphiphilic, both hydrophilic and lipophilic, characters; therefore, they have been used as emulsifying and homogenizing agents in liquid mixtures and food stuff (Linow, 1990).

Chemically, lecithins are mixtures of glycerophospholipids. They may be derived from phosphatidylcholine, phosphatidylserine, phosphatidylethanolamine, phosphatidylinositol, phosphatidylserine, and phosphatidic acid (Fig. 4.9.1) (Cherry et al., 1981). Soybean-derived lecithin may contain other lipid classes, such as triacylglycerols, sterols, and fatty acids.

4.9.4 Beneficial and detrimental effects on health

Lecithin or phosphatidylcholine is a source of choline which is essential of the proper function of the liver, nerves as well as movement of the muscles (Canty and Zeisel, 1994). Furthermore, lecithin is essential in fetus development, thus pregnant women always use lecithin supplements as a prenatal vitamin regimen (Malcolm et al., 2003). Lecithin also plays a crucial for the optimum development of the brain; this is why egg yolk is the first important food for babies (Uauy and Dangour, 2006).

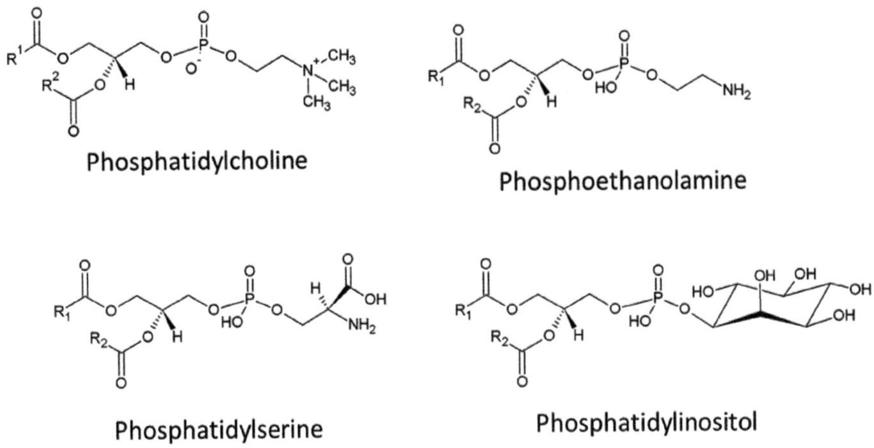

FIG. 4.9.1 Types of phospholipids in lecithin.

4.9.4.1 Benefits for the cells

Lecithin is considered is a keystone in the construction of each cell in the human body (FATIMA and SRIVASTAVA). The cell membranes need lecithin for their proper functioning, prevent their hardening and to facilitate nutrients entry into the cells. Cells protection is an important step for maintaining the body self-defense to various health problems that may attack unhealthy cells. Various body organs, such as the heart, kidney, and liver produce natural phospholipids, including lecithin in considerable amounts, but taking them as supplements also aid in health enhancement.

4.9.4.2 Benefits for the liver

Lecithin is an important source of the nutrient choline that has a vital role in liver function. Lecithin in the liver helps in preventing the scarring and cirrhosis caused by excessive alcohol use, helps in hepatic cell regeneration and recover the liver after hepatitis. Furthermore, lecithin contributes to a healthy gallbladder and heart. Lecithin being a special type of lipid so it prevents fats from accumulation in the liver (Cohn et al., 2008).

Lecithin is responsible for the regulation of fat metabolism in the liver, as it binds to proteins that lower triglycerides (TG) to increase the levels of the blood high density lipoprotein (HDL-c), which is the good cholesterol. Phosphatidylcholine also help in producing very-low-density lipoproteins (VLDL-c), which removes fats from the liver (An et al., 1997).

When phosphatidylcholine levels are low, fats accumulates in the liver, causing liver damage. In the experimental studies, liver damage was significantly reduced in mice received soybean lecithin (Baghdasaryan et al., 2008). Another study using rats found that dietary phosphatidylcholine reduced fatty buildup in the liver, reported nutrition in 2005.

4.9.4.3 Cardiovascular benefits

Soy products having lecithin additives improve cardiovascular health, in individuals suffering from hypertension or chronic cardiovascular diseases (Milsom et al., 2014).

Lecithin can also prevent and revers the damages that arising from coronary artery disease, through hindering cholesterol and other fats from sticking which is a predisposition factor to various cardiovascular diseases (Ramdath et al., 2017). Lecithin has a lubricant effect so creates a slippery lining to arteries thus prevent fatty deposits to adhere to the artery walls and carried by blood to the liver to metabolize them and produce energy. This effects improve circulation and prevent blood clots (Ramdath et al., 2017).

4.9.4.4 Benefits for the GIT

It was reported also that lecithin can improve digestion in patients with ulcerative colitis through a chain reaction that potentiate the intestinal lining mucus, hence eases the digestion process (Stremmel and Gauss, 2013). This mechanism makes it also beneficial in case of mal-digestion due to the irritable bowel syndrome (IBS) (Stremmel and Gauss, 2013).

4.9.4.5 Benefits for memory

Lecithin being a rich source of choline is thought to improve the brain's functions and memory especially in Alzheimer's patients (Kim et al., 2014). Lecithin may be beneficial to Alzheimer's patients or those suffering from memory related conditions as lecithin supplements can enhance the memorization and recalling in those patients (Moré et al., 2014).

4.9.4.6 Other health benefits

Lecithin especially the hydrogenated type also has the ability to restore skin hydration, so used as an emollient in skin care preparations (Harcharik and Emer, 2014).

Lecithin being composed of many fatty acids, it can help in processing the fat soluble vitamins and facilitate their breakdown and absorption (Wang et al., 2013).

4.9.5 The pro-oxidant activity

Food emulsions have been used widely in last decades in many formulas; however, the shelf-life of these products is largely affected by lipid peroxidation. Upon oxidation, lipids are converted to degradation products, that is, negatively affecting the taste and flavor of food (Decker et al., 2017). A successful solution to overcome this concern is to use natural emulsifiers with antioxidant activity. Several reports have shown that using lecithin as an emulsifying agent significantly improves the emulsions oxidative stability of (Choe et al., 2014; Cui et al., 2018).

Lecithin antioxidant activity can be attributed to its ability to regenerate natural antioxidants as tocopherols as well as its ability to chelate pro-oxidative metals (Cui and Decker, 2016).

4.9.6 *In-vitro* studies

Studies have demonstrated that lecithin antioxidant activity can be affected by combination with drugs. One study demonstrated antioxidant effect of soy lecithin in triolein models when used alone or in combination with quercetin (1:1, w/w) and it was observed that the effect of quercetin–lecithin mixture was stronger than each one alone, and the mixture also showed a concentration dependent effect, this can be attributed to the synergistic effect between polar lipids and quercetin (Ramadan, 2008).

Pan et al. (2013) made a comparative study between the antioxidant properties of lecithin and tween 20 in emulsions of encapsulated curcumin. Lecithin showed significantly higher antioxidant activity when compared to tween 20 as it decreased the rate of permeation of peroxyl radicals across the emulsion interface, accordingly encapsulated curcumin showed higher oxidative stability in lecithin stabilized emulsions. Dihydromyricetin–lecithin complex was prepared by Liu et al. (2009) combined by noncovalent bond, to enhance the hydrophobicity of dihydromyricetin, The complex was evaluated using the Rancimat antioxidant test using lard oil as substrate and it showed free radical scavenging activity against DPPH (IC_{50} 22.60 μg/mL).

Lecithin has also been reported to enhance the bioavailability of several drugs. Lecithin-drug hybrid nanoparticles were prepared by Javed et al. (2013) using amphotericin B as model drug (Lec-AmB NPs) and the results demonstrated that an enhancement of the antileishmanial activity and oral bioavailability of Lec-AmB NPs compared to the deoxycholate complex of AmB (injectable market product) attributed to the emulsifying properties of lecithin in the mucus of the gut walls which enhanced the absorption and thereby bioavailability. PLGA was used with lecithin to develop polymeric micelles for delivery of methotrexate (MTX) to cancer cells, which showed enhanced cytotoxicity of MTX by 2.13 folds on MDA-MB-231 cells with intracellular delivery as well as enhanced bioavailability (Singh et al., 2017). Lecithin-based nano-emulsification was investigated by Heo et al. (2016) and it showed enhancement of the *in vitro* bioavailability when tested on Caco-2 human intestinal cells using conjugated linoleic acid (CLA) in different free fatty acid (FFA) and triglyceride (TG) forms.

The effect of lecithin on neuronal development was studied by Latifi et al. (2016). Natural lecithin was formulated into nanoliposomes have been shown to have increased neurite outgrowth, network complexity and neural activity of cortical rat neurons in vitro as well as an upregulation of synapsin I (SYN1), which supports the positive role of lecithin in synaptogenesis, synaptic development and maturation. This can be attributed to lecithin being rich in poly unsaturated fatty acids (PUFAs) which are critical to nervous system function and structure.

Ramadan and Asker (2009) reported that lecithin had a synergistic effect on antimicrobial activity of quercetin in quercetin-enriched-lecithin formulation. It also increased the antibacterial activity against gram-positive bacteria with minimum inhibitory concentration between 750 and 1000 mg/mL and also an inhibitory effect on the biosynthesis of DNA and RNA protein in cells of Bacillus subtilis. Zhang et al. (2017) observed that significant synergistic antimicrobial occurred in *E. coli* cultures upon exposure to a constant concentration of eugenol (0.043%–0.050% [wt/wt]) plus lecithin (0.5–1 mg/100 mL), however, it was also observed that increasing the concentration of lecithin above 1 mg/100 mL stop its synergistic effect. So, the synergistic effect may be due to the formation of nanoscale aggregates (<100 nm) at the lecithin concentration below 1 mg/Ml, which is responsible for improving eugenol antimicrobial effects.

4.9.7 Animal studies

Raj et al. (2010) investigated the hepatoprotective activity of lecithin against D-galactosamine induced hepatotoxicity in rat hepatocytes and animal models. Lecithin showed significant antihepatotoxic effect at 100 μg/mL, its effect was comparable to the standard silymarin. Pretreatment with lecithin was also shown to inhibit increases in liver enzyme levels confirmed by histopathological observation (Raj et al., 2011).

4.9.8 Mechanism of action

Lecithin (phosphatidylcholine) works through increasing the synthesis of the neurotransmitter acetylcholine as it is considered as a precursor for choline that included in acetylcholine synthesis. It also works through the regulation of the permeability of the cell membrane as it has a crucial role in normal metabolism and homeostatic regulation of membrane fluidity (Kraus et al., 2010).

Lecithin (high concentrations of phospholipids) has antioxidant protection against reactive oxygen species from different chemical toxins (Kidd, 2000; Lieber et al., 1997). It is considered as an obligatory micellizing constituent of bile due to its amphipathic properties (Thistle and Schoenfield, 1968; Toouli et al., 1975). Also, it can protect the epithelial-luminal membrane of GI tract and lungs and for its surfactant properties (Dunjic et al., 1993; Lloyd et al., 1999).

4.9.9 Clinical trials

Lecithin for many centuries has a great interest from scientists to discover its properties and benefits of it for human beings, in brief will show some of these trials.

Patients with bipolar disorder showed low choline levels and alternation in membrane phospholipid metabolism, so usage of lecithin supplementation obviously changes action potential and stabilizes the membrane (Kraus et al., 2010).

By enhancing acetylcholine synthesis, lecithin is considered one of the early cholinergic medications used for Tardive dyskinesia (TD) which is caused by high dose and long-term usage of antipsychotic drugs (Wurtman et al., 1977; Wurtman and Growdon, 1978), and there are many studies used lecithin in doses ranging from 50–60 g/day to treat tardive dyskinesia (Tammenmaa-Aho et al., 2018).

It is known that phospholipids increase the solubility of biliary cholesterol and patients with gallstones showed low levels of phospholipids, so lecithin (high concentrations of phospholipids) has the ability to decrease the bile lithogenicity and that led to decrease formation of the gallstones, whereas some studies showed no relationship between the phospholipid content of lithogenicity of the bile (Anderson and Bouchier, 1969).

In a study where eight gallstone patients receiving 100 mg t.i.d of lecithin for two years showing significant decrease in biliary cholesterol levels, while biliary phospholipid content was significantly increased, in one patient of them, the gallstones size was decreased and changed in its shape while other patients showed no change (Tuzhilin et al., 1976).

In another study, 4.5 g of soybean lecithin taken daily for weeks resulted in a non-significant change in the cholesterol saturation index of bile (Holan et al., 1979).

While in another study as patients were Taking 100 mg lecithin t.i.d. led to increase solubility of cholesterol, while using lecithin alone showed no significant effects on gallstone dissolution (Pizzorno et al., 2016).

In another study for assessing fetal pulmonary maturity found that the lecithin concentration in amniotic fluid increases dramatically at about 35 weeks of gestation which considered as a function of gestational age for normal pregnancies (Gluck et al., 1971).

Many studies used lecithin/quercetin preparation to treat permanent focal ischemia showing that mixing lecithin with quercetin increased possibility of crossing the blood brain barrier and the mixture have a neuroprotective action and beneficial effect against oxidative damage (Dajas et al., 2002), as a (30 mg/kg, ip) of a lecithin/quercetin mixture was administered 30 min after artery occlusion in rats, showing that the volume of the ischemic lesion was significantly passive diffusion decreased by 56% (Swanson et al., 1990).

Another combination with lecithin in one of the NSAIDs where, indomethacin is connected to lecithin via a 5-carbon length linker. This combination appears better GI and renal safety and does not cause local gastric toxicity because it selectively inhibits COX-2 and not COX-1 (Dvir et al., 2007).

In the pharmaceutical field, lecithin organogels (LOs) plays an important role as it can form reverse micelles that can be used in topical drug delivery. LOs are clear, thermodynamically stable, viscoelastic, and biocompatible jelly-like phases (Scartazzini and Luisi, 1988).

4.9.10 Bioavailability

Lecithin is surface-active agent due to its amphiphilic character that has a glycerol backbone, where hydrophobic tail groups are esterified with fatty acids in carbon number one and two and with phosphoric acid in carbon number three, then the hydrophilic headgroup undergo esterification with an additional alcohol (van Hoogevest and Wendel, 2014). Lecithin properties differ according to phospholipid concentrations as it is classified to liquid lecithins (regular and hydrolysed) and deoiled lecithin powder (Additives and Feed, 2016).

Regular and hydrolyzed liquid lecithins are brown viscous fluids (viscosity: 10 Pa.s at 25°C) with unknown bioavailability, dispersible in water and with a high solubility in fats, their density of about 1.04 g/cm^3 (technical dossier) (dossier).

Lecithin undergoes lipid digestion in the oral cavity and buccal cavity by lingual lipases and Ebner's glands on the tongue, followed by gastric digestion in the stomach by lingual and gastric enzymes, considering Lecithin as a complex mixture of acetone insoluble phosphatides (less than 80% phospholipids) in combination with other substances as carbohydrates, triglycerides, and fatty acids (van Hoogevest, 2017), so it undergoes emulsification by stomach's peristalsis and changed to fine lipid droplets in the duodenum and mixed with bile and pancreatic juice where undergoes many reactions to complete its emulsification (Iqbal and Hussain, 2009).

After enzymatic hydrolysis of phospholipids and by passive diffusion the enterocytes take the fatty acids where monoacyl-phospholipids undergo acylation by acyl-CoA to diacyl-phospholipids and further hydrolyzed to glycerophosphocholine (GPC) and analogues and fatty acids by lysolecithinase (Scow et al., 1967).

Lecithin as a pharmaceutical dosage form if there is no gastric coating it will undergo hydrolysis by gastric fluids with a pH ranging from 1.5 to 3.5 in the stomach. The hydrolysis rate of lecithin will decrease by increasing pH ranging from pH (1–3) (Ottesen et al., 2004).

There is no recommended daily allowance (RDA) for lecithin as it is not considered an essential nutrient. In general, dietary intake of lecithin ranges from 1–5 g/d (Potter et al., 2009). Lecithin is generally recognized as safe (GRAS) pharmaceutical excipient, described in Pharmacopoeias (American, Chinese, Japanese) and relevant regulatory guidance documentation of the Food and Drug Administration (FDA) and European Medicines Agency (EMA) (van Hoogevest and Wendel, 2014).

Conclusion

Lecithin the most common phosphatidylcholine that constitutes the membranes of every living cell has many health benefits as antioxidant, to improve the functions in Alzheimer's patients and in case of gallstones and liver diseases. Although some animal and human studies proved its importance, more researches are still required to study its bioavailability and the most suitable dosage form.

References

Additives, E.P.o., Feed, P.o.S.u.i.A., 2016. Safety and efficacy of lecithins for all animal species. EFSA J. 14, e04561.

An, B., Nishiyama, H., Tanaka, K., Ohtani, S., Iwata, T., Tsutsumi, K., Kasai, M., 1997. Dietary safflower phospholipid reduces liver lipids in laying hens. Poult. Sci. 76, 689–695.

Anderson, F., Bouchier, I.A., 1969. Phospholipids in human lithogenic gall bladder bile. Nature 221, 372.

Baghdasaryan, A., Fickert, P., Fuchsbichler, A., Silbert, D., Gumhold, J., Hörl, G., Langner, C., Moustafa, T., Halilbasic, E., Claudel, T., 2008. Role of hepatic phospholipids in development of liver injury in Mdr2 (Abcb4) knockout mice. Liver Int. 28, 948–958.

Canty, D.J., Zeisel, S.H., 1994. Lecithin and choline in human health and disease. Nutr. Rev. 52, 327–339.

Cherry, J.P., Gray, M.S., Jones, L.A., 1981. A review of lecithin chemistry and glandless cottonseed as a potential commercial source. J. Am. Oil Chem. Soc. 58, 903–913.

Choe, J., Oh, B., Choe, E., 2014. Effect of soybean lecithin on iron-catalyzed or chlorophyll-photosensitized oxidation of canola oil emulsion. J. Food Sci. 79, C2203–C2208.

Cohn, J.S., Wat, E., Kamili, A., Tandy, S., 2008. Dietary phospholipids, hepatic lipid metabolism and cardiovascular disease. Curr. Opin. Lipidol. 19, 257–262.

Cui, L., Decker, E.A., 2016. Phospholipids in foods: prooxidants or antioxidants? J. Sci. Food Agric. 96, 18–31.

Cui, L., Fan, J., Sun, Y., Zhu, Z., Yi, J., 2018. The prooxidant activity of salts on the lipid oxidation of lecithin-stabilized oil-in-water emulsions. Food Chem. 252, 28–32.

Dajas, F., Rivera, F., Blasina, F., Urbanavicius, J., Arredondo, F., Lafon, L., Costa, G., Echeverry, C., Ferreira, M., Morquio, A., 2002. Mechanisms of neuroprotection. The contribution of quercetin in focal ischemia and cell culture, XVII Annual Meeting of the Federação de Sociedades Brasileiras de Biologia Experimental (FESBE), pp. 28–31.

Decker, E.A., McClements, D.J., Bourlieu-Lacanal, C., Durand, E., Figueroa-Espinoza, M.C., Lecomte, J., Villeneuve, P., 2017. Hurdles in predicting antioxidant efficacy in oil-in-water emulsions. Trends Food Sci. Technol. 67, 183–194.

Dunjic, B., Axelson, J., Ar'Rajab, A., Larsson, K., Bengmark, S., 1993. Gastroprotective capability of exogenous phosphatidylcholine in experimentally induced chronic gastric ulcers in rats. Scand. J. Gastroenterol. 28, 89–94.

Dvir, E., Elman, A., Simmons, D., Shapiro, I., Duvdevani, R., Dahan, A., Hoffman, A., Friedman, J.E., 2007. DP-155, a lecithin derivative of indomethacin, is a novel nonsteroidal antiinflammatory drug for analgesia and Alzheimer's disease therapy. CNS Drug Rev. 13, 260–277.

Gluck, L., Kulovich, M.V., Borer Jr, R.C., Brenner, P.H., Anderson, G.G., Spellacy, W.N., 1971. Diagnosis of the respiratory distress syndrome by amniocentesis. Am. J. Obstet. Gynecol. 109, 440–445.

Harcharik, S., Emer, J., 2014. Steroid-sparing properties of emollients in dermatology. Skin Ther. Lett. 19, 5–10.

Heo, W., Kim, J.H., Pan, J.H., Kim, Y.J., 2016. Lecithin-based nano-emulsification improves the bioavailability of conjugated linoleic acid. J. Agric. Food Chem. 64, 1355–1360.

Holan, K.R., Holzbach, T., Hsieh, J.Y., Welch, D.K., Turcotte, J.G., 1979. Effect of oral administration of 'essential' phospholipid, β-glycerophosphate, and linoleic acid on biliary lipids in patients with cholelithiasis. Digestion 19, 251–258.

Iqbal, J., Hussain, M.M., 2009. Intestinal lipid absorption. Am. J. Physiol. Endocrinol. Metab. 296, E1183–E1194.

Javed, I., Hussain, S.Z., Ullah, I., Khan, I., Ateeq, M., Shahnaze, G., Rehmana, H.u., Razi, M.T., Shah, M.R., Hussain, I., 2013. Synthesis, characterization and evaluation of lecithin based nanocarriers for enhanced pharmacological and oral pharmacokinetic profile of amphotericin B. J. Mater. Chem. B.

Juneja, L., 1996. Egg Yolk Lipids. CRC press, Boca Raton.

Kidd, P., 2000. Dietary phospholipids as anti-aging nutraceuticals. Anti-aging medical therapeutics 4, 282–300.

Kim, H.-Y., Huang, B.X., Spector, A.A., 2014. Phosphatidylserine in the brain: metabolism and function. Prog. Lipid Res. 56, 1–18.

Kralova, I., Sjöblom, J., 2009. Surfactants used in food industry: a review. J. Dispersion Sci. Technol. 30, 1363–1383.

Kraus, R., Stricker, G., Speyer, C., 2010. Online Counseling: A Handbook for Mental Health Professionals. Academic Press, United States.

Küllenberg, D., Taylor, L.A., Schneider, M., Massing, U., 2012. Health effects of dietary phospholipids. Lipids Health Dis. 11, 3.

Latifi, S., Tamayol, A., Habibey, R., Sabzevari, R., Kahn, C., Geny, D., Eftekharpour, E., Annabi, N., Blau, A., Linder, M., Arab-Tehrany, E., 2016. Natural lecithin promotes neural network complexity and activity. Sci. Rep. 6, 25777.

Lieber, C.S., Leo, M.A., Aleynik, S.I., Aleynik, M.K., DeCarli, L.M., 1997. Polyenylphosphatidylcholine decreases alcohol-induced oxidative stress in the baboon. Alcohol. Clin. Exp. Res. 21, 375–379.

Linow, F., 1990. Lecithins: Sources, Manufacture & Uses. Herausgegeben von B. F. Szuhaj. 283 Seiten, zahlr. Abb. und Tab. American Oil Chemists' Society, Champaign, Illinois, 1989. Food /Nahrung 34, 403.

Liu, B., Du, J., Chen, C., Zeng, J., Niu, S., 2009. Characterization and antioxidant activity of dihydromyricetin–lecithin complex. Eur. Food Res. Technol. 230, 325–331.

Lloyd, J., Todd, D.A., John, E., 1999. Serial phospholipid analysis in preterm infants: comparison of Exosurf and Survanta. Early Hum. Dev. 54, 157–168.

Ma, J.K., Drake, P.M., Christou, P., 2003. Genetic modification: the production of recombinant pharmaceutical proteins in plants. Nat. Rev. Genet. 4, 794.

Malcolm, C., McCulloch, D., Montgomery, C., Shepherd, A., Weaver, L., 2003. Maternal docosahexaenoic acid supplementation during pregnancy and visual evoked potential development in term infants: a double blind, prospective, randomised trial. Arch. Dis. Child 88, F383–F390.

Milsom, V.A., Malcolm, R.J., Johnson, G.C., Pechon, S.M., Gray, K.M., Miller-Kovach, K., Rost, S.L., O'Neil, P.M., 2014. Changes in cardiovascular risk factors with participation in a 12-week weight loss trial using a commercial format. Eat. Behav. 15, 68–71.

Moré, M.I., Freitas, U., Rutenberg, D., 2014. Positive effects of soy lecithin-derived phosphatidylserine plus phosphatidic acid on memory, cognition, daily functioning, and mood in elderly patients with Alzheimer's disease and dementia. Adv. Ther. 31, 1247–1262.

Ottesen, J.T., Olufsen, M.S., Larsen, J.K., 2004. Applied Mathematical Models in Human Physiology. SIAM.

Pan, Y., Tikekar, R.V., Nitin, N., 2013. Effect of antioxidant properties of lecithin emulsifier on oxidative stability of encapsulated bioactive compounds. Int. J. Pharm. 450, 129–137.

Pizzorno, J., Murray, M., Joiner-Bey, H., 2016. The Clinician's Handbook of Natural Medicine E-Book. Elsevier Health Sciences Publisher, Amsterdam.

Potter, M., Moses, A., Wozniak, J., 2009. Alternative treatments in pediatric bipolar disorder. Child Adolesc. Psychiatr. Clin. N. Am. 18, 483–514.

Raj, P.V., Nitesh, K., Chandrashekhar, H.R., Rao, C.M., Rao, J.V., Udupa, N., 2010. Effect of lecithin and silymarin on d-galactosamine induced toxicity in isolated hepatocytes and rats. Indian J. Clin. Biochem. 25, 169–174.

Raj, P.V., Nitesh, K., Prateek, J., Sankhe, M.N., Rao, J.V., Rao, C.M., Udupa, N., 2011. Effect of lecithin on d-galactosamine induced hepatotoxicity through mitochondrial pathway involving Bcl-2 and Bax. Indian J. Clin. Biochem. 26, 378–384.

Ramadan, M.F., 2008. Quercetin increases antioxidant activity of soy lecithin in a triolein model system. LWT - Food Sci. Technol. 41, 581–587.

Ramadan, M.F., Asker, M.M.S., 2009. Antimicrobical and antivirial impact of novel quercetin-enriched lecithin. J. Food Biochem. 33 (2009), 557–571 33.

Ramdath, D., Padhi, E., Sarfaraz, S., Renwick, S., Duncan, A., 2017. Beyond the cholesterol-lowering effect of soy protein: a review of the effects of dietary soy and its constituents on risk factors for cardiovascular disease. Nutrients 9, 324.

Scartazzini, R., Luisi, P.L., 1988. Organogels from lecithins. J. Phys. Chem. 92, 829–833.

Scow, R.O., Stein, Y., Stein, O., 1967. Incorporation of dietary lecithin and lysolecithin into lymph chylomicrons in the rat. J. Biol. Chem. 242, 4919–4924.

Singh, A., Thotakura, N., Kumar, R., Singh, B., Sharma, G., Katare, O.P., Raza, K., 2017. PLGA-soya lecithin based micelles for enhanced delivery of methotrexate: Cellular uptake, cytotoxic and pharmacokinetic evidences. Int. J. Biol. Macromol. 95, 750–756.

Singh, H., Ye, A., Horne, D., 2009. Structuring food emulsions in the gastrointestinal tract to modify lipid digestion. Prog. Lipid Res. 48, 92–100.

Stremmel, W., Gauss, A., 2013. Lecithin as a therapeutic agent in ulcerative colitis. Dig. Dis. 31, 388–390.

Swanson, R.A., Morton, M.T., Tsao-Wu, G., Savalos, R.A., Davidson, C., Sharp, F.R., 1990. A semiautomated method for measuring brain infarct volume. J. Cereb. Blood Flow Metab. 10, 290–293.

Szuhaj, B.F., 1989. Lecithins: Sources, Manufacture & Uses. The American Oil Chemists Society.

Szuhaj, B.F., JuDong Y., Fereidoon S., 2005. "Lecithins." Bailey's industrial oil and fat products, 1–86.

Tammenmaa-Aho, I., Asher, R., Soares-Weiser, K., Bergman, H., 2018. Cholinergic medication for antipsychotic-induced tardive dyskinesia. Cochrane Database Syst. Rev 3.

Thistle, J., Schoenfield, L., 1968. Bile acid, lecithin, and cholesterol in repeated human duodenal biliary drainage: effect of lecithin feeding. Clin. Res. 16, 450.

Toouli, J., Jablonski, P., Watts, J.M., 1975. Gallstone dissolution in man using cholic acid and lecithin. Lancet North Am. Ed. 306, 1124–1126.

Tuzhilin, S., Dreiling, D., Narodetskaja, R., Lukash, L., 1976. The treatment of patients with gallstones by lecithin. Am. J. Gastroenterol. 65, 231–235.

Uauy, R., Dangour, A.D., 2006. Nutrition in brain development and aging: role of essential fatty acids. Nutr. Rev. 64, S24–S33.

van Hoogevest, P., 2017. Review–An update on the use of oral phospholipid excipients. Eur. J. Pharm. Sci. 108, 1–12.

van Hoogevest, P., Wendel, A., 2014. The use of natural and synthetic phospholipids as pharmaceutical excipients. Eur. J. Lipid Sci. Technol. 116, 1088–1107.

van Nieuwenhuyzen, W., Tomás, M.C., 2008. Update on vegetable lecithin and phospholipid technologies. Eur. J. Lipid Sci. Technol. 110, 472–486.

Wang, T.Y., Liu, M., Portincasa, P., Wang, D.Q.H., 2013. New insights into the molecular mechanism of intestinal fatty acid absorption. Eur. J. Clin. Invest. 43, 1203–1223.

Wilde, P., Chu, B., 2011. Interfacial & colloidal aspects of lipid digestion. Adv. Colloid Interface Sci. 165, 14–22.

Williams, R., Chapman, D., 1971. Phospholipids, liquids crystals and cell membranes. Prog. Chem. Fats Other Lipids 11, 1–79.

Wurtman, R., Growdon, J., Hirsch, M., 1977. Lecithin consumption raises serum-free-choline levels. Lancet North Am. Ed. 310, 68–69.

Wurtman, R.J., Growdon, J.H., 1978. Dietary enhancement of CNS neurotransmitters. Hosp. Pract. 13, 71–77.

Zhang, H., Dudley, E.G., Davidson, P.M., Harte, F., 2017. Critical concentration of lecithin enhances the antimicrobial activity of eugenol against Escherichia coli. Appl. Environ. Microbiol. 83, e03467–16.

CHAPTER

Lignans

4.10

Juan M. Álvarez-Caballero[a], **Ericsson Coy-Barrera**[b]

[a]*Chemistry and Bioprospecting of Natural Products Group, Universidad del Magdalena, Santa Marta, Colombia*

[b]*Bioorganic Chemistry Laboratory, Faculty of Basic and Applied Science, Universidad Militar Nueva Granada, Cajicá, Colombia*

4.10.1 Background

Lignans belong to the phenylpropanoid class and are recognized as biologically active antioxidants, which are further subdivided into lignans and neolignans (Teponno et al., 2016). For centuries, (neo)lignans are included in both herbal remedies and edible plants as diet. Predominant sources of edible (neo)lignans comprise some seeds (e.g., *Linum usitatissimum* L., commonly known as flaxseed), and whole grains (Milder et al., 2005). Therefore, phytochemists, food chemists and pharmacologists are interested on the structures, functionality and biological activity, respectively, of various dietary and nondietary lignans. For instance, podophyllotoxin, honokiol and secoisolariciresinol (SECO)-type (neo)lignans have motivated important attention (Teponno et al., 2016). Within the last 50 years, numerous studies focused on the health benefits and biological properties of different lignans has raised, even those transformed by gut microbiota. In this sense, investigations about (neo)lignan bioactivity on cancer, cardiovascular, and metabolic conditions/disorders has mostly revealed beneficial actions, in spite of some inconclusive or negative effects are also reported (Zálešák et al., 2019). Gene suppression, hormonal metabolism, angiogenesis, and antioxidant activity/capacity are generally the plausible mechanistic descriptions for the declared promissory biological activity of (neo)lignans (Durazzo et al., 2018). In the available literature, the join terms "natural" and "antioxidants" are extensively disseminated. From this fact, some interesting reviews describing some important points associated to these terms have been recently published (Guo et al., 2019; Hunyadi, 2019; Lourenço et al., 2019). In the case of antioxidant actions of (neo) lignans, such properties are justified owing to their phenolic nature (Hosseinian et al., 2007), so a valuable information prevails but a smaller dimension is reached in comparison to other phenolics.

In this context, reactive oxygen (ROS) and reactive nitrogen (RNS) species are the main targets (although not exclusively) of antioxidants in biological systems

Antioxidants Effects in Health. DOI: https://doi.org/10.1016/B978-0-12-819096-8.00050-1

(Apak et al., 2013; Hunyadi, 2019). ROS/RNS are free radicals that are created from by-products of redox processes at cellular level, in order to satisfy various functions related to cell signaling and immune roles. However, if these reactive species reach elevated concentrations, oxidative stress is promoted, whose occurrence might alter important structures and functionality in cells (Pham-Huy et al., 2008; Manivannan et al., 2016). An unfortunate effect of this distortion is the metabolic syndrome (MetS) (Guo et al., 2019). Hence, antioxidants arise to counteract the imbalance of those ROS/RNS (Hunyadi, 2019). In general terms, antioxidants constitute chemical entities that prolong auto-oxidation through inhibition of the free radical formation and/or disruption of the spread of oxidant chemical species. Such actions are accomplished by means of scavenging of peroxidation-initiating species, chelation of metal ions to impede their ability to generate reactive species, reduction of localized O_2 contents, prevention of peroxide formation, and/or breaking of the auto-oxidative chain reaction (Nawar, 1996).

On the other hand, the data obtained from methods/assays to measure/evaluate the antioxidative properties of compounds of natural origin are other part of the huge information about it. However, the information of those methods related to mechanisms and ways to express results and scope is highly diversified and usually difficult to be compared (Haida and Hakiman, 2019). Therefore, as a contribution to the rationale for understanding this specific condition, explicitly the case of (neo)lignans, several concepts deserve to be clarified. In the first place, a single method/assay to measure/evaluate the antioxidant properties of a particular compound has not been developed so far (Apak, 2019). However, the current reported methods/assays can be generally categorized into the mechanisms based on electron transfer (ET) and hydrogen atom transfer (HAT). Hence, the discrepancies and incomparable results from different methods can be rationalized depending on these mechanisms and other variables such as pH, temperature, solubility, redox and chemical potentials, presence of other ions and interferences, among others (Bunaciu et al., 2016). Secondly, antioxidant capacity and antioxidant activity have distinctive connotations as previously described in the technical report from the International Union for Pure and Applied Chemistry (IUPAC) (Apak et al., 2013). Thus, the first term comprises the reaction kinetics of the free radical-antioxidant pair while the second one assesses the efficiency of the thermodynamic transformation of an oxidant agent (as probe) when reacts with the antioxidant.

From the above-mentioned background, the current compilation gathers and discusses the chemistry and biological activity of (neo)lignans, making emphasis on antioxidant actions. Its assembly was achieved from the available published information in literature according to the *in-vitro* and/or *in-vivo* evidences, in order to connect some ideas to clarify the potential benefits of these compounds on human health.

4.10.2 Sources of lignans

Lignans are secondary metabolites of wide distribution in nature. These compounds are present in the plant kingdom from pteridophytes, gymnosperms, and angiosperms. From the chemotaxonomic point of view, its distribution throughout plant organisms appears to be related to its structural complexity. In this sense, the simplest structures occurred in pteridophytes, but increasing in number and complexity in gymnosperms and especially in the angiosperm group (Ríos et al., 2002). The number and levels is another differential factor between sources. In pteridophytes, they have been identified in ferns with a reduced number of compounds. In gymnosperms, lignans are more widely distributed throughout the various families and become more structurally complex (Castro et al., 1996). For angiosperms, they present a large structural variety and a wide distribution mainly in dicotyledons (Lewis and Davin, 1994).

Lignans are present in various plant structures such as leaves, bark, wood, fruits (Ayres and Loike, 2008) and are part of the nature of various food sources, including seeds (e.g., flax, pumpkin, and sesame), grains (e.g., barley, wheat, and oats), legumes (e.g., soybeans, lentils, and beans), and vegetables (e.g., carrots, garlic, broccoli, and asparagus), among others (Murphy and Hendrich, 2002; Tham et al., 1998). As mentioned, occurrence and levels of particular lignans can vary depending on the plant source (Zálešák et al., 2019; Zhang et al., 2014). For instance, SECO can be found ranging from 0.1 (vegetables and cereals) to 3,699 (flaxseed) mg/kg dry weight (DW), whereas secoisolariciresinol diglucoside (SDG) and matairesinol (MAT) showed levels of 11,900–25,900 mg/kg DW (flaxseed) and 0.2 (fruits) to 28.5 (flaxseed) mg/kg DW, respectively (Touré and Xueming, 2010). Sesamin and sesamolin occurred at 1,547–8,852 and 1,235–4,765 mg/kg DW, respectively, in sesame seeds (Touré and Xueming, 2010).

Lignans are not only present in plants. These metabolites have also been identified in different secretions and fluids (e.g., urine, serum, bile) of mammals, including humans (Setchell et al., 1981, 1980). The presence of lignans in vertebrates is determined by the metabolic transformation developed by the intestinal microbiota from a diet rich in integral and vegetable products. The main microbiota-transformed lignans that have been identified and studied are enterolactone (ENL) and enterodiol (END). Such lignan-derived end-products are generated from the bacterial transformation of MAT and SECO, which are produced/accumulated in various food sources (Borriello et al., 1985).

4.10.3 Chemistry

The label "lignan" was firstly used to categorize a type of compounds structurally formed by phenylpropanoid units (comprising C_6-C_3 building blocks), which are linked by means of an 8-8′ bond (Haworth, 1936). Subsequently, Gottlieb uses the term neolignan to reference those metabolites that are similarly conformed by two C_6-C_3 units, but linked differently to an 8-8′ bond (Gottlieb, 1972). Hitherto, lignans have been defined as compounds that result from the radical pairing of one-electron-oxidized cinnamic alcohols (even acids), whereas allyl/propenyl phenols are the monomers of neolignans (Gottlieb, 1978). However, under a strict differentiation, lignans exhibited more reported studies (possibly by the abundance in edible sources) than neolignans. Therefore, in order to avoid confusions, "(neo)lignan" was considered as a general term to involve lignans and neolignans in the present compilation. Otherwise, it is specified.

4.10.3.1 (Neo)lignans classification: A summary

Due to the high structural variety that this type of compound has exhibited, various subgroups have been raised that attempt to explain the great diversity of these molecules in nature. Therefore, a brief summary of such a classification is compiled in this section, but detailed reviews are then referred (Moss, 2000; Teponno et al., 2016; Whiting, 1987; Zhang et al., 2014). Briefly, lignans have been classified into eight subgroups, taking into account their cyclization assembly and the position of an oxygen atom between two C_6-C_3 units. Such subgroups comprise arylnaphthalenes, aryltetralins, dibenzylbutanes, dibenzocyclooctadienes, dibenzylbutyrolactols dibenzylbutyrolactones, furofurans and furans (diaryltetrahydrofurans) (Fig. 4.10.1) (Whiting, 1987; Lewis and Davin, 1994).

In the case of neolignans, around fifteen groups have been demarcated such as benzofurans, dihydrobenzofurans, diarylethanes, benzodioxines, alkylaryleters, biphenyls, bicycle[3.2.1]octanes, among others. Owing to some of NL has been assigned no special names, some authors designate them with the term "NL" (i.e., neolignans) from 1 to 15 (Fig. 4.10.2) (Teponno et al., 2016; Ríos et al., 2002; Whiting, 1987). In addition, hybrid lignans are also referenced, which correspond to structures formed by phenylpropanoids linked to other types of compounds such as coumarins, flavonoids or triterpenes, forming flavonolignans, coumarinolignans, among others (Fig. 4.10.3) (Ríos et al., 2002; Whiting, 1987). On the other hand, there is also occurrence of oligomeric (neo)lignans and norlignans. Oligomeric (neo) lignans are formed by those compounds that have more than two C_6-C_3 units (in some cases 4 or 5 phenylpropanoids) forming the molecule. In the case of norlignans, they are defined as phenolic compounds with a diphenylpentane moiety (C_6-C_5-C_6),

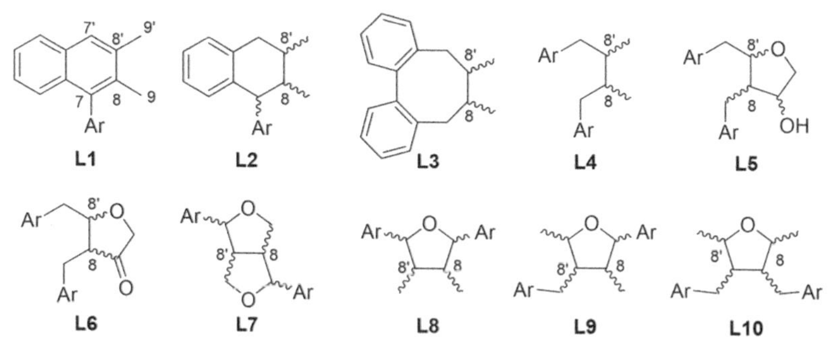

FIG. 4.10.1 Structures of various subgroups of lignans (L1–L10).

L1, arylnaphthalene; L2, aryltetralin; L3, dibenzocyclooctadiene; L4, dibenzylbutane; L5, dibenzylbutyrolactol; L6, dibenzylbutyrolactone, L7, furofuran; L8, 2,5-diaryltetrahydrofuran; L9, 2-aryl-4-benzyltetrahydrofuran; L10, 3,4-dibenzyltetrahydrofuran.

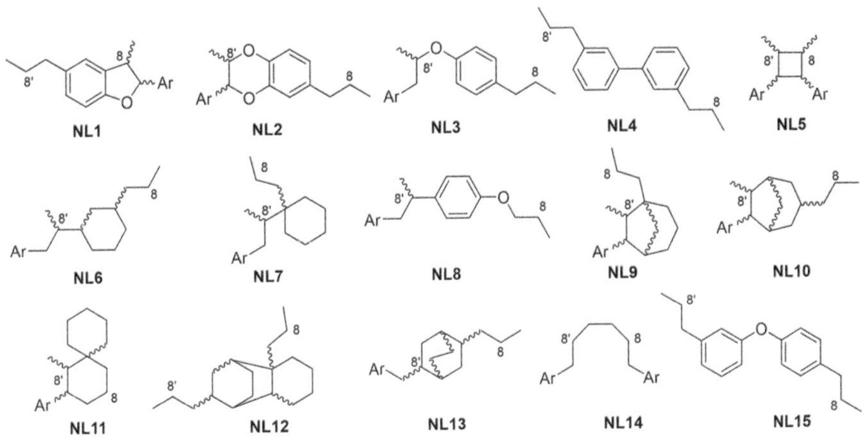

FIG. 4.10.2 Structures of various subgroups of neolignans (NL1–NL15).

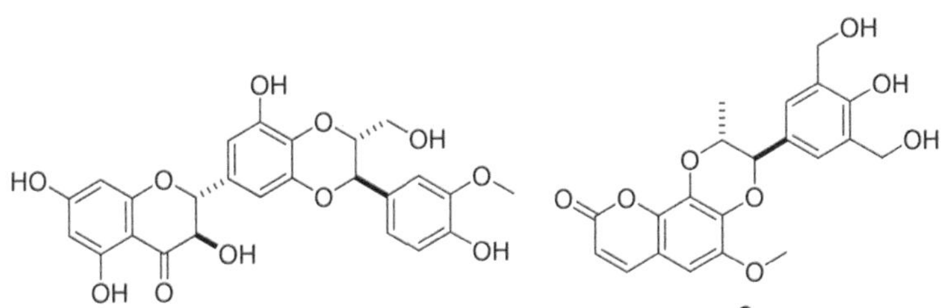

1 2

FIG. 4.10.3

Two examples of hybrid lignans: silybin (1) (i.e., flavolignan) and jatrocin B (2) (i.e., coumarolignan)

which have been isolated mainly in conifers and monocotyledons (Suzuki et al., 2001; Ríos et al., 2002; Whiting, 1987).

In order to have greater clarity into the classification and numbering of the huge lignan chemodiversity, IUPAC recommended some important issues to organize them. In 2000, Moss states that there are two basic structures for these compounds: lignans and neolignans (under the same definitions mentioned above). In addition, there is a subgroup, called oxineolignans, whose compounds comprise two C_6-C_3 units linked together by an ether bridge. In the same way, this classification gathers the superior (neo)lignans analogs (i.e., those structures formed by three or more C_6-C_3 units (oligomers)) and, similarly to that of terpenoid organization, (neo)lignans have been called sesquineolignans (C_6-C_3 trimers), dineolignans (C_6-C_3 tetramers) and sesterneolignans (C_6-C_3 pentamers) (Moss, 2000) (see Fig. 4.10.4). In the case of norlignans, they have been defined as derivatives of lignans or neolignans whose structural core lost carbon atoms (one or more). In other words, they represent dimeric structures of phenylpropanoids comprising C15, C16 or C17 moieties (Moss, 2000; Suzuki et al., 2001) (See Fig. 4.10.5).

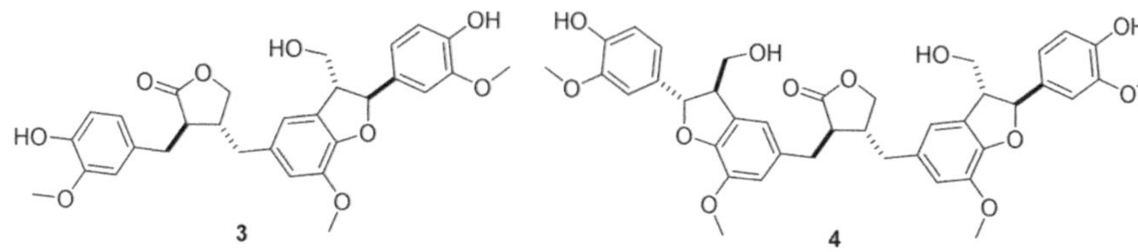

FIG. 4.10.4

Two examples of neolignan oligomers: lappaol A (3) (i.e., sesquineolignan) and lappaol F (4) (i.e., dineolignan).

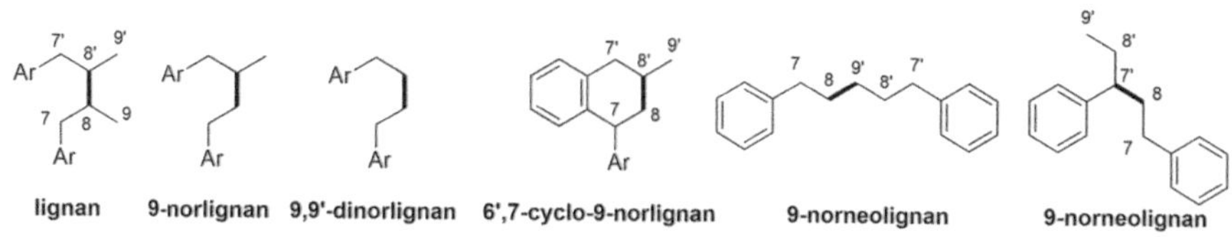

FIG. 4.10.5 Some examples of norlignan-like structures.

4.10.3.2 Biosynthesis of lignans: An overview

The biosynthesis of the basic structures of lignans has been proposed and described. However, there is a lack of enough information to verify many of the proposals established for several subgroups of these metabolites (Suzuki et al., 2001). Many of such investigations have currently focused to explain the enzyme-mediated transformations, especially those related to regioselectivity, stereoselectivity and enantiospecificity (Ríos et al., 2002). Thus, the present compilation describes a general overview on those processes that characterize the biosynthetic aspects of the basic structures of (neo)lignans. Some interesting reviews on biosynthetic insights of lignans were previously reported (Satake et al., 2015; Solyomváry et al., 2017; Suzuki and Umezawa, 2007).

Lignans and neolignans are originated biosynthetically via shikimic acid pathway. These two kind of metabolites are generated by two phenylpropanoid units that are obtained from L-phenylalanine (Fig. 4.10.6). In this process, this amino acid undergoes a deamination reaction due to the action of phenylalanine ammonia lyase (PAL) to produce cinnamic acid. A subsequent hydroxylation (by P450) produces the *p*-coumaric acid, which can be also converted to caffeic acid. This dihydroxylated analog is then transformed by methylation (through *O*-methyltransferases) to produce *p*-coumaric acid derivatives such as ferulic and/or synapic acids. These derivatives are reduced to aldehydes until its corresponding alcohol, such as coniferyl or synapilic alcohols (Vogt, 2010; Zálešák et al., 2019). These phenolic C_6-C_3 units are the basic structures that are implicated in the radical coupling towards the formation of lignans and/or neolignans through dimerization processes. The coupling mechanisms of the involved monomeric C_6-C_3 units are mediated by specific enzyme structures and accompanied by auxiliary oxidase or peroxidase (Ward, 2000; Suzuki and Umezawa, 2007). Knowledge about the enzymes responsible for such transformation processes has been achieved in several

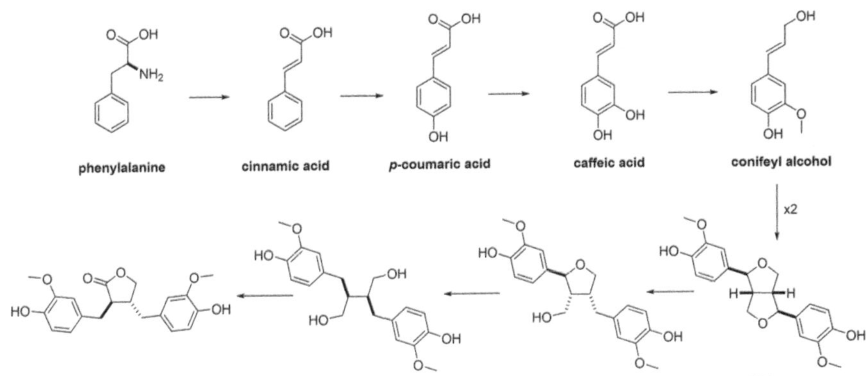

FIG. 4.10.6 Scheme of the biosynthetic process of representative lignans occurred in some food sources.

(+)-PIN, (+)-pinoresinol; (+)-LAR, (+)-lariciresinol; (-)-SECO, secoisolariciresinol; (-)-MAT, (-)-matairesinol.

experiments in various plant species, mainly those studies conducted on *Forsythia intermedia* and *F. suspensa*. These experiments allowed the first evidence of the transformation of pinoresinol (PIN) to MAT. Previous information on radical coupling process for phenylpropanoid units was based mainly on the action of the H_20_2-dependent peroxidases, 0_2-dependent lacases or phenol oxidases, that generate racemic-type products (Lewis and Davin, 1994). However, studies on *Forsythia* plants, for the first time, let to observe the pairing of *E*-coniferyl radicals to generate stereoselectively (+)-PIN and (-)-MAT (Ward, 2000). In recent years, a larger number of experiments elucidates this route deeply, revealing those enzymes responsible for this transformation, including their isolation and identification, such as PIN synthase, PIN/lariciresinol (LAR) reductase, and finally, SECO dehydrogenase (Suzuki et al., 2001; Umezawa, 2003; Suzuki and Umezawa, 2007).

In summary, it could be established that the monomer structures of coniferyl alcohol follow a transformation sequence of (+)-PIN, (+)-LAR, (-)-SECO and (-)-MAT (Fig. 4.10.6). It has been demonstrated that, for this sequence, two units of coniferyl alcohol undergo a stereoselective radical-mediated coupling to produce (+)-PIN under the action of PIN synthase. This lignan is transformed, by an enantiospecific reduction, to (+)-LAR and (-)-SECO, mediated by PIN reductase (Umezawa, 2003; Suzuki and Umezawa, 2007). In the last phase, due to the participation of SECO dehydrogenase, (-)-SECO is oxidized to produce (-)-MAT. These steps, in addition to explaining the biosynthesis of these metabolites, describe a proposal for the formation of different types of lignan (even neolignan) moieties. Thus, from a furanofuran (i.e., (+)-PIN) moiety, a furan-type (i.e., (+)-LAR) is generated, which in turn is transformed into a dibenzylbutane-type (i.e., (-)-SECO) moiety and, finally, a dibenzylbutyrolactone-type ((-)-MAT) is produced. The biosynthesis of lignans seems to be closely correlated with lignin production specifically through the conversion pathway from coniferyl alcohol to MAT (Suzuki and Umezawa, 2007; Lewis and Davin, 1994). In addition, other important lignans are produced from MAT after several enzyme-mediated transformations (e.g., hinokinin and podophyllotoxin-type compounds) (Satake et al., 2015). In the case of neolignans, the specific insights into the coupling conditions to produce their chemodiversity (from cinnamic-derived alcohols, and/or allyl or propenyl phenols) are not completely clear and understood due to the lack of related studies. However, an interesting fact can be considered since some neolignans are normally isolated having high diastero and enantiomeric excess, so the enzymes involved in the biosynthesis of some neolignans, apart from regiochemically-controlled transformations, employed a stereochemical restrictions during radical and/or pericyclic-mediated dimerizations (Lewis and Davin, 1992; Lourith et al., 2005).

4.10.4 Bioavailability

A particular concentration of a compound/metabolite in certain edible source does not necessarily correspond to the quantity effectively assimilated. In this process, several factors influences it, ranging from the food storage/preparation to its digestion/absorption throughout digestive system (Degerud et al., 2015). Hence, bioavailability refers to the portion/ratio of a xenobiotic compound (existing in consumed food) that can be exploited or stored by body system once previous gastrointestinal processes (i.e., digestion, absorption, distribution) have been accomplished. Some authors also define it as the systemic circulation achieved by a proportion of a certain compound/metabolite (Holst and Williamson, 2008). In general, bioavailability of a compound (in this case (neo)lignans) is determined by the chemical structure. This fact has an important influence on the rate of absorption, metabolism and therefore their biological action. Most polyphenols, such as (neo) lignans, have low intestinal absorption and are metabolized or removed very quickly. However, this bioavailability had exhibited distinguishable inter-human subjects variations (Clavel et al., 2006b), being highly-dependent on the diet style, so it might be very fluctuating (Rodríguez-García et al., 2019).

4.10.4.1 Lignans intake

For the particular case of lignans, there is a fact related to the content in food, since it is not particularly high. However, it does not necessarily indicate that their consumption in diet is low since this factor will really depend of the amount of food consumed (Peterson et al., 2010). Around 200 plant foods have been analyzed to determine their lignan content. The most-reported lignans that are consumed in higher proportion are related to MAT and SECO, ranging from 1087 to 3 and 369900 to 47 μg/100 g DW (Webb and McCullough, 2005). In a smaller proportion are found medioresinol (MED), LAR, PIN, and syringaresinol (SYR) (Webb and McCullough, 2005). The mean total intake of lignans (MTIL) is estimated from 150 to 1600 μg/day (Hedelin et al., 2008; Peterson et al., 2010). For highly-consumed metabolites from edible plants, the MAT intake is 2-74 μg/day, while SECO is reported to be within 70-992 μg/day (Ulucan et al., 2011; van der Schouw et al., 2005). Generally, a metabolite intake can vary from a geographical area to another. For instance, studies by Valsta showed that the MTIL (mostly MAT and SECO) of Finns is 434 μg/day (Valsta et al., 2003). A more recent study developed by Nurmi on the intake of dietary lignans by 100 Finn men determined that the MTIL is 1224 μg/day (Nurmi et al., 2010). Other previous studies showed that the estimated MTIL in German women was 560 μg/day, with a SECO presence around 95% (Linseisen et al., 2004), whereas the dietary lignans intake in the Dutch population was found to be around 979 μg/day (Milder et al., 2005).

4.10.4.2 Lignan metabolism

Plant lignans after consumption are transformed in digestive tract, specifically by the gut microbiota. For instance, SECO-like lignans produce demethylated and deoxygenated lignans, such as ENL and END after action of intestinal bacteria. Such a process is carried out through various metabolic reactions that lead to the lignan conversion into the respective enterolignans (Webb and McCullough, 2005). During intake, a lot of lignans are incorporated in the glycoside form, such as secoisolariciresinol diglucoside (SDG). This glycoside is hydrolyzed (by β-glucosidases derived from anaerobic organisms) and converted into SECO. This aglycone undergoes processes of dehydroxylation and demethylation by intestinal microbiota and it is transformed into END and also oxidized to afford ENL (Fig. 4.10.7) (Ford et al., 2001; Landete, 2012). *In vitro* experiments

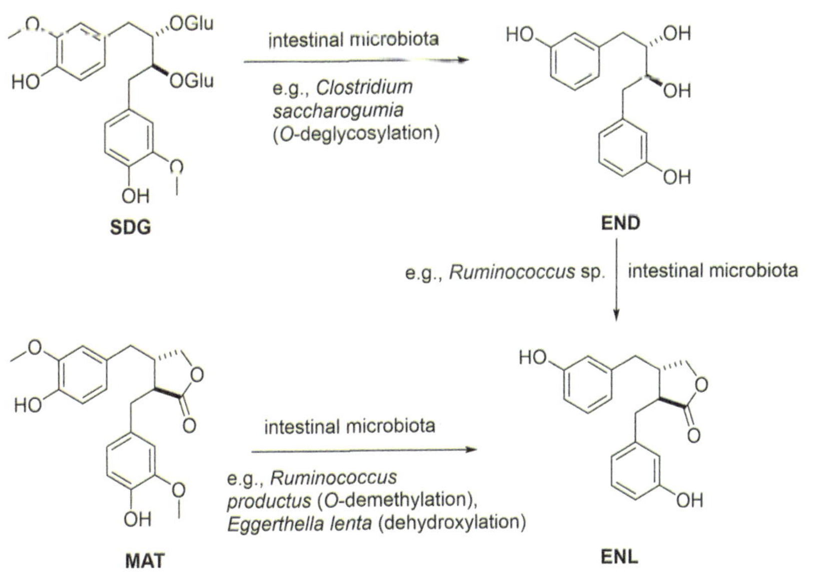

FIG. 4.10.7 Conversion of some plant lignan precursors by intestinal microbiota.

END, enterodiol; ENL, enterolactone; MAT, matairesinol; SDG, secoisolariciresinol diglycoside.

revealed the direct transformation to enterolignans (mostly ENL) of the most-common lignans existing in vegetables (e.g., MAT as the main component of the daily source of lignans) is mediated by the action of intestinal microbiota (Peterson et al., 2010). In this regard, Heinonen et al. (2001) measured the 62% conversion of MAT after its fecal incubation, being ENL mainly detected. In the case of SECO and PIN, 72% and 55% was metabolized to produce END and ENL, respectively. Only 4% of SIN was transformed. LAR was the only metabolite 100% converted into END and ENL. In general, the process of lignan transformation, their efficiency and bioavailability differs considerably from one individual to another. This fact is apparently determined, among others, by genetic factors (Low et al., 2005b, 2005a).

Several studies have identified various microorganisms responsible for intestinal action towards the metabolic conversion of plant lignans to produce enterolignans. Such microorganisms comprise *Clostridium* sp., *Clostridium cocleatum*, *Bacteroides ovatus*, *B. distasonis*, and *B. fragilis*, which are responsible for deglycosylation, while demethylation is caused by as *Peptostreptococcus productus*, *Eubacterium limosum*, *Eubacterium callanderi*, *Butyribacterium methylotrophicum*, and*Ruminococcus productus* (Clavel et al., 2006a). Some microorganisms responsible for the transformation of END to ENL correspond to *Ruminococcus* sp., END-1 and END-2 (Jin et al., 2007; Jin and Hattori, 2009). Enterolignans may be usually detected in blood of most lignan-consuming people since such lignan-transforming bacteria can be commonly presented. Hence, inter-people variations in bacterial titer can rationalize the inter-people variations of enterolignan blood content (Landete, 2012).

There is a close relationship between lignan intake and enterolignan levels in plasma and urine. The flaxseed consumption (13.5 g/day during 6 weeks) leads to μM plasma levels of END/ENL (Atkinson et al., 1993). It has also been determined that the plasma half-life of END and ENL is *ca.* 5 and 13 h, respectively. They can remain in plasma between 8 and 10 h after the initial intake (Anneleen Kuijsten et al., 2005; Tetens et al., 2013). Their excretion through urine exhibited a 3 to 285-fold increase in healthy postmenopausal women (PMW) ($n = 31$), healthy young men ($n = 6$), and healthy young women (n=18), after flaxseed intake (5-10 g/day during 6 weeks) (Hutchins et al., 2000; Lampe et al., 1994). Accordingly, several studies have also determined that a lignan-rich diet to promote higher levels of END/ENL provides health benefits, since protective or reductive effects can be favored against the risk of several negative conditions (e.g., hair loss and breast, colon, and prostate cancer) (Kuijsten et al., 2006; Lin X, Switzer BR, 2001).

Enterolignans formed from plant lignans, resume their transformation to be conjugated as *O*-glucuronides and sulfate esters, due to the enzymatic action of UDP-glucuronosyltransferases and sulfotransferases, respectively (Landete, 2012; Webb and McCullough, 2005). This process begins in the intestinal wall and ends in liver, allowing an easy excretion by urinary and biliary routes (Jansen et al., 2005; Knust et al., 2006). In urine, enterolignans are eliminated as monoglucuronides (73-94%) and another fraction as monosulfates (2%–10%), although there is a low percentage that is excreted freely (*ca.* 1%) (Adlercreutz et al., 1995). The conjugation process suffered by enterolignans is evidenced in the work developed by Jansen et al. (2005). They exposed enterolignans to colon cell cultures, so it was possible to identify ENL sulfate and END and ENL glucuronides, respectively.

In general, it should be borne in mind that there are several factors altering the levels of lignan in several fluids and, therefore, their bioavailability to generate and participate in the diversified biological roles closely related to their antioxidant properties. Among such factors, the redox state and microbial population changes of colon, the duration of intestinal transit, the half-life of END and ENL, the kind of lignans in diet and antibiotics intake can be identified (Axelson et al., 1982; Clavel et al., 2006b, 2006a; Peterson et al., 2010). Above-mentioned factors reflected the perceivable inter-individual and intra-individual variability of beneficial effects in several populations around the world after intake of plant lignans.

4.10.5 Antioxidant activity of (neo)lignans and mechanism of action

Throughout the last twenty years, naturally-occurring antioxidants (mostly phenolics and (neo)lignans as well) have been classified as important therapeutic agents (acting alone or combined with other chemical entities) against a broad range of negative conditions or diseases, including cardiovascular dysfunctions, diabetes, inflammatory diseases, cancer, even aging, among others. The beneficial effects on health of such a type of compounds are mainly attributed to their antioxidant properties, acting as free radical scavengers, cell signaling pathway interactors, redox-active metal ions chelators, and even gene expression modulators (Soobrattee et al., 2005). These facts offer the corresponding rationale for lignans related to their benefits on health owing to the formation of oxidized chemical entities (Touré and Xueming, 2010).

Such a good properties of (neo)lignans can be rationalized through the dimerization process to neutralize radicals, constituting the main antioxidative ability of lignans (Hosseinian et al., 2007). In general, antioxidative performance of (neo)lignans depends on the presence of phenolic OH group as well as the functional groups disposition on the main moiety. Conventional mechanisms for the anti-oxidative behavior of lignans showed an enhancing of the antioxidant activity by adding a further *o*-hydroxyl group to a monohydroxylated phenolic moiety. In this sense, guaiacyl-containing lignans exhibited the weakest radical scavenging capacity, while *meta*-monophenol and catechol-containing lignans showed the highest one (Eklund et al., 2005). This fact can explain the higher antioxidant activity showed by END and ENL to that of the parent lignans SDG and SECO (Kitts et al., 1999). Therefore, the antioxidant activity/capacity of (neo)lignans is mostly enhanced by the pattern and number of *H*-donor substituents (mostly phenolic OH), i.e., it is strongly related to structural factors (Cao et al., 1997). However, benzylic hydrogen atoms are also important to outline the antioxidant potential in most phenolic lignans, since they are other probable abstracting hydrogen atoms to create radicals. This formation can add likely stabilization by resonance of phenoxyl radicals (Vo et al., 2018). Thus, benzylic C-H bonds scavenge radicals through hydrogen atom transfer, while phenolic O-H bonds scavenge radicals by proton coupled electron transfer (Vo et al., 2018). These features undoubtedly constitutes the functional antioxidant properties of (neo) lignans, whose effects can be possibly feasible at *in vivo* doses (Hu et al., 2007). In this regard, particular features that some lignans have shown, explaining their antioxidant behavior, are following described.

4.10.5.1 Sesamoline and sesamol

Sesame is one of the oldest, known oilseeds, which are identified by their nutritional value and medicinal uses (Namiki, 1989). Several metabolites with antioxidant capacity have been identified from its seeds and oil, such as sesamoline (5), sesamolinol, and sesaminol (6) (Fig. 4.10.8). Other very important antioxidant component in sesame is sesamol (7). This compound is not specifically a lignan, but it is produced by a thermal reduction of sesaminol (Budowski, 1950; Kikugawa et al., 1983). Some studies have indicated that sesame oil can resist oxidative deterioration by the presence and antioxidant action of sesamol. This process seems to be related to the thermal conversion that sesamoline (5) undergoes to produce sesaminol (6) and sesamol (7) (Fukuda et al., 1994). Similarly, its ability to inhibit lipid oxidation induced by UV or Fe(II) has been demonstrated (Chen and Ahn, 1998).

Various studies have been developed over this metabolite in order to determine its antioxidant behavior. Sesamol (7) contains two groups: a phenolic and a benzodioxol (Fig. 4.10.8). The first one is mainly responsible for its antioxidant behavior (Aruoma, 1994) and the second one is associated to several biological actions

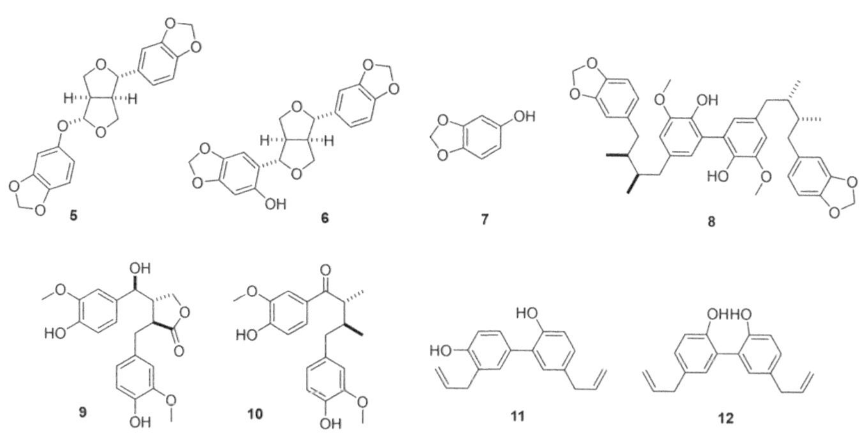

FIG. 4.10.8

Structures of some lignan-related metabolites with antioxidant activity: sesamoline (5), sesaminol (6), sesamol (7), argenteane (8), 7-hydroximatairesinol (9), cinnamophilin (10), honokiol (11), magnolol (12).

(Mazzio et al., 1998; Joshi et al., 2005). In this sense, it has been identified that its mechanism of action to scavenge hydroxyl-type radicals produces 1,2-dihydroxybenzene, by means of a demethylenation of benzodioxol. This reaction generates a catechol-type moiety which is able to trap free radicals (Kumagai et al., 1991). Sesamol is a soluble compound in aqueous and lipid phase, therefore, it can efficiently scavenge several oxidizing primary/secondary-origin radicals, including the physiologically-generated hydroxyl, tryptophyl and peroxylipid radicals. Sesamol also inhibits lipid peroxidation, deoxyribose degradation induced by hydroxyl radicals and DNA cleavage (Joshi et al., 2005). In addition, aqueous sesamol produces sesamolyl and benzoquinone ions and/or cyclohexadienyl-type and dimeric radicals under pulsed/continuous UV exposure (Nakagawa, 2000). It has also been discovered that this lignan derivative can scavenge 2,2-diphenyl-1-picrylhydrazyl (DPPH$^•$) (Kurechi et al., 1980) and imidazoquinoxaline-type radicals (Kato et al., 1996).

4.10.5.2 Argenteane

Studies on dilignan argenteane (**8**), obtained firstly from nutmeg (*Myristica argentea*, a plant commonly employed as spice) (Filleur et al., 2002), have focused on determining its antioxidant activity based on its ability to inhibit lipid peroxidation (very similar to vitamin E) and scavenge DPPH$^•$ free radicals. These results were supported by theoretical studies, which were conducted to explain the antioxidative action of argenteane. Calculations at the theory of functional density (DFT) level of dissociation enthalpies for O-H bonds correlate with the free radical-scavenging ability. This fact confirmed that the biphenolic part of the molecule is capable of providing hydrogen atoms, following the well-known reaction ArOH + R$^•$ → ArO$^•$ + RH. In other words, its antioxidant behavior is influenced by the H transfer capacity of the central OH groups. The antioxidant capacity of argenteane was compared with *meso*-dihydroguayretic acid and *erythro*-austrobailignan-6. However, this dilignan exhibited intermediate ability values to inhibit lipid peroxidation and scavenge free radicals (Calliste et al., 2010).

4.10.5.3 Enterolignans and precursors (SDG-SECO-END-ENL)

There is a high attention for determining/delineating the antioxidant actions/properties/effects of both enterolignans and their precursors, taking into account the close relationship between the free-radical scavenging capability and the anticancer activity exhibited by this type of metabolites. An example of this relationship is the case of SDG, the main constituent of flax seed, whose consumption is related to cancer prevention (Prasad, 1997). Studies have revealed the SDG's capacity to capture exogenously-generated $^•$OH radicals and prevent the $^•$OH-promoted lipid peroxidation at 319.3-2554.4 μM dose range (Prasad, 1997). This fact inferred that its antioxidant capacity may be related to its anticancer activity. In another study, in addition to the assessment of SDG, the aqueous and lipidic-phased antioxidant capacity of enterolignans (i.e., END and ENL) was also determined. All three metabolites exhibited the capacity to block the linoleic acid's peroxidation at 10–100 μM (Hu et al., 2007). Additionally, in a deoxyribose assay, the activity of Fenton reagent-induced OH capturing (non-site specific and site specific) was evaluated. ENL and END enterolignans were shown to be more effective in preventing degradation of deoxyribose in comparison to SDG at similar concentrations. This indicates the higher ability of END and ENL to react with hydroxyl radicals (Kitts et al., 1999). SDG radical scavenging activity appears to be strongly related to its ability to prevent the development of diabetes mellitus, hypercholesterolemia, and atherosclerosis (Prasad, 2000a). Antioxidative capacity of these enterolignans has also been evaluated for inhibiting polymorphonuclear leukocytes (PMNL), employing vitamin E, SDG, SECO as comparative antioxidants. Results showed a 19-94% PMNL reduction at 2.5 mg/mL. The highest activity was exhibited by END followed by SEC and ENL (Prasad, 2000a).

4.10.5.4 Other (neo)lignans

7-hydroximatairesinol (**9**) (Fig. 4.10.8) has been isolated from conifers and represents around 60% of the total lignans occurred in spruce (*Picea abies*) (Saarinen et al., 2000). Its antioxidant capacity of this lignan was measured by lipid peroxidation, elimination of superoxide and peroxyl radicals, and LDL oxidation models. Its activity was compared to known antioxidants (i.e., trolox, butylated hydroxytoluene (BHT), and butylated hydroxyanisole (BHA)). Results showed that 7-hydroximatairesinol exhibited higher antioxidant capacity to that of trolox in the different tests and was more effective in capturing superoxide and peroxyl radicals than BHA and BHT (Kangas et al., 2002). In addition, cinnamophilin (**10**) (Fig. 4.10.8) is a lignan isolated from *Cinnamomum philippinense*, whose antioxidant activity has been studied against several reactive species. It proved to be very efficient in the lipid peroxidation of various membranous systems (microsome-mitochondria) with values equal to those exhibited by trolox and α-tocopherol. The potency of cinnamophilin to scavenge peroxyl radicals was better than ascorbate, considering that this capture process was associated with the two hydrogen OH-donor groups in the molecule. In the same study, its activity was determined as a cytoprotector against oxidative stress and inhibitor of LDL oxidation. In addition, cinnamophilin reduced the generation of superoxide anion in blood vessels of animals (Hsiao et al., 2001). Finally, the neolignans honokiol (**11**) and magnolol (**12**), firstly isolated from plants of the genus Magnolia, exhibited several beneficial effects on inflammation, tumor reduction, diabetes, depression, among others. These effects have been explained by their potent antioxidant properties (Shen et al., 2010). Honokiol has been also described as a neuroprotective agent for the central nervous system to prevent/manage neurodegenerative diseases (Talarek et al., 2017).

4.10.6 Plausible pro-oxidant activity of lignans

Biological systems are regularly subjected to oxidative stress, which is promoted by different free radicals, especially ROS/RNS, which are produced by various biochemical processes. Oxidative damage is directly associated with the origin/progress of different chronic diseases (e.g., carcinogenesis, cardiovascular problems, neurodegenerative disorders, among others) (Quideau et al., 2011). In the case of lignans, the antioxidant activity depends if a phenolic moiety (or more) is presented. Therefore, they are identified as natural sources with a high antioxidant capacity. They can act as a key barrier to regulate and reduce the oxidative stress of natural systems. This fact also led to indicate that lignan-enriched foods are capable to regulate the free radical impacts and, therefore, to reduce the development of various chronic diseases. However, in recent years, several experiments have showed that antioxidant compounds can behave as prooxidants under particular conditions, that is, phenolics such as lignans could promote and generate free radicals (Decker, 2009). However, no prooxidant action had been declared for SDG, END, or ENL or any polymeric end-product (Hosseinian et al., 2007; Kitts et al., 1999).

Prooxidant activity of phenolics is supported by the fact that the presence of metal ions (e.g., Cu^{2+} and Fe^{3+}) can generate free radicals (Guardado et al., 2012). These metal ions are present in various biological processes and catalyze a feasible prooxidant feature of cathecol units, promoting the generation of hydroxyl ion and hydroxyl radicals after several redox steps under aerobic conditions, involving the formation of phenoxide/phenoxyl radicals, *o*-benzoquinone, oxygen radicals and hydrogen peroxide (Fig. 4.10.9). Hydroxyl ions/radicals interact in different ways with biological systems, causing DNA damage, enzyme inactivation, protein degradation, among others (Borg and Schaich, 1984).

Prooxidant/antioxidant capacity of lignans depends largely on the interaction of various factors, ranging from structural features (intrinsic factors) to the medium conditions (extrinsic factors) (Fig. 4.10.10). In general, if pH decreases, the iron-reducing activity heightens and the chelating capability of phenolic compounds to block the catalytic action of this metal ion is therefore reduced. Processes such as lipid oxidation are influenced by pH, so the activity can be modified from oxidant to prooxidant due to the pH variation (Moran et al. 1997; Hider et al., 2001).

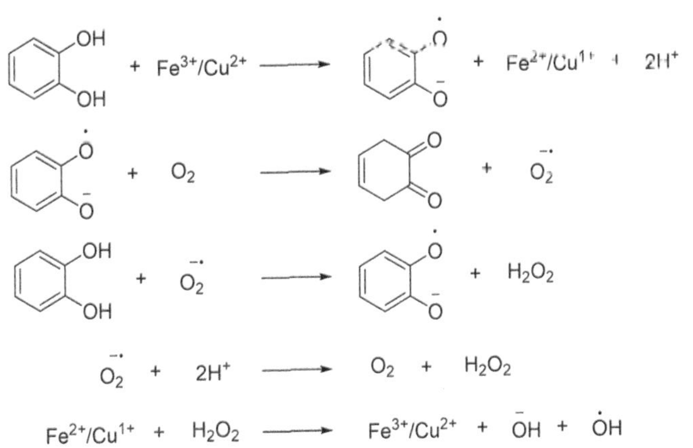

FIG. 4.10.9 Mechanism of the prooxidant action of phenolics-related compounds promoted by Cu^{2+} and Fe^{3+}.

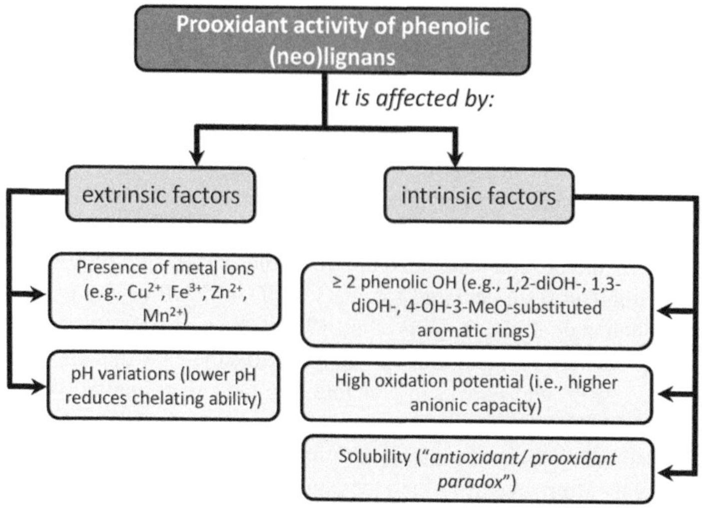

FIG. 4.10.10 Main factors that influence the pro-oxidant effect of phenolic (neo)lignans.

Solubility is another factor to take into account in the prooxidant behavior of a substance, within a process called "the antioxidant paradox". Thus, the capacity to capture hydrophilic radicals in bulk oils is more efficient than hydrophobic ones, while hydrophobic scavengers are more efficient if they are present in emulsified phases (Porter, 1993). This fact was also evidenced with hydrophilic compounds (e.g., trolox and gallic acid), which exhibited better antioxidative properties to that of α-tocopherol and propyl gallate hydrophobes in bulk oil. Hence, when an emulsified oil was used, the free radical scavenging effectiveness by α-tocopherol was better than trolox. In lipid emulsion systems, the phenolic content in emulsifier and lipid phases was higher if its hydrophobicity was enhanced and, consequently, the inhibition of lipid oxidation was also improved (Frankel et al., 1994; Huang et al., 1996). Another additional factor is the number and pattern of phenolic OH substituents (Eghbaliferiz and Iranshahi, 2016). *Ortho*-dihydroxylated or 4-hydroxy-3-methoxy-substituted compounds generated a higher prooxidant activity, reflecting a significant DNA alteration (Azmi et al., 2005). High concentrations of metal ions, such as Cu^{+2}, could influence the prooxidant effect of lignans, promoting cell apoptosis and DNA damage (Azmi et al., 2005). Other cases exhibited the formation of the protein-Cu^{+2} complex that is linked to the prooxidant capacity of some phenolics, involving ROS production, which are able to interact lasciviously with DNA (Ahmad et al., 2005; Yoshino et al., 2004). Another feature to be considered within the antioxidant/prooxidant dual behavior of (neo)lignans is the oxidation potential. Thus, lignans having a low oxidation potential (i.e., low anionic capacity) showed antioxidant activity, while those possessing high oxidation potential promote prooxidant capacity. In this regard, (neo)lignans with electron-donor substitutents (≥2) have reduced anodic potentials and enhanced antioxidative action than monosubstituted phenolic compounds, although OH groups exhibit more potent effects than methyl radicals (Simić et al., 2007). Additionally, prooxidant lignans (e.g., schisandrin B) may generate an antioxidant response through glutathione. Thus, a cytoprotection is conferred at *in-vitro* level, while a similar *in-vivo* response can vary according to different factors such as toxicity and bioavailability (Leong et al., 2012).

4.10.7 Beneficial effects of lignans on health

Occurrence of lignans is widely distributed in several daily-consuming foods involving seeds, fruits, beverages, and vegetables. Beneficial effects on health are described to several lignan-enriched edible sources, such as flaxseeds and sesame seeds, due to the lignan content (Parikh et al., 2018; Namiki, 2007). In the most cases, antioxidative properties are the main factor to be correlated to the beneficial effects on health of plant/mammalian lignans. For instance, the observed good behavior of SDG in lupus nephritis and cancer was deduced to be produced by its hydroxyl radical scavenging capacity (Touré and Xueming, 2010; Prasad, 1997). Additionally, there are evidences that some lignans can protect against osteoporosis and cancer (Bilal et al., 2014; Zheng et al., 2014; Webb and McCullough, 2005). Lignans also promote health benefits by their interaction with the estrogen metabolism (EM) due to the chemical likeness to 17β-estradiol (17βE) (Mousavi and Adlercreutz, 1992). EM can be influenced by lignans through various processes: (1) estrogen transport, (2) estradiol biosynthesis, and (3) estrogen receptor-mediated gene transcription (Rietjens et al., 2017). Indeed, part of the beneficial effect of several lignan-rich edible sources is mainly associated to the well-known impact of lignans on EM. However, such a influence can vary according to the 17βE levels, since lignans participate as estrogen antagonists if 17βE levels are normal, but they can exhibit a weak estrogenic ability at low 17βE levels (Hutchins et al., 2000). Lignans can also contribute beneficially in various cancer classes (e.g., breast, prostate, colon, ovarian, etc), diminishing angiogenesis and metastasis or preventing variations of precancerous cells (Liu et al., 2017; Saarinen et al., 2007), which had been discreetly linked to their antioxidant ability (Hosseinian et al., 2007; Kitts et al., 1999). In addition, an indirect protective effect against cancer had been related to the EM influence by the production of 2-hydroxy-estrogen (Haggans et al., 1999). In general, several experiments, performed at *in vivo* and *in vitro* levels, established that various types of lignans exhibited antimicrobial (antiviral and antibacterial), anti-inflammatory (NF-kB inhibition and nitric oxide production suppression), anticancer, and antioxidant activities, as well as cardiovascular benefit, among others (Rodríguez-García et al., 2019). Thus, the contemporary evidences of the lignan bioactivity represent the facts for cataloguing them as health-promoting naturally-occurring compounds in humans. However, other studies showed some examples having insufficient and inconclusive results or even negative effects on health (Landete, 2012). Therefore, the information and evidences of health benefits of lignans and their sources cannot be generalized and should be adopted individually depending on the lignan type.

The transformation of dietary chemicals by human intestinal bacteria (HIB) conducts to the production of other metabolites, but such compounds may exhibit beneficial or detrimental influence on human health (Russell et al., 2013). Among estrogen-like lignans, the most known naturally-occurring metabolite is SECO, presented as the diglucoside derivarive SDG, which is abundantly occurred in flaxseeds (Kezimana et al., 2018). SDG has been described to have potential to promote the incidence reduction of various chronic diseases (Herchi et al., 2014). Particularly, HIB appear to have an essential role on the beneficial health properties of bioactive lignans, since if SDG is hydrolyzed to produce SECO, this is further converted by HIB into the enterolignans END and ENL, or MAT transformed to ENL (Webb and McCullough, 2005). Numerous dietary important precursors of enterolignans are currently known, such as sesamin, LAR, medioresinol, hydroxymatairesinol, arctigenin, syringaresinol, PIN, among other similar lignans (Durazzo et al., 2018; Webb and McCullough, 2005). Several previous investigations suggested that these renowned mammalian lignans can provide beneficial effects on health owing to the evidenced antioxidant activity (Kitts et al., 1999), antiestrogenic (even weak estrogenic) effects (Mueller et al., 2004), antiproliferative and apoptotic properties (Shin et al., 2019; Mousavi and Adlercreutz, 1992), hypocholesterolemic and antiatherogenic effects (Prasad, 2000b), diabetes development interruption (Prasad, 2001), enzyme activity inhibition (Ilbeigi et al., 2017; Adlercreutz et al., 1993), and/or other actions with mechanisms not yet elucidated.

Analogously to dibenzylbutyrolactone-like dietary lignans, other (neo)lignans (such as sesamin, honokiol and magnolol, etc) can be transformed through glucuronidation and they usually exhibited no adverse effects (or weak), even if high doses are used. Therefore, different food safety authorities have recently considered harmless to various dietary and non-dietary lignans, and no detrimental effect on health had not been observed for them (Sarrica et al., 2018; Stopper et al., 2005).

4.10.8 *In-vitro* evidences of antioxidant activity of (neo)lignans

A plethora of *in vitro* assays (in both aqueous and lipid media) demonstrated the strong antioxidant capacity/activity of plant and mammalian lignans. In comparison to the most known antioxidants of natural origin, (neo)lignans have shown predominantly higher antioxidant performance, having potential as therapeutic/preventive properties (Rodríguez-García et al., 2019). However, such experimental results cannot be compared easily for structure-activity relationship (SAR) purposes due to inconsistencies and assay standardization (Xu et al., 2019; Zálešák et al., 2019). In this regard, common assays used for antioxidant activity/capacity are related to lipid peroxidation, DPPH$^{\bullet}$, 2,2′-azobis(2-amidinopropane) dihydrochloride (AAPH), 2,2′-azino-bis(3-ethylbenzothiazoline-6-sulphonic acid (ABTS) and ferric reducing antioxidant power (FRAP), among others (Runeberg et al., 2019; Teponno et al., 2016; Zálešák et al., 2019), while 3,5-di*tert*-butyl-4-hydroxytoluene (BHT), vitamin E, trolox, or α-tocopherol have been used as reference antioxidants. As a consequence, results are quite different and some of them are inconsistent (Ghafoorunissa et al., 2004; Hosseinian et al., 2007, 2006; Kyselka et al., 2017; Prasad, 2000a; Yamauchi et al., 2004). Nevertheless, as a generalization from available reported data, dibenzylbutane and furofuran lignans exhibited the highest free radical scavenging activity (DPPH IC_{50} = 6.5–98.3 μM; ABTS IC_{50} = 0.3–114.8 μM) and reducing power [FRAP = 174.1–16.3 μg/mL (at 50 μg/mL); 95.1–48.8 mg/mL (at 0.75 mg/mL)] among reported (neo)lignans (Zálešák et al., 2019). Additionally, SECO and SDG promoted a reduction of zymosan-activated polymorphonuclear leukocytes (91.2 and 23.8%, respectively), while END and ENL generated 94.2 and 81.6% reduction, respectively (Prasad, 2000a). A lipid peroxidation assay resulted in 58, 55, 39% inhibition for SECO, END and ENL, respectively (Kitts et al., 1999). 2,3-dihydrobenzofuran neolignans (i.e., chushizisins E-F) showed antioxidant activity against H_2O_2-induced impairment in PC12 cells using doses within 0.16-100 mM range (Teponno et al., 2016). Other examples were compiled by Zálešák et al. (2019), showing the antioxidant evidences at *in vitro* level for several (neo)lignans through different assays. In general, a diminished antioxidant activity can be found if (neo)lignans are glycosylated and/or methoxylated (Teponno et al., 2016; Hosseinian et al., 2007).

4.10.9 Animal and clinical studies

There is an increased interest in recognizing potential sources of health-promoting substances of natural origin. However, in the case of lignans bioactivity, human epidemiological examinations are very scarce. Association between studies in humans and animals warns that adequate intake of those (neo)lignans-enriched food sources can diminish the chance of origin/progress of chronic diseases, that is, cardiovascular disease, MetS and cancer (Rodríguez-García et al., 2019). Roles that are involved in the anticancer process may include the above-mentioned EM antagonism, antioxidant activity and modulating properties at the key control points of the cell cycle. Some lignans can reduce the growth of tumor in rats at different carcinogenesis stages, but there is a lack of satisfactory information for preventive properties against cancer in humans (Ranaware et al., 2018; Rauf et al., 2018; Majdalawieh et al., 2017; Arora et al., 2012; Mabrok et al., 2011). On the other hand, hypocholesterolemic activity and the platelet activation inhibition are some of the proposed antiatherogenic processes (Parikh et al., 2018). Lignan consumption is also related to the depletion of lipid and glucose, decrease of oxidative stress, inflammation, and blood pressure (Adolphe et al., 2010).

Lignan intake of distinct populations has been recently examined, especially in PMW, owing to the phytoestrogenic character of some lignans (Rodríguez-García et al., 2019). In general, (neo)lignans can be considered as safe chemical entities for most adult people, although exposure of pregnant women should be limited according to some animal trials (Adolphe et al., 2010). However, owing to the extensive irregularity in trial methods and results in the existing literature, there is a severe deficiency to clearly identify a global trend to the health properties of lignans. Some cases can be mentioned, such as SDG, since a dose consisting of 500 mg SDG per day during *ca.* 8 weeks can provide beneficial effects on human cardiovascular events (Adolphe et al., 2010), but showed no influence on resistance of lipoprotein oxidation in serum, and antioxidant capacity or lipid concentrations in plasma (Hallund et al., 2006). Contrarily, their metabolites (i.e., ENL and END) exhibited discrepancies in those results obtained from *in vivo* and *in vitro* experiments/assays, which can be explained by their low bioavailability (Landete, 2012). In the case of episesamin and sesamin, no accumulation was detected (using multiple doses at 50 mg) in healthy subjects, therefore, they were found to be tolerable and safe (Tomimori et al., 2013). Sesamin exhibited beneficial effects in PMW, since it improved blood levels of lipids as well as sex hormone and antioxidant conditions (Wu et al., 2006). A protective action on cardiovascular problems was also observed for sesamin in rheumatoid arthritis-affected subjects (Helli et al., 2016). Finally, an oral treatment based on nordihydroguaiaretic acid (NDGA) in patients with nonmetastatic hormone-sensitive prostate cancer demonstrated a linkage with the prostate-specific antigen doubling time. However, this compound had no potential to be used as therapeutic alternative to impede the injurious side-effects of androgen deprivation therapy to control prostate cancer (Friedlander et al., 2012).

4.10.10 Concluding remarks

Owing to the natural free radical scavenging capacity related to the beneficial effects on metabolic, gastrointestinal and cardiovascular conditions as well as anti-cancer, anti-inflammatory and chemopreventive actions, most (neo)lignans are considered as interesting chemical entities by different research and industry fields. Hence, according to the available evidences *in vitro* and *in vivo*, (neo)lignans as antioxidants have more preventive potential than curative effect on some diseases, although some evidences indicated curative properties, particularly in cardiovascular and cancer events. In addition, compiled data reinforce the attention on (neo)lignans as health-promoting components. In this sense, dietary consumption of lignan-rich edible sources can be then associated to the prevention of particular cancer conditions, such as breast and colon cancer, and the reduction of cardiovascular disease development in chronic lifestyle-related disorders. Furthermore, they are also considered inoffensive and there are no observed detrimental effects on health for some evaluated lignans, specifically for diarylbutane (e.g., SDG), diarylbutirolactone (e.g., MAT), furofurane (e.g., sesamin), and dibenzyl (e.g., honokiol) (neo)lignans. Regarding the optimal lignan dosage, the available literature mentioned that doses within 50-500 mg lignan per day during 4–8 weeks can provide beneficial effects on different conditions (estrogenic, cardiovascular and cancer events). Additionally, due the low bioavailability of lignans by HIB-mediated transformations, whose derivatives (e.g., END and ENL) may enhance the antioxidant activity of the respective parent lignan, the oral administration can be recommended as the ideal route. On the other hand, the prooxidant action of (neo)lignans depends on structural factors (e.g., presence of OH groups and diketone [as conjugated systems]) and environment (e.g., solubility, pH, metal ions concentrations [Cu^{+2} and Fe^{3+}], anionic capacity [oxidation potential]). Thus, the particular structural and environmental conditions of each lignan molecule will determine the prooxidant/oxidant/antioxidant roles at the cellular level. This fact might rationalize the observed discrepancies between *in vitro* and *in vivo* results, since bioavailability, for instance, represent an important influence to promote such roles. Therefore, further studies in humans are required to support rationally such a health-promoting features of lignans from the bioavailability point-of-view and, as previously mentioned, the beneficial and adverse properties of each (neo)lignan must be contemplated discretely, since a constant consumption doses could not be recommended without proper information on the composition and profiles of the (neo)lignan-containing source. In this regard, more examinations are required to cover and establish trends within the roles of this wide class of naturally-occurring compounds on human health. Accordingly, collaboration among researchers from distinct disciplines, being motivated to determine/delineate the bioavailability of other antioxidant lignans through systematic investigations, can stimulate additional progresses on improved applications/information and coherent promotion of lignans as antioxidants, preferably by reducing adverse but enhancing beneficial effects.

Acknowledgments

Authors thank the Universidad Militar Nueva Granada for the financial support through the project IMP-CIAS-2924, validity 2019.

References

Adlercreutz, H., Bannwart, C., Wähälä, K., Mäkelä, T., Brunow, G., Hase, T., Arosemena, P.J., Kellis, J.T., Vickery, L.E., 1993. Inhibition of human aromatase by mammalian lignans and isoflavonoid phytoestrogens. J. Steroid Biochem. Mol. Biol. 44, 147–153. https://doi.org/10.1016/0960-0760(93)90022-O.

Adlercreutz, H., van der Wildt, J., Kinzel, J., Attalla, H., Wähälä, K., Mäkelä, T., Hase, T., Fotsis, T., 1995. Lignan and isoflavonoid conjugates in human urine. J. Steroid Biochem. Mol. Biol. 52, 97–103. https://doi.org/10.1016/0960-0760(94)00146-d.

Adolphe, J.L., Whiting, S.J., Juurlink, B.H.J., Thorpe, L.U., Alcorn, J., 2010. Health effects with consumption of the flax lignan secoisolariciresinol diglucoside. Br. J. Nutr. 103, 929–938. https://doi.org/10.1017/S0007114509992753.

Ahmad, A., Syed, F.A., Singh, S., Hadi, S.M., 2005. Prooxidant activity of resveratrol in the presence of copper ions: Mutagenicity in plasmid DNA. Toxicol. Lett. 159, 1–12. https://doi.org/10.1016/j.toxlet.2005.04.001.

Apak, R., 2019. Current issues in antioxidant measurement. J. Agric. Food Chem. 67, 9187–9202. https://doi.org/10.1021/acs.jafc.9b03657.

Apak, R., Gorinstein, S., Böhm, V., Schaich, K.M., Özyürek, M., Güçlü, K., 2013. Methods of measurement and evaluation of natural antioxidant capacity/activity (IUPAC Technical Report). Pure Appl. Chem. 85, 957–998. https://doi.org/10.1351/PAC-REP-12-07-15.

Arora, S., Singh, S., Piazza, G.A., Contreras, C.M., Panyam, J., Singh, A.P., 2012. Honokiol: a novel natural agent for cancer prevention and therapy. Curr. Mol. Med. 12, 1244–1252. https://doi.org/10.2174/156652412803833508.

Aruoma, O.I., 1994. Nutrition and health aspects of free radicals and antioxidants. Food Chem. Toxicol. 32, 671–683. https://doi.org/10.1016/0278-6915(94)90011-6.

Atkinson, D.A., Hill, H.H., Shultz, T.D., 1993. Quantification of mammalian lignans in biological fluids using gas chromatography with ion mobility detection. J. Chromatogr. B Biomed. Sci. Appl. 617, 173–179. https://doi.org/10.1016/0378-4347(93)80485-M.

Axelson, M., Sjövall, J., Gustafsson, B.E., Setchell, K.D.R., 1982. Origin of lignans in mammals and identification of a precursor from plants. Nature 298, 659–660. https://doi.org/10.1038/298659a0.

Ayres, D.C., Loike, J.D., 2008. Lignans: Chemical, Biological and Clinical Properties. Cambridge University Press, Cambridge, UK.

Azmi, A.S., Bhat, S.H., Hadi, S.M., 2005. Resveratrol-Cu(II) induced DNA breakage in human peripheral lymphocytes: Implications for anticancer properties. FEBS Lett. 579, 3131–3135. https://doi.org/10.1016/j.febslet.2005.04.077.

Bilal, I., Chowdhury, A., Davidson, J., Whitehead, S., 2014. Phytoestrogens and prevention of breast cancer: The contentious debate. World J. Clin. Oncol. 5, 705–712. https://doi.org/10.5306/wjco.v5.i4.705.

Borg, D.C., Schaich, K.M., 1984. Cytotoxicity from coupled redox cycling of autoxidizing xenobiotics and metals: a selective critical review and commentary on work-in-progress. Isr. J. Chem. 24, 38–53. https://doi.org/10.1002/ijch.198400007.

Borriello, S.P., Setchell, K.D.R., Axelson, M., Lawson, A.M., 1985. Production and metabolism of lignans by the human faecal flora. J. Appl. Bacteriol. 58, 37–43. https://doi.org/10.1111/j.1365-2672.1985.tb01427.x.

Budowski, P., 1950. Sesame oil. III. Antioxidant properties of sesamol. J. Am. Oil Chem. Soc. 27, 264–267. https://doi.org/10.1007/BF02634903.

Bunaciu, A.A., Danet, A.F., Fleschin, Ş., Aboul-Enein, H.Y., 2016. Recent applications for in vitro antioxidant activity assay. Crit. Rev. Anal. Chem. 46, 389–399. https://doi.org/10.1080/10408347.2015.1101369.

Calliste, C.A., Kozlowski, D., Duroux, J.L., Champavier, Y., Chulia, A.J., Trouillas, P., 2010. A new antioxidant from wild nutmeg. Food Chem. 118, 489–496. https://doi.org/10.1016/J.FOODCHEM.2009.05.010.

Cao, G., Sofic, E., Prior, R.L., 1997. Antioxidant and prooxidant behavior of flavonoids: structure-activity relationships. Free Radic. Biol. Med. 22, 749–760. https://doi.org/10.1016/s0891-5849(96)00351-6.

Castro, M.A., Gordaliza, M., Miguel Del Corral, J.M., Feliciano, A.S., 1996. The distribution of lignanoids in the order coniferae. Phytochemistry 41, 995–1011. https://doi.org/10.1016/0031-9422(95)00512-9.

Chen, X., Ahn, D.U., 1998. Antioxidant activities of six natural phenolics against lipid oxidation induced by Fe^{2+} or ultraviolet light. J. Am. Oil Chem. Soc. 75, 1717. https://doi.org/10.1007/s11746-998-0322-2.

Clavel, T., Borrmann, D., Braune, A., Doré, J., Blaut, M., 2006a. Occurrence and activity of human intestinal bacteria involved in the conversion of dietary lignans. Anaerobe 12, 140–147. https://doi.org/10.1016/j.anaerobe.2005.11.002.

Clavel, T., Doré, J., Blaut, M., 2006b. Bioavailability of lignans in human subjects. Nutr. Res. Rev. 19, 187–196. https://doi.org/10.1017/S0954422407249704.

Decker, A., 2009. Phenolics: prooxidants or antioxidants? Nutr. Rev. 55, 396–398. https://doi.org/10.1111/j.1753-4887.1997.tb01580.x.

Degerud, E.M., Manger, M.S., Strand, T.A., Dierkes, J., 2015. Bioavailability of iron, vitamin A, zinc, and folic acid when added to condiments and seasonings. Ann. N.Y. Acad. Sci. 1357, 29–42. https://doi.org/10.1111/nyas.12947.

Durazzo, A., Lucarini, M., Camilli, E., Marconi, S., Gabrielli, P., Lisciani, S., Gambelli, L., Aguzzi, A., Novellino, E., Santini, A., Turrini, A., Marletta, L., 2018. Dietary lignans: definition, description and research trends in databases development. Molecules 23, 3251. https://doi.org/10.3390/molecules23123251.

Eghbaliferiz, S., Iranshahi, M., 2016. Prooxidant activity of polyphenols, flavonoids, anthocyanins and carotenoids: updated review of mechanisms and catalyzing metals. Phyther. Res. 30, 1379–1391. https://doi.org/10.1002/ptr.5643.

Eklund, P.C., Långvik, O.K., Wärnå, J.P., Salmi, T.O., Willför, S.M., Sjöholm, R.E., 2005. Chemical studies on antioxidant mechanisms and free radical scavenging properties of lignans. Org. Biomol. Chem. 3, 3336–3347. https://doi.org/10.1039/B506739A.

Filleur, F., Pouget, C., Allais, D.P., Kaouadji, M., Chulia, A.J., 2002. Lignans and neolignans from *Myristica Argenteal* warb. Nat. Prod. Lett. 16, 1–7. https://doi.org/10.1080/1057563029001/4764.

Ford, J.D., Huang, K.-S., Wang, H.-B., Davin, L.B., Lewis, N.G., 2001. Biosynthetic pathway to the cancer chemopreventive secoisolariciresinol diglucoside–hydroxymethyl glutaryl ester-linked lignan oligomers in flax (*Linum u sitatissimum*) Seed †. J. Nat. Prod. 64, 1388–1397. https://doi.org/10.1021/np010367x.

Frankel, E.N., Huang, S.-W., Kanner, J., German, J.B., 1994. Interfacial phenomena in the evaluation of antioxidants: bulk oils vs emulsions. J. Agric. Food Chem. 42, 1054–1059. https://doi.org/10.1021/jf00041a001.

Friedlander, T.W., Weinberg, V.K., Huang, Y., Mi, J.T., Formaker, C.G., Small, E.J., Harzstark, A.L., Lin, A.M., Fong, L., Ryan, C.J., 2012. A phase II study of insulin-like growth factor receptor inhibition with nordihydroguaiaretic acid in men with non-metastatic hormone-sensitive prostate cancer. Oncol. Rep. 27, 3–9. https://doi.org/10.3892/or.2011.1487.

Fukuda, Y., Osawa, T., Kawakishi, S., Namiki, M., 1994. Chemistry of lignan antioxidants in sesame seed and oil. Food Phytochemicals for Cancer Prevention II, chapter 27, pp. 264–274. ACS Symposium Series 547, Washington DC, USA. https://doi.org/10.1021/bk-1994-0547.ch027.

Ghafoorunissa, H,S., Rao, M.V.V., 2004. Sesame lignans enhance antioxidant activity of vitamin E in lipid peroxidation systems. Mol. Cell. Biochem. 262, 195–202. https://doi.org/10.1023/B:MCBI.0000038235.01389.a9.

Gottlieb, O.R., 1978. Neolignans. Fortschr. Chem. Org. Naturst. 35, 1–72.

Gottlieb, O.R., 1972. Chemosystematics of the lauraceae. Phytochemistry 11, 1537–1570. https://doi.org/10.1016/0031-9422(72)85001-5.

Guardado, E., Molina, E., Joo, M., Uriarte, E., 2012. Antioxidant and Pro-Oxidant Effects of Polyphenolic Compounds and Structure-Activity Relationship EvidenceNutrition, Well-Being and Health. InTech, Rijeka, Croatia. https://doi.org/10.5772/29471.

Guo, Q., Li, F., Duan, Y., Wen, C., Wang, W., Zhang, L., Huang, R., Yin, Y., 2019. Oxidative stress, nutritional antioxidants and beyond. Sci. China Life Sci 63, 866–874. https://doi.org/10.1007/s11427-019-9591-5.

Haggans, C.J., Hutchins, A.M., Olson, B.A., Thomas, W., Martini, M.C., Slavin, J.L., 1999. Effect of flaxseed consumption on urinary estrogen metabolites in postmenopausal women. Nutr. Cancer 33, 188–195. https://doi.org/10.1207/S15327914NC330211.

Haida, Z., Hakiman, M., 2019. A comprehensive review on the determination of enzymatic assay and nonenzymatic antioxidant activities. Food Sci. Nutr. 7, 1555–1563. https://doi.org/10.1002/fsn3.1012.

Hallund, J., Ravn-Haren, G., Bügel, S., Tholstrup, T., Tetens, I., 2006. A lignan complex isolated from flaxseed does not affect plasma lipid concentrations or antioxidant capacity in healthy postmenopausal women. J. Nutr. 136, 112–116. https://doi.org/10.1093/jn/136.1.112.

Haworth, R.D., 1936. Natural resins. Ann. Rep. Prog. Chem. 33, 266–279.

Hedelin, M., Löf, M., Olsson, M., Adlercreutz, H., Sandin, S., Weiderpass, E., 2008. Dietary phytoestrogens are not associated with risk of overall breast cancer but diets rich in coumestrol are inversely associated with risk of estrogen receptor and progesterone receptor negative breast tumors in Swedish women. J. Nutr. 138, 938–945. https://doi.org/10.1093/jn/138.5.938.

Heinonen, S., Nurmi, T., Liukkonen, K., Poutanen, K., Wähälä, K., Deyama, T., Nishibe, S., Adlercreutz, H., 2001. In vitro metabolism of plant lignans: new precursors of mammalian lignans enterolactone and enterodiol. J. Agric. Food Chem. 49, 3178–3186. https://doi.org/10.1021/jf010038a.

Helli, B., Mowla, K., Mohammadshahi, M., Jalali, M.T., 2016. Effect of sesamin supplementation on cardiovascular risk factors in women with rheumatoid arthritis. J. Am. Coll. Nutr. 35, 300–307. https://doi.org/10.1080/07315724.2015.1005198.

Herchi, W., Arráez-Román, D., Trabelsi, H., Bouali, I., Boukhchina, S., Kallel, H., Segura-Carretero, A., Fernández-Gutierrez, A., 2014. Phenolic compounds in flaxseed: a review of their properties and analytical methods. an overview of the last decade. J. Oleo Sci. 63, 7–14. https://doi.org/10.5650/jos.ess13135.

Hider, R.C., Liu, Z.D., Khodr, H.H., 2001. Metal chelation of polyphenols. Methods Enzymol. 335, 190–203. https://doi.org/10.1016/S0076-6879(01)35243-6.

Holst, B., Williamson, G., 2008. Nutrients and phytochemicals: from bioavailability to bioefficacy beyond antioxidants. Curr. Opin. Biotechnol. 19, 73–82. https://doi.org/10.1016/j.copbio.2008.03.003.

Hosseinian, F.S., Muir, A.D., Westcott, N.D., Krol, E.S., 2007. AAPH-mediated antioxidant reactions of secoisolariciresinol and SDG. Org. Biomol. Chem. 5, 644–654. https://doi.org/10.1039/B617426D.

Hosseinian, F.S., Muir, A.D., Westcott, N.D., Krol, E.S., 2006. Antioxidant capacity of flaxseed lignans in two model systems. J. Am. Oil Chem. Soc. 83, 835. https://doi.org/10.1007/s11746-006-5034-x.

Hsiao, G., Teng, C.M., Sheu, J.R., Cheng, Y.W., Lam, K.K., Lee, Y.M., Wu, T.S., Yen, M.H., 2001. Cinnamophilin as a novel antiperoxidative cytoprotectant and free radical scavenger. Biochim. Biophys. Acta 1525, 77–88. https://doi.org/10.1016/s0304-4165(00)00173-2.

Hu, C., Yuan, Y.V, Kitts, D.D., 2007. Antioxidant activities of the flaxseed lignan secoisolariciresinol diglucoside, its aglycone secoisolariciresinol and the mammalian lignans enterodiol and enterolactone in vitro. Food Chem. Toxicol. 45, 2219–2227. https://doi.org/10.1016/j.fct.2007.05.017.

Huang, S.-W., Hopia, A., Schwarz, K., Frankel, E.N., German, J.B., 1996. Antioxidant activity of α-tocopherol and trolox in different lipid substrates: bulk oils *vs* oil-in-water emulsions. J. Agric. Food Chem. 44, 444–452. https://doi.org/10.1021/jf9505685.

Hunyadi, A., 2019. The mechanism(s) of action of antioxidants: from scavenging reactive oxygen/nitrogen species to redox signaling and the generation of bioactive secondary metabolites. Med. Res. Rev. 39, 2505–2533. https://doi.org/10.1002/med.21592.

Hutchins, A.M., Martini, M.C., Olson, B.A., Thomas, W., Slavin, J.L., 2000. Flaxseed influences urinary lignan excretion in a dose-dependent manner in postmenopausal women. Cancer Epidemiol. Biomarkers Prev. 9, 1113–1118.

Ilbeigi, D., Nourbakhsh, M., Khaghani, S., Einollahi, N., Kheiripour, N., Gholinejad, Z., Alaee, M., Saberian, M., 2017. Enterolactone reduces telomerase activity and the level of its catalytic subunit in breast cancer cells. Cell J. 19, 37–43. https://doi.org/10.22074/cellj.2017.4705.

Jansen, G.H.E., Arts, I.C.W., Nielen, M.W.F., Müller, M., Hollman, P.C.H., Keijer, J., 2005. Uptake and metabolism of enterolactone and enterodiol by human colon epithelial cells. Arch. Biochem. Biophys. 435, 74–82. https://doi.org/10.1016/j.abb.2004.12.015.

Jin, J.-S., Hattori, M., 2009. Further studies on a human intestinal bacterium *Ruminococcus* sp. END-1 for transformation of plant lignans to mammalian lignans. J. Agric. Food Chem. 57, 7537–7542. https://doi.org/10.1021/jf900902p.

Jin, J.-S., Kakiuchi, N., Hattori, M., 2007. Enantioselective oxidation of enterodiol to enterolactone by human intestinal bacteria. Biol. Pharm. Bull. 30, 2204–2206. https://doi.org/10.1248/bpb.30.2204.

Joshi, R., Kumar, M.S., Satyamoorthy, K., Unnikrisnan, M.K., Mukherjee, T., 2005. Free radical reactions and antioxidant activities of sesamol: pulse radiolytic and biochemical studies. J. Agric. Food Chem. 53, 2696–2703. https://doi.org/10.1021/jf0489769.

Kangas, L., Saarinen, N., Mutanen, M., Ahotupa, M., Hirsinummi, R., Unkila, M., Perälä, M., Soininen, P., Laatikainen, R., Korte, H., Santti, R., 2002. Antioxidant and antitumor effects of hydroxymatairesinol (HM-3000, HMR), a lignan isolated from the knots of spruce. Eur. J. Cancer Prev. 11 (Suppl 2), S48–S57.

Kato, T., Harashima, T., Moriya, N., Kikugawa, K., Hiramoto, K., 1996. Formation of the mutagenic/carcinogenic imidazoquinoxaline-type heterocyclic amines through the unstable free radical Maillard intermediates and its inhibition by phenolic antioxidants. Carcinogenesis 17, 2469–2476. https://doi.org/10.1093/carcin/17.11.2469.

Kezimana, P., Dmitriev, A.A., Kudryavtseva, A.V., Romanova, E.V., Melnikova, N.V., 2018. Secoisolariciresinol diglucoside of flaxseed and its metabolites: biosynthesis and potential for nutraceuticals. Front. Genet. 9, 641. https://doi.org/10.3389/fgene.2018.00641.

Kikugawa, K., Arai, M., Kurechi, T., 1983. Participation of sesamol in stability of sesame oil. J. Am. Oil Chem. Soc. 60, 1528–1533. https://doi.org/10.1007/BF02666577.

Kitts, D.D., Yuan, Y.V, Wijewickreme, A.N., Thompson, L.U., 1999. Antioxidant activity of the flaxseed lignan secoisolariciresinol diglycoside and its mammalian lignan metabolites enterodiol and enterolactone. Mol. Cell. Biochem. 202, 91–100. https://doi.org/10.1023/a:1007022329660.

Knust, U., Hull, W.E., Spiegelhalder, B., Bartsch, H., Strowitzki, T., Owen, R.W., 2006. Analysis of enterolignan glucuronides in serum and urine by HPLC-ESI-MS. Food Chem. Toxicol. 44, 1038–1049. https://doi.org/10.1016/j.fct.2005.12.008.

Kuijsten, C., Arts, I.C.W., Hollman, P.C.H., van't Veer, P., Kampman, E., 2006. Plasma enterolignans are associated with lower colorectal adenoma risk. Cancer Epidemiol. Biomarkers Prev. 15, 1132–1136. https://doi.org/10.1158/1055-9965.EPI-05-0991.

Kuijsten, A., Arts, I.C.W., Vree, T.B., Hollman, P.C.H., 2005. Pharmacokinetics of enterolignans in healthy men and women consuming a single dose of secoisolariciresinol diglucoside. J. Nutr. 135, 795–801. https://doi.org/10.1093/jn/135.4.795.

Kumagai, Y., Lin, L.Y., Schmitz, D.A., Cho, A.K., 1991. Hydroxyl radical mediated demethylenation of (methylenedioxy)phenyl compounds. Chem. Res. Toxicol. 4, 330–334. https://doi.org/10.1021/tx00021a012.

Kurechi, T., Kikugawa, K., Kato, T., 1980. Studies on the antioxidants. XIII. Hydrogen donating capability of antioxidants to 2,2-diphenyl-1-picrylhydrazyl. Chem. Pharm. Bull. (Tokyo). 28, 2089–2093. https://doi.org/10.1248/cpb.28.2089.

Kyselka, J., Rabiej, D., Dragoun, M., Kreps, F., Burčová, Z., Němečková, I., Smolová, J., Bjelková, M., Szydłowska-Czerniak, A., Schmidt, Š., Šarman, L., Filip, V., 2017. Antioxidant and antimicrobial activity of linseed lignans and phenolic acids. Eur. Food Res. Technol. 243, 1633–1644. https://doi.org/10.1007/s00217-017-2871-9.

Lampe, J.W., Martini, M.C., Kurzer, M.S., Adlercreutz, H., Slavin, J.L., 1994. Urinary lignan and isoflavonoid excretion in premenopausal women consuming flaxseed powder. Am. J. Clin. Nutr. 60, 122–128. https://doi.org/10.1093/ajcn/60.1.122.

Landete, J.M., 2012. Plant and mammalian lignans: a review of source, intake, metabolism, intestinal bacteria and health. Food Res. Int. 46, 410–424. https://doi.org/10.1016/J.FOODRES.2011.12.023.

Leong, P.K., Chiu, P.Y., Ko, K.M., 2012. Prooxidant-induced glutathione antioxidant response *in vitro* and *in vivo*: a comparative study between schisandrin B and curcumin. Biol. Pharm. Bull. 35, 464–472. https://doi.org/10.1248/bpb.35.464.

Lewis, N.G., Davin, L.B., 1994. Evolution of lignan and neolignan biochemical pathways. Food Phytochemicals for Cancer Prevention II, chapter 27, pp. 202–246, ACS Symposium Series 547, Washington DC, USA. https://doi.org/10.1021/bk-1994-0562.ch010

Lewis, N.G., Davin, L.B., 1992. Stereo Selectivity in Polyphenol Biosynthesis BT - Plant Polyphenols: Synthesis, Properties, Significance, In: Hemingway, R.W., Laks, P.E. (Eds.), Springer US, Boston, MA, 73–95. https://doi.org/10.1007/978-1-4615-3476-1_5

Lin, X., Switzer, B.R., D., W.W., 2001. Effect of mammalian lignans on the growth of prostate cancer cell lines. Anticancer Res. 21, 3995–3999.

Linseisen, J., Piller, R., Hermann, S., Chang-Claude, J.German Case-Control Study, 2004. Dietary phytoestrogen intake and premenopausal breast cancer risk in a German case-control study. Int. J. Cancer 110, 284–290. https://doi.org/10.1002/ijc.20119.

Liu, H., Liu, J., Wang, S., Zeng, Z., Li, T., Liu, Y., Mastriani, E., Li, Q.-H., Bao, H.-X., Zhou, Y.-J., Wang, X., Hu, S., Gao, S., Qi, Y., Shen, Z., Wang, H., Yu, M., Gao, T., Johnston, R.N., Liu, S.-L., 2017. Enterolactone has stronger effects than enterodiol on ovarian cancer. J. Ovarian Res. 10, 49. https://doi.org/10.1186/s13048-017-0346-z.

Lourenço, S.C., Moldão-Martins, M., Alves, V.D., 2019. Antioxidants of natural plant origins: from sources to food industry applications. Molecules 24, 4132. https://doi.org/10.3390/molecules24224132.

Lourith, N., Katayama, T., Suzuki, T., 2005. Stereochemistry and biosynthesis of 8-O-4′ neolignans in Eucommia ulmoides: diastereoselective formation of guaiacylglycerol-8-O-4′-(sinapyl alcohol) ether. J. Wood Sci. 51, 370–378. https://doi.org/10.1007/s10086-004-0660-0.

Low, Y.-L., Taylor, J.I., Grace, P.B., Dowsett, M., Folkerd, E., Doody, D., Dunning, A.M., Scollen, S., Mulligan, A.A., Welch, A.A., Luben, R.N., Khaw, K.-T., Day, N.E., Wareham, N.J., Bingham, S.A., 2005a. Polymorphisms in the CYP19 gene may affect the positive correlations between serum and urine phytoestrogen metabolites and plasma androgen concentrations in men. J. Nutr. 135, 2680–2686. https://doi.org/10.1093/jn/135.11.2680.

Low, Y.-L., Taylor, J.I., Grace, P.B., Dowsett, M., Scollen, S., Dunning, A.M., Mulligan, A.A., Welch, A.A., Luben, R.N., Khaw, K.-T., Day, N.E., Wareham, N.J., Bingham, S.A., 2005b.

Phytoestrogen exposure correlation with plasma estradiol in postmenopausal women in European Prospective Investigation of Cancer and Nutrition-Norfolk may involve diet-gene interactions. Cancer Epidemiol. Biomarkers Prev. 14, 213–220.

Mabrok, H.B., Klopfleisch, R., Ghanem, K.Z., Clavel, T., Blaut, M., Loh, G., 2011. Lignan transformation by gut bacteria lowers tumor burden in a gnotobiotic rat model of breast cancer. Carcinogenesis 33, 203–208. https://doi.org/10.1093/carcin/bgr256.

Majdalawieh, A.F., Massri, M., Nasrallah, G.K., 2017. A comprehensive review on the anti-cancer properties and mechanisms of action of sesamin, a lignan in sesame seeds (Sesamum indicum). Eur. J. Pharmacol. 815, 512–521. https://doi.org/10.1016/j.ejphar.2017.10.020.

Manivannan, J., Silambarasan, T., Shanthakumar, J., Suganya, N., Kanchana, S., 2016. Role of antioxidants in human health. Omega-3 Fatty Acids. Springer International Publishing, Cham, pp. 501–512. https://doi.org/10.1007/978-3-319-40458-5_36.

Mazzio, E., Harris, N., Soliman, K., 1998. Food constituents attenuate monoamine oxidase activity and peroxide levels in c6 astrocyte cells. Planta Med. 64, 603–606. https://doi.org/10.1055/s-2006-957530.

Milder, I.E.J., Arts, I.C.W., Putte, B., Venema, D.P., Hollman, P.C.H., 2005. Lignan contents of Dutch plant foods: a database including lariciresinol, pinoresinol, secoisolariciresinol and matairesinol. Br. J. Nutr. 93, 393–402. https://doi.org/10.1079/bjn20051371.

Moran, J.F., Klucas, R.V, Grayer, R.J., Abian, J., Becana, M., 1997. Complexes of iron with phenolic compounds from soybean nodules and other legume tissues: prooxidant and antioxidant properties. Free Radic. Biol. Med. 22, 861–870. https://doi.org/10.1016/s0891-5849(96)00426-1.

Moss, G.P., 2000. Nomenclature of lignans and neolignans (IUPAC Recommendations 2000). Pure Appl. Chem. 72, 1493–1523. https://doi.org/10.1351/pac200072081493.

Mousavi, Y., Adlercreutz, H., 1992. Enterolactone and estradiol inhibit each other's proliferative effect on MCF-7 breast cancer cells in culture. J. Steroid Biochem. Mol. Biol. 41, 615–619. https://doi.org/10.1016/0960-0760(92)90393-W.

Mueller, S.O., Simon, S., Chae, K., Metzler, M., Korach, K.S., 2004. Phytoestrogens and their human metabolites show distinct agonistic and antagonistic properties on estrogen receptor α (ERα) and ERβ in human cells. Toxicol. Sci. 80, 14–25. https://doi.org/10.1093/toxsci/kfh147.

Murphy, P.A., Hendrich, S., 2002. Phytoestrogens in foods. Adv. Food Nutr. Res. 44, 195–246.

Nakagawa, K., 2000. Photochemical reactions of antioxidant sesamol in aqueous solution. J. Am. Oil Chem. Soc. 77, 1205. https://doi.org/10.1007/s11746-000-0187-1.

Namiki, M., 2007. Nutraceutical functions of sesame: a review. Crit. Rev. Food Sci. Nutr. 47, 651–673. https://doi.org/10.1080/10408390600919114.

Namiki, N., 1989. Goma no Kagaku (Science of Sesame). Asakura Shoten, Tokio.

Nawar, W., 1996. Lipids. In: Fennema, O. (Ed.), Food Chemistry. Marcel Dekker, Inc., New York, pp. 225–320.

Nurmi, T., Mursu, J., Peñalvo, J.L., Poulsen, H.E., Voutilainen, S., 2010. Dietary intake and urinary excretion of lignans in Finnish men. Br. J. Nutr. 103, 677–685. https://doi.org/10.1017/S0007114509992261.

Parikh, M., Netticadan, T., Pierce, G.N., 2018. Flaxseed: its bioactive components and their cardiovascular benefits. Am. J. Physiol. Circ. Physiol. 314, H146–H159. https://doi.org/10.1152/ajpheart.00400.2017.

Peterson, J., Dwyer, J., Adlercreutz, H., Scalbert, A., Jacques, P., McCullough, M.L., 2010. Dietary lignans: physiology and potential for cardiovascular disease risk reduction. Nutr. Rev. 68, 571–603. https://doi.org/10.1111/j.1753-4887.2010.00319.x.

Pham-Huy, L.A., He, H., Pham-Huy, C., 2008. Free radicals, antioxidants in disease and health. Int. J. Biomed. Sci. 4, 89–96.

Porter, W.L., 1993. Paradoxical behavior of antioxidants in food and biological systems. Toxicol. Ind. Health 9, 93–122.

Prasad, K., 2001. Secoisolariciresinol diglucoside from flaxseed delays the development of type 2 diabetes in Zucker rat. J. Lab. Clin. Med. 138, 32–39. https://doi.org/10.1067/mlc.2001.115717.

Prasad, K., 2000a. Antioxidant activity of secoisolariciresinol diglucoside-derived metabolites, secoisolariciresinol, enterodiol, and enterolactone. Int. J. Angiol. 9, 220–225.

Prasad, K., 2000b. Flaxseed: A source of hypocholesterolemic and antiatherogenic agents. Drug News Perspect. 13, 99–104. https://doi.org/10.1358/dnp.2000.13.2.662239.

Prasad, K., 1997. Hydroxyl radical-scavenging property of secoisolariciresinol diglucoside (SDG) isolated from flax-seed. Mol. Cell. Biochem. 168, 117–123. https://doi.org/10.1023/a:1006847310741.

Quideau, S., Deffieux, D., Douat-Casassus, C., Pouységu, L., 2011. Plant polyphenols: chemical properties, biological activities, and synthesis. Angew. Chemie Int. Ed. 50, 586–621. https://doi.org/10.1002/anie.201000044.

Ranaware, A.M., Banik, K., Deshpande, V., Padmavathi, G., Roy, N.K., Sethi, G., Fan, L., Kumar, A.P., Kunnumakkara, A.B., 2018. Magnolol: a Neolignan from the Magnolia family for the prevention and treatment of cancer. Int. J. Mol. Sci. 19, 2362. https://doi.org/10.3390/ijms19082362.

Rauf, A., Patel, S., Imran, M., Maalik, A., Arshad, M.U., Saeed, F., Mabkhot, Y.N., Al-Showiman, S.S., Ahmad, N., Elsharkawy, E., 2018. Honokiol: an anticancer lignan. Biomed. Pharmacother. 107, 555–562. https://doi.org/10.1016/j.biopha.2018.08.054.

Rietjens, I.M.C.M., Louisse, J., Beekmann, K., 2017. The potential health effects of dietary phytoestrogens. Br. J. Pharmacol. 174, 1263–1280. https://doi.org/10.1111/bph.13622.

Ríos, J.L., Giner, R.M., Prieto, J.M., 2002. New findings on the bioactivity of lignans. Stud. Nat. Prod. Chem. 26, 183–292. https://doi.org/10.1016/S1572-5995(02)80008-4.

Rodríguez-García, C., Sánchez-Quesada, C., Toledo, E., Delgado-Rodríguez, M., Gaforio, J.J., 2019. Naturally lignan-rich foods: a dietary tool for health promotion? Molecules 24, 1–13. https://doi.org/10.3390/molecules24050917.

Runeberg, P.A., Brusentsev, Y., Rendon, S.M.K., Eklund, P.C., 2019. Oxidative transformations of lignans. Molecules 24, 300. https://doi.org/10.3390/molecules24020300.

Russell, W.R., Hoyles, L., Flint, H.J., Dumas, M.-E., 2013. Colonic bacterial metabolites and human health. Curr. Opin. Microbiol. 16, 246–254. https://doi.org/10.1016/j.mib.2013.07.002.

Saarinen, N.M., Wärri, A., Airio, M., Smeds, A., Mäkelä, S., 2007. Role of dietary lignans in the reduction of breast cancer risk. Mol. Nutr. Food Res. 51, 857–866. https://doi.org/10.1002/mnfr.200600240.

Saarinen, N.M., Warri, A., Makela, S.I., Eckerman, C., Reunanen, M., Ahotupa, M., Salmi, S.M., Franke, A.A., Kangas, L., Santti, R., 2000. Hydroxymatairesinol, a novel enterolactone precursor with antitumor properties from coniferous tree (Picea abies). Nutr. Cancer 36, 207–216. https://doi.org/10.1207/S15327914NC3602_10.

Sarrica, A., Kirika, N., Romeo, M., Salmona, M., Diomede, L., 2018. Safety and toxicology of magnolol and honokiol. Planta Med. 84, 1151–1164. https://doi.org/10.1055/a-0642-1966.

Satake, H., Koyama, T., Bahabadi, S.E., Matsumoto, E., Ono, E., Murata, J., 2015. Essences in metabolic engineering of lignan biosynthesis. Metabolites 5, 270–290. https://doi.org/10.3390/metabo5020270.

Setchell, K.D., Lawson, A.M., Conway, E., Taylor, N.F., Kirk, D.N., Cooley, G., Farrant, R.D., Wynn, S., Axelson, M., 1981. The definitive identification of the lignans trans-2,3-bis(3-hydroxybenzyl)-γ-butyrolactone and 2,3-bis(3-hydroxybenzyl)butane-1,4-diol in human and animal urine. Biochem. J. 197, 447–458. https://doi.org/10.1042/bj1970447.

Setchell, K.D.R., Lawson, A.M., Mitchell, F.L., Adlercreutz, H., Kirk, D.N., Axelson, M., 1980. Lignans in man and in animal species. Nature 287, 740–742. https://doi.org/10.1038/287740a0.

Shen, J.-L., Man, K.-M., Huang, P.-H., Chen, W.-C., Chen, D.-C., Cheng, Y.-W., Liu, P.-L., Chou, M.-C., Chen, Y.-H., 2010. Honokiol and magnolol as multifunctional antioxidative molecules for dermatologic disorders. Molecules 15, 6452–6465. https://doi.org/10.3390/molecules15096452.

Shin, M.-K., Jeon, Y.-D., Jin, J.-S., 2019. Apoptotic effect of enterodiol, the final metabolite of edible lignans, in colorectal cancer cells. J. Sci. Food Agric. 99, 2411–2419. https://doi.org/10.1002/jsfa.9448.

Simić, A., Manojlović, D., Šegan, D., Todorović, M., 2007. Electrochemical behavior and antioxidant and prooxidant activity of natural phenolics. Molecules 12, 2327–2340. https://doi.org/10.3390/12102327.

Solyomváry, A., Imre, B., Beni, S., 2017. Dibenzylbutyrolactone lignans – a review of their structural diversity, biosynthesis, occurrence, identification and importance. mini-reviews. Med. Chem 17, 1053–1074. https://doi.org/10.2174/1389557516666160614005828.

Soobrattee, M.A., Neergheen, V.S., Luximon-Ramma, A., Aruoma, O.I., Bahorun, T., 2005. Phenolics as potential antioxidant therapeutic agents: mechanism and actions. Mutat. Res. Mol. Mech. Mutagen. 579, 200–213. https://doi.org/10.1016/J.MRFMMM.2005.03.023.

Stopper, H., Schmitt, E., Kobras, K., 2005. Genotoxicity of phytoestrogens. Mutat. Res. Mol. Mech. Mutagen. 574, 139–155. https://doi.org/10.1016/j.mrfmmm.2005.01.029.

Suzuki, S., Umezawa, T., 2007. Biosynthesis of lignans and norlignans. J. Wood Sci. 53, 273–284. https://doi.org/10.1007/s10086-007-0892-x.

Suzuki, S., Umezawa, T., Shimada, M., 2001. Norlignan biosynthesis in Asparagus officinalis L.: the norlignan originates from two non-identical phenylpropane units, Parts of this work were presented at the 50th annual Meeting of the Japan Wood Research Society, April 3, 2000. Kyoto, Japan the 45th L. J. Chem. Soc. Perkin Trans. 1, 3252–3257. https://doi.org/10.1039/b107949b.

Talarek, S., Listos, J., Barreca, D., Tellone, E., Sureda, A., Nabavi, S.F., Braidy, N., Nabavi, S.M., 2017. Neuroprotective effects of honokiol: from chemistry to medicine. Biofactors 43, 760–769. https://doi.org/10.1002/biof.1385.

Teponno, R.B., Kusari, S., Spiteller, M., 2016. Recent advances in research on lignans and neolignans. Nat. Prod. Rep. 33, 1044–1092. https://doi.org/10.1039/C6NP00021E.

Tetens, I., Turrini, A., Tapanainen, H., Christensen, T., Lampe, J.W., Fagt, S., Håkansson, N., Lundquist, A., Hallund, J., Valsta, L.M., 2013. Dietary intake and main sources of plant lignans in five European countries. Food Nutr. Res. 57, 19805. https://doi.org/10.3402/fnr.v57i0.19805.

Tham, D.M., Gardner, C.D., Haskell, W.L., 1998. Potential health benefits of dietary phytoestrogens: a review of the clinical, epidemiological, and mechanistic evidence. J. Clin. Endocrinol. Metab. 83, 2223–2235. https://doi.org/10.1210/jcem.83.7.4752.

Tomimori, N., Tanaka, Y., Kitagawa, Y., Fujii, W., Sakakibara, Y., Shibata, H., 2013. Pharmacokinetics and safety of the sesame lignans, sesamin and episesamin, in healthy subjects. Biopharm. Drug Dispos. 34, 462–473. https://doi.org/10.1002/bdd.1862.

Touré, A., Xueming, X., 2010. Flaxseed lignans: source, biosynthesis, metabolism, antioxidant activity, bio-active components, and health benefits. Compr. Rev. Food Sci. Food Saf. 9, 261–269. https://doi.org/10.1111/j.1541-4337.2009.00105.x.

Ulucan, O., Keskin, O., Erman, B., Gursoy, A., 2011. A comparative molecular dynamics study of methylation state specificity of JMJD2A. PLoS One 6. https://doi.org/10.1371/journal.pone.0024664.

Umezawa, T., 2003. Diversity in lignan biosynthesis. Phytochem. Rev. 2, 371–390. https://doi.org/10.1023/B:PHYT.0000045487.02836.32.

Valsta, L.M., Kilkkinen, A., Mazur, W., Nurmi, T., Lampi, A.-M., Ovaskainen, M.-L., Korhonen, T., Adlercreutz, H., Pietinen, P., 2003. Phyto-oestrogen database of foods and average intake in Finland. Br. J. Nutr. 89, S31–S38. https://doi.org/10.1079/BJN2002794.

van der Schouw, Y.T., Kreijkamp-Kaspers, S., Peeters, P.H.M., Keinan-Boker, L., Rimm, E.B., Grobbee, D.E., 2005. Prospective study on usual dietary phytoestrogen intake and cardiovascular disease risk in Western women. Circulation 111, 465–471. https://doi.org/10.1161/01.CIR.0000153814.87631.B0.

Vo, Q.V, Nam, P.C., Bay, M.V, Thong, N.M., Cuong, N.D., Mechler, A., 2018. Density functional theory study of the role of benzylic hydrogen atoms in the antioxidant properties of lignans. Sci. Rep. 8, 12361. https://doi.org/10.1038/s41598-018-30860-5.

Vogt, T., 2010. Phenylpropanoid Biosynthesis. Mol. Plant 3, 2–20. https://doi.org/10.1093/mp/ssp106.

Ward, R.S., 2000. Recent advances in the chemistry of lignans. Stud. Nat. Prod. Chem. 24, 739–798. https://doi.org/10.1016/S1572-5995(00)80054-X.

Webb, A.L., McCullough, M.L., 2005. Dietary lignans: potential role in cancer prevention. Nutr. Cancer 51, 117–131. https://doi.org/10.1207/s15327914nc5102_1.

Whiting, D.A., 1987. Lignans, neolignans, and related compounds. Nat. Prod. Rep. 4, 499–525.

Wu, W.-H., Kang, Y.-P., Wang, N.-H., Jou, H.-J., Wang, T.-A., 2006. Sesame ingestion affects sex hormones, antioxidant status, and blood lipids in postmenopausal women. J. Nutr. 136, 1270–1275. https://doi.org/10.1093/jn/136.5.1270.

Xu, W.-H., Zhao, P., Wang, M., Liang, Q., 2019. Naturally occurring furofuran lignans: structural diversity and biological activities. Nat. Prod. Res. 33, 1357–1373. https://doi.org/10.1080/14786419.2018.1474467.

Yamauchi, S., Ina, T., Kirikihira, T., Masuda, T., 2004. Synthesis and antioxidant activity of oxygenated furofuran lignans. Biosci. Biotechnol. Biochem. 68, 183–192. https://doi.org/10.1271/bbb.68.183.

Yoshino, M., Haneda, M., Naruse, M., Htay, H.H., Tsubouchi, R., Qiao, S.L., Li, W.H., Murakami, K., Yokochi, T., 2004. Prooxidant activity of curcumin: copper-dependent formation of 8-hydroxy-2′-deoxyguanosine in DNA and induction of apoptotic cell death. Toxicol. Vitr. 18, 783–789. https://doi.org/10.1016/j.tiv.2004.03.009.

Zálešák, F., Bon, D.J.-Y.D., Pospíšil, J., 2019. Lignans and neolignans: plant secondary metabolites as a reservoir of biologically active substancces. Pharmacol. Res. 146, 104284. https://doi.org/10.1016/j.phrs.2019.104284.

Zhang, J., Chen, J., Liang, Z., Zhao, C., 2014. New lignans and their biological activities. Chem. Biodivers. 11, 1–54. https://doi.org/10.1002/cbdv.201100433.

Zheng, C.-J., Zhang, X.-W., Han, T., Jiang, Y.-P., Tang, J.-Y., Brömme, D., Qin, L.-P., 2014. Anti-inflammatory and anti-osteoporotic lignans from Vitex negundo seeds. Fitoterapia 93, 31–38. https://doi.org/10.1016/j.fitote.2013.12.006.

CHAPTER

Organosulfur compounds (allyl sulfide, indoles) 4.11

Mohamad Fawzi Mahomoodally, Nouzaifa Nabee, Nawshin Baureek
Department of Health Sciences, Faculty of Medicine and Health Sciences, University of Mauritius, Réduit, Mauritius

4.11.1 Introduction

Organosulfur compounds are sulfur-containing organic compounds, often associated with foul odours (Goncharov et al., 2021). They are obtained abundantly from both plant and animal sources, including *Brassica oleracea* var. botrytis, *Brassica oleracea* var. italic, *Brassica oleracea* var. italic, *Allium sativum*, *Allium cepa*, meat, eggs and fish (Goncharov et al., 2021). Organosulfur compounds are classified according to the functional group to which sulfur is attached (Goncharov et al., 2021). Cruciferous vegetables such as cauliflower, broccoli and cabbage have become an important part in the diet due to its health benefits (Katz et al., 2018). In ancient times, extracts from these vegetables were thought to have a curative effect (Katz et al., 2018). In modern times, these vegetables have become an important part in a cancer patient's diet due to its anticarcinogenic properties (Katz et al., 2018). The phytochemical compounds responsible for this effect are the organosulfur compounds, mainly indoles (Katz et al., 2018).

Other organosulfur-containing compounds are the allium vegetables (Bianchini et al., 2001). There are approximately 500 species of the *Allium* genus, the most commonly used of which are onions (*Allium cepa*), garlic (*Allium sativum*), leeks (*Allium porrum*), chives (*Allium schoenoprasum*), and shallots (*Allium ascalonium*) (Bianchini et al., 2001). The most widely cultivated and consumed allium vegetable worldwide is garlic and its health benefits have been recognised for thousands of years due to its important compounds such as cysteine sulfoxides and γ-glutamylcysteines (Bianchini et al., 2001). Garlic and onions have a foul smell when cysteine sulfoxides are hydrolysed (Bianchini et al., 2001). The four types of cysteine sulfoxides are alliin, methiin, propiin and isoallin. Isoalliin are found abundantly in onions whereas garlic is rich in alliin (Bianchini et al., 2001). When garlic is crushed during cooking, alliin is converted into allicin by a hydrolysing enzyme alliinase, as follows,

$$2RSOCH_2CH(NH_2)COOH + H_2O \rightarrow 2NH_3 + 2CH_3COCOOH + RSSOR$$

(Bianchini et al., 2001).

Another bioactive ingredient found in garlic is allyl sulphide, which have been found in human breath after ingestion of garlic in a clinical study. (Bianchini et al., 2001).

Antioxidants Effects in Health. DOI: https://doi.org/10.1016/B978-0-128-19096-8.00055-0

The aims of this chapter is to determine whether organosulfur compounds have a more preventive than curative effect on diseases, to what extent it is detrimental to health and what are its ideal route of administration.

4.11.2 Sources, chemical structure, and bioavailability of organosulfur compounds

Indoles are found in many natural products namely alkaloids, peptides, and synthetic compounds, with biodynamic properties (Derosa, 2006). It has a benzene ring fused to a pyrrole ring at 2 & 3 positions with a hetero-atomic planar structure (Derosa, 2006).

Diallyl sulfides are produced from the decomposition of allicin in garlic (Wang et al., 2012). It is yellowish in nature and is insoluble in water with a chemical structure of $C_6H_{10}S_2$ (Wang et al., 2012).

A preliminary study conducted with garlic products revealed that no allicin was found in blood after oral ingestion (Amagase et al., 2001). The first detected compound producing strong odour after ingesting garlic was found to be allyl mercaptan (Amagase et al., 2001). After ingestion of 25 g of raw garlic (90 mg allicin), no allicin was found in urine from 1 to 24 hours in a study performed by Lawson in 1992. Allicin is thus not likely the active ingredient as it was not found in the blood or urine after garlic consumption (Amagase et al., 2001). Generally, the active ingredients in garlic have not been fully distinguished and characterised (Amagase et al., 2001). It is assumed that a biological response is dependent on the bioavailaibility of the organosulfur compounds found in various garlic preparations (Amagase et al., 2001).

Chives (*Allium schoenoprasum*),belonging to the *Alliaceae* family and having bulb-forming characteristics together with garlic, onion, shallot and leek; can be easily grown and are found fresh throughout Thailand (Rattanachaikunsopon et al., 2008). The leaves of Chives are grown for culinary purposes in traditional dishes mainly in countries like Thailand, France and Sweden (Rattanachaikunsopon et al., 2008). They are known to have several health promoting effects due to the presence of organosulfur compounds namely diallyl sulphides (diallyl monosulfide, diallyl disulfide, diallyl trisulfide, and diallyl tetrasulfides) (Rattanachaikunsopon et al., 2008).

4.11.3 Mechanisms of action

4.11.3.1 Allyl sulfides

Various organosulfur compounds are found in garlic which contribute to its biological effects. Allicin is responsible for the pungent smell of garlic and is metabolized from alliin when garlic is cut or crushed. Allicin is further metabolized - due to high instability - to other organosulfur compounds.

The possibility that allicin can penetrate different cellular compartments in biological systems and exerts its biological effects, has been raised through findings. Significance of allicin as a biological effectors' molecule is due to its:

- Reactivity with low and high molecular weight thiols
- Prominent antioxidant activity and
- Accessibility resulting from high membrane permeability (Omar and Al-Wabel, 2010).

Low cellular concentrations of glutathione (major intracellular antioxidant), and/or overproduction of reactive oxygen species (ROS) can lead to oxidative-induced stress damage. Garlic-derived allicin lowered ROS production and also increased glutathione concentration (Horev-Azaria et al., 2009). Like allicin, diallyl disulfide (DADS), and diallyl trisulfide (DATS), have been shown to stimulate antioxidant pathway (Trio et al., 2014). S-allylcysteine (SAC) is one of the most abundant organosulfur compounds in aged garlic extract (AGE) (Colin-Gonzalez et al., 2012). The latter is an odorless garlic preparation resulting from protracted extraction of fresh garlic at room temperature (Colin-Gonzalez et al., 2012). Different antioxidant mechanisms have been reported for AGE and SAC, such as their ability to (a) scavenge reactive oxygen and nitrogen species; (b) increase enzymatic and nonenzymatic antioxidant levels; (c) activate Nrf2 factor; or (d) inhibit some pro-oxidant enzymes (Colin-Gonzalez et al., 2012).

The antioxidant capacity of SAC is due to the presence of a nucleophile (thiol group) which can easily donate its proton, neutralising an electrophile or making it less reactive (Colin-Gonzalez et al., 2012). SAC readily prevents protein and lipid oxidation and nitration (Colin-Gonzalez et al., 2012). SAC can scavenge superoxide anion ($O_2^{\bullet-}$), hydrogen peroxide (H_2O_2), hydroxyl radical ($OH^{\bullet}$), peroxynitrite radical ($ONOO^{-}$), and peroxyl radical ($LOO^{\bullet}$) produced in neuronal cells, as well as hypochlorous acid (HOCl) and singlet oxygen ($^1O_2$) produced in microglial cells (Colin-Gonzalez et al., 2012). Furthermore, SAC improved catalase and glutathione peroxidase activities which are antioxidant enzymes- and activate Nrf2 (Colin-Gonzalez et al., 2012). Redox homeostasis is regulated by the transcription factor Nrf2 (nuclear factor-E2-related factor 2) (Colin-Gonzalez et al., 2012). It regulates the expression of antioxidant proteins, providing the level of protection required for normal cellular activities and against oxidative damage (Colin-Gonzalez et al., 2012).

One of the products formed when nitric oxide synthases (NOS) catalyse the conversion of L-arginine is nitric oxide (NO) (Colin-Gonzalez et al., 2012). Though increased levels of NO have been involved in health complications, NO is also imperative in cellular signaling and in a large number of physiological processes (Colin-Gonzalez et al., 2012). Thus, an imbalance in NO concentrations compromises the cell survival mechanisms (Colin-Gonzalez et al., 2012). SAC is known to inhibit NO production in lipopolysaccharide/ cytokine-stimulated macrophages and hepatocytes by suppression of inducible NOS (iNOS) gene expression. iNOS is one the isoforms of NOS and is dependent on calcium ions. Inhibition of NO production induced by SAC may be related to its ability to decrease NFκB activation, a protein that controls transcription of DNA, cytokine production and cell survival (Colin-Gonzalez et al., 2012). Thus, leading to cell protection (Colin-Gonzalez et al., 2012).

Other pro-oxidant enzymes, such as xanthine oxidase, NADPH oxidase, and cyclooxygenase, have been studied in this aspect concerning the effects AGE and SAC on them (Colin-Gonzalez et al., 2012). However, due to little information, their possible inhibitory mechanisms of action have not yet been determined (Colin-Gonzalez et al., 2012).

4.11.3.2 Indoles

Sulfur-containing compounds called glucosinolates are found in abundance in cruciferous vegetables (Ishida et al., 2014). Based on their structure, glucosinolates can be classified into three categories, one of which is indole glucosinolates. Glucosinolates and their metabolites, especially indole-3-carbidol and isothiocyanates, exhibit a diversity of biological activities by triggering anti-inflammatory and antioxidant response which contribute to cell homeostasis (Ishida et al., 2014).

4.11.3.3 Indole-3-carbinol

Biotransformation enzymes play significant roles in the metabolism and elimination of many biologically active compounds, such as estrogens, drugs, xenobiotics, carcinogens, toxins (Lampe et al., 2002). In general, phase I metabolizing enzymes catalyze reactions that escalate the reactivity of fat-soluble compounds, which prepares them for reactions catalyzed by phase II detoxifying enzymes (Lampe et al., 2002). Reactions catalyzed by phase II enzymes usually elevate water solubility and promote the elimination of these compounds (Lampe et al., 2002).

Indole-3-carbinol (I3C) and some I3C condensation products can bind to a protein in the cytoplasm of cells called the aryl hydrocarbon receptor (AhR) (Marconett et al., 2010). They appear to be important endogenous ligands for the AhR (Hubbard et al., 2015). Binding allows AhR to enter the nucleus where it forms a complex with the AhR nuclear translocator (Arnt) protein (Tsuji et al., 2014). This AhR/Arnt complex binds to specific DNA sequences, known as xenobiotic response elements (XRE), in the regulatory regions of target genes, especially those involved in xenobiotic metabolism (Safe, 2001).

I3C and DIM have been shown to induce the expression of phase II detoxifying and antioxidant enzymes via the activation of the nuclear factor E2-related factor 2 (Nrf2)-dependent pathway (Di et al., 2016).

4.11.3.4 Isothiocyanates

Isothiocyanates are derived from the hydrolysis of glucosinolates. Each glucosinolate forms a different isothiocyanate when hydrolysed (Fahey et al., 2001). Absorbed isothiocyanates are rapidly conjugated to glutathione in the liver, and then sequentially metabolised, before being excreted in the urine (Barba et al., 2016). Isothiocyanates may modulate the expression and activity of biotransformation enzymes that are involved in the metabolism and elimination of xenobiotics (e.g., carcinogens) from the body (Bryan et al., 2013).

4.11.4 Beneficial and detrimental effects on health

4.11.4.1 Beneficial effects

The potential health benefits of garlic (*Allium sativum*) have been known for 5000 years (Amagase et al., 2001). Garlic was used more often in ancient times by the Babylonians, Egyptians, Phoenicians, Vikings, Chinese, Greeks, Romans and Indian (Amagase et al., 2001). Garlic was used as a treatment against intestinal disorders, flatulence, worms, respiratory infections, skin diseases, wounds, symptoms of aging and many other diseases (Amagase et al., 2001). Grounded or sliced garlic was used to treat wounds to inhibit the spread of infectious diseases (Amagase et al., 2001). In modern times, health benefits of garlic have gained popularity worldwide through 3000 publications, including experimental and clinical studies conducted with garlic extract (Amagase et al., 2001). Health benefits of garlic include a decrease in risk factor of cardiovascular diseases and cancer, acting as an immunostimulant, enhanced foreign compound detoxification, radio-protection, restoration of physical strength, resistance to various stresses and potential anti-aging effects (Amagase et al., 2001).

Water-soluble organosulfur compounds have shown to exhibit hypolipidemic, antiplatelet and procirculatory effects (Amagase et al., 2001). Aged garlic extract has been shown to have a hepato-protective effect,anticancer and chemopreventive effects and they exhibit antioxidant activities compared to raw or heated garlic which stimulates oxidation (Amagase et al., 2001). Thus, not all garlic preparations will have the same composition and will produce the same biological responses (Amagase et al., 2001).

Garlic, due to its presence of organosulfur compounds, has shown to increase longevity and has gained importance in the scientific community due to their health benefits on the cardiovascular system by reducing atherosclerosis and its associated complications, including stroke, myocardial infarction and thrombotic disorders. It has been useful in the prevention of cancer by exhibiting its antioxidant activity through free-radical scavenging effects, and the effects on cell proliferation and tumor growth (Bianchini et al., 2001).There is a possibility that diallyl disulfide and diallyl trisulfide are important in the anticancer action of garlic (Omar et al., 2009).

Indole-3-carbinol is another organosulfur compound involved in the prevention of cancer (Katz et al., 2018). It is formed by the breakdown of a secondary metabolite in cabbage, broccoli and related vegetables; namely, glucosinolates (Katz et al., 2018). It has a potential role in cancer management and many studies have shown that it inhibits the proliferation of cancer cells, including breast, colon, prostate and endometrial cancer cells (Katz et al., 2018).

4.11.5 Detrimental effects

Garlic has shown to be detrimental to health due to the presence of an excess of oil-soluble organosulfur constituents (Amagase et al., 2001). The lipid-lowering effects of some oil-soluble sulfur compounds have shown to cause hepatoxicity, revealed by increased lactate dehydrogenase release from cells. (Amagase et al., 2001).

Another problem is the unrealistic estimated doses (10–100 cloves per day) that need to be consumed by humans to benefit from the anti-tumour effects of these organosulfur compounds when comparing the doses of the pure chemicals (40 to 100 g per day) that were found to be effective in animals (Bianchini et al., 2001). Studies have found a protective recommended intake of 18.3 g/week on average (Bianchini et al., 2001). Little is known about the toxicity of these compounds, requiring further attention if the use of garlic and other Allium vegetables are to be recommended for cancer management and/ or prevention (Bianchini et al., 2001).

4.11.5.1 In vitro evidences

Cardiac cells exposed to high glucose were treated with diallyl trisulfide (DATS) to investigate whether it inhibits hyperglycemia-induced ROS production via the Nrf2 pathway (Lui et al., 2014). It was seen that DATS protects against hyperglycemia-induced ROS-mediated apoptosis and activates Nrf2-regulated antioxidant enzymes in cardiomyocytes exposed to high glucose (Tsai et al., 2013).

Human endothelial cells were cultured to test whether aged garlic extracts have an effect on antioxidant enzymes' activity (Colin-Gonzalez et al., 2012). It was found that AGE enhanced the accumulation of Nrf2 into the nucleus in a time- and dose-dependent manner and increased the gene expression of antioxidant enzymes (Hiramatsu et al., 2016).

In a study where human primary hepatocytes were culture, the effects of naturally occurring phytochemicals on gene expression of carcinogen-metabolising enzymes were investigated (Kou et al., 2013). One of the phytochemicals was 3,3′-diindolylmethane (DIM). It was demonstrated that cytochrome P450 (CYP)1A1 was up-regulated by DIM in a dose-dependent manner (Gross-Steinmeyer et al., 2004). CYP1A2 transcription was significantly activated following DIM treatment, amongst others (Gross-Steinmeyer et al., 2004).

In a 2012 study, the mechanisms of action of broccoli-derived 3,3′-diindolylmethane (DIM) and indole-3-carbinol (I3C) were investigated using cultured models of prostate cancer cells (Wang et al., 2012). It was found that differences in efficacies and mechanisms existed between DIM and I3C and they both showed protective cancer effects (Wang et al., 2012).

Studies have shown that indole-3-carbinol (I3C) and its *in vivo* dimeric product, 3,3′-diindolylmethane (DIM), inhibit the growth of PC3 prostate cancer cells and induce apoptosis by inhibiting nuclear factor-kappaB (NFκB) and Akt pathways (Li et al., 2003). Upon analysis, it was found that both I3C and DIM up-regulated the expression of genes that are related to the Phase I and II enzymes, suggesting their increased capacity for detoxification of carcinogens or chemicals

(Banerjee et al., 2011). It was concluded that I3C and DIM affected the expression of various genes that are related to the control of carcinogenesis, cell survival and physiologic behaviours (Li et al., 2003).

4.11.5.2 Animal studies

A study shows that diallyl sulphide has an anti-hepatocarcinogenic effect in mouse model (Hayes et al., 1987). A similar study shows that tobacco carcinogen nitrogen-derived nitrosoketone was inhibited when mouse having lung carcinogenesis was treated with diallyl sulphide (Rao et al., 2015). Another research has shown that the dietary intake of diallyl sulphide inhibited formation of liver pre-neoplastic foci by blocking aflatoxin B1 and N-nitrosodiethylamine in rats (Rao et al., 2015).

Diallyl sulphide showed an increase in the activity of major antioxidant enzymes in kidney of treated Wistar rats by a study conducted by Kalayarasan et al. in 2009. Furthermore, diallyl sulphide exhibited protective antioxidant effects by decreasing immunohistochemical staining for tumor necrosis factor in renal tissues (Rao et al., 2015). This effect increased the expression of transcription nuclear factor –like 2 in Wistar rats (Rao et al., 2015).

Another study conducted showed antioxidant effects of diallyl sulphide in rat lung by up regulating the activity and transcription of major antioxidant enzymes. This effect was achieved by modulating the Nrf2 expression and nuclear translocation compared to untreated animals. Animals treated with diallyl sulphide showed an increase in enzyme activity and transcription for superoxide dismutase, glutathione peroxidase and catalase, suggesting an increase in pulmonary antioxidant capacity or a reduction in oxidative stress (Rao et al., 2015).

Based on the Fischer indole synthesis, Radulovic et al. in 2014 designed and used two indole compounds against rat peritoneal macrophages showing that these compounds exhibited a significant increase in myeloperoxidase inhibition, weak anticholinesterase activity and a high cytotoxicity (Naim et al., 2016).

A study conducted showed that 11-amino derivatives of chromeno 2, 3-b indoles prevent the proliferation of fibroblast in normal mice (Peng et al., 2012). Another-study was conducted to examine the effect of diallyl sulphide and garlic homogenates on the activities of catalase, glutathione peroxidase and superoxide dismutase in male Sprague-Dawley rats at daily doses of 50 or 200 mg/kg for eight consecutive days. This treatment caused a decrease in hepatic catalase activity, which is an enzyme involved in oxidative stress.

4.11.5.3 Clinical studies

A 2004 study tried to examine whether a large dose of allitridum and a microdose of selenium prevent gastric cancer (Li et al., 2004). A double-blind intervention study was performed on the participants aged (35–74) years, who had matched at least one of the following criteria: (1) a medical history of stomach disorder, (2) a family history of tumour, or (3) smoking and/or alcohol consumption (Li et al., 2004). A total of 2,526 and 2,507 persons were randomly enrolled into intervention group and control group, respectively, from 288 natural villages of seven communities in China (Li et al., 2004).

Those from the intervention group orally took 200 mg synthetic allitridum every day and 100 mg selenium every other day for one month of each year during November 1989 to December 1991 (Li et al., 2004). Simultaneously, people in control group were given 2 placebo capsules containing corn oil with the identical appearance to that in the intervention group (Li et al., 2004).

For all subjects the large dose of allitridum was accepted and no harmful side effects were found during the study (Li et al., 2004). In the first follow-up five years (1992-1997) after stopping the intervention, a 22% decline was observed in the intervention group with respect to the morbidity rates of malignant tumours. In comparison, a 47.3% decrease was seen in the control group (Li et al., 2004). This study proves that large doses of allitridum and microdose of selenium may effectively prevent gastric cancer, especially in men (Li et al., 2004).

A preliminary double-blind, randomized clinical trial using high-dose AGE (AGE 2.4 mL/d) as an active treatment and low-dose AGE (AGE 0.16 mL/d) as a control was performed on patients with colorectal adenomas-precancerous lesions of the large bowel. Fifty one patients who were diagnosed as carrying colorectal adenomas were enrolled. The patients were randomly assigned to the two groups after adenomas larger than 5 mm in diameter were removed by polypectomy. Thirty-seven patients (19 in the active group, 18 in the control group) completed the study and were evaluated for the efficacy of AGE. The number of adenomas increased linearly in the control group from the beginning (the baseline), but AGE significantly suppressed both the size and number of colon adenomas in patients after 12 months of high-dose treatment (P=0.04). The results suggest AGE suppresses progression of colorectal adenomas in humans. It appears that AGE has multiple pathways to reduce cancer incidence and suppress its growth and proliferation.

Conclusion

Based on the evidences found *in vitro*, *in vivo*, in clinical and animal studies, organosulfur compounds mainly allyl sulphides and indoles have shown to have more health benefits than detrimental effects. Allyl sulphides and indoles have shown to have an anticancergenic effect and decrease risk of cardiovascular diseases. Little is known about the toxicity of the organosulfur compounds, requiring further attention if the use of garlic and other Allium vegetables are to be recommended for cancer prevention.

References

Amagase, H., Petesch, B., Matsuura, H., Kasuga, S., Itakura, Y., 2001. Intake of Garlic and Its Bioactive Components. J. Nut. 131 (3), 955S–962S, https://doi.org/10.1093/jn/131.3.955S.

Banerjee, S., Kong, D., Wang, Z., Bao, B., Hillman, G.G., Sarkar, F.H., 2011. Attenuation of multi-targeted proliferation-linked signaling by 3, 3′-diindolylmethane (dim): from bench to clinic. Mutat. Res. 728 (0), 47–66.

Barba, F.J., Nikmaram, N., Roohinejad, S., Khelfa, A., Zhu, Z., Koubaa, M., 2016. Bioavailability of glucosinolates and their breakdown products: impact of processing. Front. Nutr. 3, 24.

Goncharov, N.V., Belinskaia, D.A., Ukolov, A.I., Jenkins, R.O., Avdonin, P.V., 2021. Chapter 54 - Organosulfur compounds as nutraceuticals, Editor(s): R.C., Gupta, R., Lall, A., Srivastava, Nutraceuticals (Second Edition), Academic Press, 911–924, ISBN 9780128210383, https://doi.org/10.1016/B978-0-12-821038-3.00054-9.

Bianchini, F., Vainio, H., 2001. Allium vegetables and organosulfur compounds: do they help prevent cancer? Environ. Health Perspect. 109, 893.

Bryan, H.K., Olayanju, A., Goldring, C.E., Park, B.K., 2013. The Nrf2 cell defence pathway: keap1-dependent and -independent mechanisms of regulation. Biochem. Pharmacol. 85 (6), 705–717.

Colin-Gonzalez, A.L., Santana, R.A., Silva-Islas, C.A., Chanez-Cardenas, M.E., Santamarıa, A., Maldonado, P.D., 2012. The antioxidant mechanisms underlying the aged garlic extract- and S-allylcysteine-induced protection. Oxid. Med. Cell. Longev. 2012, 907162. doi: 10.1155/2012/907162.

Derosa, T.F., 2006. Indoles. Advances in Synthetic Organic Chemistry and Methods Reported in US Patents, 326–336.

Di, W., Shi, X., Lv, H., Liu, J., Zhang, H., Li, Z., Fang, Y., 2016. Activation of the nuclear factor E2-related factor 2/antioxidant response element alleviates the nitroglycerin-induced hyperalgesia in rats. J. Headache Pain 17 (1), 99.

Fahey, J.W., Zalcmann, A.T., Talalay, P., 2001. The chemical diversity and distribution of glucosinolates and isothiocyanates among plants. Phytochemistry 56 (1), 5–51.

Gross-Steinmeyer, K., Stapleton, P.L., Liu, F., Tracy, J.H., Bammler, T.K., Quigley, S.D., Farin, F.M., Buhler, D.R., Safe, S.H., Strom, S.C., Eaton, D.L., 2004. Phytochemical-induced changes in gene expression of carcinogen-metabolizing enzymes in cultured human primary hepatocytes. Xenobiotica 34 (7), 619–632.

Hayes, MA., Rushmore, TH., Goldberg, MT., 1987. Inhibition of hepatocarcinogenic responses to 1,2-dimethylhydrazine by diallyl sulfide, a component of garlic oil. Carcinogenesis 8 (8), 1155–1157.

Hiramatsu, K., Tsuneyoshi, T., Ogawa, T., Morihara, N., 2016. Aged garlic extract enhances heme oxygenase-1 and glutamate-cysteine ligase modifier subunit expression via the nuclear factor erythroid 2-related factor 2-antioxidant response element signaling pathway in human endothelial cells. Nutr. Res. 36 (2), 143–149.

Horev-Azaria, L., Eliav, S., Izigov, N., Pri-Chen, S., Mirelman, D., Miron, T., Rabinkov, A., Wilchek, M., Jacob-Hersch, J., Amariglio, N., Savion, N., 2009. Allicin up-regulates cellular glutathione level in vascular endothelial cells. Eur. J. Nutr. 48 (2), 67–74.

Hubbard, T.D., Murray, I.A., Perdew, G.H., 2015. Indole and tryptophan metabolism: endogenous and dietary routes to Ah receptor activation. Drug Metab. Dispos. 43 (10), 1522–1535.

Ishida, M., Hara, M., Fukino, N., Kakizaki, T., Morimitsu, Y., 2014. Glucosinolate metabolism, functionality and breeding for the improvement of Brassicaceae vegetables. Breed. Sci. 64 (1), 48–59.

Katz, E., Nisani, S., Chamovitz, D.A., 2018. Indole-3-carbinol: a plant hormone combatting cancer. F1000 Res. 7, 689.

Kou, X., Kirberger, M., Yang, Y., Chen, N., 2013. Natural products for cancer prevention associated with Nrf2–ARE pathway. Food Sci. Hum. Wellness 2 (1), 22–28.

Lampe, J.W., Peterson, S., 2002. Brassica, biotransformation and cancer risk: genetic polymorphisms alter the preventive effects of cruciferous vegetables. J. Nutr. 132 (10), 2991–2994.

Li, H., Li, H.Q., Wang, Y., Xu, H.X., Fan, W.T., Wang, M.L., Sun, P.H., Xie, X.Y., 2004. An intervention study to prevent gastric cancer by micro-selenium and large dose of allitridum. Chin. Med. J. 117, 1155–1160.

Li, Y., Li, X., Sarkar, F.H., 2003. Gene expression profiles of I3C- and DIM-treated PC3 human prostate cancer cells determined by cDNA microarray analysis. J. Nutr. 133 (4), 1011–1019.

Lui, Q., Wang, S., Cai, L., 2014. Diabetic cardiomyopathy and its mechanisms: role of oxidative stress and damage. J. Diabetis Investig. 5 (6), 623–634.

Marconett, C.N., Sundar, S.N., Poindexter, K.M., Stueve, T.R., Bjeldanes, L.F., Firestone, G.L., 2010. Indole-3-carbinol triggers aryl hydrocarbon receptor-dependent estrogen receptor (ER) alpha protein degradation in breast cancer cells disrupting an ERalpha-GATA3 transcriptional cross-regulatory loop. Mol. Biol. Cell 21 (7), 1166–1177.

Naim, M.J., Alam, O., Bano, F., 2016. Recent review on indole: a privileged structure scaffold. Int. J. Pharm. Sci. Res. 7, 51–62.

Omar, S.H, Al-Wabel, N., 2009. Organosulfur compounds and possible mechanism of garlic in cancer. Saudi Pharm. J. 18, 51–58.

Peng, W., Switalska, M., Wang, L., Mei, Z., Edazawa, Y., Pang, C., El-Sayed, I.E., Wietrzyk, J., Inokuchi, T., 2012. Synthesis and in vitroantiproliferative activity of new 11-aminoalkylaminosubstituted chromeno[2,3-b]indoles. Eur. J. Med. 58 (1), 441–451.

Rattanachaikunsopon, P., Phumkhachorn, P., 2008. Diallyl sulphide content and antimicrobial activity against food-borne pathogenic bacteria of chives (*Allium schoenoprasum*). Biosci. Biotechnol. Biochem. 72, 2987–2991.

Rao, P., Midde, N., Miller, D., Chauhan, S., Kumar, A., Kumar, S., 2015. Diallyl sulphide: potential use in novel therapeutic interventions in alcohol, drugs, and disease mediated cellular toxicity by targeting cytochrome P450 2E1. Curr. Drug Metab. 16, 486–503.

Radulovic, N.S., Zlatkovic, D.B., Mitic, K.V., Randjelovic, P.J., Stojanovic, N.M., 2014. Synthesis, spectral characterization,cytotoxicity and enzyme-inhibiting activity of new ferrocene–indole hybrid. Polyhedron 14 (80), 134–141.

Safe, S., 2001. Molecular biology of the Ah receptor and its role in carcinogenesis. Toxicol. Lett. 120 (1-3), 1–7.

Trio, P.Z., You, S., He, X., He, J., Sakao, K., Hou, D.X., 2014. Chemopreventive functions and molecular mechanisms of garlic organosulfur compounds. Funct. Foods 5 (5), 833–844.

Tsai, C.Y., Wang, C.C., Lai, T.Y., Tsu, H.N., Wang, C.H., Liang, H.Y., Kuo, W.W., 2013. Antioxidant effects of diallyl trisulfide on high glucose-induced apoptosis are mediated by the PI3K/Akt-dependent activation of Nrf2 in cardiomyocytes. Int. J. Cardiol. 168 (2), 1286–1297.

Tsuji, N., Fukuda, K., Nagata, Y., Okada, H., Haga, A., Hatakeyama, S., Yoshida, S., Okamoto, T., Hosaka, M., Sekine, K., Ohtaka, K., Yamamoto, S., Otaka, M., Grave, E., Itoha, H., 2014. The activation mechanism of the aryl hydrocarbon receptor (AhR) by molecular chaperone HSP90. FEBS Open Bio 4, 796–803.

Wang, H.-C., Pao, J., Lin, S.-Y., Sheen, L.-Y., 2012. Molecular mechanisms of garlic-derived allyl sulfides in the inhibition of skin cancer progression. Ann. N. Y. Acad. Sci. 1271, 44–52.

Wang, T.T., Schoene, N.W., Milner, J.A., Kim, Y.S., 2012. Broccoli-derived phytochemicals indole-3-carbinol and 3, 3′-diindolylmethane exerts concentration-dependent pleiotropic effects on prostate cancer cells: comparison with other cancer preventive phytochemicals. Mol. Carcinog. 51 (3), 244–256.

CHAPTER

Phenolic acids 4.12

Hari P. Devkota, Anjana Adhikari-Devkota

Graduate School of Pharmaceutical Sciences, Kumamoto University, Kumamoto, Japan

4.12.1 Introduction

Polyphenolic compounds comprise one of the largest groups of plant secondary metabolites that are made of one or more aromatic rings with one or more hydroxyl groups. Phenolic acids, a chemical class belonging to polyphenolic compounds are widely found in plants and specially in vegetables, berries, fruits, and beverages (Barros et al., 2009; Mattila et al., 2006; Nile and Park, 2014; Zadernowski et al., 2005). Phenolic acids are synthesized in plants during normal growth and development, as well as in response to stress conditions and against adverse factors (drought, UV radiation, wounding, infections or physical damage, etc.) (Carl, 2000). Phenolic acids can be mainly divided into two groups: (1) hydroxybenzoic acid derivatives and (2) hydroxycinnamic acid derivatives. Gallic acid, protocatechuic acid, vanillic acid, and syringic acid (Fig. 4.12.1) are the most common hydroxybenzoic acid derivatives. These hydroxybenzoic acids occur in plants as free or conjugated (bound) such as hydrolysable tannins and lignins. Some of these hydroxybenzoic acid derivatives are also bound to flavonoid aglycones or flavonoid glycosides such as in case of epigallocatechingallate (EGCG) or galloyl derivatives of flavonoid glycosides. p-Coumaric acid, caffeic acid, ferulic acid, and sinapic acid (Fig. 4.12.2) are the main hydroxycinnamic acid derivatives (Barros et al., 2009; Mattila et al., 2006). Hydroxycinnamic acids are usually found in bound form with quinic acid, tartaric acid or sugars. The most common examples are mono-caffeoyl quinic acid (e.g., chlorogenic acid (5-O-caffeoyl quinic acid), di-caffeoyl quinic acids, and tri-caffeoyl quinic acid. Mono-, di-, tri-, or tetra- galloyl quinic acid derivatives are also reported from various plant sources (Dirar et al., 2019). Some of the common sources of these phenolic acids are given in Table 4.12.1. The content of phenolic acids in fruits, berries and vegetables are affected by many genetic, physical and environmental factors such as geographic region, altitude, weather conditions, and stages of ripeness (Benvenuti et al., 2004; Biosci et al., 2013; Castrejón et al., 2008; Connor et al., 2002; Deighton et al., 2000; Ehala et al., 2005; Ha and To, 2000; Haffner et al., 2002; Hakala et al., 2003; Skupień and Oszmiański, 2004). Due to their common natural abundance, phenolic acids are well studied for their antioxidative, anticancer and other health beneficial activities. A Scopus database search with the term "phenolic

Antioxidants Effects in Health. DOI: https://doi.org/10.1016/B978-0-12-819096-8.00014-8

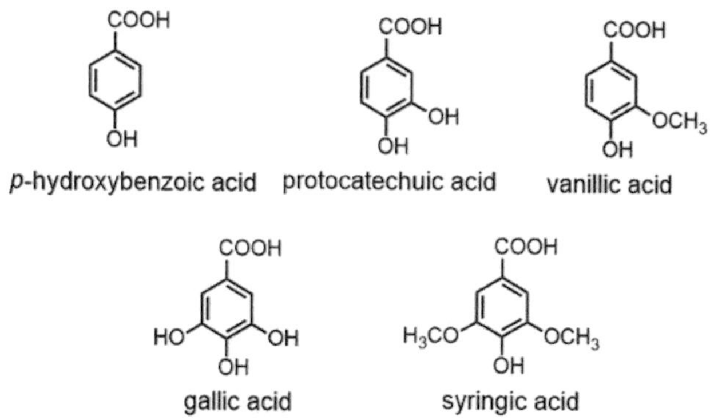

FIG. 4.12.1 Structures of common hydroxybenzoic acid derivatives.

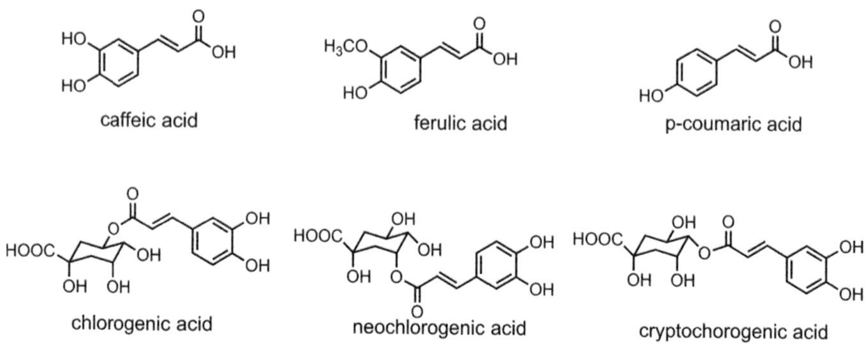

FIG. 4.12.2 Structures of common hydroxycinnamic acid derivatives.

Table 4.12.1 List of some common sources of phenolic acids.

Phenolic acids	Sources	References
Gallic acid	Tea, emblic myrobalan, grape skin	(Brewer, 2011; Yilmaz and Toledo, 2004)
p-Coumaric acid	White grapes, tomato, cabbage, asparagus	(Rice-Evans et al., 1997)
Caffeic acid	White grapes, olive, cabbage, asparagus, coffee	(Rice-Evans et al., 1997) (Zduńska et al., 2018)
Ferulic acid	Grains, tomato, asparagus, cabbage	(Rice-Evans et al., 1997)
Chlorogenic acid	Apple, cherry, coffee, pear, tomato	(Rice-Evans et al., 1997)

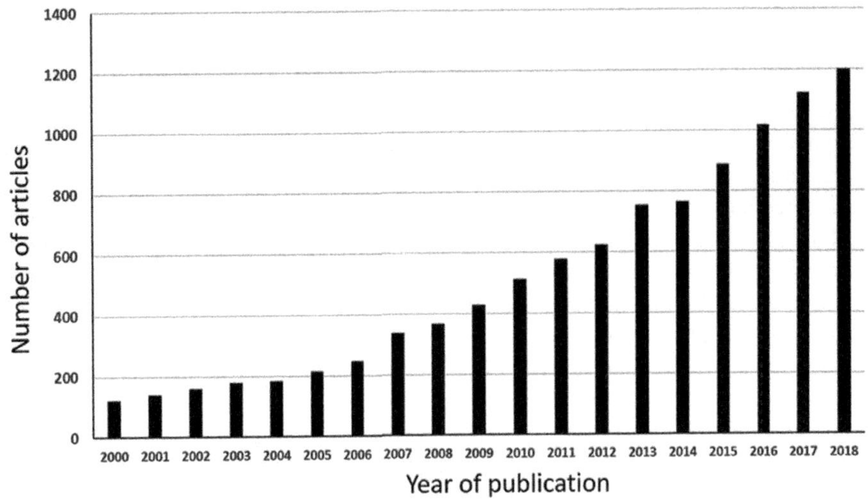

FIG. 4.12.3 Trends in scientific publications regarding curcumin from year 2000 to 2018 (Retrieved on August 10, 2019).

acid" for the past 20 years (from 2000–2018) showed constant growth in number of publications each year (Fig. 4.12.3). Many of these publications are also focused on the *in vitro* or *in vivo* antioxidant activity of phenolic acids.

4.12.2 Antioxidant activity of phenolic acids

There are hundreds of publications that report the *in vitro* antioxidant activities and metal chelating activities of phenolic acids. Most of these studies reported the free radical scavenging activities of these compounds where the number of free hydroxyl groups play significant role. Rice-Evans et al. reviewed the antioxidant capacity of different phenolic acids and expressed their antioxidant activities relative to Trolox. Monohydroxy phenolic acids had very low antioxidant activity, but dihydroxy phenolic acids (e.g., protocatechuic acid and resorcylic acid) have higher antioxidant activity. Trihydroxy substituted phenolic acid (gallic acid) showed strongest activity. However, the methoxy-substituted phenolic acids, for example, vanillic acid and syringic acid had lower activity as compared to their unsubstituted analogues, that is, protocatechuic acid and gallic acid, respectively (Rice-Evans et al., 1996). Thus, the number of unsubstituted hydroxyl groups was reported as an essential requirement for the *in vitro* antioxidant activity. Many other articles have also reported the weaker antioxidant/free radical scavenging activity for protocatechuic acid as compared to gallic acid (Joshi et al., 2014). Similar pattern was also observed for the hydroxycinnamic acid derivatives. Ortho-monohydroxy substituted hydroxycinnamic acid (*O*-coumaric acid) has no activity however its meta and para substituted analogues had antioxidant activity. In case of caffeic acid, the esterification of 3-hydroxyl group enhanced the antioxidant activity (Rice-Evans et al., 1996). Ferulic acid is reported to exert its antioxidant activity not only by scavenging the free radicals and chelating metals but also by inhibiting the different enzymes that are involved in the generation of free radicals (Zduńska et al., 2018). Lenzi et al. (2015) reported that the treatment with ferulic acid (1 mg/kg) showed antidepressant-like effect in male Swiss mice as evaluated by tail suspension test and forced swimming test. Administration of ferulic acid increased superoxide dismutase (SOD), catalase (CAT) and glutathione peroxidase (GSH-Px) activities and decreased the thiobarbituric acid-reactive substance (TBA-RS) level in blood, hippocampus and cerebral cortex.

The *in vitro* antioxidant activity of phenolic acids is highly influenced by the solvent and pH of the reaction mixtures as reported by Amorati et al. (2006). Marinova and Yanishlieva (2003) reported the antioxidant activity of protocatechuic, syringic, sinapic and caffeic acid at room temperature and high temperature (90°C). They reported that the antioxidant activity of cinnamic derivatives was higher than that of hydroxybenzoic acid derivatives at room temperature towards the oxidation of sunflower oil. While the increase in temperature increased the antioxidant activity of hydroxycinnamic acids but that of hydroxybenzoic acids was not changed.

Nair and Nair (2013) investigated the radioprotective effect of gallic acid in Swiss albino mice exposed to gamma radiation. Pre-administration of gallic acid (100 mg/kg body weight, oral) 1 h before the gamma radiation exposure prevented the radiation induced decrease in cellular antioxidant enzymes and DNA damage. Similarly the antioxidant activity of gallic acid has been reported in lead (Pb) induced toxicity in mice (Reckziegel et al., 2016), sodium fluride induced oxidative stress in rat erythrocytes (Nabavi et al., 2013), tramadol-induced hepato- and nephrotoxicity in rats (Sheweita et al., 2018), mercuric chloride-induced liver damage in rats (Goudarzi et al., 2017), among others. A recent review by Dludla et al. (2019) highlighted the antioxidative activity of gallic acid or gallic acid rich fruits based on the literature review of *in vitro*, *in vivo* and human studies and its protective role in inflammation and obesity.

Adedara et al. (2019) reported that the protocatechuic acid restored the antioxidant status, inhibited lipid peroxidation and suppressed the level of pro-inflammatory biomarkers in tests and epididymis of diabetic rats. Caffeic acid has been reported to exhibit antioxidant activity against oxytetracycline induced lipid peroxidation in albino rats (Jayanthi and Subash, 2010), methamphetamine induced tissue toxicity in Sprague Dawley rats (Koriem et al., 2013), among others. Similar results are also reported for other hydroxycinnamic acid derivatives.

4.12.3 Pro-oxidant activity of phenolic acids

Although phenolic acids are strong antioxidants, phenoxyl radicals are produced during the enzymatic process or direct radical scavenging mechanisms. These phenoxy radicals can exhibit prooxidant activities (Fukumoto and Mazza, 2000; Sakihama et al., 2002). Pro-oxidant activity of polyphenols is also reported to be beneficial for cytotoxic activities as they can show selective cytotoxic activity towards cancer cells by generating reactive oxygen species (León-González et al., 2015). Antioxidant/pro-oxidant behavior of phenolic acids is reported to be dependent on many factors such as structural characteristics, concentrations, presence of metals, etc. (Yang et al., 2012; Zeraik et al., 2014). Yoshino et al. (2002) reported that the gallic acid exerted prooxidant activity in presence of copper ion and caused DNA damage and formation of 8-hydroxy-2′-deoxyguanoside in DNA, which was reduced by the addition of catalase (CAT), suggesting the role of hydroxyl radicals in DNA damage. Iwasaki et al. (2011) also reported that phenolic compounds show pro-oxidant activity in presence of cupper ions. Utrera and Estevez (2013) reported that gallic acid showed dose dependent antioxidant/pro-oxidant activity. Gallic acid at lower concentration (10 μM) acted as antioxidant and at higher concentration (200 μM) acted as pro-oxidant against formation of protein oxidation products. Zeraik et al. (2014) reported that the pro-oxidant activity of protocatechuic acid was enhanced by esterification, specially with the alkyl ester having 7 carbon chain alkyl group, which suggested that lipophilicity may play important role in pro-oxidant activity. However, gallic acid and butyl gallate did not show such activities which also implicated that the number of hydroxyl groups is also an important factor.

Maurya and Devasagayam (2010) reported that both caffeic acid and ferulic acid showed antioxidant activity against nitric oxide, superoxide and 2,2′-azino-bis(3-ethylbenzothiazoline-6-sulfonic acid) (ABTS) radical. However, the activity against 1,1-diphenyl-2-picrylhydrazyl (DPPH) radical was decreased above 20 μM possibly due to formation of an adduct. But in the Fenton reaction, both of these compounds showed pro-oxidant activity above 20 μM probably due to their ferric reducing capacity. However, Paciello et al. (2020) reported that ferulic acid acted as pro-oxidant at lower concentration and antioxidant at higher concentration. At higher concentration, ferulic acid promoted chemoresistance and suggested that such concentration dependent activities may limit the clinical use of ferulic acid. Yang et al. (2012) reported that the ortho-methoxy substitution of p-coumaric acid decreased the pro-oxidant activity.

4.12.4 Bioavailability and metabolism of phenolic acids

Various studies have been performed regarding the bioavailability of phenolic acids in different experimental animals. Lafay and Gil-Izquierdo had compiled the available information on bioavailability of different phenolic acids, including gallic acid, caffeic acid, ferulic acid, chlorogenic acid, among others. Studies have been performed for both pure form of these acids or as constituents of different foods and beverages such as cereals, wine, tea, etc. Gallic acid, a hydroxybenzoic acid was reported to be readily absorbed in rats, however in humans, it was predominantly absorbed as 4-*O*-methylated or *O*-glucuronidated forms. In case of hydroxycinnamic acids, free forms are reported to be absorbed readily and then converted to their conjugated forms

(Lafay and Gil-Izquierdo, 2008). Caffeic acid is reported to undergo hydrogenation of side chain, dihydroxylation, and methylation of phenolic hydroxyl group, β-oxidation and finally conversion to its conjugates (Rice-Evans et al., 1996). Piazzon et al. has synthesized various hydroxycinnamic acid derivatives, that is, sulfate and acyl glucuronide derivatives of ferulic acid and sulphate derivatives of caffeic acid and evaluated their antioxidant activity by FRAP and ABTS free radical scavenging assay methods. Substitution of phenolic hydroxyl groups by sulphate or glucuronide moieties reduced the antioxidant activity however the acyl glucuronide derivative of ferulic acid showed strong antioxidant activity (Piazzon et al., 2012).

Conclusions

Phenolic acids, members of plant polyphenols, are one of the most common secondary metabolites in plants. They can be mainly divided into hydroxybenzoic acid and hydroxycinnamic acid derivatives. Phenolic acids are reported to have various health beneficial effects including potent free radical scavenging and antioxidant activities. The number of free hydroxyl groups is one of the main determining factor for their antioxidant activities. Many phenolic acids are also reported to exhibit pro-oxidant activity through their phenoxy radicals which can be both beneficial and harmful to human body depending upon various factors. However, more detailed studies are necessary to access the exact mechanism of antioxidant and pro-oxidant activities of different phenolic acids and how they affect human body in protective manner or make more harmful effects depending upon various physiological conditions.

References

Adedara, I.A., Okpara, E.S., Busari, E.O., Omole, O., Owumi, S.E., Farombi, E.O., 2019. Dietary protocatechuic acid abrogates male reproductive dysfunction in streptozotocin-induced diabetic rats via suppression of oxidative damage, inflammation and caspase-3 activity. Eur. J. Pharmacol. 849, 30–42. https://doi.org/10.1016/j.ejphar.2019.01.033.

Amorati, R., Pedulli, G.F., Cabrini, L., Zambonin, L., Landi, L., 2006. Solvent and pH effects on the antioxidant activity of caffeic and other phenolic acids. J. Agric. Food Chem. 54, 2932–2937. https://doi.org/10.1021/jf053159+.

Barros, L., Dueñas, M., Ferreira, I.C.F.R., Baptista, P., Santos-Buelga, C., 2009. Phenolic acids determination by HPLC-DAD-ESI/MS in sixteen different Portuguese wild mushrooms species. Food Chem. Toxicol. 47, 1076–1079. https://doi.org/10.1016/j.fct.2009.01.039.

Benvenuti, S., Pellati, F., Melegari, M., Bertelli, D., 2004. Polyphenols, anthocyanins, ascorbic acid, and radical scavenging activity of rubus, ribes, and aronia. Food Chem. 69, 164–169. https://doi.org/10.1111/j.1365-2621.2004.tb13352.x.

Biosci, I.J., Kivi, A.R., Sartipnia, N., 2013. Environmental and genetic variation of phenolic compounds and antioxidant capacity in raspberry. J. Food Compos. Anal. 6655, 153–159.

Brewer, M.S., 2011. Natural antioxidants: sources, compounds, mechanisms of action, and potential applications. Compr. Rev. Food Sci. Food Saf. 10, 221–247. https://doi.org/10.1111/j.1541-4337.2011.00156.x.

Carl, H.B., 2000. Phenolic-storing cells: keys to programmed cell death and periderm formation in wilt disease resistance and in general defence responses in plants? Physiol. Mol. Plant Pathol. 57, 101–110. https://doi.org/10.1006/pmpp.2000.0287.

Castrejón, A.D.R., Eichholz, I., Rohn, S., Kroh, L.W., Huyskens-Keil, S., 2008. Phenolic profile and antioxidant activity of highbush blueberry (Vaccinium corymbosum L.) during fruit maturation and ripening. Food Chem. 109, 564–572. https://doi.org/10.1016/j.foodchem.2008.01.007.

Connor, A.M., Luby, J.J., Hancock, J.F., Berkheimer, S., Hanson, E.J., 2002. Changes in fruit antioxidant activity among blueberry cultivars during cold-temperature storage. J. Agric. Food Chem. 50, 893–898. https://doi.org/10.1021/jf011212y.

Deighton, N., Brennan, R., Finn, C., Davies, H.V., 2000. Antioxidant properties of domesticated and wild Rubus species. J. Sci. Food Agric. 80, 1307–1313. https://doi.org/10.1002/1097-0010(200007)80:9<1307::AID-JSFA638>3.0.CO;2.

Dirar, A.I., Adhikari-Devkota, A., Mahadi Hassan, M., Wada, M., Watanabe, T., Devkota, H.P., 2019. Phenolic compounds as potent free radical scavenging and enzyme inhibitory components from the leaves of guiera senegalensis. Nat. Prod. Commun. 14. https://doi.org/10.1177/1934578X19857364.

Dludla, P.V., Nkambule, B.B., Jack, B., Mkandla, Z., Mutize, T., Silvestri, S., Orlando, P., Tiano, L., Louw, J., Mazibuko-Mbeje, S.E., 2019. Inflammation and oxidative stress in an obese state and the protective effects of gallic acid. Nutrients 11 (1), 23. https://doi.org/10.3390/nu11010023.

Ehala, S., Vaher, M., Kaljurand, M., 2005. Characterization of phenolic profiles of Northern European berries by capillary electrophoresis and determination of their antioxidant activity. J. Agric. Food Chem. 53, 6484–6490. https://doi.org/10.1021/jf050397w.

Fukumoto, L.R., Mazza, G., 2000. Assessing antioxidant and prooxidant activities of phenolic compounds. J. Agric. Food Chem. 48, 3597–3604. https://doi.org/10.1021/jf000220w.

Goudarzi, M., Kalantar, M., Kalantar, H., 2017. The Hepatoprotective effect of gallic acid on mercuric chloride-induced liver damage in rats. Jundishapur J. Nat. Pharm. Prod. 12. https://doi.org/10.5812/jjnpp.12345.

Ha, S.H., To, A.R., 2000. Content of flavonols and selected phenolic acids in strawberries and Vaccinium species: influence of cultivar, cultivation site and technique. Food Res. Int. 33, 517–524.

Hattner, K., Rosenfeld, H.J., Skrede, G., Wang, L., 2002. Quality of red raspberry Rubus idaeus L. cultivars after storage in controlled and normal atmospheres. Postharvest Biol. Technol. 24, 279–289. https://doi.org/10.1016/S0925-5214(01)00147-8.

Hakala, M., Lapveteläinen, A., Huopalahti, R., Kallio, H., Tahvonen, R., 2003. Effects of varieties and cultivation conditions on the composition of strawberries. J. Food Compos. Anal. 16, 67–80. https://doi.org/10.1016/S0889-1575(02)00165-5.

Iwasaki, Y., Hirasawa, T., Maruyama, Y., Ishii, Y., Ito, R., Saito, K., Umemura, T., Nishikawa, A., Nakazawa, H., 2011. Effect of interaction between phenolic compounds and copper ion on antioxidant and pro-oxidant activities. Toxicol. Vitr. 25, 1320–1327. https://doi.org/10.1016/j.tiv.2011.04.024.

Jayanthi, R., Subash, P., 2010. Antioxidant effect of caffeic acid on oxytetracycline induced lipid peroxidation in albino rats. Indian J. Clin. Biochem. 25, 371–375. https://doi.org/10.1007/s12291-010-0052-8.

Joshi, K.R., Devkota, H.P., Watanabe, T., Yahara, S., 2014. Phenolic compounds from the flowers of Nepalese medicinal plant Aconogonon molle and their DPPH free radical-scavenging activities. Nat. Prod. Res. 28, 2208–2210. https://doi.org/10.1080/14786419.2014.915829.

Koriem, K.M.M., Abdelhamid, A.Z., Younes, H.F., 2013. Caffeic acid protects tissue antioxidants and DNA content in methamphetamine induced tissue toxicity in Sprague Dawley rats. Toxicol. Mech. Methods 23, 134–143. https://doi.org/10.3109/15376516.2012.730561.

Lafay, S., Gil-Izquierdo, A., 2008. Bioavailability of phenolic acids. Phytochem. Rev. 7, 301–311. https://doi.org/10.1007/s11101-007-9077-x.

Lenzi, J., Rodrigues, A.F., Rós, A.de S., de Castro, B.B., de Lima, D.D., Magro, D.D.D., Zeni, A.L.B., 2015. Ferulic acid chronic treatment exerts antidepressant-like effect: role of antioxidant defense system. Metab. Brain Dis. 30, 1453–1463. https://doi.org/10.1007/s11011-015-9725-6.

León-González, A.J., Auger, C., Schini-Kerth, V.B., 2015. Pro-oxidant activity of polyphenols and its implication on cancer chemoprevention and chemotherapy. Biochem. Pharmacol 98 (3), 371–380. https://doi.org/10.1016/j.bcp.2015.07.017.

Marinova, E.M., Yanishlieva, N.V., 2003. Antioxidant activity and mechanism of action of some phenolic acids at ambient and high temperatures. Food Chem. 81, 189–197. https://doi.org/10.1016/S0308-8146(02)00411-9.

Mattila, P., Hellström, J., Törrönen, R., 2006. Phenolic acids in berries, fruits, and beverages. J. Agric. Food Chem. 54, 7193–7199. https://doi.org/10.1021/jf0615247.

Maurya, D.K., Devasagayam, T.P.A., 2010. Antioxidant and prooxidant nature of hydroxycinnamic acid derivatives ferulic and caffeic acids. Food Chem. Toxicol. 48, 3369–3373. https://doi.org/10.1016/j.fct.2010.09.006.

Nabavi, S.F., Habtemariam, S., Sureda, A., Hajizadeh Moghaddam, A., Daglia, M., Nabavi, S.M., 2013. In vivo protective effects of gallic acid isolated from peltiphyllum peltatum against sodium fluoride-induced oxidative stress in rat erythrocytes. Arh. Hig. Rada Toksikol. 64, 553–559. https://doi.org/10.2478/10004-1254-64-2013-2353.

Nair, G.G., Nair, C.K.K., 2013. Radioprotective effects of gallic acid in mice. Biomed Res. Int. 2013, 953079. https://doi.org/10.1155/2013/953079.

Nile, S.H., Park, S.W., 2014. Edible berries: bioactive components and their effect on human health. Nutrition 30 (2), 134–144. https://doi.org/10.1016/j.nut.2013.04.007.

Paciello, F., Rita Fetoni, A., Mezzogori, D., Rolesi, R., Di Pino, A., Paludetti, G., Grassi, C., Troiani, D., 2020. The dual role of curcumin and ferulic acid in counteracting chemoresistance and cisplatin-induced ototoxicity. Sci. Rep. 10, 1063. https://doi.org/10.1038/s41598-020-57965-0.

Piazzon, A., Vrhovsek, U., Masuero, D., Mattivi, F., Mandoj, F., Nardini, M., 2012. Antioxidant activity of phenolic acids and their metabolites: Synthesis and antioxidant properties of the sulfate derivatives of ferulic and caffeic acids and of the acyl glucuronide of ferulic acid. J. Agric. Food Chem. 60, 12312–12323. https://doi.org/10.1021/jf304076z.

Reckziegel, P., Dias, V.T., Benvegnú, D.M., Boufleur, N., Barcelos, R.C.S., Segat, H.J., Pase, C.S., dos Santos, C.M.M., Flores, É.M.M., Bürger, M.E., 2016. Antioxidant protection of gallic acid against toxicity induced by Pb in blood, liver and kidney of rats. Toxicol. Reports 3, 351–356. https://doi.org/10.1016/j.toxrep.2016.02.005.

Rice-Evans, C.A., Miller, N.J., Paganga, G., 1997. Antioxidant properties of phenolic compounds. Trends Plant Sci 2 (4), 152–159. https://doi.org/10.1016/S1360-1385(97)01018-2.

Rice-Evans, C.A., Miller, N.J., Paganga, G., 1996. Structure-antioxidant activity relationships of flavonoids and phenolic acids. Free Radic. Biol. Med. 20, 933–956. https://doi.org/10.1016/0891-5849(95)02227-9.

Sakihama, Y., Cohen, M.F., Grace, S.C., Yamasaki, H., 2002. Plant phenolic antioxidant and prooxidant activities: phenolics-induced oxidative damage mediated by metals in plants. Toxicology 177, 67–80. https://doi.org/10.1016/S0300-483X(02)00196-8.

Sheweita, S.A., Almasmari, A.A., El-Banna, S.G., 2018. Tramadol-induced hepato- and nephrotoxicity in rats: role of curcumin and gallic acid as antioxidants. PLoS One 13. https://doi.org/10.1371/journal.pone.0202110.

Skupień, K., Oszmiański, J., 2004. Comparison of six cultivars of strawberries (Fragaria x ananassa Duch.) grown in northwest Poland. Eur. Food Res. Technol. 219, 66–70. https://doi.org/10.1007/s00217-004-0918-1.

Utrera, M., Estevez, M., 2013. Impact of trolox, quercetin, genistein and gallic acid on the oxidative damage to myofibrillar proteins: the carbonylation pathway. Food Chem. 141, 4000–4009. https://doi.org/10.1016/j.foodchem.2013.06.107.

Yang, B., Chen, F., Hua, Y., Huang, S.S., Lin, S., Wen, L., Jiang, Y., 2012. Prooxidant activities of quercetin, p-courmaric acid and their derivatives analysed by quantitative structure-activity relationship. Food Chem. 131, 508–512. https://doi.org/10.1016/j.foodchem.2011.09.014.

Yilmaz, Y., Toledo, R.T., 2004. Major flavonoids in grape seeds and skins: antioxidant capacity of catechin, epicatechin, and gallic acid. J. Agric. Food Chem. 52, 255–260. https://doi.org/10.1021/jf030117h.

Yoshino, M., Haneda, M., Naruse, M., Htay, H.H., Iwata, S., Tsubouchi, R., Murakami, K., 2002. Prooxidant action of gallic acid compounds: copper-dependent strand breaks and the formation of 8-hydroxy-2′-deoxyguanosine in DNA. Toxicol. Vitr. 16, 705–709. https://doi.org/10.1016/S0887-2333(02)00061-9.

Zadernowski, R., Naczk, M., Nesterowicz, J., 2005. Phenolic acid profiles in some small berries. J. Agric. Food Chem. 53, 2118–2124. https://doi.org/10.1021/jf040411p.

Zduńska, K., Dana, A., Kolodziejczak, A., Rotsztejn, H., 2018. Antioxidant properties of ferulic acid and its possible application. Skin Pharmacol. Physiol. https://doi.org/10.1159/000491755.

Zeraik, M.L., Petrônio, M.S., Coelho, D., Regasini, L.O., Silva, D.H.S., Da Fonseca, L.M., Machado, S.A.S., Bolzani, V.S., Ximenes, V.F., 2014. Improvement of pro-oxidant capacity of protocatechuic acid by esterification. PLoS One 9 (10), e110277. https://doi.org/10.1371/journal.pone.0110277.

CHAPTER

Phytic acid: As a natural antioxidant 4.13

Jyoti Upadhyay[a], Nidhi Tiwari[b], Sumit Durgapal[b], Arvind Jantwal[b], Aadesh Kumar[b]

[a]*School of Health Science and Technology, Department of Pharmaceutical Sciences, University of Petroleum and Energy Studies, Dehradun, Uttarakhand, India*

[b]*Department of Pharmaceutical Sciences, Kumaun University, Campus Bhimtal, Bhimtal, Uttarakhand, India*

4.13.1 Introduction

Reactive oxygen species generated in living organisms play a vital role both by their beneficial as well as toxic effects in normal physiological and pathological conditions. When generated in excess, they induce oxidative damage and lipid peroxidation of biomolecules. A number of unsaturated and saturated carbonyl compounds like aldehydes produced by lipid peroxidation, contribute to cell damage by inhibiting DNA, RNA, and synthesis of proteins, depleting glutathione pool and blocking respiration. Excessive lipid peroxidation generates free radicals which are involved in advancement of several pathological conditions like malignancies, infections and injuries. Antioxidants play a major role in the prevention of free radical generation by inhibiting lipid peroxidation of proteins, DNA, and RNA. Phytic acid is the storage form of phosphate and inositol, found in many plant foods, like grains, legumes, seeds and nuts. Phosphorus is the mineral ion responsible for energy production and in the structural formation of cell membranes (Jacela et al., 2010). Phytic acid content acts as an anti-nutrient because it has the ability to bind to minerals which are essential like iron, calcium, zinc and magnesium in the stomach and inhibits their absorption (Weaver and Kannan, 2002). Recent research studies show that despite of having anti-nutrient activity, actually PA has some beneficial property like it can be used as an antioxidant agent.

Phytic acid is myoinositol, 1, 2, 3, 4, 5, 6 hexa kis-dihydrogen phosphate and is an abundant constituent found in plants (Hidvegi and Lasztity, 2002). Phytic acid is also named as myoinositol hexaphosphoric acid, IP 6 and its molecular is 660.3 and molecular formula is $C_6H_{18}O_{24}P_6$.

Fig. 4.13.1, represents the structure of phytic acid. Phosphorus mineral is tightly bound in a snowflake form. Depending upon its total weight, it constitutes 1% to 5% of the edible oil seeds, legumes, cereals, and nuts and serve as a major source of animal and human sustenance (Reddy et al., 1982, Jenab and Thompson, 2002) suggesting the beneficial role of phytate as an antioxidant, anticancer, and more.

Antioxidants Effects in Health. DOI: https://doi.org/10.1016/B978-0-12-819096-8.00015-X

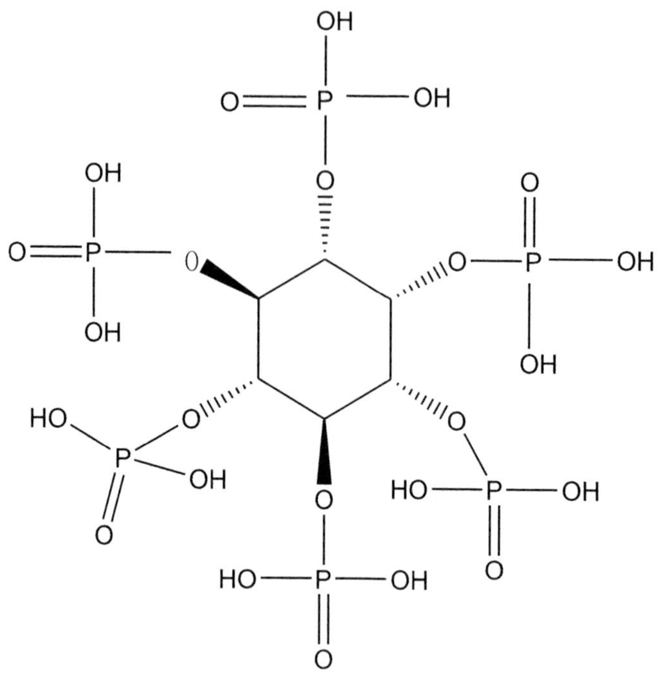

FIG. 4.13.1 Structure of phytic acid.

4.13.2 Sources of phytic acid

Phytic acid is present abundantly in oilseeds, legumes and whole grains. It is also found in roots and tubers, but to a lesser extent. In most of the grains, phytic acid is segregated from aleurone layer, by which most of the concentration of phytic acid found in the bran. Cotyledon layer is reservoir for phytic acid in case of legumes. The maximum concentrated form of phytic acid is present in sesame seeds (5.36% dry weight), whole bran cereals (3.29% dry weight), soya beans (1.00–2.22% dry weight), pinato beans (0.60–2.38% dry weight), parboiled brown rice (1.60%), and oats (1.37%) (Source https://wholegrainscouncil.org/sites/default/files/atoms/files/PhytateProsCons_0910_DK-WGC.pdf).

4.13.3 Mechanism of action of phytic acid as an antioxidant

The mechanism behind its antioxidant activity is due to its metal chelating properties. Also, excessive phytic acid content prevents putrefaction and browning of vegetables and fruits by inhibiting the activity of enzyme polyphenol oxidase (Graf et al., 1987). Phytic acid has the unique property of chelating ions that suppress the iron-catalyzed redox reactions and acts as a potent antioxidant and helps in seeds preservation. This same mechanism of phytic acid would also help in lowering the incidences of colon cancer and prevents inflammatory bowel disease (Graf and Eaton, 1990). Fig. 4.13.1 shows the interacting nature of the phytic acid with different possibilities of reacting with metal cations as well as protein residues. It forms complexes with several divalent and trivalent cations and their solubility and stability depends upon the cation-phytate complex, their pH, molar ratio of phytate is to cation and other compounds which are present (Oberales et al., 1983). The presence of six reactive phosphate groups in phytate fulfills the criteria of chelating agent. Phytates tend to be more soluble in lower pH than higher pH (Torre et al., 1991). Their solubility increases at the following pH values:- 5.5–6.0 with calcium ion (Ca^{2+}); 7.2–8.0 with magnesium ion (Mg^{2+}) and 4.3–4.5 with zinc ion (Zn^{2+}) as the counter ion. Ferric (Fe^{3+}) phytate was found insoluble with pH length of 1–3.5 and the solubility of this complex gets increases above pH 4 (Askar et al., 1983). Interaction of phytic acid with protein is pH dependent (Cheryan, 1980). The phosphate group of phytic acid binds with cationic group present in protein and forms insoluble complexes which get dissolved at pH 3.5. This property of phytic acid plays an important role in the prevention of oxidative stress produced by mineral ion generated free radicals.

Phytic acid act as an antioxidant by forming chelates with mineral ion so that the mineral ion generated free radical production is inhibited, especially in case of Fenton reaction of free radical production. In Fenton reaction, the ferrous or ferric ion is thought to facilitate the hydrogen peroxide (H_2O_2) decomposition. Ferrous ions behave as a catalyst in this reaction, if there is reduction in ferric ion. Some of the ferric ion is reduced by H_2O_2 back to ferrous ion (Schoonen et al., 2006).

$$Fe^{3+}(aq) + H_2O_2(aq) = Fe2^{+} + HO_2^{*} + H^{*}$$

Phytate form complex with iron and hinders the formation of ternary ligand metal hydrogen peroxide complex (Graf et al., 1984). Chronic inflammation related with mineral ion generated free radical production linked to several diseases including cancer, cardiovascular disorder and lung diseases like silicosis (Kamp et al., 1992; Brook et al., 2003; Knaapan et al., 2004). Therefore it is important to investigate the impact of mineral ion exposures on the cellular responses i.e. impairment of antioxidant defense mechanism and the exact role of phytic acid in preventing mineral ion generated oxidative stress.

4.13.4 Possible pro-oxidant activity

Several previous evidences showed that phytic acid is not working as pro-oxidant, dissimilar with other antioxidants like vitamins A and vitamin E. It only inhibits the generation of highly reactive oxygen species, thus resulting in the prevention of a range of diseases (Midorikawa et al., 2001). Another studies reported that phytic acid inhibits cellular growth in the progression of cancer and there is no proven data regarding its ability to formation of chelate and inactivate pro-oxidant metals (Higuchi, 2014).

4.13.5 Role of phytic acid as antioxidant in health and disease

Phytic acid is a natural antioxidant found in plant seeds and often referred as anti-nutrient due to its effects on mineral absorption. Moreover, metabolic effects of phytate and its degradation products have a number of beneficial effects against kidney stone formation (Grases et al., 2000), diabetes mellitus dental caries (Kaufman et al., 1971), atherosclerosis, coronary heart disease (Jariwalla et al., 1990), and in a certain type of cancers (Vaucenik et al., 2003). Some disease conditions are being discussed, in which phytic acid plays a major role for management of disease extent.

4.13.5.1 Diabetes mellitus (DM)

DM is a metabolic disorders characterized by hyperglycemia over a large period of time. It is also referred as lifestyle diseases when the pancreatic β-cells unable to produce adequate amount of insulin, a glucose regulatory hormone that regulates the level of blood glucose (Prasad et al., 2019). Oxidative stress plays an important role in the pathophysiology of diabetes. Hyperglycemia induced oxidative stress resulting in enhance formation of free radical and decreases the activity of antioxidant system (Bajaj and Khan, 2012). So, appropriate antioxidants either by supplementation or by diet decreases the symptoms- associated with diabetes. Previous studies reported PA involved in a secretion of insulin, inhibits the formation of plaque (Colosia et al., 2013) and decreases the level of lipid in serum. Sanchis et al. reported beneficial and detrimental effects of phytate supplements in the treatment of STZ (Streptozotocin)-induced diabetes prone rats (Sanchis et al., 2018). Diabetes mellitus is also considered as nutrition dependent disease caused by diet rich in calories and carbohydrates. Phytic acid rich source diet valuable in this regard for example, grains, legumes, and cereals. Few preclinical studies showed diabetic KK mice treated with phytate rich diet decreases the level of blood glucose when compared with the diabetic control rats (Lee et al., 2006). Nowadays, research is being focusing on tradition medicine system (natural products) over allopathic drugs. Therefore, choosing a proper target having less adverse effects plays a crucial role in treatment of diabetes.

4.13.5.2 Nephrolithiasis (Calculi)

Renal lithiasis is common diseases in the treatment of urinary tract system. Previous reported evidences have shown formation of oxidative free radicals due to interaction of calcium oxalate crystals with renal tubular-epithelial cells (Ceban et al., 2016). Exploring the potential of suitable antioxidant and antiradical prevents the urinary calculi. Several preclinical and clinical studies showed that phytic acid prevents formation of renal calculi by lowering the level of calcium oxalate and calcium phosphate crystals (Grases et al., 2004). Further, it has been documented that dietary phytate is essential for maintaining sufficient urinary level by inhibiting calcium salts, thus prevents renal stone formation (Nissar et al., 2017). A of lot studies are needed for the beneficial role of phytate as antioxidants in the treatment of renal calculi.

4.13.5.3 Cardiovascular diseases (CHD)

Heart disease is one of the major causes of death due to increase in the level of plasma cholesterol (low density lipoprotein; LDL). Reactive oxygen species (ROS) linked with CHD development, contributes to oxidative stress, directly relates to changes in sub-cellular organelles and stimulate intracellular Ca^{2+} overload. ROS induced free radicals associated with various diseases, such as hypertension, atherosclerosis, myocardial infarction, heart failure, etc. (Dhalla et al., 2000). It has been documented that dietary fiber antioxidants supplementation such as phytate controls the etiology of heart diseases (Potter, 1995). Preclinical studies showed daily intake of phytate supplementation lowers the triglyceride and serum cholesterol levels (Jariwalla et al., 1990). Another study showed the presence of antioxidants (PA) in red wine declines the symptoms of heart disease (Omoruyi et al., 2013).

4.13.5.4 Dental caries

Dental caries is the most prevalent infectious/inflammation oral diseases worldwide. Oxidative stress plays a significant role in the prevalence of dental caries and defects in enamel are due to decrease in dietary/nutrition components (Ahmadi-Motamayel et al., 2018). Diet plays a major role in the development of enamel and tooth for eg. PA reduces the solubility of various ions like calcium, fluoride and phosphate (Kaufman et al., 1971). Moreover, phytate is considered as valuable anticaries agent due to strong fibrous food that protects the teeth because of mechanically stimulates salivary flow (Moynihan and Petersen, 2004). Moreover, phytate for hydroxyl apatite inhibit the growth of plague and tartar (Nissar et al., 2017).

4.13.5.5 Cancer

Oxidative stress involves in all cancer by two possible mechanisms, one is mutations in gene resulting from cell injury/damage second one is transduction and transcription signaling factors. Preventive measures of oxidative stress induced ROS, capable of reducing symptoms linked with certain types of cancer (Noda and Wakasugi, 2001). Phytic acid is a broad-spectrum antineoplastic and agent having antioxidant property affecting several cells and tissue system (Vaucenik and Shamsuddin, 2003). It also inhibits the growth of various human cells lines like leukemic hematopoietic K-562 cell line (Lambertenghi et al., 2002), HT-29 cell line responsible for colon cancer (Shamsuddin et al., 1992), cell lines for breast cancer (Shamsuddin et al., 1995), cervical cancer cell lines (Ferry et al., 2002), prostate cancer cell lines (Zi et al., 2000), and HepG2 haepatoma cell line (Vucenik et al., 1998a). Previous preclinical studies reported phytic acid is beneficial in a number of carcinoma like mammary carcinoma (Hirose et al., 1994), papillomas of skin (Ishikawa et al., 1999), tumor volume of metastatic fibrosarcoma, and preliminary lung metastases (Vucenik et al., 1992), expansion of rhabdomyosarcoma cells or tissue and relapse of previous liver cancer cell (Vucenik et al., 1998b). Other studies have showed dietary supplements with phytic acid decreases the size and number of tumors in colon cancer (Nissar et al., 2017).

4.13.6 *In-vivo* studies

4.13.6.1 Anticarcinogenic effect

Several *In-vivo* studies have been investigated to determine the effect of phytic acid on various types of cancers. Singh et al., 2004 performed *in-vivo* mouse model study of prostate cancer and it was observed that phytic acid decrease cellular proliferation and inhibits vascular endothelial growth factor (VEGF). *In-vivo* experimental study of colon cancer on rat and mouse model was investigated by Challa et al., 1997; Zhang et al., 2005; Norazalina et al., 2010. Their study shows that phytic acid causes inhibition of cell proliferation, aberrant crypts formation, tumoral development and increase natural killer activity. Another study on hepatocellular carcinoma on animal model shows that phytic acid promotes inhibition of tumoral growth and development and induction of tumoral regression (Vucenik et al., 1998a; Lee et al., 2005). A study of phytic acid on carcinoma of mammary gland in rats demonstrated its inhibitory effect on tumoral development, incidences and multiplicity (Hirose et al., 1994; Vucenik et al., 1993; 1997). All these observed anti-carcinogenic effects are linked to the ability of phytic acid in modulating differentiation, apoptosis and proliferation of neoplastic cells. This effects conferred by phytic acid is associated with inhibition of the formation of free radicals via prevention of iron chelation in Fenton reaction (Graf and Eaton, 1985).

4.13.6.2 Alzheimer's disease

Alzheimer's disease is the neurodegenerative disease identified by progressive decline in cognitive functions. It was implicated by the presence of amyloid beta peptides in the brain (Polito et al., 2018). Abe and Taniguchi, 2014 hypothesized phytic acid in rice grains has the capacity of preventing amyloid beta (Aβ) accumulation in the brain. According to their study Aβ is excised from precursor protein amyloid-beta by sequential cleavage with the help of enzyme aspartic beta secretase 1 (BACE 1) and G secretase. It also inhibits Aβ synthesis in neuroblastoma cells causing no harm to normal cells recommending phytic acid as potent therapeutic agents in the treatment of Alzheimer's disease.

4.13.6.3 Parkinson's disease

Parkinson's disease is caused by selective degradation of dopaminergic neuron in the substantia nigra. It is the second type of neurodegenerative disease after alzheimer's disease. It is caused by multiple factors and also excess iron accumulation in the brain can lead to parkinson's disease. Phytic acid being an antinutrient compound can inhibit the accumulation of excess iron in the brain (Xu et al., 2011). The preventive effect of phytic acid on neuron and inflammatory activity in 1-methyl-4-phenyl-1,2,3,6-tetrahydropyridine (MPTP) induced parkinson's disease mice model was investigated by Lv et al. (2015). They also study the motor behavior test, activity of microglia, tyrosine hydroxylase (TH) immunohistochemistry, inducible nitric oxide synthase action, and quantitative polymerase chain reaction were markedly suppressed by phytic acid. They concluded that phytic acid has neuroptotective effect in MPTP induced Parkinson's model and this activity is correlated with anti-inflammatory effect of phytic acid (Lv et al., 2015).

4.13.6.4 **Diabetes mellitus**

Omoruyi et al. (2013) studied the effect of phytic acid supplement in diabetic rats induced by streptozotocin. Their results suggested that phytic acid supplement was found beneficial in the management of diabetes mellitus. Another study reported by Yoon et al., analyzed the blood glucose lowering effect of phytate-enriched diet in humans. Kuppuswamy et al. (2011) studied the alpha glucosidase and alpha amylase inhibitory activity of phytic acid in a dose dependent manner and also shows its antidiabetic activity against streptozotocin nicotinamide induced type-2 diabetes in rat model. Phytic acid has the capacity of binding with transition metal ions and used in the management of protein glycation (metal catalyzed) that contributes in regulating diabetes related disorders (Sanchis et al., 2018).

4.13.6.5 **Hypolipidemia**

Phytic acid shows lipid lowering properties of unregulated lipase enzyme activity in the intestine, increasing the cholesterol output in fecal matter thus lowering the cholesterol level (Dilworth et al., 2005). This may result in reduced incidences of developing cardiovascular complications and other related disorders. Some studies suggests that oral intake of 1% and 1.5% phytic acid supplementation reduces lipid profile of blood as well as liver in case of diabetic mice. Another study evaluated the efficacy of phytic acid (0.02%) on liver and blood lipid profile of albino rats fed with high sucrose diet. They suggested that supplementation with phytic acid lowers the total cholesterol level (TC), triglycerides (TG) level (Onomi et al., 2004).

4.13.7 ***In-vitro* studies**

To get evidence that PA is protective in Alzheimer's disease Anekonda et al., 2011 performed experimental study in MC65 cells and outcomes of their study revealed that PA in a concentration of 100 μl is highly effective in its neuroprotective effect as it showed remarkable results against amyloid-β-peptide (Aβ) whose accumulation in the brain is the main cause of this disease. In the same study they also found that in same concentration PA plays pivotal role against Aβ cytotoxicity via the reduction of hydrogen ion concentration (Anekonda et al., 2011). Parkinsonism comes after AD as a neurodegenerative disorder and numbers of studies have been conducted by scientists working in this arena throughout the globe. One such *in-vitro* study to know the role of PA in Parkinsonism against 1-methyl-4-phenylpyridinium (MPP+) induced neurodegeneration was conducted by Xu et al., in a cell culture medium. Data obtained with this experimental study clearly showed that PA in 30 μM concentration was found to be highly effective against the neurodegenerative effect of MPP+ (Rhodes et al., 2008). Other study to know the protective role of PA in Parkinsonism was conducted by the same group of scientists to know the protective role of PA in a normal or high iron concentration exposed cell culture model against 6-hydroxydopamine induced cell apoptosis based on the fact obtained with several experimental studies which claimed high accumulation of iron in the brain of patients suffering from Parkinsonism disease is one of the major causes of occurrence of this disease thus lowering the deposition of excess iron from the brain is one strategy to combat this disease (Kaur D et al., 2003; Xu Q et al., 2011). Results of this study clearly showed the protective effect of PA against 6-hydroxydopamine induced cell apoptosis in high iron exposed cell culture model (Xu Q et al., 2008). PA also play significant role in the treatment of cancer also as one of the recently conducted study showed the effectiveness of nano drug delivery system of PA in the treatment of human colorectal cancer (HT-29) cell line (Taitt, 2018). Few research studies showed the role of PA in prostate cancer as well. One such recent study was done by Jagadeesh et al., in which they determined the effect of PA in prostate cancer cells by studying the role of PA in telemorase activity and found a dose dependent effect of PA in the reduction of telemorase activity for the prevention of prostate cancer (Jagdeesh S et al., 2006) hepatocarcinoma which is responsible for number of deaths throughout the globe and stood third in a rank of cancer induced deaths can also be treated by PA as researchers showed its vital role in the pathogenesis of this cancer type. Al-Fatlawi et al. carried their research studies to demonstrate the role of PA in hepatocarcinoma by using $HepG_2$ cells. Results of the study which was based on the expression analysis of genes responsible for apoptosis showed significant reduction of cancer cells by PA (Al-fatlawi et al., 2014). Wawszczyk J et al. in one of their experimental work determined the role of PA in the treatment of skin cancer using human melanoma cell lines for their study and found dose dependent decrease in the excessive growth of melanoma cells (Wawszczyk et al., 2015). PA was also investigated by some researchers for its antibacterial activity. Kim and Rhee conducted an experimental study to know the potential of PA as an antibacterial agent against acid resistant enterohemorrhagic *Escherichia coli* and found strong antibacterial effects of PA compared to other organic acids used in the study in equivalent concentration (Kim et al., 2016). Similarly, Yadav et al. determined the antibacterial effect of PA in combination with methanolic extract of *Syzygium cumini* and sodium chloride using *Bacillus subtilis* and obtained data supporting the marked antibacterial activity of PA (Yadav et al., 2018). Otake et al. in one of their research work demonstrated the antiviral effect of PA on human immunodeficiency virus (HIV) by investigating its effects against MT-4 cells and revealed about its high potential as results obtained with this study clearly showed that PA in a concentration of 1.67 mg/ml greately inhibited the cytopathic effect of HIV (Otake et al., 1989). Same group of researchers in their other similar study showed the antiviral effect of PA in T cell lines and found the considerable inhibition of HIV-1 cells by PA (Otake et al., 1999).

Conclusion

Phytic acid is present abundantly in oilseeds, legumes and whole grains. Phytic acid act as an anti-nutrient because of its complex structure which forms chelates with the mineral ion. This property of phytic acid prevents the free radical generation caused by mineral ion like iron. This chapter is compilation of beneficial and detrimental effects of phytic acid in human body. Lack of data is available of phytic acid as antioxidants, a lot research to be done in further investigations with proven preclinical *in-vitro* and *in-vivo* effects.

References

Abe, T.K., Taniguichi, M., 2014. Identification of myo-inositol hexakisphosphate (IP6) as aβ-secretase 1 (BACE 1) inhibitory molecule in rice grain extract and digest. FEBs Open Bio. 4, 162–167.

Ahmadi-Motamayel, F., Goodarzi, M.T., Mahdavinezhad, A., Jamshidi, Z., Darvishi, M., 2018. Salivary and serum antioxidant and oxidative stress markers in dental caries. Caries Res. 52 (6), 565–569.

Al-Fatlawi, A.A., Rizvi, M.M., Ahmad, A.Y., 2014. Anticarcinogenic activity of rice bran phytic acid against human breast cancer cell line (MCF-7). Asian J. Pharm. Clin. Res. 7 (1), 151–155.

Anekonda, T.S., Wadsworth, T.L., Sabin, R., Frahler, K., Harris, C., Petriko, B., et al., 2011. Phytic acid as a potential treatment for Alzheimer's pathology: evidence from animal and in vitro models. J. Alzheimer's Dis. 23 (1), 21–35. https://doi.org/10.3233/JAD-2010-101287.

Askar, A., El-Samahy, S.K., Abd El-Fadeel, M.G., 1983. Phytinsaure in Lebensmittel. Alimenta 22, 131–137.

Bajaj, S., Khan, A., 2012. Antioxidants and diabetes. Indian J. Endocrinol. Metab. 16 (2), S267–S271.

Brook, R.D., Brook, J.R., Rajagopalan, S., 2003. Air pollution: the "heart" of the problem. Curr. Hypertension Rep. 5, 32–39.

Ceban, E., Banov, P., Galescu, A., Botnari, V., 2016. Oxidative stress and antioxidant status in patients with complicated urolithiasis. J. Med. Life 9 (3), 259.

Challa, A., Rao, D.R., Reddy, B.S., 1997. Interactive suppression of aberrant crypt foci induced by azoxymethane in rat colon by phytic acid and green tea. Carcinog 18, 2012–2016.

Cheryan, M., 1980. Phytic acid interactions in food systems. Crit. Rev. Food Sci. Nutr. 13, 297–335.

Colosia, A.D., Palencia, R., Khan, S., 2013. Prevalence of hypertension and obesity in patients with type 2 diabetes mellitus in observational studies: a systematic literature review. Diabetes Metab. Syndr. Obes. 6, 327.

Dilworth, L.L., Omoruyi, F.O., Simon, O.R., Marison, F.Y.S.A., Asemota, H.N., 2005. The effect of phytic acid on the levels of blood glucoseand some enzymes of carbohydrate and lipid metabolism. West Indian Med. J. 5 (2), 102e6.

Dhalla, N.S., Temsah, R.M., Netticadan, T., 2000. Role of oxidative stress in cardiovascular diseases. J. Hypertens. 18 (6), 655–673.

Ferry, S., Matsuda, M., Yoshida, H., Hirata, M., 2002. Inositol hexakisphosphate blocks tumor cell growth by activating apoptotic machinery as well as by inhibiting the Akt/NFKB-mediated cell survival pathway. Carcinogenesis 23, 2031–2041.

Graf, E., Mahoney, J.R., Bryant, R.G., Eaton, J.W., 1984. Iron-catalyzed hydroxyl radical formation. Stringent requirement for free iron coordination site. J. Biol. Chem. 259, 3620–3624.

Graf, E., Eaton, J.W., 1990. Antioxidant functions of phytic acid. Free Radic. Biol. Med. 8 (1), 61–69.

Graf, E., Eaton, J.W., 1985. Dietary suppression of colonic cancer. Fiber or phytate? Cancer 56, 717–718.

Graf, E., Empson, K.L., Eaton, J.W., 1987. Phytic acid. A natural antioxidant. J. Biol. Chem. 262, 11647–11650.

Grases, F., March, J.G., Prieto, R.M., Simonet, B.M., CostaBauzá, A., García-Raja, A., 2000. Urinary phytate in calcium oxalatestones formers and healthy people. Scand. J. Urol. Nephrol. 34, 162–164.

Grases, F., Perello, J., Prieto, R.M., Simonet, B.M., Torres, J.J., 2004. Dietary myo-inositol hexaphosphate prevents dystrophic calcifications in soft tissues: a pilot study in Wistar rats. Life Sci. 75, 11–19.

Hidvegi, M., Lasztity, R., 2002. Phytic acid content of cereals and legumes and interaction with proteins. Periodica Polytechnica Ser. Chem. Eng. 46 (1–2), 59–64.

Higuchi, M., 2014. Antioxidant properties of wheat bran against oxidative stress. Wheat and Rice in Disease Prevention and Health. Academic Press, Elseiver, pp. 181–199.

Hirose, M., Hoshiya, T., Akagi, K., Futakuchi, M., Ito, N., 1994. Inhibition of mammary gland carcinogenesis by green tea catechins and other naturally occurring antioxidants in female Spargue- Dawley rats pretreated with 7,12dimethylbenz(a)anthracene. Cancer Lett. 83, 149156.

Ishikawa, T., Nakatsuru, Y., Zarkovic, M., Shamsuddin, A.M., 1999. Inhibition of skin cancer by IP6 in vivo: Initiationpromotion model. Anticancer Res. 19, 3749–3752.

Jacela, J.Y., DeRouchey, J.M., Tokach, M.D., Goodband, R.D., Nelssen, J.L., Renter, D.G., 2010. Feed additives for swine: Fact sheets- prebiotics and probiotics and phytogenesis. J. Swine Health food 18, 87–91.

Jagadeesh, S., Banerjee, P.P., 2006. Inositol hexaphosphate represses telomerase activity and translocates TERT from the nucleus in mouse and human prostate cancer cells via the deactivation of Akt and PKCα. Biochem. Biophys. Res. Commun. 349 (4), 1361–1367.

Jariwalla, R.J., Sabin, R., Lawson, S., Herman, Z.S., 1990. Lowering of serum cholesterol and triglycerides and modulation of divalent cations by dietary phytate. J. Appl. Nutr. 42, 18–28.

Jenab, M., Thompson, L.U., 2002. Role of phytic acid in cancer and other diseases. In: Reddy, N.R., Sathe, S.K. (Eds.), Food Phytates. CRC Press, Boca Raton, FL, pp. 225–248.

Kim, N.H., Rhee, M.S., 2016. Phytic acid and sodium chloride show marked synergistic bactericidal effects against nonadapted and acid-adapted Escherichia coli O157: H7 strains. Appl. Environ. Microbiol. 82 (4), 1040–1049.

Kamp, D.W., Graceffa, P., Pryor, W.A., Weitzman, S.A., 1992. The role of free radicals in asbestos-induced diseases. Free Radic. Biol. Med. 12, 293–315.

Kaufman, H.W., Kleinberg, I., 1971. Effect of pH on calcium binding by phytic acid and its inositol phosphoric acid derivatives and on the solubility of their calcium salts. Archs Oral Biol. 16, 445–460.

Kaur, D., Yantiri, F., Rajagopalan, S., Kumar, J., Mo, J.Q., Boonplueang, R., et al., 2003. Genetic or pharmacological iron chelation prevents MPTP-induced neurotoxicity in vivo: a novel therapy for Parkinson's disease. Neuron 37 (6), 899–909.

Knaapen, A.M., Borm, P.J.A., Albrecht, C., Schins, R.P.F., 2004. Inhaled particles and lung cancer. Part A: mechanisms. Int. J. Cancer 109, 799–809.

Kuppusamy, A., Muthusamy, U., Thirumalaisamy, S.A., Varadharajan, S., et al., 2011. *In-vitro* (α-glucosidase and α-amylase inhibition) and in vivo antidiabetic property of phytic acid (IP6) in streptozotocin-nicotinamide-induced Type 2 diabetes mellitus (NIDDM) in rats. J. Complement. Integr. Med. 8, 1483.

Lambertenghi, D.G., Servida, F., Fracchiola, N.S., Ricci, C., Borsotti, C., Colombo, G., 2002. Effect of inositol hexaphosphate (IP6) on human mormal and leukaemic haematopoietic cells. Br. J. Haematol. 117, 577–587.

Lee, S.H., Park, H.J., Cho, S.Y., Jung, H.J., et al., 2005. Effects of dietary phytic acid on serum and hepatic lipid levels in diabetic KK mice. Nutr. Res. 25 (869), e76.

Lee, S.H., Park, H.J., Chun, H.K., Cho, S.Y., Cho, S.M., Lillehoj, H.S., 2006. Dietary phytic acid lowers the blood glucose level in diabetic KK mice. Nutr. Res. 26 (9), 474–479.

Lv, Y., Zhang, Z., Hou, L., 2015. Phytic acid attenuates inflammatory responses and the levels of NFκB and p-ERK in MPTP-induced Parkinson's disease model of mice. Neurosci. Lett. 597, 132–136. https://doi.org/10.1016/j.neulet.2015.04.040.

Midorikawa, K., Murata, M., Oikawa, S., Hiraku, Y., Kawanishi, S., 2001. Protective effect of phytic acid on oxidative DNA damage with reference to cancer chemoprevention. Biochem. Biophys. Res. Commun. 288 (3), 552–557.

Moynihan, P., Petersen, P.E., 2004. Diet, nutrition and the prevention of dental diseases. Public Health Nutr. 7 (1), 201–226.

Nissar, J., Ahad, T., Naik, H.R., Hussain, S.Z., 2017. A review phytic acid: as antinutrient or nutraceutical. J. Pharmacogn. Phytochem. 6 (6), 1554–1560.

Noda, N., Wakasugi, H., 2001. Cancer and oxidative stress. JMAJ 44 (12), 535–539.

Norazalina, S., Norhaizan, M.E., Hairuszah, I., Norashareena, M.S., 2010. Anticarcinogenic efficacy of phytic acid extracted from rice bran on azoxymethane induced colon carcinogenesis in rats. Exp. Toxicol. Pathol. 62, 259–268.

Oberleas, D., 1983. The role of phytate in zinc bioavailability and homeostasis. In: Inglett, G.E. (Ed.), Nutritional Bioavailability of Zinc. American Chemical Society, Washington DC. 145–158.

Omoruyi, F.O., Budiaman, A., Eng, Y., Olumese, F.E., Hoesel, J.L., Ejilemele, A., Okorodudu, A.O., 2013. The potential benefits and adverse effects of phytic acid supplement in streptozotocin-induced diabetic rats. Adv. Pharmacol. Sci. 2013, 172494. doi:10.1155/2013/172494. Epub 2013 Dec 22.

Onomi, S., Okazaki, Y., Katayami, T., 2004. Effect of dietary level of phytic acid on hepatic and serum lipid status in rats fed a high sucrose diet. Bio. Sci. Biotech. Biochem. 68 (6), 1379–1381.

Otake, T., Shimonaka, H., Kanai, M., 1989. *In-vitro* Inhibitory effect of inositol hexasulfate and inositol hexaphosphoric acid (phytic acid) on the proliferation of the human immuno deficiency virus (HIV). Kansenshogaku zasshi 63 (7), 676–683.

Otake, T., Mori, H., Morimoto, M., Miyano, K., 1999. Anti-HIV-1 activity of myo-inositol hexaphosphoric acid (IP6) and myo-inositol hexasulfate (IS6). Anticancer Res. 19 (5A), 3723–3726.

Polito, C.A., Cai, Z., Shi, Y, Li, X., Yang, R., Shi, M., et al., 2018. Association of tea consumption with risk of Alzheimer's disease and antibeta-amyloid effects of tea. Nutrients 10, 655. https://doi.org/10.3390/nu10050655.

Potter, S.M., 1995. Overview of proposed mechanisms for the hypocholesterolemic effect of soy. J. Nutr. 125, 606S–611S.

Prasad, R.B., Groop, L., 2019. Precision medicine in type 2 diabetes. J. Intern. Med. 285 (1), 40–48.

Reddy, N.R., Sathe, S.K., Salunkhe, D.K., 1982. Phytates in legumes and cereals. Adv. Food Res 28, 1–92.

Rhodes, S.L., Ritz, B., 2008. Genetics of iron regulation and the possible role of iron in Parkinson's disease. Neurobiol. Dis. 32 (2), 183–195.

Sanchis, P., Rivera, R., Berga, F., Fortuny, R., Adrover, M., Costa-Bauza, A., Grases, F., Masmiquel, L., 2018. Phytate decreases formation of advanced glycation end-products in patients with type II diabetes: randomized crossover trial. Sci. Rep. 8 (1), 9619.

Schoonen, M.A.A., Cohn, C.A., Roemer, E., Laffers, R., Simon, S.R., O'Riordan, T, 2006. Mineral-induced formation of reactive oxygen species. Rev. Mineral. Geochem. (64), 30–35.

Shamsuddin, A.M., Baten, A., Lalwani, N.D., 1992. Effects of inositol hexaphosphate on growth and differentiation in K-562 erythroleukemia cell line. Cancer Lett. 64, 195–202.

Shamsuddin, A.M., Yang, G.Y., 1995. Inositol hexaphosphate inhibits growth and induces differentiation of PC-3 human prostate cancer cells. Carcinogenesis 16, 1975–1979.

Singh, R.P., Sharma, G., Mallikarjuna, G.U., Dhanalakshmi, S., 2004. *In-vivo* suppression of hormone refractory prostate cancer growth by inositol hexaphosphate: induction of insulin like growth factor binding protein 3 and inhibition of vascular endothelial growth factor. Clin. Cancer Res. 10, 244–250.

Taitt, H.E., 2018. Global trends and prostate cancer: a review of incidence, detection, and mortality as influenced by race, ethnicity, and geographic location. Am. J. Mens Health 12 (6), 1807–1823.

Torre, M., Rodriguez, A.R., Saura-Calixto, F., 1991. Effects of dietary fiber and phytic acid on mineral bioavailability. Crit. Rev. Food Sci. Nutr. 1, 1–22.

Vucenik, I., Kalebic, T., Tantivejkul, K., Shamsuddin, A.M., 1998a. Novel anticancer function of inositol hexaphosphate: inhibition of human rhabdomyosarcoma *in-vitro* and *in-vivo*. Anticancer Res. 18, 1377–1384.

Vucenik, I., Sakamoto, K., Bansal, M., Shamsuddin, A.M., 1993a. Inhibition of mammary carcinogenesis by inositol hexaphosphate (phytic acid): a pilot study. Cancer Lett. 75, 95–102.

Vucenik, I., Shamsuddin, A.M., 2003. Cancer inhibition by inositol hexaphosphate (IP6) and inositol: Ffrom laboratory to clinic. J. Nutr. 133, 3778S–3784S.

Vucenik, I., Yang, G., Shamsuddin, A.M., 1997. Comparison of pure inositol hexaphosphate (IP6) and high bran diet in the prevention of DMBA induced rat mammary carcinogenesis. Nutr. Cancer 28, 7–13.

Vucenik, I., Tantivejkul, K., Zhang, Z.S., Cole, K.E., Saied, I., Shamsuddin, A.M., 1998b. IP6 in treatment of liver cancer. I. IP6 inhibits growth and reverses transformed phenotype in HepG2 hunman liver cancer cell line. Anticancer Res. 18, 4083–4090.

Vucenik, I., Tomazic, V.J., Fabian, D., Shamsuddin, A.M., 1992. Antitumor activity of phytic acid (inositol hexaphosphate) in murine transplanted and metastatic fibrosarcoma, a pilot study. Cancer Lett. 65, 9–13.

Wawszczyk, J., Kapral, M., Lodowska, J., Jesse, K., 2015. Antiproliferative effect of inositol hexaphosphate on human skin melanoma cells in vitro. Acta Poloniae Pharm-Drug Res. 72 (5), 895–900.

Weaver, C.M., Kannan, S., 2002. Phytate and mineral bioavailability. Food Phytates, 211–223.

Xu, Q., Kanthasamy, A.G., Reddy, M.B., 2011. Phytic acid protects against 6-hydroxydopamine induced dopaminergic neuron apoptosis in normal and iron excess conditions in a cell culture model. Parkinsons Dis. 2011, 1–6. doi: https://doi.org/10/4061/2011/431068.

Xu, Q., Kanthasamy, A.G., Reddy, M.B., 2008. Neuroprotective effect of the natural iron chelator, phytic acid in a cell culture model of Parkinson's disease. Toxicology 245 (1-2), 101–108.

Yadav, A.K, Sirohi, P., Saraswat, S., Rani, M., 2018. Inhibitory mechanism on combination of phytic acid with methanolic seed extract of Syzygium cumini and sodium chloride over Bacillus subtilis. Curr. Microbiol. 75 (7), 849–856.

Zhang, Z., Song, Y., Wang, X.L., 2005. Inositol hexaphosphate-induced enhancement of natural killer cell activity correlates with suppression of colon carcinogenesis in rats. World J. Gastroenterol. 11, 5044–5046.

Zi, X., Singh, R.P., Agarwal, R., 2000. Impairment of erbB1 receptor and fluid-phase endocytosis and associated mitogenic signaling by inositol hexaphosphate in human prostate carcinoma DU145 cells. Carcinogenesis 21, 2225–2235.

CHAPTER

Protein hydrolysates 4.14

Mohamad Fawzi Mahomoodally, Meeajan M. Irfaan
Department of Health Sciences, Faculty of Medicine and Health Sciences, University of Mauritius, Réduit, Mauritius

4.14.1 Introduction

Oxidative stress is a discrepancy between oxidative species and endogenous antioxidant systems (Karamać et al., 2016). Such stress is capable of altering biomolecules (nucleic acids, protein, carbohydrates and polyunsaturated lipids) when living organism is persistently exposed to reactive species and it exerts a role in the advancement of tissue damage. Thus, causing cellular injuries, apoptosis, aging process (Rizzo et al., 2010) and diseases, such as inflammatory illnesses, tumor, diabetes mellitus, heart diseases, and hepatic diseases (Yang et al., 2019).

In order to overcome these devastating consequences the use of antioxidants is on rise as it shows a main part as a prophylactic features and it is used for food conservation to hinder off unwanted smells, tastes and discolouration caused by lipid oxidation (Yarnpakdee et al., 2014). Conversely, synthetic antioxidants have displayed side effects such as liver damage and carcinogenesis. This intron has resulted in consumer anxiety and gloomily affecting their potential application. Thus, there has been a foremost attention in probing for antioxidants from natural sources as substitutes to artificial antioxidants for opposing these contrary effects (Yang et al., 2019).

4.14.2 Sources, chemistry, and bioavailability

In the past few years numerous studies have conveyed that protein hydrolysates from numerous food sources, besides their nutritional properties, displayed several biotic function including antioxidant properties (Sila and Bougatef, 2016). Antioxidant hydrolysates of different kinds were isolated from cereal, fish, seed, and egg white protein and also proteins from other food sources (Zhang et al., 2016).

Using exogenous proteases numerous protein bioactive peptides were produced from plant and animal origin and some studies have been conducted on protein hydrolysates using microbial fermentation. Concerning hydrolysis of parent protein and during fermentation of food such as meat and milk products, proteolytic systems of lactic acid bacteria have been castoff (Havenaar et al., 1992). Lactic acid bacteria

Antioxidants Effects in Health. DOI: https://doi.org/10.1016/B978-0-128-19096-8.00059-8

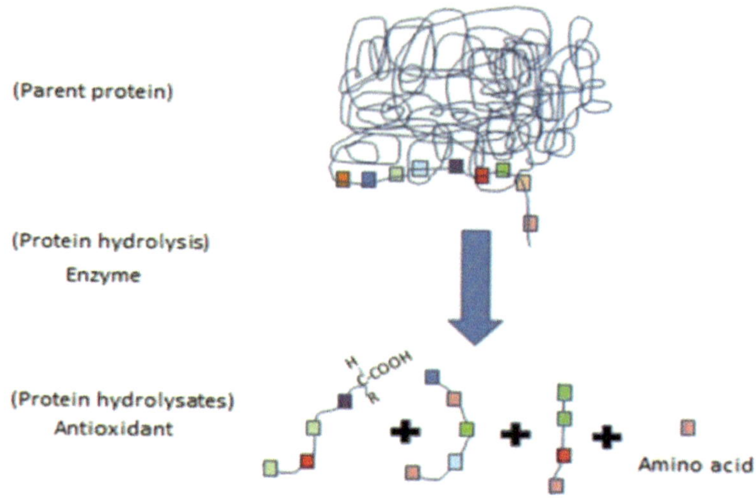

FIG. 4.14.1 Formation of hydrolysates via enzymatic hydrolysis of parent protein.

are used as the bacteria and its products are considered safe (Generally Recognized as Safe, GRAS) status by the US Food and Drug Administration (Zambrowicz et al., 2012). The antioxidant hydrolysates are liberated from the parent protein by secreted microbial proteases (Sila and Bougatef, 2016).

Protein hydrolysates comprise of 2 to 20 amino acid residues, these protein sequences may be released from their parent protein by enzymatic hydrolysis either when processing the food (e.g., cheese ripening and milk fermentation) or during gastrointestinal digestion in the body. Bioactive peptides liberated in the body possibly will operate as regulatory compounds with hormone-like action (Udenigwe and Aluko, 2011).

Hydrolysis of protein product *In vitro* using proper proteolytic enzymes, generates bioactive peptides and is depicted in Fig. 4.14.1. Depending on the reaction environment, high level of certain amino acid residues can hypothetically either increase or decrease the antioxidant action of protein hydrolysates (Udenigwe and Aluko, 2011).

The amino acid composition of the protein hydrolysates, the molecular weight and the biological function are prominently influenced by the class of the protein substrate, the specificity of the enzyme utilized for the proteolysis, the environments used during hydrolysis (temperature and time) along with the enzyme/substrate ratio (Sila and Bougatef, 2016; van der Ven et al., 2002).

4.14.3 Mechanism of protein hydrolysates as antioxidants

Other than their chemical composition, the hydrophobicity of the hydrolysates affects the antioxidant action of the bioactive peptides (Karamać et al., 2016). Moreover, various investigations are being done to determine the correlation between the antioxidant activity of the hydrolysates and the aforementioned factors. The matter becomes progressively more intricate because of numerous modes of activity of hydrolysates as antioxidants (Karamać et al., 2016).

Firstly, the antioxidant action of peptides derived from food are capable of scavenging or quenching free radicals and reactive oxygen species (ROS) (Karamać et al., 2016) mainly due to proton-coupled single electron or hydrogen atom transfer mechanism (Udenigwe and Aluko, 2011). Additionally, bioactive peptides have significant reducing power, hence exhibits capacity to chelates pro-oxidative metal ion (Karamać et al., 2016). For instance, iron (Fe^{2+}) and copper (Cu^{2+}), inhibit the oxidation of unsaturated fatty acids and cellular regulation of gene expression of antioxidant proteins, e.g., heme oxygenase-1 and ferritin (Udenigwe and Aluko, 2011). In the same line, pipeline, various of amino acids including Tryptophan, Tyrosine, Methionine, Cysteine, Histidine, Phenylalanine and Proline have been suggested to donate positively to the antioxidant properties of protein hydrolysates (Udenigwe and Aluko, 2011).

Following several studies, there is a tend to a general agreement that hydrolysates have better antioxidant activities than their parent protein and integral amino acids and this may be due to the increased ease of access of the functional side chain (R-group) to the reactive species, and the electron-dense peptide bonds. In addition, some researches have pointed out that some amino acids may be more dynamic than their parent peptides (Udenigwe and Aluko, 2011).

Additionally, there are many other mechanisms reported that tend to suggest that protein hydrolysates could act as antioxidants (Yarnpakdee et al., 2014). Likewise, antioxidant activities could be increased if there is a higher contact of the functional groups of peptides (Yarnpakdee et al., 2014). Wu et al. (2003) have put forward that phenolic hydroxyl group found in aromatic amino acids may contribute significantly to scavenge free radicals *via* acting as electron donor. Furthermore, in one study by Wang et al. (2008), it was suggested that hydrolysates containing tyrosine residues at the C-terminus, lysine or phenylalanine residues at the N-terminus and tyrosine in their sequence, had strong capacity to scavenge free radical (Yarnpakdee et al., 2014).

4.14.4 Degree of hydrolysis on protein hydrolysates activity

When the extraction of antioxidant peptides was studied it has been found that other methods can leave toxic chemicals or residual organic solvents, hence, the method which is the most preferred in pharmaceutical and food industries is the enzymatic hydrolysis method (Sila and Bougatef 2016).

In addition, it has been reported that through hydrolysis of proteins with certain enzymes, the antioxidant activities of protein hydrolysates can be increased and some peptides or portions possess stronger antioxidant potential than others (Chen et al., 1995; Sila and Bougatef, 2016).

In a recent study carried out on hairtail (*Tichiurus japonicas*), muscle protein was hydrolyzed using five different proteases (alcalase, trypsin, neutrase, pepsin, and papain). The hydrolysates formed from papain and alcalase showed higher 2,2-di-phenyl-1picryhydrazyl radicals (DPPH•) and hydroxyl radical (HO•) scavenging activity than the other hydrolysates from the other three proteases. Hence, the amino acid sequences of the protein hydrolysates of hairtail muscle were identified as Gln-Asn-Asp-Glu-Arg (TJP1), Lys-Ser (TJP2), Lys-Ala (TJP3), Ala-Lys-Gly (TJP4), Thr-Lys-Ala (TJP5), Val-Lys (TJP6), Met-Lys (TJP7), and Ile-Tyr-Gly (TJP8) with molecular weights of 660.3, 233.0, 217.1, 274.1, 318.0, 245.1, 277.0, and 351.0 Da, respectively (Yang et al., 2019). In addition to this, TJP3, TJP4, and TJP8 exhibited higher reducing power and capacity to inhibits lipid peroxidation in a linoleic acid model system, and TJP3, TJP4, and TJP8 showed robust scavenging activities (Yang et al., 2019).

Therefore, according to some of these studies it can be deduced that the activity of endogenous antioxidant hydrolysates may depend on the areas (stomach and intestine) in which hydrolysates are produced by fermentation and by which endogenous enzymes the antioxidant hydrolysates are produced in the body.

4.14.5 Therapeutic action of protein hydrolysates

Protein hydrolysates have a broad spectrum of therapeutic action, compared to synthetic drugs as they have lesser side effects and are better absorbed in the gastrointestinal tract (Agyei and Danquah, 2011). Yoshie-Stark et al. (2006) stated that bioactive peptide generated from rapeseed owning strong angiotensin converting enzyme (ACE) inhibitory and DPPH radical scavenging activity. Additionally, protein hydrolysates produced by egg yolk protein when it is hydrolyzed by orientase and protease from *Bacillus* sp, as found by lecithin extraction, displayed DPPH and hydroxyl radical scavenging activity (Zambrowicz et al., 2012). In the same context, ovokinin, a hydrolysate liberated from egg white, reduce blood pressure, without induction of hypotension or side effects. Thus, patient with regular blood pressure or moderate hypertension, ovokinin may be administered (Zambrowicz et al., 2012).

Various hydrolysates derived from food exert actions against thrombotic, cancer, hypertension, oxidant, and have effect such as immunomodulation, mineral-binding, antimicrobial or antioxidant properties (Hartman and Miesel, 2007). For instance, hydrolysates having antimicrobial properties are cysteine-rich hydrolysates obtained from oysters when hydrolyzed with alcalase and bromelain prevents the development of both Gram-positive and negative bacteria (Zambrowicz et al., 2012).

Furthermore, many products available on the market have therapeutic effects. For example, fermented milks products such as Calpis AMEEL S (Japan) and Evolus (Finland) or the energy drink CholesteBlock (Japan), drink which exhibit anti-hypercholesterolemic and antihypertensive effects (Zambrowicz et al., 2012). Pure bioactive peptides application is used only in pharmacy. For instance, LKPN peptide called "*Katshuobushi Oligopeptide*" liberated upon Thermolysin (enzyme) treatment of Bonito fish and it is the bioactive ingredient of the antihypertensive drugs Vasotensin 120, PeptACE, and Peptides 90 (Zambrowicz et al., 2012; Hartman and Miesel, 2007).

4.14.6 In vitro test for the appraisal of antioxidant potential of protein hydrolysates

To determine the antioxidant properties of bioactive peptides or a mixture of peptides, specific method has not yet been standardized. Thus, assays that are mostly used for measuring antioxidant activity of non-peptic antioxidants have been used in the literature to measure the antioxidant activity of protein hydrolysates (Sila and Bougatef, 2016). In assonance, many *in vitro* assays have been used to determine the antioxidant activity of protein hydrolysates for instance, assays such as DPPH radical-scavenging activity, reducing power assays, total antioxidant capacity, lipid peroxidation inhibition in rat liver homogenate, β-carotene bleaching assay, superoxide radical scavenging assay, hydroxyl radical scavenging activity assay, ferrous ion chelating activity and inhibition of linoleic acid autoxidation (Sila and Bougatef, 2016).

Furthermore, protein hydrolysates of fenugreek was assessed on the human colonic adenocarcinoma Caco2 cell line TC7 clone and it was revealed that the bioactive peptides have a selective anti-proliferative property on colorectal cancer cells, by augmenting intrinsic apoptosis rather than necrosis on Caco2/TC7, and by blocking the cell cycle in the G1 phase (Allaoui et al., 2019). The peptides hydrolysates caused modification in mitochondrial membrane permeability, provoked cytochrome C release to the cytoplasm, and encouraged caspase-3 activation (Allaoui et al., 2019). In addition, the hydrolysates exhibited an antioxidant activity by inhibiting ROS. In consonance with the results of the study bioactive peptides of fenugreek could be considered as a potential nutraceutical in the management of colon cancer (Allaoui et al., 2019).

Additionally, among the *Pseudosciaena crocea* protein hydrolysates (PCPH), two antioxidant peptides were isolated and identified by LC-MS/MS and the sequences of the hydrolysates are Serine-Arginine-Cysteine-Histidine-Valine and Proline-Glutamine-Histidine-Tryptophan (Zhang et al., 2016). Increasing the peptides concentration bring about a dose-dependent increase in antioxidant ability (Zhang et al., 2016). In the same context, human hepatoma cells (HepG2 cell line) were used as a model to explore the antioxidant activity and reviewing the mechanism of defense counter to oxidative stress. Up to a concentration of 300 μg/mL the protein hydrolysates did not exhibit cytotoxicity in HepG2 cells, nevertheless, when the HepG2 exposure to H_2O_2 were pre-treated with different concentration of hydrolysates, the level of SOD, CAT and GSH-Px were expressively up-regulated in a dose-dependent manner (Zhang et al., 2016).

Further to this, the first lines of defense system against ROS-mediated oxidative stress are antioxidant enzymes such as SOD, CAT and GSH-Px (Zhang et al., 2016). SOD is a vital cellular antioxidant enzyme that has the ability of transforming superoxide anions to H_2O_2 and water, hindering the signaling process convinced by superoxide anions. CAT and GSH-Px are capable of degrading H_2O_2 into H_2O and O_2, hence, thwarting the configuration of reactive species such as peroxynitrite or hydroxyl radicals (Zhang et al., 2016).

4.14.7 An appraisal of protein hydrolysates activity in vivo

Bashir and colleagues (2018) used mice with alcoholic fatty acid liver disease and fed them with *S. japonicus* (mackerel) protein hydrolysates to determine the antioxidant properties. Mackerel muscle protein hydrolysates (MMPH) were prepared using Protamex. MMPH operated in the cytoplasm of the liver cells and amplified antioxidant activity by influencing the expression of xanthine oxidase (XO) and the detoxifying enzyme such as SOD and CAT (Bashir et al., 2018). The bioactive peptides also suppressed alcohol-induced oxidative stress by inhibiting lipid peroxidation. MMPH may be a strong antioxidant, as the protein hydrolysates normalized mechanisms involving the glutathione family, which are recognized to be secondary enzymes during detoxification (Bashir et al., 2018). Glutathione peroxide-mediated glutathione or CAT converts H_2O_2 to water (Bashir et al., 2018). In addition, oral administration of the bioactive peptides lead to decreased and controlled serum triglyceride levels and XO activity. Also, the expression of SOD protein was up-regulated when administered with MMPH. In assonance, higher SOD activity was observed after treatment with MMPH at 100 mg kg^{-1} (Bashir et al., 2018).

Additionally, three protein hydrolysates from potato (phenylalanine-glycine-glutamine-arginine, phenylalanine-asparagine-arginine-arginine and phenylalanine-glycine-glutamine-arginine-arginine) were isolated and they have an inhibitory effect on linoleic acid oxidation at 55.3, 58.5 and 61.7%, respectively, using a β-carotene bleaching test. When these hydrolysates were orally administrated in Wistar rats at doses of 100 mg/kg, decreased ethanol-induced damage to the stomach lining by 67.9%, 57.0%, and 60.3%, respectively (Martínez Leo et al., 2018).

Further to this, studies have reported that protein hydrolysates liberated from oysters (*Crassostrea rivularis*), with simulated gastrointestinal digestion treatment or with other enzymatic reactions, have the ability to enhance free radical-scavenging activities *in vivo* and the capacity to enhance sexual behavior in normal male mice (Zhang et al., 2019).

In the same context, the hydrolysates of the oysters increased the activities of the antioxidant enzymes (GSH-Px, SOD and CAT) and the levels of T-AOC, as well as reduced the MDA level (Zhang et al., 2019). T-AOC is another index that indicates the capability of the non-enzymatic antioxidant defense system and a decreased in MDA levels demonstrates a good antioxidant activity. The increase of reactive free radicals can quickly inactivate nitric oxide (NO) and its capacity to relax cavernosa smooth muscles, bring injury to the body by attacking large molecules and cell organs and result in fatigue, thus thwarting sexual behavior (Zhang et al., 2019). The MDA levels are decrease by increasing the activities of antioxidant enzymes which, in turn, promote the erection mechanism (Zhang et al., 2019). Consequently, the protein hydrolysates obtained from oysters have antioxidant properties, may provide an enhancement in sexual behavior, thus, results in an aphrodisiac property (Zhang et al., 2019).

4.14.8 An appraisal of protein hydrolysates activity in human trial

A study investigated if γ-GC (hydrolysate) at single doses of (2 and 4 g) which are orally ingested can transitorily increase the GSH content of lymphocytes in human. The protein hydrolysate which is a tripeptide (γ-L-glutamyl-L-cysteinylglycine) is known as reduced glutathione (GSH) and is habitually referred as the "master antioxidant". The peptide is formed within cytoplasm of all cell types at concentrations up to 10 mM. Cytosolic GSH *de novo* synthesis occurs by two consecutive ATP dependent enzyme catalyzed reactions. Firstly, glutamate cysteine ligase (GCL) forms a rare γ-peptide bond between L-glutamic acid and L-cysteine to make γ-glutamylcysteine (γ-GC). Then glycine is added to γ-GC by glutathione synthetase (GS) to generate GSH.

Furthermore, GSH level gradually deteriorate with aging, leaving the people to many age-related diseases. Likewise, γ-GC has been proved to enhance oxidative injury in neurons and astrocytes *in vitro* and enhances brain glutathione *in vivo* (Le et al., 2011). Moreover, γ-GC was used in the human trial as the immediate precursor to GSH, as a means to increase cellular GSH. No adverse effects were conveyed by any of the study participants. The administration of γ-GC resulted in a momentous increase in lymphocyte GSH, the GSH level was increased by 2.5 fold higher for 4 g dose than that observed for the 2 g and GSH levels were 2–3 fold higher than basal GSH level.

Furthermore, no deaths or unfavorable effects were observed in Wistar rats following the acute oral (gavage) administration of 2000 mg sodium γ-GC /kg body weight and γ-GC can be assigned as not acutely toxic at 2000 mg/kg, with a no-observed-adverse-effect level (NOAEL) of at least 1000 mg/kg/day for systemic toxicology from repeated dose oral gavage administration (Chandler et al., 2012).

4.14.9 Safety, regulation, and application

There is a need to address whether antioxidant protein hydrolysates have health benefits and if it has to what extent in order not to cause harmful health effect. Therefore, to address these problems there is a general principle and it is to focus on their intended use. For instance, if a product is taken as a food (i.e., the primary use is to gain nourishment), it should be considered a food. Alternatively, if the primary use is to reduce the incidence of a disease or improve a bodily function, then it is a drug (Chakrabarti et al., 2018). Also, the favorable effect of protein hydrolysates on human health can be regarded as a function claim or disease risk reduction claim. In assonance, due to insufficient data from human studies and the data from *in vitro* and *in vivo* cannot be extrapolate to humans due to variability, health benefits claim should not be made (Chalamaiah et al., 2019).

Nitric oxide (NO), superoxide anion, and related ROS at reasonable concentrations, however, play an important role in the proper functioning of human. They function as regulatory mediators in signaling processes and T lymphocytes are strongly enhanced by ROS (Dröge, 2002). Hence, large amount of antioxidants as protein hydrolysates can disrupt these life maintaining events. In addition, the problem for the use of antioxidants is the lack of clinical evidence. It is also argued that antioxidant supplementation could disturb the normal homeostasis (Sarangarajan et al., 2017), such as the "redox homeostasis" (Dröge, 2002).

Furthermore, oral administration of therapeutic antioxidant protein hydrolysates has been an intimidating job due to the low pH denaturation of the peptides in the stomach, enzymatic instability, and low transport across the tight junctions subsequent in a very low bioavailability (Tyagi et al., 2018).

Nevertheless, a large body evidence is emerging toward the fact that scavenging of ROS is injurious to cancer patients rather than inhibiting the risk (Sarangarajan et al., 2017). In other research, increasing the glutathione levels, which is a powerful antioxidant that scavenges peroxyl radical was shown to speed up the tumour progression in a B-RAF and K-RAS induced lung cancer in murine models (Sarangarajan et al., 2017). Also, the study showed that these antioxidants endorse tumour progression by dropping ROS levels and in sequence diminish p53 expression levels that could have induced apoptosis heading to the cell death. Therefore, using antioxidant from protein hydrolysates can be deleterious in cancer patients through the ROS-p53 axis (Sarangarajan et al., 2017).

Conclusion

Protein hydrolysates can have antioxidant properties if the parent protein is hydrolyzed in an optimum environment. In addition, the bioactive peptide can have many other health benefits such as anticancer, antihypertensive, immunomodulation, mineral-binding and antimicrobial properties. The bioactive peptides have shown strong antioxidant properties both in vitro and in vivo study. However, due to lack of human trials focused on the antioxidant properties of protein hydrolysates and variability, the available data cannot be extrapolated to human hence intake of high amounts of antioxidant hydrolysates should be avoided as it may disrupt the "redox homeostasis." Moreover, it is believed that some antioxidants are detrimental to cancer patients when taking into consideration the ROS-p53 axis. Therefore, the dark side of antioxidant hydrolysates is still unknown even if they have shown favorable results in non-human studies.

References

Agyei, D., Danquah, M., 2011. Industrial-scale manufacturing of pharmaceutical-grade bioactive peptides. Biotechnol. Adv. 29 (3), 272–277.

Allaoui, A., Gascón, S., Benomar, S., Quero, J., Osada, J., Nasri, M., Rodríguez-Yoldi, M., Boualga, A., 2019. Protein hydrolysates from fenugreek (*Trigonella foenum graecum*) as nutraceutical molecules in colon cancer treatment. Nutrients 11 (4), 724.

Bashir, K., Mohibbullah, M., An, J., Choi, J., Hong, Y., Sohn, J., Kim, J., Choi, J., 2018. In vivo antioxidant activity of mackerel (*Scomber japonicus*) muscle protein hydrolysate. PeerJ 6, e6181.

Chakrabarti, S., Guha, S., Majumder, K., 2018. Food-derived bioactive peptides in human health: challenges and opportunities. Nutrients 10 (11), 1738.

Chalamaiah, M., Keskin Ulug, S., Hong, H., Wu, J., 2019. Regulatory requirements of bioactive peptides (protein hydrolysates) from food proteins. J. Funct. Food 58, 123–129.

Chandler, S., Zarka, M., Vinaya Babu, S., Suhas, Y., Raghunatha Reddy, K., Bridge, W., 2012. Safety assessment of gamma-glutamylcysteine sodium salt. Regul. Toxicol. Pharmacol. 64 (1), 17–25.

Chen, H., Muramoto, K., Yamauchi, F., 1995. Structural analysis of antioxidative peptides from soybean .beta. conglycinin. J. Agric. Food Chem. 43 (3), 574–578.

Dröge, W., 2002. Free radicals in the physiological control of cell function. Physiol. Rev. 82 (1), 47–95.

Hartmann, R., Meisel, H., 2007. Food-derived peptides with biological activity: from research to food applications. Curr. Opin. Biotechnol. 18 (2), 163–169.

Havenaar, R., Brink, B., Huis In't Veld, J., 1992. Selection of strains for probiotic use. Probiotics, 209–224.

Karamać, M., Kosińska-Cagnazzo, A., Kulczyk, A., 2016. Use of different proteases to obtain flaxseed protein hydrolysates with antioxidant activity. Int. J. Mol. Sci. 17 (7), 1027.

Le, T., Jiang, H., Cunningham, G., Magarik, J., Barge, W., Cato, M., Farina, M., Rocha, J., Milatovic, D., Lee, E., Aschner, M., Summar, M., 2011. γ-Glutamylcysteine ameliorates oxidative injury in neurons and astrocytes *in vitro* and increases brain glutathione *in vivo*. Neurotoxicology 32 (5), 518–525.

Martínez Leo, E., Martín Ortega, A., Segura Campos, M., 2018. Bioactive peptides impact in cancer therapy. Therapeutic, Probiotic, and Unconventional Foods, 157–166. https://doi.org/10.1016/B978-0-12-814625-5.00009-1.

Rizzo, A., Berselli, P., Zava, S., Montorfano, G., Negroni, M., Corsetto, P., Berra, B., 2010. Endogenous antioxidants and radical scavengers. Adv. Exp. Med. Biol., 52–67.

Sarangarajan, R., Meera, S., Rukkumani, R., Sankar, P., Anuradha, G., 2017. Antioxidants: friend or foe? Asian Pac. J. Trop. Med. 10 (12), 1111–1116.

Sila, A., Bougatef, A., 2016. Antioxidant peptides from marine by-products: isolation, identification and application in food systems. A review. J. Funct. Foods 21, 10–26.

Tyagi, P., Pechenov, S., Anand Subramony, J., 2018. Oral peptide delivery: translational challenges due to physiological effects. J. Control. Release 287, 167–176.

Udenigwe, C., Aluko, R., 2011. Chemometric analysis of the amino acid requirements of antioxidant food protein hydrolysates. Int. J. Mol. Sci. 12 (5), 3148–3161.

Van der Ven, C., Gruppen, H., de Bont, D., Voragen, A., 2002. Optimisation of the angiotensin converting enzyme inhibition by whey protein hydrolysates using response surface methodology. Int. Dairy J. 12 (10), 813–820.

Wang, S., Zhu, F., Han, F., Wang, H., 2008. Purification and characterization of antioxidative peptides from salmon protamine hydrolysates. J. Food Biochem. 32 (5), 654–671.

Wu, H., Chen, H., Shiau, C., 2003. Free amino acids and peptides as related to antioxidant properties in protein hydrolysates of mackerel (*Scomber austriasicus*). Food Res. Int. 36 (9-10), 949–957.

Yang, X., Zhang, L., Ding, D., Chi, C., Wang, B., Huo, J., 2019. Preparation, Identification, and Activity Evaluation of eight antioxidant peptides from protein hydrolysate of Hairtail (*Trichiurus japonicas*) Muscle. Mar. Drug 17 (1), 23.

Yarnpakdee, S., Benjakul, S., Kristinsson, H., Kishimura, H., 2014. Antioxidant and sensory properties of protein hydrolysate derived from Nile tilapia (*Oreochromis niloticus*) by one- and two-step hydrolysis. J. Food Sci. Technol. 52 (6), 3336–3349.

Yoshie-Stark, Y., Wada, Y., Schott, M., Wäsche, A., 2006. Functional and bioactive properties of rapeseed protein concentrates and sensory analysis of food application with rapeseed protein concentrates. LWT - Food Sci. Technol. 39 (5), 503–512.

Zambrowicz, A., Timmer, M., Polanowski, A., Lubec, G., Trziszka, T., 2012. Manufacturing of peptides exhibiting biological activity. Amino Acids 44 (2), 315–320.

Zhang, N., Zhang, C., Chen, Y., Zheng, B., 2016. Purification and characterization of antioxidant peptides of *Pseudosciaena crocea* Protein Hydrolysates. Molecules 22 (1), 57.

Zhang, Z., Su, G., Zhou, F., Lin, L., Liu, X., Zhao, M., 2019. Alcalase-hydrolyzed oyster (*Crassostrea rivularis*) meat enhances antioxidant and aphrodisiac activities in normal male mice. Food Res. Int. 120, 178–187.

CHAPTER

Selenium

4.15

Tanuj Joshi, Sumit Durgapal, Vijay Juyal, Arvind Jantwal, Mahendra Rana, Aadesh Kumar

Department of Pharmaceutical Sciences, Kumaun University, Campus Bhimtal, Bhimtal, Uttarakhand, India

4.15.1 Introduction

Antioxidants may be defined as the compounds which decrease, delay or retard the oxidative reactions. Antioxidants inhibit the formation of reactive oxygen species (ROS) and reactive nitrogen species (RNS) and thus protect the cell from the harmful effects produced by ROS and RNS (Saxena et al., 2012). There are various types of antioxidants like vitamin E, β- carotene, water soluble vitamin C, butylated hydroxyanisole (BHA), butylated hydroxyl toluene (BHT), and propyl gallate (Sheikh Arshad Saeed, 2005). Some antioxidants are found within the human body and some antioxidants are obtained by humans from natural sources. Also many antioxidants are now synthesized in laboratories and industries. An epidemiologic study suggested that a beneficial effect on a number of age related diseases is produced by consumption of antioxidant vitamins (Maxwell et al., 1999). Other than antioxidant vitamins there are other naturally occurring antioxidants like phenolic acids and flavonoids that are useful in the prevention of various human diseases. They are synthesized by plants and are present in the foods we eat. Different antioxidants may have different molecular weight, composition, physical, chemical properties, and mechanism of action (Anbudhasan et al., 2014). Flavonoids and tannins are the main phenolic compounds which are obtained from various medicinal plants.

Oxygen is inevitable for the survival of all the living beings. Energy is generated by cells through utilization of oxygen. Mitochondrion plays an important role in the generation of free radicals, the Table 4.15.1 below represents the list of oxidative stress related diseases in humans (Pham-Huy et al., 2008). Free radicals generation occurs through homolytic bond fission so that each fragment has one unpaired electron. Examples of some free radicals are: superoxide anion radical ($O_2{\cdot}^-$), nitric oxide ($NO{\cdot}$), hydroxyl ($OH{\cdot}$), nitrogen dioxide ($NO_2{\cdot}$), peroxyl ($ROO{\cdot}$) and lipid peroxyl ($LOO{\cdot}$). Singlet oxygen ($^1O_2$), Hydrogen peroxide (H_2O_2), ozone (O_3), peroxynitrite ($ONOO^-$), nitrous acid (HNO_2), dinitrogentrioxide (N_2O_3), lipid peroxide (LOOH), hypochlorous acid (HOCl) are referred to as oxidants (Aruoma et al., 1998). ROS and RNS species behave both as toxic and beneficial compounds. When the concentration

Antioxidants Effects in Health. DOI: https://doi.org/10.1016/B978-0-12-819096-8.00017-3

Table 4.15.1 List of oxidative stress induced diseases in humans (Pham-Huy et al., 2008).

S.No.	Body part	Disease/complications
1	Lungs	In Lungs Asthma and chronic bronchitis are the complications
2	Joints	Arthritis, rheumatism are the complications in joints
3	Brain	Alzheimer's, Parkinson's, memory loss, depression, stroke are problems associated with the brain
4	Kidney	Glomerulonephritis, chronic renal failure are problems associated with the kidney
5	Fetus	Preeclampsia, IU growth restriction
6	Eyes	Cataract, retinal diseases are associated with eyes
7	Heart - vessels	Arteriosclerosis, hypertension, ischemia, cardiomyopathy, heart failure are problems related to heart
8	Multiorgans	Cancer, Aging, diabetes, inflammation, infection

of ROS and RNS are low they positively affect responses and functions related to cell and immunity but at increased concentrations oxidative stress is produced by them (Halliwell et al., 2007; Bahorun et al., 2006; Valko et al., 2004; Valko et al., 2007; Droge et al., 2002; Willcox et al., 2004; Pacher et al., 2007; Genestra et al., 2007; Halliwell et al., 2007; Young et al., 2001).

Oxidative stress is a detrimental process that produces imbalance between activities of the antioxidant defenses and the production of free radicals. It is a severe condition that can damage all the cell structures (Aruoma et al., 1998). It plays an important role in development of diseases, which are chronic and degenerative in nature. Abnormal production of ROS can cause cancer, cardiovascular diseases (CVD), a number of neurological complications, pulmonary diseases, rheumatoid arthritis, nephropathy and metabolic diseases.

4.15.2 Selenium and its role as an antioxidant

The discovery of selenium occurred in 1817 by chemist Jöns Jacob Berzelius of Swedish origin. Selenium (Se) derived its meaning from Greek word Selene, meaning moon. Selenium occurs as Se II in organic form and as selenite (Se IV) or selenate (Se VI) in inorganic form. Cereals, eggs, dairy products, nuts, meat, and fish are important sources of selenium. Selenium is essential as a cofactor to activate various enzyme systems in humans. Fifty five to 70 micrograms is the recommended dietary allowance of selenium (Morris and Crane, 2013). Selenium possesses several pharmacological activities like it acts as an anti-inflammatory, antioxidant and immunological agent. Selenium mainly occurs in organic form as selenomethionine in animal and human tissues. Around 28% to 46% of total pool of selenium found in the body occurs in the skeletal muscle, making it a major site for selenium storage. Various multivitamin/multimineral formulations containing selenium are available in the market.

For the proper functioning of the body selenium is very necessary. Selenium has a very important role in metabolism and synthesis of thyroid hormones (Ragozzino et al., 2017). Selenium is involved as a cofactor in both autoimmune diseases and physiological functions of thyroid glands. Scientists have put forward the theory that even small deficiency of selenium can lead to autoimmune diseases. Prevention of cancer is also another role ascribed to selenium. Selenium plays an important function in apoptosis, deoxy ribonucleic acid (DNA) repair, immune functions and endocrine functions. Thus proper amount of selenium is very important to prevent diseases like cancer. There is an association of polythiols with proteins of cancer cell membrane. These polythiols show their characteristic presence in reducing conditions associated with hypoxic tissues of tumor. Sodium selenite can oxidize these polythiols to their corresponding disulfides. Polythiols also form a coat outside tumor cells. This coat prevents the detection and killing of antigens associated with tumor by immune system components like natural killer cells (Ragozzino et al., 2017). By their oxidizing effect on thiols they can inhibit the generation of this coat surrounding the tumor cells. Deficiencies in selenium levels lead to infertility in both females and males (Ragozzino et al., 2017). Reduced levels of selenium are also involved in development of depression. Studies in elderly and adult population have linked depression with deficient selenium consumption. Studies have also indicated a link between low levels of selenium and symptoms of depression in young adult population. Supplementation with selenium has been found to be useful in enhancement of mood and relief in postpartum depression in a study. Still a large clinical trial conducted on elderly people showed no benefit of selenium supplementation on mood. This may be due to the fact that years of exposure to free radicals might have reduced the beneficial effects of selenium in the elderly population (Ragozzino et al., 2017).

Though normal amount of selenium is important for proper functioning of the body, selenium in high amounts can cause significant adverse effects that can be harmful to the body. Metallic taste in the mouth and breath with a smell of garlic are early signs of excess selenium intake. Consumption of high amount of selenium, a condition called selenosis is associated with loss of nail and hair. Skin rashes, diarrhea, lesions of skin, nervous system abnormalities, nausea, irritability, and mottled teeth are some other symptoms associated with high levels of selenium (Ragozzino et al., 2017).

4.15.3 Evidence of beneficial effects of selenium from *in-vitro* and preclinical studies

In a model of rats fed with cholesterol, selenium has shown beneficial effects by lowering the levels of total cholesterol (Dhingra and Bansal, 2005). Inhibition of oxidation of LDL *in-vitro* was shown by diphenyl diselenide (an organoselenium compound). The compound showed the above activity by its thiol-peroxidase activity (Bem et al., 2008). Diphenyl diselenide also showed beneficial effect on total cholesterol levels in serum and various parameters of oxidation in rabbits that were hypercholesterolaemic. Also this compound effectively lowered the concentration of lipid in plasma in hyperlipidaemic mice (Sanmartin et al., 2011). Selenium is a constituent of glutathione peroxidase 1 (GPX-1). Studies on transgenic and knockout mice have thrown light on the fact that vascular function is modulated by GPX-1. Also selenium has shown beneficial effect in animal models of Parkinson's disease. Low doses of selenium along with insulin administration are useful in regulation of blood glucose level. Combination of selenium and insulin is also effective in controlling levels of glucose in blood and improving the distribution of transporter associated with glucose (GLUT4) in rats that were made diabetic. Selenium was found to be useful in cardiomyopathy associated with Chagas disease in albino mice and digestive disorders (Sanmartin et al., 2011). It is known that the distribution of selenium in yeast and garlic occurs as selenomethionine (85%) and γ-glutamyl-Se-methylselenocysteine (73%) respectively. In a study it was found that selenium found in garlic effectively inhibited development of premalignant lesions and adenocarcinomas in rat mammary glands, when these rats were administered with carcinogenic substance (Barciela et al., 2008). Ebselen is an organic compound containing selenium. A feeble activity resembling glutathione peroxidase was exhibited by this organic compound and it can act as a cardioprotective agent. Animal studies have proven that ebselen given before ischaemia reperfusion can provide beneficial effect on heart. The action responsible for this effect of ebselen is the protection of glutathione levels and induction of HSP27, a protein involved with stress. In an embolic stroke model using rabbits, ebselen has shown a neuroprotective effect. Co-administration of a plasminogen activator with ebselen has shown to improve behavior in rabbits. Newer research has also produced an important finding that ebselen might be beneficial against COVID-19. Attenuation of cytokines, inflammatory antioxidants and other mechanisms might be involved in the efficacy of ebselen against COVID-19. Much research is required in this area and ebselen can prove to be a beneficial agent against COVID-19.

In a study anticancer effects of selenium on prostate cancer were studied on male beagle dogs. It was found that when selenium in the diet of these dogs was increased, the damage to DNA was reduced, denoting the beneficial effect of selenium in prostate cancer. Destruction of antigens by macrophages and neutrophils is decreased due to deficiencies in selenium levels as demonstrated by animal studies. These studies also point out the fact that optimum selenium levels are very necessary for proper functioning of the immune system (Tinggi, 2008). The incidence of skin cancers can increase due to deficiencies of selenium in the body as shown by reduced levels of antioxidant selenoenzymes like glutathione peroxidases. Animal models have shown that tumors in skin induced by phorbol ester and ultraviolet irradiation can increase due to low levels of selenium and decrease due to increased levels of selenium. Selenium methionine possesses radioprotective properties and in mice it has shown protective effect on skin against damage induced by ultraviolet light. Selenium has an inhibitory effect on mammary carcinogenesis induced by chemical carcinogens and inhibits both post initiation and as well as initiation phases of mammary carcinogenesis. Studies have shown that consumption of broccoli with enriched selenium content provided protection against cancer in rats and mice. In a study mammary tumor was induced in rats by 7,12-Dimethylbenzathracene (DMBA). It was found in the study that an organoselenium compound had beneficial effect on mammary tumors in both initiation as well as progression phase.1,4 phenylenebis(methylene) selenocyanate a compound containing selenium showed beneficial effects in lung tumor induced in mice (Tapiero et al., 2003). It was observed in a study that rats administered a diet low in vitamin E, selenium and sulfur amino acids can lead to development of necrosis (Savitha, 2014). In an animal experimental the antioxidant defense system of rats was exposed to the toxic effect of cadmium. Selenium administration to rats provided protection against toxicity induced by selenium. The study demonstrates that selenium being an antioxidant has a protective effect on antioxidant defense system of the body (Pavlovic et al., 2001).

4.15.4 Evidence of beneficial effect of selenium from clinical studies

A clinical trial was conducted in Germany to assess ascorbic acid, coenzyme Q10, alpha-tocopherol, beta-carotene and selenium in elderly women in Germany. It was concluded by the study that there was deficiency of alpha-tocopherol and selenium in the elderly women, although ascorbic acid and beta-carotene levels were found in sufficient amount. Prevention of different diseases is an important role played by antioxidants. Thus, deficiencies in levels of antioxidants like selenium can lead to various diseases (Wolters et al., 2006). Researchers have demonstrated that selenium is useful in various cardiovascular diseases. There are evidences of relations between reactive oxygen species and cardiovascular diseases. Thus, selenium being an antioxidant possesses the potential to provide beneficial effects in cardiovascular diseases. Increase in oxidation of lipoproteins, tissue and vascular damage and vascular endothelial complications can be produced by oxidative stress.

Hypertension which is produced by unhealthy lifestyle and stress is important risk factors that can lead to cardiovascular diseases. A clinical study was carried out by researchers on 710 people that were randomly recruited from 6 rural districts of Belgium. The results indicated that there was a relation between deficient selenium levels and blood pressure recordings in men whereas no such association existed in women involved in the study. Thus, the role of selenium in prevention of cardiovascular diseases is highlighted by this study (Alsafwah et al., 2007). A clinical trial that was double blind, found that the patients involved in the study showed an increase in GPX-1 activity and expression when they were given sodium selenite supplementation. The increase in GPX-1 levels by sodium selenite outlines its importance in cardiovascular diseases. Different formulations of selenium have different effects with respect to cardiovascular complications. Selenium-methyl-selenocysteine hydrochloride gave similar results as sodium selenite, whereas selenomethionine did not produce cardioprotective effect similar to sodium selenite (Sanmartin et al., 2011). Scientists have found that decrease in selenium levels are involved with death in acute coronary syndrome patients (Lubos et al., 2010). Another clinical study on Americans has found the fact that there is a lack of association between levels of selenium and development of atherosclerosis (Xun et al., 2010). Selenium also has a protective role in rheumatoid arthritis. It was found in a study that in comparison to the control group, patients suffering from rheumatoid arthritis had lower levels of selenium (Onal et al., 2011; Kose et al., 1996). Researchers have found that the relation between consumption of selenium and development of rheumatoid arthritis is inverse in nature. This supports the claim that selenium can be helpful in retarding the development of rheumatoid arthritis (Helgeland et al., 2000). Kashin-Beck disease (KBD) is a disease with features similar to osteoarthritis. It can lead to impairment of skeletal formation and chronic secondary osteoarthritis. People living in selenium deficient areas are especially prone to this disease. Researchers conducted clinical trials to study the role of deficiencies in the level of selenium and genetic variability in the development of this disease. The clinical study utilized 129 patients of KBD that were divided into groups based on the fact whether they belonged to KBD endemic area or whether they belonged to an area were KBD was not endemic. There was also detection of modifications in chromosomes 11 and 2 in patients of KBD belonging to the area where KBD was not endemic. All the patients were treated with selenium supplementation at various concentrations and the results concluded that genetics play a more important role than environmental conditions (selenium deficiency in different areas) in development of KBD. Since selenium shows its actions mainly as a constituent of proteins (selenoproteins), an experiment was conducted to evaluate the association of genetic polymorphism of four selenoproteins and risk of developing KBD in Chinese patients. The selenoproteins mentioned above are associated with thioredoxin reductases(TrxRs),selenoprotein P(SelP) and glutathione peroxidases (GPXs). The results found that genetic modification in GPXs can prove to be a risk factor in production of KBD. However, there is no concrete evidence available that selenium is useful in arthritis or related disorders (Sanmartin et al., 2011).

In the thyroid glands selenium plays a very important role. Thyroid glands contain highest amount of selenium than any other organ. Selenium is very important in the normal functioning of thyroid glands just like iodine. The role of selenium in the functioning of thyroid glands lies in the fact that deficiency in selenium can lead to myxedematous cretinism. Selenium plays an important role in the thyroid glands by being a component of selenocysteine (Sanmartin et al., 2011). Three important selenoenzymes that are of importance to the thyroid gland are: Type-I ID (iodothyronine deiodinase), GPX and TrxR. Type-I ID (iodothyronine deiodinase) plays an important role by enhancing the rate of thyroxine deiodination and leads to activation of hormones of the thyroid glands. GPX by reducing H_2O_2 protects thyroid cell from destruction caused by oxidative species and TrxR also protects the thyroid gland from reactive oxygen species. Selenium supplementation has provided beneficial effects by increasing activity of GPX and other selenoproteins in different pathophysiological conditions associated with the thyroid glands. Selenium's important role in maintenance of thyroid function can also be outlined by the fact that supplementation with selenium reduces the levels of antibody against thyroid peroxidase enzyme in adults suffering from autoimmune thyroiditis. However, this effect was not observed in adolescents and children (Sanmartin et al., 2011). Selenoenzymes belonging to ID type are important targets that can be utilized in the diagnosis and treatment of tumors of thyroid gland. Researchers have found that selenium supplementation can be administered along levothyroxine for the treatment of Hashimoto's thyroiditis. In pregnant women detection of thyroid peroxidase antibodies is a sign that they can develop thyroid dysfunction, however pregnant women who were given supplementation with selenium during the time of pregnancy and in postpartum period demonstrated decrease in inflammatory activity of thyroid. Selenium can also be used as coadjuvant therapy to levothyroxine. Thus, all the research data that has been obtained by various sources lead to the common finding that for prevention of disorders of thyroid gland, optimal levels of selenium should be maintained within the thyroid gland by sufficient intake of selenium.

Selenium supplementation also has a role in treatment of cancers (Sanmartin et al., 2011). Selenium has shown beneficial effects in patients with advanced prostate cancer and in healthy men it reduces the risk of prostate cancer. Corcoran and coworkers have determined the maximum dose of sodium selenate that can be tolerated in order to study its pharmacokinetic profile by administration of oral capsule to patients. The anti-angiogenic efficacy of sodium selenate was found to be modest when studies were carried out on it in a way like single-agent studies of anti-angiogenic drugs such as sorafenib, sunitinib, and bevacizumab. However, there was a reduction in the levels of prostate-specific antigen (PSA) in patients suffering from cancer of prostate. Exploring the combination of selenium with anticancer drugs like docetaxel and others can lead to a new path in the treatment of prostate cancer. The combination of genistein with selenium can be a new approach towards the treatment of cancer of prostate and can become a new adjuvant to the standard therapy of cancer of prostate. Also, researchers have found that selenium can also provide a protective effect during radiotherapy of prostate cancer. A study was conducted in patients undergoing radical prostatectomy and it was found that silymarin in combination with selenium methionine exerted beneficial effects in these patients by enhancing the levels of blood selenium and ameliorating the lipid parameters. Also, the above combination is helpful in decreasing the progress of cancer of prostate in patients. However, in some studies in men it was also indicated that selenium or vitamin E alone or their combination was not beneficial in reducing the progress of cancer of prostate (Sanmartin et al., 2011). Another study was carried out in which the participants were placed in three groups. In one group placebo was given and the other two groups were treated with different amount of selenium. It was shown in the study that there was no beneficial effect of selenium in these participants, who were suffering from cancer of prostate which was localized. The lack of effectiveness of selenium in cancer of prostate by some studies like mentioned above can be due to lack of proper *in-vivo* studies and use of right formulations and doses. However overall researchers believe that selenium supplementation can be of use in the treatment of cancer of prostate. Selenium can act synergistically with other drugs used in cancer and provide a protective effect in reducing adverse effects associated with treatment of cancer with hormonal drugs or radiotherapy.

Selenium has also proven to be beneficial in treatment of HIV-1. A decrease in antioxidant status, HIV-1progression and high rates of mortality are associated with low selenium levels in HIV-1 patients. Different studies have been carried out that provide evidence that selenium can exert an anti-HIV effect. In a study morbidity associated with diarrhea in pregnant women from Tanzania was decreased by selenium methionine. Also, in two different small scale trials carried out in American men, it was found that selenium methionine decreased the load of HIV virus in the body. Inhibition of transcription factor leading to inhibition of production of HIV was also seen by researchers by using a combination of aspirin with selenium (Sanmartin et al., 2011).

Selenium has provided beneficial effects in neurodegenerative diseases. Decreased antioxidants and levels of trace elements were found in the plasma of patients with Alzheimer's disease (AD). Researchers have found that the levels of selenium were found to be low in erythrocytes, nails and plasma of AD patients in comparison to the control group. Evaluation of elderly patients by neurological tests showed that the performance was decreased in the elderly with respect to coordination due to low levels of selenium. As mentioned earlier selenium performs its various functions in the body in the form of selenoproteins and there is a link between low expressions of selenoproteins and disorders of the nervous system, particularly damage to neurons produced by oxidative stress. Selenoprotein P acts as an important protein that prevents development of oxidative stress. Its expressions are found to be decreased in diseases like Parkinson's disease, AD, epilepsy etc. Dietary supplementation consisting of selenium can lead to increase in its levels and decrease the incidence of neurodegenerative diseases. Beneficial effects of a combination consisting of donepezil and formula F (consisting of antioxidants including selenium) was observed by researchers in treatment of AD. Selenium has assumed an important role in treatment of depression as evidenced by various studies. It is known to reduce the adverse effects associated with use of lithium used in bipolar disorder. Also it provides protective effect on neurons. Selenium can be useful in adolescent depression and depression associated problems in alcoholics (Sanmartin et al., 2011). Perinatal depression can result due to inadequate intake of selenium by pregnant women. This problem can be treated by providing adequate selenium in the diet of pregnant women. Selenium binding protein-1 (SELENBP-1) is a biomarker that is associated with schizophrenia. A study was carried out by researchers in which patients suffering from schizophrenia, depression and normal people were used. It was found in the study that the levels of (SELENBP-1) were found to increase in patients suffering from schizophrenia. Reduction of antioxidants in serum has been observed in patients suffering from acute pancreatitis. It was found in a study that in comparison to the control group, selenium levels were low in patients suffering from chronic pancreatitis. Thus, in patients with severe acute pancreatitis selenium supplementation can ameliorate the antioxidant status. Selenium supplementation can thus provide benefit in acute pancreatitis (Sanmartin et al., 2011).

4.15.5 Dark side of selenium: Adverse and toxic effects

Humans can get exposed to selenium toxicity from various sources like water, food, soil and air. Use of fertilizers containing excess selenium can pollute the soil and contaminate the food grown in it. Also, water that passes through waste areas and farmlands can get polluted with selenium. Thus, there are multiple ways by which selenium can enter the body and cause toxicity. Industrial waste containing selenium can lead to contamination of water and food with selenium. Workers of paint, metal and selenium related industries are exposed to selenium through breathing. Exposure to air that contains high amount of selenium can lead to dizziness, irritation of mucous membrane and fatigue. Bronchitis and collection of fluid in the lungs is a consequence of excess administration of selenium to humans. Experimentally it has been proven that 1000 μg/day of selenium given continuously for a very long period of time can cause selenosis in humans (Kuila et al., 2012).

Selenium toxicity depends on the pharmacokinetic profile of particular selenium compound. Selenium sulphide which is insoluble in nature is less toxic than selenite, selenate and selenomethionine. Generally, the inorganic forms of selenium like selenate and selenite have a higher toxicity causing potential than the organic forms like selenocysteine, selenocystine and selenomethionine. Damaged nails, hypochromic anaemia and leucopenia have been observed in workers who are involved with production of selenium rectifiers. There have been incidences of selenium poisoning due to accidental consumption of selenic acid (30 g/l) and vitamin tablets that contain high amounts of selenium. Decrease in cognitive functions, weakness, convulsions, acroparesthesias, hair and nail changes, gastrointestinal disturbances are consequences of selenium toxicity. The characteristic features of acute selenium toxicity are emesis, hypersalivation, garlic breath (due to elimination of selenium compounds that are volatile (Kuila et al., 2012). Neurological disturbances, fatigue and hair loss also accompany the symptoms of acute toxicity of selenium. Selenosis or chronic selenium poisoning involves pain, hypoaesthesia, hyperreflexia and development of convulsions, numbness and paralysis (Kuila et al., 2012).

China, Venezuela and South Dakota, United States of America (USA) are areas where there is a high prevalence of selenium toxicity (Agarwal et al., 2016). In USA and China selenosis is usually observed when its intake is greater than 0.91 mg/day. In mice and rats selenium sulphide has been found to be carcinogenic. Overall selenium compounds have tested negative in *in vivo* models of mutagenicity. Lethal dose of sodium selenite has produced complications in bone marrow of hamster in experimental studies. In some areas of India like Punjab (Nawanshahar and Hoshiarpur district) joint swelling, hair loss and other types of deformities have been observed due to selenium toxicity. The high occurrence of selenium in these areas is responsible for toxicity in cattle also (Kuila et al., 2012). Though selenium has anticancer properties, excess of selenium can itself cause carcinogenesis, genotoxicity and cytotoxicity. Cardiovascular diseases and amyotrophic lateral sclerosis are other complications associated with selenium. Diabetes has been associated with high selenium levels (Sun et al., 2014). Research has shown that doses of selenium which are 900 μg per day can cause irritability, nausea, fatigue, hair loss, gastrointestinal upset and mild nerve damage. In many cases dietary habits may not contribute to selenium toxicity but excess of selenium levels in soil may contribute to selenium toxicity. In China soils of certain regions contained selenium levels in very high quantity which matched with dietary intake levels of 5000 μg of selenium per day. People in these areas developed selenosis and were treated immediately by removing these people from their respective regions. Also, dietary changes were done to fight selenosis in these people (Barceloux, 1999). Many efforts are needed to understand the exact mechanism by which selenium produces toxicity. However, it has been implicated that formation of selenotrisulfides by interaction of selenium with glutathione and generation of superoxide and hydrogen peroxide can lead to toxic effects of selenium by oxidizing cell membrane and macromolecules. This leads to damage in integrity of cells, apoptosis and necrosis. The toxic effect of selenium in human cells was evaluated in a study. It was proved in the study that selenium can decrease the levels of proteins and lipids. It can also produce DNA damage. In mice treated with 5 mg/L of selenium it was observed that DNA adducts are formed in their keratinocytes due to oxidation produced by generation of hydroxyl radicals by selenium. Loss of hair and nails can occur due to selenium toxicity. Loss of hairs can occur from any area of the body. In selenium toxicity there is generation of skin lesions on the backside of the neck and forearm. Redness and swelling are observed in the skin due to selenium toxicity which ultimately leads to ulceration and blistering.

Dental problems may also arise due to selenium toxicity. Various abnormalities associated with the nervous system such as pain in extremities, hyperreflexia associated with the tendons, motor disturbances and peripheral anesthesia can occur due to selenium toxicity. Selenium toxicity is associated with reduction in glutathione levels and prothrombin time (Agarwal et al., 2016). Caustic burns which are painful in nature can result from exposure of the skin to selenious acid, selenium oxychloride and selenium dioxide. Accumulation of selenium under nails of fingers can lead to pain. Accidental exposure of selenium dioxide to face can result in injury to the eyes in the form of conjunctival edema, severe pain and lacrimation. Chronic exposure to eyes can lead to a condition called palpebral conjunctivitis and red discoloration of the eyes. Selenium compounds can lead to respiratory tract irritancy. Industrial gas like hydrogen selenide is especially toxic to the respiratory system. Also, eye pain, throat pain, rhinorrhea, wheezing and pneumomediastinum can occur due to exposure with hydrogen selenide. In an incidence it was reported that twenty-eight workers came in contact with enhanced levels of selenium oxide in a selenium rectifier plant. Burning and irritation in the respiratory tract, bronchospasm, tachycardia, tachypnea and hypotension were some early symptoms of these workers exposed to smoke of selenium oxide. Later they developed chills, fever, chemical pneumonitis, vomiting, diarrhea and headache. Five workers were hospitalized for requirement of artificial respiratory support and complication of leukocytosis, bilateral infiltrates and fever. Another gas called selenium hexafluoride used in industries as electrical insulator can lead to burning sensation and pain in the skin, eyes, respiratory tract. Severe toxicity with inorganic selenium can lead to circulatory failure. The symptoms of severe selenium toxicity include chest pain, dypnea, tachycardia and hypotension. Ventricular dysrhythmias, mesenteric, myocardial infarction and metabolic acidosis associated with selenium toxicity make the situation more complex. Coagulopathy, hyperkalemia, hypokalemia, leukocytosis, thrombocytopenia, hemolysis, metabolic acidosis and enhanced lactate levels are other complications associated with selenium toxicity (See et al., 2006; Spiller et al., 2007).

Selenium toxicity comprising of ocular pain, nail bed burns and painful skin can be treated with an ointment or solution of sodium thiosulfate (10%). Reduction of selenium dioxide to elemental selenium is responsible for the selenium toxicity treating effect of sodium thiosulfate. Interestingly sponges soaked in ammonium hydroxide provided relief to industrial workers exposed to selenium dioxide. The above-mentioned use of ammonium hydroxide in selenium dioxide toxicity has not been proven experimentally. Exposure with selenium hexafluoride gas may be treated with application of calcium gluconate gel. The patient suffering from selenium toxicity should be immediately removed from the site of exposure. In case of dermal toxicity, the skin should be washed thoroughly. Selenium compounds that are systemically absorbed should be removed from the body with the help of active charcoal or gastric lavage. Not much data is available on the availability of chelating agents to inhibit selenium toxicity. Theoretically it has been put forward that ascorbic acid might counteract the oxidative damage produced by selenium but this has not been proved by experimental studies. Bromobenzene has been found to be useful in enhancing the excretion of selenium in urine but the toxicity associated with bromobenzene limits its use. Hemodialysis and hemofiltration are two techniques that can lower the levels of selenium in serum (Kise et al., 2004; Koppel et al., 1986). However in spite of various techniques being used in counteracting the toxicity of selenium, ultimate protection against selenium toxicity can only be provided by avoiding food and water containing high levels of selenium.

Conclusion

Selenium is an important essential nutrient that is found in our food and water. For proper functioning of the body normal amounts of selenium are very essential. The importance of selenium lies in the fact that selenium acts as a cofactor for activation of several enzymes. Numerous beneficial pharmacological actions have been associated with normal levels of selenium. Selenium can be classified as an anti-inflammatory, antioxidant, anticancer and immunological agent. Low levels of selenium can lead to various complications inside the body. In the cardiovascular system and the nervous system selenium produces a beneficial effect. Evidence of beneficial effect of selenium on cardiovascular system, neurodegenerative diseases, diabetes, cancer, arthritis etc. has been found by various experimental studies and clinical trials. A useful role in the human body is played by selenium yet excess of selenium is harmful to the human body. Excess selenium can reach our food and water from farmlands and industries. Selenium in high amounts can be harmful to the body in a variety of ways. Selenium can produce cancer, neurological complications, mottling of teeth, skin rashes, diarrhea, irritability, lesions of the skin, nausea, garlic breath and abnormalities of the nervous system etc. Thus, normal amounts of selenium are very essential for the body but high amounts lead to damaging effects in the body comprising of the dark side of selenium. In depth analysis of selenium in future research is essential to establish its beneficial and harmful effects accurately so that its use can benefit human health in a significant way.

References

Agarwal, P., Sharma, S., Agarwal, U.S., 2016. Selenium toxicity: a rare diagnosis. Indian J. Dermatol. Venereol. Leprol 82, 690–693.

Alsafwah, S., Laguardia, S.P., Arroyo, M., Dockery, B., Bhattacharya, S.K., Ahokas, R.A., Newman, K.P., 2007. Congestive heart failure is a systemic illness: arole for minerals and micronutrients. Clin. Med. Res. 5, 238–243.

Anbudhasan, P., Surendraraj, A., Karkuzhali, S., Sathishkumaran, S., 2014. Natural antioxidants and its benefits. Int. J. Food. Nutr. Sci. 3 (6), 225–232.

Bahorun, T., Soobrattee, M.A., Luximon-Ramma, V., Aruoma, O.I., 2006. Free radicals and antioxidants in cardiovascular health and disease. Internet J. Med. Update **1**, 1–17.

Barceloux, D.G., 1999. Selenium. J. Toxicol. Clin. Toxicol. 37 (2), 145–172.

Barciela, J., Herrero, C., García, M.S., Pena, R.M., 2008. A brief study of the role of selenium as antioxidant. 7 (8), 3151–3155.

De Bem, A.F., Farina, M., Portella, R.D., Nogueira, C.W., Dinis, T.C., Laranjinha, J.A., Almeida, L.M., Rocha, J.B., 2008. Diphenyl diselenide, a simple glutathione peroxidase mimetic, inhibits human LDL oxidation *in-vitro*. Atherosclerosis 201, 92–100.

Dhingra, S., Bansal, M.P., 2005. Hypercholesterolemia and apolipoprotein B expression: regulation by selenium status. Lipids Health Dis. 4, 28–41.

Droge, W., 2002. Free radicals in the physiological control of cell function. Review. Physiol Rev. 82, 47–95.

Genestra, M., 2007. Oxyl radicals, redox-sensitive signalling cascades and antioxidants. Review. Cell Signal. 19, 1807–1819.

Halliwell, B., Gutteridge, J.M.C., 2007. Free Radicals in Biology and Medicine, 4th ed. Clarendon Press, Oxford, UK.

Halliwell, B., 2007. Biochemistry of oxidative stress. Biochem. Soc. Trans. 35, 1147–1150.

Helgeland, M., Svendsen, E., Førre, O., Haugen, M., 2000. Dietary intake and serum concentrations of antioxidants in children with juvenile arthritis. Clin. Exp. Rheumatol. 18, 637–641.

Kise, Y., Yoshimura, S., Akieda, K., 2004. Acute oral selenium intoxication with ten times the lethal dose resulting in deep gastric ulcer. J. Emerg. Med. 26, 183–187.

Koppel, C., Baudisch, H., Veter, K.H., 1986. Fatal poisoning with selenium dioxide. J. Toxicol. Clin. Toxicol. 24, 21–35.

Kose, K., Dogan, P., Kardas, Y., 1996. Saraymen, R. Plasma selenium levels in rheumatoid arthritis. Biol. Trace Elem. Res. 53, 51–56.

Kuila, A., Mukhopadhyay, M., Banerjee, R., 2012. Selenium: Beneficial and toxic effect on Human. Human and Animal Health: Environmental Perspectives. In: Garg, S.R. (Ed.), Satish Serial Publishing House, India, pp. 111–120.

Lubos, E., Sinning, C.R., Schnabel, R.B., Wild, P.S., Zeller, T., Rupprecht, H.J., Bickel, C., Lackner, K.J., Peetz, D., Loscalzo, J., Münzel, T., Blankenberg, S., 2010. Serum selenium and prognosis in cardiovascular disease: results from the AtheroGene study. Atherosclerosis 209, 271–277.

Maxwell, S.R., 1999. Antioxidant vitamin supplements: update of their potential benefits and possible risks. Drug Saf. 21 (4), 253–266.

Morris, J.S., Crane, S.B., 2013. Selenium toxicity from a misformulated dietary supplement, adverse health effects, and the temporal response in the nail biologic monitor. Nutrients 5, 1024–1057.

Aruoma, Okezie I., 1998. Free radicals, oxidative stress, and antioxidantsin human health and disease. J. Am. Oil Chem. Soc. 75 (2), 199–212.

Onal, S., Naziroglu, M., Colak, M., Bulut, V., Flores-Arce, M.F., 2011. Effects of different medical treatments on serum copper, selenium and zinc levels in patients with rheumatoid arthritis. Biol. Trace Elem. Res. 142 (3), 447–455.

Pacher, P., Beckman, J.S., Liaudet, L., 2007. Nitric oxide and peroxynitrite in health and disease. Physiol. Rev. 87, 315–424.

Pham-Huy, L.A., He, H., Pham-Huy, C., 2008. Free radicals, antioxidants in disease and health. Int. J. Biomed. Sci. 4, 89–96.

Ragozzino, G., Tanzilli, V., Cristiano, N.M.V., Guarino, G., Mattera, E., 2017. Selenium: properties and clinical applications. a systematic review. Int. J. Clin. Endocrinol. 1 (1), 8–17.

Sanmartín, C., Plano, D., Font, M., Palop, J.A., 2011. Selenium and clinical trials: new therapeutic evidence for multiple diseases. Curr. Med. Chem. 18, 4635–4650.

Savitha, P., 2014. Role of selenium. J. Pharm. Sci. Res. 6 (1), 56–59.

Saxena, M., Saxena, J., Pradhan, A., 2012. Flavonoids and phenolic acids as antioxidants in plants and human health. Int. J. Pharm. Sci. Rev. Res. 16 (2), 130–134.

See, K.A., Lavercombe, P.A., Dillon, J., 2006. Accidental death from acute selenium poisoning. Med. J. Aust. 185, 388–389.

Sheikh, A.S., Mian, Z.S., Urfy, Talaha, M.A., Farhad, W.K., Anwar-ul-Hassan, G., 2005. Antioxidants: their role in health and disease. Int. J. Pharmacol. 1, 226–233.

Sladan, Z., Pavlovic, B.I., Radoslav, Ognjanovic, Andras, V., ikic, Stajn, S., Saicicand, Z.S., Vojislav, M.P., 2001. The effect of selenium on antioxidant defense system in the blood of rats chronically treated with cadmium. Kragujevac J. Sci. 23, 105–114.

Spiller, H.A., 2007. Two fatal cases of selenium toxicity. Forensic Sci. Int. 171, 67–72.

Sun, H.J., Rathinasabapathi, B., Wu, B., Luo, J, Pu, L.P., Ma, L.Q., 2014. Arsenic and selenium toxicity and their interactive effects in humans. Environ. Int. 69, 148–158.

Tapiero, H., Townsend, D.M., Tew, K.D., 2003. The antioxidant role of selenium and seleno-compounds. Biomed. Pharmacother. 57, 134–144.

Tinggi, U., 2008. Selenium: its role as antioxidant in human health. Environ. Health Prev. Med. 13, 102–108.

Valko, M., Izakovic, M., Mazur, M., Rhodes, C.J., Telser, J., 2004. Role of oxygen radicals in DNA damage and cancer incidence. Mol. Cell. Biochem. 266, 37–56.

Valko, M., Leibfritz, D., Moncola, J., Cronin, M.D., 2007. Free radicals and antioxidants in normal physiological functions and human disease. Review. Int. J. Biochem. Cell Biol. 39, 44–84.

Willcox, J.K., Ash, S.L., Catignani, G.L., 2004. Antioxidants and prevention of chronic disease. Review. Crit. Rev. Food Sci. Nutr. 44, 275–295.

Wolters, M., Hermann, S., Golf, S., Katz, N., Hahn, A., 2006. Selenium and antioxidant vitamin status of elderly German women. Eur. J. Clin. Nutr. 60 (1), 85–91.

Xun, P., Liu, K., Morris, J.S., Daviglus, M.L., He, K., 2010. Longitudinal association between toenail selenium levels and measures of subclinical atherosclerosis: the CARDIA trace element study. Atherosclerosis 210, 662–667.

Young, I., Woodside, J., 2001. Antioxidants in health and disease. J. Clin. Pathol 54, 176–186.

CHAPTER

Sterols: benificial or detrimental for human helath 4.16

Tanuj Joshi[a], Kiran Patni[b], Tarun Belwal[c], Harikesh Maurya[d], Aadesh Kumar[a]

[a]*Department of Pharmaceutical Sciences, Kumaun University, Campus Bhimtal, Bhimtal, Uttarakhand, India*

[b]*Graphic Era Hill University Bhimtal, Nainital, Uttarakhand, India*

[c]*College of Biosystem Engineering and Food Science, Zhejiang Key Laboratory for Agri-Food Processing, Zhejiang University, Hangzhou, People's Republic of China*

[d]*M.G.B. Rajat College of Pharmacy. Gohila, Hanswar, Ambedkar Nagar, U.P., India*

4.16.1 Sterols and their role as antioxidants

Oxidative processes taking place inside the human body result in formation of reactive oxygen species. In normal circumstances formation of free radicals is not harmful but excess formation of free radicals through chain reactions can lead to harmful consequences inside the body. Reactive oxygen species and reactive nitrogen species are scavenged by endogenous as well as externally supplied antioxidants. Antioxidants themselves get oxidized and save the body from harmful consequences of free radicals and oxidants. Thus, supplementation with antioxidants can protect the body from harmful consequences of environmental stress, physiological stress, atherosclerosis, aging and cancer. Antioxidants are of several types. Superoxide dismutase, catalase, cofactor Q10, glutathione peroxidise, vitamin C, vitamin E, vitamin A, minerals and peptides are some endogenous antioxidants found inside the body. Plants also serve as an important source of natural antioxidants. Various parts of the plants like seeds, leaves, fruits, roots, and bark have important phytoconstituents which have antioxidant activity (Anwar et al., 2018). The antioxidants obtained from vegetable sources are generally polyphenolic in nature. Broccoli, spinach, blueberries, strawberries, orange, lemons, citrus fruits, red beans, prunes, plums, spinach, kale, alfa-alfa sprouts, etc., are some sources of natural antioxidants. Gingerol, gallic acid, curcumin, flavonoids, ellagic acid, vitamin C, vitamin E, quercetin, p-coumaric acid are important examples of phytochemicals with antioxidant potential. *Allium sativum* (garlic), *Curcuma longa* (turmeric), *Eugenia caryophyllus* (clove), etc., are some important medicinal plants with tremendous antioxidant potential. Antioxidants from marine sources are also nowadays attracting significant interest. Marine biotechnologists are making a lot of effort towards the utilization of marine diversity for making useful pharmaceutical products. Sponges and seaweeds have very potent antioxidant and antimicrobial activities. Sponges and seaweeds with their antioxidant potential and the various

Antioxidants Effects in Health. DOI: https://doi.org/10.1016/B978-0-12-819096-8.00018-5

beneficial bacteria's contained in them can provide numerous benefits to human health. Polysaccharides, phenolic acids, and organic acids are the important constituents of seaweeds and sponges that are responsible for their various bioactivities (Anwar et al., 2018). These constituents also provide significant protection against environment pollution and radiation. Various types of algae found in marine environment have been studied in detail for their bioactive constituents so that useful super food can be manufactured from these constituents. Algae grow in a very harsh environment which is full of light, salinity and temperature. Thus algae have developed various metabolites to cope with these conditions. Carotenoids, vitamins, phycobilins, and polyphenols are some of the constituents of algae with antioxidant activities. People have used different types of seaweeds in fresh as well as dried form as part of their diet. These seaweeds are abundant in minerals, vitamins and proteins. Various constituents of algae have anticancer activities. Seaweeds are gaining considerable attention for their antioxidant activity. Bacteria's present in the algae protect the algae from pathogens by producing various bioactive compounds. The exopolysaccharides produced from these bacteria's are used in various pharmaceutical preparations and food products. Alaska pollock, tuna, mackerel, and yellowfin sole provide important fish protein hydrosylate. These fish protein hydrosylate possess significant antioxidant activity. Various peptides are obtained from fish muscle, skin, bone and other tissues. The amino acids found in fish can show significant antioxidant activity. The most prominent among these are cystine, methionine, tyrosine, tryptophan, and phenylalanine. Phycocyanin a pigment that is obtained from blue-green algae possesses antioxidant activity. Antioxidants obtained from algae are also used in cosmeceutical industries. They are used in antiaging preparations (Anwar et al., 2018).

Sterols are distributed throughout the world in different life forms. They play diverse role in various organisms (Galea and Brown, 2009). In vertebrates the major sterol is cholesterol, in yeast and fungus it is ergosterol and in plants sterols occur in the form of phytosterols (Galea and Brown, 2009). Sterols are an essential constituent of the membrane of all eukaryotic organisms. Control of permeability and fluidity of membrane is an important function ascribed to sterols. Modulation produced in the functions of enzymes bound to membranes is an important role played by sterols in plants. Sterols also play an important role in cell proliferation and signal transduction in plants. Besides animals and plants, sterols are also distributed in microorganisms (Volkman, 2003).

The sterols that are gaining special attention nowadays are phytosterols. Phytosterols include both plant derived sterols and stanols (Trautwein and Demonty, 2007). Sterols are unsaturated compounds and have double blond in their sterol ring and stanols are saturated compounds and lack double bond in their sterol ring. Campesterols and sitosterols form the most abundant plant sterols. Stanols form only around 10% of the phytosterols found in the diet (Ogbe et al., 2015). In biological function and structure they resemble cholesterol. Greater than thirty enzymatic reactions that occur in the plant cell membrane are held responsible for the synthesis of phytosterols. More than 200 compounds found in various marine sources and plants are termed as phytosterols. Stanols are saturated forms of plant sterols. They are different from sterols due to the fact that they lack alkyl side chain and presence of double bond in the steroid nucleus. Some examples of phytosterols are campesterol, brassicasterol, sitosterol, stigmasterol, etc. Campestanol, sitostanol are important plant stanols. Sterols can be obtained from different dietary sources. Vegetable oil can be considered as a good source of sterols. Products produced from vegetable oil like spreads and margarine contains plant sterols. Legumes, cereal grains, other products based on cereals, vegetables, fruits and nuts are also important sources of plant sterols. Non hydrogenated oils obtained from vegetables, corn, rye wheat are sources of plant stanols. Spruce and pine also contain stanols. The dietary intake of plant sterols range from 150 to 400 mg/day and that of plant stanols is around 25 mg/day (Trautwein and Demonty, 2007). There are various important pharmacological activities associated with phytosterols. The most important properties associated with phytosterols are their property to lower the levels of cholesterol through partial inhibition of cholesterol absorption from intestine. Stimulation of immune system, anti-inflammatory activity and antiatherogenic activities are some other activities that have been associated with phytosterols (Trautwein and Demonty, 2007). Also sterols possess antibacterial, antiulcerative, and antioxidative activities. Prevention of benign prostatic hyperplasia is another role that has been assigned to phytosterols. Phytosterols also have beneficial effects on various types of cancer like colon, prostate, stomach, lung, breast and ovarian. Phytosterols demonstrate their anticancer activities by inhibiting cancer cell growth, inhibition of production of carcinogens, inhibition of metastasis, invasion of cancer cells and enhancement of apoptosis of cells that have become cancerous (Azadmard-Damirchi and Nasirpour-Tabrizi, 2017). β-sitosterol has been studied extensively and has shown anti-inflammatory, antipyretic, immunomodulating, and antineoplastic activities (Cherif, 2012). Products containing phytosterols have been marketed since the nineties for their health benefits. Phytosterols are not only used for their therapeutic properties but they also serve as important precursor molecules for synthesis of important substances. Ergosterol is used as a precursor for vitamin D2. Also it is involved in the production of flavones and cortisone. Since the demand of phytosterols are rising day per day. Researchers are exploring for newer and high yielding sources of phytosterols (Luo et al., 2015).

4.16.2 Bright side of sterols

Sterols are essential components of human diet and are required for proper functioning of the body. Antioxidant, Anti-inflammatory, anti-ulcerative and antibacterial activities are some important properties of sterols. There are numerous benefits of sterol on human health and it comprises of the bright side of sterols. The various benefits of sterols are supported by *in-vitro*, *in-vivo* and clinical trials.

4.16.2.1 Beneficial effect of sterols as evidenced by *in vitro* and animals studies

Three different trials were conducted by a group of scientists using different animals to study the effect of phytosterol on atherosclerosis. The animals used in the study were knockout mice, chicken, rabbit and hamsters. The results of these studies showed that phytosterols were beneficial in decreasing the accumulation of lipids in the arteries. This also led to a decrease in the development of atherosclerosis. Decrease in size of lesion or plaques, inhibition of progression and formation of lesions and also regression of the existing lesions were consequences of phytosterol administration to the animals used in the study. In an *in vitro* study it was found that phytosterols showed good activity against atherosclerosis. Isolation of vascular smooth muscle cells (VSMC) from rats was performed in this study. Prostacyclin were found to be released from VSMC under the influence of phytosterols and this suggests that phytosterols have antihyperproliferative potential that could be of benefit in atherosclerosis. It was found that phytosterols by decreasing the release of prostaglandins produced vasodilation and an antiaggregatory effect beneficial in atherosclerosis. Antihypercholesterolemic and antioxidant effect was shown by methanolic extract (rich in sterol) of stem of *Musa sapientum* (banana) in Wistar rats fed with cholesterol. It was found in the study that the extract showed significant antioxidant and antihypercholesterolemic activity (Trautwein and Demonty, 2007; Dikshit et al., 2016). Sterols have also demonstrated potent anti-inflammatory activities. They have shown to inhibit the release of markers of inflammation like tumor necrosis factor alpha and interleukin-6 from monocytes. The inflammation of the lungs related to eosinophil and leukocytosis infiltration in asthmatic mice was also decreased by intraperitoneal administration of beta-sitosterol. In a murine model of inflammation, topical application and oral consumption of a mixture of phytosterols mainly consisting of beta-sitosterol was shown to decrease or inhibit oedema. However many other animal models do not support the claim of beta-sitosterol in preventing and reducing inflammation. Anticancer activities of phytosterols have also been demonstrated in various animal studies. It has been found in these studies that administration of beta-sitosterol or a mixture consisting of phytosterols to animals who were administered carcinogenic stimuli/substances and those who were injected or implanted with cancer cells decreased the occurrence of tumours in these animals. Also phytosterols decreased the number of metastasis of prostate, colon or breast cancer and slowed proliferation of cells (Trautwein and Demonty, 2007). The anticancer activity of phytosterols can be due to various mechanisms. Sterols can promote apoptosis of cells; they inhibit cell cycle progression; down regulate cholesterol synthesis; inhibit adhesion, invasion and migration of cells and stimulate functions of immune system. Estrogenic activity of phytosterols might also account for its anticancer effect but the data on this is not very firm. In another study phytosterols were studied for their anti-inflammatory potential. The animals used in the study were male balb/c mice. Colitis model was used in the study to evaluate the anti-inflammatory potential of phytosterols. Colitis was induced in the animals by administration of dextran sodium sulphate. Phytosterol treatment reduced inflammation induced by dextran sodium sulphate in mice. Phytosterols produced their effects by decreasing inflammatory cell infiltration and accelerating mucosal healing. Antioxidant effect might be responsible for the anti-inflammatory activity of phytosterols. Also a regulatory effect on the microflora in the intestine is responsible for their action (Aldini et al., 2014). In apo E-KO mice phytosterols have shown antiatherogenic effects. In the presence or absence of 0.15% w/w cholesterol in diet, phytosterols have reduced the lesions of atherosclerosis by 50%. The antiatherogenic potential is co-relatable to the cholesterol decreasing effect of phytosterols. Not only phytosterols obtained from plants lower the size of atherosclerotic lesions but they also decrease various other components of the atherosclerotic lesions. They decrease cholesterol clefts, foam cells, extracellular matrix and smooth muscle cells that show proliferation. Besides cholesterol lowering actions, phytosterols may produce anti-atherogenic action by actions on antioxidant system, coagulation system and lipoprotein lipase and hepatic activities. *In vitro* cholesterol lowering actions of phytosterols have also been investigated. Incubation of a colon tumor cell line with β-sitosterol resulted in decrease in cholesterol synthesis, cholesterol uptake from the incubation medium, HMG-CoA activity and levels of mRNA. Antixanthomatosis effect of phytosterols found in diet has also been observed in apo E-KO mice. Phytosterols also show numerous anti-metabolic effects other than that mentioned above. Red blood cells have shown increased resistance to osmotic haemolysis in E-KO mice in comparison to the control group when these animals were administered phytosterols. *In vitro* red blood cells are known to take phytosterols. The administration of phytosterols altered the composition of the membrane of red blood cells and thus reduced the susceptibility of these cells towards osmotic haemolysis (Moghadasian, 2000).

Animal studies have also demonstrated that phytosterols are effective in Alzheimer's diseases. Mice that have been fed with diet containing stigmasterol have shown beneficial effect in Alzheimer's disease. Also combination of phytosterols mainly containing stigmasterol can prove to be of tremendous value in Alzheimer's disease (Burg et al., 2013). In a study anti-HIV (human immunodeficiency virus) activity of different extracts of *Aerva lanata* roots and isolated phytosterol from its chloroform extract was evaluated. Different extracts were prepared of the above plant using solvents like chloroform, ethylacetate, acetone, hexane, and methanol. All extracts as well as the isolated active compound showed significant inhibitory activity against the HIV reverse transcriptase enzyme. Among all the extracts chloroform extract showed the highest inhibition of HIV reverse transcriptase. Thus, inhibitory activity of phytosterols against HIV can lead to the development of drugs that might cure AIDS. Also, beneficial compounds can be developed for cure of other microbial infections (Gujjeti et al., 2017).

4.16.2.2 Beneficial effect of phytosterols as evidenced by clinical studies

The effects of phytosterols on human health have been assessed by various clinical studies. Effect of administration of a beverage enriched by sterol and containing milk fat globule membrane was observed on cytokines and serum lipid profile in post menopausal women. There was a decrease in total and LDL (low density lipoprotein) cholesterol in post menopausal women who consumed the beverage. Also markers of phytosterol intake like campesterol, β sitosterol and precursors of cholesterol synthesis like lanosterol decreased significantly. The intake of the beverage produced an increase in anti-inflammatory cytokines levels like IL-10 whereas a decrease was produced in pro-inflammatory cytokine IL-1β levels. Thus the intake of the beverage by postmenopausal women showed that it can provide beneficial effects on the cardiovascular and anti-inflammatory status of these women (Alvarez-Sala et al., 2018). Plant stanols and plant sterols were studied for their effects on markers of oxidative stress, antioxidant status, endothelial dysfunction, low grade inflammation and lipid metabolism in participants on stable statin therapy. Sixteen weeks of treatment with stanols and sterols in participant who were already on statin produced no significant alterations in markers of antioxidant status, oxidative stress, endothelial dysfunction and low grade inflammation. Also no significant effect was produced on C-reactive protein, soluble adhesion molecules and monocyte chemotactic protein-1 concentrations. However significant reduction was found in the levels of LDL cholesterol in these patients (Dejong et al., 2008). In another study a non fat orange juice fortified with plant sterols was studied for evaluating its cholesterol lowering action in healthy individuals with mild hypercholesterolemia. Administration of this orange juice significantly lowered LDL, non-high density lipoprotein and total cholesterol in comparison to placebo group and to baseline levels. The orange juice also significantly reduced the levels of apolipoprotein B. The levels of HDL cholesterol, triglycerides and homocyteine were found to be unchanged. Folate and levels of vitamin B12 were found to be increased significantly (Devaraj et al., 2004).

4.16.3 Dark side of sterols: Adverse effects and toxicity

Though sterols provide numerous benefits to human health still they can cause various types of adverse effects in humans. Several studies have thrown light on the adverse effects of sterols. High amounts of phytosterols in the erythrocytic membrane can increase the fragility of erythrocytes. Phytosterolemia can result in increased haemolysis in patients suffering from this disease. It was also found that rat liver microsomes enriched with campesterol and β-sitosterol can lead to increased rigidity of the microsomal membrane. *In vitro* experiments have found that high levels of β-sitosterol can lead to contraction of endothelial cells of human umbilical cord vein. Also cytotoxicity and interference with cellular functions have been observed with very high plasma levels of β-sitosterol. Reproductive organ toxicity has been related to elevated levels of phytosterols in plasma of laboratory animals. The sperm count and weight of testes were decreased in albino rats by subcutaneous administration of 0.5 to 5 mg/kg body weight of β-sitosterol per day. It was also seen in an experiment that when sitosteryl ester was applied to the vagina of rabbits their pregnancy rate was found to be decreased. Also studies in goldfish have demonstrated a decrease in the levels of 17β estradiol and testosterone by intraperitoneal administration of β-sitosterol in female and male goldfish respectively. High doses of stanol ester (5%) in diet for 13 weeks have led to a decrease in the plasma levels of vitamin E, vitamin D, vitamin K, and total protein in rats (Moghadasian, 2000). Sitosterolemia or phytosterolemia is a genetic disease that occurs in people who receive mutation in both the copies of ABC G5 or ABC G8 gene. In this disease people have an elevated level of phytosterols due to enhanced intestinal absorption and reduced biliary excretion of phytosterols. Even though the amount of cholesterol in the plasma may be normal or only slightly elevated still people suffering from this disease are at risk of developing premature atherosclerosis. One solution to this problem can be that people with this disease should avoid foods rich in phytosterols. In humans phytosterols have been known to produce constipation, nausea, indigestion, diarrhoea etc. Food supplements containing plant sterols and plant stanols are not recommended for use in pregnancy since there is no sufficient data available for their safety profile. However there is also not enough data available that consumption of high amounts of phytosterols by vegetarian women could be harmful in pregnancy and lactation. There can also occur an interaction between phytosterols and HMG CoA (3-hydroxy-3-methyl-glutaryl-coenzyme A) reductase inhibitors (Statins). The interaction will occur in the form that an additional effect will be produced on the the lowering of LDL cholesterol by phytosterols when given with statins. Controlled clinical studies have shown that consumption of plant sterols and stanols in an amount of 2–3 g/day by patients on statin therapy can lead to decrease in LDL cholesterol by 7%–11%. This interaction between phytosterol and statins can produce an effect similar to double dose of statin. Consumption of foods that are rich in plant sterols and stanols on a long term basis has led to the reduction in plant carotenoids by 10%–20%. Thus it is recommended that the use of plant sterols and stanols should not be done in amounts exceeding 3 g/day so that humans and animals can benefit from them without the incidences of adverse effects (Ogbe et al., 2015). Phytosterols though have many health benefits still they too have a dark side which should be explored thoroughly if humans want to use them wisely without the incidences of toxicity and adverse reactions. Table 4.16.1 shows the bright side as well as the dark side of phytosterols.

Table 4.16.1 Bright and dark aspects of phytosterols on human health.

Bright side of phytosterols	Dark side of phytosterols
Antioxidant	Hemolysis
Anti-inflammatory	Reproductive system toxicity
Antiulcerative	Carcinogenic potential
Antiatherogenic	Potential to decrease levels of vitamins and carotenoids
Antibacterial	Phytosterolemia
Antihypercholesterolemic	Interaction with other drugs specially lipid lowering agents
Anticancer	Lack of teratogenic data

Conclusion

Sterols are beneficial for humans in a variety of ways. Plant sterols and stanols also provide numerous health benefits like they have anticancer, antioxidant and antiatherogenic potentials. Also consumption of fruits and vegetables containing phytosterols is essential for the body because of the various functions of phytosterols in human body. Still lot of research is required on humans to clearly denote what benefits sterols and phytosterols can play in the human body. Phytosterols have a bright side on human health but still we have to look at the dark side of the sterols. Sterols can be used safely for human health only if extensive research is performed on their benefits as well as adverse effects and toxicity in future.

References

Aldini, R., Micucci, M., Cevenini, M., Fato, R., Bergamini, C., Nanni, C., Cont, M., Camborata, C., Spinozzi, S., Montagnani, M., Roda, G., D'Errico-Grigioni, A., Rosini, F., Roda, A., Mazzella, G., Chiarini, A., Budriesi, R., 2014. Antiinflammatory effect of phytosterols in experimental murine colitis model: Prevention, induction, remission study. PLoS One 9, e108112.

Alvarez-Sala, A., Blanco-Morales, V., Cilla, A., Silvestre, R.Á., Hernández-Álvarez, E., Granado-Lorencio, F., Barberá, R., Garcia-Llatas, G., 2018. A positive impact on the serum lipid profile and cytokines after the consumption of a plant sterol-enriched beverage with a milk fat globule membrane: a clinical study. Food Funct. 95, 209–5219.

Anwar, H., Hussain, G., Mustafa, I., 2018. Antioxidants from natural sources. In: Shalaby, E., Azzam, G.M. (Eds.), Antioxidants in food and its application. Intech Open, United Kingdom, pp. 3–28.

Azadmard-Damirchi, S., Nasirpour-Tabrizi, P., 2017. Phytosterols and phytostanols effect on health. EC Nutr. 12 (1), 01–02.

Burg, V.K., Grimm, H.S., Rothhaar, T.L., Grösgen, S., Hundsdörfer, B., Haupenthal, V.J., Zimmer, V.C., Mett, J., Weingärtner, O., Laufs, U., Broersen, L.M., Tanila, H., Vanmierlo, T., Lütjohann, D., Hartmann, T., Grimm, M.O., 2013. Plant sterols the better cholesterol in Alzheimer's disease? A mechanistical study. J. Neurosci 33, 16072–16087.

Cherif, A.O., 2012. Phytochemicals Components as Bioactive Foods. In: Rasooli, I. (Ed.), Bioactive Compounds in Phytomedicine. Intech, Crotia, pp. 113–124.

De Jong, A., Plat, J., Bast, A., Godschalk, R.W.L., Basu, S., Mensink, R.P., 2008. Effects of plant sterol and stanol ester consumption on lipid metabolism, antioxidant status and markers of oxidative stress, endothelial function and low-grade inflammation in patients on current statin treatment. Eur. J. Clin. Nutr. 62, 263–273.

Devaraj, S., Jialal, I., Vega-López, S., 2004. Plant sterol-fortified orange juice effectively lowers cholesterol levels in mildly hypercholesterolemic healthy individuals. Arterioscler. Thromb. Vasc. Biol. 24, e25–e28.

Dikshit, P., Tyagi, M.K., Shukla, K., Gambhir, J.K., Shukla, R., 2016. Antihypercholesterolemic and antioxidant effect of sterol rich methanol extract of stem of Musa sapientum (banana) in cholesterol fed wistar rats. J. Food Sci. Technol. 53, 1690–1697.

Galea, A.M., Brown, A.J., 2009. Special relationship between sterols and oxygen: were sterols an adaptation to aerobic life? Free Radic. Biol. Med. 47, 880–889.

Gujjeti, R.P., Mamidala, E., 2017. Anti-HIV activity of phytosterol isolated from *Aerva lanata* roots. Pharmacogn. J. 9, 112–116.

Luo, X., Su, P., Zhang, W., 2015. Advances in microalgae-derived phytosterols for functional food and pharmaceutical applications. Mar. Drugs 13, 4231–4254.

Moghadasian, M.H., 2000. Pharmacological properties of plant sterols in vivo and in vitro observations. Life Sci. 67, 605–615.

Ogbe, R.J., Ochalefu, D.O., Mafulul, S.G., Olaniru, O.B., 2015. A review on dietary phytosterols: their occurrence, metabolism and health benefits. Asian J. Plant Sci. Res. 5, 10–21.

Trautwein, E.A., Demonty, I., 2007. Phytosterols: natural compounds with established and emerging health benefits. OCL 14, 259–266.

Volkman, J., 2003. Sterols in microorganism. Appl. Microbiol. Biotechnol. 60, 495–506.

CHAPTER

Tartaric acid

4.17

Arvind Jantwal[a], Sumit Durgapal[a], Jyoti Upadhyay[b], Tanuj Joshi[a], Aadesh Kumar[a]

[a]*Department of Pharmaceutical Sciences, Kumaun University, Campus Bhimtal, Bhimtal, Uttarakhand, India*

[b]*School of Health Science and Technology, Department of Pharmaceutical Sciences, University of Petroleum and Energy Studies, Dehradun, Uttarakhand, India*

4.17.1 Introduction

In the recent past the interest of the human civilization has shifted towards the use of antioxidants in food and as drugs for maintaining the normal wellbeing, due to the fact that these biologically active phytochemicals possess the potential to reduce oxidative stress levels. A large number of studies suggest that intake of antioxidant rich food reduces the oxidative stress moderately and helps in the prevention of a large number of degenerative conditions. Protection against oxidation can be carried out by the use of compounds that inhibit oxidation; these compounds are called oxidation inhibitors commonly known as antioxidants.

Antioxidants represent a very wide variety of chemical classes and varied mechanisms of action. The mechanism basically involves reaction with free radicals (peroxy/alkoxy) that are released by decomposition of lipid hydroperoxides or prevent lipid hydroperoxides from decomposing and forming free radicals. Oxidation inhibitory activity of an antioxidant compound depends on a number of factors such as the lipid composition, concentration, temperature, pressure of oxygen, presence of other antioxidants, and other food compounds (Anwar et al., 2018).

Antioxidants can be either natural or synthetic compounds. Commonly used natural antioxidants are readily available in food, especially fruits and vegetables.

4.17.2 Classification of natural antioxidants

Antioxidants obtained from Natural sources can be identified as following classes:

1. Enzymatic
   - Primary
   - Secondary

Antioxidants Effects in Health. DOI: https://doi.org/10.1016/B978-0-12-819096-8.00019-7

2. Nonenzymatic
   - **Minerals** → for example, iodine, zinc
   - **Vitamins** → for example, vit A, vit C, vit E
   - **Carotenoids**
   - **Polyphenols**
     - Flavonoids → for example, flavonols, flavones
     - Gingerols
     - Polyphenolic acids
       - Hydroxycinnamic acid → for example, coumaric acid
       - Hydrobenzoic acid → for example, gallic acid
     - Curcumin
   - **Others** → for example, bilirubin, albumin

Table 4.17.1 shows the class of antioxidant and possible mode of action.

Tartaric acid occurs naturally in a wide diversity of plants and is produced by microorganisms. It is most abundant in tamarind, grapes, banana, apple, and some citrus fruits. TA is a white crystalline powder with melting point 171°C possessing acidic taste. TA is readily soluble in water and alcohol bur sparingly soluble in ether and shows optical isomerism. Fig 4.17.1 shows the structure of tartaric acid.

Tartaric acid readily oxidizes to tartonic acid when it is oxidized. When oxidized in presence of HNO_3, tartaric acid converts to oxalic acid in another oxidation reaction. TA oxidizes to dihydroxymaleic acid in presence of Fenton reagent (Fe^{3+}/H_2O_2) (Arun Bahal and B.S. Bahal). This property of readily oxidizing to dihydroxymaleic, oxalic acid and tartonic acid indicates that TA has good antioxidant property.

TA is one of the most abundant phytochemical present in *Tamarindus indica* commonly known as tamarind. It is the main bioactive component of tamarind pulp. On testing aqueous and alcoholic extract of tamarind fruit pulp it was observed that both the extracts exhibited excellent antioxidant capacity (Ugwuona and Onweluzo, 2013). In another study from 2013, it was reported that organic acids in tamarind,

Table 4.17.1 Possible mechanisms of antioxidant activity (Pokorný, 2001).

SN	Antioxidant class	Mechanism of action	Example
1	Proper antioxidants	Inactivating lipid free radicals	Phenolic compounds
2	Hydroperoxide stabilizers	Preventing decomposition of hydroperoxides into free radicals	Phenolic compounds
3	Synergists	Promoting activity of proper antioxidants	Ascorbic acid
4	Metal chelators	Binding heavy metals into inactive compounds	Phosphoric acid
5	Singlet oxygen quenchers	Transforming singlet oxygen	Carotenes

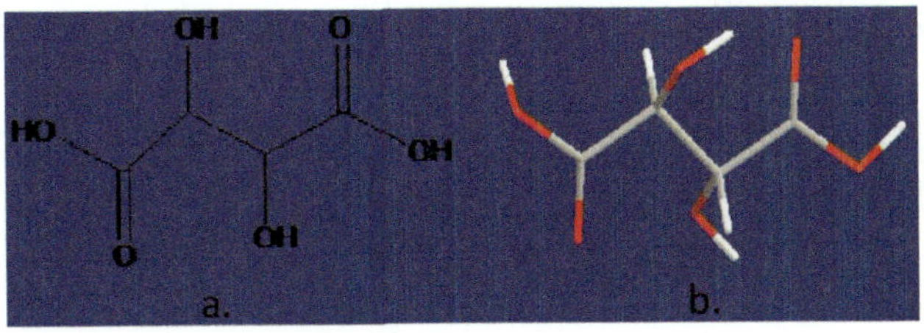

FIG. 4.17.1

Tartaric acid structure (A) 2D structure and (B) 3D structure.

including tartaric, malic, and citric acids, along with polyphenols and flavonoids are responsible for the strong antioxidant capacity of the tamarind fruit (Mbunde, 2013).

The phenolics present in wine are generally categorized in two groups (1) flavonoids and (2) nonflavonoids. Both of which are responsible for the antioxidant activity. Hydroxycinnamates arenon-flavonoids present in wine. They are present in form of esters of tartaric acid in grapes.

TA along with phosphoric, citric, malic, and ascorbic acids possesses pronounced chelating activities. These chelating agents have a very pronounced effect on increasing the oxidation stability by blocking the pro-oxidant metal ions, and henceforth limiting the formation of chain initiators by preventing metal-assisted homolysis of hydroperoxides (Yanishlieva-Maslarova, 2001).

4.17.3 Source of tartaric acid

Tartaric acid occurs naturally in a wide diversity of plants and microorganisms. It is most abundant in tamarind, grapes banana apple, *Gymnema sylvestre,* and some citrus fruits. Tamarind in India is basically used in form of tamarind pulp. TA is one of the most abundant phytochemical present in tamarind, *Tamarindus* sp., as it is the principle component of tamarind pulp. It is the constituent responsible for its acidic taste. TA occurs in both free and combined form. An estimated amount of about 50% of TA in the fruit pulp is in combined form as potassium bitartarate and as calcium tartarate (low amounts). TA content differed with variety of tamarind. Other constituents of tamarind include organic acids (malic acid [MA], lactic acid, oxalic acid, succinic acid, and citric acid) proteins, fats, fiber, carbohydrates, vitamins, and minerals. The detail of the chemical composition of tamarind is given in Table 4.17.2, and the difference between chemical composition between red and green varieties of tamarind is listed in Table 4.17.3.

Tamarind pulp is used mainly as a carminative laxative, it administered as an infusion in bilious disorders. It is used as gargle for sore throat. The pulp also possesses antiseptic properties and is therefore used as poultice for inflammatory swelling (The Wealth of India, 2009).

Vitis sp. (family Vitaceae) commonly known as grape is another major source for TA. There are more than 50 species of grapes in India including *V. indica*,

Table 4.17.2 Chemical composition of tamarind (The Wealth of India, 2009).

Phytoconstituent	Amount (per 100 g)
Moisture	20.9 g
Proteins	3.1 g
Fats	0.1 g
Carbohydrates	67.4 g
Minerals	2.9 g
Calcium	170 mg
Phosphorus	110 mg
Iron	10.9 mg
Riboflavin	0.07 mg
Niacin	0.7 mg
Vitamin C	3 mg

Table 4.17.3 Content (%) variation in red and green variety of tamarind (The Wealth of India, 2009).

Parameter	Red variety (%)	Green variety (%)
Moisture	20.1	18.2
Tartaric acid free	6.6	9.8
Tartaric acid combined	11.4	6.7
Invert sugar	36.4	38.2
Pectin	4.4	2.4

V. latifolia, *V. vinifera*. The fruit is characterized by the presence of large amounts of organic acids mainly TA and MA. These constitute about 90% of the total acids present in grapes. Other acids present are succinic, fumaric, caffeic, and glyceric acids. Citric acid is generally present in very low amounts. It has been noted that as the fruit matures, MA content decreases but TA content remains the same. TA occurs in free and combined form (as cream of tartar, i.e., potassium hydro tartrate), it is found in large amounts near the skin (sometimes in form of crystalline deposits) (The Wealth of India, 2009) There are a number of uses of *Vitis*. It is used as laxative and purgative, as a diuretic, an aphrodisiac, as an appetizer, for treating asthma. The fruit is also used for the treatment of jaundice and as hepatoprotective. Fruit extract has cardioprotective effect (The Wealth of India, 2009).

Banana (*Musa* sp.) is another fruit that is a rich source of TA. The principal organic acid present in banana is MA and TA. Other organic acids are citric acid, oxalic acid, boric acid. Some species have been reported to contain acetic acid and butyric acid (The Wealth of India, 2009).

4.17.4 Pharmacological activity of tartaric acid

Calculi formation is a problem which a number of people suffer with. It is the formation of brushite and calcium hydrogen phosphate dehydrate (CHPD) crystals that constitute the urinary stones. In an *in-vitro* study conducted to see the effect of TA of urinary type CHPD crystals, it was observed that CHPD crystal inhibition increased with increased concentrations of TA. Crystal size and number also decreased with increased concentrations of TA. Increased TA concentration also facilitated the dissolution of CHPD crystals (Joseph et al., 2005).

The content of TA in grapes is very high, this TA enters the large intestines where it is fermented by bacteria to produce short-chain fatty acids (SCFA), this increases the acidity of fecal matter, which in turn decreases the risk of colon cancer (Bell, 2011). In a similar study it was observed that consumption of TA (as cream-of-tartar) increased faecal weight, decreased intestinal transit time by eleven hrs, and there was softening of stool resulting in easier stool elimination. The lithocholic acid (LC) and deoxycholic acid (DC) ratio (LC:DC) was reduced. This reduction in LC:DC ratio is an indication of protective effect of TA against colo-rectal cancer (Spiller et al., 2003).

Effect of tartaric acid-induced cough on pulmonary function in normal and asthmatic individuals was conducted on 40 subjects (20 healthy and 20 asthmatic volunteers). For the reflex cough test (RCT), 20% TA solution in 0.15 M NaCl solution was inhaled as microaerosol. The results showed that there was no decrease in pulmonary function parameters after inhalation of tartaric acid in normal and asthmatic subjects. All the subjects had normal reflex response to RCT, consisting of laryngeal cough expiratory reflex (LCER) efforts followed by laryngeal cough reflex (LCR), without any adverse effects. This may have resulted due to CNS mediated bronchodilatation, due to TA induced cough (Addington et al., 2003).

4.17.5 Toxicity studies of tartaric acid

The acceptable daily intake (ADI) level of TA is 0–30 mg/kg body weight (Kamila Miková, 2001). In 1947, Fitzhugh demonstrated that TA was not toxic up to a concentration of 1.2% (Fitzhugh et al., 1947).

In a dental study conducted on 15 girls it was observed that the girls that had been indulged in mixing activities of TA, sucrose, sodium bicarbonate and magnesium sulfate product. The percent composition of dust is as follows: free tartaric acid:7.7%; Combined tartaric acid:22.5%; sodium bicarbonate:16.5%; magnesium sulfate:37.2%; sucrose:14.3%; insoluble residue:9.5%. It was observed that teeth of all girls except one were affected by the dust, which resulted in dental erosion (Elsbury et al., 1951). In 2011, it was reported metabolic acidosis due to TA, where renal functions are reduced (Emmett, 2011).

TA was identified as one of the major pain inducing toxin along with oxalic acid in *Urticathunbergiana* using Formaline test. In the HPLC analysis it was observed that formic acid was not responsible for the long lasting pain inducing toxin as it was not present in significant amount in the plant (Fu et al., 2006).

Toxicity studies of TA, diacetyl derivative and its esters (TEM) were carried out. It was observed that TA and its diacetyl derivative were toxic to dogs whereas large dose of TEM (27 gm/kg) was nontoxic. The rate of urinary excretion of TA and its diacetyl derivative is very high at nonlethal dose (0.5 gm/kg and 0.72 gm/kg, respectively). It was observed that nearly half of TA was eliminated within in urine (Sourkes and Koppanyi, 1950).

Conclusion

In the past decade or two a significant increase in the interest towards antioxidants has been observed. The masses are focusing on alternatives other than the synthetic sources that are easily available and convenient to consume in daily diet.

TA is an antioxidant that is readily available from fruits sources (with no pro-oxidant activity). Besides acting as an antioxidant agent, it is also a synergist to other antioxidants. Its consumption prevents kidney stone formation and it is an anti-cancer agent too. Tamarind, rich sources of TA, is used as carminative laxative, in bilious disorders and to treat sore throat. Although TA has a lot of potential and health benefits but as it is said that "too much of anything is not beneficial" the same goes for TA also. If it exceeds its ADI it causes a number of problems to the body like acidity, dental erosion, among others. Despite of all these pros and cons TA as an antioxidant has not yet been fully explored and an extensive study in this field is required.

References

Addington, W.R., Stephens, R.E., Widdicombe, J.G., Anderson, J.W., Rekab, K., 2003. Effect of tartaric acid–induced cough on pulmonary function in normal and asthmatic humans. Am. J. Phys. Med. Rehabil. 82 (5), 374–378.

Anwar, H., Hussain, G., Mustafa, I., 2018. Antioxidants from natural sources. Antioxidants in Foods and Its Applications, pp. 3–28. doi:10.5772/intechopen.75961.

Bell, S.J., 2011. A review of dietary fiber and health: focus on raisins. J. Med. Food 14 (9), 877–883.

Elsbury, W.B., Browne, R.C., Boyes, J., 1951. Erosion of teeth due to tartaric acid dust. Br. J. Ind. Med. 8 (3), 179.

Emmett, M., 2011. Recurrent tartaric acid acidosis? Kidney Int. 79 (2), 258–259.

Fitzhugh, O.G., Nelson, A.A., 1947. The comparative chronic toxicities of fumaric, tartaric, oxalic, and maleic acids. J. Am. Pharm. Assoc. 36 (7), 217–219.

Fu, H.Y., Chen, S.J., Chen, R.F., Ding, W.H., Kuo-Huang, L.L., Huang, RN., 2006. Identification of oxalic acid and tartaric acid as major persistent pain-inducing toxins in the stinging hairs of the nettle, Urticathunbergiana. Ann. Bot. 98 (1), 57–65.

Joseph, K.C., Parekh, B.B., Joshi, M.J., 2005. Inhibition of growth of urinary type calcium hydrogen phosphate dihydrate crystals by tartaric acid and tamarind. Curr. Sci. 88 (8), 1232–1238.

Mbunde, M., 2013. Quantification of phenolics, flavanoids and antioxidant activity of tamarindusindica L. from selected areas in Tanzania (Doctoral dissertation. Sokoine University of Agriculture.

Miková, K., 2001. The regulation of antioxidants in food. Antioxidant in Foods: Practical Applications. CRC Press, pp. 267–284.

Pokorný, J., 2001. Introduction. In: Pokorný, J., Yanishlieva, N., Gordon, M. (Eds.), Antioxidants in Food: Practical Applications. CRC Press, pp. 1–3.

Sourkes, T.L., Koppanyi, T., 1950. Correlation between the acute toxicity and the rate of elimination of tartaric acid and certain of its esters. J. Am. Pharm. Assoc. 39 (5), 275–276.

Spiller, G.A., Story, J.A., Furumoto, E.J., Chezem, J.C., Spiller, M., 2003. Effect of tartaric acid and dietary fibre from sun-dried raisins on colonic function and on bile acid and volatile fatty acid excretion in healthy adults. Br. J. Nutr. 90 (4), 803–807.

The Wealth of India, 2009. A dictionary of Indian raw materials and industrial products, Raw Materials, Vol X: Sp-W. CSIR, New Delhi, pp. 114–122.

The Wealth of India, 2009. A dictionary of Indian raw materials and industrial products, Raw Materials, Vol X: Sp-W. CSIR, New Delhi, pp. 527–559.

The Wealth of India, 2009. Raw materials, Vol VI L-M CSIR, New Delhi, pp. 449–470.

Ugwuona, F.U., Onweluzo, J.C., 2013. Assessment of antioxidant properties of tamarind fruit pulp and its effect on storage stability of African bread fruit seed dhal and flour. Niger. Food J. 31 (2), 41–47.

Yanishlieva, N., Pokorný, J., Gordon, M., 2001. Antioxidants in Food: Practical Applications. CRC Press.

CHAPTER

Turmeric 4.18

Filippo Maggi
School of Pharmacy, University of Camerino, Camerino, Italy

4.18.1 Introduction

An increasing interest of the scientific academia towards medicinal and aromatic plants used in the Ayurvedic and Traditional Chinese Medicine, driven by the huge demand of pharmaceutical and nutraceutical companies for healthier products, has produced a huge amount of research disclosing the pharmacological properties of herbs, spices, and aromatic plants (Li et al., 2019). Among them, turmeric, nowadays known as "the golden spice" and "the spice of life," represents the milestone allowing the merging of the tribal medicine with the advanced explorative technologies applied in laboratories of western countries (Jager, 1997).

Turmeric is among the most famous medicinal plants used in India by the whole population as medicine, food and color. From 1950 onward, the turmeric phytochemical and pharmacological traits and its coloring substance, curcumin, have been the subject of many studies highlighting a plethora of biological activities (Ammon and Wahl, 1991; Chattopadhyay et al., 2004; Khanna, 1999).

This chapter attempts to collect the main information on this golden spice, including different aspects, such as etymology, systematics, distribution, ethnobotany, traditional and commercial uses, phytochemistry, and nutraceutical and pharmaceutical uses.

4.18.2 Etymology

The term "turmeric" comes from the medieval latin "*terra merita*" – deserving earth – for the resemblance of the colored yellow powder with that of a mineral pigment (Ravindran et al., 2007). Alike, the term "*Curcuma*" derives from the Arabic "*kurkum*" meaning saffron, alluding to the yellow-orange color of rhizome (Purseglove et al., 1981; Aggarwal et al., 2013).

Since ancient times this spice, called *Oushadhi*, meaning "the medicinal herb," has been used by Indians as a multipurpose remedy, for instance to cure wounds, digestive problems, poisoning, as well as to dye clothes and yarns and as talisman

Antioxidants Effects in Health. DOI: https://doi.org/10.1016/B978-0-12-819096-8.00007-0

(Ravindran et al., 2007). However, several names are used to indicate turmeric, each of them describing a particular property of the spice (e.g., *Mahaghni, Anestha, Haridra, Varna-datri, Hemaragi, Bhadra, Pavitra, Hridayavilasini, Shobhna*, and others) (Ravindran et al., 2007). In Europe, turmeric is also known as Indian saffron or yellow root.

4.18.3 Systematics

The term "turmeric" refers to the species *Curcuma longa* L. (formerly *C. domestica* Valeton) belonging to the Zingiberaceae or ginger family. For sake of completeness, under this term are often included other species such as *C. phaeocaulis* Valeton, *C. zanthorrhiza* Roxb., *C. zedoaria* (Christm.) Roscoe, and *C. aromatica* Salisb., but they are of secondary importance compared with *C. longa* on a pharmacological and industrial level. More precisely, what we named "the golden spice" has to be referred exclusively to *C. longa*. The Zingiberaceae family encompasses mainly herbs which are famous all over the world for the culinary uses as spices and flavourings. Most of them are endowed with internal secretory cells spread in all organs and exuding volatile oils giving their typical spicy fragrance. Zingiberaceae are characterized by simple, pennate, alternate leaves that are disposed onto two lines and equipped with petiole and sheath at the base. Flowers are grouped in cymes and spikes; they are hermaphrodite and zygomorphic and remain functional just for a day. They are pollinated by bees, moths, butterflies and birds. These plants have a stout rhizome that is a rich source of starch. Zingiberaceae are distributed mainly in tropical regions, in shaded underwood. Besides the genus *Curcuma*, other commercially important genera are *Zingiber* Mill., *Amomum* Roxb., and *Elettaria* Maton.

The genus *Curcuma* L. encompasses about 110 species occurring in the Asia-Pacific area, having the greatest diversity in India and Thailand. The genus comprises many species of economic importance, with *C. longa*, known as the true turmeric, as the most famous one. Other economically important species are *C. aromatica*, *C. zedoaria*, *C. caesia* Roxb., *C. roscoeana* Wall., *C. amada* Roxb., and *C. aeruginosa* Roxb.

4.18.4 Distribution

Turmeric is native to southern Asia although its area of origin is uncertain (probably Vietnam or China). Some authors consider China as the turmeric's home from which it was brought to India by tribal people or Buddhist monks (Ravindran et al., 2007). Currently, turmeric is cultivated in other tropical countries such as Madagascar, Antilles, Polynesia, Sri Lanka, Indonesia, Cina, Giamaica, and Brazil. India is the leading country in the cultivation and production of turmeric, with more than 160,000 ha committed to turmeric cultivation and more than 500,000 tons/year of produced rhizome, of which only 10% is exported worldwide (COMTRADE, 2009).

4.18.5 Botanical description

Turmeric is a perennial herb with a stout rhizome from which 5–10 elliptic-oblong and long petioled big leaves emerge and surround a dense spike with little yellow or white flowers.

The part used as drug is an orange-yellow rhizome which is composed of a bigger, central ovoid part bearing various cylindrical tubercles and rootlets in the lower part, whereas scars of the jets of the leaves are on the upper part. On a commercial level, there are two varieties: i) *rhizoma Curcumae rotunda*, with ovoid, sharp tubercles, and ii) *rhizoma Curcumae longa*, with cylindrical-fusiform rhizome. The breaking surface of rhizome is of a bright reddish yellow color, aromatic odor, hot taste, slightly bitter. The rhizome is full of secretory structures such as oil cells, curcumin cells and ducts; their density is depending on the variety (Remashree, 2003).

4.18.6 Turmeric preparations

Turmeric is found on the market under disks of a few mm tick or powder of a bright yellow color. Rhizomes are collected once the aerial parts dry. After collection, the rhizome is boiled for a short time then dried; this procedure allows to convert starch in solid in order to facilitate drying. It is used under different preparations such as powder, decoction, infusion, extract, and tincture. India is the greatest producer of turmeric which is exported to more than 100 countries around the world. The production is attested by 636 t per year from an area of 175 ha (data 2002–2003) and is increasing at an annual rate of about 8% (Ravindran et al., 2007). The Indian turmeric is qualitatively considered as the best one due to the high content of curcumin, the marker bioactive compound. In Europe, turmeric extracts (ethanolic) and powder encapsulated in capsules and tablets are indicated as immunomodulatory as well as to relief joint and digestive problems. Other preparations include ointments, herbal teas, energy drinks, soaps and cosmetics (Gupta et al., 2013). The global market for curcumin is steadily increasing by a rate of 12% in the period 2015–2022 with an overall income of $84 million. The two main markets for curcumin are North America and Europe. India is the main supplier of turmeric rhizome worldwide. Prices for turmeric depends on the type of product, for example, dried rhizome, powder, extract, curcumin amount, organic production, etc. Generally, the price for 1 kg of crude powdered rhizome ranges from $1.9 to $4.4. Dry extracts are quoted to approximately €150/kg.

4.18.7 Uses as dye

Turmeric and its extracts are used from ancient time as a dye in food and cosmetics. It is possible that these uses originated from the habit of adding turmeric to the food for preservative purposes (Ravindran et al., 2007). In the past they were also used to dye clothes. Turmeric gives a yellow color that turns red under the action of alkali. The main use in food consists in the preparation of curry together with cinnamon, cardamom, cumin, coriander, mustard, nutmeg, cloves, ginger and red pepper. Turmeric and its derivatives are classified as E100 food additive giving a yellow color, similar to that of saffron, to foodstuffs (e.g., cheese, mustard, soups, yogurt, ice creams, cereals etc.). Nowadays, they are recognized as a safe substance (Generally Recognized as Safe, GRAS) by FDA (Liju et al., 2013; Prasad and Aggarwal, 2011). Indeed, literature data proved that curcumin is safe to humans even at 8 g/day (Cheng, 2001).

4.18.8 Ethnobotany

The first data on the medical uses of turmeric trace back to 6000 years B.C. and refer to the treatment of jaundice and leprosy. It is reported that the ancient travelers bringing turmeric from China to India used it to treat wounds and stomach problems (Ravindran et al., 2007). Turmeric was also believed to possess magic properties; thus, it was used in religious practices and marriage ceremonies (Remadevi and Ravindran, 2005). The rhizome held in the hand was believed to help against bad spirits. The day of marriage the bride was anoint with turmeric and oil. The fourth day of marriage women used to smear their body with turmeric paste in oil and have bath with turmeric water (Ravindran et al., 2007). Together with coconut milk, turmeric constituted the first meal of the newborn. Based on these customs, turmeric may be considered one of the first cosmetics used in India. This practice was exported to Rome together with the spice during the Roma age. Its cultivation is believed to be started thousands of years ago, first in Moluccas and Polynesia, then in Madagascar (Ravindran et al., 2007). At the time of the explorer Marco Polo (1280 A.D.), turmeric was known as a famous dye (Rosengarten, 1969). Later, in the middle ages, due to its resemblance with saffron, it received the name of "*Crocus indicus*". At the end of 19th century turmeric was present in all grocery stores of England (Ravindran et al., 2007). When smallpox and chickenpox were frequent in India, turmeric was used to heal skin eruptions due to its antiseptic properties (Ravindran et al., 2007). Turmeric is well used in Ayurveda, Unani, and Chinese traditional medicines to heal a plethora of disorders including inflammations, allergies, itching, joint pain, cold, cough, fever, jaundice, liver diseases, gastrointestinal, gynecological and hepatic problems, infections, and skin diseases (Gilani et al., 2005; Aggarwal et al., 2013). Most of these uses have been confirmed in modern medicine by many research studies since 1971.

4.18.9 Turmeric metabolites

Turmeric contains macronutrients (polysaccharides, protein, fat, minerals), an oleoresin containing curcuminoids (3–8%), with curcumin as the most abundant one (up to 80%), and an essential oil (1.5–5.5%), made up mainly of sesquiterpenes,

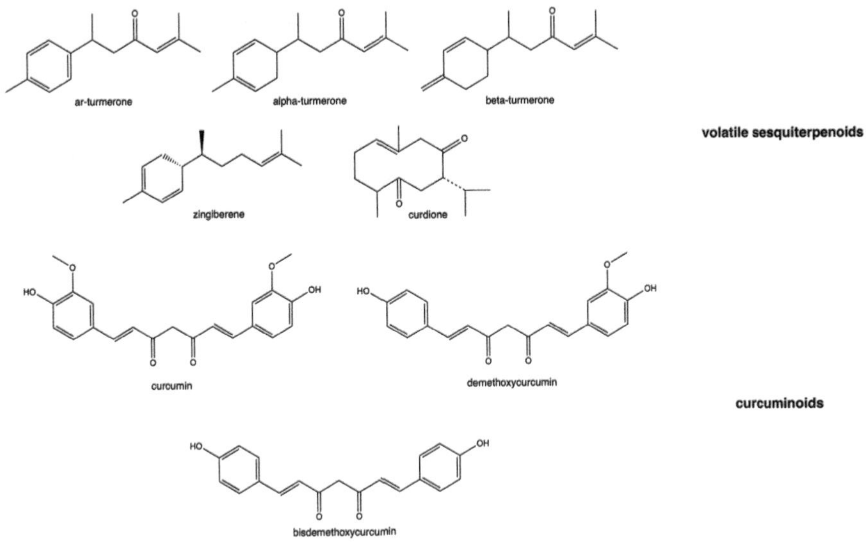

FIG. 4.18.1

Chemical structures of the main secondary metabolites in turmeric.

most of them having bisabolane skeleton, such as *α*- and *β*-turmerone, *ar*-turmerone, zingiberene, and curdione (Fig. 4.18.1) (Amalraj et al., 2017). These compounds affect the aroma of turmeric when used as seasoning. Besides bisabolanes, turmeric essential oil may also contain high amounts of monoterpenoids such as *α*-phellandrene and carvacrol, or furanosesquiterpenoids such as furanodiene and its degradation product curzerene (Chempakam et al., 2008). Curcumin is chemically defined as diferuloylmethane (molecular formula: $C_{21}H_{20}O_6$), and displays methoxy and phenolic groups on the aromatic rings along with an *α*,*β*-unsaturated *β*-diketo linker and keto-enol tautomerism, with the keto form predominant in acidic environment and the enol form occurring under basic conditions (Li et al., 2019). Pharmacological activities displayed by this molecule seem to be correlated to the methoxy and hydroxyl groups on the phenyl ring and 1, 3-diketone systems. The wealth of functional groups makes curcumin able to bind to many molecular targets (Gupta et al., 2013). This molecule was firstly isolated in the 19th century and its structure was elucidated by Roughley and Whitin in 1973 (Roughley and Whitin, 1973). Curcumin occurs in turmeric together with its two main derivatives, demethoxycurcumin ($C_{20}H_{18}O_5$, ~18% of curcuminoid fraction) and bisdemethoxycurcumin ($C_{19}H_{16}O_4$, ~2%). These compounds occur in their *trans-trans* keto-enol form. Different extraction techniques have been employed for the separation of curcuminoids from turmeric, with supercritical carbon dioxide extraction, microwave assisted extraction with solvent, and hydrotropy based extraction giving the highest yields (Jayaprakasha et al., 2005). Generally, 70% ethanol is preferred to extract curcuminoids from turmeric (Aggarwal et al., 2013). Curcuminoids are poorly soluble in water and have low absorption and bioavailability, thus being quickly metabolized and escreted. These drawbacks have recently overcome through the development of nanocarriers such as liposomes, micelles, nanoparticles and metal complexes (Amalraj et al., 2017). Other metabolites isolated from rhizome of turmeric are quinoline alkaloids, bisabolane sesquiterpenes, diterpenes, triterpenoids and steroids (Aggarwal et al., 2013). In addition, a heat stable protein (maximum absorbance at 280 nm, molecular weight of approximately 24,000 Da), named turmerin, was isolated from turmeric aqueous extract and revealed to be a strong inhibitor of lipid peroxidation in brain and cod liver oil (Selvam et al., 1995).

4.18.10 Nutraceutical and medicinal uses

Many uses are reported for turmeric and curcumin, ranging from cosmetics to drugs (e.g, treatment of Alzheimer disease). Turmeric is also qualified as the queen of natural COX-2 inhibitors (Duke, 2003). Most of pharmacological properties such as anti-inflammatory and antioxidant ones are attributed to curcumin whose concentration in the rhizomes ranges from 3 to 9%. However, it is important to underline that also curcumin-free extracts and preparations have been reported biologically active on different diseases such as tumors and diabetes (Gupta et al., 2013). In this respect, it has been found that turmeronoids (i.e. *ar*-, α- and β-turmerone), the volatile bisabolane sesquiterpenes, may give an important contribution to the plethora of pharmacological activities attributed to turmeric (Del Prete et al., 2016). In Europe, turmeric and its essential oil have been listed among botanicals allowed to be used in food supplements. Turmeric and curcumin are also listed by the FDA among the generally recognized as safe (GRAS) substances for use as a dye and flavoring in food (Yallapu et al., 2015). A monograph of turmeric is also included in the European Pharmacopoeia (*Curcuma longa* rhizome). Nutraceutics containing turmeric and its major bioactive compound curcumin are mainly indicated to support the immune system and to treat joint problems, and menstrual and digestive disorders. When used in herbal medicinal products (HMP), turmeric is mainly indicated to relief digestive problems, mainly for its capability to increase the bile flow. For the purpose, this medical use is reported in the European Scientific Cooperative on Phytotherapy (ESCOP) and World Health Organization (WHO) monographs (ESCOP, 2003). European Medicines Agency (EMEA) reports a turmeric monograph as traditional herbal medicinal product (THMP) including several preparations such as powder, herbal tea, and 1:10 tincture, for the treatment of dyspepsia (EMEA, 2008). Other medical indications reported in literature concern the treatment of liver disorders, improvement of brain disfunctions and cardiovascular diseases and relief of musculoskeletal pain (Amalraj et al., 2017; Eke-Okoro et al., 2018; Li et al., 2019). Curcumin is endowed with important anti-inflammatory action, being able to inhibit lipoxygenase, phospholipase A2 and COX-2, whereas it is ineffective on COX-1, thus having no side effects compared with non-selective analgesic drugs (Neha et al., 2009). At CNS level, curcumin seems to act on opioid and serotonin receptors giving antinociceptive effects (Motaghinejad et al., 2015). Finally, curcumin showed potential as anticancer drug, being capable of inducing autophagy, growth suppression and death in cancer cells (Gupta et al., 2013). However, to support the use of curcumin in the treatment of cancer, further clinical trials on humans remain to be performed in the next years. Curcumin, after oral administration, is converted in glucuronide and glucuronide-sulfate conjugates which are responsible for most of pharmacological effects into the body (Asai and Miyazawa, 2000). Bioavailability of curcumin can be increased by co-administration with adjuvants such as antioxidants, carriers (nanoparticles, micelles, phospholipid complexes etc.) and cellulosic derivatives (Eke-Okoro et al., 2018). Recently, it has been found that the alkaloid piperine, which is an inhibitor of glucuronidation, may be used in combination with curcumin to enhance significantly its plasma concentration (Eke-Okoro et al., 2018). Also, co-administration of turmerones, the sesquiterpenes occurring in the turmeric's essential oil, may enhance bioavailability of curcumin (Yue et al., 2012). Turmeric and its derivatives (extracts, curcumin) are devoid of side effects and toxicity and can be assumed even at high doses (0.5–8 g/day/person) (Chainani-Wu, 2003; Niranjan and Prakash, 2008), although there are no studies on children and pregnant women for which recommended doses are still to be determined. Also, long-term toxicity studies are desirable to be performed. Turmeric and its curcuminoids own several interactions with hepatic enzymes such as cytochromes P450 and transferases, thus possible contraindications may occur when they are co-administered with some drugs such as anticoagulants, antibiotics, anticancer, cardiovascular and antidepressant (Soleimani et al., 2018).

4.18.11 Uses as antioxidant

The main advantages of using turmeric as a food additive as well as phytomedicine rely on the significant antioxidant activity exerted by its extracts and the main constituent curcumin. When given to broiler chickens as a food supplement, the turmeric powder reversed the detrimental effects of the aflatoxin B1 diet due to its antioxidant power (Gowda et al., 2009). Notably, the turmeric supplementation improved the chicken performance by protecting liver from aflatoxin damage, and enhancing serum biochemical parameters and liver antioxidant status. The administration of turmeric extract in Wenchang broiler chickens led to enhancement of antioxidant enzymes such as superoxide dismutase and gluthatione peroxidase, and decreased the content of malondialdehyde (Wang et al., 2015). Administration of turmeric extracts to rabbits fed with a high fat diet reduced the membrane oxidation damage in erythrocyte and liver microsome and reduced the cholesterol level in blood and liver (Mesa et al., 2003). Supplementation of turmeric powder in pigs led to an improvement of meat quality by decreasing lipid oxidation (Mancini et al., 2017). Supplementation with curcumin (15 g) in common carp led to enhancement of antioxidant genes (e.g., superoxide dismutase and catalase) and reduction of oxidative stress in terms of concentration of malondialdehyde (Giri et al., 2019).

Turmeric supplementation (10 g/kg) in common carp after copper exposure reduced the negative effects of copper by increasing the activity of superoxide dismutase, catalase and glutathione peroxidase (Rajabiesterabadi et al., 2020). Several turmeric supplements exhibited antioxidant capacity in Wistar rats, notably by decreasing the level of malondialdehyde and displaying radical scavenging activity (Guerrero-Romero et al., 2020).

Curcuminoids, particularly curcumin, are able to inhibit the formation of thiobarbituric acid reactive substances (TBARs) during lipid peroxidation in a concentration dependent manner. Notably, curcumin showed an IC_{50} value of 2.7 μM (Chaniad et al., 2018). In addition, curcumin protects from vitamin E depletion during oxidative stress. It has been demonstrated that curcuminoids retain their antioxidant capacity even after exposition at high temperatures (e.g., boiled or roasted) (Sun et al., 2019). As a food additive, turmeric powder (8%) added to fried rice snack improved the oxidative deterioration (Lim and Han, 2016).

The total phenolic content (TPC) of turmeric has been evaluated by the Folin-Ciocalteu method and resulted to be in the range 4.52–16.07 g gallic acid eq/100 g of sample depending on the method of extract preparation and variety (Tanvir et al., 2017). The radical scavenging properties of turmeric extracts as determined by DPPH and ABTS assays and expressed as Trolox Equivalent Antioxidant capacity (TEAC) and Ascorbic Acid Antioxidant Capacity (AEAC) for ethanolic and aqueous extracts, respectively, were in the ranges 0.36–0.74 and 0.04–0.15, and 1.03–1.44 and 0.27–0.63, respectively (Tilak et al., 2004). When different varieties (Mura and Chora) were compared for radical scavenging activity in the DPPH assay, they showed IC_{50} values from 1.08 to 16.55 μg/mL (Tanvir et al., 2017). In addition, turmeric extracts were able to reduce iron complex as determined by the FRAP assay, with TEAC and AEAC in the ranges of 0.79–0.85 and 0.16–0.41, respectively. The major contribution for these properties is given by curcumin. This phenolic was reported as a quencher of reactive oxygen species (ROS) generated during cell metabolism (Das and Das, 2002). Curcumin, at a concentration of 42 μM, was able to inhibit 100% lipid peroxidation in rat liver mitochondria (Tilak et al., 2004). Generally, the extracts obtained from young tissues of turmeric exhibited a higher antioxidant activity than the aged ones (Cousins et al., 2007). Other constituents of turmeric extracts exhibiting antioxidant activity besides curcuminoids are protocatechuic acid and ferulic acid. These phenolics have radical scavenging and reducing power and protect DNA from oxidative damage (Kumar et al., 2006). Also, the essential oil and oleoresin from fresh turmeric rhizomes, rich in α- and β-turmerones, displayed notable antioxidant activity in lipid peroxidation assays and radical scavenging (DPPH) and Iron chelating activity (Singh et al., 2010).

It is important to mention that curcumin can also behave as a pro-oxidant agent. Indeed, in presence of copper, curcumin may generate hydroxyl radicals that can produce DNA cleavage (Ahsan et al., 1999). This is caused by three linkages for Cu(II) on the curcumin scaffold, at the phenolic and methoxy groups on the aromatic rings, and at the 1,3-diketone moiety. Thus, curcumin is also capable of triggering apoptosis in cells through generation of radicals, thus acting as a pro-oxidant drug.

It has been shown that in combination with Trolox, a vitamin E-like hydrosoluble compound, curcumin triggers apoptosis in several cancer cells through PARP cleavage, caspase-3 activation, and upregulation of proapoptotic proteins (Zheng et al., 2012). These results provide evidence that the anticancer activity of curcumin may be potentiated by combination with some antioxidant drugs.

Conclusions

Turmeric, the golden spice, is an excellent example of how the traditional medical knowledge brought by the Asian communities has been nowadays perfectly integrated into the western world finding application in different fields, from culinary uses and natural dyes to food additives and herbal remedies for the treatment of digestive and inflammatory disorders. Most of pharmacological properties of turmeric, such as antioxidant and anti-inflammatory, have been correlated to the presence of curcuminoids although other metabolites, for instance bisabolane sesquiterpenes, may contribute to the overall effect through synergistic effects and improvement of the curcuminoids bioavailability. Different extraction techniques have been recently developed to optimize the yield of turmeric active constituents from the plant matrix, with new eco-friendly approaches aimed at limiting the issues related to the waste disposal. Turmeric appears to be safe since it is classified as a GRAS; indeed, it is well tolerated for food and medical purposes. However, special attention should be paid when concomitant administration of other therapeutics occurs as well as potential toxicity in children and pregnant women.

References

Aggarwal, B.B., Yuan, W., Li, S., Gupta, S.C., 2013. Curcumin-free turmeric exhibits anti-inflammatory and anticancer activities: identification of novel components of turmeric. Mol. Nutr. Food Res. 57, 1529–1542.

Ahsan, H., Parveen, N., Khan, N.U., Hadi, S.M., 1999. Pro oxidant, anti oxidant and cleavage activities on DNA of curcumin and its derivatives demethoxycurcumin and bisdemethoxycurcumin. Chem. Biol. Interact. 121, 161–175.

Amalraj, A., Pius, A., Gopi, S., Gopi, S., 2017. Biological activities of curcuminoids, other biomolecules from turmeric and their derivatives – a review. J. Tradit. Complement. Med. 7, 205–233.

Ammon, H.P.T., Wahl, M.A., 1991. Pharmacology of Curcuma longa. Planta Med. 57, 1–7.

Asai, A., Miyazawa, T., 2000. Occurrence of orally administered curcuminoid as glucuronide and glucuronide/sulfate conjugates in rat plasma. Life Sci. 67, 2785–2793.

Chainani-Wu, N., 2003. Safety and anti-inflammatory activity of curcumin: a component of tumeric (Curcuma longa). J. Altern. Complement. Med. 9, 161–168.

Chaniad, P., Morales, N.P., Rojsitthisak, P., Luechapudiporn, R., 2018. Effects of turmeric extract on hemin-induced low-density lipoprotein oxidation. J. Food Biochem. 42, e12507.

Chattopadhyay, I., Biswas, K., Bandyopadhyay, U., Banerjee, R.K., 2004. Turmeric and curcumin: biological actions and medicinal applications. Curr. Sci. 87, 44–53.

Chempakam, B., Parthasarathy, V. A. Chemistry of spices In: Parthasarathy, V.A., Chempakam, N., Zachariah, T.J. (Eds.), Cabi Wallingford, UK, 2008, pp. 97–123.

Cheng, A.L., 2001. Phase I clinical trial of Curcumin, a chemopreventive agent in patients with high-risk or pre-malignant lesions. Anticancer Res. 21, 2895–2900.

Cousins, M., Adelberg, J., Chen, F., Rieck, J., 2007. Antioxidant capacity of fresh and dried rhizomes from four clones of turmeric (*Curcuma longa* L.) grown in vitro. Ind. Crops Prod. 25, 129–135.

Das, K.C., Das, C.K., 2002. Curcumin (diferuloylmethane), a singlet oxygen ($^1O_2$) quencher. Biochem. Biophys. Res. Commun. 295, 62–66.

Del Prete, D., Millán, E., Pollastro, F., Chianese, G., Luciano, P., Collado, J.A., Munoz, E., Appendino, G., Taglialatela-Scafati, O., 2016. Turmeric sesquiterpenoids: Expeditious resolution, comparative bioactivity, and a new bicyclic turmeronoid. J. Nat. Prod. 79, 267–273.

Duke, J.A., 2003. CRC Handbook of Medicinal Spices. CRC Press, Boca Raton, FL.

Eke-Okoro, U.J., Raffa, R.B., Jr, Pergolizzi, J., V., Breve, F., Jr, Taylor, R., NEMA Research Group, 2018. Curcumin in turmeric: basic and clinical evidence for a potential role in analgesia. J. Clin. Pharm. Ther. 43, 460–466.

European Medicines Agency (EMEA) Committee on Herbal Medicinal Products (HMPC). Draft Community Herbal Monograph on Curcuma longa L., Rhizoma. London, UK EMEA November 6, 2008. http://www.emea.europa.eu/pdfs/human/hmpc/curcuma_longa/45684508en.pdf. (Accessed 2 December 2021).

ESCOP, 2003. European scientific cooperative on phytotherapy, Monographs, 2nd ed. Thieme, New York.

Gilani, A.H., Shah, A.J., Ghayur, M.N., Majeed, K., 2005. Pharmacological basis for the use of turmeric in gastrointestinal and respiratory disorders. Life Sci. 76, 3089–3105.

Giri, S.S., Sukumaran, V., Park, S.C., 2019. Effects of bioactive substance from turmeric on growth, skin mucosal immunity and antioxidant factors in common carp, *Cyprinus carpio*. Fish Shellfish. Immunol. 92, 612–620.

Gowda, N.K., Ledoux, D.R., Rottinghaus, G.E., Bermudez, A.J., Chen, Y.C., 2009. Antioxidant efficacy of curcuminoids from turmeric (*Curcuma longa* L.) powder in broiler chickens fed diets containing aflatoxin B1. Br. J. Nutr. 102, 1629–1634.

Guerrero-Romero, F., Simental-Mendía, L.E., Martínez-Aguilar, G., Sánchez-Meraz, M.A., Gamboa-Gómez, C.I., 2020. Hypoglycemic and antioxidant effects of five commercial turmeric (*Curcuma longa*) supplements. J. Food Biochem. 44, e13389.

Gupta, S.C., Kismali, G., Aggarwal, B.B., 2013. Curcumin, a component of turmeric: from farm to pharmacy. Biofactors 39, 2–13.

Jager, P.D., 1997. Turmeric. Vidyasagar Pub, California, USA, p. 67.

Jayaprakasha, G.K., Rao, J.M., Sakariah, K.K., 2005. Chemistry and biological activities of C. longa. Trends Food Sci. Technol. 16, 533–548.

Khanna, N.M., 1999. Turmeric–nature's precious gift. Curr. Sci. 76, 1351–1356.

Kumar, G.S., Nayaka, H., Dharmesh, S.M., Salimath, P.V., 2006. Free and bound phenolic antioxidants in amla (*Emblica officinalis*) and turmeric (*Curcuma longa*). J Food Comp. Anal. 19, 446–452.

Li, H., Sureda, A., Devkota, H.P., Pittalà, V., Barreca, D., Silva, A.S., Tewari, D., Xu, S., Nabavi, S.M., 2019. Curcumin, the golden spice in treating cardiovascular diseases. Biotechnol. Adv. 38, 107343. doi:10.1016/j.biotechadv.2019.01.010.

Liju, V.B., Jeena, K., Kuttan, R., 2013. Acute and subchronic toxicity as well as mutagenic evaluation of essential oil from turmeric (Curcuma longa L). Food Chem. Toxicol. 53, 52–61.

Lim, S.T., Han, J.A., 2016. Improvement in antioxidant functionality and shelf life of yukwa (fried rice snack) by turmeric (*Curcuma longa* L.) powder addition. Food Chem. 199, 590–596.

Mancini, S., Paci, G., Pisseri, F., Preziuso, G., 2017. Effect of turmeric (*Curcuma longa* L.) powder as dietary antioxidant supplementation on pig meat quality. J. Food Process. Preserv. 41, e12878.

Mesa, M.D., Aguilera, C.M., Ramírez-Tortosa, C.L., Ramírez-Tortosa, M.C., Quiles, J.L., Baró, L., Gil, A., 2003. Oral administration of a turmeric extract inhibits erythrocyte and liver microsome membrane oxidation in rabbits fed with an atherogenic diet. Nutrition 19, 800–804.

Motaghinejad, M., Bangash, M.Y., Hosseini, P., Karimian, S.M., Motaghinejad, O., 2015. Attenuation of morphine withdrawal syndrome by various dosages of curcumin in comparison with clonidine in mouse: possible mechanism. Iran J. Med. Sci. 40, 125.

Neha, S., Ranvir, G.D., Jangade, C.R., 2009. Analgesic and antipyretic activities of Curcuma longa rhizome extracts in Wister Rats. Vet World 2 (8), 304–306.

Niranjan, A., Prakash, D., 2008. Chemical constituents and biological activities of turmeric (*Curcuma longa* L.) – a review. J. Food Sci. Technol. 45, 109–116.

Prasad, S., Aggarwal, B.B., 2011. Turmeric, the golden spice: from traditional medicine to modern medicine. Herbal Medicine. CRC Press, Boca Raton, Florida, US, pp. 273–298.

Purseglove, J.W., Brown, E.G., Green, C.L., Robbins, S.R.J., 1981. Spices 2. Longman, London.

Rajabiesterabadi, H., Hoseini, S.M., Fazelan, Z., Hoseinifar, S.H., Doan, H.V., 2020. Effects of dietary turmeric administration on stress, immune, antioxidant and inflammatory responses of common carp (*Cyprinus carpio*) during copper exposure. Aquac. Nutr. 26, 1143–1153.

Ravindran, P.N., Babu, K.N., Sivaraman, K., 2007. Turmeric: The Genus *Curcuma*. CRC press.

Remadevi, R., Ravindran, P.N., 2005. Turmeric: myths and traditions. Spice India 18, 11–17.

Remashree, A.B., 2003. Ontogeny of oil cells and ducts in turmeric (Curcuma longa L.). Phytomorphol 53, 261–268.

Rosengarten Jr., F., 1969. The Book of Spices. Livingston Pub. Co, Philadelphia, PA.

Roughley, P.J., Whiting, D.A., 1973. Experiments in the biosynthesis of curcumin. J. Chem. Soc. 20, 2379–2388.

Selvam, R., Subramanian, L., Gayathri, R., Angayarkanni, N., 1995. The anti-oxidant activity of turmeric (Curcuma longa). J. Ethnopharmacol. 47 (2), 59–67.

Singh, G., Kapoor, I.P.S., Singh, P., De Heluani, C.S., De Lampasona, M.P., Catalan, C.A., 2010. Comparative study of chemical composition and antioxidant activity of fresh and dry rhizomes of turmeric (*Curcuma longa* Linn). Food Chem. Toxicol. 48, 1026–1031.

Soleimani, V., Sahebkar, A., Hosseinzadeh, H., 2018. Turmeric (Curcuma longa) and its major constituent (curcumin) as nontoxic and safe substances. Phytother. Res. 32, 985–995.

Sun, J.L., Ji, H.F., Shen, L., 2019. Impact of cooking on the antioxidant activity of spice turmeric. Food Nutr. Res. 63.

Tanvir, E.M., Hossen, M., Hossain, M., Afroz, R., Gan, S.H., Khalil, M., Karim, N., 2017. Antioxidant properties of popular turmeric (*Curcuma longa*) varieties from Bangladesh. J. Food Qual. 2017, 8471785.

Tilak, J.C., Banerjee, M., Mohan, H., Devasagayam, T.P.A., 2004. Antioxidant availability of turmeric in relation to its medicinal and culinary uses. Phytother. Res. 18, 798–804.

United Nations Commodity Trade Statistics Database (COMTRADE). Accessed September 22, 2009.

Wang, D., Huang, H., Zhou, L., Li, W., Zhou, H., Hou, G., Hu, L., 2015. Effects of dietary supplementation with turmeric rhizome extract on growth performance, carcass characteristics, antioxidant capability, and meat quality of Wenchang broiler chickens. Ital. J. Anim. Sci. 14, 3870.

Yallapu, M.M., Nagesh, P.K.B., Jaggi, M., Chauhan, S.C., 2015. Therapeutic applications of curcumin nanoformulations. AAPS J. 17, 1341–1356.

Yue, G.G., Cheng, S.W., Yu, H., Xu, Z.S., Lee, J.K., Hon, P.M., Lee, M.Y.H., Kennelly, E.J., Deng, G., Yeun, S.K., Cassileth, B.R., Fung, K.P., Leung, P.C., Lau, C.B.S., 2012. The role of turmerones on curcumin transportation and P-glycoprotein activities in intestinal Caco-2 cells. J. Med. Food 15, 242–252.

Zheng, J., Payne, K., Taggart, J.E., Jiang, H., Lind, S.E., Ding, W.Q., 2012. Trolox enhances curcumin's cytotoxicity through induction of oxidative stress. Cell. Physiol. Biochem. 29, 353–360.

CHAPTER

Uric acid

4.19

Sumit Durgapal[a], Arvind Jantwal[a], Jyoti Upadhyay[b], Tanuj Joshi[a], Aadesh Kumar[a]

[a]*Department of Pharmaceutical Sciences, Kumaun University, Campus Bhimtal, Bhimtal, Uttarakhand, India*

[b]*School of Health Science and Technology, Department of Pharmaceutical Sciences, University of Petroleum and Energy Studies, Dehradun, Uttarakhand, India*

4.19.1 Introduction

Uric acid (UA), a nitrogen containing heterocyclic compound (Hafez RM et al., 2017; Karwur and Pujiastuti, 2017), is one of the major antioxidants found in the human body. Surprisingly almost the half of the antioxidant activity of blood comes from UA (El Ridi and Tallima, 2017). From the chemical point of view, UA is known to possess a molecular mass of 168Da with a molecular formulae of $C_5H_4N_4O_3$ which is 7,9-dihydro-1H-purine-2,6,8(3H)-trione in IUPAC (Maiuolo et al., 2016). Studies showed that UA is present in most of the organisms from bacteria to humans, birds, snakes, insects and reptiles. Mammals are known to contain uric acid mainly in blood and urine. UA is an ultimate product of purine metabolism which occurs mainly in liver, kidneys, intestine, and vascular endothelium (Cortese et al., 2019; El Ridi and Tallima, 2017) by using purine from various endogenous and exogenous sources. The endogenous sources of purine are mainly constituted by nucleic acids derived from living and dying cells while dietary sources, fructose catabolism, and animal proteins are the source of exogenous purine (Cortese et al., 2019). The pathway of production of UA via the conversion of nucleic acids such as adenine and guanine involve several different enzymes to catalyze the process of conversion. Two basic reactions for the production of UA are conversion of hypoxanthine to xanthine and further conversion of xanthine to UA. The enzyme which plays most significant role in this conversion is xanthine oxidoreductase, a flavoenzyme (Harrison, 2002). Other than humans, most of the organisms possess an enzyme uricase which is often referred as ureate oxidase which converts UA, which is practically very insoluble in water, into its more soluble form called allantoin. Further, allantoin gets converted into allantonate and then it results in the production of gloxylate and urea, both of which are more soluble in water as compared to UA. In humans and closely related primates uricase gene has evolved and structurally changed into a gene which cannot convert UA into allantoin and other soluble compounds and this left human beings with highest plasma concentration of UA among mammals and make them highly susceptible

Antioxidants Effects in Health. DOI: https://doi.org/10.1016/B978-0-12-819096-8.00008-2

and prone towards the diseases which are associated with increased amount of UA in blood such as gout, urolithiasis, nephropathy and hyperuricaemia and related diseases which are associated with the deposition of urate crystals. Recently several research studies reported that there occurs a great association between prevalence of hyperuricaemia and other severe disorders like cardiovascular, renal and metabolic disorder (Borghi et al., 2015). UA comes under the category of weak acids and possess a PKa value of 5.8 and mainly exist as salt from of UA which is called urate. Blood urate levels are directly related to the increase in UA crystals. Studies showed that the level of UA in men ranges from 2.5 to 7.0 mg/dl whereas women possess somewhat lower amounts of UA in the range of 1.5 to 6.0 mg/dl and compared to mammals in which UA levels ranges from 10–20 ug/ml, humans and apes due to the loss of uricase activity are known to have 3 to 10 times higher UA levels. The chief eliminating organ for the elimination of UA through body in urine is kidney. The process of elimination solely depends on the normal functioning of this organ and any kind of impairment in normal functioning of kidney interferes with the process of elimination and reflects lower rates of elimination. Diet also affects the process of elimination and one classic examples to support this fact is the increase in the process of reabsorption of UA in the proximal tubule with low sodium diet which leads high UA concentrations in blood (Wang et al., 2018; Lei and Wang, 2018; Todd, 2017).

4.19.2 Uric acid as an antioxidant

The antioxidant effect of UA is well known and well established since decades. Kellogg and Fridovich were the scientists who first explained the antioxidant potential of UA and credited this finding under their names (Sautin et al., 2008). Later Ames et al., conducted numerous experiments and described UA as one of the most powerful antioxidants present in humans with immense potential to scavenge a vast number of free radicals like hydroxyl, peroxyl, nitric oxide and singlet oxygen thus prevents processes of lipid peroxidation and nitration of proteins and is highly crucial for the antioxidant activity of plasma (Nery et al., 2015; Cutler et al., 2019). Several research studies supported the active participation of UA in various physiological and pathological conditions (Settle and Klandorf, 2014). Based on this fact Ames and his co-workers reported the loss of uricase activity in humans during Miocene age due to gene mutation is more a boon than a curse in increasing the vitality and life span of human beings because with the loss of this activity humans are not able to convert UA into other compounds. Thus, increased levels of UA in plasma acts like key players in combating a vast number of serious disorders such as cancer, aging (Ames et al., 1981), depression, anxiety (Black et al., 2018) and many more by managing excessive oxidative stress which is the main culprit and is one of the underlying causes for most of these life threatening disorders. Some researchers also believe that the increased levels of UA due to the loss of uricase activity are related to the loss of vitamin C synthesis ability in human beings with the mutation of L-gulono-lactone oxidase because of excessively rich vitamin C diet consumed by our primates (Johnson et al., 2005). UA is also very important for providing protective activities to vitamins like C and E via their stabilization (Batta, 2016). UA is a potent antioxidant because of its electron donor nature and protects from the oxidative effects and damages caused by H_2O_2 and harm proteins by acting as an oxidizable agent towards them. Other than its free radical deactivating activity it has got immense potential for chelating metal ions such as iron and copper so that they become inactive to accelerate free radical reactions (Glantzounis et al., 2005). Experiments showed that on reaction with ROS and several other free radicals, UA converts into allantoin and other products thus acts as an important and deciding parameter to determine the extent of OS. The evidence of UA as the most potent antioxidant in human body comes from the fact that it effectively inhibits and inactivates the radicals comes into existence after peroxynitrite decomposition. Being a strong oxidizing agent, this is known to be involved in the oxidation of almost all cells and results in severe cell injuries for the pathogenesis of OS induced diseases. Peroxynitrite also found to alter the structures and functions of protein molecules via the nitration of tyrosine residues. Research studies demonstrated that the rate of reaction of peroxynitrite with carbon dioxide (CO_2) is much faster than the reaction of UA with CO_2. Reaction between peroxynitrite and CO_2 produces main free radicals such as $CO_3^{\cdot -}$ and $NO_2^{\cdot}$ which are the key inducers of OS related diseases. As UA cannot compete with peroxynitrite for CO_2 therefore, instead of directly scavenging peroxynitrite to produce these radicals, UA scavenges $CO_3^{\cdot -}$ and NO_2 to produce its antioxidant effect (Whiteman et al., 2002). Urate also plays a significant role in the maintenance of blood pressure through renin-angiotensin system and via the deactivation of peroxynitrite in central nervous system, which is of great importance for the prevention of multiple sclerosis (Hediger et al., 2005). UA along with other antioxidants, such as vitamin C and reduced glutathione which is an endogenous antioxidant helps in strengthening natural antioxidant defenses mechanism to protect the body from the adverse effects of increasing OS. Recently, studies have revealed the profound role of presence of high concentration of UA in epithelial lining fluid in reducing the higher levels of OS in severe diseases like chronic obstructive pulmonary disease, lung cancer and asthma. OS is one of the leading causes of cardiovascular diseases (CVD). It has been found through extensive research studies that reactive oxygen species (ROS) is a major participant in the normal cell signaling for the smooth running and regulation of vascular functions such as endothelial tone and reactivity along with cellular redox homeostasis maintenance (Cervantes et al., 2017). Excessive increase in the levels of ROS from its normal physiological requirements results in the progression of number of cardiovascular diseases including atherosclerosis, myocardial infarction, cardiac hypertrophy, cardiomyopathy, heart failure due to the failure of endogenous systems to maintain normal physiology. OS also plays a significant role in the pathogenesis of cardiovascular and general vascular dysfunction induced diabetes complications (Glantzounis et al., 2005).

Antioxidants, the key role players in scavenging the adverse effects of ROS and other free radicals, are therefore highly required to overcome the problems related to various CVD (Sridevi et al., 2018). UA, as an antioxidant, helps in reducing excessively high levels of ROS and prevents OS during CVD. Number of studies conducted in this arena so far demonstrated that the high levels of UA due to low salt diet can maintain blood pressure acutely as well as chronically to its optimum levels. This maintenance of low blood pressure due to high UA levels enables hominoids of Miocene age to overcome the problems of low blood pressure due to environmental stress and also in the maintenance of vertical position (Johnson et al., 2005). Experimental studies conducted on the perfused heart of guinea pig in order to access the antioxidant effect of UA showed improved ability of heart to perform normal functions under OS (Becker et al., 1989). Brain is highly vulnerable towards OS because of its weak antioxidant defense mechanism, high lipid content and excessively high oxygen demand. Several studies revealed that one fifth of the overall amount of oxygen inhaled daily is used by the brain to perform its complex physiological functions to fulfil the demand of the body. As, brain cells always remain exposed to high amount of oxygen which results in overproduction of free radicals such as ROS and other major classes. Particularly proteins, Deoxy Ribonucleic acid (DNA) and phospholipids are the major targets of ROS mediated neuronal damage in brain (Salim, 2017). This damage to the neurons leads various life threatening neurodegenerative diseases such as Alzheimer's disease, Parkinson's diseases, multiple sclerosis and brain strokes and research studies gave number of evidences about the protective role of UA as an antioxidant against all these diseases (Passi et al., 2006). Patients suffering from these neurodegenerative diseases have been found to possess lower levels of UA compared to normal persons and this fact supports the use of UA as an effective therapy for their management. Formation of free radical such as peroxynitrite in the neurons due to the reaction between superoxide and nitric oxide gets prevented by UA. UA being an antioxidant plays a significant role in the neutralization of cellular superoxide and render it ineffective for the reaction with nitric oxide thus stops the formation of CO_3 and NO_2 from the breakdown of peroxynitrite. This function of UA is associated with its high plasma concentration. It has also been found that the OS in neurons cannot be prevented at low concentration of UA because of inefficiency of antioxidant defense system of the body to prevent the action of peroxynitrite. There are experimental evidences of protective effect of UA in the lab cultured neuronal cells of hippocampal region of rats impaired with OS which further supports the antioxidant effect of UA (Yu et al., 1998). Higher concentrations of UA have been found to be associated with the lower risk of Parkinsonism disease (PD) because of its antioxidant action and the plausible mechanism is the role of UA against free radical theory of PD (Batta, 2016). This remarkable involvement of UA in PD has now been exploited as important biomarkers in the diagnosis of PD by considering its serum levels in various clinical manifestations (Schirinzi et al., 2018). Recently one experimental study revealed that the low concentration of serum UA levels is associated with the loss of proper functioning of URAT1 a proximal tubular cell transporter in kidney and SLC22A12 which encodes blood vessels leads in vivo endothelial dysfunctioning. Being a main initiator of inflammatory process UA also helps in repairing of tissues (El Ridi and Tallima, 2017).

Paradoxically, some of researchers have put forth their theories, hypothesis and established with their studies that the results of increased UA levels in plasma are just opposite to that of their beneficial roles. Several studies showed that the patients with the history of hyperuricemia are at increased risk of death because of various disorders such as gout, hypertension, diabetes mellitus etc and people who possess lower levels of UA are in the safer side (Johnson et al., 2003). Hyperuricaemia is a condition in which the level of serum uric acid increases than the normal level of uric acid which is required for its antioxidant activity. Condition in which the level of uric acid is ≥7 mg/dL for men and ≥6.0 mg/dL for women is defined as hyperuricemia (de Oliveira and Burini, 2012). Principal cause of hyperuricemia is mainly related to the low solubility of uric acid in extravascular region which ultimately results in the formation of uric acid crystals and their accumulation to specific tissues. The amount of uric acid in blood increases with age and one of the classic examples of this is the lower level of uric acid in women of young age as compared to women of post menopausal age. Studies showed that the normal daily turnover of UA production with the metabolism of purine obtained through endogenous and exogenous sources ranges from 600–700 mg/day with a maximum limit of 1000 mg/day and further its exit to maintain its optimum levels in blood is mainly controlled by several processes such as excretion through kidneys, complex urine forming physiological processes and renal blood flow for the maintenance of equilibrium between UA production and its exit (Mount et al., 2006). Problems associated with the imbalance between UA production and its exit through the body results in the higher accumulation of UA. Sometimes, diet particularly rich in purine also has a significant impact on increasing the levels of UA. These elevated levels of UA have been found to be associated with severe life-threatening diseases which if not managed or treated timely results in the serious health related consequences (de Oliveira and Burini, 2012).

4.19.3 Risk factors associated with the high concentration of uric acid

4.19.3.1 Gout

Gout has long been associated with the high concentration of serum UA levels in body and characterized by the deposition of monosodium ureate crystals mainly in the joints and periarticular tissues. Risk factor increases when the level of UA increases from 7 mg/dL. Studies demonstrated that for the interaction of monosodium ureate crystals with articular cells membranes to show some inflammatory responses, these crystals first needs to be coated with serum proteins. This interaction ultimately results in the activation of NLRP3 inflammasome release of IL-1 moiety and neutrophils accumulation in the joints (El Ridi and Tallima, 2017; de Oliveira and Burini, 2012).

4.19.3.2 Cardiovascular disease

Frederick A. Mohamed was the scientist who explained the association between higher levels of UA and cardiovascular disease (CVD) almost 140 years ago. Some of the well understood and main mechanisms by which UA plays a profound role in the pathogenesis of CVS are depletion of nitric oxide (NO), pro-oxidant activity, endothelial dysfunction, promotion of inflammation and potentiation of vasoconstrictor and proliferative vascular stimuli. Though, UA acts as a powerful antioxidant and scavenges several free radicals but the episodes of conversion of UA to pro-oxidant and further to low density lipoproteins in the presence of transition metals have great role in the pathogenesis of CVD. Formation of free radicals such as aminocarbonyl and triuretcarbonyl when UA reacts with peroxynitrites again strengthens the pro-oxidant state of UA. UA directly reacts with NO and results in the depletion of NO and production of 6-aminouracil via this reaction. UA also blocks insulin and vascular endothelial growth factor regulated endothelial nitric oxide synthase and NO release, uptake of amino acid L-arginine and accelerates its degradation. This ultimately results in the improper functioning of endothelium and increased burden of vascular risk as NO derived from endothelium is highly essential for prevention of platelets aggregation, reduction of intima proliferation and control of vascular tone. UA is also known to involve greatly in the progress of atherosclerotic lesions as it produces various factors such as IL-1ß, IL-6 and TNF-α and smooth muscle cells to produce monocyte chemoattractant protein-1 via the triggering of mononuclear cells. UA also stimulates innate immune response which is a principal cause of atherosclerosis, arterial hypertension and CHF (Ndrepepa, 2018).

4.19.3.3 Renal disorder

The relationship between elevated levels of UA and kidney diseases has been known since time immemorial. Kidney is the major organ which regulates the exit of UA from the body for the maintenance of UA levels through processes like filtration from glomeruli and then reabsorption and secretion in the proximal tubule. With these mechanisms almost 90% of UA is absorbed into the blood capillaries (El Ridi and Tallima, 2017). The renal tubular process for UA is controlled mainly by proteins which come under the category of organic anion transporter class. Urate transporter 1 (URAT1) encoded by SLC22A12 gene present on the apical membrane of renal proximal tubule and OAT1 and OAT3, encoded by the SLC22A6 and SLC22A8 genes present on the basolateral membrane of the renal proximal tubules forms principal pathway for the luminal excretion of UA. The imbalance between increased production of UA and improper exit of UA from the body due to impaired kidney functions results in the higher levels of UA in the body. Due to this increased amount of UA the chances of acute kidney injury increase and intraglomerular mesangial cells starts losing their contractile activity. Hyperuricemia also interferes with the normal functioning of mesangial and proximal tubule epithelial cells and damages them through the upregulation of NLRP3 and IL -1b via TLR4 dependent mechanism. Experimental studies revealed that hyperuricemia lead severe injury to endothelial cells and induce pro-inflammatory and chemotactic cytokines and stimulate inflammasome which results in chronic kidney disease in patients suffering from type 2 diabetes. UA also effects kidney by the formation and deposition of stones due to its accumulation. UA stones constitute almost 10% of total kidney stones and come second to the calcium oxalate stones which forms the major percentage of the kidney stones. Low urine pH because of improper urinary uric acid excretion, that is, <5.5 is the most alarming and notable reason for the crystallization of UA into crystals. Some other reasons for low urinary pH are severe dehydration, chronic diarrhea and diabetic ketoacidosis. Recently one experimental study conducted in rat models revealed that elevated levels of UA are associated with the start of glomerular hypertension and renal diseases addressed by the tubulointerstitial fibrosis and glomerular injury. Similarly some pilot studies conduct in patients with chronic kidney disease also showed decrease in the rate of renal diseases with the decrease in the concentration of UA (El Ridi and Tallima, 2017; Johnson et al., 2013).

4.19.4 Experimental studies conducted to elucidate the risk of uric acid

Haig and Oxon in 1889 revealed in one of their study that the amount of UA in blood has a great impact over arterial tension. Gerteler et al., in 1951 proposed the significant role of UA in coronary heart disease (CHD) after considering the results of their study which showed markedly increased levels of UA in subjects suffering from CHD as compared to normal subjects. This finding resulted in the addition of UA as one of the serious parameters to be identified for CVD in Framingham study design. Since then numerous epidemiological studies have been conducted till date to investigate the role of UA in the pathogenesis of CVD and other diseases and most of the studies clearly suggested the close association between increased levels of UA and CVD (Ndrepepa, 2018). Same results of increased risk of CVD with increasing levels of UA were obtained with the study conducted in 15,773 subjects in Third National Health and Nutrition Examination Survey. Krishnan et al., conducted a study in 12,866 men and found a close association between increased levels of UA and myocardial infarction (Krishnan et al., 2006). A study conducted by Bos et al., in 4385 mid-age volunteers in Rotterdam clearly demonstrated the independent relationship between UA and myocardial infarction and stroke (Bos et al., 2006). One study conducted in 13,338 participants from Atherosclerosis Risks in Communities and the Cardiovascular Health Study revealed that the small change in 1mg/dl UA is significantly related with 7 to 11% increased risk for the development of CVD (Weiner et al., 2008). A cross-sectional population-based study conducted in the First National Health and Nutrition Examination survey by using 5926 volunteers of age ranging from 24-75 years gave strong evidence of association of UA and increased risk of CVD. Another, Apolipoprotein Mortality Risk Study conducted in 417,734 men and women in Stockholm with the aim of determination of association of UA and the risk for myocardial infarction showed strong association of UA and myocardial infarction or ischemia stroke in women than men (Holme et al., 2009). A prospective coherent study conducted by taking data from MJ Health Screening Centers in Taiwan in 41,879 men and 48,514 women of ages above 35 years old to determine association between hyperuricemia and cardiovascular problems revealed that hyperuricemia play significant role and is major independent parameter in CVD and mortality (Fang and Alderman, 2000). Chuang et al., carried out a study to know the association of UA with CVD by determining the role of UA in ischemic heart disease (IHD). In this study National Health Insurance and the Death Certification Registry databases were used to gather overall hospital records to figure out IHD events as per ICD-9-CM codes 410-414. Hazard ratios were estimated using Cox proportional hazard model and it was determined that men were suffered more than women and results showed association between hyperuricemia and IHD (Chuang et al., 2012). Lin et al., performed an experimental study to determine UA as independent mortality marker in 1054 patients who were angiographically diagnosed with obstructive coronary artery disease (CAD). The study was conducted for an average time of 3.2 years using Cox regression analysis in both men and women and during this tenure it was found that the mean level of UA in all participants was 410.4 mol/L and with the results of the study it was concluded that UA may be an independent mortality parameter in CAD (Lin et al., 2013). Majority of studies conducted in order to find out the relationship between increased levels of UA and heart related problems such as hypertension and CVD, clearly revealed great association between UA and these problems and gave strong evidences of prevalence of this association among women and young patients (Bombelli et al., 2014). Li et al., performed a dose response meta analysis to know the association between hyperuricemia and Coronary heart disease (CHD) by using random effect model and concluded that hyperuricemia may increase the CHD mortality, particularly in females (Li et al., 2016). Junnan et al., conducted a study to identify the role of hyperuricemia in CAD in 2142 elderly patients of age 65 years or greater for 5 years. Subjects who were suffered from diseases like hypertension, obesity, diabetes, hyperlipidemia, chronic kidney diseases, gout, any cardiovascular disorder, hyperuricemia etc were strictly excluded from the studies. To determine hazard ratio for CAD between hyperuricemic and patients with normal UA levels Poisson regression analysis was used and with the results of the studies it has been found that hyperuricemia was associated with the increased risk of CAD and acts like a valuable biomarker for the prediction of development of CAD (Wu et al., 2017).

Studies other than CVD were also conducted by researchers working in this arena to know the association of UA with diseases like hypertension, renal disorders, diabetes mellitus, metabolic syndrome etc.

Billa et al., determined the prevalence of hyperuricemia in patients suffering from hypertension (HTN) and type 2 diabetes mellitus (T2DM) in Indian subjects. In this retrospective study patients getting diagnosis for hyperuricemia in various health centers during the months from April to May 2017 were included and data related to levels of UA, T2DM and HTN were taken with the patients of mean age 47.9 years and found sufficiently high prevalence of hyperuricemia in both T2DM and HTN. They have concluded age wise increase in the levels of UA and increase in the prevalence of hyperuricemia with increasing duration of T2DM and HTN (Billa et al., 2018). Wang et al., conducted a review and meta-analysis to know the association between levels of UA and hypertension (HTN) through extensive literature review with the help of highly informative databases such as MEDLINE, EMBASE and CBM in September 2013 to gather articles based on the studies related to levels of UA and HTN and accessed quality of these articles using Newcastle-Ottawa Scale. Random effect model was employed to access the relationship between hyperuricemia and HTN by including total 25 studies with 97,842 participants and based on the result of the meta-analysis they concluded that hyperuricemia is high risk factor for the development of HTN and play critical role in the prevalence of HTN (Wang et al., 2014). Buzas et al., in one of his research studies evaluated the association of UA with arterial hypertension, blood pressure, kidney function and carotid intima membrane thickness in Romanian adults. In this study a survey called Study for the Evaluation of Prevalence of Hypertension and Cardiovascular Risk in Romania (SEPHAR III) was conducted by taking 1920 adults including both males and females with mean age 48.63 years and 56.76% females to collect data related to UA levels, blood pressure measurement, kidney function through glomerular filtration rate and carotid IMT. Results of the study revealed that increased levels of UA are greatly associated with the arterial hypertension, decrease in kidney function and marked increase IMT values (Buzas et al., 2018). Stein et al., conducted one study to determine the association between increased UA levels and damage to the nucleosides due to oxidation with the assessment of urinary 8-hydroxydeoxyguanosine (8-OHdG) in patients with type 2 diabetes (T2D) and in healthy individuals. In this study 61 subjects were pooled into two groups on the basis of median UA levels < 5.3 mg/dl and their UA levels. Results of the studies showed marked increase in the levels of 8-OHdG in patients suffering from T2D and UA levels ≥ 5.3 mg/dL than patients having UA levels < 5.3 mg/dl. A very recent study conducted by Schirinzi et al. in the year 2018 related to the role of UA in an inherited neurodegenerative disorder Friedreich Ataxia (FRDA) to access UA as novel biomarker for this disease. Results of this study found elevated levels of UA under OS in the patients suffering from this disease which are completely opposite to that of other neurodegenerative disorders due to OS where the levels of UA were found to be excessively low in patients (Schirinzi et al., 2018).

Conclusion

Antioxidants are highly essential for the maintenance of proper functioning of various physiological processes by impeding oxidative stress and hence reduce the risk of most of the oxidative stress induced life-threatening diseases such as cancer, ageing, depression, diabetes, metabolic syndrome, cardiovascular disorders etc. Uric acid, a natural antioxidant which is responsible for half of the antioxidant activity of blood is known to play a crucial role in the normal functioning of human body. Since several decades it has been a vexing question that increased levels of UA in body are actually responsible for the prophylaxis of severe diseases or these increased levels of UA in blood are the main underlying causes for the development of these diseases because there are various studies which support both facts. There is a great controversy in the fact that whether increased levels of UA are beneficial, or these are harmful for the human body. Therefore, extensive research studies are desirable to know the actual facts related to the increased levels of UA.

References

Ames, B.N., Cathcart, R., Schwiers, E., Hochstein, P., 1981. Uric acid provides an antioxidant defense in humans against oxidant-and radical-caused aging and cancer: a hypothesis. Proc. Natl. Acad. Sci. 78 (11), 6858–6862.

Batta, A., 2016. Uric acid is a significant antioxidant. Eur. J. Pharm. Med. Res. 3 (8), 242–248.

Becker, B.F., Reinholz, N., Ozcelik, T., Leipert, B., Gerlach, E., 1989. Uric acid as radical scavenger and antioxidant in the heart. Pflugers Arch. 415, 127–135.

Billa, G., Dargad, R., Mehta, A., 2018. Prevalence of hyperuricemia in Indian subjects attending hyperuricemia screening programs: a retrospective study. J. Assoc. Physicians India 66 (4), 43–46.

Black, C.N., Bot, M., Scheffer, P.G., Snieder, H., Penninx, B.W., 2018. Uric acid in major depressive and anxiety disorders. J. Affect. Disord 1 (225), 684–690.

Bombelli, M., Ronchi, I., Volpe, M., Facchetti, R., Carugo, S., Dell'Oro, R., Cuspidi, C., Grassi, G., Mancia, G., 2014. Prognostic value of serum uric acid: new-onset in and out-of-office hypertension and long-term mortality. J. Hypertes 32 (6), 1237–1244.

Borghi, C., Rosei, E.A., Bardin, T., Dawson, J., Dominiczak, A., Kielstein, J.T., Manolis, A.J., Perez-Ruiz, F., Mancia, G., 2015. Serum uric acid and the risk of cardiovascular and renal disease. J. Hypertens. 33 (9), 1729–1741.

Bos, M.J., Koudstaal, P.J., Hofman, A., Witteman, J.C., Breteler, M.M., 2006. Uric acid is a risk factor for myocardial infarction and stroke: the Rotterdam study. Stroke 37 (6), 1503–1507.

Buzas, R., Tautu, O.F., Dorobantu, M., Ivan, V., Lighezan, D., 2018. Serum uric acid and arterial hypertension, data from Sephar III survey. PLoS One 13 (7), e0199865.

Cervantes, G.K., Llanas, C.D., Husi, H., 2017. CVD and oxidative stress. J. Clin. Med 6 (2), 22.

Chuang, S.Y., Chen, J.H., Yeh, W.T., Wu, C.C., Pan, W.H., 2012. Hyperuricemia and increased risk of ischemic heart disease in a large Chinese cohort. Int. J. Cardiol. 154 (3), 316–321 9.

Cortese, F., Giordano, P., Scicchitano, P., Faienza, M.F., De Pergola, G., Calculli, G., Meliota, G., Ciccone, M.M., 2019. Uric acid: from a biological advantage to a potential danger. A focus on cardiovascular effects. Vasc. Pharmacol. 29, 106565.

Cutler, R.G., Camandola, S., Feldman, N.H., Yoon, J.S., Haran, J.B., Arguelles, S., Mattson, M.P., 2019. Uric acid enhances longevity and endurance and protects the brain against ischemia. Neurobiol. Aging 75, 159–168.

de Oliveira, E.P., Burini, R.C., 2012. High plasma uric acid concentration: causes and consequences. Diabetol. Metab. Syndr 4 (1), 12.

El Ridi, R., Tallima, H., 2017. Physiological functions and pathogenic potential of uric acid: a review. J. Adv. Res. 8 (5), 487–493.

Fang, J., Alderman, M.H., 2000. Serum uric acid and cardiovascular mortality: the NHANES I epidemiologic follow-up study, 1971–1992. JAMA 283 (18), 2404–2410.

Glantzounis, G.K., Tsimoyiannis, E.C., Kappas, A.M., Galaris, D.A., 2005. Uric acid and oxidative stress. Curr. Pharm. Des. 11 (32), 4145–4151.

Hediger, M.A., Johnson, R.J., Miyazaki, H., Endou, H., 2005. Molecular physiology of urate transport. Physiology 20 (2), 125–133.

Hafez, R.M., Abdel-Rahman, T.M., Naguib, R.M., 2017. Uric acid in plants and microorganisms: Biological applications and genetics-a review. J. Adv. Res. 8 (5), 475–486.

Harrison, R., 2002. Structure and function of xanthine oxidoreductase: where are we now? Free Radic. Biol. Med. 33, 774–797.

Holme, I., Aastveit, A.H., Hammar, N., Jungner, I., Walldius, G., 2009. Uric acid and risk of myocardial infarction, stroke and congestive heart failure in 417, 734 men and women in the Apolipoprotein MOrtality RISk study (AMORIS). J. Intern. Med. 266 (6), 558–570.

Johnson, R.J., Kang, D.H., Feig, D., 2003. Is there a pathogenetic role for uric acid in hypertension and cardiovascular and renal disease? Hypertension 41, 1183–1190.

Johnson, R.J., Titte, S., Cade, J.R., Rideout, B.A., Oliver, W.J., 2005. Uric acid, evolution and primitive cultures. Semin. Nephrol. 25 (1), 3–8.

Johnson, R.J., Nakagawa, T., Jalal, D., Sanchez-Lozada, L.G., Kang, D.H., Ritz, E., 2013. Uric acid and chronic kidney disease: which is chasing which? Nephrol. Dial. Transplant 28 (9), 2221–2228.

Karwur, F.F., Pujiastuti, D.R., 2017. Uric acid homeostasis and disturbances. Folia Med. Indonesiana 53 (4), 292–298.

Krishnan, E., Baker, J.F., Furst, D.E., Schumacher, H.R., 2006. Gout and the risk of acute myocardial infarction. Arthritis & Rheumatism 54 (8), 2688–2696.

Lei, L., Wang, J.G., 2018. Dietary sodium intake and serum uric acid: a mini-review. Pulse. 6 (1-2), 124–130.

Li, M., Hu, X., Fan, Y., Li, K., Zhang, X., Hou, W., Tang, Z., 2016. Hyperuricemia and the risk for coronary heart disease morbidity and mortality a systematic review and dose-response meta-analysis. Sci. Rep. 27 (6), 19520.

Lin, G.M., Li, Y.H., Zheng, N.C., Lai, C.P., Lin, C.L., Wang, J.H., Jaiteh, L.E., Han, C.L., 2013. Serum uric acid as an independent predictor of mortality in high-risk patients with obstructive coronary artery disease: a prospective observational cohort study from the ET-CHD registry, 1997–2003. J. Cardiol 61 (2), 122–127.

Maiuolo, J., Oppedisano, F., Gratteri, S., Muscoli, C., Mollace, V., 2016. Regulation of uric acid metabolism and excretion. Int. J. Cardiol 15 (213), 8–14.

Mount, D.B., Kwon, C.Y., Zandi, N.K., 2006. Renal urate transport. Rheum. Dis. Clin. 32 (2), 313–331.

Ndrepepa, G., 2018. Uric acid and cardiovascular disease. Clin. Chim. Acta 1 (484), 150–163.

Nery, R.A., Kahlow, B.S., Skare, T.L., Tabushi, F.I., Castro, A.D., 2015. Uric acid and tissue repair. ABCD. Arquivos Brasileiros de Cirurgia Digestiva (Sao Paulo) 28 (4), 290–292.

Passi, S., Gianni, G., Cocchi, M., 2006. Oxidative stress in brain: neurodegenerative diseases and possible treatment (part III). Prog. Nutr 8 (4), 241.

Salim, S., 2017. Oxidative stress and the central nervous system. J. Pharmacol. Exp. Ther. 360 (1), 201–205.

Sautin, Y.Y., Johnson, R.J., 2008. Uric acid: the oxidant-antioxidant paradox. Nucleosides. Nucleotides. Nucleic. Acids. 27 (6-7), 608–619.

Schirinzi, T., Vasco, G., Zanni, G., Petrillo, S., Piemonte, F., Castelli, E., Bertini, E.S., 2018. Serum uric acid in Friedreich Ataxia. Clin. Biochem. 54, 139–141.

Settle, T., Klandorf, H., 2014. The role of uric acid as an antioxidant in selected neurodegenerative disease pathogenesis: a short review. Brain Disord. Ther 3, 129.

Sridevi, P., Budde, S., Adapa, D., 2018. Anti-oxidants and their role in disease management. Int. J. Med. Res. Health Sci. 7 (3), 175–190.

Todd, A.S., Walker, R.J., MacGinley, R.J., Kelly, J., Merriman, T.R., Major, T.J., Johnson, R.J., 2017. Dietary sodium modifies serum uric acid concentrations in humans. Am. J. Hypertens 30 (12), 1196–1202.

Wang, Y., Chu, C., Wang, K.K., Hu, J.W., Yan, Y., Lv, Y.B., Cao, Y.M., Zheng, W.L., Dang, X.L., Xu, J.T., Chen, W., 2018. Effect of salt intake on plasma and urinary uric acid levels in Chinese adults: an interventional trial. Sci. Rep. 8 (1), 1434.

Wang, J., Qin, T., Chen, J., Li, Y., Wang, L., Huang, H., Li, J., 2014. Hyperuricemia and risk of incident hypertension: a systematic review and meta-analysis of observational studies. PLoS One 9 (12), e114259.

Weiner, D.E., Tighiouart, H., Elsayed, E.F., Griffith, J.L., Salem, D.N., Levey, A.S., 2008. Uric acid and incident kidney disease in the community. J. Am. Soc. Nephrol. 19 (6), 1204–1211.

Whiteman, M., Ketsawatsakul, U., Halliwell, B., 2002. A reassessment of the peroxynitrite scavenging activity of uric acid. Ann. N. Y. Acad. Sci. 962, 242–259.

Wu, J., Lei, G., Wang, X., Tang, Y., Cheng, H., Jian, G., Wu, X., Wang, N., 2017. Asymptomatic hyperuricemia and coronary artery disease in elderly patients without comorbidities. Oncotarget 8 (46), 80688.

Yu, Z.F., Bruce-Keller, A.J., Goodman, Y., Mattson, M.P., 1998. Uric acid protects neurons against excitotoxic and metabolic insults in cell culture, and against focal ischemic brain injury in vivo. J. Neurosci. Res. 53, 613–625.

CHAPTER

Vanillin 4.20

Gökçe Ş. Karatoprak

Department of Pharmacognosy, Faculty of Pharmacy, Erciyes University, Kayseri, Turkey

4.20.1 Introduction

Vanillin is one of the most popular and most widely used flavoring natural products worldwide, found as a glycoside in the fruit of natural vanilla. It is known that vanilla has been used as a flavoring agent by the Aztecs since the 14th century and Europe has been introduced to the use of vanilla via Cofiezin's discovery of the vanilla pods in the Aztec kingdom of Montezuma (Deveci et al., 2016; Clark, 1990). The main source for obtaining natural vanillin is cured *Vanilla planifolia* pods. In addition, *Vanilla pompona* and *Vanilla tahitensis*, which have less vanillin content, have also commercial significance (Kundu, 2017). It has been reported that vanilla extract contains many volatile compounds such as guaiacol, *p*-anisaldehyde and methyl cinnamate, and these compounds contributes to the flavor but vanillin is the main flavor content of vanilla (Anuradha et al., 2013). Due to the high demand of vanillin as a flavoring agent, and its use in the food, pharmaceutical and cosmetic industries has made vanillin synthesis the focus of interest for researchers (Kundu, 2017). Currently, about 50% of the synthetic vanillin production in the world is used as an intermediate in the chemical and pharmaceutical industries for the production of herbicides, antifoam agents or drugs (papaverine, 1-methyldopa, 1-dopa, and trimethoprim) (Walton et al., 2003).

Vanillin is chemically known as 4-hydroxy-3-methoxybenzaldehyde. This compound has been known as a naturally occurring hindered phenolic compound and can be found in several species of genus *Vanilla* (family: Orchidaceae) (Sinha et al., 2007) (Figs. 4.20.1, and 4.20.2). Studies of its biosynthesis revealed several pathways but the general approach is that the vanillin is synthesized from L-phenylalanine, therefore it is considered to be a product of the phenylpropanoid pathway, and that the hydroxyl group at the 4-position of the aromatic ring (para to the side chain) is produced by the action of the cytochrome P450 enzyme cinnamate 4-hydroxylase (Havkin-Frenkel and Belanger, 2011).

Antioxidants act by neutralizing the reactive oxygen species (ROS), avoiding the oxidizing effects on lipids, proteins and DNA bases. Therefore, compounds of natural origin with antioxidant power have the potential of preventing several diseases. But most of the phenolic compounds which are naturally occurring display

Antioxidants Effects in Health. DOI: https://doi.org/10.1016/B978-0-12-819096-8.00009-4

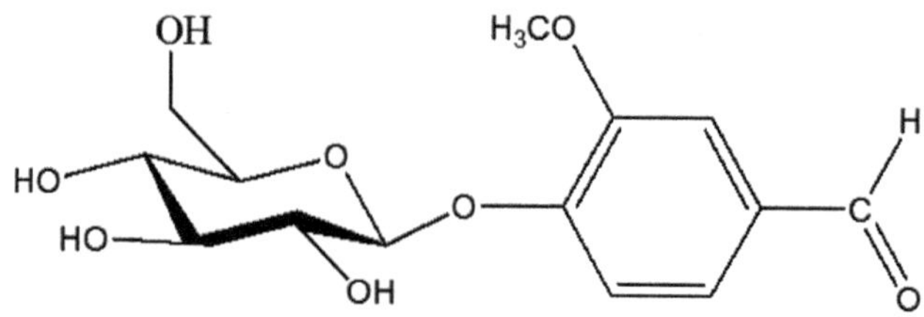

FIG. 4.20.1 The glucoside containing vanillin.

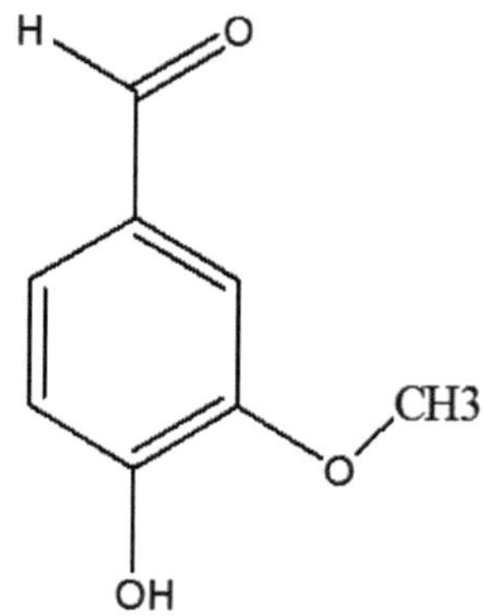

FIG. 4.20.2 Vanillin.

antioxidant and pro-oxidant activity, therefore mixed results regarding the antioxidant effect of vanillin have been declared. Although it is known that the vanillin can react in three different types due to the aldehyde group, phenolic hydroxyl group and aromatic nucleus, the researches proved that it offers interesting antioxidant properties depending on the method used (Chethanaa, 2012).

4.20.2 *In-vitro* antioxidant activity of vanillin

Vanillin, isocurcumin, diacetylcurcumin, and bisdesmethoxycurcumin all had low ability in radical scavenging activity and EC50 values were found higher than100 μM in the study of Deters et al. (2008).

In the research of Tai et al. (2011), the scavenging effect of vanillin on 1,1-diphenyl-2-picrylhydrazyl (DPPH) radical and galvinoxyl radicals was found in very small amounts but in the 2,2′-Azino-bis(3-ethylbenzthiazoline-6-sulfonic acid) ($ABTS^{\bullet +}$) test system vanillin showed significant radical scavenging activity. Because of vanillin's slowly and continuously reaction with $ABTS^{\bullet +}$, it is reported that in a period of 120 min, one molecule of vanillin scavenges 2.6 mol of ABTS radical. The oxidative self-dimerization of vanillin increased the potency of vanillin to neutralize the ABTS radical. Also, vanillin was found to have the potent antioxidant capacity in the oxygen radical absorbance capacity (ORAC) test system and the ranking of peroxyl radical inhibition was mentioned as vanillin > vanillic acid ≥ ferulic acid > Trolox > ascorbic acid. In the oxidation of erythrocyte membranes via AAPH (2,2′-azobis(2-methylpropionamidine) dihydrochloride) assay, the inhibition order in oxidation of erythrocyte membranes was found as ferulic acid (25 μM) > vanillin (25 μM) ≥ vanillic acid (25 μM) ≈ Trolox (50 μM) > ascorbic acid (50 μM) (Tai et al., 2011).

Vanillin and vanillic acid could not be found strong enough to inhibit 50% of beta carotene linoleic acid oxidation, while the values that inhibit 50% of DPPH radical were reported as 480 and 216 ppm, respectively (Shymala et al., 2007).

The activity of vanillin against the degradation of the structure of linoleic acid and cholesterol was examined by Rosa et al. (2005). The researchers have suggested that the phenolic hydroxyl in the vanillyl part of the vanillin compound has an important role in protecting these biomolecules against attack by ROO^- radicals.

Harish et al. (2005) assessed that vanillin showed the lowest activity in superoxide radical scavenging with 2.33 ± 0.16 μmol/mL among the isolated secondary metabolites from *Decalepis hamiltonii*. It also demonstrated strong inhibition of lipid peroxidation and hydroxyl radical with 0.37 ± 0.04 μmol/mL, 0.16 ± 0.01 μmol/mL, respectively.

In a study with plasmid pBR322, human and mouse peripheral blood leucocytes and splenic lymphocytes, vanillin was found to inhibit DNA damage to γ-radiation. Possible mechanism has been reported as the removal of radicals that occur during radiation, except for the modulation of DNA repair observed previously (Maurya et al., 2007).

Pretreatment of vanillin in rotenone induced SH-SY5Y neuroblastoma cells was found to attenuate mitochondrial dysfunction, oxidative stress (decreasing intracellular ROS generation), and apoptosis (Dhanalakshmi et al., 2015).

Murcia and Martinez-Tome (2001) reported that vanillin had an inhibition of lipid peroxidation of 21%–26%, inhibition of desoxybirose attack (OH·) 11 to 16%, and no hydrogen peroxide and HOCl scavenging activity.

4.20.3 *In-vivo* antioxidant activity of vanillin

In an *in-vivo* antioxidant test system, mouse were orally given 25 mg/mL (100 mg/kg) concentrated vanillin and plasma samples were evaluated with plasma ORAC assay. Metabolites of vanillin were determined as vanillic acid and protocatechic acid. Plasma vanillin level and the plasma antioxidant activity increased after oral administration. Plasma ORAC test showed the highest activity in the 5th minute when vanillin and its metabolites were in the highest amounts (Tai et al., 2011).

In the study, in which mRNA expressions of kidney glutathione peroxidase, superoxide dismutase, metallothionein (MT1 and MT2) were significantly enhanced in mice treated with potassium bromate ($KBrO_3$), vanillin was administered intraperitoneally (i.p) to the treatment group (100 mg/kg). In the treatment group, vanillin brought the mRNA levels closer to the values in the control group and showed that it has radical scavenging activity (Ben Saad et al., 2016).

Pre-treatment with vanillin in rat (daily injection, 150 mg/kg bw), prior to carbon tetrachloride (CCL_4), decreased the hepatic lipid peroxidation (MDA) and carbonyl (PCO) protein formation. It prevented the decrease in antioxidant enzymes like catalase and superoxide dismutase (SOD) and also glutathione (GSH) levels caused by CCI_4 (Makni et al., 2011).

In mouse liver microsomes, vanillin showed little or no antioxidant activity against lipid peroxidation and weak superoxide anion scavenging (IC50, 2.945 ± 247 μM) (Yan-Chun and Rong-Liang, 1991).

Saad et al. (2017), designed a study to estimate the cardioprotective efficacy of vanillin in $KBrO_3$ stimulated cardiotoxicity in adult mice. Effects on lipid peroxidation, enzymatic and non-enzymatic antioxidants levels and histopathological evidences were evaluated. For a period of 15 days $KBrO_3$ induced mice were administered daily vanillin (100 mg/kg by i.p injection). A significant reduction in lipid peroxidation and protein oxidation levels and increased levels of antioxidant enzymes and non-enzymatic antioxidants were found in the vanillin-treated group.

In rats treated with metribuzine, while MDA levels increased in all tissues compared to the control group, it was observed that GPX enzyme and GSH levels fluctuated. However, treatment with vanillin (150 mg/kg) for three weeks ensured normalization of the hematological and serum biochemical profile and reversed the evaluated oxidative stress parameters (Kadeche et al., 2016).

Antioxidant activity of many phenolic compounds similar to vanillin's structure has been elucidated. The hydroxyl in the aromatic ring has an active role in exhibiting the antioxidant activity through the homolytic cleavage of the O-H bond. It has been shown in many studies that vanillin has the ability to scavenge radicals such as hydroxyl and $ABTS^{\cdot +}$. The reason for the different results obtained in in vitro chemical reactions may be due to the use of different radical sources, the conditions of the reactions and different mechanisms. 1 mole of vanillin has only one oxidizable hydroxyl group in the aromatic ring, so, it can scavenge multiple free radical equivalents through the formation of adducts with radical, and the formation of oxidative dimer increases the ability of radical scavenging activity. Thus, phenolic substances may have antioxidant activity as demonstrated by vanillin. Three reaction mechanisms for the antioxidant activity of vanillin are as follows: hydrogen transfer from phenolic OH, single electron transfer and sequential proton electron transfer mechanism (Pereira Bezerra et al., 2016; Zhao et al., 2017).

4.20.4 Prooxidant activity of vanillin

The prooxidant effects of free radicals derived from vanillin by oxidation mediated by horseradish peroxidasc/hydrogen peroxide were evaluated on biomolecules such as cysteine, GSH, ovalbumin and coenzyme NADPH. Castor et al. (2010) evaluated that transient free radicals created by vanillin may have a pro-oxidant role when produced intracellularly. Vanillin (10 μM) was found responsible for the oxidation of GSH, sulfhydryl groups and NADPH.

Aruoma et al. (1990) observed that propyl gallate (PG) and vanillin slightly stimulate ·OH production when tested in the $FeCl_3$/EDTA/H_2O_2/ ascorbate assay. This stimulation is particularly pronounced when ascorbate is removed from the reaction mixture. It was determined that the formation of OH in the reaction was only induced by vanillin or PG when iron ions were added to the mixture as a complex with ethylenediaminetetraacetic acid (EDTA). Deoxyribose degradation was also inhibited by PG and vanillin if added to the mixture as an iron salt in the form of $FeCl_3$. It has been stated that the supplementation of iron-EDTA and vanillin aroma to the foods is possible to produce pro-oxidant effects on the proteins and DNA in the food and cause the formation of toxic products (Aruoma et al., 1990).

Pro-oxidant effect of vanillin on Fe(III) superoxide induced damage to benzoate, deoxyribose, amino acids and benzoate hydroxylation was reported by Liu and Mori (1993). According to the results of the experiment, vanillin was found to induce bleomycin-iron-induced damage of DNA (237.9%) at only a high concentration of 10 mM.

In the work by Nguyen et al. (2014), researchers declared that at increasing concentrations vanillin also increases the levels of oxidation, and that the effect of treatment with 8 mM vanillin was almost identical on intracellular oxidation with 0.4 mM H_2O_2 in yeast cells. According to the results, after the application of vanillin, the accumulated ROS increased and more oxidation occurred in the intracellular environment.

4.20.5 Vanillin formulations and their antioxidant activities

The antioxidant properties of the vanillin loaded PLA (poly-lactic acid) nanoparticles in the ABTS radical scavenging activity were found lower than the free vanillin and showed both time and concentration dependent profile due to its prolonged release, while the activity of free vanillin was concentration dependent. This research revealed that nanoscale particles prepared with PLA are potential carriers for the delivery of vanillin (Dalmolin et al., 2016).

In order to increase the antioxidant and cytotoxic effects, different self-nano emulsifying drug delivery systems (SNEDDS) of vanillin were developed by Shakeel et al. (2016). In the DPPH scavenging assay, it was shown that the optimized vanillin SNEDDS has a more notable antioxidant effect compared to free vanillin. Also it has been stated that antioxidant potential of vanillin SNEDDS was analogous with standard ascorbic acid. The outcomes of this study showed that SNEDDS can be used efficiently to increase the therapeutic efficacy of vanillin.

4.20.6 Evaluation of vanillin in terms of human health

Industrial vanillin is mainly consumed as food and beverage. The amount of vanillin is low in topical products such as perfumes and products for skin care. Worldwide use of vanillin in food products means that almost everyone in the world is exposed to very little vanillin. The 10 mg/kg dose was approved by the FAO/WHO and the EU as the acceptable daily intake (ADI).

For a 70 kg person, 700 mg of vanillin is designated as an ADI. If this amount is given as an example of chocolate and ice cream, it corresponds to 700 and 7000 grams, respectively. It is supposed that even those who consume food and drink containing high amounts of vanillin do not have a vanillin intake higher than ADI (OECD, 1996).

Occupational exposure to vanillin is defined as the inhalation of vanillin powder by employee in the factory packaging area and this exposure is not more than 0.5 mg/kg/day. It is likely that this level will never be achieved as the amount of vanillin in the over-all dust is predicted to be only 10%. In addition, only less than 1% of this compound has enough particle size to get through to the lungs (OECD, 1996).

In the closed patch test applied to human dermis, vanillin was found to be non-irritating when examined at concentrations of 20% on 29 normal volunteers, 2% on 30 normal volunteers and 0.4% on 35 volunteers with dermatosis. Results of maximization tests on groups of 25 subjects reported that vanillin, examined at doses of 2% and 5% in petrolatum did not give sensitivity reaction. Sensitivity is only found in patients susceptible to vanillin, isogenol and coniferyl benzoate, suggesting that vanillin is a secondary allergen. Drug interactions between drugs metabolized by CYP2E1/CYP1A2 and vanillin may be possible (Opdyke, 1979).

4.20.7 Acute and repeated dose toxicity

In the studies of acute toxicity, LD50 values for rats were found to be 3.5–4 g/kg, which corresponds to a safety margin of 350 to 400 times for humans, which is considered to be the acceptable limit.

In oral repeated dose toxicity test, mice were observed to have 2.5 g/kg/day NOEL after oral administration. This result is also considered acceptable, giving a safety margin of 250 for consumer use (OECD, 1996).

4.20.8 Bioavailability

Pharmacokinetics of vanillin was determined in a study on mechanical hypersensitivity in a rat model of neuropathic pain (n = 30 rats). 20 or 100 mg/kg of vanillin was applied i.v. and p.o. to the rats. In this *in vivo* research, it was reported that baseline levels were taken for hyperalgesia and allodynia 5 days before surgery for pharmacodynamic study. After p.o. administration, the pharmacokinetic results were reported as follows: C (max) 290.24 ng/mL, T (max) 4 h, relative gap 62.17 L/h/kg and T (1/2) 10.3 h, bioavailability is 7.6% (Beaudry et al., 2010).

4.20.9 Clinical trials

In a double-blind, placebo-controlled study using 1 g of vanillin per day for 40 days in 30 patients with sickle cell disease, binding to HbS and reducing the polymerization and sickling by both an allosteric shifting of oxygen affinity and a stereospecific inhibition of polymer assembly by vanillin was reported (Archer et al., 2015).

In a single-blind randomized clinical study, 40 premature neonates with Apnea of prematurity (AOP) were divided into two groups as experimental (n = 20) and control (n = 20) groups. Cotton in which 2 mL of vanillin extract was impregnated exposed for 24 hours to the experimental group. When the results were examined, it was seen that the number of apnea attacks did not show a significant difference between the two groups ($p > 0.05$). The results of the experimental group were found to be more significant from the standpoint of the mean number of attacks, average heart rate and SaO_2 levels on the first and second days (Yaghoubi et al., 2017).

Conclusion

As discussed in this chapter, the antioxidant and prooxidant properties of vanillin, which is used as a flavor and fragrance agent, have been proven by different studies. It is likely that vanillin is effectively metabolized and excreted from the body, demonstrating its safety by exhibiting non-toxic effects in animal models. Although nanocarrier systems prepared with vanillin are not subject to enough research, a limited number of studies have shown that these systems can increase the stability, bioavailability and bioactivity of vanillin. Therefore, some promising advances in this area have been associated with enhanced antioxidant activity with extended release of vanillin, and this advantage will allow us to further evaluate its usability as an effective biopharmaceutical or industrial ingredient.

References

Antonia Murcia, M., Martínez-Tomé, M., 2001. Antioxidant activity of resveratrol compared with common food additives. J. Food Prot. 64 (3), 379–384.

Anuradha, K., Shyamala, B.N., Naidu, M.M., 2013. Vanilla–its science of cultivation, curing, chemistry, and nutraceutical properties. Crit. Rev. Food Sci. Nutr. 53 (12), 1250–1276.

Archer, N., Galacteros, F., Brugnara, C., 2015. Clinical trials update in sickle cell anemia. Am. J. Hematol. 90 (10), 934–950.

Aruoma, O.I., Evans, P.J., Kauri, H., Sutcliffe, L., Halliwell, B., 1990. An evaluation of the antioxidant and food additives and of trolox C, vitamin E and probucol. Free Radic. Res. Commun. 10, 143–157.

Beaudry, F., Ross, A., Lema, P.P., Vachon, P., 2010. Pharmacokinetics of vanillin and its effects on mechanical hypersensitivity in a rat model of neuropathic pain. Phytother. Res. 24 (4), 525–530.

Ben Saad, H., Driss, D., Ellouz Chaabouni, S., Boudawara, T., Zeghal, K.M., Hakim, A., Ben Amara, I., 2016. Vanillin miti–gates potassium bromate–induced molecular, biochemical and histopathological changes in the kidney of adult mice. Chem. Biol. Interact. 252, 102–113.

Castor, L.R.G., Locatelli, K.A., Ximenes, V.F., 2010. Pro-oxidant activity of apocynin radical. Free Radic. Biol. Med. 48, 1636–1643.

Chethanaa, B.K., Basavannab, S., Arthoba Naik, Y., 2012. Determination of vanillin in real samples using Lysine modified carbon paste electrode. J. Chem. Pharm. Res. 4 (1), 538–545.

Clark, G.S., 1990. Vanillin. Perfum. Flavor. 15, 45–54.

Dalmolin, L.F., Khalil, N.M., Mainardes, R.M., 2016. Delivery of vanillin by poly(lactic-acid) nanoparticles: development, characterization and *in vitro* evaluation of antioxidant activity. Mater. Sci. Eng. C 62, 1–8.

Deters, M., Knochenwefel, H., Lindhorst, D., Koal, T., Meyer, H.H., Hänsel, W., Kaever, V., 2008. Different curcuminoids inhibit T-lymphocyte proliferation independently of their radical scavenging activities. Pharm. Res. 25 (8), 1822–1827.

Deveci, B., Türkmen, S., Avcıkurt, C., 2016. Vanilla spice and investigation on areas of usage. J. Hum. Sci. 13, 5340–5350.

Dhanalakshmi, C., Manivasagam, T., Nataraj, J., Thenmozhi, A.J., Essa, M.M., 2015. Neurosupportive role of vanillin, a natural phenolic compound, on rotenone induced neurotoxicity in SHSY5Y neuroblastoma cells. Evid. Based Complement. Alternat. Med. 2015, 626028. doi:10.1155/2015/626028.

Harish, R., Divakar, S., Srivastava, A., Shivanandappa, T., 2005. Isolation of antioxidant compounds from the methanolic extract of the roots of *Decalepis hamiltonii* (Wight and Arn.). J. Agr. Food Chem. 53 (20), 7709–7714.

Havkin-Frenkel, D., Belanger, F.C., 2011. Handbook of Vanilla Science and Technology. Blackwell Publishing, India.

Kadeche, L., Bourogaa, E., Saoudi, M., Boumendjel, A., Djeffal, A., Feki, A., Messarah, M., 2016. Ameliorative effects of vanillin against metribuzin-induced oxidative stress and toxicity in rats. Int. J. Pharm. Pharm. Sci. 9 (1), 56–62.

Kundu, A., 2017. Vanillin biosynthetic pathways in plants. Planta 245 (6), 1069–1078.

Liu, J., Mori, A., 1993. Antioxidant and pro-oxidant activities of p-hydroxybenzyl alcohol and vanillin: effects on free radicals, brain peroxidation and degradation of benzoate, deoxyribose, amino acids and DNA. Neuropharmacology 32 (7), 659–669.

Makni, M., Chtourou, Y., Fetoui, H., Garoui, E.M., Boudawara, T., Zeghal, N., 2011. Evaluation of the antioxidant, anti-inflammatory and hepatoprotective properties of vanillin in carbon tetrachloride treated rats. Eur. J. Pharmacol. 668, 133–139.

Maurya, D.K., Adhikari, S., Nair, C.K.K., Devasagayam, T.P.A., 2007. DNA protective properties of vanillin against γ-radiation under different conditions: possible mechanisms. Mutat. Res. 634, 69–80.

Nguyen, T.T., Iwaki, A., Ohya, Y., Izawa, S., 2014. Vanillin causes the activation of Yap1 and mitochondrial fragmentation in *Saccharomyces cerevisiae*. J. Biosci. Bioeng. 117 (1), 33–38.

OECD SIDS initial assessment profile, MDI. 1996. http://www.inchem.org/documents/sids/sids/121335.pdf.

Opdyke, D.L.J., 1979. Monographs on fragrance raw materials: vanillin. Food Cosmet. Toxicol. 15, 633–638.

Pereira Bezerra, D., Nascimento Soares, A.K., a de Sousa, D.P., 2016. Overview of the role of vanillin on redox status and cancer development oxidative medicine and cellular longevity. 2016, 9, Article ID 9734816. https://doi.org/10.1155/2016/9734816.

Rosa, A., Deiana, M., Corona, G., Atzeri, A., Incani, A., Appendino, G., Dessì, M.A., 2005. Protective effect of capsinoid on lipid peroxidation in rat tissues induced by FeNTA. Free Radical Res. 39 (11), 1155–1162.

Saad, H.B., Amara, I.B., Boudawara, T., 2017. Preventive effect of vanillin on lipid peroxides and antioxidants in potassium bromateinduced cardiotoxicity in adult mice: biochemical and histopathological evidences. Res. Rev. Biosci. 12 (1), 113.

Shakeel, F., Haq, N., Raish, M.A., Siddiqui, N.K., Alanazi, F.A., Alsarra, I., 2016. Antioxidant and cytotoxic effects of vanillin via eucalyptus oil containing self-nanoemulsifying drug delivery system. J. Mol. Liq. 218, 233–239.

Shyamala, B.N., Madhava Naidu, M., Sulochanamma, G., Srinivas, P., 2007. Studies on the antioxidant activities of natural vanilla extract and its constituent compounds through *in vitro* models. J. Agric. Food Chem. 55 (19), 7738–7743.

Sinha, A.K., Sharma, U.K., Sharma, N., 2007. A comprehensive review on vanilla flavor: extraction, isolation and quantification of vanillin and others constituents. Int. J. Food Sci. Nutr. 59 (4), 299–326.

Tai, A., Sawano, T., Yazama, F., Ito, H., 2011. Evaluation of antioxidant activity of vanillin by using multiple antioxidant assays. BBA Gen Subjects 1810 (2), 170–177.

Walton, N.J., Mayer, M.J., Narbad, A., 2003. Vanillin. Phytochemistry 63 (5), 505–515.

Yaghoubi, S., Salmani, N., Dehghani, K.D., Davoodi Zadeh Jolgeh, H., 2017. Investigating effect of olfactory stimulation by vanilla on the rate of apnea attacks in neonates with apnea of prematurity: a randomized clinical trial. Int. J. Pediatr. 5, 6221–6229.

Yan-Chun, Z., Rong-Liang, Z., 1991. Phenolic compounds and an analog as superoxide anion scavengers and antioxidants. Biochem. Pharmacol. 42 (6), 1177–1179.

Zhao, D., Sun, J., Sun, B., Zhao, M., Zheng, F., Huang, M., Sun, X., Li, H., 2017. Intracellular antioxidant effect of vanillin, 4-methylguaiacol and 4-ethylguaiacol: three components in Chinese Baijiu. RSC Adv. 7, 46395–46405.

CHAPTER

Vitamin A

4.21

Ömer F. Yakıncı[a,b], Ipek Süntar[b]
[a]National Poisons Information Service, Ministry of Health, Ankara, Turkey
[b]Department of Pharmacognosy, Faculty of Pharmacy, Gazi University, Ankara, Turkey

4.21.1 Introduction

A free radical is a molecule that contains one or more unpaired electrons. Reactive oxygen species (ROS) are oxygen radicals which cause oxidative stress leading to potential cellular damage. Oxidative stress is caused by the deterioration of the balance between the formation of ROS and antioxidant defense mechanisms. Several markers have been developed to assess oxygen-induced ROS damage. Most of these markers identify changes in oxidative-sensitive tissue components such as proteins, lipids, carbohydrates, and DNA. Although these markers are not specific for ROS damage, they may help diagnose by measuring changes in the presence of antioxidants. ROS can be produced either from metabolic processes or external sources. In conditions where ROS is active, cellular macromolecules become susceptible to oxidation. ROS is highly toxic and interacts with lipids, proteins, carbohydrates, and DNA, resulting in the pathogenesis of many chronic diseases, such as cancer, cardiovascular, autoimmune diseases (including rheumatoid and ankylosing spondylitis), various respiratory diseases, an increased risk of infectious diseases, diabetes (both noninsulin dependent and insulin-dependent), eye disease (cataract and retinal injury and age-related macular degeneration) and aging (Zadak et al., 2009; Temple, 2000; Mazlum, 2012; Goodarzi et al., 2018). Furthermore, it is also responsible for the pathogenesis of mental disorders such as schizophrenia, mood disorders, anxiety disorders, autism and attention deficit hyperactivity disorder (ADHD), and Alzheimer's disease (Mazlum, 2012).

An increase in oxidative stress can even lead to cell death by necrosis or apoptosis. Most cells can tolerate mild oxidative stress as they have adequate antioxidant defense capacity and repair systems that recognize and eliminate the molecules damaged by oxidation (Zadak et al., 2009). ROS can be removed by antioxidant defence system via enzymatic, metal chelating, and free radical scavenging activities. Moreover, intake of dietary antioxidants such as food or dietary supplements (including vitamins A, C, E, minerals, polyphenols, etc.) may help to maintain an adequate antioxidant status by inactivating toxic oxygen radicals (Lü et al., 2010; Goodarzi

Antioxidants Effects in Health. DOI: https://doi.org/10.1016/B978-0-12-819096-8.00064-1

et al., 2018). Indeed, dietary antioxidants can reduce the damaging effects of ROS, thus are suggested to be used for health promotion. Antioxidant vitamins such as vitamins A, C and E are considered as nonenzymatic defense against oxidative stress (Ayeleso et al., 2016). In this chapter we aimed to review the role of vitamin A in the antioxidative processes.

4.21.2 Vitamin A: Its functions and chemistry

The chemical structure of Vitamin A was first described in 1930s. It has an isoprenoid structure with a six-membered ring and an eleven-carbon side chain. It is a lipid-soluble compund. Provitamin A carotenoids are α-carotene, β-carotene, and β-cryptoxanthin; and among them the most abundant is β-carotene is converted to vitamin A, by an oxygenase in the intestine (Ayeleso et al., 2016). Vitamin A is a general term that refers to a retinoid form containing retinol, retinal, and retinoic acid, which is a mixture of the original. The chemical structures of provitamin A are shown in Fig. 4.21.1. Retinol, retinal, and retinoic acid are physiologically active forms of vitamin A. Retinol is the free alcohol form of vitamin A and can be converted to the retinal which is the active form of vitamin A in various tissues through enzymatic activity. The retinal can also be irreversibly converted to the strong transcription factor, retinoic acid. Apart from the retina, due to enzyme kinetics retinal levels are rather low in all tissues that support conversion to retinol or conversion to retinoic acid (Palace et al., 1999).

Vitamin A is an essential micronutrient that cannot be biosynthesized in the body and therefore, must be obtained from dietary sources (Kawaguchi et al., 2007). Vitamin A is mainly found as retinyl esters and provitamin A carotenoids in diets. Yellow and green vegetables (carrots, spinach, tomatoes, sweet potatoes, broccoli, zucchini), dairy products (cheese, milk, butter, margarine), fish, eggs, and organ meats (liver, kidney) are all important dietary sources. The recommended daily intake of vitamin A is 1000 retinol equivalents (RE) for adults and 375 for infants and 700 RE for children. Serum concentrations and diet intake of vitamin A have been shown to be significantly different between individuals on a daily, daily and seasonal basis, geographical regions and sex, age and alcohol consumption (Palace et al., 1999).

Vitamin A undergoes metabolic transformations as retinoid binding proteins as coligands. Thanks to these specific protein complexes, vitamin A is protected from oxidation in the β-barrel in the protein. Vitamin A is generally substantially protected *in vivo* by firstly lipid antioxidants and tocopherols in chylomicra and then their combination with certain carrier proteins. Therefore, vitamin A is a sequestered molecule whose transport and metabolism are regulated (Shils et al., 1994).

Depending on thc obtaining source (whether is an animal or a plant), vitamin A can be classified into two categories. Provitamin A is chiefly found in the plant sources, while retinoids are found chiefly in animal sources. Vitamin A is present as retinyl esters in foods of animal origin and is broken down into retinol and free fatty acids in the intestinal lumen. Retinyl esters are regenerated from retinol,

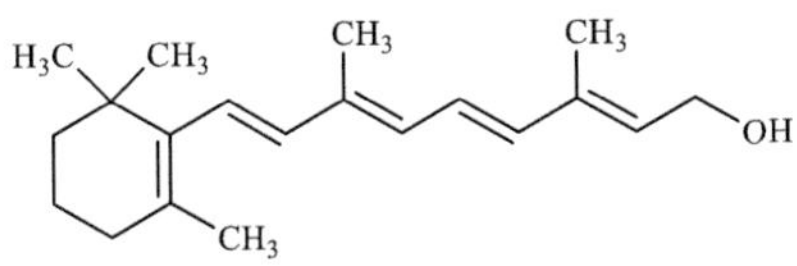

1

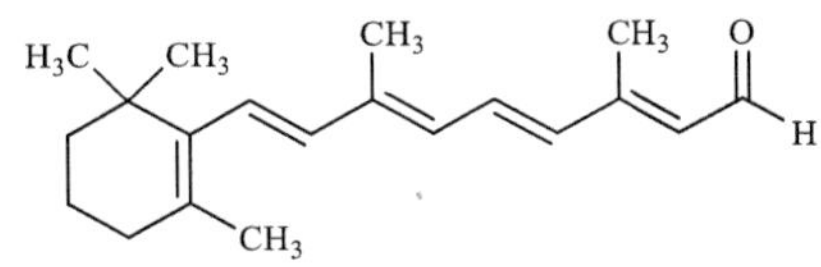

2

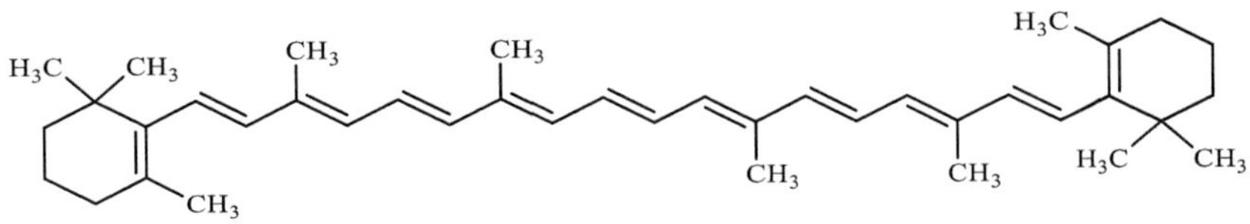

3

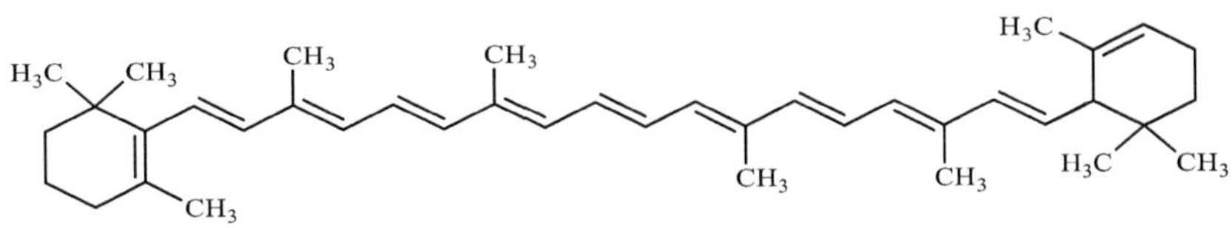

4

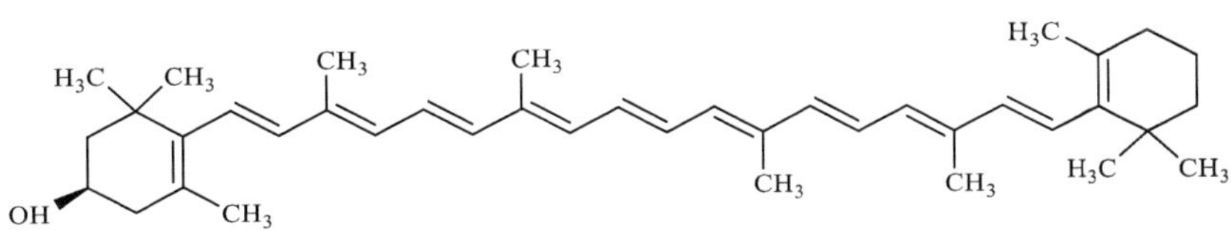

5

FIG. 4.21.1 Chemical structures of provitamin A carotenoids.

1: Retinol; 2: All- *trans*-retinal; 3: β-Carotene; 4: α-Carotene; 5: β-Cryptoxanthin.

which is introduced into the intestinal cell. β-carotene in the intestinal lumen is introduced into the intestinal cell by means of SR-BI (scavenger receptor class B Type 1) and retinoic acid or retinol with a number of enzymes, primarily β, β-carotene 15,15′-monoxygenase-1 (BCMO1). Retinyl esters and provitamin A carotenoids are incorporated into the structure of chylomicrons and released into the lymphatics. Most of vitamin A is stored in the liver. Vitamin A is carried in the blood by retinol binding protein (RBP). They have identified a specific membrane receptor (STRA6) for RBP. STRA6 not only binds to the retinol binding protein but also catalyzes the separation of vitamin A from the vitamin A-RBP complex (Mazlum, 2012).

Vitamin A is a regulator of important functions such as embryogenesis, cell growth differentiation, immune and reproductive functions and vision. In addition, this micronutrient is a very important antioxidant which contributes for protection against oxidative damage caused by free radicals in the human body. Vitamin A deficiency results in several problems in vision, fertility, and increased susceptibility to cancer. When plasma concentrations are low, retinyl ester reserves are converted to retinol by hydrolytic enzymes. On the other hand, vitamin A should be provided in balanced amounts, otherwise, excess portion of vitamin A is stored when taken at high concentrations. Long chain fatty acid esters of vitamins produced in the liver are the primary form of storage. Retinyl palmitate, oleate, stearate, myristate, and linoleate are generally isolated from special storage cells known as stellate cells in the liver, intestine, kidney, and lung. As a consequence of the administration of large amounts of vitamin A preparations, hypervitaminosis A can cause different serious problems including headache, vomiting, nausea, vertigo, visual disorientation, peeling of the skin, and chronic liver disease through exertion of its oxidative properties (Kawaguchi et al., 2007).

4.21.3 Antioxidant effect of vitamin A

There are about 600 different carotenoids and the majority of them has antioxidant activity, hence contribute physiological functions by increasing the responses immune system, intermittent junction communication, and carcinogen-metabolizing enzyme effect. Nonetheless, there are only about 50 carotenoid displaying provitamin A activity. Lutein, canthaxanthin, zeaxanthin, and lycopene have little or no provitamin A activity, whereas the common dietary carotenoids α- and β-carotene and β-cryptoxanthin have. Antioxidant effect has been reported for many provitamin A compounds, including vitamin A1 (retinol) and A2 (dehydroretinol) as well as α- and β-carotenes. Other similar carotenoid molecules found in significant amounts in human diet and tissues (lycopene, lutein, zeaxanthin, cataxanthin, neoxanthin, olaxanthin, astaxanthin, etc.) were also reported to possess antioxidant activity (Palace et al., 1999).

The antioxidant potential of vitamin A and carotenoids was first described in 1930s. Furthermore, vitamin A displays an important antioxidant potential in protecting human LDL against copper-stimulated oxidation (Nimse and Pal, 2015). The singlet oxygen is scavenged by carotenoids was then reported in 1968. Afterwards, in 1984, the mechanism by which carotenoids quench lipid radicals in biological membranes was first proposed. In recent years, a number of studies have examined the role of vitamin A and carotenoids as biological antioxidants. Today, more than 600 carotenoid type constituents are known to be present and numerous forms of vitamin A have been isolated and identified (Palace et al., 1999).

β-carotene displays antioxidant activity by neutralizing free radicals. It is the main source of vitamin A in the organism and displays a vital role in maintaining the cell health (Ayeleso et al., 2016). *In-vitro* studies have shown that carotenoids also inhibit the oxidation of fats under certain conditions and have antiatherosclerotic potential. Vitamin A reveals antioxidant activity in different ways. Vitamin A acts as an antioxidant by preventing lipid peroxidation of hydroperoxides in the cell (Yadav et al., 2016).

Retinol has been shown to inhibit iron-dependent peroxidation of rat liver microsomes by adriamycin. It has also been shown to inhibit the production of arachidonic acid, prostaglandin and hydroxyeicosatetraenoic acid from bovine seminal vesicles and kidneys (Hiramatsu and Packer, 1990). Retinol was also shown to possess peroxyl, hydroxyl, and superoxide anion radicals scavenging effect and singlet oxygen quenching activity as well as glutathione production inducing effect (Ayeleso et al., 2016). In tests, retinol was found to be more effective than tocopherol in the removal of peroxyl radicals by binding to a higher mobility and shorter polyene chain. On the other hand, the aqueous interacting chroman head on the surface of the retinol membrane cannot sweep aqueous radicals as effectively as tocopherol. Vitamin A is also directly oxidized by radical species, producing a 5,6-retinoid epoxide and thus stabilizing the lipid radical. Retinol was reported to be used in pulse radiolysis techniques to remove potentially damaging glutathione radicals. However, retinoic acid concentration, which is considerably lower than vitamin A in modulating cellular oxidative stress, is not a possible factor due to its relatively low antioxidant activity. Similarly, the antioxidant effect of retinyl esters is much less than that of free alcohol forms of vitamin A. Using the *in vitro* peroxidation system, the antioxidant effects of retinoid compounds were classified as retinol ≥ retinal ≥ retinyl palmitate > retinoic acid (Palace et al., 1999). All-trans-retinol was reported to exert chain breaking antioxidant effect in scavenging the lypoperoxyl radical. Retinol's cyclohexenyl ring was pointed out to be a predominant site when compared to the hydrogen atom transfer (HAT) ones in terms of radical adduct formation (RAF) reaction. Achieved radical formation may be decomposed to yield an alkoxyl radical (LO•) from the lipoperoxyl (LOO•) and 2,3-retinoid epoxide. Other way is that the RAF reacts with singlet oxygen to produce retinoid-derivative peroxyl radical (RetOO•), or it may undergo a second addition of the lipoperoxyl radical. Epoxide formation in consuming LOO• radical forms another radical (LO•). Retinol can display radical trapping antioxidant effect in adding a maximum of six LOO• radicals. Moreover, RetOO• radical could become a pro-oxidant for lipid system by reacting with other retinoids and causing a self-oxidation (Dao et al., 2017).

In an *in vivo* study on rats, feeding of 100000 I.U. of vitamin A on alternate days, five times resulted in a significant reduction in lipid peroxidation of the tissue hemogenates. Vitamin A enhanced the antioxygenic activity of the tissues, and it was suggested that retinol also could be used as a potential antioxidant as tocopherol in animal nutrition (Kartha and Krishnamurthy, 1977).

Meydani et al. (1994) demonstrated an enhanced antioxidant capacity in elderly women after receiving β-carotene containing nutritional supplements. Positive correlation was also detected between vitamin A consumption and the antioxidant capacity in trolox equivalent antioxidant capacity and oxygen radical absorbance capacity assays reported by Arredondo et al. (2016). However, the results were not statistically significant (Meydani et al., 1994, Ojeda Arredondo et al., 2016).

Low doses of vitamin A or β-carotene were shown to contribute for cancer prevention. The anti- cancer effects was attributed to the ROS scavenging ability, thus improvement of immune function as well as anti-proliferative effect through the

RAR and RXR receptors (Ross et al., 2000). On the other hand, high consumption of β-carotene was reported to display carcinogenic effects and was considered to be via the asymmetric β-carotene cleavage pathway induction and breakdown products which may destroy RA through cytochrome P450 activation and therefore, stimulation of cell proliferation (Mueller and Boehm, 2011).

According to a systematic review by Bjelakovic, the favorable and negative effects of antioxidant supplementation on adult mortality prevention were investigated. For this purpose, primary and secondary prevention randomized clinical trials on antioxidant supplements (β-carotene, vitamin A, vitamin C, vitamin E, and selenium) were included. In trials vitamin A did not significantly affect mortality (Bjelakovic et al., 2012).

Conclusion

Oxidative stress in cells is known to occur owing to the normal physiological processes and environmental interactions. There is great evidence that the complex network of antioxidant defense systems has an essential role in protecting against oxidative damage. According to epidemiological studies, high antioxidant rich food intake positively influences antioxidant capacity of the plasma. These processes appear to be irregular and complex in many cases. Indeed, some studies suggest that antioxidant supplements may prolong life, while other observational research demonstrates harmful effects. To guide future therapeutic improvements to combat oxidative damage, a deeper knowledge of biochemical activities occurring at the cellular level is required. Based on the above mentioned reports, it is clear that vitamin A plays an important role in human health and its antioxidant mechanism is particularly important to reveal its free radical scavenging properties. Nevertheless, vitamin A intake should be balanced, otherwise this micronutrient and its derivatives can act as oxidants rather than behaving as antioxidants.

References

Arredondo, O., Lucia, M., Betancourt, M.C.P., Yoshida, M.L.B., Herrera, V.M.C., Vega, Á.S.G., Rodríguez, J.C.R., Sequeda, G., Diez, O., Lucci, P., 2016. Relationship between vitamin intake and total antioxidant capacity in elderly adults. Univ. Sci. 21 (2), 167–177.

Ayeleso, A., Brooks, N., Oguntibeju, O., Mukwevho, E., 2016. Natural antioxidant vitamins: a review of their beneficial roles in management of diabetes mellitus and its complications. Trop. J. Pharm. Res. 15 (6), 1341–1348.

Bjelakovic, G., Nikolova, D., Gluud, L.L., Simonetti, R.G., Gluud, C., 2012. Antioxidant supplements for prevention of mortality in healthy participants and patients with various diseases. Cochrane Database Syst. Rev. 3, 1–247.

Dao, D.Q., Ngo, T.C., Thong, N.M., Nam, P.C., 2017. Is vitamin a an antioxidant or a pro-oxidant? J. Phys. Chem. B 121 (40), 9348–9357.

Goodarzi, S., Rafiei, S., Javadi, M., Haghighian, H.K., Noroozi, S., 2018. A review on antioxidants and their health effects. J. Nutr. Food Secur. 3 (2), 106–112.

Hiramatsu, M., Packer, L., 1990. Antioxidant activity of retinoids. Methods in Enzymology. Elsevier, Academic Press, United States, pp. 273–280.

Kartha, V.N., Krishnamurthy, S., 1977. Antioxidant function of vitamin A. Int. J. Vitam. Nutr. Res. 47 (4), 394–401.

Kawaguchi, R., Yu, J., Honda, J., Hu, J., Whitelegge, J., Ping, P., Wiita, P., Bok, D., Sun, H.S., 2007. A membrane receptor for retinol binding protein mediates cellular uptake of vitamin A. Science 315 (5813), 820–825.

Lü, J.-M., Lin, P.H., Yao, Q., Chen, C., 2010. Chemical and molecular mechanisms of antioxidants: experimental approaches and model systems. J. Cell. Mol. Med. 14 (4), 840–860.

Mazlum, B., 2012. Antioksidan Vitaminler ve Psikiyatride Kullanımı. Psikiyatride Güncel Yaklaşımlar 4 (4), 486–505.

Meydani, M., Martin, A., Ribaya-Mercado, J.D., Gong, J., Blumberg, J.B., Russell, R.M., 1994. β-carotene supplementation increases antioxidant capacity of plasma in older women. J. Nutr. 124 (12), 2397–2403.

Mueller, L., Boehm, V., 2011. Antioxidant activity of β-carotene compounds in different in vitro assays. Molecules 16 (2), 1055–1069.

Nimse, S.B., Pal, D., 2015. Free radicals, natural antioxidants, and their reaction mechanisms. RSC Adv. 5 (35), 27986–28006.

Palace, V.P., Khaper, N., Qin, Q., Singal, P.K., 1999. Antioxidant potentials of vitamin A and carotenoids and their relevance to heart disease. Free Radic. Biol. Med. 26 (5-6), 746–761.

Ross, S.A., McCaffery, P.J., Drager, U.C., Luca, LMDe, 2000. Retinoids in embryonal development. Physiol. Rev. 80 (3), 1021–1054.

Shils, M.E., Olson, J.A., Shike, M., 1994. Modern nutrition in health and disease, 8th ed., Lea & Febiger, Philadelphia, (OCoLC)624171609.

Temple, N.J., 2000. Antioxidants and disease: more questions than answers. Nutr. Res. 20 (3), 449–459.

Yadav, A., Kumari, R., Yadav, A., Mishra, J.P., Srivatva, S., Prabha, S., 2016. Antioxidants and its functions in human body-A Review. Res. Environ. Life Sci. 9 (11), 1328–1331.

Zadak, Z., Hyspler, R., Ticha, A., Hronek, M., Fikrova, P., Rathouska, J., Hrnciariková, D., Stetina, R., 2009. Antioxidants and vitamins in clinical conditions. Physiol. Res. 58 (Suppl 1), S13–S17.

CHAPTER

Vitamin C

4.22

Manish Kumar[a], Vinay Pratap[a], Jalaj K. Gour[a], Manoj K. Singh[b]

[a]*Department of Biochemistry, Faculty of Science, University of Allahabad, Prayagraj, India*

[b]*Centre for Non Communicable Diseases (NCD), National Centre for Disease Control (NCDC), Ministry of Health & Family Welfare-Government of India, Delhi, India*

4.22.1 Introduction

Humankind has experienced much earlier the importance of absence essential nutrition. Scurvy is one of the several instances of poor health condition due to deficiency of nutrition. The biochemical name of ascorbic acid is L-ascorbic acid, which is generally termed as vitamin C (vit C). Vit C is more essential in human health system as a protective agent against a number of diseases. Ascorbic acid is water soluble and flexible crystalline nutrient (Carr et al., 2017), obtained from plant-based diets and dietary supplements. It is utilized for influencing many metabolic reactions in the living organisms (Kuhn et al., 2018). Vit C play a key role in the maintenance of human well-being by mediating various biological activities such as scavenging of free radicals, neutralization of xenobiotics, reduction of cataract formation, collagen formation, synthesis of immunoglobulin's, bone formation, tryptophan metabolism, tyrosine metabolism, cholesterol metabolism, synthesis of corticosteroid, and cellular respiration. Along with defensive role against scurvy vit C also possess preventive immune- modulatory action, pro-oxidant activity and anti-inflammatory action (Iqbal et al., 2004). The daily allowance of the vit C is most significant in human health, but the high dose of vit C can cause disorders such as gastrointestinal disorders, sleeping disorder, skin problem, headache, kidney damage, cardiovascular disease, and cataract, and in some cases excess dose of vit C can also cause cancer. In addition, ascorbic acid also plays a substantial role in repairing tissue damage and produces many enzymes that participate in a certain neurotransmitter reaction mechanism (Covarrubias-Pinto et al., 2017). The defensive mechanisms of the immune system and anti-oxidation reaction is influenced by Vit C. Recently many researchers have suggested that ascorbic acid is not used in the remedial action of common cold for complete treatment (Maggini et al., 2012).

Antioxidants Effects in Health. DOI: https://doi.org/10.1016/B978-0-12-819096-8.00042-2

4.22.2 History

The derivation of the ascorbic acid or vit C is started with the knowing of scurvy disease, which was caused by the deficiency of dietary supplementation of vit C for centuries. It was the first disease found to be associated with various diets (Carpenter, 2012). In the sixteenth century about 10,000 mariners died of deficiency of vegetables in their diet. In 1753 James Lind established that scurvy could be cured by citrus fruit. Albert Szent-Gyorgyi in 1928 first sequestered ascorbic acid and in 1932 S zent-Gyorgyi and King validated vit C as an antiscorbutic factor for the first time (Padayatty and Levine, 2016). In 1907 Axel Holst and Alfred Frohlich proposed the existence of vit C. In 1920s Szent Gyorgyi was reviewing the chemical changes that occur when cells utilizes fats, carbohydrates, and proteins. During this course of study Szent Gyorgyi sequestered a molecule from adrenal glands that serve to allocation of the hydrogen atom. This molecule which was acting as a carrier for hydrogen atom containing six carbon atoms and showed the behavior of both the sugar and acid. Szent Gyorgyi named this molecule hexuronic acid and this hexuronic acid now known as vit C. Waugh and King in 1932 first isolated vit C in crystalline form from lemon juices. For this contribution Albert Szent Gyorgyi was granted the Nobel prizes in Physiology and Medicine separately (8, Hoyle and Santos, 2010).

4.22.3 Sources and daily allowance of vitamin C

Vit C is obtained basically from the diets based on plants and as well as some animal sources (Devaki and Raveendran, 2017). The best sources of the vit C include citrus fruits, kiwifruit and many vegetables like broccoli, brussels sprouts, strawberries etc. (Fact Sheet for Health Professionals, 2016). Various animals (synthesize from glucose in liver) and plants (synthesize from D-Gluccose and D-Galactose) can synthesize vit C but human beings cannot synthesize vit C because they lack the enzyme gulonolactone oxidase (that help the synthesis of ascorbic acid). Oranges, mangoes, lemons, watermelons, strawberries, papaya, pineapple, cherries and raspberries, and leafy vegetables, such as broccoli, tomatoes, red and green peppers, cabbage, and cauliflower are the chief source of vit C. Plant with major sources and availability of vit C are shown in Table 4.22.1. For adult male the daily supplement of the vit C in the diet range about 90 mg to 75 mg for an adult female. 100 mg per day of vitamin intake is accompanied with lowering the risk of death from stroke, cardiovascular disease and cancer. Smokers must have consumed an extra 40 mg for every day in to prevent from oxidative stress from tobacco smoke. Observational information supports the theory that high dietary intakes and supplementation of vit C may minimize the risk of hip breaks in postmenopausal woman (Naidu, 2003).

4.22.4 Chemical structure and biochemistry of vitamin C

Vit C is a five membered carbon compound known as lactone sugar acid. In 1933 Micheel and Kraft after investigated the chemical properties of the compound and suggested that 2-(4,5-dihydo-3,4-dihydroxy-5-hydroxymethyl) furanyl-carboxylic

Table 4.22.1 Plant as bioavailable scurces of vitamin C.

S. N.	Plant name (Sources)	Usable part	Family	Amount (mg/100 g)	S. N.	Plant name (sources)	Usable part	Family	Amount (mg/100g)
1	*Rosa moyesii* (Rose hips)	Fruits	Rosaceae	426	16	*Fragaria ananassa* (strawberry)	Fruits	Rosaceae	58.8
2	Capsicum annuum (Green Chili)	Fruits	Solanaceae	242	17	*Citrus sinensis* (orange)	Fruits	Rutaceae	53.2
3	*Psidium guajava* (Guava)	Fruits	Myrtaceae	228.3	18	*Citrus limon* (lemon)	Fruits	Rutaceae	53
4	Capsicum annuum (Yellow Bell Pepper)	Fruits	Solanaceae	183	19	*Citrus latifolia* (lime)	Fruits	Rutaceae	29.1
5	*Petroselinum crispum* (garden parsley)	Leaves	Apiaceae	133	20	*Citrus clementina* (clementine)	Fruits	Rutaceae	48.8
6	Capsicum annuum (bell pepper)	Fruits	Solanaceae	128	21	*Ananas comosus* (pineapple)	Stem	Bromeliaceae	47.8
7	Brassica oleracea (leaf cabbage)	Leaf	Brassicaceae	120	22	Brassica oleracea (cauliflower)	Flower	Brassicaceae	46.4
8	Chinese gooseberry (Kiwi)	Fruits	Actidiaceae	92.7	23	*Brassica rapa* (Chinese cabbage)	Flower	Brassicaceae	45
9	Brassica oleracea (Brussels Sprouts)	Flower	Brassicaceae	85	24	*Nasturtium officinale (watercress)*	Leaves	Brassicaceae	43
10	*Syzygium aromaticum* (Cloves)	Flower buds	Myrtaceae	80.8	25	*Cucumis melo (muskmelon)*	Fruits	Cucurbitaceae	36.7
11	*Chenopodium album* (lamb's quarters)	Leaves	Amarantha-ceae	80	26	*Citrus paradise (grapefruit)*	Fruits	Rutaceae	31.2
12	*Litchi chinensis* (Lychee)	Fruits	Sapindaceae	71.5	27	*(Beta vulgaris (Swiss chard)*	Leaves	Amaranthaceae	18
13	*Brassica juncea* (brown mustard)	Leaves	Brassicaceae	70	28	*Spinacia oleracea (spinach)*	Leaves	Amaranthaceae	28
14	*Brassica oleracea* (Musturd green)	Green leaves	Brassicaceae	62	29	*Ribes uva-crispa (gooseberry)*	Fruits	Grossulariaceae	27.7
15	*Carica papaya* (Papaya)	Fruits	Caricaceae	62	30	*Mangifera* species (mango)	Fruits	Anacardiaceae	36.4

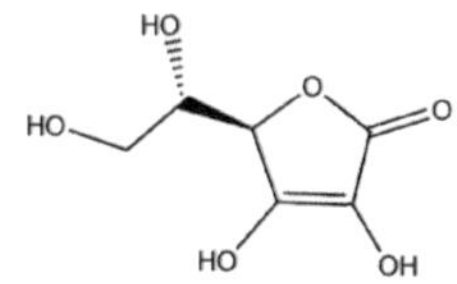

FIG. 4.22.1 Basic structure of ascorbic acid (vitamin C).

acid as the constitutional formula of ascorbic acid (Olabisi and Wimalasena, 2009). The basic structure of vit C shown in Fig. 4.22.1. Many authors have reported that the plasma and tissues predominantly contain ascorbic acid in reduced form in living organisms (Padayatty and Levine, 2016). At neutral pH DHA is unstable and if it is not reduced to regenerate reduced ascorbate by thioredoxin or glutathione, DHA is broken down into threonic acids, oxalic and diketogulonic acids by kidney. Consequently, ascorbate is present in the form of DHA in very small fraction in vivo at concentrations that is not probably to be present above 1–2 μM. Ascorbic acid can chelate and reduce Fe^{3+} and Cu^{2+} and this property is responsible for the ability of ascorbic acid to promote iron absorption present in the diet (Vissers and Das, 2018). Ascorbic acid is a ketolactone that is soluble in water and contain two ionizable hydroxyl groups in which the $_{p}K_1$ is 4.2 and $_{p}K_2$ is 11.6 of OH groups respectively. Ascorbate acts as an excellent reducing agent and can donate one electron two times consecutively and is converted to dehydroascorbic acid (DHA) and ascorbate radical ($Asc^{\bullet -}$).

Vit C found in ionized form, and this property is responsible for solubility of this compound in water. Pure ascorbic acid is a white crystalline and colourless solution. The reducing properties of vit C are very significant for the scavenging of free radical in an organism. The reducing properties of ascorbic acid exhibited in two forms one is, oxidation of electron to form ascorbate radical that represented by ($Asc^{\circ -}$) and another is dehydroascorbic acid form. The unpaired electrons of the ascorbate stabilized because of the resonance and make the ascorbate relatively unreactive and ascorbate radical is converted to DHA and ascorbate.

$$2\text{Asc}^- + \text{H}^+ \leftrightarrow \text{AscH}^- + \text{Dihydroascorbic acid (DHA)}$$

Because of these properties ascorbate behaves as an excellent antioxidant (Du et al., 2012).

4.22.5 Biosynthesis of vitamin C in the plants and animals

Vit C is synthesized in plants and in many vertebrates including fish, amphibians, and reptiles (in kidney) and mammals (in liver). The biosynthetic pathway of ascorbic acid is different in animals and plants. In animals UDP-glucuronate is converted to d-glucuronate which is then reduced to l-gulonate that leads to change in functional groups and hence change in the configuration. L-gulonate is transformed to its lactone and l-gulonolactone oxidase oxidize l-gulonate to l-ascorbate. In plants l-galactonolactone is derived from GDP—D mannose, which is eventually transformed to ascorbic acid with help of l-galactonolactone dehydrogenase (Linster and Schaftingen, 2006). Albert Lehninger in 1957 studied biosynthesis of vitamin C in animals, and found that humans are unable to synthesize vit C unlike many species, such as dogs and cats, which are capable of biosynthesizing vitamin C for themselves. Humans are incapable to synthesize vit C because they cannot perform the step of converting l-gulono-g-lactone to ascorbic acid that are catalyzed by the gulonolactone oxidase enzyme. This occurs because of the inactive form of the gene that encode the gulonolactone oxidase enzyme determined by the observation of Nishikimi and coworkers. Not only in human beings, this inactive gene is also present in orangutans, chimps, gorillas and some monkeys (De Tullio, 2010). The biosynthetic pathway of ascorbic acid in plants and animals is shown in Fig. 4.22.4.

4.22.6 Beneficial effects of vitamin C on health

The previous research determined that ascorbic acid (vit C) play an efficacious role in the prevention of many types of disorders in human beings such as scurvy, cancer, diabetes, neurodegenerative disease, atherosclerosis, inflammation, etc. (Chambial et al., 2013). The ascorbic acid shows a structural part in the incorporation of collagen molecules synthesis. The most important properties of the ascorbic acid are antioxidant activities that are responsible for the scavenging and reducing mechanism of the free radicals. Free radicals, one of the molecules that contain a number of unpaired electrons, which leads to damaging the protein, lipids, nucleic acid. Ascorbic acid diminishes the risk of free radicals (Tai et al., 2017). Vit C acts, for instance, an enzyme cofactor in many hydroxylation reactions for example enzymes like: prolyl 4-hydroxylase, prolyl 3-hydroxylase and lysyl hydroxylase enzymes which are involved in collagen biosynthesis uses vit C as cofactor. Certain neurotransmitters and carnitine and tyrosine are also synthesized by vit C. During tyrosine biosynthesis vit C help iron and homogentisate dioxygenase enzyme interaction and cuporous (c^{2+}) copper interaction. These two enzymes are involved in the process of converting phenylalanine to tyrosine (Schlueter and Johnston, 2011). The most important beneficial role and detrimental effects of vit C on health are shown in the Fig. 4.22.2.

4.22.7 Anticancer activity of ascorbic acid

Cancer has become worldwide problem with a lot of treatment and preventive measures developed globally. The disease is characterized by uncontrolled and continuous division of the cells (Greenwell and Rahman, 2015). In 1950s it was proposed that ascorbic acid has anticancer property. Scottish surgeon Ewan Cameron and Allan Campbell in 1970 tried for the first time to treat cancer by oral or intravenous dosing and observed various responses. Various preclinical data suggest that intravenous dosing of vit C can act as cancer therapeutic (Nauman et al., 2018). Ascorbic acid one of the greatest significant reducing agent and it also involved in the prevention of

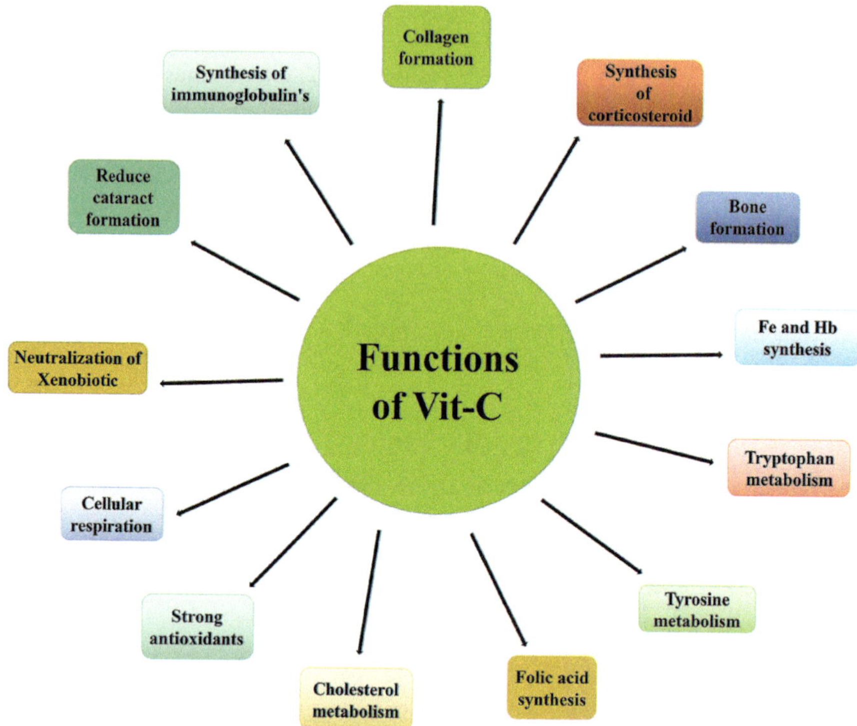

FIG. 4.22.2 Vitamin C in beneficial effects on health.

many types of cancer in the organisms (Rangarajan et al., 2014). Many researchers investigated the cytotoxicity or pharmacologic activity in selected cancer cells line in vitro and determined that vit C is responsible for anticancer activity (Alam et al., 2017). Vit C containing foods are allied with lowering the threat of certain sorts of cancer particularly gastric and esophageal cancer (Williams, 2013). The balancing action of the vit C is shown in Fig. 4.22.3.

4.22.8 Antioxidant activity of vitamin C

Vit C plays a key role in antioxidant activity. Antioxidants are the most important natural substances that are responsible for reducing free radical molecules or oxidative stress (Kasote et al., 2015). Ascorbic acid is a vital physiological antioxidant, very solid radical scavenger and has been shown to rejuvenate other antioxidants within the body. Vit C reduces unstable oxygen, sulfur radicals and nitrogen radicals (Lü et al., 2010). Aqueous peroxyl radicals induces per oxidative damage of plasma lipids and studies with human plasma have shown that vit C can prevent this damage. Dietary phytochemicals have antioxidant actions and can guard our body against oxidative

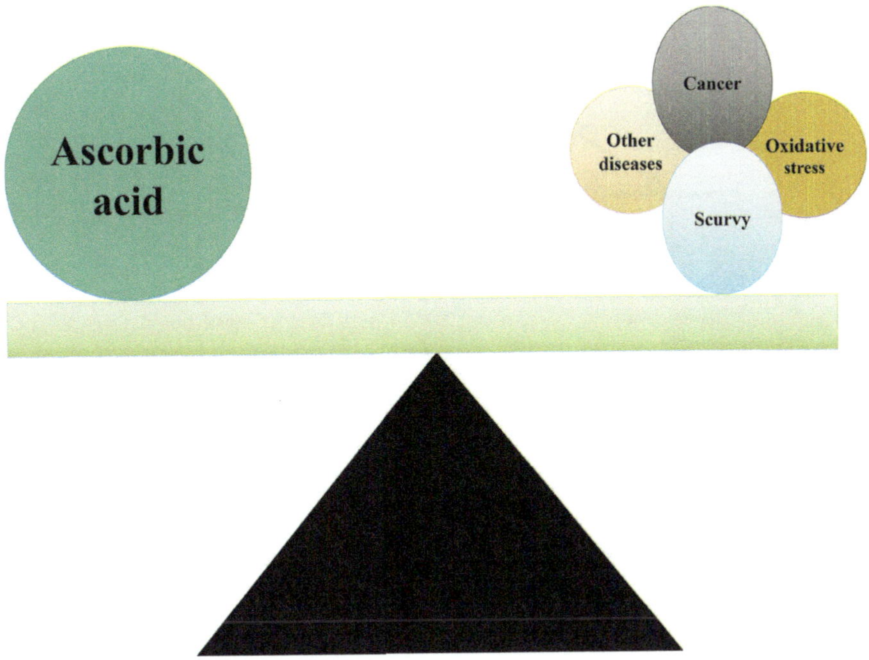

FIG. 4.22.3 Balancing action of ascorbic acid to several diseases.

abuse to cells. They can go through redox reactions and scavenge reactive species, such as unstable nitrogen, oxygen, and sulfur radicals (Aprioku, 2013). Usually, antioxidants prevent irreversible ROS injuries to DNA and proteins. Long-term use of antioxidants is associated with inhibition of mitochondrial F_1–F_0 ATP synthesis complex which may lead to the overpowering of NADH feed to the respiratory chain with oxidative phosphorylation (Traber and Stevens, 2011).

4.22.9 Detrimental effects of ascorbic acid on health

Vitamin C plays a significant role in lowering the risk of incidence of certain disease in animals but there are number of cytotoxicity that is associated with higher intake of its dose. Clinically many pieces of evidence support that the high intakes of the vit C causes, diarrhea, nausea, flushing, vomiting, headache, disturb sleep and fatigue both in adults and infants. Skin rashes are produced in infants due to reactions participating in collagen synthesis and enzymatic reaction with vit C. It is described that high intake of vit C leads to the expanded risk of cardiovascular infection and mortality in postmenopausal ladies with diabetes. Vit C enhances the absorption of iron therefore it can cause iron poisoning in patients suffering from a rare iron overload disease called such as hemochromatosis. Very large doses of vit

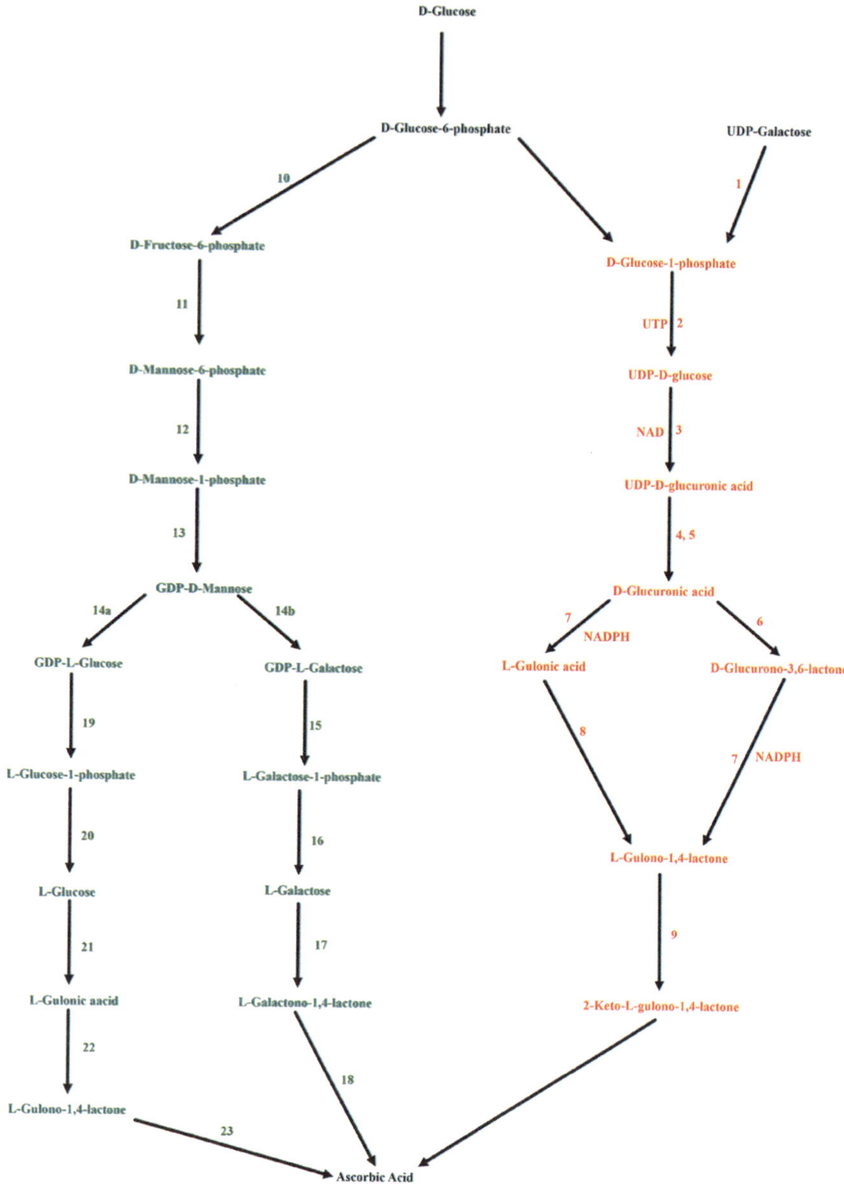

FIG. 4.22.4 Biosynthesis of ascorbic acid in plants and animals. Reactions 1–9 represent the pathway in animals and reactions 10–23 represents the pathways in plants.

1, UDP-glucose-4-epimerase; 2, UDP-glucose pyrophosphorylase; 3, UDP-glucose dehydrogenase; 4, UDPglucuronic acid pyrophosphatase; 5, glucuronic acid-l-P phosphatase; 6, D-glucuronolactonase; 7, L-glucuronic acid reductase; 8, gulono-l,4-1actone hydrolase; 9, a-gulono-l,4-1actone oxidase; 10, glucose-6-phosphate isomerase; 11, mannose-6-phosphate isomerase; 12, phosphomannose mutase; 13, GDP-mannose pyrophosphorylase; 14, GDP-mannose–epimerase; 15, GDP-L-galactose phosphorylase; 16, L-galactose-1-phosphate phosphatase; 17, L-galactose dehydrogenase; 18, L-galactono-1,4-lactone dehydrogenase; 19, phosphodiesterase; 20, sugar phosphatase; 21, L-gulose dehydrogenase; 22, aldonolactonase; 23, gulono-1,4-lactone oxidase.

C can cause hemolytic anemia by oxidizing glucose-6-phosphate dehydrogenase. This condition arise due to genetic alteration where the level of glucose-6-phosphate dehydrogenase in the cell. High dose of vit C that is 2000 mg/day may cause skin rashes or gastrointestinal upset however there are some evidences which prove that 4000 mg/day intake of vit C in general population is well tolerated. In a special condition if high intake of vit C is received intravenously or it is given to the patient suffering from chronic renal failure, it leads to the increase in uric acid and urinary oxalate production which could add to the formation of kidney stones, particularly in people with renal disorder (Korah et al., 2017).

4.22.10 Pro-oxidant activity of ascorbic acid

Pro-oxidants have the opposite action of the antioxidants. At higher concentration ascorbic acid in organisms it exerts a cytotoxic effect (Putchala et al., 2003). Many Authors reported that at higher dose of vit C put forth a cytotoxic special effects on cancer cells in vitro and in vivo, that leads to the damages to many organs in the body, thus it is determined that the ascorbic acid acts as a pro-oxidative drug, of which reaction is catalyzed by production of hydrogen peroxide in tissues in its place of acting as a radical scavenger (Phaniendra et al., 2015). In addition, ascorbic acid can also generate hydrogen peroxide by an oxidation reaction and which is improved by divalent cations, such as iron and copper (Fritz et al., 2014). Hydrogen peroxide reaction and its derivatives compound lead damaging of cell membranes and mitochondria. Many scientists reported that the oxidative reaction is very limited in mammalian health systems, because the most metal particles are associated to proteins in serum, which makes them inaccessible to improve the pro-oxidant action of vit C, while in danger, these particles are promptly accessible (Rahal et al., 2014). Many types of reactive species, such as reactive nitrogen species and oxygen species are also capable for inducing many types of negative effects on cell components, such as protein, lipid, DNA damage and strand breaks, disruption of membrane function via lipid peroxidation, and destruction of cellular ATP are also included in pro oxidant activity in the cell. The author examined that vit C facilitated hydrogen peroxide generated is repressed by serum because of specific proteins, for example, egg whites and glutathione as cancer preventive agent or catalase, which breaks down hydrogen peroxide (Das et al., 2014).

Conclusion

This chapter has explained that diet has played very crucial role in the life of human beings. Vit C is one of the many nutritious components that help combat a number of disease such as cancer, scurvy, and oxidative stress. The history of discovering vit C states that how does the researchers were able to find the biological role of the vit C. Ascorbic acid is a hydrophilic ketolactone containing two ionizable hydroxyl groups with $_pK_1$ 4.2 and $_pK_2$ 11.6 of OH groups respectively. Ascorbate can donate one electron two times consecutively so that it acts as an excellent reducing agent and and is converted to DHA and Ascorbate radical $Asc^{\bullet-}$. Vit C is synthesized in plants and while in animals it is limited to many vertebrates with different biosynthetic pathway except human beings because humans have inactive genes for the specific enzyme involved in vit C biosynthesis. Along with pharmacological advantage excessive dose of vit C can produce negative impacts on the subject. In conclusion Vit C have great importance in life by boosting the immune system of the body and protect against many dangerous impacts on body.

References

Alam, P., Beg, A.Z., Siddiqi, M.K., Chaturvedi, S.K., Rajpoot, R.K., Ajmal, M.R., Zaman, M., Abdelhameed, A.S., Khan, R.H., 2017. Ascorbic acid inhibits human insulin aggregation and protects against amyloid induced cytotoxicity. Arch. Biochem. Biophys. 621, 54–62.

Aprioku, J.S., 2013. Pharmacology of free radicals and the impact of reactive oxygen species on the testis. J. Reprod. Infertil. 14 (4), 158–172.

Carpenter, K.J., 2012. The discovery of vitamin C. Ann. Nutr. Metab. 61 (3), 259–264.

Carr, A.C., Rosengrave, P.C., Bayer, S., Chambers, S., Mehetens, J., Shaw, G.M., 2017. Hypovitaminosis-c and vitamin-C deficiency in critically ill patients despite recommended enteral and parenteral intakes. Crit. Care 21 (1), 300.

Chambial, S., Dwivedi, S., Shukla, K.K., John, P.J., Sharma, P., 2013. Vitamin C in disease prevention and cure: an overview. Ind. J. Clin. Biochem. 28 (4), 314–328.

Covarrubias-Pinto, A., Acuña, A.I., Beltrán, F.A., Torres-Díaz, L., Castro, M.A., 2017. Old things new view: ascorbic acid protects the brain in neurodegenerative disorders. Int. J. Mol. Sci. 16 (12), 28194–28217.

Das, H., Lateef, A., Parsani, H.R., Abidali, D., 2014. Role of dietary antioxidant vitamins and minerals in ruminants. IJAVMS 8 (1), 21–30.

De Tullio, M.C., 2010. The mystery of vitamin C. Nat. Educ. 3 (9), 48.

Dedicated May 11, 2002, at the University of Szeged Albert Szent-Györgyi Medical Faculty in Szeged, Hungary.

Devaki S.J., Raveendran L.R., 2017. Vitamin C: sources, functions, sensing and analysis. doi.10.5772/intechopen.70162.

Du, J., Cullen, J.J., Buettner, G.R., 2012. Ascorbic acid: chemistry, biology and the treatment of cancer. Biochim. Biophys. Acta 1826 (2), 443–457.

Fact Sheet for Health Professionals, 2016. Vitamin C. Office of Dietary Supplements. US National Institutes of Health February.

Fritz, H., Flower, G., Weeks, L., Cooley, K., Callachan, M., McGowan, J., Skidmore, B., Kirchner, L., Seely, D., 2014. Intravenous vitamin C and cancer: a systematic review. Integr. Cancer Ther 13 (4), 280–300.

Greenwell, M., Rahman, P.K., 2015. Medicinal plants: their use in anticancer treatment. Int. J. Pharm. Sci. Res. 6 (10), 4103–4112.

Hoyle, C.H.V., Santos, J.H., 2010. Cyclic voltammetric analysis of antioxidant activity in citrus fruits from Southeast Asia. Int. Food Res. J. 17, 937–946.

Iqbal, K., Khan, A., Khattak, M., 2004. Biological significance of ascorbic acid (vitamin C) in human health – a review. PJN 3 (1), 5–13.

Kasote, D.M., Katyare, S.S., Hegde, M.V., Bae, H., 2015. Significance of antioxidant potential of plants and its relevance to therapeutic applications. Int. J. Biol. Sci. 11 (8), 982–991.

Korah, M.C., PvJ, R., Rajeswari, R., Behanan, A., Elizabeth, P.P., Sivakumar, T., 2017. Adverse effects and side effects on vitamin therapy: a review. Asian J. Pharm. Clin. Res. 10 (5), 19–26.

Kuhn, S.O., Meissner, K., Mayer, L.M., Bartels, K., 2018. Vitamin-c in sepsis. Curr. Opin. Anaesthesiol. (1), 55–60.

Linster, C.L., Schaftingen, E.V., 2006. Vitamin C biosynthesis, recycling and degradation in mammals. FEBS J. 274 (1).

Lü, J.M., Lin, P.H., Yao, Q., Chen, C., 2010. Chemical and molecular mechanisms of antioxidants: experimental approaches and model systems. J. Cell. Mol. Med. 14 (4), 840–860.

Maggini, S., Beveridge, S., Suter, M., 2012. A combination of high-dose vitamin C plus zinc for the common cold. J. Int. Med. Res. 40 (1), 28–42.

Naidu, 2003. Vitamin C in human health and disease is still a mystery? An overview. Nutr. J. 2, 7.

Nauman, G., Gray, J.C., Parkinson, R., Levine, M., Paller, C.J., 2018. Systematic review of intravenous ascorbate in cancer clinical trials. Antioxidants (Basel, Switzerland) 7 (7), 89.

Olabisi, A., Advisor Wimalasena, K., 2009. The chemistry of L-ascorbic acid derivatives in the asymmetric synthesis of C2- and C3-substituted aldono-gamma-lactones [electronic resource].

Padayatty, S.J., Levine, M., 2016. Vitamin C physiology: the known and the unknown and Goldilocks. Oral Dis. 22 (6), 463–493.

Phaniendra, A., Jestadi, D.B., Periyasamy, L., 2015. Free radicals: properties, sources, targets, and their implication in various diseases. Indian J. Clin. Biochem. 30 (1), 11–26.

Putchala, M.C., Ramani, P., Sherlin, H.J., Premkumar, P., Natesan, A., 2003. Ascorbic acid and its pro-oxidant activity as a therapy for tumours of oral cavity – a systematic review. Arch. Oral Biol. 58 (6), 563–574.

Rahal, A., Kumar, A., Singh, V., Yadav, B., Tiwari, R., Chakraborty, S., Dhama, K., 2014. Oxidative stress, prooxidants, and antioxidants: the interplay. Biomed. Res. Int. 2014, 761264.

Rangarajan, S., Bhuvana, S., Lisa, M.C., 2014. Ascorbic acid in cancer: a renewed hope? J. Cancer Sci. Ther. 6, 9.

Schlueter, A.K., Johnston, C.S., 2011. Vitamin C: overview and update. J. Evid.-Based Complement. Altern. Med. 16 (1), 49–57.

Tai, A., Iomori, A., Ito, H., 2017. Structural evidence for the DPPH radical-scavenging mechanism of 2-O-α-d-glucopyranosyl-l-ascorbic acid. Bioorg. Med. Chem. 25 (20), 5303–5310.

Traber, M.G., Stevens, J.F., 2011. Vitamins C and E: beneficial effects from a mechanistic perspective. Free Radic. Biol. Med. 51 (5), 1000–1013.

Vissers, M.C.M., Das, A.B., 2018. Potential mechanisms of action for vitamin C in cancer: reviewing the evidence. Front Physiol. 9, 809.

Williams, C.D., 2013. Antioxidants and prevention of gastrointestinal cancers. Curr. Opin. Gastroenterol. 29 (2), 195–200.

CHAPTER

4.23 Vitamin E (tocopherols and tocotrienols) (natural-occurring antioxidant; bright and dark side)

Ziyad Khan[a], Salman Ahmad[b], Marya[c,d], Hammad Ullah[c], Haroon Khan[d]

[a]*Department of Pharmacy, University of Swabi, Swabi, Pakistan*

[b]*Department of Pharmacognosy, Faculty of Pharmacy and Pharmaceutical Sciences, University of Karachi, Karachi, Pakistan*

[c]*Department of Pharmacy, School of Medicine and Surgery, University of Naples Federico II, Via Domenico Montesano, Naples, Italy*

[d]*Department of Pharmacy, Abdul Wali Khan University Mardan, Mardan, Pakistan*

4.23.1 Introduction

About 60 years back, the nutritional importance of vitamin E was acknowledged for the first time. Rats in 1992 were given vitamin free semi-synthetic diet and they failed to reproduce (Evans and Bishop, 1922). It was later found that the absence of this factor in the diet caused fetal resorption. This factor was confirmed as Vitamin E and the richest sources of vitamin was germ oil and lettuce. The significance of vitamin E was confirmed by (Pappenheimer and Goettsch, 1934) when they gave processed diet to chicks which leaded to nutritional encephalomalacia (a central nervous system defect, clinically characterized by spasms, ataxia and paralysis). In 1936, Evans and Bishop isolated an alcohol which was fat soluble in nature and later it was named as vitamin E. Its activity was reported as α-tocopherol. During the course of time, a year later, another derivative of vitamin E was isolated from soya bean oil and it was named as β-tocopherol. Since these discoveries frequent researchers, using numerous species, have established nutritionally important interrelationships between vitamin E and several dietary components (Dam, 1962; Diplock et al., 1971).

Recently it has been realized that biological systems can be highly affected due to adverse effects of free radicals processing (Henderson and Tyagi, 2006). This realization paved the way to produce interest in understanding the functions of vitamin E as the most potent antioxidant and biological form among other vitamins

Antioxidants Effects in Health. DOI: https://doi.org/10.1016/B978-0-12-819096-8.00066-5

(Burton et al., 1985). Now to determine the role of vitamin E in reducing ischemic-reperfusion injury reflects the interest (Massey and Burton, 1989) in exploring the way it impede the production of low density lipoproteins (LDL) and their harmful effects like atherosclerosis (Steinberg et al., 1989).

To completely identify the potential of vitamin E, one has to determine not only its antioxidant properties on molecular level but also the kinetics, its distribution mechanism in tissues and organs and bioavailability (Burton and Traber, 1990). Evidence suggests that the main causative phenomenon of most of the disorders like cancer and aging is because of the oxidative stress of DNA, lipids and proteins in the biological systems (Niki, 2000). That is the reason antioxidants have got much more attention. Human beings consists of strong guard system which is comprised of diverse natural antioxidants like vitamin E and C, which help in the free radical scavenging and defend the body from harmful effects of oxidative stress (Niki and Noguchi, 2004). Vitamin E is usually converted into vitamin E-radical inside the body, but in the presence of vitamin C it is again reverted to its original form of vitamin E (Buettner, 1993). In a study it was reported that the patients who took vitamin E supplements have a low risk of stroke as compared to others who did not used it. Also this vitamin is proved to reduce the threat of stroke in smokers (Schürks et al., 2010).

Vitamin E is found to have considerable part in preventing muscles weakness in children suffering from cystic fibrosis and neuropathy in infants with biliary atresia. The recommended dose of vitamin E for infants is found to be 0.7 IU/Kcal of energy intake. Vitamin E reduces the risk of lung damage in neonates with broncho-pulmonary dysplasia because of its antioxidant activity. Vitamin E is also helpful in preventing retinopathy in premature infants with ischemic-reperfusion injury and it may has an advantage in dropping the occurrence of intra-ventricular hemorrhage (Herman et al., 2011). In some disorders like cataract Packer (Packer, 1991) recommends the use of vitamin E at dose of 1000 to 1200 IU daily. While the recommended vitamin E daily intake is 10 mg for males and 8 mg for females. Due to bearing strong radical scavenging properties, vitamin E finds its role in the prevention of failure in glaucoma surgery. Complications in this type of surgery are due to development of fibro cellular scar derived from Tenon's capsule fibroblasts. Vitamin E is found to prevent proliferation of Tenon's capsule fibroblasts (Haas et al., 1996). In 2002 a heart antioxidant study was conducted in Cambridge and it revealed that the patients who received vitamin E at a dose of 400–800 IU/day were 77 percent less prone to myocardial infarction during a follow up period of 1.3 years (Stephens et al., 1996). However, further detail studies including intervention through different signaling pathways, clinical effects are highly recommending ascertaining its safe uses in different human illness.

4.23.2 Sources

Tocopherols are extensively distributed all over the plant kingdom, α-tocopherol is the most extensive, being found essentially in the chloroplast. Other tocopherols are located outside the chloroplast. The subcellular location of the non-α-tocopherols is not fully implicit. It has been observed that on green leaves and found that γ-tocopherol was situated outside the chloroplast (Traber and Atkinson, 2007). Other co-researchers investigated plastid fractions from green leaves, have found small amounts of β- and γ -tocopherols (Bucke, 1968) and 5-tocopherolquinone (Barr and Arntzen, 1969). In the brown algae *Fucus spiralis* 3-, γ- and *S*-tocopherols are located outside the plastid (Traber and Atkinson, 2007). No tocopherols were found in the nucleus; cell wall or ribosomes of the cell. 8-Tocopherol is found in the microsomal and soluble fractions whilst γ-tocopherol is associated with the cell fraction containing organelles such as mitochondria and Golgi (Traber and Atkinson, 2007). Vitamin E is a fat soluble vitamin, having proficient antioxidant properties are mainly occur naturally in various foods (Shils and Shike, 2006).

4.23.3 Chemistry

The diverse nature of tocopherols isolated from oils showed that these were analogues of α-tocopherol differing in arrangement of methyl groups and numbers of methyl group. The term vitamin E is used as a generic description for all tocol and tocotrienol derivatives qualitatively showing the biological activity of α-tocopherol. The natural stereo-isomer of α-tocopherol is (2 R, 4 R′, 8′R)-α-tocopherol or RRR α-tocopherol. Later it was shown that some of these substances contained three double bonds in the side chain (Pennock et al., 1964). Collectively, vitamin E illustrated in eight various chemical forms, that is, α, β, γ, and δ-tocopherol (Fig. 4.23.1) and α, β, γ, and δ-tocotrienol (Brigelius-Flohe and Traber, 1999; Kamal-Eldin and Appelqvist, 1996) (Fig. 4.23.2). Among them, the α-tocopherol type is most the effective and the only form having an exceptional antioxidant activity and is renowned to fulfill the human needs.

Furthermore, α-tocopherol type is the most potent fat soluble vitamin, having efficient antioxidant properties, reacts with lipid radicals that are produced in lipid peroxidation chain reactions and protects oxidation of cell membranes, thus prevent the radical chain by generating a low reactivity derivative which is not capable of attacking the lipid substrate (Rigotti, 2007).

4.23.4 Absorption and metabolism

Absorption of vitamin E is not that much efficient in the body. At routine oral doses the approximate absorption of vitamin E is 20% to 40% in healthy subjects. The percentage of absorption is gradually decreased with increase in the dose (Traber et al., 1993). Vitamin E is a liposoluble vitamin therefore both bile secretions and pancreatic juices are necessary for the absorption of vitamin E. Concomitant intake and digestion of dietary fats increased the absorption of vitamin E. After absorption the vitamin E molecules are taken up by enterocytes (columnar epithelial absorptive cells lining the inner surface of small and large intestine) and released via chylomicra (microscopic particles of triglycerides synthesized in intestine during digestion

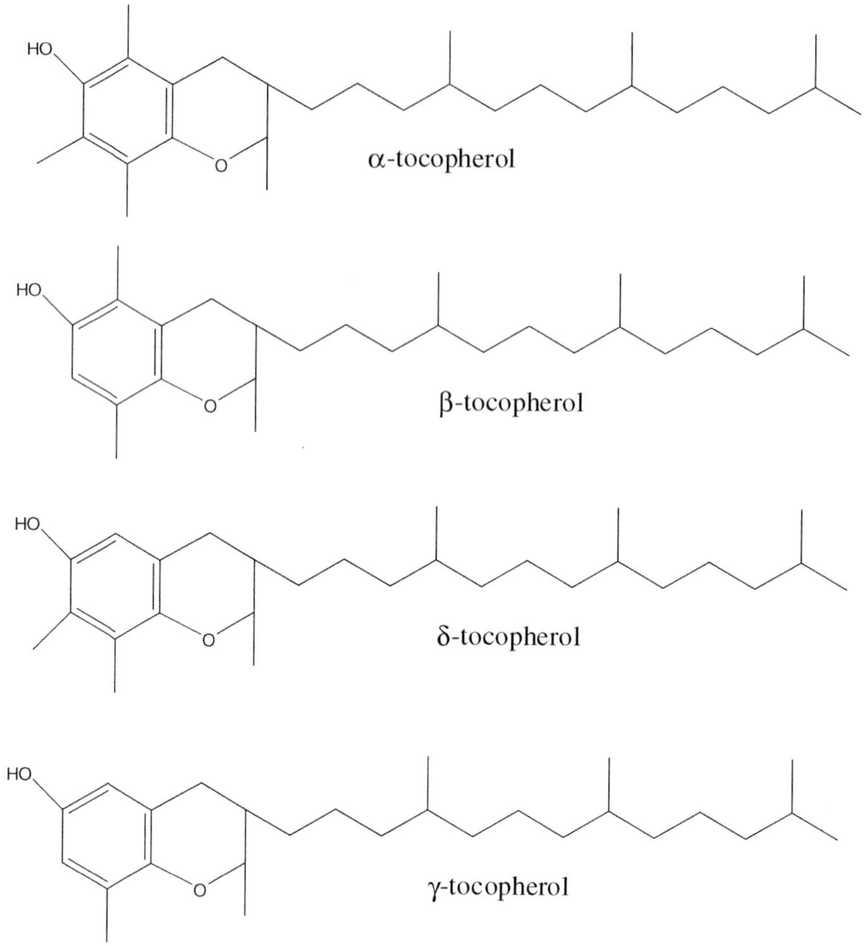

FIG. 4.23.1 Showing chemical structures of α, β, δ, and γ-tocopherol.

which release fatty acids into the blood). These chylomicra containing vitamin E molecules are taken up by the liver and then released back to the blood stream with LDL (see Fig. 4.23.3) (Machlin, 1991).

It has been learned from past few years that the liver is the important site of regulation of vitamin E in the mammalian physiological systems (Yoshida et al., 1992). Liver contains a specific preference for the regulation of α-form of vitamin E (Traber and Kayden, 1989), although upper tissues layer has the ability to retain the RRR-isomer as well (Traber et al., 1990). To best of our knowledge, from all isomers of vitamin E that are obtained and isolated form food sources only the α and γ-tocopherols are best absorbed and retained in human tissues. Studies also suggest that if we increase the intake of α-isomer of tocopherol, it decreases the concentration of γ-tocopherol at the tissues level (Handelman et al., 1994).

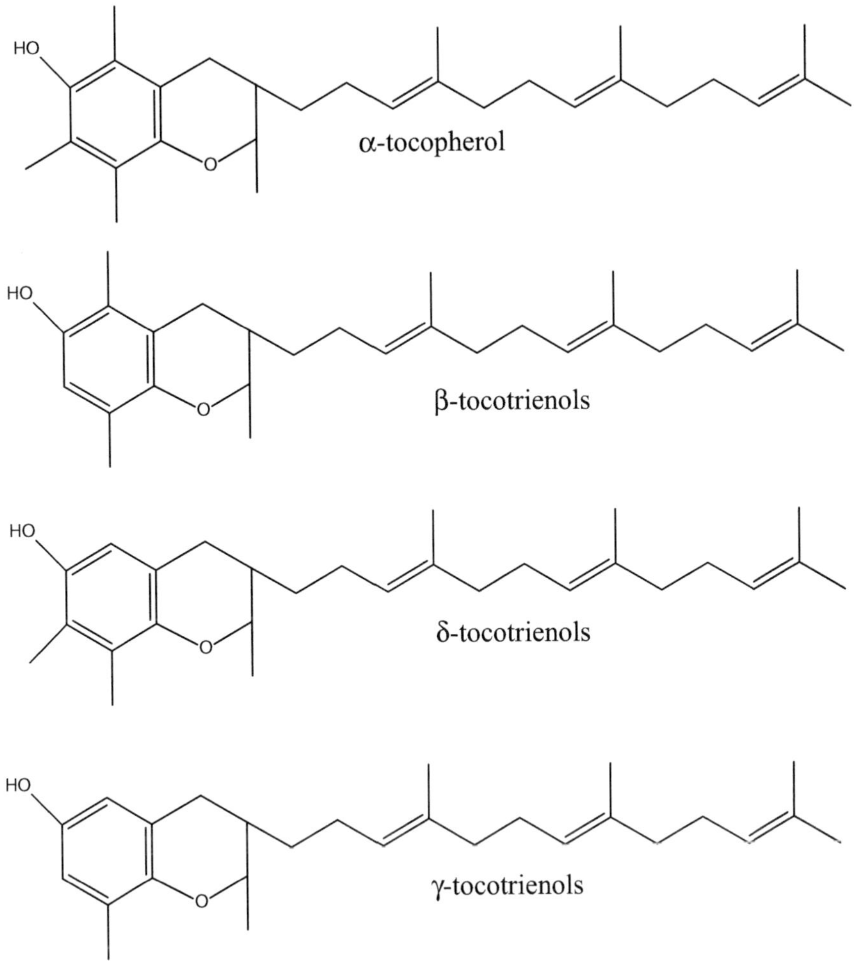

FIG. 4.23.2 Showing chemical structures of α, β, δ, and γ-tocotrienols

4.23.5 Bioavailability

The primary period of the assimilation retention procedure is the disintegration of vitamin E in the lipid phase of the meal. No digestion of vitamin E seems to exist in the stomach. In the duodenum, vitamin E is joined, alongside lipid processing items, in blended micelles, structures that are hypothetically fundamental for its retention by the enterocyte. Significant factor deciding vitamin E bio-accessibility is the measure of fat given in the feast, as fat likely encourages vitamin E extraction from its food framework, invigorates biliary emission, and advances micelle development. Its retention is really interceded, in any event to some extent, by cholesterol layer transporters including the scrounger receptor class B type I (SR-BI), CD36 particle

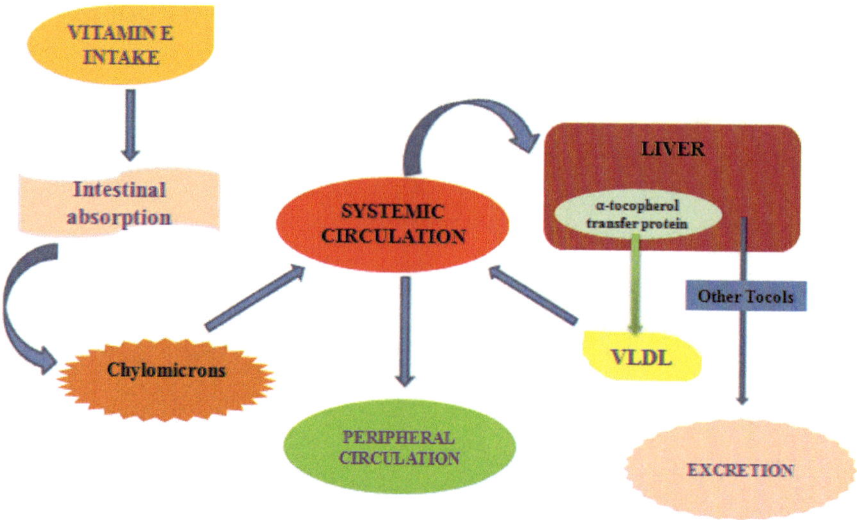

FIG. 4.23.3 Pathway of vitamin E absorption and metabolism.

(CD36), NPC1-like transporter 1 (NPC1L1), and ATP-binding tapes A1 and G1 (Reboul, 2017; Reboul, 2019). A study was conducted on rats to compare the bioavailability and absorption of α-tocopherol and γ-tocopherol. After administration of these two isomers it was found that the bioavailability of α-isomer was higher (36%) as compared to γ-isomer (9%). This significant increase in absorption of α-isomer is attributed to its higher permeability across intestinal membranes. So the bioavailability of tocopherols can be increased by enhancing its permeability by different techniques like nanoformulations (Abuasal et al., 2012). Another study was reported in which tocopherols were co-administered with lipids and it resulted in higher bioavailability. The mechanism behind this was lipids slowed the gastric emptying and stimulated the bile salts to dissolve tocopherols and ultimately increased its absorption and bioavailability (Pouton, 2006).

4.23.6 Mechanism of action

Vitamin E (α-tocopherol) is a radical-scavenging antioxidant. At the point when it searches peroxyl radicals, it is changed over into Vitamin E (α-tocopheroxyl) radical, which might be additionally oxidized into α-tocopheryl quinone or reduced by vitamin C (asborbic acid) or other reducing compounds to recover vitamin E (Fig. 4.23.4). α-Tocopheryl quinone is a biomarker of the cancer prevention agent and active form of Vitamin E. Strangely, a high amount of α-tocopheryl quinone is found in human atherosclerotic plaque (Niki, 2015).

From epidemiological studies vitamin E is proved to have antiatherogenic agent properties and its intake reduced the risk of heart diseases in human. The mechanism of

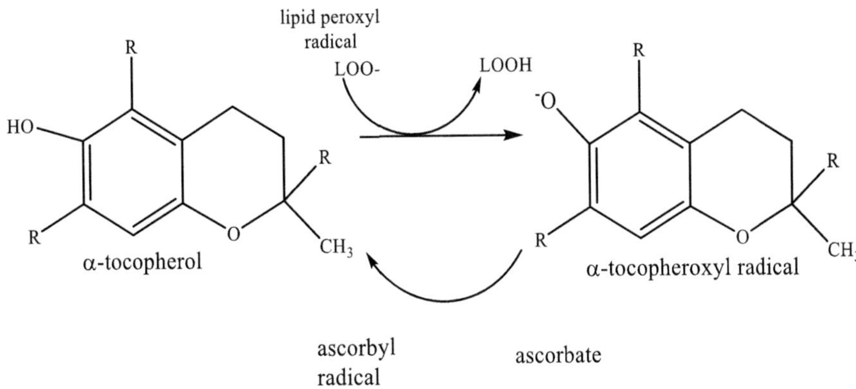

FIG. 4.23.4 The antioxidant mechanism of vitamin E (α-tocopherol).

antiatherogenesis of vitamin E is due to increase in the synthesis and release of selectin which is an adhesion molecule and hence prevents the binding of monocytes with endothelial cells (Rimm et al., 1993). Another suggested mechanism of vitamin E is to prevent the development of atheromatous plaque by reducing the oxidation of LDL cholesterol which is an important step in the pathogenesis of fatty streak (Faruqi et al., 1994).

4.23.7 Possible pro-oxidant activity

Vitamin E is the real lipid solvent antioxidant, giving hydrogen to yield a phenoxyl radical (PhO) in any case, under some particular conditions, the PhO of vitamin E catalyzed prooxidant action, for example, lipoprotein oxidation. Which is clarify in following equation (See Fig. 4.23.5) (Tafazoli et al., 2005).

High-portion of vitamin E enhancements might be related with expanded mortality. For this reason, a ferrous oxidation xylenol test was utilized to evaluate plasma oxidation action levels in tests from a randomized, placebo treatment controlled, 6-week clinical trial of day-by-day vitamin E supplementation in grown-ups with asthma (n = 72). 27% expansion in plasma oxidation action levels was seen in patients getting vitamin E. This shows a pro-oxidant impact of high-portion vitamin E supplementation that may clarify the expansion in mortality saw in mediation studies utilizing this supplement (Pearson et al., 2006).

4.23.8 Beneficial effects of vitamin E on health

Vitamin E is a physiologically basic micronutrient and has been used in different fields including medicine, pharmaceutics, beauty care products, and by food industry. The beneficial effects of Vitamin E on health by reducing Oxidative Stress is presented in

Peroxidase

$PhOH + H_2O_2$ -------------------> $PhO^* + H_2O_2$

$PhO^* + GSH$ -------------------> $PhO^* + GS^*$

$GS^* + GS^-$ -------------------> $GSSG^{*-}$

$GSSG^{*-} + O_2$ -------------------> $GSSG + O_2{}^{*-}$

$2O_2 + 2H+$ -------------------> $H_2O_2 + O_2$

FIG. 4.23.5 The pro-oxidant activity of vitamin E.

Fig. 4.23.6. Vitamin E is accepted to assume a significant role in the advancement of wellbeing and counteractive action and additionally treatment of certain ailments and disorders. The everyday prescribed admission is 15 mg (22.4 IU, International Unit) for adults. Numerous elements of vitamin E have been shown or proposed, including (1) antioxidant by scavenging free radicals, particularly peroxyl radicals, and singlet oxygen, (2) membrane stabilization by forming complexes with destabilizing molecules in order to avoid aggravation of the amphipathic balance inside the structure, (3) physiological controller of enzyme activity, cell flagging, cell multiplication, and gene articulation, which isn't directly related to antioxidant activity, (4) restraint of platelet coagulation, (5) aversion of infections including neurological issue, cardiovascular ailments, age-related eye and skin harm, and barrenness, and (6) biocompatible modifier of biomaterials and therapeutic gadgets, for instance in high molecular weight polyethylene utilized in hip and knee implants. It has been contended that tocotrienols have extra positive wellbeing impacts past those of tocopherols including, for instance, enlistment of immune reactions and bringing down of serum cholesterol levels (Niki, 2019). The Alzheimer's Disease Cooperative Study in 1997 demonstrated that vitamin E may moderate illness progression in patients with modestly extreme AD. High portions of vitamin E deferred the loss of the patient's capacity to do everyday exercises and their resulting situation in private consideration for a while. Other than the previously mentioned infections, vitamin E has additionally been found to assume a valuable job in different sicknesses, for example, photo dermatitis, menstrual agony/dysmenorrhoea, pre-eclampsia and tardive dyskinesia, when brought with vitamin C (Rizvi et al., 2014). Insufficiency of vitamin E is extraordinary as average eating regimens seem to give adequate sums, in spite of the fact that lack of healthy sustenance and hereditary issue may result in vitamin E inadequacy. Untimely infants of exceptionally low birth weight may be inadequate in vitamin E. Further, individuals with fat malabsorption issue and acquired issue in which the liver's α-tocopherol transfer protein (αTTP) is flawed or absents' or there are diminished dimensions of selenoproteins are bound to progress toward becoming vitamin E inadequate and require high portions of supplemental vitamin E. The potential job of vitamin E against periodontal illness, nonalcholic steatohepatitis,

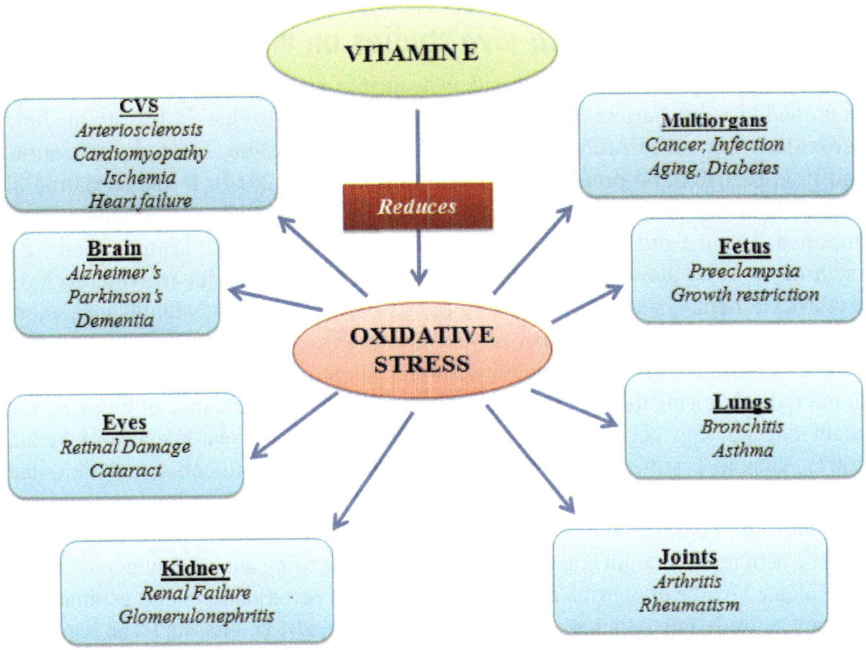

FIG. 4.23.6 Beneficial effects of vitamin E on health by reducing oxidative stress.

and sarcopenia has gotten consideration. Vitamin E is likewise utilized for biomedical materials. α-tocopherol is mixed to a dialysis layer for clinical administration of interminable hemodialysis patients to improve stability and functions. Vitamin E mixed ultrahigh molecular weight polyethylene (UHMWPE) has been created as a material for use in all out knee and hip replacements (Niki, 2019).

4.23.9 Side effects/unwanted of vitamin E

One examination found an expanded danger of death at dosages of 400 IU/day, in spite of the fact that the hazard started to increment at 150 IU (Bjelakovic et al., 2007). Results from the same of late distributed, huge SELECT clinical trial demonstrate that vitamin E supplements (400 IU/day) may hurt grown-up men in the overall public by expanding their danger of prostate cancer (Klein et al., 2011). Follow-up studies are surveying whether the malignancy hazard was related with baseline blood dimensions of vitamin E and selenium before supplementation just as whether changes in one or more genes may expand a man's danger of creating prostate disease while taking vitamin E. The mechanism underlying the expanded danger of high-portion vitamin E supplementation is obscure. The phenomenon might be because of the enlistment of cytochrome P450, which quickens digestion of different medications (Traber, 2013).

4.23.10 *In-vitro* and *in-vivo* studies on vitamin E

The activity of vitamin E as an antioxidant against *in-vitro* lipid peroxidation has been examined broadly. Various examinations show plainly that vitamin E restrains lipid peroxidation in the test tube as evaluated by oxygen take-up, substrate utilization, and lipid peroxidation products formation (Ayala et al., 2014). In the oxidation of entire blood actuated by free radicals created in aqueous phase, the antioxidants diminished in the order of vitamin C > bilirubin > uric acid, plasma vitamin E> erythrocyte thios and vitamin E. Another model is the restraint by vitamin E of erythrocyte hemolysis initiated by free radicals. Vitamin E was significantly more viable at hindering hemolysis than Trolox, a water-dissolvable vitamin E homologue, which has similar substance reactivity toward oxygen radicals yet restricted access to the radicals inside the membrane, recommending the significance of breaking the chain engendering occurring in the membranes (Krajcovicova-Kudackova et al., 2004). Vitamin E stifled ROS creation, lipid peroxidation, and cell demise initiated by selenium lack in Jurkat cells and furthermore by glutamate in juvenile essential cortical neuron cultures (Niki, 2014).

The activity of vitamin E as a peroxyl radical-scavenging antioxidant *in vivo* might be evaluated by the dimension and circulation of lipid peroxidation items estimated in biological fluids and tissues as a marker (Yoshida et al., 2013). Vitamin E can scavenge lipid peroxyl radicals to restrain lipid peroxidation (Min et al., 2016). Its supplementation diminished the dimensions of the lipid peroxidation products and ameliorated the cognitive deficits, with corresponding increment in α-tocopherylquinone, unequivocally recommending that vitamin E hindered lipid peroxidation by scavenging lipid peroxyl radicals (Chen et al., 2016). This vitamin and selenium act in cooperation to detoxify lipid hydro peroxides (Niki, 2014).

By and large, it might be expressed that vitamin E, applies antioxidant impacts by scavenging lipid peroxyl radicals *in vivo* just as *in vitro* frameworks. It ought to be included that vitamin E isn't a proficient scrounger of hydroxyl radical, alkoxyl radical, nitrogen dioxide, thiyl radical, ozone, hypochlorite, and likely singlet oxygen *in vivo*. It might be noticed that the impacts of antioxidant supplementation including vitamin E to well-sustained subjects are frequently little and some huge scale mediation studies have demonstrated disillusioning outcomes (Niki, 2014).

4.23.11 Clinical trial studies

Among a gathering of 5133 people pursued for a mean of 14 years, higher vitamin E admissions from nourishment were related with diminished mortality from CHD. Notwithstanding, randomized clinical trials provide reason to feel doubt about the adequacy of vitamin E supplementation to avoid CHD. For instance, the Heart Outcomes Prevention Evaluation think about, which pursued just about 10,000 patients at high danger of heart assault or stroke for 4.5 years, found that members taking 400 IU/day of common vitamin E encountered no less cardiovascular occasions or hospitalizations for heart disappointment or chest torment than members taking a placebo treatment. Two randomized controlled trials in which members took supplements of vitamin E or a placebo treatment failed to demonstrate a defensive impact for vitamin E on AMD. The Age-Related Eye Disease Study, a huge randomized clinical trial, found that members at high danger of creating propelled AMD diminished their danger of creating advanced AMD by 25% by taking an everyday supplement containing vitamin E, beta-carotene, vitamin C, zinc, and copper contrasted with members taking a placebo treatment more than 5 years. Another study where 769 people with mellow intellectual debilitation were arbitrarily relegated to get 2000 IU/day vitamin E, a cholinesterase inhibitor, or placebo treatment found no noteworthy contrasts in the progression rate of Alzheimer's sickness between the vitamin E and placebo treatment bunches. In synopsis, most research results don't bolster the utilization of vitamin E supplements by healthy or mildly impaired individuals to keep up psychological execution or moderate its decrease with typical aging (N.I.H., 2018). Nonetheless, there are a few purposes behind the frustrating and conflicting side effects of vitamin E treatment. The side effects of inexhaustible *in-vitro* and *in-vivo* tests and the nearby relationship between the dimensions of oxidative pressure and the sickness state detailed in numerous examinations emphatically bolster the thought that free radical-intervened oxidation of biological molecules assumes a causative job in the pathogenesis of certain ailments. Oxidative pressure might be significant for the commencement of illnesses however turns out to be continuously less significant during the later phases of chronic ailments, including atherosclerosis and neurodegenerative conditions.

Vitamin E is potent radical-scavenging antioxidant, however it is insufficient against non-radical oxidants, for example, lipoxygenase, cyclooxygenase, cytochrome P450, and hypochlorite. These oxidants might be engaged with the pathogenesis of illnesses, for example, atherosclerosis, and their relative commitments may change contingent upon the conditions. This recommends the impacts of vitamin E against oxidative pressure incited maladies are constrained and that even enormous scale, randomized clinical studies of vitamin E may not give clear outcomes (Niki, 2015).

Conclusions

In short, it is concluded that vitamin E has a strong antioxidant potential that is considered as the basis of its therapeutic potential. The explicit role of this vitamin has been observed in several human disorders, such as cardiovascular diseases, traumatic brain injury, renal system, metabolic disorders, and inflammations. The different roles of vitamin E as antioxidant agent both as a preventive and causative hazard factor in different human illnesses propose a huge number of variables must be considered in the prescription of antioxidants as a preventive measure to diminish the danger of specific infections. The negative cases watched for its high doses and pro-oxidant action additionally make it confusing and lemmatized its wide spread use. In addition, for the treatment different disorders, it needed to be studied in combination with other agents. Hence, further details studies are required to include vitamin E as an integral component of any prescription.

References

Abuasal, B.S., Qosa, H., Sylvester, P.W., Kaddoumi, A., 2012. Comparison of the intestinal absorption and bioavailability of γ-tocotrienol and α-tocopherol: in vitro, in situ and in vivo studies. Biopharm. Drug Dispos 33, 246–256.

Ayala, A., Muñoz, M.F., Argüelles, S., 2014. Lipid peroxidation: production, metabolism, and signaling mechanisms of malondialdehyde and 4-hydroxy-2-nonenal. Oxid. Med. Cell Longev. 2014, 360438.

Barr, R., Arntzen, C., 1969. The occurence of δ-tocopherylquinone in higher plants and its relation to senescence. Plant Physiol. 44, 591–598.

Bjelakovic, G., Nikolova, D., Gluud, L.L., Simonetti, R.G., Gluud, C., 2007. Mortality in randomized trials of antioxidant supplements for primary and secondary prevention: systematic review and meta-analysis. JAMA 297, 842–857.

Brigelius-Flohe, R., Traber, M.G., 1999. Vitamin E: function and metabolism. FASEB J. 13, 1145–1155.

Bucke, C., 1968. The distribution and stability of α-tocopherol in subcellular fractions of broad bean leaves. Phytochemistry 7, 693–700.

Buettner, G.R., 1993. The pecking order of free radicals and antioxidants: lipid peroxidation, -tocopherol, and ascorbate. Arch. Biochem. Biophys. 300, 535–543.

Burton, G.W., Doba, T., Gabe, E., Hughes, L., Lee, F., Prasad, L., Ingold, K.U., 1985. Autoxidation of biological molecules. 4. Maximizing the antioxidant activity of phenols. J. Am. Chem. Soc. 107, 7053–7065.

Burton, G.W., Traber, M.G., 1990. Vitamin E: antioxidant activity, biokinetics, and bioavailability. Annu. Rev. Nutr. 10, 357–382.

Chen, B., Rao, J., Ding, Y., McClements, D.J., Decker, E.A., 2016. Lipid oxidation in base algae oil and water-in-algae oil emulsion: impact of natural antioxidants and emulsifiers. Food Res. Int. 85, 162–169.

Dam, H., 1962. Interrelations Between Vitamin E and Polyunsaturated Fatty Acids in Animals, Vitamins and Hormones. Elsevier, United States, pp. 527–540.

Diplock, A., Baum, H., Lucy, J., 1971. The effect of vitamin E on the oxidation state of selenium in rat liver. Biochem. J. 123, 721–729.

Evans, H.M., Bishop, K.S., 1922. On the existence of a hitherto unrecognized dietary factor essential for reproduction. Science 56, 650–651.

Faruqi, R., de la Motte, C., DiCorleto, P.E., 1994. Alpha-tocopherol inhibits agonist-induced monocytic cell adhesion to cultured human endothelial cells. J. Clin. Invest. 94, 592–600.

Haas, A.L., Boscoboinik, D., Mojon, D.S., Böhnke, M., Azzi, A., 1996. Vitamin E inhibits proliferation of human Tenon's capsule fibroblasts in vitro. Ophthalmic Res. 28, 171–175.

Handelman, G.J., Epstein, W.L., Peerson, J., Spiegelman, D., Machlin, L.J., Dratz, E.A., 1994. Human adipose α-tocopherol and γ-tocopherol kinetics during and after 1 y of α-tocopherol supplementation. Am. J. Clin. Nutr. 59, 1025–1032.

Henderson, B.C., Tyagi, S.C., 2006. Oxidative Mechanism and Homeostasis of Proteinase/ Antiproteinase in Congestive Heart Failure. Elsevier, United States.

Herman, R., Btaiche, I., Teitelbaum, D.H., 2011. Nutrition support in the pediatric surgical patient. Surg. Clin. 91, 511–541.

Kamal-Eldin, A., Appelqvist, L.Å., 1996. The chemistry and antioxidant properties of tocopherols and tocotrienols. Lipids 31, 671–701.

Klein, E.A., Thompson Jr., I.M., Tangen, C.M., Crowley, J.J., Lucia, M.S., Goodman, P.J., Minasian, L.M., Ford, L.G., Parnes, H.L., Gaziano, J.M., Karp, D.D., Lieber, M.M., Walther, P.J., Klotz, L., Parsons, J.K., Chin, J.L., Darke, A.K., Lippman, S.M., Goodman, G.E., Meyskens Jr., F.L., Baker, L.H., 2011. Vitamin E and the risk of prostate cancer: the Selenium and Vitamin E Cancer Prevention Trial (SELECT). JAMA 306, 1549–1556.

Krajcovicova-Kudackova, M., Paukova, V., Bacekova, M., Dusinska, M., 2004. Lipid peroxidation in relation to vitamin C and vitamin E levels. Cent. Eur. J. Public Health 12, 46–48.

Machlin, L.J., 1991. Handbook of Vitamins. In: Dekker, M. (Ed.). Marcel Dekker, Inc., New York.

Massey, K.D., Burton, K.P., 1989. Alpha-tocopherol attenuates myocardial membrane-related alterations resulting from ischemia and reperfusion. Am. J. Physiol. Heart Circ. Physiol. 256, H1192–H1199.

Min, Y., Sun, T., Niu, Z., Liu, F., 2016. Vitamin C and vitamin E supplementation alleviates oxidative stress induced by dexamethasone and improves fertility of breeder roosters. Anim. Reprod. Sci. 171, 1–6.

N.I.H., 2018. Vitamin E-Fact Sheet for Health Professionals. National Institute of Health, Office of Dietary Supplements. U.S. Department of Health & Human Services. https://ods.od.nih.gov/factsheets/VitaminE-HealthProfessional/.

Niki, E., 2000. Free radicals in the 1900's: from in vitro to in vivo. Free Radic. Res. 33, 693–704.

Niki, E., 2014. Role of vitamin E as a lipid-soluble peroxyl radical scavenger: in vitro and in vivo evidence. Free Radic. Biol. Med. 66, 3–12.

Niki, E., 2015. Evidence for beneficial effects of vitamin E. Korean J. Intern. Med. 30, 571–579.

Niki, E., 2019. Vitamin E: Chemistry and Nutritional Benefits. Royal Society of Chemistry, United Kingdom.

Niki, E., Noguchi, N., 2004. Dynamics of antioxidant action of vitamin E. Acc. Chem. Res. 37, 45–51.

Packer, L., 1991. Protective role of vitamin E in biological systems. Am. J. Clin. Nutr. 53, 1050S–1055S.

Pappenheimer, A.M., Goettsch, M., 1934. Nutritional myopathy in ducklings. J. Exp. Med. 59, 35–42.

Pearson, P., Lewis, S.A., Britton, J., Young, I.S., Fogarty, A., 2006. The pro-oxidant activity of high-dose vitamin E supplements in vivo. BioDrugs 20, 271–273.

Pennock, J., Hemming, F., Kerr, J.D., 1964. A reassessment of tocopherol chemistry. Biochem. Biophys. Res. Commun. 17, 542–548.

Pouton, C.W., 2006. Formulation of poorly water-soluble drugs for oral administration: physicochemical and physiological issues and the lipid formulation classification system. Eur. J. Pharm. Sci. 29, 278–287.

Reboul, E., 2017. Vitamin E Bioavailability: Mechanisms of Intestinal Absorption in the Spotlight. Antioxidants (Basel) 6, 95.

Reboul, E., 2019. Vitamin E intestinal absorption: Regulation of membrane transport across the enterocyte. IUBMB Life 71, 416–423.

Rigotti, A., 2007. Absorption, transport, and tissue delivery of vitamin E. Mol. Aspects Med. 28, 423–436.

Rimm, E.B., Stampfer, M.J., Ascherio, A., Giovannucci, E., Colditz, G.A., Willett, W.C., 1993. Vitamin E consumption and the risk of coronary heart disease in men. N. Engl. J. Med. 328, 1450–1456.

Rizvi, S., Raza, S.T., Ahmed, F., Ahmad, A., Abbas, S., Mahdi, F., 2014. The role of vitamin e in human health and some diseases. Sultan Qaboos Univ. Med. J. 14, e157–e165.

Schürks, M., Glynn, R.J., Rist, P.M., Tzourio, C., Kurth, T., 2010. Effects of vitamin E on stroke subtypes: meta-analysis of randomised controlled trials. BMJ 341, c5702.

Shils, M.E., Shike, M., 2006. Modern Nutrition in Health and Disease. Lippincott Williams & Wilkins, UK.

Steinberg, D., Parthasarathy, S., Carew, T., Khoo, J., 1989. Witztum Beyond cholesterol JL. Modifications of low-density lipoprotein that increase its atherogenicity. N. Engl. J. Med. 320, 915–924.

Stephens, N.G., Parsons, A., Brown, M., Schofield, P., Kelly, F., Cheeseman, K., Mitchinson, M., 1996. Randomised controlled trial of vitamin E in patients with coronary disease: Cambridge Heart Antioxidant Study (CHAOS). Lancet North Am. Ed. 347, 781–786.

Tafazoli, S., Wright, J.S., O'Brien, P.J., 2005. Prooxidant and antioxidant activity of vitamin E analogues and troglitazone. Chem. Res. Toxicol. 18, 1567–1574.

Traber, M., Cohn, W., Muller, D., 1993. Absorption, transport and delivery to tissues. Vitamin E in health and disease, 35–51.

Traber, M.G., 2013. Mechanisms for the prevention of vitamin E excess. J. Lipid Res. 54, 2295–2306.

Traber, M.G., Atkinson, J., 2007. Vitamin E, antioxidant and nothing more. Free Radic. Biol. Med. 43, 4–15.

Traber, M.G., Burton, G., Ingold, K., Kayden, H., 1990. RRR-and SRR-alpha-tocopherols are secreted without discrimination in human chylomicrons, but RRR-alpha-tocopherol is preferentially secreted in very low density lipoproteins. J. Lipid Res. 31, 675–685.

Traber, M.G., Kayden, H.J., 1989. Preferential incorporation of alpha-tocopherol vs gamma-tocopherol in human lipoproteins. Am. J. Clin. Nutr. 49, 517–526.

Yoshida, H., Yusin, M., Ren, I., Kuhlenkamp, J., Hirano, T., Stolz, A., Kaplowitz, N., 1992. Identification, purification, and immunochemical characterization of a tocopherol-binding protein in rat liver cytosol. J. Lipid Res. 33, 343–350.

Yoshida, Y., Umeno, A., Shichiri, M., 2013. Lipid peroxidation biomarkers for evaluating oxidative stress and assessing antioxidant capacity in vivo. J. Clini. Biochem. Nutr. 52, 9–16.

CHAPTER

Vitamin K

4.24

Abhishek K. Das, Sumit Ghosh, Parames C. Sil

Division of Molecular Medicine, Bose Institute, Kolkata, India

4.24.1 Introduction

Vitamin K was discovered in 1929 by the Danish nutritional biochemist Henrik Dam. When he was working on sterol metabolism, he observed that the cholesterol-free diet developed hemorrhage (Ferland, 2012). Initially, he thought that it is due to vitamin C deficiency (Holst and Halbrook, 1933) but the severe bleeding was not cured by administration of a considerable amount of vitamin C. Based on further studies Dam suggested that there should be an active, fat-soluble compound apart from other fat-soluble vitamins and cholesterol (Dam, 1934). In 1935, he named this anti-hemorrhagic factor as vitamin K, after the German word 'Koagulation' (Dam, 1935). Subtypes of vitamin K, vitamin K1 was first isolated from alfalfa (Binkley et al., 1939) and vitamin K2 was isolated from putrefied fish (McKee et al., 1939). The structure of both the vitamin K subtype was derived by Edward Doisy and colleagues (Binkley et al., 1940; MacCorquodale et al., 1939). Both Dam and Doisy were awarded the Nobel Prize on physiology in 1943 for their contribution. Since Vitamins are needed in very minute amount for their biological activity; they were termed as micronutrients (Shearer et al., 2012). The role of micronutrients is to act as a cofactor or coenzyme on different metabolic reactions (Shenkin, 2006). Vitamin K, which is a cofactor of enzymes, is developing as a vital micronutrient with various actions including bone health, vascular calcification, etc. Vitamin K is not a traditional antioxidant and its role as an antioxidant is not extensively reported. In this book chapter we will be focusing on the absorption, mechanism of action of vitamin K, deficiency of the vitamin, vitamin K dependent proteins along with its limited antioxidant power.

4.24.2 General informations about vitamin K

4.24.2.1 Types of vitamin K and structures

There are mainly three types of biologically active forms of vitamin K (Fig. 4.24.1), two of which are obtained from natural sources and one is from laboratory synthesis.

- *Vitamin K1 (phylloquinone/phytonadione):* The phylloquinone is structurally a polycyclic ketone, 2-methyl-1,4-naphthoquinone with a 3-phytyl substituent.

Antioxidants Effects in Health. DOI: https://doi.org/10.1016/B978-0-12-819096-8.00067-7

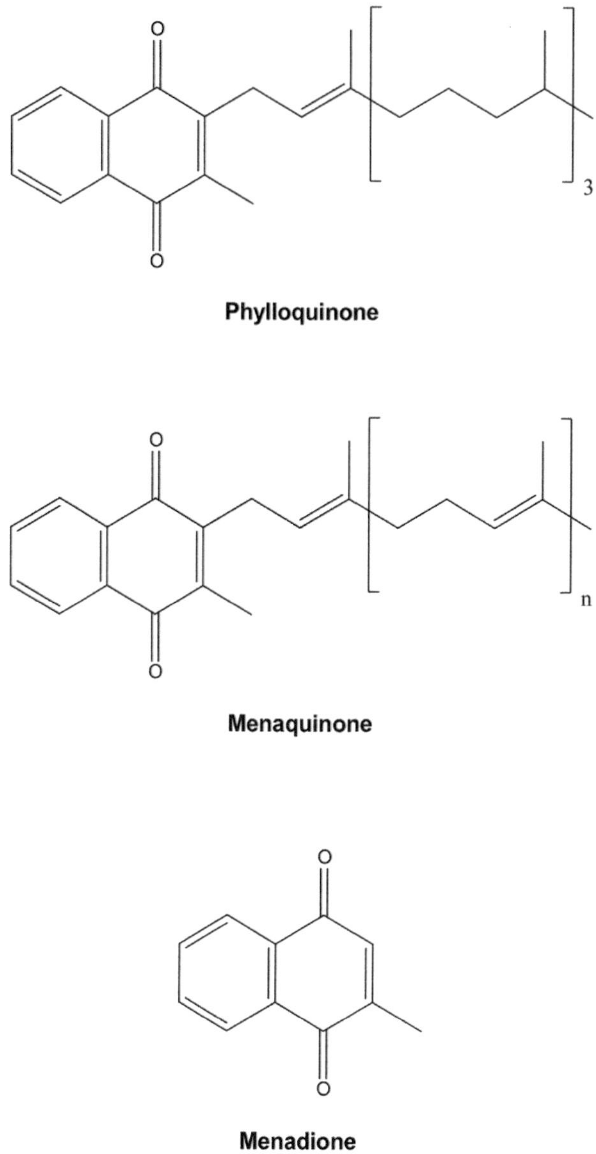

FIG. 4.24.1 Three subtypes of vitamin K.

- *Vitamin K2 (menaquinones, MKs):* MKs are structurally similar to vitamin K1 except that having multiple isoprenoid units (ranging from 4-11) as a side chain.
- *Vitamin K3 (menadione):* Remarkably, the core structure of phylloquinone and MKs, 2-methyl-1,4-naphthoquinone, is not naturally available but is a synthetic compound known as menadione. Two water-soluble forms of menadione, namely menadione sodium bisulfate and menadione dimethyl pyrimidinol

bisulfate, are commercially available. The later one was found to be more effective in promoting plasma prothrombin level (Griminger, 1965).

4.24.2.2 Dietary sources of vitamin K

There are two dietary sources of vitamin K, plant source and endogenous source. Phylloquinone, the prime dietary source of vitamin K, is available in all photosynthetic plants (Gross et al., 2006). Since phylloquinone is a standard component of the chloroplast, green and leafy vegetables (Gross et al., 2006) contain the highest amount of the compound. In a study, it was found that vegetables contribute almost 60% of total phylloquinone intake whereas cooked vegetables contribute almost 28% of total phylloquinone intake (Thane et al., 2002). Dark green colored vegetables, like spinach, cabbage, broccoli etc. have a higher concentration of phylloquinone since they contain a larger amount of chlorophyll than other vegetables. Plant-derived oils like soybean oil, canola oil, etc. also contain phylloquinone. Table 4.24.1 listed the major sources of phylloquinone along with menaquinone sources.

Table 4.24.1 Vitamin K content (μg/100 g or μg/100 mL) in different food items (Schurgers and Vermeer, 2000).

Nature of food	Phylloquinone content	MK-4 content	Other menaquinone content
Fruits and vegetables			
Spinach	387	-	-
Broccoli	156	-	-
Kale	817	-	-
Natto	34.7	-	998 (MK-7), 84.1 (MK-8)
Oils			
Margarine	93.2	-	-
Butter	14.9	15.0	-
Olive oil	53.7	-	-
Sunflower oil	5.7	-	-
Dairy produce			
Whole milk	0.5	0.8	-
Hard cheeses	10.4	4.7	16.9 (MK-8), 51.1 (MK-9)
Soft cheese	2.6	3.7	11.4 (MK-8), 39.6 (MK-9)
Egg yolk	2.1	31.4	-
Animal protein			
Goose leg	4.1	31.0	-
Goose liver paste	10.9	369	-
Chicken breast	-	8.9	-
Chicken leg	-	8.5	-

Table 4.24.2 Synthesis of subtypes of menaquinone by gut microbiota (Mathers et al., 1990).

Bacterial species	Major menaquinones	Minor menaquinones
Enterobacter sp.	MK-8	DemethylatedMK-8
Veillonella sp.	MK-7	MK-6
Enterococcus sp.	MK-9	Demethylated forms of MK-6, MK-7, MK-8
Bacteroides fragilis	MK-11, MK-10, MK-12	MK-9, MK-8, MK-7
Bacteroides vulgatus	MK-11, MK-10, MK-12	MK-9, MK-8, MK-7

Menaquinones are mainly synthesized by different bacteria present in gut microflora. About 50% of our daily vitamin K requirement comes from the endogenous source. *Enterobacter, Veillonella, Enterococci, Bacteroides fragilis* and other *Bacteroides* species are the primary organisms that synthesize MKs (Table 4.24.2) (Duda-Chodak et al., 2015). Natto, a traditional Japanese food, contains a massive amount of MK-7 as a fermented product of *Bacillus subtilis natto*. Natto also contains a considerable amount of MK-6, MK-8, and phylloquinone. Fermented products like a few types of cheese were found to contain menaquinones (Schurgers and Vermeer, 2000). MK-4, the only menaquinone that is not synthesized by bacteria, can be obtained from different types of meat, butter and dairy products, egg yolk (Schurgers and Vermeer, 2000). It is to be noted that breast milk is deficient in vitamin K (Table 4.24.3) (Erick, 2018).

MK-4 is a specific menaquinone, which is not present in the diet but is synthesized in our body from phylloquinone. A large portion of the scientific community primarily misunderstood that generation of MK-4 is from menaquinone. However, later several studies broke this concept. Feeding menadione free diet of experimental chicks showed the presence of MK-4 in serum, liver and kidney (Will et al., 1992). Double radiolabeled phylloquinone (tritium in the 2-methyl group in the core and ^{14}C in the side chain) consumption on chick or pigeon model results in several organs containing MK-4 radiolabeled only in the core (Shearer and Newman, 2008). Vitamin K free diet depletes the phylloquinone and MK-4 levels in tissues but subsequent supplementation of phylloquinone results in increasing both phylloquinone and MK-4 levels (Shearer and Newman, 2008).

Table 4.24.3 Vitamin K content in human breast milk (Canfield et al., 1991).

Sample	Vitamin K concentration (nmol/L)
Colostrum (30–81 h)	7.52 ± 5.90
Mature Milk – 1 month	6.98 ± 6.36
Mature Milk – 3 months	5.14 ± 4.52
Mature Milk – 6 months	5.76 ± 4.48

4.24.2.3 Bioavailability

Surprisingly, very few data are available regarding the bioavailability of different forms of vitamin K. Phylloquinone in its free form has 80% absorption rate but phylloquinone from plant origin has very poor bioavailability. The body absorbs only about 4% from spinach regarding phylloquinone supplements although the absorption increases when spinach was consumed with butter, which is not surprising as vitamin K is lipophilic (Garber et al., 1999). Since phylloquinone is tightly bound with chloroplast, the bioavailability is lower in the case of green vegetables than plant-derived oils (L. Booth, 2012). Bioavailability of food derived menaquinones is higher than phylloquinone. The postprandial serum concentration of MK-7 from natto was found to be 10 fold higher than phylloquinone from spinach (Schurgers and Vermeer, 2000).

4.24.2.4 Some important information at a glance

- The LD50 value for a single parenteral dose of vitamin K2 is in the range of 75 to 200 mg per kg of body weight while the LD50 value for a single oral dose is 600 to 800 mg per kg of body weight for mice, rats, and dogs (Shapiro and Richards, 1945).
- Vitamin K1 is usually dissolved in 99.9% ethanol for use in vitro (Wei et al., 2010).
- In 2001, the IOM Food and Nutrition Board prescribed the oral dose of vitamin K to be 120 mcg/day in adult male and 90 mcg/day in adult female humans (Shearer et al., 2012).

4.24.3 Cellular metabolism of vitamin K

4.24.3.1 Absorption of vitamin K

Since vitamin K is fat soluble, its absorption is similar to lipid absorption. MKs with longer side chains are mainly synthesized by gut bacteria, but because of the presence of a cell wall, the bioavailability is very poor or insignificant. Inside the intestinal lumen, phylloquinone and MK-7 from dietary source form mixed micelles with bile salts, triacylglycerols, free fatty acids and phospholipids. These mixed micelles enter into the intestinal enterocytes present in intestinal villi. Study with tritium-labeled phylloquinone revealed that phylloquinone associates with nascent chylomicrons after it is being absorbed by intestinal enterocytes (Blomstrand and Forsgren, 1968). These nascent chylomicrons, comprised of apoA and apoB48, are subsequently released into lacteal and eventually enter into the bloodstream through the thoracic duct. During circulation in the bloodstream, chylomicrons get apoC and apoE bind with it. Upon entering capillaries of tissues, triacylglycerols are stripped from chylomicrons by the action of lipases present in the surface of the endothelium. The resulting chylomicron remnants maintain the lipophilic core and vitamin K although the loss of apoA and apoC occurs (Shearer and Newman, 2008). Chylomicron remnants are rapidly engulfed by the liver. The liver is the place of synthesis of vitamin K dependent coagulation factors. Chylomicron remnants bind to the low-density lipoprotein receptor (LDLR) and low-density lipoprotein receptor-related protein (LRP) receptors present in hepatic tissue and different constituents of chylomicron remnants are engulfed by endocytosis including vitamin K and apoE (Herz and Strickland, 2001). Chylomicron remnants are reconstituted into very low-density lipoprotein (VLDL) having apoB100 and come back to circulation where it acquires apoE and apoC. Gradually triacylglycerols are removed from VLDL by lipoprotein lipase and loss of apoproteins, making it intermediate density lipoprotein (IDL) and low-density lipoprotein (LDL). This LDL and chylomicron remnants transport lipids to bone matrix cells, known as osteoblasts. Receptor-mediated endocytosis of vitamin K along with other lipid molecules takes place in osteoblasts as they express LRP and LDLR receptors. It has been observed that the major portion of phylloquinone is transported by chylomicron remnants whereas most of the MK-7 is transported by LDL (Shearer and Newman, 2008).

4.24.3.2 Storage of vitamin K

The liver is the primary storage of various types of vitamin K. Menaquinones (MK7-13) are major vitamer found in liver comprising almost 80%–90% of total vitamin K storage; the rest is made up of phylloquinone. The mean concentration of vitamin K in the liver is around 200 pmol/g (wet weight) (Usui et al., 1990). Extrahepatic tissues are mostly storage place of phylloquinone and MK-4. Heart and pancreas have a higher level of phylloquinone whereas kidney and brain contains a much higher concentration of MK-4 (Thijssen, 1996).

4.24.3.3 Redox cycling of vitamin K

The oxidation and subsequent reduction of vitamin K coupled with carboxylation of glutamate residues of peptides forming γ-carboxylated proteins is known as vitamin cycle (Fig. 4.24.2). As the main enzymatic reaction involved in this process is oxidoreductase type, the cycle is also known as redox cycle. The redox cycle is consisting of mainly three reactions in the lumen of the endoplasmic reticulum (Ivanova et al., 2018):

- *Vitamin K epoxide formation*: The oxidation of dihydroxy vitamin K (the hydroquinone form, often denoted as KH_2) by γ-glutamyl carboxylase (GGCX) to vitamin K 2,3-epoxide (denoted as KO) is the key step of the pathway. This step is a part of the post-translational modification of glutamate-containing proteins. Vitamin K acts as a cofactor and GGCX carboxylates vitamin K dependent proteins using O_2 and CO_2 (Fig. 4.24.2). Interestingly, GGCX itself is a vitamin K dependent protein and in this process, it also gets carboxylated (Berkner and Pudota, 1998).
- *Formation of vitamin K quinone*: Vitamin K epoxide reductase (VKOR), a microsomal enzyme, converts vitamin K epoxide to vitamin K quinone in an active site

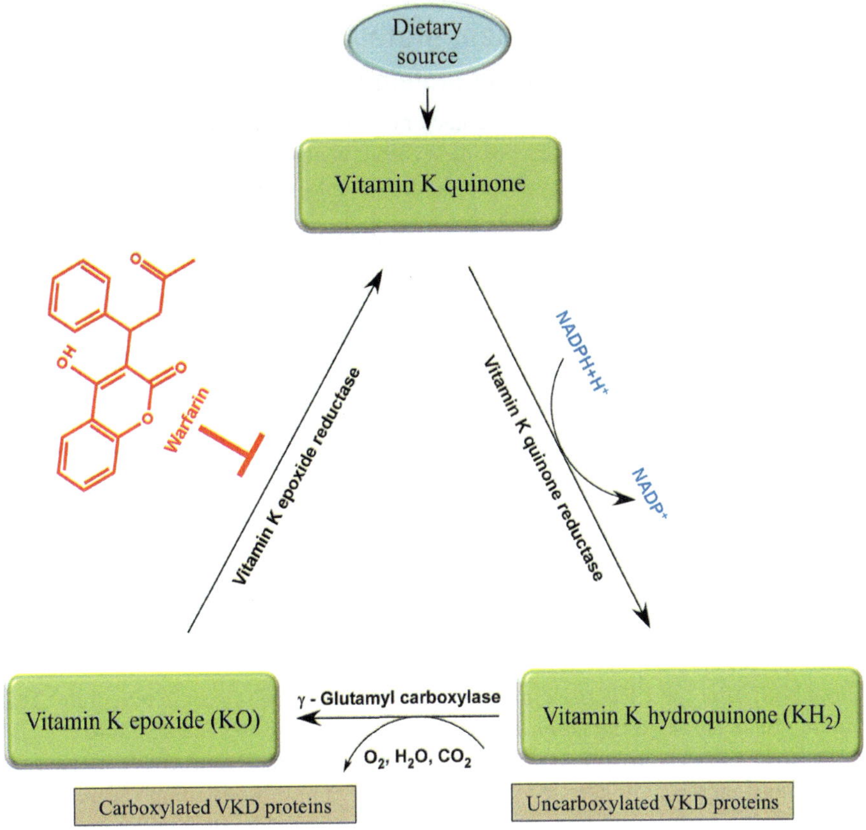

FIG. 4.24.2 The vitamin K redox cycle.

Hydroquinone form is the active form of vitamin K which acts as a cofactor of γ-glutamyl carboxylase converting uncarboxylated VKD proteins to carboxylated VKD proteins. The resulting vitamin K epoxide is converted to active hydroquinone form by the action of vitamin K reductases.

membrane localized, dithiol dependent reaction. The catalytic activity lies within the VKOR1 subunit of the enzyme which consists of three cysteine residues and one serine/threonine residue (Goodstadt and Ponting, 2004). VKOR is inhibited by 4-hydroxycoumarin analog compounds like warfarin, causing different complications and would be described in section 4.24.5.6 (Fasco et al., 1982). Quione form is the most prevalent form of vitamin K in the diet; the active form of vitamin K i.e. KH_2 is regenerated from KO by VKOR (Fig. 4.24.2).

- *Reproduction of active dihydroxy vitamin K*: The reduction of vitamin K quinone to vitamin K hydroquinone is catalyzed by two types of enzymes, namely vitamin K quinine reductase having the same active site structure and the same mechanism of action as VKOR, a dithiol dependent enzyme (Gardill and Suttie, 1990) and DT-diaphorase, an NADPH dehydrogenase (Wallin et al., 1978).

4.24.4 Vitamin K dependent proteins and their functions

4.24.4.1 Vitamin K dependent protein carboxylation

Vitamin K hydroquinone serves as a cofactor of the enzymatic reaction conducted by γ-carboxylase. The enzyme γ-carboxylase carboxylates Glu residues and forms Gla of specific proteins, known as vitamin K dependent (VKD) proteins. The reaction (Fig. 4.24.3) is initiated by Lys218 (Rishavy et al., 2006) present in the active site of the carboxylase enzyme that deprotonates vitamin K hydroxyquinone or reduced vitamin K (KH_2). Subsequent reaction with O_2 converts the deprotonated form to a strong base K-. K- in terms deprotonates Glu residue present in proteins and protonates itself to produce vitamin K epoxide (KO) and a carbanion intermediate. The pKa of the carbanion intermediate was found to be 25~28 (Berkner, 2008) which shows the need for a robust base other than the carboxylate residues. In order to react with O_2, firstly the active site Lys218, deprotonates KH_2. Subsequent reaction with O_2 makes a strong vitamin K intermediate K-. This mechanism where a weak carboxylase base produces a robust vitamin K base is named as 'base amplification model' (Dowd et al., 1995).

The binding of carboxylase to VKD proteins is tightly regulated. The carboxylase has three binding sites with VKD; mature VKD protein binding site, Glu-binding domain and propeptide-binding domain. Propeptide part of the VKD proteins binds with residues 495-513 of carboxylase enzyme and is subjected to proteolysis in Glogi body after the VKD protein is carboxylated in the endoplasmic reticulum. The binding of substrate to the Glu binding site of carboxylase (residues 393-404) is allosterically enhanced by propeptide binding (Knobloch and Suttie, 1987). These two interactions increase the affinity of the cofactor (vitamin K) binding (Soute et al., 1992). Thus carboxylase limits its reaction from generating vitamin K intermediates when there is lack of Glu residues to be carboxylated. The conversion of multiple Glu residues, generally 3-13 for a single VKD, to Gla residues is a single binding event (processive mechanism); although partial carboxylation of Glu residues in one single VKD protein is observed in case of low vitamin K level (Berkner, 2008).

4.24.4.2 The basic structure of a VKD protein

The general structure of VKD proteins consists of 4 parts (Fig. 4.24.4A). (1) The N-terminal signal peptide for its targeted delivery to subcellular organs, (2) followed by propeptide region for targeting the VKD protein to the carboxylase present in endoplasmic reticulum and binding with the enzyme. Two conserved residues, phenylalanine at -16 position and alanine at -10 position are present in the propeptide of various VKD proteins (Furie et al., 1999). These two parts are proteolytically cleaved after their requirement is fulfilled. (3) The Gla domain contains multiple Glu residues that are carboxylated followed by d) a protein domain with a specific function (Berkner and Runge, 2004). Specific VKD protein structure is listed in Table 4.24.4.

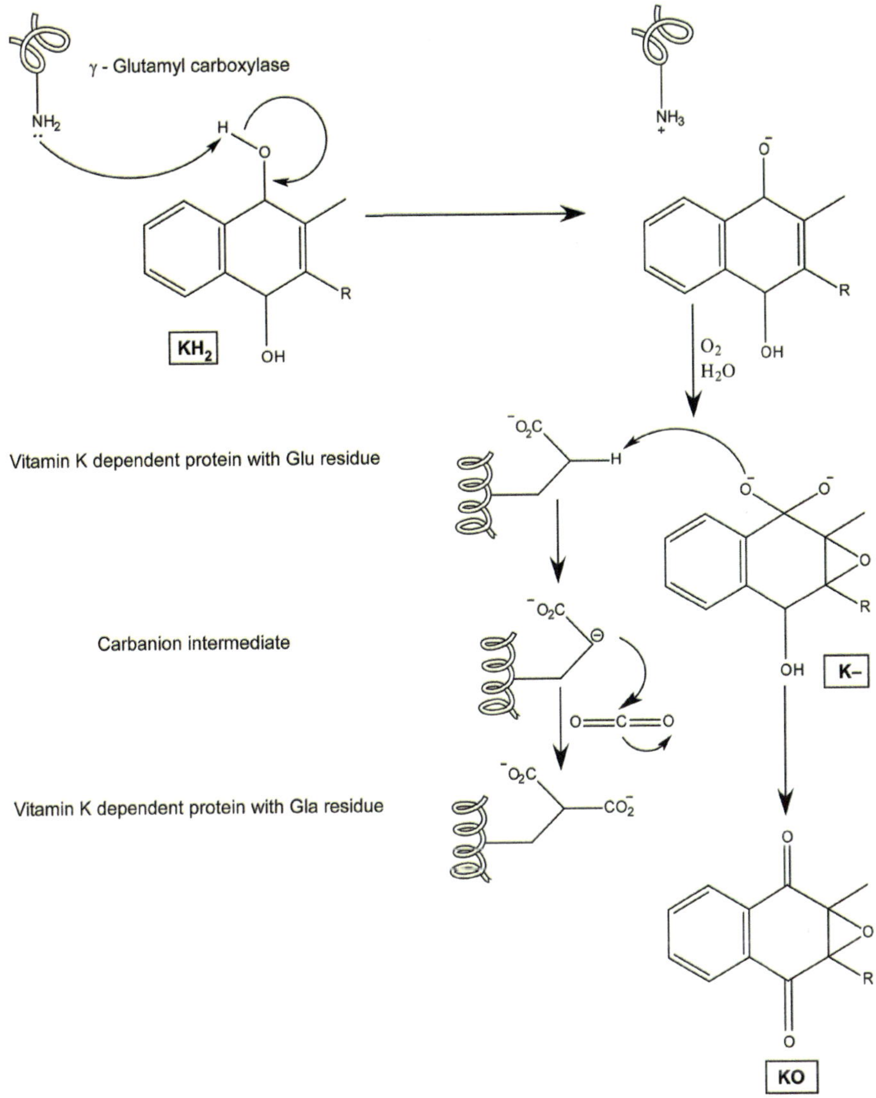

FIG. 4.24.3 The mechanism of the chemical reaction catalyzed by γ-carboxylase.

The reaction is initiated by Lys218 of γ-carboxylase attacking KH_2 followed by reaction with O_2 and H_2O, generating a strong dialkoxide base K–. The intermediate base K– subsequently deprotonates Glu residue of a VKD protein, forming KO and a carbanion intermediate. Subsequent reaction with CO_2 creates carboxylated VKD protein with Gla residues.

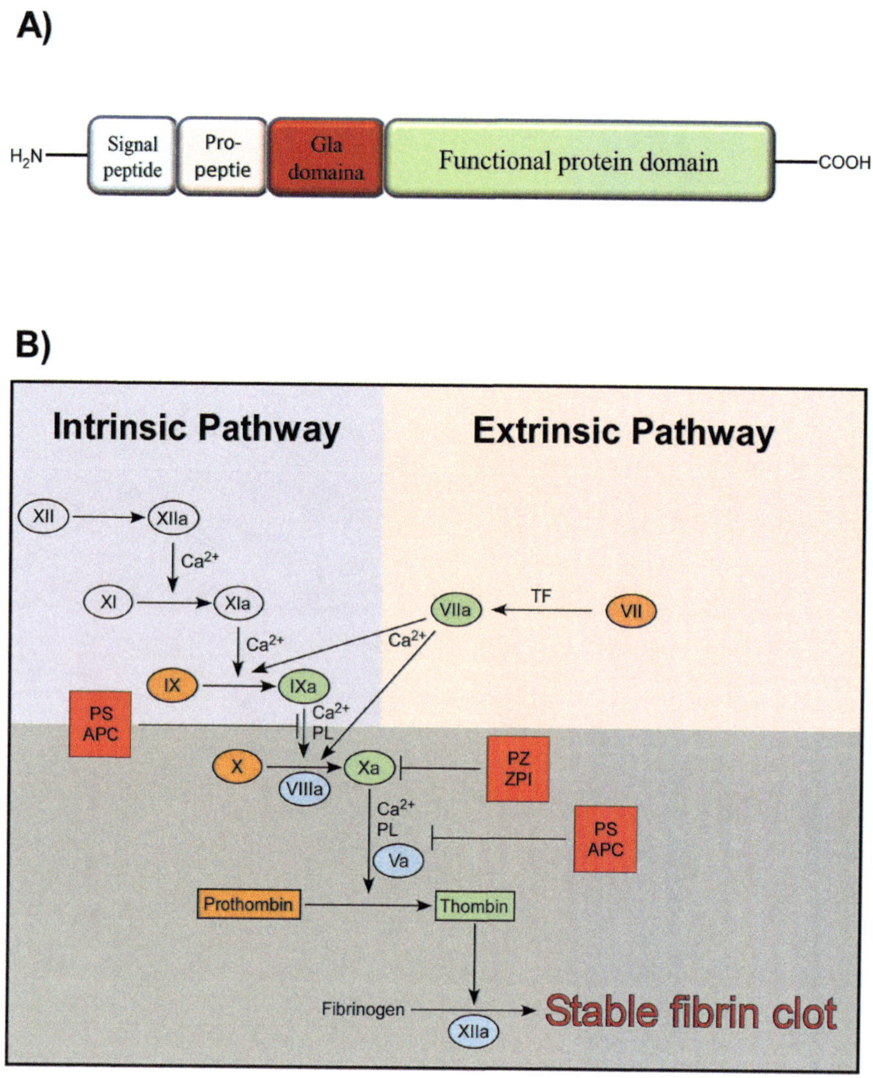

FIG. 4.24.4

(A) Basic structure of VKD proteins. The general primary structure of vitamin K dependent protein consists of four parts, namely signal peptide, propeptide, Gla domain, and functional domain. (B) Blood coagulation cascade (details in the text). APC, activated protein C; PS, protein S; PZ, protein Z; ZPI, protein Z related protease inhibitor. Factors are represented in Roman numerical where subscript "a" means activated form.

4.24.4.3 Blood coagulation and vitamin K

Blood transports oxygen and other nutrients to the cells and carries away the waste product through its circulatory anatomy with the help of the blood vessels and heart. Rupture of blood vessels leads to bleeding and can be dangerous. In order to stop

Table 4.24.4 Domain structure of vitamin K dependent proteins.

Name of VKD protein	Primary domain structure	References
Factor VII (FVII)	H_2N— PRO | GLA | AS | EGF | Serine protease —COOH	(Hagen et al., 1986)
Factor IX (FIX)	H_2N— PRO | GLA | AS | EGF | AP | Serine protease —COOH	(Yoshitake et al., 1985)
Factor X (FX)	H_2N— PRO | GLA | AS | EGF | AP | Serine protease —COOH	(Leytus et al., 1986)
Prothrombin	H_2N— PRO | GLA | AS | K1 | K2 | Serine protease —COOH	(Degen and Davie, 1987)
Protein C (PC)	H_2N— PRO | GLA | AS | EGF | AP | Serine protease —COOH	(Foster and Davie, 1984)
Protein Z (PZ)	H_2N— PRO | GLA | TS | EGF | Pseudo serine protease —COOH	(Berkner and Runge, 2004)
Protein S (PS)	H_2N— PRO | GLA | TS | EGF | SHBD —COOH	(Lundwall et al., 1986)
Osteocalcin	H_2N— PRO | GLA | —COOH	(Pan and Price, 1985)
Matrix Gla protein (MGP)	H_2N— GLA | CRS | GLA | —COOH	(Laizé et al., 2006)
Growth arrest specific 6 (Gas6)	H_2N— PRO | GLA | TS | EGF | SHBD —COOH	(Manfioletti et al., 1993)
Proline-rich Gla protein 1/2 (PRGP1/2)	H_2N— PRO | GLA | | TM | CYT —COOH	(Kulman et al., 1997)
TransmembaneGla protein 3/4 (TMG3/4)	H_2N— PRO | GLA | | TM | CYT —COOH	(Kulman et al., 2001)

Factor VII – PS: involve in blood coagulation; PS – MGP: involve in bone mineralization; Gas6 – TMG3/4: involve in other physiological activates.
AS, aromatic stack; AP, activation peptide; CRS, carboxylase recognition signal; CYT, cytoplasmic domain; EGF, epithermal growth factor domain; GLA, Gla domain; K1/K2, kringle domain; PRO, propeptide; SHBD, steroid hormone binding domain; TS, thrombin sensitive domain; TM, single pass transmembrane domain (Berkner and Runge, 2004; Furie and Furie, 1992).
Relative sizes of a single domain are only an approximation. There may be multiple numbers of domains (EGF, SHBD) present, which is not mentioned.

it, by the process of hemostasis, our body naturally heals blood vessels. Hemostasis has two major phases viz. primary hemostasis and secondary hemostasis. Primary hemostasis includes vasoconstriction, that is, reduced blood pressure near the ruptured area by endothelial derived factors initiating construction of ruptured blood vessels and platelet aggregation, that is, the interaction of platelet with collagen matrix to form platelet aggregation known as a plug. Secondary hemostasis refers to blood coagulation or clotting (Gale, 2011). Clotting is a consecutive procedure which includes the interaction of various blood components known as clotting factors. There are a total of 13 clotting factors involved in the blood clotting cascade. Moreover, there are two mechanisms of blood clotting namely extrinsic pathway and intrinsic pathway. The extrinsic pathway is activated when the rupture is outside the vessel which leads to blood escaping from the vascular system. This pathway is triggered faster than the intrinsic pathway and comprises factor VII. On the other hand, the intrinsic pathway is triggered when the rupture is inside the vascular system and includes factors XII, XI and IX. The net result of both the pathways is to activation of factor X and subsequent activation of other proteins to form a fibrin clot (Davie, 2003).

The blood clotting factors are present in an inactive zymogen form. In order to be activated, they need to be proteolytically cleaved. Several blood clotting factors are VKD proteins in nature, namely prothrombin (Factor II), Factor IX, Factor X and Factor VII. These factors circulate through the blood flow as serine protease zymogens and upon exposure to tissue factors, they are proteolytically cleaved (with cofactor Ca^{2+}) and activate coagulation cascade. Factor VII is involved in initiating extrinsic pathway. Activated Factor VII, known as FVIIa along with tissue factor and FX forms tenase complex which converts FX to FXa. On the other hand, FIX is associated with the intrinsic pathway (Fig. 4.24.4B). FIX and FVIIIa converts FX to FXa. FX is present at the junction where both the intrinsic and the extinct pathways meet. FIX can be activated by FVIIa by a crosstalk mechanism. FXa, FVa and prothrombin form prothrombinase complex that converts prothrombin to active thrombin. Soluble fibrinogen is converted to insoluble fibrin by the action of activated thrombin. This insoluble fibrin forms a mesh that is integrated into the platelet plug stabilizing the blood clot (Davie, 2003). This coagulation cascade is highly controlled by other vitamin K dependent proteins. FXa is degraded by the protein Z and protein Z related protease inhibitor (ZPI), proteins localized in the cell surface during coagulation. Protein Z is a VKD protein with similar structure with FIX, FX and FVII (Broze Jr, 2001). Protein C dependent regulation is another anticoagulant reaction that controls the coagulation cascade. When free thrombin is attached to thrombomodulin, the complex exerts a feedback loop inhibition, inhibiting prothrombinase complex. The thrombin-thrombomodulin complex proteolyzes protein C making it active and inhibits FVa and FVIIIa. Protein C also has structural similarity with FIX, FX and FVII (Rezaie, 2003). Protein S is a VKD protein which acts as a cofactor of protein C (Andersson et al., 2010). Protein S exists in two forms in blood circulation; C4BP bound form and free form. The free form acts as a cofactor in coagulation cascade whereas the bound form is involved in the activation of the complement system (Webb et al., 2002).

4.24.4.4 Bone mineralization

The deposition of minerals crystals into the extracellular matrix is known as bone mineralization. Three major VKD proteins are identified to be present in this process.

- *Osteocalcin*: the most crucial VKD protein that has been characterized in the calcified tissues is the bone Gla protein or osteocalcin. Osteocalcin is made up of 49 amino acids and is subject to γ–carboxylation. Although there is no sequential homology of osteocalcin with the other VKD proteins of coagulating cascade (Pan and Price, 1985). The sequence homology of osteocalcin is highly maintained within species (Nielsen-Marsh et al., 2005). Mature osteocalcins are secreted from osteoblast cells into the bone microenvironment. Subsequent conformational change occurs to attain proper secondary structure allowing it to bind with Ca^{2+} and hydroxyapatite and inhibits to the formation of hydroxyapatite crystals (Price et al., 1976; Zoch et al., 2016). It is unclear whether osteocalcin helps in bone formation or disruption. It is secreted from osteoblast cells, the bone-forming cells but there are several conflicting reports. In osteocalcin knockout mice the bone formation was enhanced unaffecting the bone resorption (Ducy et al., 1996). Osteocalcin is secreted in the late stage of bone formation. The very minute amount of most of the osteocalcin incorporated in bone (the undercarboxylated proteins ranging from 10%–30%) is released into the bloodstream. For this reason, osteocalcin is considered to-be bone formation/remodeling marker (Szulc et al., 1993).
- *Protein S* (effect in bone health): Protein S has active participation in bone formation along with its anticoagulatory effect. The protein is synthesized by osteoblast cells and associated with bone turnover. It has been reported that a low level of protein S is associated with severe osteopenia, fractures and low bone mineral density (Maillard et al., 1992).
- *Matrix Gla protein (MGP)*: This VKD protein is expressed in soft tissues inhibiting calcification. It needs two-step post translational modification before it can demineralize bone matrix; carboxylation and serine phosphorylation (Epstein, 2016). The inactive form of MGP, known as dephospho-uncarboxylated MGP (dp ucMGP), is measured in plasma as a cardiovascular marker (Liu et al., 2015).

Both osteopenia and osteoporosis, characterized by the loss of minerals from bone tissue, is a leading cause of the bone fracture. Administration of phylloquinone causes a reduction in fractures of osteopenic postmenopausal women (Cheung et al., 2008). Menaquinones also exhibit reduced chances of the hip, vertebral and nonvertebral fractures (Cockayne et al., 2006).

4.24.4.5 Other VKD proteins

- *Growth arrest specific* (Dugas et al., 1998): Gas6 is a VKD protein having extensive sequence similarity with protein S (Table 4.24.4) (Manfioletti et al., 1993). Gas6 acts as a ligand for three specific receptor tyrosine kinases namely, axl (Varnum et al., 1995) sky (Matsubara et al., 1996) and mer (Chen et al.,

2001). Gas6 is involved in various cellular processes, like phagocytosis, cell adhesion, proliferation, etc. (Berkner and Runge, 2004).

- *Proline rich Gla proteins (PRGP)*: PRGP1 and PRGP2 are two single-pass transmembrane proteins that have been identified and cloned (Kulman et al., 1997). Their function is still unknown but they act as transcriptional coactivator proteins (Kulman et al., 2007).
- *Transmembrane Gla Proteins (TMG)*: TMG3 and TMG4 are two transmembrane proteins similar to PRGPs (Table 4.24.4). Both PRGPs and TMGs have a similar primary structure with a single pass transmembrane domain and a cytosolic domain. Their functions are not clearly known but since both of them contain the cytosolic domain, they might have a role in signal transduction. Both proteins have diverse tissue distribution (Kulman et al., 2001).

4.24.5 Reported *in vivo, in vitro*, and clinical effect of vitamin K in the mammalian system

4.24.5.1 Vitamin K and cardiovascular system

Ischemic heart disease, the narrowing of arteries due to the accumulation of fats on the inner arterial wall or vascular calcification, is one of the most common cardiovascular diseases. Vitamin K2 subtypes especially MK-7, MK-8 and MK-9 significantly decrease chances of coronary heart disease in clinical studies (Gast et al., 2009). Matrix Gla protein (MGP), a VKD protein, is expressed in smooth muscle cells and maintains arterial calcification level by means of inhibition. It has been reported that MGP knockout mice grow enormous vascular calcification causing premature death (Luo et al., 1997).

4.24.5.2 Vitamin K and diabetes

In a study, it was found that vitamin K helps to improve insulin sensitization on older men (Yoshida et al., 2008). Vitamin K1 was found to be inhibiting insulin resistance in type 2 diabetes (Dihingia et al., 2018) The improvement of insulin sensitization is linked with osteocalcin metabolism (Choi et al., 2011) although the exact mechanism of action is not known.

4.24.5.3 Oxidative stress and vitamin K

Vitamin K is responsible for the production of hydroxyl and hydroperoxyl radicals, thereby inducing oxidative stress (Frei, 1994). Superoxide radical mediated one-electron reduction of vitamin K leads to the generation of intermediary semi-quinone radicals (Gant et al., 1988). These radicals, in turn, can convert transition metal ions from M^{3+} to M^{2+} state, thereby inducing Fenton's reaction and the production of toxic hydroxyl and hydroperoxyl radicals (Chen and Pignatello, 1997; Ivanova et al., 2018; Plaza, 2003). Thus, the decrease of superoxide radicals (oncogenic ROS) in the course of the vitamin K cycle can be regarded as its potential anti-cancer effect (Pervaiz and Clement, 2007). On the other hand, pro-vitamin K3 elicits oxidative insult in cancer cells by augmenting the production of hydroxyl radicals and the generation of breaks in the DNA strands (Nutter et al., 1992).

However, the vitamin K cycle exhibits potent antioxidant effects. Vitamin K acts as a cofactor of carboxylase which converts glutamic acid to γ-carboxy glutamic acid. Vitamin K has reduced to vitamin K hydroquinone, which in turn, gets oxidized to vitamin K epoxide and provides the necessary energy for the carboxylation of glutamic acid. Vitamin K epoxide gets reduced to regenerate vitamin K and the cycle continues (Vervoort et al., 1997). Vitamin K hydroquinone is a potential radical scavenger (Mukai et al., 1993).

In two studies, the antioxidant effects of the vitamin K cycle were investigated in the microsomes of the murine liver and lecithin liposomes. It was observed that in the presence of dithiothreitol or ascorbic acid/Fe^{2+}, a cofactor of vitamin K epoxide reductase, vitamin K1 & K2 suppressed lipid peroxidation. Lipid peroxidation was found to be inversely related to the amount of vitamin K hydroquinone (Vervoort et al., 1997). In a particular study, it was observed that vitamin K blocked oxidative injury of oligodendrocyte precursor and fetal cortical neuron primary cultures. On the other hand, vitamin K could not prevent the depletion of intracellular glutathione caused by cystine deprivation. It, however, completely blocked the accumulation of free radicals in vitro. Thus, vitamin K can be used as a potential therapeutic agent against oxidative damage of undifferentiated oligodendrocytes in perinatal hypoxic/ischemic brain injury (Li et al., 2003).

4.24.5.4 Vitamin K toxicity, catabolism and excretion in mammals

Phylloquinone and menaquinones exhibit no adverse effect upon excessive administration by any route. Although the synthetic vitamin K menadione exerts toxicity when administered in high doses causing anemia, mortality in rats and other organ toxicities in animals (Gundberg et al., 2012).

The catabolic mechanism of phylloquinone and menaquinones lies on a common pathway where the polyisoprenoid subunits are cut to 5 7 carbon chain containing carboxylic acids. The reaction involves omega oxidation of vitamin K followed by β-oxidation of side chains. The resulting metabolites are then conjugated with glucuronic acid and subsequently expelled in the bile and urine (Olson, 1984).

4.24.5.5 Vitamin K deficiency in mammals

Although the daily need for vitamin K is very low, sometimes deficiency of vitamin K is observed among adult individuals. The absorption of vitamin K is hampered by different gastrointestinal infections (Vervoort et al., 1997), liver disease (Strople et al., 2009), cystic fibrosis (Conway et al., 2005), biliary stasis (Papadopoulos et al., 2007), etc. Warfarin and other 4-hydroxycoumarin molecules act as vitamin K inhibitor by inhibiting the VCOR enzyme in the redox cycle.

Several broad spectrum antibiotics can practically sterilize the intestine. Thereby, the body loses the production site of different menaquinones, the gut bacteria. Studies with penicillin, cephalosporin, and other β-lactam antibiotics demonstrated hypoprothrombinemia (Smith and Lipsky, 1983). Most of the cases, the vitamin K status can be revived by vitamin K treatment. On the other hand, neonates are prone to vitamin K deficiency which can be severe. The deficiency may arise due to poor placental transport of the vitamin, the amount of vitamin K in breast milk is remarkably low and absent of gut microbiota of new-borns (Lippi and Franchini, 2011). As a result, clotting factors including plasma prothrombin levels are very low leading to hemorrhage. This state is known as vitamin K deficiency bleeding (VKDB) and seen on newborns extensively relied on breast milk as the food source (Sutor et al., 1999). For this reason, the introduction of phylloquinone is a common practice. VKDs are usually used as markers of vitamin K deficiency (El Asmar et al., 2014).

4.24.5.6 Vitamin K antagonist

4-hydroxycoumarins are well known vitamin K antagonist. It contains an enolic benzopyran structure that is responsible for its effect (Au and Rettie, 2008). Warfarin, the most prominent vitamin K antagonist, inhibits the liver microsomal enzyme VKOR (Shen et al., 2017). Warfarin is a racemic mixture of two optically active R and S enantiomers (Breckenridge, 1978; O'Reilly, 1976). The S enantiomer is more potent than R and is metabolized by CYP2C9 enzyme of the cytochrome P450 system (Miners and Birkett, 1998). Carboxylation is necessary for calcium ion mediated conformational change of coagulation proteins by promoting binding to cofactors on phospholipid surfaces (Nelsestuen, 1976). The S enantiomer of warfarin effectively inhibits the activity of the enzyme Vitamin K epoxide reductase, thereby decreasing blood clotting (Whitlon et al., 1978). Thus, warfarin acts as a potent anticoagulant and warfarin administration leads to vitamin K deficiency.

Warfarin administration leads to the interference in the production of new competent coagulation factors (X, IX, VII and II) along with others, like Factor C. However, the already synthesized coagulation factors remain unaffected. Therefore, the coagulation factors that are already present in the blood will remain entirely unaffected and will be able to function correctly until their half-lives are over. Factor C has the shortest half-life, so it drops first while other factors like II, IX and X are still able to function correctly and needs some more time to disappear completely. So, from the time Factor C disappears until the time the rest of the coagulation factors disappear, the risk of transient hypercoagulation increases (Ansell et al., 2008; Lurie et al., 2010).

Vitamin K epoxide reductase is sensitive to warfarin but vitamin K reductase is less sensitive. Therefore, the anticoagulant effect of warfarin can be overcome by low doses of vitamin K (phytonadione) (Choonara et al., 1988; Whitlon et al., 1978).

Conclusion

Current advances in vitamin K research indicate that is has a plethora of potential implications. Historically the research of vitamin K was delayed due to lack of knowledge about its bioavailability and absence of stable isotope studies. Isotope labeling has helped to understand the absorption and fate of vitamin K in the body. Several functions of vitamin K including its role in bone mineralization and in blood coagulation have been extensively studied in past decades. Although its effects on insulin sensitivity, oxidative stress, signal transduction mechanisms are poorly understood or unknown in *in vivo*, *in vitro*, and clinical trials. Vitamin K is in its emerging stage as a potent bioactive micronutrient.

References

Andersson, H.M., Arantes, M.J., Crawley, J.T., Luken, B.M., Tran, S., Dahlbäck, B., Lane, D.A., Rezende, S.M., 2010. Activated protein C cofactor function of protein S: a critical role for Asp95 in the EGF1-like domain. Blood 115, 4878–4885.

Ansell, J., Hirsh, J., Hylek, E., Jacobson, A., Crowther, M., Palareti, G., 2008. Pharmacology and management of the vitamin K antagonists: American College of Chest Physicians evidence-based clinical practice guidelines. Chest 133, 160S–198S.

Au, N., Rettie, A.E., 2008. Pharmacogenomics of 4-hydroxycoumarin anticoagulants. Drug Metab. Rev. 40, 355–375.

Berkner, K., Runge, K., 2004. The physiology of vitamin K nutriture and vitamin K-dependent protein function in atherosclerosis. J. Thromb. Haemost. 2, 2118–2132.

Berkner, K.L., 2008. Vitamin K-dependent carboxylation. Vitam. Horm. 78, 131–156.

Berkner, K.L., Pudota, B.N., 1998. Vitamin K-dependent carboxylation of the carboxylase. Proc. Natl. Acad. Sci. 95, 466–471.

Binkley, S., Maccobquodale, D., Thayer, S.A., Doisy, E.A., 1939. The isolation of vitamin K1. J. Biol. Chem. 130, 219–234.

Binkley, S., McKee, R., Thayer, S.A., Doisy, E.A., 1940. The constitution of vitamin K2. J. Biol. Chem. 133, 721–729.

Blomstrand, R., Forsgren, L., 1968. Vitamin K1-3H in man. Its intestinal absorption and transport in the thoracic duct lymph. Int. Z. Vitaminforsch. 38, 45–64.

Breckenridge, A., 1978. Oral anticoagulant drugs: pharmacokinetic aspects. Semin. Hematol. 15, 19–26.

Broze Jr, G.J., 2001. Protein Z-dependent regulation of coagulation. Thromb. Haemost. 86, 08–13.

Canfield, L., Hopkinson, J., Lima, A., Silva, B., Garza, C., 1991. Vitamin K in colostrum and mature human milk over the lactation period—a cross-sectional study. Am. J. Clin. Nutr. 53, 730–735.

Chen, J., Hammonds, R.G., Godowski, P.J., Mark, M.R., Mather, J.P., Li, R., 2001. Mer receptor activation by gas6. Google Patents.

Chen, R., Pignatello, J.J., 1997. Role of quinone intermediates as electron shuttles in Fenton and photoassisted Fenton oxidations of aromatic compounds. Environ. Sci. Technol. 31, 2399–2406.

Cheung, A.M., Tile, L., Lee, Y., Tomlinson, G., Hawker, G., Scher, J., Hu, H., Vieth, R., Thompson, L., Jamal, S., 2008. Vitamin K supplementation in postmenopausal women with osteopenia (ECKO trial): a randomized controlled trial. PLoS Med. 5, e196.

Choi, H.J., Yu, J., Choi, H., An, J.H., Kim, S.W., Park, K.S., Jang, H.C., Kim, S.Y., Shin, C.S., 2011. Vitamin K2 supplementation improves insulin sensitivity via osteocalcin metabolism: a placebo-controlled trial. Diabetes Care. 34, e147.

Choonara, I., Malia, R., Haynes, B., Hay, C., Cholerton, S., Breckenridge, A., Preston, F., Park, B., 1988. The relationship between inhibition of vitamin K1 2, 3-epoxide reductase and reduction of clotting factor activity with warfarin. Br. J. Clin. Pharmacol. 25, 1–7.

Cockayne, S., Adamson, J., Lanham-New, S., Shearer, M.J., Gilbody, S., Torgerson, D.J., 2006. Vitamin K and the prevention of fractures: systematic review and meta-analysis of randomized controlled trials. Arch. Intern. Med. 166, 1256–1261.

Conway, S.P., Wolfe, S.P., Brownlee, K.G., White, H., Oldroyd, B., Truscott, J.G., Harvey, J.M., Shearer, M.J., 2005. Vitamin K status among children with cystic fibrosis and its relationship to bone mineral density and bone turnover. Pediatrics 115, 1325–1331.

Dam, H., 1934. Haemorrhages in chicks reared on artificial diets: a new deficiency disease. Nature 133, 909.

Dam, H., 1935. The antihæmorrhagic vitamin of the chick.: occurrence and chemical nature. Nature 135, 652.

Davie, E.W., 2003. A brief historical review of the waterfall/cascade of blood coagulation. J. Biol. Chem. 278, 50819–50832.

Degen, S.J.F., Davie, E.W., 1987. Nucleotide sequence of the gene for human prothrombin. Biochemistry 26, 6165–6177.

Dihingia, A., Ozah, D., Ghosh, S., Sarkar, A., Baruah, P.K., Kalita, J., Sil, P.C., Manna, P., 2018. Vitamin K1 inversely correlates with glycemia and insulin resistance in patients with type 2 diabetes (T2D) and positively regulates SIRT1/AMPK pathway of glucose metabolism in liver of T2D mice and hepatocytes cultured in high glucose. J. Nutr. Biochem. 52, 103–114.

Dowd, P., Hershline, R., Ham, S.W., Naganathan, S., 1995. Vitamin K and energy transduction: a base strength amplification mechanism. Science 269, 1684–1691.

Ducy, P., Desbois, C., Boyce, B., Pinero, G., Story, B., Dunstan, C., Smith, E., Bonadio, J., Goldstein, S., Gundberg, C., 1996. Increased bone formation in osteocalcin-deficient mice. Nature 382, 448.

Duda-Chodak, A., Tarko, T., Satora, P., Sroka, P., 2015. Interaction of dietary compounds, especially polyphenols, with the intestinal microbiota: a review. Eur. J. Nutr. 54, 325–341.

Dugas, T.R., Morel, D.W., Harrison, E.H., 1998. Impact of LDL carotenoid and α-tocopherol content on LDL oxidation by endothelial cells in culture. J. Lipid Res. 39, 999–1007.

El Asmar, M.S., Naoum, J.J., Arbid, E.J., 2014. Vitamin K dependent proteins and the role of vitamin K2 in the modulation of vascular calcification: a review. Oman Med. J. 29, 172.

Epstein, M., 2016. Matrix Gla-protein (MGP) not only inhibits calcification in large arteries but also may be renoprotective: connecting the dots. EBioMedicine 4, 16–17.

Erick, M., 2018. Breast milk is conditionally perfect. Med. Hypotheses 111, 82–89.

Fasco, M., Hildebrandt, E., Suttie, J., 1982. Evidence that warfarin anticoagulant action involves two distinct reductase activities. J. Biol. Chem. 257, 11210–11212.

Ferland, G., 2012. The discovery of vitamin K and its clinical applications. Ann. Nutr. Metab. 61, 213–218.

Foster, D., Davie, E.W., 1984. Characterization of a cDNA coding for human protein C. Proc. Natl. Acad. Sci. 81, 4766–4770.

Frei, B., 1994. Reactive oxygen species and antioxidant vitamins: mechanisms of action. Am. J. Med. 97, S5–S13.

Furie, B., Bouchard, B.A., Furie, B.C., 1999. Vitamin K-dependent biosynthesis of γ-carboxyglutamic acid. Blood 93, 1798–1808.

Furie, B., Furie, B.C., 1992. Molecular and cellular biology of blood coagulation. N. Engl. J. Med. 326, 800–806.

Gale, A.J., 2011. Continuing education course# 2: current understanding of hemostasis. Toxicol. Pathol. 39, 273–280.

Gant, T.W., Rao, D.R., Mason, R.P., Cohen, G.M., 1988. Redox cycling and sulphydryl arylation; their relative importance in the mechanism of quinone cytotoxicity to isolated hepatocytes. Chem. Biol. Interact. 65, 157–173.

Garber, A.K., Binkley, N., Krueger, D.C., Suttie, J., 1999. Comparison of phylloquinone bioavailability from food sources or a supplement in human subjects. J. Nutr. 129, 1201–1203.

Gardill, S.L., Suttie, J., 1990. Vitamin K epoxide and quinone reductase activities: evidence for reduction by a common enzyme. Biochem. Pharmacol. 40, 1055–1061.

Gast, G.-C.M., de Roos, N.M., Sluijs, I., Bots, M.L., Beulens, J.W., Geleijnse, J.M., Witteman, J.C., Grobbee, D.E., Peeters, P.H., van der Schouw, Y.T., 2009. A high menaquinone intake reduces the incidence of coronary heart disease. Nutr. Metab. Cardiovasc. Dis. 19, 504–510.

Goodstadt, L., Ponting, C.P., 2004. Vitamin K epoxide reductase: homology, active site and catalytic mechanism. Trends Biochem. Sci. 29, 289–292.

Griminger, P., 1965. Relative vitamin K potency of two water-soluble menadione analogues. Poult. Sci. 44, 210–213.

Gross, J., Cho, W.K., Lezhneva, L., Falk, J., Krupinska, K., Shinozaki, K., Seki, M., Herrmann, R.G., Meurer, J., 2006. A plant locus essential for phylloquinone (vitamin K1) biosynthesis originated from a fusion of four eubacterial genes. J. Biol. Chem. 281, 17189–17196.

Gundberg, C.M., Lian, J.B., Booth, S.L., 2012. Vitamin K-dependent carboxylation of osteocalcin: friend or foe? Adv. Nutr. 3, 149–157.

Hagen, F.S., Gray, C.L., O'Hara, P., Grant, F.J., Saari, G.C., Woodbury, R.G., Hart, C.E., Insley, M., Kisiel, W., Kurachi, K., 1986. Characterization of a cDNA coding for human factor VII. Proc. Natl. Acad. Sci. 83, 2412–2416.

Herz, J., Strickland, D.K., 2001. LRP: a multifunctional scavenger and signaling receptor. J. Clin. Invest. 108, 779–784.

Holst, W.F., Halbrook, E.R., 1933. A "scurvy-like" disease in chicks. Science (Washington) 77, 354.

Ivanova, D., Zhelev, Z., Getsov, P., Nikolova, B., Aoki, I., Higashi, T., Bakalova, R., 2018. Vitamin K: redox-modulation, prevention of mitochondrial dysfunction and anticancer effect. Redox. Biol. 16, 352–358.

Knobloch, J.E., Suttie, J., 1987. Vitamin K-dependent carboxylase. Control of enzyme activity by the" propeptide" region of factor X. J. Biol. Chem. 262, 15334–15337.

Kulman, J.D., Harris, J.E., Haldeman, B.A., Davie, E.W., 1997. Primary structure and tissue distribution of two novel proline-rich γ-carboxyglutamic acid proteins. Proc. Natl. Acad. Sci. 94, 9058–9062.

Kulman, J.D., Harris, J.E., Xie, L., Davie, E.W., 2001. Identification of two novel transmembrane γ-carboxyglutamic acid proteins expressed broadly in fetal and adult tissues. Proc. Natl. Acad. Sci. 98, 1370–1375.

Kulman, J.D., Harris, J.E., Xie, L., Davie, E.W., 2007. Proline-rich Gla protein 2 is a cell-surface vitamin K-dependent protein that binds to the transcriptional coactivator yes-associated protein. Proc. Natl. Acad. Sci. 104, 8767–8772.

Booth, S.L., 2012. Vitamin K: food composition and dietary intakes. Food Nutr. Res. 56, 5505.

Laizé, V., Viegas, C.S., Price, P.A., Cancela, M.L., 2006. Identification of an osteocalcin isoform in fish with a large acidic prodomain. J. Biol. Chem. 281, 15037–15043.

Leytus, S.P., Foster, D.C., Kurachi, K., Davie, E.W., 1986. Gene for human factor X: a blood coagulation factor whose gene organization is essentially identical with that of factor IX and protein C. Biochemistry 25, 5098–5102.

Li, J., Lin, J.C., Wang, H., Peterson, J.W., Furie, B.C., Furie, B., Booth, S.L., Volpe, J.J., Rosenberg, P.A., 2003. Novel role of vitamin k in preventing oxidative injury to developing oligodendrocytes and neurons. J. Neurosci. 23, 5816–5826.

Lippi, G., Franchini, M., 2011. Vitamin K in neonates: facts and myths. Blood Transfus. 9, 4.

Liu, Y.-P., Gu, Y.-M., Thijs, L., Knapen, M.H., Salvi, E., Citterio, L., Petit, T., Carpini, S.D., Zhang, Z., Jacobs, L., 2015. Inactive matrix Gla protein is causally related to adverse health outcomes: a Mendelian randomization study in a Flemish population. Hypertension 65, 463–470.

Lundwall, A., Dackowski, W., Cohen, E., Shaffer, M., Mahr, A., Dahlbäck, B., Stenflo, J., Wydro, R., 1986. Isolation and sequence of the cDNA for human protein S, a regulator of blood coagulation. Proc. Natl. Acad. Sci. 83, 6716–6720.

Luo, G., Ducy, P., McKee, M.D., Pinero, G.J., Loyer, E., Behringer, R.R., Karsenty, G., 1997. Spontaneous calcification of arteries and cartilage in mice lacking matrix GLA protein. Nature 386, 78.

Lurie, Y., Loebstein, R., Kurnik, D., Almog, S., Halkin, H., 2010. Warfarin and vitamin K intake in the era of pharmacogenetics. Br. J. Clin. Pharmacol. 70, 164–170.

MacCorquodale, D., Binkley, S., Thayer, S., Doisy, E., 1939. On the constitution of vitamin K1. J. Am. Chem. Soc. 61, 1928–1929.

Maillard, C., Berruyer, M., Serre, C.M., Dechavanne, M., Delmas, P.D., 1992. Protein-S, a vitamin K-dependent protein, is a bone matrix component synthesized and secreted by osteoblasts. Endocrinology 130, 1599–1604.

Manfioletti, G., Brancolini, C., Avanzi, G., Schneider, C., 1993. The protein encoded by a growth arrest-specific gene (gas6) is a new member of the vitamin K-dependent proteins related to protein S, a negative coregulator in the blood coagulation cascade. Mol. Cell. Biol. 13, 4976–4985.

Mathers, J., Fernandez, F., Hill, M., McCarthy, P., Shearer, M., Oxley, A., 1990. Dietary modification of potential vitamin K supply from enteric bacterial menaquinones in rats. Br. J. Nutr. 63, 639–652.

Matsubara, N., Takahashi, Y., Nishina, Y., Mukouyama, Y., Yanagisawa, M., Watanabe, T., Nakano, T., Nomura, K., Arita, H., Nishimune, Y., 1996. A receptor tyrosine kinase, Sky, and its ligand Gas 6 are expressed in gonads and support primordial germ cell growth or survival in culture. Dev. Biol. 180, 499–510.

McKee, R., Binkley, S., Thayer, S.A., MacCorquodale, D., Doisy, E.A., 1939. The isolation of vitamin K2. J. Biol. Chem. 131, 327–344.

Miners, J.O., Birkett, D.J., 1998. Cytochrome P4502C9: an enzyme of major importance in human drug metabolism. Br. J. Clin. Pharmacol. 45, 525–538.

Mukai, K., Morimoto, H., Kikuchi, S., Nagaoka, S.-i., 1993. Kinetic study of free-radical-scavenging action of biological hydroquinones (reduced forms of ubiquinone, vitamin K and tocopherol quinone) in solution. Biochim. Biophys. Acta Gen. Subj. 1157, 313–317.

Nelsestuen, G., 1976. Role of gamma-carboxyglutamic acid. An unusual protein transition required for the calcium-dependent binding of prothrombin to phospholipid. J. Biol. Chem. 251, 5648–5656.

Nielsen-Marsh, C.M., Richards, M.P., Hauschka, P.V., Thomas-Oates, J.E., Trinkaus, E., Pettitt, P.B., Karavanić, I., Poinar, H., Collins, M.J., 2005. Osteocalcin protein sequences of Neanderthals and modern primates. Proc. Natl. Acad. Sci. 102, 4409–4413.

Nutter, L.M., Ngo, E., Fisher, G., Gutierrez, P., 1992. DNA strand scission and free radical production in menadione-treated cells. Correlation with cytotoxicity and role of NADPH quinone acceptor oxidoreductase. J. Biol. Chem. 267, 2474–2479.

Olson, R.E., 1984. The function and metabolism of vitamin K. Annu. Rev. Nutr. 4, 281–337.

O'Reilly, R.A., 1976. Vitamin K and the oral anticoagulant drugs. Annu. Rev. Med. 27, 245–261.

Pan, L.C., Price, P.A., 1985. The propeptide of rat bone gamma-carboxyglutamic acid protein shares homology with other vitamin K-dependent protein precursors. Proc. Natl. Acad. Sci. 82, 6109–6113.

Papadopoulos, V., Filippou, D., Manolis, E., Mimidis, K., 2007. Haemostasis impairment in patients with obstructive jaundice. J. Gastrointestin. Liver Dis. 16, 177.

Pervaiz, S., Clement, M.-V., 2007. Superoxide anion: oncogenic reactive oxygen species? Int. J. Biochem. Cell Biol. 39, 1297–1304.

Plaza, S.M., 2003. The anticancer effects of vitamin K. Altern. Med. Rev. 8, 303–318.

Price, P.A., Otsuka, A., Poser, J.W., Kristaponis, J., Raman, N., 1976. Characterization of a gamma-carboxyglutamic acid-containing protein from bone. Proc. Natl. Acad. Sci. 73, 1447–1451.

Rezaie, A.R., 2003. Exosite-dependent regulation of the protein C anticoagulant pathway. Trends Cardiovasc. Med. 13, 8–15.

Rishavy, M.A., Hallgren, K.W., Yakubenko, A.V., Shtofman, R.L., Runge, K.W., Berkner, K.L., 2006. Brønsted analysis reveals Lys218 as the carboxylase active site base that deprotonates vitamin K hydroquinone to initiate vitamin K-dependent protein carboxylation. Biochemistry 45, 13239–13248.

Schurgers, L.J., Vermeer, C., 2000. Determination of phylloquinone and menaquinones in food. Pathophysiol. Haemost. Thromb. 30, 298–307.

Shapiro, S., Richards, R.K., 1945. The prothrombin response to large doses of synthetic vitamin K in liver disease. Ann. Intern. Med. 22, 841–851.

Shearer, M.J., Fu, X., Booth, S.L., 2012. Vitamin K nutrition, metabolism, and requirements: current concepts and future research. Adv. Nutr. 3, 182–195.

Shearer, M.J., Newman, P., 2008. Metabolism and cell biology of vitamin K. Thromb. Haemost. 100, 530–547.

Shen, G., Cui, W., Zhang, H., Zhou, F., Huang, W., Liu, Q., Yang, Y., Li, S., Bowman, G.R., Sadler, J.E., 2017. Warfarin traps human vitamin K epoxide reductase in an intermediate state during electron transfer. Nat. Struct. Mol. Biol. 24, 69.

Shenkin, A., 2006. Micronutrients in health and disease. Postgrad. Med. J. 82, 559–567.

Smith, C.R., Lipsky, J.J., 1983. Hypoprothrombinemia and platelet dysfunction caused by cephalosporin and oxalactam antibiotics. J. Antimicrob. Chemother. 11, 496–498.

Soute, B.A., Ulrich, M.M., Watson, A.D.J., Maddison, J.E., Ebberink, R.H., Vermeer, C., 1992. Congenital deficiency of all vitamin K-dependent blood coagulation factors due to a defective vitamin K-dependent carboxylase in Devon Rex cats. Thromb. Haemost. 68, 521–525.

Strople, J., Lovell, G., Heubi, J., 2009. Prevalence of subclinical vitamin K deficiency in cholestatic liver disease. J. Pediatr. Gastroenterol. Nutr. 49, 78–84.

Sutor, A.H., von Kries, R., Cornelissen, M.E., McNinch, A.W., Andrew, M., 1999. Vitamin K deficiency bleeding (VKDB) in infancy. Thromb. Haemost. 81, 456–461.

Szulc, P., Chapuy, M., Meunier, P., Delmas, P., 1993. Serum undercarboxylated osteocalcin is a marker of the risk of hip fracture in elderly women. J. Clin. Invest. 91, 1769–1774.

Thane, C., Paul, A., Bates, C., Bolton-Smith, C., Prentice, A., Shearer, M., 2002. Intake and sources of phylloquinone (vitamin K 1): variation with socio-demographic and lifestyle factors in a national sample of British elderly people. Br. J. Nutr. 87, 605–613.

Thijssen, H., 1996. Vitamin K status in human tissues: tissue-specific accumulation of phylloquinone and menaquinone-4. Br. J. Nutr. 75, 121–127.

Usui, Y., Tanimura, H., Nishimura, N., Kobayashi, N., Okanoue, T., Ozawa, K., 1990. Vitamin K concentrations in the plasma and liver of surgical patients. Am. J. Clin. Nutr. 51, 846–852.

Varnum, B.C., Young, C., Elliott, G., Garcia, A., Bartley, T.D., Fridell, Y.-W., Hunt, R.W., Trail, G., Clogston, C., Toso, R.J., 1995. Axl receptor tyrosine kinase stimulated by the vitamin K-dependent protein encoded by growth-arrest-specific gene 6. Nature 373, 623.

Vervoort, L.M., Ronden, J.E., Thijssen, H.H., 1997. The potent antioxidant activity of the vitamin K cycle in microsomal lipid peroxidation. Biochem. Pharmacol. 54, 871–876.

Wallin, R., Gebhardt, O., Prydz, H., 1978. NAD (P) H dehydrogenase and its role in the vitamin K (2-methyl-3-phytyl-1, 4-naphthaquinone)-dependent carboxylation reaction. Biochem. J. 169, 95–101.

Webb, J.H., Blom, A.M., Dahlbäck, B., 2002. Vitamin K-dependent protein S localizing complement regulator C4b-binding protein to the surface of apoptotic cells. J. Immunol. 169, 2580–2586.

Wei, G., Wang, M., Hyslop, T., Wang, Z., Carr, B.I., 2010. Vitamin K enhancement of sorafenib-mediated HCC cell growth inhibition in vitro and in vivo. Int. J. Cancer 127, 2949–2958.

Whitlon, D., Sadowski, J., Suttie, J., 1978. Mechanism of coumarin action: significance of vitamin K epoxide reductase inhibition. Biochemistry 17, 1371–1377.

Will, B.H., Usui, Y., Suttie, J., 1992. Comparative metabolism and requirement of vitamin K in chicks and rats. J. Nutr. 122, 2354–2360.

Yoshida, M., Jacques, P.F., Meigs, J.B., Saltzman, E., Shea, M.K., Gundberg, C., Dawson-Hughes, B., Dallal, G., Booth, S.L., 2008. Effect of vitamin K supplementation on insulin resistance in older men and women. Diabetes Care. 31, 2092–2096.

Yoshitake, S., Schach, B.G., Foster, D.C., Davie, E.W., Kurachi, K., 1985. Complete nucleotide sequences of the gene for human factor IX (antihemophilic factor B). Biochemistry 24, 3736–3750.

Zoch, M.L., Clemens, T.L., Riddle, R.C., 2016. New insights into the biology of osteocalcin. Bone 82, 42–49.

CHAPTER

Zinc

4.25

Toorabally B. Zaynab, Subratty A. Hussein, Mohamad Fawzi Mahomoodally
Department of Health Sciences, Faculty of Medicine and Health Sciences, University of Mauritius, Réduit, Mauritius

4.25.1 Sources

Zinc is a ubiquitous nutrient that is available in small quantity in a wide range of foods. In the human body, zinc is present in the rough amount of 2–4 g while the plasma concentration of the mobile pool of zinc ions is barely 12–16 μM (Jansen et al., 2012). Recommended amount of dietary zinc can be obtained by consuming appreciable amounts of the following food items: (1) Oysters which are recognized as the ideal source of zinc, (2) poultry, red meat, certain seafood such as crab and lobsters as well as fortified breakfast cereals, which are also referred to as good sources of zinc, and (3) beans, nuts, whole grains, and dairy products, which offer a comparatively smaller amount zinc (National Institutes of Health, 2016).

Foods which are nutritively higher in protein observe a comparatively higher content of zinc content, while diets which are rich in carbohydrates have been found to be poor in zinc (Osis et al., 1972). Meat and meat products contains a fairly larger amount of zinc (0.40 to 6.77 mg/100 g) as compared to the other food groups such as grain group which has 0.30 to 2.54 mg/100 g, dairy products which have 0.36 to 0.49 mg/100g, vegetables which have 0.12 to 0.60 mg per 100 g, and fruits which observe the least amount of zinc ranging from 0.02 to 0.26 mg/100 g (Haeflein and Rasmussen, 1977).

4.25.2 Chemistry

The name for zinc is of German backgrounds, zink. The terrestrial chemistry of Zn is referred to as Zn (II) rather than Zn (0). This chemical metallic element has an atomic number of 30 and stable isotopes with mass of 66, 67, 68, and 70, with an average of 65.8 a.m.u. The electron configuration of Zn (II) $1s^2$, $2s^2$, $2p^6$, $3d^{10}$, hence consequently lacks unoccupied *d* subshells within the recognized oxidation state, the essential norm for true transition metals (Barak and Helmke, 1993). Zinc is among one of the alleged trace elements existing in the human body, but naturally it is available in the Earth's crust to the level of approximately 0.02% and is the 23rd most abundant element (Jackson, 1989).

Antioxidants Effects in Health. DOI: https://doi.org/10.1016/B978-0-12-819096-8.00068-9

4.25.3 Bioavailability of zinc

Zinc deficiency continues to be a significant health issue in developing countries, chiefly due to inadequate dietary consumption. This problem is accounted by a large intake of cereals and cereal products having small concentrations and poor bioavailability of Zn (Cakmak and Kutman, 2018). Approximately, 33% of the earth's population is assessed as being at risk of Zn deficiency, with particular predominance in the paediatric population for children below the age of 5 years owing to their relatively high requirement for Zn to sustain growth and development (Wessells and Brown, 2012).

Bioavailability refers to the proportion of nutrient intake, which can be absorbed into the blood system and utilized for the physiological functions of the human body. In healthy individuals, zinc bioavaibility is determined by three factors which are: (1) the zinc status of an individual, (2) dietary zinc, and (3) the availability of soluble zinc from food constituents in the diet (Lonnerdal, 2000). Zinc status refers to the consumption of dietary zinc over an extended period of time. If the zinc status of an individual is low, the zinc absorption relies mainly its solubility in the intestinal lumen. This is sequentially influenced by the chemical form of zinc as well as by the presence of inhibitors and enhancers specific for zinc absorption (Roohani, 2013)

The main inhibitor of dietary zinc absorption is phytic acid. The latter is present in legumes, nuts, seeds and unrefined cereals by chelating with zinc (Adams et al., 2002). Phytate forms an insoluble complex, thereby hindering absorption (Türk et al., 1996). However, the inhibitory effect may be eliminated by food processing techniques, which make use of thermal processing or enzymes to hydrolyse phytic acid (Samman, 2007). For instance, the heat released in the baking process eliminates more than 50% of the phytate in yeast-leavened whole meal breads and phytase present in wheat grains break down phytate in yeast fermentation (Türk et al., 1996). Additionally, the soaking and sprouting of seeds, beans and, grains also decrease the phytate content.

A diet with a higher level of protein will augment absorption, since zinc binds to proteins. However, it is to be noted that different types of protein may either positively or negatively influence zinc absorption. Casein, which is a protein present in milk inhibits zinc absorption, while soy protein does not (Lonnerdal, 2000). Compared to human milk, cows milk has more casein thereby influencing zinc absorption. This micronutrient is available in a diluted form in breast milk while its bioavailability in milk is high. Hence, infants fed with breast-milk will have better zinc absorption in contrast to those who are formula-fed (Institute of Medicine (US) Committee on the Evaluation of the Addition of Ingredients New to Infant Formula, 2004).

Among vegetarians whose protein intake is mostly about pulses, incidence of zinc deficiency is likely to be higher. Additionally, the consumption of foods like cereals, starchy vegetables and legumes, which are high in phytate may contribute to a reduced bioavailability of zinc (Hunt, 2003).

4.25.4 Zinc as an antioxidant

The role of zinc within the antioxidant defense system has been extensively researched. Oxidative stress underlies the molecular mechanisms which are responsible for the development of many inflammatory diseases, such as atherosclerosis, diabetes mellitus, rheumatoid arthritis, cancer, and neurodegeneration (Valko et al., 2007).

Zinc operates as an antioxidant through diverse mechanisms, which can be either acute or chronic. Zinc delays oxidative processes by two acute mechanisms, one being the stabilization of protein sulfhydryls against oxidation (Powell, 2000). Zinc reduces sulfhydryl reactivity via the following three ways: Zinc firstly binds directly to the thiol group. Secondly, it forms steric hindrance while binding closely to the sulfhydryl group of the protein. Thirdly, it alters the bonding arrangement of the protein, by binding to the other position site on the protein. The enzyme δ-aminolevulinate dehydratase which is the most widely studied enzyme for sulfhydryl protection by zinc is, accelerates the creation of the pyrrole porphobilinogen. The presence of the metal inhibits enzyme thiol oxidation and disulfide formation (Jarosz et al., 2017).

Another antioxidant effect of zinc comprise of providing an antagonist effect in transition metal-catalyzed reactions, when zinc competes with iron (Fe) and copper (Cu) ions for binding to cell membranes and proteins, displacing these redox active metals, such as reduction of •OH formation from H_2O_2 and O_2^- (Powell, 2000) In general, redox-active transition metals coupled with cellular components create a site for the cyclic formation of •OH radicals. Only high affinity chelators or some redox-inactive agents can antagonize the formation of •OH or shift the formation site to less critical one. Consequently, the metal has the capacity of reducing post-ischemic injury in an array of tissues and organs, such as stomach, kidney, intestine, retina, and brain (Powell, 2000; Tapiero and Tew, 2003).

Zinc also intensifies the activation of antioxidant proteins, molecules, and enzymes for example glutathione (GSH), catalase, and SOD and furthermore diminishes the activities of oxidant-promoting enzymes such as inducible nitric acid synthase (iNOS) and NADPH oxidase and hinders the generation of lipid peroxidation products (Bao et al., 2013).

Zinc is considered as a strong metallothionein inducer and has the ability to bind to metallothionein when exposed to normal physiological conditions. This micronutrient is released from its complex with metallothionein under oxidative stress conditions and reallocated in the cells to effect antioxidant mechanisms (Marreiro, 2017). A study carried out by Liang et al. (2015), showed that an elevated metallothionein expression in the liver of rats supplemented with zinc at a dose of 5 mg/kg of body weight per day, promoted antioxidant and anti-inflammatory effects mediated by metallothionein.

Studies have shown that zinc deficiency instigate oxidative damage and subsequently, endothelial dysfunction. This micronutrient deficiency within the endothelial cells augments the inflammatory response mediated by cytokines and lipids, probably through mechanisms related to elevated cellular oxidative stress. However, a supplementation of zinc applies a defensive role against harm to the vascular system. The mechanisms by which zinc defends the blood vessels comprise of the regulation of Nrf2 which is a transcription factor essential for the expression of genes encoding antioxidant enzymes as well as for the induction and expression of metallothionein (Biagiotti et al., 2016).

Zinc, as an inhibitor of N-methyl-d-aspartate (NMDA) receptors, contributing to the movement of calcium from the extracellular environment to the cytosol has also been largely studied. Hence, any deficiency in zinc will support the stimulation of NMDA receptors thereby elevating intracellular calcium concentration. With an increase in intracellular calcium concentration, release of Phosphate will be released by neuronal cells, consequently activating leukocytes and macrophages, promoting the release of inflammatory cytokines and the production of free radicals, thereby enhancing the expression of oxidative stress (Biagiotti et al., 2016).

4.25.5 Beneficial and detrimental effects on health

The significance of zinc as an essential element in health and nutrition is categorical, as its importance in health is progressively being recognised and its deficiency may play an essential part in the manifestation of ailments. Currently, the World Health Organization (WHO) estimates that approximately two billion individuals residing in the developing world may be zinc deficient.

Zinc is vital for life on earth because it is needed whether as a structural component or as a reaction site in many proteins and participates in innumerable cellular functions. It is anticipated that around 10% of each of the proteins in the human organism, equivalent to almost 3000 proteins, are Zn-dependent (Andreini et al., 2006; Krezel and Maret, 2016). Two of the most important amino acids synchronized with trace metal ions are cysteine and histidine, which have a significant role in living bodies, though, in some cases, are causative to the happening of neurodegeneration (Kozlowski, 2013).

The significance of zinc for humans was recognized in the Middle East (Iran, Egypt), in the early 1960s, among patients with delayed growth, severe iron deficiency anemia, hypogonadism, hepatomegaly, splenomegaly and dry and wrinkled skin (Prasad, 2008). In a study carried out by Prasad (2008), patients with Zn deficiency had severe immune dysfunction owing to which they died from opportunistic infections before attaining 25 years old. Further indicators of zinc deficiency narrated successively, comprise of high incidence of infection like pneumonia and diarrhea, because of an immune deficiency, various forms of skin lesions, weakened wound healing capacity, and night blindness (Samman, 2007; King and Cousins, 2006). In developing countries, deficiencies of Zn and other micronutrients contributes to the economic burden and have a substantial consequence on the overall national productivity by the reducing output while facing a rise in health care costs (Darnton-Hill et al., 2005; Stein, 2014).

Fortunately, in the last few decades, there has been more research on the impacts of Zn deficiency and its supplementation vis à vis human health disorders. Zn deficiency triggers substantial weakening of the inborn and adaptive immune responses and stimulates systemic inflammation (Wong and Ho, 2012). The occurrence of metal ions in our organism is certainly indispensible, nonetheless there is a small margin between their valuable and unpleasant consequence on cellular processes.

Zinc – excess	Zinc – deficiency
Brain	**Brain**
• Lethargy • Focal neuronal deficits	• Decreased nerve conduction • Neuropsychiatric disorders • Neurosensory disorders • Mental lethargy
Respiratory tract	**Thymus**
• Respiratory disorder after inhalation of zinc smoke • Metal fume fever	• Thymic actrophy
Gastrointestinal tract	**Skin**
• Nausea/vomiting • Epigastric pain • Diarrhea	• Skin lesions • Decreased wound healing • Acrodermatitis
Prostate	**Reproductive systems**
• Elevated risk of prostate cancer	• Infertility • Retarded genital development • Hypoganadism

4.25.6 Zinc supplementation in humans

Preserving acceptable concentrations of zinc in the cell compartments is vital for the appropriate performance of the antioxidant defense system. In stimulated mononuclear cells it has been observed that a daily supplementation with ≥45 mg of zinc has been described to reduce ex vivo generated levels of pro-inflammatory cytokine mRNA and proteins (Bao et al., 2008; Prasad et al., 2007). Contrariwise, an elevated cytokine concentrations have been reported in stimulated mononuclear cells isolated from human subjects supplemented with ≤20 mg zinc/day, implying a zinc dose-response (Sandstead et al., 2008; Aydemir et al., 2006; Kahmann et al., 2008). During the first year of life, the daily nutritional recommendation of elemental zinc in infants having zinc deficiency is generally 3 mg/d for first half and 5 mg/d for second half. Successively, zinc supplement 10 mg/day can be given from 1 year of age to 10 years, 15 mg/day for adolescents and adults while 20–25 mg/d during gestation and breastfeeding. The WHO suggested that the use of zinc supplementation or fortification of complementary foods to be given infants who are still being breastfed as from six months of age as a strategy to meet their nutritional recommendation for zinc. Additionally, it has been discussed that the incorporation of meat in complementary feeding will meet zinc requirements and mitigate the demand for supplementation in developed and developing countries (Krebs and Hambidge, 2007).

Besides of the daily diet, zinc can be obtained through dietary supplements as most of them contain zinc whether as the sole ingredient or in combination with calcium, magnesium or other nutrients. Additionally, Zinc is present in some oral over-the-counter products, as well as those considered as homeopathic medications used for flus. Nowadays, on the market, there are different types of zinc supplements that are available as follows:

- **Zinc gluconate:** Available as over-the-counter forms of zinc and frequently often used in cold medications, for example lozenges and nasal sprays.
- **Zinc acetate:** Similar to zinc gluconate in its use and addition cold lozenges to decrease symptoms and accelerate recovery. A study carried out by Hemilä et al. (2017) reported that patients administered with zinc lozenges recovered faster as compared to the control group by rate ratio 3:1.
- **Zinc sulfate:** Besides in aiding in the prevention of zinc deficiency, zinc sulfate has been reported to be beneficial to alleviate the severity of acne.
- **Zinc picolinate:** Some studies has suggested that the human body has the ability to absorb this form better as compared to other types of zinc, comprising of zinc gluconate and zinc citrate. It is available in tablet, capsule or lozenge form.
- **Zinc orotate:** This form of zinc is bound to orotic acid and it is one of the most common types of zinc supplements on the market.
- **Zinc citrate:** One study showed that zinc citrate as a supplement is absorbed efficiently as zinc gluconate while however having more appealing and less bitter taste.

4.25.7 *In-vitro* studies in human cells

Zinc regulates immune responses and displays antioxidant and anti-inflammatory activities. The mineral delays oxidative processes continually by prompting the expression of metallothioneins. The latter are metal-binding cysteine-rich proteins which are accountable for preserving zinc-related cell homeostasis. In addition, they serve as strong electrophilic scavengers and cytoprotective agents (Jarosz et al., 2017).

Inflammatory cytokine production following zinc depletion or supplementation has been assessed in stimulated and unstimulated cells through numerous *in vitro* studies. Studies carried out by Bao et al. (2011) and Hayashi et al. (2008) indicated that in inflammatory conditions IFN-γ gene expression is diminished by a deficiency in zinc and zinc supplementation.

4.25.8 Animal studies and clinical studies

The incidence of mild to moderate zinc deficiency in human beings, apparently influence the generation of a range of cytokines, including IL-1β, IL-2, IL-6, and TNF-α. A study carried out by Prasad et al. (2007) reported that following zinc supplementation in elderly subjects, it was observed that there were improvements in biochemical markers in contrast to the placebo group with an elevation in plasma zinc, reduction in oxidative stress markers and reduced production of inflammatory cytokine.

A deficiency in zinc has been reported to augment endothelial cell apoptosis and intensifies the unfavorable consequences of oxidized LDL in atherosclerosis (Foster and Samman, 2010). A study conducted by Beattie et al. (2012) on suboptimal dietary zinc intake within a mouse model of atherosclerosis, observed that a deprivation of zinc encouraged vascular inflammation and arterial plaque formation. Other studies showed that negligible dietary zinc deficiency *in vivo* stimulated vascular smooth muscle cell apoptosis; whereas long-term zinc deficiency, on the hand, accelerated rat vascular smooth muscle cell proliferation by suppressing the c-Jun N-terminal kinase 1/2 (JNK1/2) expression in mitogen-activated protein kinase (MAPK) signaling pathway (Allen-Redpath et al., 2013; Alcantara et al., 2013).

Such studies stipulate that there is a strong association in terms of reduced zinc consumption with amplified plasma lipoprotein levels, enhanced smooth muscle remodeling, higher vascular inflammation, and plaques development in mice consuming dietary zinc.

In humans, there are strong epidemiological evidences which demonstrated that the progression of atherosclerosis is altered by nutrition-related factors involving zinc (Kritchevsky, 2001). However, many factors should be considered such as the interaction between other nutrients and zinc, as well as the real bioavailability of zinc, which varies according to an individual's genetic profile and dietary habits.

Zinc increases insulin sensitivity and glycemic control since it assists in diminishing the synthesis of ROS under hyperglycemic environments, thus restraining the activation of oxidative stress pathways (Lima and Sampaio, 2011). In type 2 diabetes mellitus, chronic hyperglycemia has been correlated with lipid peroxidation and oxidative damage in cells, thereby weakening the antioxidant defense in type 2 diabetes patients.

Conclusion

Zinc functions as a co-factor for certain vital enzymes that assist in the correct performance of the antioxidant defense system. Nonetheless, studies have still not fully clarified the function of zinc in terms of the molecular mechanisms related to the pathophysiology of certain chronic diseases, such as cancer, obesity and diabetes. Therefore, more in-depth clinical studies should be carried out in terms of zinc supplementation to the actual daily requirement as a protective and therapeutic mediator for human health.

References

Adams, C.L., Hambidge, M., Raboy, V., Dorsch, J.A., Sian, L., Westcott, J.L., Krebs, N.F., 2002. Zinc absorption from a low–phytic acid maize. Am. J. Clin. Nutr. 76, 556–559.

Alcantara, E.H., Shin, M.Y., Feldmann, J., Nixon, G.F., Beattie, J.H., Kwun, I.S., 2013. Long-term zinc deprivation accelerates rat vascular smooth muscle cell proliferation involving the down-regulation of JNK1/2 expression in MAPK signaling. Atherosclerosis 228, 46–52.

Allen-Redpath, K., Ou, O., Beattie, J.H., Kwun, I.-S., Feldmann, J., Nixon, G.F., 2013. Marginal dietary zinc deficiency in vivo induces vascular smooth muscle cell apoptosis in large arteries. Cardiovasc. Res. 99, 525–534.

Andreini, C., Banci, L., Bertini, I., Rosato, A., 2006. Zinc through the three domains of life. J. Proteome Res. 5, 3173–3178.

Aydemir, T.B., Blanchard, R.K., Cousins, R.J., 2006. Zinc supplementation of young men alters metallothionein, zinc transporter, and cytokine gene expression in leukocyte populations. Proc. Natl. Acad. Sci. 103, 1699–1704.

Bao, B., Ahmad, A., Azmi, A., Li, Y., Prasad, A.S., Sarkar, F.H., 2013. The biological significance of zinc in inflammation and aging. In: Rahman, I., Bagchi, D. (Eds.), Inflammation, Advancing and Nutrition. Elsevier Inc, New York, NY, pp. 15–27.

Bao, B., Prasad, A.S., Beck, F.W., Bao, G.W., Singh, T., Ali, S., Sarkar, F.H., 2011. Intracellular free zinc up-regulates IFN-γ and T-bet essential for Th1 differentiation in Con-A stimulated HUT-78 cells. Biochem. Biophys. Res. Commun. 407, 703–707.

Bao, B., Prasad, A.S., Beck, F.W., Snell, D., Suneja, A., Sarkar, F.H., Doshi, N., Fitzgerald, J.T., Swerdlow, P., 2008. Zinc supplementation decreases oxidative stress, incidence of infection, and generation of inflammatory cytokines in sickle cell disease patients. Transl. Res. 152, 67–80.

Barak, P., Helmke, P.A., 1993. The Chemistry of Zinc. In: Robson, A.D. (Ed.), Zinc in Soils and Plants. Developments in Plant and Soil Sciences, 55. Springer, Dordrecht.

Beattie, J.H., Gordon, M.J., Duthie, S.J., McNeil, C.J., Horgan, G.W., Nixon, G.F., Feldmann, J., Kwun, I.S., 2012. Suboptimal dietary zinc intake promotes vascular inflammation and atherogenesis in a mouse model of atherosclerosis. Mol. Nutr. Food Res. 56, 1097–1105.

Biagiotti, S., Menotta, M., Orazi, S., Spapperi, C., Brundu, S., Fraternale, A., Bianchi, M., Rossi, L., Chessa, L., Magnani, M., 2016. Dexamethasone improves redox state in ataxia telangiectasia cells by promoting an NRF2-mediated antioxidant response. FEBS J. 283, 3962–3978.

Cakmak, I., Kutman, U.á., 2018. Agronomic biofortification of cereals with zinc: a review. Eur. J. Soil Sci. 69, 172–180.

Darnton-Hill, I., Webb, P., Harvey, P.W., Hunt, J.M., Dalmiya, N., Chopra, M., Ball, M.J., Bloem, M.W., De Benoist, B., 2005. Micronutrient deficiencies and gender: social and economic costs. Am. J. Clin. Nutr. 81, 1198S–1205S.

Foster, M., Samman, S., 2010. Zinc and redox signaling: perturbations associated with cardiovascular disease and diabetes mellitus. Antioxid. Redox Signal. 13, 1549–1573.

Haeflein, K., Rasmussen, A., 1977. Zinc content of selected foods. J. Am. Diet. Assoc. 70, 610–616.

Hayashi, K., Ishizuka, S., Yokoyama, C., Hatae, T., 2008. Attenuation of interferon-γ mRNA expression in activated Jurkat T cells by exogenous zinc via down-regulation of the calcium-independent PKC–AP-1 signaling pathway. Life Sci. 83, 6–11.

Hemilä, H., Fitzgerald, J.T., Petrus, E.J., Prasad, A., 2017. Zinc acetate lozenges may improve the recovery rate of common cold patients: an individual patient data meta-analysis. Open Forum Infect Dis Spring; 4 (2), ofx059.

Hunt, J.R., 2003. Bioavailability of iron, zinc, and other trace minerals from vegetarian diets. Am. J. Clin. Nutr. 78, 633S–639S.

Institute of Medicine (US) Committee on the Evaluation of the Addition of Ingredients New to Infant Formula, 2004. Infant formula: evaluating the safety of new ingredientsComparing Infant Formulas with Human Milk. National Academies Press (US), Washington (DC). https://www.ncbi.nlm.nih.gov/books/NBK215837/.

Jackson, M.J., 1989. Physiology of Zinc: General Aspects. In: Mills, C.F. (Ed.), Zinc in Human Biology. ILSI Human Nutrition Reviews. Springer, London.

Jansen, J., Rosenkranz, E., Overbeck, S., Warmuth, S., Mocchegiani, E., Giacconi, R., Weiskirchen, R., Karges, W., Rink, L., 2012. Disturbed zinc homeostasis in diabetic patients by in vitro and in vivo analysis of insulinomimetic activity of zinc. J. Nutr. Biochem. 23, 1458–1466.

Jarosz, M., Olbert, M., Wyszogrodzka, G., Młyniec, K., Librowski, T., 2017. Antioxidant and anti-inflammatory effects of zinc. Zinc-dependent NF-κB signaling. Inflammopharmacology 25, 11–24.

Kahmann, L., Uciechowski, P., Warmuth, S., Plümäkers, B., Gressner, A.M., Malavolta, M., Mocchegiani, E., Rink, L., 2008. Zinc supplementation in the elderly reduces spontaneous inflammatory cytokine release and restores T cell functions. Rejuvenation Res. 11, 227–237.

King, J.C., Cousins, R.J., 2006. Zinc. In: Shils, M.E., Shike, M., Ross, A.C., Caballero, B., Cousins, R.J. (Eds.), Modern Nutrition in Health and Disease, 10th ed. Lippincott Williams & Wilkins, Philadelphia, PA, USA, pp. 271–285.

Kozlowski, H., Potocki, S., Remelli, M., Rowinska-Zyrek, M., Valensin, D., 2013. Specific metal ion binding sites in unstructured regions of proteins. In: Krebs, N.F., Hambidge, K.M. (Eds.), Coordination Chemistry Reviews, 257, 2625–2638.

Krebs, N.F., Hambidge, K.M., 2007. Complementary feeding: clinically relevant factors affecting timing and composition. Am. J. Clin. Nutr. 85, 639S–645S.

Krężel, A., Maret, W., 2016. The biological inorganic chemistry of zinc ions. Arch. Biochem. Biophys. 611, 3–19.

Kritchevsky, D., 2001. Diet and atherosclerosis. J. Nutr. Health Aging 5, 155–159.

Liang, T., Zhang, Q., Sun, W., Xin, Y., Zhang, Z., Tan, Y., Zhou, S., Zhang, C., Cai, L., Lu, X., 2015. Zinc treatment prevents type 1 diabetes-induced hepatic oxidative damage, endoplasmic reticulum stress, and cell death, and even prevents possible steatohepatitis in the OVE26 mouse model: Important role of metallothionein. Toxicol. Lett. 233, 114–124.

Lima, V.B.d.S., Sampaio, F.d.A., Bezerra, D.L.C., Moita Neto, J.M., Marreiro, D.d.N., 2011. Parameters of glycemic control and their relationship with zinc concentrations in blood and with superoxide dismutase enzyme activity in type 2 diabetes patients. Arq Bras Endocrinol Metabol 55, 701–707.

Lonnerdal, B., 2000. Dietary factors influencing zinc absorption. J. Nutr. 130, 1378S–1383S.

Marreiro, D.D.N., Cruz, K.J.C., Morais, J.B.S., Beserra, J.B., Severo, J.S., De Oliveira, A.R.S., 2017. Zinc and oxidative stress: current mechanisms. Antioxidants 6(2), 24.

National Institutes of Health, 2016. Zinc. https://ods.od.nih.gov/factsheets/Zinc-HealthProfessional/ (Accessed 10 June 2019).

Osis, D., Kramer, L., Wiatrowski, E., Spencer, H., 1972. Dietary zinc intake in man. Am. J. Clin. Nutr. (United States) 25.

Powell, S.R., 2000. The antioxidant properties of zinc. J. Nutr. 130, 1447S–1454S.

Prasad, A.S., 2008. Clinical, immunological, anti-inflammatory and antioxidant roles of zinc. Exp. Gerontol. 43, 370–377.

Prasad, A.S., 2008. Zinc in human health: effect of zinc on immune cells. Mol. Med. 14, 353–357.

Prasad, A.S., Beck, F.W., Bao, B., Fitzgerald, J.T., Snell, D.C., Steinberg, J.D., Cardozo, L.J., 2007. Zinc supplementation decreases incidence of infections in the elderly: effect of zinc on generation of cytokines and oxidative stress. Am. J. Clin. Nutr. 85, 837–844.

Roohani, N., Hurrell, R., Kelishadi, R., Schulin, R., 2013. Zinc and its importance for human health: An integrative review. J. Res. Med. Sci. 18, 144.

Samman, S., 2007. Zinc. Nutr. Diet. 64, S131–S134.

Sandstead, H.H., Prasad, A.S., Penland, J.G., Beck, F.W., Kaplan, J., Egger, N.G., Alcock, N.W., Carroll, R.M., Ramanujam, V., Dayal, H.H., 2008. Zinc deficiency in Mexican American children: influence of zinc and other micronutrients on T cells, cytokines, and antiinflammatory plasma proteins. Am. J. Clin. Nutr. 88, 1067–1073.

Stein, A.J., 2014. Rethinking the measurement of undernutrition in a broader health context: should we look at possible causes or actual effects? Glob. Food Secur. 3, 193–199.

Tapiero, H., Tew, K.D., 2003. Trace elements in human physiology and pathology: zinc and metallothioneins. Biomed. Pharmacother. 57, 399–411.

Türk, M., Carlsson, N.-G., Sandberg, A.-S., 1996. Reduction in the levels of phytate during wholemeal bread making; effect of yeast and wheat phytases. J. Cereal Sci. 23, 257–264.

Valko, M., Leibfritz, D., Moncol, J., Cronin, M.T., Mazur, M., Telser, J., 2007. Free radicals and antioxidants in normal physiological functions and human disease. Int. J. Biochem. Cell Biol. 39, 44–84.

Wessells, K.R., Brown, K.H., 2012. Estimating the global prevalence of zinc deficiency: results based on zinc availability in national food supplies and the prevalence of stunting. PLoS One 7, e50568.

Wong, C.P., Ho, E., 2012. Zinc and its role in age-related inflammation and immune dysfunction. Mol. Nutr. Food Res. 56, 77–87.

PART

5

Antioxidants and diseases: beneficial and detrimental effects

CHAPTER

Beneficial and detrimental effects of antioxidants in cancer

5.1

Saeideh Momtaz[a,b,c,*], Shokoufeh Hassani[b,d,*], Amir Hossein Abdolghaffari[a,b,c,e]

[a]*Medicinal Plants Research Center, Institute of Medicinal Plants, ACECR, Karaj, Iran*

[b]*Toxicology and Diseases Group (TDG), Pharmaceutical Sciences Research Center (PSRC), The Institute of Pharmaceutical Sciences (TIPS), and Faculty of Pharmacy, Tehran University of Medical Sciences, Tehran, Iran*

[c]*GI Pharmacology Interest Group (GPIG), Universal Scientific Education and Research Network (USERN), Tehran, Iran*

[d]*Department of Toxicology and Pharmacology, School of Pharmacy, Tehran University of Medical Sciences, Tehran, Iran*

[e]*Department of Toxicology & Pharmacology, Faculty of Pharmacy, Tehran Medical Sciences, Islamic Azad University, Tehran, Iran*

**Both authors contributed equally*

5.1.1 Introduction

Epidemiological data demonstrated that cancer is spreading at a dreadful rate (Ferlay et al., 2015). Cancer is a complex of events caused by a variety of stimuli, which initially occurs at molecular level. Cancer progression involves cellular transformation, proliferation, apoptosis, angiogenesis, and survival (Zhang et al., 2012). Natural antioxidants have extensively studied in cellular models as therapeutic agents, for their impressive anticarcinogenic effects and low toxicity. Generally, cancer cells are resistant to chemotherapeutic drugs, generating multidrug resistant cells. In this regard, antioxidants can enhance the sensitization to chemotherapeutic drugs and have become interesting approach to overcome chemoresistance and increase the efficacy of anticancer drugs (Guestini et al., 2016). This chapter study evaluates the literature on the oxidative stress function in cancer development and introduces promising antioxidants favorable for cancer prevention or management.

5.1.2 Antioxidants in cancer development and treatment

Exogenous or endogenous changes in organisms may generate one of the cellular intermediate called reactive oxygen species (ROS) (Choudhari et al., 2014). Interestingly, at high level, ROS leads to many disorders (i.e., carcinogenesis), however, at low level,

Antioxidants Effects in Health. DOI: https://doi.org/10.1016/B978-0-12-819096-8.00071-9

it may even have some beneficial effects (Acharya et al., 2010). In order to protect itself, the human body has a proper system named as antioxidant system. A complex of enzymatic and nonenzymatic system provides defense against free radical damage (Prasad et al., 2017). The nonenzymatic antioxidants are classified as metabolic and nutrient antioxidants. Metabolic antioxidants are byproducts of the body metabolism such as lipoid acid, glutathione, L-ariginine, coenzyme Q10, melatonin, uric acid, bilirubin, metal-chelating proteins, transferrin, etc. Nutrient antioxidants are externally supplied through foods and/or supplements such as vitamin E, vitamin C, carotenoids, trace metals (selenium, manganese, and zinc), flavonoids, and omega-3 and omega-6 fatty acids. The imbalance between the ROS production and neutralization may lead to carcinogenesis (Prasad et al., 2017). Numerous studies have shown ROS level increases in cancer cells. Exposure to ROS results in oxidative damage to DNA, proteins, and lipids develops DNA instability and oncogenic mutations. Oxidative stress within cells induces the free radicals productions. Free radicals disrupt the activities of many antioxidant enzymes such as glutathione reductase (GR), superoxide dismutase (SOD), glutathione peroxidase (GPx), glutathione S-transferase (GST), catalase (CAT), and glutathione (GSH), thereby, this whole mechanism prevails events of carcinogenesis (Mariadoss et al., 2019). In between, the main role of antioxidants is to neutralize ROS or mitigate oxidative stress, protect DNA and tumor suppressor genes from being mutated (Athreya and Xavier, 2017), consecutively moderate the stability of DNA and finally to prevent carcinogenesis.

Many studies provide a clear evidence that antioxidants are capable of treating cancer in vitro, primarily attributed as prophylactic agents. Since, ROS is a preliminary factor for progression of many cancers and oncogenic mutations, antioxidants supplementation may assist to prevent and/or treat cancer through encoding or modulating the genes involved in processes such as oncogenic transformation of normal cells, as well as tumor growth, progression, angiogenesis, and metastasis (Gupta et al., 2010). Antioxidants were shown to suppress the oncogenic survival kinases such as phosphoinositide 3-kinase (PI3K) and Akt (Wang et al., 2014); cell proliferation regulators including extracellular signal regulated kinase (ERK)1/2, D-type cyclins, and cyclin-dependent kinases (Kim et al., 2006); transcription factors such as nuclear factor (NF)-kB, nuclear factor-erythroid 2 p45 subunit-related factor-2 (Nrf2), and signal transducer and activator of transcription (STAT); and angiogenic factors including vascular endothelial growth factor (VEGF), matrix metalloproteinase (MMPs), epidermal growth factor receptor 1 (EGFR1), and fibroblast growth factor (FGF) (Kunnumakkara et al., 2008). Moreover, antioxidants can downregulate the expression of tumor suppressor proteins such as p53, phosphatase and tensin homolog (PTEN), p21, and p27 (Singh et al., 2017).

5.1.3 Antioxidant foods and cancer prevention

Previous investigations have demonstrated that lifestyle modifications (such as regular fruits and vegetables intake) are beneficial for neutralizing ROS effects. Regular intake of diet rich in antioxidants can be a protective barrier against oxidative stress, also reduces the incidence and cancer mortality rate (Sharma et al., 2017). In a clinical study, the consumption of antioxidant reduced the incidence of pancreatic cancer (Han et al., 2013). Individuals consuming fresh fruits and vegetables regularly are less prone to cancer risk, thus, such antioxidant-rich substances may act as chemoprotective nutrition (Prasad et al., 2017). Fat-free, fiber-rich, and antioxidant-rich nutrition in concomitant with some lifestyle modification can serve as chemoprotective therapy (Society, 2013). Recently, it has been proposed that dietary intake of antioxidants for lengthy period leads to chemoprotective effects and reduces the risk of cancer (Yeung et al., 2019).

Bioactive molecules and minerals found in fresh fruits, herbs, and vegetables including flavonoids, carotenoids (β-carotene), vitamins (E, C), polyphenols, quercetin, phytochemicals, anthocyanin, selenium, exhibited beneficial effects toward reducing oxidative stress (Georgiev et al., 2014). A recent prospective investigation revealed that a Mediterranean plant-based regime abundant in antioxidants consisted of fruits and vegetables, legumes, olive, and canola oil reduced the incidence of cancer as well as the mortality rate. The chemoprotective actions of this diet attributed to the complex interaction between bioactive molecules of this diet (Schwingshackl and Hoffmann, 2016).

5.1.4 Antioxidants; clinical trials in cancer prevention and treatment

Dietary intake of antioxidants showed variable outcomes in different clinical studies involving cancerous patients. It was documented that a diet enriched with fruits and vegetables or antioxidants, lowered the risk of mortality in cancer patients (Genkinger et al., 2004). Similar results were obtained in a cohort study, which was associated with high content of vitamin C, provitamin A, carotenoids, and lycopene (Agudo et al., 2007). A comprehensive review study showed that a low-fat diet supplemented with high antioxidants improved the breast cancer in patients (Hauner et al., 2011).

In a study by the Nutritional Prevention of Cancer Study Group, it was shown that dietary supplementation with selenium reduced the incidence of skin basal and/or squamous cancers in patients (Clark et al., 1996), with no effect on other skin cancers. Selenium supplementation reduced the cancer incidence by 40%, particularly prostate cancer. The Linxian General Population Nutrition Intervention Trial reported that a complex of selenium, vitamin E, and β-carotene reduced the overall mortality and cancer rates (Qiao et al., 2009).

Controversially, several other studies revealed that antioxidants may even stimulate the cancer onset and/or development. For instance, in the SELECT controlled trial on old males supplemented with vitamin E, selenium, both, or neither for 7–12 years, the participants who received vitamin E alone, were more prone to develop prostate cancer, although, there was no significant different in overall cancer incidence between groups. Selenium at high concentration increased the risk of high-grade prostate carcinoma, while at low doses, it was not effective (Kristal et al., 2014).

It is worth mentioning that the inverse association between dietary total antioxidant capacity (DTAC) and cancer was observed in studies with large sample size. For example, out of seven studies indicating a positive correlation between DTAC and cancer, five studies were prospective (Mekary et al., 2010; Pantavos et al., 2015; Russnes et al., 2014; Serafini et al., 2012; Zamora-Ros et al., 2013), and four studies included more than 10,000 participants (Mekary et al., 2010; Russnes et al., 2014; Serafini et al., 2012; Zamora-Ros et al., 2013). Based on the US Preventative Services Task Force (USPSTF) report, 2014, there is a lack of adequate scientific data on the majority of nutrient supplements for cancer prevention (Moyer, 2014), although, there are two exceptions for vitamin E and β-carotene. Data obtained from many trials and meta-analysis studies indicated that vitamin E is not effective for cancer prevention, while β-carotene may even induce the risk of lung cancer in smoker subjects.

In addition to cancer prevention, antioxidants were considered as adjuvant therapy in cancer treatment trials. Modalities used in cancer treatment are majorly including immunotherapy, radiation therapy, and chemotherapy. It has been demonstrated that antioxidants are able to assist chemotherapeutical agents in treating cancer. In the other hand, clinical trials have proved that subjects with high antioxidant concentration in their plasma can resist radiotherapy more than those with less antioxidant content (Athreya and Xavier, 2017). Another study confirmed the efficacy and improvement of patient survival rate following the use of adjuvant antioxidants during chemotherapy or radiotherapy (Yasueda et al., 2016). In head and neck cancer patients receiving radiotherapy, vitamin E, and β-carotene supplementation enhanced the risk of cancer recurrence and mortality, particularly in smoker population (Meyer et al., 2008). Overall, antioxidants can significantly play both preventive and/or therapeutic roles in a number of cancers in some population; however, the antioxidant use may enhance the cancer incidence, specifically in individuals at high risk.

5.1.5 Common dietary antioxidants in cancer prevention

5.1.5.1 Tea polyphenols and catechin

Catechin is the main flavonal in tea, of which epigallactocatechin-3-gallate (EGCG) is the most active and abundant catechin. In vitro and in vivo studies indicated that EGCG can inhibit the cancer cell growth and development. According to data published by the Scientific Panel on Food Additives and Nutrient Sources added to Food (ANS), the intake of doses equal or above 800 mg EGCG/day could induce hepatic toxicity by increasing of serum transaminases (EFSA Panel on Food Additives and Nutrient Sources added to Food (ANS) et al., 2018).

Yang et al. (2011) showed that green tea and black tea have protective and therapeutic effects on esophageal tumorigenesis in rats. A population-based case–control study in Shanghai reported that the incidence of colorectal cancer reduced considerably following tea consumption (Yang et al., 2007). It was shown that EGCG suppressed the NF-κB activity in various human cancer cells (Peng et al., 2006). Several studies indicated that the chemopreventive properties of EGCG were mainly mediated through inhibiting the MMPs induction. Onoda et al. (2011) found that EGCG treatment inhibited the human gastric adenosquamous carcinoma (i.e., MKN-1, MKN-28, MKN-45, NUGC-3, and TMK-1) cell growth, although the NUGC-3 was the most sensitive cell line. EGCG was also shown to downregulate the Met signaling in human colorectal HCT116 cancer cells (Larsen and Dashwood, 2010). In endothelial cells, EGCG inhibited the MMP-2 and -9 activities in a direct and indirect manners, repressed the VEGF binding to its receptors, and inhibited the cancer invasion and metastasis (Cheng et al., 2005). As a chemopreventive agent, EGCG can modulate angiogenesis and apoptosis by downregulation of multiple signaling pathways such as VEGF, uPA, insulin-like growth factor (IGF)-1, EGFR, cell cycle regulatory proteins, NF-κB, PI3-K/Akt, ERK, JNK, Ras/Raf/mitogen-activated protein kinase (MAPK) and AP-1 (Shankar et al., 2007). It was also reported that black tea polyphenols (i.e., catechin) can prevent cancer invasion and metastasis through regulating the genes and pathways involved in tumor cells migration such as MMP-2, -7, and -9 or the Wnt/β-catenin pathway (El Bedoui et al., 2005; Patel et al., 2008).

5.1.5.2 Curcumin

According to the United State Food and Drug Administration (FDA), curcumin is generally safe (Gupta et al., 2013). Curcumin was shown to possess chemopreventive effect in a number of cancers such as colon, stomach, esophageal, breast, liver, prostate, lung, and oral carcinogenesis in a dose-dependent manner (Aggarwal et al., 2003; Rajasekaran, 2011). Curcumin nearly disrupts all the cancer steps including tumor promotion, growth, and invasion (Nagpal and Sood, 2013) by targeting cancer-related transcription, growth, adhesion, and angiogenesis regulators; apoptotic genes; and cellular signaling molecules, mainly correlated with its antioxidant properties (Aggarwal et al., 2003). Curcumin reduced carcinogen-induced tumorigenesis and N-ethyl-N′-nitro-nitrosoguanidine-induced tumors.

In benzopyrene-induced forestomach tumors mice, curcumin reduced the incidence and the intensity of the disease, representing its preventive and therapeutic effects (Singh et al., 1998). It was reported that curcumin reduced the intestinal adenoma in $Min^{-/-}$ mice (Mahmoud et al., 2000). Curcumin induced apoptosis and prevented adenoma development in mice (Anand et al., 2008). A combination of curcumin and celecoxib inhibited the cancer progression in DMH-induced rat model (Shpitz et al., 2006). Several in vitro and in vivo studies reported that curcumin suppressed tumor invasion and metastasis through downregulation of the expression of MMP-9 (Bimonte et al., 2013; Kunnumakkara et al., 2008). Curcumin also downregulated the MMPs expression by reducing the NF-κB activity and the AP-1 transcription (Kunnumakkara et al., 2008). In human colon cancer cells, the compound restricted the cancer cell migration and invasion through inhibiting NF-κB/p65 and suppression of the cyclooxygenase (COX)-2 and MMP-2 expressions (Su et al., 2006). It was demonstrated that curcumin inhibited invasion of human carcinoma cells through suppressing the MMP-2 promoter activity and inhibiting the formation of lipid raft-associated Rac1/PI3K/Akt signaling pathway (Meng et al., 2015).

5.1.5.3 Resveratrol

Clinical trials confirmed that resveratrol and *trans*-resveratrol supplementation is safe and well tolerated at different doses, although some gastrointestinal complications were reported, at doses 2.5–5 g of resveratrol (Ramírez-Garza et al., 2018). Suppression of metastasis-associated factors seems to be the main mechanism underlying the anticancer potential of resveratrol. Resveratrol acted as a potent chemopreventive compound in various experimentally tumor models, besides, resveratrol inhibited the growth and proliferation of different cancer cells. For instance, resveratrol possessed chemopreventive effects in HCT116 CRC, mediated by Wnt/β-catenin signaling pathway through inhibiting MMP-7 (Ji et al., 2013). Similarly, this stilbene potentiated the chemopreventive activity of gemcitabine in pancreatic cancer models (Harikumar et al., 2010). In a colon cancer model, oral administration of resveratrol for 6 weeks increased the antioxidant enzymes expressions (i.e., heme oxygenase-1 (HO-1) and GR), through upregulation of the Nrf2 signaling (Chiou et al., 2011).

In APC-/+ mice, resveratrol inhibited the intestinal tumorigenesis initiation (Schneider et al., 2001). In human hepatocellular carcinoma cells, resveratrol inhibited the MMP-9 expression and modulated the MAP kinases; ERK1/2, JNK, and p38 MAPK. The compound also down regulated the expression of epithelial-mesenchymal transition (EMT) and inhibited the cell proliferation, invasion, and metastasis (Li et al., 2013). In accordance, resveratrol and its methoxy analog MR-3 exerted same anticancer activity in hepatoma cells via suppression of MMPs (Weng et al., 2010). Recently, the potential application of resveratrol in suppressing the initiation and progression of cancers has been proven through modulation of miRNAs and lncRNAs (Jiang et al., 2019). The antineoplastic potential of resveratrol against cervical cancer cells HPV18 and HPV16 was attributed to down regulation of E6 and VEGF proteins, due to the growth arrest indicated by suppressing the proliferating cell nuclear antigen (PCNA) expression (Chatterjee, 2019).

5.1.5.4 Quercetin

Data on safety profile of quercetin supplements are controversial, indicating that the duration and the dosage of the compound intake are determinative, however, certain potential risk groups have been identified, for example, patients with a kidney dysfunction (Andres et al., 2018). Quercetin is known to ameliorate ROS-mediated cellular damages, resulting in remarkable protection against progression of different cancer types (Ghanaatian et al., 2018). In an extensive review by Khan et al., the antitumorigenesis activity of quercetin was ascribed to its ability to bind stress-activated protein kinase (SAPK)/ERK kinase 1 (SEK1), c-Jun N-terminal kinase 1/2 (JNK1/2), MEK1, and ERK1/2 pathways; induction of cellular senescence and suppression of telomerase activity; induction of cell death; cell cycle arrest; antioxidant activity; interaction with cellular receptors (i.e., Raf and MEK, aryl hydrocarbon receptor (AhR)); and modification of signal transduction (i.e., ERK and p38 MAPK) (Khan et al., 2016; Darband et al., 2018).

In addition, intensive attention has been paid to the chemotherapeutic properties of this flavonoid, either alone or in combination with other natural antioxidants, that is, resveratrol or curcumin, and even in combination with chemotherapeutic agents such as cisplatin (Rauf et al., 2018). The efficacy of quercetin to inhibit the progress of numerous human cancers such as breast (Nguyen et al., 2017), colon (Zhao et al., 2017), lung (Zhao and Zhang, 2015), prostate (Zhao et al., 2016), skin (Zhu et al., 2016), ovarian (Zhou et al., 2015), and many other type of carcinogenesis has been reported broadly.

5.1.5.5 Flavonoids

Flavonoids were found favorable for human health by means of their antioxidant power and chelating ability, owing to their diverse frameworks (Heim et al., 2002). The presence of multiple hydroxyl groups and high affinity to pair delocalized electrons, amplifies the free radical scavenging ability of flavonoids in a way to interact with the redox system of the cell (Singh et al., 2014). Pharmacokinetic and the bioavailability of flavonoids in the body are highly affecting their chemopreventive efficacy (Thilakarathna and Rupasinghe, 2013).

It has been evidenced that flavonoids are effectual for both cancer treatment and prevention. Flavonoids reduce the risk of developing certain types of cancer through inhibition of cell proliferation and growth or by apoptosis induction in a ROS-dependent (i.e., genistein, quercetin, silymarin, luteolin, apigenin, kaempferol, and EGCG) (Antosiak et al., 2017; Liao et al., 2016; Yildiz et al., 2017) or independent manner. In vitro models, flavonoids may modulate cancer either through antioxidant-dependent or -independent pathways (Arango et al., 2012).

Overall, the anticarcinogenesis effects of flavonoids are ascribed to induction of apoptosis (i.e., by apigenin, luteolin, silibinin) (Lin et al., 2008, 2015; Kim et al., 2014); cell cycle arrest by inhibiting the key cell cycle regulators such as cyclin-dependent kinases (CDKs) (i.e., by apigenin, luteolin) (Lin et al., 2015); suppression of metabolizing enzymes such as cytochromes P450 and P3A involved in the activation of carcinogenic compounds (i.e., by quercetin) (Ekstrand et al., 2015); reduction of ROS production (i.e., by luteolin); inactivation of various intracellular signaling pathways; and inhibition of angiogenesis (i.e., by nobiletin) (Amawi et al., 2017).

Several flavonoids have also been identified to inhibit ROS generation in macrophages, thereby, reducing the immunosuppressive conditions of the tumor microenvironment (TME) (Pérez-Cano and Castell, 2016). In mouse macrophage ANA-1 cells, apigenin increased ROS and induced apoptosis by activating the caspase-3 and MAPK pathway (Liao et al., 2014). Increased number of tumor-associated macrophages (TAMs) in the TME has also been correlated with defective immune-surveillance, increased metastasis and poor clinical prognosis of cancers (Medrek et al., 2012), however, it was documented that TAMs may decline the efficacy of chemotherapy (De Palma and Lewis, 2013).

5.1.5.6 Other promising dietary antioxidants

Beyond dietary antioxidants, some other antioxidants such as melatonin, lycopene, retinoic acid, vitamin C, and vitamin E play potential preventive role against various cancers. Melatonin is an indigenous hormone with free radical scavenging, anti-inflammatory, and antioxidant activities, which is synthesized by the pineal gland (Reiter et al., 2007). It was evidenced that the consumption of tropical fruits enhanced the serum level of melatonin and the antioxidant capacity of the blood serum (Rudra et al., 2013). Dietary intake of melatonin for a long time reduced the tumor number and size in aged mice through inhibiting the growth of carcinoma cells via p21 and Bax overexpression, and Bcl-2 suppression (Xu et al., 2013). In HepG2 cells, melatonin induced apoptosis and reduced metastasis through TIMP-1 upregulation and downregulation of the MMP-9 activity via Nf-κB signal pathway (Nemeth et al., 2011).

Lycopene was shown to have 11 conjugated double bonds, thereby, is accounted as the most potent antioxidant among carotenoids, warranting its DNA protective effect against oxidative damage. The anticancer effect of lycopene has been implicated to regulation of oxidative enzymes (i.e., COX-2, 5-lipoxygenase (5-LOX)); induction of antioxidant/detoxifying phase II enzymes modulated by the Nrf2/antioxidant responsive element (ARE) system; inhibition of tumor growth and progression; induction of apoptosis; modulation of cell cycle regulatory proteins; modulation of immune function; and regulation of growth factor signaling. Moreover, lycopene reduces the risk of breast, lung, gastric, liver, head and neck, ovarian, pancreatic, prostate, skin, and many other type of cancers (Kim and Kim, 2015; Ono et al., 2015; Sahin et al., 2010; Tang et al., 2012).

The retinoid derivative all-trans retinoic acid, retinoic acid, and its derivatives (retinoids) are accounted as potential chemotherapeutic or chemopreventive agents due to their antiproliferative, proapoptotic, antiangiogenesis, antimetastasis, and antioxidant properties. Retinoic acid involves in a set of pleiotropic effects, regulating the survival and differentiation of tumor-initiating cells or cancer stem cells. In combination with other therapeutics, retinoic acid is beneficial for resensitization of resistant cancer cells (Di Masi et al., 2015; Tripathi et al., 2019). Retinoic acid also attained the FDA approval for the treatment of lymphoma and leukemia (Tripathi et al., 2019).

Dietary supplementation of ascorbic acid may lower the risk of cancer either way; as an antioxidant by neutralizing free radicals, hence, suppressing cancer progression; or at higher doses it may act as a pro-oxidant, which induces cytotoxicity in cancer cells. In addition, inhibition of MMPs, apoptosis-inducing activity, antiangiogenic effect, and inhibition of tumorigenesis and cancer cell growth are implicated as other mechanisms involved in its cancer prevention and treatment effects (Sunil et al., 2017).

Vitamin E (γ-tocotrienol) and its derivatives are known to increase the sensitization of cancer cells against chemotherapeutic agents. In a recent review, it was stated that two isoforms of vitamin E, gamma-tocopherol and delta-tocopherol, are promising scavengers of ROS and reactive nitrogen species. Moreover, inhibition of 3-hydroxy-3-methylglutaryl-coenzyme A reductase (HMG CoA) and suppression of NF-κB pathway were shown responsible for anticancer effect of tocotrienols (chemicals that are part of the vitamin E family), representative of its potent antiangiogenic effect compared with other vitamin E isoforms (Abraham et al., 2018).

5.1.6 Antioxidants for cancer management; cure or threat

In the context of cancer therapy or prevention, antioxidants play either beneficial or detrimental. Antioxidants may be considered as a treatment strategy for cancer patients (Godic et al., 2014) and at the same time are implicated with deleterious effects of increasing the risk of cancer progression (Goodman et al., 2011). Indeed, the majority of studies signify that ROS elimination may be deleterious to cancer patients rather than preventing the risk. Under certain condition, some of antioxidants act as pro-oxidants (Valko et al., 2004). For example, in cigarette smokers diagnosed with lung cancer, β-carotene supplementation, worsened the disease condition due to the instability of β-carotene in the oxygen-rich environment (Wang and Russell, 1999). Another study showed that the consumption of β-carotene should be cautiously, since concomitant use of vitamin C and β-carotene could have provoking effects on ultra violet (UV)-induced skin carcinogenesis (Black, 2004).

In another study, combination of vitamin E and N-acetylcysteine (NAC) led to the acceleration of tumor progression in B-RAF- and K-RAS-induced lung cancer in murine models. It was concluded that these antioxidants promoted the tumor progression by reducing the ROS level and the p53 expression, leading to apoptosis and cell death. The supplementation of NAC and vitamin E caused p53 inactivation, promoted tumor progression through the ROS-p53 axis (Sayin et al., 2014). It has been demonstrated that the GSH synthesis is a critical step for cancer initiation, but only prior to the cancer onset. When tumor is progressed, the alternative antioxidant pathways are replaced with GSH (Harris et al., 2015). Following an intensive literature review, Yasueda and colleagues concluded that the deteriorate/beneficial effect of antioxidant supplementation remains doubtful for patients during cancer therapy, except for smokers undergoing radiotherapy (Yasueda et al., 2016).

Overall, it seems once antioxidants are consumed by healthy individuals, confer favorable effects; while under tumorigenesis state, high dosage of antioxidants is not recommended as results in exaggerated proliferation of tumor cells (Saeidnia and Abdollahi, 2013). It is also noteworthy that topical application of antioxidants could turn unstable due to the reactions associated with UV-irradiation, thereby, causing harmful consequences. Such facts raise considerable questions regarding the application of antioxidant in cancer therapy. Considering the disease statues, selection of certain antioxidants and consumption of a regular diet, containing a balanced amount of antioxidants might be a proper strategy, rather than the intake of direct supplements (Sarangarajan et al., 2017).

5.1.7 Antioxidants in combination therapy in combating cancer

Today, combination therapy is deliberated as a cornerstone strategy for cancer management, since enhances the efficacy of therapeutic agents in a synergistic or additive manner, rather than mono-therapy. Regarding multidimensional nature of cancer, such approach potentially reduces the anticancer drugs resistance, the tumor growth and metastasis potential, besides provides a way to specifically target cancer-inducing or cell-sustaining pathways (Mokhtari et al., 2017). There are evidence that antioxidants alone in combination with other chemo preventive drugs seem effective (Del Follo-Martinez et al., 2013; Li et al., 2014; Panchuk et al., 2014; Srivastava et al., 2010), although this is still debated and needs more studies specially in the clinic level.

Conclusion

Oxidative stress is a core factor in the pathogenesis of many cancers. Antioxidants provide chemopreventive options to either prevent or treat cancer. It is worth mentioning that antioxidants play bilateral roles during tumorigenesis either by increasing the survival via metabolic rescue or by preventing oxidative DNA damage. Thereby, before or at the early onset of the disease the consumption of antioxidants seems beneficial to neutralize the DNA oxidation, while during the late stages of carcinogenesis the intake of antioxidants is better to be restricted, allowing ROS production and destroying the malignant cells. A balanced diet containing both oxidants and antioxidants may help to maintain optimum health condition. Overall, use of antioxidants for cancer therapy or prevention depends on stage of the disease, the type of antioxidant compounds, and to the health condition of each individual. Based on the experimental and preclinical outcomes, antioxidants can be a promising platform for cancer prevention or treatment, however, many clinical trials come across with controversial results. Various clinical studies indicated that selenium, vitamin E, and carotenoids have no anticancer effects, while several flavonoids and resveratrol with a diet full of antioxidants, could lower the incidence of certain cancers. Combinatorial treatments of antioxidants and chemotherapeutics may lead to synergistic effects on tumor cells survival, growth and death, offering the possibility of the usage of chemotherapeutics at lower concentrations, enhancing their therapeutic efficacy and reducing their adverse effects. Further explorations should focus on larger and prolonged clinical trials to estimate the efficacy, safety, optimal dosage, and to indicate a proper formulation with adequate antioxidant and anticarcinogenic effects.

Abbreviations

ROS	reactive oxygen species
GR	glutathione reductase
SOD	superoxide dismutase
GPx	glutathione peroxidase
GST	glutathione S-transferase
CAT	catalase
GSH	glutathione
PI3K	phosphoinositide 3-kinase
ERK	extracellular signal regulated kinase
NF-κB	nuclear factor kappa β
NRF2	nuclear factor-erythroid 2 p45 subunit-related factor 2
STAT	signal transducer and activator of transcription
VEGF	vascular endothelial growth factor
MMPs	matrix metalloproteinase (MMPs)
EGFR	epidermal growth factor receptor 1
FGF	fibroblast growth factor
DTAC	dietary total antioxidant capacity
USPSTF	US Preventative Services Task Force
EGCG	epigallactocatechin-3-gallate
FDA	Food and Drug Administration
HO-1	heme oxygenase-1
EMT	epithelial-mesenchymal transition
JNK	c-Jun N-terminal kinase
AhR	aryl hydrocarbon receptor
TME	tumor microenvironment
TAMs	HMG CoA, tumor-associated macro hydroxy-3-methylglutaryl-coenzyme A
UV	ultra violet
NAC	N-acetylcysteine
PTEN	phosphatase and tensin homolog
IGF	insulin-like growth factor
SAPK	stress-activated protein kinase
MAPK	mitogen-activated protein kinase
5-LOX	5-lipoxygenase
ARE	antioxidant responsive element
CDKs	cyclin-dependent kinases

References

Abraham, A., Kattoor, A.J., Saldeen, T., Mehta, J.L., 2018. Vitamin E and its anticancer effects. Crit. Rev. Food Sci. Nutr. 59 (17), 1–8.

Acharya, A., Das, I., Chandhok, D., Saha, T., 2010. Redox regulation in cancer: a double-edged sword with therapeutic potential. Oxidative Med. Cell. Longevity 3 (1), 23–34.

Aggarwal, B.B., Kumar, A., Bharti, A.C., 2003. Anticancer potential of curcumin: preclinical and clinical studies. Anticancer Res. 23 (1/A), 363–398.

Agudo, A., Cabrera, L., Amiano, P., Ardanaz, E., Barricarte, A., Berenguer, T., Larranaga, N., 2007. Fruit and vegetable intakes, dietary antioxidant nutrients, and total mortality in Spanish adults: findings from the Spanish cohort of the European Prospective Investigation into Cancer and Nutrition (EPIC-Spain). Am. J. Clin. Nutr. 85 (6), 1634–1642.

Amawi, H., Ashby Jr, C.R., and Tiwari, A.K. 2017. Cancer chemoprevention through dietary flavonoids: what. 癌症 (英文版) (2017 年 10), 455–467.

Anand, P., Sundaram, C., Jhurani, S., Kunnumakkara, A.B., Aggarwal, B.B., 2008. Curcumin and cancer: an "old-age" disease with an "age-old" solution. Cancer Lett. 267 (1), 133–164.

Andres, S., Pevny, S., Ziegenhagen, R., Bakhiya, N., Schäfer, B., Hirsch-Ernst, K.I., Lampen, A., 2018. Safety aspects of the use of quercetin as a dietary supplement. Mol. Nutr. Food Res. 62 (1), 1700447.

Antosiak, A., Milowska, K., Maczynska, K., Rozalska, S., Gabryelak, T., 2017. Cytotoxic activity of genistein-8-C-glucoside form *Lupinus luteus* L. and genistein against human SK-OV-3 ovarian carcinoma cell line. Med. Chem. Res. 26 (1), 64–73.

Arango, D., Parihar, A., Villamena, F.A., Wang, L., Freitas, M.A., Grotewold, E., Doseff, A.I., 2012. Apigenin induces DNA damage through the PKCδ-dependent activation of ATM and H2AX causing down-regulation of genes involved in cell cycle control and DNA repair. Biochem. Pharmacol. 84 (12), 1571–1580.

Athreya, K., Xavier, M.F., 2017. Antioxidants in the treatment of cancer. Nutr. Cancer 69 (8), 1099–1104.

Bimonte, S., Barbieri, A., Palma, G., Luciano, A., Rea, D., Arra, C., 2013. Curcumin inhibits tumor growth and angiogenesis in an orthotopic mouse model of human pancreatic cancer. Biomed. Res. Int. 2013, 810423–810423. doi:10.1155/2013/810423.

Black, H.S., 2004. Pro-carcinogenic activity of β-carotene, a putative systemic photoprotectant. Photochem. Photobiol. Sci. 3 (8), 753–758.

Chatterjee, K., 2019. An In Vitro and In Vivo Evaluation of the Anticancer Potential of Resveratrol and Pterostilbene Against HPV-E6 Positive Cancers. CUNY Academic Works, City University of New York.

Cheng, X.W., Kuzuya, M., Nakamura, K., Liu, Z., Di, Q., Hasegawa, J., Iguchi, A., 2005. Mechanisms of the inhibitory effect of epigallocatechin-3-gallate on cultured human vascular smooth muscle cell invasion. Arterioscler. Thromb. Vasc. Biol. 25 (9), 1864–1870.

Chiou, Y.-S., Tsai, M.-L., Nagabhushanam, K., Wang, Y.-J., Wu, C.-H., Ho, C.-T., Pan, M.-H., 2011. Pterostilbene is more potent than resveratrol in preventing azoxymethane (AOM)-induced colon tumorigenesis via activation of the NF-E2-related factor 2 (Nrf2)-mediated antioxidant signaling pathway. J. Agric. Food Chem. 59 (6), 2725–2733.

Choudhari, S.K., Chaudhary, M., Gadbail, A.R., Sharma, A., Tekade, S., 2014. Oxidative and antioxidative mechanisms in oral cancer and precancer: a review. Oral Oncol. 50 (1), 10–18.

Clark, L.C., Combs, G.F., Turnbull, B.W., Slate, E.H., Chalker, D.K., Chow, J., Gross, E.G., 1996. Effects of selenium supplementation for cancer prevention in patients with carcinoma of the skin: a randomized controlled trial. JAMA 276 (24), 1957–1963.

Darband, S.G., Kaviani, M., Yousefi, B., Sadighparvar, S., Pakdel, F.G., Attari, J.A., Majidinia, M., 2018. Quercetin: a functional dietary flavonoid with potential chemo-preventive properties in colorectal cancer. J. Cell. Physiol. 233 (9), 6544–6560.

De Palma, M., Lewis, C.E., 2013. Macrophage regulation of tumor responses to anticancer therapies. Cancer Cell 23 (3), 277–286.

Del Follo-Martinez, A., Banerjee, N., Li, X., Safe, S., Mertens-Talcott, S., 2013. Resveratrol and quercetin in combination have anticancer activity in colon cancer cells and repress oncogenic microRNA-27a. Nutr. Cancer 65 (3), 494–504.

Di Masi, A., Leboffe, L., De Marinis, E., Pagano, F., Cicconi, L., Rochette-Egly, C., Nervi, C., 2015. Retinoic acid receptors: from molecular mechanisms to cancer therapy. Mol. Aspects Med. 41, 1–115.

EFSA Panel on Food Additives and Nutrient Sources added to Food (ANS)Younes, M., Aggett, P., Aguilar, F, Crebelli, R., Dusemund, B., Filipič, M., Frutos, M.J., Galtier, P., Gott, D., Gundert-Remy, U., 2018. Scientific opinion on the safety of green tea catechins. EFSA J. 16 (4), e05239.

Ekstrand, B., Rasmussen, M.K., Woll, F., Zlabek, V., Zamaratskaia, G., 2015. In vitro gender-dependent inhibition of porcine cytochrome p450 activity by selected flavonoids and phenolic acids. Biomed. Res. Int. 2015, 387918–387918. doi:10.1155/2015/387918.

El Bedoui, J., Oak, M.-H., Anglard, P., Schini-Kerth, V.B., 2005. Catechins prevent vascular smooth muscle cell invasion by inhibiting MT1-MMP activity and MMP-2 expression. Cardiovasc. Res. 67 (2), 317–325.

Ferlay, J., Soerjomataram, I., Dikshit, R., Eser, S., Mathers, C., Rebelo, M., Bray, F., 2015. Cancer incidence and mortality worldwide: sources, methods and major patterns in GLOBOCAN 2012. Int. J. Cancer 136 (5), E359–E386.

Genkinger, J.M., Platz, E.A., Hoffman, S.C., Comstock, G.W., Helzlsouer, K.J., 2004. Fruit, vegetable, and antioxidant intake and all-cause, cancer, and cardiovascular disease mortality in a community-dwelling population in Washington County, Maryland. Am. J. Epidemiol. 160 (12), 1223–1233.

Georgiev, V., Ananga, A., Tsolova, V., 2014. Recent advances and uses of grape flavonoids as nutraceuticals. Nutrients 6 (1), 391–415.

Ghanaatian, N., Lashgari, N.A., Abdolghaffari, A.H., Rajaee, S.M., Panahi, Y., Barreto, G.E., Sahebkar, A., 2018. Curcumin as a therapeutic candidate for multiple sclerosis: molecular mechanisms and targets. J. Cell. Physiol. 234 (8), 12237–12248. doi:10.1002/jcp.27965.

Godic, A., Poljšak, B., Adamic, M., Dahmane, R., 2014. The role of antioxidants in skin cancer prevention and treatment. Oxidative Med. Cell. Longevity 2014, 860479–860479. doi:10.1155/2014/860479.

Goodman, M., Bostick, R.M., Kucuk, O., Jones, D.P., 2011. Clinical trials of antioxidants as cancer prevention agents: past, present, and future. Free Radic. Biol. Med. 51 (5), 1068–1084.

Guestini, F., McNamara, K.M., Ishida, T., Sasano, H., 2016. Triple negative breast cancer chemosensitivity and chemoresistance: current advances in biomarkers indentification. Expert Opin. Ther. Targets 20 (6), 705–720.

Gupta, S.C., Kim, J.H., Prasad, S., Aggarwal, B.B., 2010. Regulation of survival, proliferation, invasion, angiogenesis, and metastasis of tumor cells through modulation of inflammatory pathways by nutraceuticals. Cancer Metastasis Rev. 29 (3), 405–434.

Gupta, S.C., Patchva, S., Aggarwal, B.B., 2013. Therapeutic roles of curcumin: lessons learned from clinical trials. AAPS J. 15, 195–218.

Han, X., Li, J., Brasky, T.M., Xun, P., Stevens, J., White, E., He, K., 2013. Antioxidant intake and pancreatic cancer risk: the Vitamins and Lifestyle (VITAL) Study. Cancer 119 (7), 1314–1320.

Harikumar, K.B., Kunnumakkara, A.B., Sethi, G., Diagaradjane, P., Anand, P., Pandey, M.K., Aggarwal, B.B., 2010. Resveratrol, a multitargeted agent, can enhance antitumor activity of gemcitabine in vitro and in orthotopic mouse model of human pancreatic cancer. Int. J. Cancer 127 (2), 257–268.

Harris, I.S., Treloar, A.E., Inoue, S., Sasaki, M., Gorrini, C., Lee, K.C., Cox, M.A., 2015. Glutathione and thioredoxin antioxidant pathways synergize to drive cancer initiation and progression. Cancer Cell 27 (2), 211–222.

Hauner, D., Janni, W., Rack, B., Hauner, H., 2011. The effect of overweight and nutrition on prognosis in breast cancer. Deuts. Ärzteblatt Int. 108 (47), 795.

Heim, K.E., Tagliaferro, A.R., Bobilya, D.J., 2002. Flavonoid antioxidants: chemistry, metabolism and structure-activity relationships. J. Nutr. Biochem. 13 (10), 572–584.

Ji, Q., Liu, X., Fu, X., Zhang, L., Sui, H., Zhou, L., Ren, J., 2013. Resveratrol inhibits invasion and metastasis of colorectal cancer cells via MALAT1 mediated Wnt/β-catenin signal pathway. PLoS One 8 (11), e78700.

Jiang, S., Yu, F., Zhou, L., Wang, K., 2019. Noncoding RNAs as molecular targets of resveratrol underlying its anticancer effects. J. Agric. Food Chem 67 (17), 4709–4719.

Khan, F., Niaz, K., Maqbool, F., Ismail Hassan, F., Abdollahi, M., Nagulapalli Venkata, K., Bishayee, A., 2016. Molecular targets underlying the anticancer effects of quercetin: an update. Nutrients 8 (9), 529.

Kim, A.L., Zhu, Y., Zhu, H., Han, L., Kopelovich, L., Bickers, D.R., Athar, M., 2006. Resveratrol inhibits proliferation of human epidermoid carcinoma A431 cells by modulating MEK1 and AP-1 signalling pathways. Exp. Dermatol. 15 (7), 538–546.

Kim, M.J., Kim, H., 2015. Anticancer effect of lycopene in gastric carcinogenesis. J. Cancer Prevent. 20 (2), 92.

Kim, T.H., Woo, J.S., Kim, Y.K., Kim, K.H., 2014. Silibinin induces cell death through reactive oxygen species-dependent downregulation of notch-1/ERK/Akt signaling in human breast cancer cells. J. Pharmacol. Exp. Ther. 349 (2), 268–278.

Kristal, A.R., Darke, A.K., Morris, J.S., Tangen, C.M., Goodman, P.J., Thompson, I.M., Parnes, H.L., 2014. Baseline selenium status and effects of selenium and vitamin E supplementation on prostate cancer risk. J. Natl. Cancer Inst. 106 (3), djt456.

Kunnumakkara, A.B., Anand, P., Aggarwal, B.B., 2008. Curcumin inhibits proliferation, invasion, angiogenesis and metastasis of different cancers through interaction with multiple cell signaling proteins. Cancer Lett. 269 (2), 199–225.

Kunnumakkara, A.B., Diagaradjane, P., Guha, S., Deorukhkar, A., Shentu, S., Aggarwal, B.B., Krishnan, S., 2008. Curcumin sensitizes human colorectal cancer xenografts in nude mice to γ-radiation by targeting nuclear factor-κB-regulated gene products. Clin. Cancer Res. 14 (7), 2128–2136.

Larsen, C.A., Dashwood, R.H., 2010. (−)-Epigallocatechin-3-gallate inhibits Met signaling, proliferation, and invasiveness in human colon cancer cells. Arch. Biochem. Biophys. 501 (1), 52–57.

Li, W., Liu, M., Xu, Y.-F., Feng, Y., Che, J.-P., Wang, G.-C., Zheng, J.-H., 2014. Combination of quercetin and hyperoside has anticancer effects on renal cancer cells through inhibition of oncogenic microRNA-27a. Oncol. Rep. 31 (1), 117–124.

Li, W., Ma, J., Ma, Q., Li, B., Han, L., Liu, J., Wang, F., 2013. Resveratrol inhibits the epithelial-mesenchymal transition of pancreatic cancer cells via suppression of the PI-3K/Akt/NF-κB pathway. Curr. Med. Chem. 20 (33), 4185–4194.

Liao, W., Chen, L., Ma, X., Jiao, R., Li, X., Wang, Y., 2016. Protective effects of kaempferol against reactive oxygen species-induced hemolysis and its antiproliferative activity on human cancer cells. Eur. J. Med. Chem. 114, 24–32.

Liao, Y., Shen, W., Kong, G., Lv, H., Tao, W., Bo, P., 2014. Apigenin induces the apoptosis and regulates MAPK signaling pathways in mouse macrophage ANA-1 cells. PLoS One 9 (3), e92007.

Lin, C.-H., Chang, C.-Y., Lee, K.-R., Lin, H.-J., Chen, T.-H., Wan, L., 2015. Flavones inhibit breast cancer proliferation through the Akt/FOXO3a signaling pathway. BMC Cancer 15 (1), 958.

Lin, Y., Shi, R., Wang, X., Shen, H.-M., 2008. Luteolin, a flavonoid with potential for cancer prevention and therapy. Curr. Cancer Drug Targets 8 (7), 634–646.

Mahmoud, N.N., Carothers, A.M., Grunberger, D., Bilinski, R.T., Churchill, M.R., Martucci, C., Bertagnolli, M.M., 2000. Plant phenolics decrease intestinal tumors in an animal model of familial adenomatous polyposis. Carcinogenesis 21 (5), 921–927.

Mariadoss, A.V.A., Vinayagam, R., Xu, B., Venkatachalam, K., Sankaran, V., Vijayakumar, S., David, E., 2019. Phloretin loaded chitosan nanoparticles enhance the antioxidants and apoptotic mechanisms in DMBA induced experimental carcinogenesis. Chem. Biol. Interact. 308, 11–19.

Medrek, C., Pontén, F., Jirström, K., Leandersson, K., 2012. The presence of tumor associated macrophages in tumor stroma as a prognostic marker for breast cancer patients. BMC Cancer 12 (1), 306.

Mekary, R.A., Wu, K., Giovannucci, E., Sampson, L., Fuchs, C., Spiegelman, D., Smith-Warner, S.A., 2010. Total antioxidant capacity intake and colorectal cancer risk in the Health Professionals Follow-up Study. Cancer Causes Control. 21 (8), 1315–1321.

Meng, X., Huang, Z., Teng, F., Xing, L., Yu, J., 2015. Predictive biomarkers in PD-1/PD-L1 checkpoint blockade immunotherapy. Cancer Treat. Rev. 41 (10), 868–876.

Meyer, F., Bairati, I., Fortin, A., Gélinas, M., Nabid, A., Brochet, F., Têtu, B., 2008. Interaction between antioxidant vitamin supplementation and cigarette smoking during radiation therapy in relation to long-term effects on recurrence and mortality: a randomized trial among head and neck cancer patients. Int. J. Cancer 122 (7), 1679–1683.

Mokhtari, R.B., Homayouni, T.S., Baluch, N., Morgatskaya, E., Kumar, S., Das, B., Yeger, H., 2017. Combination therapy in combating cancer. Oncotarget 8 (23), 38022.

Moyer, V.A., 2014. Vitamin, mineral, and multivitamin supplements for the primary prevention of cardiovascular disease and cancer: US Preventive Services Task Force recommendation statement. Ann. Intern. Med. 160 (8), 558–564.

Nagpal, M., Sood, S., 2013. Role of curcumin in systemic and oral health: an overview. J. Nat. Sci. Biol. Med. 4 (1), 3.

Nemeth, C., Humpeler, S., Kallay, E., Mesteri, I., Svoboda, M., Rögelsperger, O., Ekmekcioglu, C., 2011. Decreased expression of the melatonin receptor 1 in human colorectal adenocarcinomas. J. Biol. Regul. Homeost. Agents 25 (4), 531–542.

Nguyen, L.T., Lee, Y.-H., Sharma, A.R., Park, J.-B., Jagga, S., Sharma, G., Nam, J.-S., 2017. Quercetin induces apoptosis and cell cycle arrest in triple-negative breast cancer cells through modulation of Foxo3a activity. Korean J. Physiol. Pharmacol. 21 (2), 205–213.

Ono, M., Takeshima, M., Nakano, S., 2015. Mechanism of the anticancer effect of lycopene (tetraterpenoids). In: Bathaie, S. Zahra, Tamanoi, Fuyuhiko (Eds.), The Enzymes 37. Elsevier, UK, pp. 139–166.

Onoda, C., Kuribayashi, K., Nirasawa, S., Tsuji, N., Tanaka, M., Kobayashi, D., Watanabe, N., 2011. (-)-Epigallocatechin-3-gallate induces apoptosis in gastric cancer cell lines by down-regulating survivin expression. Int. J. Oncol. 38 (5), 1403–1408.

Panchuk, R., Skorokhyd, N., Chumak, V., Lehka, L., Omelyanchik, S., Gurinovich, V., Stoika, R., 2014. Specific antioxidant compounds differentially modulate cytotoxic activity of doxorubicin and cisplatin: in vitro and in vivo study. Croat. Med. J. 55 (3), 206–217.

Pantavos, A., Ruiter, R., Feskens, E.F., de Keyser, C.E., Hofman, A., Stricker, B.H., Kiefte-de Jong, J.C., 2015. Total dietary antioxidant capacity, individual antioxidant intake and breast cancer risk: the R otterdam study. Int. J. Cancer 136 (9), 2178–2186.

Patel, R., Ingle, A., Maru, G.B., 2008. Polymeric black tea polyphenols inhibit 1, 2-dimethylhydrazine induced colorectal carcinogenesis by inhibiting cell proliferation via Wnt/β-catenin pathway. Toxicol. Appl. Pharmacol. 227 (1), 136–146.

Peng, G., Dixon, D.A., Muga, S.J., Smith, T.J., Wargovich, M.J., 2006. Green tea polyphenol (−)-epigallocatechin-3-gallate inhibits cyclooxygenase-2 expression in colon carcinogenesis. Mol. Carcinog. 45 (5), 309–319.

Pérez-Cano, F., Castell, M., 2016. Flavonoids, Inflammation and Immune System. Multidisciplinary Digital Publishing Institute, Switzerland.

Prasad, S., Gupta, S.C., Tyagi, A.K., 2017. Reactive oxygen species (ROS) and cancer: role of antioxidative nutraceuticals. Cancer Lett. 387, 95–105.

Qiao, Y.-L., Dawsey, S.M., Kamangar, F., Fan, J.-H., Abnet, C.C., Sun, X.-D., Yu, B., 2009. Total and cancer mortality after supplementation with vitamins and minerals: follow-up of the Linxian General Population Nutrition Intervention Trial. J. Natl. Cancer Inst. 101 (7), 507–518.

Rajasekaran, S.A., 2011. Therapeutic potential of curcumin in gastrointestinal diseases. World J. Gastrointest. Pathophysiol. 2 (1), 1.

Ramírez-Garza, S.L., Laveriano-Santos, E.P., Marhuenda-Muñoz, M., Storniolo, C.E., Tresserra-Rimbau, A., Vallverdú-Queralt, A., Lamuela-Raventós, R.M., 2018. Health effects of resveratrol: results from human intervention trials. Nutrients 10 (12), 1892.

Rauf, A., Imran, M., Khan, I.A., ur-Rehman, M., Gilani, S.A., Mehmood, Z., Mubarak, M.S., 2018. Anticancer potential of quercetin: a comprehensive review. Phytother. Res. 32 (11), 2109–2130.

Reiter, R.J., Tan, D.-x., Manchester, L.C., Simopoulos, A.P., Maldonado, M.D., Flores, L.J., Terron, M.P., 2007. Melatonin in edible plants (phytomelatonin): identification, concentrations, bioavailability and proposed functions. World Rev. Nutr. Diet. 97, 211–230.

Rudra, D.S., Pal, U., Maiti, N.C., Reiter, R.J., Swarnakar, S., 2013. Melatonin inhibits matrix metalloproteinase-9 activity by binding to its active site. J. Pineal Res. 54 (4), 398–405.

Russnes, K.M., Wilson, K.M., Epstein, M.M., Kasperzyk, J.L., Stampfer, M.J., Kenfield, S.A., Willett, W.C., 2014. Total antioxidant intake in relation to prostate cancer incidence in the Health Professionals Follow-Up Study. Int. J. Cancer 134 (5), 1156–1165.

Saeidnia, S., Abdollahi, M., 2013. Antioxidants: friends or foe in prevention or treatment of cancer: the debate of the century. Toxicol. Appl. Pharmacol. 271 (1), 49–63.

Sahin, K., Sahin, N., Kucuk, O., 2010. Lycopene and chemotherapy toxicity. Nutr. Cancer 62 (7), 988–995.

Sarangarajan, R., Meera, S., Rukkumani, R., Sankar, P., Anuradha, G., 2017. Antioxidants: friend or foe? Asian Pac. J. Trop. Med. 10 (12), 1111–1116.

Sayin, V.I., Ibrahim, M.X., Larsson, E., Nilsson, J.A., Lindahl, P., Bergo, M.O., 2014. Antioxidants accelerate lung cancer progression in mice. Sci. Transl. Med. 6 (221), 221ra215.

Schneider, Y., Duranton, B., Goss, F., Schleiffer, R., Seiler, N., Raul, F., 2001. Resveratrol inhibits intestinal tumorigenesis and modulates host-defense-related gene expression in an animal model of human familial adenomatous polyposis. Nutr. Cancer 39 (1), 102–107.

Schwingshackl, L., Hoffmann, G., 2016. Does a Mediterranean-type diet reduce cancer risk? Curr. Nutr. Rep. 5 (1), 9–17.

Serafini, M., Jakszyn, P., Luján-Barroso, L., Agudo, A., Bueno-de-Mesquita, H.B., Van Duijnhoven, F.J., Boeing, H., 2012. Dietary total antioxidant capacity and gastric cancer risk in the European prospective investigation into cancer and nutrition study. Int. J. Cancer 131 (4), E544–E554.

Shankar, S., Ganapathy, S., Srivastava, R.K., 2007. Green tea polyphenols: biology and therapeutic implications in cancer. Front. Biosci. 12 (4881), 99.

Sharma, P., McClees, S., Afaq, F., 2017. Pomegranate for prevention and treatment of cancer: an update. Molecules 22 (1), 177.

Shpitz, B., Giladi, N., Sagiv, E., Lev-Ari, S., Liberman, E., Kazanov, D., Arber, N., 2006. Celecoxib and curcumin additively inhibit the growth of colorectal cancer in a rat model. Digestion 74 (3-4), 140–144.

Singh, M., Kaur, M., Silakari, O., 2014. Flavones: an important scaffold for medicinal chemistry. Eur. J. Med. Chem. 84, 206–239.

Singh, S.K., Banerjee, S., Acosta, E.P., Lillard, J.W., Singh, R., 2017. Resveratrol induces cell cycle arrest and apoptosis with docetaxel in prostate cancer cells via a p53/p21WAF1/CIP1 and p27KIP1 pathway. Oncotarget 8 (10), 17216.

Singh, S.V., Hu, X., Srivastava, S.K., Singh, M., Xia, H., Orchard, J.L., Zaren, H.A., 1998. Mechanism of inhibition of benzo [a] pyrene-induced forestomach cancer in mice by dietary curcumin. Carcinogenesis 19 (8), 1357–1360.

Society, A.C., 2013. Cancer Facts and Figures 2015. American Cancer Society Atlanta, United States.

Srivastava, A.N., Gupta, A., Srivastava, S., Natu, S., Mittal, B., Negi, M., Prasad, R., 2010. Cisplatin combination chemotherapy induces oxidative stress in advance non small cell lung cancer patients. Asian Pac. J. Cancer Prev. 11 (2), 465–471.

Su, C.-C., Chen, G.-W., Lin, J.-G., WU, L.-T., Chung, J.-G., 2006. Curcumin inhibits cell migration of human colon cancer colo 205 cells through the inhibition of nuclear factor kappa B/p65 and down-regulates cyclooxygenase-2 and matrix metalloproteinase-2 expressions. Anticancer Res. 26 (2A), 1281–1288.

Sunil Kumar, B., Singh, S., Verma, R, 2017. Anticancer potential of dietary vitamin D and ascorbic acid: a review. Crit. Rev. Food Sci. Nutr. 57 (12), 2623–2635.

Tang, F.Y., Pai, M.H., Kuo, Y.H., Wang, X.D., 2012. Concomitant consumption of lycopene and fish oil inhibits tumor growth and progression in a mouse xenograft model of colon cancer. Mol. Nutr. Food Res. 56 (10), 1520–1531.

Thilakarathna, S., Rupasinghe, H., 2013. Flavonoid bioavailability and attempts for bioavailability enhancement. Nutrients 5 (9), 3367–3387.

Tripathi, S.K., Pandey, K., Panda, M., Spinella, M.J., Rengasamy, K., Biswal, B.K., 2019. The potential of retinoids for combination therapy of lung cancer: updates and future directions. Pharmacol. Res. 147, 104331. doi:10.1016/j.phrs.2019.104331.

Valko, M., Izakovic, M., Mazur, M., Rhodes, C.J., Telser, J., 2004. Role of oxygen radicals in DNA damage and cancer incidence. Mol. Cell. Biochem. 266 (1-2), 37–56.

Wang, C., Zhang, X., Teng, Z., Zhang, T., Li, Y., 2014. Downregulation of PI3K/Akt/mTOR signaling pathway in curcumin-induced autophagy in APP/PS1 double transgenic mice. Eur. J. Pharmacol. 740, 312–320.

Wang, X.-D., Russell, R.M., 1999. Procarcinogenic and anticarcinogenic effects of β-carotene. Nutr. Rev. 57 (9), 263–272.

Weng, C.-J., Wu, C.-F., Huang, H.-W., Wu, C.-H., Ho, C.-T., Yen, G.-C., 2010. Evaluation of anti-invasion effect of resveratrol and related methoxy analogues on human hepatocarcinoma cells. J. Agric. Food Chem. 58 (5), 2886–2894.

Xu, L., Liu, H., Zhang, H., Wang, R.X., Song, J., Zhou, R.X., 2013. Growth-inhibitory activity of melatonin on murine foregastric carcinoma cells in vitro and the underlying molecular mechanism. Anat. Rec. 296 (6), 914–920.

Yang, G., Shu, X.-O., Li, H., Chow, W.-H., Ji, B.-T., Zhang, X., Zheng, W., 2007. Prospective cohort study of green tea consumption and colorectal cancer risk in women. Cancer Epidemiol. Biomarkers Prev. 16 (6), 1219–1223.

Yang, G., Zheng, W., Xiang, Y.-B., Gao, J., Li, H.-L., Zhang, X., Shu, X.-O., 2011. Green tea consumption and colorectal cancer risk: a report from the Shanghai Men's Health Study. Carcinogenesis 32 (11), 1684–1688.

Yasueda, A., Urushima, H., Ito, T., 2016. Efficacy and interaction of antioxidant supplements as adjuvant therapy in cancer treatment: a systematic review. Integr. Cancer Ther. 15 (1), 17–39.

Yeung, A.W.K., Tzvetkov, N.T., El-Tawil, O.S., Bungău, S.G., Abdel-Daim, M.M., Atanasov, A.G., 2019. Antioxidants: scientific literature landscape analysis. Oxidative Med. Cell. Longevity 2019, 8278454–8278454. doi:10.1155/2019/8278454.

Yildiz, I., Sen, O., Erenler, R., Demirtas, I., Behcet, L., 2017. Bioactivity-guided isolation of flavonoids from *Cynanchum acutum* L. subsp. sibiricum (willd.) Rech. f. and investigation of their antiproliferative activity. Nat. Prod. Res. 31 (22), 2629–2633.

Zamora-Ros, R., Fedirko, V., Trichopoulou, A., González, C.A., Bamia, C., Trepo, E., Bredsdorff, L., 2013. Dietary flavonoid, lignan and antioxidant capacity and risk of hepatocellular carcinoma in the European prospective investigation into cancer and nutrition study. Int. J. Cancer 133 (10), 2429–2443.

Zhang, T., Qi, Y., Liao, M., Xu, M., Bower, K.A., Frank, J.A., Chen, G., 2012. Autophagy is a cell self-protective mechanism against arsenic-induced cell transformation. Toxicol. Sci. 130 (2), 298–308.

Zhao, J., Liu, J., Wei, T., Ma, X., Cheng, Q., Huo, S., Liang, X.-J., 2016. Quercetin-loaded nanomicelles to circumvent human castration-resistant prostate cancer in vitro and in vivo. Nanoscale 8 (9), 5126–5138.

Zhao, X., Zhang, J., 2015. Mechanisms for quercetin in prevention of lung cancer cell growth and metastasis. Zhong nan da xue xue bao. Yi xue ban= J. Central South Univ. Med. Sci. 40 (6), 592–597.

Zhao, Y., Fan, D., Zheng, Z.P., Li, E.T., Chen, F., Cheng, K.W., Wang, M., 2017. 8-C-(E-phenylethenyl) quercetin from onion/beef soup induces autophagic cell death in colon cancer cells through ERK activation. Mol. Nutr. Food Res. 61 (2), 1600437.

Zhou, J., Gong, J., Ding, C., Chen, G., 2015. Quercetin induces the apoptosis of human ovarian carcinoma cells by upregulating the expression of microRNA-145. Mol. Med. Rep. 12 (2), 3127–3131.

Zhu, X., Zeng, X., Zhang, X., Cao, W., Wang, Y., Chen, H., Chang, D., 2016. The effects of quercetin-loaded PLGA-TPGS nanoparticles on ultraviolet B-induced skin damages in vivo. Nanomed. Nanotechnol. Biol. Med. 12 (3), 623–632.

CHAPTER

Antioxidants and cardiovascular diseases 5.2

Ankita Mandal*, Uday Hossain*, Parames C. Sil
Division of Molecular Medicine, Bose Institute, Kolkata, India
**Both the authors contributed equally to the work.*

5.2.1 Introduction

Cardiovascular diseases (CVD) are multifactorial abnormalities including atherosclerosis, coronary heart disease, cardiac ischemia, hypertension, cardiomyopathies, cardiac hypertrophy, and heart failure. Despite of different therapeutic approaches to combat CVD and its related disorders, it accounts for almost quarter of all deaths worldwide and according to World Health Organization (WHO), CVDs are the leading cause of mortality globally (Mendis et al., 2011). Various cardiovascular diseases cause deaths of around 17.6 million people per year worldwide, which is predicted to be 25 million (approx.) by 2020 (Dahlöf, 2010). The predominant cardiac disorders namely cardiac ischemia and strokes accounts >80% of all death caused due to CVD. According to a survey conducted by the Global Burden of Disease (2010), about 24.8% of all deaths in India are caused due to CVD (Kassebaum et al., 2017). Moreover, the age standard death due to CVD is 272 people per 100,000 populations in India, which is higher than average death worldwide (i.e., 235 people per 100,000 populations) (Prabhakaran et al., 2016).

On the basis of existing evidences and ongoing investigations, it is becoming clear that oxidative stress is the leading cause of different cardiovascular diseases. Excessive production of reactive oxygen species (ROS) occurs in various conditions such as diabetes, obesity, metabolic diseases, excessive smoking, etc. and can lead to oxidative stress-induced cardiovascular abnormalities (Katagiri et al., 2007). Sources of ROS include mitochondrial electron transport chain, nicotinamide-adenine dinucleotide phosphate (NADPH) oxidase, lipoxygenases, xanthine oxidase, uncoupling of endothelial NO synthase (eNOS), etc. (Lakshmi et al., 2009). Oxidation of low-density lipoprotein (ox-LDL), overload, and desensitization of Ca^{2+} and uncoupling of eNOS are the major deleterious effects of oxidative stress that results in endothelium dysfunction, atherosclerosis, hypertension, ischemic heart disorders, and subsequently other cardiovascular abnormalities.

In regard of the role of oxidative stress in the onset and progression of various cardiovascular diseases, antioxidants are the most efficient and beneficial therapeutic agents for amelioration of cardiovascular disorders. For last two decades, association

Antioxidants Effects in Health. DOI: https://doi.org/10.1016/B978-0-12-819096-8.00072-0

of CVD and antioxidant has been intensively investigated in cell culture, animal models as well as in clinical studies. In this book chapter, we focused on the most important natural or dietary antioxidants, namely, vitamins and polyphenolic compounds that have been demonstrated to significantly inhibit oxidative stress-mediated cellular damages which results in cardiovascular disorders. The mechanism of actions of these antioxidants in relation to its role in CVD has also been described with in vitro, in vivo, and clinical trials evidences.

5.2.2 Oxidative stress and its role in cardiovascular disease: a brief idea

Oxidative stress can be broadly defined as a condition in which the balance between ROS and antioxidants in the cell is disturbed (Sinha et al., 2013). Immoderate generation of ROS, exceeding the neutralizing capacity of antioxidant defense system, has been suggested to be the cause of oxidation of biological macromolecules, such as lipids, proteins, DNA, and carbohydrates. Currently, these circumstances are regarded as oxidative stress (Cai and Harrison, 2000). The main source of ROS generation is mitochondrial electron transport chain and in addition to that various enzymes including, (eNOS), cyclooxygenase, (NADPH) oxidase, lipooxygenase, glucose oxidase, xanthine oxidase, lypoxidases (Dahlöf), and myeloperoxidases (MPO) leads to the production of reactive oxygen species (Higashi et al., 2009; Cervantes Gracia et al., 2017). ROS can be of two classes: free radicals, such as superoxide anion (O_2^-), hydrogen peroxide (H_2O_2), hydroxyl radical (Podrez et al., 2000), nitric oxide (NO), and peroxynitrite ($ONOO^-$); and nonradical derivatives, such as hydrogen peroxide (H_2O_2), hypochlorous acid (HOCl), peroxynitrite ($ONOO^-$), and singlet oxygen ($^1O_2$) (Sinha et al., 2013; Higashi et al., 2009; Dhalla et al., 2000). Unpaired electrons of free radicals make them more reactive than nonradicals but can oxidize biomacromolecules and to the state of cellular oxidative stress (Fig. 5.2.1).

Collective evidences suggested that acute and chronic over production of ROS is essential in the onset of cardiovascular diseases including ischemic heart disease, atherosclerosis, cardiomyopathies, congestive heart failure, hypertension, and cardiac hypertrophy. Complications related to atherosclerosis are the leading causes of most of the cardiovascular disease. Atherosclerosis is characterized by the accumulation of lipids, foam cell formation, hardening, and narrowing of arteries (Cervantes Gracia et al., 2017; Raggi, 2016). Oxidation of low-density lipoprotein (LDL) by reactive oxidation species leads to the initiation of ROS-induced atherosclerosis (Ross, 1999). Oxidized LDL (Ox-LDL) induces vascular cell adhesion molecule (VCAM) expression on endothelial cells resulting in recruitment of monocytes into the intima, which matures and proliferates to form macrophages (Zwaka et al., 2001). These macrophages engulf ox-LDL with the help of CD36 (ox-LDL receptor) and transforms into foam cells (Podrez et al., 2000). Chemokines secreted by endothelial cells and macrophages recruits T cells, which in turn secrete proinflammatory cytokines (Tumor necrosis factor-α, TNF-α and Interferon-γ, IFN-γ) and leads to amplification of inflammation in blood vessels (Pirillo et al., 2013). In addition, fibroblast growth

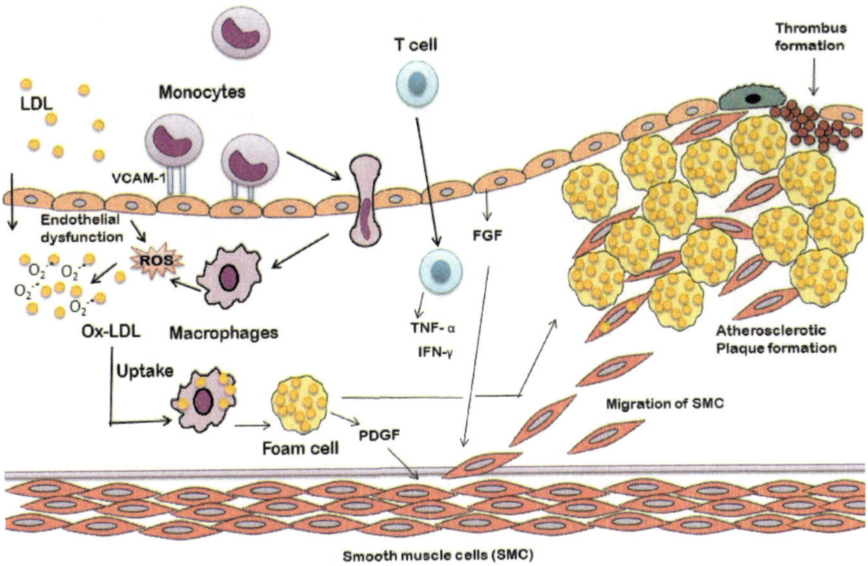

FIG. 5.2.1 Role of reactive oxygen species and immune cells in atherosclerotic plaque formation.

factor (FGF) and platelet-derived growth factor (PDGF) secreted from endothelium and foam cells respectively stimulates migration and proliferation of smooth muscle cells, which can also contribute in oxidized lipid accumulation (Raggi, 2016). All together, these events result in the formation atherosclerotic plaque characterized by lump of necrotic foam cells surrounded by smooth muscle cells and a fibrous cap of collagen. The adverse consequence of these events is the reduction of blood supply to the heart, leading to myocardial ischemia and hypertension (Dhalla et al., 1999).

Endothelium dysfunction (Dhalla et al., 1999) is referred to as an early clinical manifestation of atherosclerosis and is the indicator of future cardiovascular pathological events. Superoxide anion (O_2^{*-}) leads to oxidative degradation of tetrahydrobiopterin (BH_4), an essential cofactor of NO synthase, resulting in uncoupling of eNOS (endothelial NO synthase). As a result superoxide anion (O_2^{*-}) is generated instead of NO, leading to reduction of NO bioavailability (Kelly and Fussell, 2017). In addition, O_2^{*-} reacts with NO and generates peroxynitrite, leading to the NO loss (Lakshmi et al., 2009). Abnormal endothelial function due to minimal NO availability results in a series of disorders including accumulation of platelets, fibrinolytic imbalance, thrombogenesis, and atheroma formation (Anderson et al., 1995). As a consequence, endothelium-dependent vasodilation is prohibited and ultimately leads to various cardiovascular disorders like hypertension.

Ample amount of ROS generation is associated with myocardial ischemia/reperfusion. ROS can directly impose myocardial cell injury or it effect myocardium via inflammatory consequences. The critical phenomenon related to ROS-induced myocardial injury is protein and lipid peroxidation in myocardial cells, which can cause damage to sarcolemma and contractile proteins (such as troponin, myosin, and α-tropomyosin). As a result of these, overload and desensitization of Ca^{2+} occur in the region of ischemia/reperfusion of cardiac myocytes, ultimately lead to reduced contraction ability of myocardium and heart failure (Lefer and Granger, 2000).

5.2.3 Antioxidants and cardiovascular diseases

Antioxidants are the substances that possess ROS neutralizing properties and can prevent or slow down the deleterious effect of oxidative stress in various pathophysiological conditions. Antioxidant defense system of human body comprises proteins, enzymatic, and nonenzymatic antioxidants that continuously combat with ROS and neutralize them before they become deleterious to health. The major antioxidant enzymes that act as intracellular armor are superoxide dismutase (SOD), catalase, and glutathione peroxidase (GPx) that collectively reduce hydrogen peroxide/superoxide to H_2O (Goszcz et al., 2015; Nojiri et al., 2004). Nonenzymatic antioxidants include vitamins (A, C, and E), uric acid, bilirubin, ubiquinol-10, carotenoids, α-tocopherol as well as polyphenolic flavonoids compounds and minerals (Nojiri et al., 2004; Halliwell, 2007a). Other antioxidants are proteins such as transferrin, ceruloplasmin, hemopexin, haptoglobin, and albumin (Halliwell, 1988). In addition, some therapeutic agents like quercetin, resveratrol, N-acetylcysteineare, epicatechin, and allopurinol are under clinical trials for their antioxidant effects in relation to CVD (Goszcz et al., 2015). If the intrinsic or endogenous antioxidant defense system fails to conquer the over produced ROS in pathophysiological disorders such as diabetes, cardiovascular disease, etc. than dietary and therapeutic antioxidant supplements are provided to get rid of the deleterious effects of oxidative stress. Collective evidences suggested that oxidation of LDL, uncoupled endothelial NO synthase, and Ca^{2+} overload resulting due to oxidative stress are the major cause for various CVD, so antioxidants can be an inexpensive of preventing onset or progression of cardiovascular diseases (Table 5.2.1). Some of the major dietary and naturally occurring antioxidants in relation to CVD are discussed in this chapter. In addition, the roles of some other antioxidants that have gained attention for its antioxidative role in cardiovascular disorders are included in Table 5.2.2.

Table 5.2.1 Endogenous antioxidants and their ROS neutralizing reactions.

Endogenous antioxidants	ROS neutralizing reactions
Superoxide dismutase (SOD)	$O_2^{\cdot -} + O_2^{\cdot -} + 2H^+ \xrightarrow{\text{SOD}} H_2O_2$
Catalase	$2H_2O_2 \xrightarrow{\text{Catalase}} 2H_2O + O_2$
Glutathione (GSH)	$H_2O_2 + 2GSH \xrightarrow{\text{GPx}} GSSG + 2H_2O$ $GSSG + NADPH + H^+ \xrightarrow{\text{GSH Reductase}} 2GSH + NADP^+$

Table 5.2.2 Dietary antioxidants and its role in cardiovascular diseases.

Compounds	Sources	Type	Antioxidative role in cardiovascular diseases			References
			In vitro studies	Animal model studies	Clinical trials	
Quercetin	Tea, red wine fruits, and vegetables	Polyphenol (flavonol)	• Scavenges per-oxynitrite – ↓ ox-LDL • Inhibits xanthine oxidase • ↑ NO bioavailability	• Reduce hypertension in rats • Alters gene expression related to oxidative stress	• Reduce systolic BP in patients with hypertension • ↑ HDL and ↓ ox-LDL	Han et al. (2009), Duarte et al. (2001), Haenen et al. (1997), Egert et al. (2009)
Coenzyme Q10	Meats, eggs, fruits, vegetable, cereals. Nuts, pulses, and dairy products	Quinone	• Prevents protein and DNA • Prevents lipid peroxidation • Reduce ROS generation	• Functional recovery after cardiac arrest • Prevents lipid peroxidation in myocardial cell wall • Functional recovery after reperfusion	• ↓ Blood pressure • Prevents myocardial infarction • Prevents congestive heart failure	Pravst et al. (2010), Yamamura et al. (1967), Singh et al. (1998), Turunen et al. (2004)
Selenium	Brazil nuts, fish, pork, beef, cheese, eggs, mushroom, etc.	Mineral	• Enhance antioxidant capacity • Inhibits NF-κB and ROS generation	• Preservation of glutathione (GSH) • Prevents ischemia-reperfusion(I/R) injury	• Reverse coronary atherosclerosis	Baljinnyam et al. (2006), Benstoem et al. (2015)
Naringenin	Grapefruit, orange, and tomatoes	Polyphenol (flavanones)	• Scavenges superoxide radicals • Inhibits xanthine oxidase • Inhibits oxidation of LDL	• Enhance activity of SOD, GPx, and catalase • Prevents lipid peroxidation	• Reduced risk of ischemic stroke • Prevented cardiomyopathy	Cavia-Saiz et al. (2010), Russo et al. (2000), Jeon et al. (2002), Gorinstein et al. (2006)

5.2.3.1 Vitamins and CVD

SOD, GPx, and catalase are some of the antioxidative enzymes present in the body and provide defense against free radicals. Some vitamins, like β-carotene, vitamin C, and vitamin D, cooperate or sometimes work synergistically to protect the body from oxidative stress. In atherosclerosis, modified LDL accumulates in the intima of the arterial wall and induces plaque progression, which results into different cardiovascular events (Finking and Hanke, 1997; Witztum and Steinberg, 1991).

5.2.3.2 Vitamin E

A group of fat-soluble compounds are collectively termed as vitamin E. These compounds show significant antioxidant properties which is beneficial for our health (Niki and Traber, 2012). The fat-soluble property of the vitamin helps them to get stored within the fatty tissues of humans and other animals. The vitamin E group is divided into two groups, tocotrienols and tocopherols, together termed as tocochromanols. Each group shows four forms, alpha(α), beta(β), gamma(ϒ), and delta (δ) (Traber, 2007). Edible vegetable oil is a rich source of vitamin E. Various seed, nuts, green vegetables contain significant amount of alpha-tocopherol.

It has already been established that atherosclerosis is a chronic inflammatory disease (Diaz et al., 1997) and vitamin E has observed to mediate antioxidant as well as anti-inflammatory properties (Devaraj et al., 1996), which extends its protective role against CVD. Vitamin E decreases monocyte recruitment by reducing foam cell formation(Devaraj et al., 1996; Wu et al., 1999), chemokine secretion(Munteanu et al., 2006), and expression of scavenger receptors on macrophages (CD36; Zapolska-Downar et al., 2000). Vitamin E has the ability to reduce expression of adhesion molecule (Wu et al., 1999; Keaney Jr. et al., 1999), inhibit proliferation of smooth muscle cell (Keaney et al., 1993), and enhance bioavailability of NO (Murohara et al., 2002), which lowers the progression of atherosclerosis. These above-mentioned effects are partially mediated by nonantioxidant properties of vitamin E. It also inhibits the signaling pathway, mainly protein kinase C, activated by oxidized LDL (Sugiyama et al., 1998). It has been observed that vitamin E prevents activation of NF-κB, induced by oxidized LDL, via suppression of protein kinase C (Li et al., 2000) and inhibits IkB degradation (Goya et al., 2006), which further reduces the CVD-mediated inflammatory response. Vitamin E exerts its anti-atherogenic property by controlling the gene expression. Expression of endothelial NO synthase mRNA is upregulated by vitamin E, therefore NO level is also elevated (Keaney et al., 1996) and protects the endothelium. Vitamin E protects the endothelium against oxidized LDL and reactive oxygen species (Ulrich-Merzenich et al., 2002). It stimulates endothelial cell proliferation and reduces endothelial apoptosis (Kuzuya et al., 1991; Haendeler et al., 1996). These help in prevention of endothelial dysfunction. These effects are possibly facilitated via modulation of apoptosis-related proteins of Bcl-2 family (Uemura et al., 2002), caspase-3 inhibition, phosphatase PP2A activation, and inhibition of mRNA and protein upregulation of angiotensin II receptor (AT1R), which is induced by oxidized LDL (Azzi et al., 1998; Koga et al., 2004; Jialal and Grundy, 1991).

5.2.3.2.1 *In vitro and animal studies*

Different experiments on animals, in vitro and in vivo, lead us toward this option of using vitamin E, and especially α-tocopherol as a protective against cardiovascular disease. It is known that, inflammation, oxidative stress and endothelial dysfunction play key role in the development of atherosclerotic plaque. In some studies, it has been observed that α-tocopherol can prevent above-mentioned processes. In different experiments it is seen that α-tocopherol can reduce endothelium to monocyte adhesion in primary human monocytes (Devaraj et al., 1996) and cultured monocytes. It has been discovered that vitamin E can prevent expression of different adhesion molecules, like intercellular adhesion molecule (ICAM-1) and VCAM-1, which is induced by oxidized LDL in cultured endothelial cell (Cominacini et al., 1997). Alpha-tocopherol can also prevent proliferation of vascular smooth muscle cell by inhibiting activity of protein kinase C (Tasinato et al., 1995). But the protective roles of vitamin E in *in vivo* animal models are not well understood properly. α-tocopherol can remarkably reduce circulating C reactive protein (CRP), which is a marker of inflammation, linked with atherosclerosis (Patrick and Uzick, 2001). On the other hand, if we look from antioxidant viewpoint, peroxidation of LDL can be inhibited by supplements of vitamin E, which ultimately slows down the atherosclerotic plaque formation (Reaven et al., 1993). α-tocopherol can also scavenge ROS which results into prevention of peroxidation of LDL (Sagach et al., 2002). In an experiment, diabetic mouse model, which is prone to develop atherosclerosis, is treated with vitamin E supplements. Those animals were observed to show lower macrophage activation and intravascular fat deposition (Otero et al., 2005). But on the contrary, in many studies, vitamin E has failed to reduce occurrence of atherosclerosis.

5.2.3.2.2 *Clinical studies and trials*

The Heart Outcomes Prevention Evaluation Study (HOPE) and the Cambridge Heart Antioxidant Study (CHAOS) are the two major clinical trials conducted on vitamin E. In CHAOS trial, α-tocopherol has succeeded in lowering the occurrence of nonfatal myocardial infarction in coronary artery disease patients, but failed to reduce the mortality rate in cardiovascular disease (Stephens et al., 1996). But the results of the HOPE trial have been considered as failure (Yusuf et al., 2000). It is still not fully understood, why one trial failed and another one succeeded. Despite the unsatisfactory results, scientists are still planning to conduct further clinical studies on this. US Preventative Task Force still has not recommended not using vitamin E supplementation for CVD primary prevention.

5.2.3.3 Vitamin C

Vitamin C or ascorbic acid is a simple carbohydrate with low-molecular weight with an ene-diol structure. Humans lack L-glucono-ϒ-lactone oxidase enzyme, which is needed to synthesize vitamin C. For this reason, it is not synthesized in human body.

Because of this, dietary ascorbate is essential. Citrus fruits and leafy vegetables are rich source of vitamin C.

The independent protective role of vitamin C has not been widely studied in clinical trials. Vitamin C can perform synergistically with other vitamins, which might enhance their beneficial role (Myasnikova, 1947). Vitamin C also has some nonantioxidant properties. Administration of vitamin C has been observed to lower the cholesterol level in the hypercholesterolemic patients (Rössig et al., 2001). Vitamin C suppresses apoptosis of endothelium which is mediated by oxidized LDL and inflammatory cytokines (Rayment et al., 2003). It has been observed that vitamin C promotes endothelial cell proliferation and inhibits growth of vascular smooth muscle (Kuzuya et al., 1991) via the extracellular signal-regulated kinase-signaling pathway. Vitamin C is also able to modulate gene expression. It reduces monocyte adherence to endothelium by downregulating the expression of intercellular adhesion molecule-1 gene (Heller et al., 1999). Studies have shown that vitamin C elevates the synthesis of NO in endothelial cells (Gokce et al., 1999). An in vivo study has shown that vitamin C exerts sustained beneficial effects on endothelial-derived NO-dependent flow-mediated dilation (Siow et al., 1999). Apoptosis of vascular smooth muscle cell can be reduced by vitamin C supplementation, which prevents plaque instability seen in late-stage atherosclerosis. Vitamin C gives protection oxidative changes in LDL induced by different types of oxidative stress including metal-induced oxidative stress (Lehr et al., 1994).

5.2.3.3.1 ***In vivo studies***

In one study, Lehr et al. induced atherosclerosis in Syrian Golden hamsters by cigarette smoke. They have observed that vitamin C inhibits the inflammatory response by stopping leucocyte adhesion and aggregation to the endothelium (Padayatty et al., 2003). Different studies have shown that vitamin C has the ability to stop peroxidation of lipid by preventing LDL oxidation and subsequent ox-LDL uptake (Frei, 1991; Maeda et al., 2000). Maeda et al. have shown in their study that L-glucono-ϒ-lactone oxidase enzyme lacking mice shows extensive vascular damage, which includes disruption of elastic lamina, proliferation of smooth muscle cell, epithelial cell desquamation, when their dietary vitamin C had been removed. This indicates the importance of vitamin C in vascular function and development (Nakata and Maeda, 2002).

5.2.3.3.2 ***Clinical trials***

As vitamin C is synthesized in huge number of animal species, animal studies of this vitamin is difficult. Few studies have been held to discover the role of vitamin C in vascular health-related factors. It has been observed that proper dose of vitamin C (3 g) helped to improve endothelium-dependent vasodilation of epicardial coronary artery in hypertension patients (Solzbach et al., 1997). Vitamin C has major effect on endothelial dysfunction. Patients with NO-mediated or hyperglycemia-induced vasodilation showed good result with intra-arterial infusion (10 min of 24 mg/min) or oral dose (6 g over 2 days) administration of vitamin C (Levine et al., 1996).

Reduction of arterial stiffness and aggregation platelet in healthy person by oral administration of vitamin C (2 g) has been observed, though the exact mechanism behind this is not known properly (Wilkinson et al., 1999). A study by the European Prospective Investigation into Cancer and Nutrition (EPIC) has shown that vitamin C concentration in plasma has inverse relationship with mortality due to cardiovascular disease (Sargeant et al., 2001).

5.2.3.4 β-Carotene and vitamin A

The carotenoids are brightly colored pigments (red, yellow, and orange) found largely in vegetables and fruits. This compound is also lipid soluble. Synthesis of retinol or vitamin A in gut can occur from β-carotene, before as well as after absorption (Norum and Blomhoff, 1992). Lycopene is the precursor of β-carotene. It is an efficient scavenger of singlet oxygen and a second-line antioxidative defense for LDL compound after vitamin E is utilized (Jessup et al., 1990). Carotenoids as an antioxidant are pretty unpredictable. Studies show that it can act as neutral, pro-, and antioxidant depending on the situation (Princen et al., 1992; Tsuchihashi et al., 1995). Dietary consumption of β-carotene in high amount is linked with lowered mortality in CVD as well as in all-cause (Buijsse et al., 2005), but this effect is limited to elderly persons. Whether β-carotene can prevent LDL oxidation or not, has not been explored yet. A number of studies have been performed to examine LDL oxidation inhibition by β-carotene, but the results are inconclusive.

5.2.3.4.1 In vitro studies

In vitro studies show conflicting results regarding the protective role of β-carotene in CVD. In one study, β-carotene prevents peroxidation of LDL which is mediated by Cu^{2+} (Romanchik et al., 1995). But in another study of similar kind, β-carotene shows no beneficial effect upon peroxidation of lipid (Dugas et al., 1998). 3-hydroxy-3-methyl-glutaryl-CoA (HMG-CoA) reductase is a rate-limiting enzyme helps in synthesis of cholesterol. In vitro studies show that β-carotene can reduce plasma cholesterol level by inhibiting HMG-CoA reductase. It has been studied that β-carotene and fluvastatin reduce HMG-CoA activity to same extent, which suggests that carotenoids can be used as therapeutic drug. Activity of LDL receptor of macrophage is increased with the help of carotenoids which reduces circulating LDL. From this it might be concluded that carotenoids have protective effect on CVD (Rao, 2002).

5.2.3.4.2 Clinical studies and trials

A protective property of carotenoids against CVD has been proven in clinical studies. In a report of Physicians Health Study, it has been observed that people consuming vegetables containing high carotenoids are less prevalent to heart disease (Liu et al., 2001). Some current studies like the young adult longitudinal trends in antioxidants (YALTA) and coronary artery risk development in young adults (CARDIA) have shown that high concentrations of plasma carotenoid are related with lowered inflammation, endothelial dysfunction, and oxidative stress, three significant features of atherosclerosis (Hozawa et al., 2007). On the other hand, reduced occurrence of atherosclerosis has been observed in Bruneck study with higher plasma β-carotene as well as α-carotene (D'Odorico et al., 2000). Endothelial dysfunction markers, like ICAM-1, as well as inflammation (CRP) are reduced due to more vegetables and fruit consumption containing high carotenoids (Watzl et al., 2005). On the contrary, in the β-carotene and retinol efficacy trial (CARET), inverse relation has been noticed between risk cardiovascular disease and combinatorial supplementation of retinol and β-carotene (Omenn et al., 1996). Strong correlation has been observed between lowered risk of MI and serum β-carotene (Karppi et al., 2011). Similarly, vitamin E has found to be more effective than β-carotene in reducing the oxidation of LDL (Reaven et al., 1993). On the other hand, lycopene, present tomatoes in large amount, the precursor of β-carotene is receiving increasing interest as an antioxidant. EURAMIC study has shown that high concentration of adipose lycopene can correlated with lowered mortality in CVD (Kohlmeier et al., 1997). Epidemiological studies have revealed that supplement- and diet-derived lycopene lowers oxidative stress and inflammation in healthy as well as overweight individuals, which results into reduced CVD occurrence (O'Kennedy et al., 2006).

5.2.4 Polyphenols

Polyphenol is a group of compounds that consists of two or more phenol groups and can be categories into flavonoids, stilbenes, phenolics acids, and lignans (Halliwell, 2007b; Manach et al., 2004). The relative position and number of OH and catechol groups in chemical structures are responsible for their antioxidant property (Quideau et al., 2011). The mechanism of action involves direct interaction with reactive oxygen species, metal ions chelating, upregulation of endogenous antioxidants, etc. (Goszcz et al., 2015). Herein we discussed about the major polyphenols that are investigated in animal studies, in vitro and clinical trials and their role in CVD are mostly understood.

5.2.4.1 Anthocyanins

Anthocyanins are group of water-soluble polyphenol flavonoid constituents that occurs abundantly in plant realm forming red–orange to blue–violet pigments of vegetables, fruits, grains, flowers, and other edible parts of plants (Wallace, 2011). Anthocyanin possess various pharmacological properties owing to their chemical structures, most importantly anti-inflammatory and antioxidant activity (Castaneda-Ovando et al., 2009). Collective evidences from several epidemiological studies suggested that regular intake of flavonoid-rich food materials is associated with decrease chance of CVD onset and progression as well as reduced mortality due to cardiac disorders. Regular intake of anthocyanin is estimated to be nine folds higher (approx. 180–215 mg/d) than other dietary flavonoids (approx. 20–25 mg/d) (Hertog et al., 1993). The mechanism of action of anthocyanin includes inhibition of oxidative stress via free radical scavenging, endothelial dysfunction, and inflammation, as well

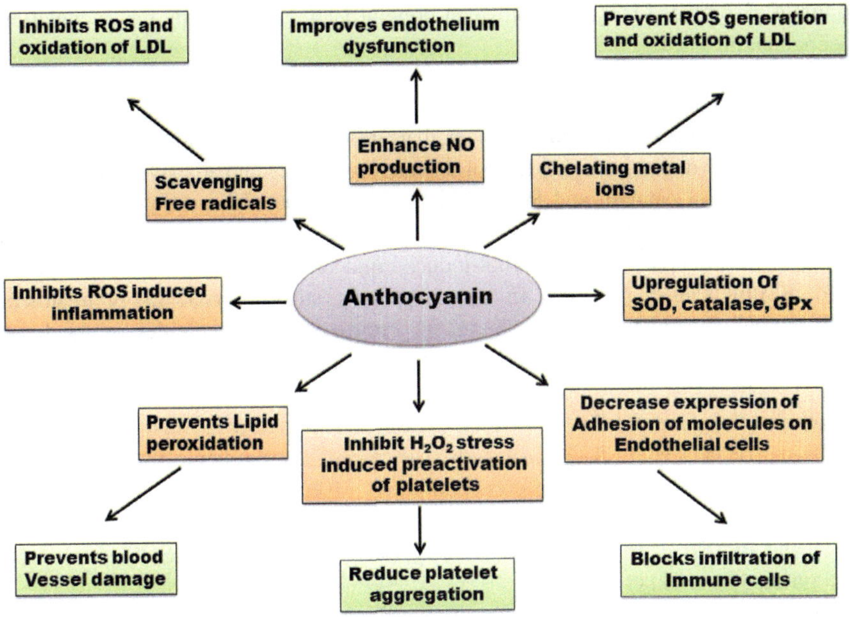

FIG. 5.2.2 Antioxidative role of anthocyanins and its beneficiary effects in cardiovascular diseases.

as metal ions chelating. The antioxidant role of anthocyanin regarding CVD has been demonstrated in various in vitro, in vivo (animal models) as well as clinical trials and human epidemiological studies (Fig. 5.2.2).

5.2.4.1.1 *In vitro studies*

Youdim et al. (2002) isolated four types of anthocyanins from elder berries and incorporated them into plasma, cytoplasm, and lemma of endothelial cells in vitro and examine their role directly. The result of their study indicated that anthocyanins can protect endothelial cells from oxidative stress significantly via direct incorporation into them (Youdim et al., 2002). Anthocyanins and proanthocyanidins, found in red–black rice grains and grape seeds, can scavenge hydroxyl radical and superoxide radicals *in vitro* (Da Silva et al., 1991; Walter and Marchesan, 2011). Delphinidin, a type of anthocyanin, prevents apoptosis of endothelial cells *in vitro* in presence of NOS inhibitors, via regulation of NO-signaling pathways and maintaining calcium ion homeostasis (Martin et al., 2003). In another study, it was demonstrated that anthocyanin prevents thrombosis or atherosclerotic plaque formation by inhibiting platelet activation via inhibition of P-selectin expression and thrombin receptor activating peptides (TRAP) and H_2O_2 stress-induced preactivation of platelets (Rechner and Kroner, 2005). Moreover, anthocyanin from purple grape extract can prevent activation of platelets in both in vitro and in vivo, by reducing ROS generation and enhancement of NO production (Goszcz et al., 2015; Freedman et al., 2001).

Viana et al. reported that flavonoids can prevent LDL oxidation by metal chelating mechanism in vitro.

5.2.4.1.2 In vivo studies

There are evidences supporting the fact that anthocyanins can increase antioxidant capacity *in vivo* due to its higher radical scavenging activity and adaptive improvement in the activity of catalase and SOD (Chiang et al., 2006). In a study conducted by Karthikeyan et al. (2007), significant improvement of SOD, catalase, GSH, and GPx activity was observed in isoproterenol-induced injected rat model (Karthikeyan et al., 2007; Karthikeyan et al., 2009). In this study, it was also observed that pretreatment with proanthocyanidins leads to significant improvement of activities of the heart mitochondrial enzymes involved in electron transport system. Rodriguez et al. (2013) showed that supplementation of blueberries containing anthocyanin improves endothelial function and diminish blood pressure in rats fed with high-fat diet(HFD; Rodriguez-Mateos et al., 2013). In a study with hypercholesterolemic rabbits, Sozanski et al. demonstrated that a diet of cornelian cherry (100 mg/kg) for 60 days decreases triglycerides, LDL levels, and atherogenic index in blood. It was also evaluated that GSH and PPARα expression levels were elevated while GPx and SOD showed no differences (Sozański et al., 2014).

5.2.4.1.3 Clinical trials

In a study, Mazza et al. found that when adult male human subjects were supplemented with anthocyanin (1.2 g), the serum antioxidant capacity is directly proportional to the serum anthocyanins concentration (Mazza et al., 2002). This event suggests that anthocyanins may play a crucial role in reducing superoxide production by NADPH oxidase because diminishing activity NADPH oxidase may cause increase in serum antioxidant potential (Wallace, 2011). In a study conducted by Alvarez et al. (2014), they showed that a diet of fresh fruits (strawberry) containing anthocyanin for 30 days can lower cholesterol, LDL, and triglycerides level and partially prevent the pathogenesis of CVD (Alvarez-Suarez et al., 2014). Qin et al. (2009) studied the role of 17 purified anthocyanins from blackcurrant and blueberry, in ameliorating CVD abnormalities in 120 dyslipidemic subjects and they observed significant decrease in LDL and (CETP) cholesteryl ester protein and increase in HDL concentration in blood (Qin et al., 2009). In another study, 66 patients with obesity and metabolic syndrome were suggested to have blueberries (containing anthocyanin) in their diet for 8 weeks and after evaluation at the end of this period, it was found that serum level of ox-LDL and MDA were significantly reduced with almost normal blood pressure (Basu et al., 2010). This result suggests a close association of blueberries and improvement of CVD.

5.2.4.2 Catechins

Catechins are the major group of flavanols found in green tea and epidemiological, *in vitro* and *in vivo* studies postulated a strong association between cardiovascular conditions and green tea consumption. The most active biologically active catechin in green tea is epigallocatechin-3-gallate (EGCG) and others catechins are gallocatechin

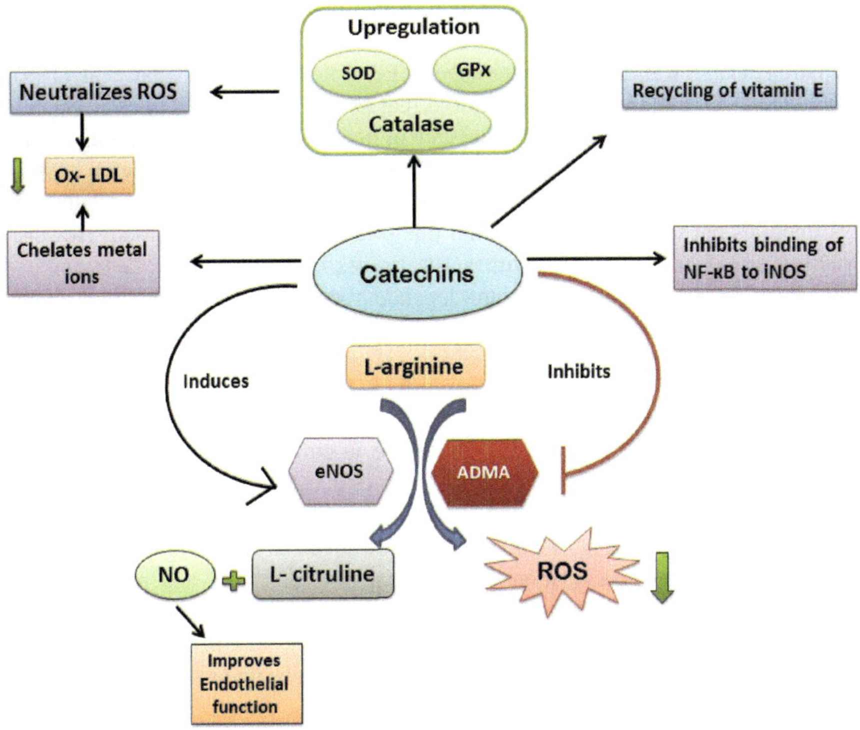

FIG. 5.2.3 Antioxidative role of catechins and its beneficiary effects in cardiovascular diseases.

Catechins Inhibit asymmetric dimethylarginine (ADMA) that used to block eNOS activity and leads to ROS generation.

and epicatechin (Islam, 2012). Catechins exhibit cardiovascular protective role via various mechanisms such as antihypertensive, antithrombogenic, anti-inflammation, antioxidative, and lipid minimizing effects (Balentine et al., 1997). Catechins exert antioxidant activities by metal ion chelating, free radical scavenging, inactivation of pro-oxidant enzymes activity (iNOS), promoting activity of antioxidant enzymes (SOD, catalase, and GPx) and inactivation of transcription factors (NF-κB) involve in redox reactions (Babu and Liu, 2008). The mechanism behind prevention of oxidative injury by catechin via scavenging reactive oxygen species may include hydrogen bond formation, electron delocalization, molecular rearrangements, and metal ion chelating (Babu and Liu, 2008) (Fig. 5.2.3).

5.2.4.2.1 *In vitro studies*

There are evidences regarding antioxidant activity of green tea extracts or catechins in *in vitro* studies, indicating metal ion chelating and superoxide scavenging properties of catechins (Deka and Vita, 2011). In another study, it has been reported that catechins in tea extract can prevent lipid peroxidation (Miura et al., 1994). In an *ex vivo* study, Pearson et al. (1998) demonstrated that when aortic endothelial cells and human LDL was incubated with green tea extract (5 and 0.08 ppm, respectively), oxidation of LDL was inhibited significantly (Pearson et al., 1998). Chan et al. (1997) conducted a cell culture study and showed that EGCG prevents iNOS expression in LPS-treated macrophages in a dose-dependent manner by inhibiting binding of NF-κB to iNOS and lowering the production of harmful nitric oxide (Chan et al., 1997). In many experimental studies, catechins have been able to improve endothelium dysfunctions but the mechanism behind this role is not fully understood. However in some recent *ex vivo* studies with aortic tissues of rat and rabbit demonstrated that catechins can induce endothelium-dependent vasorelation via increment in NO bioavailability (Schroeter et al., 2006).

5.2.4.2.2 Animal studies

In a mice model for atherosclerosis, Miura et al. (1994) demonstrated that treatment of EGCG increases apolipoprotein E in circulation and improvement of antioxidant capacity in vascular tissues (Chyu et al., 2004). Several investigations with animal model confirmed that catechin can enhance antioxidant enzymes activity. For instance, when 0.2% catechins are given to mice via drinking water results in significant upregulation of SOD, catalase, and GPx and promote ROS scavenging (Khan et al., 1992). In another study, 2 weeks of green tea consumption by hypertensive rats, catechins enhanced catalase expression in aorta (Babu et al., 2006). Moreover catechins also take part in recycling of vitamin E and supplement the action of glutathione (Zhu et al., 1999).

5.2.4.2.3 Clinical studies

In a recent study, Wang et al. (2011) reported that consumption of one or more cup of green tea per day may prevent the risk of coronary artery disorders development (Wang et al., 2011). In a clinical trial, it has been demonstrated that green tea consumption reversed endothelial dysfunction and significantly reduces 8-iso-prostaglandin-F2α which is an oxidative stress index (Bhardwaj and Khanna, 2013). Consumption of 600 mL green tea (with 5.2 g solid tea) regularly for 4 weeks results in significant reduction of LDL oxidation (Sung et al., 2005). It was reported that regular consumption of 120 ml per day of green tea prevented hypertension development in Chinese population (Deka and Vita, 2011). Long-term ingestion of high-dose of catechins significantly increases bioavailability NO and decreases malondialdehyde (MDA) and chemotactic protein-1 (MCP-1; Oyama et al., 2010). Daily consumption of catechins or green tea improves endothelial function in patients affected with coronary heart disorder (Islam, 2012). In human studies with EDCG (10 μM) reduces oxidation by 68% in dose-dependent manner (Liao et al., 2001).

5.2.4.3 Resveratrol

Resveratrol (3,4',5-trihydroxystilbene) is a natural compound which is produced by some spermatophytes, like grapevines. It has been observed that resveratrol is the main active compound of phytoalexins and it might have a beneficial role upon

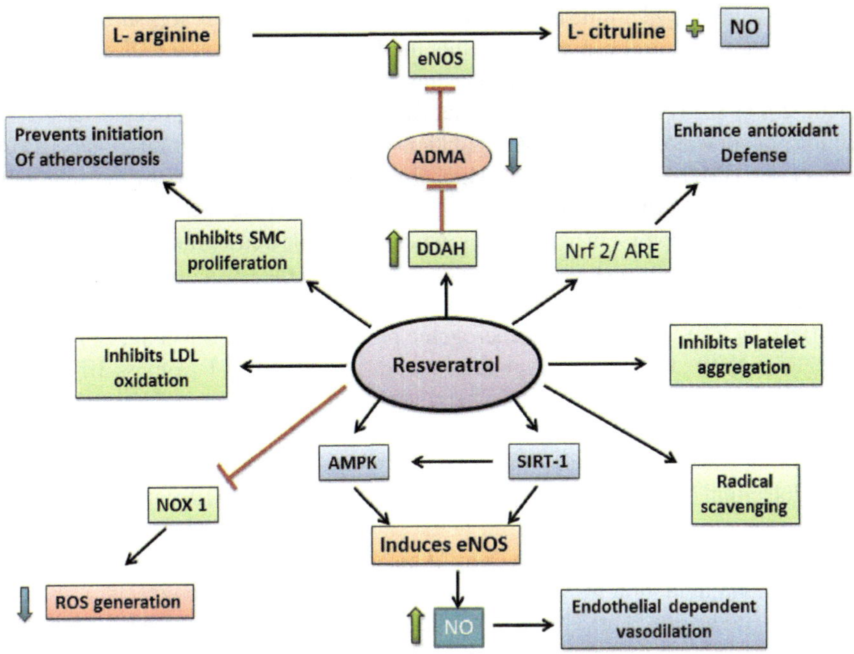

FIG. 5.2.4 Antioxidative role of resveratrol and its beneficiary effects in cardiovascular diseases.

Resveratrol upregulates dimethylaminohydroase (DDAH) which inhibits asymmetric dimethylarginine (ADMA) and restores the activity of eNOS and ultimately improves endothelium-dependent vasodilation. *AMPK*, AMP-activated protein kinase; *ARE*, antioxidant response elements; *SIRT*, silent information regulator 2/sirtuin 1.

human health. Resveratrol is also used in traditional oriental medicine. As skin of grapeberry contains resveratrol, not flesh, red wine has larger amount of resveratrol than that of white wine. Resveratrol shows antioxidant properties. It has been noticed that resveratrol can modulate lipid metabolism, inhibit LDL oxidation, and platelets aggregation. Also, resveratrol, being a phytoestrogen, might play a role in cardiovascular disease prevention. Besides being an antioxidant, resveratrol also contains anticancer and anti-inflammatory properties (Fig. 5.2.4).

5.2.4.3.1 ***Antiatherosclerotic effects of RES***

The intimal layer of arterial vessel wall gets affected in atherosclerosis. It can be characterized by extracellular lipid deposition, local smooth muscle cell proliferation and migration, and a chronic type inflammation. It results into liminal narrowing, might accompanied with formation of thrombus. All of these eventually lead to clinical proceedings like peripheral arterial disease, coronary artery disease, or stroke (Glass and Witztum, 2001). As lipids, LDLs mainly, are involved in atherosclerosis;

improvement of lipid profile might be considered to be beneficial. It is shown in some studies that resveratrol can modulate this lipid profile by reducing LDL-cholesterol level and plasma triglyceride, and elevating the level of HDL-cholesterol (Gocmen et al., 2011). Cho et al. have reported that RES can increase the hypocholesterolemia action of pravastatin via downregulation of 3-hydroxy-3-methyl-glutaryl-CoA reductase (HMG-CoA reductase), which is an enzyme takes part in the first step of cholesterol biosynthesis (Cho et al., 2008). Also, RES elevates the expression of LDL-R (LDL receptors) in hepatocytes, which further decreases the LDL-cholesterol level in blood (Yashiro et al., 2012). RES also lowers the oxidation of LDL, which contributes to the antioxidant properties of RES. Migration of smooth muscle cell is inhibited by RES, this is the antiatherogenic property of RES (Lin et al., 2014). RES activates eNOS, antioxidant response element (ARE), SIRT-1 (a class III histone deacetylase), Nrf2 and on the other hand decreases production of TNFα. All these above-mentioned actions of RES lead to activation of endothelium, decrease in endothelial apoptosis, vascular inflammation, which in a way improves the function of endothelium (Haskó et al., 2010). It has been observed that RES can downregulate the expression of different adhesion molecules (ICAM-1, VCAM-1) by inhibiting the activation of NF-κB pathway (Deng et al., 2011). RES can also reduce foam cell formation via inhibition of NADPH oxidase 1 and monocyte chemotactic protein-1 (MCP-1) production (Park et al., 2009). RES exerts its anti-inflammatory action by activating Nrf2 and suppressing the production of proinflammatory cytokines (Hao et al., 2013).

5.2.4.3.2 *Antihypertensive effects of RES*

Hypertension is one of the main risk factors of CVDs. Several studies have shown that RES has an antihypertensive property. Endothelium-dependent pathways with the involvement of AMPK (an energy metabolism regulator), Nrf2, and SIRT-1 are some of the involved mechanisms by which RES exerts its antihypertensive properties (Zordoky et al., 2015). RES elevates the expression as well as activity of eNOS, which improves the availability of NO, resulting into vasodilation (Leikert et al., 2002). This mechanism is linked with the antioxidant role of RES. RES induces the activation of SIRT-1 which increases expression as well as activity of eNOS (Arunachalam et al., 2010). RES can also induce activation of AMPK, results into increased NO production (Dolinsky et al., 2013). In several studies, it has been also reported that RES perform its antihypertensive role via endothelium-independent mechanism. RES induces activation of AMPK, which inhibits angiotensin II (AngII)-induced phosphorylation of myosin light chain and myosin phosphatase-targeting subunit 1. This results into inhibition of contractility of smooth muscle cell. Also, It has been observed that RES can inhibit contraction of aorta, induced by AngII (Cao et al., 2014).

5.2.4.3.3 *In vitro and in vivo study of RES*

Many studies have revealed that RES can cure contractile dysfunction and cardiac hypertrophy, one is functional and another one is structural abnormalities linked with hypertension (Chan et al., 2011). Fukuda et al. have shown that RES pretreatment has elevated the expression of inducible and endothelial NOS (iNOS and eNOS, respectively) in rats with ischemic myocardium. Many evidences have shown that the involved mechanism behind the beneficial role of RES against heart failure and cardiac hypertrophy is decrease in oxidative stress. In in vitro study, Tanno et al. have observed that increased level of Mn-superoxide dismutase (SOD2), a mitochondrial antioxidant enzyme, is the reason behind decreased oxidative stress (Tanno et al., 2010). In several *in vivo* studies, RES has shown significant antihypertensive effect upon different animals in 10–320 mg RES/kg body weight per day (Zordoky et al., 2015).

5.2.4.3.4 Clinical study of RES

High CV risk patients, who are treated with statin for primary prevention, have shown decrease in LDL-cholesterol by 4.5% and decrease in oxidized LDL by 20% with additional RES treatment (350 mg/day RES extract containing 8 mg RES; Timmers et al., 2011). A clinical study was performed on six random group in controlled environment (247 patients), and it showed that higher dose of RES (more than 150 mg/day) is able to lower the blood pressure significantly but lower dose than that has failed to do so (Liu et al., 2015). In order to find out the role of RES regarding secondary prevention, another clinical trial was performed upon patients who already had myocardial infarction. This study was a randomized, placebo-controlled, double-blind 3-month trial, which included 40 Caucasian postinfarction patients (14 women and 26 men). They were treated with either 10 mg/day RES with other additional medication or with placebo. It has been noticed that RES has helped to improve systolic as well as endothelial function, decrease LDL-cholesterol, and lowers aggregation of platelets. From this result the author has derived the conclusion that RES treatment, combined with other standard medication, has a significant beneficial role in lowering the risk of secondary MI in post-MI patients (Magyar et al., 2012).

Conclusion

Oxidative stress plays a critical role in onset and progression anthcroclcrotic plaque and various other cardiovascular diseases. This leads to the assumption that antioxidant would be an efficient therapeutic agent in this regard. Multiple evidences from in vitro, animal studies, and clinical trials have demonstrated various mechanisms by which antioxidants are capable of regulating oxidative stress and diminish its effects in cardiovascular systems. Among naturally occurring antioxidants, those that belongs to dietary sources have proven to significantly improve the ROS-induced cardiovascular diseases. Vitamins E, C, A, β-carotene, and polyphenols like resveratrol, catechins, anthocyanin, and quercetin are the major dietary antioxidants that have been demonstrated to ameliorate CVD by neutralizing ROS, minimizing lipid peroxidation, upregulation of endogenous antioxidants, reducing expression of adhesion molecules, preventing platelet aggregation, and diminishing oxidation of LDL, proteins, and DNA. In spite of these multiple role of dietary antioxidants in relation to cardiovascular diseases, there are some evidences where they have failed to give significant results. However, if we consider only the positive evidences, dietary antioxidants can be an efficient therapeutic approach for the prevention of cardiovascular diseases. In future more clinical studies should be carried out with these antioxidants in patients with different disorders related to cardiovascular diseases. Investigations must carry on discovering other dietary sources of antioxidants with relevant mechanism of action related to the prevention of ROS-induced atherosclerosis, endothelial dysfunctions, heart failure, and other deleterious CVD.

Abbreviations

ROS	reactive oxygen species
CVD	cardiovascular diseases
LDL	low-density lipoprotein
HDL	high-density lipoprotein
eNOS	endothelium nitric oxide synthase
MPO	myeloperoxidases
FGF	fibroblast growth factor
PDGF	platelet-derived growth factor
SOD	superoxide dismutase
GPx	glutathione peroxidase
CRP	circulating C reactive protein
EPIC	European Prospective Investigation into Cancer and Nutrition
HOPE	Heart Outcomes Prevention Evaluation study
EURAMIC	European Community Multicenter Study on Antioxidants Myocardial infarction and Breast Cancer
TRAP	thrombin receptor activating peptides
CETP	cholesteryl ester protein
EGCG	epigallocatechin-3-gallate
MDA	malondialdehyde
MCP-1	chemotactic protein-1

References

Alvarez-Suarez, J.M., Giampieri, F., Tulipani, S., Casoli, T., Di Stefano, G., González-Paramás, A.M., Santos-Buelga, C., Busco, F., Quiles, J.L., Cordero, M.D., 2014. One-month strawberry-rich anthocyanin supplementation ameliorates cardiovascular risk, oxidative stress markers and platelet activation in humans. J. Nutr. Biochem. 25, 289–294.

Anderson, T.J., Uehata, A., Gerhard, M.D., Meredith, I.T., Knab, S., Delagrange, D., Lieberman, E.H., Ganz, P., Creager, M.A., Yeung, A.C., 1995. Close relation of endothelial function in the human coronary and peripheral circulations. J. Am. Coll. Cardiol. 26, 1235–1241.

Arunachalam, G., Yao, H., Sundar, I.K., Caito, S., Rahman, I., 2010. SIRT1 regulates oxidant- and cigarette smoke-induced eNOS acetylation in endothelial cells: role of resveratrol. Biochem. Biophys. Res. Commun. 393, 66–72.

Azzi, A., Aratri, E., Boscoboinik, D., Clément, S., Özer, N.K., Ricciarelli, R., Spycher, S., 1998. Molecular basis of alpha-tocopherol control of smooth muscle cell proliferation. Biofactors 7, 3–14.

Babu, P.V., Liu, D., 2008. Green tea catechins and cardiovascular health: an update. Curr. Med. Chem. 15, 1840–1850.

Babu, P.V., Sabitha, K.E., Shyamaladevi, C.S., 2006. Therapeutic effect of green tea extract on oxidative stress in aorta and heart of streptozotocin diabetic rats. Chem. Biol. Interact. 162, 114–120.

Balentine, D.A., Wiseman, S.A., Bouwens, L.C., 1997. The chemistry of tea flavonoids. Crit. Rev. Food Sci. Nutr. 37, 693–704.

Baljinnyam, E., Hasebe, N., Morihira, M., Sumitomo, K., Matsusaka, T., Fujino, T., Fukuzawa, J., Ushikubi, F., Kikuchi, K., 2006. Oral pretreatment with ebselen enhances heat shock protein 72 expression and reduces myocardial infarct size. Hypertens. Res. 29, 905–913.

Basu, A., Du, M., Leyva, M.J., Sanchez, K., Betts, N.M., Wu, M., Aston, C.E., Lyons, T.J., 2010. Blueberries decrease cardiovascular risk factors in obese men and women with metabolic syndrome. J. Nutr. 140, 1582–1587.

Benstoem, C., Goetzenich, A., Kraemer, S., Borosch, S., Manzanares, W., Hardy, G., Stoppe, C., 2015. Selenium and its supplementation in cardiovascular disease–what do we know? Nutrients 7, 3094–3118.

Bhardwaj, P., Khanna, D., 2013. Green tea catechins: defensive role in cardiovascular disorders. Chin. J. Nat. Med. 11, 345–353.

Buijsse, B., Feskens, E., Schlettwien-Gsell, D., 2005. for the SENECA investigators: plasma carotene and α-tocopherol in relation to 10-y all-cause and cause-specific mortality in European elderly: the Survey in Europe on Nutrition and the Elderly, a Concerted Action (SENECA). Am. J. Clin. Nutr. 82, 879–886.

Cai, H., Harrison, D.G., 2000. Endothelial dysfunction in cardiovascular diseases: the role of oxidant stress. Circ. Res. 87, 840–844.

Cao, X., Luo, T., Luo, X., Tang, Z., 2014. Resveratrol prevents AngII-induced hypertension via AMPK activation and RhoA/ROCK suppression in mice. Hypertens. Res. 37, 803–810.

Castaneda-Ovando, A., de Pacheco-Hernández, L.M., Páez-Hernández, M.E., Rodríguez, J.A., Galán-Vidal, C.A., 2009. Chemical studies of anthocyanins: a review. Food Chem. 113, 859–871.

Cavia-Saiz, M., Busto, M.D., Pilar-Izquierdo, M.C., Ortega, N., Perez-Mateos, M., Muniz, P., 2010. Antioxidant properties, radical scavenging activity and biomolecule protection capacity of flavonoid naringenin and its glycoside naringin: a comparative study. J. Sci. Food Agric. 90, 1238–1244.

Cervantes Gracia, K., Llanas-Cornejo, D., Husi, H., 2017. CVD and oxidative stress. J. Clin. Med. 6, 22.

Chan, M.M., Fong, D., Ho, C.T., Huang, H.I., 1997. Inhibition of inducible nitric oxide synthase gene expression and enzyme activity by epigallocatechin gallate, a natural product from green tea. Biochem. Pharmacol. 54, 1281–1286.

Chan, V., Fenning, A., Iyer, A., Hoey, A., Brown, L., 2011. Resveratrol improves cardiovascular function in DOCA-salt hypertensive rats. Curr. Pharm. Biotechnol. 12, 429–436.

Chiang, A.N., Wu, H.L., Yeh, H.I., Chu, C.S., Lin, H.C., Lee, W.C., 2006. Antioxidant effects of black rice extract through the induction of superoxide dismutase and catalase activities. Lipids 41, 797–803.

Cho, I.J., Ahn, J.Y., Kim, S., Choi, M.S., Ha, T.Y., 2008. Resveratrol attenuates the expression of HMG-CoA reductase mRNA in hamsters. Biochem. Biophys. Res. Commun. 367, 190–194.

Chyu, K.Y., Babbidge, S.M., Zhao, X., Dandillaya, R., Rietveld, A.G., Yano, J., Dimayuga, P., Cercek, B., Shah, P.K., 2004. Differential effects of green tea-derived catechin on developing versus established atherosclerosis in apolipoprotein E-null mice. Circulation 109, 2448–2453.

Cominacini, L., Garbin, U., Pasini, A.F., Davoli, A., Campagnola, M., Contessi, G.B., Pastorino, A.M., Cascio, V.L., 1997. Antioxidants inhibit the expression of intercellular cell adhesion molecule-1 and vascular cell adhesion molecule-1 induced by oxidized LDL on human umbilical vein endothelial cells. Free Radic. Biol. Med. 22, 117–127.

Dahlöf, B., 2010. Cardiovascular disease risk factors: epidemiology and risk assessment. Am. J. Cardiol. 105, 3A–9A.

Da Silva, J.M.R., Darmon, N., Fernandez, Y., Mitjavila, S., 1991. Oxygen free radical scavenger capacity in aqueous models of different procyanidins from grape seeds. J. Agric. Food Chem. 39, 1549–1552.

Deka, A., Vita, J.A., 2011. Tea and cardiovascular disease. Pharmacol. Res. 64, 136–145.

Deng, Y.H., Alex, D., Huang, H.Q., Wang, N., Yu, N., Wang, Y.T., Leung, G.P., Lee, S.M., 2011. Inhibition of TNF-alpha-mediated endothelial cell-monocyte cell adhesion and adhesion molecules expression by the resveratrol derivative, trans-3,5,4′-trimethoxystilbene. Phytother. Res. 25, 451–457.

Devaraj, S., Li, D., Jialal, I., 1996. The effects of alpha tocopherol supplementation on monocyte function. Decreased lipid oxidation, interleukin 1 beta secretion, and monocyte adhesion to endothelium. J. Clin. Invest. 98, 756–763.

Dhalla, N.S., Golfman, L., Takeda, S., Takeda, N., Nagano, M., 1999. Evidence for the role of oxidative stress in acute ischemic heart disease: a brief review. Can. J. Cardiol. 15, 587–593.

Dhalla, N.S., Temsah, R.M., Netticadan, T., 2000. Role of oxidative stress in cardiovascular diseases. J. Hypertens. 18, 655–673.

Diaz, M.N., Frei, B., Vita, J.A., Keaney Jr., J.F., 1997. Antioxidants and atherosclerotic heart disease. N. Engl. J. Med. 337, 408–416.

D'Odorico, A., Martines, D., Kiechl, S., Egger, G., Oberhollenzer, F., Bonvicini, P., Sturniolo, G.C., Naccarato, R., Willeit, J., 2000. High plasma levels of α-and β-carotene are associated with a lower risk of atherosclerosis: results from the Bruneck study. Atherosclerosis 153, 231–239.

Dolinsky, V.W., Chakrabarti, S., Pereira, T.J., Oka, T., Levasseur, J., Beker, D., Zordoky, B.N., Morton, J.S., Nagendran, J., Lopaschuk, G.D., Davidge, S.T., Dyck, J.R., 2013. Resveratrol prevents hypertension and cardiac hypertrophy in hypertensive rats and mice. Biochim. Biophys. Acta 1832, 1723–1733.

Duarte, J., Pérez-Palencia, R., Vargas, F., Angeles Ocete, M., Pérez-Vizcaino, F., Zarzuelo, A., Tamargo, J., 2001. Antihypertensive effects of the flavonoid quercetin in spontaneously hypertensive rats. Br. J. Pharmacol. 133 (1), 117–124.

Dugas, T.R., Morel, D.W., Harrison, E.H., 1998. Impact of LDL carotenoid and α-tocopherol content on LDL oxidation by endothelial cells in culture. J. Lipid Res. 39, 999–1007.

Egert, S., Bosy-Westphal, A., Seiberl, J., Kurbitz, C., Settler, U., Plachta-Danielzik, S., Wagner, A.E., Frank, J., Schrezenmeir, J., Rimbach, G., Wolffram, S., Muller, M.J., 2009. Quercetin reduces systolic blood pressure and plasma oxidised low-density lipoprotein concentrations in overweight subjects with a high-cardiovascular disease risk phenotype: a double-blinded, placebo-controlled cross-over study. Br. J. Nutr. 102, 1065–1074.

Finking, G., Hanke, H., 1997. Nikolaj Nikolajewitsch Anitschkow (1885–1964) established the cholesterol-fed rabbit as a model for atherosclerosis research. Atherosclerosis 135, 1–7.

Freedman, J.E., Parker Iii, C., Li, L., Perlman, J.A., Frei, B., Ivanov, V., Deak, L.R., Iafrati, M.D., Folts, J.D., 2001. Select flavonoids and whole juice from purple grapes inhibit platelet function and enhance nitric oxide release. Circulation 103, 2792–2798.

Frei, B., 1991. Ascorbic acid protects lipids in human plasma and low-density lipoprotein against oxidative damage. Am. J. Clin. Nutr. 54, 1113S–1118S.

Glass, C.K., Witztum, J.L., 2001. Atherosclerosis. The road ahead. Cell 104, 503–516.

Gocmen, A.Y., Burgucu, D., Gumuslu, S., 2011. Effect of resveratrol on platelet activation in hypercholesterolemic rats: CD40-CD40L system as a potential target. Appl. Physiol. Nutr. Metab. 36, 323–330.

Gokce, N., Keaney Jr., J.F., Frei, B., Holbrook, M., Olesiak, M., Zachariah, B.J., Leeuwenburgh, C., Heinecke, J.W., Vita, J.A., 1999. Long-term ascorbic acid administration reverses endothelial vasomotor dysfunction in patients with coronary artery disease. Circulation 99, 3234–3240.

Gorinstein, S., Caspi, A., Libman, I., Lerner, H.T., Huang, D., Leontowicz, H., Leontowicz, M., Tashma, Z., Katrich, E., Feng, S., Trakhtenberg, S., 2006. Red grapefruit positively influences serum triglyceride level in patients suffering from coronary atherosclerosis: studies in vitro and in humans. J. Agric. Food Chem. 54, 1887–1892.

Goszcz, K., Deakin, S.J., Duthie, G.G., Stewart, D., Leslie, S.J., Megson, I.L., 2015. Antioxidants in cardiovascular therapy: panacea or false hope? Front. Cardiovasc. Med. 2, 29.

Goya, K., Sumitani, S., Otsuki, M., Xu, X., Yamamoto, H., Kurebayashi, S., Saito, H., Kouhara, H., Kasayama, S., 2006. The thiazolidinedione drug troglitazone up-regulates nitric oxide synthase expression in vascular endothelial cells. J. Diab. Complic. 20, 336–342.

Haendeler, J., Zeiher, A.M., Dimmeler, S., 1996. Vitamin C and E prevent lipopolysaccharide-induced apoptosis in human endothelial cells by modulation of Bcl-2 and Bax. Eur. J. Pharmacol. 317, 407–411.

Haenen, G.R., Paquay, J.B., Korthouwer, R.E., Bast, A., 1997. Peroxynitrite scavenging by flavonoids. Biochem. Biophys. Res. Commun. 236, 591–593.

Halliwell, B., 1988. Albumin—an important extracellular antioxidant? Biochem. Pharmacol. 37, 569–571.

Halliwell, B., 2007a. Biochemistry of Oxidative Stress. Biochem. Soc. Trans. 35 (5), 1147–1150.

Halliwell, B., 2007b. Dietary polyphenols: good, bad, or indifferent for your health? Cardiovasc. Res. 73, 341–347.

Han, J.J., Hao, J., Kim, C.H., Hong, J.S., Ahn, H.Y., Lee, Y.S., 2009. Quercetin prevents cardiac hypertrophy induced by pressure overload in rats. J. Vet. Med. Sci. 71, 737–743.

Hao, E., Lang, F., Chen, Y., Zhang, H., Cong, X., Shen, X., Su, G., 2013. Resveratrol alleviates endotoxin-induced myocardial toxicity via the Nrf2 transcription factor. PLoS One 8, e69452.

Haskó, G., Pacher, P., 2010. Endothelial Nrf2 activation: a new target for resveratrol? Am. J. Physiol. Heart Circ. Physiol. 299 (1), H10–H12.

Heller, R., Münscher-Paulig, F., Gräbner, R., Till, U., 1999. L-ascorbic acid potentiates nitric oxide synthesis in endothelial cells. J. Biol. Chem. 274, 8254–8260.

Hertog, M.G., Hollman, P.C., Katan, M.B., Kromhout, D., 1993. Intake of potentially anticarcinogenic flavonoids and their determinants in adults in The Netherlands.

Higashi, Y., Noma, K., Yoshizumi, M., Kihara, Y., 2009. Endothelial function and oxidative stress in cardiovascular diseases. Circ. J. 73, 411–418.

Hozawa, A., Jacobs, D.R., Steffes, M.W., Gross, M.D., Steffen, L.M., Lee, D.-H., 2007. Relationships of circulating carotenoid concentrations with several markers of inflammation, oxidative stress, and endothelial dysfunction: the Coronary Artery Risk Development in Young Adults (CARDIA)/Young Adult Longitudinal Trends in Antioxidants (YALTA) study. Clin. Chem. 53, 447–455.

Islam, M.A., 2012. Cardiovascular effects of green tea catechins: progress and promise. Recent Pat. Cardiovasc. Drug Discov. 7, 88–99.

Jeon, S.M., Bok, S.H., Jang, M.K., Kim, Y.H., Nam, K.T., Jeong, T.S., Park, Y.B., Choi, M.S., 2002. Comparison of antioxidant effects of naringin and probucol in cholesterol-fed rabbits. Clin. Chim. Acta 317, 181–190.

Jessup, W., Rankin, S., De Whalley, C., Hoult, J., Scott, J., Leake, D., 1990. α-Tocopherol consumption during low-density-lipoprotein oxidation. Biochem. J. 265, 399–405.

Jialal, I., Grundy, S.M., 1991. Preservation of the endogenous antioxidants in low density lipoprotein by ascorbate but not probucol during oxidative modification. J. Clin. Invest. 87, 597–601.

Karppi, J., Laukkanen, J.A., Mäkikallio, T.H., Kurl, S., 2011. Low serum lycopene and β-carotene increase risk of acute myocardial infarction in men. Eur. J. Public Health 22, 835–840.

Karthikeyan, K., Bai, B.S., Devaraj, S.N., 2007. Cardioprotective effect of grape seed proanthocyanidins on isoproterenol-induced myocardial injury in rats. Int. J. Cardiol. 115, 326–333.

Karthikeyan, K., Bai, B.S., Devaraj, S.N., 2009. Efficacy of grape seed proanthocyanidins on cardioprotection during isoproterenol-induced myocardial injury in rats. J. Cardiovasc. Pharmacol. 53, 109–115.

Kassebaum, N., Smith, A., Bernabé, E., Fleming, T., Reynolds, A., Vos, T., Murray, C., Marcenes, W., Collaborators, G.O.H., 2017. Global, regional, and national prevalence, incidence, and disability-adjusted life years for oral conditions for 195 countries, 1990–2015: a systematic analysis for the global burden of diseases, injuries, and risk factors. J. Dent. Res. 96, 380–387.

Katagiri, H., Yamada, T., Oka, Y., 2007. Adiposity and cardiovascular disorders: disturbance of the regulatory system consisting of humoral and neuronal signals. Circ. Res. 101, 27–39.

Keaney, J.F., Gaziano, J.M., Xu, A., Frei, B., Curran-Celentano, J., Shwaery, G.T., LoScalzo, J., Vita, J.A., 1993. Dietary antioxidants preserve endothelium-dependent vessel relaxation in cholesterol-fed rabbits. Proc. Natl. Acad. Sci. 90, 11880–11884.

Keaney, J.F., Guo, Y., Cunningham, D., Shwaery, G.T., Xu, A., Vita, J.A., 1996. Vascular incorporation of alpha-tocopherol prevents endothelial dysfunction due to oxidized LDL by inhibiting protein kinase C stimulation. J. Clin. Invest. 98, 386–394.

Keaney Jr., J.F., Simon, D.I., Freedman, J.E., 1999. Vitamin E and vascular homeostasis: implications for atherosclerosis. FASEB J. 13, 965–975.

Kelly, F.J., Fussell, J.C., 2017. Role of oxidative stress in cardiovascular disease outcomes following exposure to ambient air pollution. Free Radic. Biol. Med. 110, 345–367.

Khan, S.G., Katiyar, S.K., Agarwal, R., Mukhtar, H., 1992. Enhancement of antioxidant and phase II enzymes by oral feeding of green tea polyphenols in drinking water to SKH-1 hairless mice: possible role in cancer chemoprevention. Cancer Res. 52, 4050–4052.

Koga, T., Kwan, P., Zubik, L., Ameho, C., Smith, D., Meydani, M., 2004. Vitamin E supplementation suppresses macrophage accumulation and endothelial cell expression of adhesion molecules in the aorta of hypercholesterolemic rabbits. Atherosclerosis 176, 265–272.

Kohlmeier, L., Kark, J.D., Gomez-Gracia, E., Martin, B.C., Steck, S.E., Kardinaal, A.F., Ringstad, J., Thamm, M., Masaev, V., Riemersma, R., 1997. Lycopene and myocardial infarction risk in the EURAMIC Study. Am. J. Epidemiol. 146, 618–626.

Kuzuya, M., Naito, M., Funaki, C., Hayashi, T., Yamada, K., Asai, K., Kuzuya, F., 1991. Antioxidants stimulate endothelial cell proliferation in culture. Artery 18, 115–124.

Lakshmi, S.V., Padmaja, G., Kuppusamy, P., Kutala, V.K., 2009. Oxidative stress in cardiovascular disease.

Lefer, D.J., Granger, D.N., 2000. Oxidative stress and cardiac disease. Am. J. Med. 109, 315–323.

Lehr, H.-A., Frei, B., Arfors, K.-E., 1994. Vitamin C prevents cigarette smoke-induced leukocyte aggregation and adhesion to endothelium in vivo. Proc. Natl. Acad. Sci. 91, 7688–7692.

Leikert, J.F., Rathel, T.R., Wohlfart, P., Cheynier, V., Vollmar, A.M., Dirsch, V.M., 2002. Red wine polyphenols enhance endothelial nitric oxide synthase expression and subsequent nitric oxide release from endothelial cells. Circulation 106, 1614–1617.

Levine, G.N., Frei, B., Koulouris, S.N., Gerhard, M.D., Keaney Jr., J.F., Vita, J.A., 1996. Ascorbic acid reverses endothelial vasomotor dysfunction in patients with coronary artery disease. Circulation 93, 1107–1113.

Li, D., Saldeen, T., Mehta, J.L., 2000. Effects of α-tocopherol on ox-LDL-mediated degradation of IκB and apoptosis in cultured human coronary artery endothelial cells. J. Cardiovasc. Pharmacol. 36, 297–301.

Liao, S., Kao, Y.H., Hiipakka, R.A., 2001. Green tea: biochemical and biological basis for health benefits. Vitam. Horm. 62, 1–94.

Lin, Y.C., Chen, L.H., Varadharajan, T., Tsai, M.J., Chia, Y.C., Yuan, T.C., Sung, P.J., Weng, C.F., 2014. Resveratrol inhibits glucose-induced migration of vascular smooth muscle cells mediated by focal adhesion kinase. Mol. Nutr. Food Res. 58, 1389–1401.

Liu, S., Lee, I.-M., Ajani, U., Cole, S.R., Buring, J.E., Manson, J.E., 2001. Intake of vegetables rich in carotenoids and risk of coronary heart disease in men: the Physicians' Health Study. Int. J. Epidemiol. 30, 130–135.

Liu, Y., Ma, W., Zhang, P., He, S., Huang, D., 2015. Effect of resveratrol on blood pressure: a meta-analysis of randomized controlled trials. Clin. Nutr. 34, 27–34.

Maeda, N., Hagihara, H., Nakata, Y., Hiller, S., Wilder, J., Reddick, R., 2000. Aortic wall damage in mice unable to synthesize ascorbic acid. Proc. Natl. Acad. Sci. 97, 841–846.

Magyar, K., Halmosi, R., Palfi, A., Feher, G., Czopf, L., Fulop, A., Battyany, I., Sumegi, B., Toth, K., Szabados, E., 2012. Cardioprotection by resveratrol: A human clinical trial in patients with stable coronary artery disease. Clin. Hemorheol. Microcirc. 50 (3), 179–187.

Manach, C., Scalbert, A., Morand, C., Rémésy, C., Jiménez, L., 2004. Polyphenols: food sources and bioavailability. Am. J. Clin. Nutr. 79, 727–747.

Martin, S., Giannone, G., Andriantsitohaina, R., Carmen Martinez, M., 2003. Delphinidin, an active compound of red wine, inhibits endothelial cell apoptosis via nitric oxide pathway and regulation of calcium homeostasis. Br. J. Pharmacol. 139, 1095–1102.

Mazza, G., Kay, C.D., Cottrell, T., Holub, B.J., 2002. Absorption of anthocyanins from blueberries and serum antioxidant status in human subjects. J. Agric. Food Chem. 50, 7731–7737.

Mendis, S., Puska, P., Norrving, B.Organization, W.H., 2011. Global Atlas on Cardiovascular Disease Prevention and Control. World Health Organization, Geneva.

Miura, S., Watanabe, J., Tomita, T., Sano, M., Tomita, I., 1994. The inhibitory effects of tea polyphenols (flavan-3-ol derivatives) on Cu^{2+} mediated oxidative modification of low density lipoprotein. Biol. Pharm. Bull. 17, 1567–1572.

Munteanu, A., Taddei, M., Tamburini, I., Bergamini, E., Azzi, A., Zingg, J.-M., 2006. Antagonistic effects of oxidized low density lipoprotein and α-tocopherol on CD36 scavenger receptor expression in monocytes involvement of protein kinase B and peroxisome proliferator-activated receptor-γ. J. Biol. Chem. 281, 6489–6497.

Murray, C.J., Ezzati, M., Flaxman, A.D., Lim, S., Lozano, R., Michaud, C., Naghavi, M., Salomon, J.A., Shibuya, K., Vos, T., Wikler, D., 2012. GBD 2010: design, definitions, and metrics. The Lancet 380 (9859), 2063–2066.

Murohara, T., Ikeda, H., Katoh, A., Takajo, Y., Otsuka, Y., Haramaki, N., Imaizumi, T., 2002. Vitamin E inhibits lysophosphatidylcholine-induced endothelial dysfunction and platelet activation. Antioxidants Redox Signal. 4, 791–798.

Myasnikova, I., 1947. Effect of ascorbic acid, nikotinic acid, and thiamine on cholesterolemia. Voenno-Morskoi Med. Akad. Leningrado 8, 140.

Nakata, Y., Maeda, N., 2002. Vulnerable atherosclerotic plaque morphology in apolipoprotein E-deficient mice unable to make ascorbic acid. Circulation 105, 1485–1490.

Niki, E., Traber, M.G., 2012. A history of vitamin E. Ann. Nutr. Metab. 61, 207–212.

Nojiri, S., Daida, H., Inaba, Y., 2004. Antioxidants and cardiovascular disease: still a topic of interest. Environ. Health Prevent. Med. 9, 200–213.

Norum, K.R., Blomhoff, R., 1992. McCollum Award Lecture, 1992: Vitamin A Absorption, Transport, Cellular Uptake, and Storage. Oxford University Press.

O'Kennedy, N., Crosbie, L., Whelan, S., Luther, V., Horgan, G., Broom, J.I., Webb, D.J., Duttaroy, A.K., 2006. Effects of tomato extract on platelet function: a double-blinded crossover study in healthy humans. Am. J. Clin. Nutr. 84, 561–569.

Omenn, G.S., Goodman, G.E., Thornquist, M.D., Balmes, J., Cullen, M.R., Glass, A., Keogh, J.P., Meyskens Jr, F.L., Valanis, B., Williams Jr, J.H., Barnhart, S., 1996. Effects of a combination of beta carotene and vitamin A on lung cancer and cardiovascular disease. N. Engl. J. Med. 334 (18), 1150–1155.

Otero, P., Bonet, B., Herrera, E., Rabano, A., 2005. Development of atherosclerosis in the diabetic BALB/c mice: prevention with vitamin E administration. Atherosclerosis 182, 259–265.

Oyama, J., Maeda, T., Kouzuma, K., Ochiai, R., Tokimitsu, I., Higuchi, Y., Sugano, M., Makino, N., 2010. Green tea catechins improve human forearm endothelial dysfunction and have antiatherosclerotic effects in smokers. Circ. J. 74, 578–588.

Padayatty, S.J., Katz, A., Wang, Y., Eck, P., Kwon, O., Lee, J.-H., Chen, S., Corpe, C., Dutta, A., Dutta, S.K., 2003. Vitamin C as an antioxidant: evaluation of its role in disease prevention. J. Am. Coll. Nutr. 22, 18–35.

Park, D.W., Baek, K., Kim, J.R., Lee, J.J., Ryu, S.H., Chin, B.R., Baek, S.H., 2009. Resveratrol inhibits foam cell formation via NADPH oxidase 1- mediated reactive oxygen species and monocyte chemotactic protein-1. Exp. Mol. Med. 41, 171–179.

Patrick, L., Uzick, M., 2001. Cardiovascular disease: C-reactive protein and the inflammatory disease paradigm: HMG-CoA reductase inhibitors, alpha-tocopherol, red yeast rice, and olive oil polyphenols. A review of the literature. Altern. Med. Rev. 6, 248.

Pearson, D.A., Frankel, E.N., Aeschbach, R., German, J.B., 1998. Inhibition of endothelial cell mediated low-density lipoprotein oxidation by green tea extracts. J. Agric. Food Chem. 46 (4), 1445–1449.

Pirillo, A., Norata, G.D., Catapano, A.L., 2013. LOX-1, OxLDL, and atherosclerosis. Mediators Inflamm. 2013, 1–12.

Podrez, E.A., Febbraio, M., Sheibani, N., Schmitt, D., Silverstein, R.L., Hajjar, D.P., Cohen, P.A., Frazier, W.A., Hoff, H.F., Hazen, S.L., 2000. Macrophage scavenger receptor CD36 is the major receptor for LDL modified by monocyte-generated reactive nitrogen species. J. Clin. Invest. 105, 1095–1108.

Prabhakaran, D., Jeemon, P., Roy, A., 2016. Cardiovascular diseases in India: current epidemiology and future directions. Circulation 133, 1605–1620.

Pravst, I., Zmitek, K., Zmitek, J., 2010. Coenzyme Q10 contents in foods and fortification strategies. Crit. Rev. Food Sci. Nutr. 50, 269–280.

Princen, H., van Poppel, G., Vogelezang, C., Buytenhek, R., Kok, F.J., 1992. Supplementation with vitamin E but not beta-carotene in vivo protects low density lipoprotein from lipid peroxidation in vitro. Effect of cigarette smoking. Arterioscler. Thromb. 12, 554–562.

Qin, Y., Xia, M., Ma, J., Hao, Y., Liu, J., Mou, H., Cao, L., Ling, W., 2009. Anthocyanin supplementation improves serum LDL-and HDL-cholesterol concentrations associated with the inhibition of cholesteryl ester transfer protein in dyslipidemic subjects. Am. J. Clin. Nutr. 90, 485–492.

Quideau, S., Deffieux, D., Douat-Casassus, C., Pouysegu, L., 2011. Plant polyphenols: chemical properties, biological activities and synthesis angew. Chem. Int. Ed. 50, 586–621.

Raggi, P., 2016. Inflammation, depression and atherosclerosis or depression, inflammation and atherosclerosis? Atherosclerosis 251, 542–543.

Rao, A., 2002. Lycopene, tomatoes, and the prevention of coronary heart disease. Exp. Biol. Med. 227, 908–913.

Rayment, S.J., Shaw, J., Woollard, K.J., Lunec, J., Griffiths, H.R., 2003. Vitamin C supplementation in normal subjects reduces constitutive ICAM-1 expression. Biochem. Biophys. Res. Commun. 308, 339–345.

Reaven, P.D., Khouw, A., Beltz, W.F., Parthasarathy, S., Witztum, J.L., 1993. Effect of dietary antioxidant combinations in humans. Protection of LDL by vitamin E but not by beta-carotene. Arterioscler. Thromb. 13, 590–600.

Rechner, A.R., Kroner, C., 2005. Anthocyanins and colonic metabolites of dietary polyphenols inhibit platelet function. Thromb. Res. 116, 327–334.

Rodriguez-Mateos, A., Ishisaka, A., Mawatari, K., Vidal-Diez, A., Spencer, J.P., Terao, J., 2013. Blueberry intervention improves vascular reactivity and lowers blood pressure in high-fat-, high-cholesterol-fed rats. Br. J. Nutr. 109, 1746–1754.

Romanchik, J.E., Morel, D.W., Harrison, E.H., 1995. Distributions of carotenoids and α-tocopherol among lipoproteins do not change when human plasma is incubated in vitro. J. Nutr. 125, 2610–2617.

Ross, R., 1999. Atherosclerosis—an inflammatory disease. N. Engl. J. Med. 340, 115–126.

Rössig, L., Hoffmann, J., Hugel, B., Mallat, Z., Haase, A., Freyssinet, J.-M., Tedgui, A., Aicher, A., Zeiher, A.M., Dimmeler, S., 2001. Vitamin C inhibits endothelial cell apoptosis in congestive heart failure. Circulation 104, 2182–2187.

Russo, A., Acquaviva, R., Campisi, A., Sorrenti, V., Di Giacomo, C., Virgata, G., Barcellona, M.L., Vanella, A., 2000. Bioflavonoids as antiradicals, antioxidants and DNA cleavage protectors. Cell Biol. Toxicol. 16, 91–98.

Sagach, V.F., Scrosati, M., Fielding, J., Rossoni, G., Galli, C., Visioli, F., 2002. The water-soluble vitamin E analogue Trolox protects against ischaemia/reperfusion damage in vitro and ex vivo. A comparison with vitamin E. Pharmacol. Res. 45, 435–439.

Sargeant, L., Khaw, K., Bingham, S., Day, N., Luben, R., Oakes, S., Welch, A., Wareham, N., 2001. Fruit and vegetable intake and population glycosylated haemoglobin levels: the EPIC-Norfolk Study. Eur. J. Clin. Nutr. 55, 342.

Schroeter, H., Heiss, C., Balzer, J., Kleinbongard, P., Keen, C.L., Hollenberg, N.K., Sies, H., Kwik-Uribe, C., Schmitz, H.H., Kelm, M., 2006. (-)-Epicatechin mediates beneficial effects of flavanol-rich cocoa on vascular function in humans. Proc. Natl. Acad. Sci. U S A 103, 1024–1029.

Singh, R.B., Wander, G.S., Rastogi, A., Shukla, P.K., Mittal, A., Sharma, J.P., Mehrotra, S.K., Kapoor, R., Chopra, R.K., 1998. Randomized, double-blind placebo-controlled trial of coenzyme Q10 in patients with acute myocardial infarction. Cardiovasc. Drugs Ther. 12, 347–353.

Sinha, K., Das, J., Pal, P.B., Sil, P.C., 2013. Oxidative stress: the mitochondria-dependent and mitochondria-independent pathways of apoptosis. Arch. Toxicol. 87, 1157–1180.

Siow, R.C., Richards, J.P., Pedley, K.C., Leake, D.S., Mann, G.E., 1999. Vitamin C protects human vascular smooth muscle cells against apoptosis induced by moderately oxidized LDL containing high levels of lipid hydroperoxides. Arterioscler. Thromb. Vasc. Biol. 19, 2387–2394.

Solzbach, U., Hornig, B., Jeserich, M., Just, H., 1997. Vitamin C improves endothelial dysfunction of epicardial coronary arteries in hypertensive patients. Circulation 96, 1513–1519.

Sozański, T., Kucharska, A., Szumny, A., Magdalan, J., Bielska, K., Merwid-Ląd, A., Woźniak, A., Dzimira, S., Piórecki, N., Trocha, M., 2014. The protective effect of the *Cornus mas* fruits (cornelian cherry) on hypertriglyceridemia and atherosclerosis through PPARα activation in hypercholesterolemic rabbits. Phytomedicine 21, 1774–1784.

Stephens, N.G., Parsons, A., Brown, M., Schofield, P., Kelly, F., Cheeseman, K., Mitchinson, M., 1996. Randomised controlled trial of vitamin E in patients with coronary disease: Cambridge Heart Antioxidant Study (CHAOS). Lancet North Am. Ed. 347, 781–786.

Sugiyama, S., Kugiyama, K., Ogata, N., Doi, H., Ota, Y., Ohgushi, M., Matsumura, T., Oka, H., Yasue, H., 1998. Biphasic regulation of transcription factor nuclear factor-κB activity in human endothelial cells by lysophosphatidylcholine through protein kinase C-mediated pathway. Arterioscler. Thromb. Vasc. Biol. 18, 568–576.

Sung, H., Min, W.K., Lee, W., Chun, S., Park, H., Lee, Y.W., Jang, S., Lee, D.H., 2005. The effects of green tea ingestion over four weeks on atherosclerotic markers. Ann. Clin. Biochem. 42, 292–297.

Tanno, M., Kuno, A., Yano, T., Miura, T., Hisahara, S., Ishikawa, S., Shimamoto, K., Horio, Y., 2010. Induction of manganese superoxide dismutase by nuclear translocation and activation of SIRT1 promotes cell survival in chronic heart failure. J. Biol. Chem. 285, 8375–8382.

Tasinato, A., Boscoboinik, D., Bartoli, G., Maroni, P., Azzi, A., 1995. d-alpha-tocopherol inhibition of vascular smooth muscle cell proliferation occurs at physiological concentrations, correlates with protein kinase C inhibition, and is independent of its antioxidant properties. Proc. Natl. Acad. Sci. 92, 12190–12194.

Timmers, S., Konings, E., Bilet, L., Houtkooper, R.H., van de Weijer, T., Goossens, G.H., Hoeks, J., van der Krieken, S., Ryu, D., Kersten, S., Moonen-Kornips, E., Hesselink, M.K.C., Kunz, I., Schrauwen-Hinderling, V.B., Blaak, E., Auwerx, J., Schrauwen, P., 2011. Calorie restriction-like effects of 30 days of resveratrol supplementation on energy metabolism and metabolic profile in obese humans. Cell Metab. 14, 612–622.

Traber, M.G., 2007. Vitamin E regulatory mechanisms. Annu. Rev. Nutr. 27, 347–362.

Tsuchihashi, H., Kigoshi, M., Iwatsuki, M., Niki, E., 1995. Action of β-carotene as an antioxidant against lipid peroxidation. Arch. Biochem. Biophys. 323, 137–147.

Turunen, M., Olsson, J., Dallner, G., 2004. Metabolism and function of coenzyme Q. Biochim. Biophys. Acta 1660, 171–199.

Uemura, M., Manabe, H., Yoshida, N., Fujita, N., Ochiai, J., Matsumoto, N., Takagi, T., Naito, Y., Yoshikawa, T., 2002. α-Tocopherol prevents apoptosis of vascular endothelial cells via a mechanism exceeding that of mere antioxidation. Eur. J. Pharmacol. 456, 29–37.

Ulrich-Merzenich, G., Metzner, C., Schiermeyer, B., Vetter, H., 2002. Vitamin C and vitamin E antagonistically modulate human vascular endothelial and smooth muscle cell DNA synthesis and proliferation. Eur. J. Nutr. 41, 27–34.

Wallace, T.C., 2011. Anthocyanins in cardiovascular disease. Adv. Nutr. 2, 1–7.

Walter, M., Marchesan, E., 2011. Phenolic compounds and antioxidant activity of rice. Braz. Arch. Biol. Technol. 54, 371–377.

Wang, Z.M., Zhou, B., Wang, Y.S., Gong, Q.Y., Wang, Q.M., Yan, J.J., Gao, W., Wang, L.S., 2011. Black and green tea consumption and the risk of coronary artery disease: a meta-analysis. Am. J. Clin. Nutr. 93, 506–515.

Watzl, B., Kulling, S.E., Möseneder, J., Barth, S.W., Bub, A., 2005. A 4-wk intervention with high intake of carotenoid-rich vegetables and fruit reduces plasma C-reactive protein in healthy, nonsmoking men. Am. J. Clin. Nutr. 82, 1052–1058.

Wilkinson, I., Megson, I., MacCallum, H., Sogo, N., Cockcroft, J., Webb, D., 1999. Oral vitamin C reduces arterial stiffness and platelet aggregation in humans. J. Cardiovasc. Pharmacol. 34, 690–693.

Witztum, J.L., Steinberg, D., 1991. Role of oxidized low density lipoprotein in atherogenesis. J. Clin. Invest. 88, 1785–1792.

Wu, D., Koga, T., Martin, K.R., Meydani, M., 1999. Effect of vitamin E on human aortic endothelial cell production of chemokines and adhesion to monocytes. Atherosclerosis 147, 297–307.

Yamamura, Y., Ishiyama, T., Yamagami, T., Morita, Y., Ishio, S., Kashiwamura, S., Terada, A., Tsukamoto, N., Toyama, S., Nakajima, Y., Wada, N., 1967. Clinical use of coenzyme-Q for treatment of cardiovascular diseases. Jpn. Circ. J. 31 (1), 168.

Yashiro, T., Nanmoku, M., Shimizu, M., Inoue, J., Sato, R., 2012. Resveratrol increases the expression and activity of the low density lipoprotein receptor in hepatocytes by the proteolytic activation of the sterol regulatory element-binding proteins. Atherosclerosis 220, 369–374.

Youdim, K.A., McDonald, J., Kalt, W., Joseph, J.A., 2002. Potential role of dietary flavonoids in reducing microvascular endothelium vulnerability to oxidative and inflammatory insults. J. Nutr. Biochem. 13, 282–288.

Yusuf, S., Dagenais, G., Pogue, J., Bosch, J., Sleight, P., 2000. Vitamin E supplementation and cardiovascular events in high-risk patients. N. Engl. J. Med. 342, 154–160.

Zapolska-Downar, D., Zapolski-Downar, A., Markiewski, M., Ciechanowicz, A., Kaczmarczyk, M., Naruszewicz, M., 2000. Selective inhibition by α-tocopherol of vascular cell adhesion molecule-1 expression in human vascular endothelial cells. Biochem. Biophys. Res. Commun. 274, 609–615.

Zhu, Q.Y., Huang, Y., Tsang, D., Chen, Z.Y., 1999. Regeneration of alpha-tocopherol in human low-density lipoprotein by green tea catechin. J. Agric. Food Chem. 47, 2020–2025.

Zordoky, B.N., Robertson, I.M., Dyck, J.R., 2015. Preclinical and clinical evidence for the role of resveratrol in the treatment of cardiovascular diseases. Biochim. Biophys. Acta 1852, 1155–1177.

Zwaka, T.P., Hombach, V., Torzewski, J., 2001. C-reactive protein-mediated low density lipoprotein uptake by macrophages: implications for atherosclerosis. Circulation 103, 1194–1197.

CHAPTER

Antioxidants and cataracts/age-related macular degeneration

5.3

Laxman Singh[a], Pooja Patni[b], Archana N. Sah[c], Devesh Tewari[d]

[a]*Centre of Biodiversity Conservation & Management, G.B. Pant National Institute of Himalayan Environment, Kosi-Katarmal, Almora, Uttarakhand, India*

[b]*School of Pharmaceutical Sciences, Lovely Professional University, Phagwara, Punjab, India*

[c]*Department of Pharmaceutical Sciences, Faculty of Technology, Bhimtal Campus, Kumaun University, Nainital, Uttarakhand, India*

[d]*Department of Pharmacognosy, School of Pharmaceutical Sciences, Lovely Professional University, Phagwara, Punjab, India*

5.3.1 Introduction

Ocular abnormalities, that is, cataract, glaucoma, diabetic retinopathy, and macular degeneration, are serious health concerns internationally, with cataract alone being accountable for more than 50% of the world blindness. Cataractogenesis is influenced by multiple factors, for instance, age-related factors, genetic disorders, oxidative stress and inflammation, diabetes mellitus, substance abuse, toxins, other ocular diseases, UV light exposure, osmotic stress, and abnormal glucose metabolism. Of these, oxidative stress is identified as an initiating factor in the development of cataractogenesis. This further causes modification of the ocular lens proteins via protein–protein aggregation, crosslinking, and finally precipitation (Reddy et al., 1998). Free radical species, mainly the toxic aldehyde, are generated as a result of oxidative stress which causes lens epithelium membrane peroxidation that makes the retina vulnerable and finally, damaged lens protein results into opacity (Altomare et al., 1997). Thus, targeting or reduction of oxidative stress produced by reactive oxygen species (ROS) is the first-hand strategy as no drugs have been approved to prevent this disorder so far.

Given this, natural compounds possessing antioxidant or anti-inflammatory effects are seen as a positive controller since they can eliminate the already formed ROS by trapping, scavenging, or by quenching activity. A critical first step to delineate the possible antioxidant potential is to study a brief mechanism of action against the genesis of cataract disorder (Fig. 5.3.1).

Several studies involving in vitro, in vivo, and clinical investigations have been conducted with proposing anticataract potential; even then a promising nutraceutical's

Antioxidants Effects in Health. DOI: https://doi.org/10.1016/B978-0-12-819096-8.00073-2

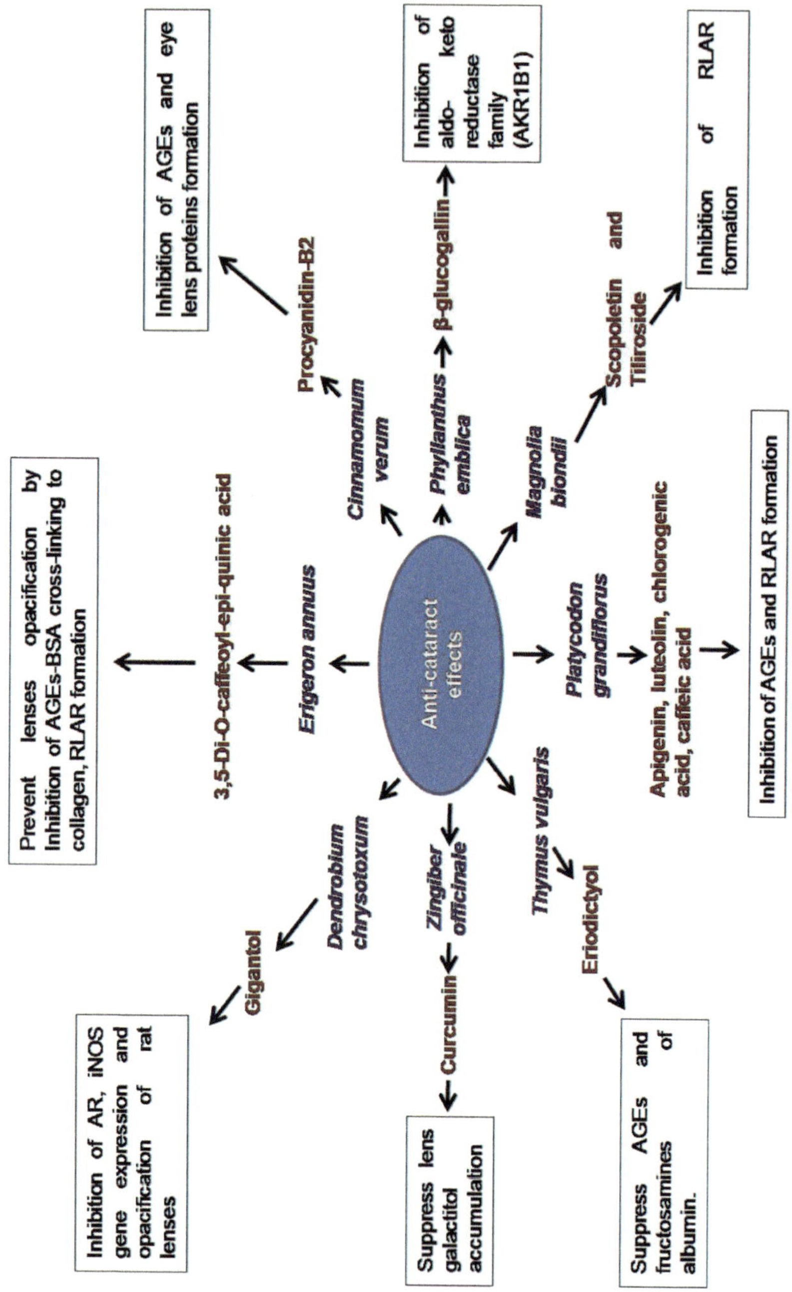

FIG. 5.3.1 Mechanism of plants active constituents and its role against cataract progression.

supplement entrusted for carving out cataract effects is yet to be established. Taking this into consideration, the present chapter congregates the major finding of previous reports and envisage future course of actions needed to improve the existing knowledge in terms of preclinical and clinical studies.

5.3.2 Cataract, a global problem

According to a global assessment, around 51% of global blindness is due to cataract which represents around 20 million people. Though surgical removal of cataracts is an option, in several countries, patients are still unable to access cataract surgery due to different reasons, and thus cataract remnants as a major cause of blindness. Due to increasing life expectancy, cataract prevalence globally is expected to increase. Apart from blindness, cataract is a major reason for low vision in the industrially developed and developing countries (https://www.who.int/blindness/causes/priority/en/index1.html accessed on 20.06.2019). It is an estimate that, in 2010, approximately 32 million blind people and 191 million people with poor vision were affected by cataract and the same study showed that out of three blind people, one blind due to cataract (Khairallah et al., 2015).

The statistics by the World Health Organization (WHO) anticipates that around 90 million global populations will be blind by 2020 (Khairallah et al., 2015; Taylor, 2016). The approach to fight this challenge is expensive and requires very well-trained human resources, development of infrastructure, and operative disease control. Cataract prevalence is directly associated with age progression, and it is interesting to note that in-between age group 52–62 years the prevalence is 5% while the prevalence in the age group over 70 years is around 64% (Prokofyeva et al., 2013). Age progression of the rapidly growing population is an alarming issue as the age is a nonmodifiable risk factor for cataract pathogenesis. It is vital to figure out the modifiable risk factors of cataract that might help in the establishment of preventive measures (Tewari et al., 2019).

5.3.3 Pathophysiology of cataract

The age-related cataract pathogenesis is associated with many mechanisms. Opacity in lens occurs because of alteration in the microarchitecture, sometimes by mutations, physical, or biomechanical changes (Tewari et al., 2019). Lens opacity is a straight outcome of oxidative stress. Depending upon the opacification in the lens, the age-related cataracts are categorized into the following three types: nuclear, cortical, and posterior subcapsular cataracts. The epithelial cells of the lens are extremely active cells in metabolism that undergo oxidation, insolubilization, and crosslinking. These cells migrate to the center of a lens and outline lens fibers, in opacity disorder. These fibers are gradually compressed and finally results in lens nuclear sclerosis (Alshamrani, 2018).

Table 5.3.1 Potential role of nutraceuticals involving clinical, in vivo and in vitro studies, and anticataract effect.

S. no.	Objectives	Interventions/ formulation	Experimentation mode	Possible outcomes	References
1	Assessment of serum atherosclerotic low-density lipoprotein levels mild COPD patients	Curcumin capsule (25 mg)	Clinical trial	Reduced the levels of AT-LDL that leads to atherosclerotic events in COPD patients	Funamoto et al. (2016)
2	Estimation of IMA levels concerning oxidant–antioxidant profiles of cataract patients	-	Clinical trial	Cataract patients had higher levels of oxidative markers, i.e., serum MDA and IMA whereas, lower level of antioxidant properties in terms of lower levels of serum catalase (CAT) and super oxidase dismutase (SOD) than control (healthy individuals without cataract symptoms) group	Elmazar et al. (2018)
3	Examine folic acid, vitamins B_6, and B_{12} infusion on age-related anticataract effect	Infusion folic acid (2.5 mg/day), vitamin B6 (50 mg/day), and vitamin B12 (1 mg/day)	Clinical trial	Daily infusion supplementation had no significant effect on cataract	Christen et al. (2016)
4	Assess the efficacy and safety of oral saffron, on AMD	20 mg saffron capsule	Clinical trial	Saffron supplementation moderately improved visual function; longer-term supplementation was suggested beneficial in long-term usage of saffron	Broadhead et al. (2019)
5	Quantifying ascorbic acid concentration after Vit. C supplementation in aqueous humor of patients with cataract	Administration ratio of Vit. C: oral (2 g), intravenous (20 g)	Clinical trial	In aqueous humor ascorbic acid concentration increased, with intravenous Vit. C administration being more effective than oral intake	Hah et al. (2017)

6	Investigate *Ginkgo biloba* extract on selenite-induced cataractogenesis	100 mg/ kg body weight	In vivo	Effective against oxidative stress markers, one among the initiators of cataractogenesis, perhaps by preventing reduction of antioxidant enzymes, i.e., CAT, SOD, GPx, GST, and GR and by inhibiting lipid peroxidation (MDA)	Cao et al. (2015)
7	Curcumin permeability on coronary artery in rat coronary atherosclerosis disease model	Curcumin (100 mg/ kg/day)	In vivo	Curcumin inhibit expression of MMP-9, CD40L, TNF-α, and CRP to improve the permeability of coronary artery	Li et al. (2015)
8	To examine probucol on the progression of cataracts	1 mL/kg probucol	In vivo	Probucol prevented the progression of cataracts in rats with severe hyperglycemia, the beneficial effects were attributed to antioxidative effect in lenses	Higashi et al. (2018)
9	Effect of roasting coffee beans on cataract formation	0.2 mL/day (100% coffee)	In vivo	Concentrations of glutathione (GSH) and ascorbic acid (AA) in cataract-induced rats reduced substantially. Pyrocatechol (antioxidant agent), generated during the roasting process, was credited to prevent cataract formation.	Ishimori et al. (2017)

(*Continued*)

Table 5.3.1 Potential role of nutraceuticals involving clinical, in vivo and in vitro studies, and anticataract effect. Continued

S. no.	Objectives	Interventions/ formulation	Experimentation mode	Possible outcomes	References
10	Investigate effect of EGCG (from green tea) on αA(66–80) (peptide fragment) derived from αA crystallin and anticataract activity	1–50 mM EGCG	In vitro	EGCG at 50 mM efficiently prevented and disaggregate preformed αA(66–80) aggregates, thus blocking cataract formation	Kumar et al. (2017)
11	Investigate effect of lutein (L) and fatty acids (LA, EPA, DHA, and OA) on oxidative stress leading to cataract formation	Lutein (1.3 mmol/ kg body weight), and 7.5 mM LA, or 7.5 mM EPA + DHA or 7.5 mM OA	In vivo	NO·, MDA, and protein carbonyls were higher in cataract rats than control and experimental groups, while, CAT, SOD, GPx, and GST in serum and lens of cataract group were decreased significantly Lutein with EPA + DHA anticataract was higher compared to LA or OA	Padmanabha and Vallikannan (2018)
12	To screen root of *Leucas aspera* against anticataractogenic potential	300 µg/mL extract	In vitro	Lenticular opacification was prevented by action of enzymatic antioxidants (GR and MDA), while, gene expression encoding αA- and βB1-crystallins gene, and crystallin proteins was near about normal levels, thus prevented cataractogenesis	Sundararajan et al. (2017)

13	To investigate anticataract protective effects of HDAC is on HLECs following UV-B exposure	1 µmol/L, 2 µmol/L SAHA	In vitro	Low concentrations of HDACis (1 µmol/L SAHA) placidly inhibit oxidative stress, thus protecting HLECs from oxidation suggesting anticataract property	Qiu et al. (2019)
14	To investigate gigantol usefulness against cataractogenesis	1µg/mL gigantol	In vitro	Gigantol inhibited AR, along with AR gene expression, thus affective against diabetic cataract	Wu et al. (2017)
15	The potential of hydrocortisone, dexamethasone, and deoxycorticosterone (steroidal drugs) as a promising anticataract agent	20–30 mM dexamethasone	In vitro	The oxidative stress induced by glucose was significantly reduced highest in dexamethasone (30 mM) in the treated lens when compared to other steroids drugs	Rana et al. (2018)

In congenital cataracts, the lens opacity manifests at the time of birth. On the other hand, infantile cataracts infer to lens opacity progression after a year of life. Based on the cause, pediatric cataracts may be unilateral or bilateral. Moreover, the hereditary reason is a major cause of cataract and around a third of pediatric cataracts are of the hereditary type. One-third of these, coexist with various ocular anomalies indicative of a kind of multisystem syndrome, and the other one-third have undetermined reasons (Alshamrani, 2018).

5.3.4 Role of antioxidants in cataract

Since oxidative stress is one of the main mechanisms for the progression of age-related cataract, the importance of antioxidants in the prevention of cataract is tremendous. In this section, we discussed about various in vitro, in vivo, and preclinical studies on various antioxidants against cataract (Table 5.3.1).

Conclusion

Preclinical and clinical studies revealed that antioxidants are tremendously potential for prevention of cataractogenesis because of the significant involvement of ROS in the formation of cataract. Limited antioxidants have been evaluated for their anticataract potency; hence, it is highly recommended to find out novel antioxidant compounds from various natural sources and to evaluate their potential against cataract.

Abbreviations

CAT	catalase
SOD	superoxide dismutase
GPx	glutathione peroxidase
GST	glutathione S-transferase
MDA	malondialdehyde
GR	glutathione reductase
NO	nitric oxide
EGCG	epigallocatechin-3-gallate
iNOS	inducible nitric oxide synthase
AGEs	advanced glycation end product
BSA	bovine serum albumin
RLAR	rat lens aldose reductase
AKR1B1	aldo-keto reductase family
MMP-9	matrix metalloproteinase-9
ELISA	enzyme-linked immunosorbent assay

TNF-α	tumor necrosis factor-α
CRP	C-reaction protein
HDACis	histone deacetylase inhibitors
HLECs	human lens epithelial cells
UV-B	ultraviolet-B
SAHA	suberoylanilide hydroxamic acid
AR	aldose reductase
AA	ascorbic acid
AMD	age-related macular degeneration
Vit. C	vitamin C
IMA	ischemia-modified albumin
AT	α1-antitrypsin
LDL	low-density lipoprotein

References

Alshamrani, A.Z., 2018. Cataracts pathophysiology and managements. Egypt. J. Hosp. Med. 70, 151–155.

Altomare, E., Grattagliano, I., Vendemaile, G., Micelli-Ferrari, T., Signorile, A., Cardia, L., 1997. Oxidative protein damage in human diabetic eye: evidence of a retinal participation. Eur. J. Clin. Invest. 27, 141–147.

Broadhead, G.K., Grigg, J.R., McCluskey, P., Hong, T., Schlub, T.E., Chang, A.A., 2019. Saffron therapy for the treatment of mild/moderate age-related macular degeneration: a randomised clinical trial. Graefe's Arch. Clin. Exp. Ophthalmol. = Albr. von Graefes Arch. Klin. Exp. Ophthalmol. 257, 31–40. https://doi.org/10.1007/s00417-018-4163-x.

Cao, S., Gao, M., Wang, N., Liu, N., Du, G., Lu, J., 2015. Prevention of selenite-induced cataratogenesis by *Ginkgo biloba* extract (Egb761) in Wistar rats. Curr. Eye Res. 40, 1028–1033. https://doi.org/10.3109/02713683.2014.980005.

Christen, W.G., Glynn, R.J., Chew, E.Y., Albert, C.M., Manson, J.E., 2016. Folic acid, vitamin B6, and vitamin B12 in combination and age-related cataract in a randomized trial of women. Ophthalmic Epidemiol. 23, 32–39. https://doi.org/10.3109/09286586.2015.1130845.

Elmazar, H.M., Elmadbouh, I., Mandour, S.S., Al Ariny, G.M., Ibrahim, A.M., 2018. Association between cataract progression and ischemia-modified albumin in relation to oxidant-antioxidant profiles in the serum, aqueous humor, and lens. J. Cataract Refract. Surg. 44, 134–139. https://doi.org/10.1016/j.jcrs.2017.10.051.

Funamoto, M., Sunagawa, Y., Katanasaka, Y., Miyazaki, Y., Imaizumi, A., Kakeya, H., Yamakage, H., Satoh-Asahara, N., Komiyama, M., Wada, H., Hasegawa, K., 2016. Highly absorptive curcumin reduces serum atherosclerotic low-density lipoprotein levels in patients with mild COPD. Int. J. COPD 11, 2029. https://doi.org/10.2147/COPD.S104490.

Hah, Y.-S., Chung, H.J., Sontakke, S.B., Chung, I.-Y., Ju, S., Seo, S.-W., Yoo, J.-M., Kim, S.-J., 2017. Ascorbic acid concentrations in aqueous humor after systemic vitamin C supplementation in patients with cataract: pilot study. BMC Ophthalmol. 17, 121. https://doi.org/10.1186/s12886-017-0515-2.

Higashi, Y., Higashi, K., Mori, A., Sakamoto, K., Ishii, K., Nakahara, T., 2018. Anti-cataract effect of resveratrol in high-glucose-treated streptozotocin-induced diabetic rats. Biol. Pharm. Bull. 41, 1586–1592. https://doi.org/10.1248/bpb.b18-00328.

Ishimori, N., Oguchi, J., Nakazawa, Y., Kobata, K., Funakoshi-Tago, M., Tamura, H., 2017. Roasting enhances the anti-cataract effect of coffee beans: ameliorating selenite-induced cataracts in rats. Curr. Eye Res 42 (6), 864–870. https://doi.org/10.1080/02713683.2016.1262877.

Khairallah, M., Kahloun, R., Bourne, R., Limburg, H., Flaxman, S.R., Jonas, J.B., Keeffe, J., Leasher, J., Naidoo, K., Pesudovs, K., Price, H., White, R.A., Wong, T.Y., Resnikoff, S., Taylor, H.R., 2015. Number of people blind or visually impaired by cataract worldwide and in world regions, 1990 to 2010. Invest. Ophthalmol. Vis. Sci. 56, 6762–6769. https://doi.org/10.1167/iovs.15-17201.

Kumar, V., Gour, S., Peter, O.S., Gandhi, S., Goyal, P., Pandey, J., Harsolia, R.S., Yadav, J.K., 2017. Effect of green tea polyphenol epigallocatechin-3-gallate on the aggregation of αA(66-80) peptide, a major fragment of αA-crystallin involved in cataract development. Curr. Eye Res 42 (10), 1368–1377. https://doi.org/10.1080/02713683.2017.1324628.

Li, X., Lu, Y., Sun, Y., Zhang, Q., 2015. Effect of curcumin on permeability of coronary artery and expression of related proteins in rat coronary atherosclerosis heart disease model. Int. J. Clin. Exp. Pathol 8 (6), 7247.

Padmanabha, S., Vallikannan, B., 2018. Fatty acids modulate the efficacy of lutein in cataract prevention: assessment of oxidative and inflammatory parameters in rats. Biochem. Biophys. Res. Commun 500 (2), 435–442. https://doi.org/10.1016/j.bbrc.2018.04.098.

Prokofyeva, E., Wegener, A., Zrenner, E., 2013. Cataract prevalence and prevention in Europe: a literature review. Acta Ophthalmol. 91, 395–405. https://doi.org/10.1111/j.1755-3768.2012.02444.x.

Qiu, X., Rong, X., Yang, J., Lu, Y., 2019. Evaluation of the antioxidant effects of different histone deacetylase inhibitors (HDACis) on human lens epithelial cells (HLECs) after UVB exposure. BMC Ophthalmol. 19, 42. https://doi.org/10.1186/s12886-019-1056-7.

Rana, D., Sharma, R., Kumar, A., 2018. Comparative potential of hydrocortisone, deoxycorticosterone and dexamethasone in the prevention of cataract: an in silico and in vitro study. Orient. Pharm. Exp. Med 18 (4), 403–422. https://doi.org/10.1007/s13596-018-0332-z.

Reddy, V.N., Giblin, F.J., Lin, L.R., Chakrapani, B., 1998. The effect of aqueous humor ascorbate on ultraviolet-B-induced DNA damage in lens epithelium. Invest. Ophthalmol. Vis. Sci. 39, 344–350.

Sundararajan, M., Thomas, P.A., Babyshalini, K., Geraldine, P., 2017. Identification of phytoconstituents and in-vitro evaluation of the putative anticataractogenic effect of an ethanolic root extract of *Leucas aspera*. Biomed. Pharmacother. 85, 87–101. https://doi.org/10.1016/j.biopha.2016.11.126.

Taylor, H.R., 2016. The global issue of vision loss and what we can do about it: José Rizal Medal 2015. Asia Pacific J. Ophthalmol. 5, 95–96.

Tewari, D., Samoilă, O., Gocan, D., Mocan, A., Moldovan, C., Devkota, H.P., Atanasov, A.G., Zengin, G., Echeverría, J., Vodnar, D., Szabo, B., Crişan, G., 2019. Medicinal plants and natural products used in cataract management. Front. Pharmacol 10, 466.

Wu, J., Li, X., Wan, W., Yang, Q., Ma, W., Chen, D., Hu, J., Chen, C.-Y.O., Wei, X., 2017. Gigantol from *Dendrobium chrysotoxum* Lindl. binds and inhibits aldose reductase gene to exert its anti-cataract activity: an in vitro mechanistic study. J. Ethnopharmacol 198, 255–261. https://doi.org/10.1016/j.jep.2017.01.026.

CHAPTER

5.4 Antioxidants and cognitive decline in elderly

Devina Lobine, Mohamad Fawzi Mahomoodally

Department of Health Sciences, Faculty of Medicine and Health Sciences, University of Mauritius, Réduit, Mauritius

5.4.1 Introduction

Age-related cognitive decline is emerging as one of the greatest health challenges of the twenty-first century. With the projected trends in population ageing and population growth, the prevalence of cognitive decline and risk of dementia is constantly increasing since cognitive impairment results from the physiological process of brain aging (Cardoso et al., 2014; Ogawa, 2014). It is estimated that by 2050, the number of > 60-year-olds will be approximately 2 billion and will account for 22% of the world's population, with the majority living in developing countries. Consequently, the imminent growth in elderly population living with cognitive impairment will placed significantly greater demands of healthcare facilities, leading to increased economic burden (Mavrodaris, 2013).

Cognitive decline is characterized by a decline in memory and other cognitive processes, changes in health behaviours and impaired their everyday functional abilities (Cunningham and Hennessy, 2015; Murman, 2015). Some are within the spectrum of normal aging, whereas others exceed expected decline for corresponding age and are referred as mild cognitive impairment (MCI), which represents an intermediate stage between the expected cognitive decline of normal aging and the more serious decline associated with dementia. Although MCI causes major problems with everyday living, it is associated with greater risk of developing dementia such as Alzheimer's disease (AD) and Parkinson disease (PD) (Plassman et al., 2010; Cardoso et al., 2014). Findings from numerous epidemiological and clinical studies have suggested that onset of cognitive decline is influenced by multiple biological, psychological, social, and environmental factors (Deary, 2009; Ferry and Rousse, 2011; Lipnick et al., 2013). Reports have showed that lifestyle and nutritional status as well as health behaviours, frailty, disability, functionality status and chronic diseases, such as hypertension, cardiovascular diseases and diabetes mellitus contribute to cognitive decline and dementia (Agüero-Torres et al., 2002; Gillette-Guyonnet et al., 2007; Lee et al., 2010; Solfrizzi et al., 2011; Mitnitski et al., 2011).

Currently, available treatments for cognitive impairment are of limited value, with none slowing the progression or attenuating cognitive decline in the elderly.

Antioxidants Effects in Health. DOI: https://doi.org/10.1016/B978-0-128-19096-8.00054-9

Studies are focusing on potential benefits of various pharmacological treatments for optimizing the quality of life of elderly. Antioxidants have received much attention, in particular, because of the increasing evidence that cognitive decline is accompanied by elevated oxidative stress (Amieva et al., 2013).

5.4.2 Oxidative stress and brain aging

Long-term oxidative stress is believed to play a key pathological role in the decline of cognitive functions. The brain is particularly highly vulnerable to oxidative damage: (1) as it is consumes approximately 20% of the body's total oxygen; (2) has a high content of polyunsaturated fatty acids, so it is strongly sensitive to peroxidation; (3) accumulation of redox metals (iron, copper and zinc) which are capable of catalysing the formation of ROS; (4) and lower levels of endogenous antioxidant activity relative to other tissue (Head, 2009; Cardoso et al., 2014; Collin, 2019). Normal metabolic processes result in the generation of reactive oxygen species (ROS) and reactive nitrogen species (RNS), which can lead to accumulated oxidative damage to of critical biomolecules (lipids, proteins, DNA), especially coupled with the inability of the biological systems' defence mechanisms to detoxify those species and repair the resulting cytological damage (Baierle et al., 2015; Usyal et al., 2018). To defence against ROS, the antioxidant system consists of several enzymes, such as catalase, superoxide dismutase (SOD), glutathione peroxidase, and numerous nonenzymatic endogenous antioxidants like glutathione (GSH), or nutritional components, like vitamins A, C, and E, carotenoid and flavonoids (Mehta and Gowder, 2015).

Studies on both normal and pathological brain aging have showed higher levels of oxidative damage to biomolecules occurred (Cini and Moretti, 1995; Montine et al., 2002) in aged as compared to young human brain. Accumulation of several markers of oxidative damage in the aging brain, such as malondialdehyde, 4-hydroxynonenal (4-HNE), fluorescent lipid peroxidation products, protein carbonyls and protein nitrotyrosines have been reported in several rodents' models (Negre-Salvayre et al., 2008; Wilhelm et al., 2016; Zhang and Forman, 2017; Poddar et al., 2019). Furthermore, oxidative stress also results in the inactivation of soluble and membrane-bound enzymes (Na^+/K^+-ATPase and Ca^{2+}-ATPase) and ion channels as well as alteration of fluidity and permeability of membranes in the aging brain (Head, 2009; Zaidi et al., 2015).

Oxidative damage may also play a role in age-associated neurodegenerative diseases, such as Parkinson's disease (PD) and Alzheimers disease (AD). ROS are highly active in the brain and neuronal tissue as excitatory amino acids and neurotransmitters, whose metabolism produces ROS, serving as sources of oxidative stress. ROS attack glial cells and neurons, which are post-mitotic cells and are predominantly susceptible to free radicals; hence, ROS lead to neuronal damage and death (Gilgun-Sherki et al., 2001; Uttara et al., 2009). To combat ROS and eliminate damaged macromolecules, neurons utilize highly effective antioxidant defences and efficient repair, and removal mechanism but in aged brain this defence mechanism goes weak due to the reduced level of antioxidants and low regenerative capacity of aged brain (Uttara et al., 2009; Garbarino et al., 2015).

5.4.3 Effects of antioxidants on cognitive decline

Antioxidants are exogenous (natural or synthetic) or endogenous molecules that scavenge ROS or their precursors, indirectly act to up-regulate antioxidant defences or inhibit ROS generation. Epidemiological studies have demonstrated that antioxidants reduce the rate of cognitive decline by limiting the negative effects of free radicals (Requejo et al., 2003; Uttara et al., 2009). A summary of antioxidants which improve cognitive performance are provided in Table 5.4.1, which while some major antioxidants are explored in detail below.

5.4.3.1 Vitamins and carotenoids

Clinical studies have signified the protective effect of vitamins on brain health. Vitamin E, the most potent lipophilic chain-breaking antioxidant, is naturally found in several foods, including green leafy vegetables, nuts and cereals (Stocker, 2007). Study has showed that Vitamin E is able to cross brain barrier and reach a therapeutic level in the central nervous system, where it suppresses lipid peroxidation and significantly reduces β-amyloid deposition (Sung et al., 2004). Vitamin E is also reported for repairing neuronal damage and reducing β-amyloid deposition in the brain by decreasing isoprostane levels (Praticó et al., 2002). However, some observational studies on dietary or supplemented Vitamin E have showed that intake of Vitamin C and E has no benefit or do not significantly reduce the risk of AD (Zandi et al., 2004; Luchsinger et al., 2003; Yaffe et al., 2004). In other hand, the Chicago health and aging study demonstrated that higher Vitamin E intake, either from diet or supplement use, was associated with improved cognitive function in a dose-dependent manner, over an 18-month follow-up. In the same study, it was observed that carotene was not protective against cognitive decline, while there was limited evidence for an association between cognitive decline and vitamin C (Morris et al., 2009).

Among the carotenoids, lutein (L) and zeaxanthin (Z) are the only two antioxidants that exclusively accumulate in the macula of the primate retina forming macular pigment. Macular pigment filters short-wavelength blue light and protect the retina from photo-oxidative damage (Vishwanathan et al., 2014). They also accumulate in at least four brain areas (cerebellum, frontal cortices, occipital cortices and pons) (Johnson et al., 2012; Ajana et al., 2018). Macular pigment optical density (MPOD) is a non-invasive way of measuring of L+Z concentration. Akbaraly et al. (2007) have reported that existence of L+Z in plasma has showed to improve cognitive function in healthy elderly. Furthermore, L supplementation has been associated with improved measures of executive function, learning and short-term memory in healthy elderly women (Johnson et al., 2008). The work by (Hammond et al., 2017) has demonstrated that elderly adults (62 years) receiving active L and Z supplement has showed significant improvement in cognitive functions as compared elderly placebo. Experimental data from the works carried by Vishwanathan et al. (2014) and Ajana et al. (2018) have suggested that L+Z concentrations in the plasma and MPOD could both be associated with cognitive function.

Table 5.4.1 Overview of clinical studies that showed antioxidants improved cognitive performance in elderly.

Antioxidant (extracts/compounds derived from natural sources/marketed formulations)	Number of subjects	Administered dose/intervention period	Findings	References
Chlorogenic acid	Eight healthy elderly men and women complaining of subjective memory loss	Intake of a test beverage containing 330 mg of chlorogenic acid before bedtime, over a period of 6 months	The Cogstate and CNS Vital Signs test batteries employed in the study showed chlorogenic acid enhanced attentional, executive, and memory functions in the subjects.	Kato et al., 2018
Cocoa flavanols	90 elderly subjects with no clinical evidence of cognitive dysfunction.	Drink containing 993 mg high flavanol], 520 mg [intermediate flavanol], or 48 mg [low flavanol] cocoa flavanols over 6 months.	Regular consumption of cocoa flavanol reduced rate of age-related cognitive decline.	Mastroiacovo et al., 2015
	RCT comprising of 50–69-year-old healthy subjects	Consumed either a high or low cocoa flavanol–containing diet for 3 months.	Improved dentate gyrus function, a region in the hippocampal formation, which reduces in function with increasing age.	Brickman et al., 2014
Bacopa monnieri	107 healthy participants (between ages 18 and 60 years)	Subjects are given extract of *B. monniera* (2 × 150 mg KeenMind) or placebo	Showed cognitive enhancing effects in subjects; improved performance on the 'Working Memory' factor	Stough et al., 2008
	RCT consisting of 54 subjects of 65 years or older with clinical signs of dementia	Orally intake of standardized *B. monnieri* extract (300 mg/day) or a similar placebo tablet for 12 weeks.	Enhancement in performance in a restraint recall and Stroop Task, that is, evaluating the capability to bypass unnecessary input.	Calabrese et al., 2008
	RCT composing of two groups of age 40- 65 years; *B. monniera* (Keenmind extract) group (n = 37) and a placebo group (n-9)	Subject were given capsules at the dosage recommended by the manufacturers; 300 mg for persons under 90 kg, and 450 mg for persons over 90 kg, equivalent to 6g and 9 g dried rhizome, respectively.	Enhanced learning ability.	Pase et al., 2012

Antioxidant (extracts/ compounds derived from natural sources/ marketed formulations)	Number of subjects	Administered dose/ intervention period	Findings	References
Combined nutraceuticals based on *Bacopa monnieri*, L-theanine, *Crocus sativus*, copper, folate and vitamins of B and D group	30 elderly subjects with MMSE score between 20 and 27 and self-perceived cognitive decline.	Orally intake *B. monnieri* dry extract (320 mg), L-theanine (100 mg), *C. sativus* (30 mg), copper (2 mg), folate (400 µg), and vitamins of B (450 µg -9 mg) and D (25 µg) over a period of 2 months	Significant improvement of the cognitive functions tested with the MMSE, PSQ-Index and SRDS score	Cicero et al., 2016
Sideritis scardica and *Bacopa monnieri* extract	10 mild cognitive impairment subjects (mean age: 61.88 ± 6.69 years)	*S. scardica* (500 mg) or *B. monnieri* (320 mg) and three combinations thereof	*Sideritis* extract alone or in combination with a low amount of Bacopa extract lead showed significant improvement during performance of the d2-concentration test.	Dimpfel et al., 2016
Crocus Sativus (saffron)	Single blind, parallel RCT consisting of 35 participants	Oral intake of powered saffron at a dosage 125 mg/ day for 48 weeks	As compared to control, saffron was superior in enhancing cognitive functions. High MMSE score was observed. At dose less than 1.5 g/daily Saffron is considered relatively safe in healthy humans, however, toxic effects are reported with doses 5 g/daily and above with a lethal dose of about 20 g/daily.	Tsolaki et al., 2016; Moshiri et al., 2015
Pomegranate juice – rich in polyphenols	261 subjects of aged 50-75 years with normal aging or MIC	Oral intake of pomegranate juice [8 oz (236.5 mL) per day] or a placebo drink (8 oz, matched constituents of pomegranate juice except for pomegranate polyphenols) over 12 months	Visual memory and visual learning and recall were evaluated at baseline and after 6 and 12 months. Visual memory performance, specifically the ability to learn visual information was stabilised in individuals who consumed pomegranate, whereas the placebo group showed a significant decline ($p = 0.002$).	Siddarth et al., 2020

(*Continued*)

Table 5.4.1 Overview of clinical studies that showed antioxidants improved cognitive performance in elderly. *Continued*

Antioxidant (extracts/compounds derived from natural sources/marketed formulations)	Number of subjects	Administered dose/intervention period	Findings	References
Pyrroloquinoline quinone (PQQ)	41 elderly healthy subjects	Oral intake of 20 mg of BioPQQ™ per day or placebo, for 12 weeks	Prevent decline in cognitive functions in aged persons, especially in attention and working memory.	Itoh et al., 2016
Pycnogenol® (PYC; *Pinus pinaster* ssp. *Atlantica*	150 healthy elderly people of age 55-70 years. Control: Subjects with comparable oxidative stress	Pycnogenol(® supplement was taken daily (100mg/day)	Enhanced cognitive function and reduced oxidative stress in normal subjects between 55 and 70 years of age.	Belcaro et al., 2015
Sailuotong (SLT) – a herbal formulation consisting of 10 bioactive components from *G. biloba* (ginkgo), *P. ginseng* (ginseng), and *C. sativus* (saffron),	RCT consisting of 60 older subjects with MCI	180 mg (4 × 45-mg capsules, 2 in the morning and 2 at night) of SLT per day for 12 weeks.	Enhanced cognition functions in particpants with vascular dementia and improved neuronal activity and cerebral perfusion in healthy adults and in seniors with vascular dementia, respectively	Steiner et al., 2018

CNS, central nervous system; PSQ, perceived stress questionnaire; SRDS, self-rating depression scale.

5.4.3.2 Selenium

Selenium (Se) is an essential trace metal with beneficial properties and is principally known for playing antioxidant role because it is the main constituent of antioxidant enzymes that are expressed in different tissues, including the brain (Nogales et al., 2013). In 2007, a cross-sectional survey conducted by (Gao et al. 2007) has revealed that higher selenium levels were associated with better cognition performance in a dose-dependent manner. Cardoso et al. (2014) have carried out a study to assess the nutritional status of selenium in AD and MCI in elderly and compared them with healthy older adults. The clinical studies carried Yan et al. (2020) and Gerardo et al. (2020) concluded that higher blood selenium is linked with higher cognitive scores in elderly subjects. However, there is report highlighting that high selenium may play a harmful role in the development of hypertension in elderly.

5.4.3.3 Polyphenolics from berries

There is a growing body of studies, although in some cases not conclusive, showing promise for polyphenolics from berries use as a therapeutic for cognitive decline. Shukitt-Hale et al. (2005), showed that diet enriched with high amounts of blueberries reduced age-related behavioural deficits in rats. Duffy et al. (2008) demonstrated that dietary supplementation with high amounts of blueberries can protect the brain against oxidative stress and associated learning deficits. Blueberry appears to have a marked effect on short-term (Ramirez et al., 2005) and long-term memory (Casadesus et al., 2004). A RCT was performed to investigate the effect of two proprietary blueberry formulations (a whole wild blueberry powder at 500 mg and 1000 mg and a purified extract at 100 mg) on cognitive performance in a population of 120 adults of 65–80 years. The subjects were randomly given to a 6-month, daily regimen of either placebo or one of the three interventions and the participants were tested at baseline, 3, and 6 months on a range of cognitive tasks. The results showed that 3 months intervention with purified extract can improve episodic memory performance in the elderly population (Whyte et al., 2018). Recently, (Millin and Rickert, 2018) have investigated the effect of a high antioxidant diet supplemented by freeze-dried spinach and strawberries on spatial learning in early and late middle-aged female rats. The results showed that the supplemented diets of relatively short duration mitigated the mild cognitive decline which was observed in control animals during the developmental period of late middle-age.

5.4.3.4 Polyphenolics from concord grape juice

Concord grape juice is one of the richest sources of phenolic compounds, particularly flavonoids, such as proanthocyanidins and anthocyanins, which are strong antioxidants (Mansouri et al., 2015; Lamport et al., 2016; Yang et al., 2018). Preliminary *in vivo* studies have showed that grape juice supplementation improved memory and motor function, suggesting potential for cognitive benefits in ageing humans (Shukitt-Hale et al., 2006; Siahmard et al., 2012). A RCT was performed to assess the effect of moderate-term supplementation with 100% concord grape juice supplementation in older individuals with memory decline but not dementia. The data revealed memory function was enhanced, and fasting insulin was increased in individuals who consumed grape juice regularly (Krikorian et al., 2010). Lamport et al. (2016) reported that intake of concord grape juice for 12 weeks was associated with slight improve in immediate spatial memory and safer driving behavior relative to the placebo in a cohort of healthy working mothers (aged 40–50 years) of preteen aged children.

5.4.3.5 Polyunsaturated fatty acids from walnuts

Walnuts are well known for their high levels of the polyunsaturated fatty acids (PUFAs), specifically α-linolenic acid and linoleic acid, which are dietary sources of polyphenols, antioxidants and lipids. Improvement in cognitive and motor performance has been demonstrated in both animal and human studies associated with rich walnuts-diets. Willis et al. (2008) investigated the effect of walnut supplementation on motor and cognitive ability in aged rats. Fischer rats (344), aged 19 months, were fed with 2%, 6%, or 9% walnut diet for 8 weeks before motor and cognitive testing were performed. The results indicated that 2 and 6 % walnut were able to improved age-related motor and cognitive deficits as assessed by rod walk, plant walk and Morris water maze (MWM) tests. Rats fed with 2% walnut diet showed significant improvement in performance on rod walking, while animals fed with 6% walnut diet exhibited improved performance on the medium plank walk. A 9% dietary supplementation resulted in large plank impaired reference memory. All of the walnut diets exhibited improved working memory on MWM. The Walnuts and Healthy Aging Study has demonstrated that regular walnut consumption delayed cognitive impairment and retinal pathology in the elderly cohort following daily ingestion of walnuts for 2 years (Rajaram et al., 2017). Although several reports have pointed out that PUFAs supplementations may improve cognitive function (Bo et al., 2017; Jackson et al., 2016; Stavrinou et al., 2020; Zhang et al., 2015), there are studies that show no beneficial effects (Balachandar et al., 2020; Baleztena et al., 2018; Philips et al., 2015; Sydenham et al., 2012). Similarly, some studies have revealed that supplementing older adults with fish oil do not prevent cognitive decline (van de rest et al., 2008; Gao et al., 2011; Danthiir et al., 2018;). No serious adverse effects were reported in conducted studies; however, it is postulated that increased cancer risks may be associated with omega-3 PUFAs supplementation, possibly due to the PUFA oxidation products or added vitamin E (Lange and Nakamura, 2020).

5.4.3.6 Ginkgo biloba

Extracts from *Ginkgo biloba* leaves have a long history in the Chinese traditional medicine, and currently standardized *G. biloba* extract (EGb 761® - Tanakan®) is one of the most marketed herbal medicinal products in Europe as a medication for memory impairment and in the United States as a dietary supplement with health claims (Heinonen and Gaus, 2015). One of well-known therapeutic effects of *G. biloba* extract is protection of neuronal cell membranes from ROS damage; however, the EGb 761® is found to exert its beneficial properties through different mechanisms (Tomino et al., 2021). Studies have showed that EGb 761® reduce amyloid-induced toxicity (Luo et al., 2002; Wu et al., 2006), stabilize mitochondrial function (Abdel-Kader et al., 2007) and enhance hippocampal neurogenesis (Tchantchou et al., 2007). Investigations on rodent models have demonstrated that EGb 761® enhances dopaminergic and cholinergic neurotransmission (Kehr et al., 2012; Yoshitake et al., 2010). Therefore, EGb 761® is considered as a multi-target drug. Randomised controlled trials concluded that EGb 761® is effective in the treatment of patients with Alzheimer's disease (Janssen et al., 2010; Yang et al., 2016). However, Ginkgo Evaluation of Memory study conducted on 3069 participants of age ≥ 75 years with mild cognitive impairment (DeKosky et al., 2008) and the Guid Age study conducted on 2854 participants of age ≥70 years with memory complaints (Vellas et al., 2012) failed to show memory improvement and reduced risk up population-based study (the PAQUID study) to assess the association between EGb 761 intake and cognitive function in elderly adults. The study involved 3612 patients, aged over 65, highlighted a slower progression of cognitive impairment in the EGb 761® group of patients than in Piracetam (drug marketed as a treatment for myoclonus and a cognitive enhancer) ones. Mini-Mental State Examination (MMSE), verbal fluency and visual memory were analysed with a multivariate mixed linear effects model to assess the cognitive decline. A significant difference in MMSE decline was noted in EGb761® and piracetam treatment groups compared to the 'neither treatment' group. It was observed that cognitive decline in the EGb761® group was less rapid than the 'neither treatment' group, whereas the piracetam group showed rapid decline in cognitive decline. These studies affirm the studies affirm a beneficial effect of *G. biloba* in delaying cognitive impairment and enhancing memory performance in elderly people.

Reports have highlighted that a relatively low risk is with the intake of Ginkgo leaf products. Occasional side effects such headaches, dizziness, gastrointestinal disturbances and allergic skin reactions have been observed when Ginkgo leaf extract is consumed in excess. Besides, the long-term safety needs of Ginkgo leaf products to be properly addressed (Amin et al., 2013).

5.4.3.7 Curcumin

Curcumin is a low molecular mass polyphenol that is concentrated in the spice turmeric (*Curcuma longa*) and is credited antioxidant, anti-inflammatory, anti-mutagenic anti-microbial and anti-cancer activities among others. The molecular structure of curcumin and its ability to cross the blood-brain barrier make it a promising potential therapeutic (Mishra and Palanivelu, 2008; Shehzad et al., 2013; Sarkar and Franks, 2018). Turmeric has a long history for its use for an ingredient in curry in Middle east and Asian countries. Various studies have indicated a lower incidence and prevalence of dementia in India and epidemiological studies has attributed lower dementia rates to higher curry consumption (Mishra and Palanivelu, 2008;

Sarkar and Franks, 2018). In 2006, Ng et al., studied the association between curry consumption and cognitive benefits in elderly Asian adults of age 60-93 years. The results revealed that individuals with higher amount of curry consumption performed significantly better on the MMSE of cognitive function compared to those who never or rarely consumed curry. The cross-sectional study carried by Seen et al., 2017 has demonstrated that there is strong association between consumption of curcumin in the form of curry and improved cognition. Although reports support the beneficial health effects of curcumin, its effectiveness and usefulness are highly limited due its low bioavailability. Curcumin is characterized by low water solubility and high instability in most body fluids; thus, it is poorly absorbed by the gastrointestinal tract (Berry et al., 2021).

Conclusion

In humans, increased oxidative damage has been observed in aging brain. Epidemiological studies have suggested that antioxidants intake is beneficial for improving cognitive functions. However, in contrast to strong in vivo positive outcomes, several observational studies in humans have shown no significant reduction in cognitive decline associated with the intake of dietary antioxidants. The contrasting observations may be partly explained by taking inadequate amounts of antioxidants supplements, their form and source, their duration and regularity of use, and the challenges of determining the required optimum amount of dietary intake of antioxidants and the individual's lifestyle. Additionally, the duration of the study is another factor resulting in inconclusive data; therefore, longer studies are required, during which greater changes in cognitive function may occur, to enable researchers to identify possible benefits of antioxidants in preventing cognitive decline.

Abbreviations:

AD Alzheimer's disease
MCI Mild cognitive impairment
MMSE Mini-mental state examination
RCT Randomized controlled trial
ROS Reactive oxidative species

References

Abdel-Kader, R., Hauptmann, S., Keil, U., Scherping, I., Leuner, K., Eckert, A., Müller, W.E., 2007. Stabilization of mitochondrial function by Ginkgo biloba extract (EGb 761). Pharmacoll. Res. 56 (6), 493–502.

Agüero-Torres, H., Thomas, V.S., Winblad, B., Fratiglioni, L., 2002. The impact of somatic and cognitive disorders on the functional status of the elderly. J. Clin. Epidemiol. 55 (10), 1007–1012.

Ajana, S., Weber, D., Helmer, C., Merle, B.M., Stuetz, W., Dartigues, J.F., Rougier, M.B., Korobelnik, J.F., Grune, T., Delcourt, C., Féart, C., 2018. Plasma concentrations of lutein and zeaxanthin, macular pigment optical density, and their associations with cognitive performances among older adults. Invest. Ophthalmol. Vis. Sci. 59 (5), 1828–1835.

Akbaraly, N.T., Faure, H., Gourlet, V., Favier, A., Berr, C., 2007. Plasma carotenoid levels and cognitive performance in an elderly population: results of the EVA Study. J. Gerontol. A Biol. Sci. Med. Sci. 62 (3), 308–316.

Amieva, H., Meillon, C., Helmer, C., Barberger-Gateau, P., Dartigues, J.F., 2013. Ginkgo Biloba extract and long-term cognitive decline: A 20-Year Follow-Up Population-Based Study. PLoS One 8 (1), e52755.

Amin, T., Bhat, S.V., Singh, P., 2013. Ginkgo (Ginkgo biloba)-A promising tree for a nutraceutical industry. Indian Food Industry Mag. 32 (1), 27–34.

Baierle, M., Nascimento, S.N., Moro, A.M., Brucker, N., Freitas, F., Gauer, B., Durgante, J., Bordignon, S., Zibetti, M., Trentini, C.M., Duarte, M.M.F., Grune, T., Breusing, N., Garcia, S.C., 2015. Relationship between inflammation and oxidative stress and cognitive decline in the institutionalized elderly. Oxid. Med. Cell. Longev. 2015, 1–12.

Balachandar, R., Soundararajan, S., Bagepally, B.S., 2020. Docosahexaenoic acid supplementation in age-related cognitive decline: a systematic review and meta-analysis. Eur. J. Clin. Pharmacol. 76, 639–648.

Baleztena, J., Ruiz-Canela, M., Sayon-Orea, C., Pardo, M., Anorbe, T., Gost, J.I., Gomez, C., Ilarregui, B., Bes-Rastrollo, M., 2018. Association between cognitive function and supplementation with omega-3 PUFAs and other nutrients in ≥ 75 years old patients: A randomized multicenter study. PLOS One 13 (3), e0193568.

Belcaro, G., Dugall, M., Ippolito, E., Hu, S., Saggino, A., Feragalli, B., 2015. The COFU3 study. Improvement in cognitive function, attention, mental performance with Pycnogenol® in healthy subjects (55-70) with high oxidative stress. J. Neurosurg. Sci. 59 (4), 437–446.

Berry, A., Collacchi, B., Masella, R., Varì, R., Cirulli, F., 2021. Curcuma longa, the "golden spice" to counteract neuroinflammaging and cognitive decline—what have we learned and what needs to be done. Nutrients 13 (5), 1519.

Bo, Y., Zhang, X., Wang, Y., You, J., Cui, J., Zhu, Y., Pang, W., Liu, W., Jiang, Y., Lu, Q., 2017. The n-3 polyunsaturated fatty acids supplementation improved the cognitive function in the chinese elderly with mild cognitive impairment: A double-blind randomized controlled trial. Nutr. 9 (1), 54.

Brickman, A.M., Khan, U.A., Provenzano, F.A., Yeung, L.K., Suzuki, W., Schroeter, H., Wall, M., Sloan, R.P., Small, S.A., 2014. Enhancing dentate gyrus function with dietary flavanols improves cognition in older adults. Nat. Neurosci *17* (12), 1798–1803.

Calabrese, C., Gregory, W.L., Leo, M., Kraemer, D., Bone, K., Oken, B., 2008. Effects of a standardized *Bacopa monnieri* extract on cognitive performance, anxiety, and depression in the elderly: a randomized, double-blind, placebo-controlled trial. J. Altern. Complement. Med. 14, 707–713.

Casadesus, G., Shukitt-Hale, B., Stellwagen, H.M., Zhu, H.M., Lee, H.G., Smith, M.A., Joseph, J.A., 2004. Modulation of hippocampal plasticity and cognitive behavior by short-term blueberry supplementation in aged Rats. Nutr. Neurosci. 7 (5–6), 306–316.

Cicero, A., Bove, M., Colletti, A., 2016. Short-term impact of a combined nutraceutical on cognitive function, perceived stress and depression in young elderly with cognitive impairment: a pilot, double-blind, randomized clinical trial. J. Prev. Alz. Dis 4, 12–15.

Cini, M., Moretti, A., 1995. Studies on lipid peroxidation and protein oxidation in the aging brain. Neurobiol. Aging 16 (1), 53–57.

Collin, F., 2019. Chemical basis of reactive oxygen species reactivity and involvement in neurodegenerative diseases. Int. J. Mol. Sci. 20 (10), 2407.

Cunningham, C., Hennessy, E., 2015. Co-morbidity and systemic inflammation as drivers of cognitive decline: new experimental models adopting a broader paradigm in dementia research. Alzheimers Res. Ther. 7 (1), 33.

Danthiir, V., Hosking, D.E., Nettelbeck, T., Vincent, A.D., Wilson, C., O'Callaghan, N., Calvaresi, E., Clifton, P., Wittert, G.A., 2018. An 18-mo randomized, double-blind, placebo-controlled trial of DHA-rich fish oil to prevent age-related cognitive decline in cognitively normal older adults. Am. J. Clin. Nutr. 107 (5), 754–762.

Deary, I.J., Corley, J., Gow, A.J., Harris, S.E., Houlihan, L.M., Marioni, R.E., Penke, L., Rafnsson, S.B, Starr, J.M., 2009. Age-associated cognitive decline. Br. Med. Bull. 92 (1), 135–152.

DeKosky, S.T., Williamson, J.D., Fitzpatrick, A.L., Kronmal, R.A., Ives, D.G., Saxton, J.A., Lopez, O.L., Burke, G., Carlson, M.C., Fried, L.P., Kuller, L.H., 2008. Ginkgo biloba for prevention of dementia: a randomized controlled trial. JAMA 300 (19), 2253–2262.

Dimpfel, W., Schombert, L., Biller, A., 2016. Psychophysiological effects of Sideritis and Bacopa extract and three combinations thereof—a quantitative EEG study in subjects suffering from mild cognitive impairment (MCI). Adv. Alzheimer's Dis. 5 (01), 1.

Duffy, K.B., Spangler, E.L., Devan, B.D., Guo, Z., Bowker, J.L., Janas, A.M., Hagepanos, A., Minor, R.L., DeCabo, R., Mouton, P.R., Shukitt-Hale, B., Joseph, J.A., Ingram, D.K., 2008. A blueberry-enriched diet provides cellular protection against oxidative stress and reduces a kainate-induced learning impairment in rats. Neurobiol. Aging 29 (11), 1680–1689.

Ferry, M., Roussel, A.-M., 2011. Micronutrient status and cognitive decline in ageing. Eur. Geriatr. Med. 2 (1), 15–21.

Gao, S., Jin, Y., Hall, K.S., Liang, C., Unverzagt, F.W., Ji, R., Murrell, J.R., Cao, J., Shen, J., Ma, F., Matesan, J., Ying, B., Cheng, Y., Bian, J., Li, P., Hendrie, H.C., 2007. Selenium level and cognitive function in rural elderly Chinese. Am. J. Epidemiol. 165 (8), 955–965.

Gao, Q., Niti, M., Feng, L., Yap, K.B., Ng, T.P., 2011. Omega-3 polyunsaturated fatty acid supplements and cognitive decline: Singapore longitudinal aging studies. J. Nutr. Health Aging 15, 32–35.

Garbarino, V.R., Orr, M.E., Rodriguez, K.A., Buffenstein, R., 2015. Mechanisms of oxidative stress resistance in the brain: Lessons learned from hypoxia tolerant extremophilic vertebrates. Arch. Biochem. Biophys 576, 8–16.

Gerardo, B., Cabral Pinto, M., Nogueira, J., Pinto, P., Almeida, A., Pinto, E., Marinho-Reis, P., Diniz, L., Moreira, P.I., Simões, M.R., Freitas, S., 2020. Associations between trace elements and cognitive decline: an exploratory 5-year follow-up study of an elderly cohort. nt. J. Environ. Res. Public Health. 17 (17), 6051.

Gilgun-Sherki, Y., Melamed, E., Offen, D., 2001. Oxidative stress induced-neurodegenerative diseases: the need for antioxidants that penetrate the blood brain barrier. Neuropharmacol 40 (8), 959–975.

Gillette-Guyonnet, S., Van Kan, G.A., Andrieu, S., Barberger-Gateau, P., Berr, C., Bonnefoy, M., Dartigues, J.F., De Groot, L., Ferry, M., Galan, P., Hercberg, S., 2007. IANA task force on nutrition and cognitive decline with aging. J. Nutr. Health Aging 11 (2), 132.

Hammond Jr, B.R., Miller, L.S., Bello, M.O., Lindbergh, C.A., Mewborn, C., Renzi-Hammond, L.M., 2017. Effects of lutein/zeaxanthin supplementation on the cognitive function of community dwelling older adults: A randomized, double-masked, placebo-controlled trial. Front. Aging Neurosci. 9, 254.

Head, E., 2009. Oxidative damage and cognitive dysfunction: antioxidant treatments to promote healthy brain aging. Neurochem. Res. 34 (4), 670–678.

Heinonen, T., Gaus, W., 2015. Cross matching observations on toxicological and clinical data for the assessment of tolerability and safety of Ginkgo biloba leaf extract. Toxicol. 327, 95–115.

Itoh, Y., Hine, K., Miura, H., Uetake, T., Nakano, M., Takemura, N., Sakatani, K., 2016. Effect of the Antioxidant Supplement Pyrroloquinoline Quinone Disodium Salt (BioPQQ^{TM}) on Cognitive Functions. In: Elwell, C.E., Leung, T.S., Harrison, D.K. (Eds.), Oxygen Transport to Tissue XXXVII. Advances in Experimental Medicine and Biology, 876. Springer, New York, pp. 319–325.

Jackson, P.A., Forster, J.S., Bell, J.G., Dick, J.R., Younger, I., Kennedy, D.O., 2016. DHA supplementation alone or in combination with other nutrients does not dodulate cerebral hemodynamics or cognitive function in healthy older adults. Nutr. 8 (2), 86.

Janßen, I.M., Sturtz, S., Skipka, G., Zentner, A., Garrido, MV, Busse, R., 2010. Ginkgo biloba in Alzheimer's disease: a systematic review. Vienna Medical Weekly 160 (21-22), 539–546.

Johnson, E.J., 2012. A possible role for lutein and zeaxanthin in cognitive function in the elderly. Am. J. Clin. Nutr. 96 (5), 1161S–1165S.

Johnson, E.J., Chung, H.Y., Caldarella, S.M., Snodderly, D.M., 2008. The influence of supplemental lutein and docosahexaenoic acid on serum, lipoproteins, and macular pigmentation. Am. J. Clin. Nutr. 87 (5), 1521–1529.

Kato, M., Ochiai, R., Kozuma, K., Sato, H., Katsuragi, Y., 2018. Effect of chlorogenic acid intake on cognitive function in the elderly: a pilot study. Evid. Based Complement. Alternat. Med. 2018. Doi: https://doi.org/10.1155/2018/8608497.

Kehr, J., Yoshitake, S., Ijiri, S., Koch, E., Nöldner, M., Yoshitake, T., 2012. Ginkgo biloba leaf extract (EGb 761®) and its specific acylated flavonol constituents increase dopamine and acetylcholine levels in the rat medial prefrontal cortex: possible implications for the cognitive enhancing properties of EGb 761®. Int. Psychogeriatr. 24 (S1), S25–S34.

Krikorian, R., Nash, T.A., Shidler, M.D., Shukitt-Hale, B., Joseph, J.A., 2010. Concord grape juice supplementation improves memory function in older adults with mild cognitive impairment. Bri. J. Nutr. 103 (5), 730–734.

Lamport, D.J., Lawton, C.L., Merat, N., Jamson, H., Myrissa, K., Hofman, D., Chadwick, H.K., Quadt, F., Wightman, J.D., Dye, L., 2016. Concord grape juice, cognitive function, and driving performance: a 12-wk, placebo-controlled, randomized crossover trial in mothers of preteen children. Am. J. Clin. Nutr. 103 (3), 775–783. doi:10.3945/ajcn.115.114553.

Lange, K.W., Nakamura, Y., 2020. The role of omega-3 polyunsaturated fatty acids in mental disorders. J. Dis. Prev. Health Promot. 4, 73–82.

Lee, Y., Back, J.H., Kim, J., Kim, S.H., Na, D.L., Cheong, H.K., Hong, C.H., Kim, Y.G., 2010. Systematic review of health behavioral risks and cognitive health in older adults. Int. Psychogeriatr. 22 (2), 174–187.

Lipnicki, D.M., Sachdev, P.S., Crawford, J., Reppermund, S., Kochan, N.A., Trollor, J.N., Draper, B., Slavin, M.J., Kang, K., Lux, O., Mather, K.A., Brodaty, H., 2013. Risk factors for late-life cognitive decline and variation with age and sex in the sydney memory and ageing study. PLoS One 8 (6), e65841.

Luchsinger, J.A., Tang, M.-X., Shea, S., Mayeux, R., 2003. Antioxidant vitamin intake and risk of alzheimer disease. Arch. Neurol. 60 (2), 203.

Luo, Y., Smith, J.V., Paramasivam, V., Burdick, A., Curry, K.J., Buford, J.P., Khan, I., Netzer, W.J., Xu, H., Butko, P., 2002. Inhibition of amyloid-β aggregation and caspase-3 activation by the Ginkgo biloba extract EGb761Proc. Natl. Acad. Sci. 99 (19), 12197–12202.

Mansouri, E., Khorsandi, L., Fard, A.A., 2015. Protective role of grape seed proanthocyanidin antioxidant properties on heart of streptozotocin-induced diabetic rats. Vet. Res. Forum 6 (2), 119–124.

Mastroiacovo, D., Kwik-Uribe, C., Grassi, D., Necozione, S., Raffaele, A., Pistacchio, L., Righetti, R., Bocale, R., Lechiara, M.C., Marini, C., Ferri, C., 2015. Cocoa flavanol consumption improves cognitive function, blood pressure control, and metabolic profile in elderly subjects: the cocoa, cognition, and aging (CoCoA) study—a randomized controlled trial. Am. J. Clin. Nutr. *101* (3), 538–548.

Mavrodaris, A., Powell, J., Thorogood, M., 2013. Prevalences of dementia and cognitive impairment among older people in sub-Saharan Africa: a systematic review. Bull. World Health Organ. 91 (10), 773–783.

Mehta, S.K., Gowder, S.J.T., 2015. Members of antioxidant machinery and their functionsBasic Principles and Clinical Significance of Oxidative Stress. BoD-Book on Demand, UK, 55–59. doi:10.5772/61884.

Millin, P.M., Rickert, G.T., 2018. Effect of a strawberry and spinach dietary supplement on spatial learning in early and late middle-aged female rats. Antioxidants 8 (1), 1–13.

Mishra, S., Palanivelu, K., 2008. The effect of curcumin (turmeric) on Alzheimer's disease: an overview. Ann. Indian Acad. Neurol. 11 (1), 1–13.

Mitnitski, A., Fallah, N., Rockwood, M.R.H., Rockwood, K., 2011. Transitions in cognitive status in relation to frailty in older adults: a comparison of three frailty measures. J. Nutr. Health Aging 15 (10), 863–867.

Montine, T.J., Neely, M.D., Quinn, J.F., Beal, M.F., Markesbery, W.R., Roberts, L.J., Morrow, J.D., 2002. Lipid peroxidation in aging brain and Alzheimer's disease. Free Radic. Biol. Med. 33 (5), 620–626.

Morris, M.C., 2009. The role of nutrition in Alzheimer's disease: epidemiological evidence. Eur. J. Neurol. 16, 1–7.

Moshiri, M., Vahabzadeh, M., Hosseinzadeh, H., 2015. Clinical applications of saffron (*Crocus sativus*) and its constituents: a review. Drug. Res. 65 (6), 287–289.

Murman, D.L., 2015. The impact of age on cognition. Semin Hear 36 (3), 111–121.

Negre-Salvayre, A., Coatrieux, C., Ingueneau, C., Salvayre, R., 2008. Advanced lipid peroxidation end products in oxidative damage to proteins. Potential role in diseases and therapeutic prospects for the inhibitors. Br. J. Pharmacol. 153 (1), 6–20.

Ng, T.P., Chiam, P.C., Lee, T., Chua, H.C., Lim, L., Kua, E.H., 2006. Curry consumption and cognitive function in the elderly. Am. J. Epidemiol. 164 (9), 898–906.

Nogales, F., Ojeda, M.L., Fenutría, M., Murillo, M.L., Carreras, O., 2013. Role of selenium and glutathione peroxidase on development, growth, and oxidative balance in rat offspring. Reproduction 146 (6), 659–667.

Ogawa, S., 2014. Nutritional management of older adults with cognitive decline and dementia. Geriatr. Gerontol. Int. 14, 17–22.

Pase, M.P., Kean, J., Sarris, J., Neale, C., Scholey, A.B., Stough, C., 2012. The cognitive-enhancing effects of Bacopa monnieri: a systematic review of randomized, controlled human clinical trials. J. Altern. Complement. Med *18* (7), 647–652.

Philips, M.A., Childs, C.E., Calder, P.C., Rogers, P.J., 2015. No effect of Omega-3 fatty acid supplementation on cognition and mood in individuals with cognitive impairment and probable Alzheimer's Disease: A randomised controlled trial. Int. J. Mol. Sci. 16 (10), 24600–24613.

Plassman, B.L., Williams, J.W., Burke, J.R., Holsinger, T., Benjamin, S., 2010. Systematic review: factors associated with risk for and possible prevention of cognitive decline in later life. Ann. Intern. Med 153 (3), 182–193.

Poddar, J., Pradhan, M., Ganguly, G., Chakrabarti, S., 2019. Biochemical deficits and cognitive decline in brain aging: intervention by dietary supplements. J. Chem. Neuroanat. 95, 70–80.
Praticò, D., Clark, C.M., Liun, F., Lee, V.Y.-M., Trojanowski, J.Q., 2002. Increase of brain oxidative stress in mild cognitive impairment. Arch. Neurol. 59 (6), 972.
Rajaram, S., Valls-Pedret, C., Cofán, M., Sabaté, J., Serra-Mir, M., Pérez-Heras, A.M., Arechiga, A., Casaroli-Marano, R.P., Alforja, S., Sala-Vila, A., Doménech, M., Roth, I., Freitas-Simoes, T.M., Calvo, C., López-Illamola, A., Haddad, E., Bitok, E., Kazzi, N., Huey, L., Fan., Ros, E., 2017. The walnuts and healthy aging study (WAHA): protocol for a nutritional intervention trial with walnuts on brain aging. Front. Aging Neurosci. 8, 1–12.
Ramirez, M.R., Izquierdo, I., Raseira, M.B.C.B., Zuanazzi, J.A., Barros, D., Henriques, A.T., 2005. Effect of lyophilised Vaccinium berries on memory, anxiety and locomotion in adult rats. Pharmacol. Res. 52 (6), 457–462.
Requejo, A.M., Ortega, R.M., Robles, F., Navia, B., Faci, M., Aparicio, A., 2003. Influence of nutrition on cognitive function in a group of elderly, independently living people. Eur. J. Clin. Nutr. 57 (S1), S54.
Rita Cardoso, B., Silva Bandeira, V., Jacob-Filho, W., Franciscato-Cozzolino, S.M., 2014. Selenium status in elderly: Relation to cognitive decline. J. Trace Elem. Med. Biol. 28 (4), 422–426.
Sarker, M.R., Franks, S.F., 2018. Efficacy of curcumin for age-associated cognitive decline: a narrative review of preclinical and clinical studies. Geroscience 40 (2), 73–95.
Seen, W.P., Mun, T.Y., Mohanty, B.K., Ebenezer, E., Sugathan, S., 2017. Curcumin consumption and cognitive function in elderly. Int. J. Pharma. Sci. Res. 8 (12), 5367–5372.
Shehzad, A., Rehman, G., Lee, Y.S., 2013. Curcumin in inflammatory diseases. Biofactors 39 (1), 69–77.
Shukitt-Hale, B., Carey, A., Simon, L., Mark, D.A., Joseph, J.A., 2006. Effects of Concord grape juice on cognitive and motor deficits in aging. Nutr. 22 (3), 295–302.
Shukitt-Hale, B., Galli, R.L., Meterko, V., Carey, A., Bielinski, D.F., McGhie, T., Joseph, J.A., 2005. Dietary supplementation with fruit polyphenolics ameliorates age-related deficits in behavior and neuronal markers of inflammation and oxidative stress. AGE 27 (1), 49–57.
Siahmard, Z., Alaci, H., Reisi, P., Pilehvarian, A.A., 2012. The effect of red grape juice on Alzheimer's disease in rats. Adv. Biomed. Res. 1, 63.
Siddarth, P, Li, Z, Miller, K.J., Ercoli, L.M., Merril, D.A., Henning, S.M., Heber, D, Small, G.W., 2020. Randomized placebo-controlled study of the memory effects of pomegranate juice in middle-aged and older adults. Am. J. Clin. Nutr 1 (1), 170–177 111.
Solfrizzi, V., Panza, F., Frisardi, V., Seripa, D., Logroscino, G., Imbimbo, B.P., Pilotto, A., 2011. Diet and Alzheimer's disease risk factors or prevention: the current evidence. Exp. Rev. Neurother. 11 (5), 677–708.
Stavrinou, P.S., Eleni, A., Aphamis, G., Pantzaris, M., Loannou, M., Patrikios, I.S., Giannaki, C.D., 2020. The effects of a 6-month high dose omega-3 and omega-6 polyunsaturated fatty acids and antioxidant vitamins supplementation on cognitive function and functional capacity in older adults with mild cognitive impairment. Nutr. 12 (2), 325.
Steiner, G.Z., Bensoussan, A., Liu, J., Hihenberg, M.I., Chang, D.H., 2018. Study protocol for a randomised, double-blind, placebo-controlled 12-week pilot phase II trial of Sailuotong (SLT) for cognitive function in older adults with mild cognitive impairment. Trials 19, 522.
Stocker, R., 2007. Vitamin E. In: Bock, G., Goode, J. (Eds.), Dietary Supplements and Health: Novartis Foundation Symposia, 282. John Wiley & Sons, pp. 77–87.
Stough, Con, Downey, Luke A., Lloyd, Jenny, Silber, Beata, Redman, Stephanie, Hutchison, Chris, Wesnes, Keith, Nathan, Pradeep J., 2008. Examining the nootropic effects of a

special extract of Bacopa monniera on human cognitive functioning: 90 day double-blind placebo-controlled randomized trial. Phytother. Res. 22 (12), 1629–1634.

Sung, S., Yao, Y., Uryu, K., Yang, H., Lee, V.M., Trojanowski, J., Praticò, D., 2004. Early vitamin E supplementation in young but not aged mice reduces Aβ levels and amyloid deposition in a transgenic model of Alzheimer's disease. FASEB J. 18 (2), 323–325.

Sydenham, E., Danger, A.D., Lim, W.S., 2012. Omega 3 fatty acid for the prevention of cognitive decline and dementia. Cochrane Database Syst. Rev. 1–42. doi:https://doi.org/10.1002/14651858.CD005379.pub3.

Tchantchou, F., Xu, Y., Wu, Y., Christen, Y., Luo, Y., 2007. EGb 761 enhances adult hippocampal neurogenesis and phosphorylation of CREB in transgenic mouse model of Alzheimer's disease. FASEB J. 21 (10), 2400–2408.

Tomino, C., Ilari, S., Solfrizzi, V., Malafoglia, V., Zilio, G., Russo, P., Proietti, S., Marcolongo, F., Scapagnini, G., Muscoli, C., Rossini, P.M., 2021. Mild cognitive impairment and mild dementia: the role of Ginkgo biloba (EGb 761®). Pharmaceuticals 14 (4), 305.

Tsolaki, M., Karathanasi, E., Lazarou, I., Dovas, K., Verykouki, E., Karacostas, A., Georgiadis, K., Tsolaki, A., Adam, K., Kompatsiaris, I., Sinakos., Z., 2016. Efficacy and safety of *crocus sativus* L. in patients with mild cognitive impairment: one year single-blind randomized, with parallel groups, clinical trial. J. Alzheimers Dis. 54 (1), 129–133.

Uttara, B., Singh, A., Zamboni, P., Mahajan, R., 2009. Oxidative stress and neurodegenerative diseases: a review of upstream and downstream antioxidant therapeutic options. Curr. Neuropharmacol. 7 (1), 65–74.

Uysal, S., Senkardes, I., Mollica, A., Zengin, G., Bulut, G., Dogan, A., Glamočlija, J., Soković, M., Lobine, D., Mahomoodally, F.M., 2018. Biologically active compounds from two members of the Asteraceae family: Tragopogon dubius Scop. and Tussilago farfara L. J. Biomol. Struct. Dyn. 37 (12), 1–13.

van de Rest, O., Geleijnse, J.M., Kok, F.J., van Staveren, W.A., Dullemeijer, C., OldeRikkert, M.G., Beekman, A.T., De Groot, C.P.G.M., 2008. Effect of fish oil on cognitive performance in older subjects: a randomized, controlled trial. Neurology *71* (6), 430–438.

Vellas, B., Coley, N., Ousset, P.J., Berrut, G., Dartigues, J.F., Dubois, B., Grandjean, H., Pasquier, F., Piette, F., Robert, P., Touchon, J., 2012. Long-term use of standardised Ginkgo biloba extract for the prevention of Alzheimer's disease (GuidAge): a randomised placebo-controlled trial. Lancet Neurol. 11 (10), 851–859.

Vishwanathan, R., Iannaccone, A., Scott, T.M., Kritchevsky, S.B., Jennings, B.J., Carboni, G., Forma, G., Satterfield, S., Harris, T., Johnson, K.C., Schalch, W., 2014. Macular pigment optical density is related to cognitive function in older people. Age Ageing 43 (2), 271–275.

Whyte, A., Cheng, N., Fromentin, E., Williams, C., 2018. A randomized, double-blinded, placebo-controlled study to compare the safety and efficacy of low dose enhanced wild blueberry powder and wild blueberry extract (ThinkBlue™) in maintenance of episodic and working memory in older adults. Nutrients 10 (6), 660.

Wilhelm, J., Vytášek, R., Uhlík, J., Vajner, L., 2016. Oxidative stress in the developing rat brain due to production of reactive oxygen and nitrogen species. Oxid. Med. Cell Longev. 2016.

Willis, L.M., Shukitt-Hale, B., Cheng, V., Joseph, J.A., 2008. Dose-dependent effects of walnuts on motor and cognitive function in aged rats. Br. J. Nutr. 101 (8), 1140–1144.

Wu, Y., Wu, Z., Butko, P., Christen, Y., Lambert, M.P., Klein, W.L., Link, C.D., Luo, Y., 2006. Amyloid-β-induced pathological behaviors are suppressed by Ginkgo biloba extract EGb 761 and ginkgolides in transgenic Caenorhabditis elegans. J. Neurosci. 26 (50), 13102–13113.

Yaffe, L., Clemons, T.E., McBee, W.L., Lindblad, A.SAge-related Eye Disease Study Research Group, 2004. Impact of antioxidants, zinc and copper on cognition in the elderly: a randomized, controlled trial. Neurol 63 (9), 1705–1707.

Yan, X., Liu, K., Sun, X., Qin, S., Wu, M., Qin, L., Wang, Y., Li, Z., Zhong, X., Wei, X., 2020. A cross-sectional study of blood selenium concentration and cognitive function in elderly Americans: National Health and Nutrition Examination Survey 2011–2014. Ann. Human Biol. 47 (7-8), 610–619.

Yang, G., Wang, Y., Sun, J., Zhang, K., Liu, J., 2016. Ginkgo biloba for mild cognitive impairment and Alzheimer's disease: a systematic review and meta-analysis of randomized controlled trials. Curr. Top Med. Chem 16 (5), 520–528.

Yang, L., Xian, D., Xiong, X., Lai, R., Song, J., Zhong, J., 2018. Proanthocyanidins against oxidative stress: from molecular mechanisms to clinical applications. Biomed. Res. Int. 2018, 1–11.

Yoshitake, T., Yoshitake, S., Kehr, J., 2010. The Ginkgo biloba extract EGb 761® and its main constituent flavonoids and ginkgolides increase extracellular dopamine levels in the rat prefrontal cortex. Br. J. Pharmacol. 159 (3), 659–668.

Zaidi, A., Sheikh, S., Williams, S., Ramlow, P., 2015. Defective calcium pumps in neurons in the aging brain and in parkinson's disease. Eur. Sci. J. 11 (10).

Zandi, P.P., Anthony, J.C., Khachaturian, A.S., Stone, S.V., Gustafson, D., Tschanz, J.T., Norton, M.C., Welsh-Bohmer, K.A., Breitner, J.C.Cache County Study Group, 2004. Reduced risk of Alzheimer disease in users of antioxidant vitamin supplements: the Cache County Study. Arch. Neurol 61 (1), 82–88.

Zhang, X.W., Hou, W.S., Li, M., Tang, Z.Y., 2015. Omega-3 fatty acids and risk of cognitive decline in the elderly: a meta-analysis of randomized controlled trials. Aging Clin. Exp. Res. 28, 165–166.

Zhang, H., Forman, H.J., 2017. 4-hydroxynonenal-mediated signaling and aging. Free Radic. Biol. Med. 111, 219–225.

CHAPTER

Antioxidant and dentistry

5.5

Govind Rajpal[a], Lokesh Patni[b], Arasana Dhariwal[a], Ankit Kumar[a], Sandeep Visvarma[c], Aadesh Kumar[d]

[a]*Department of Pharmaceutical Sciences, Sir J.C. Bose Technical Campus, Nainital, Uttarakhand, India*

[b]*Shree Dev Dental Clinic, Nainital, Uttarakhand, India*

[c]*Deep Dental Clinic, Nainital, Uttarakhand, India*

[d]*Department of Pharmaceutical Sciences, Kumaun University, Campus Bhimtal, Bhimtal, Uttarakhand, India*

5.5.1 Introduction

One of the major and leading concerns about molecules available in natural sources are antioxidants. These molecules are useful for reducing as well as preventing the oxidation of some alternative molecules. When oxidation occurs free radicals are produced, these complimentary radicals are free to bind with another molecule and they cause detrimental series reactions that cause cell injury or cell death. This was observed that these molecules are responsible for the oxidation of low-density lipoprotein (LDL) in cardiovascular diseases (CVS) and carcinogenesis. Ageing is one of the factors that occur due to free radicals release. Health-related problems also occur when low levels of antioxidants are taken with diet they may be gout, osteoporosis, skin problems, hair loss, and obesity. Free radicals are neutralized by the antioxidants as they donate their free electrons and helps in chain termination which ends the electron taking reaction (Bhuvaneswari, 2014; Cadenas, 1989).

Free radicals are known to be chemical ions or active ions that hold some charge which is generated due to surplus or inferior count of electrons. They are categorized into two types according to their species; one is a potent reactive chemical known as reactive oxygen species (ROS) and the other one is the group of nitric oxide derivative compound (Cadenas, 1989). Oxygen-free radicals damage the biological mechanism is known as ROS. X-rays, γ-rays, and UV light irradiation may produce damaging free radicals. Free radicals are produced and remain dispersed in the atmosphere as pollutants after the catalyzation of metals. Macrophages and neutrophils also produce free radicals during the Inflammation cycle in the human body. Free radicals also developed in the process of energy generation, when mitochondrial cells utilize oxygen (O_2) to create power in the form of adenosine triphosphate (ATP) molecule (Valko et al., 2004; Tiwari, 2004). The oxidative stress would be increased

Antioxidants Effects in Health. DOI: https://doi.org/10.1016/B978-0-12-819096-8.00055-0

by an inequality in connecting the free radical management and antioxidant count to rummage ROS (Shivana and Gupta, 2013). Based on lipophilic and hydrophilic nature antioxidant can be divided into two subcategories.

Hydrophilic antioxidants: These are responsible for reaction with oxidants in blood plasma.
Lipophilic antioxidants: These types of antioxidants help to preserve cell laminate through lipid catabolism.

There are various types of antioxidants available within nature, which could be in the form of enzymes, vitamin, and phytochemicals. Green algae, plants, and fresh leafy veggies are a great source of antioxidants, plants developed antioxidants with the help of UV light present in Sun exposure (Bhuvaneswari, 2014). Attack of free radicals on the human body could be prohibited with the help of antioxidants by creating an extremely composite antioxidant complex that may be enzymic or non-enzymic. Antioxidants work harmonious and in sequence with one to prohibit the oxidation of cells as well as organs (Shivana and Gupta, 2013).

Based on occurrence antioxidants can be classified as:

1. Exogenous
2. Endogenous or internal

Exogenous antioxidants are such type of antioxidants which we grab from our diet and supplements that enriched with antioxidant (e.g., retinol, ascorbic acid, and tocopherols). It seems impractical to grab enough antioxidants from our daily diet and food habits to compensate for the daily scavenge of free radicals automatically. So there is the necessity to complete the requirement with a supplement. Supplementation could be of dosage form like capsule, liquid, or tablet that meet the need of exogenous antioxidants. In dentistry, the source of antioxidants may be the kinds of toothpaste, mouth bathe, or oral sprinkler that containing antioxidant additives. The majority of oral additives include green tea, propolis from beehives, dried grape seed, and extract of pine bark (Bhuvaneswari, 2014).

Endogenous antioxidants are those which are produced by our body. Internal antioxidants are responsible for the repair of all damage occurred by the free radical while commencing cell reclamation from the innards on away, instead of it exogenous antioxidants helps only in overhaul a few of the free radical destruction from the surface by challenging (not initiating) cell reconstruction. Examples of five highly dominant internal antioxidants include glutathione (gamma-glutamylcysteinylglycine, GSH receive from yeast and animal tissue extract), alpha lipoic acid (gain from yeast, liver, kidney, and spinach), superoxide dismutase (gain from melon), catalase (present in living tissue), and coenzyme Q10 (CoQ10) (Bhuvaneswari, 2014).

5.5.2 Enzymes as antioxidant

Different enzymes are synthesized inside the human body and work by scavenging free radicals inside the body. These enzymes are superoxide dismutase (SOD), glutathione peroxidase, glutathione reductase, and catalases. Those are the good sample of antioxidants that drop in from the protein and minerals present in our daily food intake. Generally, enzymes required some co-factors to produce maximum antioxidant activity. These co-factors are iron, copper, selenium, magnesium, and zinc (Valko et al., 2004).

1. *Superoxide dismutase*: SOD plays a decisive aspect in the scavenging of ROS. It abolishes anions of superoxide that comes from extracellular origin (Tosun et al., 2019). Superoxide dismutase is the association of stimulant or enzymes that facilitate cellular protection form the ROS by converting free radicals to oxygen (O_2) and hydrogen peroxide (H_2O_2) molecule. Extensive research on SOD conclude that therapeutically SOD is used as anti-inflammatory, anticarcinoma, and protection from radiation (Wang and Zhang, 2015).
2. *Glutathione peroxidase*: Glutathione is a tri-peptide molecule containing cystine, glycine, and glutamic acid, developed inside the body by cellular metabolism. It helps in biotransformation of xenobiotics materials, and protect human body from oxidation and reducing agent. Glutathione transferase enzymes family helps by scavenging of electrophile from the body, which is responsible for cellular deterioration and mutations (Gad, 2014).
3. *Glutathione reductase*: The role in preserving the cellular capacity of antioxidants is Glutathione reductase (GR) (Cardoso et al., 2008). The supply of reduced glutathione is maintained by glutathione reductase; in most cells, one of the most abundant thiol reducers. Glutathione plays an essential character by reducing the count of ROS cells. Species of reactive oxygen act as intracellular and extracellular signaling agents, and complicated chat between ROS heights, glutathione, and other thiols oxidized and levels are decreased also providing the most suited redox control conditions within the cell or the activation of programmed cell deaths by antioxidant enzymes such as glutathione reductase (Couto et al., 2016).
4. *Catalases*: One of the most significant enzymes of antioxidants is the catalase. When hydrogen peroxide is broken down into harmless products like water and oxygen. Catalase is employed as a treatment for several stress-related oxidative illnesses. It is still difficult to implement the catalase enzyme in acceptable concentrations at the suitable place. Nanoparticles for the delivery of catalase to human neural cells and protection against oxidative stress were assessed with these catalase-loaded nanoparticles (Nandi et al., 2019).

5.5.3 Antioxidants and dental caries (antioxidants have a more preventive than a curative effect on following oral problems)

The creation of cavities in the teeth is one of the most common oral conditions. Dental cavities are developed because of lack of available antioxidant, free radical generation, and ROS in saliva (Battino et al., 2002). Uric acid, glutathione, ascorbic acid, albumin, and antioxidant enzymes such as salivary peroxidase are the main antioxidants. Salivary peroxidase system, salivary antioxidant systems, and salivary peroxidase catalyze the peroxidation of thiocyanate ion (–SCN) to generate oxidation products (more stable OSCN–); to suppresses the growth and metabolism, hence inhibiting caries, of several microorganisms. The uric acid constitutes 70% of the total antioxidant potential of the salivary antioxidant system (Cadenas, 1989; Shivana and Gupta, 2013).

Sources of free radical insult in dental therapy (Chapple et al., 2007).

In dentistry, many commonly used dental materials may form free radicals like.

1. Materials used as bleaching agents.
2. Some components of composite fillings.
3. Materials associated with dental cements.
4. Materials used for ceramic restoration.
5. Metals used in restoration of tooth.
6. Tooth implants.
7. Drugs used for intracanal medicament.

There are various disorders in the oral cavity which have more preventive than therapeutic effects when studying antioxidants.

5.5.4 Periodontology

The inflammatory response between the bacterial infection and the inflame reaction of the host is periodontal disorders. The main cause of inflammatory response is freely available radicals and ROS. ROS overproduction is evaluated by periodontal pathogens, leading to collagen and periodontal tissue collapse. Collagen breakdown by scavenging ROS generation reverses with the help of antioxidants. Support for antioxidants against periodontal diseases is favored. Ascorbic acid insufficiency is one of the conditions causing gingivitis (Cadenas, 1989; Valko et al., 2004).

Free radical chain response can be damaged by plant oil and green leafy vegetables. Thus, periodontal inflammation can be reduced by these natural sources. Flavonoids from natural sources acquiring antioxidants and anti-inflammatory characteristics, which reduce inflammatory molecule remark in immune system fighters. For example, monocytes found in the gingival connective tissues. In addition to this, the fraction of the cranberry can prevent the production of biofilm by the main pathogen of chronic periodontitis called *Phorphyromonas gingivalis*. Vitamin E, enriched with the capacity that reduce periodontitis oxidative damage.

Many researchers believe that poor nutrition is more rapid and serious in population with poor nutritional diets. The compromised host response not only because of periodontal disease but also characterized by poor nutrition (Dahiya et al., 2013; Abebe, 2003).

The risk of gum disease is considerably enhanced by low intake of vitamins A, C, β-carotene, and β-cryptoxanthin as well. Low antioxidant content is a risk factor for periodontal and infectious diseases. As a result of bacterial clearance and death, free radicals are released. Periodontal tissue is dependent upon endogenous antioxidants to counteract this oxidative damage and to maintain hemostasis. Inflammation decreases the systemic glutathione (GSH). GSH has antioxidant protection and immunological modulation. Pyridox (vitamin B6) and riboflavin (vitamin B2) are playing vital role in preserving GSH position. Selenium has significant oxidation and selenium dependent GSH enzymes are responsible for changes in the hydroperoxides in lipids and phospholipids to harmless products, thus neutralizing the inflammatory process at cell levels. Therefore, systemic glutathione and selenium-dependent GSH enzymes for antioxidant defense, immunological regulations, and cellular neutralization of the inflammation process are required for the maintenance of vitamins B2, B6, copper, zinc, and selenium. Micronutrients—beta-carotene and vitamins A, C, and E during inflammation may be decreased. As mitochondria (cell's power house) generate energy, responsible for ROS to release inside the cell (Labrecque et al., 2006; Moskaug et al., 2005; Nakamoto et al., 1984).

5.5.5 Clinical studies

The present study shows that the supplementation of oral lycopene and green tea extract is good for salivary uric acid and plays an essential role in gingivitis treatment. While the study focused gingivitis, the results of which can be assessed within 4–6 weeks, to authorize the performance of antioxidant remedy in periodontal disorders, further longitudinal research with bigger sample size paired with other inflammatory markers are necessary. Research into the treatment of gingivitis and periodontitis should be focused on and nutritionally supplemented with exogenous antioxidants (Nakamoto et al., 1984).

5.5.5.1 Restorative dentistry

Restorative dentistry implies how missing or damaged teeth are substituted and future dental disorders are prevented. Comparative restoration alternatives include fillings, crowns, bridges, and implants. Epigallocatechin-3-gallate molecule of green tea had a spinal effect on the avoidance of caries (Carnelio et al., 2008). Antibacterial activity was shown by Cranberries against *Streptococcus mutans* and to stop tooth decay, especially by their oligomers of type A. Pine bark and grape extracts have been useful in restoring caries and increasing bond strength. In particular to increase lower bond strength values after bleaching for restorative therapies (Schmidt et al., 2003; Berger et al., 2013).

5.5.5.2 Orthodontics

Orthodontics is the dental branch that corrected teeth and jaws that were wrongly positioned. Crooked teeth that do not fit properly together are more difficult to keep clean, risk loss early because of dental decay and periodontal disease and cause more stress on jaw muscles. Similar agents can be used in orthodontics to boost bracket strength. Ascorbic acid solutions were employed in bracket binding to boost bond strength values. In bone development, antioxidants also play an important role. Altan et al. revealed the systemic application of propolis at the suture area accelerates the creation of bones. In a recent study, pine bark extract solution was used instead of ascorbic acid solution (Vidhya et al., 2011).

5.5.5.3 Oral–maxillofacial surgery

Oral and maxillofacial surgery includes a wide variety of facial and skeletal diseases treatment, including jaws and oral caries. Common oral operative difficulties (e.g., teeth affected, dental implants), disorder on the jaw and congenital facial, facial trauma, mouth cancer, gland salivary illness, temporomandibular joint disorders, and several benign diseases (e.g., jaw tumor and cysts) (Aksakalli et al., 2013).

According to Ohnishi et al., the alveolary loss of the bone, accompanied by decreased expression of endothelial nitric oxide synthase in the mice, is due to reactive oxygen, for which hydrogen peroxide, and the production of oxidative stress is an underpinning systemic condition to enhance alveolar bone loss. Peri-implantitis caused by gram-negative, anaerobic bacteria which accumulate in subgingival areas. Supplements of antioxidants are used to treat peri-implantitis. Sheresta et al. stated that extract of grape seed has a beneficial effect on peri-implantitis treatment. For bone healing and bone creation, it was revealed that a compound found in propolis was caffeic acid phenethyl ester which considerably improved bone healing in rat models (Maxwell, 1995; Ohnishi et al., 2009).

5.5.5.4 Oral cancer

In several phases of oral carcinogenesis, antioxidants have preventative and therapeutic potential. Recently, researchers have claimed that oral cancer morphologies are inhibited after taking antioxidants. The proanthocyanidin administrations present in flavonoid structures have the ability to inhibit cell development and oral cancer proliferation. Dietary antioxidants help to prevent oxidative damage in lipids and other membrane molecules by removing oxidants before they try to destroy cells (Shrestha et al., 2012).

Antioxidants act as chemopreventive agents of cancer by reversing premalignant lesions such as oral leukoplakia reducing oral carcinogenesis. The pathogenesis of cancer which can come out of poor dietary habits and lifestyles has been identified as having an influence on oxidative damage. This process can result in DNA damage, a fundamental mechanism for the formation of cancer. The free radical defense requires adequate antioxidant state (Garewal, 1995).

The diet must be adapted to lower the hazard of oral and pharyngeal cancer, in particular oral cell carcinoma, principally to limit calorie intake through monosaturated fat and red meat. Dietary micronutrients such as vitamin A, β-carotene, lycopene, ascorbic acid, alpha-tocopherol, zinc, and selenium are antioxidant in nature should be included in food habits (Carvalho Rde et al., 2013; King et al., 2007).

The use of antioxidant nutrients inhibits cancer cell development and kills them with apoptosis (programmed cell death), stimulates cytotoxic cytokines, expression of genes, and prevents the blood supply to the tumor formation or cellular differentiation (Shirataki et al., 2000; Skibsted et al., 2006).

Retinoids are the natural and synthetic derivatives of vitamin A, emanate in the body from retinyl esters, carotenoids, and retinal obtained from diets. Retinoid receptors (RARs) and retinoid X receptors (RXRs) modulate the actions of retinoids. There are three subtypes of retinoids designated as α, β; both RARs and RXRs have been described (San Miguel et al., 2011; Baudet et al., 1991).

The lack of these dietary vitamins, systemic or mucous levels will interact with tobacco and raise the chance of precancerous mouth lesions. It would be worth continuing research of novel synthetic medicinal products for the regulation of RAs and independent receptors. It seems that the use of oral squamous cell carcinoma apoptotic potential would result in contemporary therapy; which may be less hazardous to normal cells because of their disciplined pathways of physiological survival (Swapna et al., 2014).

It is suggested that these new treatments might also treat epithelial dysplasia effectively. Ideally, chemoprevention is the basis of cancer control. A number of dietary components and micronutrients have a great potential for induction of apoptosis with the addition of the chemical therapy and chemopreventive medicines. These agents include green tea (EGCG and other) components (e.g., carotenoids (lycopene), retinoids, and many more phytochemicals). β-Carotene is a precursor of vitamin A, generally found in dark green leafy vegetables like spinach orange or yellowish vegetables like mango, carrots, oranges, papaya, and sweet potato (Shafer et al., 1993; Battino et al., 2002).

Lycopene is a major serum carotenoid, a red pigment antioxidant. This pigment contains fat-soluble red in some fruit and vegetables. Apricots, tomatoes, papaya, and other yellow fruits are the principal sources of lycopene. Lycopene and other foodstuffs rich in carotenoids are also inversely linked to neoplasms of the higher digestive tract, including mouth cancer. The protection of essential cellular biomolecules including lipids, lipoproteins, proteins, and DNA has suggested that lycopenes inhibit carcinogenesis and atherogenesis. Lycopene has the unusual characteristic to link with chemical species reacting to oxygen, therefore this is the most effective biological antioxidant (Gopinath, 2006; Hsu et al., 2003; Devasagayam et al., 2004).

5.5.6 Oral submucous fibrosis

Oral submucous fibrosis (OSF) is a well-known oral precancer disease, found mainly in South Asian ethnic populations orally (Sankaranarayanan et al., 1997).

It is characterized by a distinctive global fibrosis of oral submucosal soft tissues which results in a strong oral mucosa, which leads to gradual failure of the mouth, rigidity of the lips, and difficulties in lifting the tongue. An estimated prevalence of 0.2–1.2% in India is estimated in Oral submucus fibrosis (OSF) (Rao and Agarwal, 2000; Fig. 5.5.1) connected with the chewing of betel nut products. In India, OSF prevalence varies from 0.2% to 1.2%.

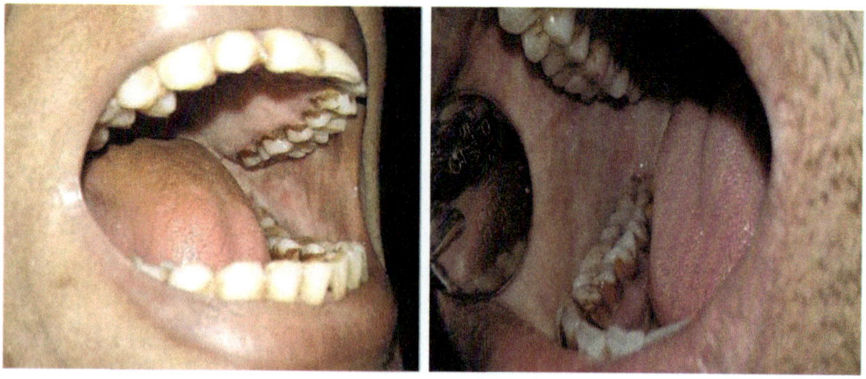

FIG. 5.5.1 Oral submucous fibrosis (A) in female and (B) in male.

Image courtesy by Dr. Sandeep Visvkarma.

It is a chronically progressive oral cavity and oropharynx scar disease, characterized by an epithelial atrophy and an inflammatory juxta-epithelial reaction with progressive lamina propria and deeper connective tissue fibrosis. The steepness of the mouth causes a gradual decrease in the opening of the mouth (Rao and Agarwal, 2000).

5.5.7 Repeal of oral leukoplakia with antioxidants

The reversal or relapse of premalignant abrasion like leukoplakia is an essential step to prevent cancer. Any agent selected for testing in premalignant lesions whose final objective is to seek the prevention of cancer should be minimized or preferable to no toxic because the intervention will expose many subjects whose sore are rare to evolution to cancer. In developing agents for general population use to reduce oral cancer incidences, antioxidants like β-carotene and vitamin E are preferred. Intervention tests on chewers of quid-tobacco and betel reveal that vitamin A delivery causes total recovery of leukoplakia (Millen et al., 2004; Fig. 5.5.2). The synthetic retinol most widely employed, 13 cis-retinoic acid is hazardous even in extremely low-dose applications. The usage of comparatively nontoxic antioxidants like beta-carotene and vitamin E is becoming increasingly important.

A study evaluating the effect of lycopene on oral cancer showed that high dosages of lycopene (8 mg/day) are good for oral health improvements (Reddy, 2011).

In several phases of oral carcinogenesis, antioxidants have preventative and therapeutic potential.

Researchers recently shown that oral cancer characteristics are inhibited with the ingestion of antioxidants. Proanthocyanidin administrations in antioxidant flavonoid structures have a capacity to lower cell development and proliferate oral carcinomas. Dietary antioxidants can prevent oxidative damage to the lipids and other membrane components by intercepting oxidants before attempting to degrade tissues (Couto et al., 2016).

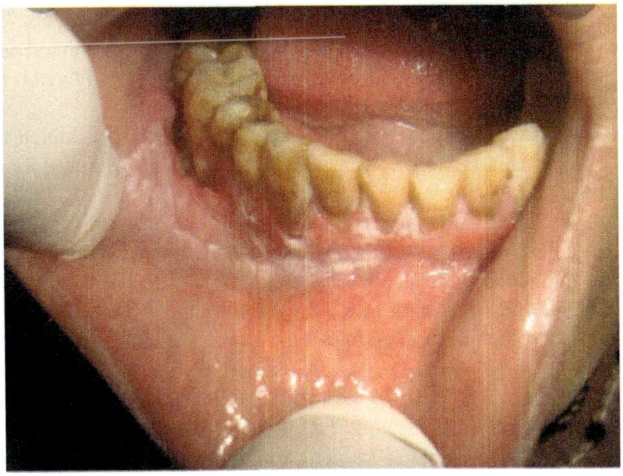

FIG. 5.5.2 Leukoplakia.

Image cortsy by Dr. Sandeep Visvakarma.

Conclusion

With the help of this chapter it may concluded that free radical poorly associated the dentistry and causes serious oral health issues but antioxidants present in different plants and other resources terminate the free radical that rectify the oral health issue caused by it and also act as preventer for free radicals by regular use of antioxidant in food habits. Oral intake of extracts of lycopene and green tea, supplementation is directly associated with salivary uric acid levels and perform an essential role in the execution of gingivitis. This study may help to overcome the oral problems by using antioxidants in routine diet.

References

Abebe, W., 2003. An overview of herbal supplement utilization with particular emphasis on possible interactions with dental drugs and oral manifestations. J. Dent. Hyg. 77, 37–46.

Aksakalli, S., Ileri, Z., Karacam, N., 2013. Effect of pine bark extract on bond strength of brackets bonded to bleached human tooth enamel. Acta Odontol. Scand. 71, 1555–1559.

Battino, M., Ferreior, M.S., Gallardo, I., Newman, H.N., Bullon, P., 2002. The antioxidant capacity of saliva. J. Clin. Periodontol. 29, 189–194.

Baudet, P.M., Janin, M.A., Souteyrand, P., 1991. Sequential immunopathologic study of oral lichen planus treated with retinoin and etretinate. Oral Surg. Oral Med. Oral Pathol. 71, 197–202.

Berger, S.B., De Souza Carreira, R.P., Guiraldo, R.D., Lopes, M.B., Pavan, S., et al., 2013. Can green tea be used to reverse compromised bond strength after bleaching? Eur. J. Oral Sci. 121, 377–381.

Bhuvaneswari, J., 2014. Antioxidants in oral healthcare Saveetha Dental College, Chennai – 77. J. Pharm. Sci. Res. 6 (4), 206–209.

Cadenas, E., 1989. Biochemistry of oxygen toxicity. Annu. Rev. Biochem. 58, 79–110.

Cardoso, L.A., Ferreira, S.T., Hermes-Lima, M., 2008. Reductive inactivation of yeast glutathione reductase by Fe(II) and NADPH. Comp. Biochem. Physiol. (Part A 151), 313–321.

Carnelio, S., Khan, S.A., Rodrigues, G., 2008. Definite, probable or dubious: antioxidants trilogy in clinical dentistry. Br. Dent. J. 204 (1), 29–32. https://doi.org/10.1038/bdj.2007.1186.

Carvalho Rde, S., de Souza, C.M., Neves, J.C., Holanda-Pinto, S.A., Pinto, L.M., et al., 2013. Vitamin E does not prevent bone loss and induced anxiety in rats with ligature-induced periodontitis. Arch. Oral. Biol. 58, 50–58.

Chapple, I.L., Brock, G.R., Milward, M.R., Ling, N., Matthews, JB., 2007. Compromised GCF total antioxidant capacity in periodontitis: cause or effect? J. Clin. Periodontol. 34, 103–110.

Couto, N., Wood, J., Barber, J., 2016. The role of glutathione reductase and related enzymes on cellular redox homeostasis network. Free Radic. Biol. Med. 95, 27–42. http://dx.doi.org/10.1016/j.freeradbiomed.2016.02.028.

Dahiya, P., Kamal, R., Gupta, R., Bhardwaj, R., Chaudhary, K., et al., 2013. Reactive oxygen species in periodontitis. J. Ind. Soc. Period 17 (4), 411–416.

Devasagayam, T., Tilak, J., Boloor, K., Sane, K., Ghaskadbi, S., Lele, R., 2004. Free radicals and antioxidants in human health: current status and future prospects. JAPI 52, 794–804.

Gad, S.C., 2014. Glutathione. In: Wexler, P. (Ed.), Encyclopedia of Toxicology. Academic Press, USA, pp. 751.

Garewal, H., 1995. Antioxidants in oral cancer pervention. Am. J. Clin. Nutr. 62 (6), 1410S–1416S. doi:10.1093/ajcn/62.6.1410S.

Gopinath, V.K., 2006. Arzreanne saliva as a diagnostic tool for assessment of dental caries. Arch. Orofac. Sci. 1, 57–59.

Hsu, S., Singh, B., Schuster, G., 2003. Induction of apoptosisin oral cancer cells: agents and mechanisms for potential therapy andprevention. Oral Oncol. 40 (5), 1–13.

King, M., Chatelain, K., Farris, D., Jensen, D., Pickup, J., et al., 2007. Oral squamous cell carcinoma proliferativephenotype is modulated by proanthocyanidins: a potential prevention and treatment alternative for oralcancer. BMC Complement. Altern. Med. 7, 22.

Labrecque, J., Bodet, C., Chandad, F., Grenier, D., 2006. Effects of a high-molecular-weight cranberry fraction on growth, biofilm formation and adherence of *Porphyromonas gingivalis*. J. Antimicrob. Chemother. 58, 439–443.

Maxwell, S.R., 1995. Prospects for the use of antioxidant therapies. Drugs 49, 345–361.

Millen, A.E., Dodd, K.W., Subar, A.F., 2004. Use of vitamin, mineral, nonvitamin, and nonmineral supplements in the United States: the 1987, 1992, and 2000 National Health Interview Survey results. J. Am. Diet. Assoc. 104, 942–950.

Moskaug, J.O., Carlsen, H., Myhrstad, M.C., Blomhoff, R., 2005. Polyphenols and glutathione synthesis regulation. Am. J. Clin. Nutr. 81, 277S–283S.

Nakamoto, T., McCroskey, M., Mallek, H.M., 1984. The role of ascorbic acid deficiency in human gingivitis--a newhypothesis. J. Theor. Biol. 108, 163–171.

Nandi, A., Yan, L., Jana, C.K., Das, N., 2019. Role of catalase in oxidative stress- and age-associated degenerative diseases. Oxidative Med. Cell. Longevity, 19.

Ohnishi, T., Bandow, K., Kakimoto, K., Machigashira, M., Matsuyama, T., et al., 2009. Oxidative stress causes alveolar bone loss in metabolic syndrome model mice with type 2 diabetes. J. Periodontal Res. 44, 43–51.

Rao, A.V., Agarwal, S., 2000. Role of antioxidant lycopene in cancer and heart disease. J. Am. Coll. Nutr. 19 (5), 563–569.

Reddy, E.S.P., 2011. Role of antioxidants in precancerous lesions. JIDA 3 (1), 99–101.

Sankaranarayanan, R., Mathew, B., Varghese, C., et al., 1997. Chemoprevention of oral leukoplakia with vitamin A and betacarotene: an assessment. Oral Oncol. 33 (4), 231–236.

San Miguel, S.M., Opperman, L.A., Allen, E.P., Svoboda, K.K., 2011. Use of antioxidants in oral healthcare. Compend. Contin. Educ. Dent. 32, E156–E159.

Schmidt, M.A., Riley, L.W., Benz, I., 2003. Sweet new world: glycoproteins in bacterial pathogens. Trends Microbiol., 554–561.

Shafer, W.G., Hine, M.K., Levy, B.M., 1993. A Text Book of Oral Pathology, fourth ed. WB Saunders & Company, Philadelphia, pp. 308–312.

Shirataki, Y., Kawase, M., Saito, S., Kurihara, T., Tanaka, W., et al., 2000. Selective cytotoxic activity of grape peel and seed extracts against oral tumor cell lines. Anticancer Res. 20, 423–426.

Shivana, V., Gupta, S., 2013. Antioxidants: stress busters in dentistry -- a review. JDPR. 1 (2), 9–19.

Shrestha, B., Theerathavaj, M.L., Thaweboon, S., Thaweboon, B., 2012. In vitro antimicrobial effects of grape seedextract on peri-implantitis microflora in craniofacial implants. Asian Pac. J. Trop. Biomed. 2, 822–825.

Skibsted, L.H., Dragsted, L.O., Dyerberg, J., Hansen, H.S., Kiens, B., et al., 2006. Antioxidants and health. Ugeskr. Laeger 168, 2787–2789.

Swapna, L.A, Pradeep, K., Reddy, P., Deepak, K., Goyal, S., 2014. Antioxidants and their implications in oral health and general health. IJCRI 5 (4), 258–263.

Tiwari, A.K., 2004. Antioxidants: new-generation therapeutic base for treatment of polygenic disorders. Curr. Sci. 86 (8), 1092–1102.

Tosun, M., Yağcı, R., Erdurmuş, M., 2019. Glaucoma and antioxidant status. In: Preedy, V.R., Watson, R.R. (Eds.), Handbook of Nutrition, Diet, and the Eye second ed. Academic Press, USA, pp. 203–219.

Valko, M., Izakovic, M., Mazur, M., Rhodes, C.J., Telser, J., 2004. Role of oxygen radicals in DNA damage and cancer incidence. Mol. Cell. Biochem. 266, 37–56.

Vidhya, S., Srinivasulu, S., Sujatha, M., Mahalaxmi, S., 2011. Effect of grape seed extract on the bond strength of bleached enamel. Oper. Dent. 36, 433–438.

Wang, F., Zhang, Y.Q., 2015. Bioconjugation of silk fibroin nanoparticles with enzyme and peptide and their characterization. In: Donev, R. (Ed.). Advances in Protein Chemistry and Structural Biology, 98. Academic Press, USA, pp. 263–291.

CHAPTER

5.6

Antioxidants and gastric lesions

Mirele da Silveira Vasconcelos[a], Tamiris de Fátima Goebel de Souza[b], Diana Célia Sousa Nunes-Pinheiro[c], Francisco Rogênio da Silva Mendes[d], Felipe Domingos de Sousa[e], Luciana de Siqueira Oliveira[f], Paulo Carvalho de Paula[g], Daniel Cordeiro Gurgel[h], Ana Sanches Silva[i,j], Seyed Mohammad Nabavi[k], Dirce Fernandes de Melo[g]

[a]*Federal Institute of Education, Science and Technology of Ceará (IFCE), Baturité, Ceará, Brazil*
[b]*Department of Physiology and Pharmacology, Nucleus of Drug Research and Development - NPDM, Federal University of Ceará, UFC, Ceará, Brazil*
[c]*Faculty of Veterinary Medicine, State University of Ceará, Fortaleza, Ceará, Brazil*
[d]*Northeast Biotechnology Network (RENORBIO), Center of Experimental Biology (Nubex), University of Fortaleza (UNIFOR), Fortaleza, Ceará, Brazil*
[e]*Department of Physics, Federal University of Ceará, Fortaleza, Ceará, Brazil*
[f]*Department of Food Technology, Federal University of Ceará, Fortaleza, Ceará, Brazil*
[g]*Department of Biochemistry and Molecular Biology (DBBM), Federal University of Ceará (UFC), Fortaleza, Ceará, Brazil*
[h]*Federal Institute of Education, Science and Technology of Ceará, Limoeiro do Norte, Ceará, Brazil*
[i]*National Institute for Agricultural and Veterinary Research (INIAV), I.P., Vairão, Vila do Conde, Portugal*
[j]*Center for Study in Animal Science (CECA), ICETA, University of Oporto, Oporto, Portugal*
[k]*Applied Biotechnology Research Center, Baqiyatallah University of Medical Sciences, Tehran, Iran*

5.6.1 Introduction

In the last years a great number of papers has been consecrated to antioxidants role in response to reactive chemical species, particularly reactive oxygen species (ROS) in order to minimize cell damage (Ali et al., 2018; Rani, 2017; Zulaikhah, 2018). It is known that ROS are responsible for aging as well as a large number of diseases and consequently it is assumed that a suitable balance between antioxidizing agents and ROS is crucial for prevention and/or cure of diseases (Pala and Gürkan, 2008; Sies, 2018).

Among several diseases, there are gastric organic disorders that could be the result of injuries stimulated by alcohol consumption, smoking, pollution, pesticides, microbes, allergens, drugs, and so on (Kalant et al., 2007). These gastric disorders present a similar symptomatology as epigastric pain, acidity, and digestive difficulty

Antioxidants Effects in Health. DOI: https://doi.org/10.1016/B978-0-12-819096-8.00056-2

but histopathological studies reveal distinct patterns according to the etiology of each one. Attention must be given to the histopathological evaluation which may signal a homeostasis imbalance potentially able to promote precancerous lesions (Katic, 2018; Kuipers and Blaser, 2012). Gastric ulcer or peptic ulcer disease is an erosion of the mucosa covering the inside of the stomach and/or duodenum produced by the corrosive action of the gastric juice (Silva and Sousa, 2011; Tovey, 2015). These and others lesions such gastritis and gastric cancer are characterized by inflammation produced by an imbalance between the aggressive and the defensive factors, in particular by a decrease of the referred protective factors (Nascimento et al., 2017). Indeed, endogenous disruptive mechanisms are mainly mediated by hypersecretion of hydrochloric acid and pepsin followed by high levels of ROS while gastric mucosal defenses consist of a protective barrier of mucus and bicarbonate, the activity of antioxidant enzymes as much as adequate blood flow (Demir et al., 2003; Mezdour et al., 2017; Priya et al., 2012).

Antioxidants can be endogenous (enzymes) and exogenous (vitamins, minerals, phytochemicals) and they function synergistically in favor of the redox balance (Valko et al., 2007). The identification and the development of new antioxidants from nature could be beneficial in limiting the deleterious effects of oxidative stress in different pathological conditions including gastric lesions such as peptic ulcer (Farzaei et al., 2013; Ramana et al., 2014).

Nowadays, several studies are focused in diets and nutraceuticals aiming to provide higher antioxidant status (Bisht, 2018; Gomes-Rochette et al., 2016; Zafra-Stone et al., 2007). This led to a rampant use of antioxidants supplements by increased marketing of these products envisaging optimal health (Khan et al., 2017). However, it must be emphasized that free radicals also play an important role as signaling several cellular processes, revealing a beneficial physiological function. Therefore, the statement that ROS is harmful, regarded as villain signal, inducing cellular damage, is not fully true (Matschke et al., 2019). Indeed, considering the cellular redox state, the double-edged effect does not only concern to ROS, but also to the antioxidants concentration in vivo which can cause different impact on ROS (Bouayed and Bohn, 2010; Martin and Barrett, 2002).

Taking into account a cellular redox state overview, exogenous antioxidants appear good targets in overall understanding about redox metabolic balance which could provide healthy life (Aslan, 2018; Pisoschi and Pop, 2015; Valko et al., 2007).

In this chapter, we highlighted the phytochemical power from natural antioxidants associated with gastric lesions, as well as its possible preventive, curative, or deleterious effects through evidence in vitro, in vivo, and clinical trials.

5.6.2 Antioxidant systems

Human biological machinery has developed a complex antioxidant network involving endogenous and exogenous systems to maintain the redox balance of cells (Fig. 5.6.1). The endogenous systems composed by enzymatic and nonenzymatic

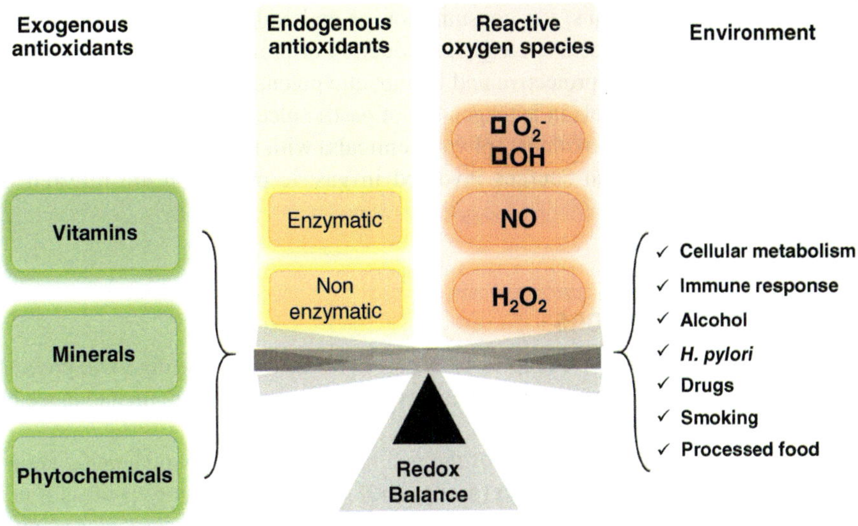

FIG. 5.6.1 Redox balance as a result of the influence of antioxidants, reactive oxygen species, and the environment.

The redox balance in the biological system is a result of adequate synergy between endogens and exogenous antioxidants to balance reactive oxygen species generated by environmental factors avoiding oxidative stress. Hydroxyl radical (OH^-), superoxide radical (O_2^{-1}), hydrogen peroxide (H_2O_2), nitric oxide (NO).

antioxidants. The most efficient enzymatic components are antioxidants enzymes such as superoxide dismutase (SOD), catalase (CAT) and glutathione peroxidase (GPx), glutathione reductase (GR), while the main nonenzymatic systems are glutathione (GSH), coenzyme Q10, lipoic acid, and thioredoxin (Bhattacharyya et al., 2014).

The exogenous antioxidants from natural sources such as vitamins C and E, minerals, phenolic compounds, and other phytochemicals also known as phytonutrients are taken from plant foods such as whole grains, beans, fruits, and herbs. Several studies reveal that the beneficial potential for human health of natural compounds can be attributed to their antioxidant activity (Batta, 2016; Belwal et al., 2017; Bouayed and Bohn, 2010; Hajialyani et al., 2019; Marchese et al., 2017; Shokoohinia et al., 2018; Zubair et al., 2017). Indeed, the Mediterranean diet rich in fruit, vegetables, and olive oils has been associated with a low cancer incidence including gastric lesions (Kuna et al., 2019). The overall components of this diet seem to have a synergism which is largely responsible for their favorable effects (Schwingshackl and Hoffmann, 2016). However, attention should be given to food processing methods, bioavailability, and bioefficacy of antioxidant compounds (Attri et al., 2017; Gayoso et al., 2016).

The gastric tissue damage has been associated with ROS generation and concomitant reduction of enzymatic and nonenzymatic antioxidant defense systems

(Suzuki et al., 2012). Thus, the consumption of antioxidant-rich foods as well as their supplements could be a strategy for reduction or prevention of gastric lesions. Several studies show the protective and therapeutic potential of natural antioxidants such as dietary polyphenols in management of gastric ulcers (Farzaei et al., 2015).

Some exogenous antioxidants (phytochemicals) with their chemical structures, main classes, and natural sources involved in gastric protection are referred in Table 5.6.1.

5.6.3 Gastric lesions

It is known that the stomach occupies a key role in orchestrating the digestive process through its control function in food intake as well as contributing to maintenance of metabolic balance (Hunt et al., 2015).

The gastrointestinal tract is one of the main sources of ROS, which generate inflammatory responses (Repetto and Llesuy, 2002). This inflammatory process if not controlled could progress leading to acute or chronic damage in mucosa, ulcerations, and even gastric cancer (Bhattacharyya et al., 2014).

The gastric mucosa uses a diversified defense system to combat endogenous and exogenous irritants, through secretion of mucus and bicarbonate, regulation of gastric mucosal blood flow, epithelial regeneration, and maintenance of epithelial homeostasis (Ock et al., 2011).

Furthermore, a reduced endogenous antioxidant enzymatic activity as superoxide dismutase (SOD) associated with low consumption of exogenous antioxidant vitamins also contributes to the development of inflammatory diseases of the gastrointestinal tract (Bhattacharyya et al., 2014; Das et al., 1997). Thus, the intake of antioxidants is important to preventing and healing gastric lesions (Repetto and Llesuy, 2002) once the approach to gastric diseases is directed to maintaining a healthy stomach (Hunt et al., 2015).

The most prevalent gastric lesions in the population around the world are gastritis, peptic ulcer as well as precancerous lesions which occur in the stomach and affect millions of people (Adinortey et al., 2013; Sipponen and Maaroos, 2015). These diseases occur in response to a wide range of injuries from alcohol, smoking, nonsteroidal anti-inflammatory drugs (AINES), *Helicobacter pylori* infection just as a rich intake in processed food and a diet poor in fruit and vegetables (Elseweidy, 2011; Sipponen and Maaroos, 2015; Wirth and Yang, 2016).

Among the gastric inflammatory processes, gastritis reveals as a predominant inflammation in the lining of the stomach. Clinically gastritis can be named as acute and chronic depending on the time of resolution and evolution of the agents that promote injuries (Rugge and Graham, 2016).

Peptic ulcer, another gastric lesion, is caused by mucosal rupture due to excessive secretion of pepsin and gastric acid, which overlap the protective effect of mucus, bicarbonate, and prostaglandins in the gastric barrier (Kulshreshtha et al., 2017; Sung et al., 2009).

Table 5.6.1 The main exogenous antioxidants (phytochemicals) with gastroprotective potential.

Phytochemical	Chemical structure	Classes flavonoids (F), polyphenolic compounds (PC), polyphenols (P), vitamins (V), and tetraterpenes (T)	Natural sources	Refs.
6-Gingerol	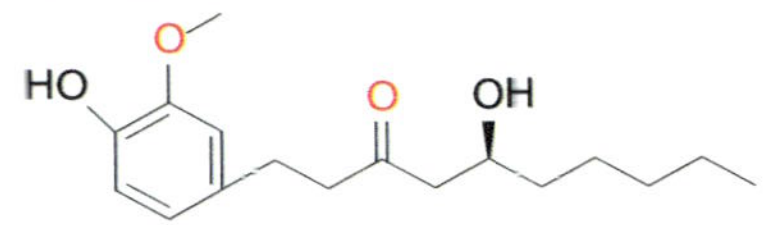	PC	Ginger	Mansingh et al., 2018
Anthocyanin	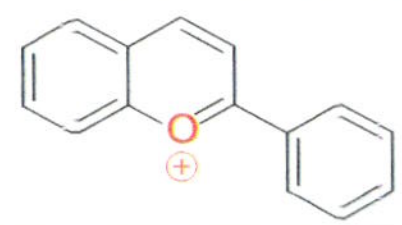	F	Grapes, berries, red-purple vegetables	Sudharameshwari and Ayshwarya, 2017.
β-Carotene	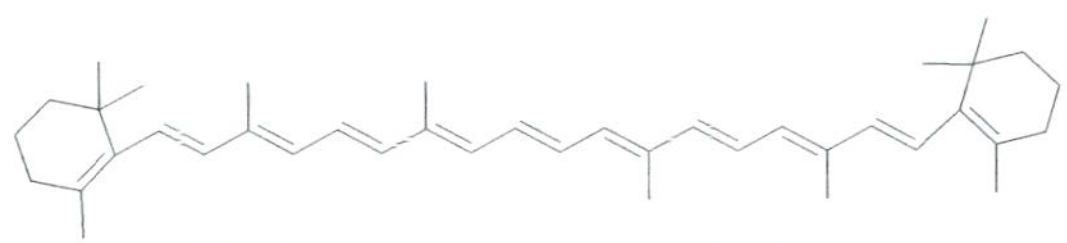	T	Pumpkin, carrot, papaya	Peiwei et al., 2014
Catechin	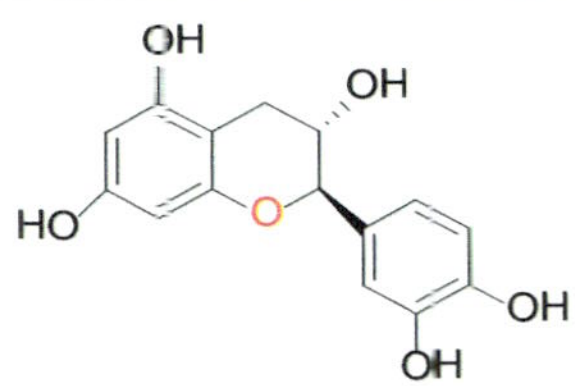	P	Green tea	Hamaishi et al., 2006
Curcumin	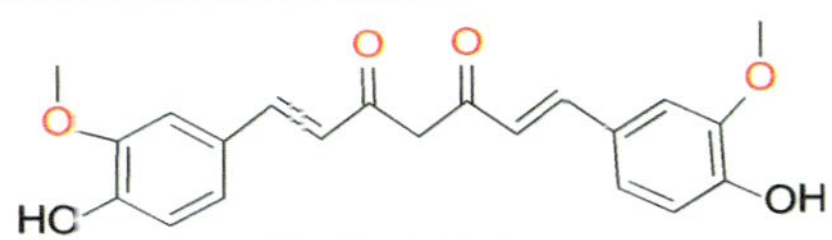	PC	Saffron, turmeric	Priyadarsini et al., 2003; Tuorkey and Karolin, 2009

(Continued)

Table 5.6.1 The main exogenous antioxidants (phytochemicals) with gastroprotective potential. *Continued*

Phytochemical	Chemical structure	Classes flavonoids (F), polyphenolic compounds (PC), polyphenols (P), vitamins (V), and tetraterpenes (T)	Natural sources	Refs.
Gallic acid	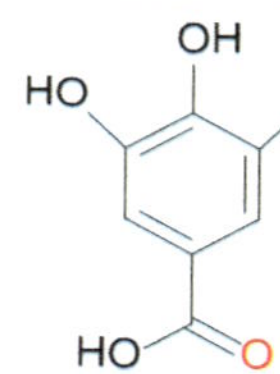	P	Pequi	Almeida et al., 2019
Genistein	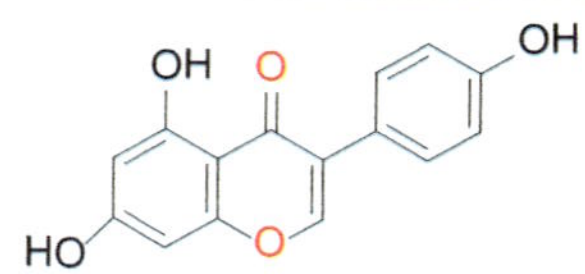	F	Soybean	Hegab et al., 2018
Kaempferol	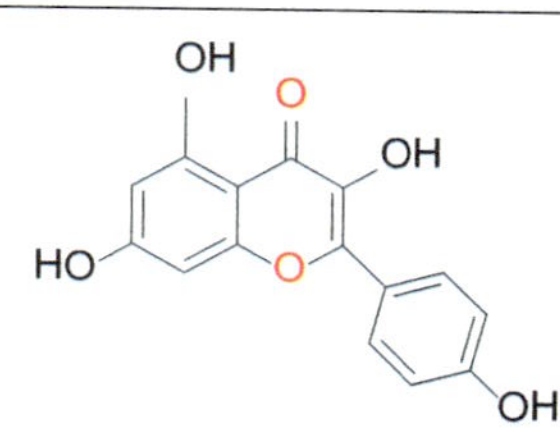	F	Strawberries, grapes, citrus fruits, onions	Li et al., 2018; Rasouli et al., 2017

Lycopene		T	Tomato, papaya, guava, watermelon	Boyacioglu et al., 2016
Quercetin	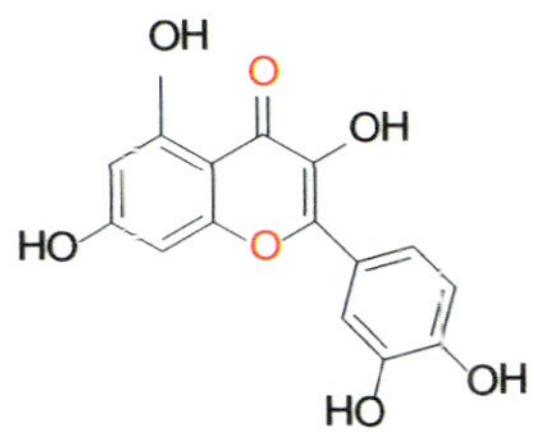	F	Apples, tea, berries, cabbage	Alkushi and Elsawy, 2017; Yeung et al., 2019
Resveratrol	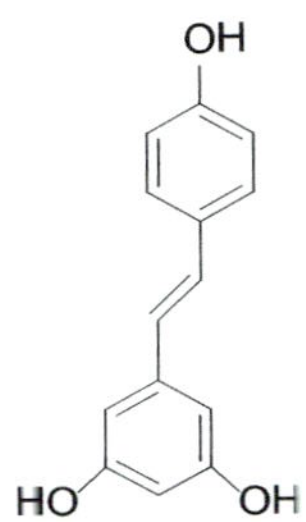	P	Grapes, peanut, berries	Hussein et al., 2016b; Solmaz et al., 2009

(Continued)

Table 5.6.1 The main exogenous antioxidants (phytochemicals) with gastroprotective potential. *Continued*

Phytochemical	Chemical structure	Classes flavonoids (F), polyphenolic compounds (PC), polyphenols (P), vitamins (V), and tetraterpenes (T)	Natural sources	Refs.
Rutin	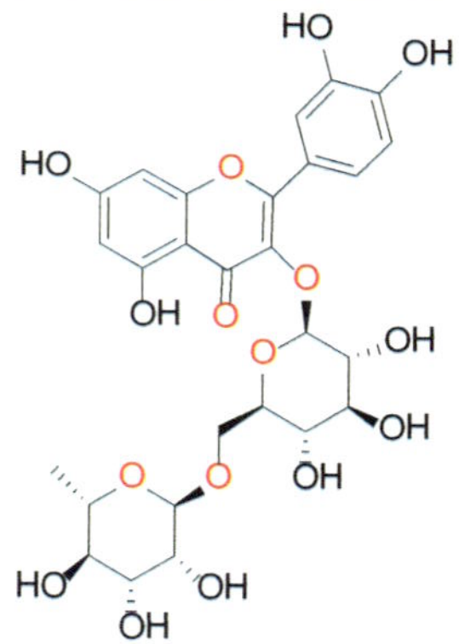	F	Onion, apple, grape, orange, tea	Abdel-Raheem, 2010; Hussein et al., 2016a
Vitamin C	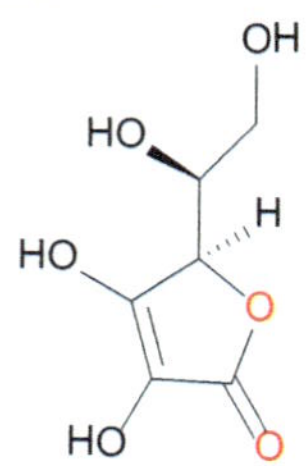	V	Citrus fruits	Seziklii et al., 2012
Vitamin E	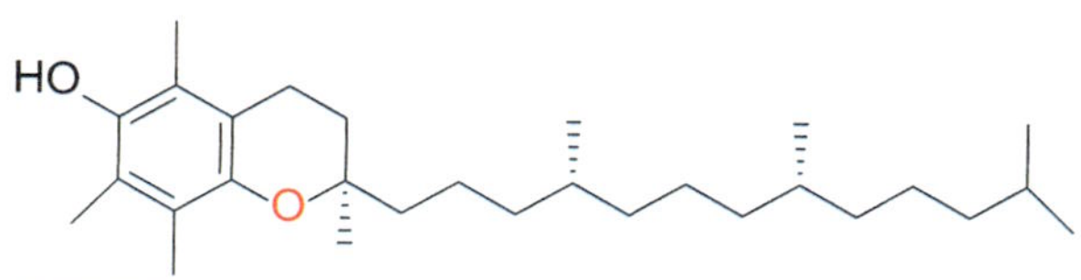	V	Vegetables oils, oil seeds, green leafy vegetables	Qiao et al., 2009; Peiwei et al., 2014

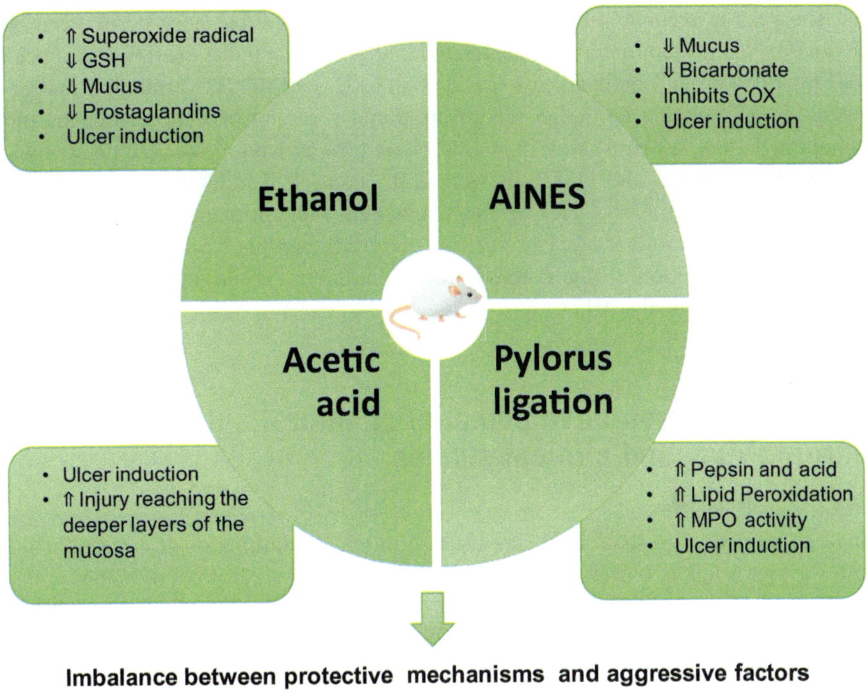

FIG. 5.6.2 Experimental models of gastric ulcers induction.

The four experimental models using irritants and erosive agents such as ethanol, acetic acid, AINES, and pylorus ligation to induce mucosal damage. *AINES*, nonsteroidal anti-inflammatory.

The gastric cancer in general originates from the gastric mucosa glands belonging to adenocarcinoma group and are heterogeneous in relation to architecture and growth, cell differentiation, histogenesis, and molecular pathogenesis (Cheng et al., 2016; Cutsem et al., 2016)

In vivo studies focused on murine models that induce ulcer in gastric mucosa have been proposed in order to evaluate the mechanisms involved in gastroprotection (Adinortey et al., 2013). Several irritant agents and models (Fig. 5.6.2) are used to induce peptic ulcer as following: absolute ethanol (Hollander et al., 1985); acidified ethanol (Mizui and Doteuchi, 1983); nonsteroidal anti-inflammatory drugs (NSAIDs), such as indomethacin (Djahanguiri, 1969) and aspirin (Brodie and Chase, 1967); erosive agents such as acetic acid (Okabe and Pfeiffer, 1972) and pylorus ligation (Debanath et al., 1974). In addition, these models are useful to investigate antisecretory, antioxidant, anti-inflammatory, and healing activity of antioxidant compounds with gastroprotective properties (Pineda-Peña et al., 2018).

The preventive or curative effect of these substances is determined by the treatment performed in the ulcer induction time. The degree of ulcer extension is an adequate index to establish the preventive level of substances. Since the treatment start after the injury indicates a curative potential, by assessing the degree of healing (Adinortey et al., 2013).

The gastroprotective mechanisms of the antioxidants substances in vivo can be divided by the pre-epithelial effects, related to the increase in the secretion of mucus and bicarbonate and reduction of the hydrochloric acid release; epithelial mechanisms, by stimulating epithelial regeneration and reducing oxidative damage; and subepithelial mechanisms, with increased blood flow and cytoprotective prostaglandin synthesis (PGE2 and PGI2), as well as the control of inflammatory mediators (Pineda-Peña et al., 2018). Gastroprotective agents with antioxidant properties may be useful in the treatment of gastric diseases, considering that oxidative stress is one of the main aggressors of the mucosa, thus modulating the tissue inflammatory response (Mezdour et al., 2017).

5.6.4 Gastroprotective potential of natural antioxidants: in vitro, in vivo, and clinical studies

5.6.4.1 In vitro

In vitro models of gastric injury are used to mimic conditions of gastric epithelial cell damage such as oxidative stress or inflammation and cytotoxicity (Lian et al., 2018). Cells are obtained from human gastric epithelial cell lines (GES-1) (Chang et al., 2015), gastric adenocarcinoma (AGS, CRL-1739) (Sangiovanni et al., 2013), or gastric tubular adenocarcinoma (MKN 28, CRL-1739) (Graziani et al., 2005).

In vitro model of gastric carcinoma using AGS cells, [6]-gingerol exhibited, in a dose-dependent manner, anticancer effects through its cytotoxic and inhibitory growth potential, stimulating the apoptosis of gastric carcinoma cells (Mansingh et al., 2018).

The aqueous extract of *Hericium erinaceus* is prevented the human gastric epithelial cell line GES-1 death due apoptosis induced by H_2O_2 (Wang et al., 2018).

The human stomach epithelial cells (MKN 28, CRL-1739), lineage derived from gastric tubular adenocarcinoma, were used to induce oxidative stress by xanthine oxidase (XO), promoting cellular damage due to ROS excessive production Apple pulp extract (APE) containing polyphenols (APE) was used to evaluate cellular cytotoxicity, phenolic compounds quantification in the cellular environment, and lipid peroxidation level. APE prevented ROS-induced cell damage, reduced lipid peroxidation, and increased intracellular antioxidants (Graziani et al., 2005).

Table 5.6.2 summarizes in vitro studies exploring the antioxidant, anti-inflammatory, and antimicrobial potential of phytochemicals.

5.6.4.2 In vivo (preclinical)

In vivo studies were performed with several phytochemicals by different experimental models in order to prove their efficiency as gastroprotective through different strategies. The obtained responses also varied according the identified compounds present in the source food and the assayed conditions.

Table 5.6.2 In vitro studies of antioxidants (phytochemicals) in gastric epithelial cells.

Phytochemical	Plant source	Treatment	Effect	Refs.
Salidroside (Sal)	*Rhodiolarosea*	Human gastric epithelial cells (GES-1) were treated with various doses of Sal 24h before H_2O_2-induced ROS production	10–40 μM of Sal significantly recovered cell viability after H_2O_2-induced stress	Chang et al., 2015
Ellagitannin-rich extracts (ETs)	*Rubus berries*	Gastric epithelial cell lines (AGS) were treated with ETs (1–25 μg/mL) after H_2O_2 or ethanol-induced ROS production	Ethanol and H_2O_2 were not toxic to AGS cells after treatment with ETs concentrations in the experiments	Sangiovanni et al., 2013
Patchouli alcohol (PA)	*Pogostemon-cablin* (Blanco) Benth. (Labiatae)	GES-1 cells were exposed to *H. pylori* and PA (5–20 μg/mL)	Treatment with 10 and 20 μg/mL PA reduced the plasma membrane damage and cell death. In addition, PA treatment significantly reduced *H. pylori*-induced cellular apoptosis; TNF-αand MCP-1 production; and, attenuated *H. pylori*-induced ROS production	Lian et al., 2018
Phenolic compounds	Freeze dried apple flesh extract (APE)	MKN 23 cells (human well-differentiated gastric tubular adenocarcinoma) were treated with APE (100 μM) for 3 h after oxidative stress induced by xanthine oxidase (XO 10–100 mU/mL) in the presence of its substrate xanthine (X 1 mM) for periods of up to 3 h	APE had a protective effect against X-XO injury by increasing the intracellular antioxidant activity	Graziani et al., 2005
Flavonoid and phenolic compounds	Mastic gum resin (MGR) from *Pistacialentiscus* var. chia	Colonic adenocarcinoma (COLO205) cells were treated with MGR at a concentration of 5.2 ± 0.8 μg/mL	MGR induced human colon cancer toxicity and apoptosis	Sulaiman Rahman, 2018
Genipin	*Gardenia jasminoides* Ellis	AGS cells were treated with genipin (1, 12.5, 25, and 50 μM) in a fresh medium for 6 h	25 μM of genipin induces antioxidant effects on AGS cell line via JNK/Nrf2/ARE signaling pathway, downregulating ROS	Kim et al., 2016
Flavonoids	*Lippia integrifolia* Hieronaqueous extract	AGS cells were infected with treated with *H. pylori* and treated with *L. integrifolia*. Hieronaqueous extracts (1–1000 mg/mL)	*L. integrifolia* extract influences *H. pylori* adhesion to stomach tissue and decreases bacteria-induced inflammation	Marcial et al., 2014
Resveratrol	*Vitis* sp.	SNU-1 cells were treated with resveratrol (1–100 μM) for 16 h	Chemoprotective effect by inducing cell cycle arrest and antioxidant action through NOS and enhanced NO production	Holian et al., 2002

Table 5.6.3 summarizes the phytochemical effects on gastric injuries as well the gastric lesion model, the experimental protocol, the mechanisms of gastroprotection, and the main effect.

Aside the phytochemical referred in Table 5.6.3 it must be highlighted other foods and extracts that show gastroprotective effects.

Studies with ginger rhizome extract revealed gastroprotective effects mediated through their antisecretory, cytoprotective, and antioxidative properties against oxidative and gastric damage, as well as prostaglandins generation (Al Mofleh, 2010; Haniadka et al., 2013). Ginger (*Zingiber officinale*) is rich in zingiberene, α-curcumene, gingerols, shogaols (da Silveira Vasconcelos et al., 2019). These phytochemicals promote gastric mucin and antioxidant enzymes generation as well as *H. pylori* growth inhibition in ulcerated ethanol stressed animals (Nanjundaiah et al., 2009). *Zingiber zerumbet* rhizome, presents Zerumbona as main constituent, promoted inhibition of the ulcerated area caused by ethanol, increased GSH activity and reduction of lipid peroxidation, as well as increased PGE2 synthesis and gastric mucus production (Sidahmed et al., 2015).

Manuka honey rich in polyphenolic flavonoids inhibited the formation of ulcers through augmented antioxidant enzymes activities (GSH, SOD, CAT, GPxe) and reduced MPO and MDA levels (Almasaudi et al., 2017). When used as pretreatment reduced lipid peroxidation (MDA), increased NO and GSH levels, stimulated GPx and SOD activities and reduced the production of TNF-α, IL-1β, and IL-6 (Almasaudi et al., 2016).

Already, *Olean europaea* methanolic extract of leaves, rich in polyphenols and flavonoids among these oleuropein, reduced MPO activity and increased levels of GSH, SOD, CAT, GPx, and GR, and reduced lipid peroxidation and NO levels (Al-quraishy et al., 2017). *Vernonia amygdalina* leaves extract attenuated the lipid peroxidation caused by aspirin-induced gastric injury in rats, by increased SOD and GSH activities (Adefisayo et al., 2017). *Moringa oleifera* Lam leaf extract increased CAT and reversion of lipid peroxidation produced in gastric ulcer models: pylorus ligation, aspirin induction, ethanol, and cold containment stress (Verma and Nripendra Singh, 2012). Aqueous extract from *Hericium erinaceusis* reduced gastric secretion, increased SOD and GPx activities, reduced TNF-α, IL-1β levels and increased PGE2, NO, and EGF (Wang et al., 2018). *Haematococcus pluvialis* extract, rich in asthaxantin, was assayed in mice with acetic acid-induced gastric lesions and *H. pylori* infected, reduced bacterial load, and gastric inflammation (Bennedsen et al., 1999).

Alpha-lipoic acid was able to protect the stomach by amelioration of MDA levels and increase in total antioxidant capacity (TAC), and reduction in TNF levels -α and iNOS, preserving the mucosa and gastric gland cells (Gomaa et al., 2018). Spirulin-containing biomass contributed to the reduction of gastric lesions induced by acetylsalicylic acid, increased GSH and CAT, GR and GPx activities, reduced MDA levels and NO and attenuates COX-2 and TNF-α levels (Mahmoud and Abd El-Ghffar, 2019). While the lycopene prevented the formation of gastric ulcer produced by indomethacin, increased SOD activity and MPO reduction and MDA level (Boyacioglu et al., 2015).

Table 5.6.3 In vivo studies of antioxidants (phytochemicals) on gastric lesions.

Phytochemical	Gastric lesion model	Experimental protocol	Mechanism of gastroprotection	Main effect	Refs.
Kaempferol (KAE)	Ethanol-induced gastric lesion	Mice, dose 40–160 mg/kg (i.g.), 1 h before induction of ulcers	Antiulcerogenic activity. Reduction of proinflammatory cytokines (TNF-α, IL-1) and pro-oxidant myeloperoxidase (MPO) levels, in addition to increase NO levels reduce the hemorrhagic lesions extensions. Also, by alleviating leucocytes infiltration and epithelium destruction in gastric mucosa	Protective	Li et al., 2018
Quercetin	Indomethacin-induced gastric lesion	Male albino rats, dose 50 mg/kg (i.g.), 15 consecutive days, 6 h after indomethacin	Reducing the extension of lesion and the volume of gastric juice; also, helping to keep the pH balance of gastric juice.	Protective	Alkushi and Elsawy, 2017
	Ethanol-induced gastric lesion	Sprague–Dawley rats, dose 200 mg/kg (i.g.), 2 h before the ethanol	Gastroprotective effects of quercetin relies on antioxidant properties, which reduce the lipid peroxidation and protein carbonyl compounds, and increase the SOD activity and total SH content as well as antihistaminic and anti-inflammatory properties	Protective	Kahraman et al., 2003
Rutin (RTN)	Ethanol-induced gastric lesion	White male albino rats, dose 200 mg/kg (orally) for 14 days prior absolute ethanol. One hour after ethanol, the animals were sacrificed or the treatment was continued with RTN for 7 days later	Significantly gastroprotection and effective treatment against ethanol-induced gastritis and oxidative damage (increased NO, SA, GSH, vitamin C concentrations), through mucus secretion (increased serum sialic acid concentration), besides prevention of DNA fragmentation. In addition to alleviate inflammatory process through by reducing NF-KB, p65, TNF-α, IL-6, IL-1β, and L-MDA concentrations	Protective and healing	Hussein et al., 2016a
	Indomethacin-induced gastric lesion and pylorus-ligation-induced gastric acid secretion	Adult female Wistar albino rats, 200 mg/kg (orally) 1 h before indomethacin. The same treatment above was applied to a group after pylorus ligation	Prevented the occurrence of ulcer and hemorrhagic lesions (reduction in the ulcer index). Pretreatment with allowed attenuation of histopathological changes and amelioration of the altered oxidative stress (reduced MPO activity and rise of NO levels) and biochemical parameters (reduced SOD activity TBARS production, increased GSH level and glutathione peroxidase activity		Abdel-Raheem, 2010

(Continued)

Table 5.6.3 In vivo studies of antioxidants (phytochemicals) on gastric lesions. *Continued*

Phytochemical	Gastric lesion model	Experimental protocol	Mechanism of gastroprotection	Main effect	Refs.
Catechin	Ethanol and acetic acid -induced gastric lesion	Male Sprague–Dawley strain SPF rats, pretreated doses 50–200 mg/kg (orally) of tea catechin 1 h prior to ethanol or for 14 consecutive days after acetic acid injection	Healing action, preventing the gastric mucosal injury through a potent antioxidant (inhibited O_2^--scavenging activity). Accelerated gastric ulcers healing by inhibiting TBARS production and increasing the gastric hexosamine content	Protective and healing	Hamaishi et al., 2006
Gallic acid and ellagic acid—*Spondiasmombin* L. ethanolic extract (SmEE)	Ethanol; acetic acid; indomethacin-induced gastric lesion and pylorus-ligation-induced gastric acid secretion	Swiss female rats, dose 50–200 mg/kg (orally), gallic acid (GA, 10 mg/kg), ellagic acid (EA, 7 mg/kg), or gallic acid + ellagic acid (GA + EA, 10 + 7 mg/kg) prior inducing the gastric lesion or for 14 days after acetic acid injection	Antiulcerogenic activity in ethanol-induced gastric lesions. Decreases the rate of lipid peroxidation (reduced levels of malondialdehyde—MDA) but also the sulfhydryl groups (–SH groups). Prevented the occurrence of injury in gastric mucosa. SmEE reduced the levels of TNF-α and increased the level of NO. After pylorus ligation, the treatment with SmEE stimulated the gastric mucus production	Protective and healing	Brito et al., 2018
Curcumin	Ethanol and indomethacin-induced gastric lesion and pylorus-ligation-induced gastric acid secretion	Sprague–Dawley rats, dose 5–60 mg/kg (i.p.), 30 min prior to ulcerogenic compound administration or laparotomy	Antiulcer effect is comparable to ranitidine. Blocks hypersecretion (reduced luminal acid content) and also prevents increased lipid peroxidation and glutathione depletion	Protective	Chattopadhyay et al., 2006

Resveratrol (RVT) Resveratrol	Acetic acid-induced gastric lesion	Wistar albino rats, dose 10 mg/kg (i.p.), pretreatment for 10 days prior to lesion induction and treatment was continued for 3 days	Systemic administration of RVT reduced the extension of gastric lesion and levels of inflammatory marker such as TNF-α, MPO, and MDA level	Protective and healing.	Solmaz et al., 2009
	Ethanol-induced gastric lesion	White male albino rats, dose 10 mg/kg (orally), pretreatment 2 days prior inducing lesion and treatment was continued for 7 days later	RVT treatment mitigated the ulceration of the tissues through increasing of NO, SA, GSH, vitamin C concentrations, enzymatic antioxidants status in addition to decreasing NF-KB p65, proinflammatory cytokines, L-MDA, DNA-fragmentation, MPO, and COX-2	Protective and healing.	Hussein et al., 2016b
	Indomethacin-induced gastric lesion	Male Swiss albino mice, dose 10 mg/kg (orally), 6 h after indomethacin	RVT aggravated indomethacin-induced gastric damage and delayed healing, through increased MPO activity and cyclooxygenase (COX)-1 inhibition with reduced synthesis of PGE2	Proulcerative	Guha et al., 2010
Genistein (GEN)	Indomethacin-induced gastric lesion	Male albino rats, dose 10 mg/kg (orally) for 7 days before induction	GEN offered protection against the induced gastric ulcer by increasing cytoprotective mediators such as NO and PGE2 levels. Also, by balancing oxidative markers to normal levels of MDA and SOD activity. Pretreatment with GEN possessed marked anti-inflammatory action by decreasing TNF-α and MPO levels, in addition to down-regulation of matrix metaloproteinase-9 (MMP-9) gene	Protective	Hegab et al., 2018
Anthocyanins flavonoids (*Brassica oleracea* var. capitata rubra methanolic extract)	Pyloric ligation	Wistar rats, dose 0.25–0.50 mg/kg (orally) after solvent evaporation. Fifteen minutes after treatment, pyloric ligation	*B. oleracea* extract reduced lesion index values with evident antiulcer activity through reducing the gastric volume, total acidity, free acidity, and increasing pH of gastric juice. The treatment inhibited the gastric lesion damage	Protective	Sudharameshwari and Ayshwarya, 2017
Proanthocyanidins (grape seed proanthocyanidin extract—GPSE, *Vitis vinifera*)	Indomethacin-induced gastric lesion and pylorus ligation	Wistar rats, dose 50–100 mg/kg (orally) for 4 days prior to induction of gastric lesion or pylorus ligation	GSPE administration has a gastroprotective (low ulcer index) and anti-inflammatory effect (low MPO activity and MDA and TBARS levels). GPSE treatment has a potent antioxidant mechanism (high levels of CAT, SOD, and GSH). Moreover, GPSE increased the gastric volume and decreased the gastric acidity	Protective	Bhardwaj et al., 2018

b/w, body weight; i.g., intragastrically; i.p., intraperitoneally.

5.6.5 Clinical studies on phytochemicals and their effect on gastric lesions

A double-blind, randomized, placebo-controlled clinical trial associated curcumin with standard treatment for *H. pylori* was able to significantly reduce the symptoms of dyspepsia, without affecting treatment *H. pylori* and demonstrating patient safety and tolerability (Khonche et al., 2016). In this work, the effect was justified by the action of curcumin on IL-8 suppression induced by *H. pylori*, through the inactivation of NF-κB (Münzenmaier et al., 1997), as well as, by the inhibition in the release of cytokines and chemokines supraregulated by the mucosa infected by *H. pylori* (Koosirirat et al., 2010).

Another pilot phase II clinical trial conducted with curcumin (*Curcuma longa*) decreased the presence of ulcer and reduced pain and abdominal discomfort early in the first and second week of curcumin use (Prucksunand et al., 2001).

A double-blind, placebo-controlled study was conducted in China to evaluate the Cranberry juice (CJ) on *H. pylori* infection into two intervention periods. The results showed that CJ presented bacteriostatic and antioxidant activity, possible due to the action of proanthocyanidins in addition to the presence of vitamin C, fructose, and bioflavonoids (Zhang et al., 2005). These data corroborate that the regular consumption of CJ can suppress *H. pylori*.

A clinical study was conducted in India with Neem bark extract (NBE) capsules in patients with gastric disease. NBE promoted healing of duodenal, esophageal, and gastric ulcer. This study indicates that NBE has both antisecretory and antiulcer effects but had no antibiotic effect on *H. pylori* infection (Bandyopadhyay et al., 2004).

The triterpenic extract from *Centella asiatica* (TECA), known as madecassol, was used in clinical trials in India in patients with historic of peptic ulcer. On patients that received treatment with madecassol, were observed subjective improvement and healing gastric ulcer effect. None alterations were observed in hematological, hepatic, and renal parameters in all patients studied (Chao et al., 1981; Shin et al., 1982). These studies reveal the antiulcer property and safety of *Centella asiatica* use.

Table 5.6.4 summarizes examples of phytochemicals and clinical trials exploring their antioxidant properties on gastric lesions.

5.6.6 Gastroprotective potential from herbs and medicinal plants

To report the gastroprotective potential of medicinal plants, taking into account the ethnopharmacological aspects, 10 species representing several countries were selected (Fig. 5.6.3).

Virola elongate (Benth.) *Warb.* (Myristicaceae) is a plant native to Brazil, Panama, Colombia, Bolivia, Ecuador, Peru, Guyana, and Suriname. The stem bark resin from *V. elongate* is used by the Yanomami Indians in rituals as a hallucinogenic substance (Viana et al., 2011). Yet, in the indigenous medicine of Brazil and Ecuador it is

Table 5.6.4 Clinical studies of antioxidants (phytochemicals) on gastric lesions.

Phytochemicals	Study design	Disease	Population	Treatment duration	Outcomes	Refs.
Neem bark extract	Placebo-controlled clinical study	Gastroduodenal or esophageal ulcers	26 patients (18–45 years)	Administration of 30–50 mg twice daily, 30 min before lunch and dinner for 6 or 10 weeks	Antisecretory and potent antiulcer effects on patients suffering from gastroduodenal and esophageal ulcers. But no antibiotic effects in duodenal ulcer with infection of *H pylori*	Bandyopadhyay et al., 2004
Triterpenic extract from *Centella asiática* (TECA)—Madecassol	Placebo-controlled clinical study	Peptic ulcer (gastric ulcer and duodenal ulcer)	40 patients (20–60 years)	60–120 mg, twice daily, 6 weeks	Improvement in subjective symptoms (93%) and healing of ulcer (73%)	Shin et al., 1982
Curcumin (Complex C3)	Double-blind randomized placebo-controlled clinical study	Peptic ulcer with *H. pylori* infection	68 patients (20–50 years)	The standard treatment for *H. pylori* was associated with curcumin (500 mg/day) for 14 days	The association of curcumin with standard treatment was able to significantly reduce the symptoms of dyspepsia without affecting treatment for *H. pylori* and demonstrating patient safety and tolerability	Khonche et al., 2016
Turmeric capsules (*Curcuma longa* rhizome)	Clinical trial phase II	Peptic ulcer	45 patients (16–60 years)	The dose of two capsules, each containing 300 mg of turmeric capsules (two units), five times a day for 4, 8, and 12 weeks	After 4, 8, and 12 weeks of treatment, of the 25 patients treated with curcumin presented absence of ulcer48%, 72%, and 76%, respectively. In addition to reducing pain and abdominal discomfort early in the first and second week of curcumin use	Prucksunand et al., 2001
Cranberry juice	Prospective, randomized, double-blind, placebo-controlled trial	*H. pylori* Infection	189 patients (48.9 ± 11.2 years)	Received two 250-mL juice boxes of cranberry juice or matching Placebo beverage daily for 90 days	The efficacy of cranberry juice on *H. pylori* infection at 35 days was similar to that at 90 days of intervention, and these results suggest that regular consumption of cranberry juice may reduce *H. pylori* infection in adults	Zhang et al., 2005

(*Continued*)

Table 5.6.4 Clinical studies of antioxidants (phytochemicals) on gastric lesions. *Continued*

Phytochemicals	Study design	Disease	Population	Treatment duration	Outcomes	Refs.
Vitamin supplement (vitamins C and E)	Intervention Double-blind study	Gastritis *with H. pylori* infection nonulcer dyspepsia	30 patients (23–64 years)	Vitamin C 500 mg BID and vitamin E 200 IU BID for 4 weeks orally	Treatment for 1 month in therapeutic doses reduces inflammation (neutrophilic activity decrease) and decrease *H. pillory rate* demonstrate that the antioxidant increases efficacy of antibiotic effects on the bacteria	Sezikli et al., 2012
Vitamin supplement (vitamins C, E, and selenium) or garlic supplement (Kyolic)	Randomized double-blind factorial trial	Precancerous lesions or gastric cancer with *H. pylori* infection	3365 subjects (35–64 years)	I. Amoxicillin and omeprazole for 2 weeks (*H. pylori* treatment) II. Vitamin supplement: vitamins C (250 mg), E (100 IU), and selenium from yeast (37.5 μg) for 7.3 years, twice a day III. Garlic supplement (400 mg): aged garlic extract and steam-distilled garlic oil (1 mg) for 7.3 years, twice a day IV. Placebo	Treatment long term with vitamin or garlic supplements were not able to reduce the prevalence of precancerous gastric lesions such as dysplasia, chronic atrophic gastritis, metaplasia intestinal, and cancerous diseases	You et al., 2006

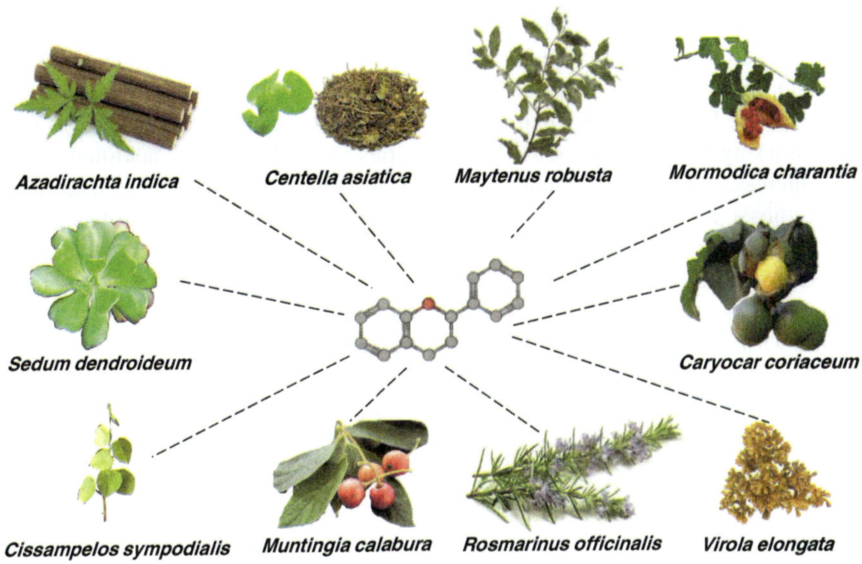

FIG. 5.6.3 A variety of species with gastroprotective potential throughout the world.

The 10 medicinal plants used to treat gastric lesions in folk medicine.

used for gastric problems from the maceration of the stem bark (Ribeiro et al., 2017). The hydroalcoholic extract of *V. elongata* stem bark (HEVe) presented flavonoids, such as gallic acid, catechin, and rutin. HEVe reduced the gastric lesions and inhibited the ulcerated area, reversed the activities of catalase and MPO and prevented the reduction of GSH-Px (Almeida et al., 2019).

Rosmarinus officinalis L (Lamiaceae), known as rosemary, native to Southern Europe and North Africa. Used for culinary, folk medicine (Borges et al., 2019; Minaiyan et al., 2011), cosmetics and perfumery (Porte and Godoy, 2001), and aromatotherapy (Filiptsova et al., 2017). *R. officinalis* presents as constituents, flavonoids and diterpenes, rosmarinic acid, carnosic acid, chlorogenic acid, rutin, quercetin, and kaempferol. The leaves extract was given orally at single doses as pretreatment reduced the gastric ulcers, prevent the lipidic peroxidation and oxidized DCS levels, increased GSH–GSSG ratio, catalase activity and Nox levels, and reduced MPO activity. Histopathological parameters revealed the protection of mucosa (Amaral et al., 2013). The essential oil of *R. officinalis* reduced the damage activities of MPO and GSH-Px and increase GSH levels. Furthermore, prevent TBARS formation, decreased SOD and LPO in tissue sample. The main constituents of essential oil were cineole, camphor, and alpha-pinene (Takayama et al., 2016).

Maytenus robusta Reissek (Celastraceae) is traditionally used in folk medicine in Brazil to treat gastric ulcer (Balbach, 1980; Mota et al., 2008). Due to the indiscriminate use of *M. ilicifolia* in Brazil, this plant is threatened with extinction. Thus, as an alternative *M. robusta* was suggested to treat gastric ulcers.

A hydroalcoholic extract of *M. robusta* dried leaves was used orally as a single dose as pretreatment which reduced the gastric lesions and gastric juice volume, total acidity, and pH (de Andrade et al., 2007; Teske and Trentini, 2001). In other research, the hydroalcoholic extract from *M. robusta* aerial parts (10 mg/kg) was used as a single dose *and* increased the activities of SOD and GSH, acidified secretion, increased secretion volume, and pepsin activity. In in vitro assays, neutralize free radicals and protect L929 cells against H_2O_2-induced toxicity but has no effect on SOD activity. This extract used orally reduced the areas of ulcer lesions but in the intraduodenal application did not alter pH, nor gastric volume neither total acidity (Mota et al., 2008).

Caryocar coriaceum Wittm (Caryocaraceae) is a native plant of Brazil, known as pequi. Its fruits are used in cooking and folk medicine (De Oliveira et al., 2015; de Oliveira et al., 2010; Matos, 1984). Hydroalcoholic extract from *C. coreaceum* leaves, orally and intraperitoneally, reduced the gastric lesions, increased the production of mucus but decreased intestinal motility. The mechanism of action of this extract, including primary receptors sensitive to capsaicin, receptor and alpha 2 adrenergic opioids, and synthesis of NO and stimulation of prostaglandin-mediated secretion. This extract present of gallic acid, chlorogenic acid, caffeic acid, rutin, and quercetin (de Lacerda Neto et al., 2017).

Momordica charantia L. (Cucurbitaceae) is a plant used in folk medicine, including to treat peptic ulcers, diabetes, and its complications (Grover and Yadav, 2004). *M. charantia* presents flavonoids, alkaloids, carotenoids among them momordicins, charantin, and beta sitosterol (Grover and Yadav, 2004; Matsuda et al., 1999). The fruit extract obtained in olive oil from *M. charantia* inhibited the formation of ulcers. Ethanolic extract was also effective in reducing lesions induced by *diethyldithiocarbamate* (Gürbüz et al., 2000), while methanolic extract, orally as pretreatment, reduced the gastric lesions, recovered the histological parameters and increased pH, total acidity, mucin content, and reduced the content of pepsin and total proteins and reduced the activity SOD and catalase (Alam et al., 2009). Hexanic extract from aerial parts of *M. charantia* has also been reported as gastroprotector effect (Leite et al., 2005).

Sedum dendroideum Moc. et Sessé ex DC (Crassulaceae) is native to South Africa. In traditional medicine in Brazil and Mexico, fresh leaves are used to treat gastric disorders (Matos, 1984). Hydroethanolic extract from *S. dendroideum* leaves, orally as pretreatment in a single dose, reduced the ulcerations, inhibit oxidation, increase of gastric mucus, reduction of gastric volume, and acidity (Carrasco et al., 2014). The pretreatment by oral via of fresh leaf tea from *S. dendroideum* reduced the area of gastric lesions, increased mucus production, and tissue GSH. Tea reduced gastric injury, protected mucus, and GSH and increased secretion and total acidity, but decreased pH. In vitro studies show dose-dependent antioxidant activity and cell viability. The substances contained in the tea are consistent with kaempferol, myrcetin, quercetin. It is worth mentioning that the authors prepared the infusion according to ethnopharmacological use (da Luz et al., 2019). Another experiment a polysaccharide isolated from tea of *S. dendroideum* leaves reduced the area of gastric lesions and was able to revert gastric mucus and tissue GSH (de Oliveira et al., 2018).

Cissampelos sympodialis Eichl (Menispermaceae) is a plant used by indigenous tribes and at folk medicine in Brazil to treat several diseases (Bezerra-Santos et al., 2012; Correa, 1984). Its main components are terpenoids, phenolic compounds, polysaccharides, and alkaloids (Cavalcanti et al., 2013), *among* them warifteine and methylwarifteine (Bezerra-Santos et al., 2012; De Sales et al., 2018). The ethanolic extract and the chloroform fraction from leaves of *C. sympodialis* were preadministered orally at a single dose reduced gastric lesions, increased tissue glutathione but did not promote change in gastric volume, pH, and total acidity. Histologically the extracts and fraction reduced mucosal ulceration, preserved the glands, altered inflammatory exudate, and tissue necrosis (De Sales et al., 2018).

Murtingia clabura L. (Muntingiaceae) is a plant used in folk medicine in Malaysia, East Asia, and South America to treat ulcer-related diseases. *M. calabura* leaves powder presents flavonoids, tannins, polyphenols, saponins, steroids, and triterpenes, and compounds (flavonoids) compatible as rutin, fisetin, quercitrin, and *dihydroquercetin* (Zakaria et al., 2014). The methanolic extract from *M. calabura* leaves used as pretreatment reduced the areas of ulcers, the volume of gastric secretion, increased mucus gastric wall, and pH. In vitro, this extract (100 mg/kg) inhibited the activity of macrophages (RAW 264.7) stimulated by interferon gamma and LPS, and did not provoke toxicity in these cells. Moreover, the extract reduced the activities of LPO and xanthine oxidase (Zakaria et al., 2014). The methanolic extract of *M. calabura* leaves, petroleum ether (PEF), ethyl acetate (EAF) fractions, orally for 7 consecutive days, prevented the gastric lesions, sequestered free radicals, and increase SOD activity. The fractions inhibited the production of NO in vitro and inhibited LPO but only EAF has action on xanthine oxidase (Balan et al., 2014).

Azadirachta indica A. Juss (Meliaceae), called Neem, it was reported as divine tree (Kumar and Navaratnam, 2013) that has been used from prehistory to the present days to treat various diseases (Atawodi and Atawodi, 2009; Bijauliya et al., 2018; Maity et al., 2009). Nimbidine, catechin, gallic acid, kaempherol, quercitrin, rutin, and chlorogenic acid are responsible for medicinal properties (Atawodi and Atawodi, 2009; Bijauliya et al., 2018; Maity et al., 2009). Neem promote antiulcer effects (Maity et al., 2009) being attributed to Nimbidine from seed oil (Pillai and Santhakumari, 1984). The aqueous extract of Neem leaves as single- or five dose as pretreatment dose-dependently reduced gastric ulcer severity and increase of adherent gastric mucus (Garg et al., 1993), inhibited pentagastrin-induced acid secretion, and inhibited proton pump activity. The extract also increased the half-life of mucosal cells, but had no effect on mucin secretion (Dorababu et al., 2006). The extract of Neem leaves orally reduced the ulcer index, increased pH, and reduced the volume of gastric contents, free acidity, and total acidity (Bhajoni et al., 2016). The bark extract inhibits acid secretion by H+–K+–ATPase and prevention of oxidative damage by S-OH, prevents adhered mucus and endogenous glutathione depletion and oxidative damage of the gastric mucosa by blocking lipid peroxidation. In vitro

Neem bark extract protected glioma cells and gastric mucosal surface epithelial cells through the oxidative damage (Bandyopadhyay et al., 2004). Another study, Neem bark extract was administered intraperitoneally and orally and reduces the ulcer index, increase the inhibition of ulceration, and increase the total gastric acidity. Neem bark extract inhibits H2 receptor, causing blockade of histamine release (Raji et al., 2004).

Centella asiatica L. Urban (Apiaceae) is a plant found throughout India and increased all over the world. It is used in traditional and alternative medicine to treat several diseases (Gohil et al., 2010). This plant presents triterpernes, mainly, asiatic acid, madecassoic acid, asiaticoside, and madecassoside (James and Dubery, 2009). The *orally* pretreatment with aqueous extract from *C. asiatica* reduced the gastric lesions and suppressed the mucosal MPO activity (Cheng and Koo, 2000). Both, *C. asiatica* water extract and asiaticoide induced effect on bFGF expression and promoted angiogenesis during ulcer healing, and also increased the number of proliferating cells and promoted attenuation of MPO activity at the ulcer tissues (Cheng et al., 2004), and attenuates of iNOS activity (Guo et al., 2004).

It was verified that the scientific research advanced a lot in the validation of the popular knowledge; however, the clinical studies are very scarce with respect to the gastroprotection. Moreover, even those available do not adequately portray a clinical approach due to the number of participants. Plants with gastroprotective activities continue to be used in the form of traditional medicine made available at natural product fairs, and often made available as herbal medicines.

5.6.7 Antioxidants in gastric cancer

Case–control studies performed in North America, Europe, and Asia point out that the intake of fruits and vegetables, especially when consumed raw, may reduce the risk of gastric cancer around 40% (International Agency for Research on Cancer, 2003; Liu and Russell, 2008; Lunet et al., 2007). However, the correlation between vegetable intake and risk of gastric cancer also suffer the influence of macronutrient composition of the diet, since it has been reported that populations with a high incidence of gastric cancer present a low-quality diet regarding proteins and carbohydrate and the lesser preference for fresh fruits and vegetables. Consequently, this profile of consumption favors catalyzed-acid nitrosation in the stomach that trigger a series of reactions that can lead to mechanical injury in the stomach lining (Berretta et al., 2012; Krejs, 2010; Tsugane and Sasazuki, 2007).

Several antioxidants compounds have shown in isolation or in synergy effects on human health related to cancer. Some antioxidants present in fruit and vegetables have been reported as anticarcinogenic, such as isothiocyanates (Ullah, 2015), catechins (Yang et al., 2015), organosulfur compounds (Ho et al., 2012), curcumin (Balstad et al., 2011; Das and Vinayak, 2015), sulforaphane (Lin et al., 2014; Russo et al., 2018), diallyl and triallyl sulfide (Kim et al., 2014), genistein (Wang et al., 2013), and also resveratrol (Wang et al., 2015).

In this context, resveratrol by being able to protect the stomach against *H. pylori*-associated gastritis by limiting oxidative stress (Wang et al., 2015), as well as reversing chemoresistance to doxorubicin by inhibiting the mesenchymal epithelial transition through the modulation of the PTEN/Akt signaling pathway (Xu et al., 2017), therefore a pivotal chemopreventive and therapeutic agent (Bishayee, 2009). In a study conducted by Petrick et al. (2015), the intake of flavonoids such as anthocyanidins was inversely correlated with decreased mortality risk for gastric cancer (Petrick et al., 2015). Yet, even though green tea contains polyphenols, powerful antioxidants that also act as chemopreventive agents, safety, and efficacy are not clear when treating cancer patients (Hou et al., 2013).

The effects generated by these substances are associated with their different mechanisms of action in cellular and molecular levels. Among them, one possible mechanism is modulates cellular homeostasis via activation of the transcription factor NrfF-E2-related factor 2 (Nrf2), which is involved in protecting the human body against endogenously produced aggression or for external agents (Huang et al., 2015). Nrf2 modulates the expression of cellular genes involved in the antioxidant response to oxidant exposure (Ma, 2013).

Likewise, the consumption of alpha- and beta-carotene and its association with the risk of developing gastric cancer seems to be conflicting; however, it is estimated that an increase of 5 mg/day in its consumption would reduce the risk of developing that cancer (Zhou et al., 2016). In the association between dietary intake/blood levels of antioxidant vitamins (vitamin C, vitamin E, β-carotene, and α-carotene) and gastric cancer risk was found that the dietary intake of vitamin C, vitamin E, β-carotene, and α-carotene was inversely associated with gastric cancer risk. On the other hand, no such association was observed for blood levels of these antioxidants (Peiwei et al., 2014).

Selenium (Se) is known to play a role in redox status as a fundamental substrate for the synthesis of glutathione peroxidases (GPxs) and this enzyme acts directly on the removal of reactive oxygen and nitrogen species (Tinggi, 2008). In patients with gastric cancer, the Se serum levels appear to be reduced (Gong et al., 2016) when associated with high serum levels of carcinoembryonic antigen (CEA), a tumor marker for prognosis clinical of gastrointestinal cancers (Charalabopoulos et al., 2009). The high Se exposure has been reported with dissimilar effects on specific types of cancer, reduced risk of cancer mortality (Gong et al., 2016), as well as gastric cancer prevention (Cai et al., 2016).

Many types of multivitamin/mineral supplements, available on the market, are recommended for health maintenance, prevention, or treatment of diseases. In this sense, in China, two large studies sought to identify the association of multivitamin consumption and garlic supplements with the incidence or prevalence of gastric cancer and observed opposite effects. A multivitamin supplement consisting of vitamin C (250 mg), vitamin E (150 mg), and selenium (37.5 μg) and long-term garlic supplementation were not able to reduce the prevalence of precancerous gastric lesions such as dysplasia, chronic atrophic gastritis, metaplasia intestinal, and cancerous diseases (You et al., 2006). However, the selenium intake (50 μg) associated with vitamin E

(30 mg) and beta-carotene (15 mg) supplementation resulted in a significant mortality reduction by gastric cancer (Qiao et al., 2009). In this context, the concept of food matrix emphasizes that the consumption of bioactive compounds in whole foods has a different synergism when consumed isolated (Thomas et al., 2018).

Another primary natural source of antioxidants is marine fish and their derivatives products. They are rich in fatty acids with an anti-inflammatory profile, such as omega-3. These nutrients reduce chemotherapy-related toxicity, improve radiotherapy response in patients with esophagogastric tumors, and further promote reducing IL-2, TNF-α, and VEGF levels, as well as favor fewer hospitalizations (Eltweri et al., 2019). It is worth mentioning that supplementation of fish oil with 1.55 g of EPA and DHA for 9 weeks in patients undergoing chemotherapy for gastrointestinal cancer improved functional performance and reduced the severity of gastrointestinal disorders (Camargo et al., 2019).

Fig. 5.6.4 summarizes some antioxidant substances (curcumin, sulforaphane, diallyl trisulfide, genistein, resveratrol) that prevented or treated gastric injuries in experimental models.

Despite controversy evidence in the literature regarding the use of some antioxidants as a curative agents or adjuvants in gastric diseases therapy, the studies reported in this chapter reveal that a wide range of antioxidants compounds from plant foods and medicinal plants are beneficial for the prevention and treatment of gastritis, peptic ulcer disease, and even precancerous lesions. However, some substances may also be associated with negative effects in the treatment of gastric cancers, due mechanisms of resistance to chemotherapy.

Conclusion

In this chapter, we comment about using antioxidants phytochemicals as gastroprotective agents and safety in preclinical studies. We also highlight underlying mechanisms that provide the beneficial role of antioxidants in gastric injuries. In vitro studies emphasize the mechanism of action of antioxidant molecules that prevent cell damage, acts to decrease inflammation, induces apoptosis of the cancer cell, and confirms its nontoxic potential once they recovers viability cell. In some situations, these molecules may also contribute to the control and elimination of *H. pylori* damage. In addition, in vivo studies show both the preventive and curative effect in gastric injuries. These results are attributed to the antioxidant, anti-inflammatory, antisecretory properties, and restorative of defense mechanisms and healing in the gastric mucosa. Although there is a shortage in clinical studies, the antioxidants studied to reveal high curative potential in gastric lesions such as gastritis, peptic ulcer, and precancerous lesions. In relation to vitamin supplement, the findings are controversial to treat gastric cancer. In conclusion, the antioxidants are able to prevent as to cure gastric injury. However, cure and prevention are associated with many factors, including genetic and environmental factors. Therefore, the efficacy and versatility of the therapeutic potential of these antioxidants need to be better assessed in clinical research that appear be crucial for this clarification.

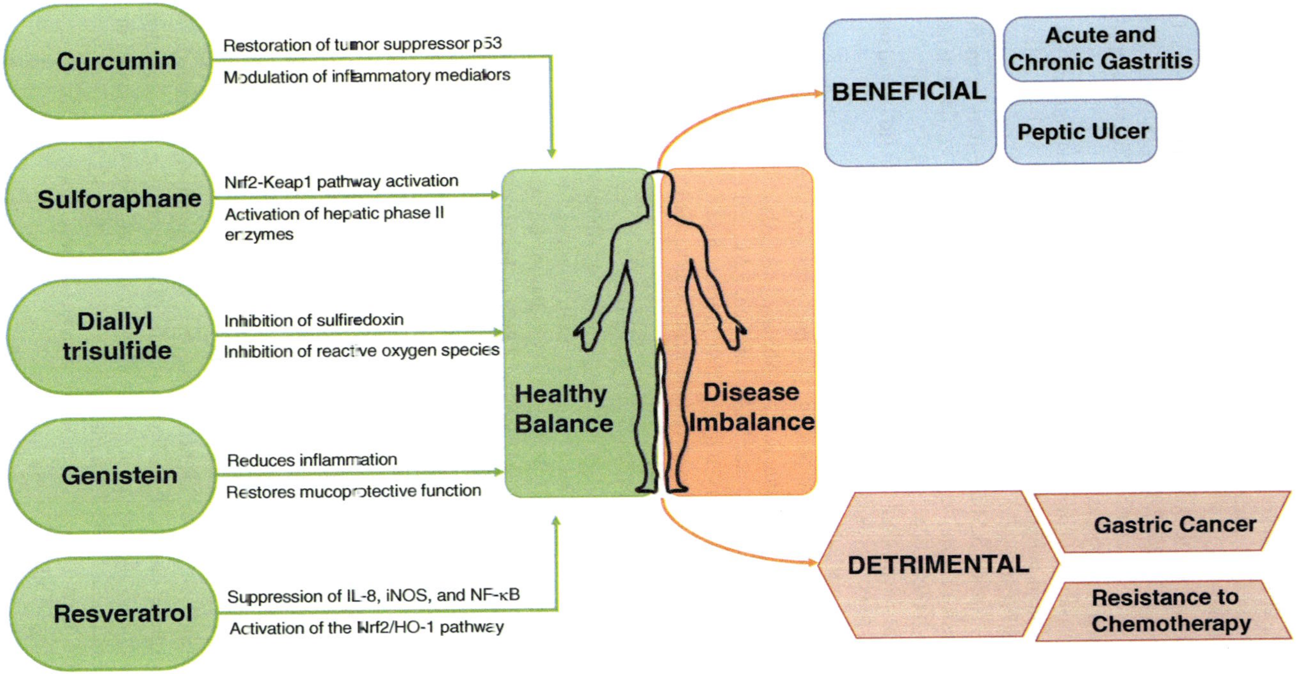

FIG. 5.6.4 Mechanism of action of some antioxidants with preventive and therapeutic potential in gastric lesions.

Several bioactive compounds in isolation or in synergy possess a beneficial effect in gastritis and ulcer peptic due favorable effects on oxidative balance. On the other hand, in gastric cancer some studies shown a detrimental effect for antioxidant therapy associated a resistance to chemotherapy in this disease. The effects generated by these substances are associated with their different mechanisms of action at cellular and molecular levels.

References

Abdel-Raheem, I.T., 2010. Gastroprotective effect of rutin against indomethacin-induced ulcers in rats. Basic Clin. Pharmacol. Toxicol. 107, 742–750. https://doi.org/10.1111/j.1742-7843.2010.00568.x.

Adefisayo, M.A., Akomolafe, R.O., Akinsomisoye, S.O., Alabi, Q.K., Ogundipe, O.L., Omole, J.G., Olamilosoye, K.P., 2017. Gastro-protective effect of methanol extract of *Vernonia amygdalina* (del.) leaf on aspirin-induced gastric ulcer in Wistar rats. Toxicol. Rep. 4, 625–633. https://doi.org/10.1016/j.toxrep.2017.11.004.

Adinortey, M.B., Ansah, C., Galyuon, I., Nyarko, A., 2013. In vivo models used for evaluation of potential antigastroduodenal ulcer agents. Ulcers 2013, 1–12. https://doi.org/10.1155/2013/796405.

Alam, S., Asad, M., Asdaq, S.M.B., Prasad, V.S., 2009. Antiulcer activity of methanolic extract of *Momordica charantia* L. in rats. J. Ethnopharmacol. 123, 464–469. https://doi.org/10.1016/j.jep.2009.03.024.

Ali, R., Sajjad, N., Hassan, S., Ganie, S.A., 2018. Reactive oxygen species (ROS) and response of antioxidants as ROS-scavengers. J. Appl. Sci. Comput. 5, 104–115.

Alkushi, A.G.R., Elsawy, N.A.M., 2017. Quercetin attenuates, indomethacin-induced acute gastric ulcer in rats. Folia Morphol 76, 252–261. https://doi.org/10.5603/FM.a2016.0067.

Almasaudi, S.B., Abbas, A.T., Al-Hindi, R.R., El-Shitany, N.A., Abdel-dayem, U.A., Ali, S.S., Saleh, R.M., Al Jaouni, S.K., Kamal, M.A., Harakeh, S.M., 2017. Manuka honey exerts antioxidant and anti-inflammatory activities that promote healing of acetic acid-induced gastric ulcer in rats. Evidence-based complement. Altern. Med. 2017, 5413917, 1–12. https://doi.org/10.1155/2017/5413917.

Almasaudi, S.B., El-shitany, N.A., Abbas, A.T., Abdel-dayem, U.A., Ali, S.S., Jaouni, S.K., Al, Harakeh, S., 2016. Antioxidant, Anti-inflammatory, and Antiulcer Potential of Manuka Honey against Gastric Ulcer in Rats. Oxid. Med. Cell. Longev. 2016, 3643824, 1–10.

Almeida, G.V.B., Arunachalam, K., Balogun, S.O., Pavan, E., Ascêncio, S.D., Soares, I.M., Zanatta, A.C., Vilegas, W., Macho, A., Oliveira Martins, D.T., 2019. Chemical characterization and evaluation of gastric antiulcer properties of the hydroethanolic extract of the stem bark of *Virola elongata* (Benth.) Warb. J. Ethnopharmacol. 231, 113–124. https://doi.org/10.1016/j.jep.2018.11.011.

Al Mofleh, I.A., 2010. Spices, herbal xenobiotics and the stomach: friends or foes? World J. Gastroenterol. 16, 2710–2719. https://doi.org/10.3748/wjg.v16.i22.2710.

Al-quraishy, S., Othman, M.S., Dkhil, M.A., Esmat, A., Moneim, A., 2017. ScienceDirect Olive (*Olea europaea*) leaf methanolic extract prevents HCl/ethanol-induced gastritis in rats by attenuating inflammation and augmenting antioxidant enzyme activities. Biomed. Pharmacother. 91, 338–349. https://doi.org/10.1016/j.biopha.2017.04.069.

Amaral, G.P., de Carvalho, N.R., Barcelos, R.P., Dobrachinski, F., Portella, R.L., da Silva, M.H., Lugokenski, T.H., Dias, G.R.M., da Luz, S.C.A., Boligon, A.A., Athayde, M.L., Villetti, M.A., Antunes Soares, F.A., Fachinetto, R., 2013. Protective action of ethanolic extract of *Rosmarinus officinalis* L. in gastric ulcer prevention induced by ethanol in rats. Food Chem. Toxicol. 55, 48–55. https://doi.org/10.1016/j.fct.2012.12.038.

Aslan, R., 2018. Antioxidant safety and antioxidant-induced stress. Kocatepe Vet. J. 11, 491–496. https://doi.org/10.30607/kvj.447022.

Atawodi, S.E., Atawodi, J.C., 2009. *Azadirachta indica* (neem): a plant of multiple biological and pharmacological activities. Phytochem. Rev. 8, 601–620. https://doi.org/10.1007/s11101-009-9144-6.

Attri, S., Singh, N., Singh, T.R., Goel, G., 2017. Effect of in vitro gastric and pancreatic digestion on antioxidant potential of fruit juices. Food Biosci 17, 1–6. https://doi.org/10.1016/j.fbio.2016.10.003.

Balan, T., Mohd. Sani, M.H., Suppaiah, V., Mohtarrudin, N., Suhaili, Z., Ahmad, Z., Zakaria, Z.A., 2014. Antiulcer activity of *Muntingia calabura* leaves involves the modulation of endogenous nitric oxide and nonprotein sulfhydryl compounds. Pharm. Biol. 52, 410–418. https://doi.org/10.3109/13880209.2013.839713.

Balbach, A., 1980. A flora nacional na medicina doméstica, 2nd ed. Vida Plena.

Balstad, T.R., Carlsen, H., Myhrstad, M.C.W., Kolberg, M., Reiersen, H., Gilen, L., Ebihara, K., Paur, I., Blomhoff, R., 2011. Coffee, broccoli and spices are strong inducers of electrophile response element-dependent transcription in vitro and in vivo -- studies in electrophile response element transgenic mice. Mol. Nutr. Food Res. 55, 185–197. https://doi.org/10.1002/mnfr.201000204.

Bandyopadhyay, U., Biswas, K., Sengupta, A., Moitra, P., Dutta, P., Sarkar, D., Debnath, P., Ganguly, C.K., Banerjee, R.K., 2004. Clinical studies on the effect of Neem (*Azadirachta indica*) bark extract on gastric secretion and gastroduodenal ulcer. Life Sci. 75, 2867–2878. https://doi.org/10.1016/j.lfs.2004.04.050.

Batta, A., 2016. A review on phytochemicals and their activities. Int. J. Curr. Res. Med. Sci. 2, 20–28.

Belwal, T., Nabavi, S.F., Nabavi, S.M., Habtemariam, S., 2017. Dietary anthocyanins and insulin resistance: when food becomes a medicine. Nutrients 9, 1111. https://doi.org/10.3390/nu9101111.

Bennedsen, M., Wang, X., Willén, R., Wadström, T., Andersen, L.P., 1999. Treatment of *H. pylori* infected mice with antioxidant astaxanthin reduces gastric inflammation, bacterial load and modulates cytokine release by splenocytes. Immunol. Lett. 70, 185–189. https://doi.org/10.1016/S0165-2478(99)00145-5.

Berretta, M., Cappellani, A., Lleshi, A., Vita, M., Menzo, E.L., Bearz, A., Galvano, F., Spina, M., Malaguarnera, M., Tirelli, U., Berretta, S., 2012. The role of diet in gastric cancer: still an open question. Front. Biosci. 17 (17), 1640–1647. https://doi.org/10.2741/4009.

Bezerra-Santos, C.R., Vieira-De-Abreu, A., Vieira, G.C., Filho, J.R., Barbosa-Filho, J.M., Pires, A.L., Martins, M.A., Souza, H.S., Bandeira-Melo, C., Bozza, P.T., Piuvezam, M.R., 2012. Effectiveness of *Cissampelos sympodialis* and its isolated alkaloid warifteine in airway hyperreactivity and lung remodeling in a mouse model of asthma. Int. Immunopharmacol. 13, 148–155. https://doi.org/10.1016/j.intimp.2012.03.014.

Bhajoni, P.S., Meshram, G.G., Lahkar, M., 2016. Evaluation of the antiulcer activity of the leaves of *Azadirachta indica*: an experimental study. Integr. Med. Int. 3, 10–16. https://doi.org/10.1159/000442750.

Bhardwaj, V., Bhardwaj, R., Reddy, B.V.K., Sharma, P.L., 2018. Investigation of gastroprotective potential of grape seed proanthocyanidin exract in experimental models of gastric ulcer in wistar rats. J. Gastrointest. Dig. Syst. 08, 561. https://doi.org/10.4172/2161-069x.1000561.

Bhattacharyya, A., Chattopadhyay, R., Mitra, S., Crowe, S.E., 2014. Oxidative stress: an essential factor in the pathogenesis of gastrointestinal mucosal diseases. Physiol. Rev. 94, 329–354. https://doi.org/10.1152/physrev.00040.2012.

Bijauliya, R.K., Alok, S., Chanchal, D.K., Sabharwal, M., Yadav, R.D., 2018. An updated review of pharmacological studies on *Azadiractha indica* (Neem). Int. J. Pharm. Sci. Res. 9, 2645–2655. https://doi.org/10.13040/IJPSR.0975-8232.9(7).2645-55.

Bishayee, A., 2009. Cancer prevention and treatment with resveratrol: from rodent studies to clinical trials. Cancer Prev. Res. 2, 409–418. https://doi.org/10.1158/1940-6207.CAPR-08-0160.

Bisht, R., 2018. Antioxidants: a brief review. J. Drug Deliv. Ther. 8, 373–376.

Borges, R.S., Ortiz, B.L.S., Pereira, A.C.M., Keita, H., Carvalho, J.C.T., 2019. *Rosmarinus officinalis* essential oil: a review of its phytochemistry, anti-inflammatory activity, and mechanisms of action involved. J. Ethnopharmacol. 229, 29–45. https://doi.org/10.1016/j.jep.2018.09.038.

Bouayed, J., Bohn, T., 2010. Exogenous antioxidants—double-edged swords in cellular redox state health beneficial effects at physiologic doses versus deleterious effects at high doses. Oxid. Med. Cell. Longev. 3, 228–237. https://doi.org/10.4161/oxim.3.4.12858.

Boyacioglu, M., Kum, C., Sekkin, S., Sultan, H., 2015. The effects of lycopene on DNA damage and oxidative stress on indomethacin-induced gastric ulcer in rats. Clin. Nutr. 35, 428–435. https://doi.org/10.1016/j.clnu.2015.03.006.

Boyacioglu, M., Kum, C., Sekkin, S., Yalinkilinc, H.S., Avci, H., Epikmen, E.T., Karademir, U., 2016. The effects of lycopene on DNA damage and oxidative stress on indomethacin-induced gastric ulcer in rats. Clin. Nutr. 35, 428–435. https://doi.org/10.1016/j.clnu.2015.03.006.

Brito, S.A., de Almeida, C.L.F., de Santana, T.I., da Silva Oliveira, A.R., do Nascimento Figueiredo, J.C.B., Souza, I.T., de Almeida, L.L., da Silva, M.V., Borges, A.S., de Medeiros, J.W., da Costa Silva Neto, J., de Cássia Ribeiro Gonçalves, R., Kitagawa, R.R., Sant'Ana, A.E.G., Rolim, L.A., de Menezes, I.R.A., da Silva, T.G., Caldas, G.F.R., Wanderley, A.G., 2018. Antiulcer activity and potential mechanism of action of the leaves of *Spondias mombin* L. Oxid. Med. Cell. Longev. 2018, 1731459. https://doi.org/10.1155/2018/1731459.

Brodie, D., Chase, B., 1967. Role of gastric acid in aspirin-induced gastric irritation in the rat. Gastroenterology 53, 604–610.

Cai, X., Wang, C., Yu, W., Fan, W., Wang, S., Shen, N., Wu, P., Li, X., Wang, F., 2016. Selenium exposure and cancer risk: an updated meta-analysis and meta-regression. Sci. Rep. 6, 1–18. https://doi.org/10.1038/srep19213.

Camargo, C.Q., Mocellin, M.C., Brunetta, H.S., Chagas, T.R., Fabre, M.E.S., Trindade, E.B.S.M., Silva, E.L., Nunes, E.A., 2019. Fish oil decreases the severity of treatment-related adverse events in gastrointestinal cancer patients undergoing chemotherapy: a randomized, placebo-controlled, triple-blind clinical trial. Clin. Nutr. ESPEN. 31, 61–70. https://doi.org/10.1016/j.clnesp.2019.02.015.

Carrasco, V., Pinto, L.A., Cordeiro, K.W., Cardoso, C.A.L., Freitas, K.D.C., 2014. Antiulcer activities of the hydroethanolic extract of *Sedum dendroideum* Moc et Sessé ex DC. (balsam). J. Ethnopharmacol. 158, 345–351. https://doi.org/10.1016/j.jep.2014.10.042.

Cavalcanti, A.C., Melo, I.C.A.R., Medeiros, A.F.D., Neves, M.V.M., Pereira, A.N., Oliveira, E.J., 2013. Studies with *Cissampelos sympodialis*: the search towards the scientific validation of a traditional Brazilian medicine used for the treatment of asthma. Braz. J. Pharmacogn. 23, 527–541. https://doi.org/10.1590/S0102-695X2013005000029.

Chang, X., Luo, F., Jiang, W., Zhu, L., Gao, J., He, H., Wei, T., Gong, S., Yan, T., 2015. Protective activity of salidroside against ethanol-induced gastric ulcer via the MAPK/NF-κB pathway in vivo and in vitro. Int. Immunopharmacol. 28, 604–615. https://doi.org/10.1016/j.intimp.2015.07.031.

Chao, K.H., Chung, T.J., Kim, S.J., Lee, T.H., Yoon, C.M., 1981. Clinical experience of madecassol (*Centella asiatica*) in the treatment of peptic ulcer. Korean J. Gastroenterol. 13, 49–86.

Charalabopoulos, K., Kotsalos, T., Batistatou, A., Charalabopoulos, A., Peschos, D., Vezyraki, P., Kalfakakou, V., Metsios, A., Charalampopoulos, A., Macheras, A., Agnantis, N., Evangelou, A., 2009. Serum and tissue selenium levels in gastric cancer patients and correlation with CEA. Anticancer Res. 29, 3465–3467.

Chattopadhyay, I., Bandyopadhyay, U., Biswas, K., Maity, P., Banerjee, R.K., 2006. Indomethacin inactivates gastric peroxidase to induce reactive-oxygen-mediated gastric mucosal injury and curcumin protects it by preventing peroxidase inactivation and scavenging reactive oxygen. Free Radic. Biol. Med. 40, 1397–1408. https://doi.org/10.1016/j.freeradbiomed.2005.12.016.

Cheng, C.L., Guo, J.S., Luk, J., Koo, M.W.L., 2004. The healing effects of *Centella* extract and asiaticoside on acetic acid induced gastric ulcers in rats. Life Sci. 74, 2237–2249. https://doi.org/10.1016/j.lfs.2003.09.055.

Cheng, C.L., Koo, M.W.L., 2000. Effects of *Centella asiatica* on ethanol induced gastric mucosal lesions in rats. Life Sci. 67, 2647–2653. https://doi.org/10.1016/S0024-3205(00)00848-1.

Cheng, X.J., Lin, J.C., Tu, S.P., 2016. Etiology and prevention of gastric cancer. Gastrointest. Tumors 3, 25–36. https://doi.org/10.1159/000443995.

Correa, M.P., 1984. Dicionário de plantas exóticas cultivadas no Brasil. Imprensa Nacional Brasilia, Rio de Janeiro.

Cutsem, E., Sagaert, X., Topal, B., Haustermans, K., Prenen, H., 2016. Gastric cancer. Lancet 388, 2654–2664. https://doi.org/10.1016/S0140-6736(16)30354-3.

da Luz, B.B., de Oliveira, A.F., Maria Ferreira, D., Dallazen, J.L., Cipriani, T.R., de Souza, L.M., Werner, M.F.P., 2019. Chemical composition, antioxidant and gastrointestinal properties of *Sedum dendroideum* Moc & Sessé ex DC leaves tea infusion. J. Ethnopharmacol. 231, 141–151. https://doi.org/10.1016/j.jep.2018.11.019.

Das, D., Bandyopadhyay, D., Bhattacharjee, M., Banerjee, R.K., 1997. Hydroxyl radical is the major causative factor in stress. Free Radic. Biol. Med. 23, 8–18.

Das, L., Vinayak, M., 2015. Long term effect of curcumin in restoration of tumour suppressor p53 and phase-II antioxidant enzymes via activation of Nrf2 signalling and modulation of inflammation in prevention of cancer. PLoS One 10, 1–22. https://doi.org/10.1371/journal.pone.0124000.

da Silveira Vasconcelos, M., Mota, E.F., Gomes-Rochette, N.F., Nunes-Pinheiro, D.C.S., Nabavi, S.M., de Melo, D.F., 2019. Chapter 3.18 - Ginger (*Zingiber officinale* Roscoe). In: Nabavi, S.M., Silva, A.S. (Eds.), Nonvitamin and Nonmineral Nutritional Supplements, Academic Press, London, pp. 235–239. https://doi.org/10.1016/B978-0-12-812491-8.00034-5.

de Andrade, S.F., Lemos, M., Comunello, E., Noldin, V.F., Filho, V.C., Niero, R., 2007. Evaluation of the antiulcerogenic activity of *Maytenus robusta* (Celastraceae) in different experimental ulcer models. J. Ethnopharmacol. 113, 252–257. https://doi.org/10.1016/j.jep.2007.06.002.

Debanath, P.K., Gode, K.D., Das, D.G., Sanyal, A.K., 1974. Effects of propranolol on gastric secretion in albino rats. Br. J. Pharmacol. 51, 213–216. https://doi.org/10.1111/j.1476-5381.1974.tb09649.x.

de Lacerda Neto, L.J., Ramos, A.G.B., Santos Sales, V., de Souza, S.D.G., dos Santos, A.T.L., de Oliveira, L.R., Kerntopf, M.R., de Albuquerque, T.R., Coutinho, H.D.M., Quintans-Júnior, L.J., Wanderley, A.G., de Menezes, I.R.A., 2017. Gastroprotective and ulcer healing effects of hydroethanolic extract of leaves of *Caryocar coriaceum*: mechanisms involved in the gastroprotective activity. Chem. Biol. Interact. 261, 56–62. https://doi.org/10.1016/j.cbi.2016.11.020.

Demir, S., Yilmaz, M., Köseoğlu, M., Akalin, N., Aslan, D., Aydin, A., 2003. Role of free radicals in peptic ulcer and gastritis *Serbest radikallerinin* peptik ülser ve gastritteki rolü. Turk. J. Gastroenterol. 14, 39–43.

de Oliveira, A.F., da Luz, B.B., Werner, M.F.P., Iacomini, M., Cordeiro, L.M.C., Cipriani, T.R., 2018. Gastroprotective activity of a pectic polysaccharide fraction obtained from infusion of *Sedum dendroideum* leaves. Phytomedicine 41, 7–12. https://doi.org/10.1016/j.phymed.2018.01.015.

De Oliveira, F.F.B., De Araújo, J.C.B., Pereira, A.F., Brito, G.A.C., Gondim, D.V., Ribeiro, R.D.A., De Menezes, I.R.A., Vale, M.L., 2015. Antinociceptive and anti-inflammatory effects of *Caryocar coriaceum* Wittm fruit pulp fixed ethyl acetate extract on zymosan-induced arthritis in rats. J. Ethnopharmacol. 174, 452–463. https://doi.org/10.1016/j.jep.2015.08.017.

de Oliveira, M.L.M., Nunes-Pinheiro, D.C.S., Tomé, A.R., Mota, É.F., Lima-Verde, I.A., Pinheiro, F.G., de, M., Campello, C.C., de Morais, S.M., 2010. In vivo topical anti-inflammatory and wound healing activities of the fixed oil of *Caryocar coriaceum* Wittm. seeds. J. Ethnopharmacol. 129, 214–219. https://doi.org/10.1016/j.jep.2010.03.014.

De Sales, I.R.P., Formiga, R.D.O., Machado, F.D.F., Nascimento, R.F., Pessoa, M.M.B., Barros, M.E.F.X., Vieira, G.C., Gadelha, F.A.A.F., Marinho, A.F., Barbosa Filho, J.M., Júnior, R.F.D.A., Antunes, A.A., Batista, L.M., 2018. Cytoprotective, antioxidant and anti-inflammatory mechanism related to antiulcer activity of *Cissampelos sympodialis* Eichl. in animal models. J. Ethnopharmacol. 222, 190–200. https://doi.org/10.1016/j.jep.2018.04.019.

Djahanguiri, B., 1969. The production of acute gastric ulceration by indomethacin in the rat. Scand. J. Gastroenterol. J Gastroenterol 4, 265–267.

Dorababu, M., Joshi, M.C., Bhawani, G., Kumar, M.M., Chaturvedi, A., Goel, R.K., 2006. Effect of aqueous extract of neem (*Azadirachta indica*) leaves on offensive and diffensive gastric mucosal factors in rats. Indian J. Physiol. Pharmacol. 50, 241–249.

Elseweidy, M.M., 2011. Papel de antioxidantes naturais na gastrite. In: Tonino, P. (Ed.), Gastritis and Gastric Cancer – New Insights in Gastroprotection, Diagnosis and Treatments. IntechOpen. Croatia. 1, 127–152. https://doi.org/10.5772/24336.

Eltweri, A.M., Thomas, A.L., Chung, W.Y., Morgan, B., Thompson, J., Dennison, A.R., Bowrey, D.J., 2019. The effect of supplementary Omegaven® on the clinical outcome of patients with advanced esophagogastric adenocarcinoma receiving palliative epirubicin, oxaliplatin, and capecitabine chemotherapy: a phase II clinical trial. Anticancer Res. 39, 853–861. https://doi.org/10.21873/anticanres.13185.

Farzaei, M.H., Abdollahi, M., Rahimi, R., 2015. Role of dietary polyphenols in the management of peptic ulcer. World J. Gastroenterol. 21, 6499–6517. https://doi.org/10.3748/wjg.v21.i21.6499.

Farzaei, M.H., Shams-Ardekani, M.R., Abbasabadi, Z., Rahimi, R., 2013. Scientific evaluation of edible fruits and spices used for the treatment of peptic ulcer in traditional iranian medicine. ISRN Gastroenterol 2013, 1–12. https://doi.org/10.1155/2013/136932.

Filiptsova, O.V., Gazzavi-Rogozina, L.V., Timoshyna, I.A., Naboka, O.I., Dyomina, Y.V., Ochkur, A.V., 2017. The essential oil of rosemary and its effect on the human image and numerical short-term memory. Egypt. J. Basic Appl. Sci. 4, 107–111. https://doi.org/10.1016/j.ejbas.2017.04.002.

Garg, G.P., Nigam, S.K., Ogle, C.W., 1993. The gastric antiulcer effects of the leaves of the neem tree. Planta Med. 59, 215–217. https://doi.org/10.1055/s-2006-959654.

Gayoso, L., Claerbout, A.S., Calvo, M.I., Cavero, R.Y., Astiasarán, I., Ansorena, D., 2016. Bioaccessibility of rutin, caffeic acid and rosmarinic acid: influence of the in vitro gastrointestinal digestion models. J. Funct. Foods 26, 428–438. https://doi.org/10.1016/j.jff.2016.08.003.

Gohil, K., Patel, J., Gajjar, A., 2010. Pharmacological review on *Centella asiatica*: a potential herbal cure-all. Indian J. Pharm. Sci. 72, 546. https://doi.org/10.4103/0250-474x.78519.

Gomaa, A.M.S., Abd El-Mottaleb, N.A., Aamer, H.A., 2018. Antioxidant and anti-inflammatory activities of alpha lipoic acid protect against indomethacin-induced gastric ulcer in rats. Biomed. Pharmacother. 101, 188–194. https://doi.org/10.1016/j.biopha.2018.02.070.

Gomes-Rochette, N.F., Vasconcelos, M.D.S., Nabavi, S.M., Mota, E.F., Nunes-Pinheiro, D.C.S., Daglia, M., Melo, D.F., 2016. Fruit as potent natural antioxidants and their biological effects. Curr. Pharm. Biotechnol. 17, 986–993.

Gong, H.-Y., He, J.-G., Li, B.-S., 2016. Meta-analysis of the association between selenium and gastric cancer risk. Oncotarget 7, 15600–15605. https://doi.org/10.18632/oncotarget.7205.

Graziani, G., D'Argenio, G., Tuccillo, C., Loguercio, C., Ritieni, A., Morisco, F., Del Vecchio Blanco, C., Fogliano, V., Romano, M., 2005. Apple polyphenol extracts prevent damage to human gastric epithelial cells in vitro and to rat gastric mucosa in vivo. Gut 54, 193–200. https://doi.org/10.1136/gut.2004.046292.

Grover, J.K., Yadav, S.P., 2004. Pharmacological actions and potential uses of *Momordica charantia*: a review. J. Ethnopharmacol. 93, 123–132. https://doi.org/10.1016/j.jep.2004.03.035.

Guha, P., Dey, A., Chatterjee, A., Chattopadhyay, S., Bandyopadhyay, S.K., 2010. Pro-ulcer effects of resveratrol in mice with indomethacin-induced gastric ulcers are reversed by l-arginine. Br. J. Pharmacol. 159, 726–734. https://doi.org/10.1111/j.1476-5381.2009.00572.x.

Guo, J.S., Cheng, C.L., Koo, M.W.L., 2004. Inhibitory effects of *Centella asiatica* water extract and asiaticoside on inducible nitric oxide synthase during gastric ulcer healing in rats. Planta Med. 70, 1150–1154. https://doi.org/10.1055/s-2004-835843.

Gürbüz, I., Akyüz, Ç., Yeşilada, E., Şener, B., 2000. Anti-ulcerogenic effect of *Momordica charantia* L. fruits on various ulcer models in rats. J. Ethnopharmacol. 71, 77–82. https://doi.org/10.1016/S0378-8741(99)00178-6.

Hajialyani, M., Farzaei, M.H., Echeverría, J., Nabavi, S.M., Uriarte, E., Eduardo, S.S., 2019. Hesperidin as a neuroprotective agent: a review of animal and clinical evidence. Molecules 24, 648. https://doi.org/10.3390/molecules24030648.

Hamaishi, K., Kojima, R., Ito, M., 2006. Anti-ulcer effect of tea catechin in rats. Biol. Pharm. Bull. 29, 2206–2213. https://doi.org/10.1248/bpb.29.2206.

Haniadka, R., Saldanha, E., Sunita, V., Palatty, P.L., Fayad, R., Baliga, M.S., 2013. A review of the gastroprotective effects of ginger (*Zingiber officinale* Roscoe). Food Funct 4, 845–855. https://doi.org/10.1039/c3fo30337c.

Hegab, I.I., Abd-Ellatif, R.N., Sadek, M.T., 2018. The gastroprotective effect of N-acetylcysteine and genistein in indomethacin-induced gastric injury in rats. Can. J. Physiol. Pharmacol. 96, 1161–1170. https://doi.org/10.1139/cjpp-2017-0730.

Ho, C.Y., Cheng, Y.T., Chau, C.F., Yen, G.C., 2012. Effect of diallyl sulfide on in vitro and in vivo Nrf2-mediated pulmonic antioxidant enzyme expression via activation ERK/p38 signaling pathway. J. Agric. Food Chem. 60, 100–107. https://doi.org/10.1021/jf203800d.

Holian, O., Wahid, S., Atten, M.J., Attar, B.M., 2002. Inhibition of gastric cancer cell proliferation by resveratrol: role of nitric oxide. Am. J. Physiol. Liver Physiol. 282, G809–G816. https://doi.org/10.1152/ajpgi.00193.2001.

Hollander, D., Tarnawski, A., Krause, W.J., Gergely, H., 1985. Protective effect of sucralfate against alcohol-induced gastric mucosal injury in the rat: macroscopic, histologic, ultrastructural, and functional time sequence analysis. Gastroenterology 88, 366–374. https://doi.org/10.1016/S0016-5085(85)80191-8.

Hou, I.C., Amarnani, S., Chong, M.T., Bishayee, A., 2013. Green tea and the risk of gastric cancer: epidemiological evidence. World J. Gastroenterol. 19, 3713–3722. https://doi.org/10.3748/wjg.v19.i24.3713.

Huang, Y., Li, W., Su, Z., Kong, A.N.T., 2015. The complexity of the Nrf2 pathway: beyond the antioxidant response. J. Nutr. Biochem. 26, 1401–1413. https://doi.org/10.1016/j.jnutbio.2015.08.001.

Hunt, R.H., Camilleri, M., Crowe, S.E., El-Omar, E.M., Fox, J.G., Kuipers, E.J., Malfertheiner, P., McColl, K.E.L., Pritchard, D.M., Rugge, M., Sonnenberg, A., Sugano, K., Tack, J., 2015. The stomach in health and disease. Gut 64, 1650–1668. https://doi.org/10.1136/gutjnl-2014-307595.

Hussein, S.A., Karousa, M.M., Amin, A., Awadalla, M.A., 2016a. Gastroprotective and antioxidant activities of rutin against ethanol-induced gastric mucosal erosion in rats via suppression of inflammation, oxidative stress and apoptosis in stomach tissues. Int. J. Pharma Sci. 6, 1658–1668.

Hussein, S.A., Karousa, M.M., Amin, A., Awadalla, M.A., 2016b. Evaluation of gastroprotective effect and anti-inflammatory role of resveratrol against gastric mucosal alterations in experimental model of gastritis in rats. Int. J. Pharma Sci. 6, 1559–1570.

International Agency for Research on Cancer, 2003. IARC Handbooks on Cancer Prevention, Vol. 8: Fruit and Vegetables. IARC, Lyon.

James, J.T., Dubery, I.A., 2009. Pentacyclic triterpenoids from the medicinal herb, *Centella asiatica* (L.) urban. Molecules 14, 3922–3941. https://doi.org/10.3390/molecules14103922.

Kahraman, A., Köken, T., Aktepe, F., Erkasap, N., Erkasap, S., Serteser, M., 2003. The antioxidative and antihistaminic properties of quercetin in ethanol-induced gastric lesions. Toxicology 183, 133–142. https://doi.org/10.1016/s0300-483x(02)00514-0.

Kalant, H., Grant, D.M., Mitchell, J., 2007. Principles of Medical Pharmacology. Saunders Elsevier, Toronto.

Katic, V.V., 2018. Study on metaplastic lesions in gastric cancerogenesis. Dig. Syst. 2, 1–6. https://doi.org/10.15761/dsj.1000116.

Khan, I., Samson, S.E., Grover, A.K., 2017. Antioxidant supplements and gastrointestinal diseases: a critical appraisal. Med. Princ. Pract. 26, 201–217. https://doi.org/10.1159/000468988.

Khonche, A., Biglarian, O., Panahi, Y., Valizadegan, G., Soflaei, S.S., Ghamarchehreh, M.E., Majeed, M., Sahebkar, A., 2016. Adjunctive therapy with curcumin for peptic ulcer: a randomized controlled trial. Drug Res. (Stuttg). 66, 444–448. https://doi.org/10.1055/s-0042-109394.

Kim, J.M., Ko, H., Kim, S.-J., Shim, S.H., Ha, C.H., Chang, H.I., 2016. Chemopreventive properties of genipin on AGS cell line via induction of JNK/Nrf2/ARE signaling pathway. J. Biochem. Mol. Toxicol. 30, 45–54. https://doi.org/10.1002/jbt.

Kim, S., Lee, H.G., Park, S.A., Kundu, J.K., Keum, Y.S., Cha, Y.N., Na, H.K., Surh, Y.J., 2014. Keap1 cysteine 288 as a potential target for diallyl trisulfide-induced Nrf2 activation. PLoS One 9, e85984. https://doi.org/10.1371/journal.pone.0085984.

Koosirirat, C., Linpisarn, S., Changsom, D., Chawansuntati, K., Wipasa, J., 2010. Investigation of the anti-inflammatory effect of *Curcuma longa* in *Helicobacter pylori*-infected patients. Int. Immunopharmacol. 10, 815–818. https://doi.org/10.1016/j.intimp.2010.04.021.

Krejs, G.J., 2010. Gastric cancer: epidemiology and risk factors. Gastric Cancer 28, 600–603. https://doi.org/10.1159/000320277.

Kuipers, E.J., Blaser, M.J., 2012. Acid peptic disease. In: Goldman, L., Schafer, A.I. (Eds.), Goldman's Cecil Medicine, 24th ed. 1. Saunders Elsevier, Philadelphia, PA, pp. 886–895 https://www.sciencedirect.com/science/article/pii/B978143771604700141X.

Kulshreshtha, M., Srivastava, G., Singh, M.P., 2017. Pathophysiological status and nutritional therapy of peptic ulcer: an update. Environ. Dis. 2, 76–86. https://doi.org/10.4103/ed.ed.

Kumar, V.S., Navaratnam, V., 2013. Neem (*Azadirachta indica*): prehistory to contemporary medicinal uses to humankind. Asian Pac. J. Trop. Biomed. 3, 505–514. https://doi.org/10.1016/S2221-1691(13)60105-7.

Kuna, L., Jakab, J., Smolic, R., Raguz-Lucic, N., Vcev, A., Smolic, M., 2019. Peptic ulcer disease: a brief review of conventional therapy and herbal treatment options. J. Clin. Med. 8, 179. https://doi.org/10.3390/jcm8020179.

Leite, K.L., Nunes-Pinheiro, D.C.S., Campello, C.C., 2005. Gastroprotective effect of hexanic extract from aerial parts of *Momordica charantia*. Ciência Anim 15, 15–20.

Li, Q., Hu, X., Xuan, Y., Ying, J., Fei, Y., Rong, J., Zhang, Y., Zhang, J., Liu, C., Liu, Z., 2018. Kaempferol protects ethanol-induced gastric ulcers in mice via pro-inflammatory cytokines and NO. Acta Biochim. Biophys. Sin. (Shanghai) 50, 246–253. https://doi.org/10.1093/abbs/gmy002.

Lian, D.W., Xu, Y.F., Ren, W.K., Fu, L.J., Chen, F.J., Tang, L.Y., Zhuang, H.L., Cao, H.Y., Huang, P., 2018. Unraveling the novel protective effect of patchouli alcohol against *Helicobacter pylori*-induced gastritis: Insights into the molecular mechanism in vitro and in vivo. Front. Pharmacol. 9, 1347. https://doi.org/10.3389/fphar.2018.01347.

Lin, H., Wei, B., Li, G., Zheng, J., Sun, J., Chu, J., Zeng, R., Niu, Y., 2014. Sulforaphane reverses glucocorticoid-induced apoptosis in osteoblastic cells through regulation of the Nrf2 pathway. Drug Des. Devel. Ther. 8, 973–982. https://doi.org/10.2147/DDDT.S65410.

Liu, C., Russell, R.M., 2008. Nutrition and gastric cancer risk: an update. Nutr. Rev. 66, 237–249. https://doi.org/10.1111/j.1753-4887.2008.00029.x.

Lunet, N., Valbuena, C., Vieira, A.L., Lopes, C., David, L., Barros, H., et al., 2007. Fruit and vegetable consumption and gastric cancer by location and histological type: case–control and meta-analysis. Eur. J. Cancer Prev. 16 (4), 312–327.

Ma, Q., 2013. Role of NRF2 in oxidative stress and toxicity. Annu. Rev. Pharmacol. Toxicol. 53, 401–426. https://doi.org/10.1146/annurev-pharmtox-011112-140320.Role.

Mahmoud, Y.I., Abd El-Ghffar, E.A., 2019. *Spirulina* ameliorates aspirin-induced gastric ulcer in albino mice by alleviating oxidative stress and inflammation. Biomed. Pharmacother. 109, 314–321. https://doi.org/10.1016/j.biopha.2018.10.118.

Maity, P., Biswas, K., Chattopadhyay, I., Banerjee, R.K., Bandyopadhyay, U., 2009. The use of neem for controlling gastric hyperacidity and ulcer. Phytother. Res. 23, 747–755. https://doi.org/10.1002/ptr.

Mansingh, D.P., Sunanda, O.J., Sali, V.K., Vasanthi, H.R., 2018. [6]-Gingerol-induced cell cycle arrest, reactive oxygen species generation, and disruption of mitochondrial membrane potential are associated with apoptosis in human gastric cancer (AGS) cells. J. Biochem. Mol. Toxicol. 32, 3–9. https://doi.org/10.1002/jbt.22206.

Marchese, A., Arciola, C.R., Barbieri, R., Silva, A.S., Nabavi, S.F., Sokeng, A.J.T., Izadi, M., Jafari, N.J., Suntar, I., Daglia, M., Nabavi, S.M., 2017. Update on monoterpenes as antimicrobial agents: a particular focus on p-cymene. Materials (Basel) 10, 1–15. https://doi.org/10.3390/ma10080947.

Marcial, G., Sendker, J., Brandt, S., De Lampasona, M.P., Catalán, C.A.N., De Valdez, G.F., Hensel, A., 2014. Gastroprotection as an example: antiadhesion against *Helicobacter pylori*, anti-inflammatory and antioxidant activities of aqueous extracts from the aerial parts of *Lippia integrifolia* Hieron. J. Ethnopharmacol. 155, 1125–1133. https://doi.org/10.1016/j.jep.2014.06.039.

Martin, K.R., Barrett, J.C., 2002. Reactive oxygen species as double-edged swords in cellular processes: low-dose cell signaling versus high-dose toxicity. Hum. Exp. Toxicol. 21, 71–75. https://doi.org/10.1191/0960327102ht213oa.

Matos, F.J.A., 1984. Plantas medicinais: guia de seleção e emprego de plantas usadas em fitoterapia no Nordeste do Brasil. Imprensa Universitária, Fortaleza.

Matschke, V., Theiss, C., Matschke, J., 2019. Oxidative stress: the lowest common denominator of multiple diseases. Neural Regen. Res. 14, 238–241. https://doi.org/10.4103/1673-5374.244780.

Matsuda, H., Li, Y., Yoshikawa, M., 1999. Roles of capsaicin-sensitive sensory nerves, endogenous nitric oxide, sulfhydryls, and prostaglandins in gastroprotection by momordin Ic, an oleanolic acid oligoglycoside, on ethanol-induced gastric mucosal lesions in rats. Life Sci. 65, 27–32.

Mezdour, H., Hanfer, M., Menad, A., Ameddah, S., 2017. Oxidative stress and its relationship with the emergence of various stomach damages. Batna J. Med. Sci. 4, 145–148.

Minaiyan, M., Ghannadi, A.R., Afsharipour, M., Mahzouni, P., 2011. Effects of extract and essential oil of *Rosmarinus officinalis* L. on TNBS-induced colitis in rats. Res. Pharm. Sci. 6, 13–21.

Mizui, T., Doteuchi, M., 1983. Effect of polyamines on acidified ethanol-induced gastric lesions in rats. Jpn. J. Pharmacol. 33, 939–945. https://doi.org/10.1254/jjp.33.939.

Mota, K.S.D.L., Pita, J.C.L.R., Estevam, E.C., Medeiros, V.M., Tavares, J.F., Agra, M.D.F., Diniz, M.D.F.F.M., Da Silva, M.S., Batista, L.M., 2008. Evaluation of the toxicity and antiulcerogenic activity of the ethanol extract of *Maytenus obtusifolia* Mart. leaves. Braz. J. Pharmacogn. 18, 441–446. https://doi.org/10.1590/S0102-695X2008000300019.

Münzenmaier, A., Lange, C., Glocker, E., Covacci, A., Moran, A., Baeuerle, P.A., Kist, M., Pahl, H.L., Munzenmaier, A., Lange, C., Clocker, E., Covacci, A., Moran, A., Bereswill, S., Baeuerle, P.A., Kist, M., Pah, H.L., 1997. A secreted/shed product of *Helicobacter pylori* activates transcription factor nuclear A Secretedhhed Product of *Helicobacter pylori* activates transcription factor nuclear factor-KB. J. Immunol. 159, 6140–6147.

Nanjundaiah, S.M., Annaiah, H.N.M., Dharmesh, S.M., 2009. Gastroprotective effect of ginger rhizome (*Zingiber officinale*) extract: role of gallic acid and cinnamic acid in H+, K+-ATPase/*H. pylori* inhibition and anti-oxidative mechanism. Evidence-based complement. Altern. Med. 2011, 1–13. https://doi.org/10.1093/ecam/nep060.

Nascimento, A.M., Maria-Ferreira, D., de Souza, E.F.J., de Souza, L.M., Sassaki, G.L., Iacomini, M., Maria, M.F., Cipriani, T.R., 2017. Gastroprotective effect and chemical characterization of a polysaccharide fraction from leaves of *Croton cajucara*. Int. J. Biol. Macromol. 95, 153–159. https://doi.org/10.1016/j.ijbiomac.2016.11.044.

Ock, C.Y., Hong, K.S., Kim, J.H., Hahm, K.-B., 2011. Free radicals and gastric mucosal injury. Front. Gastrointest. Res. 29, 97–110.

Okabe, S., Pfeiffer, C.J., 1972. Chronicity of acetic acid ulcer in the rat stomach. Am. J. Dig. Dis. 17, 619–629. https://doi.org/10.1007/BF02231748.

Pala, F.S., Gürkan, H., 2008. The role of free radicals in ethiopathogenesis of diseases. Adv. Mol. Biol. 1, 1–9.

Peiwei, L., Zhang, H., Chen, J., Shi, Y., Cai, J., Yang, J., Wu, Y., 2014. Association between dietary antioxidant vitamins intake/blood level and risk of gastric cancer. Int. J. Cancer. 135, 1444–1453. https://doi.org/10.1002/ijc.28777.

Petrick, J.L., Steck, S.E., Bradshaw, P.T., Trivers, K.F., Abrahamson, P.E., Engel, L.S., He, K., Chow, W.H., Mayne, S.T., Risch, H.A., Vaughan, T.L., Gammon, M.D., 2015. Dietary intake of flavonoids and oesophageal and gastric cancer: incidence and survival in the

United States of America (USA). Br. J. Cancer 112, 1291–1300. https://doi.org/10.1038/bjc.2015.25.

Pillai, N.R., Santhakumari, G., 1984. Effects of nimbidin on acute and chronic gastro-duodenal ulcer models in experimental animals. Planta Med. 50, 143–146. https://doi.org/10.1055/s-2007-969654.

Pineda-Peña, E.A., Velázquez-Moyado, J.A., Orona-Ortíz, A., Tavares-Carvalho, J.C., Navarrete, A., 2018. Phytotherapy for the Gastric Ulcer Treatment. Ulcer Treatment, 1st ed. 1. Avid Science, Berlin, Germany, pp. 2–61.

Pisoschi, A.M., Pop, A., 2015. The role of antioxidants in the chemistry of oxidative stress: a review. Eur. J. Med. Chem. 97, 55–74. https://doi.org/10.1016/j.ejmech.2015.04.040.

Porte, A., Godoy, R.L.O., 2001. Alecrim (*Rosmarinus officinalis* L.): propriedades antimicrobiana e química do óleo essencial. Bol. Cent. Pesqui. Process. Aliment. 19, 193–210. https://doi.org/10.5380/cep.v19i2.1233.

Priya, G., Parminder, N., Jaspreet, S., 2012. Oxidative stress induced ulcer protected by natural antioxidants: a review. Int. Res. J. Pharm. 3, 76–81.

Priyadarsini, K.I., Maity, D.K., Naik, G.H., Kumar, M.S., Unnikrishnan, M.K., Satav, J.G., Mohan, H., 2003. Role of phenolic O-H and methylene hydrogen on the free radical reactions and antioxidant activity of curcumin. Free Radic. Biol. Med. 35, 475–484. https://doi.org/10.1016/S0891-5849(03)00325-3.

Prucksunand, C., Indrasukhsri, B., Leethochawalit, M., Hungspreugs, K., 2001. Phase II clinical trial on effect of the long turmeric (*Curcuma longa* Linn) on healing of peptic ulcer. Southeast Asian J. Trop. Med. Public Health 32, 208–215.

Qiao, Y.L., Dawsey, S.M., Kamangar, F., Fan, J.H., Abnet, C.C., Sun, X., Johnson, L.L., Gail, M.H., Dong, Z.W., Yu, B., Mark, S.D., Taylor, P.R., 2009. Total and cancer mortality after supplementation with vitamins and minerals: follow-up of the Linxian general population nutrition intervention trial. J. Natl. Cancer Inst. 101, 507–518. https://doi.org/10.1093/jnci/djp037.

Raji, Y., Ogunwande, I.A., Osadebe, C.A., John, G., 2004. Effects of *Azadirachta indica* extract on gastric ulceration and acid secretion in rats. J. Ethnopharmacol. 90, 167–170. https://doi.org/10.1016/j.jep.2003.09.020.

Ramana, K.V., Singhal, S.S., Reddy, A.B., 2014. Therapeutic potential of natural pharmacological agents in the treatment of human diseases. Biomed. Res. Int. 2014, 1–4. https://doi.org/10.1155/2014/573452.

Rani, K., 2017. Role of antioxidants in prevention of diseases. J. Appl. Biotechnol. Bioeng. 4, 495–496. https://doi.org/10.15406/jabb.2017.04.00091.

Rasouli, H., Farzaei, M.H., Khodarahmi, R., 2017. Polyphenols and their benefits: A review. Int. J. Food Prop. 20, 1700–1741. doi:10.1080/10942912.2017.1354017.

Repetto, M.G., Llesuy, S.F., 2002. Antioxidant properties of natural compounds used in popular medicine for gastric ulcers. Braz. J. Med. Biol. Res. 35, 523–534. https://doi.org/10.1590/S0100-879X2002000500003.

Ribeiro, R.V., Bieski, I.G.C., Balogun, S.O., Martins, D.T.O., 2017. Ethnobotanical study of medicinal plants used by Ribeirinhos in the North Araguaia microregion, Mato Grosso, Brazil. J. Ethnopharmacol. 205, 69–102. https://doi.org/10.1016/j.jep.2017.04.023.

Rugge, M., Graham, D.Y., 2016. Gastritis. Pract. Gastroenterol. Hepatol. Board Rev. Toolkit 2, 126–131.

Russo, M., Spagnuolo, C., Russo, G.L., Skalicka-Woźniak 2, K., Daglia, M., Sobarzo-Sánchez, E., Nabavi, S.F., Nabavi, S.M., 2018. Nrf2 targeting by sulforaphane: a potential therapy for cancer treatment. Crit. Rev. Food Sci. Nutr. 58, 1391–1405. https://doi.org/10.1016/j.poly.2012.02.008.

Sangiovanni, E., Vrhovsek, U., Rossoni, G., Colombo, E., Brunelli, C., Brembati, L., Trivulzio, S., Gasperotti, M., Mattivi, F., Bosisio, E., Dell'Agli, M., 2013. Ellagitannins from *Rubus berries* for the control of gastric inflammation: in vitro and in vivo studies. PLoS One 8, 1–12. https://doi.org/10.1371/journal.pone.0071762.

Schwingshackl, L., Hoffmann, G., 2016. Does a Mediterranean-type diet reduce cancer risk? Curr. Nutr. Rep. 5, 9–17. https://doi.org/10.1007/s13668-015-0141-7.

Sezikli, M., Çetinkaya, Z.A., Güzelbulut, F., Çimen, B., Özcan, Ö., Özkara, S., Yeşil, A., Gümrükçü, G., İpçioğlu, O.M., Sezikli, H., Övünç, A.O.K., 2012. Effects of alpha tocopherol and ascorbic acid on *Helicobacter pylori* colonization and the severity of gastric inflammation. Helicobacter 17, 127–132. https://doi.org/10.1111/j.1523-5378.2011.00925.x.

Shin, H.S., Choi, I.G., Lee, M.H., Park, K.N., 1982. Clinical trials of Madecassol (*Centella asiatica*) on gastrointestinal ulcer patient. Korean J. Gastroenterol. 14, 49–56.

Shokoohinia, Y., Jafari, F., Mohammadi, Z., Bazvandi, L., Hosseinzadeh, L., Chow, N., Bhattacharyya, P., Farzaei, M.H., Farooqi, A.A., Nabavi, S.M., Yerer, M.B., Bishayee, A., 2018. Potential anticancer properties of osthol: a comprehensive mechanistic review. Nutrients 10, 1–16. https://doi.org/10.3390/nu10010036.

Sidahmed, H.M.A., Hashim, N.M., Abdulla, M.A., Ali, H.M., Mohan, S., Abdelwahab, S.I., Taha, M.M.E., Fai, L.M., Vadivelu, J., 2015. Antisecretory, gastroprotective, antioxidant and anti-helicobcter pylori activity of zerumbone from *Zingiber zerumbet* (L.) smith. PLoS One 10, 1–21. https://doi.org/10.1371/journal.pone.0121060.

Sies, H., 2018. On the history of oxidative stress: concept and some aspects of current development. Curr. Opin. Toxicol. 7, 122–126. https://doi.org/10.1016/j.cotox.2018.01.002.

Silva, M.I.G., Sousa, F.C.F., 2011. Gastric ulcer etiology. In: Chai, J. (Ed.), Peptic Ulcer Disease. IntechOpen, London, UK, pp. 1–28. https://doi.org/10.5772/20796.

Sipponen, P., Maaroos, H.I., 2015. Chronic gastritis. Scand. J. Gastroenterol. 50, 657–667. https://doi.org/10.3109/00365521.2015.1019918.

Solmaz, A., Şener, G., Çetinel, Ş., Yüksel, M., Yeğen, C., Yeğen, B.Ç., 2009. Protective and therapeutic effects of resveratrol on acetic acid-induced gastric ulcer. Free Radic. Res. 43, 594–603. https://doi.org/10.1080/10715760902977424.

Sudharameshwari, K., Ayshwarya, M., 2017. Evaluation of antiulcerogenic activity of methanol extracts of *Brassica olerace*a Var. Capitata Rubra on albino rat gastric ulceration. Asian J. Pharm. Clin. Res. 10, 314. https://doi.org/10.22159/ajpcr.2017.v10i3.16217.

Sulaiman Rahman, H., 2018. Phytochemical analysis and antioxidant and anticancer activities of mastic gum resin from *Pistacia atlantica* subspecies kurdica. Oncol. Targets. Ther. 2018 (11), 4559–4572. https://doi.org/10.2147/OTT.S170827.

Sung, J.J.Y., Kuipers, E.J., El-Serag, H.B., 2009. Systematic review: the global incidence and prevalence of peptic ulcer disease. Aliment. Pharmacol. Ther. 29, 938–946. https://doi.org/10.1111/j.1365-2036.2009.03960.x.

Suzuki, H., Nishizawa, T., Tsugawa, H., Mogami, S., Hibi, T., 2012. Roles of oxidative stress in stomach disorders. J. Clin. Biochem. Nutr. 50, 35–39. https://doi.org/10.3164/jcbn.11.

Takayama, C., de-Faria, F.M., de Almeida, A.C.A., Dunder, R.J., Manzo, L.P., Socca, E.A.R., Batista, L.M., Salvador, M.J., Souza-Brito, A.R.M., Luiz-Ferreira, A., 2016. Chemical composition of *Rosmarinus officinalis* essential oil and antioxidant action against gastric damage induced by absolute ethanol in the rat. Asian Pac. J. Trop. Biomed. 6, 677–681. https://doi.org/10.1016/j.apjtb.2015.09.027.

Teske, M., Trentini, A.M.M., 2001. Herbarium Compêndio de Fitoterapia, 4ª. ed. Herbarium Laboratório Botânico, Curitiba, Curitiba.

Thomas, G., Kalla, A.M., Rajunaik, B., Kumar, A., 2018. Food matrix: a new tool to enhance nutritional quality of food. J. Pharmacogn. Phytochem. 7, 1011–1014.

Tinggi, U., 2008. Selenium: its role as antioxidant in human health. Environ. Health Prev. Med. 13, 102–108. https://doi.org/10.1007/s12199-007-0019-4.

Tovey, F.I., 2015. Role of dietary phospholipids and phytosterols in protection against peptic ulceration as shown by experiments on rats. World J. Gastroenterol. 21, 1377–1384. https://doi.org/10.3748/wjg.v21.i5.1377.

Tsugane, S., Sasazuki, S., 2007. Diet and the risk of gastric cancer: review of epidemiological evidence. Gastric Cancer 10, 75–83. https://doi.org/10.1007/s10120-007-0420-0.

Tuorkey, M., Karolin, K., 2009. Anti-ulcer activity of curcumin on experimental gastric ulcer in rats and its effect on oxidative stress/antioxidant, IL-6 and enzyme activities. Biomed. Environ. Sci. 22, 488–495. https://doi.org/10.1016/S0895-3988(10)60006-2.

Ullah, M., 2015. Sulforaphane (SFN): an isothiocyanate in a Cancer Chemoprevention Paradigm. Medicines 2, 141–156. https://doi.org/10.3390/medicines2030141.

Valko, M., Leibfritz, D., Moncol, J., Cronin, M.T.D., Mazur, M., Telser, J., 2007. Free radicals and antioxidants in normal physiological functions and human disease. Int. J. Biochem. Cell Biol. 39, 44–84. https://doi.org/10.1016/j.biocel.2006.07.001.

Verma, V.K., Nripendra Singh, P.S.R.S., 2012. Anti-ulcer and antioxidant activity of *Moringa oleifera* (Lam) leaves against aspirin and ethanol induced gastric ulcer in rats. J. Pharm. 02, 46–57. https://doi.org/10.3390/ijms19082244.

Viana, C.A.S., Paiva, A.O., Jardim, C.V., Rios, M.N.S., Rocha, N.M.S., Pinagé, G.R., Arimoro, O.A.S., Suganuma, E., Guerra, C.D., Alvez, M.M., Pastore, J.F., 2011. Plantas da Amazônia: 450 espécies de uso geral, 2nd ed. Brasilia.

Wang, J., Jiang, A., Dou, J., Zhang, X., Ma, Z., Qi, B., Xiong, Y., 2015. Resveratrol protects against *Helicobacter pylori*-associated gastritis by combating oxidative stress. Int. J. Mol. Sci. 16, 27757–27769. https://doi.org/10.3390/ijms161126061.

Wang, R., Tu, J., Zhang, Q., Zhang, X., Zhu, Y., Ma, W., Cheng, C., Brann, D.W., Yang, F., 2013. Genistein attenuates ischemic oxidative damage and behavioral deficits via eNOS/Nrf2/HO-1 signaling. Hippocampus 23, 634–647. https://doi.org/10.1002/hipo.22126.

Wang, X., Yin, J., Zhao, M., Liu, S., Nie, S., Xie, M., 2018. Gastroprotective activity of polysaccharide from *Hericium erinaceus* against ethanol-induced gastric mucosal lesion and pylorus ligation-induced gastric ulcer, and its antioxidant activities. Carbohydr. Polym. 186, 100–109. https://doi.org/10.1016/j.carbpol.2018.01.004.

Wirth, H.-P., Yang, M., 2016. Different pathophysiology of gastritis in east and west? A western perspective. Inflamm. Intest. Dis. 1, 113–122. https://doi.org/10.1159/000446300.

Xu, J., Liu, D., Niu, H., Zhu, G., Xu, Y., Ye, D., Li, J., Zhang, Q., 2017. Resveratrol reverses Doxorubicin resistance by inhibiting epithelial-mesenchymal transition (EMT) through modulating PTEN/Akt signaling pathway in gastric cancer. J. Exp. Clin. Cancer Res. 36, 1–14. https://doi.org/10.1186/s13046-016-0487-8.

Yang, G.Z., Wang, Z.J., Bai, F., Qin, X.J., Cao, J., Lv, J.Y., Zhang, M.S., 2015. Epigallocatechin-3-gallate protects HUVECs from $PM_{2.5}$-induced oxidative stress injury by activating critical antioxidant pathways. Molecules 20, 6626–6639. https://doi.org/10.3390/molecules20046626.

Yeung, A.W.K., Tzvetkov, N., Zengin, G., Wang, D., Xu, S., Mitrović, G., Brncic, M., Dall'Acqua, S., Pirgozliev, V., Kijjoa, A., Georgiev, M., Atanasov, A., 2019. The berries on the top. J. Berry Res. 9, 1–15. 10.3233/JBR-180357. https://doi.org/10.3233/JBR-180357.

You, W.C., Brown, L.M., Zhang, L., Li, J.Y., Jin, M.L., Chang, Y.S., Ma, J.L., Pan, K.F., Liu, W.D., Hu, Y., Crystal-Mansour, S., Pee, D., Blot, W.J., Fraumeni, J.F., Xu, G.W., Gail,

M.H., 2006. Randomized double-blind factorial trial of three treatments to reduce the prevalence of precancerous gastric lesions. J. Natl. Cancer Inst. 98, 974–983. https://doi.org/10.1093/jnci/djj264.

Zafra-Stone, S., Yasmin, T., Bagchi, M., Chatterjee, A., Vinson, J.A., Bagchi, D., 2007. Berry anthocyanins as novel antioxidants in human health and disease prevention. Mol. Nutr. Food Res. 51, 675–683. https://doi.org/10.1002/mnfr.200700002.

Zakaria, Z.A., Balan, T., Suppaiah, V., Ahmad, S., Jamaludin, F., 2014. Mechanism(s) of action involved in the gastroprotective activity of *Muntingia calabura*. J. Ethnopharmacol. 151, 1184–1193. https://doi.org/10.1016/j.jep.2013.12.045.

Zhang, L., Ma, J., Pan, K., Go, V.L., Chen, J., You, W., 2005. Efficacy of Cranberry juice on *Helicobacter pylori* infection. Helicobacter 10, 139–145.

Zhou, Y., Wang, T., Meng, Q., Zhai, S., 2016. Association of carotenoids with risk of gastric cancer: a meta-analysis. Clin. Nutr. 35, 109–116. https://doi.org/10.1016/j.clnu.2015.02.003.

Zubair, H., Azim, S., Ahmad, A., Khan, M.A., Patel, G.K., Singh, S., Singh, A.P., 2017. Cancer chemoprevention by phytochemicals: nature's healing touch. Molecules 22, 1–24. https://doi.org/10.3390/molecules22030395.

Zulaikhah, S.T., 2018. The role of antioxidant to prevent free radicals in the body. Sains Med 8, 39. https://doi.org/10.26532/sainsmed.v8i1.1012.

CHAPTER

Antioxidants and immune functions

5.7

Kasturi Sarkar[a], **Parames C. Sil**[b]

[a]*Department of Microbiology, St. Xavier's College, Kolkata, India*

[b]*Division of Molecular Medicine, Bose Institute, Kolkata, India*

5.7.1 Introduction

The short-lived and highly reactive oxygen and nitrogen species (ROS/RNS) take part in many important physiological functions like progression in cell cycle and proliferation, differentiation, migration, cell death, immune system-mediated killing of pathogens, maintenance of the redox balance, etc. at a low to moderate dose. However, production of excess ROS and RNS has detrimental effects on host cellular biomolecules, which ultimately lead to cellular death. The amount of ROS/RNS in a cell is a result of the equilibrium between the generation of these species and antioxidant defenses. Along with the cellular antioxidative system present in all animals, dietary antioxidants also help to remove the free radicals. This chapter discusses on the generation of ROS/RNS in the immune system and the effects of antioxidant supplements, both beneficial and harmful, on different branches and cells of the immune system.

5.7.2 Origin and role of ROS

5.7.2.1 Beneficial roles of reactive species

Lower amount of ROS and RNS is essential for life to sustain and the amount, duration, and localization of ROS inside a cell affect the redox regulatory network necessary for many biological processes (Brieger et al., 2012). Reactive species and redox signaling play an important role in the regulation of root architecture, polar growth, organ senescence, and long-distance communications in plants (Dietz et al., 2016). ROS generated inside chloroplast, mitochondria have been found to affect processes at molecular level, for example, gene expression, cellular damage, and death.

5.7.2.2 Generation of ROS/RNS in our immune system and their effect on foreign molecules

Innate immunity, provided by neutrophils, dendritic cells (DCs), macrophages, mainly depends on the generation of huge amounts of ROS and RNS. Upon engulfing

Antioxidants Effects in Health. DOI: https://doi.org/10.1016/B978-0-12-819096-8.00057-4

a microorganism or an immunogen, a metabolic process called the respiratory burst, is initiated by the enzyme complex NADPH oxidase inside the phagosome. The inactive components of NADPH oxidase (Grumping and Rittinger, 2005), which are found on the plasma membrane as well as on the membranes of phagosomes, assemble and activated in response to phagocytosis. The active NADPH oxidase complex then transfers one electron from NADPH to O_2 creating superoxide radical ($O_2^{\cdot-}$) within the phagosomal lumen (Peterhans, 1997). These superoxide radicals are then converted to H_2O_2 and O_2 spontaneously or through the action of specific enzymes. The enzyme myeloperoxidase catalyzes the reaction between H_2O_2 and halides and produces reactive hypohalous acid. On the other hand, $O_2^{\cdot-}$ interacts with nitric oxide (NO) and produces peroxynitrite (ONOO–). All these ROS/RNS can form adducts with unsaturated bonds present in biomolecules and damage enzymes, unsaturated fatty acids or can oxidize solvent exposed iron–sulfur clusters in enzymes or proteins resulting in metabolic defects. Iron released from damaged Fe–S clusters undergo Fenton reaction with H_2O_2 to yield hydroxyl radical (·OH), which leads to more damage. H_2O_2 can directly react with cysteine residues present in proteins and enzymes. The bacterial membrane is not damaged directly by superoxides as bacteria lack polyunsaturated fatty acids (Fig. 5.7.1).

5.7.2.3 Diseases associated with ROS

ROS has been found to be associated with the development of many diseases. Some diseases are associated with low ROS concentration, for example, chronic granulomatous disease, an autoimmune disorder while excess ROS can lead to cardiovascular, neurodegenerative, and many other diseases. Neurons are prone to oxidative stress (OS)-induced damages because of their high demand for oxygen but weakened antioxidant defense mechanism, and presence of abundant polyunsaturated fatty acids in their cell membranes (Rego and Oliveira, 2003). Scientists have found a strong relation between OS and neurodegenerative disorders like Parkinson's disease, Alzheimer's disease, amyotrophic lateral sclerosis (ALS), spinocerebellar ataxia etc. (Dias et al., 2013; Liu et al., 2018).

5.7.2.4 Sources of ROS

5.7.2.4.1 Endogenous sources of ROS

The main source of endogenous or intracellular ROS is the electron transport chain (ETC) in mitochondrion. During the transfer of electrons from complex I or II to ubiquinone, an unstable intermediate semiquinone anion ($^{\cdot}Q^{-}$) is formed which immediately transfers electrons to O_2 leading to the formation of superoxide radical. This ROS generation is nonenzymatic and increases with increased metabolic rate inside the cells (Finkel et al., 2000).

Another site for generation of various ROS is the peroxisome. The enzyme acyl CoA oxidase produces H_2O_2 during β-oxidation of fatty acids. The other enzymes involved in ROS production are D-amino acid oxidase, L-α-hydroxy oxidase, urate oxidase, xanthine oxidase, D-aspartate oxidase, etc. (Schrader and Fahimi, 2006).

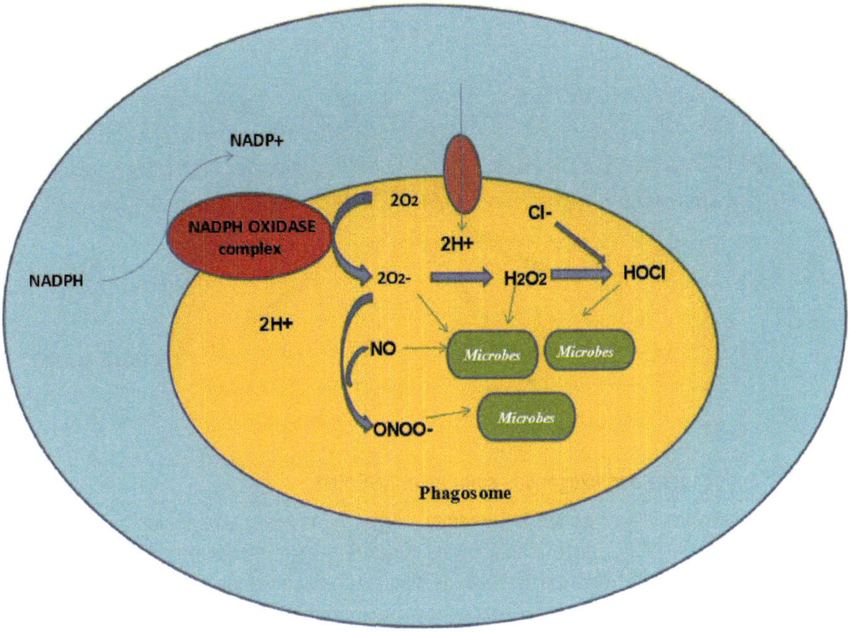

FIG. 5.7.1 Generation of ROS/RNS inside the phagocytic cells.

Oxygen-dependent killing of the phagocytosed microorganisms through respiratory burst has been shown in this figure. The process requires increased amount of oxygen which is consumed by the enzyme NADPH oxidase present on the surface of phagocytic cells. Oxygen is converted to various ROS/RNS by the action of specific enzymes. The reactive species interact with the biomolecules present in the microorganism.

ROS generation also takes place inside the endoplasmic reticulum by cytochrome p-450 oxidase, b5 enzymes, and diamine oxidase (Cheeseman and Slater, 1993). Another important enzyme, endoplasmic reticulum oxidoreductin 1p, catalyzes the formation of H_2O_2 by transferring electrons from dithiols to O_2 (Gross et al., 2006).

ROS are also generated during different biosynthetic pathways like synthesis of prostaglandin, auto-oxidation of adrenalin, synthesis of riboflavin, etc. ROS have been found to generate in conditions like mental stress, excessive exercise, certain infections, cancer, aging, ischemia, etc. (Cheeseman and Slater, 1993).

5.7.2.4.2 Exogenous sources of reactive species

Reactive oxygen and nitrogen species can be generated through the pollution of air and water, heavy metal toxicity like Fe, Cu, Co, Cr, transition metals like Cd, Hg, Pb, As, etc. Solvents used in the industries, pesticides used in farming, certain drugs like halothane, paracetamol, bleomycine, doxorubicin, metrenidazole and radiation, ultraviolet light, alcohol, tobacco smoke, used oil, and fat can also lead to the development of ROS/RNS (Krumova and Cosa, 2016; Fig. 5.7.2).

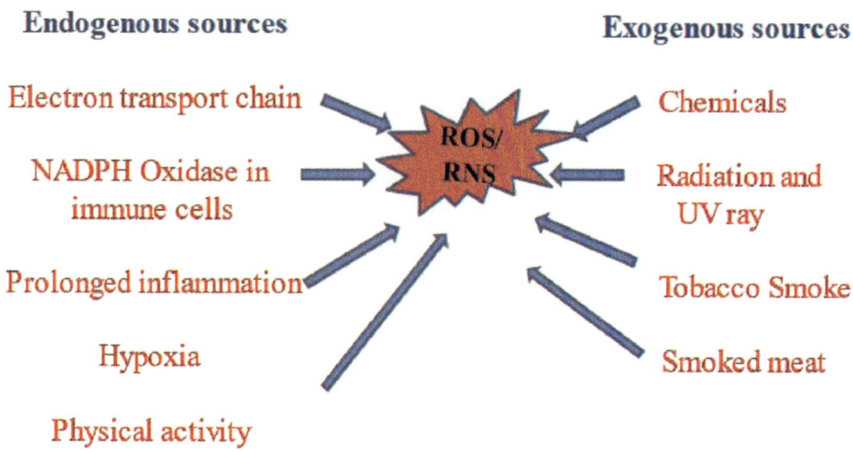

FIG. 5.7.2 Endogenous and exogenous sources of ROS/RNS.

The sources responsible for generating reactive oxygen and nitrogen species have been shown in this figure. The endogenous sources include electron leakage from mitochondrial electron transport chain, the enzyme NADPH oxidase, chronic or prolonged inflammation, strenuous physical activities, etc. The exogenous sources of ROS and RNS are radiations, chemicals, smoking, activated carbons in foods, etc.

5.7.3 Oxidative stress and endogenous antioxidative systems

The antioxidant detoxification systems present in our cells try to keep the ROS/RNS levels low. An imbalance between the generation and detoxification of ROS would lead to the accumulation of excess ROS and oxidative stress. No ROS sensor has been found in mammals till now, which can provide an indication of the existence or accumulation of excess ROS. Transcription factor, Nrf2 (NF-E2-related factor 2), regulates the expression of enzymes required for detoxification and antioxidants in response to environmental stresses. Another enzyme, Keap1 or Kelch-like ECH-associated protein 1, can detect the presence of various stress stimuli, including oxidative and electrophilic stresses and regulates the Nrf2 activity. This Nrf2-Keap 1 system controls the expression of phase II detoxifying enzymes through the antioxidant/electrophile response element (Dinkova-Kostova et al., 2002; Itoh et al., 1999) though no demonstrated direct role has been found in ROS sensing per se (Lee et al., 2003). Similarly, several stress-responsive regulators, like the c-Jun N-terminal kinase, stress-activated MAP kinase, and p53 can be activated by H_2O_2. These pathways help cells cope with OS-induced damage through induction of DNA and protein repair systems and apoptosis. Sirtuins, the NAD deacetylases, play an important role in the control of ROS and are associated with the development of age-associated degenerative conditions. Sirt3, linked to several stress-related problems, is expressed more in response to caloric restriction and activates MnSOD, the ROS scavenger (Qiu et al., 2010).

5.7.3.1 Enzymes that help in the removal of ROS/RNS

Our endogenous antioxidant system has several antioxidant enzymes like superoxide dismutase (SOD), catalase, glutathione peroxidase, NADPH oxidase, and antioxidant molecules like GSH, etc. SOD is present in all aerobic cells and extracellular fluids; catalase is present in peroxisomes and catalyzes hydrogen peroxide to molecular oxygen. The antioxidant thioredoxin system is composed of the enzymes thioredoxin and thioredoxin reductase.

5.7.3.2 Antioxidants that donate electrons to free radicals

Extensive research has shown that dietary antioxidants or supplements play a major role against the ROS/RNS generated in the immune system. We obtain many antioxidants like ascorbic acid, alpha-tocopherol, carotenoids, and polyphenolic flavonoids from our food directly. Antioxidants like vit A, E, C protect the lipid membrane of the cells while B vitamins promote wound healing by stimulation of synthesis of FAs and collagen (Hughs, 1999).

5.7.3.3 Proteins that help in minimizing OS

There are some proteins such as transferrin, heptoglobins, hemopexin, and metallothionein, which can help in minimizing the availability of prooxidants. Heat Shock Protein 70 has been found to elevate the activities of antioxidant enzymes like SOD, GSH peroxidase, and total amount of antioxidants. It also inhibits lipid peroxidation and hence OS in the intestinal mucosa significantly (Gu and Wang, 2012; Fig. 5.7.3).

5.7.4 Effect of antioxidants on cell-mediated immunity

Cell-mediated immunity is provided by phagocytic cells like neutrophils, macrophages, DCs, and nonphagocytes like T lymphocytes, natural killer cells. We have discussed the roles of antioxidants, supplemented and dietary, on the immune effects mediated by these cells.

5.7.4.1 Effect of antioxidant supplements on neutrophils

Neutrophils, the first cells to reach the site of an infection, mediate their antimicrobial effects by phagocytosis of the pathogen and releasing ROS/RNS and antimicrobial molecules. In addition to it, they have been found to release DNA and bactericidal proteins like elastase, cathepsin G, histones, etc., which form an extracellular matrix in response to bacteria, pathogenic fungi like *C. albicans*, etc. This matrix, called

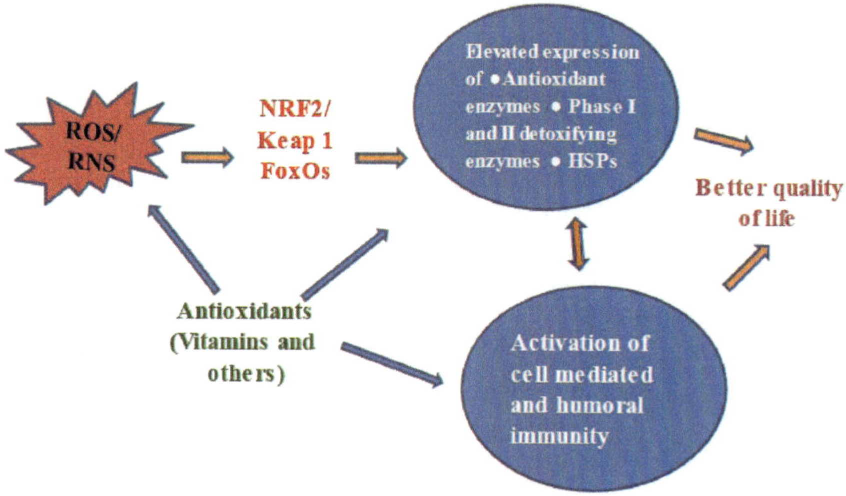

FIG. 5.7.3 Effect of antioxidants on the immune system.

Production of ROS/RNS and killing of microorganisms are the major events in the innate immunity while excess ROS/RNS is associated with oxidative stress-related disorders. Thus, the balance between the amount of ROS/RNS and antioxidants plays a critical role in a natural infection. Dietary supplementation of antioxidants in proper dosage can activate the immune system and reduce the pathophysiological conditions raised due to OS.

the neutrophil extracellular trap or NETs, is also involved in the pathogenesis of various autoimmune and inflammatory diseases. The formation of this matrix starts with the activation of NADPH oxidase and ROS and thus susceptible to all the ROS scavengers. In a recent study, Vorobjeva and Pinegin showed that antioxidants like Trolox, Tiron, and Tempol can inhibit the formation of NET induced by phorbol 12-myristate 13-acetate *in vitro*. Trolox has an inhibitory role on ROS-dependent NET release while Tempol has a suppressive effect on the formation of NET. Tiron was not found to have any inhibitory activity on the release of NET. Hence, Trolox and Tempol can be used for the treatment of autoimmune and inflammatory diseases that involve the generation of ROS and formation of NET (Vorobjeva and Pinegin, 2016).

In another study, Vorobjeva and Pinegin showed the regulatory role of ROS in neutrophil activation and release of the primary (azurophil) and secondary (specific) granules. Use of the antioxidant SkQ1, which targets mtROS, showed increased apoptosis both spontaneous and stimulated, of fMLP-activated neutrophils. Hence, it can be said that mtROS is involved in the activation of neutrophils and modulates the signal transduction pathway (Vorobjeva et al., 2017).

Antoni Sureda and his group studied the effects of antioxidant supplements on the inflammatory response generated during intense exercises, which leads to tissue damage due to thermal stress and ischemia. They measured the effects of vit C and E on the mRNA expression of neutrophil tocopherol-associated protein before (following a specific regimen) and after finishing a half-marathon race. The authors showed through different experiments that antioxidant supplementation prevented exercise-induced plasma damage without having any effect on the inflammatory process (Sureda et al., 2007). In another study, the authors tried to investigate the relation between the activation of neutrophils and OS generated during an exercise. The observations suggest that exercise promotes the release of neutrophils into the circulation having enhanced capacity to generate some forms of ROS. Antioxidant supplements like N-acetylcysteine (NAC) are effective at attenuating the generation of ROS in neutrophils when stimulated in vitro, whereas vitamin E reduces tissue infiltration of neutrophils during exercise (Jonathan and Katsuhiko, 2004; Peake and Suzuki, 2004).

It has been found that bovine serum albumin (BSA) acts as an antioxidant and the IC_{50} value of BSA was found to be 33.5 mg/mL, 6.5 mg/mL, and 6.85 mg/mL for $O-_2$, H_2O_2, and HOCl, respectively, after activation of neutrophils with phorbol-12-myristate-13-acetate *in vitro*. Washing of the neutrophils after incubation with BSA reduces the chances of dismutation of ROS production, providing evidence of direct interaction between BSA and ROS (Kouoh et al., 1999).

Estragole, a major component of basil oil, has been found to decrease the infiltration of peritoneal exudate leukocytes in vivo and prevented migration of neutrophils toward the chemotactic tripeptide fMLP in vitro. Estragole also prevented the phagocytosis of macrophages and nitric oxide (NO) production (Francielli Maria, 2014).

The role of a polyherbal preparation, Septilin, was evaluated on the immune system at low and high dosage in a study. Both the dosages stimulated humoral immunity and increased the number of circulating neutrophils without having any effect on weight gain, total lymphocyte counts, or host resistance against *E. coli* sepsis. Septilin at a higher dosage was found to reduce the phagocytic activity of the PMN cells/reticuloendothelial system. Thus, Septilin acts in a dose-dependent manner, with lower dosage showing greater stimulant of the immune system (Daswani et al., 2002).

5.7.4.2 Effects of antioxidants on macrophages

The tissue resident macrophages (Mphi) provide the first line of defense against the foreign molecules entering our body. In a study, the authors compared the effects of antioxidants, both intracellular and supplemented, on the Mphi response to endotoxins. Cells were pretreated with antioxidants and cell supernatants were analyzed for TNF, F2 isoprostane production and cellular monolayers for procoagulant activity after 18 h treatment of the cells with LPS. It was found that pretreatment with NAC and butylated hydroxyanisole inhibited both TNF and procoagulant activity production, but no such effect was found with vit C, Trolox, or SOD whereas only butylated hydroxyanisole and Trolox pretreatment inhibited membrane lipid peroxidation. Thus, specific antioxidants can modulate the proinflammatory response of Mphi by modulating the signal transduction pathways (Bulgar et al., 2002).

Excess ROS generated during hyperoxia may result in reduced phagocytic ability in macrophages. Morrow et al. tested this hypothesis in cultured macrophage-like RAW 264.7 cells and alveolar macrophages from mice exposed to hyperoxia. Results showed impaired phagocytosis of *P. aeruginosa*, both the mucoid and the nonmucoid form, in alveolar macrophages and RAW cells exposed to hyperoxia. The changes appeared in cytoskeleton, actin, and phagocytosis due to oxidation were restored by the treatment with antioxidants (Morrow et al., 2007).

Another problem created by secreted ROS/RNS from the inflammatory cells is on the biomaterial implantation and its failure. Incorporation of antioxidants into implanted biomaterials has been found to scavenge reactive species, preventing the oxidative degradation of the bioimplant (Lurier, 2015).

5.7.4.3 Effect of antioxidants on dendritic cells

DCs are the sentinels of immune system and act as the bridge between innate and adaptive immunity. Francischetti et al. found the effect of Tempol, a SOD mimetic, on cerebral malaria induced by *Plasmodium berghei* Anka infection. Tempol had been found to prevent the expression of procoagulant tissue factor in endothelial cells, aggregation of platelets, and oxidative burst stimulated by LPS. Tempol also inhibited the production of different cytokines like TNF-α, IL-6, IL-12p70, and downregulated expression of co-stimulatory molecules on DCs stimulated by LPS and thus increased the rate of survival of mice (Francischetti et al., 2014).

Vitamins C and E were found to have no effect on the morphology and function of DCs but reduced the levels of intracellular ROS and regulated several signaling pathways by modulating protein kinase C, NF-kB, and p38 MAPK pathways. The concurrent use of the two vitamins had no additive effects. The allogeneic T cells became anergic when exposed to vitamin-treated DCs and were found to secrete higher levels of Th2 cytokines and IL-10 than the T cells incubated with control DCs. The T cells acted as regulatory T cells in a contact-dependent manner and were not dependent on IL-4, IL-5, IL-10, IL-13, etc. Hence, the combined effect of vitamins C and E on the DCs might be useful for inducing tolerance to allo- or autoantigens (Peng et al., 2005).

Tan Qin and group had shown the protective effect of bursopentine (BP5), a novel pentapeptide isolated from the bursa of fabricius of chicken, against LPS-induced OS in DCs by upregulating heme oxygenase 1 (HO-1) expression, elevating the level of reduced GSH and the activities of glutathione peroxidase, catalase, and SOD. BP5 also suppressed submucosal DC maturation in the LPS-stimulated intestinal epithelial cells/DCs coculture system. These results suggested that BP5 might be useful in DC-related inflammatory responses (Qin et al., 2015).

5.7.4.4 Effect of antioxidants on T-cells

In a study, the number of antigen-specific $CD8^+$ T cells was enumerated in mice after infection with lymphocytic choriomeningitis virus (LCMV) and post-treatment with Mn (III) tetrakis (4-benzoic acid) porphyrin chloride (MnTBAP) for 8 days. The numbers of Ag-specific $CD8^+$-T-cells were significantly lower in MnTBAP-treated mice than in control mice on day 8 and on days 8–30 (10-fold and 25-fold, respectively). MnTBAP had no effect on the number of Ag-specific $CD8^+$ effector T cells during the secondary viral infection and the number of memory T cells was equivalent in both the groups. Hence, ROI has a major role to play in the expansion and control of Ag-specific effector $CD8^+$ T cells in primary viral infection (Nathan and Jason, 2004).

Chaudhri et al. checked the possibility of reactive species having a role in the lymphoproliferative response to alloantigens in the mixed lymphocyte culture. Three classes of antioxidant agents, for example, the nonpermeant electron acceptor ferricyanide, iron chelators, and pyridoxal isonicotinoyl hydrazone, were tested. These molecules prevented the expansion of reactive cells in the mixed lymphocyte culture in a dose-dependent manner if added early (less than 40 h) while they had no effects on the proliferation of CTL-2 cells dependent on IL-2 or on IL-1 production by peritoneal exudate cells. The expression of IL 2 receptors was found to be inhibited on stimulated T cells in the presence of these drugs. Hence, free radicals have effects on the early developmental stages of T lymphocyte before the expression of IL 2 receptors (Chaudhri et al., 1986).

Supplementation with the antioxidants Ebselen and NAC was found to result in a significantly higher intracellular GSH:GSSG ratio and less oxidative DNA damage in CD4+ human T cell clones (Shiva et al., 2013).

The intracellular GSH:GSSG ratio was found to decrease in PBMC, CD4+, and CD8+ lymphocytes in HIV-1 patients. Administration of NAC has been found to have many beneficial effects on CD4 lymphocyte in the survival of HIV-1-infected individuals (high antioxidant responder). Both NAC and 2 mercaptoethanol have been found to inhibit TNF-α and phorbol myristate acetate-induced cell death in HIV-infected U937 cell lines (Alfonso et al., 1996; Scheffel et al., 2016).

5.7.5 Effect of antioxidants on humoral immunity

The effect of MnTBAP was evaluated in mice infected with LCMV. After 8 days post-treatment with the virus, the splenic response was at its peak and the virus-specific IgM and IgG secreting plasma cells were decreased 22- and 457-fold in MnTBAP-treated animals compared to control. The number of virus-specific $CD4^+$ T cells were also decreased, though there was no change in LCMV-specific IgG memory B cell count. Hence, these data suggest the involvement of ROS in regulating antiviral B cell expansion and control of the infection (Crump et al., 2013).

Vitamin E, known to quench-free radicals, has shown contradictory results in many clinical studies. The effect of dl-alpha-tocopherol acetate supplementation decreased the level of TBARS in the plasma, whereas the radical scavenger activities of RBCs were significantly increased in all age groups. It had no effect on the humoral immune response (Park et al., 2003).

5.7.6 Effect of antioxidants on hypersensitivity and inflammation

In a study, the authors investigated the effects of MitoVitE, the mitochondria targeted antioxidant, on mitochondrial damage induced by paclitaxel and hypersensitivity in rat dorsal root ganglion cells and compared its effects with a common vitamin E, Trolox. MitovitE was more effective than Trolox in preventing loss of membrane activity and potential in dorsal root ganglion cells without decreasing cancer cell cytotoxicity. The mechanical hypersensitivity induced by paclitaxel was also ameliorated by the treatment of MitoVitE like duloxetine, the well-known antihypersensitivity drug. These data suggest the use of mitochondria targeted antioxidants for prevention of mitochondrial damages (McCormick et al., 2016).

Many natural products have been found to reduce inflammatory responses which are the main concern of many diseases. Two such antioxidant compounds, cannabidiol, a constituent of *Cannabis sativa*, and moringin, an isothiocyanate, found in *Moringa oleifera* seeds, have anti-inflammatory activity, though the mechanisms of action of the compounds are different. The results showed that the combined effect of these two molecules was more than the individual constituents and had positive effects on inflammatory response stimulated by LPS in murine macrophages as measured by the levels of TNF-α, interleukin-10, iNOS, nuclear factor erythroid 2-related factor 2, nitrotyrosine. The combination also has antiapoptotic property and upregulates the levels of Bcl-2, downregulates the Bax, and cleaves caspase-3 (Rajan et al., 2016).

Oxidative stress both pulmonary and systemic increases inflammatory responses relevant to asthma and allergy. Hence, dietary or vitamin supplementation with antioxidant activity might be an approach in reducing asthma incidence or morbidity. However, there are not much data on clinical trials to support the use of dietary antioxidants or supplements to prevent asthma or allergy (Moreno-Macias et al., 2014).

The flavonoids and phenolic compounds exert both antioxidant and anti-inflammatory properties. They block two major signaling pathways such as NF-κB and MAPKs which are involved in the production of various proinflammatory mediators. The dichloromethane soluble fraction of *Acanthopanax senticosus* Harms showed the strongest anti-inflammatory activities through the inhibition of expression of inducible NO synthase (iNOS), COX-2, TNF-α, IL-1β, and mRNAs and the generation of ROS in LPS-induced RAW 264.7 cells (Jiang et al., 2015). Chloroform extract of *Actinidia arguta* exerted anti-inflammatory effects by suppression of NF-κB pathway and MAPK phosphorylation. The extract also inhibits NO production and iNOS mRNA expression in RAW 264.7 cells on LPS-stimulated macrophages (Kim et al., 2014).

The two major active phenolic compounds of *Ainsliaea fragrans* Champ., 3,5-dicaffeoylquinic acid and 4,5-dicaffeoylquinic acid have been found to reduce the expression of inflammatory mediators through impact on NF-κB signaling pathway (Chen et al., 2015).

5.7.7 Effect of dietary or supplemented antioxidants on age-related tumor immunity

El-Senousey and group conducted a comparative study to compare the antioxidant capacity among the antioxidants vitamin C, vitamin E, and alpha-lipoic acid (ALA) on OS induced by dexamethasone in broilers. They observed highest antioxidant activity and lowest mRNA gene expression levels of IFN-γ, IL-1b, IL-6, and LPS-induced TNF-α in the birds fed with ALA-supplemented diet compared to vit C and E (El-Senousey et al., 2018).

Park et al. showed in their study that supplementation of alpha- and gamma-tocopherol has a major role to play in the transcriptional alterations of the genes related to aging in the heart and brain (neocortex) in mouse model (Park et al., 2008). They found out that both the tocopherols were effective in inhibiting the expression of genes associated with cardiomyocyte hypertrophy and preventing expression of genes encoding proteins of ribosomes and ATP biosynthesis associated with aging.

Consumption of green tea or its active ingredient, epigallocatechin-3-gallate (EGCG), has been found to have lots of health benefits including immunomodulation. EGCG has a pronounced effect on the activation, proliferation, differentiation of T cells and on the production of cytokines. Hence, EGCG can be used as a potential therapeutic agent in T cell-mediated autoimmune disorders (Pae et al., 2013).

5.7.8 Antioxidants shield immune cells from environmental damage

Post-treatment with the ethanolic extract of Ji-Xue-Teng (*Spatholobus suberectus*) was found to have a positive effect on blood profile with reduced leucopenia, thrombocytopenia, RBC, and hemoglobin levels, etc. compared to the control where mice were exposed to γ-radiation for the whole body at the dose of 6.0 Gy resulting in a typical hematopoietic syndrome (Dong et al., 2018). The elevated levels of ROS, SOD, and GSH-PX were found to decrease and the expressions of pJAK2/JAK2, pSTAT5a/STAT5a, and Bcl-2 were increased in bone marrow tissue in Ji-Xue-Teng-treated mice compared to control.

The addition of antioxidants in the nutrition of livestocks is considered the key in improving animal production (Elwan et al., 2019). Consumption of dietary tomato powder supplementation provided positive effects on hemato-immunological, biochemical, and antioxidant parameters and enhanced immune responses through a significant increase in phagocytosis, chemotaxis, and levels of immunoglobulins in rabbits in New Zealand.

A water-soluble curcumin lysinate incorporated into hydroxypropyl-β-cyclodextrin (NDS27) has been found to show anti-inflammatory properties and its effect was compared with curcumin lysinate and curcumin on the release of superoxide anion by PMNs using a chemiluminescence assay and on the enzymatic activity of MPO (Franck et al., 2019). It was shown that curcumin and NDS27 exhibit similar inhibition activities on superoxide anion release by stimulated PMNs but also on MPO peroxidase and halogenation activities. They demonstrated that both curcumin and NDS27 are reversible inhibitors of the MPO by acting as excellent electron donors for redox intermediate compound I ($\sim10^7$ M/s) but not for compound II ($\sim10^3$ M/s) in the peroxidase cycle of the enzyme, thereby trapping the enzyme in the compound II state. In conclusion, the hydroxypropyl-β-cyclodextrin of NDS27 enhances curcumin solubilization without affecting its antioxidant capacity and inhibitory activity on MPO.

Cadmium (Cd) pollution is an important issue as it impairs the immune system and affects the food safety of aquatic products. Exogenous metallothionein and vitamin E have been found to exhibit immunoprotection activities in Cd poisoned grass carp, *Ctenopharyngodon idellus* by decreasing Cd contents and hence lessening histological damage, reducing the percentage of apoptosis, and recovering immune-related mRNA transcript expression. Metallothionein had a more pronounced effect than vit E and can be to protect against Cd poisoning (Huang et al., 2019).

5.7.9 Effects of exogenous or dietary antioxidants against ROS/RNS generated in an immune response

Dietary carotenoids like lutein, canthaxanthin, lycopene, and astaxanthin have been found to be more active than beta-carotene in activating the immune system by reducing the toxic effects of ROS which are responsible for causing diseases like cancer, cardiovascular and neurodegenerative disorders, and aging (Chew et al., 2004).

Selenium, found in the selenocysteine-containing proteins like glutathione peroxidase (EC 1.11.1.9), has an important role to play. Glutathione peroxidase protects neutrophils from excess ROS that are produced to kill ingested foreign organisms (Arthur et al., 2003).

Dietary arginine (1.62%) has been found to enhance antioxidant and immune status in the intestine of juvenile blunt snout bream (Liang et al., 2018). The counts of WBC, RBC, hemoglobin, plasma albumin (Liang et al., 2018) content and activities of antioxidant enzymes were found to increase significantly and ALT, MDA levels were decreased in fish fed with dietary arginine level. The relative expression of Nrf2 was significantly increased and the expression of Keap1 found to be decreased. The relative expressions of IL-1β, TNF-α and NF-κB were lowered whereas the relative expression of TGF-β was increased.

5.7.10 Role of antioxidants in combating diseases

The levels of antioxidants have been found to decrease in tuberculosis patients compared to control and come back to normal again during and after treatment with conventional antituberculosis treatment (ATT; Wiid et al., 2004). Several studies have shown the beneficial role of antioxidant supplements along with ATT in the management of TB (Reddy et al., 2004). Micronutrients have been found to increase the concentration of IL-2 and increased proliferation of T-cells and to decrease the production of prostaglandin-2 leading to T cell suppression (Lamsal et al., 2007). Vitamin C and E supplementation along with ATT had been found to have positive results on TB treatment (Kowalski et al., 2004). The antioxidants might be used to prevent and treat the toxicity of drugs, particularly hepatotoxicity, caused by the antitubercular drugs (Verma et al., 2014).

Antioxidant curcumin has been found to be beneficial against influenza A virus. Xu and Liu showed in their study that curcumin downregulates the production of cytokines in a dose-dependent manner and inhibits the NF-κB-mediated signaling pathway in IAV infection (Yiming et al., 2017).

Administration of antioxidants such as vit E, C has been found to have positive effects on oxidative biomarkers in *in vivo* and *in vitro* models of community-acquired pneumonia (Ilias et al., 2008) by inhibiting ROS, DNA damage, autophagy, and reducing the concentrations of TNF-α, IL-6 in LPS-stimulated macrophages (Rodrigo et al., 2013) (Yuanyuan et al., 2014).

Oxidants are very important in removing microorganisms like bacteria, viruses, etc. ROS and antioxidant micronutrients are involved in viral inactivation, including measles, influenza, HIV, and virus-related hepatitis. Vitamin A deficiency is associated with diarrhea, severe measles, HIV, etc. (Villamor and Fawzi, 2005). Vitamin E deficiency has been found to be associated with increased virulence of coxsackievirus B3 (Beck, 1997). De-Flora et al. reported that administration of NAC significantly attenuates the symptoms of influenza in high-risk patients (De Flora et al., 1997).

In vitro studies have shown that HIV replication is enhanced in the presence of ROS, possibly by activation of NF-KB, as well as TNF-α. Coinfection of HIV with mycoplasmas, influenza, or paramyxoviruses is also able to enhance HIV replication, by directly activating macrophages to produce ROS. HIV infection increases the synthesis of IFN-γ, which in turn enhances ROS production by macrophages and neopterin and 7,8-dihydroneopterin. Thus, individuals with HIV infection are under chronic oxidative stress. Hence, it might be beneficial to have vitamin E supplementation in the treatment of patients with AIDS (Wang and Watson, 1993).

5.7.11 Role of antioxidants on autoimmunity and oxidative stress

To find out the connection between autoimmunity and OS, Weimann and Weiser studied the effects of several antioxidants on a disorder that mimics systemic lupus erythematosus (SLE) in mice. The use of antioxidants was found to show reduced symptoms like absence of extensive lymphoproliferation, significantly lower IgG levels and reduced elevation in anti-dsDNA antibody titer compared to control. Hence, the authors concluded that diets rich in antioxidants might have a major role to play in decreasing lupus-like syndromes (Weimann and Weiser, 1992).

Effect of the OS is very prominent in another autoimmune disorder, rheumatoid arthritis (RA). The ROS and RNS produced by the neutrophils activate transcription factors NF-κB, IL-2, and TNF-α. The concentrations of antioxidants like vitamins E, C, beta-carotene, selenium, and zinc as well as antioxidant enzymes like catalase, glutathione peroxidase, SOD were found to be less in the serum of RA patients compared to healthy individuals. However, there was a high concentration of iron in the fluids of RA patients, indicating the possibility of the presence of hydroxyl, hydrogen peroxide or superoxide radical production by Fenton and/or Haber–Weiss chemistry which stimulate the release of iron from proteins like hemoglobin (Knight, 2000). ROS, especially superoxide also produces in RA patients by xanthine oxidase, in ischemic and inflammatory sites. Weinberg et al. reported increased levels of NO in MRL-lpr/lpr mice, which leads to the development of arthritis, vasculitis, and immune complex glomerulonephritis (Weinberg et al., 1994). Similar reports show that the conc of S-nitrosoproteins, which are produced by the combination of NO and thiol-containing proteins, increases in the synovial fluid of patients with RA (Mahajan and Tandon, 2004). NSAIDs intervention and micronutrients administration in RA showed fewer swollen and painful joints with improvement of general health (Vugt et al., 2008; Khanna et al., 2017). Administration of vitamin E in RA patients seems to reduce joint inflammation and destruction in the transgenic KRN/NOD mouse model of RA18. Hadi Abdollahzad showed that coenzyme Q_{10}, a fat-soluble antioxidant, has a decreasing effect on serum MDA and TNF-α and improved the serum LDL concentration in RA patients (Hadi, 2016).

In a pilot study, the authors showed that consumption of antioxidant-enriched margarine increases the antioxidant status in blood and reduces the oxidative stress markers and thus provide significant relief of clinical symptoms in RA patients (Richard et al., 2008).

5.7.12 Possibility of exerting harmful or no effects of antioxidants on immune system

It is well known that consumption of fruits, vegetables containing antioxidants help in fighting with ROS and decreases the risk of cancer. However, detailed studies on the mechanism of action of different antioxidants are providing contradictory results regarding their beneficial role in immunity or carcinogenesis.

Physicians Health Study II which was carried out for a period of 10 years (between 1997 and 2007), evaluated the effects of vitamin E and C on cardiovascular events in a randomized, double-blind trial. It was found that no vitamin (C [500 mg daily]; E [400 IU daily]) could reduce the risk of major cardiovascular events in 14,641 US male physicians with an initially age of 50 years or more and a large number of (1245) confirmed major cardiovascular events were observed in the follow-up years (Hajhashemi et al., 2010).

The well-known antioxidants, naringenin and quercetin, inhibit the HOCl production through different systems and the inhibition was more pronounced for quercetin, even in the cell-free systems. The microbicidal activity, that is, killing of *Staphylococcus aureus* by neutrophils was completely inhibited both by these two antioxidants as they controlled the production and effector mechanisms of ROS (Francielli et al., 2013).

Retinol is known for its ability to quench free radicals and enhance the immune responses and surveillance against tumorigenesis. For a long time, it was known that a combination of β-carotene, vitamin E, and selenium decreases the risks of lung cancer in healthy individuals and reduce stomach cancer mortality by combating ROS (Neuhouser, 2003). However, recent studies have confirmed an increased incidence of cancer among smokers taking β-carotene. Selenium has been found to reduce total cancer incidence predominantly among males (Bardia et al., 2008). β-Carotene has been found to stimulate the production of free radicals at high concentration, whereas it exerts antioxidant activity at low concentration. β-Carotene is cleaved into many unstable derivatives in the presence of free radicals derived from cigarette smoke and undergo further oxidation (Palozza et al., 2004; Baker et al., 1999). The expression of the enzyme, heme oxigenase 1 has been found to be repressed in the presence of β-carotene alone or in combination with cigarette smoke condensate both in rat fibroblasts and human lung cancer cells (Palozza et al., 2006).

Supplementation of vitamin E has shown contradictory results on immune system. An association has been established between vit E treatment and an increased incidence of mortality related to neoplastic diseases in some studies (Bjelakovic et al., 2007). On the other hand, vitamin E supplementation has been found to stimulate both cell mediated and humoral immunity, protect the polyunsaturated fatty acids in the membrane from oxidation, etc. in both experimental animals and in humans (Lee and Han, 2018).

Polyphenols and NSAIDs have been found to act on different neurodegenerative diseases (Calabrese et al., 2003). However, the major problem in the application of polyphenols is that only a negligible amount crosses the blood–brain barrier and reaches the brain tissue (Calabrese et al., 2008).

ALS, Amyotrophic Lateral Sclerosis, a systemic motor neuron disease (Neary et al., 2000), has been found to be linked to point mutation in the Cu/Zn SOD gene and administration of α-tocopherol with riluzole, did not have any effect on the survival and motor functions in ALS patients compared to the control group treated with riluzole alone (Desnuelle et al., 2001). Treatment of ALS with a novel compound, AEOL-10150 (Aeolus), which is analogous to the SOD catalytic site, has a positive effect on it (Daria et al., 2008; Orrell, 2006).

Goswami et al. evaluated the effects of therapeutically relevant antioxidants like ascorbic acid, glutathione, and NAC on the *E. coli*-induced acute bacterial peritonitis. They observed reduced numbers of immune cells, for example, macrophages, B-cells, and DCs at the primary site of infection and decreased level of TNF-α in the serum of antioxidant-treated peritoneal mice. The treatment with antioxidants reduced the phagocytic efficacy and oxidative burst in peritoneal macrophages in response to *E. coli* in vitro though infiltration of neutrophils was found to increase in response to antioxidants. Based on the results, the authors proposed that antioxidant supplementation during bacterial infections should not be recommended as it could suppress the immune system and delay the recovery (Goswami et al., 2014).

Conclusion

The balance between the amount of reactive oxygen and nitrogen species and antioxidants plays a critical role in a natural infection, on the activity of immune system, in autoimmunity, hypersensitivity, etc. ROS/RNS takes part directly in the killing of microorganisms in the innate immunity and administration of excess antioxidants has been found to interfere with the immune responses. However, increased ROS/RNS and hence the OS has been found to be associated with neurogenerative disorders. The amount of ROS/RNS increases considerably in some pathological conditions like tuberculosis, HIV which need prolonged medications. Similarly, autoimmune disorders like RA show significantly low levels of endogenous antioxidants and increased OS. Hence, dietary supplementation of antioxidants in proper dosage can be used therapeutically to reduce the pathophysiological conditions raised due to oxidative stress in these cases.

References

Alfonso, C., Franqoise, V., Gustavo, G., Guillaume, D., 1999. *In vitro* antioxidant treatment recovers proliferative responses of anergic CD4+ lymphocytes from human immunodeficiency virus-infected individuals. Blood 87 (11), 4746–4753.

Arthur, J.R., McKenzie, R.C., Beckett, G.J., 2003. Selenium in the immune system. J. Nutr. 133, 1457S–1459S.

Baker, D.L., Krol, E.S., Jacobsen, N., Liebler, D.C., 1999. Reactions of beta-carotene with cigarette smoke oxidants. Identification of carotenoid oxidation products and evaluation of the prooxidant/antioxidant effect. Chem. Res. Toxicol. 12, 535–543.

Bardia, A., Tleyjeh, I.M., Cerhan, J.R., Sood, A.K., Limburg, P.J., Erwin, P.J., Montori, V.M., 2008. Efficacy of antioxidant supplementation in reducing primary cancer incidence and mortality: systematic review and meta-analysis. Mayo Clin. Proc. 83, 23–34.

Beck, M.A., 1997. Increased virulence of coxsackievirus B3 in mice due to vitamin E or selenium deficiency. J. Nutr. 127 (5), 966S–970S. doi:10.1093/jn/127.5.966S.

Bjelakovic, G., Nikolova, D., Gluud, L.L., Simonetti, R.G., Gluud, C., 2007. Mortality in randomized trials of antioxidant supplements for primary and secondary prevention: systematic review and meta-analysis. JAMA 297, 842–857. doi:10.1001/jama.297.8.842.

Brieger, K., Schiavone, S., Miller Jr., F.J., Krause, K.H., 2012. Reactive oxygen species: from health to disease. Swiss Med. Wkly. 142, w13659.

Bulger, E.M., Garcia, I., Maier, R.V., 2002. Intracellular antioxidant activity is necessary to modulate the macrophage response to endotoxin. Shock 18 (1), 58–63.

Calabrese, V., Butterfield, D.A., Stella, A.M., 2003. Nutritional antioxidants and the heme oxygenase pathway of stress tolerance: novel targets for neuroprotection in Alzheimer's disease. Ital. J. Biochem. 52, 177–181.

Calabrese, V., Guagliano, E., Sapienza, M., Panebianco, M., Calafato, S., Puleo, E., Pennisi, G., Mancuso, C., Butterfield, D.A., Stella, A.G., 2008. Redox regulation of cellular stress response in aging and neurodegenerative disorders: role of vitagenes. Neurochem. Res. 32, 757–773.

Chaudhri, G., Clark, I.A., Hunt, N.H., Cowden, W.B., Ceredig, R., 1986. Effect of antioxidants on primary alloantigen-induced T cell activation and proliferation. J. Immunol. 137 (8), 2646–2652.

Cheeseman, K.H., Slater, T.F., 1993. An introduction to free radical biochemistry. Br. Med. Bull. 49 (3), 481–493.

Chen, X., Miao, J., Wang, H., et al., 2015. The anti-inflammatory activities of *Ainsliaea fragrans* Champ. extract and its components in lipopolysaccharide-stimulated RAW264.7 macrophages through inhibition of NF-κB pathway. J. Ethnopharmacol. 170, 72–80. doi:10.1016/j.jep.2015.05.004.

Chew, B.P., Park, J.S., 2004. Carotenoid action on the immune response. J. Nutr. 134 (1), 257S–261S.

Crump, K., Langston, P.K., Rajkarnikar, S., Grayson, J.M., 2013. Antioxidant treatment regulates the humoral immune response during acute viral infection. J. Virol. 87 (5), 2577–2586.

Daria, B., Cesare, M., Mariagrazia, R.S., Paolo, B., Giuseppina, C., Laurence, L., Giulia Di, B., Salvatore, P., Renato, B., 2008. The role of antioxidant supplement in immune system, neoplastic, and neurodegenerative disorders: a point of view for an assessment of the risk/benefit profile. Nutr J. 7, 29.

Daswani, B.R., Yegnanarayan, R., 2002. Immunomodulatory activity of septilin, a polyherbal preparation. Phytother. Res. 16 (2), 162–165.

De Flora, S., Grassi, C., Carati, L., 1997. Attenuation of influenza-like symptomatology and improvement of cell-mediated immunity with long-term N-acetylcysteine treatment. Eur. Respir. J. 10 (7), 1535–1541.

Desnuelle, C., Dib, M., Garrel, C., Favier, A., 2001. A double-blind, placebo-controlled randomized clinical trial of alpha-tocopherol (vitamin E) in the treatment of amyotrophic lateral sclerosis. ALS Riluzole-Tocopherol Study Group. Amyotroph Lateral Scler. Other Motor Neuron. Disord. 2, 9–18.

Dias, V., Junn, E., Mouradian, M.M., 2013. The role of oxidative stress in Parkinson's disease. J. Parkinsons Dis. 3, 461–491.

Dietz, K.J., Mittler, R., Noctor, G., 2016. Recent progress in understanding the role of reactive oxygen species in plant cell signaling. Plant Physiol. 171, 1535–1539.

Dinkova-Kostova, A.T., Holtzclaw, W.D., Cole, R.N., Itoh, K., Wakabayashi, N., Katoh, Y., Yamamoto, M., Talalay, P., 2002. Direct evidence that sulfhydryl groups of Keap1 are the sensors regulating induction of phase 2 enzymes that protect against carcinogens and oxidants. Proc. Natl. Acad. Sci. 99 (18), 11908–11913.

Dong, X.Z., Wang, Y.N., Tan, X., Liu, P., Guo, D.H., Yan, C., 2018. Protective effect of JXT ethanol extract on radiation-induced hematopoietic alteration and oxidative stress in the liver. Oxid. Med. Cell Longev. 2018, 9017835. doi:10.1155/2018/9017835.

El-Senousey, H.K., Chen, B., Wang, J.Y., Atta, A.M., Mohamed, F.R., Nie, Q.H., 2018. Effects of dietary vitamin C, vitamin E, and alpha-lipoic acid supplementation on the antioxidant defense system and immune-related gene expression in broilers exposed to oxidative stress by dexamethasone. Poult. Sci. 97 (1), 30–38.

Elwan, H.A.M., Elnesr, S.S., Mohany, M., Al-Rejaie, S.S., 2019. The effects of dietary tomato powder (*Solanum lycopersicum* L.) supplementation on the haematological, immunological, serum biochemical and antioxidant parameters of growing rabbits. J. Anim. Physiol. Anim. Nutr. (Berl.) 103 (2), 534–546.

Emily, L., 2015, Conferences >2015 41st Annual Northeast Bi Effects of Radical Oxygen Species and Antioxidants on Macrophage Polarization.

Finkel, T., Holbrook, N.J., 2000. Oxidants, oxidative stress and the biology of ageing. Nature 408, 239–247.

Francielli, C., Nishimura, Y., Almeida, A.C., Ratti, B.A., Ueda-Nakamura, T., Nakamura, C.V., Ximenes, V.F., Silva, S.O., 2013. Antioxidant effects of quercetin and naringenin are associated with impaired neutrophil microbicidal activity. Evidence-Based Complement. Altern. Med. 2013, 795916 Article ID7 pages.

Francielli Maria, S.S.-C., et al., 2014. Effect of estragole on leukocyte behavior and phagocytic activity of macrophages. Evidence-Based Complement. Altern. Med. 2014, 784689, 7 pages.

Francischetti, I.M.B., et al., 2014. Tempol, an intracellular antioxidant, inhibits tissue factor expression, attenuates dendritic cell function, and is partially protective in a murine model of cerebral malaria. PLoS One 9 (2), e87140.

Franck, T., Aldib, I., Zouaoui, B.K., Furtmüller, P.G., Obinger., C., Neven, P., Prévost, M., Soubhye, J., Van, A.P., Mouithys-Mickalad., A., Serteyn, D., 2019. The soluble curcumin derivative NDS27 inhibits superoxide anion production by neutrophils and acts as substrate and reversible inhibitor of myeloperoxidase. Chem. Biol. Interact. 5 (297), 34–43.

Goswami, M., Deepak, S., Nazir, M.K., Rahul, C., Santosh, K.S., Narendra, J., 2014. Antioxidant supplementation enhances bacterial peritonitis in mice by inhibiting phagocytosis. J. Med. Microbiol. 63, 355–366.

Groemping, Y., Rittinger, K., 2005. Activation and assembly of the NADPH oxidase: a structural perspective. Biochem. J. 386 (Pt 3), 401–416.

Gross, E., Sevier, C.S., Heldman, N., Vitu, E., Bentzur, M., Kaiser, C.A., et al., 2006. Generating disulfides enzymatically: reaction products and electron acceptors of the endoplasmic reticulum thiol oxidase Ero1p. Proc. Natl. Acad. Sci. USA. 103 (2), 299–304.

Gu, X.H., Hao, Y., Wang, X.L., 2012. Overexpression of heat shock protein 70 and its relationship to intestine under acute heat stress in broilers: 2. Intestinal oxidative stress. Poult. Sci. 91 (4), 790–799.

Hadi, A., 2016. Importance of antioxidants in rheumatoid arthritiss. Austin Arthritis 1 (1), 1005.

Hajhashemi, V., Vaseghi, G., Pourfarzam, M., Abdollahi, A., 2010. Are antioxidants helpful for disease prevention? Res. Pharm. Sci. 5 (1), 1–8.

Huang, X., Feng, Y., Fan, W., Duan, J., Duan, Y., Xiong, G., Wang, K., Deng, Y., Geng, Y., Ouyang, P., Chen, D., Yang, S., 2019. Potential ability for metallothionein and vitamin E protection against cadmium immunotoxicity in head kidney and spleen of grass carp (*Ctenopharyngodon idellus*). Ecotoxicol. Environ. Saf. 159 (170), 246–252.

Hughes, D.A., 1999. Effects of dietary antioxidants on the immune function of middle-aged adults. Proc. Nutr. Soc. 58 (1), 79–84.

Ilias, I., Siempos, K.Z., Vardakas, P.K., Falagas, M.E., 2008. Adjunctive therapies for community-acquired pneumonia: a systematic review. J. Antimicrob. Chemother. 62 (4), 661–668. https://doi.org/10.1093/jac/dkn283.

Itoh, K., Ishii, T., Wakabayashi, N., Yamamoto, M., 1999. Regulatory mechanisms of cellular response to oxidative stress. Free Radic. Res. 31 (4), 319–324.

Jiang, Y., Wang, M.H., 2015. Different solvent fractions of acanthopanax senticosus harms exert antioxidant and anti-inflammatory activities and inhibit the human Kv1.3 channel. J. Med. Food 8 (4), 468–475. doi:10.1089/jmf.2014.3182.

Jonathan, P., Katsuhiko, S., 2004. Neutrophil activation, antioxidant supplements and exercise-induced oxidative stress. Exerc. Immunol. Rev. 10 (1), 129–141.

Khanna, S., Jaiswal, K.S., Gupta, B., 2017. Managing rheumatoid arthritis with dietary interventions. Front. Nutr. 4, 52. doi:10.3389/fnut.2017.00052. Published online 2017 Nov 8.

Kim, H.Y., Hwang, K.W., Park, S.Y., 2014. Extracts of *Actinidia arguta* stems inhibited LPS-induced inflammatory responses through nuclear factor-κB pathway in Raw 264.7 cells. Nutr. Res. 34 (11), 1008–1016.

Knight, J.A., 2000. Review: free radicals, antioxidants, and the immune system. Ann. Clin. Lab. Sci. 30 (2), 145.

Kouoh, F., et al., 1999. Antioxidant properties of albumin: effect on oxidative metabolism of human neutrophil granulocytes. Farmaco 54 (10), 695–699.

Kowalski, J., Janiszewska-Drobinska, B., Pawlicki, L., Ceglinski, T., Irzmanski, R., 2004. Plasma antioxidative activity in patients with pulmonary tuberculosis. Pol. Merkur. Lekarski. 16 (92), 119–122.

Krumova, K., Cosa, G., 2016. Chapter 1: Overview of reactive oxygen species. In: Nonell, Santi, Flors, Cristina (Eds.), Singlet Oxygen: Applications in Biosciences and Nanosciences. Vol. 1, Royal society of chemistry, pp. 1–21. From Book Series: Comprehensive Series in Photochemical & Photobiological Sciences.

Lamsal, M., Gautam, N., Bhatta, N., Toora, B.D., Bhattacharya, S.K., Baral, N., 2007. Evaluation of lipid peroxidation product, nitrite and antioxidant levels in newly diagnosed and two months follow up patients with pulmonary tuberculosis. Southeast Asian J. Trop. Med. Public Health 38 (4), 695–703.

Lee, J.M., Calkins, M.J., Chan, K., Kan, W.Y., and Johnson, J.A., 2003. Identification of the NF-E2-related factor-2-dependent genes conferring protection against oxidative stress in primary cortical astrocytes using oligonucleotide microarray analysis. 278(14), 12029–12038.

Lee, G.Y., Han, S.N., 2018. The role of vitamin E in immunity. Nutrients 10 (11), 1614.

Liang, H., Mokrani, A., Ji, Ke, Ge, X., Ren, M., Pan, L., Sun, A., 2018. Effects of dietary arginine on intestinal antioxidant status and immunity involved in Nrf2 and NF-κB signaling pathway in juvenile blunt snout bream, Megalobrama amblycephala. Fish Shellfish Immunol. 82, 243–249.

Liu, Z., Ren, Z., Zhang, J., Kandaswamy, E., Zhou, T., Zuo, L., 2018. Role of ROS and nutritional antioxidants in human diseases. Front. Physiol. 9, 477. doi:10.3389/fphys.2018.00477.

Mahajan, A., Tandon, V.R., 2004. Antioxidants and rheumatoid arthritis. J. Indian Rheumatol. Assoc. 12, 139–142.

McCormick, B., Lowes, D.A., Colvin, L., Torsney, C., Galley, H.F., 2016. MitoVitE, a mitochondria-targeted antioxidant, limits paclitaxel-induced oxidative stress and mitochondrial damage *in vitro*, and paclitaxel-induced mechanical hypersensitivity in a rat pain model. BJA: Br. J. Anaesthesia 117 (5), 659–666.

Moreno-Macias, H., Romieu, I., 2014. Effects of antioxidant supplements and nutrients on patients with asthma and allergies. J. Allergy Clin. Immunol. 133 (5), 1237–1244.

Morrow, D.M., Entezari-Zaher, T., Romashko 3rd, J., Azghani, A.O., Javdan, M., Ulloa, L., Miller, E.J., Mantell, L.L., 2007. Antioxidants preserve macrophage phagocytosis of *Pseudomonas aeruginosa* during hyperoxia. Free Radic. Biol. Med. 42 (9), 1338–1349.

Nathan, G.L., Jason, M.G., 2004. Antioxidant treatment reduces expansion and contraction of antigen-specific CD8 T cells during primary but not secondary viral infection. J. Virol. 78 (20), 11246–11257.

Neary, D., Snowden, J.S., Mann, D.M., 2000. Cognitive change in motor neurone disease/amyotrophic lateral sclerosis (MND/ALS). J. Neurol. Sci. 180, 15–20.

Neuhouser, M.L., Patterson, R.E., Thornquist, M.D., Omenn, G.S., King, I.B., Goodman, G.E., 2003. Fruits and vegetables are associated with lower lung cancer risk only in the placebo arm of the beta-carotene and retinol efficacy trial (CARET). Cancer Epidemiol. Biomarkers Prev. 12, 350–358.

Orrell, R.W., 2006. AEOL-10150 (Aeolus). Curr. Opin. Investig. Drugs 7, 70–80.

Pae, M., Wu, D., 2013. Immunomodulating effects of epigallocatechin-3-gallate from green tea: mechanisms and applications. Food Funct 4 (9), 1287–1303.

Palozza, P., Serini, S., Currò, D., Calviello, G., Igarashi, K., Mancuso, C., 2006. Beta-carotene and cigarette smoke condensate regulate heme oxygenase-1 and its repressor factor Bach1: relationship with cell growth. Antioxid. Redox. Signal. 8, 1069–1080.

Palozza., P., Serini, S., Di Nicuolo, F., Boninsegna, A., Torsello, A., Maggiano, N., Ranelletti, F.O., Wolf, F.I., Calviello, G., Cittadini, A., 2004. Beta-carotene exacerbates DNA oxidative damage and modifies p53-related pathways of cell proliferation and apoptosis in cultured cells exposed to tobacco smoke condensate. Carcinogenesis 25, 1315–1325.

Park, O.J., Kim, H.Y., Kim, W.K., Kim, Y.J., Kim, S.H., 2003. Effect of vitamin E supplementation on antioxidant defense systems and humoral immune responses in young, middle-aged and elderly Korean women. J. Nutr. Sci. Vitaminol. 49 (2), 94–99.

Park, S.K., Page, G.P., Kim, K., Allison, D.B., Meydani, M., Weindruch, R., Prolla, T.A., 2008. Alpha- and gamma-tocopherol prevent age-related transcriptional alterations in the heart and brain of mice. J. Nutr. 138 (6), 1010–1018.

Peake, J., Suzuki, K., 2004. Neutrophil activation, antioxidant supplements and exercise-induced oxidative stress. Exerc. Immunol. Rev. 10, 129–141.

Peng, H.T., Sagoo., P., Cliburn, C., Yates., J.B., Campbell., J., Beutelspacher., S.C., Foxwell., B.M.J., Giovanna, L., George, A.J.T., 2005. Inhibition of NF-B and oxidative pathways in human dendritic cells by antioxidative vitamins generates regulatory T cells. J. Immunol. 174, 7633–7644.

Peterhans, E., 1997. Oxidants and antioxidants in viral diseases: disease mechanisms and metabolic regulation. J. Nutr. 127 (5), 962S–965S.

Qin, T., Yin, Y., Yu, Q., Yang, Q., 2015. Bursopentin (BP5) protects dendritic cells from lipopolysaccharide-induced oxidative stress for immunosuppression. PLoS One 10 (2), e0117477.

Qiu, X., Brown, K., Hirschey, M.D., Verdin, E., Chen, D., 2010. Calorie restriction reduces oxidative stress by SIRT3-mediated SOD2 activation. Cell Metab. 12, 662–667.

Rajan, T.S., Giacoppo, S., Iori, R., De Nicola, G.R., Grassi, G., Pollastro, F., Bramanti, P., Mazzon, E., 2016. Anti-inflammatory and antioxidant effects of a combination of cannabidiol and moringin in LPS-stimulated macrophages. Fitoterapia 112, 104–115.

Reddy, Y.N., Murthy, S.V., Krishna, D.R., Prabhakar, M.C., 2004. Role of free radicals and antioxidants in tuberculosis patients. Indian J. Tuberc. 51, 213–218.

Rego, A.C., Oliveira, C.R., 2003. Mitochondrial dysfunction and reactive oxygen species in excitotoxicity and apoptosis: implications for the pathogenesis of neurodegenerative diseases. Neurochem. Res. 28 (10), 1563–1574.

Richard, M., Van, V., Rijken, P.J., Rietveld, A.G., Vugt, A.C., Dijkmans, B.A.C., 2008. Antioxidant intervention in rheumatoid arthritis: results of an open pilot study. Clin. Rheumatol. 27 (6), 771–775.

Rodrigo, L.C., Rodrigo, A.C., Pedro, I.A., Mauricio, R., Vivian, R., Enna, Z., María, A.M., Luis, F.M., 2013. Relationship between severity of adult community-acquired pneumonia and impairment of the antioxidant defense system. Biol. Res. 46 (2), 207–213.

Scheffel, M.J., Scurti, G., Simms, P., Garrett-Mayer, E., Mehrotra, S., Nishimura, M.I., Voelkel-Johnson, C., 2016. Efficacy of adoptive T-cell therapy is improved by treatment with the antioxidant N-Acetyl cysteine, which limits activation-induced T-cell death. Cancer Res. 76 (20), 6006–6016.

Schrader, M., Fahimi, H.D., 2006. Review peroxisomes and oxidative stress. Biochim. Biophys. Acta 1763 (12), 1755–1766.

Shiva, M., Paul, H., Graham, P., Yvonne, B., 2013. An investigation of the effects of the antioxidants, ebselen or N-acetyl cysteine on human peripheral blood mononuclear cells and T cells. Immunity Ageing 10, 7.

Sureda, A., Tauler, P., Aguil´o., A., Cases, N., Llompart, I., Tur, J.A., Pons, A., 2007. Antioxidant supplementation influences the neutrophil tocopherol associated protein expression, but not the inflammatory response to exercise. CEJB 2 (1), 56–70.

Verma, I., Jindal, S., Ganguly, N.K., 2014. Oxidative stress in tuberculosis. Studies on Respiratory Disorders 2197-7224, Springer Link, pp. 101–114.

Villamor, E., Fawzi, W.W., 2005. Effects of vitamin A supplementation on immune responses and correlation with clinical outcomes. Clin. Microbiol. Rev. 18 (3), 446–464.

Vorobjeva, N., Prikhodko, A., Galkin, I., Pletjushkina, O., Zinovkin, R., Sud'ina, G., Chernyak, B., Pinegin, B., 2017. Mitochondrial reactive oxygen species are involved in chemoattractant-induced oxidative burst and degranulation of human neutrophils in vitro. Eur. J. Cell Biol. 96 (3), 254–265.

Vorobjeva, N.V., Pinegin, B.V., 2016. Effects of the antioxidants Trolox, Tiron and Tempol on neutrophil extracellular trap formation. Immunobiology 221 (2), 208–219.

Vugt, R.M., Rijken, P.J., Rietveld, A.G., Vugt, A.C., Dijkmans, B.A.C., 2008. Antioxidant intervention in rheumatoid arthritis: results of an open pilot study. Clin. Rheumatol. 27 (6), 771–775.

Wang, Y., Watson, R.R., 1993. Is vitamin E supplementation a useful agent in AIDS therapy? Prog. Food Nutr. Sci. 17 (4), 351–375.

Weimann, B.J., Weiser, H., 1992. Effects of antioxidant vitamins C, E, and beta-carotene on immune functions in MRL/lpr mice and rats. Ann. NY Acad. Ser. 669, 390–392.

Weinberg, J.B., Granger, D.L., Pisetsky, D.S., et al., 1994. The role of nitric oxide in the pathogenesis of spontaneous murine autoimmune disease: increased nitric oxide production and nitric oxide synthase expression in MRL-lpr/lpr mice, and reduction of spontaneous glomerulonephritis and arthritis by orally administered NG-monoethyl-L-arginine. J. Exp. Med. 179, 651–660.

Wiid, I., Seaman, T., Hoal, E.G., Benade, A.J., Van, H.P.D., 2004. Total antioxidant levels are low during active TB and rise with antituberculosis therapy. IUBMB Life 56 (2), 101–106.

Yiming, X., Ling, L., 2017. Curcumin alleviates macrophage activation and lung inflammation induced by influenza virus infection through inhibiting the NF-κB signaling pathway. Influenza Other Respir. Viruses 11 (5), 457–463.

Yuanyuan, C., Guangyan, L., Jiao, Y., Wang, Y., Yang, X., Wang, X., Li, G., Liu, G., Zhong, N., 2014. Vitamin C mitigates oxidative stress and tumor necrosis factor-alpha in severe community-acquired pneumonia and LPS-induced macrophages. Mediators Inflamm. 2014, 426740. 11 pages. http://dx.doi.org/10.1155/2014/426740.

CHAPTER

Antioxidants and infertility

5.8

Harvesh Kumar Rana, Amit Kumar Singh, Abhay K. Pandey
Department of Biochemistry, University of Allahabad, Prayagraj, India

5.8.1 Introduction

The success of pregnancy is the consequence of a complex interaction between males and females physiological systems. Any disruption in this interactive system, whether in a man or a woman, has the potential to prevent the birth of a biological child. Infertility is described as the inability to conceive after a year of frequent unprotected intercourse. Infertility affects around 15–20% of couples of reproductive age, and it can be caused by both male and female causes.

Oxidative stress (OS) has been linked to infertility in both men and women. It is a situation that occurs when the generation or buildup of reactive oxygen species (ROS) exceeds the biological system's antioxidant capability. John MacLeod was the first to notice that OS might be a significant cause of male infertility in 1943 (MacLeod, 1943). The major sources of ROS in the male reproductive system are neutrophils, macrophages, and immature sperm. The generation of ROS in leukocytes after the beginning of infection and inflammation might result in an increase in ROS levels in the surrounding tissues (Sharma et al., 1999). Depending on the amount of the OS, ROS can cause varying levels of sperm problem. ROS can cause DNA fragmentation in sperm from infertile men in two ways: first, it can cause single or double stranded DNA breaks, which can lead to infertility (Kodama et al., 1997). Aitken et al. (1995) demonstrated that low levels of H_2O_2 had no effect on sperm motility but did inhibit sperm–egg fusion. Second, higher levels of ROS may cause damage by causing lipid peroxidation of the sperm plasma membrane through a sequence of chemical processes (Alvarez, 1987). Lack of membrane fluidity, which is crucial for sperm motility and sperm–oocyte fusion, is caused by lipid peroxidation. According to Griveau and Le Lannou (1997), the acrosome response in human spermatozoa is susceptible to ROS. Several studies have shown that the levels of ROS are inversely related to sperm motility.

OS because of free radicals load impacts the exceptional of gametes and affects oocytes, spermatozoa, embryos, and their situations. The microenvironments of follicular fluid, hydrosalpinges fluid, and peritoneal fluid have an immediate impact on oocyte quality, sperm–oocyte contact, sperm-mediated oocyte activation,

Antioxidants Effects in Health. DOI: https://doi.org/10.1016/B978-0-12-819096-8.00058-6

implantation, and early embryo development. OS has an effect on early embryo development and implantation, which has an effect on the likelihood of becoming pregnant (Agarwall et al., 2005).

Antioxidants (vitamins C and E, folic acid, zinc, selenium, carnitine, and carotenoids) are ROS scavengers, and their usage as a therapy has been demonstrated to counteract the negative effects of high ROS levels on sperm and blood parameters in female infertility. Those with a high dietary intake of antioxidants had a reduced incidence of sperm aneuploidy and higher sperm quality than men with a lower consumption, according to studies (Silver et al., 2005; Young et al., 2008). Women with unexplained infertility had greater ROS levels than their fertile counterparts (Ruder et al., 2008). The normal buildup of free radicals with ageing may explain why older women's egg cells are of lower quality (Agarwal et al., 2012). The current understanding of the role of ROS in the normal functioning of male and female fertility, as well as their role in infertility, is summarized in this chapter.

5.8.2 Infertility

Infertility is defined as the inability to achieve a medical pregnancy after 12 months of typical unprotected intercourse or decline in a person's fertility, either as an individual or with their partner, and it is an essential factor of pregnancy beginning. Infertility, according to the most recent WHO definition, is a condition that results in impairment due to a decline in function (Zegers-Hochschild et al., 2017). Infertility is a major issue. OS may play a role in endometriosis, tubal, and peritoneal issues, and unexplained infertility. Infertility affects 15% of couples of childbearing age. Only male factors can account for a quarter of these occurrences, whereas male and female factors can account for up to half of them (Sharlip et al., 2002).

The male physiological condition has traditionally been defined as an anomaly in one or more of the commonly studied spermatozoa characteristics, namely, concentration, morphology, and motility. Normal sperm analysis, on the other hand, has been proven to be ineffective in distinguishing between men with high and low fertility (Bonde et al., 1998). Furthermore, there appears to be a considerable overlap in observed sperm concentration, morphology, and motility between infertile individuals and men with confirmed fertility (Guzick et al., 2001). The considerable intra-individual variability of these measures, even during a spermatogenic cycle (about 10 weeks), might explain why conventional sperm parameters have a limited predictive value for diagnosing male infertility (Alvarez et al., 2003). It is commonly recommended that an abnormal humor analysis be performed at least once. Another possible causal factor is that regular spermatozoon analysis is often done manually using light research on a spermatozoa sample distribution (usually 200–400). This results in a high level of perspicacity as well as the possibility of significant inter-laboratory and intralaboratory variance.

Aitken et al. (1989) presented that the ability of human sperm to fuse with egg cells with increased OS decreased in a dose-dependent manner, which can be inverted in the addition of vitamin E. Hughes et al. (1998) found that the addition of antioxidants to Percoll sperm preparation for an aided copy improved the DNA integrity of the sperm. However, the majority of those studies include uncontrolled, fertilized males, and pregnancy is rarely used as a prognosis indication (Kefer et al., 2009; Lewis and Agbaje, 2008).

Infertility in reproductive-elderly women is expected to affect one out of every seven couples in the Western world, and one out of every four couples in developing countries. Infertility rates may reach 30% in several parts of the world, including South Asia, a few countries in Sub-Saharan Africa, Middle East, North Africa, Central and Eastern Europe, and Central Asia (Mascarenhas et al., 2012). Males are shown to be entirely responsible for 20–30% of infertility cases, while they contribute to 50% of cases overall. However, such numbers no longer accurately represent all parts of the globe. Male infertility rates were highest in Africa and Central/Eastern Europe, according to Agarwal et al. (2015), whereas rates in North America, Australia, Central, and Eastern Europe ranged from 4.5 to 6%, 9%, and 8–12%, respectively.

Infertility can also be classified as primary or secondary. The main infertile female is a woman who has never been diagnosed with a clinical pregnancy yet satisfies the criteria for being classified as infertile. Secondary female infertility refers to a woman who is unable to have a clinical pregnancy but has previously been diagnosed with a clinical pregnancy (Zegers-Hochschild et al., 2017). The same classification is likely to apply to the male in terms of his involvement in the conception of a child.

To improve sperm quality and, as a result, male reproductive potential, a type of medication was developed (Kamischke and Nieschlag, 1999). In the age of evidence-based medicine, accurate control of infertility must be based on identifying reversible causes of infertility and treating them with suitable medications. This, however, will be a difficulty because, despite extensive research, no specific explanation for infertility in over 25% of infertile guys can be found (March and Isidori, 2002). More couples are seeking medical treatment for infertility, including pharmacological therapies, as infertility treatment continues to improve in the United States. According to the National Survey of Family Growth from 2002, 12% (7.3 million) of reproductive-elderly women (15–44 years) in the United States said they have used infertility treatments. According to the Centers for Disease Control and Prevention and the American Pregnancy Association, around 6 million women aged 15–44 have difficulty becoming or remaining pregnant. Because age is a significant risk factor for infertility, the requirements for an infertility analysis are based on the patient's age.

5.8.3 Male infertility

In more than 90% of cases, infertility in men is caused by low sperm counts, poor sperm quality, or both. Anatomical issues, hormone abnormalities, and genetic anomalies are among the other causes. Hypothalamic hypophyseal tract, testicular

Table 5.8.1 Common cause of male and female infertility.

Causes of male infertility	
Hypothalamic hypophyseal	Pituitary insufficiency, hyperprolactinemia, Kallmann syndrome, hemochromatosis, pituitary tumors, chronic illness (Dohle et al., 2002)
Testicular disorders	Infection, Klinefelter syndrome, testicular atrophy, Y chromosome deletions, systemic disorders (Dohle et al., 2002)
Seminal tract disorders	Retrograde ejaculation, obstructive azoospermia (Quinn et al., 1996)
Immunological disorders	Autoimmunity to sperm. This antibody hobby may be detected through some of the methods, inclusive of agglutination, immobilization, cytotoxicity, and immunofluorescence (Shulman, 1972).
Psychosomatic disorders	Psychological attitudes associated with neuroticism, melancholy, etc. Clinical observations advocate that remedy of subjective anxiety or strain can enhance fertility (Urry, 1977)
Causes of female infertility	
Chromosome abnormalities	Within the phenotypic lady, sex chromosome defects include gonadal dysgenesis (including Turner's disease) and androgen insensitivity (David et al., 1994)
Ovulatory disorders	Hirsutism, obesity, and endometrial cancer are all linked to anovulatory disorders (David et al., 1994)
Oocyte factor	Weakening of oocyte first-rate inflicting a decrease being pregnant price and an expanded abortion price can be chargeable for massive age-associated lower in woman fecundity (Shoham et al., 1993)
Tubal infertility	Tubal infertility is caused by pelvic inflammatory disease (PID) caused by sexually transmitted bacteria such as gonococci, chlamydia, or other infections (David et al., 1994)
Implantation failure	The role of implantation failure as an infertility cause. The common belief is that insufficient progesterone secretion causes a nonreceptive endometrium in some women (Klentzeris et al., 1993)
Endometriosis	With the help of pelvic adhesions, altered anatomy, and ovarian or tubal damage, severe endometriosis can impair fertility (David et al., 1994)
Recurrent miscarriage	Parental chromosomal abnormalities, antiphospholipid antibodies, and uterine hollow space abnormalities are all linked to recurrent abortion. Polycystic ovaries may be the single most significant reason (Clifford and Regan, 1993)
Life style, physiological, and occupational factors	Tobacco smoking in ladies will increase the threat of infertility, the eggs of people who smoke have reduced IVF capacity. In females, occupational publicity to fabric dyes, lead, mercury, and cadmium were related to infertility (Rosevar et al., 1992)

issues of the seminal tract, immunological, psychosomatic, prior cancer treatments (chemotherapy), and various drugs like testosterone supplements, anabolic steroids, antifungals like ketoconazole, and a few antihypertensives are all factors that contribute to male infertility (Table 5.8.1).

5.8.4 Female infertility

The reasons for female infertility may also range from one topographical and community place to another. A WHO task force identified the following reasons for female infertility: tubal component 36%, ovulatory problems 33%, endometriosis 6%, and no apparent explanation 40%. In Asia, Latin America, and the Middle East, a similar pattern emerges, but in Africa, tubal infertility was the most common cause of infertility. Unprecedented infertility (considering each companion) has been documented in 8–28% of couples. Causes of infertility in females are also given in Table 5.8.1.

5.8.5 Role of oxidative stress in male infertility

Within the male duplicate system, physiological levels of ROS play an important function (Agarwal et al., 2008). Moderately higher ROS concentrations cause sperm immobility by depleting intracellular ATP and then lowering axonemal phosphorylation, whereas ROS concentrations above physiological levels cause lipid peroxidation and sperm cell death (Misro et al., 2004). Spermatozoa membranes are vulnerable to free radical damage because they are rich in polyunsaturated fatty acids and have limited antioxidant enzyme activity (Maneesh and Jayalekshmi, 2006). Furthermore, spermatogenesis in the testes is a highly energetic replicative mechanism for rapidly producing sperm. This rapid pace of cell division is accompanied with an increase in the production of free radicals as a result of an increase in mitochondrial oxygen intake through the usage of germinal epithelium (Aitken and Roman, 2008). As a result, an imbalance of free radical production and detoxification in sperm and testicular tissues causes injury to cell lipids, proteins, amino acids, sugars, nucleic acids, and mid-portions, resulting in next-to-worst semen characteristics. In male animals, poor semen characteristics are responsible for more than 80% of fertilization and embryogenesis failures, miscarriage, and infertility (Gadea and Matas, 2000).

Spermatozoa are liable to oxidative harm for some of the reasons. First, plasma membranes of sperm comprise a huge range of polyunsaturated fatty acids, compounds that might be intrinsically liable to OS. Second, the bulk of the cytoplasm of the cell is eliminated throughout spermatogenesis alongside cytoplasmic enzymes which include catalase and glutathione peroxidase, which commonly serve to scavenge loose radicals. Third, excessive ranges of ROS are related each with sperm DNA harm (Agarwal et al., 2003) and reduced capacity to restore such harm. Damaged sperm in flip produce extra ROS, thereby perpetuating this cycle.

5.8.6 Role of oxidative stress in female infertility

Free radicals are produced in the female reproductive system in a variety of ways, according to in vivo and in vitro studies. ROS are produced directly from oocytes and embryos, as well as their environment, and they mediate embryonic development strategies (Guérin et al., 2001). RNS are involved in oocyte meiotic maturation in rats, pigs (Chmelková et al., 2009), cattle (Matta et al., 2009), and sheep, in addition to ROS generation (Amale et al., 2011). In the female replica system, free radicals play two roles, one of which is ovulation (Shkolnik et al., 2011). In addition the everyday technology of free radicals in vivo, many in vitro elements reason OS and harm to replica system. For example, better ambient temperature and numerous pollution are the most important methods inducing OS. Changes in membrane properties, chromatin structure, and the meiotic spindle are among the reactions of oocytes or embryos to heat shock (Ju, 2005). Animal embryonic and larval stages are often the most vulnerable stages of the living cycle to heavy metals and pollution (Daka, 2002). There is a link between high blood and milk lead concentrations and ovarian dysfunction in cattle (Ahmed et al., 2010), and OS has been linked to the development of lead- and cadmium-related reproductive illnesses in animals (Patra et al., 2011).

Free radicals, like males, have a dual role in the pathophysiology of preeclampsia (Buhimschi et al., 1998), endometriosis (Uchiide et al., 2002), hydatid shape mole, delivery abnormalities (Williams, 2010), infertility (Celi, 2011), and abortion in the female reproductive system. Vandaele et al. (2010) discovered that short-term exposure to H_2O_2 during oocyte maturation accelerated bovine embryo development. Extra free radicals, on the other hand, had negative consequences. Furthermore, research in cattle has found that early stages of embryonic development, such as the two-cell and four-cell stage, are more susceptible to free radical-induced strain than oocytes, morulas, and blastocysts, due to the presence of more active mitochondria (Tarazona et al., 2006).

If the antioxidant device has been depleted due to excessive ROS production, female genital tract features may be altered. Oocyte maturation, steroidogenesis, and ovulation can all be affected by this. Furthermore, it has the potential to increase granulosa cell apoptosis, which is a well-known phenomena in terms of programmed cell death. In vitro studies have shown that an ovarian glutathione deficit accelerates antral follicle atresia, a condition in which antral follicles are too sensitive to OS (Mulla et al., 2018).

5.8.7 Role of antioxidant in infertility

Humans have devised a complex antioxidant defense system to protect the body's cells and organ structures from ROS (Kumar and Pandey, 2015). It consists of a variety of chemicals, both endogenous and foreign in origin, that work together to neutralize free radicals in an interactive and synergistic manner. These additives include: (a) antioxidants derived from food, such as ascorbic acid (vitamin C), tocopherols and tocotrienols (vitamin E), carotenoids, and various low molecular weight compounds, as well as glutathione and lipoic acid; (b) antioxidant enzymes, such as superoxide dismutase (SOD), glutathione peroxidase, and glutathione reductase; (c) ferrous binding proteins, including as ferritin, lactoferrin, albumin, and ceruloplasmin, which sequester loose iron and copper ions that may catalyze oxidative reactions; and (d) a variety of antioxidant phytonutrients found in a variety of plant foods (Kumar and Pandey, 2014).

While O_2 percent and H_2O_2 combine with ferric ion (Fe^{3+}), SOD inhibits the production of hydroxyl radical (HO) (Kumar et al., 2015). O'Flaherty (2014) found that the HO is highly reactive with lipids, promoting lipid peroxidation in human sperm membranes. According to one study, the best stage for keeping sperm alive is 50 units of SOD per milliliter. SOD supplementation has also been shown to increase the vitality of spermatozoa in other studies (Perumal, 2014). Catalase increases the price of sperm acrosome response and lowers the price of spermatozoa DNA fragmentation (Chi et al., 2008). Glutathione peroxidase (GPX), in combination with selenium as selenocysteine, protects cells when glutathione (GSH) is present. Within the reproductive system, GSH is the most important nonenzymatic antioxidant. After giving 600 mg of GSH per day intramuscularly in a 2-month pilot study, it was shown that patients' sperm characteristics improved (Lenzi et al., 1993). In humans, the failure to express a GPX inside the spermatozoa is linked to infertility (Imai et al., 2001). GPX becomes the leading defense against H_2O_2-mediated attack on spermatozoa and male genital tract tissues when catalase expression is compromised or absent entirely (Drevet, 2006). Peroxiredoxins (PRDXs) are the first line of defense against OS damage in spermatozoa, regulating the local movement of ROS to maintain sperm function (Manandhar et al., 2009). In human spermatozoa, they are crucial defenders against H_2O_2 and other ROS (hydroperoxides, peroxynitrite; Zini et al., 1993). Coenzyme Q10 (CoQ10), or its reduced metabolite ubiquinol, is an antioxidant that is abundant in the mitochondria of the sperm midpiece (Lewin and Lavon, 1997). CoQ10 levels in seminal plasma and spermatozoa of infertile men with idiopathic- and varicocele-related asthenospermia were shown to be lower in several studies (Balercia et al., 2002).

Carnitine is an antioxidant that accumulates explicitly in the epididymis (Tang et al., 2008) and protects the cell DNA harm brought on via way of means of free oxygen radical precipitated male infertility (Dokmeci, 2005). L-carnitine elements strength to the sperms (Ruiz-Pesini et al., 2001) via way of means of transferring and breaking or oxidizing the fatty acids into the mitochondria, ensuing in power generation. In the treatment of individuals with prostate-vesicular-epididymitis, carnitine is a third-line medication having antibacterial and anti-inflammatory properties (Vicari and Calogero, 2001). Following copper poisoning, a recent animal study discovered that L-carnitine reduced ROS production and improved sperm quality in Wistar rats (Khushboo et al., 2017). Vitamin E (α-tocopherol) is a lipid-soluble antioxidant that acts as a sequence breaker rather than a ROS scavenger (Sharma et al., 2001). Vitamin E has been shown to be effective in reducing sperm dependency and infertility in several studies (Eskenazi et al., 2005). The use of oral vitamin E was reported to improve in vitro fertilization and semen characteristics (Suleiman et al., 1996). Ascorbic acid (vitamin C) is a water-soluble antioxidant that protects spermatozoa DNA from oxidative damage (Fraga et al., 1991). It increases sperm quality and characteristics such as attentiveness, motility, and dependability. In low quantities, ascorbic acid is a strong antioxidant, but a higher dose of vitamin C may cause autooxidation (Kumar and Pandey, 2014). For 2 months, 64 infertile guys were given 1 g of vitamin C and 1 g of vitamin E, together with faster attention to unusual spermatozoa in their ejaculates, resulting in reduced DNA-fragmented spermatozoa (Greco et al., 2005). In a group of individuals, co-administering vitamin E with selenium resulted in similar results and improved sperm motility (Keskes-Ammar et al., 2003). A recent study discovered that combining vitamin E with flaxseed oil reduced ROS and improved the semen quality of cloned goats (Kargar et al., 2017).

Because of their antioxidant properties, isoflavones are phytoestrogens that include genistein and equol antioxidants and help with male fertility. Adding dose-structured amounts of genistein or equol to sperm has shown to have protective effects on sperm. Cai and Wei (1996) reported that genistein increased antioxidant interest in mice and suppressed H_2O_2 production in vitro and in vivo (Wei et al., 1993). Pentoxifylline is a methylxanthine by-product that improves the cAMP level in human sperm and functions as a sperm motility activator. In a dose-based awareness, it is a fantastic superoxide anion radical inhibitor (Gavella et al., 1991). In vitro treatment with pentoxifylline resulted in reduced ROS production in 15 patients' spermatozoa (Okada et al., 1997). It was recently discovered that inexperienced tea has the capacity to prevent OS inside the male reproductive apparatus (Roychoudhary et al., 2017). Polyphenols (catechins) and flavonoids are the two primary bioactive components in inexperienced tea. Catechins are 20 times more powerful antioxidants than vitamin C, and they help to reduce OS by quenching free radicals and chelating transition metals (Saeed et al., 2017). Catechins in green tea increase sperm awareness inside the epididymis, reduce lipid peroxidation and DNA damage, and improve enzymatic activity in the testis of mice. In individual rats, however, a higher dosage of inexperienced tea causes spermatogenesis suppression. Quercetin is a bioflavonoid with strong antioxidant effects. It boosts antioxidant enzymes while lowering NADPH oxidase and NADH-structured oxidoreductase activity (Kumar et al., 2019). Tempol is a SOD mimetic antioxidant and transforms superoxide into much less poisonous H_2O_2 (Santiani et al., 2013). Tempol protects OS from DNA breakage and lipid peroxidation caused by cryopreservation. Tempol and N-acetyl-cysteine were shown to reduce sperm motility in thawed ram spermatozoa in one study. Tempol is also used to protect cells against OS in molecular cultures techniques. When administered alone, 10 M Quercetin or 5 M Tempol increased the motility and viability of cryopreserved sperm, however adjunct therapy revealed no significant changes in semen parameters (Azadi et al., 2017).

Conclusion

OS has been suggested to be one of the important causes of infertility in males and females affecting varying degrees of sperm dysfunction and adversely affects fertilization, implantation and embryo viability. Antioxidant supplementation of natural origin (phenolic compounds and vitamins) has shown marked effect in reducing infertility in both the sexes.

Acknowledgments

H.K.R. acknowledges UGC New Delhi for providing financial support in the form of UGC-CRET fellowship. A.K.S. acknowledges the CSIR New Delhi for providing Senior Research Fellowship. All the authors gratefully acknowledge DST-FIST and UGC-SAP facilities of the Department of Biochemistry, University of Allahabad, India.

References

Agarwal, A., Aponte-Mellado, A., Premkumar, B.J., Shaman, A., Gupta, S., 2012. The effects of oxidative stress on female reproduction: a review. Reprod. Biol. Endocrinol. 10, 49.

Agarwall, A., Gupta, S., Sharma, R., 2005. Oxidative stress and its implications in female infertility – a clinician's perspective. Reprod. BioMed. 11 (5), 641–650.

Agarwal, A., Makker, K., Sharma, R., 2008. Clinical relevance of oxidative stress in male factor infertility: an update. Am. J. Reprod. Immunol. 59, 2–11.

Agarwal, A., Mulgund, A., Hamada, A., Chyatte, M.R., 2015. A unique view on male infertility around the globe. Reprod. Biol. Endocrinol. 13, 37.

Agarwal, A., Saleh, R.A., Bedaiwy, M.A., 2003. Role of reactive oxygen species in the pathophysiology of human reproduction. Fertil. Steril. 79, 829–843.

Ahmed, W.M., Zaabal, M.M., Hameed, A.R.A.E., 2010. Relationship between ovarian activity and blood lead concentration in cows and buffaloes with emphases on gene frequencies of hemoglobin. Global J. Biotechnol. Biochem. 5, 1–5.

Aitken, A.J., Roman, S.D., 2008. Antioxidant systems and oxidative stress in the testes. Oxida. Med. Cell Longev. 1, 15–24.

Aitken, R., Paterson, M., Fischer, H., Buckingham, D.W., van Duin, M., 1995. Redox regulation of tyrosine phosphorylation in human spermatozoa and its role on the control of human sperm function. J. Cell Sci. 108, 2017–2025.

Aitken, R.J., Clarkson, J.S., Fishel, S., 1989. Generation of reactive oxygen species, lipid peroxidation, and human sperm function. Biol. Reprod. 41, 183–197.

Alvarez, C., Castilla, J.A., Martinez, L., Ramirez, J.P., Vergara, F., Gaforio, J.J., 2003. Biological variation of seminal parameters in healthy subjects. Hum. Reprod. 18, 2082–2088.

Alvarez, J.G., Touchstone, J.C., Blasco, L., Storey, B.T., 1987. Spontaneous lipid peroxidation and production of hydrogen peroxide and superoxide in human spermatozoa: superoxide dismutase as major enzyme protectant against oxygen toxicity. J. Androl. 8, 338–348.

Amale, M.H., Shahne, A.Z., Abavisani, A., Nasrollahi, S., 2011. Effects of inhibiting nitric oxide synthase on cumulus expansion and nuclear maturation of sheep oocytes. Czech J. Anim. Sci. 56, 284–291.

Azadi, L., Tavalaee, M., Deemeh, M.R., Arbabian, M., Nasr-Esfahani, M.H., 2017. Effects of tempol and quercetin on human sperm function after cryopreservation. Cryo Lett. 38 (1), 29–36.

Balercia, G., Arnaldi, G., Fazioli, F., Serresi, M., Alleva, R., Mancini, A., Mosca, F., Lamonica, G.R., Mantero, F., Littarru, G.P., 2002. Coenzyme Q10 levels in idiopathic and varicocele-associated asthenozoospermia. Andrologia 34 (2), 107–111.

Bonde, J.P., Ernst, E., Jensen, T.K., Hjollund, N.H., Kolstad, H., Henriksen, T.B., Scheike, T., Giwercman, A., Olsen, J., Skakkebaek, N.E., 1998. Relation between semen quality and fertility: a population-based study of 430 first-pregnancy planners. Lancet 352, 1172–1177.

Buhimschi, I.A., Saade, G.R., Chwalisz, K., Garfield, R.E., 1998. The nitric oxide pathway in pre-eclampsia: pathophysiological implications. Hum. Reprod. Update 4, 25–42.

Cai, Q., Wei, H., 1996. Effect of dietary genistein on antioxidant enzyme activities in SENCAR mice. Nutr. Cancer 25 (1), 1–7.

Celi, P., 2011. Biomarkers of oxidative stress in ruminant medicine. Immunopharmacol. Immunotoxicol. 33, 233–240.

Chi, H.J., Kim, J.H., Ryu, C.S., Lee, J.Y., Park, J.S., Chung, D.Y., Choi, S.Y., Kim, M.H., Chun, E.K., Roh, S.I., 2008. Protective effect of antioxidant supplementation in sperm-preparation medium against oxidative stress in human spermatozoa. Hum. Reprod. 23 (5), 1023–1028.

Chmelíková, E., Sedmíková, M., Petr, J., Kott, T., Lánská, V., Tùmová, L., Tichovská, H., Ješeta, M., 2009. Expression and localization of nitric oxide synthase isoforms during porcine oocyte growth and acquisition of meiotic competence. Czech J. Anim. Sci. 54, 137–149.

Clifford, K.A., Regan, L., 1993. A systematic approach to the clinical investigation of recurrent miscarriage: experience of 500 cases. Hum. Reprod. 8 (suppl 1), 388 (abstr).

Daka, E.R., 2002. Reproduction effects of heavy metals on the rough periwinkle *Littorina saxatilis* (Mollusca: Prosobranchia). Indian J. Mainer Sci. 31, 218–224.

David, L.H., Trounson, A.O., Andersen, A.N., 1994. Female infertility: Causes and Treatment, The lancet 343, 1539–1544.

Dohle, G.R., Halley, D.J., Van Hemel, J.O., van den Ouwel, A.M., Pieters, M.H., Weber, R.F., Govaerts, L.C., 2002. Genetic risk factors in infertile men with severe oligozoospermia and azoospermia. Hum. Reprod. 17, 13–16.

Dokmeci, D., 2005. Oxidative stress, male infertility and the role of carnitines. Folia Med. (Plovdiv) 47 (1), 26–30.

Drevet, J.R., 2006. The antioxidant glutathione peroxidase family and spermatozoa: a complex story. Mol. Cell. Endocrinol. 250 (1–2), 70–79.

Eskenazi, B., Kidd, S.A., Marks, A.R., Sloter, E., Block, G., Wyrobek, A.J., 2005. Antioxidant intake is associated with semen quality in healthy men. Hum. Reprod. 20 (4), 1006–1012.

Fraga, C.G., Motchnik, P.A., Shigenaga, M.K., Helbock, H.J., Jacob, R.A., Ames, B.N., 1991. Ascorbic acid protects against endogenous oxidative DNA damage in human sperm. Proc. Natl. Acad. Sci. USA 88, 11003–11006.

Gadea, J., Matas, C., 2000. Sperm factors related to in vitro penetration of porcine oocytes. Theriogenology 54, 1343–1357.

Gavella, M., Lipovac, V., Marotti, T., 1991. Effect of pentoxifylline on superoxide anion production by human sperm. Int. J. Androl. 14 (5), 320–327.

Greco, E., Iacobelli, M., Rienzi, L., Ubaldi, F., Ferrero, S., Tesarik, J., 2005. Reduction of the incidence of sperm DNA fragmentation by oral antioxidant treatment. J. Androl. 26 (3), 349–353.

Griveau, J.F., Le Lannou, D., 1997. Reactive oxygen species and human spermatozoa: physiology and pathology. Int. J. Androl. 20, 61–69.

Guérin, G.E.I., Mouatassim, S., Ménézo, Y., 2001. Oxidative stress and protection against reactive oxygen species in the pro-implantation embryo and its surroundings. Hum. Reprod. Update 7, 175–189.

Guzick, D.S., Overstreet, J.W., Factor-Litvak, P., Brazil, C.K., Nakajima, S.T., Coutifaris, C., Carson, S.A., Cisneros, P., Steinkampf, M.P., Hill, J.A., Xu, D., Vogel, D.L., 2001. Sperm morphology, motility and concentration in fertile and infertile men. New Engl. J. Med. 345, 1388–1393.

Hughes, C., Lewis, S., McKelvey-Martin, V., Thompson, W., 1998. The effects of antioxidant supplementation during Percoll preparation on human sperm DNA integrity. Hum. Reprod. 13, 1240–1247.

Imai, H., Suzuki, K., Ishizaka, K., Ichinose, S., Oshima, H., Okayasu, I., Emoto, K., Umeda, M., Nakagawa, Y., 2001. Failure of the expression of phospholipid hydroperoxide glutathione peroxidase in the spermatozoa of human infertile males. Biol. Reprod. 64 (2), 674–683.

Ju, J.C., 2005. Cellular responses of oocytes and embryos under thermal stress: hints to molecular signaling. Anim. Reprod. 2, 79–90.

Kamischke, A., Nieschlag, E., 1999. Analysis of medical treatment of male infertility. Hum. Reprod. 14 (suppl. 1), 1–23.

Kargar, R., Forouzanfar, M., Ghalamkari, G., Nasr Esfahani, M.H., 2017. Dietary flax seed oil and/or vitamin E improve sperm parameters of cloned goats following freezing-thawing. Cryobiology 74, 110–114.

Kefer, J.C., Agarwal, A., Sabanegh, E., 2009. Role of antioxidants in the treatment of male infertility. Int. J. Urol. 16, 449–457.

Keskes-Ammar, L., Feki-Chakroun, N., Rebai, T., Sahnoun, Z., Ghozzi, H., Hammami, S., Zghal, K., Fki, H., Damak, J., Bahloul, A., 2003. Sperm oxidative stress and the effect of an oral vitamin E and selenium supplement on semen quality in infertile men. Arch. Androl. 49 (2), 83–94.

Khushboo, M., Murthy, M.K., Devi, M.S., Sanjeev, S., Ibrahim, K.S., Kumar, N.S., Roy, V.K., Gurusubramanian, G., 2017. Testicular toxicity and sperm quality following copper exposure in Wistar albino rats: ameliorative potentials of L-carnitine. Environ. Sci. Pollut. Res. 25 (2), 1837–1862.

Klentzeris, L.D., Bulmer, J.N., Trejdosiewicz, L.K., Morrison, L., Cooke, I.D., 1993. Beta-1 integrin cell adhesion molecules in the endometrium of fertile and infertile women. Hum. Reprod. 8, 1223–1230.

Kodama, H., Yamaguchi, R., Fukuda, J., Kasai, H., Tanaka, T., 1997. Increased oxidative deoxyribonucleic acid damage in the spermatozoa of infertile male patients. Fertil. Steril. 68, 519–524.

Kumar, R., Singh, A.K., Gupta, A., Bishayee, A., Pandey, A.K., 2019. Therapeutic potential of *Aloe vera*--a miracle gift of nature. Phytomedicine 2019, 152996. doi:10.1016/j.phymed.2019.152996

Kumar, S., Dwivedi, A., Kumar, R., Pandey, A.K., 2015. Preliminary evaluation of biological activities and phytochemical analysis of *Syngonium podophyllum* leaf. Natl. Acad. Sci. Lett. 38 (2), 143–146.

Kumar, S., Pandey, A.K., 2014. Pharmacological activities of some common Indian weeds: a review. Mintage J. Pharm. Med. Sci. 3, 12–17.

Kumar, S., Pandey, A.K., 2015. Free radicals: health implications and their mitigation by herbals. Br. J. Med. Med. Res. 7 (6), 438–457.

Lenzi, A., Culasso, F., Gandini, L., Lombardo, F., Dondero, F., 1993. Placebo-controlled, double-blind, cross-over trial of glutathione therapy in male infertility. Hum. Reprod. 8 (10), 1657–1662.

Lewin, A., Lavon, H., 1997. The effect of coenzyme Q10 on sperm motility and function. Mol. Aspects Med. 18 (Suppl.), S213–S219.

Lewis, S., Agbaje, I., 2008. Using the alkaline comet assay in prognostic tests for male infertility and assisted reproductive technology outcomes. Mutagenesis 23, 163–170.

MacLeod, J., 1943. The role of oxygen in the metabolism and motility of human spermatozoa. Am. J. Physiol. 138, 512–518.

Manandhar, G., Miranda-Vizuete, A., Pedrajas, J.R., Krause, W.J., Zimmerman, S., Sutovsky, M., Sutovsky, P., 2009. Peroxiredoxin 2 and peroxidase enzymatic activity of mammalian spermatozoa. Biol. Reprod. 80 (6), 1168–1177.

Maneesh, M., Jayalekshmi, H., 2006. Role of reactive oxygen species and antioxidants on pathophysiology of male reproduction. Indian J. Clin. Biochem. 21, 80–89.

March, M.R., Isidori, A., 2002. New frontiers in the treatment of male sterility. Contraception 65, 279–281.

Mascarenhas, M.N., Flaxman, S.R., Boerma, T., Vanderpoel, S., Stevens, G.A., 2012. National, regional, and global trends in infertility prevalence since 1990: a systematic analysis of 277 health surveys. PLoS Med. 9 (12), e1001356.

Matta, S.G., Caldas-Bussiere, M.C., Viana, K.S., Faes, M.R., Paes de Carvalho, C.S., Dias, B.L., Quirino, C.R., 2009. Effect of inhibition of synthesis of inducible nitric oxide synthase-derived nitric oxide by aminoguanidine on the maturation of oocyte-cumulus complexes of cattle. Anim. Reprod. Sci. 111, 189–201.

Misro, M.M., Choudhury, L., Upreti, K., Gautam, D., Chaki, S.P., Mahajan, A.S., Babbar, R., 2004. Use of hydrogen peroxide to assess the sperm susceptibility to oxidative stress in subjects presenting a normal semen profile. Int. J. Androl. 27, 82–87.

Mulla, A.A., Fazari, A.B.E., Elkhouly, M., Moghaddam, N., 2018. Role of antioxidants in female fertility. Open J. Obstet. Gynecol. 8, 85–91.

O'Flaherty, C., 2014. The enzymatic antioxidant system of human spermatozoa. Adv. Androl. 15, 626374.

Okada, H., Tatsumi, N., Kanzaki, M., Fujisawa, M., Arakawa, S., Kamidono, S., 1997. Formation of reactive oxygen species by spermatozoa from asthenospermic patients: response to treatment with pentoxifylline. J. Urol. 157 (6), 2140–2146.

Patra, R.C., Rautray, A.K., Swarup, D., 2011. Oxidative stress in lead and cadmium toxicity and its amelioration. Vet. Med. Int. 9.

Perumal, P., 2014. Effect of superoxide dismutase on semen parameters and antioxidant enzyme activities of liquid stored (5°C) mithun (Bos frontalis) semen. J. Anim. 9.

Quinn, T.C., Gaydos, C., Shepherd, M., Bobo, L., Hook, E.W., Viscidi, R., Rompalo, A., 1996. Epidemiologic and microbiologic correlates of *Chlamydia trachomatis* infection in sexual partnerships. J. Am. Med. Assoc. 276, 1737–1742.

Rosevar, S.K., Holt, D.W., Lee, T.D., Ford, C.L, Wardle, P.G., Hull, M.G.R., 1992. Smoking and decreased fertilization rates in vitro. Lancet 340, 1195–1196.

Roychoudhury, S., Agarwal, A., Virk, G., Cho, CL., 2017. Potential role of green tea catechins in the management of oxidative stress-associated infertility. Reprod. Biomed. Online 34 (5), 487–498.

Ruder, E.H., Hartman, T.J., Blumberg, J., Goldman, M.B., 2008. Oxidative stress and antioxidants: exposure and impact on female fertility. Hum. Reprod. Update 14, 345–357.

Ruiz-Pesini, E., Alvarez, E., Enríquez, J.A., López-Pérez, M.J., 2001. Association between seminal plasma carnitine and sperm mitochondrial enzymatic activities. Int. J. Androl. 24 (6), 335–340.

Saeed, M., Naveed, M., Arif, M., Kakar, M.U., Manzoor, R., Abd El-Hack, M.E., Alagawany, M., Tiwari, R., Khandia, R., Munjal, A., Karthik, K., Dhama, K., Iqbal, H.M.N., Dadar, M., Sun, C., 2017. Green tea (*Camellia sinensis*) and l-theanine: medicinal values and beneficial applications in humans—a comprehensive review. Biomed. Pharmacother. 95, 1260–1275.

Santiani, A., Evangelista, S., Valdivia, M., Risopatrón, J., Sánchez, R., 2013. Effect of the addition of two superoxide dismutase analogues (Tempo and Tempol) to alpaca semen extender for cryopreservation. Theriogenology 79 (5), 842–846.

Sharlip, I.D., Jarow, J.P., Belker, A.M., Lipshultz, L.I., Sigman, M., Thomas, A.J., Schlegel, P.N., Howards, S.S., Nehra, A., Damewood, M.D., Overstreet, J.W., Sadovsky, R., 2002. Best practice policies for male infertility. Fertil. Steril. 77, 873–882.

Sharma, R.K., Pasqualotto, F.F., Nelson, D.R., Thomas Jr., A.J., Agarwal, A., 1999. The reactive oxygen species-total antioxidant capacity score is a new measure of oxidative stress to predict male infertility. Hum. Reprod. 14, 2801–2807.

Sharma, R.K., Pasqualotto, A.E., Nelson, D.R., Thomas Jr., A.J., Agarwal, A., 2001. Relationship between seminal white blood cell counts and oxidative stress in men treated at an infertility clinic. J. Androl. 22 (4), 575–583.

Shkolnik, K., Tadmor, A., Ben-Dor, S., Nevo, N., Galiani, D., Dekel, N., 2011. Reactive oxygen species are indispensable in ovulation. Proc. Natl Acad. Sci. 108, 1462–1467.

Shoham, Z., Jacobs, H.S., Insler, V., 1993. Luteinizing hormone: its role, mechanism of action, and detrimental effects when hypersecreted during the follicular phase. Fertil. Steril. 59, 1153.

Shulman, S., 1972. Immunologic barriers to fertility. Obst. Gynec. Survey 27, 553.

Silver, E.W., Eskenazi, B., Evenson, D.P., Block, G., Young, S., Wyrobek, A.J., 2005. Effect of antioxidant intake on sperm chromatin stability in healthy nonsmoking men. J. Androl. 26, 550–556.

Suleiman, S.A., Ali, M.E., Zaki, Z.M., el-Malik, E.M., Nasr, M.A., 1996. Lipid peroxidation and human sperm motility: protective role of vitamin E. J. Androl. 17 (5), 530–537.

Tang, L.F., Jiang, H., Shang, X.J., Zhao, L.M., Bai, Q., Hong, K., Liu, D.F., Liu, J.M., Yuan, R.P., Chen, Q., Ma, LL, 2008. Seminal plasma levocarnitine significantly correlated with semen quality. Zhonghua Nan Ke Xue 14 (8), 704–708.

Tarazona, A.M., Rodriguez, J.I., Restrepo, L.F., Olivera-Angel, M., 2006. Mitochondrial activity, distribution and segregation in bovine oocytes and in embryos produced in vitro. Reprod. Domestic Anim. 41, 5–11.

Uchiide, I., Ihara, T., Sugamata, M., 2002. Pathological evaluation of the rat endometriosis model. Fertil. Steril. 78, 782–786.

Urry, R.L., 1977. Stress and infertility, Male Infertility. Gruner & Stratton, New York, pp. 145–162.

Vandaele, L., Thys, M., Bijttebier, J., Langendonckt, A.V., Donnay, I., Maes, D., Meyer, E., Soom, A.V., 2010. Short-term exposure to hydrogen peroxide during oocyte maturation improves bovine embryo development. Reproduction 139, 505–511.

Vicari, E., Calogero, A.E., 2011. Effects of treatment with carnitines in infertile patients with prostato-vesiculo epididymitis. Hum. Reprod. 16 (11), 2338–2342.

Wei, H., Wei, L., Frenkel, K., Bowen, R., Barnes, S., 1993. Inhibition of tumor promoter-induced hydrogen peroxide formation in vitro and in vivo by genistein. Nutr. Cancer 20 (1), 1–12.

Williams, D.L., 2010. Congenital abnormalities in production animals. Vet. Clin.: Food Anim. 26, 477–486.

Young, S.S., Eskenazi, B., Marchetti, F.M., Block, G., Wyrobek, A.J., 2008. The association of folate, zinc, and antioxidant intake with sperm aneuploidy in healthy non-smoking men. Hum. Reprod. 23, 1014–1022.

Zegers-Hochschild, F., Adamson, G.D., Dyer, S., Racowsky, C., de Mouzon, J., Sokol, R., Rienzi, L., Sunde, A., Schmidt, L., Cooke, I.D., Simpson, J.L., van der Poel, S., 2017. The international glossary on infertility and fertility care. Fertil. Steril. 108 (3), 393–406.

Zini, A., de Lamirande, C.E., 1993. Gagnon, reactive oxygen species in semen of infertile patients: levels of superoxide dismutase- and catalase-like activities in seminal plasma and spermatozoa. Int. J. Androl. 16 (3), 183–188.

CHAPTER

Antioxidants and kidney diseases

5.9

Roberta Jeane Bezerra Jorge[a], Aline Diogo Marinho[a], João Alison de Moraes Silveira[a], Márcia Maria Vieira Ramos[b], Jacqueline Ramos Machado Braga[c], Renata de Sousa Alves[d], Francisco Assis Nogueira-Junior[a], Mirele da Silveira Vasconcelos[e], Ana Sanches Silva[f,g], Seyed Mohammad Nabavi[h], Dirce Fernandes de Melo[i]

[a]*Department of Physiology and Pharmacology, Drug Research and Development Center (NPDM), Federal University of Ceará (UFC), Fortaleza, Ceará, Brazil*

[b]*University Center Estácio of Ceará, Fortaleza, Ceará, Brazil*

[c]*Federal University of Recôncavo of Bahia (UFRB), Cruz das Almas, Bahia, Brazil*

[d]*Department of Pharmacy and Clinical Analysis, Federal University of Ceará (UFC), Fortaleza, Ceará, Brazil*

[e]*Federal Institute of Education, Science and Technology of Ceará (IFCE), Baturité, Ceará, Brazil*

[f]*National Institute for Agricultural and Veterinary Research (INIAV), I.P., Vairão, Vila do Conde, Portugal*

[g]*Center for Study in Animal Science (CECA), ICETA University of Oporto, Oporto, Portugal*

[h]*Applied Biotechnology Research Center, Baqiyatallah University of Medical Sciences, Tehran, Iran*

[i]*Department of Biochemistry and Molecular Biology (DBBM), Federal University of Ceará (UFC), Fortaleza, Ceará, Brazil*

5.9.1 Introduction

Acute and chronic kidney diseases (CKDs) occur partly due to the imbalance between the molecular mechanisms that leads to oxidative stress (OS), inflammation, autophagy, and cell death (Sureshbabu et al., 2015).

Under physiological conditions, there is a balance between the pro-oxidant and antioxidant defense systems (Ozbec, 2012). Oxidative compounds are produced as part of tissue repair, inflammation, and defense reactions (Modaresi et al., 2015). However, a chronic activation process of them can contribute to the tissue injury creating pathological conditions (Locatelli et al., 2003).

OS results from the cumulative effects of highly reactive oxidizing molecules which lead to biomolecules oxidation (lipids, proteins, complex carbohydrates, or nucleic acids), and promoting biological functions loss and homeostatic imbalance, resulting in oxidative damage against cells and tissues (Gyurászová et al., 2020; Sureshbabu et al., 2015). Thus, OS can be defined as an imbalance state generated by

Antioxidants Effects in Health. DOI: https://doi.org/10.1016/B978-0-12-819096-8.00059-8

excessive production of highly reactive oxygen and nitrogen species (ROS and RNS) as well as their insufficient removal (Daenen et al., 2019).

It is well known that OS and autophagy are involved in health or kidney disease. Autophagic flow changes can regulate ROS production and redox signaling which also act as autophagy-inducing modulators (Sureshbabu et al., 2015; Scherz-Shouval and Elazar, 2007). Kidney disease OS has been reported as ROS increased production in detriment of antioxidants depletion (Daenen et al., 2019). Indeed, the kidney organ has intense metabolism since the proximal renal tubules demand a great energy quantity for water and solutes reabsorption requiring a large mitochondria number which are the major cellular producers of ROS (Gyurásová et al., 2020). Besides, under pathological conditions, the mitochondrial membrane loses its integrity, excessively producing reactive oxygen species (superoxide, peroxynitrite and nitric oxide) that trigger mitochondrial permeability changes and release proapoptotic proteins as cytochrome *C*-inducing cell death (Barnett and Cummings, 2019).

Another kidney feature is the long chain polyunsaturated fatty acids abundance in the composition of renal lipids, which are potent generators of alkoxyl (LO •) and peroxyl (LO2 •) free radicals (Daenen et al., 2019). These characteristics increase renal vulnerability to OS-induced effects, accelerating damage progression (Eirin et al., 2016; Himmelfarb, 2005). Advanced stages of CKD involve inflammation process and cardiovascular alteration (Popolo et al., 2013; Locatelli et al., 2003).

Once kidney has been failed there are three important clinical managements: hemodialysis (HD), peritoneal dialysis, or transplantation. Several interventions have been suggested to ameliorate OS in renal patients such as exogenous administration of antioxidants (synthetic or from natural sources) to prevent OS in kidney disease treatment to mitigate damages in this organ (Chen and Siriki, 2015; Daenen et al., 2019). Natural compounds due their antioxidant, anti-inflammatory, and antiapoptotic activities could be explored as drug candidates as nephroprotectors (Kpemissi et al., 2019; Liakopoulos et al., 2019). Evidences have shown that antioxidant supplementation associated with selenium, zinc, vitamin E, flavonol, polyphenols, and L-arginine, promote renal and endothelial functions as well as cardioprotective effects (Daenen et al., 2019). On the other hand, it has been shown conflicting results involving antioxidant use. Nonetheless, this integrating approach stimulates the understanding search concerning this subject (Tamadon et al., 2015).

The focus of the present chapter is to highlight the potential role of synthetic drugs and pharmacological agents isolated from natural compounds used as antioxidants in renal disease treatment revealed through in vitro, in vivo, and clinical trial studies.

5.9.2 Kidney diseases

According to the International Society of Nephrology 850 million people worldwide have kidney diseases of which about 5 million people experience kidney failure. These people require dialysis or a kidney transplant to survive, because kidney is extremely sensitive to hypoxia and ischemia (ISN, 2020).

Renal failure occurs when the kidney loses the ability to filter waste, salts, and fluids from the blood. Diseases such as diabetes and hypertension promoting protein loss involving oxidative/inflammatory pathways and damage to small vessels, arteriolitis, and nephrosclerosis, respectively. A self-perpetuating cycle of these events can contribute to progressive renal deterioration until irreversible CKD that needs a kidney replacement therapy as HD or transplantation (Amorim et al., 2019; Yang et al., 2020).

Kidney diseases may originate from renal glomerular or tubular cell injury. Glomerular diseases such as nephrotic or nephritic syndrome are characterized by proteinuria ≥3 g/day. They are classified according to urinary pattern changes that involve fatty cylinders, fatty oval bodies, and hematuria, usually with dysmorphic cells or erythrocyte cylinders. They may have an exogenous cause such as poststreptococcal glomerulonephritis or autoimmune one such as lupus nephritis (Modaresi et al., 2015).

The renal disease has been classified as two distinct syndromes, known as acute kidney disease (AKD) and CKDs according Kidney Disease Improving Global Outcomes. Therefore, in accordance with Kidney Disease Improving Global Outcomes standards: AKD has glomerular filtration rate (GFR) <60 mL/min/1.73 m^2 or markers of kidney damage for ≤3 months and acute kidney injury (AKI) which is a subcategory of AKD: oliguria for >6 h, rise in serum creatinine levels (SCr) by >0.3 mg/dL in 2 days or by >50% in 1 week. AKI is reserved for duration ≤3 months (Levey et al., 2020). Some diseases show increased circulating proinflammatory biomarkers, pro-oxidative mediators, and cytokines due to protein and lipid oxidation, showing correlation with other ones as AKI or CKD (Dennis and Witting, 2017; Modaresi et al., 2015).

However, studies suggest that the two syndromes are not distinct entities but rather are closely interconnected being a risk factor for each one and both are risk factor for cardiovascular disease (Chawla et al., 2014). They are delineated according to the serum creatinine concentration or the GFR.

Kidney failure could be further classified according to patient-reported outcomes (symptoms) while CKD (duration >3 months) is a general term for heterogeneous disorders affecting kidney structure and function. GFR could be directly determined or estimated using equations. Another criterion for definition of CKD is the albumin to creatinine ratio >2000 mg/g in urine accompanied by signs and symptoms of nephrotic syndrome such as low serum albumin, edema, and high serum cholesterol (Levey and Coresh, 2012).

Kidney biopsy samples can show CKD definitive evidence, through common changes such as glomerular sclerosis, tubular atrophy, and interstitial fibrosis (Webster et al., 2016) or more recently by bioinformatic analysis (Wang et al., 2020).

Several tests are utilized from kidney disease diagnostics and an ideal biomarker must be noninvasive, sensitive, specific, reliably to kidney disease response, it means, corresponding to kidney injury. Indeed, it must be applicable in different populations identifying the injury mechanisms such as prerenal, intrarenal, postrenal, if possible (Wasung et al., 2015). Table 5.9.1 describes the most important classical biomarkers as well as the new ones.

Table 5.9.1 Classical and new renal dysfunction biomarkers.

	Biomarkers	References
Classical	Urea, creatinine, GFR estimates utilizing serum creatinine, cystatin C, proteinuria, microalbuminuria, dysmorphic red blood cells, low molecular weight proteins like β2-microglobulin or beta trace protein, urine microscopy with evaluation of the urine sediment	Perazella, 2015 Sodré et al., 2007 Wasung et al., 2015
New	Endothelial markers—asymmetric dimethylarginine (ADMA) and symmetric dimethylarginine (SDMA)—liver-type fatty acid-binding protein (L-FABP) and interleukin-18 (IL-18); fibroblast growth factors-23; tubular proteins; uromodulin; neutrophil gelatinase-associated lipocalin (NGAL) and kidney injury molecule-1 (KIM-1); N-acetyl-β-D-glucosaminidase (NAG) and microRNAs (miRNAs) or metabolomic studies; connective tissue growth factor; collagen IV; podocalyxin and nephrin	Wasung et al., 2015 Georgea and Goundenb, 2019 Hu et al., 2020 Veiga et al., 2020

It is clear that experimental models of renal ischemia and/or nephrotoxicity reveal increased oxidative damage as well as decreased tissue antioxidant status. In this scenario, it would be useful to administrate antioxidant vitamins for ROS removing as well as damage preventing in tubular cells and renal tissue.

5.9.3 Natural antioxidants and kidney diseases

Natural compounds such as carotenoids, alkaloids, flavonoids among others derived from herbs or medicinal plants have gained significant attention from the scientific community as potent therapeutic and preventive agents applied in kidney disease, considering their antioxidant inhibitory effect on several oxidizing molecular mechanisms (Kanlaya and Thongboonkerd, 2019; Yaribeygi et al., 2018a).

Green tea and olive oil improved kidney function when tested in clinical trials. The renoprotective action of antioxidants crocin, quercetin, berberine, ginger, cinnamon, and garlic was also demonstrated in experimental assays, however, the absence of clinical studies of them is still a limiting factor for their therapeutic use (Yaribeygi et al., 2018a).

The renoprotective and therapeutic antioxidant effects could be evidenced by several mechanisms such as the crocin that reduce IL-18, Nox-4 and p53 and prevent DN (Yaribeygi et al., 2018b; Altinoz et al., 2015), astaxanthin (ATX) that improves mitochondrial function (Jiang et al., 2020), berberine that delays the progression of DN by improving the thickening of the glomerular basement membrane (Yaribeygi et al., 2018a; Wu et al., 2012), quercetin that mitigates renal OS in DN (Anjaneyulu,

2004; Tong et al., 2017) and cinnamaldehyde that is capable of to reduce OS in kidney tissues (Gowder and Devaraj, 2006; Zheng et al., 2011).

Leaves, roots, and seeds extracts could present renoprotective activities. Particularly, garlic extract improves renal function and prevents diabetic complications (Ahmad and Ahmed, 2006; Atkin et al., 2016). The ginger (*Zingiber officinale*) already prevents diabetes-induced nephropathy through AMPK-dependent pathways (Tzeng et al., 2013).

The nephroprotective activity of lotus seed extracts has already been demonstrated in vivo and in vitro models (Moghaddam et al., 2012). Other extract originally from mint leaves (*Mentha aquatic* L.) exert antidiabetic and nephroprotective activities (Konda et al., 2020). Mangosteen xanthones (*Garcinia mangostana* L.) protect lead-induced CKD by activating Nrf-2 which modulate NF-κB via MAPK (Rana et al., 2019). Furthermore, the bioactive IF peptide of the potato protein hydrolysate exerts a protective effect against ROS-mediated kidney damage coupled to hypertension (Tsai et al., 2020).

Nonetheless, considering that animal models cannot adequately replace human ones, the most considerable difficulty is the scarcity of randomized controlled trials in humans.

Green tea leaves are rich in polyphenolic compounds known as "catechins" that have potent antioxidant activity in several human tissues (Tavafi, 2014). Green tea polyphenols can prevent OS disorders as well as diabetic side effects (Mustata et al., 2005; Sabu et al, 2002; Tavafi, 2014). It is known that olive oil has a potent renoprotective effect (Hoile et al., 2014). Thus, the Mediterranean diet, rich in olive oil consumption, improves kidney function preventing DN (Chauveau et al., 2017) and diabetic retinopathy (Díaz-López et al., 2018).

The development of nanoparticles as drug delivery molecules appears as new perspective for kidney diseases treatment. It is known that gold nanoparticles are carriers highly effective from the antioxidative properties of *Ficus carica* L. extract transport which reduce oxidative toxicity induced by cisplatin (El-Sayed et al., 2019).

Natural products can be used as a complementary therapy in kidney injury treatment associated with allopathic therapies. It is known that homeopathy is an alternative and effective method for treating various diseases with low cost. However, it is necessary to know the risks, interactions, and toxicity of medicinal plants (Kuba and Vattimo, 2015).

5.9.4 Drugs or isolated natural antioxidant potential products in kidney diseases

As it is well known, an antioxidant might be defined as any substance that delays, prevents, or removes oxidative damage to a target molecule (Daenen et al., 2019; Su et al., 2019). The OS is a pathological mechanism shared by many diseases. Briefly, it can involve disturbance on endogenous antioxidants enzymes such as catalase, superoxide dismutase, glutathione peroxidase, peroxiredoxin, epoxide hydrolase, glicose-6-phosphate dehydrogenase (G6PD), glutathione-S-transferase, and glutathione reductase as well as nicotinamide adenine dinucleotide phosphate cofactor. Antioxidants and metal chelators also offer resistance against ROS and RNS (e.g., uric acid, glutathione, α-tocopherol, ascorbic acid, β-carotene, coenzyme Q10, dihydrolipoic acid, melatonin, bilirubin, albumin, transferrin, ceruloplasmin, lactoferrin; Su et al., 2019).

5.9.4.1 In vitro studies

The experimental in vitro assays using exogenous compounds (xenobiotics), or in hypoxia/ischemia conditions, among others, for simulate alterations on these systems aiming to understand the specific cellular targets as well as molecular biomarkers to the mechanism of protection in the prospection of new antioxidant agents. The OS caused by exposure kidney cells to these situations, depending upon intensity and/or length, induces inflammation, which increase proinflammatory cytokines and chemokines (both processes are deeply interrelated in kidney diseases), DNA damage, mitochondrial dysfunction, ATP depletion, alteration of calcium homeostasis, and leads to cell death (Xu et al., 2015; Kandemir et al., 2018; Vera et al., 2018). The mitochondria are the major common sources of ROS, but there are others ROS sites generation that include the endoplasmic reticulum, peroxisomes, and lysosomes (Zhang et al., 2018; Samiei et al., 2019). Since kidney is rich in mitochondria it means that it is highly vulnerable to damage caused by OS.

The administration of antioxidants, in both acute and chronic kidney injuries (CKIs) models, has shown therapeutic potential, acting on several pathways that increase success chances. The current knowledge reveals vitamin E as promising agent against OS since it presents high affinity for unpaired electrons. In addition, it could be incorporated into the phospholipid bilayer of the cell membrane avoiding lipid peroxidation chain reactions. Vitamin E also suppresses ferroptosis, ω-3 fatty acids displace arachidonic acid in the cell membrane decreasing arachidonic-derived ROS. The inflammation process is significantly reduced. Indeed, the cysteine residue of N-acetylcysteine (NAC) is a precursor for glutathione synthesis and the thiol group scavenge ROS directly. Allopurinol as an antioxidant on renal disorders can decrease OS through inhibits xanthine oxidase derived ROS (Kajarabille and Latunde-Dada, 2019; Small et al., 2012).

Nicotinamide adenine dinucleotide phosphate oxidases (Nox) are enzymatic sources of ROS production, which are coupled to electrons transfer across biological membranes. This process is associated with kidney diseases, such as diabetic nephropathy, Iodinated contrast medium-induced AKI, hypoxia, ischemia, and renal tumorigenesis (He et al., 2016; Jha et al., 2016; Jeong et al., 2018; Tsuchiya et al., 2018). It is known that ROS trigger apoptosis by multiple mechanisms, such as phosphorylation suppress of the anti-apoptotic B-cell lymphoma-2 (Bcl-2) protein as well as mitochondrial membrane potential loss (ΔΨm). The potential membrane rupture leads to mitochondrial permeability transition releasing the pro-apoptotic factor cytochrome *C* (CytC). Several studies show that ROS and RNS regulate cell signaling involved in programmed cell death by apoptosis and/or necrosis, but also act as upstream modulators of autophagy induction (Navarro-Yepes et al., 2014; Sureshbabu et al., 2015; Avila-Rojas et al., 2019).

Cisplatin-induced nephrotoxicity by ROS ($O_2^{\bullet-}$; H_2O_2 and $OH^{\bullet}$), decreasing antioxidant defense, increasing products of lipid peroxidation (malondialdehyde—MDA), which lead cellular alterations (Chirino et al., 2009). Using this model, curcumin (CUR) showed protects kidney cells, cleans $O_2^{\bullet-}$, $OH^{\bullet}$, and $NO_2^{\bullet}$ radicals and inhibit lipid peroxidation and cytolysis (Ugur et al., 2015). Gentamycin enhances the generation of $OH^{\bullet}$, $O_2^{\bullet-}$, and H_2O_2, release of iron from renal cortical mitochondria and causes lipid peroxidation. CUR showed protective effect antioxidant against gentamycin, alleviated lipid peroxidation, increasing glutathione, and SOD levels, increased expression Bcl-2 (this protein protects against OS), reduced Bax (a pro-apoptotic protein) and Caspase 3, that reflect their antiapoptotic and renoprotective effect (Galaly et al., 2014).

Table 5.9.2 summarizes the main antioxidants effects on renal studies using several oxidant conditions in vitro.

5.9.4.2 In vivo studies

There is a great interest in preventive/prophylactic studies of antioxidants in vivo envisaging kidney injury through acute and/or chronic models. More recently, there are several researches focused in AKI induced mainly by drugs antineoplastic (cisplatin or doxorubicin), antibiotic (gentamicin or imipenem), nonsteroidal anti-inflammatories (diclofenac), nephrotoxic agents (snake venom, glycerol, nicotine, thioacetamide), and ischemia/reperfusion (Table 5.9.3). Concerning CKI in vivo, adenine has been an inducing agent widely used mainly for its ease in induction and reproducibility compared to other chronic models (Ali et al., 2018a). However, there are still few studies with antioxidant agents acting in the prevention or protection in CKI (Ali et al., 2018b; He et al., 2019).

Noteworthy that all drugs obtained from natural products as (-)-α-bisabolol, alpha-lipoic acid (ALA), allicin, betanin, 6-gingerol, hesperidin, resveratrol, and rutin showed antioxidant and anti-inflammatory effects improving renal function. Furthermore, it was observed in a great assays number an attenuation in morphological changes (Table 5.9.3).

It must be mentioned here that several drugs used against renal injuries act also on the cardiovascular system as AT1 receptor for angiotensin II blockers (irbesartan, angiotensin 1-7, and losartan), selective β-1 adrenergic receptor blocker (Nebivolol), cholesterol regulator blocker inhibiting HMG-CoA reductase (atorvastatin), phosphodiesterase 5 inhibitor (sildenafil) used to heart disease and erectile dysfunction. Indeed, the renin–angiotensin–aldosterone system (RAAS) nonbalance is involved in several kidney diseases. According to clinical practice guidelines, RAAS inhibitors are drugs recommended for cardiovascular and renal diseases treatments (Mirabito Colafella et al., 2019). Experimental assays showed that RAAS blockers can slow down renal function deterioration, reducing inflammation, and OS (Al-Kuraishy et al., 2019; Ramalingam et al., 2019; Safari et al., 2019).

Table 5.9.2 In vitro renal studies of antioxidants.

Antioxidant agent	Pharmacological class	Oxidant/model used	Effector mechanism	References
Apocynin	Acetovanilone	Cisplatin (alkylating agent, chemotherapy drug)	Scavenging ROS by reduced Nox2 and Nox4 expression; reduced mRNA levels of TNF-α, IL-1β, and IL-6; downregulated necroptosis and apoptosis.	Meng et al., 2018
Beberine	Alkaloid	Methotrexate (antimetabolite and folate antagonist)	Alleviated oxidative stress, restored the content of GSH and enzymatic activity of SOD, modulated the expression of Nrf2 (cytoprotective pathways in oxidative stress) and NF-κB transcription factors.	Hassanein et al., 2019
Cilastatin	Inhibitor of dehydropeptidase I	Tacrolimus (immunosuppressant)	Decreased oxidative stress, recovered the antioxidant (MnSOD).	Luo et al., 2019
Curcumin (free and hyaluronic acid-curcumin)	Polyphenol and polymeric prodrug	H_2O_2 exposure	Prevent against oxidative stress, reduced expression of caspase-3 and caspase-9, promoted autophagy (enhanced autophagic flux), inhibiting the PtdIns3K-AKT-mTOR signaling pathway against oxidative stress.	Hu et al., 2018
Lentinan	Dextran	Cisplatin (alkylating agent, chemotherapy drug)	Downregulation of ROS, by activation of an endogenous antioxidant system, Nrf2 activation, Block cleaved caspase-3.	Chen et al., 2016
Madecassoside	Triterpenoidsaponin	Doxorubicin (antitumor anthracycline antibiotic)	Attenuated oxidative stress and inflammatory response, inhibit p-ERK1/2 and NF-κB p65 expression, suppressions of cleaved caspase-3 expression, apoptosome formation (consisting of Apaf-1, Cyto-c, and Cleaved caspase-9) and reduced the ratio of Bax/Bcl-2.	Su et al., 2015
N-acetylcysteine (NAC)	Amino acid (L-cysteine)	Cadmium (Cd) (chemical element, heavy metal)	ROS scavenger, alleviated of p62 and LC3-II (autophagy markers proteins), inhibit of autophagosome-lysosome fusion.	Liu et al., 2017
		Nicotine (parasympathomimetic alkaloid, addictive drug)	Decreased ROS levels, modulates the expression of EMT-related marker genes (inhibited the expressions of DNMT3b and HMT1), restored, partially, EMT marker protein ZO-1 (involved in dedifferentiation, proliferation, and tumor development)	Chang and Singh, 2019

Omeprazole	Proton pump inhibitor	Cisplatin (alkylating agent, chemotherapy drug)	Suppressing oxidative stress and inflammation, protect mitochondrial damage, increased SOD and decreased MDA, inhibit NF-κB activation, reduced accumulation of Cisplatin by modulating the expression levels of uptake and efflux transporters (inhibitors of OCT2 and increased the expression levels of P-gp).	Gao et al., 2019
Panax Ginseng		Gentamycin (aminoglycoside)	ROS scavenger, NOX inhibitor, decreased of Bax, release of Cyto c, and cleaved caspase-3 and -9, inhibited a decrease of Bcl2 expression.	Shin et al., 2014
Polydatin	Polyphenolic compound (a glucoside of resveratrol)	Ischemia/reperfusion (oxygen–glucose deprivation followed by reoxygenation)	Antioxidative effect, at least in part, by upregulated the expression of Shh (involved in differentiation, proliferation, and apoptosis and regeneration), increased GSH, decreased MDA.	Meng et al., 2016
Resveratrol	Polyphenol	High glucose	Blocked ROS overproduction, upregulation of NOX1 and NOX4, inhibited the downregulation of cytokeratin and upregulation of vimentin, protect against EMT.	He et al., 2015

Note: P-gP, P-glycoprotein; ROS, reactive oxygen species; Nox1, NADPH oxidase isoform 1; Nox2, NADPH oxidase isoform 2; Nox4, NADPH oxidase isoform 4; NF-κB, nuclear factor κB; MDA, malondialdehyde; GSH, reduced glutathione; TNF-α, tumor necrosis factor-α; IL-6 interleukin 6; IL-8, interleukin 8; IL-1β, interleukin 1 beta; SOD, superoxide dismutase; Bax, apoptosis regulator; Bcl-2, antiapoptotic protein; Nrf2, nuclear factor erythroid 2-related factor 2; p-ERK1/2, extracellular signal-regulated protein kinases 1 and 2; Apaf-1, apoptotic protease-activating factor 1; Cyto-c, cytochrome C; PtdIns3K-AKT-mTOR, phosphatidylinositol 3-kinase Akt-mTOR; MnSOD, manganese superoxide dismutase; ZO-1, zonula occludens-1; Shh, Sonic Hedgehog; OCT2, organic cation transporter 2.

Table 5.9.3 In vivo renal studies of antioxidants.

Drug/synthetic compound	Pharmacological class/chemical classification	Dose/route/frequency and duration of treatment	Model used/acute or chronic injury	Main findings	Prophylactic or therapeutic	References
Ascorbic acid	Vitamin C	20 mg/kg daily (orally) for 14 days	Cisplatin-induced nephrotoxicity	Improved renal function (decreased serum urea, creatinine, uric acid, serum electrolyte levels) Antioxidant effects (decreased MDA, NO increased GSH, SOD, GPx, CAT); anti-inflammatory effects (decreased serum and renal TNF-α)	Prophylactic	Abdel-Daim et al., 2019a
(-)-α-Bisabolol	Sesquiterpene	100 mg/kg/24 h (treated orally)	Ischemia/reperfusion (I/R) (AKI)	Ameliorated renal function (biochemical and metabolic parameters, urinary proteins); decreased urinary KIM-1; morphological changes attenuated; reduced oxidative stress (decreased MDA and increased GSH)	Prophylactic	Sampaio et al., 2016
α-Tocopherol	Vitamin E	100 mg/kg daily for 10 days (orally)	Cisplatin-induced nephrotoxicity	Improved renal function (serum uric acid, urea, and creatinine); anti-inflammatory effect (decreased serum and renal TNF-α); antioxidant effects (increased CAT, GPx, GSH, ROS, SOD, TAC, and decreased MDA and NO)	Prophylactic	Abdel-Daim et al., 2019b
Allicin	Diallylthiosulfinate	10 mg/kg daily (orally) for 14 days	Cisplatin-induced nephrotoxicity	Improved renal function (decreased serum urea, creatinine, uric acid, serum electrolyte levels) Antioxidant effects (decreased MDA, NO increased GSH, SOD, GPx, CAT); anti-inflammatory effects (decreased serum and renal TNF-α)	Prophylactic	Abdel-Daim et al., 2019a

Alpha-lipoic acid	Dithiol	10 mg/kg daily for 14 days (treated orally)	Cisplatin (antineoplastic)-induced nephrotoxicity (AKI)	Improved renal parameters (serum urea, creatinine, and albumin); reduction inflammatory markers (NF-κB and IL-6 and P53 levels); hematological level elevation (Hb and RBC count); antioxidant effect renal (increased levels of GST, GSR and CAT in the renal tissue and decreased MDA); Morphological changes attenuated	Prophylactic and therapeutic	Zaazaa et al., 2019
Allopurinol	Xanthine oxidase (XO) inhibitor	300 mg/L of water for 7 days (drinking water) or 25 mg/Kg (intraperitoneal; once time)	*Bothrops* snake venom-induced nephrotoxicity (AKI)	Reduced oxidative stress (increased GSH and reduced nitrotyrosin and F2-IsoP); improved renal hemodynamics (GFR, RBF, RVR)	Prophylactic and therapeutic	Gois et al., 2017
		300 mg/L of water for 7 days (drinking water) or 50 mg/Kg (intravenous; once time)	Glycerol-induced rhabdomyolysis (AKI)	Reduced oxidative stress (systemic, renal, and muscular); reduced F2-IsoP renal expression; attenuated inflammation (inflammasome cascade and uric acid levels); decreased apoptosis (Bax/Bcl-2 ratio reduced); improved renal hemodynamics (GFR, RBF, RVR)	Prophylactic and therapeutic	Gois et al., 2016
		50 mg/kg for 14 days before I/R (intraperitoneal)	Renal ischemia/reperfusion (I/R) (AKI)	Improved renal functional (reduced levels of urea nitrogen and serum creatine and morphological changed); decreased apoptosis (caspase-3 and Bax/Bcl-2 ratio reduced); reduced oxidative damage (decreased MDA and increased SOD); inhibition of the renal inflammatory marker (HMGB1)	Prophylactic	Zhou et al., 2016

(*Continued*)

Table 5.9.3 In vivo renal studies of antioxidants. *Continued*

Drug/synthetic compound	Pharmacological class/ chemical classification	Dose/route/ frequency and duration of treatment	Model used/ acute or chronic injury	Main findings	Prophylactic or therapeutic	References
Atorvastatin	Statin/inhibition of HMG-CoA reductase	440 mg/kg/day (orally)	Imipenem (antibiotic)-induced nephrotoxicity (AKI)	Improved kidney function (normalized values of GFR, decreased levels serum of BUN, Cr, and levels urinary of protein, glucose, creatinine); antioxidant effects (increased levels of CAT, SOD, and GPx).	Prophylactic	Rababa'h et al., 2018
Betanin (betanidin-5-Ob-glycoside)	Betacyanin	25 mg/kg daily for 28 days (treated orally)	Diclofenac (nonsteroidal anti-inflammatories)-induced renal damage (AKI)	Improved function renal (reduction of KIM-1, urea, creatinine, uric acid, DNA fragmentation, and histopathological alterations) Decreased inflammatory markers (VCAM-1 and ICAM-1) and oxidative stress (reduced MDA and increased GSH) Increased of endogenous antioxidants (POX-1, HOX-1, CAT, and SOD).	Prophylactic	Motawi et al., 2020
6-Gingerol	Gingerol	100 mg/kg daily for 10 days (intraperitoneal)	Gentamicin (antibiotic)-induced renal damage(AKI)	Improved renal function (urea nitrogen, serum creatinine, and renal morphology); antiapoptotic effect (decreased caspase-3); antioxidant effect (increased GSH, decreased MDA, and HSP47)	Prophylactic	Hegazy et al., 2016
Hesperidin	Flavanone-glycoside (bioflavonoid)	100 mg/kg daily for 7 days (orally)	Renal ischemia/ reperfusion (I/R) (AKI)	Reduced kidney damage (decreased levels urea and creatinine) antioxidant effects (increased levels SOD, CAT, GPx, NO)	Prophylactic	Park et al., 2019

Irbesartan	Antagonist of AT1 receptor for angiotensin II	10 mg/kg for 12 days (intraperitoneal)	Gentamicin-induced nephrotoxicity (AKI)	Reduced kidney injury markers (blood urea, serum creatinine, NGAL, KIM-1,Cys-c, serum levels); antioxidant effect (decreased MDA, increased SOD serum levels)	Prophylactic	Al-Kuraishy et al., 2019
		10 mg/kg for 28 days (intraperitoneal)	Nicotine-induced nephrotoxicity (AKI)	Improved renal parameters (serum urea, creatinine, renal morphology); antioxidant effect renal (increased GSH, SOD decreased MDA).	Prophylactic	Ramalingam et al., 2019
Angiotensin 1-7 and losartan	Angiotensin peptide/ antagonist of AT1 receptor for angiotensin II	50 μg/kg of angiotensin 1-7 and/ or 10 mg/kg of losartan 15 min before induction of I/R or immediately after induction of I/R (intraperitoneally)	Renal ischemia/ reperfusion (I/R) (AKI)	Both drugs improved renal function (normal levels of BUN, Cr, nitrite in serum and MDA, LDH in serum and kidneys) These drugs showed synergistic effect in the evaluated parameters	Prophylactic and therapeutic	Safari et al., 2019
Melatonin (N-acetyl-5-methoxytryptamine)	Indolaminergic hormone	10 mg/kg daily for 30 days (intraperitoneal)	Renal I/R in diabetic rats (AKI)	Improved renal parameters (serum urea, creatinine, renal morphology); antiapoptotic effect (reduction of TUNEL-positive cells); antioxidant effect renal (GSH increased, MDA decreased, expression decreased of SIRT1 and activated of Nrf2/HO-1)	Prophylactic	Shi et al., 2019

(Continued)

Table 5.9.3 In vivo renal studies of antioxidants. *Continued*

Drug/synthetic compound	Pharmacological class/chemical classification	Dose/route/frequency and duration of treatment	Model used/acute or chronic injury	Main findings	Prophylactic or therapeutic	References
N-acetylcysteine	Mucolytic agent	20 mg/kg/day for 6 weeks (orally)	Imipenem (antibiotic)-induced nephrotoxicity (AKI)	Improved kidney function (normalized values of GFR, decreased levels serum of BUN, Cr and levels urinary of protein, glucose, creatinine); Antioxidant effects (increased levels of CAT, SOD, and GPx)	Prophylactic	Rababa'h et al., 2018
N-Acetylglucosamine	Glucosamine derivative (amino sugars)	50 mg/kg day for 3 weeks (intramuscular or orally)	Doxorubicin (antineoplastic/anthracyclines)-induced nephropathy (CKI)	Intramuscular injection had best results than intraperitoneal and oral administration Improved renal function (recovered diuresis, GFR, and kidney weight coefficient; decreased serum creatinine and urea, increased urea clearance); antioxidant effect (increased endogenous GlcNAc and decreased MDA)	Therapeutic	Shebeko et al., 2019a, Shebeko et al., 2019b
Nebivolol	Selective β-1 adrenergic receptor blocker	10 mg/kg daily for 14 days (orally)	Gentamicin-induced nephrotoxicity (AKI)	Improved renal function (reduced plasma urea, creatinine, and histological changes, normalized 24-h urine volume levels); antioxidant effect (decreased total nitrite, MDA increased GSH levels)	Prophylactic	Dursun et al., 2018
Resveratrol	Polyphenol	8 mg/kg for 28 days (intraperitoneal)	Nicotine-induced nephrotoxicity (AKI)	Improved renal parameters (serum urea, creatinine, renal morphology); antioxidant effect renal (increased GSH, SOD decreased MDA)	Prophylactic	Ramalingam et al., 2019
		10 mg/kg for 14 days (orally)	Thioacetamide (organosulfur)-induced nephrotoxicity	Improved function renal (reduced serum creatinine, urea, creatine kinase); anti-inflammatory effect (reduced TNF-α, IL-4, and IFN-γ); antioxidant effect (reduced 8-OHdG, MDA, and increased CAT)	Prophylactic	Zargar et al., 2019

Rutin	Flavonoid	150 mg/kg daily for 2 weeks (orally)	Cisplatin (antineoplastic)-induced nephrotoxicity (AKI)	Improved renal parameters (serum urea, creatinine, and albumin); reduction inflammatory markers (NF-κB and IL-6 and P53 levels); hematological level elevation (Hb and RBC count); antioxidant effect renal (increased levels of GST, GSR, and catalase in the renal tissue and decreased MDA); morphological changes attenuated	Prophylactic and therapeutic	Zaazaa et al., 2019
Sildenafil	Phosphodiesterase 5 (PDE-5) inhibitor	0.1, 0.5, and 2.5 mg/kg for 5 weeks (orally)	Adenine (purine derivative)-induced chronic kidney disease (CKI)	The results were dose-dependently Improved renal function (decreased plasma concentrations of urea, creatinine, uric acid, phosphorous, NGAL, cystatin C, nitrite, and indoxyl sulfate, albumin and albumin/creatine concentrations and NAG activity, increased osmolality and creatinine clearance and reduced histological changes); anti-inflammatory and antiapoptotic effects (decreased TNF-α, IL-1β, caspase 3, MAPK); antioxidants effects (increased GR, SOD, CAT, and TAC, decreased MDA)	Prophylactic	Ali et al., 2018b

Note: CAT, catalase; GPx, glutathione peroxidase; GSH, reduced glutathione; MDA, malondialdehyde; NO, nitric oxide; ROS, reactive oxygen species; SOD, superoxide dismutase; TAC, total antioxidant capacity; GST, glutathione-s-transferase; GSR, glutathione reductase; POX-1, paraoxonase-1; HOX-1, heme oxygenase-1; VCAM-1, vascular cell adhesion molecule; ICAM-1, intracellular cell adhesion molecule; SIRT1, silent information regulator 2 associated protein 1; Nrf2, nuclear factor erythroid 2-related factor 2; AKI, acute kidney injury; F2-IsoP, phospholipid-bound F2-isoprostane; HMGB1, high mobility group box 1; HSP47, anti-heat shock protein 47; NGAL, neutrophil gelatinase-associated lipocalin; KIM-1, kidney injury molecule-1; 8-OHdG, 8-hydroxy-2-deoxyguanosine; Cys-c, cystatin c; IL-10, interleukin-10; IL-1β, interleukin 1-beta sclerostin; TNF-α, tumor necrosis factor alpha; MAPK, mitogen-activated protein kinase; NF-κB, nuclear factor kappa B; BUN, urea nitrogen; Cr, creatinine; Na, sodium; K, potassium; P, phosphorus; LDH, lactate dehydrogenase; RBC, red blood cells; Hb, hemoglobin; GFR, glomerular filtration rate; RBF, renal blood flow; RVR, renal vascular resistance; GlcNAc, N-acetylglucosamine; Bax, apoptosis regulator; Bcl-2, antiapoptotic protein.

Allopurinol, NAC, and N-acetylglucosamine are known for their antioxidant properties (Alirezaei et al., 2017; Saleem et al., 2018). Allopurinol has been tested in animal models revealing protective effects on kidney, intestinal, and heart functions. However, it must be assayed in clinical trials to test these functions (Zhou et al., 2016; Gois et al., 2016, 2017). NAC appears to prevent AKI through OS biomarkers reduction as well as antioxidants biomarkers enhancement followed by proinflammatory mediator reduced production (Rababa'h et al., 2018; Ozturk et al., 2018).

Melatonin is an indolaminergic hormone produced by the pineal gland, in the dark, derived from tryptophan. Endogenous and exogenous melatonin have antioxidative properties and act as a potent anti-inflammatory (Reiter et al., 2016). A great number of melatonin assays is promising to avoid Alzheimer's disease, Parkinson's disease, multiple sclerosis, osteoporosis, diabetes, metabolic syndrome, sepsis, cancer, tropical diseases, snake, nematocyst venom toxicity, and renal diseases (Hrenak et al., 2015; Reiter et al., 2016). More recently, Shi et al. (2019) showed that melatonin attenuated the acute kidney ischemia/reperfusion injury in diabetic rats by activation of the SIRT1/Nrf2/HO-1 signaling pathway. The transcription factor Nrf2 (nuclear factor erythroid 2-related factor 2) is a major OS regulator through antioxidant proteins transcription (Ma, 2013). Nrf2 activation showed beneficial effects on several pathophysiological processes as type-2 diabetes, Alzheimer's and Parkinson's diseases (Francisqueti-Ferron et al., 2019) as well as various tissue injuries (lung, brain, liver, cardiac, and kidney injuries; Chen and Maltagliati, 2018).

Both vitamins C and E are powerful antioxidant agents that attenuate oxidative damage and/or inflammation on acute and chronic renal injuries in animal models. Moreover, with AKI assays it was shown ROS reduction, lipid peroxidation protection, inflammation and tubular damage decrease, antioxidant status improvement, benefiting microcirculation, endothelial and renal functions (Fryer, 2000; Thabet and Chan, 2006; Dennis and Witting, 2017; Liakopoulos et al., 2019).

5.9.4.3 Clinical studies

Some drugs previously used in vitro and in vivo assays (such as NAC) or components of the diet (such as ascorbic acid and α-tocopherol) were evaluated in clinical trials. The focus was to investigate the preventive or protective role of these drugs in acute (mainly contrast-induced) and/or CKD (diabetic nephropathy or patients undergoing HD). Nonetheless, the results obtained were different from those in vitro and in vivo without potential effects in AKD and CKD (Table 5.9.4).

The ALA is fat and water soluble and occurs naturally as potent antioxidant. It has the ability to regenerate other factors, such as ascorbic acid and tocopherol (vitamins C and E, in addition to raising glutathione intracellularly). Patients undergoing simultaneous kidney and pancreatic transplantation were evaluated. They revealed a preventive effect with inflammatory markers reduction (serum IL-8 and IL-6, secretory leukocyte protease inhibitor), an early renal dysfunction and clinical post-transplant pancreatitis (islet-derived protein 3β/pancreatitis-associated protein; Ambrosi et al., 2016). Another study with patients undergoing HD that were fed with

Table 5.9.4 Clinical studies renal studies of antioxidants.

Drug/synthetic compound	Dose/route/frequency and duration of treatment	Study design	Population	Disease/AKI or CKI	Effector mechanism/main findings/outcomes	Protective or preventive effect	References
α-Lipoic acid (ALA)	(1) No ALA treatment; (2) 600 mg to the recipients only immediately before the surgical procedure; (3) 600 mg to the deceased donor at the time of procurement and to the recipients immediately before the surgical procedure.	Prospective clinical trial.	26 kidney-pancreas transplant patients (11 men and 15 women; age range: 21–57 years).	Kidney–pancreas (SKP) transplantation.	Reduced inflammatory markers (serum IL-8, IL-6, secretory leukocyte protease inhibitor). Decreased early kidney dysfunction and clinical post-transplant pancreatitis (islet-derived protein 3 β/pancreatitis-associated protein).	Preventive	Ambrosi et al., 2016
	(1) ALA (600 mg/d) + mixed tocopherols (666 IU/d) for 6 months. (2) Placebos for 6 months.	Prospective, randomized, placebo-controlled, double-blind clinical trial.	238 patients undergoing maintenance hemodialysis therapy (MHD).	Maintenance hemodialysis therapy (MHD).	Safe and well tolerated. Did not influence biomarkers of inflammation and oxidative stress or the erythropoietic response.	Not effective	Himmelfarb et al., 2014
Bardoxolone methyl	(1) Placebo or (2) Bardoxolone methyl (20 mg), oral, once daily for 52 weeks.	Phase 3, randomized, double-blind, parallel-group, international, multicenter clinical trial.	2185 patients with chronic kidney disease (CKD) and type 2 diabetes (T2D) randomized into 2 groups.	CKD associated with T2D.	Decreased serum magnesium that is not associated with changes in intracellular and urinary magnesium levels or with adverse effects on QT interval.	Protective	Rizk et al., 2019

(Continued)

Table 5.9.4 Clinical studies renal studies of antioxidants. *Continued*

Drug/ synthetic compound	Dose/route/frequency and duration of treatment	Study design	Population	Disease/AKI or CKI	Effector mechanism/main findings/outcomes	Protective or preventive effect	References
	(1) Placebo or (2) Bardoxolone methyl (20 mg), oral, once daily for 52 weeks.	Phase 3, randomized, double-blind, parallel-group, international, multicenter clinical trial.	2185 patients with chronic kidney disease (CKD) and type 2 diabetes (T2D) randomized into 2 groups.	CKD associated with T2D.	Increased eGFR sustained through study week 48 and sustained 4 weeks after cessation of treatment. Preservation of kidney function and may delay the onset of ESRD in patients with T2D and stage 4 CKD.	Preventive and protective	Chin et al., 2018
	(1) Placebo or (2) Bardoxolone 25, 75, or 150 mg, oral, once daily for 52 weeks.	Phase 2, randomized, double-blind, placebo-controlled trial.	227 adults with chronic kidney disease (CKD).	CKD associated with type 2 diabetes.	Improvement in GFR in patients with advanced CKD and type 2 diabetes at 24 weeks. Improvement persisted at 52 weeks.	Protective	Pergola et al., 2011
Coenzyme Q10 (CoQ10)	Serum CoQ10 levels measured.	Prospective clinical trial.	41 CAPD patients (21 men, 20 women), 38 HD patients (20 men, 18 women), and 43 healthy control subjects (23 men, 20 women).	Hemodialysis (HD) and continuous ambulatory peritoneal dialysis (CAPD).	Increased oxidative stress in HD and CAPD patients (reduced TAS and adjusted CoQ10 levels and increased TOS and IMA levels). Antioxidant supplementation suggested.	Preventive or protective	Mehmetoglu et al., 2012

	200 mg, oral, once daily during the week before ESWL and for 1 week after.	Prospective, randomized, double-blind, placebo-controlled clinical trial.	100 patients with renal lithiasis treated with Extracorporeal shockwave lithotripsy (ESWL).	Renal lithiasis treated with ESWL.	Improvement in vasoactive hormone parameters, VRI and interleukin levels maintained until the end of the follow-up period. Not associated with significant changes in the oxidative stress parameters.	Protective	Carrasco et al., 2014
Curcumin (CUR)	(1) 500 mg turmeric (22.1 mg CUR); or (2) Placebo, oral, once daily for 12 weeks.	Randomized, double-blind, placebo-controlled clinical trial.	71 chronic hemodialysis (CH) patients.	Chronic hemodialysis patients.	No adverse effects. Increased albumin levels in turmeric group and no meaningful changes in potassium or liver function tests neither within nor between groups. Effective anti-inflammatory supplement in hemodialysis patients (reduced plasma level of hsCRP, IL-6 and TNF-α).	Preventive	Samadian et al., 2017

(Continued)

Table 5.9.4 Clinical studies renal studies of antioxidants. *Continued*

Drug/synthetic compound	Dose/route/frequency and duration of treatment	Study design	Population	Disease/AKI or CKI	Effector mechanism/main findings/outcomes	Protective or preventive effect	References
	CUR 320 mg, oral, once daily for 8 weeks.	Randomized, double-blind, placebo-controlled clinical trial.	101 patients with nondiabetic or diabetic proteinuric chronic kidney disease (CKD).	Chronic kidney disease (CKD).	No improvements in proteinuria, estimated GFR, or lipid profile. Attenuated lipid peroxidation in individuals with nondiabetic proteinuric CKD and enhanced the antioxidant capacity in subjects with diabetic proteinuric CKD. CUR was observed on the antioxidant enzymes activities or Nrf2 activation.	Protective	Jiménez-Osorio et al., 2016
	(1) 500 mg turmeric (22.1 mg CUR); or (2) Placebo, oral, once daily for 8 weeks.	Prospective, randomized, double-blind clinical trial.	183 hemodialysis (HD) patients (>18 years).	Renal disease, hemodialysis (HD).	No adverse effects. Reduced plasma MDA levels, RBC count and plasma albumin levels in HD patients. Increased GPX, GR, and CAT levels in both groups.	Protective	Pakfetrat et al., 2015

	(1) 824 mg purified turmeric extract (95% CUR) + 516 mg *Boswellia serrata* extract (10% 3-acetyl-11-keto-β-boswellic acid) or (2) Placebo (roasted rice powder), oral, twice/day, for 8 weeks.	Prospective, randomized, double-blinded, placebo-controlled trial.	Patients >18 years of age in chronic kidney disease (CKD) stages 1 through 5.	Chronic kidney disease (CKD).	Reduced inflammation as measured by IL-6. No changes observed in any other inflammatory or antioxidant.	Protective	Moreillon et al., 2013
Elamipretide	(1) Elamipretide (0.05 mg/kg/h – IV-infusion, n=6) before and during PTRA. (2) Placebo (n = 8) before and during PTRA.	Phase 2a, randomized, double-blinded, placebo-controlled pilot trial.	14 patients with severe atherosclerotic renal artery stenosis (ARAS).	ARAS patients scheduled for revascularization with percutaneous transluminal renal angioplasty and stenting (PTRA).	Attenuated postprocedural hypoxia. Increased renal blood flow (RBF). Improved kidney function (decreased serum creatinine).	Preventive	Saad et al., 2017
Green propolis	(1) EPP-AF propolis 500 mg, oral, twice/day. (2) 500 mg of placebo, oral, twice/day.	Randomized, double-blind, placebo-controlled trial.	32 patients ranging from 18 to 90 years of age with CKD caused by diabetes or of another etiology.	Chronic kidney disease (CKD).	Safe and well tolerated. Reduced proteinuria in patients with diabetic and nondiabetic CKD.	Protective	Silveira et al., 2019

(*Continued*)

Table 5.9.4 Clinical studies renal studies of antioxidants. *Continued*

Drug/ synthetic compound	Dose/route/frequency and duration of treatment	Study design	Population	Disease/AKI or CKI	Effector mechanism/main findings/outcomes	Protective or preventive effect	References
N-Acetylcysteine (NAC) with or without ascorbic acid (AA, vitamin C)	(1) NAC 600 mg, oral, twice/day, from 24 h before the operation until 2 postoperative days. (2) NAC 1500 mg, oral, twice/day, from 24 h before the operation until 2 postoperative days.	Open-label, parallel group, randomized trial.	291 patients randomized into 3 groups: selenium (n = 72), vitamin C (n = 73), NAC (n = 73), or nothing (n = 73).	Acute kidney injury (AKI) after off-pump coronary bypass graft surgery (CABG).	No significant differences in the incidence, the time of occurrence, the severity, or the duration of AKI.	Not effective	Amini et al., 2018
	(1) NAC 600 mg, oral, twice/day for 5 days. (2) Placebo, oral, twice/day for 5 days.	Double-blind, randomized, placebo-controlled trial.	70 deceased-donor KT recipients.	Acute kidney injury (AKI) after kidney transplantation (KT).	Reduced u-NGAL levels. Increased early-phase eGFR.	Preventive	Modarresi et al., 2018
	NAC (1200 mg) + AA (2000 mg), IV, 2 h before and at 10 h and 18 h following the infusion of contrast agent.	One-center, two-arm, prospective, randomized, open-label, controlled trial.	124 patients, with 64 patients in the Control Group and 60 patients in the NAC and AA group.	Contrast-induced nephropathy (CIN) in critical care patients.	Failed to reduce the incidence of CIN associated with increased baseline values of the serum urea/creatinine serum ratio and the concomitant use of nephrotoxic medications. May have reduced oxidative stress (indicated by 8-isoprostane serum levels).	Not effective	Palli et al., 2017

NAC 1200 mg, oral, twice/day (2400 mg in total), for 3 months.	Open-label, parallel group, randomized, multicenter trial.	54 patients undergoing hemodialysis for at least 3 months and with a residual urine volume [100 mL/24 h.	Hemodialysis (HD) in end-stage renal disease (ESRD).	Increased GFR. Increased 24 h-urine volume. Increased Kt/V (not statistically significant).	Protective	Ahmadi et al. 2017
(1) NAC 1200 mg, oral, before angiography + NAC 1200 mg, oral, twice/day for 3 doses + good hydration; (2) NAC 600 mg, oral, before angiography + NAC 600 mg twice/day for 3 doses + AA 3000 mg, one dose, before angiography and AA 2000 mg, 2 doses after angiography + good hydration; (3) Hydration with 0.9% saline.	Randomized, prospective, placebo-controlled trial.	105 patients with ischemic heart disease or peripheral vascular disease undergoing elective cardiac catheterization, randomized in 3 groups: NAC (n = 30); NAC + AA (n = 45) and Control (n = 30).	Contrast-induced nephropathy (CIN) in high-risk patients undergoing elective cardiac catheterization.	NAC + hydration provide better protection against CIN than NAC + AA + hydration, or hydration alone.	Protective	Habib et al., 2016

(Continued)

Table 5.9.4 Clinical studies renal studies of antioxidants. *Continued*

Drug/synthetic compound	Dose/route/frequency and duration of treatment	Study design	Population	Disease/AKI or CKI	Effector mechanism/main findings/outcomes	Protective or preventive effect	References
	NAC 600 mg, oral, twice/day, before meals for 6 weeks.	Randomized, double-blind, controlled clinical trial.	40 chronic HD patients.	Chronic hemodialysis (HD) in chronic kidney disease (CKD) patients.	Increased total antioxidant capacity (TAC).	Protective	Shahbazian et al., 2016
	(1) NAC 600 mg, oral, twice/day + Saline (100 mL/h); (2) AA 250 mg, oral, twice/day) + saline (100 ml/h); (3) Saline (100 mL/h).	Randomized clinical trial.	120 patients who scheduled for elective coronary angiography divided in 3 groups (n = 40, each).	Contrast-induced nephropathy (CIN) in the patient's undergone coronary angiography.	NAC or AA + saline infusion has more beneficial effect than saline infusion.	Preventive	Khaledifar et al., 2015
	NAC 600 mg, twice/day from one day before surgery to 5 days after surgery through esophagus catheter.	Blind, randomized, clinical trial.	50 candidate patients for coronary artery bypass graft surgery (CABG).	Chronic kidney disease (CKD) patients undergoing CABG.	Reduced serum creatinine and BUN levels. Increased GFR. Not significant effect on Na^+, K^+, and HCO_3^- concentrations.	Protective	Firoozabadi and Ebadi, 2014

(1) NAC 150 mg/kg + 50 mg/kg for 6 h in 0.9% saline. (2) 0.9% saline.	Prospective, double-blind, placebo-controlled clinical trial.	70 chronic kidney disease (CKD) patients, stage 3 or 4, who underwent coronary artery bypass graft surgery (CABG).	CKD patients submitted to CABG.	Reduced incidence of AKI in patients with CKD. Abolished oxidative stress. Mitigated negative effects of cardiopulmonary bypass (CPB) on renal function.	Protective	Santana-Santos et al., 2014
(1) NAC 600 mg + conventional therapy; or (2) Placebo + conventional therapy; or (3) AA 500 mg + conventional therapy, IV, administered intravenously the day before and the day of contrast dye exposure.	Prospective, randomized, double-blind, placebo-controlled, single-center clinical trial.	520 patients randomized to 1 of 3 groups: NAC group (n = 208), AA group (n = 104), and placebo group (n = 208).	Contrast-induced nephropathy (CIN) in patients with chronically impaired renal function undergoing elective cardiac catheterization.	Nonsignificant benefit of the use of NAC, AA, or a combination of both drugs.	Not effective	Brueck et al., 2013

(Continued)

Table 5.9.4 Clinical studies renal studies of antioxidants. *Continued*

Drug/synthetic compound	Dose/route/frequency and duration of treatment	Study design	Population	Disease/AKI or CKI	Effector mechanism/main findings/outcomes	Protective or preventive effect	References
	(1) NAC 600 mg, oral, twice/day for 2 days starting the evening before the procedure. (2) AA 3000 mg, oral, 2 h before the angiogram, 2000 mg after the angiogram, and 2000 mg 24 h after the angiogram. (3) Combination of both drugs. (4) Standard hydration.	Randomized, open-label, single-center clinical trial.	243 patients randomized to 1 of 4 Groups: NAC (n = 62), AA (n = 57), Both Drugs (n = 58), Placebo (n = 66).	Contrast-induced nephropathy (CIN) who were referred for coronary angiography or PCI.	Nonsignificant benefit of the use of NAC, AA, or a combination of both drugs.	Not effective	Al-babtain et al., 2013
Resveratrol	(1) "Placebo first" (4 weeks placebo; 8 weeks washout, 4 weeks 500 mg of resveratrol/day). (2) "Resveratrol first" (4 weeks 500 mg of resveratrol/day, 8 weeks washout, 4 weeks placebo).	Randomized, double-blind, placebo-controlled, crossover clinical trial.	26 eligible nondialyzed patients (>40 years); CKD stages 3 and 4; and normal liver function.	Nondialyzed chronic kidney disease (CKD) patients.	Nrf2 supplementation and NF-kB as not significant. No difference in proinflammatory biomarkers or antioxidant biomarkers after resveratrol supplementation. No antioxidant and anti-inflammatory effect in nondialyzed CKD patients.	Not effective	Saldanha et al., 2016

Selenium	0.5 mg, oral, twice/day, from 24 h before the operation until 2 postoperative days.	Open-label, parallel group, randomized clinical trial.	291 patients randomized into 3 groups: selenium (n = 72), AA (n = 73), NAC (n = 73), or nothing (n = 73).	Acute kidney injury (AKI) after off-pump coronary bypass graft surgery (CABG).	No significant differences in the incidence, the time of occurrence, the severity, and the duration of AKI.	Not effective	Amini et al., 2018
Spironolactone	(1) 50 or 100 mg/day, oral, of spironolactone; or (2) Placebo group (caplets with corn starch) 3 days before kidney transplant (KT) and up to 5 days post-transplantation KT.	Pilot, randomized, double-blind, placebo-controlled clinical trial.	77 patients randomized in 3 groups: placebo (n = 27), spironolactone 50 mg (n = 25) or spironolactone 100 mg groups (n = 25).	Acute kidney injury induced by ischemia/reperfusion in KT.	No significant effect on urinary NGAL levels. Reduced the acute increase in urinary oxidative stress in living-donor KT recipients.	Preventive	Morales-Buen-rostro et al., 2019
Tocopherol (vitamin E)	(1) 0.9% saline infusion 12 h prior to and after intervention + 600 mg vitamin E 12 h before + 400 mg vitamin E 2 h before coronary angiography. (2) Placebo before coronary angiography.	Randomized, double-blind, placebo-controlled, 2-center, clinical trial.	300 patients randomized into 2 groups.	Contrast medium-induced acute kidney injury (CIAKI) in patients undergoing elective coronary angiography.	Reduced CIAKI incidence. Reduced, but not significantly, creatinine levels.	Preventive	Rezaei et al., 2016

(Continued)

Table 5.9.4 Clinical studies renal studies of antioxidants. *Continued*

Drug/synthetic compound	Dose/route/frequency and duration of treatment	Study design	Population	Disease/AKI or CKI	Effector mechanism/main findings/outcomes	Protective or preventive effect	References
	(1) Tocotrienol-rich vitamin E from palm oil 200 mg, twice/day for 8 weeks; (2) Placebo twice/day for 8 weeks.	Prospective, randomized, double-blinded, placebo-controlled clinical trial.	52 patients randomized into 2 groups.	Diabetic nephropathy (DN) in patients with type 2 diabetes.	Reduced serum creatinine. No effects in eGFR, UACR, HbA1c, blood pressure, and serum biomarkers.	Protective	Tan et al., 2018
	(1) Vitamin E 1500 IU/day for 12 weeks; (2) Placebo for 12 weeks.	Randomized, double-blind, placebo-controlled clinical trial.	60 patients randomized into 2 groups.	Diabetic nephropathy.	Reduced urine protein, protein-to-creatinine ratio, serum TNF-α, matrix metalloproteinase-2, matrix metalloproteinase-9, MDA, advanced glycation end products, and insulin concentrations.	Protective	Khatami et al., 2016

Note: NAC, N-acetylcysteine; AA, ascorbic acid; CUR, curcumin; Nrf2, nuclear factor erythroid 2-related factor 2; NF-kB, nuclear factor κB; u-NGAL, urine neutrophil gelatinase-associated lipocalin; GFR, glomerular filtration rate; eGFR, estimated GFR; CIN, contrast-induced nephropathy; BUN, serum urea nitrogen; KT, kidney transplantation; CIAKI, contrast-induced acute renal injury; Kt/V, K-dialyzer clearance of urea, t, dialysis time, V, volume of distribution of urea; TAC, total antioxidant capacity; GSH, reduced glutathione; GR, glutathione reductase; MDA, malondialdehyde; CAT, catalase; GPx, glutathione peroxidase; ROS, reactive oxygen species; TNF-α, tumor necrosis factor-α; IL-6, interleukin 6; IL-8, interleukin 8; HbA1c, glycated hemoglobin; UACR, urine albumin creatinine ratio; T2D, type 2 diabetes; RBC, red blood cells; hsCRP, high sensitivity C-reactive protein; TAS, total antioxidant status; TOS, total oxidant status; IMA, ischemia-modified albumin; VRI, vascular resistance index.

ALA and tocopherols, not revealed changes in inflammation biomarkers and OS or the erythropoietic response (Himmelfarb et al., 2014).

Bardoxolone methyl (BARD) a natural product obtained from oleanolic acid is an inflammation modulator antioxidant that activates the Keap1–Nrf2 pathway, which plays an important role in maintaining kidney function and structure (Yates et al., 2007; Sporn et al., 2011; Pergola et al. 2011; Kania et al., 2013). BARD structure and activity profile resemble those of the cyclopentenone prostaglandins and exerts anti-inflammatory effects by inhibiting the proinflammatory nuclear factor κB pathway (Levonen et al., 2004; Serhan et al., 2008). A randomized, double-blind, placebo study by Pergola et al. (2011) in diabetic nephropathy model revealed patients GFR improvement with advanced CKD and type 2 diabetes for 24 weeks, which shows a relevant protective effect. Similar results were obtained by Chin et al. (2018) who showed that BARD preserves kidney function and may delay the ESRD onset in patients with T2D and stage 4 CKD. Rizk et al. (2019) showed that BARD decreases serum magnesium which is not associated with changes in intracellular and urinary levels or with adverse effects on QT interval by this electrolyte.

Coenzyme Q10 (CoQ10) is a fat-soluble vitamin-like quinone, also known as ubiquinone, that exerts antioxidative functions (Xu et al., 2019). CoQ10 treatment decreases superoxide production in endothelial cells and can reduce major adverse cardiovascular events, improving cardiac capacity in patients with heart failure (DiNicolantonio et al., 2015; Sharma et al., 2016). Mehmetoglu et al. (2012) showed that plasma CoQ10 concentration is depressed in patients with nondialysis CKD, and in that undergoing dialysis. A prospective, randomized, double-blind, placebo-controlled clinical study by Carrasco et al. (2014) in renal lithiasis models showed a protective effect, vasoactive hormonal parameters improvement, in the VRI levels and interleukin, maintained until the end of the follow-up period. CoQ10, however, was not associated with significant changes in OS parameters.

CUR, a polyphenol derived from the Curcuma rhizome has a wide range of biomedical and pharmacological properties (Samadian et al., 2017). Several clinical trials conducted on CKD models were carried out with CUR which was observed inflammation-reducing effects at the level of IL-6 and TNF-α (Moreillon et al., 2013; Samadian et al., 2017). A protective role was also purposed by Jiménez-Osorio et al. (2016), based on attenuated lipid peroxidation in individuals with nondiabetic CKD proteinuria and increased antioxidant capacity in individuals with CKD and diabetic proteinuria, in addition to enzymatic antioxidant action on activation of Nrf2. In another study, reduced plasma MDA levels, red blood cell count and plasma levels of albumin were observed in HD patients and increased levels of GPX, GR, and CAT in both groups (Pakfetrat et al., 2015).

Elamipretide (also known as MTP-131 or Bendavia) is a small peptide that targets the mitochondrial matrix independently of membrane potential, preventing cardiolipin peroxidation (Nilakantan et al., 2007; Szeto et al., 2015). It presents a potential role for attenuating IRI in AKI experimental model improving renal function after PTRA in experimental atherosclerotic renal artery stenosis (Eirin et al., 2012). Saad et al. (2017) showed through an atherosclerotic renal artery stenosis model in a clinical trial that elamipretide promoted a preventive effect postprocedure increasing renal blood flow as well as renal function (reduced serum creatinine). Another natural compound is propolis which is derived from plant exudates as resins (Silveira et al., 2019). Its composition varies from geographic region, flora, and environmental conditions. Propolis is used by bees to protect the hive against macro- and microinvaders (Zaccaria et al., 2017). Furthermore, it presents antimicrobial, anti-inflammatory, immunomodulatory, antioxidant, and anticancer properties (MacHado et al., 2012). Silveira et al. (2019), in a randomized, placebo-controlled study, in a CKD model, found propolis protective effect with reduced proteinuria in patients with diabetic and nondiabetic CKD.

NAC is usually combined with ascorbic acid (AA, vitamin C), one of the most ubiquitous hydrosoluble antioxidants endowed with enhanced activity, being readily oxidized to dehydroascorbic acid (Pisoschi and Pop, 2015). Briefly, in randomized clinical trials, with CKD models, protective findings regarding increased total antioxidant capacity were observed (Shahbazian et al., 2016). Reduced incidence of AKI in patients with CKD, abolished OS and mitigated negative effects of cardiopulmonary bypass on renal function (Santana-Santos et al., 2014).

However, in contrast-induced nephropathy (CIN) models, hydration + NAC provided better protection against CIN than hydration + NAC + AA or hydration alone (Habib et al., 2016). On the other hand, the use of NAC, AA or both of them not revealed significant benefits in CIN models (Albabtain et al., 2013; Brueck et al., 2013).

Resveratrol has been characterized as an antioxidant, cyclooxygenase inhibitor, as phytoalexin, peroxisome proliferator-activated receptor stimulator, endothelial nitric oxide synthase inducer, as well as silent mating type information regulation 2 homolog 1 activator (Nakata et al., 2012; Pisoschi and Pop, 2015). Contrary, Saldanha et al. (2016), with a nondialyzed CKD model, in a clinical trial not detected antioxidant or anti-inflammatory effects.

Selenium is not a direct reactive oxygen/nitrogen species scavenger itself. Indeed, selenium compounds antioxidant capacity is not relevant for its function in biological media (Pisoschi and Pop, 2015). The selenium primordial role is as glutathione peroxidase cofactor promoting hydrogen peroxide and peroxides reduction (Huang et al., 2005). Selenium is also present in the structure of selenoprotein P and thioredoxin reductase. Amini et al. (2018) in a randomized clinical trial with patients suffering AKI after coronary artery bypass graft surgery without cardiopulmonary bypass (CABG), treated with NAC or AA and selenium, observed that there were no significant differences in incidence, time of occurrence, severity, and length of the AKI when selenium was used.

Several studies involving experimental models have showed that aldosterone has an important role in renal injury physiopathology induced by ischemic process (Mejia-Vilet et al., 2007). Spironolactone administration at different postischemia intervals prevents or reduces functional/structural renal injuries, suggesting that aldosterone is a key molecule in mediating ischemic renal injury (Sanchez-Pozos et al., 2012; Barrera-Chimal et al., 2013). In a clinical trial conducted by Morales-Buenrostro et al. (2019), using a kidney transplantation (KT) ischemia and reperfusion model, it was observed that spironolactone reduced the acute increase in urinary OS in KT recipients from living donors, demonstrating a preventive effect. Morales-Buenrostro et al. (2019) used KT ischemia and reperfusion model (clinical trial) to assay Spironolactone which reduced the acute increase in urinary OS in KT recipients from living donors, demonstrating a preventive effect.

α-Tocopherol (vitamin E, VIT E) prevents lipid peroxidation of cell membranes (Descamps-Latscha et al., 2001). VIT E could positively alter the OS biomarkers, improving erythropoiesis (Himmelfarb and Hakim, 2003; Pisoschi and Pop, 2015). Clinical trials with VIT E in diabetic nephropathy models showed a protective effect followed by reduction in serum creatinine level, without effects on eGFR, UACR, HbA1c, blood pressure, or serum biomarkers (Tan et al., 2018). In addition, there was reduction in proteinuria, protein/creatinine ratio, serum TNF-α, matrix metalloproteinase-2, matrix metalloproteinase-9, and MDA (Khatami et al. 2016).

VIT E also presented preventive effect in a model of contrast-induced acute renal injury (CIAKI) in patients undergoing elective coronary angiography revealing reduced incidence of CIAKI. However, the low creatinine level was not significant (Rezaei et al., 2016).

Conclusion

The kidney is a highly metabolic organ, rich in oxidation reactions in mitochondria that indicates a highly vulnerable organ to damage caused by ROS. Since OS contributes to the development of AKI and CKD, the exogenous administration of antioxidants (synthetic or from natural sources) could be candidates nephroprotective drugs. Currently, alpha-tocopherol (vitamin E), CUR, NAC combined with ascorbic acid (vitamin C), RAAS inhibitors are drugs more recommended for prevent or protect renal diseases. Thus, this chapter reunites pharmacological agents and synthetic drugs isolated from natural compounds most promising in kidney diseases and their renoprotective abilities in the OS suppressing revealed in vitro, in vivo, and clinical trial studies. Among the most recent studies with antioxidants or substances that reveal nephroprotective activity in vitro are apocynin, berberine, cilastatin, lentinan, madecassoside, omeprazole, panax ginseng, and polydatin. All the synthetic agents obtained from natural products as (-)-α-bisabolol, ALA, allicin, betanin, 6-gingerol, melatonin, hesperidin, resveratrol, and rutin, drugs as allopurinol and N-acetylglucosamine, vitamins C and E, RAAS inhibitors showed antioxidant and anti-inflammatory effects with improving renal function in vivo. In clinical studies, BARD, coenzyme Q10, CUR, elamipretide, NAC, and propolis have a protective or preventive effect in the renal system while resveratrol or selenium had not effective activity.

References

Abdel-Daim, M.M., Abushouk, A.I., Donia, T., Alarifi, S., Alkahtani, S., Aleya, L., Bungau, S.G., 2019a. The nephroprotective effects of allicin and ascorbic acid against cisplatin-induced toxicity in rats. Environ. Sci. Pollut. Res. Int. 26, 13502–13509. https://doi.org/10.1007/s11356-019-04780-4.

Abdel-Daim, M.M., Aleya, L., El-Bialy, B.E., Abushouk, A.I., Alkahtani, S., Alarifi, S., Alkahtane, A.A., Al Basher, G., Ali, D., Almeer, R.S., Al-Sultan, N.K., Alghamdi, J., Alahmari, A., Bungau, S.G., 2019b. The ameliorative effects of ceftriaxone and vitamin E against cisplatin-induced nephrotoxicity. Environ. Sci. Pollut. Res. Int. 26, 15248–15254. https://doi.org/10.1007/s11356-019-04801-2.

Ahmad, M.S., Ahmed, N., 2006. Antiglycation properties of aged garlic extract: possible role in prevention of diabetic complications. J. Nut. 136, 796S–799S. https://doi.org/10.1093/jn/136.3.796S.

Ahmadi, F., Abbaszadeh, M., Razeghi, E., Maziar, S., Khoidaki, S.D., Najafi, M.T., Lessan-Pezeshki, M., 2017. Effectiveness of N-acetylcysteine for preserving residual renal function in patients undergoing maintenance hemodialysis: multicenter randomized clinical trial. Clin. Exp. Nephrol. 21, 342–349. https://doi.org/10.1007/s10157-016-1277-5.

Albabtain, M.A., Almasood, A., Alshurafah, H., Alamri, H., Tamim, H., 2013. Efficacy of ascorbic acid, N-acetylcysteine, or combination of both on top of saline hydration versus saline hydration alone on prevention of contrast-Induced nephropathy: a prospective randomized study. J. Interv. Cardiol. 26, 90–96. https://doi.org/10.1111/j.1540-8183.2012.00767.x.

Ali, B.H., Al-Salam, S., Al Suleimani, Y., et al., 2018b. Curcumin ameliorates kidney function and oxidative stress in experimental chronic kidney disease. Basic Clin. Pharmacol. Toxicol. 122, 65–73 10.1111/bcpt.12817.

Ali, B.H., Al Za'abi, M., Adham, S.A., Suleimani, Y.A., Karaca, T., Manoj, P., Kalbani, J.A., Ysin, J., Nemmar, A., 2018a. The effect of sildenafil on rats with adenine-Induced chronic kidney disease. Biomed. Pharmacother. 108, 391–402. https://doi.org/10.1016/j.biopha.2018.09.061.

Alirezaei, A., Argani, H., Asgharpour, M., Bahadorimonfared, A., Bakhtiyari, M., 2017. An update on allopurinol and kidney failure; new trend for an old drug. J. Renal Inj. Prev. 6, 297–302. https://doi.org/10.15171/jrip.2017.57.

Al-Kuraishy, H.M., Al-Gareeb, A.I., Al-Naimi, M.S., 2019. Renoprotective effect of irbesartan in a rat model of gentamicin-induced nephrotoxicity: role of oxidative stress. J. Lab. Phys. 11, 200–205. https://doi.org/10.4103/JLP.JLP_136_18.

Altinoz, E., Oner, Z., Elbe, H., Cigremis, Y., Turkoz, Y., 2015. Protective effects of saffron (its active constituent, crocin) on nephropathy in streptozotocin-induced diabetic rats. Hum. Exp. Toxicol. 34, 127–134. https://doi.org/10.1177/0960327114538989.

Ambrosi, N., Arrosagaray, V., Guerrieri, D., Uva, P.D., Petroni, J., Herrera, M.B., Iovanna, J.L., León, L., Incardona, C., Chuluyan, H.E., Casadei, D.H., 2016. α-Lipoic acid protects against ischemia-reperfusion injury in simultaneous kidney-pancreas transplantation. Transplantation, 100, 908–915. https://doi.org/10.1097/TP.0000000000000981.

Amini, S., Robabi, H.N., Tashnizi, M.A., Vakili, V., 2018. Selenium, vitamin C and N-acetylcysteine do not reduce the risk of acute kidney injury after off-pump CABG: a randomized clinical trial. Braz. J. Cardiovasc. Surg. 33, 129–134. https://doi.org/10.21470/1678-9741-2017-0071.

Amorim, R.G., Guedes, G.S., Vasconcelos, S.M.L., Santos, J.C.F, 2019. Kidney disease in diabetes mellitus: cross-linking between hyperglycemia, redox imbalance and inflammation. Arq. Bras. Cardiol. 112 (5), 577–587. https://doi.org/10.5935/abc.20190077.

Anjaneyulu, M.C.K., 2004. Quercetin, an anti-oxidant bioflavonoid, attenuates diabetic nephropathy in rats. Clin. Exp. Pharmacol. Physiol. 31, 244–248. https://doi.org/10.1111/j.1440-1681.2004.03982.x.

Atkin, M., Laight, D., Cummings, M.H., 2016. The effects of garlic extract upon endothelial function, vascular inflammation, oxidative stress and insulin resistance in adults with type 2 diabetes at high cardiovascular risk. A pilot double blind randomized placebo controlled trial. J. Diab. Complic. 30, 723–727. https://doi.org/10.1016/j.jdiacomp.2016.01.003.

Avila-Rojas, S.H., Lira-León, A., Aparicio-Trejo, O.E., Reyes-Fermín, L.M., Pedraza-Chaverri, J., 2019. Role of autophagy on heavy metal-induced renal damage and the protective effects of curcumin in autophagy and kidney preservation. Medicina (Kaunas). 55, 360. https://doi.org/10.3390/medicina55070360.

Barnett, L.M.A., Cummings, B.S., 2019. Cellular and molecular mechanisms of kidney toxicity. Semin. Nephrol. 39, 141–151. https://doi.org/10.1016/j.semnephrol.2018.12.004.

Barrera-Chimal, J., Pérez-Villalva, R., Rodríguez-Romo, R., Reyna, J., Uribe, N., Gamba, G., Bobadilla, N.A., 2013. Spironolactone prevents chronic kidney disease caused by ischemic acute kidney injury. Kidney Int. 83, 93–103. https://doi.org/10.1038/ki.2012.352.

Brueck, M., Cengiz, H., Hoeltgen, R., Wieczorek, M., Boedeker, R.H., Scheibelhut, C., Boening, A., 2013. Usefulness of N-acetylcysteine or ascorbic acid versus placebo to prevent contrast-induced acute kidney injury in patients undergoing elective cardiac catheterization: a single-center, prospective, randomized, double-blind, placebo-controlled trial. J. Invasive Cardiol. 25, 276–283.

Carrasco, J., Anglada, F.J., Campos, J.P., Muntané, J., Requena, M.J., Padillo, J., 2014. The protective role of coenzyme Q10 in renal injury associated with extracorporeal shockwave lithotripsy: a randomised, placebo-controlled clinical trial. BJU Int. 113, 942–950. https://doi.org/10.1111/bju.12485.

Chang, Y.W., Singh, K.P., 2019. Nicotine-induced oxidative stress contributes to EMT and stemness during neoplastic transformation through epigenetic modifications in human kidney epithelial cells. Toxicol. Appl. Pharmacol. 374, 65–76. https://doi.org/10.1016/j.taap.2019.04.023.

Chauveau, P., Aparicio, M., Bellizzi, V., Campbell, K., Hong, X., Johansson, L., Kolko, A., Molina, P., Sezer, S., Wanner, C., Ter Wee, P.M., Teta, D., Fouque, D., Carrero, J.J., 2017. Mediterranean diet as the diet of choice for patients with chronic kidney disease. Nephrol. Dial. Transplant. 33, 725–735. http://doi.org/10.1093/ndt/gfx085.

Chawla, L.S., Eggers, P.W., Star, R.A., Kimmel, P.L, 2014. Acute kidney injury and chronic kidney disease as interconnected syndromes. N. Engl. J. Med. 371, 58–66. https://doi.org/10.1056/nejmra1214243.

Chen, J., Siriki, R., 2015. Antioxidants therapy for patients with chronic kidney disease: a question of balance. Am. J. Nephrol. 42, 318–319 10.1159/000441628.

Chen, Q., Peng, H., Dong, L., Chen, L., Ma, X., Peng, Y., Dai, S., Liu, Q., 2016. Activation of the NRF2-ARE signalling pathway by the lentinula edodes polysaccharose LNT alleviates ROS-mediated cisplatin nephrotoxicity. Int. Immunopharmacol. 36, 1–8. https://doi.org/10.1016/j.intimp.2016.04.007.

Chen, Q.M., Maltagliati, A.J., 2018. Nrf2 at the heart of oxidative stress and cardiac protection. Physiol. Genom. 50, 77–97. https://doi.org/10.1152/physiolgenomics.00041.2017.

Chin, M.P., Bakris, G.L., Block, G.A., Chertow, G.M., Goldsberry, A., Inker, L.A., Heerspink, H., O'Grady, M., Pergola, P.E., Wanner, C., Warnock, D.G., Meyer, C.J., 2018. Bardoxolone methyl improves kidney function in patients with chronic kidney disease stage 4 and type 2 diabetes: post-hoc analyses from bardoxolone methyl evaluation in patients with chronic kidney disease and type 2 diabetes study. Am. J. Nephrol. 47, 40–47. https://doi.org/10.1159/000486398.

Chirino, Y.I., Pedraza-Chaverri, J., 2009. Role of oxidative and nitrosative stress in cisplatin-induced nephrotoxicity. Exp. Toxicol. Pathol. 61, 223–242. https://doi.org/10.1016/j.etp.2008.09.003.

Daenen, K., Andries, A., Mekahli, D., Schepdael, A.V., Jouret, F., Bammens, B., 2019. Oxidative stress in chronic kidney disease. Pediatr. Nephrol. 34, 975–991. https://doi.org/10.1007/s00467-018-4005-4.

Dennis, J.M., Witting, P.K, 2017. Review protective role for antioxidants in acute kidney disease. Nutrients 9, 718. https://doi.org/10.3390/nu9070718.

Descamps-Latscha, B., Drüeke, T., Witko-Sarsat, V., 2001. Dialysis-induced oxidative stress: biological aspects, clinical consequences, and therapy. Semin. Dial. 14, 193–199. https://doi.org/10.1046/j.1525-139x.2001.00052.x.

Díaz-López, A., Babio, N., Martínez-González, M.A., Corella, D., Amor, A.J., Fitó, M., Estruch, R., Arós, F., Gómez-Gracia, E., Fiol, M., Lapetra, J., Serra-Majem, L., Basora, J., Basterra-Gortari, F.J., Zanon-Moreno, V., Muñoz, M.Á., Salas-Salvadó, J., 2018. Erratum. Mediterranean diet, retinopathy, nephropathy, and microvascular diabetes complications: a post hoc analysis of a randomized trial. Diabetes Care. 38, 2134–2141. 2015Diabetes Care, 41(10), 2260–2261. https://doi.org/10.2337/dc18-er10a.

Di Nicolantonio, J.J., Bhutani, J., McCarty, M.F., O'Keefe, J.H., 2015. Coenzyme Q10 for the treatment of heart failure: a review of the literature. Open. Heart. 2, e000326. https://doi.org/10.1136/openhrt-2015-000326.

Dursun, M., Sahin, S., Besiroglu, H., Otunctemur, A., Ozbek, E., Cakir, S.S., Cekmen, M., Somay, A., 2018. Protective effect of nebivolol on gentamicin-induced nephrotoxicity in rats. Bratisl. Lek. Listy. 119, 718–725. https://doi.org/10.4149/BLL_2018_128.

Eirin, A., Lerman, A., Lerman, L.O., 2016. The emerging role of mitochondrial targeting in kidney diseasePharmacology of Mitochondria. Springer, pp. 229–250. https://doi.org/10.1007/164_2016_6.

Eirin, A., Li, Z., Zhang, X., Krier, J.D., Woollard, J.R., Zhu, X.Y., Tang, H., Herrmann, S.M., Lerman, A., Textor, S.C., Lerman, L.O., 2012. A mitochondrial permeability transition pore inhibitor improves renal outcomes after revascularization in experimental atherosclerotic renal artery stenosis. Hypertension 60, 1242–1249. https://doi.org/10.1161/HYPERTENSIONAHA.112.199919.

El-Sayed, S.M., El-Naggar, M.E., Hussein, J., Medhat, D., El-Banna, M., 2019. Effect of *Ficus carica* L. leaves extract loaded gold nanoparticles against cisplatin-induced acute kidney injury. Colloids Surf. B Biointerfaces 1, 110465. http://doi.org/10.1016/j.colsurfb.2019.110465.

Firoozabadi, M.D., Ebadi, A., 2014. The effect of oral N-acetylcysteine on serum creatinine in chronic kidney diseases patients under CABG surgery. Life Sci. J. 11, 5s.

Francisqueti-Ferron, F.V., Ferron, A.J.T., Garcia, J.L., Silva, C.C.V.A., Costa, M.R., Gregolin, C.S., Moreto, F., Ferreira, A.L.A., Minatel, I.O., Correa, C.R., 2019. Basic concepts on the role of nuclear factor erythroid-derived 2-like 2 (Nrf2) in age-related diseases. Int. J. Mol. Sci. 20, 3208. https://doi.org/10.3390/ijms20133208.

Fryer, M.J., 2000. Vitamin E as a protective antioxidant in progressive renal failure. Nephrology 5, 17. https://doi.org/10.1046/j.1440-1797.2000.00504.x.

Galaly, S.R., Ahmed, O.M., Mahmoud, A.M., 2014. Thymoquinone and curcumin prevent gentamicin-induced liver injury by attenuating oxidative stress, inflammation and apoptosis. J. Physiol. Pharmacol. 65, 823–832.

Gao, H., Zhang, S., Hu, T., Qu, X., Zhai, J., Zhang, Y., Tao, L., Yin, J., Song, Y., 2019. Omeprazole protects against cisplatin-induced nephrotoxicity by alleviating oxidative stress, inflammation, and transporter-mediated cisplatin accumulation in rats and HK-2 cells. Chem. Biol. Interact. 297, 130–140. https://doi.org/10.1016/j.cbi.2018.11.008.

Georgea, J.A., Goundenb, V., 2019. Novel glomerular filtration markers. Adv. Clin. Chem. 88, 91–119. https://doi.org/10.1016/bs.acc.2018.10.005.

Gois, P.H.F., Canale, D., Volpini, R.A., Ferreira, D., Veras, M.M., Andrade-Oliveira, V., Câmara, N.O.S., Shimizu, M.H.M., Seguro, A.C., 2016. Allopurinol attenuates rhabdomyolysis-associated acute kidney injury: renal and muscular protection. Free Radic. Biol. Med. 101, 176–189. https://doi.org/10.1016/j.freeradbiomed.2016.10.012.

Gois, P.H.F., Martines, M.S., Ferreira, D., Volpini, R., Canale, D., Malaque, C., Crajoinas, R., Girardi, A.C.C., Shimizu, M.H.M., Seguro, A.C., 2017. Allopurinol attenuates acute kidney injury following *Bothrops jararaca* envenomation. PLoS Negl. Trop Dis. 11, e0006024. https://doi.org/10.1371/journal.pntd.0006024.

Gowder, S.J., Devaraj, H., 2006. Effect of the food flavor cinnamaldehyde on the antioxidant status of rat kidney. Basic Clin. Pharmacol. Toxicol. 99, 379–382. https://doi.org/10.1111/j.1742-7843.2006.pto_560.x.

Gyurászová, M., Gurecká, R., Bábíčková, J., Tóthová, L., 2020. Oxidative stress in the pathophysiology of kidney disease: implications for noninvasive monitoring and identification of biomarkers. Oxidative Med. Cell Longev., 5478708. https://doi.org/10.1155/2020/5478708.

Habib, M., Hillis, A., Hammad, A., 2016. N-acetylcysteine and/or ascorbic acid versus placebo to prevent contrast-induced nephropathy in patients undergoing elective cardiac catheterization: the NAPCIN trial; a single-center, prospective, randomized trial. Saudi J. Kidney Dis. Transpl. 27, 55–61. https://doi.org/10.4103/1319-2442.174072.

Hassanein, E.H.M., Shalkami, A.G.S., Khalaf, M.M., Mohamed, W.R., Hemeida, R.A.M., 2019. The impact of Keap1/Nrf2, P 38 MAPK/NF-κB and Bax/Bcl2/caspase-3 signaling pathways in the protective effects of berberine against methotrexate-induced nephrotoxicity. Biomed. Pharmacother. 109, 47–56. https://doi.org/10.1016/j.biopha.2018.10.088.

He, T., Guan, X., Wang, S., Xiao, T., Yang, K., Xu, X., Wang, J., Zhao, J., 2015. Resveratrol prevents high glucose-induced epithelial-mesenchymal transition in renal tubular epithelial cells by inhibiting NADPH oxidase/ROS/ERK pathway. Mol. Cell. Endocrinol. 402, 13–20. https://doi.org/10.1016/j.fct.2016.07.032.

He, T., Xiong, J., Nie, L., Yu, Y., Guan, X., Xu, X., Xiao, T., Yang, K., Liu, L., Zhang, D., Huang, Y., Zhang, J., Wang, J., Sharma, K., Zhao, J., 2016. Resveratrol inhibits renal interstitial fibrosis in diabetic nephropathy by regulating AMPK/NOX4/ROS pathway. J. Mol. Med. 94, 1359–1371. https://doi.org/10.1007/s00109-016-1451-y.

He, Y., Lang, X., Cheng, D., Yang, Z., 2019. Curcumin ameliorates chronic renal failure in 5/6 nephrectomized rats by regulation of the mTOR/HIF-1α/VEGF signaling pathway. Biol. Pharm. Bull. 42, 886–891 10.1248/bpb.b18-00787.

Hegazy, A.M., Mosaed, M.M., Elshafey, S.H., Bayomy, N.A., 2016. 6-Gingerol ameliorates gentamicin induced renal cortex oxidative stress and apoptosis in adult male albino rats. Tissue Cell 48, 208–216. https://doi.org/10.1016/j.tice.2016.03.006.

Himmelfarb, J., 2005. Relevance of oxidative pathways in the pathophysiology of chronic kidney disease. Cardiol. Clin. 23, 319–330. https://doi.org/10.1016/j.ccl.2005.03.005.

Himmelfarb, J., Hakim, R.M., 2003. Oxidative stress in uremia. Curr. Opin. Nephrol. Hypertens 12, 593–598. https://doi.org/10.1097/00041552-200311000-00004.

Himmelfarb, J., Ikizler, T.A., Ellis, C., Wu, P., Shintani, A., Dalal, S., Kaplan, M., Chonchol, M., Hakim, R.M., 2014. Provision of antioxidant therapy in hemodialysis (PATH): a randomized clinical trial. J. Am. Soc. Nephrol. 25, 623–633. https://doi.org/10.1681/ASN.2013050545.

Hoile, S.P., Clarke-Harris, R., Huang, R.-C., Calder, P.C., Mori, T.A., Beilin, L.J., Lillycrop, K.A., Burdge, G.C., 2014. Supplementation with N-3 long-chain polyunsaturated fatty acids or olive oil in men and women with renal disease induces differential changes in the DNA methylation of FADS2 and ELOVL5 in peripheral blood mononuclear cells. PLoS One 9, e109896. https://doi.org/10.1371/journal.pone.0109896.

Hrenak, J., Paulis, L., Repova, K., Aziriova, S., Nagtegaal, E.J., Reiter, R.J., Simko, F., 2015. Melatonin and renal protection: novel perspectives from animal experiments and human studies (review). Curr. Pharm. Des. 21, 936–949. https://doi.org/10.2174/1381612820666140929092929.

Hu, J.B., Lib, S.J., Kanga, X.Q., Qia, J., Wua, J.H., Wanga, X.J., Xua, X.L., Yinga, X.Y., Jiangc, S.P., Youa, J., Dua, Y.Z., 2018. CD44-targeted hyaluronic acid-curcumin prodrug protects renal tubular epithelial cell survival from oxidative stress damage. Carbohydr. Polym. 193, 268–280. https://doi.org/10.1016/j.carbpol.2018.04.011.

Hu, Y., Yu, X.A., Zhang, Y., Zhang, R., Bai, X., Lu, M., Li, J., Gu, L., Liu, J.H., Yu, B.Y., Tian, J., 2020. Rapid and sensitive detection of NGAL for the prediction of acute kidney injury via a polydopamine nanosphere/aptamer nanocomplex coupled with DNase I-assisted recycling amplification. Analyst 145, 3620–3625. https://doi.org/10.1039/d0an00474j.

Huang, D., Ou, B., Prior, R.L., 2005. The chemistry behind antioxidant capacity assays. J. Agric. Food Chem. 53, 1841–1856. https://doi.org/10.1021/jf030723c.

International Society of Nephrology (ISN). World Kidney Day 2020: The onset and progression of CKD are often preventable. 2020. Internet [cited 2020 Apr 16]. Available at: https://www.healthawareness.co.uk/kidney/world-kidney-day-2020-the-onset-and-progression-of-ckd-are-often-preventable/

Jeong, B.Y., Lee, H.Y., Park, C.G., Kang, J., Yu, S.L., Du-RiChoi, Han, S.Y., Park, M., H., Cho, S., Lee, S.Y., Hwang, W.M., Yun, S.R., Ryu, H.M., Oh, E.J., Park, S.H., Kim, Y.L., Yoon, S.H., 2018. Oxidative stress caused by activation of NADPH oxidase 4 promotes contrast-induced acute kidney injury. PLoS One 13, e0191034. https://doi.org/10.1371/journal.pone.0191034.

Jha, J.C., Banal, C., Chow, B.S.M., Cooper, M.E., Jandeleit-Dahm, K., 2016. Diabetes and kidney disease: role of oxidative stress. Antioxid. Redox. Signal. 25, 657–684. https://doi.org/10.1089/ars.2016.6664.

Jiang, W., Zhao, H., Zhang, L., Wu, B., Zha, Z., 2020. Maintenance of mitochondrial function by astaxanthin protects against bisphenol A-induced kidney toxicity in rats. Biomed. Pharmacother. 121, 109629. https://doi.org/10.1016/j.biopha.2019.109629.

Jiménez-Osorio, A.S., García-Niño, W.R., González-Reyes, S., Álvarez-Mejía, A.E., Guerra-León, S., Salazar-Segovia, J., Falcón, I., Montes de Oca-Solano, H., Madero, M., Pedraza-Chaverri, J., 2016. The effect of dietary supplementation with curcumin on redox status and Nrf2 activation in patients with nondiabetic or diabetic proteinuric chronic kidney disease: a pilot study. J. Ren. Nutr. 26, 237–244. https://doi.org/10.1053/j.jrn.2016.01.013.

Kajarabille, N., Latunde-Dada, G.O., 2019. Programmed cell-death by ferroptosis: antioxidants as mitigators. Int. J. Mol. Sci. 20, 4968. https://doi.org/10.3390/ijms20194968.

Kandemir, F.M., Yildirim, S., Kucukler, S., Caglayan, C., Mahamadu, A., Muhammet Dortbudak, M.B., 2018. Therapeutic efficacy of zingerone against vancomycin-induced oxidative stress, inflammation, apoptosis and aquaporin 1 permeability in rat kidney. Biomed. Pharmacother. 105, 981–991. https://doi.org/10.1016/j.biopha.2018.06.048.

Kania, D.S., Smith, C.T., Nash, C.L., Gonzalvo, J.D., Bittner, A., Shepler, B.M., 2013. Potential new treatments for diabetic kidney disease. Med. Clin. North. Am. 97, 115–134. https://doi.org/10.1016/j.mcna.2012.10.004.

Kanlaya, R., Thongboonkerd, V., 2019. Protective effects of epigallocatechin-3-gallate from green tea in various kidney diseases. Adv. Nutr. 10, 112–121. https://doi.org/10.1093/advances/nmy077.

Khaledifar, A., Momeni, A., Ebrahimi, A., Kheiri, S., Mokhtari, A., 2015. Comparison of N-acetylcysteine, ascorbic acid, and normal saline effect in prevention of contrast-induced nephropathy. ARYA Atheroscler. 11, 228–232.

Khatami, P.G., Soleimani, A., Sharifi, N., Aghadavod, E., Asemi, Z., 2016. The effects of high-dose vitamin E supplementation on biomarkers of kidney injury, inflammation, and oxidative stress in patients with diabetic nephropathy: a randomized, double-blind, placebo-controlled trial. J. Clin. Lipidol. 10, 922–929. https://doi.org/10.1016/j.jacl.2016.02.021.

Konda, P.Y., Egi, J.Y., Dasari, S., Katepogu, R., Jaiswal, K.K., Nagarajan, P., 2020. Ameliorative effects of *Mentha aquatica* on diabetic and nephroprotective potential activities in STZ-induced renal injury. Comp. Clin. Pathol. 29, 189–199. https://doi.org/10.1007/s00580-019-03042-6.

Kpemissi, M., Eklu-Gadegbeku, K., Veerapur, V.P., Potârniche, A.V., Adi, K., Vijayakumar, S., Banakar, S.M., Thimmaiah, N.V., Metowogo, K., Aklikokou, K., 2019. Antioxidant and nephroprotection activities of *Combretum micranthum*: a phytochemical, in-vitro and ex-vivo studies. Heliyon 5, e01365. https://doi.org/10.1016/j.heliyon.2019.e01365.

Kuba, G., Vattimo, M.F.F., 2015. O uso de fitoterápicos orientais nas lesões renais: revisão integrativa. Rev. Bras. Pl. Med. Campinas 17, 1177–1181. https://doi.org/10.1590/1983-084x/14_149.

Levey, A.S., Coresh, J., 2012. Chronic kidney disease. Lancet; 379, 165–180. doi:10.1016/S0140-6736(11)60178-5.

Levey, A.S., Eckardt, K., Dorman, N.M., Christiansen, S.L., Cheung, M., Jadoul, M., Winkelmayer, W.C, 2020. Nomenclature for kidney function and disease: executive summary and glossary from a kidney disease: improving global outcomes (KDIGO) consensus conference. Transpl. Int. 2020. https://doi.org/10.1111/tri.13627.

Levonen, A.L., Landar, A., Ramachandran, A., Ceaser, E.K., Dickinson, D.A., Zanoni, G., Morrow, J.D., Darley-Usmar, V.M., 2004. Cellular mechanisms of redox cell signalling: role of cysteine modification in controlling antioxidant defences in response to electrophilic lipid oxidation products. Biochem. J. 378, 373–382. https://doi.org/10.1042/BJ20031049.

Liakopoulos, V., Roumeliotis, S., Bozikas, A., Eleftheriadis, T., Dounousi, E., 2019. Antioxidant supplementation in renal replacement therapy patients: is there evidence? Oxid. Med. Cell Longev., 9109473. https://doi.org/10.1155/2019/9109473.

Liu, F., Wang, X.Y., Zhou, X.P., Liu, Z.P., Song, X.B., Wang, Z.Y., Wang, L., 2017. Cadmium disrupts autophagic flux by inhibiting cytosolic Ca^{2+}-dependent autophagosome-lysosome fusion in primary rat proximal tubular cells. Toxicology 383, 13–23. https://doi.org/10.1016/j.tox.2017.03.016.

Locatelli, F., Canaud, B., Eckardt, K-U., Stenvinkel, P., Wanner, C., Zoccali, C., 2003. Oxidative stress in end-stage renal disease: an emerging threat to patient outcome. Nephrol. Dial. Transplant. 18, 1272–1280. https://doi.org/10.1093/ndt/gfg074.

Luo, C.J., Luo, F., Bu, Q.D., Jiang, W., Zhang, W., Liu, X.M., Che, L., Luan, H., Zhang, H., Ma, R.X., Luo, K., Lim, S.W., Jin, J., Jin, L., Gil, H.W., Im, D.S., Hwang, H.S., Yang, C.W., 2019. Cilastatin protects against tacrolimus induced nephrotoxicity via anti-oxidative and anti-apoptotic properties. BMC Nephrol. 20, 221. https://doi.org/10.1186/s12882-019-1399-6.

Ma, Q., 2013. Role of nrf2 in oxidative stress and toxicity. Annu. Rev. Pharmacol. Toxicol. 53, 401–426 10.1146/annurev-pharmtox-011112-140320.

Machado, J.L., Assunção, A.K., da Silva, M.C., Dos Reis, A.S., Costa, G.C., Arruda, D., Rocha, B.A., Vaz, M.M., Paes, A.M., Guerra, R.N., Berretta, A.A., do Nascimento, F.R., 2012. Brazilian green propolis: anti-inflammatory property by an immunomodulatory activity. Evid. Based Complement Altern. Med., 157652. https://doi.org/10.1155/2012/157652.

Mehmetoglu, I., Yerlikaya, F.H., Kurban, S., Erdem, S.S., Tonbul, Z., 2012. Oxidative stress markers in hemodialysis and peritoneal dialysis patients, including coenzyme Q10 and ischemia-modified albumin. Int. J. Artif. Organs. 35, 226–232. https://doi.org/10.5301/ijao.5000078.

Mejía-Vilet, J.M., Ramírez, V., Cruz, C., Uribe, N., Gamba, G., Bobadilla, N.A., 2007. Renal ischemia-reperfusion injury is prevented by the mineralocorticoid receptor blocker spironolactone. Am. J. Physiol. Renal 293, F78–F86. https://doi.org/10.1152/ajprenal.00077.2007.

Meng, Q.H., Liu, H.B., Wang, J.B., 2016. Polydatin ameliorates renal ischemia/reperfusion injury by decreasing apoptosis and oxidative stress through activating sonic hedgehog signaling pathway. Food Chem. Toxicol. 96, 215–225. https://doi.org/10.1016/j.fct.2016.07.032.

Meng, X.M., Ren, G.L., Gao, L., Yang, Q., Li, H.D., Wu, W.F., Huang, C., Zhang, L., Lv, X.W., Li, J., 2018. NADPH oxidase 4 promotes cisplatin-induced acute kidney injury via ROS-mediated programmed cell death and inflammation. Lab. Invest. 98, 63–78. https://doi.org/10.1038/labinvest.2017.120.

Mirabito Colafella, K.M., Bovée, D.M., Danser, A.H.J., 2019. The renin-angiotensin-aldosterone system and its therapeutic targets. Exp. Eye Res. 186, 107680. https://doi.org/10.1016/j.exer.2019.05.020.

Modaresi, A., Nafar, M., Sahrae, Z., 2015. Oxidative stress in chronic kidney disease. Iran. J. Kidney Dis. 9, 165–179.

Modarresi, A., Nafar, M., Sahraei, Z., Salamzadeh, J., Chaibakhsh, S., Ziaie, S., Parvin, M., Panahi, Y., Einollahi, B., 2018. N-acetylcysteine decreases urinary level of neutrophil gelatinase-associated lipocalin in deceased-donor renal transplant recipients: a randomized clinical trial. Biomarkers 23, 589–596. https://doi.org/10.1080/1354750X.2018.1468823.

Moghaddam, A.H., Nabavi, S.M., Nabavi, S.F., Bigdellou, R., Mohammadzadeh, S., Ebrahimzadeh, M.A., 2012. Antioxidant, antihemolytic and nephroprotective activity of aqueous extract of diospyros lotus seeds. Acta. Pol. Pharm. 69, 687–692.

Morales-Buenrostro, L.E., Ortega-Trejo, J.A., Pérez-Villalva, R., Marino, L.A., González-Bobadilla, Y., Juárez, H., Zamora-Mejía, F.M., González, N., Espinoza, R., Barrera-Chimal, J., Bobadilla, N.A., 2019. Spironolactone reduces oxidative stress in living donor kidney transplantation: a randomized controlled trial. Am. J. Physiol. Renal 317, F519–F528. https://doi.org/10.1152/ajprenal.00606.2018.

Moreillon, J.J., Bowden, R.G., Deike, E., Griggs, J., Wilson, R., Shelmadine, B., Cooke, M., Beaujean, A., 2013. The use of an anti-inflammatory supplement in patients with chronic kidney disease. Complement Integr. Med. 10. /j/jcim.2013.10.issue-1/jcim-2012-0011/jcim-2012-0011.xml. https://doi.org/10.1515/jcim-2012-0011.

Motawi, T.K., Ahmed, S.A., El-Boghdady, N.A., Metwally, N.S., Nasr, N.N., 2020. Impact of betanin against paracetamol and diclofenac induced hepato-renal damage in rats. Biomarkers 25, 86–93. https://doi.org/10.1080/1354750x.2019.1697365.

Mustata, G.T., Rosca, M., Biemel, K.M., Reihl, O., Smith, M.A., Viswanathan, A., Strauch, C., Du, Y., Tang, J., Kern, T.S., Lederer, M.O., Brownlee, M., Weiss, M.F., Monnier, V.M., 2005. Paradoxical effects of green tea (*Camellia sinensis*) and antioxidant vitamins in diabetic rats: improved retinopathy and renal mitochondrial defects but deterioration of collagen matrix glycoxidation and cross-linking. Diabetes 54, 517–526. https://doi.org/10.1016/j.foodres.2019.108862.

Nakata, R., Takahashi, S., Inoue, H., 2012. Recent advances in the study on resveratrol. Biol. Pharm. Bull. 35, 273–279. https://doi.org/10.1248/bpb.35.273.

Navarro-Yepes, J., Burns, M., Anandhan, A., Khalimonchuk, O., del Razo, L.M., Quintanilla-Vega, B., Pappa, A., Panayiotidis, M.I., Franco, R., 2014. Oxidative stress, redox signaling, and autophagy: cell death versus survival. Antioxid. Redox Signal. 21, 66–85. https://doi.org/10.1089/ars.2014.5837.

Nilakantan, V., Hilton, G., Maenpaa, C., Van Why, S.K., Pieper, G.M., Johnson, C.P., Shames, B.D., 2007. Favorable balance of anti-oxidant/pro-oxidant systems and ablated oxidative stress in Brown Norway rats in renal ischemia-reperfusion injury. Mol. Cell. Biochem. 304, 1–11. https://doi.org/10.1007/s11010-007-9480-z.

Ozbek, E., 2012. Induction of oxidative stress in kidney. Int. J. Nephrol., 465897. https://doi.org/10.1155/2012/465897.

Ozturk, O., Eroglu, H.A., Ustebay, S., Kuzucu, M., Adali, Y., 2018. An experimental study on the preventive effects of N-acetyl cysteine and ozone treatment against contrast-induced nephropathy. Acta Cir. Bras. 33, 508–517. https://doi.org/10.1590/s0102-865020180060000005.

Pakfetrat, M., Akmali, M., Malekmakan, L., Dabaghimanesh, M., Khorsand, M., 2015. Role of turmeric in oxidative modulation in end-stage renal disease patients. Hemodial. Int. 19, 124–131. https://doi.org/10.1111/hdi.12204.

Palli, E., Makris, D., Papanikolaou, J., Garoufalis, G., Tsilioni, I., Zygoulis, P., Zakynthinos, E., 2017. The impact of N-acetylcysteine and ascorbic acid in contrast-induced nephropathy in critical care patients: an open-label randomized controlled study. Crit. Care 21, 269. https://doi.org/10.1186/s13054-017-1862-3.

Park, W.S., Park, M.S., Kang, S.W., Jin, S.A., Jeon, Y., Hwang, J., Kim, S.K., 2019. Hesperidin shows protective effects on renal function in ischemia-induced acute kidney injury (Sprague-Dawley rats). Transplant. Proc. 51, 2838–2841. https://doi.org/10.1016/j.transproceed.2019.02.055.

Perazella, M.A., 2015. The urine sediment as a biomarker of kidney disease. Am. J. Kidney Dis. 66, 748–755. https://doi.org/10.1053/j.ajkd.2015.02.342.

Pergola, P.E., Raskin, P., Toto, R.D., Meyer, C.J., Huff, J.W., Grossman, E.B., Krauth, M., Ruiz, S., Audhya, P., Christ-Schmidt, H., Wittes, J., Warnock, D.G.BEAM Study Investigators, 2011. Bardoxolone methyl and kidney function in CKD with type 2 diabetes. N. Engl. J. Med., 365, 327–336. https://doi.org/10.1056/NEJMoa1105351.

Pisoschi, A.M., Pop, A., 2015. The role of antioxidants in the chemistry of oxidative stress: a review. Eur. J. Med. Chem. 97, 55–74. https://doi.org/10.1016/j.ejmech.2015.04.040.

Popolo, A., Autore, G., Pinto, A., Marzocco, S., 2013. Oxidative stress in patients with cardiovascular disease and chronic renal failure. Free Radic. Res. 47, 346–356. https://doi.org/10.3109/10715762.2013.779373.

Rababa'h, A.M., Hijjawi, T.B., Alzoubi, K.H., Al Demour, S., Ababneh, M.A., 2018. The nephroprotective effect of N-acetyl-L-cysteine and atorvastatin against imipenem induced nephrotoxicity. Curr. Mol. Pharmacol. 11, 155–161. https://doi.org/10.2174/1874467210666170731115928.

Ramalingam, A., Santhanathas, T., Shaukat Ali, S., Zainalabidin, S., 2019. Resveratrol supplementation protects against nicotine-induced kidney injury. Int. J. Environ. Res. Public Health 16, 4445. https://doi.org/10.3390/ijerph16224445.

Rana, M.N., Tangponga, J., Rahman, M.A., 2019. Xanthones protects lead-induced chronic kidney disease (CKD) via activating Nrf-2 and modulating NF-kB, MAPK pathway. Biochem. Biophys. Rep. 10, 100718. https://doi.org/10.1016/j.bbrep.2019.100718.

Reiter, R.J., Mayo, J.C., Tan, D.X., Sainz, R.M., Alatorre-Jimenez, M., Qin, L., 2016. Melatonin as an antioxidant: under promises but over delivers. J. Pineal Res. 61, 253–278. https://doi.org/10.1111/jpi.12360.

Rezaei, Y., Khademvatani, K., Rahimi, B., Khoshfetrat, M., Arjmand, N., Seyyed-Mohammadzad, M.H., 2016. Short-term high-dose vitamin E to prevent contrast medium-induced acute kidney injury in patients with chronic kidney disease undergoing elective coronary angiography: a randomized placebo-controlled trial. J. Am. Heart Assoc. 5, e002919. https://doi.org/10.1161/JAHA.115.002919.

Rizk, D.V., Silva, A.L., Pergola, P.E., Toto, R., Warnock, D.G., Chin, M.P., Goldsberry, A., O'Grady, M., Meyer, C.J., McCullough, P.A., 2019. Effects of bardoxolone methyl on magnesium in patients with type 2 diabetes mellitus and chronic kidney disease. Cardiorenal. Med. 9, 316–325. https://doi.org/10.1159/000500612.

Saad, A., Herrmann, S., Eirin, A., Ferguson, C.M., Glockner, J.F., Bjarnason, H., McKusick, M.A., Misra, S., Lerman, L.O., Textor, S.C., 2017. Phase 2a clinical trial of mitochondrial protection (elamipretide) during stent revascularization in patients with atherosclerotic renal artery stenosis. Circ. Cardiovasc. Interv. 10, e005487. https://doi.org/10.1161/CIRCINTERVENTIONS.117.005487.

Sabu, M., Smitha, K., Kuttan, R., 2002. Anti-diabetic activity of green tea polyphenols and their role in reducing oxidative stress in experimental diabetes. J. Ethnopharmacol. 83, 109–116. https://doi.org/10.1016/s0378-8741(02)00217-9.

Safari, T., Shahraki, M.R., Miri, S., Mirakzehi Bakhshani, N., Niazi, A.A., Komeili, G.R., Bagheri, H., 2019. The effect of angiotensin 1-7 and losartan on renal ischemic/reperfusion injury in male rats. Res. Pharm. Sci. 14, 441–447. https://doi.org/10.4103/1735-5362.268205.

Saldanha, J.F., Leal, V.O., Rizzetto, F., Grimmer, G.H., Ribeiro-Alves, M., Daleprane, J.B., Carraro-Eduardo, J.C., Mafra, D., 2016. Effects of resveratrol supplementation in Nrf2 and NF-κB expressions in nondialyzed chronic kidney disease patients: a randomized, double-blind, placebo-controlled, crossover clinical trial. J. Ren. Nutr. 26, 401–406. https://doi.org/10.1053/j.jrn.2016.06.005.

Saleem, T.H., Abo El-Maali, N., Hassan, M.H., Mohamed, N.A., Mostafa, N.A.M., Abdel-Kahaar, E., Tammam, A.S., 2018. Comparative protective effects of N-acetylcysteine, N-acetyl methionine, and N-acetyl glucosamine against paracetamol and phenacetin therapeutic doses-induced hepatotoxicity in rats. Int. J. Hepatol. 2018, 7603437. https://doi.org/10.1155/2018/7603437.

Samadian, F., Dalili, N., Poor-Reza Gholi, F., Fattah, M., Malih, N., Nafar, M., Firoozan, A., Ahmadpoor, P., Samavat, S., Ziaie, S., 2017. Evaluation of curcumin's effect on inflammation in hemodialysis patients. Clin. Nutr. ESPEN 22, 19–23. https://doi.org/10.1016/j.clnesp.2017.09.006.

Samiei, F., Sajjadi, H., Jamshidzadeh, A., Seydi, E., Pourahmad, J., 2019. Contrasting role of concentration in rivaroxaban induced toxicity and oxidative stress in isolated kidney mitochondria. Drug Res. (Stuttg). 69, 523–527. https://doi.org/10.1055/a-1001-2154.

Sampaio, T.L., Menezes, R.R., da Costa, M.F., Meneses, G.C., Arrieta, M.C., Chaves Filho, A.J., Morais, G.B., Libório, A.B., Alves, R.S., Evangelista, J.S., Martins, A.M., 2016.

Nephroprotective effects of (-)-α-bisabolol against ischemic-reperfusion acute kidney injury. Phytomedicine 23, 1843–1852. https://doi.org/10.1016/j.phymed.2016.11.008.

Sánchez-Pozos, K., Barrera-Chimal, J., Garzón-Muvdi, J., Pérez-Villalva, R., Rodríguez-Romo, R., Cruz, C., Gamba, G., Bobadilla, N.A., 2012. Recovery from ischemic acute kidney injury by spironolactone administration. Nephrol. Dial. Transplant. 27, 3160–3169. https://doi.org/10.1093/ndt/gfs014.

Santana-Santos, E., Gowdak, L.H., Gaiotto, F.A., Puig, L.B., Hajjar, L.A., Zeferino, S.P., Drager, L.F., Shimizu, M.H., Bortolotto, L.A., De Lima, J.J., 2014. High dose of N-acetylcystein prevents acute kidney injury in chronic kidney disease patients undergoing myocardial revascularization. Ann. Thorac. Surg. 97, 1617–1623. https://doi.org/10.1016/j.athoracsur.2014.01.056.

Scherz-Shouval, R., Elazar, Z., 2007. ROS, mitochondria and the regulation of autophagy. Trends Cell Biol. 17, 422–427. https://doi.org/10.1016/j.tcb.2007.07.009.

Serhan, C.N., Chiang, N., Van Dyke, T.E., 2008. Resolving inflammation: dual anti-inflammatory and pro-resolution lipid mediators. Nat. Rev. Immunol. 8, 349–361. https://doi.org/10.1038/nri2294.

Shahbazian, H., Shayanpour, S., Ghorbani, A., 2016. Evaluation of administration of oral N-acetylcysteine to reduce oxidative stress in chronic hemodialysis patients: a double-blind, randomized, controlled clinical trial. Saudi. J. Kidney Dis. Transpl. 27, 88–93. https://doi.org/10.4103/1319-2442.174084.

Sharma, A., Fonarow, G.C., Butler, J., Ezekowitz, J.A., Felker, G.M., 2016. Coenzyme Q10 and heart failure: a state-of-the-art review. Circ. Heart Fail. 9, e002639. https://doi.org/10.1161/CIRCHEARTFAILURE.115.002639.

Shebeko, S.K., Zupanets, I.A., Popov, O.S., 2019a. N-acetylglucosamine is the most effective glucosamine derivative for the treatment of membranous nephropathy in rats. Pharmazie 74, 667–670. https://doi.org/10.1691/ph.2019.9733.

Shebeko, S.K., Zupanets, I.A., Tarasenko, O.O., 2019b. Nephroprotective effect of N-acetylglucosamine in rats with acute kidney injury. Ceska Slov. Farm. 68, 173–179.

Shi, S., Lei, S., Tang, C., Wang, K., Xia, Z., 2019. Melatonin attenuates acute kidney ischemia/reperfusion injury in diabetic rats by activation of the SIRT1/Nrf2/HO-1 signaling pathway. Biosci. Rep. 39, BSR20181614. https://doi.org/10.1042/BSR20181614.

Shin, H.S., Yu, M., Kim, M., Choi, H.S., Kang, D.H., 2014. Renoprotective effect of red ginseng in gentamicin-induced acute kidney injury. Lab. Invest. 94, 1147–1160. https://doi.org/10.1038/labinvest.2014.101.

Silveira, M., Teles, F., Berretta, A.A., Sanches, T.R., Rodrigues, C.E., Seguro, A.C., Andrade, L., 2019. Effects of Brazilian green propolis on proteinuria and renal function in patients with chronic kidney disease: a randomized, double-blind, placebo-controlled trial. BMC Nephrol. 20, 140. https://doi.org/10.1186/s12882-019-1337-7.

Small, D.M., Coombes, J.S., Bennett, N., Johnson, D.W., Gobe, G.C., 2012. Oxidative stress, anti-oxidant therapies and chronic kidney disease. Nephrology (Carlton). 17, 311–321. https://doi.org/10.1111/j.1440-1797.2012.01572.x.

Sodré, F.L., Costa, J.C.B., Lima, J.C.C., 2007. Avaliação da função e da lesão renal: um desafio laboratorial. J. Bras. Pato. Med. Lab. 43, 329–337. https://doi.org/10.1590/s1676-24442007000500005.

Sporn, M.B., Liby, K.T., Yore, M.M., Fu, L., Lopchuk, J.M., Gribble, G.W., 2011. New synthetic triterpenoids: potent agents for prevention and treatment of tissue injury caused by inflammatory and oxidative stress. J. Nat. Prod. 74, 537–545. https://doi.org/10.1021/np100826q.

Su, L.J., Zhang, J.H., Gomez, H., Murugan, R., Hong, X., Xu, D., Jiang, F., Peng, Z.Y., 2019. Reactive oxygen species-induced lipid peroxidation in apoptosis, autophagy, and ferroptosis. Oxid. Med. Cell Longev., 5080843. https://doi.org/10.1155/2019/5080843.

Su, Z., Ye, J., Qin, Z., Ding, X., 2015. Protective effects of madecassoside against doxorubicin induced nephrotoxicity in vivo and in vitro. Sci. Rep. 5, 18314. https://doi.org/10.1038/srep18314.

Sureshbabu, A., Ryter, S.W., Choi, M.E., 2015. Oxidative stress and autophagy: crucial modulators of kidney injury. Redox Biol. 4, 208–214. https://doi.org/10.1016/j.redox.2015.01.001.

Szeto, H.H., Liu, S., Soong, Y., Birk, A.V., 2015. Improving mitochondrial bioenergetics under ischemic conditions increases warm ischemia tolerance in the kidney. Am. J. Physiol. Renal Physiol. 308, F11–F21. https://doi.org/10.1152/ajprenal.00366.2014.

Tamadon, M.R., Zahmatkesh, M., Beladi Mousavi, S.S., 2015. Administration of antioxidants in chronic kidney disease. J. Nephropharmacol. 4, 9–11.

Tan, S.M.Q., Chiew, Y., Ahmad, B., Kadir, K.A., Tan, S., Chiew, Y., Ahmad, B., Kadir, K.A., 2018. Tocotrienol-rich vitamin E from palm oil (Tocovid) and its effects in diabetes and diabetic nephropathy: a pilot phase II clinical trial. Nutrients 10, 1315. https://doi.org/10.3390/nu10091315.

Tavafi, M., 2014. Antioxidants against contrast media induced nephrotoxicity. J. Ren. Inj. Prev. 3, 55–56. https://doi.org/10.12861/jrip.2014.18.

Thabet, M.A., Chan, J.C, 2006. Vitamin E in renal therapeutic regiments. Pediatr. Nephrol. 21, 1790–1801. https://doi.org/10.1007/s00467-006-0211-6.

Tong, F., Liu, S., Yan, B., Li, X., Ruan, S., Yang, S., 2017. Quercetin nanoparticle complex attenuated diabetic nephropathy via regulating the expression level of ICAM-1 on endothelium. Intl. J. Nanomed. 12, 7799–7813. https://doi.org/10.2147/IJN.S146978.

Tsai, B.C.K., Hsieh, D.J.Y., Lind, W.T., Tamilselvi, S., Day, C.H., Hog, T.J., Chang, R.L., Viswanadha, V.P., Kuo, C.H., Huan, C.Y, 2020. Functional potato bioactive peptide intensifies Nrf2-dependent antioxidant defense against renal damage in hypertensive rats. Food Res. Int. 129, 108862. https://doi.org/10.1016/j.foodres.2019.108862.

Tsuchiya, T., Kijima, A., Ishii, Y., Takasu, S., Yokoo, Y., Nishikawa, A., Yanai, T., Umemura, T., 2018. Mechanisms of oxidative stress-induced in vivo mutagenicity by potassium bromate and nitrofurantoin. J. Toxicol. Pathol. 31, 179–188. https://doi.org/10.1293/tox.2018-0024.

Tzeng, T.-F., Liou, S.-S., Chang, C.J., Liu, I.-M., 2013. The ethanol extract of *Zingiber zerumbet* attenuates streptozotocin-induced diabetic nephropathy in rats. Evid. Based Complement. Altern. Med., 340645. https://doi.org/10.1155/2013/340645.

Ugur, S., Ulu, R., Dogukan, A., Gurel, A., Yigit, I.P., Gozel, N., Aygen, B., Ilhan, N., 2015. The renoprotective effect of curcumin in cisplatin-induced nephrotoxicity. Ren. Fail. 37, 332–336. https://doi.org/10.3109/0886022X.2014.986005.

Veiga, G., Alves, B., Perez, M., Alcantara, L.V., Raimundo, J., Zambrano, L., Encina, J., Pereira, E.C., Bacci, M., Murad, N., Fonseca, F., 2020. NGAL and SMAD1 gene expression in the early detection of diabetic nephropathy by liquid biopsy. J. Clin. Pathol., 206494. https://doi.org/10.1136/jclinpath-2020-206494.

Vera, M., Torramade-Moix, S., Martin-Rodriguez, S., Cases, A., Cruzado, J.M., Rivera, J., Escolar, G., Palomo, M., Diaz-Ricart, M., 2018. Antioxidant and anti-inflammatory strategies based on the potentiation of glutathione peroxidase activity prevent endothelial dysfunction in chronic kidney disease. Cell Physiol. Biochem. 51, 1287–1300. https://doi.org/10.1159/000495540.

Wang, W., Shen, J., Qi, C., Pu, J., Chen, H., Zuo, Z., 2020. The key candidate genes in tubulointerstitial injury of chronic kidney diseases patients as determined by bioinformatic analysis. Cell. Biochem. Funct. https://doi.org/10.1002/cbf.3545.

Wasung, M.E., Chawla, L.S., Madero, M., 2015. Biomarkers of renal function, which and when? Clin. Chim. Acta 438, 350–357. https://doi.org/10.1016/j.cca.2014.08.039.

Webster, A.C., Nagler, E.V., Morton, R.L., Masson, P., 2016. Chronic kidney disease. Lancet North Am. Ed. 389, 1238–1252. https://doi.org/10.1016/s0140-6736(16)32064-5.

Wu, D., Wen, W., Qi, C.L., Zhao, R.X., Lü, J.H., Zhong, C.Y., Chen, Y.Y., 2012. Ameliorative effect of berberine on renal damage in rats with diabetes induced by high-fat diet and streptozotocin. Phytomedicine 19, 712–718. https://doi.org/10.1016/j.phymed.2012.03.003.

Xu, S., Nam, S.M., Kim, J.H., Das, R., Choi, S.K., Nguyen, T.T., Quan, X., Choi, S.J., Chung, C.H., Lee, E.Y., Lee, I.K., Wiederkehr, A., Wollheim, C.B., Cha, S.K., Park, K.S., 2015. Palmitate induces ER calcium depletion and apoptosis in mouse podocytes subsequent to mitochondrial oxidative stress. Cell. Death Dis. 6, e1976. https://doi.org/10.1038/cddis.2015.331.

Xu, Y., Liu, J., Han, E., Wang, Y., Gao, J., 2019. Efficacy of coenzyme Q10 in patients with chronic kidney disease: protocol for a systematic review. BMJ Open 9, e029053. https://doi.org/10.1136/bmjopen-2019-029053.

Yang, C.W., Harris, D.C., Luyckx, V.A., Nangaku, M., Hou, F.F., Garcia, G.G., Abu-Aisha, H, Niang, A., Sola, L., Bunnag, S., Eiam-Ong, S., Tungsanga, K., Richards, M., Richards, N., Goh, B.L., Dreyer, G., Evans, R., Mzingajira, H., Twahir, A., McCulloch, M.I, Ahn, C., Osafo, C., Hsu, H.H., Barnieh, L., Donner, JA, Tonelli, M., 2020. Global case studies for chronic kidney disease/end-stage kidney disease care. Kidney Int. Suppl. *10* (1), e24–e48. https://doi.org/10.1016/j.kisu.2019.11.010.

Yaribeygi, H., Mohammadi, M.T., Rezaee, R., Sahebkar, A., 2018b. Crocin improves renal function by declining Nox-4, IL-18, and p53 expression levels in an experimental model of diabetic nephropathy. J. Cell. Biochem. 119, 6080–6093. https://doi.org/10.1002/jcb.26806.

Yaribeygi, H., Simental Mendía, L.E., Butler, A.E., Sahebkar, A., 2018a. Protective effects of plant-derived natural products on renal complications. J. Cell. Physiol. 234, 12161–12172. https://doi.org/10.1002/jcp.27950.

Yates, M.S., Tauchi, M., Katsuoka, F., Flanders, K.C., Liby, K.T., Honda, T., Gribble, G.W., Johnson, D.A., Johnson, J.A., Burton, N.C., Guilarte, T.R., Yamamoto, M., Sporn, M.B., Kensler, T.W., 2007. Pharmacodynamic characterization of chemo preventive triterpenoids as exceptionally potent inducers of Nrf2-regulated genes. Mol. Cancer Ther. 6, 154–162. https://doi.org/10.1158/1535-7163.MCT-06-0516.

Zaazaa, A.M., Motelp, B.A.A.E., Aniss, N.N., 2019. Potential protective role of rutin and alpha-lipoic acid against cisplatin-induced nephrotoxicity in rats. Pak. J. Biol. Sci. 22, 361–371. https://doi.org/10.3923/pjbs.2019.361.371.

Zaccaria, V., Curti, V., Di Lorenzo, A., Baldi, A., Maccario, C., Sommatis, S., Mocchi, R., Daglia, M., 2017. Effect of green and brown propolis extracts on the expression levels of microRNAs, mRNAs and proteins, related to oxidative stress and inflammation. Nutrients 9, 1090. https://doi.org/10.3390/nu9101090.

Zargar, S., Alonazi, M., Rizwana, H., Wani, T.A., 2019. Resveratrol reverses thioacetamide-induced renal assault with respect to oxidative stress, renal function, DNA damage, and cytokine release in Wistar rats. Oxid. Med. Cell Longev. 2019, 1702959. https://doi.org/10.1155/2019/1702959.

Zhang, H., Zhang, W., Jiao, F., Li, X., Zhang, H., Wang, L., Gong., Z., 2018. The nephroprotective effect of MS-275 on lipopolysaccharide (LPS)-induced acute kidney injury by inhibiting reactive oxygen species (ROS)-oxidative stress and endoplasmic reticulum stress. Med. Sci. Monit. 24, 2620–2630. https://doi.org/10.12659/MSM.906362.

Zheng, H., Whitman, S.A., Wu, W., Wondrak, G.T., Wong, P.K., Fang, D., Zhang, D.D., 2011. Therapeutic potential of Nrf2 activators in streptozotocin-induced diabetic nephropathy. Diabetes 60, 3055–3066. https://doi.org/10.2337/db11-0807.

Zhou, J.Q., Qiu, T., Zhang, L., Chen, Z.B., Wang, Z.S., Ma, X.X., Li, D., 2016. Allopurinol preconditioning attenuates renal ischemia/reperfusion injury by inhibiting HMGB1 expression in a rat model. Acta Cir. Bras. 31, 176–182. https://doi.org/10.1590/S0102-865020160030000005.

CHAPTER

Antioxidants and liver diseases

5.10

Jyoti Upadhyay[a], Nidhi Tiwari[b], Sumit Durgapal[b,c], Mohammad Hosein Farzaei[c]

[a]*School of Health Science and Technology, Department of Pharmaceutical Sciences, University of Petroleum and Energy Studies, Dehradun, Uttarakhand, India*

[b]*Department of Pharmaceutical Sciences, Kumaun University, Campus Bhimtal, Bhimtal, Uttarakhand, India*

[c]*Pharmaceutical Sciences Research Center, Health Institute, Kermanshah University of Medical Sciences, Kermanshah, Iran*

5.10.1 Introduction

Antioxidants are natural or synthetic substances which when present at low concentration compared to that of a substrate (i.e., oxidizable), extensively inhibit or delay the oxidation of that substrate. The main physiological role of these man-made or naturally occurring plant substances is to prevent cellular damage caused by the chemical reactions including free radicals. Antioxidants work by scavenging free radicals. They act as a donor of electron and hydrogen, peroxide decomposer, enzyme inhibitor, metal chelator, singlet oxygen quencher, and synergist. Defense mechanism includes antioxidants (endogenous or exogenous) that inhibits or control generation of free radicals. They are effective because of their ability of donating their electrons to free radicals and neutralizing the undesirable effect of the latter. Intracellular enzymes are used by the cells to defend them against free radical oxidative damage and maintain the homeostasis of free radicals at low level. The level of free radicals increases during cellular dysfunction and environmental stress that significantly damages the cells in the body. This oxidative damage by free radicals involved in pathogenesis of inflammatory diseases, liver diseases, Alzheimer, cataracts, diabetes, autism, and ageing (Lu et al., 2010). Antioxidant molecule inhibits oxidative damage by their free radical scavenging activity (Lobo et al., 2010). As liver is the major organ of the body, therefore it is the target organ attacked by free radicals. The primary cells present in the liver are parenchymal cells which are supposed to get oxidatively damaged by free radicals as mitochondria, peroxisomes; microsomes are present in these cells generates free radicals, regulates nuclear receptor peroxisome proliferators-activated receptor-alpha involved in fatty acid beta oxidation gene expression (Li et al., 2015).

Antioxidants Effects in Health. DOI: https://doi.org/10.1016/B978-0-12-819096-8.00049-5

5.10.1.1 Free radicals and oxidative stress

Free radicals are molecular species that are capable of having self-governing existence and contains unpaired number of electrons in an atomic orbital (Halliwell and Gutteridge, 1989). Unpaired electrons have some common properties shared by most of the free radicals. They are weakly attracted toward magnetic field and are thought to be paramagnetic. Because of their high reactivity free radicals can either donate an electron to other molecule or extract an electron from other molecule and hence acting as oxidant or reductant. These radicals have very short half-life, that is, 10^{-6} second or less in biological systems although some radicals may survive for much longer time. The most common free radicals in various disease states are derivatives of oxygen, particularly hydroxyl and superoxide radicals. Various endogenous and environmental factors are the possible mechanisms behind the formation of free radicals in the body (Young and Woodside, 2001). The term "reactive oxygen species" (ROS) is collectively used for a group of oxidants that are either free radical species or species which are capable of producing free radicals. Nitric oxide (NO^*) and superoxide (O_2^{*-}) radicals are ROS generated intracellularly. Under normal physiological environment, approximately 2% of the oxygen consumed by the body and through cellular processes like mitochondrial respiration and phagocytosis it gets converted to O_2^{*-}. The percentages of ROS increases during exercise, infection, UV light, radiations, exposure to environmental pollutants, etc. Nitric oxide (NO^*) a neurotransmitter and an endothelial relaxing factor produced by enzyme nitric oxide synthase. Both O_2^{*-} and NO^* radicals get transformed to potent oxidizing radicals like alkoxy (RO^*), hydroxyl ($^*OH$), peroxyl (ROO^*), and singlet oxygen ($^1O_2$). Molecular oxidants like hypochlorous acid (HOCl), peroxynitrite ($ONOO^*$), and peroxides (H_2O_2) are generated from free radical conversions, and they act as a major source of ROS (Kunwar and Priyadarshini, 2011). Fig. 5.10.1 represents several pathways involved in free radical generation.

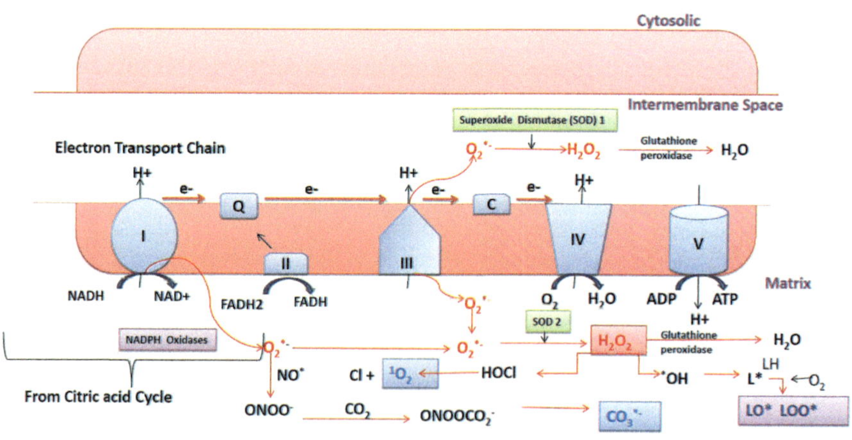

FIG. 5.10.1 Free radical production in mitochondria.

Like, at physiological concentration of carbon dioxide (CO_2), $ONOO^-$ becomes a source of anionic carbonate radical (CO_3^{*-}). Hydroxyl ($^*OH$) radical is produced from H_2O_2 by Fenton reaction and singlet oxygen ($^1O_2$) is produced from reaction of HOCl with H_2O_2 (Winterbourn, 2008).

Formation of free radicals takes place in the cells as a result of both nonenzymatic and enzymatic reactions. Enzymatic reaction source involves reactions in the respiratory chain, cytochrome system, phagocytosis, and prostaglandin synthesis (Lobo et al., 2010). Nonenzymatic reactions source includes ionizing reactions and reactions of oxygen with various organic compounds. Internally free radicals are generated from mitochondria, peroxisomes, xanthine oxidase, phagocytosis, inflammation, arachidonic pathways, ischemia/reperfusion injury, and exercise. The external sources of free radical generation are environmental pollutants, cigarette smoke, radiations, industrial solvents, pesticides, drugs, and ozone (Ebadi, 2001).

5.10.1.2 Free radicals biology

In normal physiologic condition, free radicals at low concentration plays major role in various body functions like normal cellular growth, gene expression, and immunological functions. In some of the cellular biochemical processes they also act as a stimulating agent (Droge, 2002). ROS functions through active site oxidation (reversible process) in transcription factors like activator protein (AP-1) and nuclear factor kappa B (NF-κB) responsible for gene expression and cell growth. They also activate signal transduction pathways and causes transcription factor induction indirectly for example mitogen-activated protein kinases. Free radicals also regulate embryonic development in mammals and also participate in biosynthesis of endogenous molecules like prostaglandins, thyroxin hence therefore, accelerating the developmental process (Schreck and Baeuerle, 1991). Generation of free radicals helps in induction of cell apoptosis by mediating cytotoxicity of tumor and cells infected with viruses (Schreck and Baeuerle, 1991; Droge, 2002). Reactions of free radicals produce undesirable changes which accumulate with age throughout the body. These undesirable changes are normal and common to all. However, when this normal pattern gets superimposed by the environmental and genetic factors, they modulate free radical damage and manifests diseases like cancer, liver, and cardiovascular disorders (Lobo et al., 2010).

5.10.1.3 Oxidative stress in liver caused by free radicals

Oxidative damage occurs when there is an unfavorable imbalance occurs between antioxidant defense and generation of free radicals. It is linked with damage caused by free radicals to molecular species involving proteins, lipids, and nucleic acids. Free radicals are capable of damaging cell membranes and biomolecules like DNA, lipids, proteins, and carbohydrates and also cause homeostatic disruption. Their targets include all biomolecules in the body and proteins, lipids, and nucleic acids are major targets among them (Fig. 5.10.2). The oxidative damage caused

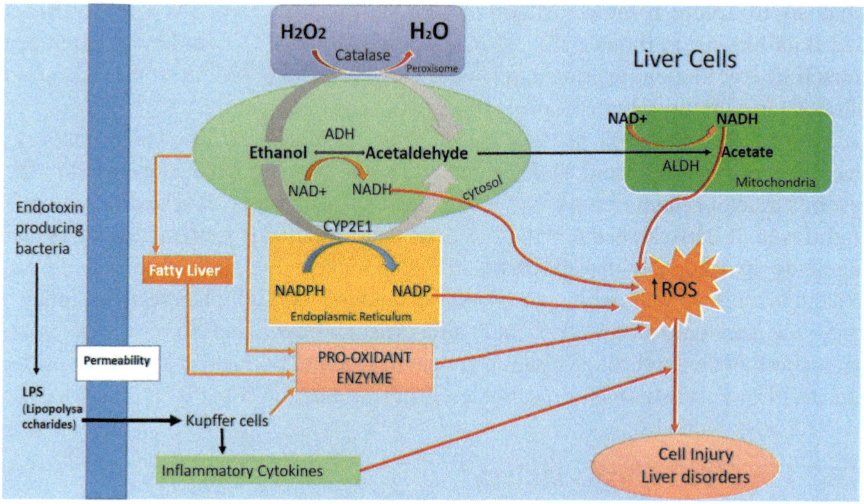

FIG. 5.10.2 Generation of free radicals contributing liver disorders.

by free radicals arises because of their high reactivity and ability to instigate free radical chain reaction. The radical which is formed by chain reaction produced a secondary radical and reacts with a nonradical. This ongoing process continues till the two radicals combined by pairing of electrons which is called as quenching of radicals (Dietz et al., 1999). Free radicals cause oxidative stress (OS) in liver by damaging the hepatic stellate cells, endothelial cells, and Kupffer cells oxidatively. It induces the production of tumor necrosis factor cytokines-alpha cytokines in Kupffer cells that causes inflammation and cell necrosis. Lipid peroxidation, caused by free radicals, triggers synthesis of collagen and cell proliferation of hepatic stellate cells (Wu and Cederbaum, 2009; Sakaguchi et al., 2011; Cichoz-Lach and Michalak, 2014). Lipid peroxidation, caused by secondary free radicals, acts either by directly reacting with biomolecules or as a second messenger causing biochemical lesions. Lipid peroxidation caused by reaction of ROS with polysaturated fatty acids situated on cell membrane and proceeds with chain reaction of free radicals. Lipid hydroperoxides (LOOH) are formed by lipid peroxidation and further gets decomposes to aldehyde like malonaldehyde or form isoprotans, cyclic endoperoxides, and hydrocarbons. These products are used as biomarkers of lipid peroxidation assay and confirmed many disease conditions like diabetes, ischemia reperfusion, and neurogenerative diseases (Lobo et al., 2010; Kunwar and Priyadarshini, 2011). OS not only causes alteration of proteins, lipids, and DNA but also causes modulation of pathways controlling biological functions, like gene expression, gene transcription, protein expression, cell necrosis, apoptosis, and activation of hepatic stellate cells. This oxidative damage is involved in pathogenesis of liver diseases like alcoholic liver disease, nonalcoholic steatohepatitis, and chronic viral hepatitis. Alcoholic liver disease progresses from steatosis to severe form of liver diseases like hepatitis and liver cirrhosis (Gao and Bataller, 2011; Banerjee et al., 2013). In this disease, ethanol metabolism is related to free radical production, steatosis, and mitochondrial injury, caused by chronic or acute alcoholic exposure (Li et al., 2015). Three different enzymatic pathways are involved in the metabolic process of ethanol. The first pathway is the primary pathway and involves dehydrogenase system, initiated by enzyme alcohol dehydrogenase, requiring NAD^+ and expressed in hepatocytes at a very high level. It causes oxidation of alcohol to acetaldehyde. Inside mitochondria, aldehyde hydrogenase acetaldehyde to acetate. Microsomal ethanol oxidizing system is the secondary pathway oxidizes ethanol, involves CYP2E1a cytochrome P450 enzyme. The third pathway requires catalase enzyme and H_2O_2 for oxidation of ethanol (Valente et al., 2012; Dey and Lakshmanan, 2013; Wang et al., 2015). During the metabolism of ethanol via primary and secondary pathway $NADP^+$ or NADH produced in excess causes increased production of ROS leading to OS resulting cell necrosis and finally triggers liver disorders (Fig. 5.10.2).

Some research studies demonstrated changes in the level of enzymes SOD (superoxide dismutase), CAT (catalase), glutathione peroxidase as well as the level of lipid peroxidation in rats treated with alcohol (Mallikarjuna et al., 2010; Shanmugam et al., 2011). Drug-induced OS in liver is an important indicator of hepatotoxicity as it increases the level of lipid peroxidation and cellular oxidants like drug sulfasalazine which is used for the treatment of inflammatory bowel disease induces hepatic damage oxidatively. This drug when administered orally it reduces SOD activity and significantly increases CAT activity (Linares, 2009). Paracetamol, an analgesic drug induces antioxidant capacity in liver during hepatic damage. This study was evaluated by Mladenovic et al. (2009), and it was observed that paracetamol causes significant increase in the level of malondialdehyde (MDA), an indicator of lipid peroxidation as well as nitrite and nitrate level in the liver with greater decline in the level of SOD activity. Environmental toxicants like heavy metals causes' OS in the liver of some animal models (Bando et al., 2005). Adegbesan and Adenuga (2007) showed in their study that mercuric chloride (0.1 mg/kg) induces significant fall in both, copper- and zinc-dependent SOD and manganese-dependent SOD activities. It also shows changes in the level of CAT enzyme and glutathione enzyme system with small increase in level of serum glutamyltransferase and alanine transaminase. Their results indicate that mercury at low dose can induce OS and finally causes liver disorders. Hepatic OS can also be induced by other factors like temperature, radiations, benzoyl peroxide, and high-fat diet. Effect of radiations from mobile phone exposure has been studied in guinea pig animal model and it was investigated that radiations produces OS by significantly increasing the level of MDA and nitric oxide in guinea pigs (Ozgur et al., 2010). Benzoyl peroxide has powerful oxidizing strength therefore it is widely used as flour bleaching agent. In liver it affects the ATPases and antioxidants as investigated in mice model (Jia et al., 2011). Zhang et al. (2011) investigated that high-fat diet when given to pregnant rats it causes increase in the level of triglycerides significantly and also increase in the lipid droplet size of rat neonates.

5.10.1.4 Preventive effect of antioxidants

5.10.1.4.1 Lipid peroxidation

Oxidative damage of lipids occurs in C=C bonds of unsaturated fatty acids, phospholipids, glycolipids, cholesterol, and esters of cholesterol. Free radicals attack unsaturated fatty acids containing methylene and several double bonds with hydrogen atoms (reactive) and instigate the chain reactions of peroxidation of radicals (Fig. 5.10.3; Lu et al., 2010). The breakdown of poly unsaturated fatty acids by peroxidation reaction has been observed in the pathogenesis of liver injuries especially liver damage caused by toxicants like carbon tetrachloride, haloalkanes, chloroform, trichlorobromomethane, halothane, ethanol, paracetamol, etc. have been shown to stimulate lipid peroxidation. In liver, the high proportions of unsaturated fatty acids are present in mitochondrial membrane and endoplasmic reticulum which are more vulnerable to peroxidation. Also, during the same time, enzymes of the electron transport chain are also present that are capable of generating free radicals and finally causing oxidative damage in liver. Fatty liver disorders comprise of liver diseases beginning with steatosis and progresses to steatohepatitis, a more advanced stage to liver cirrhosis and hepatocellular carcinoma. Steatohepatitis constitutes both alcoholic and nonalcoholic steatohepatitis, both the diseases share common clinical manifestations like inflammation, hepatocytes cell necrosis and fibrosis. In alcoholic and nonalcoholic steatohepatitis, it is believed that mitochondrial dysfunction and OS play a major role. Although other possible mechanism involves disease progression, that is, endoplasmic reticulum stress and impairment in autophagy. Therefore, nonalcoholic steatohepatitis is considered as mitochondrial disease (Garcia-Ruiz and Fernandez-Checa, 2018). ROS generated by electron transport chain interacts with lipids and produces endoperoxides and hydroperoxides, after fragmentation they generate highly reactive species such as 4-hydroxynonenal and MDA. With lipids, DNA, and proteins these intermediates form covalent adducts causing cell death. Reactive OS generally occurs in the genome of mitochondria because it is open and rounded without the protection of histone protein and also it is close to free radical species produced heavily inside the mitochondrion in comparison with nuclear region (Li et al., 2016). Antioxidants scavenge free radicals thereby preventing pathogenic conditions. Some studies on antioxidants like vitamins, flavones, ginsenosides, and polyphenols have been carried out using lipid, rat liver microsomes, human cells, and human LDL in micelles or homogenous solutions. In vitro peroxidation of lipids like linoleic acid can be initiated either by using azo initiator 2,2- azobis (2-amidinopropane) hydrochloride (AAPH) which is water soluble or initiation by Fe^{2+} or Cu^{+} with hydrogen peroxide, that is, Fenton reaction. At physiological conditions (i.e., temperature 37°C) AAPH decomposed in aqueous solution and generates alkyl radical (R); which in the presence of O_2 is converted to the subsequent peroxyl radicals (ROO^*). Due to its water solubility, nature of AAPH, the free radical generation rate can be easily measured and controlled. AAPH used as initiator for free radical generation and it is a good model for studying free radicals induced membrane damage (Gal et al., 2007). The inhibitory activity of antioxidants can be calculated as total antioxidant activity by comparing the generation kinetics

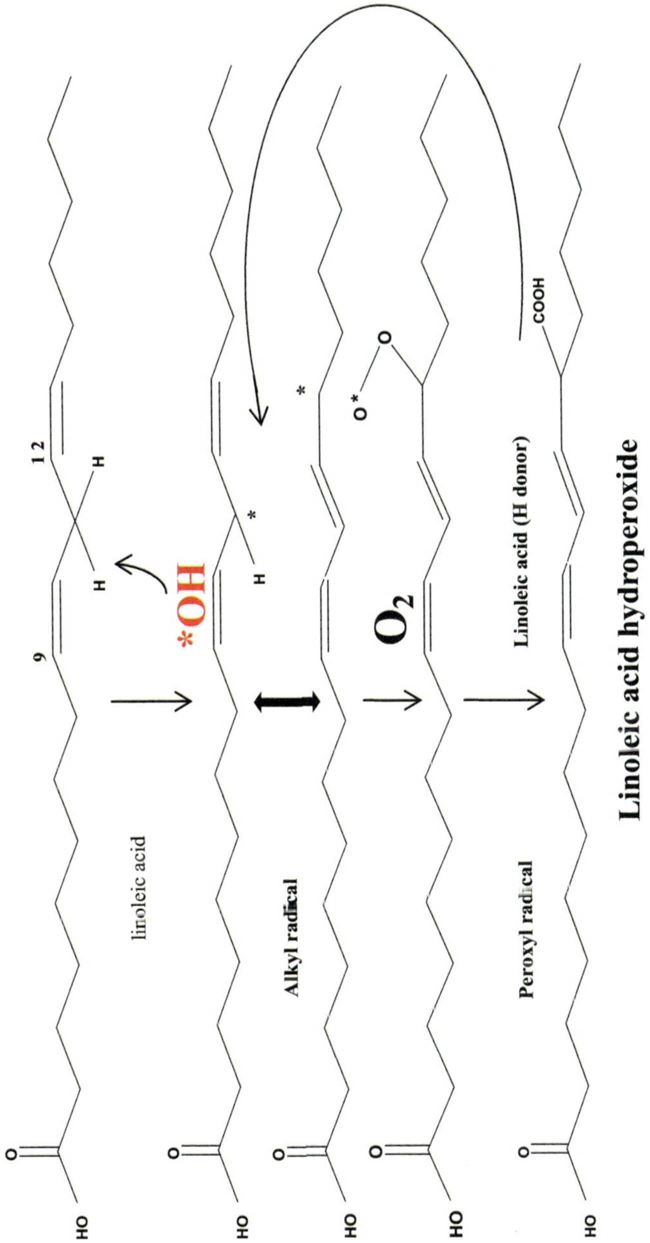

FIG. 5.10.3 Linoleic acid peroxidation initiated by •OH radical.

of H_2O_2 in the presence and absence of antioxidants. This experiment has been used to determine the important antioxidant used (Lindenmeier et al., 2002). The reduced activity of antioxidants is measured by using methyl ester of the product linoleic acid hydroperoxide as substrate and the difference calculated between total antioxidant activity and antioxidants reduced activity is measured as radical scavenging activity (Lindenmeier et al., 2007).

5.10.1.4.2 DNA damage

The free radicals ($^*OH$ and $ONOO^-$) generated from nitric oxide and O_2^{*-} can react directly with DNA macromolecule of the plasmid in vivo and they nick or cleave one strand of DNA causes oxidative damage to DNA. Hepatocellular carcinoma and primary liver cancers are the most common frequent cause of cancer death and are related with numerous risk factors involving viral hepatitis (B and C), alcohol consumption, liver infection, and obesity. Hepatocellular carcinoma arises because of chronic hepatocytes inflammation and cell injury thereby endorsing chromosomal aberrations and DNA damage which stimulates a set of signaling pathways like DNA damage response that coordinate DNA repair, cell cycle arrest, and finally cell death or senescence (Yang et al., 2014). In hepatocellular carcinoma, various types of DNA damage and their repair mechanisms have been implicated like homologous recombination of DNA replication fork (Chung and Wu, 2013), mismatch repair mechanism of nitrogenous bases (Bayram et al., 2012) and most important form of DNA damage is double strand break (Hsu et al., 2013) repair by nonhomologous end joining (Curtin, 2012). DNA damage response aberrations may damage genomic integrity triggering carcinogenesis of hepatic cells and facilitates the development of hepatocellular carcinoma. It has been recognized that genomic instability and alteration are the common characteristics of human hepatocellular carcinoma. Therefore, a better knowledge of DNA damage response pathways is required for developing strategies in the prevention and treatment of hepatocellular carcinoma (Yang et al., 2014). Plasmid or DNA damage models have been used for studying and identifying the role of antioxidants (Zhao et al., 2007). Cu^+-induced $^*OH$ radical causing DNA damage has been used as a typical research model. Cu^{2+}, H_2O_2, and ascorbic acid combine with metal free plasmid DNA at pH 7. Along with ascorbic acid Cu^{2+} is reduced to Cu^+ in situ. The Cu^+/ H_2O_2 generates $^*OH$ radical that cleaves one strand of DNA, thus unwinding the supercoiled plasmid DNA. The level of DNA damage is assessed by gel electrophoresis and separation of damaged and undamaged form of DNA occurs. Antioxidants like selenium when added inhibits the free radical $^*OH$-induced oxidative damage of DNA. Thus antioxidant activity of various compounds has been assessed, quantified, and directly compared. Carbonyls like aldehydes and ketones can also be produced during DNA damage and they react with thiobarbituric acid and form thiobarbituric acid reactive species and measured directly by thiobarbituric acid reactive species assay at 532 nm in aqueous phase (Zhao et al., 2007).

5.10.1.4.3 Protein modification

Free radical causes modification of proteins by chlorination and nitration of amino acids. Peroxynitrite (O=N–O–O$^-$) is potent oxidant and nitrating agent formed in vivo by a diffusion controlled reaction between O_2^{-*} and nitric oxide (Pacher et al., 2007). $O=N–O–O^-$ damages large number of molecules including proteins and DNA in cells. $O=N–O–O^-$ and its protonated derivative form peroxynitrous acid (ONOOH) which exert direct modification through oxidation process of one or two electron. $O=N–O–O^-$ reacts with carbon dioxide (CO_2) nucleophilically in vivo and form nitrosoperoxycarbonate ($ONOOCO_2^-$). It is the major pathway for $O=N–O–O^-$. Nitrosoperoxycarbonate ($ONOOCO_2^-$) homolyses to form carbonate (CO_3^-) and nitrogen dioxide ($^*NO_2$) free radical. $^*NO_2$ is also reactive nitrogen species that nitrate amino acid tyrosine to nitrotyrosine and thought to cause $O=N–O–O^-$ related oxidative damage (Lu et al., 2010). Protein expression modulation caused by OS mainly occurs through stimulation of redox-sensitive transcription factors like activator protein-1 (AP-1), NF-κB, G-proteins, and early growth response protein. The fate of liver cells depends upon the stimuli duration and intensity, determining the time period and degree of activation/inactivation of redox-sensitive reactions especially the associated level of AP-1, NF-κB, and early growth response protein (Bubici et al., 2006; Mari et al., 2010).

$O=N–O–O^-$ being a strong oxidant reacts directly with electron-rich groups like iron sulfur centers, sulfhydryls, zinc thiolates, and sulfhydryl group in tyrosine phosphatase which is the active site (Pacher et al., 2007). Myeloperoxidase, a heme protein generates HOCl that reacts with NO_2^- and form nitryl chloride (NO_2Cl) and decomposes spontaneously to $^*NO_2$ and Cl^*. These radicals are responsible for nitrating and chlorinating Cl-NO_2 behavior. HOCl reacts with a many biomolecules like RNA, DNA, proteins, cholesterol, and fatty acids. Nitrotyrosine or chlorotyrosine acts as a biomarker of oxidative damage by reactive nitrogen species in vivo (Mohiuddin et al., 2006). Presence of nitrotyrosine in proteins helps in identifying various pathological conditions like hypertension, atherosclerosis, diabetes, and are related with elevated OS involving high production of $O=N–O–O^-$ (Viappiani and Schulz, 2006). For attenuating the modification of proteins caused by $O=N–O–O^-$ and HOCl, nonenzymatic and enzymatic antioxidants are used. Enzymatic antioxidants like CAT eliminates H_2O_2 and inhibits HOCl formation similarly SOD or antioxidant curcumin, polyphenols scavenge O_2^{-*}, and inhibits $O=N–O–O^-$ formation (Lu et al., 2010).

5.10.2 Antioxidants in liver diseases

5.10.2.1 Liver: a crucial organ for human body

Liver is the largest glandular and resilient organ in the body and composed of several types of cells like hepatocytes, Kupffer cells, liver sinusoidal endothelial cells, pit cells, and hepatic stellate cells (Muriel and Arauz, 2012). It is responsible to perform several functions in human body, like: (i) Aids in detoxification which can eliminate massive number of toxins, breakdown of chemicals, heavy metals, and synthetic pharmaceuticals; (ii) Storage of fat soluble in the form of vitamins (A, D, E, and K), glucose (glycogen), and minerals (iron and copper); (iii) Helps in metabolism of carbohydrates, lipids, and proteins give production of energy; (iv) Production of immune factors that helps the body to challenge immune response, for example, fibrinogen; (v) Maintain the regulation of hormones such as insulin for systematic food metabolism and conversion of thyroxin (T4) into more active tri-iodothyronine (T3); (vi) Responsible for production of bile, assists in regulation of digestion and fat by controlling formation of cholesterol (blood fats) and level of triglycerides; (vii) Eradicate vitamin toxins (A, D, E, and K) from the human body by making complex with bile salts and excreting through feces; (viii) Metabolism of various xenobiotics and endogenous molecule (Tortora et al., 2008; Hirschfield and Gershwin, 2013; Zakim and Boyer, 2000).

5.10.2.2 Liver disease and oxidative stress

OS occurs due to imbalance between production of free radicals and antioxidant defenses (Betteridge, 2000). It plays a major role in the development of liver injury like fibrosis and consequently cirrhosis (Muriel, 1997). Recent studies conducted by number of scientists worldwide in this arena revealed that independent of cause of any particular disease the main common culprit which is responsible for the impairment of normal functioning in most of the cases under consideration is OS (Medina and Moreno-Otero, 2005). Presence of weak liver antioxidant defense system again makes liver highly vulnerable for the attack of ROS and other free radicals (Parola and Robino, 2001). Parenchymal cells are mostly affected cells due to OS caused liver injuries. Various components of parenchymal cells like microsomes, mitochondria, and peroxisomes are greatly involved in the production of ROS and its imbalance results in the aggravation of liver fatty acid oxidations (Li et al., 2015). Other cells such as Kupffer cells are more prone toward the attack of oxidative molecules and produces different kind of cytokines like tumor necrosis factor cytokines-alpha leads marked inflammatory responses and apoptosis. Numbers of studies have been carried out to evaluate the role of OS in various hepatic diseases and antioxidant potential of both natural and synthetic compounds and their role in combating the adverse effects due to OS. Recently one such study conducted by Camacho et al. revealed the association between ion channels and OS and concluded that this association is very important and must be considered while deciding treatment for different liver diseases. Other group of scientists working in the same area demonstrated the protective role of curcumin as a natural antioxidant in hepatic injury induced by OS with chronic alcohol consumption. A Korean, scientists group of Moon et al. evaluated the potential of oligonol a polyphenol from litchi fruit as an antioxidant in liver injury induced by CCl_4 in rats (Muriel and Gordillo, 2016). Therefore, compounds which possess antioxidant properties are just like new hopes in the horizon for all these hepatic diseases such as nonalcohol-induced fatty liver disease, hepatitis, cirrhosis for which leading cause is OS (Li et al., 2015). Various antioxidants which are proven effective for the treatment includes vitamin E, vitamins C and A, curcumin, resveratrol, quercetin, naringenin, green tea, ginseng, licorice, silymarin small amount of selenium, and zinc (Hanje, 2006; Fogden and Neuberger, 2003).

5.10.2.3 Therapeutic effect of antioxidants in liver diseases

Natural plants containing abundant antioxidant property to eliminate the liver damage induced by various models (Table 5.10.1). Several bioactive compounds like flavanoids, saponins, tannins, phenolic compounds, etc. containing enzymatic and nonenzymatic antioxidants and decreases lipid metabolism. OS is responsible for liver damage, and reduces by many plants, known for antioxidant and hepatoprotective activity (Li et al., 2015).

5.10.3 Flavonoids

Quercetin, a polyphenolic (diferuloylmethane) compound shows beneficial pharmacological profile as anti-inflammatory, antioxidant, antibacterial, and antitumor properties (Li et al., 2018; Wu et al., 2017). Preclinical studies demonstrated quercetin attenuated effects of ethanol-induced liver damage and lipid metabolism. Quercetin also increases various antioxidant enzymes like hepatic catalase, SOD, glutathione peroxidase, GR activities, and glutathione (Surapaneni and Jainu, 2014). Another animal studies showed catechin- and tannin-rich extracts obtained from pecan nut shell improving all the symptoms of liver diseases by inducing antioxidant

Table 5.10.1 Preclinical studies of various antioxidants from medicinal plants.

S.N.	Models	Plant source	Active constituents	Pharmacological effect (hepatoprotective)	References
1.	Rat treated with ethanol diet	*Camellia sinensis* (green tea), *Ziziphus mauritiana* (L.), *Hammada scoparia* (methanolic extract L.)	Epicatechin, epicatechin gallate, tannins, saponins, and phenolic compounds	↑Enzymes & ↑nonenzymatic ↓antioxidants, SOD, GSH, lipid peroxidation	Dahiru et al.,2007; Augustyniak et al., 2005
2.	Mice with acute alcohol-induced liver injury/ hepatotoxicity	Pseudofruits (*Hovenia dulcis* peduncles)	Phenolic compounds, nonstarch polysaccharides	↑SOD, GSH & ↑ALT, AST, MDA, ↑antioxidants; NO	Wang et al., 2012
3.	STZ-induced diabetic aged rats	*Terminalia glaucescens* (L.), Aloevera (L.), Resveratol, acai berries	Berberine, (-)-epicatechin, stobadine, Vit C; E	Antioxidant, hepatoprotection	Njomen et al., 2008; Sadi et al., 2014; Ramachandraiahgari et al., 2012; Guerra et al., 2011

(Continued)

Table 5.10.1 Preclinical studies of various antioxidants from medicinal plants. *Continued*

S.N.	Models	Plant source	Active constituents	Pharmacological effect (hepatoprotective)	References
4.	PCM-induced liver toxicity in mice	*Phyllanthus niruri*, *Polyalthia longifolia* (L.), *Boerhaavia diffusia* (L.), Genistein	Gallic acid, carnosic acid	Antioxidant, hepatoprotection	Sabir et al., 2008; Jothy et al., 2012; Olaleye et al., 2010; Fan et al., 2013
5.	D-Galactosamine-induced liver injury in rats	Leucasaspera, Enicostemma axillare, curcumin, betulinic acid	Combination of selenium, ascorbic acid, β-carotene, and α-tocopherol	Antioxidant, hepatoprotection	Banu et al., 2012; Jaishree et al., 2010; Cerny et al., 2011; Zheng et al., 2011
6.	CCl_4-induced hepatotoxicity	*Gareinia indica* (Kokum) fruit rind, *Hybanthus enneaspermus* (humpback flower), *Matricaria chamomilla* (chamomile), grape fruit	Anthocyanins, proanthocyanidins, artemetin	↑SOD, CAT, GSH-Px,↓ AST, ALT, ALP, lipid peroxidation, antioxidant, hepatoprotection	Panda et al., 2012; Vuda et al., 2012; Aksoy et al., 2012; Dai et al., 2014
7.	TAA-induced liver injury	Coriander, *Andrographis paniculata* leaf (green chireta)	Phenolic compounds	↓Antioxidant, ALT, AST, ALP, TBARS, MPO, NO	Moustafa et al., 2014; Abdulaziz Bardi et al., 2014
8.	Lead-induced liver injury	*Zingiber officinale* (ginger)	Gingerol	↑SOD, CAT, ↓MDA	Khaki and Khaki, 2010
9.	Cisplatin-induced liver damage in rats	*Solanum lycopersicum* (tomato juice)	Vitamin C, lycopene, carotene, quercetin, glycosides	Antioxidant, hepatoprotective	Avci et al., 2008
10.	Bile-duct-ligated rats	*Allium sativum* (garlic), *Phaseolus trilobus* (mudgaparni), *Holothuria arenicola* (thymiosycia)	Epigallocatechin-3-gallate, thymoquinone, allyl cysteine, melatonin	↑GSH, LDH, TB, ↓MDA, MPO, TNF-α, TGF-β, MMP-13	Mahmoud et al., 2014; Fursule et al., 2010; Fahmy, 2015

ALP, alkaline phosphatase; ALT, alanine transaminase; AST, aspartate aminotransferase; CAT, catalase; CCl4, carbon tetrachloride; GSH-Px, glutathione peroxidase; GSH, glutathione; LDH, lactate dehydrogenase; LDL, low-density lipoprotein; MDA, malondialdehyde; SOD, superoxide dismutases; TAA, thioacetamide; TBARS, thiobarbituric acid-reactive substances; TNF, tumor necrosis factor.

activity (Muller et al., 2013; Bharrhan et al., 2011). Alvarez-Suarez et al. reported strawberry extract, richest source of anthocyanin (flavanoids), which prevents alcohol-induced gastric lesion by stimulating gastric antioxidant enzyme (Alvarez-Suarez et al., 2011). Green tea, richest source of epicatechin and epicatechin gallate, enhances activity of enzymatic and nonenzymatic antioxidants, decreases lipid and protein oxidation in rats treated with ethanol diet (7 g/L ethanol; Lieber-DeCarli diet; Li et al., 2015). Several studies have to be done which show flavanoids provide better efficacy against liver diseases.

5.10.4 Phenolic compounds

Phenolic compounds, vital component in human diet obtained from plants and huge pharmacological profile due to their antioxidant property. Several studies showed phenolic compound have beneficial activity against liver disease. Dahiru et al. reported that *Ziziphus mauritiana* (L.), richest source of tannins, saponins, and phenolic compound, enhances activity of all antioxidant enzymes such as alanine transaminase, AST, ALP, total bilirubin, CAT; Ò GSH-Px, glutathione reductase, and SOD in rat treated with ethanol-induced liver damage. Coriander, containing highest amount of phenolic compound, prevents TAA-induced hepatotoxicity in rats, due to antioxidant effects (Moustafa et al., 2014). Numerous studies demonstrated that various models are used to induce hepatotoxicity, which can be prevented by antioxidants.

The animal models provide better prospect for study of molecular mechanism and the pathological conditions, leads to liver damage. Recognition of suitable therapeutic antioxidant is suitable for improving alcohol-inducing liver diseases. Research needs focus on selection of animal models that is suitable for the liver damage/injury.

Conclusion

Liver being the significant organ for the maintenance of hepatic homeostasis through the metabolism of several xenobiotics and endogenous molecules always remains a prime target of attack by OS due to excessive free radicals produced during metabolism. This implies how promptly liver undergoes such burdens which lay a solid foundation for the occurrence of severe chronic liver diseases. Extreme lipid peroxidation and weak antioxidant mechanism further support the overproduction of free radicals because of which the balance between antioxidant system and free radicals gets disturbed and ultimately leads enough OS among hepatic cells to damage highly essential cellular constituents such as proteins, lipids, and nucleic acids and results in severe hepatic diseases. Thus, antioxidant therapy including both natural and synthetic antioxidants is highly effective and beneficial therapeutic approach to combat the OS which is the main underlying cause of development of most of the liver diseases via the control of overwhelming effects of free radicals. However, use of antioxidants for the prevention and treatment of life-threatening liver diseases has been very prevalent since time immemorial but to get maximum therapeutic outcome extensive clinical studies and intelligent dosages forms are required.

References

Abdulaziz Bardi, D., Halabi, M.F., Hassandarvish, P., Rouhollahi, E., Paydar, M., Moghadamtousi, S.Z., Al-Wajeeh, N.S., Ablat, A., Abdullah, N.A., Abdulla, M.A., 2014. *Andrographis paniculata* leaf extract prevents thioacetamide-induced liver cirrhosis in rats. PLoS One 9, e109424.

Adegbesan, B.O., Adenuga, G.A., 2007. Effect of lead exposure on liver lipid peroxidative and antioxidant defense systems of protein-undernourished rats. Biol. Trace Element Res. 116, 219–225.

Aksoy, L., Sozbilir, N.B., 2012. Effects of *Matricaria chamomilla* L. on lipid peroxidation, antioxidant enzyme systems, and key liver enzymes in CCl_4-treated rats. Toxicol. Environ. Chem. 94, 1780–1788.

Alvarez-Suarez, J.M., Dekanski, D., Ristić, S., Radonjić, N.V., Petronijević, N.D., Giampieri, F., Astolfi, P., González-Paramás, A.M., Santos-Buelga, C., Tulipani, S., Quiles, J.L., Mezzetti, B., Battino, M., 2011. Strawberry polyphenols attenuate ethanol-induced gastric lesions in rats by activation of antioxidant enzymes and attenuation of MDA increase. PLoS One 6, e25878.

Augustyniak, A., Waszkiewicz, E., Skrzydlewska, E., 2005. Preventive action of green tea from changes in the liver antioxidant abilities of different aged rats intoxicated with ethanol. Nutrition 21, 925–932.

Avci, A., Cetin, R., Erguder, I.B., Devrim, E., Kilicoglu, B., Candir, O., Ozturk, H.S., Durak, I., 2008. Cisplatincauses oxidation in rat liver tissues: possible protective effects of antioxidant food supplementation. Turk. J. Med. Sci. 38, 117–120.

Bando, I., Reus, M.I., Andres, D., Cascales, M., 2005. Endogenous antioxidant defence system in rat liver following mercury chloride oral intoxication. J. Biochem. Mol. Toxicol. 19, 154–161.

Banerjee, P., Jana, S., Chakraborty, S., Swarnakar, S., 2013. Inflammation and MMPs in alcohol-induced liver diseases and protective action of antioxidants. Indian J. Biochem. Biol. 50, 377–386.

Banu, S., Bhaskar, B., Balasekar, P., 2012. Hepatoprotective and antioxidant activity of *Leucas aspera* against D-galactosamine induced liver damage in rats. Pharm. Biol. 50, 1592–1595.

Bayram, S., Akkiz, H., Bekar, A., Akgollu, E., Yildirim, S., 2012. The significance of Exonuclease 1 K589E polymorphism on hepatocellular carcinoma susceptibility in the Turkish population: a case-control study. Mol. Biol. Rep. 39 (5), 5943–5951.

Betteridge, D.J., 2000. What is oxidative stress? Metabolism, 49, 3–8 [PMID: 10693912 DOI: 10.1016/S0026-0495(00)80077-3].

Bharrhan, S., Koul, A., Chopra, K., Rishi, P., 2011. Catechin suppresses an array of signalling molecules and modulates alcohol-induced endotoxin mediated liver injury in a rat model. PLoS One 6, e20635.

Bubici, C., Papa, S., Pham, C.G., Zazzeroni, F., Franzoso, G., 2006. The NF-κB-mediated control of ROS and JNK signaling. Histol. Histopathol. 21 (1-3), 69–80.

Cerny, D., Lekic, N., Vanova, K., Muchova, L., Horinek, A., Kmonickova, E., Zidek, Z., Kamenikova, L., Farghali, H., 2011. Hepatoprotective effect of curcumin in lipopolysaccharide/-galactosamine model of liver injury in rats: relationship to HO-1/CO antioxidant system. Fitoterapia 82, 786–791.

Chung, Y.L., Wu, M.L., 2013. Promyelocytic leukemia protein links DNA damage response and repair to hepatitis B virus related hepatocarcinogenesis. J. Pathol. 230 (4), 377–387.

Cichoz-Lach, H., Michalak, A., 2014. Oxidative stress as a crucial factor in liver diseases. World J. Gastroenterol. 20, 8082–8091.

Curtin, N.J., 2012. DNA repair dysregulation from cancer driver to therapeutic target. Nat. Rev. Cancer 12 (12), 801–817 2012.

Dahiru, D., Obidoa, O., 2007. Evaluation of the antioxidant effects of *Ziziphus mauritiana* lam. leaf extracts against chronic ethanol-induced hepatotoxicity in rat liver. Afr. J. Tradit. Complement. Altern. Med. 5, 39–45.

Dai, N., Zou, Y., Zhu, L., Wang, H.F., Dai, M.G., 2014. Antioxidant properties of proanthocyanidins attenuate carbon tetrachloride (CCl_4)-induced steatosis and liver injury in rats via CYP2E1 regulation. J. Med. Food 17, 663–669.

Dey, A., Lakshmanan, J., 2013. The role of antioxidants and other agents in alleviating hyperglycemia mediated oxidative stress and injury in liver. Food Funct. 4, 1148–1184.

Dietz, K.J., Baier, M., Kramer, U., 1999. Free radicals and reactive oxygen species as mediators of heavy metal toxicity in plants From Biomolecules to Ecosystems. M.N.V. Prasad, Heavy Metal Stress in Plants, 2nd. ed. Springer, Berlin, Heidelberg, pp. 73–97.

Droge, W., 2002. Free radicals in the physiological control of cell function. Physiol. Rev. 82, 47–95.

Ebadi, M., 2001. Antioxidants and Free Radicals in Health and Disease: An Introduction to Reactive Oxygen Species, Oxidative Injury, Neuronal Cell Death and Therapy in Neurodegenerative Diseases. Prominent Press, Arizona.

Fahmy, S.R., 2015. Anti-fibrotic effect of *Holothuria arenicola* extract against bile duct ligation in rats. BMC Complement. Altern. Med. 15, 14.

Fan, Y.J., Rong, Y., Li, P.F., Dong, W.L., Zhang, D.Y., Zhang, L., Cui, M.J., 2013. Genistein protection against acetaminophen-induced liver injury via its potential impact on the activation of UDP-glucuronosyltransferase and antioxidant enzymes. Food Chem. Toxicol. 55, 172–181.

Fogden, E., Neuberger, J., 2003. Alternative medicines and the liver. Liver Int. 23, 213–220.

Fursule, R.A., Patil, S.D., 2010. Hepatoprotective and antioxidant activity of *Phaseolus trilobus*, Ait on bile duct ligation induced liver fibrosis in rats. J. Ethnopharmacol. 129, 416–419.

Gal, S., Lichtenberg, D., Bor, A., et al., 2007. Copper-induced peroxidation of phosphatidylserine-containing liposomes is inhibited by nanomolar concentrations of specific antioxidants. Chem. Phys. Lipids. 150, 186–203.

Gao, B., Bataller, R., 2011. Alcoholic liver disease: pathogenesis and new therapeutic targets. Gastroenterology 141, 1572–1585.

Garcıa- Ruiz, C., Fernandez-Checa, J.C., 2018. Mitochondrial oxidative stress and antioxidants balance in fatty liver disease. Hepatol. Commun. 2 (12), 1425–1439.

Guerra, J.F., Magalhaes, C.L., Costa, D.C., Silva, M.E., Pedrosa, M.L., 2011. Dietary acai modulates ROS production by neutrophils and gene expression of liver antioxidant enzymes in rats. J. Clin. Biochem. Nutr. 49, 188–194.

Halliwell, B., Gutteridge, J.M., 1989. Free Radicals in Biology and Medicine, second ed. Clarendon Press, Oxford.

Hanje, A.J., Fortune, B., Song, M., Hill, D., McClain, C., 2006. The use of selected nutrition supplements and complementary and alternative medicine in liver disease. Nutr. Clin. Pract. 21, 255–272.

Hirschfield, G.M., Gershwin, M.E., 2013. The immunobiology and pathophysiology of primary biliary cirrhosis. Ann. Rev. Pathol. 8, 303–330.

Hsu, C.M., Yang, M.D., Chang, W.S., et al., 2013. The contribution of XRCC6/Ku70 to hepatocellular carcinoma in Taiwan. Anticancer Res. 33 (2), 529–535.

Jaishree, V., Badami, S., Krishnamurthy, P.T., 2010. Antioxidant and hepatoprotective effect of the ethyl acetate extract of *Enicostemma axillare* (Lam). Raynal against CCl_4-induced liver injury in rats. Indian J. Exp. Biol. 48, 896–904.

Jia, X., Wu, Y., Liu, P., 2011. Effects of flour bleaching agent on mice liver antioxidant status and ATPases. Environ.Toxicol. Pharmacol. 31, 479–484.

Jothy, S.L., Aziz, A., Chen, Y., Sasidharan, S., 2012. Antioxidant activity and hepatoprotective potential of *Polyalthia longifolia* and *Cassia spectabilis* leaves against paracetamol-induced liver injury. Evid. Based Complement. Altern. Med. 2012, 561284.

Khaki, A.A., Khaki, A., 2010. Antioxidant effect of ginger to prevents lead-induced liver tissue apoptosis in rat. J. Med. Plants Res. 4, 1492–1495.

Kunwar, A., Priyadarshini, K.I., 2011. Free radicals, oxidative stress and importance of antioxidants in human health. J. Med. Allied Sci. 1 (2), 53–60.

Li, S., Hong, M., Tan, H.Y., Wang, N., Feng, Y., 2016. Insights into the role and interdependence of oxidative stress and inflammation in liver diseases. Oxid. Med. Cell. Longev., 1–21 Article ID 4234061.

Li, S., Tan, H.Y., Wang, N., Zhang, Z.J., Lao, L., Wong, C.W., Feng, Y., 2015. The role of oxidative stress and antioxidants in liver diseases. Int. J. Mol. Sci. 16 (11), 26087–26124.

Li, X., Jin, Q., Yao, Q., Xu, B., Li, L., Zhang, S., Tu, C., 2018. The flavonoid quercetin ameliorates liver inflammation and fibrosis by regulating hepatic macrophages activation and polarization in mice. Front. Pharmacol. 9, 72.

Linares, V., Alonso, V., Albina, M.L., Belles, M., Sirvent, J.J., Domingo, J.L., Sanchez, D.J., 2009. Lipid peroxidation and antioxidant status in kidney and liver of rats treated with sulfasalazine. Toxicology 256, 152–156.

Lindenmeier, M., Burkon, A., Somoza, V., 2007. Anovel method to measure both the reductiveand the radical scavenging activity in a linoleic acid model system. Mol. Nutr. Food Res. 51, 1441–1446.

Lindenmeier, M., Fais, t.V., Hofmann, T., 2002. Structural and functional characterization of pronyl-lysine, a novel protein modification in bread crust melanoidins showing in vitro antioxidative and phase I/II enzyme modulating activity. J. Agric. Food Chem. 50, 6997–7006.

Lobo, V., Patil, A., Phatak, A., Chandra, N., 2010. Free radicals, antioxidants and functional foods: impact on human health. Pharmacogn. Rev. 4 (8), 118–126.

Lu, J.M., Lin, P.H., Yao, Q., Chen, C., 2010. Chemical and molecular mechanisms of antioxidants: experimental approaches and model systems. J. Cell. Mol. Med. 14 (4), 840–860.

Mahmoud, M.F., Zakaria, S., Fahmy, A., 2014. Aqueous garlic extract alleviates liver fibrosis and renal dysfunction in bile-duct-ligated rats. Z. Naturforsch. C 69, 133–141.

Mallikarjuna, K., Shanmugam, K.R., Nishanth, K., Wu, M.C., Hou, C.W., Kuo, C.H., Reddy, K.S., 2010. Alcohol-induced deterioration in primary antioxidant and glutathione family enzymes reversed by exercise training in the liver of old rats. Alcohol 44, 523–529.

Mari, M., Colell, A., Morales, A., Von Montfort, C., Garcia- Ruiz, C., Fernandez-Checa, J.C., 2010. Redox control of liver function in health and disease. Antioxidants Redox Signal. 12 (11), 1295–1331.

Medina, J., Moreno-Otero, R., 2005. Pathophysiological basis for antioxidant therapy in chronic liver disease. Drugs 65 (17), 2445–2461.

Mladenovic, D., Radosavljevic, T., Ninkovic, M., Vucevic, D., Jesic-Vukicevic, R., Todorovic, V., 2009. Liver antioxidant capacity in the early phase of acute paracetamol-induced liver injury in mice. Food Chem. Toxicol. 47, 866–870.

Mohiuddin, I., Chai, H., Lin, P.H., et al., 2006. Nitrotyrosine and chlorotyrosine: clinical significance and biological functions in the vascular system. J. Surg. Res. 133, 143–149.

Moustafa, A.H., Ali, E.M., Moselhey, S.S., Tousson, E., El-Said, K.S., 2014. Effect of coriander on thioacetamide-induced hepatotoxicity in rats. Toxicol. Ind. Health 30, 621–629.

Müller, L.G., Pase, C.S., Reckziegel, P., Barcelos, R.C., Boufleur, N., Prado, A.C., Fett, R., Block, J.M., Pavanato, M.A., Bauermann, L.F., da Rocha, J.B., Burger, M.E., 2013. Hepatoprotective effects of pecan nut shells on ethanolinduced liver damage. Exp. Toxicol. Pathol. 65, 165–171.

Muriel, P., 1997. Peroxidation of lipids and liver damage. In: Baskin, S.I., Salem, H. (Eds.), Oxidants, Antioxidants and Free Radicals. Taylor & Francis, Washington, DC, USA, pp. 237–357.

Muriel, P., Arauz, J., 2012. Coffee and liver health. In: Chu, Y. (Ed.), Coffee Emerging Health Effects and Disease Prevention. IFT Press/Wiley-Blackwell, West Sussex, UK, pp. 123–139.

Muriel, P., Gordillo, K.R., 2016. Role of oxidative stress in liver health and disease. Oxid. Med. Cell. Longev 2016, 1–2.

Njomen, G.B., Kamgang, R., Oyono, J.L., Njikam, N., 2008. Antioxidant potential of the methanol-methylene chloride extract of Terminalia glaucescens leaves on mice liver in streptozotocin-induced stress. Indian J. Pharmacol. 40, 266–270.

Olaleye, M.T., Akinmoladun, A.C., Ogunboye, A.A., Akindahunsi, A.A, 2010. Antioxidant activity and hepatoprotective property of leaf extracts of *Boerhaavia diffusa* Linn against acetaminophen-induced liver damage in rats. Food Chem. Toxicol. 48, 2200–2205.

Ozgur, E., Guler, G., Seyhan, N., 2010. Mobile phone radiation-induced free radical damage in the liver is inhibited by the antioxidants N-acetyl cysteine and epigallocatechin-gallate. Int. J. Radiat. Biol. 86, 935–945.

Pacher, P., Beckman, J.S., Liaudet, L., 2007. Nitric oxide and peroxynitrite in health and disease. Physiol. Rev. 87, 315–424.

Panda, V.S., Ashar, H.D., 2012. Antioxidant and hepatoprotective effects of *Garcinia indica* choisy fruits in carbon tetrachloride-induced liver injury in rats. J. Food Biochem. 36, 240–247.

Parola, M., Robino, G., 2001. Oxidative stress-related molecules and liver fibrosis. J. Hepatol. 35 (2), 297–306.

Ramachandraiahgari, R.M.Y., Somesula, S.R., Adi, P.J., Mannur, I.S., Enamala, M., Matcha, B., 2012. Protective role of ethanolic extract of aloe vera antioxidant properties on liver and kidney of streptozotocin-induced diabetic rats. Dig. J. Nanomater. Biostruct. 7, 175–184.

Sabir, S.M., Rocha, J.B.T., 2008. Water-extractable phytochemicals from *Phyllanthus niruri* exhibit distinct in vitro antioxidant and in vivo hepatoprotective activity against paracetamol-induced liver damage in mice. Food Chem. 111, 845–851.

Sadi, G., Bozan, D., Yildiz, H.B., 2014. Redox regulation of antioxidant enzymes: post-translational modulation of catalase and glutathione peroxidase activity by resveratrol in diabetic rat liver. Mol. Cell. Biochem. 393, 111–122.

Sakaguchi, S., Takahashi, S., Sasaki, T., Kumagai, T., Nagata, K., 2011. Progression of alcoholic and non-alcoholic steatohepatitis: common metabolic aspects of innate immune system and oxidative stress. Drug Metab. Pharmacokinet. 26, 30–46.

Schreck, R., Baeuerle, P.A., 1991. A role for oxygen radicals as se-cond messengers. Trends Cell Biol. 1, 39–42.

Shanmugam, K.R., Mallikarjuna, K., Reddy, K.S., 2011. Effect of alcohol on blood glucose and antioxidant enzymes in the liver and kidney of diabetic rats. Indian J. Pharmacol. 43, 330–335.

Surapaneni, K.M., Jainu, M., 2014. Comparative effect of pioglitazone, quercetin and hydroxy citric acid on the status of lipid peroxidation and antioxidants in experimental non-alcoholic steatohepatitis. J. Physiol. Pharmacol. 65, 67–74.

Tortora, G.J., Derrickson, B.H., 2008. Principles of Anatomy and Physiology, 12th ed. John Wiley and Sons, p. 945 ISBN 978-0-470-08471-7.

Valente, M.J., Carvalho, F., Bastos, M., de Pinho, P.G., Carvalho, M., 2012. Contribution of oxidative metabolism to cocaine-induced liver and kidney damage. Curr. Med. Chem. 19, 5601–5606.

Viappiani, S., Schulz, R., 2006. Detection of specific nitrotyrosine-modified proteins as a marker of oxidative stress in cardiovascular disease. Am. J. Physiol. Heart Circ. Physiol. 290, H2167–H2168.

Vuda, M., D'Souza, R., Upadhya, S., Kumar, V., Rao, N., Kumar, V., Boillat, C., Mungli, P., 2012. Hepatoprotective and antioxidant activity of aqueous extract of *Hybanthus enneaspermus* against CCl_4-induced liver injury in rats. Exp. Toxicol. Pathol. 64, 855–859.

Wang, H., Chen, X., Su, Y., Paueksakon, P., Hu, W., Zhang, M.Z., Harris, R.C., Blackwell, T.S., Zent, R., Pozzi, A., 2015. p47phox contributes to albuminuria and kidney fibrosis in mice. Kidney Int. 87, 948–962.

Wang, M., Zhu, P., Jiang, C., Ma, L., Zhang, Z., Zeng, X., 2012. Preliminary characterization, antioxidant activity in vitro and hepatoprotective effect on acute alcohol-induced liver injury in mice of polysaccharides from the peduncles of *Hovenia dulcis*. Food Chem. Toxicol. 50, 2964–2970.

Winterbourn, C.C., 2008. Reconciling the chemistry and biology of reactive oxygen species. Nat. Chem. Biol. 4, 278–286.

Wu, D., Cederbaum, A.I., 2009. Oxidative stress and alcoholic liver disease. Semin. Liver Dis. 29, 141–154.

Wu, L., Zhang, Q., Mo, W., Feng, J., Li, S., Li, J., et al., 2017. Quercetin prevents hepatic fibrosis by inhibiting hepatic stellate cell activation and reducing autophagy via the TGF-β1/Smads and PI3K/Akt pathways. Sci. Rep. 7, 9289.

Yang, S.F., Chien-Wei, C., Ren-Jie, W., Yow-Ling, S., 2014. Involvement of DNA Damage Response Pathways in Hepatocellular Carcinoma. Hindawi Publishing Corporation BioMed Research International, p. 1–18. Article ID 153867.

Young, I.S., Woodside, J.V., 2001. Antioxidants in health and disease. J. Clin. Pathol. 54, 176–186.

Zakim, D., Boyer, T.D., 2000, fourth ed. Hepatology: A Textbook of Liver Disease 2, Saunders.

Zhang, X., Strakovsky, R., Zhou, D., Zhang, Y., Pan, Y.X, 2011. A maternal high-fat diet represses the expression of antioxidant defense genes and induces the cellular senescence pathway in the liver of male offspring rats. J. Nutr. 141, 1254–1259.

Zhao, F., Tang, Y.Z., Liu, Z.Q., 2007. Protective effect of icariin on DNA against radical-inducedProtective effect of icariin on DNA against radical-induced oxidative damage. J. Pharm. Pharmacol. 59, 1729–1732.

Zheng, Z.W., Song, S.Z., Wu, Y.L., Lian, L.H., Wan, Y., Nan, J.X., 2011. Betulinic acid prevention of D-galactosamine/lipopolysaccharide liver toxicity is triggered by activation of Bcl-2 and antioxidant mechanisms. J. Pharm. Pharmacol. 63, 572–578.

PART

Actual and future perspectives on antioxidants

6

CHAPTER

Antioxidants effects in health: The bright and the dark sides

6.1

Sajad Fakhri, Mohammad Hosein Farzaei

Pharmaceutical Sciences Research Center, Health Institute, Kermanshah University of Medical Sciences, Kermanshah, Iran

6.1.1 Introduction

Human exposure to a variety of environmental and biological risk factors in agriculture and food industry causes oxidative stress (OS). OS is triggered by an extra release of reactive oxygen species (ROS) or reactive nitrogen species (RNS) as other pro-oxidative factors to overwhelm the detoxification potential of antioxidants (Sies, 2015). This condition seems to be involved in several health complications, including cancer, cardiovascular diseases, diabetes mellitus, hypertension, aging, and neurodegenerative disorders (Suzuki, 2009). Therefore, studying the questionable involvement of OS in a large number of pathologies and antioxidants, as well as their main ability to interact with OS is of great importance for health maintenance (Upadhayay et al., 2012). Oxygen, nitrogen, and other free radicals consisting of unpaired electrons are reactive and unstable compounds which rapidly interact with proteins, lipids, and DNA and cause oxidative damage. However, as the final acceptor of electron in the respiration system, oxygen plays both the role of a friend and a foe inside the biological systems. Indeed, reactive species play a key constructive role through implicating in the regulation of the cell function, gene expression, cytotoxic activity of the immune system, and oxidative metabolism to produce heat and locomotion energy, alongside a destructive role in diseases. So, it is crucial to keep an appropriate balance between the tissue/cell damage and the benefits of radicals needed for healthy signaling pathways (Dewaele et al., 2010, Fransen and Lismont, 2018; Saeidnia and Abdollahi, 2013).

Antioxidants are vital for metabolic and signaling mechanisms (Ahuja et al., 2012), scavenging of free radicals, increasing the redox potential or decreasing the complexing of catalytic trace metal ions. Being supported by advanced detecting methods, several antioxidant compounds have been found in plant supplements and functional food like fruits, nuts, vegetables, oilseeds, cocoa, teas, spices, coffee, cereals, and meat (Hall, 2001). The protective role of food enriched by antioxidants could also be applied through targeting specific compartment of human cells; hence, preventing a variety of diseases (Eastwood, 1999; Frei, 2004). In spite of the existing

Antioxidants Effects in Health. DOI: https://doi.org/10.1016/B978-0-12-819096-8.00061-6

evidence and increase of antioxidants use over the last two decades, not only chronic diseases have not been diminished but also increased. This fact highlights the probable harmful role of antioxidants and shows that other factors need to be unrevealed. It also introduces antioxidants as a double-edged sword (Bouayed and Bohn, 2010; Seifirad et al., 2014). It is currently clear that OS is not our enemy and that antioxidants are not essentially good. Altogether, it is essential to maintain a balance between the oxidants and antioxidative agents. This chapter focuses on the bright and dark sides of antioxidant effects on health.

6.1.2 Oxidative stress: sources and the pathophysiology

As the main cause of OS, reactive species are produced both from endogenous and exogenous sources and they basically rely on nonenzymatic reactions. Inflammation, immune cells, infection, ischemia, cancer, aging, and mental stress are all the pathophysiological conditions responsible for the production of free radicals from endogenous sources. Besides, as the main endogenous components and enzymes involved in the ROS/RNS production, mitochondria, endoplasmic reticulum, the cytochrome P450, peroxisomes, cyclooxygenases and lipoxygenases, endothelial phagocytic and inflammatory cells, as well as the nitric oxide synthases (NOSs) and nicotinamide adenine dinucleotide phosphate (NADPH) oxidases, all play key roles in both the physiological and pathophysiological situations (Al-Gubory et al., 2012; Kumar and Pandey, 2015; Nathan and Ding, 2010).

In addition to endogenous enzymatic reactions, nonenzymatic reactions are also responsible for the generation of free radicals, where the cells react with reactive species (Genestra, 2007). Moreover, exogenous-free radicals are also produced following the exposure to environmental heavy metals, chemicals/toxicants/pollutants, cigarette smoke, ozone, ionizing radiation, hyperoxia, chemical solvents, alcohol, cooking (used oil/fat, and smoked meat), as well as specific drugs (e.g., tacrolimus, cyclosporine, bleomycin, and gentamycin) (Halliwell, 2007; Valko et al., 2005). Upon the penetration of these exogenous agents to human body, free radicals are produced as by-products following their degradation or metabolization.

These sources produce different species of ROS/RNS, including hydroxyl radical ($OH^{\bullet}$), superoxide ($O_2^{\bullet -}$), and nitric oxide radical ($NO^{\bullet}$) (Genestra, 2007). Besides, $O_2^{\bullet -}$, $OH^{\bullet}$, and hydrogen peroxide (H_2O_2) also react to produce new free radicals, including organic hydroperoxides (ROOH), alkyloxyl (ROO^-), peroxynitrite ($ONOO^-$), as well as carbon-, nitrogen-, and sulfur-containing radicals through several reactions (Sies, 2017).

From a pathophysiological point of view, several studies have supposed that increased levels of ROS should cause severe diseases and that OS contributes to the development of several diseases, including multiple sclerosis, Parkinson's disease, Alzheimer's disease, Huntington's disease, and amyotrophic lateral sclerosis like neurological diseases. OS also plays a key role in the pathogenesis of rheumatic, cardiovascular, pulmonary, psychiatric, and chronic kidney diseases, as well as cancer, ischemia–reperfusion injuries, ischemic stroke, metabolic disorders, and diabetes. Controversially, it has been confirmed that though the high level of ROS may harm nucleic acids, proteins, and lipids (Wu et al., 2013) and thereby, responsible for the onset and progression of a variety of diseases, low levels of ROS/RNS do not cause permanent damages to cellular nucleic acids, proteins, and lipids and serve as the secondary messengers of normal signaling processes (Pizzino et al., 2017; Taniyama and Griendling, 2003). In this regard, the low produced level of reactive species is essential to handle the cell function, differentiation, apoptosis, protein phosphorylation, and gene expression (Rajendran et al., 2014). For instance, while H_2O_2 starts the apoptosis at a high level, it regulates the signaling pathways and cell proliferation at low doses. The type, location, concentration, and kinetics of reactive species determine their possible involvement in the oxidative damages or signaling pathways (Auten and Davis, 2009).

Free radicals are produced by NADPH oxidase, stored and released by phagocytes of the host defense system in order to destroy pathogenic microbes (Droge, 2002; Young and Woodside, 2001). For this reason, a deficiency in the production of free radicals is possible to create the risk of multiple and persistent infections like what happens in granulomatous disease. The patients with such diseases lack $O_2^{\bullet-}$ due to a deficiency in NADPH oxidase (Droge, 2002). In addition, free radicals play a crucial regulatory role in intracellular signaling pathways of several cell types, including endothelial, cardiac, and vascular smooth muscle cells, as well as the fibroblasts and thyroid tissue. They also induce mitogenic responses to combat diseases (Genestra, 2007; Pacher et al., 2007).

In general, to be helpful for the organisms, free radicals must be kept at low/moderate concentrations. For instance, low levels of $NO^{\bullet}$ and H_2O_2 as moderate oxidants may cause reversible oxidation of definite amino acid side chains, especially cysteine thiol groups [Cys-SH] (Sies, 2017). Sulfinylation (Cys-SO_2H), S-nitrosylation (Cys-S-NO), sulfenylation (Cys-S-OH), and S-glutathionylation (Cys-S-SG) are among other reversible oxidative cysteine modifications which all display the crucial role of cysteine group in the structure, function, and localization of proteins. On the other hand, exposure to high levels of severe oxidative insults carry out irreversible and negative modifications on protein folding, stability, and function, for example, carbonylation of arginine and lysine, sulfonylation of cysteine, nitration of tyrosine and tryptophan, formation of dityrosine, and protein–protein crosslinking (Ghezzi and Bonetto, 2003).

As an imbalance between antioxidants and pro-oxidants, OS causes molecular damage via the disturbance of redox signaling (Gào and Schöttker, 2017; Sies, 2017) and has been considered as the key contributor to several diseases. In this situation, the ROS/RNS pathological role overwhelms their physiological mechanisms, as well as the antioxidants beneficial roles (Selim et al., 2017).

6.1.3 Antioxidants: sources and related mechanisms

An antioxidant is defined as any substrate that delays, prevents, or inhibits the oxidative damage (Halliwell, 2007) in particular concentrations (Halliwell and Gutteridge, 1995) by scavenging or inhibiting ROS (Khlebnikov et al., 2007) and/or creating a new stable radical via intramolecular hydrogen bonding against further oxidations (Halliwell, 1990). Over the years, antioxidants have been demonetized from miracles to marvelous and then physiological molecules (Berger et al., 2012; Singh et al., 2010).

Antioxidants consist of both synthetic and natural sources, as well as exogenous and endogenous compounds (Kancheva, 2009; Pokorný, 2007). The natural endogenous antioxidant system is classified into nonenzymatic and enzymatic compounds (Hanschmann et al., 2013; Marengo et al., 2016). The later is itself divided into primary and secondary compounds (Carocho and Ferreira, 2013). While the primary enzymatic antioxidants (e.g., superoxide dismutase [SOD], catalase (CAT), and glutathione peroxidase) neutralize and prevent the formation of free radicals and also convert $O_2^{\bullet-}$ to H_2O_2 and H_2O_2 into the water and peroxide elimination (Rahman, 2007; Nordgren and Fransen, 2014; Szabó et al., 2007), the secondary types (e.g., glucose-6-phosphate dehydrogenase and glutathione reductase) are not able to neutralize free radicals directly and play a supporting role for other endogenous antioxidants (Ratnam et al., 2006).

Along with enzymatic natural endogenous antioxidants, there are several nonenzymatic ones, including vitamins, coenzyme Q10, glutathione as a tripeptide, and uric acid as a nitrogen compound. As a carotenoid produced by liver or through β-carotene breakdown, vitamin A prevents lipid peroxidation (LPO) through the combination with peroxyl radicals and thereby, shows potential positive effects in health (Palace et al., 1999). Coenzyme Q10 neutralizes and prevents the formation of lipid peroxyl radicals throughout the body (Turunen et al., 2004). As other endogenous nonenzymatic antioxidants, glutathione donates an electron or a hydrogen atom and uric acid scavenges singlet oxygen and $OH^{\bullet}$, through which overwhelms OS (Kand'ár et al., 2006).

In addition to natural endogenous antioxidants, exogenous antioxidants are also needed to keep the balance between OS and antioxidants which could be obtained from fruits, vegetables, chocolates, and green tea (Pietta, 2000). Flavonoids, including flavanols, flavonols, isoflavonoids, flavanones, flavones, and anthocyanins, are all considered as food materials with antioxidant potentials which possess the same diphenyl propane skeleton. Besides activating the antioxidant enzymes and increasing the uric acid level, the phenolic hydroxyl groups attached to ring skeleton of flavonoids act as singlet oxygen quenchers, hydrogen donators, metal chelators, and superoxide radical scavengers. However, their biotransformation by gut microorganisms in addition to their chemical structure features reduces their absorption (Hassan et al., 2017; Marchesi et al., 2016; Procházková et al., 2011). Nevertheless, the low plasma concentrations of polyphenols, as well as their rapid oxidation and metabolization, make many of these results artifactual (Halliwell, 2008; Halliwell et al., 2005; Rechner et al., 2002). In recent years, the potential use of aromatic compounds taken from plants and food has been developed against cancer and several clinical trials with an attempt to find their mechanisms of action are in progress. Hydroxybenzoic and hydroxycinnamic acids as phenolic acids scavenge-free radicals, especially $O_2^{\bullet-}$ and $ONOO^-$ anions, respectively (Krimmel et al., 2010; Terpinc et al., 2011). As another exogenous group of natural pigments with antioxidants activity, carotenoids are produced by a variety of plants. The conjugated unsaturated bonds of carotenoids remove free radicals, neutralize singlet oxygen, scavenge radicals to prevent chain reactions, react with them to produce harmless products, and also inhibit LPO (Fakhri et al., 2018; Heidari Khoei et al., 2019; Paiva and Russell, 1999). Other supplementary nutrients like vitamin C scavenges $OH^{\bullet}$, $O_2^{\bullet -}$, H_2O_2, and reactive nitrogen oxide (Barros et al., 2011) and vitamin E donates its phenolic hydrogen atom to prevent LPO, through which it protects the membranes integrity (Burton and Traber, 1990). Vitamin K, a fat-soluble compound, also shows antioxidant property via its 1,4-naphthoquinone structure (Vervoort et al., 1997). Moreover, selenium and zinc are key minerals with indirect antioxidant activity. Selenium plays a crucial role in the antioxidant activity of glutathione peroxidase, thioredoxin reductase, and selenocysteine, without which no antioxidant effect would have been possible (Tabassum et al., 2010). Whereas, zinc is an inhibitor of NADPH oxidases and activates SOD, induces metallothionein, and competes with copper in binding to the cell wall, which results in the scavenging of hydroxyl radical.

Besides the natural endogenous and exogenous antioxidants, butylated hydroxyanisole and butylated hydroxytoluene have also been developed as synthetic antioxidants to be incorporated into food.

Considering both exogenous and endogenous sources of antioxidants, there is a need to keep a balance between exogenous antioxidants and antioxidants production. Indeed, dosing with an exogenous antioxidant decreases the synthesis and rate of endogenous antioxidant as a compensatory event and may change the complex endogenous antioxidant system in order to maintain the overall balance (Yoshihara et al., 2010). Cutler *et al.* presented "The oxidative stress compensation model" to explain this matter (Cutler and Rodriguez, 2003; Poljsak et al., 2013).

6.1.4 The interplay of antioxidants and pro-oxidants

There is no doubt that food is the safe source of antioxidants and plays a crucial protective and physiological role. However, some reports do not confirm the negative relationship between the intake of dietary antioxidant and risks of disease progression. Conflicting results have forced the researchers to dig deeper into the role of pro-oxidants and antioxidants. They have found that some supplemental antioxidants may be regarded as a double-edged sword (Bouayed and Bohn, 2010; Saso and Firuzi, 2014). According to Gilgun-Sherki *et al.*, there are several conflicting results on the protective effects of antioxidants (Gilgun-Sherki et al., 2001) and there are yet typical arguments in favor of the benefits of antioxidants in the prevention of disease progression (Bains and Hall, 2012). For instance, an increased level of the free radicals induced by deletion of SOD could not accelerate aging, *in vivo*. In two studies on adenocarcinoma and healthy professional men, dietary antioxidant supplementation did not reduce the risk of endometrial and rectal cancer, respectively (Cui et al., 2011; Mekary et al., 2010). Another study also indicated that while the intake of total phenolics from food reduced the risk of endometrial cancer, the supplementation of the subjects with phenolic compound did not (Gifkins et al., 2012). In line with this and to evaluate the correlation of antioxidant intake in women with non-Hodgkin lymphoma, Thompson *et al.* showed that α-carotene, proanthocyanidins, vitamin C, and manganese found in food materials reduced the risk for non-Hodgkin lymphoma, but artificial vitamins C and E, copper, zinc, selenium, and manganese did not decrease the risk of this disease (Thompson et al., 2010). As Miller *et al.* reported, vitamin E even enhanced the mortality rate in the patients with head/neck cancers, cardiovascular diseases, and diabetes (Miller III and Guallar, 2009). This implies the role of exogenous antioxidant in cancer initiation and cancer cell survival (Berger et al., 2012; Fakhri et al., 2020).

In a population-based case–control study, antioxidant did not reduce the osteoporotic hip fracture among long-term smokers (Zhang et al., 2005). A prospective community-based study also revealed the nonbeneficial role of flavonoids, β-carotene, and vitamins C and E in mid-life and showed that they were not able to decrease the risk of dementia in late life (Laurin et al., 2004).

> In a clinical trial by Murphy *et al.*, vitamin C decreased the risk of esophageal adenocarcinoma, but it could not decrease the risk of reflux esophagitis and Barrett's esophagus (Murphy et al., 2010). It has also been reported that supplementation with oral vitamin C could not achieve the serum level in order to play the antioxidant role (Talaulikar and Manyonda, 2011). Daryani et al. reported that cardiovascular risk and metabolic disorder in immigrant women from the Middle East was not due to the low antioxidant intake (Daryani et al., 2007). Hart *et al.* even warned against the use of some antioxidants for cancers (Hart et al., 2012). Therefore, antioxidants may have also a negative effect on some diseases (Gomez-Cabrera et al., 2012).

Chemical analyses may reveal the functions of some antioxidants. Since an antioxidant donates its electron or hydrogen atom to a free radical or a ROS (Litwinienko and Ingold, 2007), it is oxidized to form a radical or an oxidant. What makes an antioxidant real depends on the related derived product (Liu, 2014). If the product is stable enough (not initiating the propagation of free radicals or not combining with other free radicals leading to their functional termination), it is considered a real antioxidant. Conversely, if the product is so active and initiates secondary radical propagation, it is considered a pro-oxidant. So, antioxidants can also mimic the behavior of pro-oxidants when they combine with copper and iron which in turn produce $OH^{\bullet}$ from H_2O_2 (Duarte and Lunec, 2005). To suppress this additional radical propagation, simultaneous consumption of antioxidants is recommended to prevent the conversion of single antioxidant to a pro-oxidant. For instance, α-tocopherol may produce α-tocopherol radical after donating its hydrogen atom in the hydroxyl group which in turn propagates low-density lipoprotein oxidation in the absence of other antioxidants (Bowry and Stocker, 1993). Also, while vitamin C is a potent antioxidant, it could be a pro-oxidant too. The long-lasting exposure to OS unquestionably elevates the risk of disease development (Minelli and Gögele, 2011). Phenolic acids could also be considered as a pro-oxidants in certain situations (Yordi et al., 2012). In order to protect the body against free radicals, antioxidants must pass through an appropriate pathway to be considered as valid complying with theoretical expectations (Perera and Bardeesy, 2011).

Though the results of some clinical trials suggest little or no benefits for antioxidants, the emphasis is on a balanced status of antioxidants (Schafer et al., 2009). Thus, the idea of good antioxidants and bad ROS depends on both the context and the extent, as well as epigenetic, genetic, and microenvironmental variations (Walton, 2016). It is also necessary to reveal other conditions for converting an antioxidant to a pro-oxidant in order to know the associated risks of antioxidants that depend on their concentration and neighboring molecules (Villanueva and Kross, 2012).

Due to the importance of the pathway through which the antioxidants pass, further studies are crucial to provide the involved signaling pathway of the antioxidant as resilience pathways.

6.1.5 Resilience pathways

Increasing evidence introduce ROS as one of the key factors in both resilience and oxidative damage pathways and declining the resilience pathway to oxidative damage causes disease development. While high levels of ROS cause harmful results and mitochondrial dysfunction (Fakhri et al., 2021; Forbes-Hernández et al., 2014), low levels play a helpful role through resilience pathways (Martindale and Holbrook, 2002). Several sources of antioxidants are involved in the metabolization of ROS to the extent that an appropriate balance is kept between the oxidative damage and the benefits of radicals needed for healthy resilience pathways (Halliwell, 2008). Two main molecular mechanisms and signaling pathways of the resilience pathways are the heat shock factor (HSF) and Kelch-like ECH-associated protein 1 (Keap1)-nuclear factor erythroid 2-related factor 2 (Nrf2)-antioxidant response element (ARE).

Keap1-Nrf2/ARE is the major signaling pathway against OS (Fakhri et al., 2020; Wu et al., 2010). In nonstressed conditions, Nrf2 protein is constantly degraded in cytoplasm by proteasome via Keap1. Upon oxidative situation, Keap1 as a cytoplasmic and cysteine-rich protein is inactivated and then dissociated from Nrf2. This modification leads to cytoplasmic accumulation and the nucleus translocation of newly synthesized Nrf2 in order to bind to ARE (Wu et al., 2010). Nrf2/ARE complex, in turn, adjusts the expression of NAD(P)H quinone oxidoreductase-1, antioxidant enzymes, and other key regulator enzymes to activate the defense system (Kansanen et al., 2012). These antioxidants/pro-oxidants are believed to decrease/increase ROS concentrations to the extent that activate Keap1-Nrf2/ARE or HSF (de Roos and Duthie, 2015; Rodrigo et al., 2013). Although the major activation pathway of HSF1 is not completely known, it is clear that heat shock proteins (HSPs) are expressed at low levels under normal physiological conditions and OS take the inhibitory effects of HSP90 and HSP70 on HSF1 and then dramatically increases the expression of HSPs. Upon activation, HSF1 undergoes multistep processing in order to bind to heat shock elements (Akerfeldt et al., 2010; de Roos and Duthie, 2015; Gorrini et al., 2013; Lu et al., 2016).

Considering the role of Nrf2 and HSF in cellular responses to OS, their activators could be therapeutic candidates against oxidative and inflammatory diseases. Conversely, their inhibitors increase the oxidative damage, as well as the efficacy of chemotherapeutic drugs (Akerfeldt et al., 2010; de Roos and Duthie, 2015; Gorrini et al., 2013; Lu et al., 2016).

In fact, the resilience response could be regulated not only by stress but also by external factors such as dietary pro-oxidant compounds. Moderate exposure to pro-oxidants obtained through diet could potentiate the resilience pathways which in turn enhance the endogenous protection and protective enzymes through regulation of Keap1-Nrf2/ARE and HSF. In this regard, novel therapeutic agents are being developed to attenuate the resilience response in patients suffering from a debility in cellular antioxidant defenses (Rochette et al., 2014).

6.1.6 The signaling pathways of antioxidants

Antioxidants could directly affect free radicals (Dominguez-Rodriguez et al., 2012) through scavenging ROS/RNS (Aldini et al., 2010; Borges Bubols et al., 2013; Doeppner and Hermann, 2010) and chelation of metals (Kupershmidt et al., 2012; Mladěnka et al., 2011). The direct effect of antioxidants has been suggested as the important but not the only mechanism in certain conditions (Hollman et al., 2011; Lu et al., 2013; Sies, 2010). For instance, considering polyphenols poor bioavailability, they directly act against free radicals in the gastrointestinal tract (Hu, 2011). Skin also benefits from the direct antioxidant activity of topical antioxidants (Stahl and Sies, 2012).

Antioxidants could also act through inducing endothelial NOS (eNOS) and heme oxygenase-1 (HO-1). In this regard, inducing eNOS increases the production of NO and decreases $O_2^{\bullet-}$ production, thereby play a key therapeutic role (Kietadisorn et al., 2012). The induction of HO-1, as a HSP, also showed protective effects as well (Hall, 2011). As described above, antioxidants act through Nrf2 in addition to HSP. Therefore, activation of Nrf2 as a master regulator of the antioxidant defense system causes the activity of antioxidant enzymes like CAT, SOD, glutathione S-transferase, glutathione peroxidase (GPx), and HO-1 (Joshi and Johnson, 2012).

Antioxidants can also increase the activity of sirtuins (SIRTs), as a family of enzymes with SIRT1–SIRT7 members, with different functions and distributions (Kincaid and Bossy-Wetzel, 2013) which are linked with stress-induced signaling pathways. For instance, SIRT1 with a nucleus and cytosol dominancy mediates the antioxidant effects through the activation of nuclear factor-κB, and peroxisome proliferator-activated receptor gamma (Radak et al., 2013). SIRT3 is more located in mitochondria (Baur et al., 2012) and activates manganese-dependent SOD, isocitrate dehydrogenase 2 (Kincaid and Bossy-Wetzel, 2013), CAT, and NADPH which play an antioxidant effect in the prevention of diseases (Jacobs et al., 2008). Increasing the levels of uric acid (Wahlqvist, 2013), suppressing the formation of free lipid radicals, and initiating the resilience pathway by setting the pro-oxidant tone are among the other ways observe the effects of antioxidants (Kancheva, 2009; Pokorný, 2007).

6.1.7 Antioxidant therapy: novel approaches

Mitochondria is the key target in biology and the main source of reactive species. Actually, about 1–2% of the consumed O_2 converts to $O_2^{\bullet -}$ in human body (Tahara et al., 2009) and mitochondria plays a crucial role in the production of ROS/RNS (Andriantsitohaina et al., 2012; Cenini et al., 2019; Dodson et al., 2013). So, the involvement of mitochondrial dysfunction in the development of diseases associated with OS has been revealed (Hernandez-Resendiz et al., 2013; Murphy, 2009; Smith et al., 2012; Wallace et al., 2010). Altogether mitochondria could be introduced as a novel approaches to antioxidant therapy. Some compounds have been revealed to target mitochondria in a more specific manner including triphenylphosphonium derivatives (Murphy and Smith, 2007), ss-peptides (Smith and Murphy, 2011), dimebon (Su et al., 2010), and thymoquinone (Severina et al., 2013).

In addition to the mitochondria, targeting peroxisome as another compartment being a source of antioxidant enzymes to protect against different diseases where the balance between ROS/RNS and antioxidants has been disrupted. As the peroxisome antioxidant enzymes, CAT neutralizes ROS, while CAT-serine-lysine-leucine (SKL), as a CAT derivative, has been designed to specifically target peroxisome which in turn decreases the level of H_2O_2 (Fransen and Lismont, 2018; Koepke et al., 2007). Besides, peroxisome-targeted therapy would be an auspicious method to attenuate the level of OS through which protect against diseases (Undyala et al., 2011). In addition to mitochondria and peroxisome, NADPH oxidase 4 (NOX4) localization to endoplasmic reticulum could attenuate the ROS-induced pathogenesis (Chen et al., 2008; Hassan et al., 2017).

Antioxidant gene therapy is another approach toward antioxidant therapy, in order to maintain the balance between ROS/RNS production and the antioxidant defense. As mentioned earlier, OS is a pathological situation of an imbalance between the activity of ROS/RNS and endogenous antioxidant defense systems. Thus, as an alternative therapy to the use of natural and synthetic antioxidants, antioxidant gene therapy could potentiate the expression of SOD and CAT as well as inducing HO-1 (Chan et al., 2011) and Nrf2 (Joshi and Johnson, 2012) in the resilience pathways. Antioxidant gene therapy would be of promising approaches to combat against OS-induced dysfunctions including cardiovascular (Liu et al., 2012; Navarro-Yepes et al., 2014) diseases, ischemia/reperfusion injuries (Li et al., 1998, 2001). Furthermore, covalent lipophilic modification of the antioxidants could potentiate the cell uptake, as well as their membrane penetration and mitochondrial targeting (Laguerre et al., 2013). In this line, rosmarinic acid and hydroxycinnamic acid were lyophilized by the addition of aliphatic chains and alkyl esters, respectively, to attain these goals (Bayrasy et al., 2013; Garrido et al., 2012; Fakhri et al., 2022).

Conclusion

OS is a double-edge sword, with a complex physiological and pathological role. While long-lasting exposure to OS correlates with disease development, OS is also a natural defense system in the human body working against malignancies and diseases. Increasing evidence revealed that right pro-oxidant compound, in the right concentrations, and the right method of administration enhanced resilience pathways to protect diseases. On the other hand, the benefits of antioxidants to display useful role in human homeostasis are also confirmed. However, a high dose of antioxidant could be harmful instead of beneficial for health. So, it is critical to keep a balance between OS and antioxidants in a healthy system. Recent conflicting evidence has revealed a double-edged sword role for pro- and antioxidants, also forced the researchers to dig deeper into their role.

Developing new antioxidant therapies with related mechanisms of action as well as the situation affect their effects are seemed to be necessary.

References

Ahuja, I., Kissen, R., Bones, A.M., 2012. Phytoalexins in defense against pathogens. Trends Plant Sci. 17, 73–90.

Akerfeldt, M., Morimoto, R., Sistonen, L., 2010. Heat shock proteins: integrators of cell stress, development and lifespan. Nat. Rev. Mol. Biol. 7, 1–10.

Aldini, G., Vistoli, G., Regazzoni, L., Benfatto, M.C., Bettinelli, I., Carini, M., 2010. Edaravone inhibits protein carbonylation by a direct carbonyl-scavenging mechanism: focus on reactivity, selectivity, and reaction mechanisms. Antioxid. Redox Signal. 12, 381–392.

Al-Gubory, K.H., Garrel, C., Faure, P., Sugino, N., 2012. Roles of antioxidant enzymes in corpus luteum rescue from reactive oxygen species-induced oxidative stress. Reprod. Biomed. Online 25, 551–560.

Andriantsitohaina, R., Duluc, L., Garcia-Rodriguez, J.C., Gil-del Valle, L., Guevara-Garcia, M., Simard, G., Soleti, R., Su, D.-F., Velásquez-Pérez, L., Wilson, J.X., 2012. Systems biology of antioxidants. Clin. Sci. 123, 173–192.

Auten, R.L., Davis, J.M., 2009. Oxygen toxicity and reactive oxygen species: the devil is in the details. Pediatr. Res. 66, 121.

Bains, M., Hall, E.D., 2012. Antioxidant therapies in traumatic brain and spinal cord injury. Biochim. Biophys. Acta (BBA)-Mol. Basis Dis. 1822, 675–684.

Barros, A.I., Nunes, F.M., Gonçalves, B., Bennett, R.N., Silva, A.P., 2011. Effect of cooking on total vitamin C contents and antioxidant activity of sweet chestnuts (*Castanea sativa* Mill.). Food Chem. 128, 165–172.

Baur, J.A., Ungvari, Z., Minor, R.K., Le Couteur, D.G., De Cabo, R., 2012. Are sirtuins viable targets for improving healthspan and lifespan? Nat. Rev. Drug Discov. 11, 443.

Bayrasy, C., Chabi, B., Laguerre, M., Lecomte, J., Jublanc, É., Villeneuve, P., Wrutniak-Cabello, C., Cabello, G., 2013. Boosting antioxidants by lipophilization: a strategy to increase cell uptake and target mitochondria. Pharm. Res. 30, 1979–1989.

Berger, R., Lunkenbein, S., Ströhle, A., Hahn, A., 2012. Antioxidants in food: mere myth or magic medicine? Crit. Rev. Food Sci. Nutr. 52, 162–171.

Borges Bubols, G., da Rocha Vianna, D., Medina-Remon, A., von Poser, G., Maria Lamuela-Raventos, R., Lucia Eifler-Lima, V., Cristina Garcia, S., 2013. The antioxidant activity of coumarins and flavonoids. Mini Rev. Med. Chem. 13, 318–334.

Bouayed, J., Bohn, T., 2010. Exogenous antioxidants—double-edged swords in cellular redox state: health beneficial effects at physiologic doses versus deleterious effects at high doses. Oxid. Med. Cell. Longev. 3, 228–237.

Bowry, V.W., Stocker, R., 1993. Tocopherol-mediated peroxidation. The prooxidant effect of vitamin E on the radical-initiated oxidation of human low-density lipoprotein. J. Am. Chem. Soc. 115, 6029–6044.

Burton, G.W., Traber, M.G., 1990. Vitamin E: antioxidant activity, biokinetics, and bioavailability. Annu. Rev. Nutr. 10, 357–382.

Carocho, M., Ferreira, I.C., 2013. A review on antioxidants, prooxidants and related controversy: natural and synthetic compounds, screening and analysis methodologies and future perspectives. Food Chem. Toxicol. 51, 15–25.

Cenini, G., Lloret, A., Cascella, R., 2019. Oxidative stress in neurodegenerative diseases: from a mitochondrial point of view. Oxid. Med. Cell. Longev. 2019. doi:10.1155/2019/2105607.

Chan, K.H., Ng, M.K., Stocker, R., 2011. Haem oxygenase-1 and cardiovascular disease: mechanisms and therapeutic potential. Clin. Sci. 120, 493–504.

Chen, K., Kirber, M.T., Xiao, H., Yang, Y., Keaney, J.F., 2008. Regulation of ROS signal transduction by NADPH oxidase 4 localization. J. Cell Biol. 181, 1129–1139.

Cui, X., Rosner, B., Willett, W.C., Hankinson, S.E., 2011. Antioxidant intake and risk of endometrial cancer: results from the Nurses' Health Study. Int. J. Cancer 128, 1169–1178.

Cutler, R.G., Rodriguez, H., 2003. Genetic stability, dysdifferentiation, and longevity determinant genesCritical Reviews of Oxidative Stress and Aging: Advances in Basic Science, Diagnostics and Intervention. World Scientific, New Jersey, USA, pp. 1146–1235 (in 2 Volumes).

Daryani, A., Basu, S., Becker, W., Larsson, A., Risérus, U., 2007. Antioxidant intake, oxidative stress and inflammation among immigrant women from the Middle East living in Sweden: associations with cardiovascular risk factors. Nutr. Metab. Cardiovasc. Dis. 17, 748–756.

de Roos, B., Duthie, G.G., 2015. Role of dietary pro-oxidants in the maintenance of health and resilience to oxidative stress. Mol. Nutr. Food Res. 59, 1229–1248.

Dewaele, M., Maes, H., Agostinis, P., 2010. ROS-mediated mechanisms of autophagy stimulation and their relevance in cancer therapy. Autophagy 6, 838–854.

Dodson, M., Darley-Usmar, V., Zhang, J., 2013. Cellular metabolic and autophagic pathways: traffic control by redox signaling. Free Radic. Biol. Med. 63, 207–221.

Doeppner, T.R., Hermann, D.M., 2010. Free radical scavengers and spin traps—therapeutic implications for ischemic stroke. Best Pract. Res. Clin. Anaesthesiol. 24, 511–520.

Dominguez-Rodriguez, A., Abreu-Gonzalez, P., Avanzas, P., 2012. The role of melatonin in acute myocardial infarction. Front. Biosci. 17, 2433–2441.

Droge, W., 2002. Free radicals in the physiological control of cell function. Physiol. Rev. 82, 47–95.

Duarte, T.L., Lunec, J., 2005. When is an antioxidant not an antioxidant? A review of novel actions and reactions of vitamin C. Free Radic. Res. 39, 671–686.

Eastwood, M., 1999. Interaction of dietary antioxidants in vivo: how fruit and vegetables prevent disease? QJM 92, 527–530.

Fakhri, S., Abbaszadeh, F., Dargahi, L., Jorjani, M., 2018. Astaxanthin: a mechanistic review on its biological activities and health benefits. Pharmacol. Res 136, 1–20.

Fakhri, S., Abdian, S., Zarneshan, S.N., Akkol, E.K., Farzaei, M.H., Sobarzo-Sánchez, E., 2021. Targeting Mitochondria by Plant Secondary Metabolites: A Promising Strategy in Combating Parkinson's Disease. Int. J. Mol. Sci. 22 (22), 12570. doi:https://doi.org/10.3390/ijms222212570.

Fakhri, S., Khodamorady, M., Naseri, M., Farzaei, M.H., Khan, H., 2020. The ameliorating effects of anthocyanins on the cross-linked signaling pathways of cancer dysregulated metabolism. Pharmacol. Res. 159. doi:10.1016/j.phrs.2020.104895.

Fakhri, S., Pesce, M., Patruno, A., Moradi, S.Z., Iranpanah, A., Sobarzo-Sánchez, E., 2020. Attenuation of Nrf2/Keap1/ARE in Alzheimer's Disease by Plant Secondary Metabolites: A Mechanistic Review. Molecules 25 (21), 4926. doi:https://doi.org/10.3390/molecules25214926.

Fakhri, S., Tomas, M., Capanoglu, E., Hussain, Y., Abbaszadeh, F., Lu, B., Hu, X., Wu, J., Zou, L., Smeriglio, A., Simal-Gandara, J., Cao, H., Xiao, J., Khan, H., 2022. Antioxidant and anticancer potentials of edible flowers: where do we stand? Crit Rev Food Sci Nutr. doi:10.1080/10408398.2021.1931022.

Forbes-Hernández, T.Y., Giampieri, F., Gasparrini, M., Mazzoni, L., Quiles, J.L., Alvarez-Suarez, J.M., Battino, M., 2014. The effects of bioactive compounds from plant foods on mitochondrial function: a focus on apoptotic mechanisms. Food Chem. Toxicol. 68, 154–182.

Fransen, M., Lismont, C., 2018. Peroxisomes and cellular oxidant/antioxidant balance: protein redox modifications and impact on inter-organelle communication. New York, NY: Springer Singapore.

Frei, B., 2004. Efficacy of dietary antioxidants to prevent oxidative damage and inhibit chronic disease. J. Nutr. 134, 3196S–3198S.

Gào, X., Schöttker, B., 2017. Reduction–oxidation pathways involved in cancer development: a systematic review of literature reviews. Oncotarget 8, 51888.

Garrido, J., Gaspar, A., Garrido, E.M., Miri, R., Tavakkoli, M., Pourali, S., Saso, L., Borges, F., Firuzi, O., 2012. Alkyl esters of hydroxycinnamic acids with improved antioxidant activity and lipophilicity protect PC12 cells against oxidative stress. Biochimie 94, 961–967.

Genestra, M., 2007. Oxyl radicals, redox-sensitive signalling cascades and antioxidants. Cell. Signal. 19, 1807–1819.

Ghezzi, P., Bonetto, V., 2003. Redox proteomics: identification of oxidatively modified proteins. Proteomics: Int. Ed. 3, 1145–1153.

Gifkins, D., Olson, S.H., Demissie, K., Lu, S.-E., Kong, A.-N.T., Bandera, E.V., 2012. Total and individual antioxidant intake and endometrial cancer risk: results from a population-based case–control study in New Jersey. Cancer Causes Control. 23, 887–895.

Gilgun-Sherki, Y., Melamed, E., Offen, D., 2001. Oxidative stress induced-neurodegenerative diseases: the need for antioxidants that penetrate the blood brain barrier. Neuropharmacology 40, 959–975.

Gomez-Cabrera, M.C., Ristow, M., Viña, J., 2012. Antioxidant supplements in exercise: worse than useless? Am. J. Physiol.-Endocrinol. Metab. 302, E476–E477.

Gorrini, C., Harris, I.S., Mak, T.W., 2013. Modulation of oxidative stress as an anticancer strategy. Nat. Rev. Drug Discov. 12, 931.

Hall, C., 2001. Sources of Natural Antioxidants: Oilseeds, Nuts, Cereals, Legumes, Animal Products and Microbial Sources. Woodhead Publishing Limited, Cambridge, England, pp. 159–209.

Hall, E.D., 2011. Antioxidant therapies for acute spinal cord injury. Neurotherapeutics 8, 152–167.

Halliwell, B., 1990. How to characterize a biological antioxidant. Free Radic. Res. Commun. 9, 1–32.

Halliwell, B., 2007. Biochemistry of Oxidative Stress 35. Portland Press Limited, 1147–1150.

Halliwell, B., 2008. Are polyphenols antioxidants or pro-oxidants? What do we learn from cell culture and in vivo studies? Arch. Biochem. Biophys. 476, 107–112.

Halliwell, B., Gutteridge, J., 1995. The definition and measurement of antioxidants in biological systems. Free Radic. Biol. Med. 18, 125–126.

Halliwell, B., Rafter, J., Jenner, A., 2005. Health promotion by flavonoids, tocopherols, tocotrienols, and other phenols: direct or indirect effects? Antioxidant or not? Am. J. Clin. Nutr. 81, 268S–276S.

Hanschmann, E.-M., Godoy, J.R., Berndt, C., Hudemann, C., Lillig, C.H., 2013. Thioredoxins, glutaredoxins, and peroxiredoxins—molecular mechanisms and health significance: from cofactors to antioxidants to redox signaling. Antioxid. Redox Signal. 19, 1539–1605.

Hart, C., Cohen, R., Norwood, M., Stebbing, J., 2012. The emerging harm of antioxidants in carcinogenesis. Future Oncol. 8, 535–548.

Hassan, W., Noreen, H., Rehman, S., Gul, S., Amjad Kamal, M., Paul Kamdem, J., Zaman, B., da Rocha, J.B.T., 2017. Oxidative stress and antioxidant potential of one hundred medicinal plants. Curr. Top. Med. Chem. 17, 1336–1370.

Heidari Khoei, H., Fakhri, S., Parvardeh, S., Shams Mofarahe, Z., Baninameh, Z., Vardiani, M., 2019. Astaxanthin prevents the methotrexate-induced reproductive toxicity by targeting oxidative stress in male mice. Toxin Rev. 38, 248–254.

Hernandez-Resendiz, S., Buelna-Chontal, M., Correa, F., Zazueta, C., 2013. Targeting mitochondria for cardiac protection. Curr. Drug Targets 14, 586–600.

Hollman, P.C., Cassidy, A., Comte, B., Heinonen, M., Richelle, M., Richling, E., Serafini, M., Scalbert, A., Sies, H., Vidry, S., 2011. The biological relevance of direct antioxidant effects of polyphenols for cardiovascular health in humans is not established. J. Nutr. 141, 989S–1009S.

Hu, M.-L., 2011. Dietary polyphenols as antioxidants and anticancer agents: more questions than answers. Chang Gung Med. J. 34, 449–460.

Jacobs, K.M., Pennington, J.D., Bisht, K.S., Aykin-Burns, N., Kim, H.-S., Mishra, M., Sun, L., Nguyen, P., Ahn, B.-H., Leclerc, J., 2008. SIRT3 interacts with the daf-16 homolog FOXO3a in the mitochondria, as well as increases FOXO3a dependent gene expression. Int. J. Biol. Sci. 4, 291.

Joshi, G., Johnson, A.J., 2012. The Nrf2-ARE pathway: a valuable therapeutic target for the treatment of neurodegenerative diseases. Recent Patents CNS Drug Discov. 7, 218–229.

Kancheva, V.D., 2009. Phenolic antioxidants–radical-scavenging and chain-breaking activity: a comparative study. Eur. J. Lipid Sci. Technol. 111, 1072–1089.

Kand'ár, R., Žáková, P., Mužáková, V., 2006. Monitoring of antioxidant properties of uric acid in humans for a consideration measuring of levels of allantoin in plasma by liquid chromatography. Clin. Chim. Acta 365, 249–256.

Kansanen, E., Jyrkkänen, H.-K., Levonen, A.-L., 2012. Activation of stress signaling pathways by electrophilic oxidized and nitrated lipids. Free Radic. Biol. Med. 52, 973–982.

Khlebnikov, A.I., Schepetkin, I.A., Domina, N.G., Kirpotina, L.N., Quinn, M.T., 2007. Improved quantitative structure–activity relationship models to predict antioxidant activity of flavonoids in chemical, enzymatic, and cellular systems. Bioorg. Med. Chem. 15, 1749–1770.

Kietadisorn, R., Juni, R., Moens, A., 2012. Tackling endothelial dysfunction by modulating NOS uncoupling: new insights into its pathogenesis and therapeutic possibilities (vol 302, pg E481, 2012). Am. J. Physiol. Endocrinol. Metab. 303, E1505.

Kincaid, B., Bossy-Wetzel, E., 2013. Forever young: SIRT3 a shield against mitochondrial meltdown, aging, and neurodegeneration. Front. Aging Neurosci. 5, 48.

Koepke, J.I., Nakrieko, K.A., Wood, C.S., Boucher, K.K., Terlecky, L.J., Walton, P.A., Terlecky, S.R., 2007. Restoration of peroxisomal catalase import in a model of human cellular aging. Traffic 8, 1590–1600.

Krimmel, B., Swoboda, F., Solar, S., Reznicek, G., 2010. OH-radical induced degradation of hydroxybenzoic-and hydroxycinnamic acids and formation of aromatic products—a gamma radiolysis study. Radiat. Phys. Chem. 79, 1247–1254.

Kumar, S., Pandey, A.K., 2015. Free radicals: health implications and their mitigation by herbals. J. Adv. Med. Med. Res. 7, 438–457.

Kupershmidt, L., Amit, T., Bar-Am, O., Youdim, M.B., Weinreb, O., 2012. The novel multi-target iron chelating-radical scavenging compound M30 possesses beneficial effects on major hallmarks of Alzheimer's disease. Antioxid. Redox Signal. 17, 860–877.

Laguerre, M., Bayrasy, C., Lecomte, J., Chabi, B., Decker, E.A., Wrutniak-Cabello, C., Cabello, G., Villeneuve, P., 2013. How to boost antioxidants by lipophilization? Biochimie 95, 20–26.

Laurin, D., Masaki, K.H., Foley, D.J., White, L.R., Launer, L.J., 2004. Midlife dietary intake of antioxidants and risk of late-life incident dementia: the Honolulu-Asia Aging Study. Am. J. Epidemiol. 159, 959–967.

Li, Q., Bolli, R., Qiu, Y., Tang, X.-L., Guo, Y., French, B.A., 2001. Gene therapy with extracellular superoxide dismutase protects conscious rabbits against myocardial infarction. Circulation 103, 1893–1898.

Li, Q., Bolli, R., Qiu, Y., Tang, X.-L., Murphree, S.S., French, B.A., 1998. Gene therapy with extracellular superoxide dismutase attenuates myocardial stunning in conscious rabbits. Circulation 98, 1438–1448.

Litwinienko, G., Ingold, K., 2007. Solvent effects on the rates and mechanisms of reaction of phenols with free radicals. Acc. Chem. Res. 40, 222–230.

Liu, Y., Tang, L., Chen, B., 2012. Effects of antioxidant gene therapy on retinal neurons and oxidative stress in a model of retinal ischemia/reperfusion. Free Radic. Biol. Med. 52, 909–915.

Liu, Z.-Q., 2014. Antioxidants may not always be beneficial to health. Nutrition 30, 131–133.

Lu, M.C., Ji, J.A., Jiang, Z.Y., You, Q.D., 2016. The Keap1–Nrf2–ARE pathway as a potential preventive and therapeutic target: an update. Med. Res. Rev. 36, 924–963.

Lu, M.-F., Xiao, Z.-T., Zhang, H.-Y., 2013. Where do health benefits of flavonoids come from? Insights from flavonoid targets and their evolutionary history. Biochem. Biophys. Res. Commun. 434, 701–704.

Marchesi, J.R., Adams, D.H., Fava, F., Hermes, G.D., Hirschfield, G.M., Hold, G., Quraishi, M.N., Kinross, J., Smidt, H., Tuohy, K.M., 2016. The gut microbiota and host health: a new clinical frontier. Gut 65, 330–339.

Marengo, B., Nitti, M., Furfaro, A.L., Colla, R., Ciucis, C.D., Marinari, U.M., Pronzato, M.A., Traverso, N., Domenicotti, C., 2016. Redox homeostasis and cellular antioxidant systems: crucial players in cancer growth and therapy. Oxid. Med. Cell. Longev. 2016. doi:10.1155/2016/6235641.

Martindale, J.L., Holbrook, N.J., 2002. Cellular response to oxidative stress: signaling for suicide and survival. J. Cell. Physiol. 192, 1–15.

Mekary, R.A., Wu, K., Giovannucci, E., Sampson, L., Fuchs, C., Spiegelman, D., Willett, W.C., Smith-Warner, S.A., 2010. Total antioxidant capacity intake and colorectal cancer risk in the Health Professionals Follow-up Study. Cancer Causes Control. 21, 1315–1321.

Miller III, E.R., Guallar, E., 2009. Vitamin E supplementation: what's the harm in that? Clin. Trials 6, 47–49.

Minelli, C., Gögele, M., 2011. The role of antioxidant gene polymorphisms in modifying the health effects of environmental exposures causing oxidative stress: a public health perspective. Free Radic. Biol. Med. 51, 925–930.

Mladěnka, P., Macáková, K., Filipský, T., Zatloukalová, L., Jahodář, L., Bovicelli, P., Silvestri, I.P., Hrdina, R., Saso, L., 2011. In vitro analysis of iron chelating activity of flavonoids. J. Inorg. Biochem. 105, 693–701.

Murphy, M.P., 2009. Mitochondria—a neglected drug target. Curr. Opin. Investig. Drugs 10, 1022.

Murphy, M.P., Smith, R.A., 2007. Targeting antioxidants to mitochondria by conjugation to lipophilic cations. Annu. Rev. Pharmacol. Toxicol. 47, 629–656.

Murphy, S.J., Anderson, L.A., Ferguson, H.R., Johnston, B.T., Watson, P.R., McGuigan, J., Comber, H., Reynolds, J.V., Murray, L.J., Cantwell, M.M., 2010. Dietary antioxidant and mineral intake in humans is associated with reduced risk of esophageal adenocarcinoma but not reflux esophagitis or Barrett's esophagus. J. Nutr. 140, 1757–1763.

Nathan, C., Ding, A., 2010. SnapShot: reactive oxygen intermediates (ROI). Cell 140, e952 951–951.

Navarro-Yepes, J., Zavala-Flores, L., Anandhan, A., Wang, F., Skotak, M., Chandra, N., Li, M., Pappa, A., Martinez-Fong, D., Del Razo, L.M., 2014. Antioxidant gene therapy against neuronal cell death. Pharmacol. Ther. 142, 206–230.

Nordgren, M., Fransen, M., 2014. Peroxisomal metabolism and oxidative stress. Biochimie 98, 56–62.

Pacher, P., Beckman, J.S., Liaudet, L., 2007. Nitric oxide and peroxynitrite in health and disease. Physiol. Rev. 87, 315–424.

Paiva, S.A., Russell, R.M., 1999. β-Carotene and other carotenoids as antioxidants. J. Am. Coll. Nutr. 18, 426–433.

Palace, V.P., Khaper, N., Qin, Q., Singal, P.K., 1999. Antioxidant potentials of vitamin A and carotenoids and their relevance to heart disease. Free Radic. Biol. Med. 26, 746–761.

Perera, R.M., Bardeesy, N., 2011. Cancer: when antioxidants are bad. Nature 475, 43.

Pietta, P.-G., 2000. Flavonoids as antioxidants. J. Nat. Prod. 63, 1035–1042.

Pizzino, G., Irrera, N., Cucinotta, M., Pallio, G., Mannino, F., Arcoraci, V., Squadrito, F., Altavilla, D., Bitto, A., 2017. Oxidative stress: harms and benefits for human health. Oxid. Med. Cell. Longev. 2017. doi:10.1155/2017/8416763.

Pokorný, J., 2007. Are natural antioxidants better and safer—than synthetic antioxidants? Eur. J. Lipid Sci. Technol. 109, 629–642.

Poljsak, B., Šuput, D., Milisav, I., 2013. Achieving the balance between ROS and antioxidants: when to use the synthetic antioxidants. Oxid. Med. Cell. Longev. 2013. doi:10.1155/2013/956792.

Procházková, D., Boušová, I., Wilhelmová, N., 2011. Antioxidant and prooxidant properties of flavonoids. Fitoterapia 82, 513–523.

Radak, Z., Koltai, E., Taylor, A.W., Higuchi, M., Kumagai, S., Ohno, H., Goto, S., Boldogh, I., 2013. Redox-regulating sirtuins in aging, caloric restriction, and exercise. Free Radic. Biol. Med. 58, 87–97.

Rahman, K., 2007. Studies on free radicals, antioxidants, and co-factors. Clin. Interv. Aging 2, 219.

Rajendran, P., Nandakumar, N., Rengarajan, T., Palaniswami, R., Gnanadhas, E.N., Lakshminarasaiah, U., Gopas, J., Nishigaki, I., 2014. Antioxidants and human diseases. Clin. Chim. Acta 436, 332–347.

Ratnam, D.V., Ankola, D., Bhardwaj, V., Sahana, D.K., Kumar, M.R., 2006. Role of antioxidants in prophylaxis and therapy: a pharmaceutical perspective. J. Control. Release 113, 189–207.

Rechner, A.R., Kuhnle, G., Hu, H., Roedig-Penman, A., van den Braak, M.H., Moore, K.P., Rice-Evans, C.A., 2002. The metabolism of dietary polyphenols and the relevance to circulating levels of conjugated metabolites. Free Radic. Res. 36, 1229–1241.

Rochette, L., Zeller, M., Cottin, Y., Vergely, C., 2014. Diabetes, oxidative stress and therapeutic strategies. Biochim. Biophys. Acta (BBA)-Gen. Subjects 1840, 2709–2729.

Rodrigo, R., Prieto, J.C., Castillo, R., 2013. Cardioprotection against ischaemia/reperfusion by vitamins C and E plus n− 3 fatty acids: molecular mechanisms and potential clinical applications. Clin. Sci. 124, 1–15.

Saeidnia, S., Abdollahi, M., 2013. Antioxidants: friends or foe in prevention or treatment of cancer: the debate of the century. Toxicol. Appl. Pharmacol. 271, 49–63.

Saso, L., Firuzi, O., 2014. Pharmacological applications of antioxidants: lights and shadows. Curr. Drug Targets 15, 1177–1199.

Schafer, Z.T., Grassian, A.R., Song, L., Jiang, Z., Gerhart-Hines, Z., Irie, H.Y., Gao, S., Puigserver, P., Brugge, J.S., 2009. Antioxidant and oncogene rescue of metabolic defects caused by loss of matrix attachment. Nature 461, 109.

Seifirad, S., Ghaffari, A., Amoli, M.M., 2014. The antioxidants dilemma: are they potentially immunosuppressants and carcinogens? Front. Physiol. 5, 245.

Selim, K., Abdelrasoul, H., Aboelmagd, M., Tawila, A., 2017. The role of the MAPK signaling, topoisomerase and dietary bioactives in controlling cancer incidence. Diseases 5, 13.

Severina, I.I., Severin, F.F., Korshunova, G.A., Sumbatyan, N.V., Ilyasova, T.M., Simonyan, R.A., Rogov, A.G., Trendeleva, T.A., Zvyagilskaya, R.A., Dugina, V.B., 2013. In search of novel highly active mitochondria-targeted antioxidants: thymoquinone and its cationic derivatives. FEBS Lett. 587, 2018–2024.

Sies, H., 2010. Polyphenols and health: update and perspectives. Arch. Biochem. Biophys. 501, 2–5.

Sies, H., 2015. Oxidative stress: a concept in redox biology and medicine. Redox Biol. 4, 180–183.

Sies, H., 2017. Hydrogen peroxide as a central redox signaling molecule in physiological oxidative stress: oxidative eustress. Redox Biol. 11, 613–619.

Singh, P., Chandra, A., Mahdi, F., Roy, A., Sharma, P., 2010. Reconvene and reconnect the antioxidant hypothesis in human health and disease. Indian J. Clin. Biochem. 25, 225–243.

Smith, R.A., Hartley, R.C., Cocheme, H.M., Murphy, M.P., 2012. Mitochondrial pharmacology. Trends Pharmacol. Sci. 33, 341–352.

Smith, R.A., Murphy, M.P., 2011. Mitochondria-targeted antioxidants as therapies. Discov. Med. 11, 106–114.

Stahl, W., Sies, H., 2012. β-Carotene and other carotenoids in protection from sunlight. Am. J. Clin. Nutr. 96, 1179S–1184S.

Su, B., Wang, X., Bonda, D., Perry, G., Smith, M., Zhu, X., 2010. Abnormal mitochondrial dynamics—a novel therapeutic target for Alzheimer's disease? Mol. Neurobiol. 41, 87–96.

Suzuki, K., 2009. Anti-oxidants for therapeutic use: why are only a few drugs in clinical use? Adv. Drug Deliv. Rev. 61, 287–289.

Szabó, C., Ischiropoulos, H., Radi, R., 2007. Peroxynitrite: biochemistry, pathophysiology and development of therapeutics. Nat. Rev. Drug Discov. 6, 662.

Tabassum, A., Bristow, R., Venkateswaran, V., 2010. Ingestion of selenium and other antioxidants during prostate cancer radiotherapy: a good thing? Cancer Treat. Rev. 36, 230–234.

Tahara, E.B., Navarete, F.D., Kowaltowski, A.J., 2009. Tissue-, substrate-, and site-specific characteristics of mitochondrial reactive oxygen species generation. Free Radic. Biol. Med. 46, 1283–1297.

Talaulikar, V.S., Manyonda, I.T., 2011. Vitamin C as an antioxidant supplement in women's health: a myth in need of urgent burial. Eur. J. Obstet. Gynecol. Reprod. Biol. 157, 10–13.

Taniyama, Y., Griendling, K.K., 2003. Reactive oxygen species in the vasculature: molecular and cellular mechanisms. Hypertension 42, 1075–1081.

Terpinc, P., Polak, T., Šegatin, N., Hanzlowsky, A., Ulrih, N.P., Abramovič, H., 2011. Antioxidant properties of 4-vinyl derivatives of hydroxycinnamic acids. Food Chem. 128, 62–69.

Thompson, C.A., Habermann, T.M., Wang, A.H., Vierkant, R.A., Folsom, A.R., Ross, J.A., Cerhan, J.R., 2010. Antioxidant intake from fruits, vegetables and other sources and risk of non-Hodgkin's lymphoma: the Iowa Women's Health Study. Int. J. Cancer 126, 992–1003.

Turunen, M., Olsson, J., Dallner, G., 2004. Metabolism and function of coenzyme Q. Biochim. Biophys. Acta (BBA)-Biomembr. 1660, 171–199.

Undyala, V., Terlecky, S.R., Vander Heide, R.S., 2011. Targeted intracellular catalase delivery protects neonatal rat myocytes from hypoxia-reoxygenation and ischemia-reperfusion injury. Cardiovasc. Pathol. 20, 272–280.

Upadhayay, U.P.P.D.D., Ewam, P.C.V.V., Ewam, U.P.C.V.V., Sansthan, G.-A., 2012. Immunomodulatory and therapeutic potentials of herbal, traditional/indigenous and ethnoveterinary medicines" Mahima,"Anu Rahal," Rajib Deb,"Shyma K. Latheef," Hari Abdul Samad. Pak. J. Biol. Sci. 15, 754–774.

Valko, M., Morris, H., Cronin, M., 2005. Metals, toxicity and oxidative stress. Curr. Med. Chem. 12, 1161–1208.

Vervoort, L.M., Ronden, J.E., Thijssen, H.H., 1997. The potent antioxidant activity of the vitamin K cycle in microsomal lipid peroxidation. Biochem. Pharmacol. 54, 871–876.

Villanueva, C., Kross, R.D., 2012. Antioxidant-induced stress. Int. J. Mol. Sci. 13, 2091–2109.

Wahlqvist, M.L., 2013. Antioxidant relevance to human health. Asia Pac. J. Clin. Nutr. 22, 171–176.

Wallace, D.C., Fan, W., Procaccio, V., 2010. Mitochondrial energetics and therapeutics. Annu. Rev. Pathol. Mech. Dis. 5, 297–348.

Walton, E.L., 2016. The Dual Role of ROS, Antioxidants and Autophagy in Cancer 39. Elsevier, 89–92.

Wu, J.Q., Kosten, T.R., Zhang, X.Y., 2013. Free radicals, antioxidant defense systems, and schizophrenia. Prog. Neuropsychopharmacol. Biol. Psychiatry 46, 200–206.

Wu, R.P., Hayashi, T., Cottam, H.B., Jin, G., Yao, S., Wu, C.C., Rosenbach, M.D., Corr, M., Schwab, R.B., Carson, D.A., 2010. Nrf2 responses and the therapeutic selectivity of electrophilic compounds in chronic lymphocytic leukemia. Proc. Natl. Acad. Sci. 107, 7479–7484.

Yordi, E.G., Pérez, E.M., Matos, M.J., Villares, E.U., 2012. Antioxidant and pro-oxidant effects of polyphenolic compounds and structure-activity relationship evidence. Nutrition 10, 29471.

Yoshihara, D., Fujiwara, N., Suzuki, K., 2010. Antioxidants: benefits and risks for long-term health. Maturitas 67, 103–107.

Young, I., Woodside, J., 2001. Antioxidants in health and disease. J. Clin. Pathol. 54, 176–186.

Zhang, J., Munger, R.G., West, N.A., Cutler, D.R., Wengreen, H.J., Corcoran, C.D., 2005. Antioxidant intake and risk of osteoporotic hip fracture in Utah: an effect modified by smoking status. Am. J. Epidemiol. 163, 9–17.

CHAPTER

6.2 Food and food supplement antioxidants: Targets in human antioxidant system and effects on the production of endogenous antioxidants

Anna Blázovics

Department of Surgical Research and Techniques, The Heart and Vascular Center, Semmelweis University, Budapest, Hungary

6.2.1 Briefly about healthy nutrition

Numerous references can be found concerning the importance of nutrition in the Egyptian Ebers papyrus (1500 BC), in doctrine of the Far East Indian Charaka ancient Indian doctor (1000 BC), in antique Greek-Roman culture, in books of Hippocrates (460–377 BC) and Galenus (122–199 AD). Nei-Jing's Classical Internal Medicine Book of China (100 BC), in the Canon of Medicine of Avicenna, who was a prominent scholar of the golden age of Islam (1000 AD; Blázovics, 2016).

However, healthy nutrition is different in the Arctic, the Mediterranean, in temperate climate and in the tropics, or in desert areas. We also know today that healthy nutrition should be personalized due to genetic polymorphism in terms of maintaining health, preventing disease, or healing, which is unworkable.

The most common diet-related civilization diseases, cardiovascular diseases, diabetes mellitus, alcoholism, and tumors, primarily gastrointestinal tumors are now among the leading deaths on all continents (http://www.ksh.hu/szamlap/hosszuel_bet.html; KSH, 2021; https://www.who.int/healthinfo/mortality_data/en/; WHO, 2021).

It also follows that healthy nutrition is different for people with different diseases as well.

However, it is necessary to state that a healthy diet may have compounds that are essential for the body's redox homeostasis. These compounds, by their direct radical scavenging activity or indirectly, enhance the body's natural antioxidant defense. Their overdose suppresses antioxidant protection (Blázovics et al., 2016).

Antioxidants Effects in Health. DOI: https://doi.org/10.1016/B978-0-12-819096-8.00062-8

The antioxidants can be very varied in their molecular structure. Among the most significant nutritional antioxidants/scavenger compounds are: phenoloids including flavonoids (flavones, flavanones, flavonols, flavanonols, isoflavones, calcones, flavanolignans, anthocyanines, and stilbenes) and carotenoids, which strengthen the body's antioxidant protection mechanism, such as vitamins or isothiocyanates, but the importance of transmethylating agents is not questionable, such as betaine, which is indirectly linked to maintaining of the glutathione redox system. Vitamins A, C, and E are excellent scavenger molecules, but also the effects of these compounds cannot be explained by their scavenger properties, but by their activities through signal transduction for gene transcription. Current research strengthens that antioxidant compounds from natural sources are converted in various enzymatic processes and the transformed molecules get into the circulation. Their absorption depends on the activity of the intestinal bacterial flora. Their transformations depend on liver cytochrome P450 enzyme (CYP enzyme) activities as well. Therefore, they affect in different ways from the original molecules, act on the signal transduction routes differently, and by this they affect cellular functions and influence the proliferation, apoptosis, autophagy, and ubiquitin proteasome system as well (Breinholt, 1999; Blázovics et al., 2012).

There are examples of several polyphenol glycosides that are delivered to distant tissues in aglicons or phenolcarboxylic acids or hydroxylated forms. Beta-carotene is transformed into retinol or retinoic acid and then they are involved in gene transcription. Glucosinolates are transformed into isothiocyanates by the action of myrosinase enzyme, which activate the phase II enzyme synthesis through antioxidant response element (ARE; Ibrahim and Abul-Hajj, 1990; Blázovics, 2007).

6.2.2 Dietary supplements

There is no global consensus on the category of preparations for dietary supplements as food or complementary medicine as well as prescription drugs. Those are categorized as food in the European Union, giving manufacturers unjustifiable freedom and often making health professionals indecisive, given the lack of adequate evidence-based medicine data, so there is no need to carry out a clinical study to prove that it is fit for consumption, or even to issue or suggest health claims.

In the United States, for example, products of plant origin that enhance life quality are classified as dietary supplements (Dietary Supplement Health and Education Act of 1994 Public Law 103–417, 103rd Congress). This category is between food and "over the counter" preparations. The quality of the preparations is not regulated by official regulations, nor do they need to be certified for efficacy, although they may be accompanied by scientific claims.

Dietary supplements are foods, which can link to daily meals, and contain nutrients or substances with nutritional or physiological effects alone or in combination. Dietary supplements are available in the forms of tablets, capsules, pills, syrups, or other preparations (Dwyer et al., 2018).

The various formation dietary supplements are not the same as the dietetic foods (low in fat, sugar, or calories, etc., developed by dietitians for concrete case), nutriments (nutrients are elemental ingredients for the body's metabolic and physiological function) or foods for special medical purposes. Manufacturers generally follow the recommended daily diets of Recommended Dietary Allowance, but often use higher doses than balanced diets. Occasionally, amounts in excess of the tolerable upper limit are included in the formulations (https://www.ogyei.gov.hu/ETREND_LISTA/; OGYEI, 2021).

The situation is more complicated for some exotic formulations because they contain plant parts with unknown active ingredients.

Occasionally, dietary supplements may be falsified, for example, with herbs, herbal ingredients without active drug or synthetic compounds for some beneficial effects or greater material gain (Csupor et al., 2010).

Filtering out counterfeits is often cumbersome. Every year hundreds of products are withdrawn from the market.

A further problem is, if dietary supplements are used together with medications, without knowing the contents of active ingredients of that and/or the interactions between the active ingredients and medicines, which can cause severe, life-threatening conditions, such as high levels of vitamin K in curcuma preparations, which are dangerous when taken together with anticoagulants. These drugs inhibit prothrombin by reducing vitamin K utilization (Daveluy et al., 2014).

In the United States, approximately there are 85,000 (2018) varieties of dietary supplements are in the market. In Hungary 16,600 varieties (2019) are registered. From large numbers, it is clear that not only people, but also professionals, do not have adequate information on application and effectiveness (Scott et al., 2018; https://www.ogyei.gov.hu/ETREND_LISTA/).

That is why it is necessary to emphasize that dietary supplements should be used only on medical advice and under control, and rather that the diet should be modified to the desired direction.

Dietary supplements are incorrectly identified as "functional foods," which is complicated by the fact that functional food as such is still not properly defined by European Food Safety Authority (https://www.efsa.europa.eu/en/topics/topic/food-supplements; EFSA, 2021).

6.2.3 Functional foods

What is called functional food and where is it place in the nutrition? Functional foods are characterized by their traditional appearances and have special nutritional purposes and have nutritional and/or health claims. Health claims state, suggest or imply, that there are relationships between the food or one of its ingredients and health. Functional components should be quantifiable by physical and/or chemical properties and analytical measurement methods.

Functional foods, according to the American Dietetic Association, are enhanced, enriched, fortified, or altered foods that, as part of a varied diet, are consumed in optimal amounts could be advantageous to health, supported by clinical trials (Chen, 2001; van Kleef et al., 2005).

Functional foods should be well-balanced diet and safety according to the Functional Food Science Group (FUFOSE-Group; Laplace, 2006)

Functional foods have several biomolecules that influence metabolic processes and/or pathways, therefore these preserve healthy state, protect against sicknesses, and influence the outcome of diseases. Those can be consumed as part of the daily diet and consumed as needed. Functional foods may be at all levels of the nutritional pyramid.

The most important families of compounds in functional foods are the same as bioactive substances in foods consumed in a balanced diet. Many of these compounds have antioxidant properties.

The main important compounds are isoprene derivatives: carotenoids, saponins, tocopherols, tocotrienols, terpenes, plant sterols, phytoestrogens; phenolic compounds: coumarins, tannins (proanthocyanidines), lignans, isoflavones, flavones, flavonols, catechins, anthocyanidins, isoflavones; protein/amino acid components: allyl-S compounds, capsanoids, indoles, folate, choline; isothiocyanates; probiotics; prebiotics; fatty acids and structured lipids: long-chain omega-3, -6, and -9-polyunsaturated fatty acids, conjugated linoleic acid, lecithin, and carbohydrates and their derivatives; oligosaccharides, nonstarch polysaccharides: soluble β-glucans and insoluble dietary fibers, fucoidan (sulfated polysaccharide), lignins; vitamins; minerals.

These are necessary the probiotic bacterial strains such as *Bifidobacteria* or *Lactobacilli* and prebiotics (e.g., indigestible carbohydrate compounds of chicory root, garlic, onion, etc.) feeding them to function that the most important biologically active molecules with antioxidant property found in functional foods can be utilized in gut–liver axis (Abuajah et al., 2015; Egresi et al., 2017; Wan et al., 2018).

6.2.4 Why should antioxidant supplementation be considered?

In contrast to a healthy body, the patient's redox homeostasis is changed. There are diseases in which the body's antioxidant system is malfunctioning due to a genetic defect, for example, acatalasemia (catalase), Leigh syndrome (coenzyme-Q) or Gilbert's syndrome (bilirubin). In other diseases, excessive levels of ROS and RNS (reactive oxygen and reactive nitrogen species, respectively) are dominated by the environmental effects (alcoholism, smoking, and radioactivity).

In tumor patients, different free radical and antioxidant levels can be found in different stages of their diseases, which help to identify the tumor state and determine the effectiveness of the treatment. Improper nutritional habits and inadequate use of dietary supplements greatly damage redox homeostasis (Blázovics et al., 2012, 2016).

Scavengers are compounds that are capable of neutralizing ROS and RNS by their molecular structure. In their reactions, themselves also becoming radicals, but they are converted into a relative stable molecules at a much lower reaction rate than ROS, for example, alpha-tocopheroxyl radical, ascorbyl radical, phenoxy radical (Rahal et al., 2014).

According to our present knowledge, the balance of ROS and antioxidants, namely, redox homeostasis is decisive for tissue function, since the oxidative stress and the three levels of antioxidant defense mechanism, antioxidant molecules, enzymatic protection, and clearance mechanisms (autophagy, apoptosis, ubiquitin–proteasome system) play an essential role in the life processes. A healthy body can prevent the ROS/RNS overproduction.

In the tissues, the primary antioxidant defend line is enzymatic. Superoxide dismutases, catalase, peroxidases, glutathione-S-transferases, DT diaphorases, reductases regulated the free radical reactions in the coordinated action.

Enzymatic protection is supported by endogenous and exogenous antioxidants, that is, scavenger molecules. Recently, it was established that the antioxidant enzymes can regulate many redox signaling pathways and exhibit pro-oxidant functions or functions independent of their redox activities (Lei, 2016; Wang, 2013; Hershko and Ciechanover, 1998).

During the last decades, molecular biologists have demonstrated the importance of ROS during signal transduction. The relatively small amount of ROS produced and the H_2O_2 formed by different pathways, such as the activity of superoxide dismutase, play a secondary messenger role in the signal transduction process. In response to oxidative stress, NF-κB and activator protein 1 (AP-1) transcription factors trigger the synthesis of antioxidant enzymes, at the same time inhibit ROS production processes as well (Simizu et al., 1998; Ramachandiran et al., 2002; Rhodes and Campbell, 2002).

HIF1-α, REF-1, p53, NF-κB, AP-1, Nrf2, and ETS belong to the redox-sensitive transcription factors. ROS has a modulating role on the cSrc, Akt, eNOS, p38, MAPK, ERK1/2 major signaling pathways and VEGF, MMP-s, uPA, and PAI-1 are the results of redox-sensitive gene expression in tumor angiogenesis. The ROS-generating NAD(P)H-oxidases contribute to the redox signal. Potential therapeutic targets are VEGF VEGFR2, EGF, EGFR, ROS, and NAD(P) H oxidases in tumor angiogenesis (Ushio-Fukai and Nakamura, 2008; Yashuda et al., 1999; Sartippour et al., 2002; Manna et al., 2000; Kong et al., 2000).

Trace elements play an essential role in regulating signaling mechanisms and through the functions of transcription factors, they influence the expression a whole range of genes (Kudrin, 2000).

6.2.5 Some dietary antioxidants that affect endogenous antioxidant systems

Many of the thousands of biomolecules are involved in the cellular antioxidant network. During nutrition, the body should aim for a steady state of these natural agents.

The known polyphenol compounds, for example, apigenin, genistein, caffeic acid phenethyl ester, epigallocatechin gallate, resveratrol, act on the "gap junction," thereby support cellular communication (Boik, 2001).

The best known anti-inflammatory flavonoid derivatives are flavan-3-ols, flavanones, flavones, flavonols, isoflavones, and prenylated flavonoids. Some of the known agents of these compounds are epicatechin, naringenin, chrysine, apigenin, galangin, quercetin, kaempferol, morin, biochanin, and kuranidine. Their main attack points are phospholipase A2, cyclooxygenase-2, lipoxygenase, and NO-synthase enzymes. These compounds influence signaling pathways (Gábor, 2013).

Flavonoids are capable of direct binding to tyrosine kinase receptors, mitogen-activated protein kinases (MAPK), phosphoinositol-3-kinase (PI3K/Akt), inhibit NF-κB signal protein entry into the nucleus and/or inhibit proteosomal degradation of Iκ-Bα inhibitor protein. They bind directly to the AP-1 (Williams et al., 2004; Chin et al., 2013).

Polyphenols inhibit the angiogenesis connected with inflammation, reduce VEGF expression, inhibit EGFR-coupled signaling. On the one hand, flavonoids can inhibit MMP2 gene induction by capturing ROS and on the other hand inhibiting PKC activity, which result in the inhibition of cancer cell invasion (Chin et al., 2013; Cheng et al., 2003).

Major groups of phytoestrogens have different chemical structures: isoflavonoids, stilbenes, lignans, and coumestans, which can be found, for example, in *Glycine max*, *Trifolium pratense*, *Humulus lupulus* (Remport and Blázovics, 2017).

The stilbene derivative resveratrol founds in skin of red grapes has been in the center of interest, with a number of beneficial physiological effects on cardiovascular diseases (Xia et al., 2017).

It has been shown that resveratrol suppresses TNF-induced activation of NF-κB and AP-1. Resveratrol inhibits ROS-dependent VEGF-induced angiogenesis (Figueira and Cesar, 2018).

Resveratrol initiates CD95 signal-dependent apoptosis on human tumor cells (Manna et al., 2000). Recent research suggests that the in vivo antioxidant property of resveratrol is likely to be due to a gene regulatory effect. Resveratrol increases the expression of various antioxidant enzymes (McCubrey et al., 2017; Xia et al., 2017).

In high doses (20 mg/kg), resveratrol acted as pro-oxidant and caused liver injury, which was detected by liver enzyme release and diminished antioxidant enzyme activities as well as histological picture (Hassan-Khabbar et al., 2008).

Resveratrol is a Janus face molecule, because in high dose it deteriorates the transmethylation of the liver and does not protect alcoholic liver damage (Blázovics et al., 2019). Moderate consumption of glucosinolates rich cruciferous vegetables (e.g., cabbage, broccoli, cauliflower, parsnip, Brussels sprouts, radish, horseradish, turnip, kale, and watercress) is important in the early stages of tumor formation, while in the promotion and progression phases of the tumors are questionable (Langner and Rzeski, 2012).

Isothiocyanates are formed from glucosinolates by the action of myrosinase. In the body, isothiocyanates can be in cysteine conjugation, glutathione conjugation, attachment to the thiol group of proteins, or degradation to amine (Finley, 2005).

Some studies showed that isothiocyanates inhibited NF-κB and AP-1 through the MAPK signaling cascade. The defective gene expression of cyclin D1 appears in most cancers. Experimental evidences showed that phenethyl isothiocyanate and sulforaphane were able to stop the cell cycle by inhibiting cyclin D1 expression. This inhibitory effect is closely related to MAPK, AP-1, and NF-κB routes as well (Young et al., 2003; Jeong et al., 2004).

Isothiocyanates generate oxidative stress in little concentration. As a result, anti-apoptotic molecules are inhibited and proapoptotic molecules are activated. Activation of caspase cascade through caspase-3 and caspase-9 trigger apoptotic cell death. Activation of JNK and inhibition of NF-κB also lead to apoptosis. AP-1 activation occurs at low isothiocyanate concentrations (Chen and Kong, 2005).

Isothiocyanates induce the phase II enzymes (glutathione-S-transferase, NAD(P) H-quinone reductase, γ-glutamylcysteine synthase, hemoxygenase-1) through the ARE.

Nrf2 signal protein can be found in the cytosol anchored to Keap-1 as an inactive complex. Isothiocyanates can directly cleave the disulfide bridges between Nrf2 and Keap-1 and initiate the translocation of Nrf2 into the nucleus and dimerization with the Maf small protein to bind to ARE. Isothiocyanates can indirectly initiate this process through the activation of protein kinases (MAPK, PI3K, PKC, and PERK) and thus, isothiocyanates are capable of inhibiting the process of cancer formation (Keum et al., 2004).

It is also known that the excessive consumption of vegetables belonging to the Cruciferae family resulted in reduced thyroid function. These molecules have goitrogenic property. Thiocyanate-ion reduces iodine uptake and oxazolidine-2-tion interferes with thyroid hormone synthesis, therefore these molecules cause thyroid gland enlargement. In animal experiments, sinigrin and gluconapine triggered liver injury and progoitrin caused hypertrophy of the liver, kidney, and thyroid gland. Isothiocyanates caused bladder hyperplasia. Nitrile decomposition products induced liver and kidney toxicity in high doses.

Genotoxicity has also been reported. Sister chromatid exchange and mutation can be caused by the ascorbigen and neoascorbigen formed in the reaction of indole-glucosinolates and ascorbic acid, when consuming large amounts of broccoli (Ekiz et al., 2010; Latté et al., 2011; Kassie et al., 1996; Baasanjav-Gerber et al., 2011).

6.2.6 Antioxidants to fight diseases

So far numerous primary and secondary prevention studies have been performed with antioxidants with varying degrees of success. In these studies, vitamins and selenium were of particular importance, either alone or in combination. The following some examples reflect on them.

In the 1990s, major cohort studies, ATBC study (the Alpha-Tocopherol, Beta-Carotene Cancer Prevention Study, 1994) and CARET study (Beta-Carotene and Retinol Efficacy Trial, 1996), had negative results (Albanes et al., 1997).

Beta-carotene in 250–500 mg/kg doses induces CYP1A1/2, CYP3A1/2, CYP2E1, CYP2B1/2, and CYP2C11 enzymes in the liver, lung, intestine, and kidneys, resulting in ROS overproduction (Paolini et al., 2001).

Excessive intake of high β-carotene plants in Crohn's disease and ulcerative colitis are not recommended as it may trigger a returning of the disease (Tragnone et al., 1995).

Vitamin A overdose in alcoholics is due to increased induction of CYP2E1 increases vitamin toxicity and cholestasis and chronic liver disease symptoms may be formed. Vitamin C enhanced microsomal enzyme induction in alcoholics leading to the formation of ROS (Herbert, 1993).

Recently, Bjelakovic et al. (2007, 2012) have reviewed hundreds publications and randomized trials involving patients of both sexes were studied. The gender ratio was almost the same. Treatments were made in different continents.

Secondary prevention treatments were performed in gastrointestinal, cardiovascular, neurological, ophthalmic, dermatological, rheumatic, renal, and endocrinological patients.

The result was striking in terms of endpoints, which were the tumor progression and mortality in the primary prevention and in the secondary prevention studies, despite the many limitations of the meta-analysis.

Multivariate metaregression analysis showed that β-carotene, vitamin A, vitamin E alone or in combination increased mortality, but vitamin C supplementation and lifetime-related studies of selenium compounds were unclear.

Bjelakovic et al. have not been able to demonstrate the beneficial effects of antioxidant supplemental treatments in various diseases based on meta-analyses (Bjelakovic et al., 2007, 2012).

According to the EFSA, the recommended dietary reference intake value of selenium is 55 mg/day for adults. The ideal Se serum levels are between 130 and 150 μg/L (Kleiner et al., 2015).

In Nutritional Prevention of Cancer randomized double-blind study the patients were treated with 200 μg/day Se-enriched yeast for 4.5 years. Se supplementation did not have an effect on colon, prostate, and lung cancer, but increased the risk of recurrence of nonmelanoma skin cancer. Treatment at 106 μg/L initial plasma levels also increased the risk of developing squamous cell carcinomas. About 200 μg/day Se supplementation increased the incidence of type 2 diabetes. So selenium exerts its beneficial physiological effect in a very narrow concentration range (Rayman, 2012).

A dietary supplement (polyphenols: 2100 mg, vitamin C: 48 mg, carotenoids: 2.2 mg in 100 g) was applied in the treatment of colectomyzed patients in secondary prevention for 3 months. The treatment (2 × 3 g/day) did not change AFP, CEA, CA19-9, PSA tumor marker levels significantly. In each case, the tumor marker values were higher in treated groups and in few cases the normal values exceeded comparison with baseline ones. ROS reactions of patients were strengthened, as a rebound effect against antioxidant excess. The tendency was observed in healthy controls too. Beneficial effects of this dietary supplement were not clearly detected (Blázovics et al., 2016).

Biochemistry research explaining the failures has led to the recognition that the redox homeostasis is changed in diseases.

Antioxidants can alter signal transduction pathways, cell proliferation, or apoptosis, or induce tumor processes in the absence of apoptosis. Pro-oxidant effects may also occur due to overdose of antioxidants (McCay, 1985; Simopoulos and Ordovas, 2004; Terao and Matsushita, 1986). In the absence of proper knowledge, during daily meals, otherwise essential nutritive and non-nutritive nutrient components may enter the body in high concentrations from food supplements, and they due to gross shifts in the proportions of the active ingredients can lead to severe damage, especially in long-term life-threatening diseases such as inflammatory bowel diseases, alcoholic liver diseases, diabetes, cancers, and so on.

It should be pointed out, on the basis of the results of the cohort studies, that there is no need to consume a dietary supplement rich in antioxidants for healthy people. In the various diseases, it is necessary to consider what type of antioxidant or antioxidant variation is needed, because oxygen and nitrogen centered free radicals (ROS/RNS) are needed as well (Neuhouser et al., 2003). The lack of special antioxidant vitamins should not be replaced by other types of antioxidants or other vitamins, given that these compounds are signal transducers.

Flavonoids cannot be substituted by isothiocyanates because of their different mechanisms of action, but, for example, flavonoids and vitamin C together are much more effective than alone. Metal element also needed, because the Mn, Cu, Zn, and selenium are active components in the antioxidant enzymes (Blázovics, 2007).

Conclusion

Unfortunately, clinical trials do not include global assays for the control of redox homeostasis, and vitamin and elemental determinations. These would require joint testing, because only a single method is not sufficient for an accurate diagnosis (Blázovics and Sárdi, 2018; Blázovics and Kocsis, 2019).

As a result, examinations are extremely expensive and cannot be funded by the health sector at a social level.

However, the prudent physician may observe a lack of antioxidants or elements from various clinical symptoms based on the results of routine laboratory tests. Although doctors need to be very careful with laboratory tests because their patients do not tell them, what supplements or foods are being consumed at all. In many cases, these omissions mislead the doctors.

It has become known decades ago, but it is not yet reached common knowledge, that excessive consumption of antioxidants is harmful on health, that is why a wide range of correct information is needed.

References

Abuajah, C.I., Ogbonna, A.C., Osuji, C.M., 2015. Functional components and medicinal properties of food: a review. J. Food Sci. Technol. 52, 2522–2529.

Albanes, D., Virtamo, J., Taylor, P.R., Rautalahti, M., Pietinen, P., Heinonen, O.P., 1997. Effects of supplemental beta-carotene, cigarette smoking, and alcohol consumption on serum carotenoids in the alpha-tocopherol, beta-carotene cancer prevention study. Am. J. Clin. 66, 366–372.

Baasanjav-Gerber, C., Monien, B.H., Mewis, I., Schreiner, M., Barillari, J., Iori, R., Glatt, H., 2011. Identification of glucosinolate congeners able to form DNS adducts and to induce mutations upon activation by myrosinase. Mol. Nutr. Food Res. 55, 783–792.

Bjelakovic, G., Nikolova, D., Gluud, L.L., Simonetti, R.G., Gluud, C., 2007. Mortality in randomized trials of antioxidant supplements for primary and secondary prevention: systematic review and meta-analysis. JAMA 297, 842–857.

Bjelakovic, G., Nikolova, D., Gluud, L.L., Simonetti, R.G., Gluud, C., 2012. Antioxidant supplements for prevention of mortality in healthy participants and patients with various diseases. Cochrane Database Syst. Rev. 14, CD007176.

Blázovics, A., 2007. Redox homeostasis, bioactive agents and transduction therapy. Curr. Sign. Transduct. Ther. 2, 226–239.

Blázovics, A., 2016. Avicenna és a modern farmakognózia (Avicenna and the modern pharmacognosy). Kaleidoscope 7, 63–77.

Blázovics, A., Fébel, H., Bekő, G., Kleiner, D., Szentmihályi, K., Sárdi, É., 2019. Why do not polyphenols of red wine protect against the harmful effects of alcohol in alcoholism? Acta Alim. 48, 358–364.

Blázovics, A., Kocsis, I., 2018. Szabadgyök-kutatás módszerei, néhány fontosabb orvostörténeti vonatkozása. (Methods of free radical research, and its primary impact in the history of medicine). Kaleidoscope Művelődés- Tudomány- és Orvostörténeti Folyóirat 9, 159–170. http://dx.doi.org/10.17107/KH.2018.16.159-170.

Blázovics, A., Kursinszki, L., Papp, N., Szőke, É., Hegyi, G., Szilvás, Á., 2016. Is professional prescription of a commercially derived dietary supplement in colectomysed patients necessary? Eur. J. Integr. Med. 8, 219–226.

Blázovics, A., Nyirády, P., Romics, I., Szűcs, M., Horváth, A., Szilvás, Á., Székely, E., Szentmihályi, K., Bekő, G., Sárdi, É., 2012. How can cancer-associated anemia be moderated with nutritional factors and how do *Beta vulgaris* L. ssp. esculenta var. rubrass modify the transmethylation reaction in erythrocytes in cancerous patients? In: Silverberg, D.S. (Ed.), How can cancer-associated anemia be moderated with nutritional factors and how do *Beta vulgaris* L. ssp. esculenta var. rubrass modify the transmethylation reaction in erythrocytes in cancerous patients? Anemia, 1–23 InTech Open Access.

Blázovics, A, Sárdi, É., 2018. Methodological repertoire development to study the effect of dietary supplementation in cancer therapy. Microch. J. 136, 121–127.

Boik, J., 2001. Natural Compounds in Cancer Therapy. Oregon Medicinal Press, LLC, USA.

Breinholt, V., 1999. Desirable vs. harmful levels of intake of flavonoids and phenolic acids. In: Kumpulainen, J.T., Salonen, J.T. (Eds.), Natural Antioxidants and Anticarcinogens in Nutrition, Health and Disease. The Royal Society of Chemistry, Cambridge, UK, pp. 93–105.

Chen, L., 2001. Functional food science in Europe. Nutr. Metab. Cardiovasc. Dis. 4 (Suppl), 20–23.

Chen, C., Kong, A.N.T., 2005. Dietary cancer-chemopreventive compounds: from signaling and gene expression to pharmacological effects. Trends Pharmacol. Sci. 26, 318–326.

Cheng, J.J., Huang, N.K., Chen, C.F., 2003. Influence of flavonoids on regulation of MMPs activity in THP-1 cells. Chin. Pharm. J. 55, 355–360.

Chin, Y.W., Kong, J.Y., Han, S.Y., 2013. Flavonoids as receptor tyrosine kinase FLT3 inhibitors. Bioorg. Med. Chem. 23, 1768–1770.

Csupor, D., Szekeres, A., Kecskeméti, A., Vékes, E., Veres, K., Micsinay, Á., Szendrei, K., Hohmann, J., 2010. Szildenafillal és tadalafillal hamisított étrend-kiegészítők a hazai piacon (Dietary supplements on the domestic market adulterated with sildenafil and tadalafil). Orv. Hetil. 151, 1783–1789.

Daveluy, A., Geeniaux, H, Thibaud, L., Mallaret, M., Miremont-Salamee, G., Haramburu, F., 2014. Probable interaction between an oral vitamin K antagonist and turmeric (*Curcuma longa*). Therapie 69, 519–520.

Dwyer, J.T., Coates, P.M., Smith, M.J., 2018. Dietary supplements: regulatory challenges and research resources. Nutrients 10, 411–424.

Egresi, A., Kovács, Á., Szilvás, Á., Blázovics, A., 2017. Bél-máj tengely vizsgálata colitis ulcerosaban – retrospektív tanulmány (Bowel-liver axis examination in ulcerative colitis - retrospective study). Orv. Hetil. 158, 1014–1021.

Ekiz, F., Yuksel, I., Kertmen, N., Yuksel, O., 2010. Liver toxicity due to broccoli juice. Eur. J. Gastroenterol. Hepatol. 22, 898.

EFSA, 2021. https://www.efsa.europa.eu/en/topics/topic/food-supplements.

Figueira, L., Cesar Gonzalez, J., 2018. Effect of resveratrol on seric vascular endothelial growth factor concentrations during atherosclerosis. Clin. Invest. Arterioscler. 30, 209–216.

Finley, J.W., 2005. Proposed criteria for assessing the efficacy of cancer reduction by plant foods enriched in carotenoids, glucosinolates, polyphenols and selenocompounds. Ann. Bot. 95, 1075–1096.

Gábor, M., 2013. Pharmacology and Dermatopharmacology of Mouse Ear Inflammation. Akadémiai KiadóBudapest, Hungary.

Hassan-Khabbar, S., Cottart, C.H., Wendum, D., Vibert, F., Clot, J.P., Savouret, J.F., Conti, M., Nivet-Antoine, V., 2008. Postischemic treatment by trans-resveratrol in rat liver ischemia-reperfusion: a possible strategy in liver surgery. Liver Transplant. 14, 451–459.

Herbert, V., 1993. Viewpoints does mega–C do more good than harm, or more harm than good? Nutr Today 28, 126–133.

Hershko, A., Ciechanover, A., 1998. The ubiquitin system. Annu. Rev. Biochem. 67, 425–479.

Ibrahim, A.R., Abul-Hajj, Y.J., 1990. Microbiological transformation of flavone and isoflavone. Xenobiotica 20, 363–373.

Jeong, W.S., Kim, I.W., Rong, H., Kong, A.N.T., 2004. Modulatory properties of various natural chemopreventive agents on the activation of NF-κB signaling pathway. Pharm. Res. 21, 661–671.

Kassie, F., Parzefall, W., Musk, S., Johnson, I., Lamprecht, G., Sontag, G., Knasmüller, S., 1996. Genotoxic effects of crude juices from *Brassica* vegetables and juices and extracts from phytopharmaceutical preparations and spices of cruciferous plants origin in bacterial and mammalian cells. Chem. Biol. Interact. 102, 1–16.

Keum, Y.S., Jeong, W.S., Kong, A.N.T., 2004. Chemoprevention by isothiocyanates and their underlying molecular signaling mechanisms. Mutat. Res. 555, 191–202.

Kleiner, D., Mátis, E., Süle, K., Molnáoln, J., 2015. A szelén élettani szerepe és jelentősége. Gyógyszerészet 59, 148–153.

Kong, A.N., Yu, R., Chen, C., Mandlekar, S., Primiano, T., 2000. Signal transduction events elicited by natural products: role of MAPK and caspase pathways in homeostatic response and induction of apoptosis. Arch. Pharm. Res. 23, 1–16.

KSH, 2021. http://www.ksh.hu/szamlap/hosszuel_bet.html.

Kudrin, A.V., 2000. Trace elements in regulation of NF-κB activity. J. Trace Elem. Med. Biol. 14, 129–142.

Langner, E., Rzeski, W., 2012. Dietary derived compounds in cancer chemoprevention. Contemp. Oncol. (Pozn.) 16, 394–400.

Laplace, J-P., 2006. Health and nutrition claims made on food: what future? Bull. Acad. Nation Med. 190, 1663–1682.

Latté, K.P., Appel, K.E., Lampen, A., 2011. Health benefits and possible risks of broccoli – an overview. Food Chem. Toxicol. 49, 3287–3309.

Lei, X.G., Zhu, J.H., Cheng, W.H., Bao, Y., Ho, Y.S., Reddi, A.R., Holmgren, A., Arnér, E.S.J., 2016. Paradoxical roles of antioxidant enzymes: basic mechanisms and health implications. Physiol. Rev. 96, 307–364.

Manna, S.K., Mukhopadhyay, A., Aggarwal, B.B., 2000. Resveratrol suppresses TNF-induced activation of nuclear transcription factors NFkappaB, activator protein-1, and apoptosis: potential role of reactive oxygen intermediates and lipid peroxidation. J. Immunol. 164, 6509–6519.

McCay, P.B., 1985. Vitamin E: interactions with free radicals and ascorbate. Ann. Rev. Nutr. 5, 323–340.

McCubrey, A., Lertpiriyapong, K., Steelman, L.S., Abrams, S.L., Yang, L.V., Murata, R.M., Rosalen, P.L., Scalisi, A., Neri, L.M., Cocco, L., Ratti, S., Martelli, A.M., Laidler, P., Dulińska-Litewka, J., Rakus, D., Gizak, A., Lombardi, P., Nicoletti, F., Candido, S., Libra, M., Montalto, G., Cervello, M., 2017. Effects of resveratrol, curcumin, berberine and other nutraceuticals on aging, cancer development, cancer stem cells and microRNAs. Aging US 9, 1477–1536.

Neuhouser, M.L., Patterson, R.E., Thornquist, M.D., Omenn, G.S., King, I.B., Goodman, G.E., 2003. Fruits and vegetables are associated with lower lung cancer risk only in the placebo arm of the beta-carotene and retinol efficacy trial (CARET). Cancer Epidemiol. Biomark. Prevent. 12, 350–358.

OGYEI, 2021. https://www.ogyei.gov.hu/ETREND_LISTA/.

Paolini, M., Antelli, A., Pozzetti, L., Spetlova, D., Perocco, P., Valgimigli, L., Pedulli, G.F., Cantelli-Forti, G., 2001. Induction of cytochrome P450 enzymes and over-generation of oxygen radicals in beta-carotene supplemented rats. Carcinogenesis 22, 1483–1495.

Rahal, A., Kumar, A., Singh, V., Yadav, B., Tiwari, R., Chakraborty, S., Dhama, K., 2014. Oxidative stress, prooxidants, and antioxidants: the interplay. Biomed. Res. Int. 2014, 761264.

Ramachandiran, S., Huang, Q., Dong, J., Lau, S.S., Monks, T.J., 2002. Mitogen activated protein kinases contribute to reactive oxygen species-induced cell death in renal proximal tubule epithelial cells. Chem. Res. Toxicol. 15, 1635–1642.

Rayman, M.P., 2012. Selenium and human health. Lancet 379, 1256–1268.

Remport, J., Blázovics, A., 2017. Fitoösztrogének a menopauza terápiájában (Phytoestrogens in menopause therapy). Orv. Hetil. 158, 1243–1251.

Rhodes, J.M., Campbell, B.J., 2002. Inflammation and colorectal cancer: IBD-associated and sporadic cancer compared. Trends Mol. Med. 8, 10–16.

Sartippour, M.R., Shao, Z.M., Heber, D., Beatty, P., Zhang, L., Liu, C., Ellis, L., Liu, W., Go, V.L., Brooks, M.N., 2002. Green tea inhibits vascular endolthelial growth factor (VEGF) induction in human breast cancer cells. J. Nutr. 132, 2307–2311.

Scott, J.M., Lindsey, A.T., Costello, R.B., Deuster, P.A., 2018. Using the dietary supplement label database to identify potentially harmful dietary supplement ingredients. Nutr. Today 53, 229–233.

Simizu, S., Takada, M., Umezawa, K., Imoto, M., 1998. Requirement of caspase-3 (-like) protease mediated hydrogen peroxide production for apoptosis induced by various anticancer drugs. J. Biol. Chem. 273, 26900–26907.

Simopoulos, A.P., Ordovas, J.M., 2004. Nutrigenetics and nutrigenomics. World Review of Nutrition and Dieetics 93. Karger, 1–28.

Terao, J., Matsushita, S., 1986. The peroxidizing effect of α-tocopherol on autooxidation of methyl linoleate in bulk phase. Lipids 21, 255–260.

Tragnone, A., Valpiani, D., Miglio, F., Elmi, G., Bazzocchi, G., Pipitone, E, Lafranchi, G.A., 1995. Dietary habits as risk factors for inflammatory bowel disease. Eur. J. Gastroenterol. Hepatol. 7, 47–51.

Ushio-Fukai, M., Nakamura., Y., 2008. Reactive oxygen species and angiogenesis: NAD(P)H oxidase as target for cancer therapy. Cancer Lett. 266, 37–52.

van Kleef, E., van Trijp, H.C.M., Luning, P., 2005. Functional foods: health claim-food product compatibility and the impact of health claim framing on consumer evaluation. Appetite 44, 299–308.

Wan, M.L.Y., Ling, K.H., El-Nezami, H., Wang, M.F., 2018. Influence of functional food components on gut health. Crit. Rev. Food Sci. Nutr. 59, 1927–1936.

Wang, H.G., 2013. Authophagy and Cancer. Springer, New York, Heidelberg, Dordrecht, London.

WHO, 2021. https://www.who.int/healthinfo/mortality_data/en/.

Williams, R.J., Spencer, J.P.E., Rice-Evans, C., 2004. Flavonoids: antioxidants or signaling molecules? Free Radic. Biol. Med. 36, 838–849.

Xia, N., Daiber, A., Förstermann, U., Li, H., 2017. Antioxidant effects of resveratrol in the cardiovascular system. Br. J. Pharm. 174, 1633–1646.

Yashuda, M., Ohzeki, Y., Shimizu, S., Naito, S., Ohtsuru, A., Yamamoto, T., Kuroiwa, Y., 1999. Stimulation of in vitro angiogenesis by hydrogen peroxide and the relation with ETS-1 in endothelial cells. Life Sci. 64, 249–258.

Young, M.R., Yang, H.S., Colburn, N.H., 2003. Promising molecular targets for cancer prevention: AP-1, NF-kappa B and Pdcd4. Trends Mol. Med. 9, 36–41.

CHAPTER

6.3 Antioxidants effects in health: Concluding remarks and future perspectives

Ana Sanches Silva[a,b,c], **Seyed Mohammad Nabavi**[d]

[a]*National Institute for Agricultural and Veterinary Research (INIAV), I.P., Vairão, Vila do Conde, Portugal*

[b]*Center for Study in Animal Science (CECA), ICETA University of Oporto, Oporto, Portugal*

[c]*University of Coimbra, Faculty of Pharmacy, Coimbra, Portugal*

[d]*Applied Biotechnology Research Center, Baqiyatallah University of Medical Sciences, Tehran, Iran*

6.3.1 Introduction

Mediterranean diet, characterized by the consumption of olive oil, fruits, vegetables, nuts, legumes, and unprocessed cereals, is associated with health positive benefits, namely, with the reduction of the risk of chronic diseases such as cardiovascular diseases and increase of life expectancy (Trichopoulou et al., 2014; Tosti et al., 2018; Martini, 2019). One of the main responsible for these benefits is antioxidants. These compounds are able to stabilize free radicals, avoiding oxidative stress, and may have a key role in the prevention of diseases or as adjuvant therapy. According to the Regulation (EC) No 1333/200, antioxidants in food "*are substances which prolong the shelf-life of foods by protecting them against deterioration caused by oxidation, such as fat rancidity and colour changes*" (EC, 2008).

This group of compounds received considerable attention from the scientific community due to the first experiments of Dr. Denham Harman in 50s, who proposed that accumulation of reactive oxygen species (ROS)/free radical over time may induce damage in cells (Harman, 1981).

6.3.2 Antioxidants versus ROS dichotomy

One of the pervasive assumptions of consumers, in part due to the information vehiculated by food, cosmetic, and pharmaceutical industries through mass media, is that free radicals are "bad" and antioxidants are "good." Is this dichotomy true? Human body is a complex machine which requires an equilibrium to achieve a healthy

Antioxidants Effects in Health. DOI: https://doi.org/10.1016/B978-0-12-819096-8.00063-X

state. An excess or deficit of some nutrients can originate body imbalance. Also, a large concentration of the same type of antioxidants can cause an imbalance, including induce damage, and contribute for the opposite effect of a lower level of antioxidants.

There are scientific evidences regarding the benefits of antioxidants, both endogenous and exogenous compounds, in several diseases, such as cancer, cardiovascular disease, neurodegeneration, and ageing, which have been addressed along this book. Fig. 6.3.1 gives examples of endogenous and exogenous antioxidants while Fig. 6.3.2 shows the main effects of antioxidants in human health reported in the scientific literature.

For instance, in the randomized placebo-controlled trial carried out by Boaz et al. (2000) it was reported that in hemodialysis patients with cardiovascular disease, the vitamin E supplementation (800 IU/day) decreases composite cardiovascular disease endpoints as well as myocardial infarction. There are also cases where it is not possible to achieve a conclusion regarding the effect of an antioxidant in a specific disease due to studies with opposite results. Lloret, 2019 described that most studies report a reduction in the levels of vitamin E in plasma in Alzheimer's disease patients. Vitamin E supplementation showed positive results in Alzheimer's disease patients in some studies but failed to reveal good results in other studies. This evidence might be due to the difference antioxidant effect of vitamin E in each person, different nutritional status of the patients at baseline or due to the range of time required for brain compensation in each person (Lloret, 2019). Also, many antioxidants are

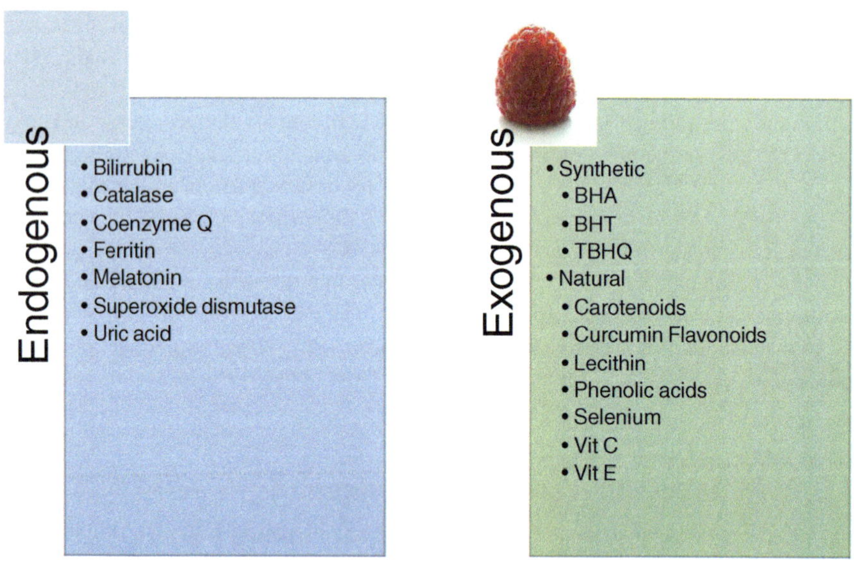

FIG. 6.3.1 Examples of endogenous and exogenous antioxidants.

BHA, butylated hydroxyanisole; *BHT*, butylated hydroxyanisole; *TBHQ*, tert-butylhydroquinone.

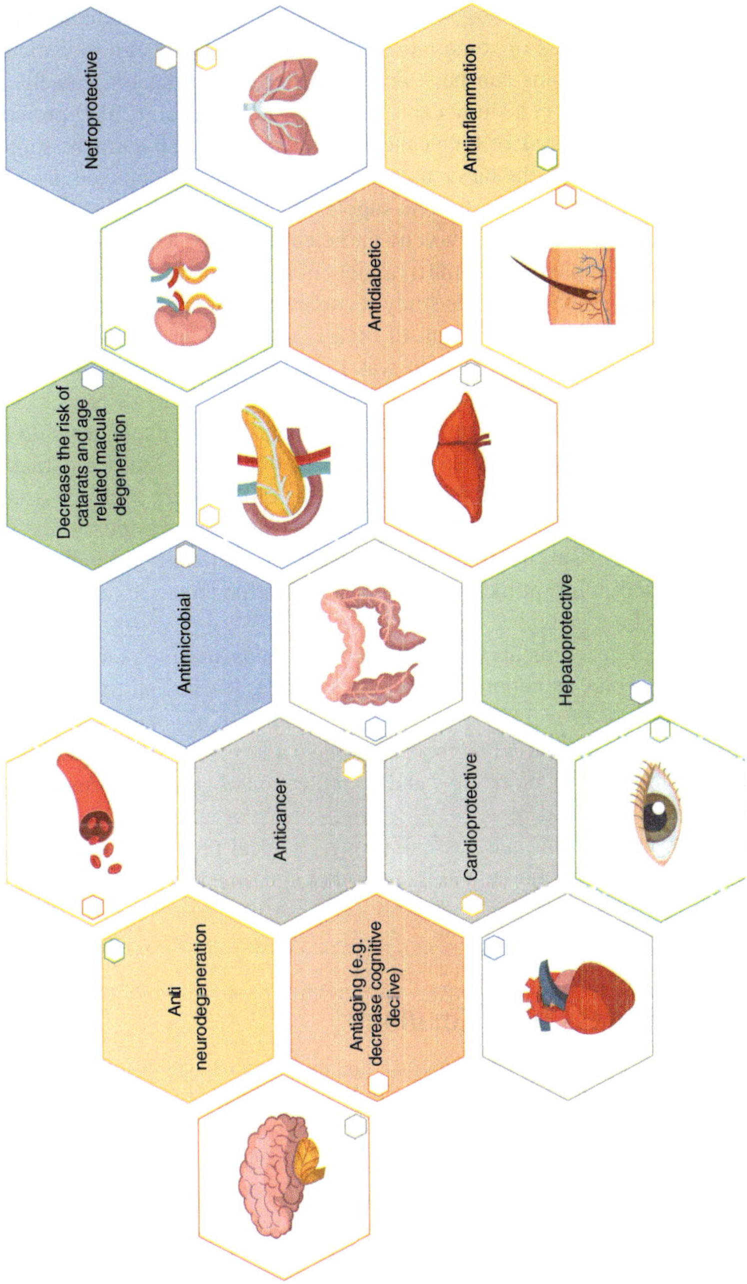

FIG. 6.3.2 Most reported benefits of antioxidants in human health in the scientific literature.

efficient-free radical quenchers in vitro studies but in in vivo this effect is not so clear.

Therefore, antioxidants shall not be regarded as a solution to all problems. There are evidences that the excess of some molecules with antioxidant capacity such as β-carotene and some vitamins can have deleterious effects on health, namely, regarding tumors. For instance, a study carried out by Middha et al. (2019) provided evidence that the increased risk of lung cancer in smokers taking β-carotene supplements does not depend upon the tar or nicotine level of cigarettes smoked and suggests that smokers should avoid β-carotene supplementation.

A meta-analysis concluded there was no sufficient evidence to sustain the effectiveness of vitamin C supplements in the treatment of breast cancer patients (Moradi-Joo, 2018). In fact, the authors of this study recommend that clinician do not highlight these supplements in the treatment of breast cancer. In an interesting review carried out by Pawlowska et al. (2019) it was reported that high doses of vitamin C may induce pro-oxidant effects, which infer detrimental effects to cancer cells. Other studies indicate that antioxidant molecules in the form of supplements do not have the same action as those obtained from the diet. In fact, in a study reported by Bouayed and Bohn (2010) the antioxidants within plant foods were considered safer and healthier than those found in food supplements. This result was due to the fact that the concentration of antioxidants in food matrices is generally low and that the mixture of compounds with antioxidant activity found in food matrices can originate additive or synergistic effects.

Two decades ago, free radicals were regarded as villains, nowadays there are evidences that they also play an important function at cellular levels. For instance, nitric oxide dilates blood vessels and favors blood flow. Together, nitric oxide/soluble guanylyl cyclase/cyclic guanosine monophosphate plays a key role in the regulation of kidney by controlling kidney blood flow and by protection of glomerular and tubular compartments (Krishnan, 2018).

ROS are also involved in many essential physiological processes, such as systemic signaling, pathogen defense, and induction of stress responses (Milisav et al., 2019).

6.3.3 Effectiveness of antioxidants

The in vitro antioxidant activity of molecules has been extensively studied; however this effect cannot be extrapolated to in vivo studies. In part this is justified due to the bioavailability of the molecules, which in the case of antioxidants is generally low or even unknown (Porrini and Riso, 2008). Bioavailability of antioxidants depends on different factors which are summarized in Fig. 6.3.3. They can mainly be divided into those related with the antioxidant molecule, the food or the host. For instance, one of the factors that affects bioavailability are the components of food besides antioxidants that can enhance (e.g., protein, lecithin, and fat) or decrease (e.g., fiber and chelating agents) the absorption of antioxidants (Porrini and Riso, 2008).

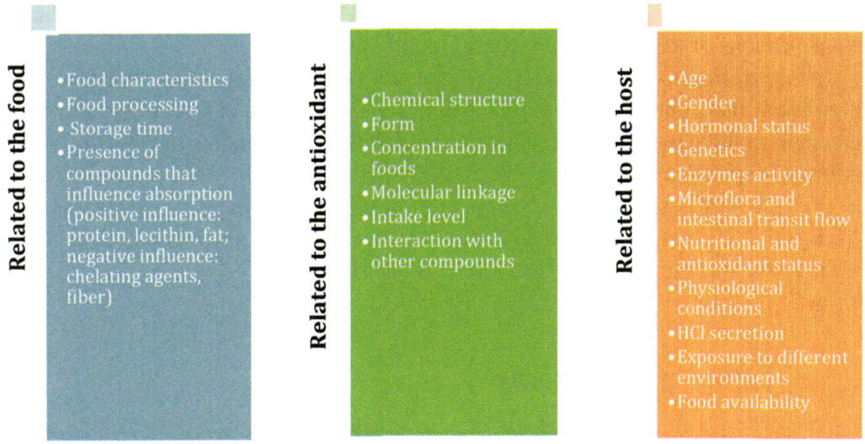

FIG. 6.3.3 Factors affecting the bioavailability of antioxidants.

The benefits of antioxidants for human health are centered in diseases prevention rather than in diseases treatment, although their effect as adjuvant therapy has also been addressed. Yasueda et al. (2016) carried out a systematic review and the only sustainable conclusions based on the results of 49 reports indicate it is difficult to demonstrate ultimately that antioxidants ameliorate therapeutic toxicities and that there is no evidence of antioxidant supplementation causing injury together with cancer therapy, except in the case of smokers undergoing radiotherapy.

A specific antioxidant shall not be indicated as the most effective in general because different antioxidants play different functions in different type of cells. Therefore, for a specific organ or group of cells, a specific molecule can better exhibit their action than in another. For instance lipophilic micronutrients like liposoluble vitamins and β-carotene modulate several processes occurring in adipocytes via the regulation of gene expression due to their nature (Landrier et al., 2012)

The therapeutic doses of antioxidants shall be investigated case by case, because a subtherapeutic dose has no benefits and an excessive amount can also have considerable drawbacks (Bouayed and Bohn, 2010). For instance, excessive vitamin C is associated with kidney stones (Sunkara et al., 2015).

6.3.4 Dietary sources of antioxidants

Diet is one of the main sources of antioxidants. In particular fruits and vegetables are rich in these compounds. Each compound has specific antioxidant properties but fruits and vegetables are generally rich in different phytochemicals with antioxidant capacity that may have an additive or synergistic effect (van Breda and de Kok, 2018).

Physiological doses of antioxidants help to maintain redox homeostasis. However, at high doses, exogenous antioxidants may disturb redox balance, inducing pro-oxidative effects (Bouayed and Bohn, 2010). They can also react with beneficial concentrations of ROS, necessary for optimal function of cells.

It is preferred to intake dietary antioxidants directly from fruits, vegetables, or other foods that contain a high levels of antioxidants. More studies are also required to conclude about the recommendation of antioxidant supplements for specific population or in specific circumstances like in the case of athletes, to evaluate the influence on exercise performance and in the prevention of exercise-induced muscle damage (Malaguti et al., 2013); during pregnancy (Mistry and Williams, 2011; Poston et al., 2011) or for specific groups such are vegans or vegetarians.

6.3.5 Future perspectives

The number of antioxidants molecules is considerably vast and the opinion on these molecules varies from those that defend that there are essential elements for a healthy state to others who consider them as irrelevant from the biological point of view or even toxic/harmful compounds. One of the causes for the hazardous effects of antioxidants can be related with their ability to block apoptosis driven by oxidants (Bast and Haenen, 2013). In this line, nutritional recommendations also interchange between advices recommending the increase of antioxidant status to those that consider the measurement of the plasma antioxidant level does not give any information about the healthy state of an individual.

Moreover, the emerging usage of antioxidants as components of food, in food packaging, in cosmetics and hygiene products or as food supplements reinforces the need for further investigations on this thematic, particularly with a focus on chronic disorders and ageing. Further research aimed to evaluate that the most relevant antioxidants for each specific disorder/disease are warranted to have evidence on the doses that can yield benefits for human health. In this line, some of the mechanisms of antioxidants observed in vitro and in animal models should be validated in clinical trials for different diseases. Therefore, there is the need of large-scale randomized placebo-controlled clinical studies subjected to rigorous scientific scrutiny to confirm whether these compounds may provide preventive or adjunct therapeutic support in different pathologies and better understand their mechanisms of action such as modulators of cell-clearing systems.

Although dietary antioxidants are regarded as safe, besides evaluating their optimal doses, it is important to study the toxicity profile (understanding their biotransformation) at doses required for obtaining health benefits. Another important question in this regard is the fact that antioxidant food and antioxidant food extracts might present variable composition due to several factors such as species, variety, edaphoclimatic conditions, and season of the year. Therefore, there is a need for standardization, especially in the case of antioxidant food supplements (Silva and Nabavi, 2018).

The excessive doses (abuse) of dietary antioxidants are a potential concern. Therefore, it is recommended that the surveillance of these compounds should be made as for the other therapeutic drugs. In the future, it is important to find biomarkers of the antioxidants effect and to monitor the antioxidant status as an indicator of disease severity.

In sum, taking into account all the above mentioned, the evaluation of potential benefits and risks emerging from the use of antioxidants, usages, and effects in different diseases/disorders shall be correctly monitored in order to obtain reliable conclusions about their potential use as prophylactic or adjunct health tools.

Acknowledgment

The work was supported by UIDB/00211/2020 with funding from FCT/MCTES through national funds.

References

Bast, A., Haenen, G.R.M.M., 2013. Ten misconceptions about antioxidants. Trends Pharmacol. Sci. 34 (8), 430–436. http://doi.org/10.1016/j.tips.2013.05.010.

Boaz, M., et al., 2000. Secondary prevention with antioxidants of cardiovascular disease in endstage renal disease (SPACE): randomised placebo-controlled trial. Lancet 356 (9237), 1213–1218. http://doi.org/10.1016/S0140-6736(00)02783-5.

Bouayed, J., Bohn, T., 2010. Exogenous antioxidants -- double-edged swords in cellular redox state: health beneficial effects at physiologic doses versus deleterious effects at high doses. Oxid. Med. Cell. Longev. 3 (4), 228–237. http://doi.org/10.4161/oxim.3.4.12858.

EC, Regulation (EC) No 1333/2008, 2008, 2008. Regulation (EC) No 1333/2008 of the European Parliament and of the Council of 16 December 2008 on food additives. Offic. J. Eur. Union L354, 16–33 L354/16-L354/33.

Harman, D., 1981. The aging process. Proc. Natl. Acad. Sci. USA 78 (11), 7124–7128. http://doi.org/10.1002/j.2326-1951.1963.tb00730.x.

Krishnan, S.M., et al., 2018. The impact of the nitric oxide (no)/soluble guanylyl cyclase (sGC) signaling cascade on kidney health and disease: a preclinical perspective. Int. J. Mol. Sci. 19 (6), 1712. http://doi.org/10.3390/ijms19061712.

Landrier, J.F., Marcotorchino, J., Tourniaire, F., 2012. Lipophilic micronutrients and adipose tissue biology. Nutrients 4 (11), 1622–1649. http://doi.org/10.3390/nu4111622.

Lloret, A., et al., 2019. The effectiveness of vitamin E treatment in Alzheimer's disease. Int. J. Mol. Sci. 20 (4), 879. http://doi.org/10.3390/ijms20040879.

Malaguti, M., Angeloni, C., Hrelia, S., 2013. Polyphenols in exercise performance and prevention of exercise-induced muscle damage. Oxid. Med. Cell. Longev. 2013, 825928. http://doi.org/10.1155/2013/825928.

Martini, D., 2019. Health benefits of Mediterranean diet. Nutrients 11 (8), 11–14. http://doi.org/10.3390/nu11081802.

Middha, P., et al., 2019. β-Carotene supplementation and lung cancer incidence in the alpha-tocopherol, beta-carotene cancer prevention study: the role of tar and nicotine. Nicotine Tobacco Res. 21 (8), 1045–1050. http://doi.org/10.1093/ntr/nty115.

Milisav, I., Ribarič, S., Poljsak, B., 2019. Antioxidant vitamins and ageing K. V. In: Harris, J. (Ed.). Biochemistry and Cell Biology of Ageing: Part I Biomedical Science. Subcellular

Biochemistry, Vol. 90. Springer, Singapore. Available at:. https://doi.org/10.1007/978-981-13-2835-0_1.

Mistry, H.D., Williams, P.J., 2011. The importance of antioxidant micronutrients in pregnancy. Oxid. Med. Cell. Longev. 2011, 841749. http://doi.org/10.1155/2011/841749.

Moradi-Joo, M., et al., 2018. Antioxidant supplements and breast cancer: a systematic review and meta-analysis. Int. J. Cancer Manage. 11 (4), e10082. http://doi.org/10.5812/ijcm.10082.

Pawlowska, E., Szczepanska, J., Blasiak, J., 2019. Pro- and antioxidant effects of vitamin C in cancer in correspondence to its dietary and pharmacological concentrations. Oxid. Med. Cell. Longev. 2019, 7286737. http://doi.org/10.1155/2019/7286737.

Porrini, M., Riso, P., 2008. Factors influencing the bioavailability of antioxidants in foods: a critical appraisal. Nutr. Metab. Cardiovasc. Dis. 18 (10), 647–650. http://doi.org/10.1016/j.numecd.2008.08.004.

Poston, L., et al., 2011. Role of oxidative stress and antioxidant supplementation in pregnancy disorders. Am. J. Clin. Nutr. 94 (6), 1980–1985. http://doi.org/10.3945/ajcn.110.001156.

Silva, A.S., Nabavi, S.M., 2019. Challenges and foresight of food supplements. Nonvitamin and Nonmineral Nutritional Supplements, Elsevier, 541–543. http://doi.org/10.1016/B978-0-12-812491-8.00070-9.

Sunkara, V., et al., 2015. Acute kidney disease due to excessive vitamin C ingestion and remote roux-en-y gastric bypass surgery superimposed on CKD. Am. J. Kidney Dis. 66 (4), 721–724. http://doi.org/10.1053/j.ajkd.2015.06.021.

Tosti, V., Bertozzi, B., Fontana, L., 2018. Health benefits of the Mediterranean diet: metabolic and molecular mechanisms. J. Gerontol. – Ser. A Biol. Sci. Med. Sci. 73 (3), 318–326. http://doi.org/10.1093/gerona/glx227.

Trichopoulou, A., et al., 2014. Definitions and potential health benefits of the Mediterranean diet: views from experts around the world. BMC Med. 12 (112), 1–16.

van Breda, S.G.J., de Kok, T.M.C.M., 2018. Smart combinations of bioactive compounds in fruits and vegetables may guide new strategies for personalized prevention of chronic diseases. Mol. Nutr. Food Res. 62 (1), 1–12. http://doi.org/10.1002/mnfr.201700597.

Yasueda, A., Urushima, H., Ito, T., 2016. Efficacy and interaction of antioxidant supplements as adjuvant therapy in cancer treatment: a systematic review. Integr. Cancer Ther. 15 (1), 17–39. http://doi.org/10.1177/1534735415610427.

Index

Page numbers followed by "*f*" and "*t*" indicate, figures and tables respectively.

G

T

U

V

9780128190968